S0-BMG-673

PROCEDURES IN NUCLEIC ACID RESEARCH

VOLUME 2

PROCEDURES IN NUCLEIC ACID RESEARCH

VOLUME 2

Edited by G. L. CANTONI
National Institute of Mental Health

DAVID R. DAVIES
National Institute of Arthritis
and Metabolic Diseases

HARPER & ROW, Publishers
New York, Evanston, San Francisco, London

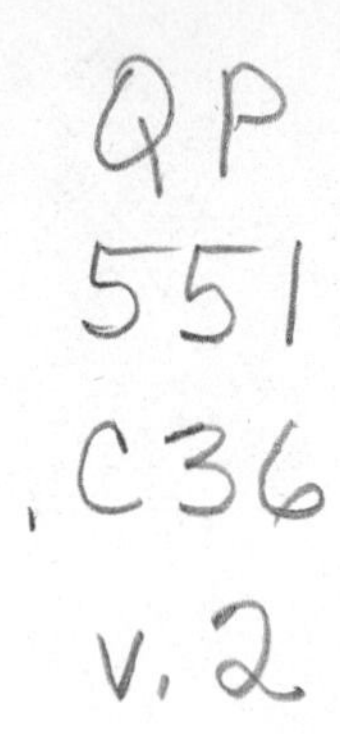

Procedures in Nucleic Acid Research, Volume 2

For information address Harper & Row, Publishers, Inc., 49 East 33rd Street, New York, N.Y. 10016.

Standard Book Number: 06-041166-X
Library of Congress Catalog Card Number: 66-14165

CONTRIBUTORS

Numbers in parentheses denote pages on which chapters begin.

J. T. AUGUST. Department of Molecular Biology, Division of Biological Sciences, Albert Einstein College of Medicine, Bronx, New York (829)

ROBERT L. BALDWIN. Department of Biochemistry, Stanford University School of Medicine, Stanford, California (355)

B. G. BARRELL. Medical Research Council, Laboratory of Molecular Biology, Cambridge, England (751)

WILLIAM BAUER. Department of Chemistry, University of Colorado, Boulder, Colorado (297)

E. K. F. BAUTZ. Molekular Genetik der Universität, Heidelberg, West Germany (743)

MICHAEL BEER. Thomas C. Jenkins Department of Biophysics, Johns Hopkins University, Baltimore, Maryland (443)

GIORGIO BERNARDI. Laboratoire de Génétique Moleculaire, Institut de Biologie Moleculaire, Faculté des Sciences, Paris, France (455)

A. BLOOM. Department of Molecular Biophysics and Biochemistry, Yale University, New Haven, Connecticut (524)

R. J. BRITTEN. Carnegie Institution of Washington, Department of Terrestrial Magnetism, Washington, D.C. (500)

RICHARD R. BURGESS. Institut de Biologie Moléculaire, Université de Genève, Genève, Switzerland (851)

CHARLES R. CANTOR. Departments of Chemistry and Biological Sciences, Columbia University, New York, New York (31)

PAUL CHADWICK. The Biological Laboratories, Harvard University, Cambridge, Massachusetts (703)

ELLIOT CHARNEY. National Institute of Arthritis and Metabolic Diseases, National Institutes of Health, Bethesda, Maryland (176)

D. M. CROTHERS. Department of Chemistry, Yale University, New Haven, Connecticut (369)

V. DANIEL. Department of Biochemistry, Weizmann Institute of Science, Rehovot, Israel (448)

DAVID R. DAVIES. National Institute of Arthritis and Metabolic Diseases, National Institutes of Health, Bethesda, Maryland (435)

C. WESLEY DINGMAN. Chemistry Branch, National Cancer Institute, National Institutes of Health, Bethesda, Maryland (623)

B. P. DOCTOR. Division of Biochemistry, Walter Reed Army Institute of Research, Walter Reed Army Medical Center, Washington, D.C. (435, 588)

J. J. DUNN. Molekular Genetik der Universität, Heidelberg, West Germany (743)

FRITZ ECKSTEIN. Max-Planck-Institut für experimentelle Medizin, Chemische Abteilung, Göttingen, Germany (665)

MARY EDMONDS. Department of Biochemistry, Faculty of Arts and Sciences, University of Pittsburgh, Pennsylvania (629)

HENRYK EISENBERG. Polymer Department, The Weizmann Institute of Science, Rehovot, Israel (137)

PAUL T. ENGLUND. Department of Physiological Chemistry, Johns Hopkins School of Medicine, Baltimore, Maryland (864)

LILLIAN EOYANG. Department of Molecular Biology, Division of Biological Sciences, Albert Einstein College of Medicine, Bronx, New York (829)

GARY FELSENFELD. Laboratory of Molecular Biology, National Institute of Arthritis and Metabolic Diseases, National Institutes of Health, Bethesda, Maryland (233)

T. FRANZE DE FERNANDEZ. Department of Molecular Biology, Division of Biological Sciences, Albert Einstein College of Medicine, Bronx, New York (840)

D. H. GAUSS. Max-Planck-Institut für experimentelle Medizin, Göttingen, Germany (643)

MARTIN GELLERT. Laboratory of Molecular Biology, National Institute of Arthritis and Metabolic Diseases, National Institutes of Health, Bethesda, Maryland (875)

W. B. GRATZER. Medical Research Council, Biophysics Unit, King's College, London, England (3)

W. S. HAYWARD Department of Molecular Biology, Division of Biological Sciences, Albert Einstein College of Medicine, Bronx, New York (840)

T. ISHIDA. Asahi Chemical Industry Co., Tokyo, Japan (608)

J. KAN. City of Hope Medical Center, Duarte, California (608)

T. KANO-SUEOKA. Departments of Biology and Biochemical Sciences, Princeton University, Princeton, New Jersey (608)

CLAUDE B. KLEE. Laboratory of Biochemical Pharmacology, National Institute of Arthritis and Metabolic Diseases, National Institutes of Health, Bethesda, Maryland (896)

D. E. KOHNE. Carnegie Institution of Washington, Department of Terrestrial Magnetism, Washington, D.C. (500)

C. H. KUO. Department of Molecular Biology, Division of Biological Sciences, Albert Einstein College of Medicine, Bronx, New York (846)

ULF LAGERKVIST. Department of Medical Biochemistry, University of Gothenburg, Gothenburg, Sweden (725)

U. Z. LITTAUER. Department of Biochemistry, Weizmann Institute of Science, Rehovot, Israel (448, 565)

BRUCE McCONNELL. Department of Biochemistry and Biophysics, University of Hawaii, Honolulu, Hawaii (389)

MATTHEW MESELSON. The Biological Laboratories, Harvard University, Cambridge, Massachusetts (889)

H. TODD MILES. Laboratory of Molecular Biology, National Institute of Arthritis and Metabolic Diseases, National Institutes of Health, Bethesda, Maryland (205)

ROGER MONIER. Centre de Biochemie et de Biologie Moléculaire, Centre National de la Recherche Scientifique, Marseilles, France (618)

KARL H. MUENCH. Departments of Medicine and Biochemistry, University of Miami School of Medicine, Miami, Florida (515)

SUSUMU NISHIMURA. Biology Division, National Cancer Center Research Institute, Tokyo, Japan (542)

ANDREW C. PEACOCK. Chemistry Branch, National Cancer Institute, National Institutes of Health, Bethesda, Maryland (623)

WARNER L. PETICOLAS. Department of Chemistry, University of Oregon, Eugene, Oregon (94)

VINCENZO PIRROTTA. The Biological Laboratories, Harvard University, Cambridge, Massachusetts (703)

MARK PTASHNE. The Biological Laboratories, Harvard University, Cambridge, Massachusetts (703)

ERIKA RANDERATH. The John Collins Warren Laboratories of the Huntington Memorial Hospital of Harvard University at the Massachusetts General Hospital, Boston, Massachusetts and Department of Biological Chemistry, Harvard Medical School, Boston, Massachusetts (796)

KURT RANDERATH. The John Collins Warren Laboratories of the Huntington Memorial Hospital of Harvard University at the Massachusetts General Hospital, Boston, Massachusetts and Department of Biological Chemistry, Harvard Medical School, Boston, Massachusetts (796)

CHARLES C. RICHARDSON. Department of Biological Chemistry, Harvard Medical School, Boston, Massachusetts (815)

PHILIP D. ROSS. Laboratory of Molecular Biology, National Institute of Arthritis and Metabolic Diseases, National Institutes of Health, Bethesda, Maryland (417)

K. L. ROY. Department of Molecular Biophysics and Biochemistry, Yale University, New Haven, Connecticut (524)

KARL-HEINZ SCHEIT. Max-Planck-Institut für experimentelle Medizin, Chemische Abteilung, Göttingen, Germany (665)

E. SCHLIMME. Max-Planck-Institut für experimentelle Medizin, Göttingen, Germany (643)

D. SÖLL. Department of Molecular Biophysics and Biochemistry, Yale University, New Haven, Connecticut (524)

ROBERT STEINBERG. The Biological Laboratories, Harvard University, Cambridge, Massachusetts (703)

R. STERN. National Institute of Dental Research, National Institutes of Health, Bethesda, Maryland (565)

TSUTOMU SUGIYAMA. Central Research Department, E. I. du Pont de Nemours and Company, Wilmington, Delaware (716)

M. SZÉKELY. Medical Research Council, Laboratory of Molecular Biology, Cambridge, England (780)

ELIZABETH H. SZYBALSKI. McArdle Laboratory, University of Wisconsin, Madison, Wisconsin (311)

WACLAW SZYBALSKI. McArdle Laboratory, University of Wisconsin, Madison, Wisconsin (311)

TERENCE TAO. Department of Molecular Biophysics and Biochemistry, Yale University, New Haven, Connecticut (31)

ANDREW A. TRAVERS. Medical Research Council, Laboratory of Molecular Biology, Cambridge, England (851)

HANS TÜRLER. Department of Molecular Biology, University of Geneva, Geneva, Switzerland (680)

JEROME VINOGRAD. Division of Chemistry and Chemical Engineering and Division of Biology, California Institute of Technology, Pasadena, California (297)

F. VON DER HAAR. Max-Planck-Institut für experimentelle Medizin, Göttingen, Germany (643)

PETER H. VON HIPPEL. Institute of Molecular Biology and Department of Chemistry, University of Oregon, Eugene, Oregon (389)

JAMES C. WANG. Chemistry Department, University of California, Berkeley, California (407)

T. YAMANE. Bell Telephone Laboratories, Murray Hill, New Jersey (262)

ROBERT YUAN. The Biological Laboratories, Harvard University, Cambridge, Massachusetts (889)

B. H. ZIMM. Department of Chemistry, Revelle College, University of California (San Diego), La Jolla, California (245)

CONTENTS

C. MISCELLANEOUS TECHNIQUES

D. PREPARATION AND PURIFICATION OF NUCLEIC ACIDS

PREFACE

Six years ago we believed that a valuable contribution could be made by bringing together in a multivolume series the wide variety of methods and techniques that are currently employed in the field of nucleic acid research. The first volume contained descriptions of methods for the isolation and characterization of enzymes of nucleic acid metabolism and degradation. In addition, techniques were described for the isolation and preparation of a variety of representative nucleic acids. The first volume was well received and we have been encouraged by the users' response to follow our original plan, which was to extend the series to cover a broader range of procedures and techniques of interest to workers in the field of nucleic acid research.

This second volume contains a variety of physical, physicochemical, and chemical techniques that are used for the analysis and characterization of nucleic acids. These include a description of various spectroscopic and hydrodynamic methods which have played such a significant role in the determination of nucleic acid conformation in solution. No volume on nucleic acid physical chemistry would be complete without a description of centrifugation techniques; in addition to chapters on the sedimentation velocity and equilibrium density gradient centrifugation we have included a chapter on a specialized technique for measuring the length of lambda DNA that may have more general application in the future. Under miscellaneous techniques we have chapters on a variety of physicochemical techniques ranging from temperature-jump methods to tRNA crystallization. We have included in this section the use of hydroxyapatite columns for examining hybridization techniques between different nucleic acids, a method which can uniquely provide information regarding fundamental interactions between nucleic acids.

This volume also contains methods for chemical modification of nucleic acids and for the study of nucleic acid–protein interactions. Six years ago only a small number of single species of RNA had been isolated; great advances have been made in this field and now a large variety of nucleic acids can be prepared in pure form. Many of these methods will have general applicability, and we have therefore included a section on the preparation and purification of nucleic acids. Paralleling this rise in the number of pure nucleic acid species has been the development of powerful new sequencing techniques that culminated in the pioneering sequence determination of a tRNA molecule by Holley and his colleagues. The techniques developed for this analysis have been developed further under the stimulus of Sanger and his colleagues in Cambridge, and the art of nucleic acid sequence determination is now comparable in power to the techniques used for protein sequence determination. In the field of DNA sequence determination, progress has been slower but significant advances are being made.

The selection of enzymes of polynucleotide metabolism for this volume was more difficult to make because of the variety of new preparations and procedures now available. In order to keep the size of the volume within reasonable limits we have restricted ourselves to a few key enzymes the description of which complements the preparations described in Volume 1, most of which are still valid.

In this volume as in the earlier one we have been extremely fortunate in securing the contributions of friends and colleagues who are all preeminent in their special fields of research. We are immeasurably indebted to them for their enthusiastic response, for the uniformly high quality of their contributions, and for the promptness and facility with which they produced them. If this series has any merit it is not due to us, the editors, but to the excellence of our contributors.

In addition we benefited enormously during the preparation of this volume, especially during the early phases, from many helpful discussions from friends and colleagues at the NIH. In particular we wish to thank Dr. Martin Gellert and Dr. Maxine Singer, who played a most important role in selection of the material to be included in this volume. In addition we wish to thank Drs. Gerald Cohen, G. Felsenfeld, M. Gellert, and Ruth McDiarmid for their critical evaluation in manuscript form of some of the chapters which we did not feel qualified to edit ourselves. Finally, we wish to thank Miss Marjorie Singer of Harper & Row, whose systematic and clearheaded approach has greatly expedited and facilitated the production of this volume.

G. L. CANTONI
DAVID R. DAVIES

SECTION

SPECTROSCOPIC AND HYDRODYNAMIC METHODS

OPTICAL ROTATORY DISPERSION AND CIRCULAR DICHROISM OF NUCLEIC ACIDS

W. B. GRATZER
MEDICAL RESEARCH COUNCIL
BIOPHYSICS UNIT
KING'S COLLEGE,
LONDON, ENGLAND

Measurements of optical activity – optical rotatory dispersion (ORD) or circular dichroism (CD) – of nucleic acids can be used either for the analysis of conformation, in terms of base-pairing and single-stranded stacking, or for following changes in conformation, most commonly melting processes. They are also of use in following interactions, whether with large molecules, such as polypeptides, or small ligands, such as antimetabolites or dyes. An important application is in the configurational analysis of nucleosides and nucleotides.

ORD vs. CD

The nature of the phenomenon of optical activity has been amply discussed, and no better introductory account is still available than that of Moscowitz (*1*). Several good surveys of the theory of optical activity exist (*2–5*). For the purposes of this discussion we will merely recall that whereas CD is confined to the region of absorption, ORD can be observed also away from the absorption bands. The simplest type of measurement is that of optical rotation in the visible and near ultraviolet. Since no light is absorbed, the measurements are relatively simple, and can be cheap and very precise at the same time. From measurements at several wavelengths in this range a Drude plot can be constructed (see below), and from the constants derived from it, certain structural inferences can be drawn (*6*). If, however, the means are on hand for extending the measurements into the absorption region, the information content becomes much higher. Between about 310 nm and the effective wavelength limit of the best modern instruments, at about 185 nm, nucleic acids show a number of Cotton effects, both positive and negative (for definitions see refs. *1, 7, 8*). The nature and magnitude of these Cotton effects reflect the composition and conformation of the polynucleotide.

The main advantage of CD over ORD is that each Cotton effect is localized, in the sense that there is no CD where there is no absorption, and is a simple function with one extremum, for which a Gaussian approximation will frequently serve. An ORD Cotton effect by contrast has two extrema and the rotation diminishes slowly toward longer and shorter wavelengths, such that each Cotton effect sits on a contribution from others at longer or shorter wavelengths or both. This makes it virtually impossible to resolve a complex ORD curve into its constituent Cotton effects, whereas CD can be resolved into a series of bands in just the same way as an absorption spectrum. The use of curve resolvers to analyze CD curves into components is already common (*9, 10*), and indeed the consequences of their overly enthusiastic application are much to be feared. From the purely spectroscopic viewpoint, and in terms of the information that can be extracted, CD is altogether superior to ORD. The advantages of ORD are: (1) Some standard data is available in terms of ORD, but not yet of CD. (2) In unfavorable conditions, for example, when it is required to work as very high concentrations of sample, such that absorbances in the ultraviolet are prohibitively high, or when solvents or cosolutes absorb so strongly that the Cotton-effect region becomes obscured, it is still possible to observe optical rotations in the visible. For example, with solvents that absorb in the ultraviolet and make it impossible to measure absorbance-melting profiles at 260 nm, the rotation at say 546 nm can be used to follow melting processes (*10*). (3) Finally, for such measurements good manual instruments are available in an altogether lower price range than scanning ORD or CD spectropolarimeters.

Definitions and Parameters

ORD

In ORD the quantity measured is a rotation in degrees, α. This is related to the standard rotation at wavelength λ by:

$$[\alpha]_\lambda = \frac{100\,\alpha}{cd} \qquad (1)$$

where c is the concentration in gm/100 ml and d the pathlength in *dm*. This quantity is sometimes used to describe the ORD. It will be noted that to be explicit this requires a knowledge of the mean residue weight of a polynucleotide, and brings with it the perennial ambiguity about whether or not to include the weight of the counterions. It is generally preferable to use instead the molar residue rotation, usually written $[m]$ or $[\phi]$, and defined as

$$[m]_\lambda = \frac{M_0}{100}[\alpha] = \frac{M_0}{cd}\alpha \tag{2}$$

where M_0 is the mean residue weight. This reduces more conveniently to

$$[m]_\lambda = \frac{100}{C}\alpha \tag{3}$$

where C is the molar concentration (per phosphorus atom), or if the molar absorptivity say at 260 nm is ϵ_p, and the absorbance in a 1-cm cell at the same wavelength is A, we have

$$[m]_\lambda = \frac{100\epsilon_p}{A}\alpha \tag{4}$$

CD

The CD is generally measured in modern instruments as an absorbance difference between left- and right-circularly polarized light, $\Delta A = A_l - A_r$. The circular dichroism is then defined as the corresponding molar absorptivity (extinction coefficient) difference, or by analogy with the definition of absorbance (optical density):

$$\epsilon_l - \epsilon_r = \Delta\epsilon_\lambda = \frac{\Delta A_\lambda}{Cl} \tag{5}$$

where C is the molar concentration and l the pathlength in *cm*. The CD is sometimes expressed in plots of $\Delta\epsilon$. Commonly, however, the *ellipticity* is the quantity plotted, for the ellipticity of plane-polarized light after passage through the optically active medium is another measure of circular dichroism (*1*). The ellipticity, written $[\theta]$, is related to the circular dichroism by

$$[\theta]_\lambda = 3300\Delta\epsilon_\lambda \tag{6}$$

and emerges in the cumbersome units of degree cm^2/decimole.

There is also a rather seldom used specific ellipticity, defined in the same way as specific rotation, written $[\psi]$, and related to the molar ellipticity by $[\theta] = M_0[\psi]/100$ (cf. Eq. (2)).

Refractive Index Correction

The optical activity depends in a secondary way on the refractive index of the medium, and this becomes a problem when it is desired to compare ORD or CD of a sample measured in different solvents. A nominal correction to give optical rotation *in vacuo* can be made by multiplying the rotation by $3/(n^2 + 2)$ where n is the refractive index of the solvent. This has been widely used in polypeptide work, where the use of a variety of solvents to control conformation is common. In the polynucleotides one is concerned in the great majority of cases with aqueous solutions, and the refractive index correction is therefore generally discarded. Some authors however do apply the correction, which is not inconsiderable: for water in the 500 nm region it is about 0.8. The correct rotation or ellipticity is denoted by a prime, e.g., $[m']$.

There still remains a further ambiguity, namely whether the correction includes refractive index dispersion. The refractive index changes with wavelength, and strictly speaking the correction factor is then different for every wavelength. It is not always obvious whether a curve has been corrected only by applying a constant multiplier, $3/(n_D{}^2 + 2)$, where n_D is the refractive index at 589 nm as normally given in tables, or by a wavelength-dependent factor. The former is often regarded as sufficient to propitiate the guardians of spectroscopic precision, and, considering the vague theoretical basis of the correction, may be regarded as adequate to give the trend. For the fastidious, however, tables of refractive indices at a series of wavelengths of various common solvents can be found in the "Handbook of Biochemistry" (*11*). These can be plotted according to a one-term Sellmeier equation, namely

$$n_\lambda{}^2 = 1 + \frac{a\lambda^2}{\lambda^2 - \lambda_0{}^2} \qquad (7)$$

where a and λ_0 are constants. For example, a plot of $(n^2 - 1)^{-1}$ against λ^{-2} will be linear, and can be used to interpolate and extrapolate to give values of n at any desired wavelength. For any measurements that are confined to aqueous solutions it is simpler and preferable that no refractive index correction be applied. The incorporation of corrections in some published work must however be borne in mind in comparing data.

For absorption intensity the correction for refractive index is $9/(n^2 + 2)^2$, and, if there is reason to express circular dichroism or ellipticity corrected to vacuum conditions ($[\theta']$), this is the factor to be used. The same strictures, however, apply as to ORD.

Other Parameters of Optical Activity

A quantity commonly encountered in discussions of optical activity is the rotational strength, which serves as a quantitative index of the optical activity associated with an absorption band. It is discussed by Moscowitz (*1*), and we note here only that it is proportional to the integrated intensity of the CD Cotton effect. If the approximation is made that the Cotton effect is

Gaussian the rotational strength is approximately

$$R_K \sim 1.23 \times 10^{-42} \; [\theta_K{}^\circ] \frac{\Delta^\circ_K}{\lambda^\circ_K} \text{(cgs units)} \tag{8}$$

where $[\theta^\circ_K]$ is the molar ellipticity at the maximum, λ°_K the corresponding wavelength, and Δ°_K the half-width of the band in the same units. (This is the half-width at the level at which $[\theta] = 0.368[\theta^\circ_K]$). The subscript K indicates that one is considering here only one (the Kth) optically active absorption band. Other parameters, such as rotational oscillator strength, are also sometimes encountered, and are defined in strict analogy with corresponding parameters of absorption spectra (*12*).

Relation Between ORD and CD

CD bears the same relation to ORD that absorbance does to refractive index dispersion. Such interrelated functions are analytically interconvertible by the general relations known as Kronig-Kramers transforms. These may be applied to single Cotton effects or to the system as a whole. In order to derive the optical rotation $[m]_i$ at a given frequency $\bar{\nu}_i$ ($\equiv 1/\lambda_i$) from the circular dichroism one must evaluate an integral (*13*):

$$[m]_i = \frac{2}{\pi} \int_0^\infty \frac{\bar{\nu}[\theta]}{\bar{\nu}^2 - \bar{\nu}_i^2} \, d\bar{\nu} \tag{9}$$

Thus to obtain a value for the molar residue rotation at any wavelength an integration must be performed over the entire wavelength range in which the circular dichroism is finite. To define the ORD curve from the CD therefore is scarcely feasible without the use of a computer. Again it is obvious that the ORD will be overwhelmingly determined by the closest-lying Cotton effects, so that a good approximation to the true ORD can in practice be obtained from the CD – even without knowledge of Cotton effects in the remote ultraviolet, where instruments cannot go. Many programs have been written to perform the calculation. A particularly thorough numerical procedure for the evaluation of the integral of equation (*9*) has been given by Emeis et al. (*13*). An example of how closely the calculated ORD can fit the observed is to be found in the work of Carver et al. (*14*) on α-helical polypeptides.

The reverse transformation can also be made:

$$[\theta]_i = -\frac{2\bar{\nu}}{\pi} \int_0^\infty \frac{[m]}{\bar{\nu}_i^2 - \bar{\nu}^2} \, d\nu \tag{10}$$

but here, since a large part of the rotational strength in ORD lies outside the Cotton effect range, the possibility of integrating effectively becomes remote.

Uses of Kronig-Kramers Transforms

If the means for measuring circular dichroism are available there is not often any need to try to derive the ORD. One can imagine useful applications only when comparison has to be made of substances for which CD data are available, with others for which there are only ORD data.

One other possible use is the attempt to divine something about the nature of the inaccessible Cotton effects in the short ultraviolet. Thus subtraction from the observed ORD of a transform derived from the measured CD gives in principle the ORD arising from the Cotton effects at shorter wavelengths, primarily of course the nearest (or biggest). A recent example exists of the use of this procedure to deduce changes in an inaccessible Cotton effect (*15*).

Approximate Relations Between ORD and CD

For a Gaussian band, the rotational strength is related to the magnitude of the Cotton effect by Eq. (8). Moscowitz (*1*) has also shown that for a similar approximation, at a wavelength λ, well away from the Cotton effect, so that $|\lambda - \lambda^\circ_K| \gg \Delta^\circ_K$, the optical rotation deriving from the *K*th Cotton effect is given by

$$[m]_K \sim 0.916 \times 10^{42} R_K \frac{\lambda_K^{\circ 2}}{\lambda^2 - \lambda_K^{\circ 2}} \tag{11}$$

and one can frequently throw in the additional assumption that the magnitude of the total rotation is dominated by the nearest prominent Cotton effect. Equations (8) and (11) can be used to obtain a useful rough estimate from the CD curve of the molar rotation at long wavelengths, e.g., a value of $[\alpha]_D$, for comparison with standard data.

ORD and CD of Nucleic Acid Components

We shall now consider the optical activity of nucleic acids and their components, and the scope of its application in the chemistry and structure of nucleic acids.

Monomers

The nucleosides are the simplest components to possess optical activity. The ORD curves of the common nucleosides and of the corresponding 5′-nucleotides, have been measured by Yang et al. (*16*). They all display apparently single Cotton effects associated with the absorption bands near 260 nm, which are smaller in magnitude than those observed in ordered polynucleotides, though by no means negligible. The common purine nucleosides and nucleotides have negative, and the pyrimidines positive Cotton effects. ORD has been a useful method in configurational studies of nucleosides, and has been used, for example, to demonstrate the prevalence of the *anti* configuration about the glycosidic link (*17*). The ORD curves of many nucleotides have been surveyed and the effects of substituents in various positions have been analyzed by Robins, Eyring, and their co-workers (*18*). This and other work on mononucleotides of various kinds is outside the scope of this survey, however, except to the extent that it relates to the optical activity of nucleic acids, as will be seen below.

Unpaired Oligo- and Polynucleotides

Theoretical Aspects

It is now well established that an unpaired polynucleotide chain is by no means structureless, and that the bases are "sticky" molecules with a strong tendency to stack on each other. The optical activity of such a chain is in fact dominated by the stacking phenomenon. Our understanding of the theoretical basis of the optical manifestations of stacking is due almost wholly to the work of Tinoco. What Tinoco has demonstrated is the following: The transitions in the different bases interact with each other to produce large Cotton effects. The interactions are of two kinds: those between identical transition moments (e.g., those corresponding to the longest-wavelength absorption band) and those between a given transition in one base with other transitions in its neighbor (e.g., longest wavelength band with all those at shorter wavelengths). This is a dispersion interaction. The first of these (exciton interaction) gives rise to a conservative pattern, in which one Cotton effect is balanced by another of equal magnitude, but opposite sign (Fig. 1, curves 2, 2a). The second produces a nonconservative pattern (Fig. 1, curves 1, 1a), where the rotational strength of the transition does not cancel out. Both these forms are found in polynucleotides, but in the general case the observed curve will be a mixture made up of both kinds of contribution. The relative magnitudes of the two depend on the strength of the two types of interaction between the transition moments, which in turn depend on the geometry, i.e., the equatorial orientation of the base relative to the helix axis, the amount of tilt, and the number of bases per turn of helix. The manner in which the circular dichroism depends on these parameters has been calculated and the results are described in the important paper by Tinoco (*19*) which should be consulted for further details. Figure 1 shows the general character of the curves generated by combination of two positive contributions, and of one positive and one negative. From the shape of observed curves it is clearly then possible to draw some inferences concerning the possible orientation of the bases to the helix axis.

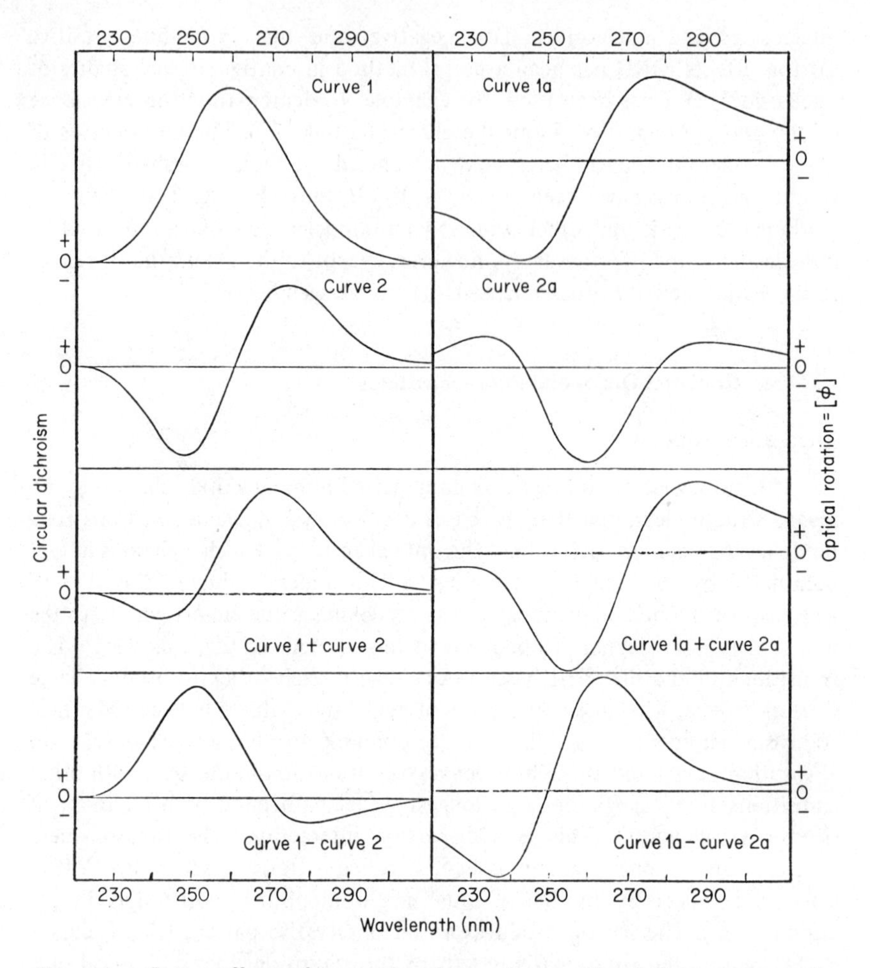

Figure 1. Cotton effects of interacting chromophores: (1) and (1a) nonconservative system in CD and ORD, respectively; (2) and (2a) conservative (exciton) interaction in CD and ORD, respectively. The remaining curves show the results of combining dispersion and exciton contributions. (From Tinoco (*19*) by courtesy of the author).

Dinucleotides

A typical exciton-type system of Cotton effects is observed in the dinucleoside phosphate, ApA, which is shown in Fig. 2. As the temperature is raised the extent of stacking diminishes, and so therefore does the optical activity, and the curve approaches that of the monomer. In CpC (Fig. 2b) there is clearly an admixture of the two types of effect. In heterodimers the situation is clearly more complicated, and the precise nature of the interaction depends on the sequence; thus ApU will not be expected to be the

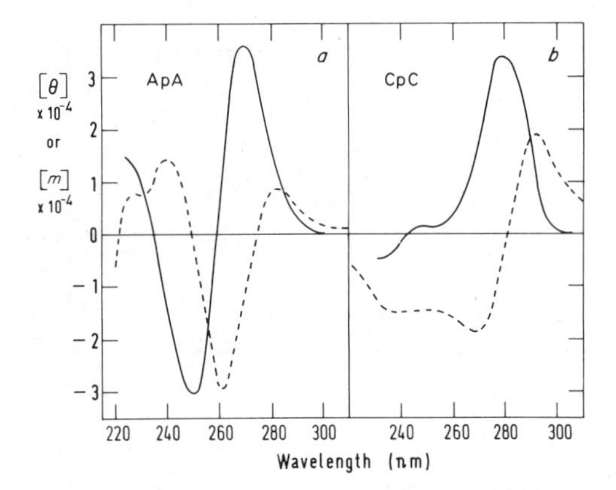

Figure 2. ORD and CD of two homodimers, ApA and CpC, showing optical activity dominated respectively by exciton (conservative) and dispersion (nonconservative) interaction. Solid lines, CD: dashed lines, ORD. The CD data are from van Holde et al. (*59*) for ApA, and Brahms et al. (*60*) for CpC, and the ORD curves from Warshaw (*20*). (Note that the conditions of measurement, in particular the temperature, are different for the CD and ORD, so that the rotational strengths are not precisely comparable.)

same as UpA (Fig. 3). The ORD curves of all the sixteen dinucleoside phosphates have been measured by Warshaw and Tinoco (*20*).

Another parameter that now has to be considered is the extent of stacking, for the observed ORD or CD curve will depend on the position of the equilibrium between the stacked and unstacked states. The stacking equilibrium is an essentially noncooperative process and so the degree of stacking decreases slowly with increasing temperature. The sixteen dinucleoside phosphates show a wide variation in the amount of stacking present at room temperature (*21*). In aqueous solutions the equilibrium in CpC for example is heavily in favor of the stacked state, whereas in UpU there is no detectable stacking (*22*). The standard curves of Warshaw and Tinoco (*20*) therefore refer to the temperature of measurement only (although the temperature dependence of stacking is sufficiently gradual that there will be no appreciable error if the temperature is standardized within 2 or 3°). We note finally that the dinucleoside phosphates are the proper standards for interpretation of polymer ORD in terms of dimers, for the presence of a charged end group, as in ApAp for example, has a considerable effect on the degree of stacking (*23*).

Unpaired Polyribonucleotides

If the stacking-unstacking equilibrium is indeed noncooperative, that is to say every stacking contact between adjacent bases is governed by its own equilibrium constant, which is independent of what goes on between any

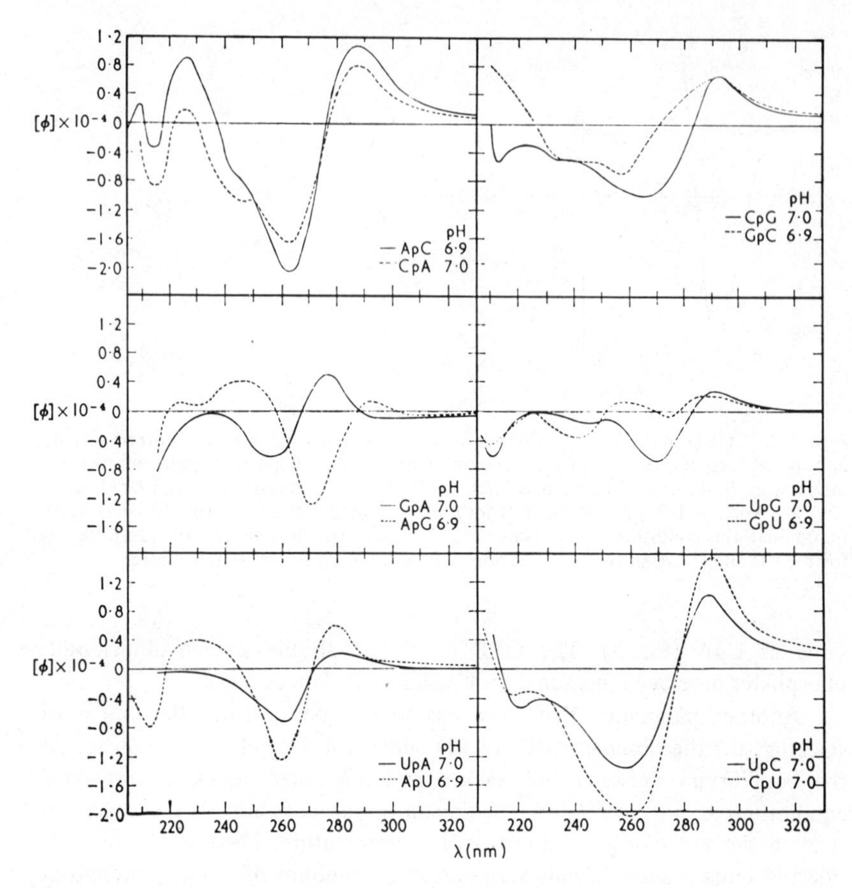

Figure 3. ORD of some heterodimers, showing how the optical activity depends on the sequence as well as the composition. (From Warshaw and Tinoco (*20*) by courtesy of the authors).

other adjacent pair in the chain, it should be possible to express the ORD or CD in terms of all the pairwise contacts along the chain. These are defined by the dinucleoside phosphate curves. The validity of this procedure is extraordinarily good. Cantor and Tinoco (*24*) showed that the ORD curves of trimers could be accurately predicted from the dimers, and so on to long RNA chains (as long of course as no base pairing is present). The key paper, in which the procedure is applied and discussed is that by Cantor et al. (*25*). An RNA species which has base-paired regions of relatively low stability, such as a viral or ribosomal RNA, can without too much difficulty be converted into a fully single-stranded form by lowering the ionic strength sufficiently, and carefully excluding divalent cations. Since the full sequence of such RNA chains is not known, it is possible only to assume a random distribution of nearest neighbors for the base composition of the material. Thus if the fractional base compositions in bases I and J are i and j, the fraction of all dinucleotide sequences IpI, JpJ, IpJ; and JpI will be, respectively, i^2, j^2, ij, and ij.

The procedure for computing the ORD of a fully unpaired RNA chain is given by Cantor et al. (*25*) and goes as follows:

$$[m] = 2\{f_{AA}[m_{AA}] + f_{AU}[m_{AU}] + f_{AG}[m_{AG}] + \ldots\} - \{f_A[m_A] + f_U[m_U] + \ldots\} \quad (12)$$

where f_{AA} is the fraction of all the dinucleotides taken pairwise along the chain that is ApA, and f_A the fraction of A in the RNA, with corresponding molar residue rotations, and so on for all sixteen dinucleotides and all four monomers. More economically

$$[m] = 2\sum_{I=1}^{4}\sum_{J=1}^{4} f_{IJ}[m_{IJ}] - \sum_{I=1}^{4} f_I[m_I] \quad (13)$$

where each sum extends over the four bases. It is important to note that all molar rotations refer as always to one mole residue in chromophores, i.e., never to a mole of dinucleotide. The derivation of Eq. (12) is straightforward and will not be given here (see ref. *24*). As already stated, most commonly the dinucleotide concentrations, f_{AA}, f_{AU} . . . are not known, and Cantor et al. have given reasons why the assumption of randomness in a chain of known base composition but unknown sequence should not in general produce serious errors – an assumption which has so far stood up very well. In this case the terms f_{IJ} are replaced by $f_I f_J$, and Eq. (13) becomes

$$[m] = 2\sum_{I=1}^{4}\sum_{J=1}^{4} f_I f_J [m_{IJ}] - \sum_{I=1}^{4} f_I[m_I] \quad (14)$$

Such a calculation may now be performed by desk calculator or computer for a sufficient number of wavelengths to define the ORD curve, for any RNA, using the values of the $[m_{IJ}]$ terms determined by Warshaw (*20*) and tabulated in Table 1.

Examples

To produce an unpaired chain in a viral or ribosomal RNA, the sample must either be very salt-free or must be dialyzed against 10^{-4} M EDTA, at neutral pH. This serves not only to remove traces of divalent metals, but also to provide marginal buffering. The problem of maintaining neutral pH under these conditions is nevertheless considerable. An example of the fit between

TABLE 1. Molar Residue Rotations of Mononucleotides and Dinucleoside Phosphates[a]

Wavelength	$[m] \times 10^{-4}$								
(nm)	ApA	ApU	ApC	ApG	UpA	UpU	UpC	UpG	CpA
215	−3.12	−0.67	−0.33	−0.55	−0.06	−0.15	−0.13	−0.35	−0.85
220	−0.75	0.20	0.49	0.08	−0.06	−0.18	−0.54	−0.06	−0.35
225	0.73	0.35	0.88	0.12	−0.05	0.10	−0.45	−0.02	0.19
230	0.75	0.39	0.72	0.08	−0.05	0.07	−0.44	−0.02	0.00
235	1.14	0.24	0.27	0.18	−0.09	−0.05	−0.55	−0.06	−0.58
240	1.43	0.22	−0.26	0.38	−0.21	−0.38	−0.73	−0.13	−0.87
245	1.04	−0.07	−0.74	0.41	−0.26	−0.79	−0.97	−0.16	−1.04
250	−0.18	−0.40	−1.00	0.36	−0.43	−1.11	−1.20	−0.12	−1.07
255	−1.77	−0.92	−1.48	0.29	−0.64	−1.44	−1.33	−0.13	−1.37
260	−2.94	−1.26	−1.95	−0.32	−0.76	−1.53	−1.36	−0.36	−1.55
265	−2.26	−0.90	−2.00	−0.89	−0.56	−1.28	−1.15	−0.58	−1.57
270	−1.14	−0.15	−1.30	−1.28	−0.12	−0.58	−0.82	−0.70	−1.09
275	0.10	0.47	−0.36	−1.05	0.16	0.07	−0.15	−0.50	−0.35
280	0.80	0.57	0.62	−0.46	0.23	0.70	0.50	−0.17	0.37
285	0.78	0.37	1.04	−0.12	0.18	0.80	0.94	0.16	0.75
290	0.47	0.17	1.02	0.14	0.13	0.60	0.99	0.28	0.77
295	0.26	0.14	0.85	0.11	0.08	0.30	0.74	0.24	0.58
300	0.16	0.11	0.59	0.05	0.04	0.20	0.52	0.18	0.39
305	0.12	0.09	0.42	−0.03	0.02	0.14	0.40	0.10	0.27
310	0.10	0.07	0.32	−0.05	0.01	0.12	0.31	0.07	0.22
315	0.08	0.05	0.26	−0.05	0	0.10	0.24	0.06	0.17
320	0.08	0.05	0.20	−0.03	0	0.06	0.20	0.04	0.15

[a] All data provided by Professor M. M. Warshaw.

calculations according to Eq. (14) and observed ORD is shown in Fig. 4, together with some curves for trimers (*24*) fitted with Eq. (13).

It will clearly be very desirable to be able to perform the same type of experiment with CD. Fortunately the standard CD curves for the sixteen dinucleoside phosphates, corresponding to the ORD data of Table 1, have now been measured (*25a*).

Polydeoxyribonucleotides

Evidently the geometry of the stacked form of the deoxypolymers differs from that of the ribopolymers, for although it is clearly present, the rotational strength of the longest wavelength Cotton effect, in poly (dA), is very small (*26*), but that at shorter wavelength, centered near 220 nm is large. This can be used to follow the thermal unstacking process. Considerable

[*m*] x 10^{-4}										
CpU	CpC	CpG	GpA	GpU	GpC	GpG	pA	pU	pC	pG
−0.31	−0.16	−0.44	−0.81	−0.28	0.57	−0.88	−0.24	0.30	0.02	−0.65
−0.34	−0.66	−0.30	−0.40	−0.15	0.35	−0.33	−0.20	0.02	−0.69	−0.24
−0.34	−0.56	−0.30	−0.19	−0.02	0.04	−0.03	−0.12	−0.17	−1.03	0.06
−0.41	−0.88	−0.39	−0.06	−0.15	−0.24	0.17	−0.06	−0.24	−1.24	0.17
−0.78	−1.26	−0.48	−0.03	−0.28	−0.48	−0.01	0.02	−0.35	−1.32	0.18
−1.14	−1.50	−0.50	−0.09	−0.36	−0.52	−0.25	0.14	−0.67	−1.36	0.09
−1.50	−1.48	−0.57	−0.21	−0.30	−0.53	−0.52	0.18	−0.94	−1.31	0.02
−1.76	−1.50	−0.72	−0.53	−0.09	−0.56	−0.68	0.17	−1.17	−1.28	−0.06
−1.93	−1.50	−0.86	−0.63	0.11	−0.68	−0.47	0.07	−1.20	−1.22	−0.10
−2.05	−1.68	−0.96	−0.57	0.11	−0.67	−0.01	−0.07	−1.01	−1.00	−0.22
−1.85	−1.80	−1.00	−0.17	0.07	−0.38	0.30	−0.20	−0.65	−0.67	−0.21
−1.33	−1.86	−0.94	0.24	−0.01	−0.08	0.37	−0.30	−0.18	−0.32	−0.20
−0.43	−1.20	−0.73	0.50	−0.06	0.13	0.36	−0.34	0.17	0.16	−0.20
0.59	−0.12	−0.37	0.40	0.09	0.37	0.15	−0.32	0.46	0.52	−0.19
1.36	1.08	0.15	0.11	0.18	0.51	0.03	−0.27	0.44	0.74	−0.18
1.52	1.88	0.57	−0.06	0.21	0.62	−0.04	−0.21	0.32	0.77	−0.16
1.17	1.66	0.60	−0.10	0.18	0.63	−0.07	−0.15	0.20	0.61	−0.14
0.81	1.16	0.45	−0.08	0.11	0.46	−0.07	−0.15	0.08	0.45	−0.12
0.58	0.80	0.29	−0.08	0.09	0.35	−0.06	−0.13	0.07	0.36	−0.11
0.47	0.62	0.19	−0.07	0.07	0.25	−0.05	−0.12	0.06	0.27	−0.11
0.39	0.50	0.14	−0.07	0.05	0.21	−0.05	−0.12	0.05	0.22	−0.10
0.33	0.40	0.11	−0.06	0.03	0.18	−0.04	−0.11	0.03	0.17	−0.09

differences in magnitude are also found in the cytidylate series (*27*). No unpaired copolymers appear to have been studied. Standard CD data for the deoxydimers have been determined by Cantor et al. (*25a*).

Double-Stranded Polynucleotides

Paired polynucleotides present an altogether more complex problem than single-stranded chains. In the first place, if the optical activity is governed by interactions between adjacent base pairs, the number of different interactions that can be generated between the two different kinds of base pairs is ten. This is because the chains run antiparallel and the interaction between a base lying on the 3′ side of another is not the same as when the two are

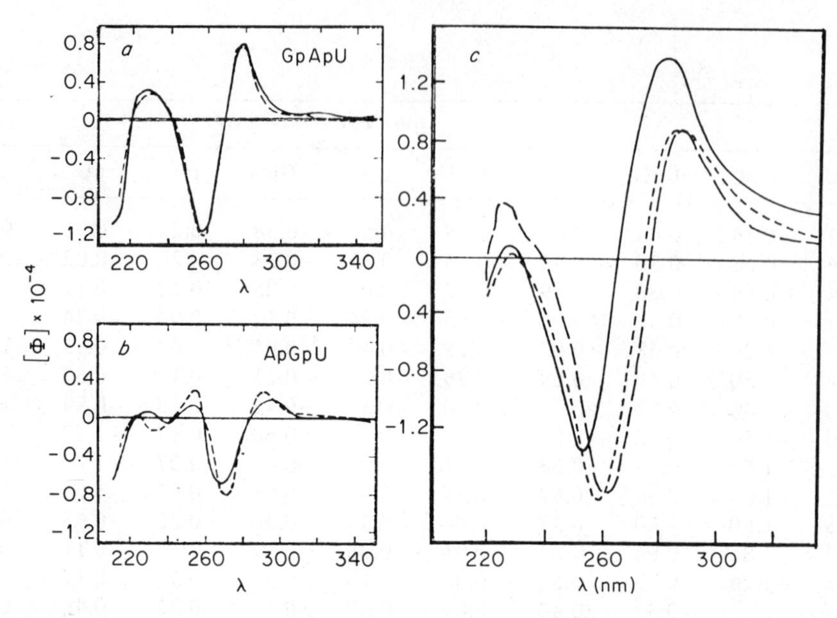

Figure 4. Observed and calculated ORD curves for systems that contain no base pairing. (a) and (b) are trinucleoside diphosphates, GpApU and ApGpU. The solid lines are experimental and the dashed lines calculated curves, computed from the data of Table 1, using Eq. (13) as described; (c) tobacco mosaic virus RNA: the dashed line is the measured ORD in the absence of salt, where there is no significant base pairing; the dotted line is the calculated curve, assuming random nearest-neighbor distribution, according to Eq. (14). The solid line is the measured ORD in 0.15 M salt, showing the effect of introducing base pairing. [(a) and (b) are from Cantor and Tinoco (*24*), and (c) is from Cantor et al. (*25*), by courtesy of Dr. I. Tinoco, Jr.]

interchanged. Thus between two adjacent GC pairs there are three possible arrangements, namely:

$$\begin{array}{ccc} 5'\,G\ldots C\,3' & 5'\,G\ldots C\,3' & 5'\,C\ldots G\,3' \\ |\quad\;\;\; | & |\quad\;\;\; | & |\quad\;\;\; | \\ 3'\,G\ldots C\,5' & 3'\,C\ldots G\,5' & 3'\,G\ldots C\,5' \end{array}$$

Data are simply not available for all the ten combinations of nearest neighbors, and in any case one cannot suppose that in a double helix all interactions beyond the nearest neighbors can be neglected. Thus in practice one assumes some sort of crude additivity of AT (AU) and GC contributions only (*25*).

The extent to which such an approximation works for DNA in CD can be seen from ref. *28* and ORD from ref. *6*. The CD curves of two DNA samples of widely different base compositions are shown in Fig. 5. Within some

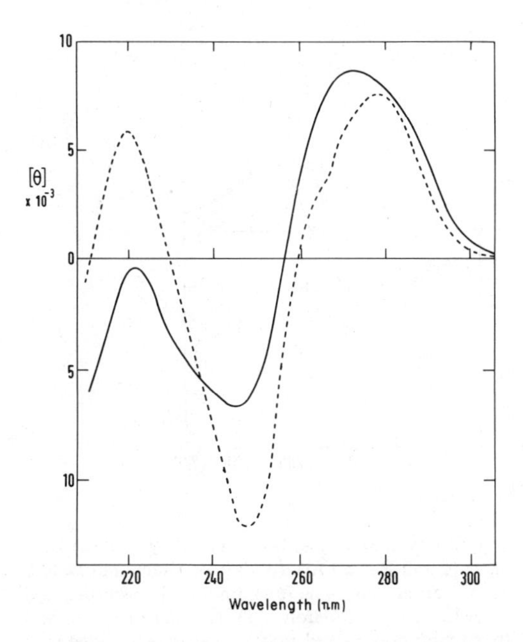

Figure 5. CD of two DNA species. Solid line; *M. lysodeikticus* (28% AT). Dashed line; *Cytophaga* sp. (67% AT). [From Gratzer et al. (*28*).]

5–10% all such curves, at any rate in the range of compositions 28–67% AT, can be expressed in terms of linear mixtures of apparent contributions of AT and GC pairs, obtained by simply extrapolating the molar ellipticities for a series of DNA species to 0 and 100% AT-content at different wavelengths. The linearity of the plots of ellipticity against AT-content is very approximate, and there are good reasons for thinking (*28*) that it will cease to apply altogether at extremes of base composition. However in the present state of knowledge this seems to be the best that can be done.

With RNA less data are available, since fully two-stranded RNA species are difficult to assemble in any number. The curves are different in shape from DNA, as has been mentioned, with a much larger nonconservative contribution, and bigger optical activities. Figure 6 shows curves for poly (A) · poly (U), poly (G) · poly (C) (*6*) and a two-stranded viral RNA (*29*). It will by now be clear that no very good fit can be expected for a two-stranded RNA of intermediate base composition by the use of the curves of Fig. 6 as standards. A rational analysis of a spectroscopic parameter determined by nearest-neighbor interactions, if random nearest-neighbor distribution is assumed, is to be found in the work of Felsenfeld and Hirschman (*30*) on absorption and hypochromicity of DNA. The simplest form that can be hoped for contains a dependence of the optical parameter

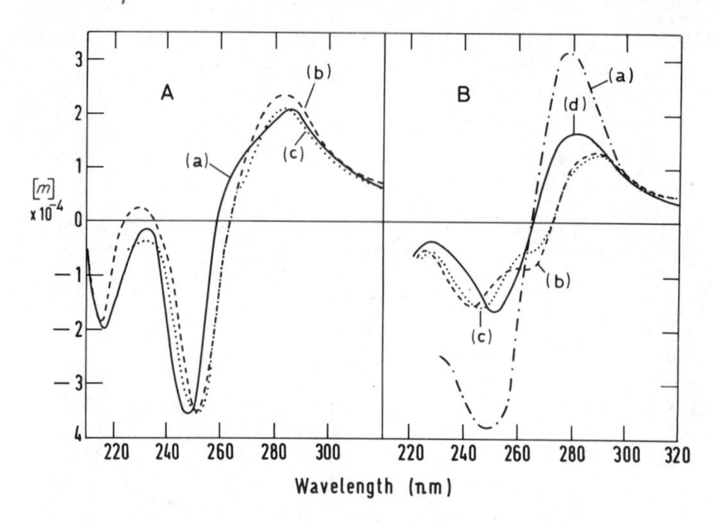

Figure 6. ORD of double-helical polyribonycleotides. A. Poly (A) · Poly (U): (a) Data of Sarkar and Yang (*61*); (b) data of Ward and Reich (*40*); (c) data of Green and Mahler (*39*). B. Poly (G) · Poly (G): (a) Data of Sarkar and Yang (*62*), on template-synthesized double helical material; (b) data of Michelson and Pochon (*63*); (c) data of Green and Mahler (*39*) on material produced by mixing poly G and poly C; (d) two-stranded viral RNA (rice dwarf virus), containing 56% AU, from Samejima et al. (*64*). Curves (a) have been used as standard data in earlier calculations, and are tabulated in Table 1.

on the square of the base composition. The use of double helices of homopolymer chains as standards with the assumption that a nucleic acid of given base composition can be simulated by a simple mixture of these, will be expected to produce dire results. With the information at present available however there is little else that can be done for RNA. The results in fact seem to be better than might have been hoped. The choice of standard data for the poly (A) · poly (U) and poly (G) · poly (C) is another problem, and it will be clear from Fig. 7 that the spread of published curves is not inconsiderable, especially for poly (G) · poly (C). The data of Yang and Samejima (*6*) were obtained on a template-synthesized double helix, and are therefore to be preferred in principle to ORD of double helices made by mixing poly G and poly C. They are the ones used by Tinoco and coworkers.

Partly-Paired RNA Chains

In order to simulate the ORD of a partly paired chain of known composition, such as a viral, ribosomal, or transfer RNA, Cantor et al. (*25*) have given the following procedure: Difference ORD curves were constructed for formation of two-stranded poly (A) · poly (U) and poly (G) · poly (C),

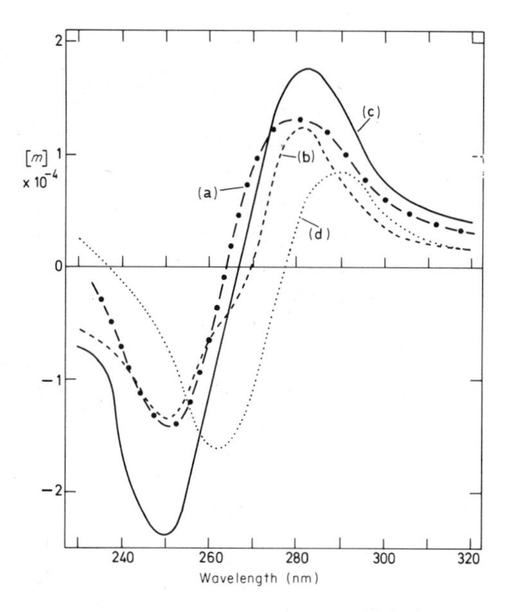

Figure 7. ORD of *E. coli* ribosomal RNA. (a) observed ORD at 0.1 M salt, (b) calculated ORD curve for 70% base pairing based on Eq. (15), (c) calculated ORD curve for the same conformation, based on separate calculations for the single-stranded and base-paired parts of the chain, (d) calculated ORD, assuming no base pairing and a random nearest-neighbor distribution, according to Eq. (14). [From Cotter and Gratzer (*31*).]

from the corresponding homopolymer strands. The ORD curve is then calculated for the given RNA from the dimer data of Table 1, and pairing is introduced by adding appropriate molar proportions of the two difference curves. The difference curves are defined as follows:

$$2[\Delta m]_{AU} = [m]_{A:U} - \tfrac{1}{2}([m]_{A_n} + [m]_{U_n})$$
$$2[\Delta m]_{GC} = [m]_{G:C} - \tfrac{1}{2}([m]_{G_n} + [m]_{C_n}) \qquad (15)$$

where $[m]_{A:U}$ and $[m]_{G:C}$ are the molar residue rotations (per phosphorus) of poly (A) · poly (U) and poly (G) · poly (C) and $[m]_{A_n}$, etc., are the molar residue rotations of the homopolymers, poly (A), etc. It may be noted that $[m]_{G_n}$ is a very dubious quantity, for poly (G) exists normally in some complex, multistranded helical form. There are arguments against the use of Eq. (15), in particular that the formation of base pairs in a real RNA from homopolymer sequences is unrealistic (*31*). In reality these will be formed from copolymeric sequences. Even assuming that the ORD of the paired chain can be expressed as a mixture of poly (A) · poly (U) and poly (G) · poly (C) the ORD of the unpaired chain cannot be seen as a mixture of poly (A), poly (U), poly (C), and poly (G). If one considers a chain containing 25% each of the four bases and calculates its ORD curve from the dimer data, assuming no pairing and a random sequence, one obtains a result which differs greatly from that obtained by mixing the curves for the homopolymers in equal proportions. The difference curves for pairing are correspondingly very different. Thus for a partially paired RNA chain, the following procedure has been suggested (*31*) as the most rational (even if it does not necessarily give a better answer): Separate calculations are performed for the unpaired part of the chain, using Eq. (14) for the appropriate base composition and for the paired part, by mixing curves for poly (A) · poly (U) and poly (G) · poly (C) (which are tabulated in Table 2). According to the fraction of bases paired, a mixture of the two components for the single- and double-stranded parts of the chain is calculated.

TABLE 2. Molar Residue Rotations of Two-Stranded Polyribonucleotides[a]

Wavelength	$[m] \times 10^{-4}$	
(nm)	Poly(A)·poly(U)	Poly(G)·poly(C)
230	-0.17	-2.45
235	-0.20	-2.65
240	-1.92	-3.31
245	-3.37	-3.75
250	-3.45	-3.75
255	-1.38	-3.54
260	0.32	-1.58
265	1.00	0.08
270	1.31	1.70
275	1.64	3.07
280	1.88	3.12
285	2.10	2.64
290	1.93	1.93
295	1.57	1.27
300	1.25	0.78
305	1.07	0.61
310	0.87	0.59
315	0.73	0.53
320	0.65	0.51

[a]Data taken from Sarkar and Yang (*61, 62*).

Figure 7 shows a set of observed and calculated curves for a ribosomal RNA (*31*), and the following observations can be made: (1) for the single-stranded stacked chain the fit with experiment is very good as before;

(2) when salt is added to the solution, and the chain assumes its normal partly paired form, the first Cotton effect peak shifts to the left and upward; (3) quantitatively the fit with the calculated curves is poor; and (4) the two ways of calculation give disparate results, that based on Eq. (15) actually giving a somewhat better fit.

A considerable number of ORD and CD curves for double-stranded RNA species have recently been appearing in the literature, and this encourages new attempts to derive suitable standards to replace those given for paired chains in Table 2. Results so far (Richards and Gratzer, unpublished data) suggest that considerable improvements in fitting the observed curves for partly paired RNA species are already possible.

Conclusions

A single-stranded RNA chain may be simulated by calculation and certainly can be recognized by ORD. In the first instance it can be compared with the calculated version, and second, if pairing can be introduced by addition of salt, as in Fig. 7, this also is strong evidence that the chain was unpaired. A simple criterion for extensive pairing is the wavelength at which the optical rotation crosses the zero line, which is much longer in an unpaired chain. The introduction of pairing into the calculation is altogether less soundly based than that of stacking only, and is at best semiquantitative. In the example given (Fig. 7) the analysis was used to show that the RNA in the ribosome could not correspond to the fully unpaired state as had at one stage been postulated. It is in this type of essentially qualitative or diagnostic sphere that the application of the calculations by Eq. (11) and (12) is often very useful.

Other Features of ORD and CD Curves

Long-Wavelength ORD

On the long-wavelength side of the Cotton effects there is of course no CD, and the ORD is a smooth curve. It can therefore be defined adequately by rotations at a few wavelengths, and before the advent of recording instruments, these were commonly the wavelengths of the lines of the mercury arc which was (and is) often used in manual instruments. For the great majority of optically active molecules the ORD in this region can be expressed by a one-term Drude equation:

$$[m] = k/(\lambda^2 - \lambda_c^2) \tag{16}$$

where k and λ_c are constants, the latter referred to as the Drude constant; λ_c is determined by the positions and magnitudes of the Cotton effects, the effect of the nearest being dominant. When the wavelength of measurement approaches the Cotton effect region, the equation breaks down. A plot of $\lambda^2 [m]$ against $[m]$ gives the value of the Drude constant. Both mononucleotides (*16*) and single- and double-stranded nucleic acids (*31a*) obey the Drude law at long wavelengths (above about 350 nm). When a two-stranded nucleic acid, such as DNA melts, there is a large drop in λ_c which is a reflection of what happens to the Cotton effects. The only likely application of the Drude analysis that one can now envisage is in monitoring samples when the Cotton effect region is for some reason inaccessible (see above). For most purposes, however, a rotation at a single wavelength, such as $[m]_D$, at least in a properly defined system, would serve equally well.

Short-Wavelength Cotton Effects

Associated with the absorption near 200 nm there are Cotton effects of considerable size, which have been examined in a number of nucleic acids, and are alleged to show greater sensitivity to conformational characteristics. Sensitivity to pH and ionic strength of ribosomal RNA Cotton effects in this region have been reported (*32*), and effects in tRNA have also been noted (*33*). This has not however been pursued, and it is likely that any increased magnitude in conformational effects in this region may be offset by the greater problems of measurement (see below).

Long-Wavelength Cotton Effects

The presence of a very small negative Cotton effect, near 310 nm in DNA, which becomes apparent only at high concentrations, was noted by Sarkar and Yang (*34*), and also a similar although somewhat larger effect in RNA at about 295 nm. Other than it disappears on melting nothing is known about its structural or spectroscopic origins.

Unusual Chromophores in Nucleic Acids

The best documented example of a nucleotide in a nucleic acid containing a chromophore with an absorption band separate from those of the common bases is 4-thiouridylic acid, which is present as a minor component in tRNA of *E. coli.* Its absorption maximum is at about 335 nm. The optical activity of this residue in single- and double-stranded situations has been studied (*34, 35*), and it has also been found that the Cotton effect associated with the long-wavelength absorption of thiouridylic acid in various tRNA species is variable in magnitude and sign, depending evidently on conformation and environment (*35*). Other resolvable optically active chromophores appear also to be present in tRNA (*36*). Many synthetic polynucleotides containing bases not found in nature have been prepared, and their optical properties examined (*37*).

Ionized States

The changes in the chromophores engendered by ionization are reflected in the optical activity. The work of Warshaw and Tinoco (*20*) on dinucleoside phosphates has demonstrated that changes in ionization state cause changes in the extent of stacking, the decrease in structure being generally much greater for acid titration. In DNA, acid- and base-induced denaturation can be followed by optical activity, and in the acid range there have been studies especially of protonation of G residues in the intact double helix (*38*) and in the denatured form.

Unusual Conformations

Optical activity has been used as a means of detecting the presence of new conformational forms of polynucleotides. An example occurs in DNA in ethylene glycol solutions, where the conformation is clearly not that of the native structure, but the enhanced optical activity in comparison with the aqueous denatured form, indicates that some different structure is present (*39*). A comparison with synthetic polynucleotides seems to exclude any explanation of this phenomenon in terms of the effects of solvation on the chromophores. This then is a different type of qualitative, diagnostic application of ORD. (Compare also the work of Tunis-Schneider and Maestre (*39a*) on CD of DNA films under varying conditions of humidity, which give rise to conformational transitions).

Another example occurs in the purine polynucleotide, polyformycin (*40*). There are a number of reasons for believing that the monomer, unlike the common nucleotides, exists preferentially in the *syn* form about the glycosidic bond. Polyformycin in the neutral single-stranded state has a system of Cotton effects at long wavelengths which are inverted compared with those of the analog, poly (A). It will however form two-stranded complexes with poly (U), and this requires that the residues be in the *anti* state. The Cotton effects of the double helices do indeed show an inversion of sign compared with the unpaired chain, which suggests that the interpretation is correct. The same kind of reasoning has been applied to dinucleoside phosphates containing enantiomers of ribose and 2′-5′-phosphodiester bonds (*40a*).

Complex Systems

One of the applications of ORD and CD has been in studying the conformation of the nucleic acid (and protein) in such systems as ribosomes, nucleohistones, nucleoprotamines, artificial nucleic acid–polybase complexes, and viruses. The use of ORD to show that the conformation of RNA in the ribosome is similar to that of the extracted material has been mentioned. Work on ribosomes has been summarized by Yang and Samejima (*6*). In nucleohistone, it appears that the DNA contribution to the CD is considerably lower than that of free DNA and it is inferred that there are changes in the geometry of the double helix when it forms the nucleohistone structure (*41*). In complexes of nucleic acids with protamines or basic polypeptides, such as polylysine, other effects are observed, and polylysine complexes especially show highly anomalous ORD and CD, the Cotton effects of the DNA being inverted and greatly increased in amplitude (*42, 43*). These systems however contain large particles, and the possibility exists that the observed phenomena are essentially optical in character, arising from internally oriented structures, possibly of a liquid crystalline type, which are known (*44*) to be capable of generating very large optical activity. Smaller changes in ORD are observed in complexes of other polynucleotides with polybases and DNA with protamines.

In spherical viruses it appears that the RNA conformation again resembles that of the free viral RNA (*45*), but in the DNA phages a completely new situation is observed. Maestre and Tinoco (*46*) have obtained very striking results: The magnitudes of the DNA Cotton effects differ greatly from those of the free DNA, and are shown to be linearly related to the concentration of DNA in the phage head. It is likely that this is related to the effect of packing on the DNA conformation, which possibly involves super-coiling (as probably occurs in nucleohistone) with consequent changes in the pitch of the primary helix. Such an effect on the tightness of super-coiling is observed in circular DNA in concentrated salt solutions, and can perhaps be correlated with diminutions in Cotton-effect amplitudes at high salt concentrations (*47*). The phenomenon may find wide applications in virus work.

Extrinsic Chromophores

When chromophoric ligands, which include many antibiotics, and mutagens as well as metal ions, are bound to nucleic acids, they in general acquire optical activity. Cases in which Cotton effects in the visible region, associated with such ligands, have been recorded are too numerous to list. Where ligands bind in different modes, e.g., intercalation, and external binding, or binding of the isolated ligand and association of ligands on adjoining sites, Cotton effects corresponding to these different modes can often be resolved, and used to analyze the different binding characteristics. A good example can be found in the work of Gardner and Mason (*48*). The binding of aminoacridine dyes and the spectroscopic manifestations has been reviewed by Blake and Peacocke (*49*). A discussion of this complex area is outside the sphere of this review.

Experimental Considerations

Two good concise accounts of the principles of measurement of ORD and CD are by Duffield and Abu-Shumays (*50*) and Woldbye (*51*). It will suffice here merely to recall that good modern high-performance instruments will measure rotations in the range of milli-degrees, or absorbance differences in CD down to about 10^{-5}. It is therefore clear that no liberties can be safely taken with the experimental conditions. The following are some critical considerations, assuming always that the instrument is operating within its specifications, and that the arc and electronics for example are both stable.

Cells

Cells must be selected for low strain. The instrument manufacturers will in general supply handpicked cells (though it is frequently much cheaper to obtain them directly from the cell manufacturers). It is usually possible to find ordinary spectrophotometer cells with low strain. A preliminary selection can be made by viewing the cell against a bright light between crossed polaroids. Thereafter the only criterion for a good cell is a low cell zero at all wavelengths. Care should be taken to mount the cell in a reproducible orientation in the cell housing. The cells may be cleaned by various methods, including when necessary, heating with a mixture of sulfuric and nitric acids.

For any work involving heating of the sample it is especially important to use cells with low strain. For this purpose and in general, it is better to avoid the use of demountable cells of any kind, which can now be regarded as obsolete. Full thermal equilibration is important, for if any refractive index gradients are set up in the sample, the measurement of meaningful rotations becomes impossible.

Solutions

It is highly desirable that solutions be free from scattering material and bubbles, for the least that can be expected if these conditions are not fulfilled is a large increase in noise. It is advisable to centrifuge or filter all solutions.

Samples

The upper absorbance limit that can be tolerated depends on the wavelength and the condition of the instrument (arc output, mirror reflectivity), but in general something in the range 1 to 1.8 is expected. The necessity to operate at the highest levels arises of course only when the ratio of $[m]$ or of $[\theta]$ to the molar absorptivity, ϵ, is low. The higher the concentration the larger the measured parameter, but the smaller the signal, since a large proportion of the light is being absorbed. A balance must therefore be found to give the best signal-to-noise ratios. Moreover, the higher the absorbance the greater the slit-width that is required, and this can lead to problems also.

Spectral Band Width

The relation between the half-band width sampled by the instrument, and the wavelength and slit width should be given in the instrument specifications. A large spectral band width causes low readings on sharp extrema and on steeply-sloping edges of Cotton effects, by exact analogy with spectrophotometry (*52*). Working at high absorbances, this can become a serious problem, and it may be advisable to measure the rotations or CD at a series of slit widths, and extrapolate to zero. This problem is most acute at long wavelengths, where the resolution of prism monochromators is low. In extreme cases it is even possible to generate completely spurious Cotton effects when the band width is large compared with the rate of change of rotation with wavelength (*53*).

Scanning Speed

It is obviously necessary to ensure that the rate of change of the parameter as the scan proceeds is well below the response time of the electronic system. Again it is advisable to ensure that the results are not dependent on the time constant of the circuit, or the scanning rate. This becomes especially relevant when it is desired to suppress noise by operating at long time constants.

Stray Light

This is in general the factor that limits the upper absorbance range that is permissible, although the inability of the system to respond to very low signal levels can also be limiting. Stray light has been much discussed, and can give rise to especially disagreeable artifacts in optical rotation. In brief, any monochromator produces some fraction of its output outside the nominal pass-band. This level is specified by the manufacturers, and can be very low especially in instruments with double monochromators. As the absorbance of the sample in the pass-band, say in the ultraviolet, is increased, so the proportion of stray light, most of which will be at longer wavelengths (depending on the energy distribution of the source output) and is therefore perhaps completely unabsorbed by the sample, increases. Thus at a sample absorbance of 2 (1% transmittance) the fraction of stray light, if unabsorbed, that reaches the detector is one hundred times greater than in the absence of a sample, and ten times greater than when the absorbance is 1. In polarimetry the rotation will increase as the scan enters the optically active absorption band. If stray light is important, it will make up an increasing part of the signal as the absorbance increases. Since the stray light is of longer wavelengths, further from the Cotton effect and therefore corresponds to lower rotation, the recorded rotation will drop, thus presenting the appearance of the first limb of a Cotton effect. Scanning through the band, the absorbance will then begin to drop beyond the peak, the fraction of stray light to diminish, and the recorded rotation therefore to rise. A demonstration of how completely spurious Cotton effects can be generated by this means, and by introducing an optically inactive absorbing material, has been given by Urnes and Doty (*54*). It would be invidious to point out any of the numerous papers in the literature based on this artifact.

Just as the simplest test for stray light effects in spectrophotometry is to examine the system for adherence to Beer's law, so it is always desirable, when working at high absorbances, or near the limits of performance of the instrument, at very short or long wavelengths, where stray light is greatest, to perform an analogous test. The rotation or CD should be measured at two or more concentrations or, more simply, pathlengths, and it should be ascertained that the ratio of the readings is also that of the pathlengths. A plot of reading against absorbance at a given wavelength will at some (it is hoped, high) value of the absorbance deviate from linearity (cf. Ref. *52*).

Measurements at Short Wavelengths

If it is desired to penetrate into the short-wavelength region below about 210 nm, a certain amount of trouble has to be taken. The instrument must be in good condition, since oxidized mirror surfaces or coated arc windows have a catastrophic effect in this region. The quality of the silica cells must also be high. Energy is low, and the danger of stray light artifacts much increased. Moreover below 200 nm it is necessary to flush oxygen-free (and oil-free) nitrogen rapidly through the entire light path to eliminate as much oxygen as possible from the beam, because oxygen has strong absorption bands below about 198 nm. The precautions that have to be taken are much the same as in spectrophotometry, and these have been fully described elsewhere (*55*). The practical limit for aqueous solutions is about 185 nm (which is in any case the limit for currently available instruments), and the water absorption band spreads toward longer wavelengths on heating. To limit solvent absorption, cells of short path should be used. Some gain can be achieved by the use of D_2O as solvent. Few nonaqueous solvents can be used below 200 nm. They include paraffins, acetonitrile, trifluoroethanol, and similar substances, trimethylphosphate, and pure sulfuric and methanesulfonic acids. Transparent anions include fluoride, perchlorate, and sulfate. A more complete discussion is given in Ref. *55*. Tests for stray light artifacts are essential in all work in this region.

Scattering Solutions and Oriented Samples

Scattering causes depolarization of light, but this does not appear in itself to be a limiting factor. The limited proportion of scattered light from turbid solutions that reaches the detector is liable to cause flattening of the absorption bands, and some red-shifts, as it may in spectrophotometry. This has been considered by Urry and Ji (*56*). Moreover trouble can arise from the "sieve" effect, in which a dense particle in the beam masks the material in the cone behind it (*56*). Of probably greater importance is the effect of differential scattering of left- and right-circulating polarized light (*57*). Some striking artifacts in the CD of scattering systems have been observed (*57, 57a*). Results obtained on turbid systems should be interpreted very circumspectly for this reason. Measurements should certainly at the very least be made at several concentrations. It should be noted that the mere introduction of scattering per se into the light path is not equivalent to measuring the optical activity of an intrinsically particulate material.

Another means of courting difficulties is to work with anisotropic samples of any kind. It is a well-known phenomenon that when birefringence is present, for example in a gel of fibrous molecules in which local orientation by shear readily occurs, large and erratic rotations or CD are often observed. Very large optical rotations have long been known to arise from cholesteric liquid crystals, and this effect has been studied in some detail in α-helical synthetic polypeptides (*44*). It seems likely that the large optical activities observed in DNA-polylysine complexes (*42, 43*) are explicable in such terms (*57b*). A simple idealized case of an oriented, uniaxial sample with its optical axis perpendicular to the beam has been treated by Disch and Sverdlik (*58*). They show that the CD in such a case can exhibit diminished intensity, red or blue shifts, and the appearance of small spurious bands.

In theoretical terms the surface of the complex problem of optical activity in oriented systems of various kinds (including locally oriented ones, where the domain dimensions are appreciable compared with the wavelength of light) has scarcely yet been penetrated. There may well be many situations in which measurements on such systems are of special interest, and the object here is only to advise caution.

Acknowledgment

I am very grateful to Professor Myron M. Warshaw for making the contents of Table 1 available.

References

1. Moscowitz, A. 1960. In C. Djerassi (Ed.), Optical rotatory dispersion, McGraw-Hill, New York. Chap 12.
2. Moscowitz, A. 1962. *Adv. Chem. Phys.*, **4**: 67.
3. Caldwell, D. W., H–C. Liu, and H. Eyring. 1968. *Chem. Rev.*, **68**: 525.
4. Mason, S. F. 1963. *Quart. Rev.*, **17**: 20.
5. Schellman, J. A. 1968. *Acc. Chem. Res.*, **1**: 144.
6. Yang, J. T. and T. Samejima. 1969. *Progr. Nucleic Acid Res.*, **9**: 224.
7. Lowry, T. M. 1935. Optical rotatory power, reissued by Dover Publications, New York, 1964.
8. Crabbé, P. 1965. Optical rotatory dispersion and circular dichroism in organic chemistry, Holden-Day, San Francisco.
9. Urry, D. W. 1968. *Ann. Rev. Phys. Chem.*, **19**: 477.
10. Ts'o, P. O. P., G. K. Helmkamp, and C. Sander. 1962. *Proc. Natl. Acad. Sci U.S.*, **48**: 686.
11. Sober, H. A. (Ed.). 1970. Handbook of biochemistry, Chemical Rubber Co., Cleveland, 2nd Edition, J-264.
12. Tinoco, I. 1965. In (B. Pullman and M. Weissbluth (Eds.)), Molecular biophysics, Academic Press, New York.
13. Emeis, C. A., L. J. Oosterhoff, and G. de Vries. 1967. *Proc. Roy. Soc., Sci. A* **287**: 54.
14. Carver, J. P., E. Schechter, and E. R. Blout. 1966. *J. Am. Chem. Soc.*, **88**: 2550.
15. Tsong, T. Y. and J. M. Sturtevant. 1969. *J. Am. Chem. Soc.*, **91**: 2382.
16. Yang, J. T., T. Samejima, and P. K. Sarkar. 1966. *Biopolymers*, **4**: 623.
17. Emerson, T. R., R. J. Swan, and T. L. V. Ulbricht. 1967. *Biochemistry*, **6**: 843.
18. Miles, D. W., R. K. Robins, and H. Eyring. 1967. *Proc. Natl. Acad. Sci. U.S.*, **57**: 1138; *J. Phys. Chem.*, **71**: 3931.
19. Tinoco, I. 1968. *J. Chim. Phys.* **65**: 91.
20. Warshaw. M. M. and I. Tinoco. 1966. *J. Mol. Biol.*, **20**: 29; Warshaw, M. M., Ph.D. Thesis, University of California, Berkeley, 1966.
21. Brahms, J., J. C. Maurizot, and A. M. Michelson. 1967. *J. Mol. Biol.*, **25**: 481.
22. Simpkins, H. and E. G. Richards. 1967. *J. Mol. Biol.*, **29**, 349.
23. Inoue, Y., S. Aoyagi, and K. Nakanishi. 1967. *J. Am. Chem. Soc.*, **89**: 5701.
24. Cantor, C. R. and I. Tinoco. 1967. *Biopolymers*, **5**: 84.
25. Cantor, C. R., S. R. Jaskunas, and I. Tinoco. 1966. *J. Mol. Biol.*, **20**: 39.

25a. Cantor, C. R., M. M. Warshaw and H. Shapiro. 1970. *Biopolymers*, **9**: 1059

26. Vournakis, J. N., D. Poland, and H. A. Scheraga. 1967. *Biopolymers*, **5**: 403.
27. Adler, A., L. Grossman, and G. D. Fasman. 1967. *Proc. Natl. Acad. Sci. U.S.*, **57**: 423.
28. Gratzer, W. B., L. R. Hill, and R. J. Owen. 1970. *European J. Biochem.*, **15**: 209.
29. McMullen, D. W., S. R. Jaskunas, and I. Tinoco. 1967. *Biopolymers*, **5**: 589.
30. Felsenfeld, G., S. Z. Hirschman. 1965. *J. Mol. Biol.*, **13**: 407.
31. Cotter, R. I. and W. B. Gratzer. 1967. *Nature*, **221**: 154.

31a. Fresco, J. R. 1961. *Tetrahedron*, **18**: 185

32. Wolfe, F. H., K. Oikawa, and C. M. Kay. 1969. *Biochemistry,* **7:** 3361.
33. Lamborg, M. R. and P. C. Zamecnik. 1965. *Biochem. Biophys. Res. Commun.,* **20:** 328.
34. Sarkar, P. K., B. Wells, and J. T. Yang. 1967. *J. Mol. Biol.,* **25:** 563.
35. Scheit, K. H. and W. Saenger. 1969. *FEBS Letters,* **2:** 305.
36. Scott, J. F. and P. Schofield. 1969. *Proc. Natl. Acad. Sci. U.S.,* **64:** 931.
37. Massoulié, J. and A. M. Michelson. 1966. *Biochim. Biophys. Acts,* **144:** 16.
38. Luck, G., C. Zimmer, and G. Snatzke. 1968. *Biochim. Biophys. Acta,* **169:** 548 Courtois, Y., P. Fromageot, and W. Guschlbauer. 1968. *European J. Biochem.,* **6:** 493.
39. Green, G. and H. R. Mahler. 1970. *Biochemistry,* **9:** 368.
39a. Tunis-Schneider, M. J. B. and M. F. Maestre. 1970. *J. Mol. Biol.* **52**: 521.
40. Ward, D. C. and E. Reich. 1968. *Proc. Natl. Acad. Sci. U.S.,* **61:** 1494.
40a. Tazawa, I., S. Tazawa, L. M. Stempel, and P. O. P. Ts'o. 1970. *Biochemistry*, **9:** 3499.
41. Permogorov, V. I., V. G. Debabov, I. A. Sladovka, and B. A. Rebentish. 1970. *Biochim. Biophys. Acta*, **199:** 556; R. T. Simpson and H. A. Sober. 1970. *Biochemistry*, **9:** 3103; T. Y. Shih and G. D. Fasman. 1970. *J. Mol. Biol.*, **52:** 125.
42. Cohen, P. and C. Kidson. 1968. *J. Mol. Biol.,* **35:** 241.
43. Shapiro, J. T., M. Leng, and G. Felsenfeld. 1969. *Biochemistry* **8:** 3219.
44. Robinson, C. 1961. *Tetrahedron,* **13:** 219.
45. Oriel, P. and J. A. Koenig. 1968. *Arch. Biochem.,* **127:** 274; H. Isenberg, R. I. Cotter, and W. B. Gratzer. 1971. *Biochim. Biophys. Acta,* **232:** 184.
46. Maestre, M. F. and I. Tinoco. 1967. *J. Mol. Biol.,* **23:** 323.
47. Tunis, M. J. B. and J. E. Hearst. 1968. *Biopolymers,* **6:** 1218.
48. Gardner, B. J. and S. F. Mason. 1967. *Biopolymers,* **5:** 79.
49. Blake, A. and A. R. Peacocke. 1968. *Biopolymers,* **6:** 1225.
50. Duffield, J. J. and A. Abu-Shumays. 1966. *Anal. Chem.,* **38:** 29A.
51. Woldbye, F. 1967. In G. Snatzke (Ed.), Optical rotatory dispersion and circular dichroism in organic chemistry, Heyden, London, Chap. 5.
52. Beaven, G. H. and E. A. Johnson. 1961. In G. H. Beaven, E. A. Johnson, H. A. Willis, and R. J. G. Miller (Eds.), Molecular spectroscopy, Heywood, London.
53. Wyss, H. R. and H. H. Günthard. 1966. *J. Opt. Soc. Am.,* **56:** 888.
54. Urnes, P. and P. Doty. 1961. *Advan. Protein Chem.,* **11:** 401.
55. Gratzer, W. B. 1968. In G. D. Fasman (Ed.), Poly-α-amino acids, Dekker, New York, Chap. 5.
56. Urry. D. W. and T. H. Ji. 1968. *Arch. Biochem.,* **128:** 802.
57. Urry. D. W. and J. Krivacic. 1970. *Proc. Natl. Acad. Sci. U.S.,* **65:** 845.
57a. Ji, T. H. and D. W. Urry. 1969. *Biochem. Biophys. Res. Commun.,* **34:** 82.
57b. Haynes, M., R. A. Garrett, and W. B. Gratzer. 1970. *Biochemistry,* **9:** 4410.
58. Disch, R. L. and D I. Sverdlik. 1969. *Anal. Chem.,* **41:** 82.
59. Van Holde, K. E., J. Brahms, and A. M. Michelson. 1965. *J. Mol. Biol.,* **12:** 726.
60. Brahms, J., J. C. Maurizot, and A. M. Michelson. 1967. *J. Mol. Biol.,* **25:** 465.
61. Sarkar, P. K. and J. T. Yang. 1965. *J. Biol. Chem.,* **240:** 2088.
62. Sarkar, P. K. and J. T. Yang. 1965. *Biochemistry,* **4:** 1238.
63. Michelson, A. M. and F. Pochon. 1969. *Biochim. Biophys. Acta,* **174:** 604.
64. Samejima, T., H. Hashizume, K. Imahori, I. Fujii, and K. I. Miura. 1968. *J. Mol. Biol.,* **34:** 39.

APPLICATION OF FLUORESCENCE TECHNIQUES TO THE STUDY OF NUCLEIC ACIDS

CHARLES R. CANTOR*
DEPARTMENTS OF CHEMISTRY AND BIOLOGICAL SCIENCES
COLUMBIA UNIVERSITY
NEW YORK, NEW YORK

TERENCE TAO†
DEPARTMENT OF MOLECULAR BIOPHYSICS AND BIOCHEMISTRY
YALE UNIVERSITY
NEW HAVEN, CONNECTICUT

INTRODUCTION

Phenomenology

Fluorescence is the radiation emitted when a molecule in an excited singlet electronic state undergoes a transition to a lower energy singlet state. For most common molecules, the radiation emitted lies in the ultraviolet or visible region of the spectrum. The initial state is almost always the lowest excited singlet state and the final state an excited vibrational level of the ground singlet state. In this chapter, attention will be restricted almost

* Alfred P. Sloan fellow.
† Helen Hay Whitney fellow.

entirely to a discussion of the ways in which this radiation can be used to obtain information about the structure and properties of nucleic acid systems. All the techniques that will be described can usually be performed on samples maintained in the range of temperatures and solvents characteristic of those in living organisms. For discussion of other luminescence techniques which are usually restricted to more severe environments, such as low temperature rigid glasses, the interested reader is referred elsewhere (*1*). For similar reasons, any references to emission spectroscopy in other regions of the spectrum, such as infrared fluorescence, have been omitted.

A number of different aspects of the fluorescent light emitted by a sample are potential sources of information for the study of biological molecules. These include: (1) The spectral distribution of the emitted light. (2) The spectral distribution of absorbed light capable of causing a subsequent emission. (3) The degree of polarization of the emitted light. (4) The overall intensity of the emitted light usually measured as that fraction of the absorbed photons which will be reemitted. (5) The time dependence of light emission. (6) The time dependence of polarized components of light emission. The ways in which each of these aspects of fluorescent light can be detected and used will be described briefly in this chapter. For more detailed information about particular techniques or for a more general overview of the field of emission spectroscopy, the interested reader is referred to a number of excellent recent books and review articles (*2–12*).

Survey of Applications

Fluorescence spectroscopy is a relatively recent addition to the arsenal of techniques that have been used to study nucleic acids. In contrast to the field of protein chemistry, where there have been a wide variety of types of applications of fluorescence techniques (*9, 12*), there is relatively little published data showing many of the possible applications of fluorescence spectroscopy to nucleic acid research. If this discussion were restricted entirely to these proven applications to nucleic acids, it would present far too narrow a view of the potential of fluorescence spectroscopy. Thus, a number of the examples we have chosen to discuss are taken from work on proteins. The past success in this field should serve to stimulate similar work on polynucleotides. In this section, we will survey a number of the different ways in which fluorescence spectroscopy can yield information about nucleic acid or protein structures. With several important exceptions, most of the types of information described below can be obtained by alternative methods which do not involve fluorescence radiation. There are, however, two general advantages of using fluorescence techniques instead of some of the alternatives. With rare exception, fluorescence spectroscopy requires smaller sample sizes than any other physical technique which has been used to study molecular structure. This is a substantial advantage in many fields of biochemistry where the difficulty of obtaining a suitable sample often ranks second to none in any of the problems encountered in research. The second real advantage of fluorescence techniques is that, quite often, many different types of data can be obtained from a single fluorescent sample, sometimes even from a single measurement. The versatility and high sensitivity of fluorescence makes it an especially useful technique for studying large, complex biological systems such as cellular organelles, membranes, and large protein aggregates. The fact that most protein and nucleic acid samples do not themselves have strong natural fluorescence in most of the visible region of the spectrum might at first seem quite a disadvantage. However, the absence of this natural background is a tremendous asset when added fluorescent labels are used to obtain information about specific regions of a complex, high-molecular-weight system.

Assay of Nucleic Acids

The natural fluorescence of most nucleic acids is so weak as to make them almost undetectable by this method. There exist, however, a number of fluorescent dyes which bind rather specifically to nucleic acids. One of these, ethidium bromide, is very weakly fluorescent in aqueous solution. When the dye is bound to nucleic acids, however, a more than 20-fold enhancement in fluorescence intensity occurs. Taking advantage of this fact, LePecq and Paoletti have devised a very sensitive method for assaying the presence of nucleic acids in complex mixtures (*13*). However, ethidium bromide binds both to RNA and DNA, thus obviating the possibility of its use in a specific assay.

Fluorescence spectroscopy has also been used to assay particular species of transfer RNA. There are a number of unusual nucleosides in various tRNA species which show weak fluorescence, as will be discussed later, but only one of these, the Y base present in the $tRNA^{Phe}$ of certain organisms, shows strong fluorescence in the visible region of the spectrum. Yoshikami et al. have reported that a very satisfactory assay for the presence of $tRNA^{Phe}$ can be developed by monitoring the visible fluorescent light when the sample is excited near 325 nm (*14*). One must be quite careful, however, to use rigidly controlled conditions of temperature and salt concentration in making these measurements since the fluorescence of the Y base is extremely sensitive to changes in environment (*15, 16*).

Binding of Small Molecules

There are a number of different ways in which fluorescence can be used to study the interaction of small molecules with nucleic acids. If the small molecule itself is fluorescent, then usually, upon binding, there will either be changes in fluorescence intensity, shifts in the absorption spectrum or the emission spectrum, or changes in fluorescence polarization. All of these can be exploited in ways quite parallel to those used for following binding by changes in visible or ultraviolet absorption spectroscopy or circular dichroism. By studying the observed fluorescence as a function of the ratio of small molecule to large molecule, information about the number of binding sites and the binding constants associated with these sites can be obtained. For example, LePecq and Paoletti have used the fluorescence of ethidium bromide to monitor the binding of this dye to various nucleic acids (*17*). Bittman has made a detailed study of the interaction of ethidium bromide with unfractionated tRNA (*18*). We have used fluorescence spectroscopy to analyze the binding of this dye to a sample of purified yeast $tRNA^{Phe}$ (*19*). There are times when fluorescence polarization can be a more effective monitor of binding than just fluorescence intensity. An extensive discussion of the use of this technique to study the binding of proflavine to DNA has been given by Ellerton and Isenberg (*20*).

The use of fluorescence offers a number of additional advantages in studying the binding of dyes to nucleic acids. Using an external quencher, the determination of free and bound dye can sometimes be greatly simplified (*21*). The fact that the shape of the fluorescence spectrum, fluorescence polarization, and quantum yield or fluorescence lifetime can be measured on the same set of samples offers a very sensitive way to distinguish the number of different types of binding sites. Fluorescence lifetime measurements are particularly useful in this regard because it is much easier to note a deviation from a single exponential decay than it might be to resolve two almost linearly dependent emission spectra.

In many cases it may be possible to get more information than just the number of bound dyes and the number of distinct classes of binding sites. The use of dyes such as ethidium bromide or 1-anilino-8-naphthalene sulfonate permits considerable information to be obtained about the nature of the environment of the bound dye (*12*). Both of these dyes are strongly fluorescent in relatively nonpolar media and their fluorescence is mostly quenched in water. Many fluorescent compounds exist with the opposite properties (*2*). If more than one dye is bound to a single macromolecule it may be possible to determine the proximity of these dyes, either by energy transfer techniques to be described below or, in an unusually favorable case, even by isolating a photodimer of two dyes that were initially bound very closely to each other (*22*). If the overall shape of the macromolecule is known, then studies of fluorescence polarization or decay of fluorescence anisotropy may permit considerable information to be learned about the orientation of the dye with respect to the macromolecule (*23*). This information is especially useful when the direction of the electronic transition moment of the dye is known.

If the macromolecule itself is fluorescent, then the binding of small molecules to the macromolecule may often be followed, since these will induce changes in the fluorescence of the macromolecule. For example, when ethidium bromide is bound to $tRNA^{Phe}$ from yeast the enhancement of the fluorescence of the dye is followed almost exactly by a corresponding quenching in the fluorescence of the Y base (*19*). Not only can the characteristics of the binding be determined from either set of data alone, but taken together they permit some information about the relative proximity of the dye and the fluorescent base, Y, to be obtained.

Determination of Molecular Conformation

Since most nucleic acids are not naturally fluorescent, we shall concentrate our attention on the types of conformational information obtainable by the use of fluorescent probes. Some of the ways in which they provide information about the macromolecule host are shown schematically in Fig. 1. As mentioned above, many fluorescent dyes can be used to give information about the nature of the environment in which they are found. Since fluorescence is quenched by many types of molecules present in

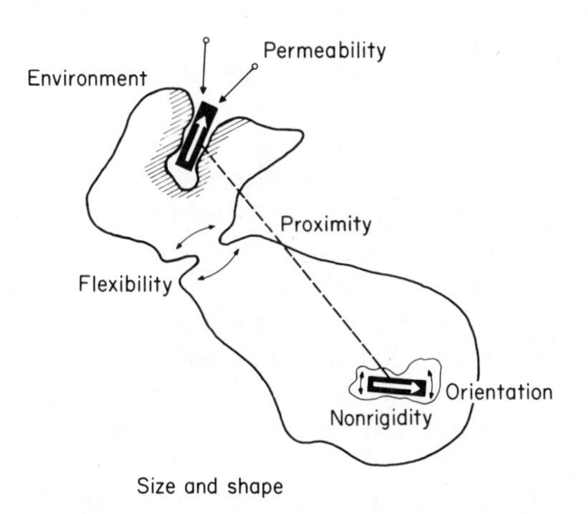

Figure 1. Schematic drawing of a number of the ways in which fluorescent dyes can be used to obtain information about the properties of macromolecules.

solution, it is often possible to assess whether a bound fluorescent probe is in a region accessible to solvent. A particularly elegant example of this approach was recently demonstrated by Vaughan and Weber who bound pyrene butyric acid to proteins and then exposed the solutions to oxygen and followed the extent to which the fluorescence of the bound hydrocarbon was quenched (*24*). Similar studies can be performed by taking advantage of the fact that D_2O often stimulates the fluorescence of many aromatic hydrocarbons (*25*). In this way, some insight about the degree of exposure of the Y base of $tRNA^{Phe}$ to the solvent could be obtained as a function of other environmental conditions (*15*). The possibility of using such subtle environmental variables as oxygen concentration and presence of deuterium instead of hydrogen offers one of the very strong advantages of fluorescence techniques.

Once a fluorescent probe has been bound to a macromolecule, it can be used to provide considerable information about the size and shape of the macromolecule. Fluorescence polarization measurements have long been used for this purpose (*26*). More recently, the advent of the technique of decay of fluorescence anisotropy has permitted far more detailed information to be obtained. This technique will be described in detail in the last section of this article. It is sufficient to say for the present that, in a very favorable case, it is possible to obtain from a single set of measurements, the axial ratio of a macromolecule containing a bound dye, the orientation of the dye relative to the principal axes of an equivalent ellipsoid of the macromolecule, and degree of hydration of the macromolecule, if the molecular weight is known (*23*). Few techniques offer such a wealth of information from one set of determinations. Furthermore, the direct study of the decay of fluorescence anisotropy eliminates what used to be the troublesome need to measure polarization as a function of solvent viscosity.

If the macromolecule on which a dye is bound is not rigid, then the results obtained by measurements of fluorescence polarization or time dependent anisotropy represent the size and shape of that part of the structure adjacent to the location of the dye which is effectively rigid. If the overall size and shape of the macromolecule are known, then placing a dye at specific places on the macromolecule can permit a detailed survey of just which parts are flexible and which parts are not. This technique has been used with great success by Stryer and his coworkers to determine the flexibility of parts of the immunoglobulin molecule (*27*). Studies with fluorescent groups located in different parts of the tRNA molecule have similarly permitted information about the flexibility of regions near the anticodon loop and near the CCA end to be obtained (*15, 19, 28, 29*).

All of the techniques mentioned thus far have their counterparts in nonfluorescent methods. But there is one aspect of fluorescence, namely resonant energy transfer (*12*), which provides information of a type usually expected only from total structure determination by x-ray crystallography. If two fluorescent groups are present in a macromolecule, it may be possible to observe singlet-singlet (*30*), triplet-singlet (*31*), or triplet-triplet (*32*) energy transfer between them. Only the former technique will be discussed in this article because it is the only one which would not normally require the presence of low temperature solid matrices for observation. Given certain assumptions, which will be discussed in detail in a later section, it may be possible from a measurement of singlet-singlet energy transfer to determine the distance between the two fluorescent groups. While other techniques exist for measuring relatively short distances such as electron paramagnetic resonance of a bound biradical, or chemical cross linking, the technique of singlet-singlet energy transfer is capable of measuring intramolecular distances of the order of 60 Å. Such information is useful, of course, not only in determining the conformation of large macromolecules but also in determining the location of bound fluorescent groups in studies of small molecule interactions with polymers.

Conformational Changes and Complex Interactions

Many techniques exist for following large conformational changes such as major alterations in secondary structure when environmental conditions are changed or when two macromolecules are allowed to interact. If the macromolecule contains a bound fluorescent label, then the change in fluorescence can be used to follow specific conformational changes in a particular region. This is especially useful if the fluorescent derivative is a modified substrate or other ligand known to be located in regions of the macromolecule where functionally significant conformational alterations are likely to take place. Since some fluorescent base analogs do exist, it is possible to synthesize fluorescent polynucleotide chains. These may be used to follow more general conformational changes such as those that take place when a double helix is formed. They should also be very useful in following the interaction of nucleic acids with many of the enzymes involved in nucleic acid metabolism. The analysis of complex conformational changes is often hindered by the lack of detailed information about what is happening in different parts of the macromolecule. Thus, the use of fluorescently labeled nucleic acids offers a considerable adjunct to the types of studies which have been carried out in the past largely by optical absorption measurements, circular dichroism and hydrodynamic studies. In fact, some of the first major successes of fluorescence spectroscopy in nucleic acid research have dealt with studies of conformational changes at the CCA end of tRNA and compared these with changes of the macromolecule as a whole (*28*).

All of the methods described thus far have dealt with samples whose properties were constant with time. However, it is perfectly possible to use any of the above named techniques to study kinetic processes. In fact, the measurement of time dependence of fluorescence affords, in principle, the possibility of following rates as fast as 10^9 sec^{-1}. Kinetics of slow processes can be followed by fluorescence spectroscopy in exactly the same way that absorption spectroscopy or circular dichroism is used. Reactions with millisecond half-lives can be observed by fluorescence stopped-flow techniques. Schechter, Chen, and Anfinsen have recently used these to follow conformational changes in micrococcal nuclease (*33*). A number of designs for workable fluorescence stopped-flow apparatus have been published (*34, 35*). Still faster processes should be observable by fluorescence temperature jump techniques, although these have not yet been well developed.

One of the great puzzles in nucleic acid biochemistry is the nature of specific interactions between nucleic acids and proteins. Fluorescence spectroscopy seems at present to be one of the most likely techniques for examining the nature of these specific interactions. Studies on the change of protein fluorescence when nucleic acids are added and on the changes in fluorescence of base analogs in the presence of proteins should enable fairly detailed information about the structure of protein-nucleic acid complexes to be determined. Some progress has already been made by Helene et al. in following, by fluorescence, the interaction between valyl-tRNA synthetase and valine-specific tRNAs (*36*). Many more studies of these types will probably develop in the near future.

From the brief summary of possible applications of fluorescence given above, it should be clear that many of these are dependent on the presence of a specifically attached fluorescent probe to overcome the fact that nucleic acids themselves do not have very useful fluorescence properties. The current state of the art in finding and covalently or noncovalently attaching such probes will be discussed in a later section. Some experiments require the presence of more than one probe and these are more difficult to perform than those involving only one. One consolation for these difficulties in the fact that, with typical fluorescence equipment, it is possible to work with only 10^{-7} M nucleic acid and still observe the presence of a single fluorescent dye bound per mole of nucleic acid.

In the sections that follow, we shall first describe in detail the precise nature of the quantities which characterize the fluorescence spectrum of a molecule. Traditional and new experimental techniques for determining these quantities will then be discussed. After a survey of the types of fluorescent chromophores that are known to be useful for the study of nucleic acids, particular attention will be given to two fluorescence techniques, singlet-singlet energy transfer and fluorescence polarization studies. These are the two techniques which differ most radically from traditional absorption spectroscopy. Most of the others that have been mentioned above can be understood simply as variations on well-established methods of spectroscopic structure determination.

PRINCIPLES OF MOLECULAR LUMINESCENCE

Molecular Energy States

For a typical organic molecule like anthracene, the energy states and the transitions between them are best summarized by the Jablonski diagram (Fig. 2). In this diagram, each horizontal line designates a discrete molecular state. The states are first classified by spin as singlet states (S) and triplet states (T). The subscripts 0, 1, 2, . . . , n designate an ascending order in electronic energy. The symbols $S_0, S_1, S_2, \ldots, S_n, T_1, T_2, \ldots, T_n$ thus

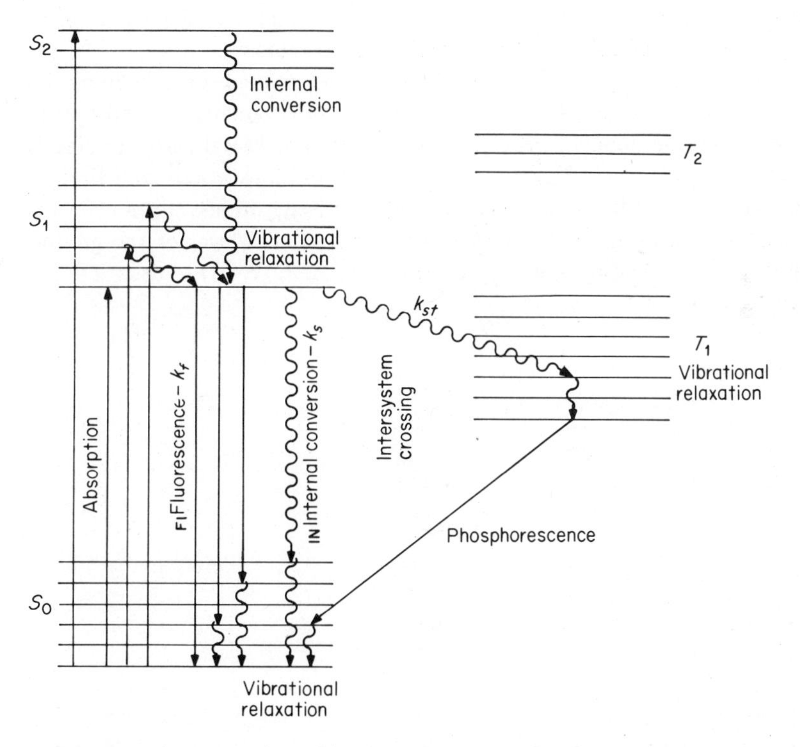

Figure 2. Jablonski diagram depicting molecular energy states. The symbols in this figure are defined in the text and in Table 1.

represent states characterized by the electronic energy and spin. Associated with each electronic state is a set of finely spaced vibrational sublevels corresponding to the discrete vibrational modes of the nuclei. In quantum mechanical terms, this assumes that a molecular wavefunction can be separated into a product of the electronic spin wavefunction, the electronic orbital wavefunction and the atomic vibrational wavefunction.

The electronic configuration depicts how the electronic orbitals are filled by the electrons of the molecule. Consider, for example, the π electron orbitals of the anthracene molecule. In the ground state (Fig. 3a) the electrons fill the lowest lying orbitals. In accordance with the Pauli exclusion principle, each orbital is filled by two electrons with their spins opposed, such that the molecule possesses no net spin. The state is therefore a singlet state, and is designated as S_0. In an excited state, a single electron is promoted to a higher energy orbital (Figs. 3b, 3c). In this situation, the promoted electron and the remaining electron can have spins either antiparallel (Fig. 3b) or parallel (Fig. 3c) to each other. The first case gives rise to an excited singlet state, and the latter to an excited triplet. Thus, in almost all cases, one would find the ground state to be singlet, and all excited states to exist in pairs of singlet and triplet states. Note that a triplet state is always slightly lower in energy than a singlet state, in accordance to Hund's rule.

The population of the vibrational levels is determined statistically by the Boltzmann distribution. At room temperature, the Boltzmann distribution heavily favors the population of the lowest vibrational level (or zeroth level) of each electronic state. Thus, under ordinary conditions (room temperature, fluid medium), and in the absence of external perturbations, the molecule is found almost entirely in the lowest energy vibrational level of the ground singlet state. We shall occasionally use the symbol $S_0(\nu = 0)$ to designate the

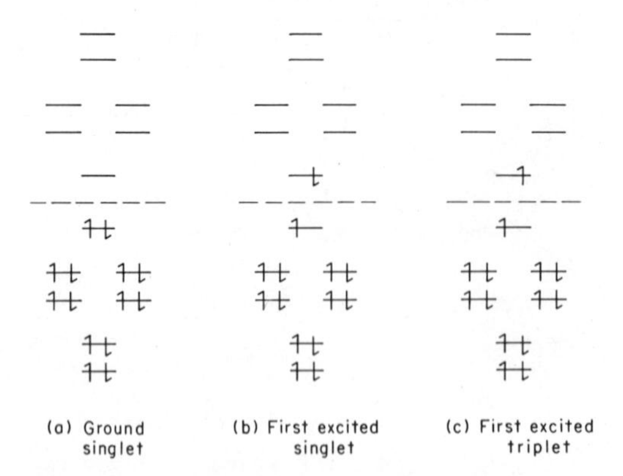

Figure 3. The electronic configuration of the anthracene molecule as predicted by Hückel molecular orbital calculations for π orbitals.

zeroth level of the ground singlet. Population of higher vibrational levels can be achieved when transitions take place between the various molecular states, as will be discussed in the next section.

Transitions Between Molecular States

The important molecular transitions are tabulated in Table 1 (see also the Jablonski diagram in Fig. 2). The first is absorption, in which light of appropriate energy induces transitions from the zeroth vibrational level of S_0 to a set of vibrational levels of an excited electronic state. For the purpose of discussion, consider S_1 as the excited electronic state. Thus, the molecule may find itself in an excited vibrational level of S_1 after excitation. However, this excess vibrational energy is rapidly dissipated into heat, via interaction with solvent molecules, such that the molecule rapidly (in 10^{-13} sec) "relaxes" to the lowest vibrational levels of S_1. This occurs before fluorescence or any other deexcitation process can take place. This process is known as "vibrational relaxation" and has been found to take place within the vibrational manifold of any electronic state.

Let us now suppose that excitation is from S_0 to a higher electronic state, say S_2. If the excitation is to an excited vibrational level of S_2, then the

TABLE 1. Transitions Between Molecular States

Process	Equation	Approximate time for completion (in seconds)
1. Absorption (or excitation)		
a. Singlet–singlet	$S_0 + h\nu \rightarrow S_1$	10^{-15}
b. Singlet–triplet	$S_0 + h\nu \rightarrow T_1$	
2. Vibrational relaxation	S_1 (excited vibrational level) $\rightarrow S_1$ (zeroth vibrational level) $+ \Delta$	10^{-13}
3. Internal conversion between excited states	$S_2 \rightarrow S_1 + \Delta$	10^{-13}–10^{-11}
4. Intersystem Crossing	$S_1 \overset{k_{st}}{\rightarrow} T_1 + \Delta$	10^{-8}–10^{-7}
5. Radiative transitions		
a. Fluorescence	$S_1 \overset{k_f}{\rightarrow} S_0 + h\nu$	10^{-8}–10^{-7}
b. Phosphorescence	$T_1 \overset{k_p}{\rightarrow} S_0 + h\nu$	10^{-2}–10^{1}
6. Internal conversion to ground state		
a. From singlet	$S_1 \overset{k_s}{\rightarrow} S_0 + \Delta$	10^{-8}–10^{-7}
b. From triplet	$T_1 \overset{k_t}{\rightarrow} S_0 + \Delta$	10^{-2}–10^{1}

molecule undergoes rapid vibrational relaxation to the zeroth level of S_2. One might expect that the zeroth level of S_2 would be depopulated via direct transitions to S_0. It has been found, however, that a more efficient pathway (in 10^{-12} sec) is "internal conversion" from S_2 to a highly excited vibrational level of S_1, from which vibrational relaxation once again brings the molecule to the zeroth level of S_1. Thus, regardless of which singlet state is excited, rapid processes like internal conversion and vibrational relaxation always bring the molecule to the zeroth level of S_1 before further deexcitation can take place. This nearly universal rule (known as Kasha's rule) is central to the interpretation of all luminescence behavior, as we shall see in the ensuing sections.

The depopulation of $S_1(\nu = 0)$ can occur via three pathways of roughly equal rate. The first of these is *fluorescence*, in which the molecule undergoes a transition from the zeroth level of S_1 to any allowed vibrational level of S_0, accompanied by the emission of a photon of the same energy as the transition. This process is also known as spontaneous emission or radiative decay. The first order rate constant is usually of the order 10^{+8} to 10^{+7} sec^{-1}. It should be obvious at this point that for emission to be observed at all, the nonradiative deexcitation processes must not be much faster than radiative pathway. One such radiationless pathway is *internal conversion* from S_1 to S_0. This process is similar to internal conversion from S_2 to S_1, but its rate is lower by several orders of magnitude. The reason for this is not perfectly understood but energy separation between the two electronic states involved appears to be an important factor. For our purposes, we may say as a qualitative rule that the larger the energy separation between two electronic states, the slower the rate of internal conversion. The energy separation between S_0 and S_1 is usually larger than that between S_1 and S_2. Consequently, the rate of internal conversion is reduced to 10^7-10^8 sec^{-1}, which is comparable to the rate of radiative decay from S_1 to S_0.

The other nonradiative pathway for deexcitation of S_1 is *intersystem crossing* from S_1 to T_1. All transitions between states of different spin multiplicity, whether radiative or nonradiative are formally forbidden by spin selection rules. It is only through small perturbations like spin-orbit coupling that such transitions occur at all at a much reduced rate than transitions between states of like multiplicity. Thus, although the singlet-triplet absorption process has been observed, it is about 10^6 times weaker than the singlet-singlet absorption. Consequently, one might expect intersystem cross from S_1 to be too slow to compete with the singlet-singlet transitions from S_1. Here again, one must take into consideration the small energy separation between S_1 and T_1. As in internal conversion, small energy separation strongly enhances the rate of intersystem crossing. The net effect of these two opposing factors, spin forbiddance and small energy gap, is to make the rate for intersystem crossing fall in the range 10^7 to 10^8 sec^{-1}, again comparable to the other excitation processes from S_1. It should be noted that this is the major pathway for the population of T_1. Direct excitation via singlet–triplet absorption is much more inefficient.

In contrast to the S_1–T_1 separation discussed above, the separation between T_1 and S_0 is considerably larger, so that the spin-forbiddance factor dominates. Thus, both the radiative decay and the nonradiative decay from T_1 to S_0 have rate constants of 10^{-1} to 10^2 sec.

In principle, S_1 and T_1 can be deexcited by processes like photochemical reactions or collisional deexcitation, in which actual contact must be made between the excited molecule and another molecule. For S_1, the internal deexcitation processes proceed at rates of the order of 10^7 to 10^8 sec^{-1}. Thus its lifetime is no more than 10^{-8} sec. In dilute solutions, not too many collisions can take place within that lifetime. In contrast, for T_1, the deexcitation processes are of the order of 10^{-1} to 10^2 sec^{-1}. Thus the lifetime of T_1 is in the millisecond–second range, such that many collisions take place within the triplet lifetime. It is for this reason that most photochemical reactions occur through the first excited triplet state.

Like any other paramagnetic species, oxygen with its triplet ground state is a potent triplet quencher upon collision. At room temperature and fluid solution there is sufficient dissolved oxygen to completely quench the triplet. Thus, phosphorescence is seldom observed in a fluid medium. Usually, it is from rigid plastics or frozen glasses that phosphorescence has been observed.

DEFINITION AND NATURE OF OBSERVABLES

In this chapter we shall discuss the kind of data that is collected in emission spectroscopy, introduce some significant defined quantities, and describe the kinds of relationships that hold between these quantities and the primary processes. Some of these quantities are listed in Table 2.

Absorption and Emission Spectra

As a typical example, consider the absorption and emission spectra of anthracene as shown in Fig. 4 (*37*). The relatively sharp bands in the absorption spectrum correspond to transitions from $S_0(\nu = 0)$ to the vibrational levels of S_1. Similarly, the lines in the fluorescence spectrum correspond to transitions from $S_1(\nu = 0)$ to the vibrational levels of S_0. Since fluorescence always occurs from the zeroth level of S_1 it is apparent that the fluorescence spectrum is always at lower energies (or longer wavelengths) than absorption. The presence of vibrational structure in the emission is only observed for those molecules with well-separated vibrational sublevels. In

TABLE 2. Definitions of Observables

Observable	Symbol	Relation to rate constants	Source
Fluorescence quantum yield	Q_f	$\frac{k_f}{k_f + k_{st} + k_s}$	Fluorescence spectrum
Phosphorescence quantum yield	Q_p	$\frac{k_p}{k_p + k_t} \times \frac{k_{st}}{k_f + k_{st} + k_s}$	Phosphorescence spectrum
Singlet lifetime	τ_s	$\frac{1}{k_f + k_{st} + k_s}$	Fluorescence decay
Triplet lifetime	τ_t	$\frac{1}{k_p + k_t}$	Phosphorescence decay
Singlet radiative lifetime	τ_0	$\frac{1}{k_f}$	Absorption spectrum

most cases, the emission spectrum is broad and structureless, because most of the vibrational bands are strongly overlapping. In the case of anthracene, the absorption spectrum and fluorescence spectrum are roughly mirror images of each other. This, too, is not a general phenomenon. It occurs only for those molecules whose vibrational levels of S_1 are spaced in the manner as those of

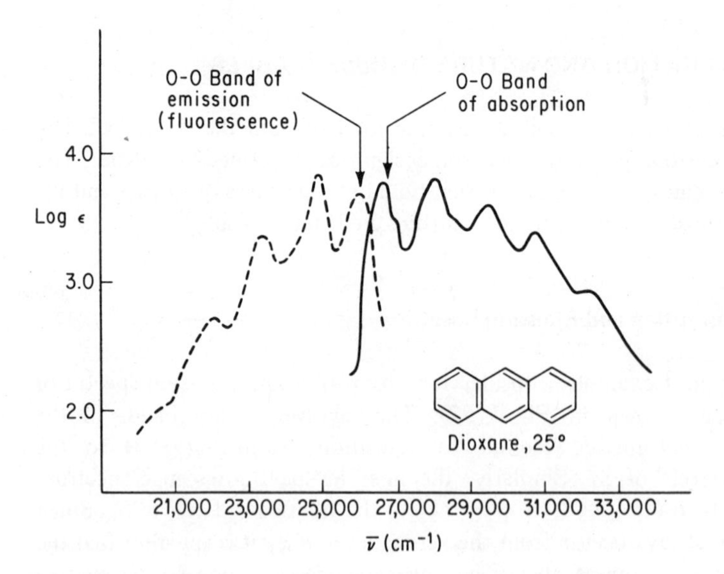

Figure 4. Absorption and fluorescence spectra of anthracene. [Reproduced from *Molecular Photochemistry* by N. J. Turro, with the permission of the publisher, W. A. Benjamin, New York (Ref. 37).]

S_0. This means that the nuclei vibrate with approximately the same modes in the ground state as in the first excited singlet state. This is more likely to be true for rigid aromatic molecules like anthracene than for molecules with flexible bonds.

The Excitation Spectrum

The excitation spectrum is obtained when fluorescence intensity is measured as a function of the excitation wavelength. In wavelength regions where the sample does not absorb, the excitation spectrum will fall to zero. Near the absorption bands, the excitation spectrum will exhibit maxima. In fact, due to the Kasha rule, the excitation spectrum should be exactly the same as the absorption spectrum. This is indeed true for all cases when the excitation spectrum is properly corrected.

The Fluorescence Quantum Yield Q_f

The fluorescence quantum yield Q_f is defined as the ratio between the number of photons emitted as fluorescence and the number of photons absorbed by the sample. Since every photon that is absorbed by the sample creates an excited molecule, Q_f also expresses the fraction of excited molecules that underwent deexcitation from S_1 to S_0 by emitting light. Still another way of looking at Q_f is that it is a measure of the efficiency of the radiative deexcitation process relative to the total rate of deexcitation. With normal first order deexcitation kinetics, Q_f can readily be written in terms of the primary rate constants:

$$Q_f = \frac{k_f}{k_f + k_s + k_{st}} \tag{1}$$

where k_f, k_s, and k_{st} are defined in Table 1.

Thus the fluorescence quantum yield depends on how well the radiative process can compete with nonradiative processes like internal conversion and intersystem crossing. When a decrease in Q_f is observed, it is generally not due to any decreases in k_f, but to either increases in the nonradiative rates, or to the presence of additional nonradiative processes like energy transfer or photochemical reactions. In other words, substances that have high quantum yields are simply good emitters (fluorescein and rhodamine B are good examples) and are therefore very easy to detect. Substances with quantum yields less than 0.01 are difficult, but not impossible to detect experimentally. As we shall see later, however, a high quantum yield is not the only criterion for the suitability of a molecule for fluorescence work.

The Singlet Lifetime, τ_s

Note that all the deexcitation pathways from S_1 follow simple first order kinetics. One can write down a rate equation as follows:

$$\frac{d[S_1]}{dt} = -(k_f + k_{st} + k_s)[S_1] \tag{2}$$

Solving for the concentration of S_1:

$$[S_1] = [S_1]_0 e^{-k_T t} \tag{3}$$

where $[S_1]_0$ is the initial concentration of S_1, and $k_T = k_f + k_s + k_{st}$.

Suppose one excites a sample with a short pulse of light at $t = 0$, and follows the fluorescence intensity $I(t)$ as a function of time. The fluorescence intensity is proportional to the concentration of S_1 at any time. Thus, $I(t)$ also decays exponentially:

$$I(t) = I_0 e^{-t/\tau_s} \tag{4}$$

where I_0 is the initial intensity, and

$$\tau_s = 1/k_T = 1/(k_f + k_s + k_{st}) \tag{5}$$

The quantity τ_s is known as the singlet lifetime, although very often called the fluorescence lifetime in the literature. τ_s is obtained from fluorescence decay measurements. It is a measure of the average amount of time that a molecule remains in the state S_1. It is proportional to the quantum yield:

$$k_f \tau_s = Q_f \tag{6}$$

The reciprocal of τ_s, k_T, represents the total rate at which S_1 is depleted and not just the rate of fluorescence. Clearly, in the presence of quenchers that provide additional pathways of deexcitation, the total deexcitation rate would be increased and the singlet lifetime, along with the fluorescence quantum yield, would be decreased.

It is clearly advantageous to work with substances that have long singlet lifetimes. It not only facilitates experimental detection, but also extends the range of measurements in energy transfer and fluorescence polarization experiments. This point will become more clear when we discuss these measurements in full detail in later sections.

The Radiative Lifetime, τ_0

The radiative lifetime is defined as follows:

$$\tau_0 = 1/k_f \tag{7}$$

where the fluorescence rate constant k_f is just the rate constant for spontaneous emission, and expresses the "allowedness" of the radiative transition from S_1 to S_0. The factor that governs this allowedness is the magnitude of the transition dipole moment between S_1 and S_0 (*38*). This is the same factor that governs the rate of absorption. Thus, we would expect a relationship to exist between the radiative lifetime and the extinction coefficient of absorption. This relation in an approximate form reads as follows (*37*):

$$\tau_0 = \frac{3.5 \times 10^8}{(\nu_m)^2 \epsilon_m (\Delta\nu_{1/2})} \tag{8}$$

In this equation, we have assumed that the absorption band is Gaussian in shape, $\Delta\nu_{1/2}$ is the half width of the band in cm^{-1}, ν_m is the frequency in cm^{-1} at the peak of the band, and ϵ_m is the molar extinction coefficient at the band maximum.

Equation (8) predicts that the more absorptive a species is, the shorter is its radiative lifetime.

The radiative lifetime represents the upper limit for the singlet lifetime. This can be seen as follows: Using the definition of τ_0 in Eq. (7), Eq. (6) now reads

$$\tau_s = \tau_0 Q_f \tag{9}$$

In the limit of total absence of radiationless transitions, $k_{st} = k_s = 0$ and $Q_f = 1$, so that $\tau_s = \tau_0$. This upper limit is often useful to know in assessing the suitability of an emittor for fluorescence work. Thus, with no other information but the absorption spectra, one can calculate the radiative lifetime of a molecule to see if it is too short for fluorescence work. Conversely, if it is calculated that a molecule has a long radiative lifetime, one knows that it has the potential of having both a long singlet lifetime and a high quantum yield.

Although Eq. (8) is often used for qualitative appraisal of a molecule's fluorescence properties, its application to quantitative measurements of τ_0 is rife with pitfalls. One has to be sure that the absorption band corresponds to the same transition as the fluorescence band. Even then, Eq. (8) becomes inaccurate when the mirror image rule between the absorption and fluorescence band does not hold (*39*). When the absorption band shape is not

Gaussian, it becomes necessary to integrate over the entire band, a tedious and often inaccurate procedure. The quantum yield is usually obtained by comparing the integrated fluorescence intensity of the unknown sample with that of a standard emittor. Intensity measurements are not trivial to make, and reliable quantum yield measurements can only be obtained under very strict experimental conditions. One should not, therefore, be surprised to find large discrepancies in reported quantum yield values. Of the three quantities appearing in Eq. (9), τ_s is by far the easiest to measure reliably. The measurement is direct and simple, and with modern day instrumentation τ_s can be measured to at least 0.5% accuracy.

Another observable in emission spectroscopy is the polarization. This will be discussed in detail in a later section.

Corresponding to Q_f, τ_s, and τ_0 analogous quantities can be defined for transitions from T_1 (see Table 2). These include the phosphorescence quantum yield, the triplet lifetime, and the triplet radiative lifetime. Except for the triplet lifetime, the other two quantities are hardly ever measured due to experimental difficulties.

EXPERIMENTAL TECHNIQUES

There are three basic types of measurements in emission spectroscopy: absorption measurements, steady state intensity measurements, and intensity decay measurements. Assuming that most researchers have ready access to a spectrophotometer, a self-sufficient luminescence laboratory would need two additional pieces of major equipment: a spectrofluorimeter, and a fluorescence lifetime apparatus. In this section we shall describe some general features of these instruments, and present some practical guides to data collection.

The Spectrofluorimeter

Numerous publications (*2, 40*) on emission spectroscopy contain detailed descriptions of spectrofluorimeters. We shall, therefore, describe the fundamental features of this instrument very briefly.

A fair number of commercial spectrofluorimeters are presently available (Ellis (*40*) gave a comprehensive list). The Aminco-Bowman model is the widely known commercial spectrofluorimeter, and has found its way into many biological laboratories. Other well known instruments are available from Turner, Perkin-Elmer, and Zeiss. These instruments are perfectly adequate for routine investigations. For versatility, added sensitivity and dependability, many workers have chosen to assemble their own spectro-

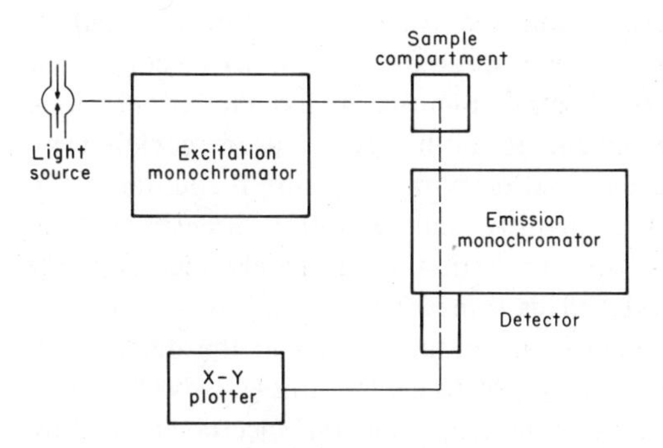

Figure 5. Schematic diagram of a simple spectrofluorimeter.

fluorimeters from the components. Whether commercial or laboratory built, a typical spectrofluorimeter will contain the following skeletal components: a light source, excitation monochromator, sample holder, emission monochromator, detector, and output devices. The common geometric arrangement of these components is shown in Fig. 5. The right angle arrangement between the excitation and emission beam is generally preferred for reasons we will discuss later. In Table 3, we have listed the manufacturers of these components in a laboratory built spectrofluorimeter that the authors have used.

To this skeletal structure, the following components may be added in order of importance: (1) Optical devices like focusing and light collecting lenses, filters, polarizers, etc. (2) Excitation spectra correction device –

TABLE 3. Components of a Typical Spectrofluorimeter [a]

1. Light Source	Osram Xenon arc lamp, 150 W
	Wild Heerbrugg Universal lamp holder
	Siemens power supply
2. Monochromators	Jarrell-Ash 0.25 meter grating, $f3.5$
3. Sample Compartment	American Instrument Company
4. Detector	E.M.I. 6256S photomultiplier
	Hewlett-Packard 6525A power supply
5. Modulation	Light chopper type 42, American Time Products
	Princeton Applied Research lock-in detector, model 121
6. Output devices	Hewlett-Packard X-Y recorder

[a] From C. K. Luk (Ref. *110*).

incorporation of this feature has two purposes: (a) To correct for fluctuations in the excitation intensity while emission spectra are taken. (b) To correct for the nonuniform distribution of spectral intensity in the excitation source while excitation spectra are taken. In both cases, this is achieved by monitoring the intensity of the excitation with a "quantum counter" (*41*). The ratio between the emission intensity and excitation intensity is then taken electronically, and it is this ratio that is ultimately recorded. (3) Emission spectra correction device. This is meant to correct for nonuniform spectral responses in the emission monochromator and in the detector. A number of techniques have been developed to achieve this (*42–44*) but none is completely satisfactory. For this reason, emission spectra are usually published without correction. We should stress that for absolute quantum yield measurements the spectra must be corrected. For relative quantum yield measurements, spectra correction is not necessary when the emission band does not change shape from sample to sample. (4) Phase sensitive detection. This feature can increase the sensitivity of the detection system by a hundredfold. It employs choppers to modulate the excitation beam at a fixed frequency. A "lock-in-amplifier" then amplifies only signals of that frequency. Random noise and other electronic interference are generally at different frequencies and would not be amplified. (5) There are, in addition, some more conventional features like thermostatted temperature control, adaptations for low temperature work, adaptations for solid samples or thin films, and so forth.

A word of prophecy: Research workers are increasingly relying on digital computers for data processing. It is worthwhile, therefore, to consider devices that digitize and interface the data for computer analysis (whether on-line or off-line). In this manner, corrected excitation spectra, corrected emission spectra, quantum yield measurements, etc., can be obtained much more efficiently.

Lifetime Apparatus

Phosphorescence occurs in the range of milliseconds to seconds. Conventional electronics can easily follow such intensity decays. Fluorescence occurs within the singlet lifetime of 10^{-9} to 10^{-8} secs. Less direct techniques must be used for reliable measurements. Currently, three techniques are in use for fluorescence lifetime work: the phase-shift method, the sampling method, and the single photon counting method. The *phase-shift method* utilizes the principle that if the excitation is a sinusoidal function with time, the emission would also be sinusoidal, but shifted by a phase lag that is related to the fluorescence lifetime. This technique has been developed to a high degree of sophistication by Weber and his co-workers (*45, 46*). The *sampling method* (*47, 48*) requires a flash lamp that provides repetitive pulses of durations short compared with the fluorescence decay. The detection system is gated in such a manner that a "window" briefly opens at a certain time t_1 after the excitation pulse to measure the intensity of the fluorescence at that time. The next excitation pulse triggers the "window" to open at time $t_2 > t_1$, the third excitation pulse opens the "window" at $t_3 > t_2$ and so on. The fluorescence decay curve can therefore be obtained from the intensities measured at these discrete times t_1, $t_2, t_3, \ldots, t_n$. The *single photon counting method* also required short excitation pulses. Its detection system, however, is a drastic departure from conventional means, in so far as it detects and records the time of flight of individual photons rather than monitoring photon fluxes.

Each method has its advantages and disadvantages. The phase-shift method is very accurate even down to subnanosecond lifetimes, but requires a strong excitation source for high sensitivity work. The sampling method suffers from both insensitivity and inaccuracy at short lifetimes. The most popular at present is the single photon counting method. Its main disadvantages are: (1) Long data collection time; at a lamp repetition rate of 4000 pulses per second, it takes at least half an hour to accumulate sufficient data to define three decades of decay. (2) Like all measurements of fast decays in real time, the finite width of the excitation pulses modifies the decay curves measured, such that decay processes with lifetimes comparable to the decay time of the excitation cannot be measured accurately. The main advantages of single photon counting are: (1) The apparatus operates only at low photon fluxes; this high sensitivity allows the use of low intensity lamps with fast decay characteristics. (2) Irregularities in the amplitude or the repetition rate of the excitation pulses do not affect the results. (3) Compared with the sampling method, the resolution time is down by an order of magnitude (about 0.2 nsec); thus, if the excitation is sufficiently short, it is theoretically possible to measure subnanosecond lifetimes. (4) The actual decay curve is obtained and readily displayed; any irregularities or deviations from simple exponential behavior become immediately obvious. (5) The assembly of a single photon counting apparatus is relatively easy because the components of such an apparatus have long been in use by experimental nuclear physicists, and are readily available from a number of excellent manufacturers. For these and other reasons, we shall give a relatively more detailed account of single photon counting. Descriptions of the method have been given by others also *(49–51)*.

Single Photon Counting

We stress again that single photon counting relies on the detection of individual photons. The time at which the photon is detected relative to the time of excitation is measured and recorded. In practice, the measurement

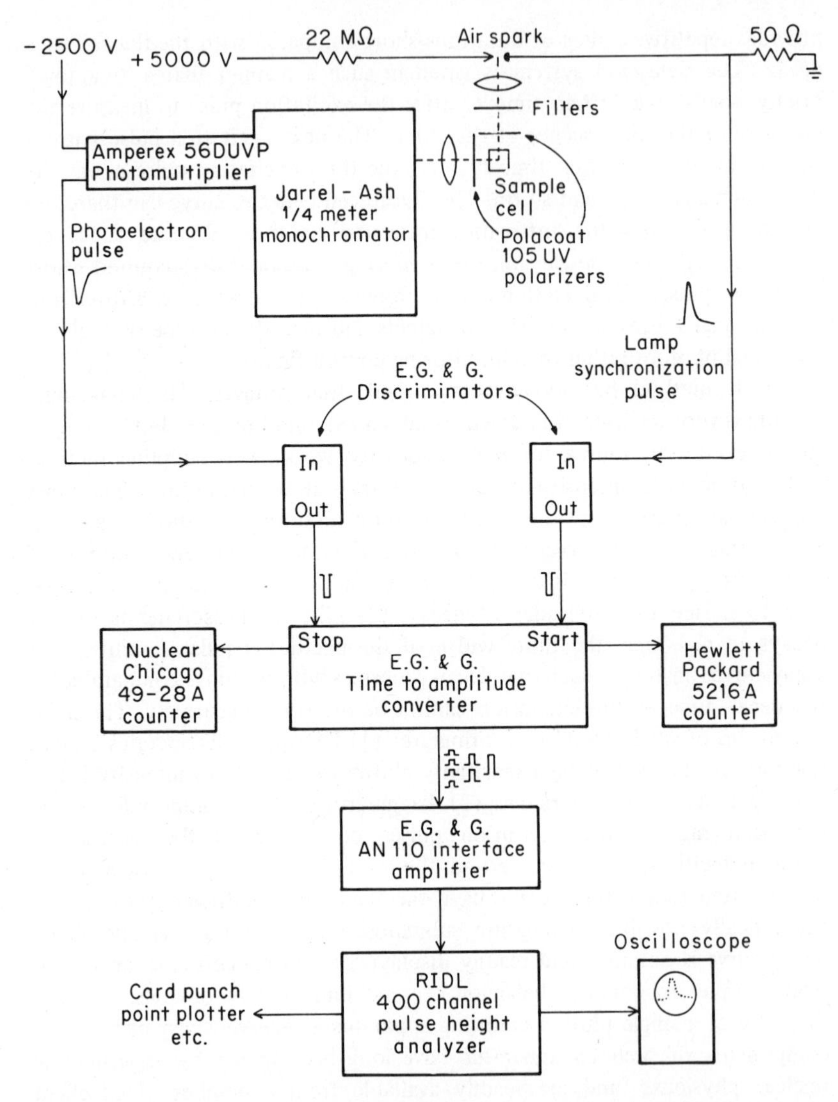

Figure 6. Schematic diagram of a single photon counting apparatus for fluorescence decay measurements. This diagram was adapted from T. Tao, *Biopolymers,* 8: 609 (1969), with permission of the publisher (*23*).

and the recording are performed rapidly and automatically by high-speed electronic modules. Figure 6 shows a block diagram of an apparatus that the authors have used. The principal components are the time-to-amplitude converter (TAC) which performs the time measurement, and the multichannel pulse height analyzer (MCPHA) which performs the recording.

Detection of single photons is achieved by high sensitivity photomultipliers that are kept at high operating voltages. To make sure that only single photons are detected, neutral density filters, small slits, or other optical attenuation devices are employed on the emission beam. Generally, the attenuation is to such an extent that less than 5 photons are detected for every 100 excitation pulses. This makes the probability of two photons being emitted after excitation extremely unlikely.

The pulse excitation source is probably the most difficult aspect of the instrumentation. Several designs are presently in use (*48, 52–54*). All designs utilize discharges between the tips of tungsten or other metallic electrodes. The discharge might be in open air, or enclosed in a gas-filled tube. We have used a design that is particularly easy to construct (*23*). We took a radiating line that was purchased from the General Radio Corporation for $13.50, and cut away the middle section of the inner conductor (see Fig. 7). Tungsten electrodes, sharpened to tips, were inserted by snug fitting. A ballast resistor of about 22 megohms was inserted between the spark gap and a 5000-volt dc power source. The ballast resistance along with stray capacitances in the

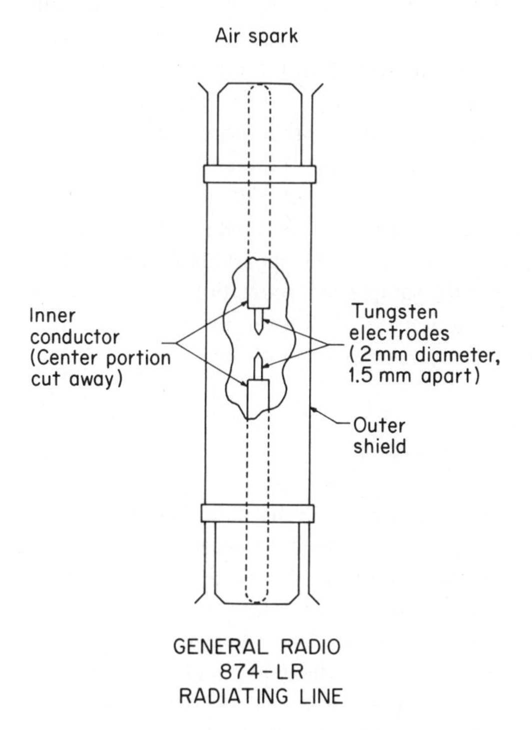

Figure 7. Sketch of a useful pulsed excitation source. From T. Tao, *Biopolymers,* **8**: 609 (1969), with permission of the publisher (*23*).

circuit builds up the voltage until breakdown occurs between the tips of the electrodes. Repetitive pulses can be obtained at a rate of about 4000 pulses/sec, with a width at half height of 3.5 nsec, and useful intensities in the wavelength region of 300–410 nm. The discharges are grounded with a 50-ohm terminator. As shown in Fig. 6, an electrical signal synchronous to the optical discharge can be drawn from the lamp circuit. After suitable attenuation, this signal is used to establish a zero time reference by activating the TAC at the "start" input. Ninety-five percent of the time, no photon is detected after the TAC was activated, the unit simply recycles without delivering any output. Once out of twenty times, a photon is detected by the photomultiplier, which delivers a photoelectron pulse. The photoelectron pulse then deactivates the TAC at the stop input, whereupon the TAC delivers an output signal whose amplitude is proportional to the time lapse between the arrival of the lamp synchronization signal and the arrival of the photoelectron pulse. This output signal is fed into the MCPHA, which "senses" the amplitude of this signal, and stores a count in a channel whose number is proportional to the amplitude. Thus the time lapse is first converted into an amplitude, and then a channel number. There will be many more counts in the initial channels than the late channels, because photons are more likely to be emitted immediately after excitation than at long times after excitation. After many counts are collected, the number of counts in each channel represents the probability that a photon is emitted at a time corresponding to the number of the channel. In other words, the display of the memory of the MCPHA is just a plot of the fluorescence decay curve. The memory of the MCPHA can also be read out to point-plotters, card-punchers, parallel printers, paper tape, etc. for further analysis of the data. The other relatively minor components of the apparatus shown in Fig. 6 include discriminators that shape the lamp synchronization signal and the photoelectron pulse and digital counters that monitor the lamp repetition rate and the rate at which photoelectron pulses arrive at the TAC.

Analysis of Fluorescence Decay Data

Figure 8 shows a typical set of fluorescence decay data. The time profile of the excitation is shown alongside the fluorescence decay of a sample of $tRNA^{Phe}$ in 0.003 M magnesium. Note that in this case as in many others, the excitation pulse is decidedly shorter than the fluorescence decay, but it is not so short that its effect on the observed decay curve can be completely ignored. In general, if the singlet lifetime is τ_s, the experimental fluorescence decay $I_e(t)$ is given by a convolution integral:

$$I_e(t) = \int_0 E(t')e^{-(t-t')/\tau_s}dt' \tag{10}$$

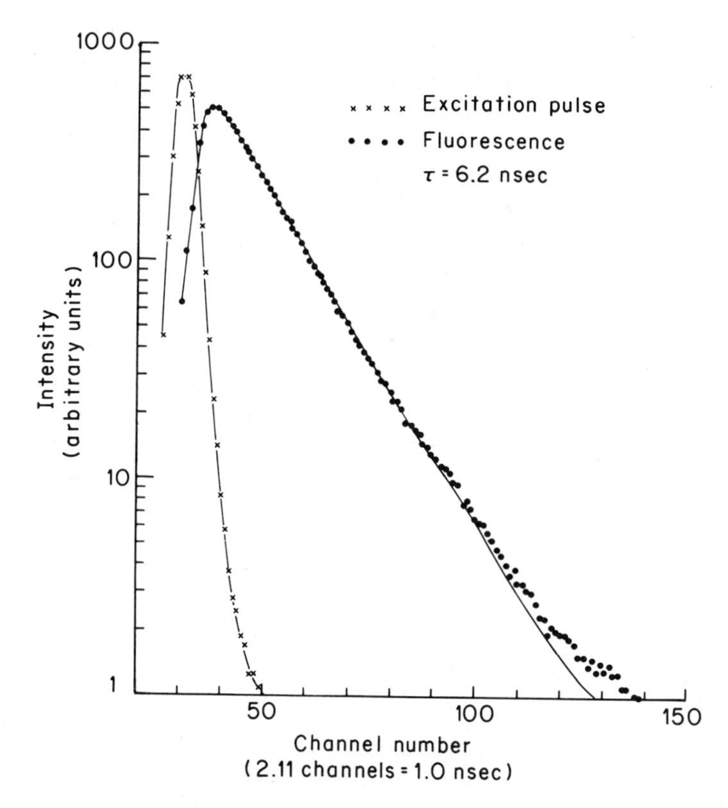

Figure 8. Fluorescence decay of the Y base of $tRNA^{Phe}$ in 2.6×10^{-3} M Mg^{2+}. Solid line passing through the filled circles is the best one exponential fit representing a decay time of 6.2 nsec.

where $E(t')$ is the observed time dependence of the excitation. In order to extract τ_s from the experimentally determined $I_e(t)$ and $E(t)$, two methods can be used: the method of moments, and a least square curve fitting technique. The application of the method of moments to this type of data analysis has recently been examined in detail by Isenberg and Dyson (*55*). Here we briefly describe the curve fitting technique.

As outlined by Hundley et al. (*54*), a computer program sets up a grid of decay times τ in steps of 0.01 nsec. For each assumed τ, the convolution integral in Eq. (10) is calculated by numerical integration. The calculated decay curve $I_c(t)$ is then compared with the experimental decay curve, and the sum of the residues squared is computed:

$$\chi^2 = \sum_{i=1}^{n} [I_c(t_i) - I_e(t_i)]^2 \tag{11}$$

This is repeated until a minimum χ^2 is found. The value of τ that gives rise to this minimum χ^2 is taken to be the fluorescence lifetime. This procedure works remarkably well for single exponential decays. The quality of the fit can readily be judged from the value of the minimized χ^2 and the overlap between the calculated decay curve and the experimental decay curve (the solid line in Fig. 8 is the calculated curve). For multiexponential decay curves, the method of moments is far superior (*55*).

Sample Handling

In this section, we shall discuss the hazards that one is likely to encounter in fluorescence work, and some techniques that can be used to overcome these difficulties. There are three principal sources of error in emission spectroscopy: (1) The inner filter effect. (2) Scattered exciting light. (3) Interference from impurities.

The Inner Filter Effect

The *inner filter effect* is a consequence of using a sample that is optically too dense. That is to say, at certain wavelengths, the sample is highly absorptive. If the excitation is at those wavelengths, it will not penetrate the entire sample uniformly. Consequently, only those molecules at the front of the incident beam are excited, causing the detected fluorescence intensity to drop drastically. In contrast, in wavelength regions of low optical density, the excitation reaches all parts of the sample cell with equal intensity, allowing the detector to view the fluorescence along the entire pathlength of the cell. Similarly, if the fluorescence is in the region of high optical density, the emitted light might be reabsorbed by the absorbing species, causing a further decrease in fluorescence intensity. Because of the inner filter effect, both the emission spectrum and the excitation spectrum might be distorted, and the fluorescence intensity will not be proportional to concentration.

In most cases, the highly absorbing species is the fluorescent molecule itself. This problem can then be overcome by simply diluting the sample. (Diluting the solute concentration also alleviates problems arising from the formation of photodimers or self-quenching via collision between the solute molecules themselves.) Generally, the inner filter effect can be ignored when the optical density of the sample does not exceed 0.05 anywhere in the absorption spectrum. When dilution is not possible, one might have to use a different optical arrangement than right angle excitation and detection. For example, a 1-mm pathlength cell may be used. When arranged in the "back surface" configuration (Fig. 9b) the optical density of the sample is reduced by a factor of 10. If this reduced optical density is less than 0.05 everywhere, the inner filter effect can again be ignored. For completely absorbing samples,

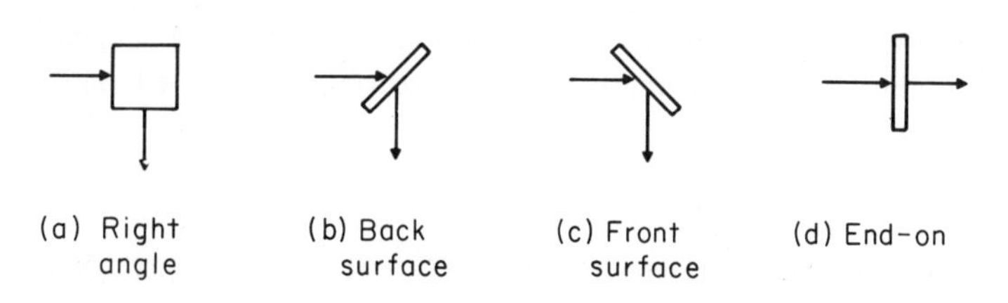

Figure 9. Four common geometries for measurements of fluorescence spectra.

the emission spectrum can be obtained by adopting either the "front surface" or the "end on" configurations of Figs. 9c and 9d. The excitation spectrum in this case will still be badly distorted.

Scattered Light

To record an emission spectrum the excitation monochromator is set at a certain wavelength, and the emission monochromator is then scanned. When the scan passes through the preset excitation wavelength, a "scattered peak" will usually be observed. Depending on the amplitude and width of this scattered peak, scattered exciting light often seriously obscures the emission spectrum.

Scattering may arise from any one of the following sources: scattering off the wall of the sample cell, Tyndall scattering off solid particles suspended in the sample, Rayleigh or Raman scattering off the sample itself. The first two are by far more serious but can be dealt with effectively. The "right angle" geometry minimizes cell wall scattering. Even then, one must make sure that the excitation beam never strikes the side walls of the cuvette. This is achieved by using appropriate slits at the sample compartment after proper collimation of the excitation beam. Millipore filtration is very effective for removing dust particles that are responsible for Tyndall scattering. Rayleigh and Raman scattering are clearly impossible to remove, but they are of much weaker intensity.

Scattering in general can be further reduced by various optical means. Thus, appropriate cut-off filters (glass or interference) can be employed on the excitation beam to reduce the scatter-to-emission ratio. In addition, the excitation beam can be polarized with the electric vector parallel to the direction of the emission beam. Since scattered light usually retains its polarization, and since a light beam cannot travel in the same direction as its electric vector, little scattered light will enter the detection system.

Finally, the slit widths of the monochromators can be reduced to enhance their resolution, thereby reducing the width of the scattered peak. If one is not certain whether a given peak is due to scattering or emission, there is a very simple test. If the excitation wavelength is changed, an emission peak will remain at the same wavelength, while a scattered peak will shift accordingly.

Impurities

Impurities are very troublesome in emission spectroscopy. Even when present in trace amounts, they can either quench or compete with the fluorescence. They are frequently difficult to identify and almost impossible to remove. The only defense is to be able to recognize the presence of impurities and even that is not always obvious. Competitive impurities are usually species of high quantum yield. The best one can do is to subject the sample to high-resolution separation procedures like column chromatography and note if the suspected impurity emission suffer any diminution. Often, shifting the excitation wavelength might reveal the presence of impurity emission when the emission spectrum changes shape. If the impurity is suspected to be present in the solvent, running a solvent blank, which is always advisable, becomes of prime importance.

Impurity quenchers might be organic molecules to which excitation transfer, or with which photochemical reaction, can take place. Or, they might be paramagnetic species like oxygen or cupric ions. Oxygen can be efficiently removed by degassing the sample (*56*), a procedure that we would not recommend for biological materials. Oxygen is not as soluble in water as in some nonpolar solvents. Thus, for lifetimes shorter than 100 nsec, oxygen quenching of fluorescence is not a serious problem. The use of deionized water in emission work carries the purpose of minimizing the amount of metal ions, paramagnetic ones included. There are, unfortunately, no general procedures for circumventing problems arising from impurities. The best approach is probably to work with strong emitters at sufficiently high concentrations.

The difficulties we have cited above are particularly acute for intensity measurements. Lifetime measurements share most of the difficulties, except for those arising from the inner filter effect, which only reduces the amplitude of the decay curve. This is another reason why singlet lifetime measurements are more reliable than quantum yield measurements.

From the foregoing discussion it is clear that the choice of a proper sample concentration is an important matter. It should be minimized to get away from the inner filter effect. On the other hand, it must not be so low that scattering and impurity interference become significant. In practice, this limits the concentration range to 10^{-3} M to 10^{-6} M. We should stress, however, that this lower limit is set not by instrumental sensitivity, but by the presence of such problems like scattering or impurities. For strong emitters like quinine sulfate or fluorescein, most instruments can easily detect

10^{-8} M of material. Roughly speaking, the fluorescence intensity I_f from a sample is given by

$$I_f \sim \epsilon c l Q_f \tag{12}$$

where ϵ is the extinction coefficient of the sample, c is the sample concentration, l is the pathlength, usually 1 cm, and Q_f is the fluorescence quantum yield. For quinine sulfate in 0.1 N H_2SO_4, $\epsilon_{348} = 15{,}000$, and $Q_f = 0.55$. Equation (12) can now be used to estimate the lower limit of detectability for any material whose absorption and fluorescence quantum yield are known.

USEFUL FLUORESCENT CHROMOPHORES

General Criteria

While many molecules show detectable fluorescence spectra, several characteristics will tend to make a small set of these much more useful for the study of nucleic acids than others. It is usually desirable to be able to selectively excite a given fluorescent chromophore with high efficiency. For nucleic acid samples, this will usually be possible only if the absorption spectrum of the fluorescent probe does not strongly overlap the principal 260 nm absorption band of natural nucleic acids. Thus, as a general rule, useful fluorescent dye probes for nucleic acids are likely to have absorption maxima at wavelengths longer than 300 nm.

The experimental detection of fluorescence becomes increasingly more difficult as the wavelength of the emitted photons becomes longer. If wavelengths greater than 650 nm are to be observed, it is usually necessary to resort to the use of red-sensitive photomultipliers which, even with cooling jackets, tend to have less favorable signal-to-noise ratios than those used for shorter wavelength regions. Thus, as a practical rule, it is usually useful to try to work with fluorescent labels which emit light at wavelengths shorter than 650 nm.

Another important criterion in the selection of a suitable fluorescent probe is that it be relatively stable to prolonged illumination. The usefulness of a number of possible fluorescent chromophores for nucleic acid studies is impaired by the fact that in the excited states these react readily with nucleic acids. While relatively useless for fluorescence studies, these photoproducts can, of course, be extremely useful sources of structural information once the site of the reaction on the macromolecule has been discovered. An extremely elegant example of this is the remarkable photo-induced cross-linking of the 4-thiouridine base of valine tRNA with a cytosine base located 5 nucleotides away in the primary structure but obviously quite close in the tertiary structure (*57*).

It is usually desirable to make measurements at the highest possible signal to noise ratios. In fluorescence spectroscopy, this means that for a constant wavelength region one attempts to maximize the intensity of the emitted light. As mentioned earlier, the fluorescence intensity of a sample depends on the product of its extinction coefficient for absorption and the quantum yield of emission. As a first approximation, this product can be used as a measure of experimental sensitivity. For more precise and absolute comparisons, the bandwidth of absorption and fluorescence and the desired resolution required must also be taken into account. These additional elaborations are discussed in detail by Parker (*2*).

Using the simplfied criterion discussed above, let us compare the relative ease of detection of an acridine dye, such as acriflavine, with the ability to detect the fluorescence of guanosine at pH 1.2 at room temperature. Guanosine is the most fluorescent of the common nucleosides, but only under extreme pH conditions. The quantum yield of guanosine is 0.009 (*5*), the ultraviolet extinction is about 10^4 (*58*). For acriflavine, these values are 0.54 (*2*) and 3×10^4 (*59*), respectively. Thus, in dilute solution, acriflavine is about 200 times easier to detect by fluorescence than guanosine. In the presence of other nucleic acid bases, this discrepancy would be even larger because it is possible to excite acriflavine with visible light where normal nucleic acid bases are transparent. There is no convenient wavelength for exciting guanosine without attenuating most of the light by absorbance of other nucleic acid bases which will fluoresce even less.

In planning fluorescence studies on nucleic acids, three different classes of chromophores should be considered. These are naturally occurring fluorescent bases, covalently attached fluorescent labels and noncovalently bound fluorescent labels. The relative merits and disadvantages of each of these techniques are complex and depend on the particular system being studied, and the type of information desired. In this section, we will survey the particular members of each class of compounds that either have been used successfully for fluorescence studies in the past or that merit consideration for use in future studies.

Naturally Fluorescent Nucleic Acid Components

The fluorescence of the common nucleosides A, C, G, and U or T at neutral pH is so weak as to be just about undetectable. While some measurable fluorescence can be observed for nucleosides of the normal bases if measurements are made at pH 1 or 11, the necessity of going to these extremes of pH probably makes these normal bases completely useless for most types of fluorescence studies. The fluorescence properties of the bases A, U, C, G, and T and their nucleosides and nucleotides have been discussed in detail by Udenfriend (*5*).

The possibility of fluorescence studies on completely unperturbed native nucleic acids would be extremely bleak indeed if it were not for the presence of a few naturally fluorescent nucleosides in several species of tRNA. The most useful of these is the base, Y, which occurs at the 3′ end of the anticodon of the phenylalanine tRNAs of yeast (*60*), wheat germ (*61*), and rat liver (*62*). The structure of this base is still not yet known. The Y base has a moderate quantum yield, 7% (*16, 59*), and has an excitation maximum at 315 nm, well to the red of the bulk of the normal nucleic acid absorption. The fluorescence peaks at 450 nm and is readily observable. Since the Y base occurs next to the anticodon, one of the most interesting regions of the tRNA structure, it will be extremely useful for many studies of the mechanism of tRNA action. The properties of the Y base have already been exploited in a number of different studies (*14–16, 59*).

There are several other fluorescent nucleosides which occur naturally in some tRNA species. These include 7-methylguanosine, 4-thiouridine, and *N*-6-acetylcytidine. These have not yet been used in many detailed studies of tRNA because the fluorescence is much weaker than that of the Y base. There have, however, been a number of studies reported on the fluorescence of these bases when incorporated into oligonucleotides (*63*). The fluorescence spectrum appears to be fairly sensitive to the conformation of the oligonucleotide. In general, the fluorescence of one of these nucleosides in a dinucleoside phosphate is considerably weaker than its fluorescence as a monomer. This strong tendency for strange bases to show fluorescence quenching as an oligonucleotide or polynucleotide attains a fairly rigid structure strongly discourages attempts to exploit these in some of the more demanding aspects of fluorescence spectroscopy.

Base Analogs

A number of synthetic nucleosides show considerable fluorescence. These include 2-aminopurine riboside, 2,6-diaminopurine riboside, formycin, 7-methyl inosine, azaguanosine, and others. The structures of some of these are shown in Fig. 10, and the fluorescence properties are summarized in Table 4. Some of these base analogs as well as some of the strange bases discussed in the previous section, are substrates for some of the enzymes of nucleic acid metabolism. Thus, it is not difficult to synthesize oligonucleotides and polynucleotides containing base analogs. Unfortunately, the fluorescence of formycin, 2-aminopurine, or 7-methylinosine in a polymer is considerably less

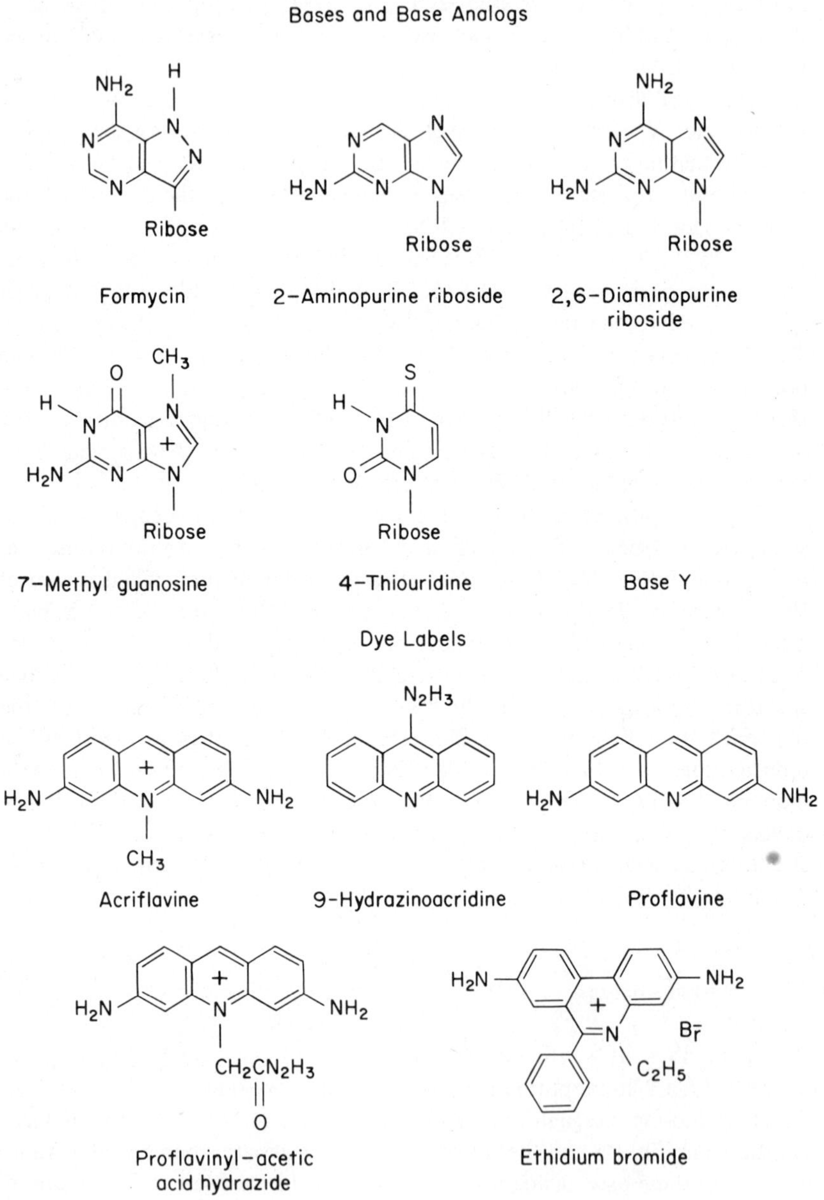

Bases and Base Analogs

Formycin

2-Aminopurine riboside

2,6-Diaminopurine riboside

7-Methyl guanosine

4-Thiouridine

Base Y

Dye Labels

Acriflavine

9-Hydrazinoacridine

Proflavine

Proflavinyl-acetic acid hydrazide

Ethidium bromide

Figure 10. Structures of some of the chromophores useful for fluorescence studies on nucleic acids.

TABLE 4. Optical Properties of Substances Useful for Fluorescence Studies on Nucleic Acids

Substance	Conditions[a]	Absorption[b]		Fluorescence			Ref.
		λ_{max}(nm)	$\epsilon_{max} \times 10^4$	Q_f	λ_{max} (nm)	τ_s(nsec)	
Y in yeast $tRNA^{Phe}$	pH 7, 0.01 M Mg^{2+}	320(e)		0.07	430	6.3	59
2-Aminopurine riboside	pH 7	303(e)		0.68	370	7.0	64
Formycin	pH 7	295(e)		0.06	340	<1	64
Diaminopurine riboside	pH 7	280(e)		0.01	350	<1	64
7-Methyl GMP	pH 5	292(e)		0.01	385		63
7-Methyl IMP	pH 7.5	280(e)		0.10	395		63
6-Acetyl CMP	pH 7	308(e)		0.03	360		63
4-Thiouridine in tRNA	pH 7	350(e)			510		111
4-Thiouridine	pH 7	338(e)			520		c
Adenosine-*n*-oxide	pH 7	295(e)			400		c
Guanosine	pH 1	285(e)		0.01	390		5
Acriflavine-poly A	covalently bound	460(e)			510		112
Acriflavine-tRNA	covalently bound	463(a)	3.31		510		59
Acriflavine	free, pH 7	440(a)	4.4	0.54			4, 112
Ethidium bromide-tRNA	covalently bound	505(a)				25.4	65
Ethidium bromide-tRNA	intercalated	515(a)	0.38	~ 1.0	600	26.5	18, 19, 65
Ethidium bromide	free, pH 7	478(a)	0.53	<0.05	628	<1.5	17, 18, 19

[a]All measurements at 25°C.
[b]e means excitation maximum, a means absorption maximum.
[c]Paul Koenig, unpublished results.

than it is in a monomer. Ward, Reich, and Stryer (*64*) have made a detailed study of the fluorescence properties of three of these base analogs. They were able to incorporate the base formycin in place of the terminal adenosine of tRNA using the terminal pyrophosphorylase. In this way, they were able to label this part of the tRNA molecule with a minimum perturbation. While formycin and 2-aminopurine riboside are reasonably fluorescent, many of the other analogs have such weak emission spectra that they are not likely to be terribly useful.

Covalently Attached Fluorescent Labels

The most effective way to overcome the disadvantages of naturally occurring fluorescent bases or their analogs is to prepare specific covalently labeled fluorescent derivatives of nucleic acids. The most widely used method for this was first described by Churchich (*29*). He specifically oxidized the 3′ end of an RNA with periodate and then coupled the resulting aldehydes with the amino groups of the dye acriflavine to form a fluorescent Schiff's base label. This has extraordinary favorable optical properties but suffers from the disadvantage that the Schiff's base link is relatively unstable and hydrolyzes with time. This problem can be overcome partially by using other fluorescent aldehyde specific reagents. Almost any amine, hydrazine, or hydrazine-containing dye should be able to be attached to the 3′ end of tRNA in this way. Recently, we have been able to covalently attach the very useful fluorescent probe, ethidium bromide, to the 3′ end of tRNA by reacting it with the oxidized aldehydes (*65*). It is unfortunate that this procedure will only work for ribonucleic acids since DNA lacks the *cis* hydroxyls needed for the specific periodate cleavage.

A second possible method exists for labeling the 3′ end of tRNA. The tRNA can be loaded with an amino acid which will then result in a free aliphatic amino group present at the 3′ end. This can be acylated with the *N*-hydroxysuccinimide adducts of fluorescent acids in an analogous way to the preparation of derivatives useful for purification or spin-labeling studies (*66, 67*). This type of a label will probably result in less perturbation of the tRNA structure than coupling a dye with oxidized aldehydes. It has not, to our knowledge, however, yet been used for fluorescence studies.

In principle, one should be able to attach a wide variety of fluorescent amines to the 5′ ends of nucleic acids. Ralph et al. have demonstrated that aniline can be coupled to the 5′ phosphate of tRNA in good yield (*68*). This technique requires, however, a very complicated solvent mixture which would probably have to be adjusted for each different dye label used. We have had only limited success thus far in attempting to use this technique to load the 5′ end of tRNAs with aminoacridines or aminonaphthalenes.

Some of the unusual bases in tRNAs are potential targets for the covalent attachment of specific fluorescent labels. While 4-thiouridine itself is only slightly fluorescent, it will react specifically with modified bromoacetamides. In this way Hara et al. were able to prepare a spin-labeled derivative of tRNA and it should be possible, in an exactly analogous way, to make fluorescently labeled derivatives (*69*). Another potential source of fluorescent labels is the nucleoside pseudouridine. Acrylonitrile reacts fairly specifically with this nucleoside (*70*), and it should be possible to design fluorescently modified acrylonitriles. There are, in addition, a number of specific reactions for the normal nucleosides. Perphthalic acid oxidation specifically converts adenosine into adenosine *n*-oxide (*71*). The product is slightly fluorescent, but not enough to be very useful. Similarly, it is possible to form a specific adduct of acetylaminofluorene on the 8-position of guanosine (*72*). Again, the adduct is fluorescent, but not really enough to be of great help.

Noncovalently Attached Fluorescent Labels

A great variety of different types of substances will bind to nucleic acids. The list includes polycyclic aromatic hydrocarbons, polyamines, alkaloids, antibiotics, proteins, and other nucleic acids (*73*). The strong affinity of many substances for nucleic acids offers an easy route to fluorescent labeling. Unfortunately, few of the small molecules that bind to nucleic acids show any great specificity. Many dyes, like ethidium bromide, will bind to double-stranded nucleic acids with much greater affinity than to single-stranded nucleic acids (*17*). These dyes often show, however, no selectivity for binding in the vicinity of certain bases, and often cannot distinguish between DNA and RNA (*17, 74*). Some of the dyes that have been used include ethidium bromide, acriflavine, proflavine, and many other acridine dyes. Such non-specifically bound fluorescent probes are probably mostly useful for determining the flexibility of DNA (*75*).

There is some reason to expect that if these dyes were bound to a polynucleotide containing a tertiary structure such as tRNA, there would be far fewer available binding sites and thus some degree of selectivity in interaction might be observed. Binding of many of these dyes occurs by intercalation which must be accompanied by an increase in the length of the double helix and rotation about the helix axis (*76, 77*). Many of the double-helical regions in the tRNA tertiary structure are probably constrained to a constant length and, therefore, it would be a relatively high barrier to dye binding. Close packing of adjacent helices would also inhibit binding. Many other double helices are probably buried and thus inaccessible to dyes. We have found, with studies on yeast $tRNA^{Phe}$ that in the absence of magnesium, only one ethidium bromide molecule will bind per molecule of tRNA (*19*).

This dye almost completely quenches the Y fluorescence, thus indicating that the binding site is quite close to the anticodon.

There is at least one possible approach which, in principle, could be used to make rather specific noncovalent fluorescent labels for nucleic acids. Complementary oligonucleotides will form quite specific complexes (*78*). If a hexa- or heptanucleotide were labeled covalently with a fluorescent dye, and then added to a large polynucleotide, it would probably bind in only a few places thus enabling fairly specific location of the fluorescent probe (*79*).

The structures of many of the noncovalent fluorescent labels that have been used in the study of nucleic acids are shown in Fig. 10. The optical properties of these compounds are summarized in Table 4. This list is only a small fraction of the molecules that are potentially of interest as noncovalent labels for nucleic acids. Several features should be borne in mind in selecting one of these for use in experiments. In most cases, an ideal label is one which is fluorescent only when bound to nucleic acid. Thus, no corrections must be made for fluorescence arising from free dye. An additional complication is introduced by the fact that a number of these dyes are known to form aggregates in aqueous solution. Since the fluorescence properties of the aggregate are different from the fluorescence properties of dye monomer, this makes it very difficult to correct for the fluorescence of dye which is not bound to nucleic acids. The dye which probably best overcomes all of these difficulties is ethidium bromide. A large number of chemical modifications of ethidium bromide exist and are potentially interesting as other possible fluorescent labels (*80*).

At the present time, there are a great many potentially useful covalent and noncovalent fluorescent labels for nucleic acids. However, it is fair to say that very few of these have been tested and found completely satisfactory. The present lack of well-established techniques for fluorescently labeling nucleic acids and the extremely weak fluorescence of most natural nucleic acids are the major obstacles to more extensive use of fluorescence techniques.

SINGLET–SINGLET ENERGY TRANSFER

Consider a system with more than one fluorescent chromophore. These might be attached to different parts of the same molecule or might simply be several different molecules in free solution. If the molecules can approach each other closely enough to interact weakly, there is a finite probability that if one is excited by a photon, the electronic excitation energy may be transferred to a second molecule, which will subsequently emit radiation. For simplicity's sake, we shall restrict our discussion to systems with two different chromophores. Suppose that the absorption and emission spectra of

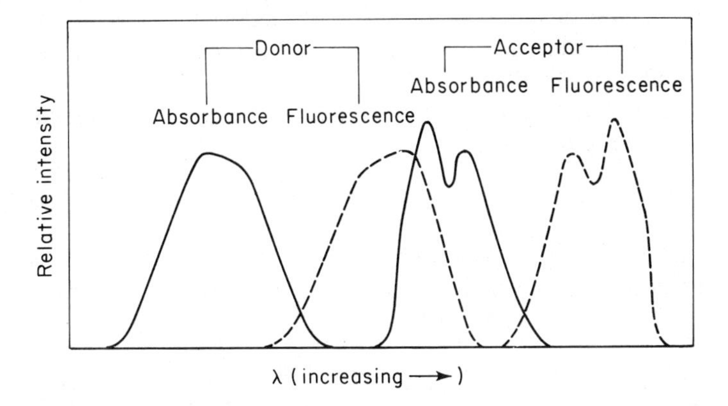

Figure 11. Schematic diagram of the spectral properties of chromophores useful for singlet–singlet energy transfer experiments.

the two interacting molecules are as shown in Fig. 11. It is possible to find a wavelength of exciting light where only molecule A absorbs and it is possible to choose a wavelength for monitoring fluorescence where the fluorescence of only molecule B can be observed. The emission spectrum of molecule A overlaps the absorption spectrum of molecule B. This is a typical situation in which the observation of energy transfer is possible.

Examination of the spectral data shown in Fig. 11 immediately suggests one possible mechanism for energy transfer. Molecule A absorbs a photon, then emits it. The photon is subsequently reabsorbed by molecule B and then reemitted by molecule B. This is the so called "trivial" mechanism of energy transfer. It is a highly improbable event in dilute solution and can take place without any direct electronic interaction between molecules A and B. According to this mechanism, molecule B in no way influences processes governing the decay of the excited singlet state of molecule A. Therefore, the lifetime of molecule A is unaltered. This is contrary to what is observed experimentally for the types of energy transfer processes we shall be considering (*81*).

A more direct mechanism for energy transfer has been proposed by Förster (*82*). He considered the case in which molecules A and B are close enough together to permit a very weak resonant interaction to occur. In this case, it is possible for a simultaneous deexcitation of molecule A and excitation of molecule B to occur without emission or absorption of the photon. Such a process will directly quench the excited singlet state of molecule A. Thus, the observed singlet lifetime of A will decrease as the rate of transfer becomes faster. To derive a theoretical expression for the rate of transfer. Förster assumed that the molecules are interacting in the very weak coupling limit. One test of very weak coupling interactions is that the shape of the absorption and emission spectra of both molecules will be unaltered by the presence of the interaction.

Using this limit, Förster was able to derive the expression for the rate of energy transfer shown below.

$$k_{tr} = 8.71 \times 10^{23}\, R^{-6}\, J\kappa^2 n^{-4} k_f \text{ sec}^{-1} \tag{13}$$

In the above equation, R is the distance between the two chromophores, n is the refractive index of the medium between the two groups participating in energy transfer, J and κ^2 are defined below.

$$J = \frac{\int f_D(\bar{\nu})\epsilon_A(\bar{\nu})\bar{\nu}^{-4}d\bar{\nu}}{\int f_D(\bar{\nu})d\bar{\nu}} \tag{14}$$

$$\kappa^2 = \langle(\cos\theta - 3\cos\omega_1\cos\omega_2)^2\rangle \tag{15}$$

J is the overlap between the normalized emission spectrum of molecule A, the donor, and the absorption spectrum of molecule B, the acceptor. κ^2 is a measure of the average orientation of the electronic transition moment of the emitting state of the donor and the absorbing state of the acceptor. It is the square of the geometric part of the dipole-dipole interaction tensor and arises from the use of the dipole approximation in describing the resonance energy of interaction between the two groups. The angles defined in κ^2 are shown in Fig. 12.

The Förster theory predicts that the rate of energy transfer will depend on the inverse sixth power of the distance between the two chromophores. This highly sensitive distance dependence makes the method, in principle, an ideal way of determining intramolecular distances if a reasonably accurate measure of the rate can be obtained. In practice, it is not the rate itself

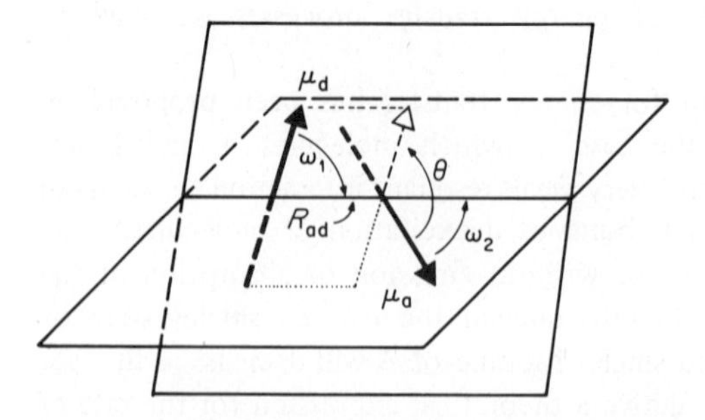

Figure 12. Definition of the angles used in defining κ^2. See Eq. (15).

which is usually measured, but other quantities which can be related to it. These must first be discussed before the methods by which the Förster theory has been tested and used can be described.

Steady-State Detection of Singlet–Singlet Energy Transfer

Since most laboratories are not yet equipped for measuring fluorescence lifetimes, most measurements of energy transfer processes have been made by following the manifestations of these events in the normal fluorescence spectra (*30, 83–85*). From the scheme described above, it should be clear that there are two ways in which an energy transfer process will alter the observed steady-state fluorescence spectra. The emission spectrum of the donor molecule will be quenched, and the excitation spectrum of the acceptor molecule will show a new excitation band corresponding to sensitized emission, that is, the light which is absorbed by the donor but emitted by the acceptor. Let us describe these observables more quantitatively. One can define the efficiency of energy transfer as the rate of transfer divided by the sum of all rate processes which can depopulate the excited singlet state of the donor. This is shown below.

$$E = \frac{k_{tr}}{k_{tr} + k_f + k_s + k_{st}} \tag{16}$$

From Eqs. (13) and (16), and the definition of the quantum yield of the donor, one can easily derive the following expression for efficiency.

$$\mathrm{E} = \frac{R_0^{\,6}}{R_0^{\,6} + R^6} \tag{17}$$

In Eq. (17), R_0 is the distance at which energy transfer is half efficient.

$$R_0 = 9.79 \times 10^3 (Jn^{-4}\kappa^2 Q_f)^{1/6} \tag{18}$$

It is thus clear that a measure of the efficiency of energy transfer will directly give the distance R between the donor and the acceptor if the numerical constant R_0 is known.

It is very easy to relate the efficiency of energy transfer to directly observable quantities. The quantum yield of the donor molecule in the absence and presence of energy transfer are shown below.

$$Q_{fd}{}^0 = \frac{k_f}{k_f + k_s + k_{st}} \qquad Q_{fd}{}^a = \frac{k_f}{k_f + k_s + k_{st} + k_{tr}} \tag{19}$$

By combining these two quantities with the above definition for the efficiency of transfer in terms of fundamental rate constants, we obtain the following simple expression.

$$\frac{Q_{fd}{}^{a}}{Q_{fd}{}^{0}} = 1 - E \qquad (20)$$

It can be seen that, as the efficiency of transfer increases from a minimum possible value of zero to a maximum possible value of 1, the fluorescence spectrum of the donor is quenched in a linear fashion to complete extinction. This equation has an important ramification because nowhere in its derivation is it necessary to consider whether or not the acceptor molecule B is capable of observable fluorescence. Thus, in fact, it is quite possible to perform energy transfer experiments with only a single fluorescent donor and a nonfluorescent acceptor molecule.

If the acceptor molecule is fluorescent, a second possible method for observing energy transfer by steady-state techniques exists. Consider what happens when the system containing one donor and one acceptor molecule is excited in the region where the absorption spectrum of the donor is a maximum and the absorption probability of the acceptor is very small. The total light emitted by the acceptor after such an excitation will occur from two sources. The first of these is simple, direct excitation of the acceptor. The fluorescence resulting from direct excitation of the acceptor will be proportional to the quantum yield of the acceptor and the extinction coefficient of the acceptor at the wavelength of excitation. In addition, energy will be absorbed by the donor and transferred to the acceptor. The number of photons absorbed by the donor will be proportional to its extinction coefficient, and the percentage of these that are transferred to the acceptor is simply the efficiency of transfer, E. Combining these two processes, we have the following expression for the fluorescence of the acceptor in the presence of the donor.

$$F_a{}^d \propto Q_{fa}\epsilon_a + Q_{fa}\epsilon_d E \qquad (21)$$

In the absence of the donor, the second term in the equation above is zero. If we consider only the ratio of the fluorescence in the two cases, then the following very simple equation results.

$$\frac{F_a{}^d}{F_a{}^0} = 1 + \frac{\epsilon_d}{\epsilon_a} E \qquad (22)$$

This equation has very important implications for energy transfer measurements. The efficiency, E, is a positive fraction greater than zero and less than one. Thus the fluorescence of the acceptor in the presence of the donor must always be greater than the fluorescence in the absence. Depending on the exact experimental system chosen, the ratio, ϵ_d/ϵ_a can vary over many orders of magnitude from a very small fraction to a very large number. Since E is the derived quantity needed to calculate distance between two chromophores and $F_a{}^d/F_a{}^0$ is the quantity measured, it is clear that for the most accurate possible determinations, the ratio of the absorption of the donor and the acceptor should be as large as possible. In general, in a well chosen system, ϵ_d/ϵ_a will be a factor of ten or more and thus determination of the efficiency of energy transfer by sensitized emission using the above equation is a much more accurate process than following the quenching of the donor. It is this reason, plus the advantage of being able to make two independent determinations of the same quantity, E, which favors the use of chromophore systems where both donor and acceptor are fluorescent.

If the shapes of the absorption spectra of the donor and the acceptor are known in great detail, then Eq. (22) can be used to follow the sensitized fluorescence as a function of wavelength and ensure that a simple singlet-singlet energy transfer process is taking place since E should be a constant independent of wavelength. Some additional complications may occur, however, if the donor is excited at the long wavelength edge of its absorption band (*86*).

Detection of Singlet-Singlet Energy Transfer by Fluorescence Decay Methods

It was mentioned earlier that fluorescence lifetime measurements offer a considerable adjunct to more commonly used static techniques. This is also true in the observation of energy transfer processes. The fluorescence lifetime of the donor in the presence and absence of energy transfer are shown below.

$$\frac{1}{\tau^0} = k_f + k_s + k_{st} \qquad \frac{1}{\tau} = k_f + k_s + k_{st} + k_{tr} \tag{23}$$

The results of these two independent measurements can be combined simply to yield directly a value for the rate of transfer or the efficiency of transfer.

$$k_{tr} = \frac{1}{\tau} - \frac{1}{\tau^0} \qquad E = 1 - \frac{\tau}{\tau^0} \tag{24}$$

One of the advantages of choosing to detect energy transfer processes by fluorescence lifetimes is that the possibility of the trivial transfer process can be immediately excluded, if results obtained by lifetime measurements are in agreement with results obtained by static fluorescence. Haugland et al. (*81*), in studies on energy transfer between two chromophores attached to a steroid nucleus have shown that the transfer efficiencies derived from kinetic measurements are in extremely good agreement with quantities derived from static fluorescence yields. This offers strong support for the Förster mechanism.

Verification of the Förster Theory

Several aspects of the Förster theory of singlet–singlet resonant energy transfer are accessible to experimental test. These include the inverse sixth distance dependence of the efficiency of energy transfer, and the dependence of the numerical constant R_0 on the spectral overlap of the donor and the acceptor. In principle, one should also be able to design experimental systems with rigid fixed values of κ^2 but, in practice, deviations from the nominal angle between two groups due to molecular vibrations and lack of exact knowledge of the directions of transition moments and the assignment of electronic states has made this difficult. However, the two first mentioned experimental implications of the Förster theory have been tested in detail by the elegant work of Stryer and his collaborators. To examine the distance dependence of energy transfer, Stryer and Haugland prepared a series of oligoprolines labeled at one end with a suitable energy donor and at the other with a suitable energy acceptor. By examining the efficiency of energy transfer as a function of length, they were able to show that the exponent of the distance in Eq. (18) is 5.9 ± 0.3. This is strong proof indeed that the square of a dipole–dipole interaction is responsible for the transfer.

A different series of experiments enabled Haugland, Yguerabide, and Stryer to examine the dependence of the efficiency of energy transfer on the spectral overlap of the fluorescence of the donor and the absorption of the acceptor (*81*). By the clever choice of a system in which the fluorescence spectrum of the donor and the absorption spectrum of the acceptor shifted in opposite directions as the solvent polarity was changed, they were able simply by varying the solvent conditions to change the spectral overlap in a precisely determinable way. In this way, they were able to show that the dependence of the transfer rate on the spectral overlap is precisely as given by the Förster theory. Both of these experiments suggest that the Förster theory is right in most of its details. The ultimate test, comparison of an experimentally determined R_0 with a theoretically calculated value, has not thus far proved quite as satisfying, but the discrepancies between calculation and experiment are not intolerably large.

Experimental Considerations

Let us consider what is involved in determining the distance between two fluorescent chromophores attached to a biological macromolecule. Regardless of whether time-dependent or static methods of detection are used, three experimental systems are needed to provide a check on the experimental results: a system containing only the donor, only the acceptor, and both the donor and the acceptor. Unless the fluorescent groups used are normally intrinsic components of the macromolecule, one must always worry whether their insertion has in any way perturbed the structure that is to be measured. In some cases, it may be possible to determine this by alternative techniques such as circular dichroism, or standard hydrodynamic methods. In other cases, it may be possible to show that the fluorescence of the donor or the acceptor alone is rather sensitive to the molecular conformation. Thus, an energy transfer experiment will provide an internal check because the results obtained by following the quenching of the donor and the sensitized emission of the acceptor will be in agreement only if the presence of one does not perturb the environment of the other directly.

The choice of donor and acceptor molecules is an important consideration in designing an experiment. The reason for this is shown clearly in Fig. 13 which displays the expected efficiency of energy transfer as a function of the distance between the two dyes in terms of the characteristic distance, R_0. If the distance to be measured is very much greater than R_0, the energy transfer will be close to zero and, furthermore, there will be little dependence of the transfer on distance. Similarly, if the distance to be measured is very much shorter than the characteristic distance, R_0, the efficiency of energy transfer will be very close to 1 and again will be relatively insensitive to distance. It is only in the region where R is approximately equal to R_0 that the possibility of a very accurate measurement of the distance R exists. Thus, if there is any outside indication of what the distance between the two groups is likely to be, the groups should be chosen such that R_0 is very close to the suspected distance. Typical values of R_0 for attainable

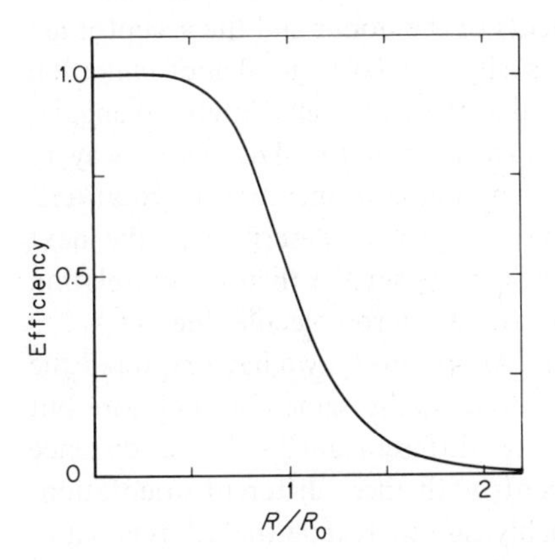

Figure 13. Efficiency of singlet–singlet energy transfer as a function of the distance between the donor and the acceptor. This is a plot of Eq. (17).

chromophore systems vary anywhere from less than 10 Å to as high as 60 Å. This means, in principle, that distances of up to 100 Å could be measured with a fair degree of accuracy. This is sufficient to provide highly useful information on a wide variety of nucleic acid systems.

To derive a distance given an experimental determination of the efficiency, E, the constant R_0 must be calculated. The quantum yield of the donor can usually be determined experimentally as described earlier in the chapter. The overlap integral, J, is directly calculable by numerical integration from experimental extinction and fluorescence data. It is important to use the corrected emission spectrum of the donor for exact measurements, although, in practice, it probably will not make too much difference if an uncorrected emission spectrum is used. The index of refraction of the medium between the donor and the acceptor is generally not known but it is usually possible to make a realistic estimate of this. In practice, we have found it convenient to use the number 1.4 for nucleic acids (*59*). This takes into account the degree of hydration and approximate refractive indices for sugar, magnesium phosphate, and a pyridine base. In practice, the distance derived depends on the two-thirds power of the refractive index chosen and thus it is not likely that a large experimental error could be made by a poor choice of refractive index.

The most variable and least easily obtainable quantity that affects the constant, R_0, is the geometrical factor, κ^2. In general, this is not known experimentally nor is there any known method to determine it in an arbitrary case. The range of possible values for κ^2 is 0 to 4. In calculating a distance from experimental data one takes the sixth root of κ^2. Thus a precise value of κ^2 is not needed but it is still important to at least place limits on the possible value of κ^2. There are two ways to do this. If there is some indication that the angles between the transition moments of the donor and the acceptor are not rigidly fixed, then it will probably usually be a fairly good approximation to assume that κ^2 is equal to 2/3, the value if there is absolutely no angular correlation between the donor and the acceptor group. The easiest way to determine whether this is a good approximation is to measure the rotational correlation times of the donor and the acceptor as described in the next section of this chapter. If one or both of these is smaller than the correlation time expected for a dye rigidly attached to the macromolecule, then $\kappa^2 = 2/3$ is probably a fairly good approximation. Alternatively, we have explored the possibility of using, as an acceptor (or donor), the same chromophore but attached to the macromolecule at several different angles. If the distance derived from energy transfer measurements with these different orientations is fairly consistent, then again it is probably safe to assume that κ^2 is equal to 2/3. There is no question at the moment, however, that uncertainties in the value of κ^2 are the major source of difficulty in energy transfer experiments.

If the macromolecule system containing the two fluorescent dyes is not rigid, then the distance between the two fluorescent groups derived from an energy transfer experiment will be an average value shown below.

$$\frac{R_a}{R_0} = \left(\left[\frac{1}{1 + \left(\frac{R}{R_0} \right)^6} \right]^{-1} - 1 \right)^{1/6} \tag{25}$$

In this equation, R_a is the apparent distance between donor and acceptor and the quantity in brackets is an average over all the actual configurations occupied by the macromolecule. This will tend to weight very short distances very heavily and thus lead to an underestimate for what would normally be considered to be the average distance between the two groups. It is quite important to consider, therefore, that the value derived for a distance from energy transfer measurements is likely to be a lower estimate if a fair amount of conformational flexibility is possible.

Until now we have considered only systems which contain precisely one donor and one acceptor. In actual experimental situations, it is often not possible to attach fluorescent dye labels to macromolecules with absolute precise stoichiometry. It is, thus, important to consider briefly the implications of having fewer donor molecules than acceptors or vice versa. Consider first a system where every macromolecule contains a donor but not all contain an acceptor. In this case, Eq. (20) must be modified. If the quenching of the donor is observed, then the efficiency can be calculated as follows.

$$\frac{Q_{fd}{}^a}{Q_{fd}{}^0} = 1 - E\chi_A \tag{26}$$

In the above equation, χ_A is the mole fraction of molecules which contain an acceptor. If sensitized emission is observed, Eq. (22) can still be used, providing that it is possible to find a wavelength where only acceptor fluorescence and not any pure donor fluorescence is detected. For time-dependent studies, the results will be more complicated. If some molecules contain an acceptor and some do not, then the fluorescence decay of the donor will be a sum of two exponentials. Given sufficiently accurate experimental data, it is possible to resolve this and determine the lifetimes of both the quenched and the unquenched state but in practice this is often not such an easy process to do.

Suppose, now, that some molecules contain an acceptor but no donor. In this case, the efficiency can still be determined from the fluorescence of the donor using Eq. (20), providing that the concentration of excess acceptor is not so great as to cause quenching by the trivial process. Equation (22) is no longer valid, however. The sensitized fluorescence of the acceptor will be given by the equation below.

$$\frac{F_0{}^d}{F_a{}^0} = 1 + \frac{\epsilon_d}{\epsilon_a} \chi_D E \tag{27}$$

In this equation, χ_D is the fraction of molecules containing a donor.

If systems are studied in which populations of molecules containing only donor, only acceptor and both are measured simultaneously, then an appropriate combination of the correction factors discussed above must be used. Even more serious complications can occur if, for example, there are macromolecules containing one donor and more than one acceptor. These systems can usually be handled by the appropriate generalizations of the equations given above. For example, suppose that there are two acceptor molecules, both at the same distance from the donor. In this case, all the equations remain the same except that the value of the overlap integral, J, used in calculating R_0 must be changed to account for twice as much absorption of the acceptor. This assumes that $\kappa^2 = 2/3$ is a good approximation for both dyes.

Typical Experimental Results

Because of the chemical difficulties of introducing two specific fluorescent chromophores into nucleic acids, there have been relatively few attempts at singlet-singlet energy transfer experiments capable of yielding distances between two fixed points in a nucleic acid molecule. The recent discovery that yeast phenylalanine tRNA contains the Y base has greatly simplified the task of making singlet-singlet energy transfer experiments on tRNA. By attaching dyes covalently to the periodate oxidized 3′ terminal aldehydes as described earlier, a number of different fluorescent chromophores have been specifically introduced into the tRNA molecule in this position (*59*). This afforded the possibility of measuring the distance between the Y base position on the anticodon loop and the CpCpA terminus which would normally contain an amino acid in aminoacyl tRNA. The spectral properties of one of the chromophores used, acriflavine, is shown along with the excitation and emission spectra of the Y base, in Fig. 14. It is clear that there is extremely good spectral overlap between the donor, Y, and the acceptor, acriflavine. The R_0 for this system was calculated to be 30 Å and the values derived for the Y-dye distance from different wavelengths of excitation, samples, and choices of dye, ranged from 45 to 59 Å. This placed a very strong constraint on possible tertiary structures for tRNA. It should serve to give strong encouragement to other attempts to measure intramolecular

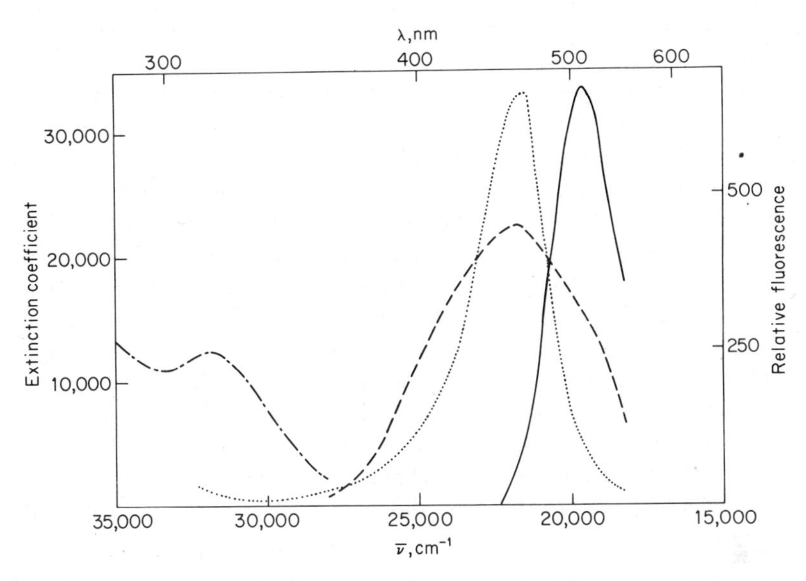

Figure 14. Optical properties of base Y of tRNA^{Phe} and covalently attached acriflavine. (—·—·) Excitation spectrum of Y base x 5. (– – –) Corrected emission spectrum of tRNA^{Phe} x 5. (. . .) Absorption spectrum of tRNA-bound acriflavine. (——). Emission spectrum of tRNA-bound acriflavine. This figure was adapted from Beardsley and Cantor (*59*) with permission of the publisher.

distances in nucleic acids by singlet–singlet energy transfer. A number of other possible distances within the tRNA molecule are potentially determinable, and if these could be determined, then a fairly definite solution conformation might be established.

All of the energy transfer experiments described in this section have been between singlets. However, the reader should be aware that it is also possible to detect triplet–singlet and triplet–triplet energy transfer. Measurements of this type are usually possible only in rigid frozen glasses. However, the extreme usefulness of the data derived makes it worthwhile to develop these techniques and a number of important advances have already been made in the protein field. There have also been numerous studies of these types of energy transfer in oligo- and polynucleotides at very low temperature (*1*). A thorough discussion of these is, however, outside the scope of this article.

POLARIZATION TECHNIQUES

Fluorescence polarization is another major observable phenomenon in emission spectroscopy. In this section, we shall discuss the principles and techniques for fluorescence polarization measurements, and cite some typical experimental results.

There are three possible kinds of fluorescence polarization measurements: (1) polarized excitation spectra; (2) steady-state fluorescence depolarization (SSFD); (3) time-dependent fluorescence depolarization (TDFD), or nanosecond depolarization. Polarized excitation spectra are primarily used to augment the analysis of absorption spectra. This technique is capable of resolving electronic transitions that are not revealed in the absorption spectra. It can be used to determine the relative orientations of transition moments (*26*). Both SSFD and TDFD have been used to probe macromolecular size, conformation, and flexibility. Of the two techniques, SSFD has had a long history of application to protein research, and numerous reviews on the topic have appeared (*26, 87–89*). We therefore will not carry a detailed discussion of SSFD, but will mention the principles behind the technique. TDFD is a relatively new technique. Its principles and applications are still under exploration. We take this opportunity to present a unified discussion of this technique.

General Considerations

Fluorescence polarization stems from the principle of photoselection, which in turn is a consequence of the electric dipole nature of excitation. Consider a single molecule whose principal transition dipole moment (for the $S_0 \rightarrow S_1$ transition) is represented by a vector $\vec{\mu}$. Much like a classical oscillating dipole, this molecule is most likely to be excited by light that is polarized with the electric vector parallel to $\vec{\mu}$. Similarly, when the molecule emits, the emitted light is most likely to be polarized parallel to $\vec{\mu}$. The probability of emission falls to zero for emitted light polarized perpendicular to $\vec{\mu}$.

Let us now consider a typical experimental geometry shown in Fig. 15. The excitation is directed along the x-axis, the emission is detected along the y-axis. The sample is placed at the origin. In this sample we have an ensemble of fluorescent molecules, each with a transition moment $\vec{\mu}$. Under ordinary conditions, these transition moments would be randomly oriented in space. We now expose the system to exciting light polarized with the electric vector along the z-axis. The intensity of the emitted light that is polarized parallel to the z-axis is defined as $I_{\parallel}$, the intensity of the emitted light that is polarized perpendicular to the z-axis is defined as $I_{\perp}$. We define the fluorescence polarization anisotropy (or simply anisotropy) as follows:

$$A = \frac{I_{\parallel} - I_{\parallel}}{I_{\perp} + 2I_{\perp}} \qquad (28)$$

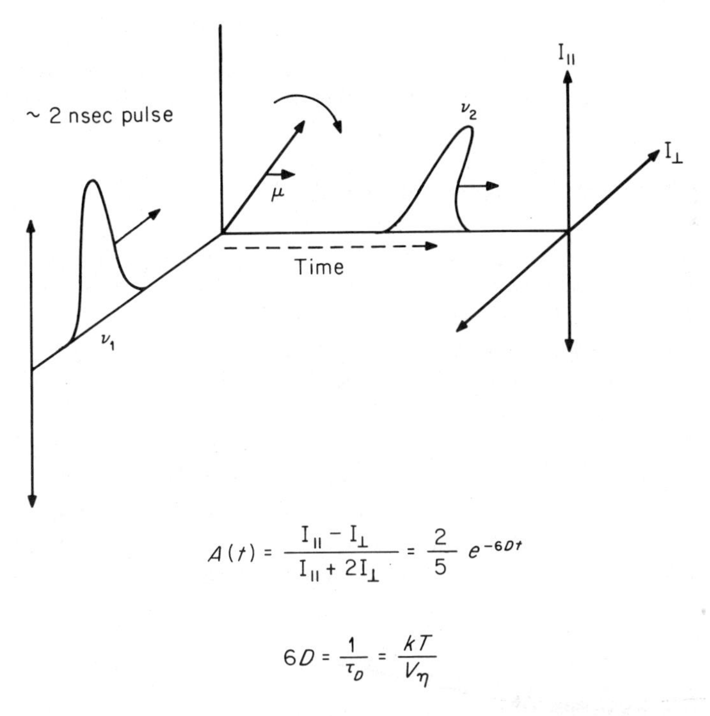

Figure 15. Schematic diagram of a typical anisotropy decay measurement. μ is the transition dipole moment of the fluorescent molecule.

Since the excitation is vertically polarized, those molecules whose transition moments are oriented along the vertical axis will be selectively excited. The light emitted by these photoselected molecules will therefore have a stronger parallel component ($I_{\parallel}$) than perpendicular component ($I_{\perp}$). This is true, however, only if the photoselected molecules retain their vertical orientations before they emit. Thus, if the molecules have moved sufficiently far away from their initial vertical orientations by the time they emit, the two components of emission would become equal. We therefore have these two extremes: (1) In a rigid or very viscous medium (such as a low-temperature glass), the photoselected molecules retain their anisotropic distribution about the vertical axis within the excited-state lifetime, such that $I_{\parallel} - I_{\perp} > 0$, and the fluorescence anisotropy (measured under steady-state conditions) attains some maximum value A_0. (2) In a fluid medium (small molecules in aqueous solution at room temperature, for example), the molecules can undergo Brownian rotational motion, such that the photoselected anisotropic distribution would be randomized within the excited-state lifetime. Thus $I_{\parallel} = I_{\perp}$, and

$A = 0$. Note that the denominator $I_{\|} + 2I_{\perp}$ can be shown to be proportional to the fluorescence decay under all conditions. Additional complications arise if the absorbing transition moment is not parallel to the emitting dipole (*26*).

Time-Dependent Fluorescence Depolarization

We now consider a case intermediate between the two extremes discussed above, namely a fluorescently labeled macromolecule at room temperature and in aqueous solution. In this case, Brownian rotational motion is neither so slow that the fluorescence is completely polarized, nor so fast that the fluorescence is completely depolarized. Suppose further that the excitation is now a short pulse of vertically polarized light, and we follow the time dependence of the anisotropy $A(t)$. The macromolecules are so large that at short times after the excitation pulse, little reorientation can take place. The situation thus resembles that of a rigid glass, and $A(0) = A_0$. At long times, much reorientation has taken place, the situation resembles that of the fluid medium extreme, and $A(\infty) = 0$. In other words, the anisotropy will in general, decay as a function of time.

Anisotropy Decay for Spherical Bodies

The qualitative ideas in the last section can be subjected to more quantitative treatment. In most cases, macromolecular reorientation is attributable to Brownian rotational motion. A diffusion equation can be derived and solved, from which the anisotropy decay can be calculated explicitly. Thus for a spherical body, the anisotropy decays as a simple exponential according to the Jablonski equation (*90*):

$$A(t) = A_0 e^{-t/\tau_D} \qquad (29)$$

where the relaxation time, τ_D is given by the Stokes-Einstein law*:

$$\frac{1}{\tau_D} = 6D = \frac{kT}{V\eta} \qquad (30)$$

where D is the diffusion coefficient, k is the Boltzmann constant, T is the temperature of the medium, η is the viscosity of the medium, and V is the volume of the body.

Thus, a single anisotropy decay measurement allows us to estimate macromolecular volumes. For example, we have measured the anisotropy

* Note that this definition of the relaxation time differs from the more common relaxation time, ρ, by a factor of three: $\rho = 3\tau_D$.

decay for ethidium bromide-labeled unfractionated yeast tRNA in the presence of 0.003 M magnesium (*91*). The decay was a single exponential with a relaxation time $\tau_D = 24.5$ nsec (see Fig. 18). Since the temperature and the viscosity were known, the molecular volume, assuming a spherical geometry, was calculated from Eq. (30) to be 114,000 Å^3. The interpretation of this result will be discussed in a later section.

Anisotropy Decay for Nonspherical Bodies

The anisotropy decay curve becomes considerably more complex when the body deviates from spherical symmetry. In the most general cases, when the macromolecule has no symmetry properties at all, five exponentials (*23*) appear in $A(t)$. Since this is far beyond the resolution of present day experimental methods, one usually has to resort to fitting the data by assuming certain well-defined shapes for the macromolecules. It is particularly convenient to treat the data as if they were true of ellipsoids of revolution, because the hydrodynamic properties of such bodies have been worked out in detail. Thus, for prolate and oblate ellipsoids, the decay of $A(t)$ is described by three exponentials (*92, 93*):

$$A(t) = A_0 [A_1(\theta)e^{-t/\tau_1} + A_2(\theta)e^{-t/\tau_2} + A_3(\theta)e^{-t/\tau_3} \tag{31}$$

where θ is the angle subtended by the transition moment of the label with respect to the symmetry axis of the body (see Fig. 16),

$$A_1(\theta) = (\tfrac{3}{2} \cos^2\theta - \tfrac{1}{2})^2$$

$$A_2(\theta) = 3 \cos^2\theta \sin^2\theta \tag{32}$$

$$A_3(\theta) = \tfrac{3}{4} \sin^4\theta$$

τ_1, τ_2, τ_3 are the three rotational relaxation times characterized by the diffusion coefficients:

$$\tau_1 = 1/6D_\perp$$

$$\tau_2 = 1/(5D_\perp + D_\parallel) \tag{33}$$

$$\tau_3 = 1/(2D_\perp + 4D_\parallel)$$

$D_\parallel$ is the diffusion coefficient for rotation about the symmetry axis, and $D_\perp$ is the diffusion coefficient for rotation about an axis that is perpendicular to the symmetry axis of the body.

Perrin's hydrodynamic equations (*94, 95*) allow us to calculate $D_\parallel$ and $D_\perp$, and hence the three relaxation times as a function of the axial ratio ρ between the longitudinal axis and the equatorial axis of the ellipsoid. The

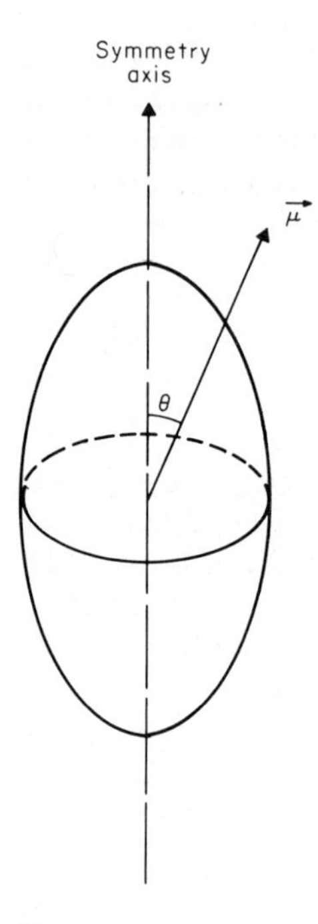

Figure 16. Definition of the orientation angle, θ, used in Eq. (32). μ is the transition dipole moment of the fluorescent molecule.

values of these relaxation times are tabulated in Table 5, normalized by the relaxation time, τ_D, of an equivolume sphere under the same conditions of temperature T and viscosity η.

The relaxation behavior associated with each of the three relaxation times is weighted by the three coefficients $A_1(\theta), A_2(\theta), A_3(\theta)$. The values of these coefficients are plotted against the orientation angle, θ, in Fig. 17.

Figure 17 together with Table 5 is very useful for the interpretation of anisotropy decay data, as we shall illustrate in a later section. Here we note some important features of the ellipsoidal model.

By inspection of Table 5 and Figure 17, the following are evident:

1. For oblate ellipsoids, although all three normalized relaxation times increase with degree of flattening, the three hardly differ from each other at any axial ratio. Equation (31) is therefore in effect a single exponential decay.
2. For prolate ellipsoids on the other hand, the three normalized relaxation times rapidly deviate from each other as the body is elongated. Note, also, that τ_3 is always close to τ_D.
3. Figure 17 shows that when $\theta = 0$, only A_1 is nonzero; Eq. (31) is once

TABLE 5. Rotational Diffusion Relaxation Times Appearing in Eq. (31) as a Function of Axial Ratios

ρ[a]	$\tau_{\parallel}/\tau_D$	τ_1/τ_D	τ_2/τ_D	τ_3/τ_D
		prolate ellipsoids		
1	1.0000	1.0000	1.0000	1.0000
2	0.8067	1.5049	1.3152	0.9543
3	0.7480	2.3408	1.7276	0.9674
4	0.7210	3.3956	2.0984	0.9777
5	0.7061	4.6405	2.4060	0.9842
6	0.6968	6.0616	2.6548	0.9884
7	0.6906	7.6505	2.8549	0.9911
8	0.6862	9.4008	3.0162	0.9930
9	0.6829	11.3080	3.1472	0.9944
10	0.6805	13.3680	3.2545	0.9954
15	0.6739	25.8607	3.5774	0.9979
20	0.6712	41.8199	3.7280	0.9988
$1/\rho$		oblate ellipsoids		
1	1.0000	1.0000	1.0000	1.0000
2	1.4100	1.1316	1.1701	1.3032
3	1.8284	1.4645	1.5147	1.6885
4	2.2495	1.8431	1.9003	2.0955
5	2.6718	2.2398	2.3018	2.5104
6	3.0948	2.6455	2.7111	2.9290
7	3.5181	3.0564	3.1247	3.3495
8	3.9418	3.4705	3.5411	3.7711
9	4.3655	3.8869	3.9593	4.1934
10	4.7894	4.3049	4.3787	4.6162
15	6.9099	6.4073	6.4859	6.7338
20	9.0312	8.5193	8.6005	8.8538

[a]ρ is the axial ratio, or ratio of the longitudinal semiaxis to the equatorial semiaxis, $\tau_{\parallel} = 1/6D_{\parallel}$, $\tau_1 = 1/6D_{\perp}$, $\tau_2 = 1/(5D_{\perp} + D_{\parallel})$, $\tau_3 = 1/(2D_{\perp} + 4D_{\parallel})$, and $\tau_D = 1/6D = V\eta/kT$ is the relaxation time under the spherical model.

again a single exponential. Note, also, that there is no region of θ in which all three amplitudes are of equal magnitude. In fact, we can roughly divide the entire range of θ into the following regions:

a. θ 0 A_1 dominates
b. θ 20° A_1 and A_2 dominate
c. θ 45° A_2 dominates
d. θ 60° A_2 and A_3 dominate
e. θ 90° A_1 and A_3 dominate

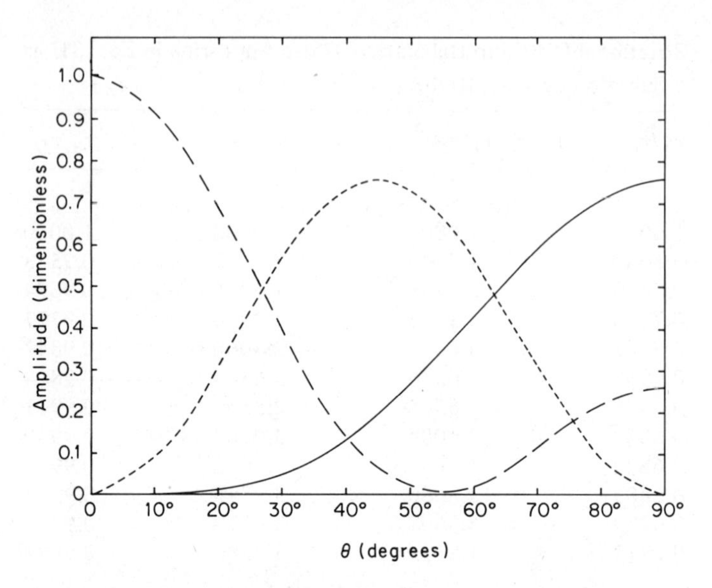

Figure 17. Plot of the three amplitudes appearing in Eq. (32). θ is defined in Figure 16. This figure was adapted from Tao, Nelson, and Cantor (*19*), with the permission of the publisher.

From these observations, we can draw the following conclusions:

1. If only one exponential decay of anisotropy is observed, one cannot immediately conclude that the body is spherical in shape. The possibilities exist that the body is an oblate ellipsoid or a prolate ellipsoid with preferential orientation on the part of the fluorescence label. Independent information such as the volume of the body helps considerably in distinguishing these possibilities. Thus, if the measured relaxation time is significantly longer than that predicted by the known volume, deviation from spherical geometry is a likely cause.
2. When more than one exponential appears in the anisotropy decay, the conclusion that the body is elongated is inescapable. If the body is assumed to be a prolate ellipsoid, then two exponentials suffice to fit the data. Again, if the volume of the body is known, then a number of consistency checks can be made; for example, the shortest possible relaxation time takes the value calculated from Eq. (30).

Steady-State Fluorescence Depolarization

All the SSFD expressions can be obtained from the time-dependent ones by taking the time average over the emission decay. Assuming that the fluorescence decay is a single exponential with lifetime τ_s, the steady-state anisotropy, $\bar{A}$, for a spherical body is calculated as follows:

$$\bar{A} = \frac{1}{\tau_s}\int_0^\infty A_0 e^{-t/\tau_D}\; e^{-t/\tau_D} = A_0 \frac{\tau_D}{\tau_D + \tau_s} \tag{34}$$

After slight rearrangement, we get an equivalent form of the Perrin equation (*96*).

$$\frac{1}{\bar{A}} = \frac{1}{A_0}\left(1 + \frac{kT}{V\eta}\tau_s\right) \tag{35}$$

Thus, the same kind of information can be obtained by making a Perrin plot. This is done by changing either the temperature or the viscosity of the medium, and measuring the steady-state anisotropy, $\bar{A}$. A plot of $1/\bar{A}$ v. T/η should be linear, and the slope contains information on the molecular volume.

For an ellipsoid of revolution, it can easily be shown that

$$\bar{A} = A_0 \sum_{i=1}^{3} A_i \frac{\tau_i}{\tau_i + \tau_s} \tag{36}$$

Knowing the values of A_i and the τ_i, one can construct the Perrin plots and compare them with the experimental Perrin plot. For small axial ratios, the Perrin plot is approximately linear, as Weber (*97*) has shown.

It should be clear that the time-dependent measurements are by far superior to the steady-state ones. They are more reliable and more direct; moreover, the necessity of altering the nature of the medium is obviated.

Depolarization Due to Energy Transfer

So far, we have attributed fluorescence depolarization to Brownian rotational motion alone. It has long been known, however, that fluorescence depolarization can take place via singlet-singlet energy transfer between like molecules (*98*). This principle can also be applied to probing macromolecular conformations (*26*). Thus, if more than one of the same fluorescence label can be loaded on the same macromolecule, the extent of depolarization reflects the proximity of the labels, and under favorable conditions, the size and shape of the macromolecule (*99, 100*).

Instrumentation for Polarization Work

This topic needs little elaboration. For polarization studies, all that is needed is the insertion of polarizers in an already existing spectrofluorimeter or lifetime apparatus. For steady-state work, designs with two photomultipliers, one for each component of fluorescence, have been described (*44, 101*).

Typical Experimental Results

Fluorescence depolarization studies can provide information on the size, conformation, and flexibility of a macromolecule. In this section we shall present a brief survey of applications of this technique to studies of tRNA. We were interested in the effect of magnesium ions on the conformation of tRNAs (*102, 103*). Circular dichroism spectra (*103, 104*) and UV hypochromism (*104*) indicate loss of hydrogen bonds and base stacking as magnesium is removed. We have used fluorescence depolarization to provide additional information on the effect of magnesium on the conformation of tRNA (*19*).

As a fluorescent probe we have used the dye ethidium bromide (*17*). In choosing this particular dye, we were influenced by the following considerations: (1) The fluorescence of ethidium bromide is much enhanced when bound to base paired regions of nucleic acids (*17*). Thus the fluorescence from free dye molecules can be largely ignored. (2) Ethidium bromide is believed to intercalate between nucleic acid base pairs (*76, 77*). The dye must, therefore, be rigidly held with respect to the macromolecule within the excited-state lifetime (*105*). This is to be contrasted with earlier SSFD studies (*28, 29*) in which flexibility on the part of the covalently attached fluorescence labels may obscure the measurements on the rotational motion of the macromolecule itself. (3) Intercalated ethidium bromide has a singlet lifetime of 23–26 nsec (*19*). This relatively long lifetime is attractive for at least two reasons: (a) The excitation pulse is so short in comparison that its effect can be ignored. (b) Although the expressions for the anisotropy $A(t)$ do not contain the fluorescence decay e^{-t/τ_s} explicitly, the actual experimental observables, $I_{\|}(t)$ and $I_{\perp}(t)$, are proportional to e^{-t/τ_s} (*23, 90*). Because of this, the resolution of the anisotropy decay measurement becomes increasingly poor as the relaxation time, τ_D, becomes much longer or much shorter than the excited-state lifetime, τ_s. Indeed, optimum conditions are reached when $\tau_s \simeq \tau_D$. The lifetime of ethidium bromide is just in the range of expected rotational relaxation times for tRNAs.

The above considerations should be taken whenever one is choosing a fluorescence label for depolarization work. The single most important consideration is probably the long singlet lifetime. Thus, the lifetimes of most traditional nucleic acid stains like molecules of the acridine family are too short for depolarization studies. At the present moment there are few

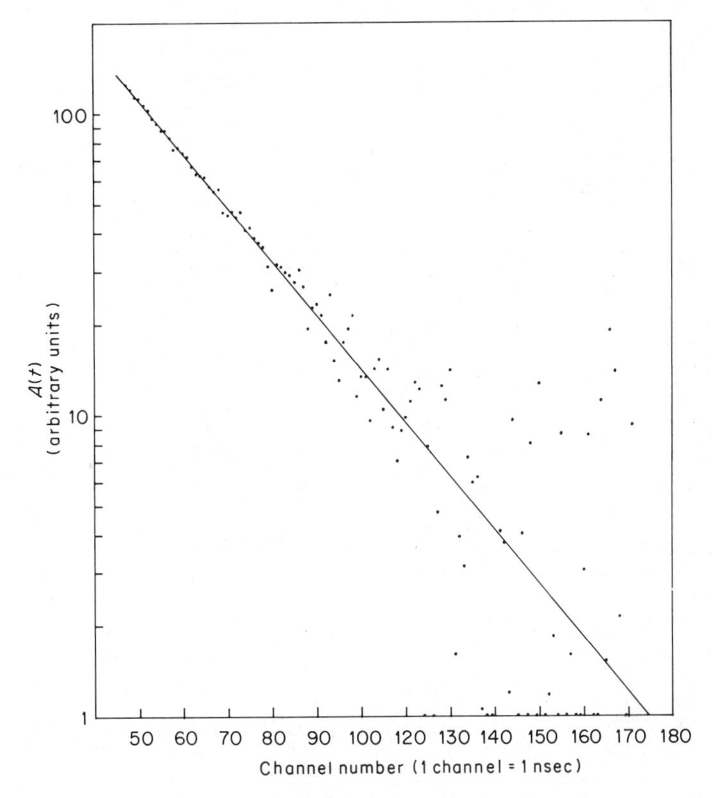

Figure 18. Decay of the fluorescence anisotropy of ethidium bromide intercalated into an unfractionated sample of yeast tRNA in the presence of magnesium ion (*91*). τ_D = 24.5 nsec. Temperature = 25.5°C.

noncovalent nucleic acid labels that have the long lifetime of ethidium bromide. Some potentially useful long-lifetime covalent labels include pyrene butyric acid, anthranilic acid, dansyl chloride, 1-anilino-8-naphthalene sulfonate, acridone, and *N*-methylacridine.

Our experimental conditions were always such that the dye concentration was less than the tRNA concentration. This is to avoid loading more than one dye on each tRNA molecule, so that depolarization due to energy transfer can be ruled out. Note that there is no requirement for stoichiometric binding either. Unlabeled tRNA molecules are simply not observed in the experiment.

Figures 18 and 19 show the anisotropy decay curves for unfractionated yeast tRNA in the presence and absence of magnesium, respectively. In the presence of magnesium the anisotropy decays with a single exponential of

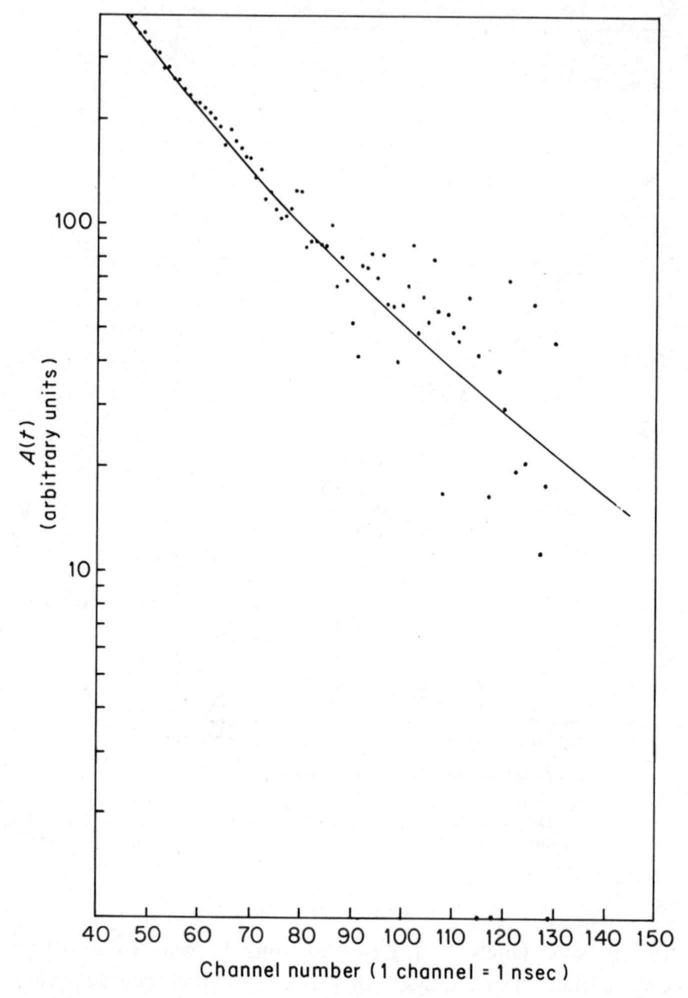

Figure 19. Decay of fluorescence anisotropy of ethidium bromide intercalated into a sample of unfractionated yeast tRNA in the absence of added magnesium ion. Best two exponential fit:

$A(t) = 2.1\, e^{-t/16.8 \text{ nsec}} + 1.7\, e^{-t/39.4 \text{ nsec}}$

Temperature is 22°C. This figure is adapted from Tao, Nelson, and Cantor (*19*) with permission of the publisher.

$\tau_D{}^{25°,\, w} = 24.8$ nsec*. As a first approximation, consider the molecule to be spherical in shape. Equation (30) is used to find a molecular volume of 114,000 $Å^3$, or a radius of 30.1 Å. In comparison, hydrodynamic work yields a Stokes' radius of 28.5 Å (*103*). The molecular volume of tRNA can be

* This relaxation time has been corrected to 25°C and the viscosity of water.

calculated to be 75,000 Å^3 by using a partial specific volume $\bar{\nu}$ = 0.503 ml/gm (*103*), mol wt = 26,400 gm/mole (*106*), and degree of hydration δ_s = 1.2 gm H_2O/gm tRNA (*107*). The excess in our measured volume may be due to deviations from spherical geometry. If the molecule is assumed to be an oblate ellipsoid, the axial ratio can be found (with the aid of Table 5) to be $1/\rho = 3.0$. If the ellipsoid is assumed prolate, the axial ratio ρ ranges from 1.8 to 3.0 when δ is taken to be 1.2 ± 0.2 gm/gm (see Ref. *19* for details).

In the absence of magnesium, the anisotropy decay is markedly nonexponential (Fig. 19), indicating that the molecule is substantially elongated. Assuming a prolate ellipsoid model, the anisotropy decay curve was fitted to two exponentials. The values to the two relaxation times, together with the ratio of the two amplitudes give rise to an axial ratio of $\rho = 4.6$, and a molecular volume of 74,000 Å^3. The calculated anhydrous volume is 23,300 Å^3 (using $\nu = 0.531$ ml/gm (*106*)), thus the degree of hydration is estimated to be 1.2 gm/gm. It should be noted that considerable uncertainty exists in the absolute values of the axial ratio and hydration reported here. However, the very fact that the anisotropy decay curve in Fig. 19 was nonexponential allowed us to firmly conclude that there is definitely a magnesium-dependent conformational change, and that the tRNA becomes more elongated as magnesium was removed.

Under favorable conditions, decay of fluorescence anisotropy also yields information on the orientation of the fluorescent label. Thus, from the data in Fig. 19, one can conclude that the orientation angle, θ, must be about 60° (*19*). This means that the principal transition moment of the ethidium bromide label is oriented at an angle of 60° to the long axis of the tRNA molecule. Since the principal transition moment of ethidium bromide is almost certainly in the plane of the molecule, and since ethidium bromide is believed to intercalate between base pairs, this result suggests that the plane of the base pairs are in turn tilted at an angle of 60° with respect to the long axis of the tRNA molecule.

Previously we have had strong reasons to believe that the fluorescent label was rigidly attached to the macromolecule, and that the macromolecule itself was a rigid hydrodynamic particle. This, of course, is not generally true, and fluorescence depolarization can be used to detect such internal flexibility. Theoretical treatment of the effect of internal flexibility on fluorescence depolarization have been given by Wallach (*108*), and by Gottlieb and Wahl (*109*). Yguerabide et al. (*27*) have inferred internal flexibility in the immunoglobulin molecule in their time dependent fluorescence depolarization work. Using the same technique, Wahl et al. found evidence for internal motion in DNA (*75*). Anisotropy decay measurements have been carried out with the fluorescence from the Y base of $tRNA^{Phe}$ (*15*). It was found that for a sample of yeast $tRNA^{Phe}$ in the presence of 0.003 M magnesium, measurements using the fluorescence of intercalated ethidium bromide yielded a relaxation time of $\tau_D{}^{25°, w} = 24.5$ nsec, almost exactly the same as that for unfractionated yeast tRNA in the presence of magnesium. Measurements with the fluorescence of Y, on the other hand, yielded $\tau_D{}^{25°, w} = 10$ nsec. The large discrepancy between these two values suggests that intercalated ethidium bromide is rigidly bound to the $tRNA^{Phe}$ molecule, while some degree of flexibility exists for either the Y base itself, or the entire anticodon loop.

Acknowledgments

This work was supported by a grant, GM-14825, from the U.S. Public Health Service. The authors have benefited greatly from conversations with Professor Richard Bersohn, Professor Lubert Stryer, D. C. K. Luk, Dr. Kenneth Beardsley, and Dr. James Nelson. We are grateful to Mr. Paul Koenig for allowing us to cite several unpublished observations.

References

1. Guéron, M., and R. G. Shulman. 1968. *Ann. Rev. Biochem.*, **37**: 571.
2. Parker, C. A. 1968. Photoluminescence of solutions. Elsevier, Amsterdam.
3. Hercules, D. M. (Ed.). 1966. Fluorescence and phosphorescence analysis. Wiley (Interscience), New York.
4. Udenfriend, S. 1962. Fluorescence assay in biology and medicine. Academic Press, New York, Vol. I.
5. Udenfriend, S. 1969. Fluorescence assay in biology and medicine. Academic Press, New York, Vol. II.
6. Lim, E. C. (Ed.). 1969. Molecular luminescence. Benjamin, New York.
7. Becker, R. S. 1969. Theory and interpretation of fluorescence and phosphorescence. Wiley (Interscience), New York.
8. Bowen, E. J. (Ed.). 1968. Luminescence in chemistry. Van Nostrand, New Jersey.
9. Chen, R. F., H. Edelhoch, and R. F. Steiner. 1969. In S. J. Leach (Ed.), Physical principles and techniques of protein chemistry. Academic Press, New York, p. 171.
10. Guilbault, G. 1967. Fluorescence, theory, instrumentation and practice. Dekker, New York.
11. Reid, C. 1957. Excited states in chemistry and biology. Butterworth, London.
12. Stryer, L. 1968. *Science*, **162**: 526.
13. LePecq, J-B. Y., and C. Paoletti. 1966. *Anal. Biochem.*, **17**: 100.
14. Yoshikami, D., G. Katz, E. B. Keller and B. S. Dudock. 1968. *Biochim. Biophys. Acta*, **166**: 714.
15. Beardsley, K., T. Tao, and C. R. Cantor. 1970. *Biochemistry*, **9**: 3524.
16. Eisinger, J., B. Feuer, and T. Yamane. 1970. *Proc. Natl. Acad. Sci. U.S.*, **65**: 638.
17. LePecq, J. B. and C. Paoletti. 1967. *J. Mol. Biol.*, **27**: 87.
18. Bittman, R. 1969. *J. Mol. Biol.*, **46**: 251.
19. Tao, T., J. Nelson, and C. R. Cantor. 1970. *Biochemistry*, **9**: 3514.
20. Ellerton, N. F. and I. Isenberg. 1969. *Biopolymers*, **8**: 767.
21. Thomes, J. C., G. Weill, and M. Daune. 1969. *Biopolymers*, **8**: 647.
22. Rosen, C. G. and G. Weber. 1969. *Biochemistry*, **8**: 3915.

23. Tao, T. 1969. *Biopolymers,* **8:** 609.
24. Vaughan, W. M. and G. Weber. 1970. *Biochemistry,* **9:** 464.
25. Stryer, L. 1966. *J. Am. Chem. Soc.,* **88:** 5708.
26. Weber, G. 1966. In D. M. Hercules (Ed.), Fluorescence and phosphorescence analysis. Wiley (Interscience), New York.
27. Yguerabide, J., H. Epstein and L. Stryer. 1970. *J. Mol. Biol.,* **51**:573.
28. Millar, D. B. and R. F. Steiner. 1966. *Biochemistry,* **5:** 2289.
29. Churchich, J. E. 1963. *Biochim. Biophys. Acta,* **75:** 274.
30. Stryer, L. and R. P. Haugland. 1967. *Proc. Natl. Acad. Sci. U.S.* **58:** 719.
31. Galley, W. C. and L. Stryer. 1969. *Biochemistry,* **8:** 1831.
32. Galley, W. C. and L. Stryer. 1966. *Proc. Natl. Acad. Sci. U.S.,* **60:** 108.
33. Schechter, A., R. F. Chen, and C. B. Anfinsen. 1970. *Science,* **167:** 886.
34. Harvey, R. 1969. *Anal. Biochem.,* **29:** 58.
35. Chen, R. F., A. N. Schechter, and R. L. Berger. 1969. *Anal. Biochem.,* **29:** 68.
36. Helene, C., F. Brun, and M. Yaniv. 1969. *Biochem. Biophys. Res. Commun.,* **37:** 393.
37. Turro, N. J. 1965. Molecular photochemistry. Benjamin, New York, Chap. 4.
38. Merzbacher, E. 1961. Quantum mechanics. Wiley, New York, Chap. 20.
39. Strickler, S. J. and R. A. Berg. 1962. *J. Chem. Phys.,* **37**: 814.
40. Ellis, D. W. 1966. In D. M. Hercules (Ed.), Fluorescence and phosphorescence analysis. Wiley (Interscience), New York.
41. Parker, C. A. 1958. *Nature,* **182:** 1002.
42. Turner, G. K. 1964. *Science,* **146:** 183.
43. Hamilton, T. D. S. 1966. *J. Sci. Instr.,* **43:** 49.
44. Witholt, B. and L. Brand. 1968. *Rev. Sci. Instr.,* **39:** 1271.
45. Spencer, R. D. and G. Weber. 1969. *Ann. N.Y. Acad. Sci.,* **158:** 361.
46. Spencer, R. D., W. M. Vaughan, and G. Weber. 1969. In E. C. Lim (Ed.), Molecular luminescence. W. A. Benjamin, New York, p. 607.
47. Bennett, R. G. 1960. *Rev. Sci. Instr.,* **81:** 1275.
48. D'Alessio, J. T., P. K. Ludwig, and M. Burton. 1964. *Rev. Sci. Instr.* **35:** 1015.
49. Bollinger, L. M. and G. E. Thomas. 1961. *Rev. Sci. Instr.* **32:** 1044.
50. Birks, J. B. and I. H. Munro. 1967. *Progr. Reaction Kinetics,* **4:** 239.
51. Amata, C. D. and P. K. Ludwig. 1967. *J. Chem. Phys.,* **47:** 3540.
52. Pfeffer, G., H. Lami, E. Lanstriat, and A. Coche. 1963. *Nucl. Instr. Methods,* **23:** 74.
53. Berlman, I. B., O. J. Steingraber, and M. J. Bensen. 1968. *Rev. Sci. Instr.* **39:** 54.
54. Hundley, L., T. Coburn, E. Garwin, and L. Stryer. 1967. *Rev. Sci. Instr.,* **38:** 488.
55. Isenberg, I. and R. D. Dyson. 1969. *Biophys. J.,* **9:** 1337.
56. Parker, C. A. 1968. Photoluminescence of solutions. Elsevier, Amsterdam, p. 239.
57. Yaniv, M., A. Favre, and B. G. Barrell. 1969. *Nature,* **223:** 1331.
58. P. L. Biochemicals, circular OR-10.
59. Beardsley, K. and C. R. Cantor. 1970. *Proc. Natl. Acad. Sci. U.S.,* **65:** 39.
60. RajBhardary, U. L., S. H. Chang, A. Stuart, R. D. Faulkner, R. M. Hoskinson, and H. G. Khorana. 1967. *Proc. Natl. Acad. Sci. U.S.,* **57:** 751.
61. Dudock, B. S., G. Katz, E. K. Taylor, and R. W. Holley. 1969. *Proc. Natl. Acad. Sci. U.S.,* **62:** 941.
62. Fink, L. M., T. Goto, F. Frankel, and I. B. Weinstein. 1968. *Biochem. Biophys. Res. Commun.,* **32:** 963.
63. Leng, M., F. Pochon, and A. M. Michelson. 1968. *Biochim. Biophys. Acta.,* **169:** 338.

64. Ward, D. C., E. Reich, and L. Stryer. 1969. *J. Biol. Chem.*, **244:** 1228.
65. Cantor, C. R., K. Beardsley, J. Nelson, T. Tao, and K. W. Chin. 1970. In F. Hahn (Ed.), Progress in molecular and subcellular biology. Springer-Verlag, Berlin, Vol. 2, in press.
66. Hoffman, B., P. Scholfield, and A. Rich. 1969. *Proc. Natl. Acad. Sci. U.S.*, **62:** 1195.
67. Gillam, I., D. Blew, R. C. Warrington, M. von Tigerstrom, and G. M. Tener. 1968. *Biochemistry,* **7:** 3459.
68. Ralph, R. K., R. J. Young, and H. G. Khorana. 1963. *J. Am. Chem. Soc.*, **85:** 2002.
69. Hara, H., T. Horiuchi, M. Saneyoshi, and S. Nishimura. 1970. *Biochem. Biophys. Res. Commun.*, **38:** 305.
70. Ofengand, J. 1965. *Biochem. Biophys. Res. Commun.*, **18:** 192.
71. Cramer, F., H. Doepner, F. v. d. Haar, E. Schlimme, and H. Seidel. 1968. *Proc. Natl. Acad. Sci. U.S.* **61:** 1384.
72. Grunberger, D., J. Nelson, C. R. Cantor, and I. B. Weinstein. 1970. *Proc. Natl. Acad. Sci. U.S.*, **66:** 488.
73. Hahn, F. (Ed.). 1970. Progress in molecular and subcellular biology. Springer-Verlag, Berlin, in press.
74. Waring, M. J. 1966. *Biochim. Biophys. Acta,* **114:** 234.
75. Wahl, P., J. Paoletti, and J-B. LePecq. 1970. *Proc. Natl. Acad. Sci. U.S.*, **65:** 417.
76. Bauer, W. and J. Vinograd. 1968. *J. Mol. Biol.*, **33:** 141.
77. Crawford, L. V. and M. J. Waring. 1967. *J. Mol. Biol.*, **25:** 23.
78. Gennis, R. and C. R. Cantor. 1970. *Biochemistry,* **9:** 4714.
79. Uhlenbeck, O. C., J. Baller and P. Doty. 1970. *Nature,* **225:** 503.
80. Hudson, B., W. B. Upholt, J. Devinny, and J. Vinograd. 1969. *Proc. Natl. Acad. Sci. U.S.*, **62:** 813.
81. Haugland, R. P., J. Yguerabide, and L. Stryer. 1969. *Proc. Natl. Acad. Sci. U.S.*, **63:** 23.
82. Förster, T. 1966. In O. Sinanoglu (Ed.), Modern quantum chemistry, Academic Press, New York, Vol. 3, p. 93.
83. Latt, S. A., H. T. Cheung, and E. R. Blout. 1965. *J. Am. Chem. Soc.* **87:** 995.
84. Badley R. A. and F. W. J. Teale. 1969. *J. Mol. Biol.* **44:** 71.
85. Bücher, H., K. H. Drexhage, M. Fleck, H. Kuhn, D. Möbius, F. P. Schäfer, J. Sondermann, W. Sperland, P. Tillmann, and J. Wiegand, 1967. *Mol. Cryst.*, **2:** 199.
86. Weber, G. and M. Shinitzky. 1970. *Proc. Natl. Acad. Sci. U.S.*, **65:** 823.
87. Weber, G. 1953. *Advan. Prot. Chem.*, **8:** 415.
88. Steiner, R. F. and H. Edelhoch. 1962. *Chem. Rev.* **62:** 462.
89. Chen. R. F. 1967. In G. Guilbault (Ed.), Fluorescence, theory, instrumentation, and practice. Dekker, New York.
90. Jablonski, A. 1961. *Z. Naturforsch.* **169:** 1.
91. Tao, T. 1969. Ph.D. Thesis. Columbia University, New York.
92. Memming, R. 1961. *Z. Physik. Chem. (Frankfurt),* **28:** 169.
93. Wahl, P. 1966. *Compt. Rend. Acad. Sci. (Paris),* **263D:** 1525.
94. Perrin, F. 1936. *J. Phys. Radium,* **7:** 1.
95. Woessner, D. E. 1962. *J. Chem. Phys.*, **37:** 647.
96. Perrin, F. 1934. *J. Phys. Radium,* **5:** 497.
97. Weber, G. 1952. *Biochem. J.*, **51:** 145.
98. Förster, T. 1951. Fluoreszenz Organischer Verbindungen. Druck Herbert, Göttingen, Chap. 9.
99. Weber, G. and L. B. Young. 1964. *J. Biol. Chem.*, **239:** 1415.
100. Weber, G. and E. Daniel. 1966. *Biochemistry,* **5:** 1900.

101. Stryer, L. 1965. *J. Mol. Biol.*, **13:** 482.
102. Lindahl, T., A. Adams, and J. R. Fresco. 1966. *Proc. Natl. Acad. Sci. U.S.*, **55:** 941.
103. Adams, A., T. Lindahl, and J. R. Fresco. 1967. *Proc. Natl. Acad. Sci. U.S.*, **57:** 1684.
104. Reeves, R. H., C. R. Cantor, and R. W. Chambers. 1970. *Biochemistry*, **9**: 3993.
105. Li, H. J. and D. M. Crothers. 1969. *J. Mol. Biol.* **39:** 461.
106. Lindahl, T., D. D. Henley, and J. R. Fresco. 1965. *J. Am. Chem. Soc.*, **87:** 4961.
107. Kuntz, I. D., T. S. Brassfield, G. D. Law, and G. V. Purcell. 1969. *Science*, **163:** 1329.
108. Wallach, D. 1967. *J. Chem. Phys.*, **47:** 5258.
109. Gottleib, Y. Y. and P. Wahl. 1963. *J. Chim. Phys.* **60:** 849.
110. Luk, C. K. 1970. Ph.D. Thesis. Columbia University, New York.
111. Favre, A., Yaniv, M. and Michelson, A. M. 1969. *Biochem. Biophys. Res. Commun.*, **37:** 266.
112. Millar, D. B. S. and Steiner, R. F. 1967. *Biochim. Biophys. Acta.*, **102:** 571.

RAMAN SPECTROSCOPY OF POLYNUCLEOTIDES AND NUCLEIC ACIDS*

WARNER L. PETICOLAS
DEPARTMENT OF CHEMISTRY
UNIVERSITY OF OREGON
EUGENE, OREGON

I. INTRODUCTION

A. Raman Processes

Raman processes using lasers comprise a whole set of new techniques for the characterization of the physical properties of polynucleotides and nucleic acids as well as other biological and synthetic polymers (*1*). Because this field is in its infancy this review will attempt to deal with some of the long-range prospects as well as with what has already been accomplished.

By the term "Raman process" we really mean any type of inelastic light scattering, i.e., an experiment in which one irradiates the sample with an incident laser beam and then looks at the intensity of the scattered light as a function of the frequency. Using three different techniques it is now possible, at least in principle, to determine the intensity of the scattered light as a function of the frequency for frequencies which differ from that of the incident light by an amount ranging from 10 to 10^{14} Hz. The Rayleigh line may be considered to fall in the range 10 to 10^5 Hz (*2*) about the central component. This light is sometimes said to be elastically or quasielastically scattered. Light scattering from thermally or mechanically produced sound

* Supported by NSF grant GB 13700.

waves in transparent solid or fluid media is called Brillouin scattering and occurs in the region 10^8 to 10^{10} Hz (*3–5*). Raman scattering generally is considered to fall in the range 10^{11} to 10^{14} Hz, i.e., 5 cm^{-1} to 3000 cm^{-1}. It will be with this last region which this review will be almost exclusively concerned.

When light from a laser is incident on a molecule there is a periodic displacement of the electrons in the molecule due to the alternating electric field of the light. This periodically induced dipole moment then becomes itself a secondary source of radiation. The induced dipole moment, p_i, in the ith direction ($i = x, y, z$) may be expanded in terms of the electric field by the equation,

$$p_i = \sum_j \alpha_{ij} E_j + \ldots \tag{I-1}$$

Thus if the polarization field of the light is applied in the jth direction, ($j = x, y, z$), there may be a component of induced electric moment p_i in any or all of the three directions, x, y, or z. Thus the components α_{ij} form a nine component tensor, α_{xx}, α_{xy}, ... α_{zz}, called the polarizability tensor. To measure one of these components, for example α_{xz}, one, in principle, orients the molecule so that the incident light is polarized along the x-axis of the molecule and observes the intensity of the light scattered at the same frequency but with its polarization in the z direction. The x-, y-, and z-axes in the molecule are taken from group theoretical convention depending upon the point symmetry group to which the molecule belongs. Thus for the polynucleotide bases the x- and y-axes may be taken as any two orthogonal axes in the plane of the base while the z-axis is taken as perpendicular to the plane. This then is the theory of simple light scattering from a single oriented molecule whose dimensions are small compared to that of the wavelength of the light so that no interference effects must be considered.

Elastic and quasielastic light scattering may be described by Eq. (I-1) if α_{ij} is simply a constant. In this case the electronic polarization is proportional to the applied field and the induced oscillating dipole acts as a source of secondary radiation with the frequency of the incident light. However, if the atoms or the molecules are moving with a translational velocity, then because of the Doppler effect the frequency of the scattered light is shifted, and if the motion is random the frequency range of the scattered light is considerably broader than that of the incident laser light. Using a technique called homodyne spectroscopy (*2, 6*) it is possible to obtain the frequency range very close to that of the laser source and information on the diffusional motion of polymers in solution has been obtained (*7*). The theory is due to Pecora (*8*).

However, in this review we will deal with the Raman effect which arises because the polarizability tensor is not simply a constant but may vary periodically in time. This is due to the fact that the nuclei of the molecule are not stationary but vibrate periodically with a set of frequencies such that each frequency corresponds to one of the normal modes of vibration. If for any molecular vibration there is a change in the polarizability tensor α_{ij} for a displacement of the normal coordinate, Q_a, i.e., $\delta\alpha_{ij}/\delta Q_a = 0$, then there will appear in the frequency spectrum of the scattered light a line shifted from that of the incident light by an amount equal to the frequency of the vibration. This can be shown in a very simple, well-known classical manner. Let us express the field of the incident light E_j by the equation,

$$E_j = E_j^0 \cos \omega_L t$$
$$\omega_L = 2\pi\nu_L \tag{I-2}$$

where ω_L is the circular frequency of the incident light and ν_L is the linear frequency.

Now let us expand the polarizability tensor, α_{ij}, in terms of the normal coordinate, Q_a, for the ath vibration,

$$\alpha_{ij} = \alpha_{ij}^0 + (\partial\alpha_{ij}/\partial Q_a)_0 Q_a \tag{I-3}$$

where α_{ij}^0 is the polarizability in the equilibrium position, $Q_a = 0$. Since the normal coordinate Q_a is periodic in time, we may write

$$Q_a = Q_a^0 \cos \Omega_a t \tag{I-4}$$

where Q_a^0 is the maximum amplitude of the displacement of the normal coordinate, Q_a, and Ω_a is the circular frequency of the molecular vibration. Putting Eq. (I-2), (I-3), and (I-4) into Eq. (I-1) we have

$$p_i = \sum_j \alpha_{ij}^0 E_j^0 \cos \omega_L t + \sum_j (\partial\alpha_{ij}/\partial Q_a) Q_a^0 E_j^0 \left[\cos(\omega_L + \Omega_a)t + \cos(\omega_L - \Omega_a)t\right] \tag{I-5}$$

The first term in Eq. (I-5) gives rise to elastic light scattering as discussed previously, but we see that the second term gives rise to frequencies at $\omega_L \pm \Omega_a$ where the minus sign denotes the Stokes line and the plus sign denotes the anti-Stokes line. For quantum statistical mechanical reasons the anti-Stokes lines are generally much weaker than the Stokes lines and go to zero as the temperature is decreased. However, we see that from the spectrum of the scattered light we can obtain the vibrational frequencies of those normal modes of vibration which change the polarizability tensor.

B. Brief Review of Work on Raman Scattering from Nucleic Acids and Related Compounds

The investigation of the Raman spectra of nucleic acids and their constituent bases began with the reporting of the Raman frequencies of the bases and some of their derivatives by Malt (*9*), followed shortly by a more comprehensive study by Lord and Thomas (*10a*). At that time they were able to give some rather general assignments for the bands they observed. Lord and Thomas were also able to obtain a spectrum of one polymer polyadenylic acid (*10b*). The Raman frequencies of all four common polyribonucleic acids were recently reported by Fanconi et al. (*11*) and preliminary Raman spectra of certain naturally occurring nucleic acids were reported by Tobin (*12*). Recently, Aylward and Koenig (*13*) also reported the Raman spectra of 20 % solution of polyadenylic acid and polycytidylic acid; Thomas has reported the Raman spectrum of rRNA (*14*).

Unfortunately all that we can say at present about the observed Raman bands of the polynucleotides is that most of the intense ones are due to vibrations of the purine and pyrimidine bases which are attached to the sugar-phosphate polymer backbone, although attempts to further characterize these vibrations have been made using infrared absorption and Raman spectroscopy of model compounds such as imidazole (*15–17*), pyrimidine (*18–20*), and purine (*21*), including some of their derivatives. However, from a brief consideration of the group theory of these vibrations something can be said about the polarization of the scattering from the vibrations. The purine and pyrimidine bases belong to the point group C_{1h} (usually called C_s) which possesses only two symmetry elements, the identity and a plane of reflection. Consequently, the vibrations of the bases belong to one of two irreducible representations, A', the totally symmetric representation in which case the vibrations are in the plane of the ring, or A'', in which case the nuclei move perpendicular to the plane of the ring. Scattering from the A' vibrations should be rather strongly although not completely polarized, while that from the A'' vibrations should be completely depolarized. Thus in-plane and out-of-plane vibrations should be distinguishable from polarization data alone. This point will be discussed in more detail later in Section IV.

Recently, Tomlinson and Peticolas (*22*) reported a temperature dependence of Raman scattering intensities in certain bands in poly A. They interpreted their results in terms of the Raman intensity theory of Peticolas et al. (*23*) as being derived from the changes in the ultraviolet absorption intensities brought about by temperature dependent conformational changes in the polymer. These changes in UV absorption intensities occur when planar molecules such as purines or pyrimidines are caused to stack face to face in a one-dimensional array by one of a number of methods, the most common being lowering the temperature. This phenomenon is generally referred to as UV hypochromism because the UV absorbance of the lower energy absorption bands of the bases are observed to decrease. Since a similar decrease in certain Raman bands of poly A was observed upon lowering the temperature it was suggested (*22*) that this phenomenon should be called Raman hypochromism, and based upon a very approximate theory, it was predicted that this phenomena should be most pronounced in the totally symmetric, i.e., the in-plane ring vibrations of the purine and pyrimidine bases. These observations will be reported in detail in Section III, where the results of Small and Peticolas (*24, 25*) will be discussed in detail.

C. Advantages and Problems in Raman Spectroscopy

There are many advantages to Raman spectroscopy as a technique for the characterization of nucleic acids. The Raman bands are very sharp, of the order 10–40 cm^{-1} and occur throughout the range 400 cm^{-1} to 3000 cm^{-1} – all of which is accessible to samples studied in both aqueous or D_2O solutions and most of it available to samples studied in water alone. Thus, studies on the configurational interactions of several different bases may be in principle carried out simultaneously. The UV absorption bands on the other hand are all very broad (of the order of 5000 cm^{-1}) and occur over approximately the same frequency range for both purines and pyrimidines. Consequently, one must resort to complicated mathematical methods to separate the effects due to specific bases. Infrared absorption bands are also very sharp but because of the high IR absorbance of water work must usually be done in D_2O over a rather narrow range of available frequencies. In spite of these limitations, much useful information has been obtained (*26*). However, there is a great deal more information in the Raman spectrum since the frequencies arise from the nuclear vibrations while the intensities arise from the change in the electronic polarizability with the nuclear vibrational displacements so that it should be possible through measurements of intensity changes and frequency shifts to distinguish many structural features.

Raman spectroscopy is not, however, without its disadvantages. If this were not so, this method would have been used more widely than it has been to date. The major disadvantages of Raman spectroscopy of nucleic acids is that Raman spectra are so hard to obtain on dilute solutions. The lower limit of concentration that can be studied is limited by the extremely broad weak Raman spectrum of water. In more recent work, concentrations of 10 to 25 mg/ml (approximately 1 to 2.5 %) have been used. Although use of higher concentrations results in greatly improved signal-to-noise ratio, it is considered of greater importance to use the lowest concentration possible. At these concentrations the Raman spectrum of water effectively supplies a background over the entire frequency range of interest, decreasing the signal-to-noise ratio. In all Raman spectra in aqueous solutions, this background comes down from the low-frequency end of the spectrum, continues to decrease across the middle, and forms a broad band at about 1635 cm^{-1}. This background is not due to fluorescent impurities since it is independent of concentration and may be simply obtained by running the

Raman spectrum of water. This background limits the concentrations we can study at the present, but we are confident that in the future improved instrumentation will enable the investigation of lower concentrations.

II. EXPERIMENTAL TECHNIQUES

Our Raman techniques have undergone considerable modifications during the last several years. This will be apparent in the quality of the data as we progress from our original work on the polynucleotides to our conformational work on the polynucleotides to our more recent studies on helical complexes. We would like to describe our own current apparatus in some detail.

The Raman apparatus consists essentially of a Coherent Radiation Model 53 argon ion laser, a Spex 1400 double grating monochromator driven by a Slo-Syn stepping motor and preset indexer, a specially selected ITT FW 130 Startracker phototube cooled to -20°C by a photoelectric cooler, and Ortec photon counting equipment with the output of a Nuclear Chicago rate meter displayed on a Speedomax recorder.

Although in our original conformational work with polynucleotides (*24*) we measured peak heights off the chart paper to plot band intensities vs. temperature, we now determine Raman intensities quickly and more accurately by the use of a RIDL model 49-25 scalar, which counts the output of our integral discriminator.

A new sample cell and holder are now used. The sample cell is 4 mm long, has a diameter of about 2 mm, and has a bottom made from the side of a quartz cuvette. The cell holder is a copper block with holes through which water of carefully controlled temperature circulates. The cell sits within the block and is covered with a brass plate with a window to allow the exit of the laser beam. A small "O"-ring fits tightly around the top of the cell and seals against the brass plate, completely eliminating the problem of solvent evaporation. A hole in the bottom of the holder permits the entry of the laser beam, and a hole on the side permits the sampling of the Raman scattered light.

Since the small cell is buried within the block, the temperature equilibrates quickly and accurately. The temperatures we report here are the temperatures of the circulating bath, and we have found them to represent the temperature in the cell to within 0.2°C. We of course do not know the temperature directly in the focused laser beam, but since we observe a T_m actually two degrees higher rather than lower for poly A • poly U double helix than that reported in the literature, the discrepancy between our measured temperature and the true temperature cannot be great.

The use of a scalar has tremendously decreased the amount of scatter in the intensity vs. temperature plots. Appropriate wavelengths are chosen to represent the background and the middle of the band of interest, and the monochromator slits are opened to 300 μ. The broad slits allow us to obtain what is essentially an integrated intensity. The temperature is adjusted, the temperature in the cell is allowed to equilibrate, the background and signal are counted in succession for one minute each, and the process is repeated. Taking the data for a typical melting plot requires from thirty minutes to an hour depending on how many points are desired. The integrated band intensity is considered to be the number of counts per minute in the broad center of the Raman band minus the number from the background. This method lends itself well to the determination of depolarization ratios by merely adding an analyzer between the sample and the monochromator, and counting on the Raman bands and background with the analyzer in both the horizontal and vertical positions. The appropriate background can then be subtracted from each signal.

The Raman spectra presented here on helical complexes were taken with a resolution of about 4 cm^{-1}, a scan speed of 25 cm^{-1}/min, and a time constant of 0.5 seconds. The wavelengths of the emitted light were inked at the top of the chart paper, the frequencies in cm^{-1} were inked below the Raman bands, and the spectra were photographed directly from the recorder chart. For the exact conditions used for the earlier work presented, we will refer the reader to the original publications (*11, 24*).

III. REVIEW OF EXPERIMENTAL RESULTS

A. Raman Spectra of Polynucleotides

We will begin this review of experimental results with a detailed comparison of the Raman spectra of the polyribonucleotides – polyadenylic acid (poly A), polyuridylic acid (poly U), polycytidylic acid (poly C), polyguanylic acid (poly G) and polyinosinic acid (poly I). The Raman bands listed in Table 1 are taken from measurements on 1.5% aqueous solutions at neutral pH and room temperature.

A few comments should be made about the Raman band assignments presented in Table 1. These assignments are based largely on work done in this laboratory, taking into account the assignments made earlier by Lord and Thomas (*10a*) in their pioneering work on the Raman spectra of nucleic acid constituents. Most of the Raman bands which are intense and highly polarized are assigned to the in-plane ring vibrations of the bases. Most of the less intense Raman bands are believed to be due to vibrations of the sugar–phosphate chain, although precise assignments can not yet be made.

One band of interest occurs at about 800 cm^{-1}. This band undoubtedly arises from the ribose–phosphate chain, and we have assigned it to the phosphate diester symmetric stretch. This assignment is, however, questionable, but the frequency of this band appears to convey information on the order of the ribose phosphate chain. Note in Table 1 that in poly A, poly C, and poly G, which all possess somewhat ordered structures under these conditions, this Raman band occurs at 810–819 cm^{-1}. In poly U and poly I however, which have considerably less ordered structures under these conditions, this band occurs at 794–800 cm^{-1}. We will see later that at higher temperatures the 811 and 810 cm^{-1} bands of poly A and poly C shift to about 800 cm^{-1}, and also, that on formation of ordered structure in poly U at low temperature in the presence of Mg^{2+}, the 800 cm^{-1} band shifts to 814 cm^{-1}. We will discuss this band later in conjunction with a number of other compounds that give rise to it under various conditions.

For more complete assignments of the Raman bands in Table 1, it would be helpful to have reliable force field calculations for the bases, ribose, and phosphate portions of the polymer, but unfortunately this is not yet possible.

B. Conformational Dependence of Raman Scattering Intensities from Polynucleotides

1. Single Chain Polymers

The conformation of polyadenylic acid has been studied previously in more detail than have any of the other synthetic polyribonucleotides and this polymer has also been examined by Raman spectroscopy in more detail than other polymers. Hydrodynamic (*27, 28*), optical (*29, 30*) and other studies on poly A and chemically modified poly A (*31–33*) have led to the conclusion that formation of structure in neutral pH aqueous solution at low temperatures is due largely to base stacking and is essentially a noncooperative process (*34–38*) involving a single chain. An excellent review of the structure of single chain polynucleotides can be found in the review article by Felsenfeld and Miles (*39*).

Figure 1 shows two Raman spectra of poly A at 20°C and 80°C taken directly off the recorder chart. The numbers at the top of each chart are the wavelengths of the emitted light and the numbers below the Raman bands give the frequency in cm^{-1} of the corresponding bands. Since the Raman bands of interest lie primarily in two regions – one from 500 to 900 cm^{-1}, and one from 1100 to 1700 cm^{-1} – only the two regions shown were obtained (*23*). One solvent and two buffer bands are evident, very broad Raman band of water at about 1600 cm^{-1}, a cacodylate band at 610 cm^{-1} labeled C, and a very weak unlabeled cacodylate band at about 640 cm^{-1}.

TABLE 1. Raman Frequencies and Assignments for the Synthetic Polyribonucleotides[a]

Poly U	Poly C	Poly A	Poly G	Poly I	Assignments[b]
			408		Ribose, phosphate, out-of-plane deformations, etc.
	429		429		
			505		
		533		529	
	545				
556				559	
			588		
	601				
639	644	635	640		
			675		G
			691		G
		705			A
		725	722	723	A, G, I
	750				C
783	790		782		U, C, G
~800				794	O–P–O diester symmetric stretch-disordered form
	810	811	819		O–P–O diester symmetric stretch-ordered form
				822	Ribose-phosphate
	866	868			Ribose-phosphate
	916	912		~912	Ribose-phosphate
			975	969	Ribose-phosphate
997	1008	1004			Ribose-phosphate
			~1026		C–O stretch
	1042	1042	1046	1050	C–O stretch
1094	1097	1095	1096	1091	$O{=}P{=}O^-$ symmetric stretch

A number of changes are evident in the Raman spectra of poly A. This is consistent with the UV data which also show changes. The most pronounced are the strongly polarized bands at 725, 1252, 1303, 1377, 1424, and 1508 cm^{-1}, which appear to increase markedly with temperature. These bands have been assigned to ring vibrations in Table 1. The band at 1483 cm^{-1} appears to decrease slightly with temperature while the band at 1576 cm^{-1} shifts slightly in frequency but does not change much in

Poly U	Poly C	Poly A	Poly G	Poly I	Assignments[b]
1144				1135	Ribose
				1158	Ribose-phosphate
	1180	1179	1183		Base external C–N stretch
			1207		G
1233		1228			U, A
			1242		G
	1256	1252			C, A
			1267	1269	G, I
	1290				C
		1303			A
				1323	I
		1336	1330		A, G
				1352	I
	1384	1377	1367	1384	C, A, G, I
1399					U
	1408				C
		1424	1419	1422	A, G, I
1466	1464	~ 1466		1470	Ribose
1476					U
		1483	1483		A, G
		1508			A
				1518	I
	1547		1543		C, G
				1556	I
		1576	1583		A, G
				1594	I
1680					U

[a] Conditions: 1.5% solutions in 0.01 N sodium cacodylate at neutral pH and room temperature.
[b] U, C, A, G, and I indicate vibrations characteristic of the uracil, cytosine, adenine, guanine, and hypoxanthine bases, respectively.

intensity. A band at approximately 810 cm^{-1} appears to shift to 794 cm^{-1} with increasing temperature. All of these changes were observed to be reversible up to about 80°C, where we began to experience trouble with solvent evaporation and possible breakdown of the polymer in the laser beam.

A plot of the height of the 1508 cm^{-1} band divided by the height of the cacodylate band at 610 cm^{-1} is given in Fig. 2. The intensity of the cacodylate band at 610 cm^{-1} was remarkably independent of temperature which

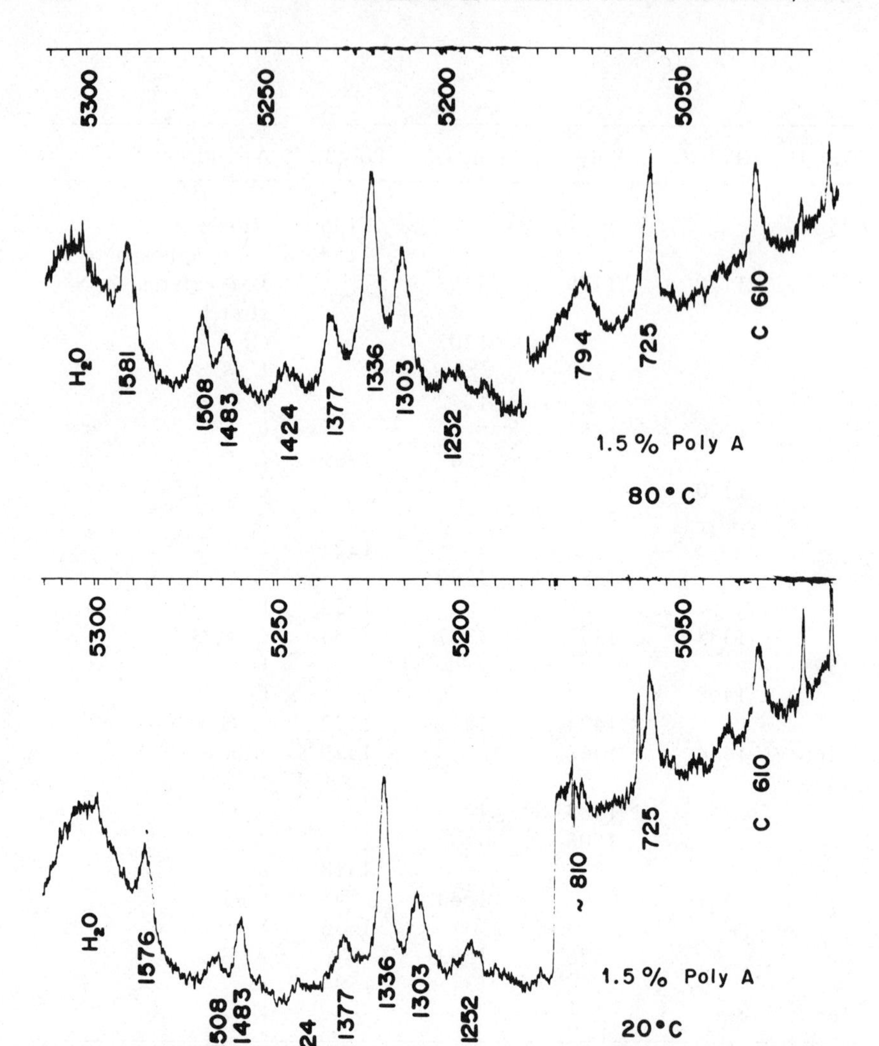

Figure 1. Raman spectra of poly A at 20° and 80°C in 0.01 N sodium cacodylate pH 7.0.

served as a very useful internal reference for intensities of the polymer bands. Figure 2 shows a marked decrease in the intensity of the Raman band when the temperature is lowered so that stacking occurs. In fact, this plot looks extremely similar to the decrease in UV absorption intensity when the temperature is lowered, and consequently we have suggested that the decrease in the Raman bands is due to the ordering of the poly A structure in some way, and that it be called Raman hypochromism (*22, 24*). A possible

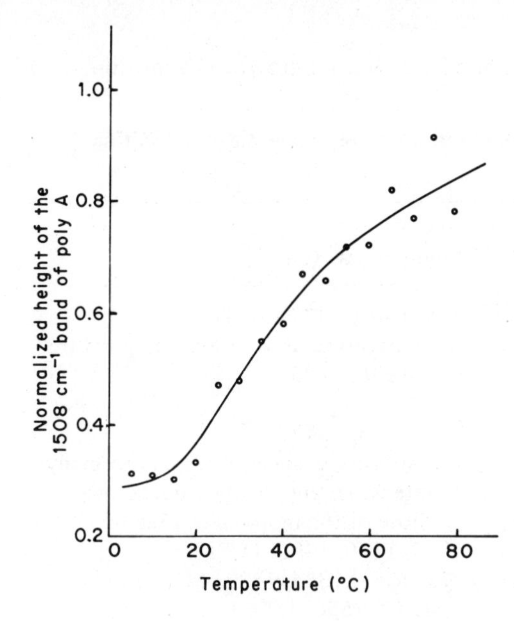

Figure 2. Plot of the normalized height of the 1508 cm^{-1} band of poly A vs. temperature.

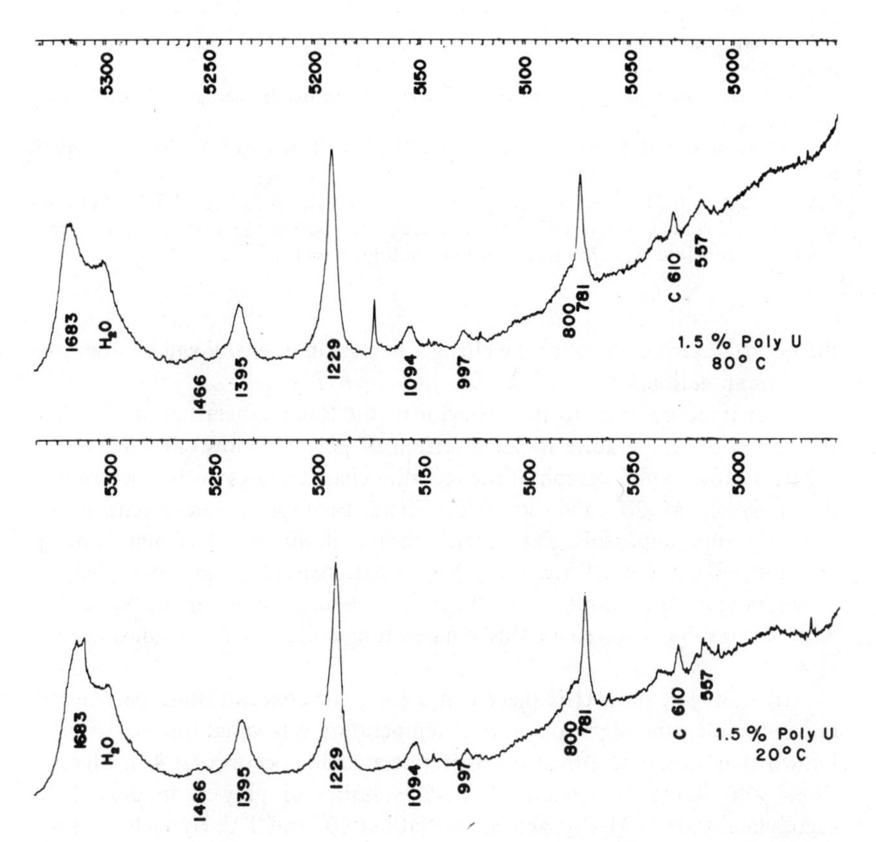

Figure 3. Raman spectra of poly U at 20° and 80°C in 0.01 N sodium cacodylate pH 6.8.

TABLE 2. Raman Bands of the Synthetic Polynucleotides Exhibiting Hypochromism

Polymer[a]	Temperature range	Frequencies[b] (cm^{-1})
Poly A	20°-80°	725, 1303, 1377, 1424, 1508
Poly U	20°-80°	None observed above room temperature.
Poly U[c] (0° with Mg^{2+})	0°-20°	791, 1236, 1403
Poly C	30°-80°	790, 1256, 1547
Poly G	25°-90°	No intensity changes observed but many intensities are greatly reduced over those of the monomers. (See text)
5′–GMP[d]	0°-30°	1325, 1370, 1490, 1580
Poly I	25°-75°	723, 1352, 1384, 1422, 1470, 1518, 1556, 1594

[a] Conditions unless otherwise noted: 1.5% solution in 0.01 N sodium cacodylate at neutral pH.
[b] Frequencies listed are from strongly polarized Raman bands characteristic of the base residues.
[c] 1% solution in 0.01 N sodium cacodylate, 0.1 M NaCl, and 0.05 M $MgCl_2$ at neutral pH.
[d] 1% solution in 0.01 N sodium acetate, Na^{2+} concentration 0.2 M at pH 5.0 (conditions used by Gellert et al. Ref. *60*) 5′–GMP, although not a polymer, was included since poly G did not show intensity changes under the conditions used.

theoretical interpretation for the changes in Raman intensity will be discussed in the next section.

In marked contrast to the behavior of the Raman spectrum of poly A as a function of temperature is the spectrum of poly U in low salt above 20°C. Figure 3 shows a photograph of the recorder chart tracings of the spectrum of this polymer at 20° and at 80°C. These two spectra are identical and perfectly superimposable. Our interpretation of this result is that between 20° and 80°C there is little if any conformational change in poly U leading to interaction or stacking of the bases. This result is in complete agreement with the results reported in this temperature range for UV studies in poly U (*40*).

If, however, poly U is placed in a high salt concentration, particularly one containing the Mg^{2+} ion and the temperature is lowered toward 0°C, it is known that structure formation with base stacking occurs (*41, 42*). Figure 4 shows the Raman spectrum of a 1% solution of poly U in 0.01 M Na cacodylate, 0.05 M $MgCl_2$, and 0.1 M NaCl at 20° and 0°C. Hypochromism is observed in the characteristic uracil ring modes at 791, 1236, and 1403 cm^{-1}.

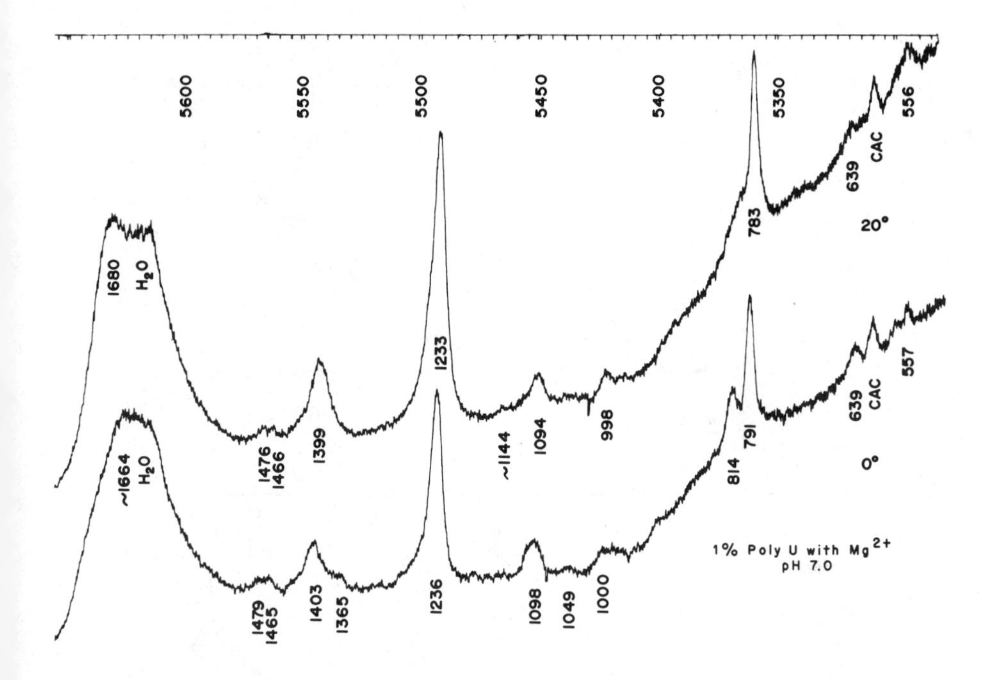

Figure 4. Raman spectra of poly U at 0° and 20°C in the presence of Mg^{2+}.

Also, the appearance of a band at 814 cm^{-1} in the spectrum taken at 0°C and the shoulder on the 783 cm^{-1} band in the spectrum taken at 20°C are of interest. This change is similar to that noted in poly A and will be discussed below.

The structure of poly C in aqueous solution is generally considered to undergo roughly the same change in configuration as does poly A (*43*). Figure 5 shows the chart spectra of poly C at 30° and 80°C. Again, there are large changes in the strong Raman bands which we have found to be completely reversible with temperature. The Raman bands observed at 790, 1256, and 1547 cm^{-1} show the largest increase with temperature. These bands appear to correspond to the bands observed by Lord and Thomas (*10a*) in CMP-5′ at 783, 1243, and 1530 cm^{-1} which they tentatively assigned as two ring vibrations and a double bond stretch, respectively.

Figure 6 shows the Raman spectra of a 1.5% solution of poly G in 0.01 M sodium cacodylate at 25° and 90°C. Since previous work showed that dilute poly G in 0.002 M Na^+ has an extremely stable ordered structure which does not melt below 100°C (*44*) – it would appear probable that under the conditions we used poly G would also form a stable structure.

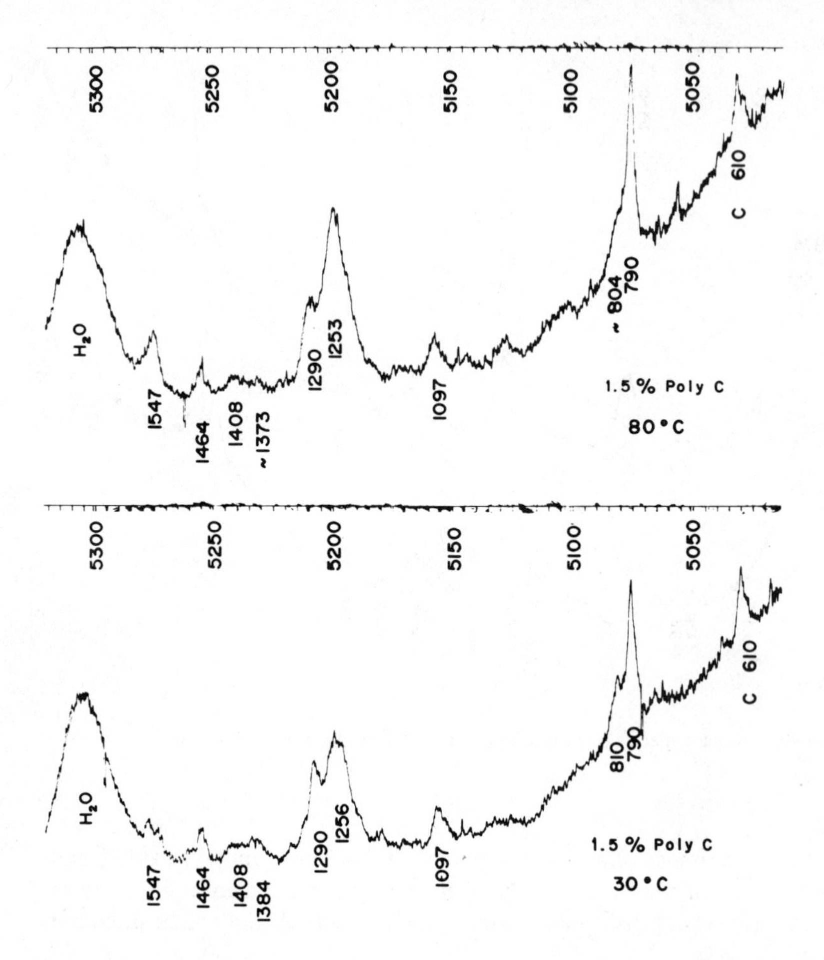

Figure 5. Raman spectra of poly C at 30° and 80°C in 0.01 N sodium cacodylate pH 6.8.

Figure 7 shows Raman spectra of poly I at 25° and 75°C. Clearly virtually every hypoxanthine ring mode shows a slight hypochromism. This is particularly true of the band at 723 cm^{-1}. The lack of any very strongly hypochromic bands we believe to be due to the fact that under these conditions at 25°C poly I possesses only a very flexible structure with partial base stacking (*45*). Table 2 gives a summary of the Raman bands of synthetic polynucleotides which show hypochromism. In addition it should be noted that there exists a band at about 670 cm^{-1} in guanine derivatives that appears to be hyperchromic. The origin of this band is uncertain.

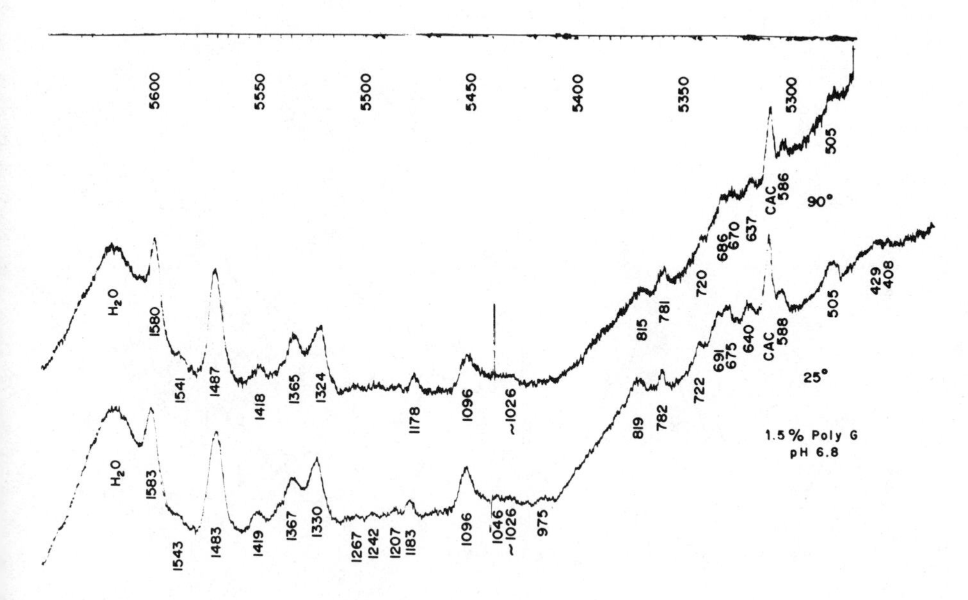

Figure 6. Raman spectra of poly G at 25° and 90°C in 0.01 N sodium cacodylate pH 6.8.

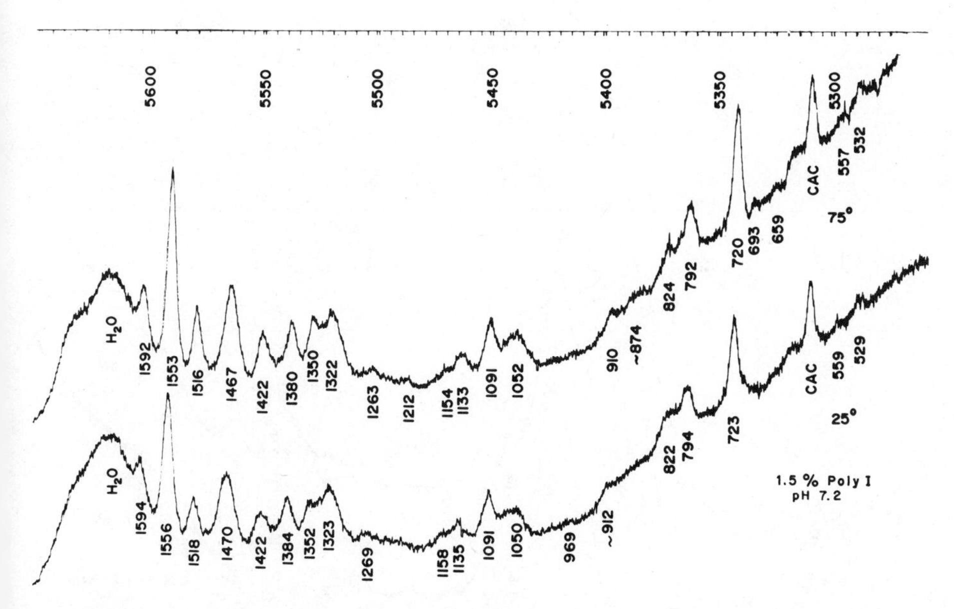

Figure 7. Raman spectra of poly I at 25° and 75°C in 0.01 N sodium cacodylate pH 7.2.

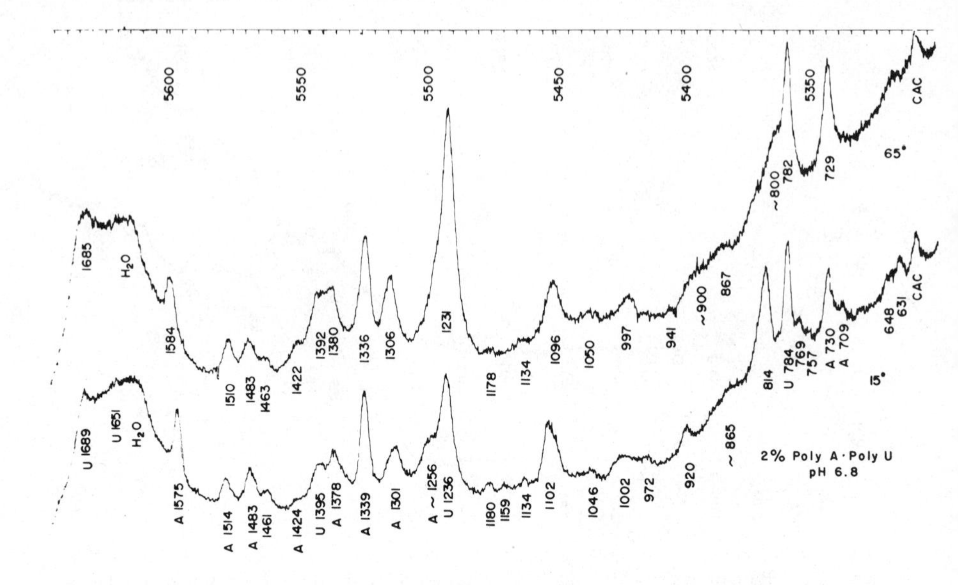

Figure 8. Raman spectra of poly A poly U at 15° and 65°C.

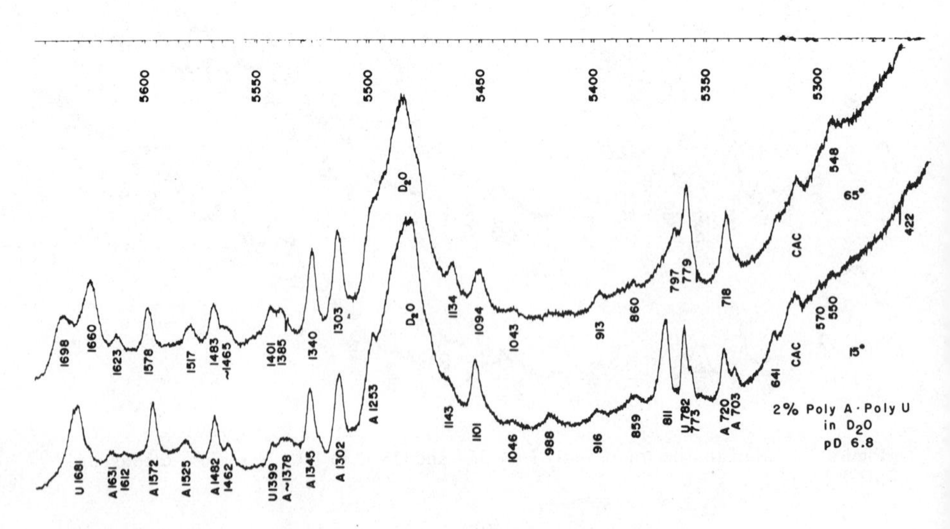

Figure 9. Raman spectra of poly A poly U at 15° and 65°C in D_2O.

2. Double Stranded Helices

The poly A · poly U complex gives a rather clear illustration of what can be learned from Raman scattering of nucleic acids in aqueous solution. With the correct ionic conditions an equimolar mixture of poly A and poly U form a double stranded helix (*46–52*). This double stranded helix has been shown to give rise to a cooperative melting transition with the transition temperature a function of the concentration and type of small ions in the polymer solution. The conditions under which the Raman spectra have been studied are the same as in the IR work of Miles et al. (*46*): a 2% solution of poly A · poly U complex (i.e., approximately 0.05 M polymer phosphate) in 0.005 N cacodylate buffer at neutral pH with sufficient NaCl to bring the total Na^+ concentration to 0.14 M. Under these conditions Miles et al. found poly A · poly U to undergo a transition at 57°C (*46*).

Figure 8 shows a Raman spectra of poly A · poly U at 15° and 65°C, i.e. temperatures both below and above the melting point. Figure 9 shows similar spectra in D_2O. With the exceptions of certain frequency shifts and variations in intensity the spectra clearly resemble a superimposition of the spectra of the component polymers of A and U. By this we mean that one can readily assign each of the Raman bands to characteristic ring modes of adenine and uracil and to vibrations of the ribose–phosphate backbone. In Figs. 8 and 9 the easily assigned ring modes are labeled either A or U for adenine or uracil.

In general, vibrations giving rise to Raman spectra which show significant changes in either intensity or frequency or both fall into one of three categories. (1) Raman bands that arise from ring vibrations of the nucleotide bases that decrease in intensity with increased stacking of the bases. This phenomenon we have called Raman hypochromism. (2) A band at 814 cm^{-1} that is absent or very broad with shifted frequency in the disordered separated chains, becomes very strong in the helical poly A · poly U complex. This band appears to be a strong indication of the helical content of the poly A · poly U complex. It apparently arises from the ribose–phosphate backbone, but is difficult to assign with complete certainty. (3) Raman bands in the 1600 to 1700 cm^{-1} region that arise from the carbonyl stretching vibrations in uracil and that change markedly on hydrogen bond formation in the helix.

Since Raman hypochromism was previously observed upon stacking of single chain polynucleotides, one would expect that similar hypochromic effects would occur upon poly A · poly U helix formation. Qualitatively this is found to be true but quantitatively there are distinct differences. Thus, although the adenine bands at 730, 1301, 1510 cm^{-1} all show a decrease in intensity upon helix formation, the band at 1510 cm^{-1} does not become as small as it does in single chain poly A solutions at low temperature. Apparently the adenine stacking interaction in poly A · poly U helix formation does not give rise to the same interactions, and does not reduce the intensity of this band as does the base stacking interaction in poly A.

As we have seen, Raman hypochromism is observed in self-stacking of poly U at 791, 1236, and 1403 cm^{-1}. These same bands show Raman hypochromism in the poly A · poly U complex. Particularly striking is the strong change in intensity of the 1236 cm^{-1} ring vibration of uracil, as can be seen from Fig. 8.

By following the intensities of the different bands as a function of temperature, one would predict that we should be able to observe separately the stacking interactions of uracil and adenine. This is readily done. Figure 10 shows a plot of the intensity of the 1236 cm^{-1} uracil band vs. temperature. This plot shows a sharp cooperative transition at 59°C, sufficiently close to the temperature observed in infrared work (*46*) (57°C) to ascribe the difference to sample variation. This curve conveys interesting information about the state of order in which the uracil finds itself both below and above the transition temperature. From an examination of this curve, it appears as though the poly U structure is weakening on approaching the transition temperature, and then suddenly undergoes unstacking to reach a constant state of disorder. The sudden leveling off of the intensity of this band with temperature is due to the fact that after dissociation of the helix, there is no subsequent change in the unstacking of the now separated poly U chain. Thus

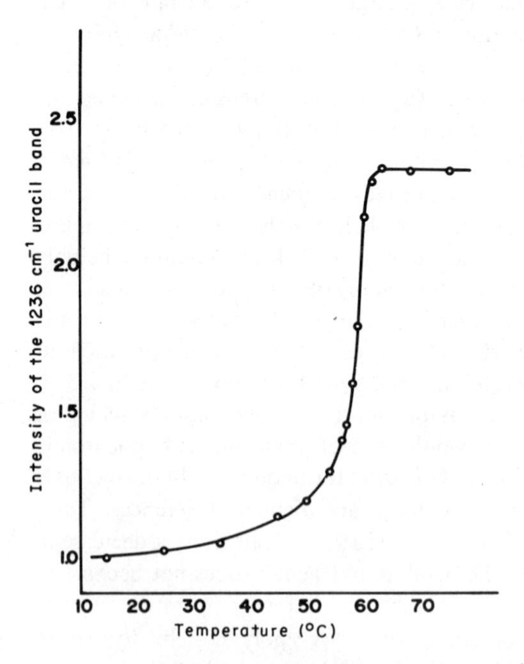

Figure 10. Plot of the intensity of the 1236 cm^{-1} uracil band of poly A·poly U vs. temperature (ratio to intensity at 15°C).

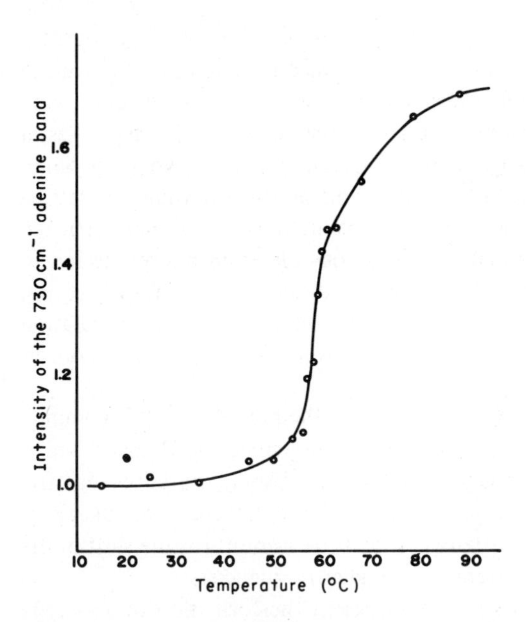

Figure 11. Plot of the intensity of the 730 cm^{-1} adenine band of poly A•poly U vs. temperature (ratio to intensity at 15°C).

the poly U chain appears to be completely disordered when it becomes dissociated from the helix.

The unstacking of adenine gives a different shaped curve. Figure 11 is a plot of the intensity of the adenine band at 730 cm^{-1} as a function of the temperature. Again the intensity is approximately constant until one reaches the transition temperature at 59°C. The intensity then suddenly rises as the helix becomes uncoiled, but above 60°C there remains a considerable amount of increase in intensity with temperature due to the continued unstacking of the now separated but partially stacked single chain of poly A. The curve then follows the melting or unstacking curve of single chain poly A which is an essentially noncooperative effect.

The spectrum of poly A · poly U (Fig. 8) possesses a strongly polarized (depolarization ratio < 0.05) Raman band at 814 cm^{-1}. Tsuboi (*53*) assigns an IR band at 814 cm^{-1} to the symmetrical stretching of the

```
C   C
 \ /
  C
  |
  O
```

group at the 2′ position of the ribose. He notes that the IR band does not shift in frequency on deuteration, and the 814 cm^{-1} Raman band which has been observed in the poly A • poly U complex also does not shift significantly on deuteration (see Fig. 9). Thomas (*14*) has also reported on this band in rRNA and singly stranded poly A, and quotes work by Yu and Lord, who assign it to the overlapping symmetric and antisymmetric phosphate diester stretching vibrations. This assignment was based on earlier work by Shimanouchi et al. (*54*), who performed calculations on dimethyl phosphate as a model compound. On the basis of their calculations the authors assigned Raman bands at 816 and 759 cm^{-1} of aqueous dimethyl phosphate ion to the antisymmetric and symmetric O–P–O diester stretching vibrations, respectively, of the C_2 gauche–gauche conformation. Their calculation showed that the frequencies of the vibrations were very sensitive to conformation.

We tend to assign the band at 814 cm^{-1} in the poly A • poly U complex to the symmetric stretch of the phosphate diester, since the Raman band is strongly polarized. It seems possible that the corresponding antisymmetric stretch may be one of the bands in the region 860 to 920 cm^{-1} which appear as weak broad bands. It is admittedly difficult to account for the shift in the symmetric and antisymmetric stretches to higher frequencies.

If the 814 cm^{-1} band does in fact represent the formation of a specific structure of the sugar–phosphate backbone, it should disappear in a sigmoidal fashion as the transition temperature is reached. Figure 12 is a plot of the height of this band as a function of the temperature. This plot is a sigmoidal

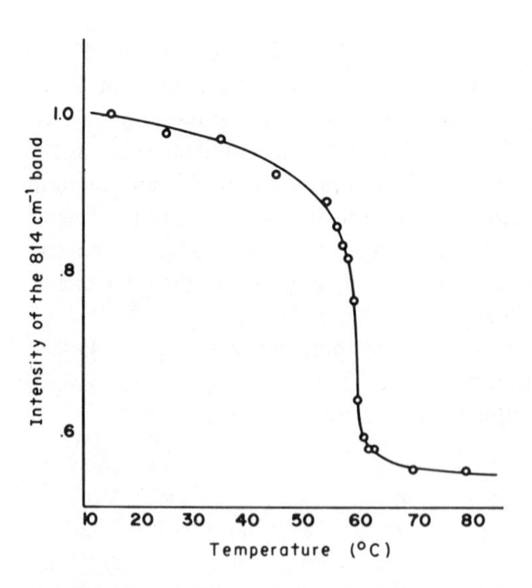

Figure 12. Plot of the intensity of the 814 cm^{-1} band of poly A • poly U vs. temperature (ratio to intensity at 15°C).

curve with a transition temperature of 59°. After the transition at 59° the curve drops off somewhat more, probably owing to the residual structure remaining in the phosphate chain of the now separated poly A.

One region of interest in the spectrum of poly A • poly U is the region 1600 to 1700 cm^{-1}. In this region double-bond stretching vibrations is observed. A considerable amount of IR work has been done on this region, since the carbonyl stretch of the uracil base in particular is extremely sensitive to interbase hydrogen bonds (*46, 51*). Using isotopic replacements Miles et al. (*55*) were able to specifically assign the IR bands at 1691 and 1672 cm^{-1} to the carbonyl stretches at the 2 and 4 positions of the uracil, respectively. On the transition to the single strands the carbonyl stretches shift, respectively, to 1692 and 1672 cm^{-1}. The IR work was done in D_2O, since the IR spectrum of H_2O completely obscures this region. Raman studies can not easily be done in H_2O in this region either, since the broad weak Raman band of water at about 1635 cm^{-1} interferes with these bands at such low concentrations. We studied this region in D_2O (see Fig. 9). One strong Raman band is present in the double helix at 1681 cm^{-1} presumably due to one of the C=O stretching vibrations of the hydrogen bonded uracil. On melting this band disappears and two other bands appear at 1698 and 1660 cm^{-1} probably corresponding to the carbonyl stretches on the 2 and 4 positions, respectively. Bernstein (*56*) has also observed a similar change in the Raman spectrum of poly A • poly U complex. The origins of the difference between the IR frequencies reported and the Raman frequencies is not entirely clear.

The appearance of the band at 1660 cm^{-1} is a clear indication of the breakup of the interbase hydrogen bonding of the double helix. Figure 13 is a plot of the intensity of this band vs. temperature. From this plot it appears that on approaching the transition temperature the hydrogen bonding begins to weaken. At 59° this structure suddenly collapses to reach a constant degree of disorder.

Table 3 summarizes the frequencies we have observed for poly A • poly U along with our Raman band assignments (*25*). The vibrations characteristic of the base residues have been assigned in a straightforward manner by comparison with the Raman spectra of poly A and poly U. Contributions to intensity from very weak bonds in the single chain spectra have been neglected.

Studies on the poly A • poly U complex have clearly shown the potential of Raman studies on nucleic acids. We have been able to follow independently the effect of temperature on adenine and uracil base stacking interactions, the breakup of the interbase hydrogen bonding, and, although not yet completely defined, the specific changes of the helical sugar–phosphate backbone.

Poly dAT, a synthetic DNA consisting of alternating deoxyadenosine and

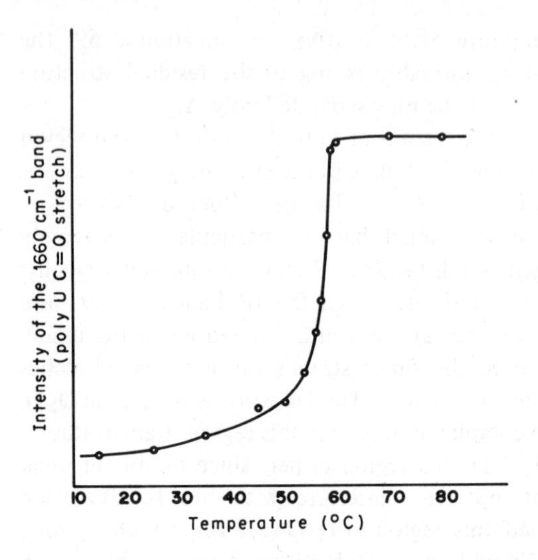

Figure 13. Plot of the intensity of the 1660 cm^{-1} uracil band of poly A•poly U vs. temperature.

deoxythymidine residues, has been shown by x-ray diffraction measurements to form a double helix of the same configuration as natural DNA (*57*). We have studied this polymer under the same conditions we used for poly A • poly U, i.e., a 2% solution in 0.005 N cacodylate buffer at neutral pH with NaCl added to bring the total Na^+ concentration to 0.14 M. Under these conditions poly dAT should undergo a sharp cooperative transition at between 60° and 65° (*58*). Figure 14 shows Raman spectra of poly dAT at 25° and 78°, below and above its transition temperature.

In Fig. 14 the readily assigned Raman bands characteristic of adenine or thymine are labeled either A or T, respectively. When a Raman band has significant contributions from both bases, the predominant contribution is listed first. It must be kept in mind that Raman spectra such as these contain the superpositions of a great many Raman bands, but fortunately the great majority are sufficiently weak to allow us to observe only a few very intense bands. Most of these bands can then be easily assigned to a specific base, and the remainder are assumed to be from the sugar–phosphate backbone.

Due to the complexity of the Raman spectra one must be careful not to come to any erroneous conclusions regarding the source of intensity changes. In our studies of the poly A • poly U complex one could diagnose the source of intensity changes in a straightforward manner. All three Raman bands studies were intense and not significantly interfered with by other bands. In poly dAT the situation is not as apparent. The adenine bands at 733, 1306, and 1485 cm^{-1} clearly show hypochromic shifts on helix formation. This is a

TABLE 3. Raman Frequencies and Assignments for Poly A · Poly U

Frequencies (cm^{-1})				
H_2O Solutions		D_2O Solutions		
15°	65°	15°	65°	Assignments[a]
		422		Ribose
		550	548	Ribose
631				
648		641		
709		703		A
730	729	720	718	A
757				
769		773		U
784	782	782	779	U
	~800		797	O–P–O Diester symmetric stretch-disordered form
814		811		O–P–O Diester symmetric stretch-helical form
~865	867	859	860	Ribose-phosphate
	~900			Ribose-phosphate
920		916	913	Ribose-phosphate
	941			Ribose-phosphate
972				Ribose-phosphate
		988		Ribose-phosphate
1002	997			
1046	1050	1046	1043	C–O Stretch
1102	1096	1101	1094	O^{-}–P–O^{-} Symmetric stretch
1134	1134	~1143	1134	Stretching at 2′ position of ribose
1159				Ribose-phosphate
1180	1178			A External C–N stretch
1236	1231			U
1256		1253		A
1301	1306	1302	1303	A
1339	1336	1345	1340	A
1378	1380	~1370	1385	A
1395	1392	1399	1401	U
1424	1422			A
1461	1463	1462	~1465	Ribose
1483	1483	1482	1483	A
1514	1510	1525	1517	A
1575	1584	1572	1578	A
		1612		
			1623	A
		1631		A
1651				U C=O Stretch
			1660	U C=O Stretch
1689	1685	1681		U C=O Stretch
			1698	U C=O Stretch

[a] U and A indicate vibrations characteristic of the uracil and adenine bases, respectively. Ribose-phosphate indicates probable origin is in the ribose-phosphate chain but cannot be readily assigned specifically to ribose or phosphate.

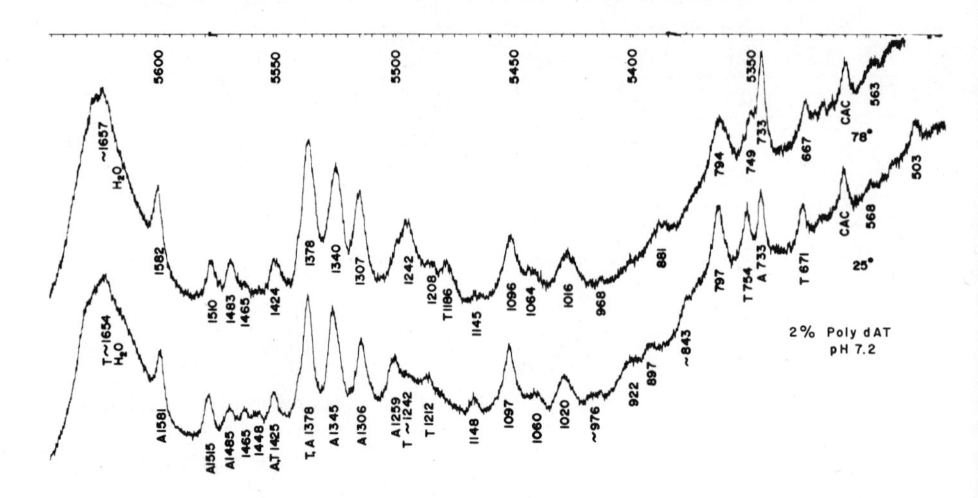

Figure 14. Raman spectra of poly dAT at 25° and 78°C.

different set of adenine bands than those observed to undergo hypochromic shifts in either single stranded poly A or poly A • poly U. This is what one might expect since in poly dAT adenine presumably undergoes stacking interactions with adjacent thymines rather than other adenines as it does in either of the other two systems. Neither the band at 1306 cm^{-1} nor the band at 1485 cm^{-1} would be a good band to study adenine stacking, since their intensity changes are not very great. The best adenine band to study would be the one at 733 cm^{-1}, but one would need to consider the shifting of the thymine band which becomes a shoulder on the adenine band at high temperatures.

Not unexpectedly, certain thymine bands also show Raman hypochromism. The characteristic thymine bands at 1186 and 1242 cm^{-1} both show intensity increases on unstacking. The band at 1378 cm^{-1}, which is predominantly due to a thymine ring mode, also increases somewhat on melting. This band however coincides with an adenine band which in single stranded poly A is also hypochromic, and it is therefore impossible to separate the effects of two bases on this band. The band at 1186 would be the preferable band for studying thymine base stacking in this system.

It is particularly interesting to note the intense strongly polarized (depolarization ratio < 0.05) band at 797 cm^{-1}, which does not change much on deuteration. This band must contain a thymine ring mode at about 787 cm^{-1}, but this leaves most of the band unaccounted for. Comparing Fig. 14 with the spectrum of poly A • poly U in Fig. 8, it can be seen that the 797 cm^{-1} band of poly dAT resembles closely in intensity and shape the band at 814 cm^{-1} in poly A • poly U. We tend therefore to assign this band also to the phosphate diester stretch of the gauche–gauche conformation around the C–O–P–O–C linkage. This linkage is also presumed to be in the gauche–gauche configuration from x-ray diffraction work (*59*). If this assignment is correct, the difference in frequency of this band between RNA (poly A • poly U) and DNA (poly dAT) must arise from some conformational difference on the backbone.

If we return again to the Raman spectra of poly A • poly U, we see that on the disappearance of the band at 814 cm^{-1} another band, which is broader but of roughly the same intensity, appears as a shoulder on the poly U band at 782 cm^{-1}. This is more clearly illustrated by the spectrum of deuterated poly A • poly U (Fig. 8) where the uracil ring mode is shifted to slightly lower frequency permitting us to observe a band in the separated chains at 797 cm^{-1}. It appears, then, as though when poly A • poly U is melted, the phosphate diester linkage becomes more like the phosphate diester linkage of poly dAT. If these assignments are correct then on formation of helical structure in RNA the frequency of the phosphate diester stretch is shifted to 814 cm^{-1} by some interaction involving directly or indirectly the $2'$ hydroxyl group of the ribose.

Table 4 summarizes the Raman frequencies and our Raman band assignments for poly dAT.

C. Raman Scattering from Nucleic Acids

One of our goals is to demonstrate the feasibility of performing structural studies with Raman spectroscopy on dilute aqueous solutions of natural nucleic acids. We have already demonstrated that certain Raman ring modes of the bases adenine, uracil, thymine, cytosine, and hypoxanthine show a decrease in intensity upon stacking. Later in the section on model compounds we will show this to also be true for certain guanine bands. Now we would like to present Raman spectra of dilute solutions of natural RNA and DNA, make preliminary Raman band assignments, and discuss a number of their interesting features.

Figure 15 shows Raman spectra of 2.5% solutions of yeast sRNA and calf thymus DNA, and Fig. 16 shows the same samples in D_2O. Although the use of higher concentrations would have resulted in a vastly improved signal-to-noise ratio, our goal is not to produce as many Raman bands as possible, but to present Raman spectra of concentrations of sufficient dilution to have some meaning in structural studies. The characteristic ring vibrations of the bases have been labeled with A,U,T,C, and G for the bases adenine, uracil, thymine, cytosine, and guanine, respectively. When more than one base makes a significant contribution to the intensity of a band, they are listed in order of the size of their contribution with the largest contribution first.

TABLE 4. Raman Frequencies and Assignments for Poly dAT

Frequencies (cm^{-1})			
H_2O Solutions		D_2O Solution	
25°	78°	25°	Assignments[a]
503		498	Deoxyribose-phosphate
		526	
568	563	571	Deoxyribose
671	667	662	T
733	733	725	A
754	749	744	T
797	794	793	O–P–O Diester symmetric stretch
843		841	O–P–O Diester antisym stretch?
		867	Deoxyribose-phosphate
	881		Deoxyribose-phosphate
897		~901	Deoxyribose-phosphate
922		932	Deoxyribose
~976	968		Deoxyribose
1020	1016	1024	C–O Stretch
1060	1064	1054	C–O Stretch
1097	1096	1095	O⁻–P–O⁻ Symmetric stretch
1148	1145	1162	Deoxyribose-phosphate
	1186		T
1212	1208		T
1242	1242		T
1259			A
1306	1307	1304	A
1345	1340	1347	A
1378	1378	1381	T,A
1425	1424	1423	A,T
1448		1444	Deoxyribose
1465	1465	1465	Deoxyribose
1485	1483	1489	A
1515	1510	1524	A
		1549	
1581	1582	1576	A
		1615	
		~1634	A
~1654	~1657		T
		1669	T

[a] T and A are for vibrations characteristic of the thymine and adenine bases with the predominant contribution listed first. Deoxyribose-phosphate indicates probable origin is in the deoxyribose-phosphate chain but cannot be readily assigned specifically to deoxyribose or phosphate.

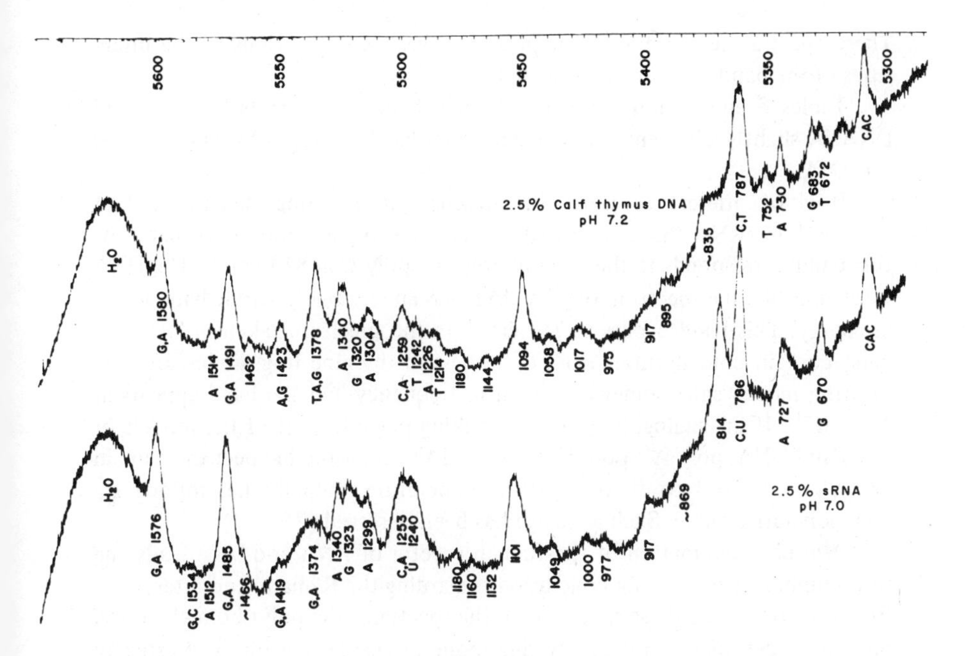

Figure 15. Raman spectra of sRNA and calf thymus DNA in 0.01 N sodium cacodylate.

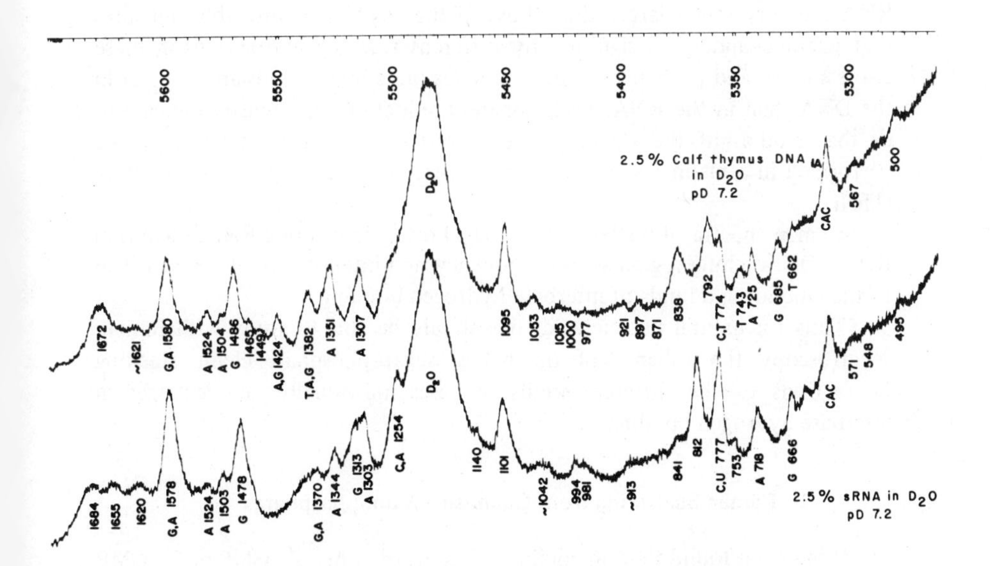

Figure 16. Raman spectra of sRNA and calf thymus DNA in D_2O.

These spectra are extremely complex and weak contributions to the intensities of the bands have been neglected.

Tables 5 and 6 summarize our Raman band assignments for RNA and DNA. A slightly different set of assignments has been given by Thomas (*14*) for RNA.

It is of interest to note the presence of a strong Raman band at 814 cm^{-1} in sRNA that is essentially unaffected by deuteration. Apparently this band corresponds to the band in poly A · poly U at 814 cm^{-1}. The DNA spectrum, like the spectrum of poly dAT, has an extremely strong band lower in energy. This band occurs at 787 cm^{-1} in the spectrum taken in H_2O and must contain contributions from cytosine and thymine ring modes. In D_2O the ring modes shift somewhat down in frequency and the band appears at 792 cm^{-1}. If an analogous process is taking place to that of the models of RNA and DNA, poly A · poly U and poly dAT, one would expect the band in sRNA at 814 cm^{-1} to disappear with temperature gradually, i.e., following a noncooperative curve. Such a change has been observed (*25*).

A visual examination of the Raman spectra of RNA and DNA leads one to a number of interesting conclusions regarding the Raman band intensities. Roughly comparing these spectra with the spectra of the polyribonucleotides, one concludes that many of the ring mode intensities have been drastically reduced over those of the disordered polymers presumably from stacking interactions. This is particularly true of the spectrum of DNA. A comparison of the spectra of DNA and RNA also indicates that, although both the DNA and RNA samples possess approximately 40% GC, the guanine bands of the RNA are very much larger than those of the DNA. This probably indicates that guanine undergoes significantly different stacking interactions in these two samples, and probably in these samples these interactions are stronger in the DNA than in the RNA. Such apparent reductions in intensity indicate to us that studies of the stacking interactions of the different bases can be performed in a manner similar to that we used in studying the poly A · poly U helix.

Raman spectra of nucleic acids in D_2O reveal interesting Raman bands in the C=O stretching region whose frequency and intensity should be sensitive to the amount and kinds of interbase hydrogen bonding.

Thus for natural nucleic acids one should be able to study with Raman spectroscopy the collapse of the helical sugar-phosphate chain, stacking interactions of the different kinds of bases individually, and changes in interbase hydrogen bonding.

D. Raman Scattering from Guanosine Monophosphates

It has been found that on cooling solutions of either 3′–GMP or 5′–GMP at pH 5 under the correct conditions, one obtains a clear viscous gel. This gel

TABLE 5. Raman Frequencies and Assignments for Yeast sRNA

Frequencies (cm^{-1})		
H_2O Solution	D_2O Solution	Assignments[a]
368		
427		Ribose
507	495	Ribose-phosphate
	548	Ribose
581	571	
601		Ribose
633		
	664	
670	666	G
727	718	A
759	753	
786	777	C,U
814	812	O–P–O Diester symmetric stretch
	841	Ribose-phosphate
~ 869		Ribose-phosphate
917	913	Ribose-phosphate
977	981	Ribose-phosphate
1000	994	
1049	1042	C–O Stretch
1101	1101	$O^{-}\!\!-\!P\!-\!O^{-}$ Symmetric stretch
1132	1140	Stretching at 2′position of ribose
1160		Ribose-phosphate
1180		Base external C–N stretch
1240		U
1253	1254	C,A
1299	1303	A
1323	1313	G
1340	1344	A
1374	1370	G,A
1419		G,A
1466		Ribose
1485	1478	G,A
	1503	
1512	1524	A
1534		G,C
1576	1578	G,A
	1620	
	1655	C=O Stretch
	1684	

[a] U, C, A, and G indicate vibrations characteristic of the uracil, cytosine, adenine, and guanine bases, respectively, listed in order of their relative contributions with the largest contribution first. Ribose-phosphate indicates probable origin is in the ribose-phosphate chain but cannot be readily assigned specifically to ribose or phosphate.

TABLE 6. Raman Frequencies and Assignments for Calf Thymus DNA

Frequencies (cm^{-1})		
H_2O Solution	D_2O Solution	Assignments[a]
	500	Deoxyribose-phosphate
	567	Deoxyribose
672	662	T
683	685	G
730	725	A
752	743	T
	774	C,T
787		O–P–O Diester symmetric stretch overlapping C,T
	792	O–P–O Diester symmetric stretch
~835	838	Deoxyribose-phosphate
	871	Deoxyribose-phosphate
895	897	Deoxyribose-phosphate
917	921	Deoxyribose
975	977	Deoxyribose
	1000	Deoxyribose
1017	1015	C–O Stretch
1058	1053	C–O Stretch
1094	1095	$O \overset{-}{-} P \overset{-}{-} O^-$ Symmetric stretch
1144		Deoxyribose-phosphate
1180		Base external C–N stretch
1214		T
1226		A
1242		T
1259		C,A
1304	1307	A
1320		G
1340	1351	A
1378	1382	T,A,G
1423	1424	A,G
1448	1449	Deoxyribose
1462	1465	Deoxyribose
1491	1486	G,A
	1504	A
1514	1524	A
1534		G,C
1580	1580	G,A
	~1621	
	1672	C=O Stretch

[a] T,C,A, and G indicate vibrations characteristic of the thymine, cytosine, adenine, and guanine bases, respectively, listed in order of their relative contributions with the largest contribution first. Deoxyribose-phosphate indicates probable origin is in the deoxyribose-phosphate chain but cannot be readily assigned specifically to deoxyribose or phosphate.

formation is accompanied by changes in the UV absorbance and optical rotation of the solution. Gellert et al. (*60*) studied these gels by a number of means including x-ray diffractions of fibers made from them. They proposed at that time a structure in which the guanine rings hydrogen bond with adjacent rings. From their x-ray data they determined that the 3′ isomer forms square planar tetramers stacked upon each other with the ribose and phosphate ends of the molecule protruding into the solvent. They suggested that the 5′ isomer also hydrogen bonds in a similar manner but that instead of forming planar tetramers, it forms a helix by hydrogen bonding continuously around.

It has been found that the characteristic guanine ring modes at about 1325, 1370, 1490, and 1580 cm^{-1} show a strong hypochromism on gel formation, probably due to a stacking interaction. Figure 17 gives melting curves of these gels plotted as the analog integrated intensity of the strong guanine ring mode at about 1490 cm^{-1}. These plots closely resemble the UV absorbance data of Gellert et al. (*60*), and are therefore further evidence that the Raman hypochromic effect is a result of the same conformational phenomenon which causes the UV hypochromic effect, i.e., presumably some type of stacking interaction.

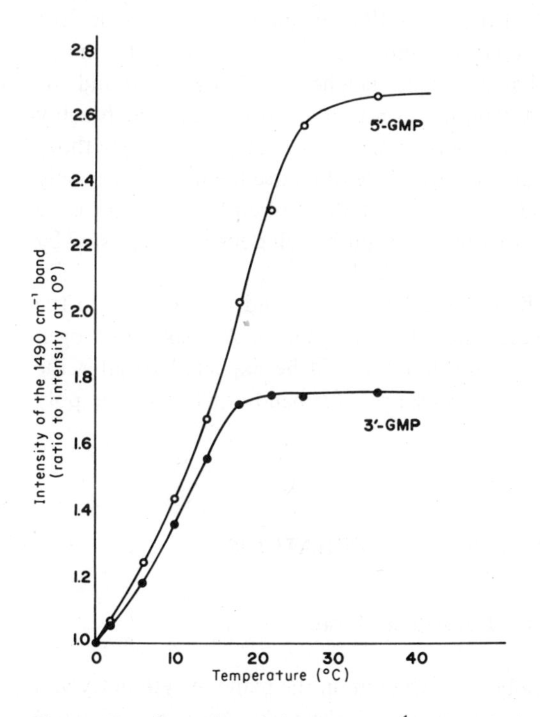

Figure 17. Intensity of the 1490 cm^{-1} band of 3′-GMP and 5′-GMP vs. temperature.

In the 5′ isomer the phosphate group has considerably more freedom to bend back and interact with the guanine rings or adjacent ribose groups. Gellert et al. (*60*) recognized this and suggested that the phosphate might stabilize the structure of the 5′ derivative by hydrogen bonding with a 2′OH group on a ribose two levels above, or with an adjacent guanine amino group. Such structures could not be formed by the 3′ isomer. Our Raman data appears to respond dramatically to this difference. There are a large number of new bands in the Raman spectrum of the 5′–GMP gel at 0°. The most spectacular new bands that appear are those at 670 and 814 cm^{-1}. As we have seen, both of these bands are present in the Raman spectrum of RNA, and we noted that the 814 cm^{-1} band appears to convey structural information about the sugar-phosphate backbone. The 814 cm^{-1} band is not present in the spectrum of DNA, and the 670 cm^{-1} band observed in the spectrum of DNA we have assigned to thymine and not guanine. We assigned instead the 683 cm^{-1} band of DNA to guanine.

If we are to retain our assignment of the 814 cm^{-1} band in polyribonucleotides to the phosphate diester symmetric stretch, we must account for its presence in the monoester 5′–GMP. The 814 cm^{-1} band in 5′–GMP clearly appears to be of the same origin, since it is of the same energy, roughly the same shape, and strongly polarized (depolarization ratio < 0.05) We constructed a CPK model of the 5′–GMP helix and hydrogen bonded the phosphate group back to an adjacent amino group. In this position in our model the phosphate group has a gauche-gauche conformation around the base . . . O–P–O–C linkage. Attempts to form other structures using roughly the helical parameters from the x-ray work were not successful without introducing noticeable strain in the model. We of course mention this as only a possible explanation for the occurrence of the 814 cm^{-1} band, since we recognize the tentative nature of the hydrogen bonding scheme suggested by the x-ray workers.

One difficulty with attributing the 814 cm^{-1} band of 5′–GMP to the gauche-gauche phosphate diester should be mentioned. A weaker hydrogen bond rather than a strong covalent bond would be expected to alter the frequency of the band, but we have observed this band at 814 cm^{-1} in poly A · poly U, sRNA, and also 5′–GMP.

IV. GROUP THEORETICAL CONSIDERATIONS

A. Group Theory of the Constituent Bases

The purine and pyrimidine bases both contain the plane of symmetry and consequently belong to the point group, C_s. The character table for this group is shown in Table 7. Both the vibrational states and the electronic

TABLE 7. Character Table for C_s Point Group

	E	σh	IR polarization	Raman tensor
A'	+1	+1	x,y	$\alpha_{xx}, \alpha_{yy}, \alpha_{zz}, \alpha_{xy}$
A''	+1	−1	z	α_{yz}, α_{zx}

states can be classified as being either in-plane and belonging to the totally symmetric representation, A', or as being out-of-plane and belonging to the irreducible representation, A''. Although all of the vibrations are both Raman and IR active, the polarization dependence of the Raman lines will be different.

From Table 7 we can construct the Raman tensors for the two types of vibrations. These are given as

$$\alpha(A') = \begin{pmatrix} a & d & 0 \\ d & b & 0 \\ 0 & 0 & c \end{pmatrix} \tag{IV-1}$$

$$\alpha(A'') = \begin{pmatrix} 0 & 0 & e \\ 0 & 0 & f \\ e & f & 0 \end{pmatrix} \tag{IV-2}$$

B. Group Theory of Helical Molecules

If the planar purine and pyrimidine molecules are stacked in a helical array with 10 residues per turn as they do in the Crick-Watson helix, then the helix itself has the linear space group or line group, C_{10} and the group table for this group is given in Table 8.

The totally symmetric A representation in the helix is polarized (in the infrared) along the z-axis, i.e., it is said to transform like the z-axis. Vibrations which belong to this irreducible representation are Raman active as well as active in the infrared. The E_1 vibrations which are polarized perpendicular to

TABLÈ 8. Character Table for the C_{10} Point Group

	E	$C_{1/10}$	$C_{2/10}$	$C_{9/10}$	IR polarization	Raman tensor elements
A	+1	+1	+1	+1	z	$\alpha_{zz}, \alpha_{xx}, {}^{+}\alpha_{yy}$
E_1	+2	2cos $\pi/5$	2cos $2\pi/5$	. . .2cos $9\pi/5$	x,y	α_{xz}, α_{yz}
E_2	+2	2cos $2\pi/5$	2cos $4\pi/5$	. . .2cos $18\pi/5$		$\alpha_{xx} - \alpha_{yy}; \alpha_{xy}$
B	+1	−1	+1	.−1		
E_3	+2	2cos $3\pi/5$	2cos $6\pi/5$	. . .2cos $27\pi/5$		
E_4	+2	2cos $4\pi/5$	2cos $2\cdot4\pi/5$	. .2cos $9\cdot4\pi/5$		

the helix axis are also both IR and Raman active while the E_2 vibrations are only Raman active. Vibrations in the helix which transform as E_3, E_4, and B are inactive both in the Raman and the infrared but the E_3 vibrations are active in the hyper-Raman effect (*37*).

Now let us consider what happens to the A' and A'' vibrations when the bases are put into a helical configuration. As is well known, Raman lines which arise from totally symmetric vibrations have a Raman tensor with a nonzero trace and tend to be much more highly polarized than unsymmetric vibrations. These latter vibrations have Raman tensor elements which are nonzero only off the diagonal and consequently they have a depolarization ratio of 3/4. The totally symmetric vibrations on the other hand have very low depolarization ratios which are of the order of 0–0.3 depending on the nature of the vibration.

It may be that polarization measurements in the Raman will become a useful technique for examining the conformation of bases because of an interesting crossover in the symmetry of the vibrations when the bases are caused to go from a random conformation to a highly ordered helical one. The out-of-plane vibrations of the bases become polarized in the helix along the z-axis, i.e., the helix axis. Hence the out-of-plane vibrations in the helix belong to the totally symmetric representation of the helix factor group, C_{10}, so that these vibrations may now become totally polarized. Similarly, the in-plane vibrations which are totally polarized in the random bases may beome short-axis polarized in the helix so that their Raman lines are totally depolarized. Whether or not such changes actually would be observed experimentally depends upon the strength of the coupling between the bases in the helical form. So far there is no experimental evidence to determine whether or not this is the case.

Table 9 gives the change in the representation upon going from a disordered array of molecules belonging to the point group C_s to an ordered array of such molecules with a linear factor group of C_{10}.

TABLE 9. Correlation Table Between C_s and C_{10} Groups

C_{10}	C_s
A	A''
E_1	A'
E_2	$A' + A''$
B	$A' + A''$
E_3	$A' + A''$
E_4	$A' + A''$

V. QUANTUM THEORY OF RAMAN SCATTERING INTENSITIES: PROPOSED RELATION BETWEEN ULTRAVIOLET AND RAMAN HYPOCHROMISM

In Section IV we reviewed the experiments of Small and Peticolas (*24*) which showed that the intensity of certain Raman bands from the ring vibrations of the bases of the polynucleotides are quite sensitive to the polymer conformation. One important qualitative observation is that virtually all the bands which show any change with conformations show a decrease in intensity with increasing ordering of the polymer configuration. In this section we will discuss the theory of Raman scattering intensities and attempt to give a qualitative explanation of some of the experimental observations. Recently, the theory of Raman scattering intensities from molecules has been given by Peticolas et al. (*23*). Unfortunately, because the theory involves infinite sums over all of the excited electronic states the theory is very difficult to use to obtain quantitative or even qualitative predictions for the ordinary Raman effect. However, much more can be said with assurance about the resonant Raman effect (RRE). When the frequency of the incident light approaches that of the first absorption band, an enormous increase in the scattering efficiency is observed and this is believed due to the appearance of one or more zeros in the energy denominator of the first term of the infinite sum. (Actually, because of damping, the denominator does not go exactly to zero but may become very small so that almost all of the intensity of the Raman scattering comes from the term involving the resonant UV excited state.)

The Raman scattering theory actually involves six different sets of infinite sums. To each of these six sets may be associated a time-ordered diagram which is a convenient mnemonic for the writing down of each of the possible mathematical combinations; in addition it also serves to give a physical flavor to the abstract terms arising from higher order time-dependent perturbation theory. In order to make some qualitative and semiquantitative estimates of Raman intensity behavior, it is convenient to oversimplify the theory and concentrate only on one set of infinite sums corresponding to one of the six time-ordered diagrams. The diagram for the sum which we shall for convenience consider in detail is given in Fig. 18. The physical meaning which may be associated with this diagram is as follows. Before the Raman scattering event occurs, there are n_1 laser photons per cubic centimeter, n_2 Raman photons per cubic centimeter and the molecule is in the νth vibrational level of the ground electronic state, $|0\rangle$. The total wave function is then given by $|n_1,n_2,\nu,0\rangle$. In the first step a laser photon is annihilated, i.e., removed from the radiations field and the molecule goes into a time-dependent electronic state which is expressed as a superposition of all of the allowed excited electronic states. Thus the effect of the perturbation by the electronic-radiation interaction Hamiltonian, H_{ER}, is to mix the excited

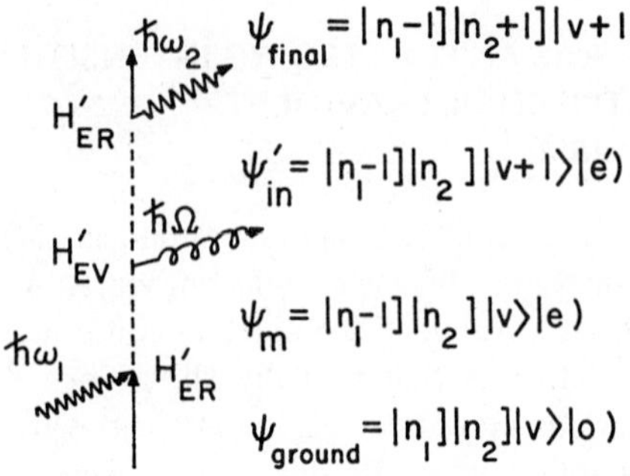

$\hbar\omega_2$ $\psi_{final} = |n_1 - 1]|n_2 + 1]|v+1\rangle|o)$

H'_{ER}

$\psi'_{in} = |n_1 - 1]|n_2]|v+1\rangle|e')$

$\hbar\Omega$

H'_{EV}

$\psi_m = |n_1 - 1]|n_2]|v\rangle|e)$

$\hbar\omega_1$ H'_{ER}

$\psi_{ground} = |n_1]|n_2]|v\rangle|o)$

Figure 18. One of the six time-ordered diagrams for Raman scattering.

electronic states of the molecule. In the second step the vibrational quantum number of the Raman active vibration is increased by one owing to the action of the Hamiltonian connecting the electronic with the vibrational motion, H_{EV}; this again mixes the excited electronic states of the molecule. The third and final step of the process involves the creation of a photon which is equal in energy to the incident photon minus the vibrational quantum, i.e., the Raman quantum, and the molecule ends up in the ground electronic state but in a vibrational state one state higher (Stokes process). In the intermediate steps the energy does not need to be conserved because the time involved for the whole process is so short that the energy uncertainty is very large during these transitions through nonstationary states. Third order time-dependent quantum theory gives for this diagram a rate of scattering events which is proportional to the measured intensity and is given by

$$I_a = K \left[\sum_{E,S} \frac{(0|\mathbf{e}_2 \cdot \mathbf{r}|E)(E|\partial H/\partial Q_a|S)(S|\mathbf{e}_1 \cdot \mathbf{r}|0)}{(E_E - E_O \pm \hbar\Omega - \hbar\omega_1)(E_S - E_O - \hbar\omega_1)} \right]^2 \tag{V-1}$$

where I_a is the intensity of scattering from the ath vibration. The sums over E and S are over all of the stationary electronic states of the molecule; $\mathbf{e}_1$ and $\mathbf{e}_2$ are the polarization vectors of the laser and Raman photons, respectively; $\mathbf{r}$ is the dipole moment operator of the electrons; $\partial H/\partial Q_a$ is the change in the electronic Hamiltonian of the molecule with the nuclear displacements which constitute the ath normal mode; the circular frequency of the incident light is ω_1 and that of the ath vibration is Ω_a; E_E, E_S, and E_O are the energies of the corresponding electronic states; and K is a constant. Note that $(E|$, $(S|$, $\mathbf{r}$, and $\partial H/\partial Q_a$ are all in electronic coordinates. The plus and minus signs in the denominator are for Stokes and anti-Stokes scattering, respectively.

As the frequency of the incident light approaches that of the first absorption band, the trace term with $|E) = |S) = |1)$ will tend to dominate where $|1)$ is the electronic wave function of the first absorption band. The reason that the trace term will tend to dominate is that for the trace term one obtains two zeros in the energy denominator while for the corresponding off-diagonal terms with $|E) \neq |S) = |1)$ will produce only one zero of energy in the denominator.

One interesting feature of this theory is that the zero $(E_E - E_O \pm \hbar\Omega_a - \hbar\omega_1)$ will be reached more quickly for the anti-Stokes than for the Stokes scattering as the frequency of the incident light approaches that of the first absorption band. Thus, the theory predicts that as one approaches the RRE there should be an anomalous increase in the relative ratio of the anti-Stokes radiation to Stokes radiation. It is of interest that this effect has recently been observed in both gaseous and liquid samples but that the anomalous increase in the anti-Stokes lines has been attributed to very large local heating which, of course, would give rise to a similar experimental observation because of the change in the Boltzmann factor. More work is obviously needed if these effects are to be definitely separated.

For the RRE where the two zeros in the denominator of the first trace term tend to make it dominate, the intensity of the scattering is given by

$$I_a = K' [(0|\mathbf{e}_2 \cdot \mathbf{r}|1)(1|\partial H/\partial Q_a|1)(1|\mathbf{e}_1 \cdot \mathbf{r}|0)]^2 \tag{V-2}$$

where K' will depend on the width of the UV absorption band and, hence, the damping which has been left out of our simple theory. Let us assume that $\mathbf{e}_1 = \mathbf{e}_2$ and that they are parallel to the transition dipole of the first excited electronic transition. In this case we obtain

$$I_a = K'(1|\partial H/\partial Q_a|1)^2 (1|\mathbf{r}|0)^4 = K''(1|\partial H/\partial Q_a|1)^2 \epsilon^2 \tag{V-3}$$

Since the UV molar absorption, ϵ, is proportional to the matrix element $(1|r|0)^2$ we see that for the RRE when the trace term is dominant, the Raman scattering intensity should be proportional to the square of the UV absorption intensity. Thus, if the UV hypochromism observed in the first UV absorption bands of the polynucleotides is in fact due to a diminution of the corresponding matrix element as is predicted by several theories then one would expect a similar effect in the RRE.

The RRE is very difficult to observe in these compounds because the UV absorption band is so high in energy that it is difficult to find a suitable exciting source. Consequently, the question naturally arises as to what can be said about the ordinary Raman effect where the incident light lies well below that of the first absorption band. In order to say anything about the ordinary Raman intensities one must make assumptions which are not strictly justifiable theoretically. However, any set of assumptions which consistently give results in agreement with experiment are useful and if we assume that some of the predictions of the RRE also carry into the ordinary Raman scattering intensities at least in a qualitative way we obtain some very interesting predictions which are in agreement with the observations reported in the previous section.

The primary prediction of the RRE theory is the dominant importance of the trace or diagonal terms in Eq. (1), i.e., those terms with $|E) = |S)$. Verlan has attempted numerical evaluation of matrix elements similar to those in Eq. (1) and has come to the conclusion that the diagonal terms give the greatest contribution to Raman intensity of symmetrical vibrations. Furthermore, it should be noted that these diagonal terms are only of importance for totally symmetric vibrations, and are identically zero for nontotally symmetric vibrations. (If the vibrations are degenerate there is a slight complication because of the self-multiplication properties of degenerate representations, but we will not discuss this here.) The reason that the diagonal terms are only of importance for symmetrical vibrations can easily be seen for the nucleic acid bases which belong to the point group, C_s. In these planar molecules both the electronic states and the vibrations may be classified as in-plane with irreducible representations, $\Gamma_{A'}$ or as out-of-plane with irreducible representations, $\Gamma_{A''}$. The same can be said about the operator, $\partial H/\partial Q_a$, which always belongs to the same representation in electronic coordinates as the vibration, Q_a does in nuclear coordinates. For the in-plane or totally symmetric vibrations, the sum over E and S in Eq. (1) must be such that both $|E)$ and $|S)$ belong to the same irreducible representations, i.e., transitions to these states from the ground state must both be in-plane or both out-of-plane transitions so that the cross product of the representations of $|E)$ and $|S)$ gives the totally symmetric representation. In this case the matrix element $(E|\partial H/\partial Q_{A'}|S)$ cannot be identically zero, since the representation of the integrand is given by

$$\Gamma_E \times \Gamma_{A'} \times \Gamma_S = \Gamma_{A'} \qquad \text{(V-4)}$$

Obviously, if $|E) = |S)$, then $\Gamma_E = \Gamma_S$ although the reverse is not necessarily true. Consequently, the diagonal terms with $E = S$ are present in the infinite sum for the in-plane vibrations. For the out-of-plane or completely unsymmetric vibrations, the diagonal terms must be identically zero since the irreducible representation of the integrand in the matrix elements $(E|\partial H/\partial Q_{A''}|E)$ must belong to the A'' representation. Hence the diagonal terms make no contribution to the intensity of the out-of-plane vibrations.

We now make the following assumptions: (1) the trace or diagonal terms are much larger than the off-diagonal terms so that the symmetric in-plane vibrations will show much more intense Raman scattering than the corresponding out-of-plane or unsymmetric vibrations, (2) certain of these in-plane vibrations will be coupled more strongly to the first or first few terms in the series than the others and therefore will derive their Raman intensity more directly from the first few UV absorption bands, (3) the hypochromism of UV transitions which is observed when aromatic rings interact with each other closely in solution is due to a decrease in the size of the corresponding matrix elements, $(0|r|1)$, $(0|r|2)$, etc. where $|1)$ and $|2)$ are the wave functions of the first and second UV absorption bands which have been observed to show hypochromism. Although these assumptions cannot be either derived or disproved theoretically because of mathematical complexities they lead to predictions which are in qualitative agreement with our experimental observations on the polynucleotides. Furthermore, since UV or visible hypochromism is known to occur in a wide variety of dyes and other organic aromatic compounds which aggregate in aqueous solution, these assumptions are subject to much further experimental investigation and should be looked upon as a set of working hypotheses for further examination of Raman intensity data in aromatic ring compounds – particularly those that aggregate. If the predictions from these assumptions prove to be borne out in a wide variety of experimental conditions then they should provide a very useful set of rules for the behavior of Raman intensities both close to and away from resonance in aromatic compounds.

The prediction which results from the first assumption is that all of the intense Raman bands should be strongly polarized and that none of the intense Raman bands should be completely depolarized. Recently, Nafie et al. have calculated what they call the depolarization ratio of all common molecular point groups. They obtain for the A' vibrations

$$\rho = \frac{a^2 + b^2 + c^2 + 3d^2 - ab - ac - bc}{3a^2 + 3b^2 + 3c^2 + 4d^2 + 2ab + 2ac + 2bc} \qquad \text{(V-5)}$$

and for the A'' vibrations they obtain the well-known result for unsymmetric vibrations,

$$\rho = 3/4 \qquad \text{(V-6)}$$

The quantities a, b, c, and d are obtained from the Raman tensor.

Just how completely polarized the A' vibrations should be is difficult to say exactly since measurements of a, b, c, and d have not yet been made independently on crystal samples. However, in our laboratory we have observed that in other liquid samples, that molecules such as benzene we obtain precisely the theoretical value of 0.75 within a few percent accuracy for vibrations which are known to be totally depolarized.

Polarized Raman spectra of all of the polynucleotides and their monomers have recently been reported from this laboratory. Referring to that paper we see immediately that not only do none of the intense bands have a depolarization ratio of 0.75 but that most of them have depolarization ratios which are very low. Only two bands of more than the two dozen intense bands in all four polymers whose polarizations were examined show a large depolarization. These are the 1580 cm^{-1} band in AMP and poly A where $\rho = 0.6$, and the 989 cm^{-1} band in UMP where $\rho = 0.7$. It is of interest that this latter band is rather weak in poly U. Thus the predictions of the first assumptions appear to be borne out by the intensity and the polarization measurements on the polynucleotides and their monomers.

The second assumption that some vibrations are more closely coupled to the first UV absorption bands than the others is necessary if we are to understand why there is such a large variation in the behavior of the intense totally polarized bands. For example, the strong, almost totally polarized band at 1336 cm^{-1} shows very little if any change while many other intense, highly polarized bands of poly A show large changes. Our explanation of this observation is that the bands which show Raman hypochromism are more strongly coupled to the first or first few UV absorption bands while those like the 1336 cm^{-1} band are not.

The assumption that even off resonance, certain Raman bands derive their intensity predominantly from the lower lying UV absorption bands is both bold and difficult to justify since it amounts to using the first or first few terms of an infinite series and ignoring the remainder. Strictly, there is absolutely no justification for this assumption except that it explains certain observed facts. As was pointed out by Tomlinson and Peticolas, a similar assumption directly underlies the theories of Bader and Pearson which have successfully accounted for a large number of reaction rates and structural features of molecules. Apparently, it does appear that for at least some vibrations, the integrals $(E|\partial H/\partial Q_a|S)$ become rapidly smaller as the energy separations between the electronic states E and S become larger so that not only do the energy denominators of the higher terms of the infinite series become larger but the matrix elements in the denominator becomes smaller thus leading to a more rapid convergence. One other point should also be mentioned; the matrix elements in the infinite sum vary in sign. Since it is the square of the sum which gives the intensity, if the terms in the sum alternate in sign they will tend to cancel each other out and resulting intensity will be affected. If the Raman intensity of certain bands derive their intensity from the first or first few absorption bands this may be due to the subsequent mutual cancelation of the higher terms.

Undoubtedly, more theoretical work needs to be done. Particularly, it would be of great use to obtain sum rules for vibrational spectroscopy. In the case of UV hypochromism, for example, if assumption (3) given above is made, that there is a decrease in the matrix elements of the first few absorption bands upon stacking, then it follows from the Kuhn-Thomas sum rule that there must be an increase in the intensity of the higher absorption bands, since the sum of the intensities of all of the bands must remain a constant. A similar rule for IR and Raman spectra would be most helpful in correlating intensity changes in vibrational spectra.

In conclusion, it can be said that with the assumption we have made, a qualitative interpretation of the changes in Raman intensities of polynucleotide ring vibrations can be given which is in agreement with the experimental observations. Quantitatively nothing can be said until experiments in the RRE can be made, since that is the only case in which there is any real theoretical justification for the use of the theory.

However, in spite of the deficiencies of the theory, it is very possible that Raman spectroscopy will become an important tool for the investigation of the conformation of genetic material since the intensity of certain bands of the bases are so conformationally dependent.

Acknowledgment

This review article is unusual in that it is taken largely from the Ph.D. dissertation of a single student – Enoch W. Small. Since at the present time there are very few Raman spectra of polynucleotides or nucleic acids in the literature Mr. Small's work provides most of what is available to the author. Furthermore the quality of Mr. Small's spectra compares favorably with the few other spectra which have been published. The author would like to express his gratitude to Mr. Small for his help in preparing this manuscript.

References

1. Peticolas, W. L., F. Fanconi, B. Tomlinson, L. A. Nafie, and W. Small. 1970. *Ann. N. Y. Acad. Sci.*, **168:** 564.
2. Cummings, H. Z., and H. L. Swinney, *Progr. Opt.*, **8:** in press.
3. Brillouin, L. 1922. *Ann. Phys.* **17**: 88.
4. Benedek, G., and T. Greytak. 1965. *Proc. IEEE*, **53**: 1623.
5. Chiao, R. Y., and B. P. Stoicheff. 1964. *J. Opt. Soc. Am.*, **54**: 1268.
6. Forrester, A. J. 1961. *J. Opt. Soc. Am.*, **51:** 253.
7. Dubin, S. B., J. H. Lunacek, and G. B. Benedek. 1967. *Proc. Natl. Acad. Sci. U.S.*, **57**: 1164.
8. Pecora, R. 1964. *J. Chem. Phys.*, **40:** 1604.
9. Malt, R. A. 1966. *Biochim, Biophys. Acta*, **120**: 461.
10a. Lord, R. C., and G. J. Thomas, Jr. 1967. *Spectrochim. Acta*, **23A:** 2551.
10b. Lord, R. C., and G. J. Thomas, Jr. 1968. In W. K. Baer, A. J. Perkins, and E. C. Grove, (Eds.), Developments in applied spectroscopy. Plenum Press, New York, Vol. 6, p. 179.
11. Fanconi, B., B. Tomlinson, L. A. Nafie, W. Small, and W. L. Peticolas. 1969. *J. Chem. Phys.*, **51**: 3993.
12. Tobin, M. C. 1969. *Spectrochim. Acta*, **25A:** 1855.
13. Aylward, N. N., and J. L. Koenig. 1970. *Macromolecules*, **3:** 590.
14. Thomas, G. J., Jr. 1970. *Biochim. Biophys. Acta*, **213**: 417.
15. Bellocq, A. M., C. Perchard, A. Novak, and M. L. Josien. 1965. *J. Chim. Phys.*, **62:** 1334.
16. Perchard, C., A. M. Bellocq, and A. Novak. 1965. *J. Chim. Phys.*, **62:** 1344.
17. Perchard, C., and A. Novak. 1968. *J. Chem. Phys.*, **48:** 3079.
18. Lord, R. C., A. L. Marston, and F. A. Miller. 1957. *Spectrochim. Acta*, **9:** 113.

19. Foglizzo, R., and A. Novak. 1967. *J. Chim. Phys.*, **64:** 1484.
20. Foglizzo, R., and A. Novak. 1969. *Spectroscopy Letters,* **2,** 165.
21. Lautie, A., and A. Novak. 1968. *J. Chem. Phys.*, **65:** 1359. (1968).
22. Tomlinson, B., and W. L. Peticolas. 1970. *J. Chem. Phys.*, **52:** 2154.
23. Peticolas, W. L., L. Nafie, B. Fanconi, and P. Stein. 1970. *J. Chem. Phys.*, **52:** 1576.
24. Small, E. W., and W. L. Peticolas. 1971. *Biopolymers,* **10:** 69.
25. Small, E. W., and W. L. Peticolas, *Biopolymers,* in press.
26. Miles, H. T. this volume, p. 205.
27. Fresco, J. R., and P. Doty. 1957. *J. Am. Chem. Soc.*, **79:** 3928.
28. Steiner, R., and R. Beers. 1957. *Biochem. Biophys. Acta,* **26:** 336.
29. Fresco, J. R., and E. Klemperer. 1959. *Ann. N. Y. Acad. Sci.*, **81:** 730.
30. Barszcz, D., and D. Shugar. 1964. *Acta Biochem. Polon.*, **11:** 481.
31. Griffin, B. E., W. J. Haslam, and C. B. Reese. 1964. *J. Mol. Biol.* **10:** 353.
32. Van Holde, K. E., J. Brahms, and A. M. Michelson, *J. Mol. Biol.* **12**: 726.
33. Holcomb, D. N., and I. Tinoco, Jr. 1965. *Biopolymers,* **3**: 121.
34. Stevens, C. L., and A. Rosenfeld. 1966. *Biochemistry,* **5:** 2714.
35. Poland, D., J. N. Vournakis, and H. A. Scheraga. 1966. *Biopolymers,* **5**: 223.
36. Brahms, J., A. M. Michelson, K. E. van Holde. 1966. *J. Mol. Biol.*, **15**: 467.
37. Leng, M., and G. Felsenfeld. 1966. *J. Mol. Biol.*, **15:** 455.
38. Applequist, J., and V. Damle. 1966. *J. Am. Chem. Soc.*, **88:** 3895.
39. Felsenfeld, G., and H. T. Miles. 1967. *Ann. Rev. Biochem.*, **36:** 407.
40. Richards, E. G., C. P. Flessel, and J. R. Fresco. 1963. *Biopolymers,* **1:** 431.
41. Lipsett, M. N., 1960. *Proc. Natl. Acad. Sci. U.S.*, **46:** 445.
42. Richards, E. G., C. P. Flessel, and J. R. Fresco. 1963. *Biopolymers,* **1:** 431.
43. Fasman, G. D., Lindblow, and L. Grossman. 1964. *Biochemistry,* **3:** 1015.
44. Pochon, F., and A. M. Michelson. 1965. *Proc. Natl. Acad. Sci. U.S.*, **53:** 1425.
45. Sarkar, P. K., and J. T. Yang. 1965. *Biochemistry,* **4:** 1238.
46. Miles, H. T., and J. Frazier. 1964. *Biochem. Biophys. Res. Commun.*, **14:** 129.
47. Warner, R. 1956. *Federation Proc.*, **15**: 379.
48. Warner, R. 1957. *Ann. N. Y. Acad. Sci.*, **69:** 314.
49. Rich, A., and D. R. Davies. 1956. *J. Am. Chem. Soc.*, **78:** 3548.
50. Felsenfeld, G., and A. Rich. 1957. *Biochim. Biophys. Acta,* **26:** 457.
51. Miles, H. T., and J. Frazier. 1964. *Biochem. Biophys. Res. Commun.*, **14:** 21.
52. Stevens, C. L., and G. Felsenfeld. 1964. *Biopolymers,* **2**: 293.
53. Tsuboi, M. 1969. *Appl. Spectroscopy. Rev.*, **3:** 45.
54. Shimanouchi, T., M. Tsuboi, and Y. Kyogoku. 1964. *Advan. Chem. Phys.*, **7:** 435.
55. Miles, H. T. 1964 *Proc. Natl. Acad. Sci. U.S.*, **51**: 1104.
56. Berstein, personal communication.
57. Davies, D. R., and R. L. Baldwin. 1963. *J. Mol. Biol.*, **6:** 251.
58. Inman, R. B., and R. L. Baldwin. 1962 *J. Mol. Biol.*, **5:** 172.
59. Sundaralingham, M. 1969. *Biopolymers,* **7**: 821.
60. Gellert, M., M. N. Lipsett, and D. R. Davies. 1962. *Proc. Natl. Acad. Sci. U.S.*, **48:** 2013.

LIGHT SCATTERING AND SOME ASPECTS OF SMALL ANGLE X-RAY SCATTERING*

HENRYK EISENBERG

POLYMER DEPARTMENT

THE WEIZMANN INSTITUTE OF SCIENCE

REHOVOT, ISRAEL

INTRODUCTION

The angular dependence of the scattering of radiation constitutes a powerful method in the study of macromolecules in solution. Scattering by visible light is limited in that the spectral region accessible to experimentation extends over a relatively narrow range of frequencies (the ultraviolet part of the spectrum is not suitable for scattering studies because DNA solutions absorb energy in this range; similar reasons, and also the fact that scattering strongly decreases with increasing wavelength, prevent the use of the low-frequency visible and near-visible spectrum). The essential complementarity of small angle x-ray scattering (SAXS) and visible-light scattering (LS) therefore provides a convenient way of extending the range accessible to either method (*1*). Elastic scattering, without a shift in frequency between the incident and scattered light, will be considered here; this eliminates Raman scattering, fluorescence, and Brillouin scattering associated with the Doppler shifts due to the molecular motion of the individual scatterers. Multiple scattering (that is, repeated scattering of the scattered beam) is

* This article was written during a tenure at the Institute of Colloid and Surface Science, Clarkson College of Technology, Potsdam, New York.

negligible in dilute systems and small scattering volumes, and will not be discussed in this work.

The theoretical aspects of LS and scattering of other electromagnetic radiation have recently been covered by Kerker in an introductory review (*2*) and in a comprehensive monograph (*3*) by the same author. A recent review which achieves an interesting insight into the complementary aspects of LS and SAXS is due to Finch and Holmes (*4*); specifically the relevance to structural studies of viruses is considered. Clear expositions on various aspects of LS are given by Tanford (*5*). In the important review of Geiduschek and Holtzer (*6*) both theoretical aspects and early studies by LS of nucleic acids and other systems of biological origin, are critically considered. A collection of reprints (and translations into English) of classical papers on LS from dilute polymer solutions has been published by McIntyre and Gornick (*7*). A recent extensive bibliography is by Kratohvil (*8*), who has also discussed (*9*) the problem of absolute calibration in LS experiments (cf. also the earlier reviews by Peterlin (*10*) and by Oster (*11*) and the comprehensive book by Stacey (*12*), which, because of errors and misprints should be used with utmost caution.

From the elastic scattering of electromagnetic radiation in solutions of high-molecular-weight materials we derive information on three parameters relevant to the solute. (1) the molecular weight of the particles, (2) the interaction between particles as manifested in the second and higher virial coefficients, and (3) information on the distribution of mass and the volume of the particles. From the shape of the angular dependence of the scattering curve it is possible to characterize the particle in terms of globular, random, or wormlike coil, or rodlike structure. Sometimes, particularly with SAXS, more detailed information may be derived (*13*) on the structure of a given macromolecule. When we refer to scattering in the following we mean aspects common to both LS and SAXS. The author is more familiar with the former method and has only slight connections with some aspects of the latter; the experimental aspects of the two methods are quite dissimilar but the common theoretical background justifies the usefulness of a discussion from a unified point of view. This will be attempted in the following section although the experimental aspects of SAXS will not be discussed. Readers interested in these aspects should refer to articles in the proceedings of a conference on SAXS, edited by Brumberger (*14*) or to the article by Kratky (*13*), devoted to substances of biological interest.

THEORETICAL BACKGROUND

For studies of dilute solutions of macromolecules, the equations for both LS and SAXS can be deduced from the Rayleigh-Debye-Gans approximation (cf. Chap. 8 of Ref. *3*) for scattering from an assembly of optically isotropic particles (*15, 16*). If the particles are noninteracting, then the total scattering from the system is the sum of the scattering from the individual particles. For interacting particles it is convenient to derive the total scattering on the basis of the Einstein-Debye treatment of fluctuations (under suitable thermodynamic restrictions) of density and concentrations. Whereas (for two component systems and short range interactions) particle–particle interactions are eliminated at high dilutions, preferential solute–solvent interactions in mixed (multicomponent) solvent systems persist in the limit of vanishing solute concentration. Nucleic acid solutions, except for rare exceptions (*17*), always contain additional low-molecular-weight salts or buffer components. They, therefore, constitute mixed solvent systems to which these considerations are applicable. The analysis of LS, as well as that of SAXS, must take proper account of the problems arising from the thermodynamics of multicomponent systems. (For a detailed review, see Casassa and Eisenberg (*18*).) The present treatment closely follows a recent discussion (*19*) on SAXS and LS of DNA solutions. Very recent developments showing that additional information may be obtained from scattering experiments (*20*) with coherent monochromatic (laser) radiation, will not be discussed in detail here.

Basic Equations

For a nonpolarized incident radiation of beam intensity I_0, the scattered intensity $I(h)$ at distance d from the sample (Fig. 1) is given by

$$I(h)/I_0 = (8\pi^4/d^2\lambda^4 V)(1 + \cos^2 2\theta)\overline{(\delta\alpha)^2}\, i_n'(h) \tag{1}$$

where λ is the wavelength (in the medium) of the radiation, 2θ is the scattering angle, $h = 4\pi \sin\theta/\lambda$, $\overline{(\delta\alpha)^2}$ is the mean-square contribution to the average polarizability α of volume element V, owing to concentration fluctuations δc, c is given in units of gm/ml, and $i_n'(h)$ is the factor representing the dependence upon scattering angle, normalized to unity at $h = 0$.

For two component systems, and assuming that pressure and temperature fluctuations are identical with those of the pure solvent, we have for the excess (over solvent) scattered intensity $\Delta I(h)$

$$\Delta I(h)/I_0 = (8\pi^4/d^2\lambda^4 V)(1 + \cos^2 2\theta)(\partial\alpha/\partial c)_{P,T}^2\overline{(\delta c)^2}\, i_n'(h) \tag{2}$$

LS changes in polarizability in the volume element V are related to changes in refractive index n by the Lorenz-Lorentz formula

$$(n^2 - n_0{}^2)/(n_2 + 2n_0{}^2) = (4\pi/3V)\alpha$$

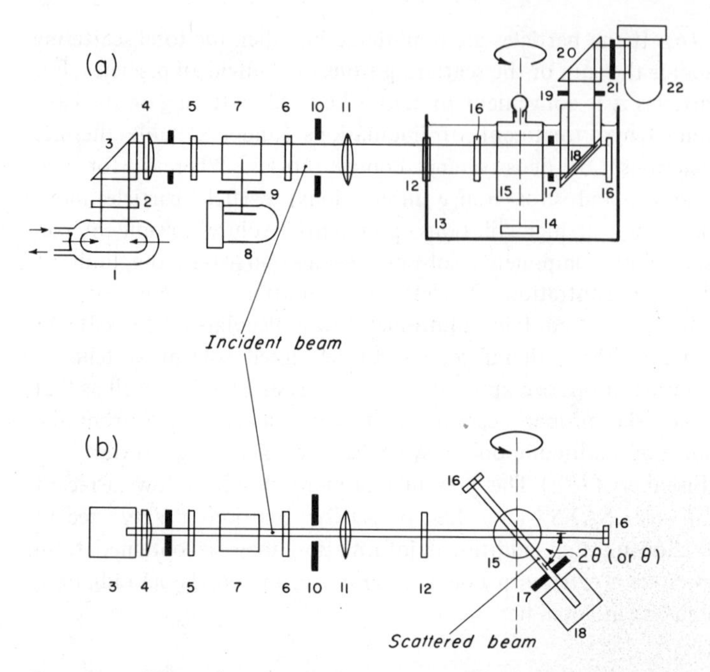

Figure 1. Schematic representation (not to scale) of Fica LS instrument. (*a*) Vertical section. (*b*) Horizontal section. (1) High pressure water cooled mercury vapor lamp. (2) Heat and ultraviolet light filter. (3) Total reflection prism. (4) Condensor. (5) Filters. (6) Polarizers. (7) Glass diffusor. (8) Reference photomultiplier. (9) Iris diaphragm. (10) Variable slit for control of light beam intensity. (11) Lens. (12) Entrance window to vat. (13) Vat filled with liquid of refractive index about 1.5. (14) Magnetic stirrer. (15) Stoppered cell. (16) Absorbing glasses. (17) Scattered beam entrance slit. (18) Air blade total reflection prism. (19) Exit slit. (20) Total reflection prism. (21) Shutter. (22) Photomultiplier tube. For details see text and references.

n_0 is the refractive index of the solvent. For small fluctuations

$$2(\partial n/\partial c)_{P,T}/n_0 = (4\pi/V)(\partial\alpha/\partial c)_{P,T}$$

The mean-square concentration fluctuation is given by

$$\overline{(\delta c)^2} = (c/VN_A)(M^{-1} + 2Bc + \ldots)^{-1}$$

where N_A is Avogadro's number, M is the molecular weight of the solute, and B is the second virial coefficient in the expansion of the osmotic pressure in powers of c.

If we now define the Rayleigh ratio $\Delta R(h) = d^2\Delta I(h)/I_0(1 + \cos^2 2\theta)$, then Eq. (2) yields the well-known expression (*21*)

$$Kc/\Delta R(h) = i_n^{-1}(h)M^{-1} + 2Bc + \ldots \tag{3}$$

where $K = (2\pi^2/N_A)(n_0{}^2/\lambda_0{}^4)(\partial n/\partial c)_{P,T}{}^2$, λ_0 is the wavelength *in vacuo*, and $i_n(h)$ is the particle scattering factor. The latter is more often designated $P(\theta)$ in polymer solution LS practice; also it is common practice in LS to define the scattering angle as θ and not as 2θ.

For low angle x-ray scattering we again refer to Eq. (2) but substitute

$$(\partial\alpha/\partial c)_{P,T} = (\partial\rho/\partial c)_{P,T}\,\alpha_{\text{el}}V$$

where α_{el} is the polarizability of an electron, and ρ is the density in electrons per milliliter. The scattering I_{el} of an electron is given by

$$I_{\text{el}}/I_0 = (8\pi^4/d^2\lambda^4)(1 + \cos^2 2\theta)\alpha_{\text{el}}^2$$

and

$$K'c/\Delta I(h) = i_n^{-1}(h)M^{-1} + 2Bc + \ldots \qquad (4)$$

where $K' = (\partial\rho/\partial c)_{\text{P,T}}{}^2 I_{\text{el}}/N_A$.

Extension to Multicomponent Systems

Equations (3) and (4) can be extended to take properly into account solute-solvent interactions in multicomponent systems by the straightforward procedure of specifying, to within a good approximation, $(\partial n/\partial c_2)_{T,\mu}$ in the constant K in Eq. (3) and $(\partial\rho/\partial c_2)_{T,\mu}$ in K' in Eq. (4), respectively (cf. Ref. *18* for more precise formulations). We use the designation 2 for the macromolecular component, 1 for the principal solvent water, and 3 and higher numbers for the added low-molecular-weight salt or neutral components. Subscript μ signifies that the increments are to be taken at constant chemical potential of all diffusible solutes. This, in essence, is the osmotic pressure condition, that is, the monodisperse macromolecular component 2 is the only component nondiffusible through a semipermeable membrane. An essentially equivalent treatment for LS has been given by Vrij and Overbeek (*22*). For charged systems, and at low-molecular-weight salt concentration of the order of 10^{-3} M NaCl, for example, and higher, the above theoretical treatment is applicable to LS with visible light at all accessible scattering angles (or values of h). Results from SAXS are at present not extensive enough to conclusively test the validity of the procedure for the highly charged DNA solutions (see Eisenberg and Cohen (*19*)).

Range of Applicability of LS and SAXS

It is interesting to compare the range of h in which LS and SAXS are applicable, as well as the information which can be obtained by both methods. This is best seen by examination of Fig. 2, taken from Finch and

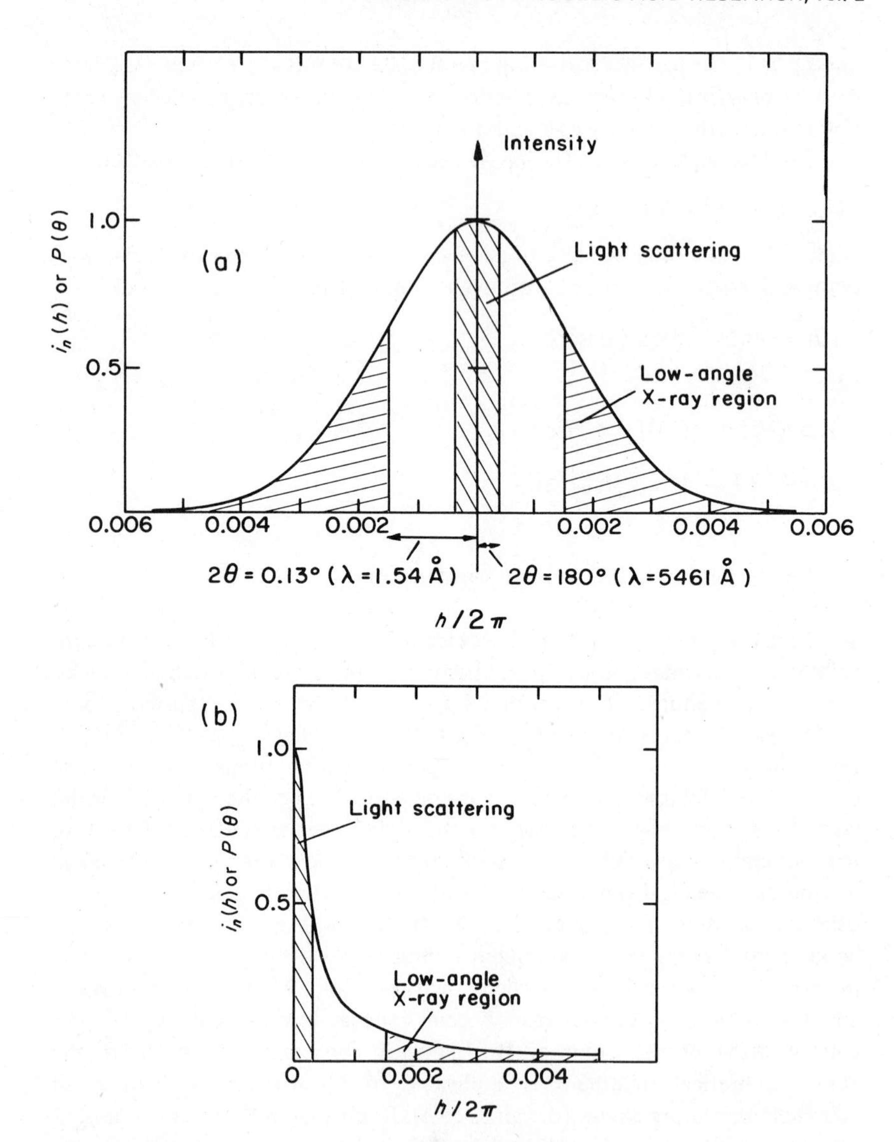

Figure 2. Comparison of LS and SAXS, according to Finch and Holmes (*4*). Particle scattering factor vs. $h/2\pi$. Scattering angle is 2θ. (a) Small spherical particle (radius 100 Å). (b) Rod particle of length 3000 Å and diameter 150 Å. This corresponds to tobacco mosaic virus particle. With modern techniques (*14*) the range of SAXS may be extended lower, to overlap with the LS range. [From J. T. Finch and K. C. Holmes, 1967. In K. Maramorosch and H. Koprowski (Eds.), Methods in virology, Vol. 3. Reprinted with permission of the publisher, Academic Press.]

Holmes (*4*). In a diffraction experiment on randomly oriented dissolved particles we record, in the limit of vanishing concentration $c \to 0$, the spherical average of the square of the Fourier transform per particle. Figure 2 shows the origin peak of the particle transform (actually the normalized function $i_n(h)$ or $P(\theta)$) for two particles: (a) a small, spherical particle of radius 100 Å and (b) a rodlike particle of length 3000 Å and diameter 150 Å, corresponding, respectively, to a globular protein and a molecule similar to tobacco mosaic virus. The regions explored by either LS or SAXS are shaded.

The wavelengths most commonly used in LS are the mercury lines* at $\lambda_0 = 4358$ Å and 5461 Å, and the scattering angles explore the range between $2\theta = 10°$ to $2\theta = 150°$; the corresponding quantities in SAXS are $\lambda_0 = 1.54$ Å for CuKα radiation, and scattering angles as low as 0.0015 to 0.00015 radians. Upon extrapolation of the scattering intensity to $h \to 0$ and $c \to 0$, the true molecular weight is obtained (cf. Eqs. (3) or (4)). The scattering intensity extrapolated to $h = 0$ is independent of the shape of the particle and depends on the molecular weight only; both $R(0)/Kc$ and $I(0)/K'c$ equal M_2 in the limit $c \to 0$.

Questions sometimes arise as to which molecular weight is determined in multicomponent systems. Is it the molecular weight of the backbone chain, or does it include some, or all of the counterions? Does it include bound molecules and is it feasible to speak of the molecular weight of the solvated, or the nonsolvated molecule? We have discussed (*18*) this problem both with respect to scattering experiments and equilibrium sedimentation. The molecular weight M_2, in the limit of vanishing polyelectrolyte concentration, is unambiguously defined in terms of the weight of material included in the definition of c_2. Equations (3) and (4) show that, in the limit $c_2 = 0$, multiplication of both c_2 and M_2 by an identical scale factor, leaves the equations unchanged. As long as the refractive index, or density, increments have been measured under the proper restrictions $(\partial n/\partial c_2)_{T,\mu}$ and $(\partial \rho/\partial c_2)_{T,\mu}$, (see p. 141), binding of any low-molecular-weight component (diffusible through a semipermeable membrane) to the macromolecular species (without change in the molarity of the latter) does not change this result. The quantities $(\partial n/\partial c_2)_{T,\mu}{}^2 c_2 M_2$ and $(\partial \rho/\partial c_2)_{T,\mu}{}^2 c_2 M_2$ which appear in this combination in Eqs. (3) and (4) easily transform into the more basic (but less practical) expressions $(\partial n/\partial C_2)_{T,\mu}{}^2 C_2$ and $(\partial \rho/\partial C_2)_{T,\mu}{}^2 C_2$, where $C_2 = c_2/M_2$ is the macromolecular concentration in moles per milliliter.

For molecular weight determinations it is necessary to have an absolute calibration (possibly in terms of a secondary standard) of the scattered intensity, and the values of the constants K or K': the incident radiation is

* It is quite likely that in the near future the use of laser beams will permit measurements at other values of λ. For example, Bünemann and Müller (*23*) describe an instrument for the study of DNA-dye complexes in which an He–Ne laser is used.

many orders of magnitude more intense than the scattered radiation and its absolute intensity can only be determined after suitable attenuation.

The shape of the origin peak around $h = 0$ is (when $c \to 0$) independent of the shape of the particle and depends on the distribution of mass only. Guinier's approximation (*24*) yields $-R_g{}^2/3$ as the limiting slope in the plot of either $\ln \Delta I(h)$ or $\ln \Delta R(h)$ against h^2; $R_g{}^2$ is the mean-square radius* and is independent of the exact nature of the particle shape. For rigid particles or for a given fixed conformation of a flexible particle

$$R_g = (\sum_i m_i r_i{}^2 / \sum_i m_i)^{1/2}$$

where we consider m_i mass elements each located a distance r_i from the center of the mass. Thus the intercept of a plot of $\ln\{\Delta R(h)/Kc_2\}$ or $\ln\{\Delta I(h)K'c_2\}$ against h^2 yields $\ln M_2$ and the limiting slope of this plot equals $-R_g{}^2/3$. (Neither absolute scattering values nor the values of K or K' are required for the determination of R_g.)

In LS it is customary to use the Zimm plot (*21*) for the double extrapolation to zero angle and zero concentration. The method is equivalent to Guinier's method, if the exponential in

$$\Delta R(\theta)/\Delta R(0) = \exp(-R_g{}^2 h^2/3)$$

is expanded for small enough values of h. It is customary to use Guinier's plot in SAXS and Zimm's plot in LS. The latter will be discussed in relation to specific LS examples below.

With increasing values of h more detailed information about the shape and the size of the particles may be obtained.† A suitable model must be assumed. For the sake of our discussion we choose a model appropriate to an idealization of DNA molecules, namely a cylinder of length L and radius r; for this model

$$R_g{}^2 = r^2/2 + L^2/12$$

Here $r^2/2$ is $r_g{}^2$, the mean-square radius of the cross section.

The complete particle scattering factor $i_n(h)$ (or $P(\theta)$) for N scattering elements was derived by Debye (*26*) and equals

$$\frac{1}{N^2} \sum_i \sum_j \frac{\sin hr_{ij}}{hr_{ij}}$$

where the summation extends over all distances r_{ij} between scattering

* Usually R is called the radius of gyration, but Hayes et al. (*25*) have objected.

† The reader is referred to Chapter 8 in the work of Kerker (*3*) for a more detailed critical discussion of size parameters which can be obtained from the analysis of Rayleigh-Debye scattering from solutions of macromolecules. Kerker also reviews LS of multicomponent solutions (Chapter 9) and problems raised by optical anisotropy (Chapter 10) in the absence and presence of external fields.

elements; the above expression correctly yields Guinier's (or Zimm's) approximation, when expanded for low values of the argument hr_{ij}. For long thin rods the particle scattering factor was obtained by Neugebauer (*27*) to be

$$\frac{2}{hL}\int_0^{hL}\frac{\sin w}{w}\,dw-\left(\frac{\sin(hL/2)}{hL/2}\right)^2$$

For values $hL >> 1$ (but $rh << 1$) we have

$$h\,\frac{\Delta I(h)}{K'c}=h\,\frac{\Delta R(h)}{Kc}=\frac{M}{L}\left(\pi-\frac{2}{Lh}\right) \tag{5}$$

which asymptotically approaches $\pi M/L$, with increasing values of h. This (see Fig. 2b) is the range in which (for elongated particles) SAXS is particularly useful, but asymptotic extrapolation of LS to high angles may also yield M/L (see p. 168). Sometimes (*28*) in LS the asymptotic behavior is plotted in the form

$$\frac{Kc}{\Delta R(h)}=\frac{2}{\pi^2 M}+\frac{L}{M}\frac{h}{\pi} \tag{6}$$

Thus the plot of $Kc/\Delta R(h)$ against h yields a straight line with the slope $L/M\pi$ and the intercept is equal to $2/\pi^2 M$. If the particle is not rigidly rodlike, or has flexible joints, the intercept may become negative and interesting information with respect to DNA structure and flexibility may be obtained (*29–31*). For the determination of M/L absolute calibration of the scattered intensity and value of the constants K or K' are required.

With h increasing even further the asymptotic expression becomes sensitive to r and for the right-hand side of Eq. (5) it is possible to write $(\pi M/L)\exp(-r_g{}^2h^2/2)$. This region is beyond the resolution of LS but is accessible to SAXS – provided that, in charged systems, the values of h correspond to volume elements which are large enough for independent electroneutral fluctuations. In the case of DNA solutions this restriction has not yet been convincingly tested. Again, for the determination of r_g, absolute calibration and the value of K', are not required.

Whenever LS and SAXS are discussed together, contributions relating to the size and shape of the particles, coming from the origin peak only, are considered. With increasing values of h, additional, much weaker side bands and maxima may appear. For systems extending over molecular and macromolecular dimensions, these are seen with SAXS only. A detailed analysis of the shape of the SAXS curve may be quite rewarding (we refer to recent studies (*32, 33*) on tRNA, (*34*) on ribosomes and, for comparison, on a rodlike associating protein system (*35*)). SAXS may provide information, for example, on the volume of the particles, or inhomogeneities in the distribution of the mass. The interested reader is referred to the specialized literature (*4, 13–16*) for more information.

Additional Features

The preceding considerations referred to homogeneous optically isotropic particles. We shall now consider extension to polydisperse systems and to optical anisotropy. We shall also briefly discuss the model for the wormlike coil which is thought to have great relevance to DNA particles, although its applicability has not been unequivocally established. Henceforth our considerations will be restricted to LS only, and the scattering angle and particle scattering factor will be denoted by θ and $P(\theta)$, respectively, to comply with general practice; therefore $h = 4\pi \sin(\theta/2)/\lambda$.

Polydispersity

Consider a collection of particles characterized by identical refractive index increment (same constant K), but dissimilar in size (for the usual DNAs, $(\partial n/\partial c)_{T,\mu}$ is not sensitive to composition – GC content – in a significant way). By Eq. (3), and in the limit $c \to 0$ and $\theta \to 0$

$$\Delta R(0) = K \Sigma c_i M_i \tag{7}$$

where the summation is over all species i, and $\sum_i c_i = c$, the total concentration. The weight average molecular weight is given by

$$M_w = \sum_i c_i M_i / \Sigma c_i$$

and substitution into Eq. (7) shows that indeed M_w is obtained

$$\{Kc/\Delta R(0)\}_{c\to 0} = 1/M_w$$

The virial coefficient B in a polydisperse system is biased toward contributions from species of higher molecular weight and exact expressions can be specified (*18*).

To average the root-mean-square radius R_g we write, at $c = 0$,

$$\Delta R(\theta) = K \sum_i c_i M_i P_i(\theta)$$

and obtain at low angles, with the use of Guinier's law, or Zimm's expansion,

$$Kc/\Delta R(\theta) = \sum_i c_i / \{\sum_i M_i c_i - (h^2/3) \sum_i M_i (R_g{}^2)_i c_i\}$$

which is transformed to

$$Kc/\Delta R(\theta) = M_w{}^{-1} (1 + \langle R_g{}^2 \rangle_z h^2/3) \tag{8}$$

where

$$\langle R_g{}^2 \rangle_z = \sum_i M_i c_i (R_g{}^2)_i / \Sigma M_i c_i$$

is the z average of the mean-square radius; $(R_g{}^2)_i$ is, over a range of chain lengths, proportional to $M_i{}^{2a}$, where a increases from $a = 0.5$ for random coils without excluded volume to $a = 1$ for rigid rods. These are two extreme cases which may be consistent with very long or very short linear DNA, respectively. Thus, an experimentally determined R_g corresponds to a molecular weight average defined by

$$\langle M \rangle_{R_g} = (\sum_i c_i M_i^{1+2a} / c_i M_i)^{1/2a} \tag{9}$$

For $a = 0.5$ this is the z average molecular weight M_z and for $a = 1$ this is $(M_z M_{z+1})^{1/2}$.

For polydisperse rodlike particles the intercept in the asymptotic plot against h (Eq. (6)), yields the number-average molecular weight M_n. Owing to deviations from this limiting model of ideal rods, it is not believed that M_n can be obtained with any degree of accuracy. In the case of polydisperse large Gaussian chains M_w and $\langle R_g^2 \rangle_z$ are obtained from the analysis at low angles ($hR_g \ll 1$), whereas at high angles ($hR_g \gg 1$) the analysis in principle leads to M_n and $\langle R_g^2 \rangle_n$ (*36, 37*).

The Wormlike Coil

A model for DNA which has been often used is the wormlike chain of Kratky and Porod (*38*). The wormlike chain is characterized by a contour length L and a persistence length A. The latter increases with increasing stiffness, but should on the basis of the model, be independent of L.* There is no universal agreement to date whether DNA is a "weakly bending rod" (*39*) or a collection of stiffer segments, connected by flexible joints (*30*) or whether the persistence length depends on chain length (*40*), the value of the persistence length, and the role of the excluded volume (*25*).

The relation between R_g and L for wormlike coils without excluded volume is (*41*)

$$12 R_g^2/L^2 = (3/2x^2)[(4x/3) - 2 + (2/x) - (1 - e^{-2x}/x^2)] \tag{10}$$

where $x = L/2A$. Equation (10) correctly yields the rigid rod relation $12 R_g^2 = L_2$ in the limit $x \to 0$ and the Gaussian coil relation $3 R_g^2 = LA$ at large values of x. It is thus possible, from measured values of R_g and L (derived from properly weighted molecular-weight measurements and the B structure of DNA – a repeat distance of 33.5 Å per 20 nucleotides, along the double helix axis) to ascribe a Kratky-Porod persistence length to DNA samples. It is also possible to express the results in terms of an elastic force constant AkT (where k is Boltzmann's constant and T the absolute temperature), directly related to the persistence length (*42, 43*). Polydispersity problems which arise in light scattering of wormlike coils can be taken into account (*25*) by ascribing a suitable value for a in Eq. (9). Although

* Numerically the persistence length A is just one-half the length of the Kuhn statistical element.

DNA may be curved over large distances, it may be approximated by a straight rod over short distances, and the asymptotic Eqs. (5) or (6) are expected to yield the mass per unit length; M/L is, of course, independent of the polydispersity of the sample. The asymptotic form of the intensity scattered by an assembly of isotropic filiform particles (with constant M/L) has been calculated by Luzzati and Benoit (*44*) for certain types of configuration and they also discuss the scattering expected from the Kratky-Porod chain, within the context of their own work.

Measurements (*45*) on short DNA fragments indicate that the persistence length formally calculated from Eq. (10) (and the assumption that the contour length L in solution corresponds to the "high humidity" B structure of DNA) markedly increases with decrease in molecular weight of the DNA (*40*). The apparent enhanced stiffness of the shorter fragments may be somewhat changed if optical anisotropy is considered in the calculation of M, R_g, and M/L (see below). In a recent study of dielectric and viscoelastic relaxation in dilute solutions of predominantly rodlike polybutylisocyanate, Dev et al. (*46*) attribute similar deviations to the inapplicability of the Kratky-Porod model to real systems for which the chain length is less than or of the order of the persistence length. Thus, although the Kratky-Porod model envisages a smooth transition from a nearly rigid rod to a Gaussian coil (with increasing values of L), the contour of "real" molecules is believed to respond more radically to increasing length. At low molecular weights the behavior is truly rodlike, in an intermediate molecular-weight range "bent rod" or semicircular arc conformations (cf. also Ref. *45*) predominate, and above this more pronounced curvature is seen leading to random coil conformations.

The problem of the conformation of long chain macromolecules with limited flexibility has also been the subject of a number of recent theoretical treatments (*47–50*).

Optical Anisotropy

Our discussion so far was restricted to optically isotropic particles. If the macromolecule is not an isotropic resonator, but is characterized by two principal polarizabilities, one in the direction of the double helix axis, the other perpendicular to it, then refinements owing to anisotropy have to be considered. This was found (*51*) to be particularly important in the case of DNA-proflavine complexes. Rodlike macromolecules may be characterized by two principal polarizabilities, one (α) in the direction of the axis, and another one (β) perpendicular to it. The anisotropy factor δ is defined by $(\alpha - \beta)/(\alpha + 2\beta)$ and can be evaluated (*52, 53*) from the value of the polarization ratio $\rho_v = H_v/V_v$, in the limit $\theta = 0$. In this case $\rho_v = 3\delta^2/(5 + 4\delta^2)$. The subscripts v indicate that the incident beam is vertically polarized and the capitals H and V refer to the horizontal and vertical components, respectively, of the scattered light; solvent corrections have been applied. As the square δ^2 is obtained from this experiment, the sign of δ has to be determined by an accessory experiment, such as streaming or electrical birefringence (for DNA Horn and Benoit (*52*) found $\delta = -0.15$ and for tobacco mosaic virus they found $\delta = +0.3$). The sign of δ is also given by the position of the minimum of H_h (the horizontal component of the scattered light with horizontally polarized incident radiation) as a function of θ; if $\theta_{min} < \pi/2$, $\delta > 0$, and vice versa. Horn and Benoit (*54*) and Benoit and Weill (*55*) showed that for anisotropic particles, in the limiting case of small angles, and for small values δ, an apparent mass $M^* = M(1 + \delta^2)$ and an apparent radius of gyration $R_g^* = R_g(1\text{-}2\delta/5)$ are obtained, from which the true quantities M and R_g may be obtained, if δ is known. Since δ is usually small, the correction is less important with regards to molecular weight than with respect to R_g. From the asymptotic behavior at high angles the apparent mass per unit length $(M/L)^*$ is connected (to within a first-order correction) to the true value by $(M/L)^* = (M/L)(1 + \delta)$. Compare also Weill et al. (*56*) for a comparative study of the optical and electrical anisotropy of DNA. Light scattering from dilute solutions of flexible chain polymers consisting of anisotropic segments has also been discussed by Utiyama and Kurata (*57*). The original publications should be consulted for details.

Coherent Incident Light

We have already mentioned the useful, but trivial application of a laser beam (*23*) in a classical study of complexes of DNA with actinomin and actinomycin.* In this case the laser beam provides a monochromatic light source with extremely small divergence of the beam. Additional information, not accessible to classical light scattering studies with incoherent light is available from a study of the shift and change in spectral distribution of the frequency of light scattered from macromolecules in solution. The fine structure of the scattered light reveals a triplet composed of a central Rayleigh band (symmetrically centered around the incident frequency) and two frequency-shifted Brillouin bands. From the broadening of the central Rayleigh band, translational diffusion coefficients (*20, 58–63*) and sometimes rotational diffusion coefficients (*60, 64*), intramolecular structural relaxation parameters, and chemical reaction rate constants of macromolecules in dilute solution may be derived (*60, 61, 65*), if such changes are accompanied by changes in polarizability of the reactants and components. From the intensity ratio of the Rayleigh unshifted band to the Brillouin, Doppler-shifted bands, it is possible to derive (*66, 67*) an absolute measure of the scattering intensity owing to concentration fluctuation without recourse to the absolute Rayleigh ratio of a standard material (such as benzene), or to an evaluation of the intensity of the incident beam. The molecular weight of standard NBS 705 polystyrene has been determined in this way (*67*).

* See footnote, p. 143.

Integrated intensities and the spectral distribution of light scattered from polydisperse rods and gaussian coils (*68, 69*) and for once-broken rods (*70*) have been calculated. These new applications provide renewed interest to LS theory and practice. Further developments may be foreseen and references given here are not exhaustive.

For the experimental study of both the shifted and unshifted scattered bands piezoelectrically driven scanning Fabry-Perot interferometers are conveniently used. For the precise and rapid analysis of the width of the central Rayleigh band new photon counting and optical self-beat techniques have recently been introduced. (It is claimed, for instance (*62*), that the coefficient of diffusion of a protein, hemocyanin, may be derived with 1% accuracy, within 30 seconds on small samples). Most studies have been undertaken with He-Ne lasers at 6329Å with laser power of 10-15 mW but shorter wavelengths will certainly be explored.

External Orienting Fields

Rigid nonspherical particles are partially oriented and nonrigid particles deformed in a flowing system or when subjected to an electrical or magnetic field. The associated optical effects are known as streaming birefringence (Maxwell effect), the Kerr effect, or the Cotton-Mouton effect. Interesting new information may be obtained when light-scattering measurements are undertaken in the presence of an electric field since both steady-state and relaxation measurements may be performed.* Recently Scheludko and Stoylov (*71*) have presented preliminary experiments on the variation in the intensity of scattered light by solutions of DNA subjected to an electric field. Stoylov and Sokerov (*72*) have described a method for the determination of rotational diffusion coefficients from light scattering in a transient electrical field.

INSTRUMENTS AND METHODS

Light Scattering Instruments

Basically a LS machine is a very simple device. An incident beam (Fig. 1) enters a cell, often cylindrical, and scattered light is quantitatively detected at various angles θ by a photomultiplier tube. A list of basic requirements, by no means exhaustive, includes (1) provision for selecting discrete spectral lines (by suitable filters) and horizontally and vertically polarized components, (2) the incident beam must be well defined and narrow, (3) reflection errors at entrance and exit faces should be either minimized or correctly accounted for, (4) the incident beam must be "trapped" efficiently after passing through

* See footnote, p. 144.

the cell, (5) the cell should be well thermostatted and constructed so that optical artifacts, parasitic reflections, or depolarized radiation are not introduced by it, and (6) the photomultiplier should be extremely sensitive and exposed only to the scattered beam. The stability of the reading is assured by continuously "sampling" the incident beam and by use of either a ratio recording device, or by adjusting the high voltage of the reading photomultiplier in proper relation to changes in intensity of the source. We shall not discuss technical details of all the optical and electrical problems which have been overcome in order to produce satisfactory instruments.

One of the most widely distributed instruments, particularly in the United States, is based on an early design (*73*) and manufactured by the Phoenix Precision Instrument Company at Gardiner, New York. They also manufacture a dual photomultiplier type instrument with ratio recorder, as well as a low-angle LS photometer. A disadvantage of the Brice-Phoenix instrument and others of its kind, is that the LS cell is surrounded by air. This necessitates reflection corrections (because the path of the optical beam encounters air-glass-medium-glass-air interfaces from source to detector) and also makes good thermostatting cumbersome. A different approach is based on a design by Carr and Zimm (*74*) who suggested that the cell (a conical "Erlenmeyer" type cell in their design) be made of thin glass and immersed in a liquid with a refractive index matching that of the glass (benzene was first used, toluene is useful for low temperatures, and silicone oil Dow Corning 702 fluid can be used at high temperatures, but it does not evaporate from cell walls and is therefore messy to handle). By this method reflection corrections are eliminated and good thermostatting is achieved. An instrument incorporating the latter principle was described by Wippler and Scheibling (*75*) and is manufactured by Fica (formerly Sofica) Le Mesnil Saint Denis, France (distributed in the United States by Bausch and Lomb, Rochester, N.Y.). It is an excellent instrument for studies on nucleic acid solutions. In the present version 15 ml of solution are required. The cylindrical cell rotates with the photomultiplier so that cells with a flat entrance face cannot be used for the study of the angular distribution of the scattered light. Because of the volatile bath liquid (benzene or toluene), well-stoppered cells should be used. Custom made cells (see Fig. 3) of thin walled (0.9-mm wall) Pyrex glass, outside diameter 28 mm, with flat bottoms and ground glass stoppers have been obtained from Fischer and Porter, Warminster, Pa.*

* It is possible to use smaller diameter (21 mm) for smaller volumes of liquid, but the curvature of the glass introduces disturbances in the measurement. We have designed and obtained (from Hellma A. G., Mullheim, Germany) octagonal cells which, with a special holder successfully permit measurements at 45°, 90°, and 135° (for dissymmetry determinations) with sample size of about 5 ml. Even smaller sample size, about 2–3 ml can be used in the Fica instrument with an ordinary optical quality glass 10 mm spectrophotometer fluorescence cell for the determination of ΔR (90).

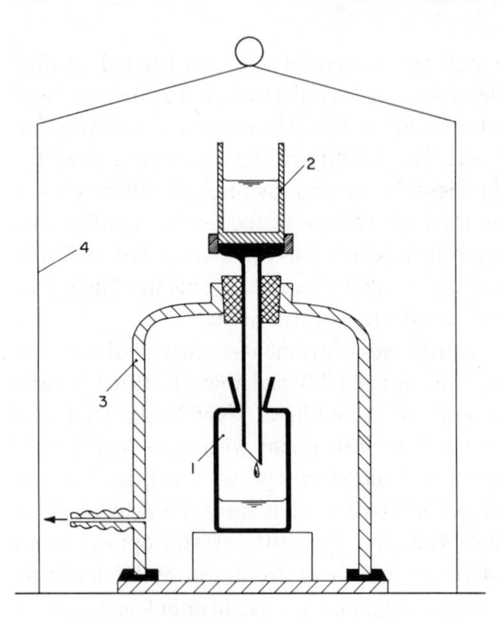

Figure 3. Filtration setup and LS cell. (1) LS cell. (2) Millipore filter holder. (3) Filtering bell jar. (4) Plastic cover.

Fica is now starting to produce a new instrument with a narrower incident beam allowing the sample size to be reduced to 5 and 1 ml; in addition the lowest scattering angle is 15° instead of 30°. Both these considerations are of extreme importance, but the new instrument has not yet been tested in nucleic acid work. According to the manufacturer, it also incorporates an analog computer to perform some basic calculations.

Reflection effects in the Brice-Phoenix and Fica LS photometers have recently been compared by Tomimatsu et al. (*76*). They refer to earlier references which deal with instrument calibration and the corrections required. A detailed comparison of a number of instruments has been made by Utiyama et al. (*77*), and they also describe (*78*) a low-angle LS photometer (lowest useful angle 20°) which is a modification of the commercial instrument manufactured by Shimadzu-Seisokusho, Kyoto, Japan. Corrections to LS measurements have also been considered by Schurz et al. (*79*) and programmed computations are described by Evans et al. (*80*). The correction for absorption and fluorescence in the determination of molecular weights by LS has been discussed by Brice et al. (*81*). The use of a laser light source in an instrument constructed for depolarization studies of simple liquids has been critically discussed by Lalanne and Bothorel (*82*).

We have already mentioned that, in the case of DNA solution, it has become increasingly important to measure at angles lower than 30°. Instruments suitable for this purpose, and studies on DNA solutions, have been undertaken by Froelich et al. (*83*) where the lowest angle is 16°* and Harpst et al. (*84, 85*) where the lowest angle is 10°. Meyerhoff et al. (*86*) describe an instrument for the angular range 6° to 170°. Recently two instruments designed for different types of work and capable of scanning the scattered beam extremely close to the incident beam have been described by Livesey and Billmeyer (*87*) (lowest angle 0.05°) and by Keijzers et al. (*88, 89*) (lowest angle 0.5°). The latter instrument may be adapted to the study of solutions and is manufactured by C. E. Bleeker, Zeist, Holland. It incorporates rotating coupled polaroids, which makes it possible to measure scattering outside the normally scanned horizontal plane; this procedure embodies certain advantages with respect to analyzing specific scattering curves. Finally, a commercial instrument, of the same type as the Fica instrument, is the Cenco-TNO turbidimeter, manufactured by Cenco, Breda, Holland.

Refractometers and Densitometer

The main accessory measurement required in LS studies is the refractive index of the solvents and the refractive index increment due to the solute ($(\partial n/\partial c_2)_{T,\mu}$ in multicomponent solutions). A useful instrument is the modified Pulfrich refractometer, manufactured by Bellingham and Stanley, London, England. This instrument as well as other refractometers and the various techniques for measuring n and dn/dc are described by Bauer et al. (*90*). The accuracy of the Pulfrich refractometer is only 2–3 units in the fifth decimal and its relatively limited usefulness is due to the fact that refractive index is linear in concentration over a wide range of n; therefore refractive index measurements may be undertaken at much higher concentrations than the LS study. The Pulfrich refractometer is a critical angle instrument which uses only a small amount of solution and, with suitable light sources and filters, the wavelength dispersion is easily determined. The manufacturers provide a split prism with which differential measurements may be performed. A differential refractometer sensitive to 3 units in the sixth decimal is based on a design by Brice and Halwer (*91*) and manufactured by the Phoenix Precision Instrument Company. The Shimadzu Seisakusho Company manufactures a similar instrument. Even higher precision and sensitivity, with some loss in convenience, are possible with a Rayleigh interference differential refractometer (see Ref. *90* for a detailed discussion). Commercial instruments are produced by Hilger and Watts, Manchester, England and by Carl Zeiss, Jena, East Germany.

* These authors (*83*) also describe a new and efficient clarification technique (cf. p. 158) for solutions of nucleic acids, one step of which involves shaking the DNA solutions with a chloroform-isoamyl alcohol mixture 5:1.

Density increments may also be determined on solutions (prepared by equilibrium dialysis against the proper solvent mixture) used for the determination of refractive index increments. Although not required for the evaluation of LS experiments, these density increments are required for the interpretation of SAXS studies (*19*) and for sedimentation studies (*92*) as well. In earlier days physical chemists shied away from time consuming pycnometric and other methods for density determinations requiring relatively large amounts of solute and solutions; this situation has been remedied by the introduction of a method (*93, 94*) in which density is determined (to an accuracy of 1.5×10^{-6} gm/ml) from the precise measurement of the change in resonating frequency of a glass swinger filled with about 0.7 ml of solution. Instruments based on this principle are manufactured by Anton Paar K. G., Graz, Austria.

AN ACTUAL EXPERIMENT

Choice of Concentration Range

The concentration ranges required for LS and for refractive index measurements do not always overlap. We consider the situation for the latter experiment first. It is usually found that, in macromolecular solutions, the refractive index is a linear function of c. Therefore we may approximate $(\partial n/\partial c_2)_{T,\mu}$ by $(\Delta n/c_2)_{T,\mu}$, where $\Delta n = n - n_0$, the difference between the refractive index of solution and solvent; n is directly measured in a differential refractometer. Since $(\partial n/\partial c_2)_{T,\mu}$ is approximately 0.15 ml/gm, and the sensitivity of the Rayleigh refractometer is about 1.5×10^{-6}, it is necessary to use a DNA concentration of at least 1 mg/ml for 1% accuracy in $(\Delta n/c_2)_{T,\mu}$. If higher accuracy is desired, then the concentration determination becomes a limiting factor. Solutions should be prepared in the concentration range 1 and 5 mg/ml. Since the refractive index is independent of molecular weight, the use of DNA broken into smaller double stranded "native" fragments (e.g., by sonic irradiation in solution) is recommended (*45, 92*). The refractive index increment should always be determined in a given solvent (buffer) system, and at a specified wavelength and temperature. On the other hand, it is not absolutely necessary to determine $(\partial n/\partial c_2)_{T,\mu}$ on the same nucleic acid sample used in the LS work, particularly if it is available in minute quantities only. The correct $(\partial n/\partial c)_{T,\mu}$ can, in most cases, be estimated (if data under identical experimental conditions are available) to a high enough accuracy, unless the nucleic acid is extremely unusual with respect to either base-composition or modification. Both refractive index and LS measurements are nondestructive operations and the solutions may be further used for other work.

In the case of LS, choice of a suitable concentration range depends on the molecular weight of the nucleic acid. For calf thymus Na-DNA of molecular weight 5×10^5 daltons in 0.2 M NaCl at $\lambda = 5461$ Å and 25°C, $\Delta R(90)/R_0$ (90) equals about unity (scattering from the solution is about twice that of the solvent mixture, R_0 *(90)*), for a concentration of about 10^{-5} gm/ml. The concentration can reasonably be raised by a factor of one hundred, and a convenient span – a five-fold dilution range – lies somewhere in between. As a rule of thumb, it can be reckoned that, over a range of molecular weights, the product *cM* remains constant for constant scattering, (this does not consider (1) reduced scattering because of interference at finite angles in the high-molecular-weight range and (2) thermodynamic non-ideality). The above considerations show the importance of reducing sample size from 15 to 1 ml (see p. 152) in the study by LS of nucleic acids available in small amounts only, such as the circular tumor virus nucleic acids (molecular weights about 3×10^6).

For the concentration determination, the optical density of suitably diluted aliquots is determined by UV absorption at 259 nm. The molar extinction coefficient, ϵ(P), for each sample should be determined by proper phosphate analysis. The use of dry-weight calibration is not recommended.* From the equivalent concentration (phosphorus, or nucleotide, equivalents per milliliter), it is possible to calculate weight concentrations in grams per milliliter by multiplying by a conversion factor. The latter for double stranded DNA of the normal type is, to a good approximation, independent of base composition. The average weight M_u of a paired nucleotide unit in the double-helix edifice is 331 daltons in the Na form and 441 daltons in the Cs form.

Dialysis

Dialysis is the required procedure in preparing solutions for the determination of $(\partial n/\partial c_2)_{T,\mu}$ (at constant chemical potential of diffusible solutes) and for LS. It is an excellent way of purifying a solution, eliminating low-molecular-weight contaminants and transferring of the nucleic acid from one solvent system to another. For $(\partial n/\partial c_2)_{T,\mu}$ equilibrium dialysis is mandatory, and most stringent precautions for successful execution must be observed; however it can sometimes be dispensed with in the preparation of dilute DNA solutions for LS, if lack of sufficient amounts of the valuable material precludes exhaustive dialysis.

* This is so because (1) it is very difficult to remove all water from a nucleic acid sample without damage or decomposition, and (2) in multicomponent systems unequal distribution of salt across a semipermeable dialysis membrane makes the interpretation of such dry-weight measurement precarious.

The importance of good dialysis for the refractive index measurements cannot be overemphasized, especially at high-salt concentrations. The value of $(\partial n/\partial c_2)_{T,\mu}$ is determined (*18*) by the value of $(\partial n/\partial c_2)_{P,T,m_3}$ (the refractive index increment at constant pressure P, temperature T, and molarity m_3 of the diffusible components) modified (*92*) by a contribution due to preferential interaction with component 3 and its own refractive index increment $(\partial n/\partial c_3)_{P,T,m_2}$ contribution (this term due to preferential interaction may amount to 20–30% of the major term in usual conditions; in the LS Eq. (3) $(\partial n/\partial c_2)_{T,\mu}$ is squared!). The concentration of the DNA component is usually much lower than that of component 3; thus even a slight shift in the concentration of component 3 may have deleterious effects on the Δn determination.

For good dialysis it is essential to mix both *inside* and *outside* dialysis solutions. We have been taught by W. R. Carroll to introduce the DNA solution into a dialysis bag, in which a little air space is left after the bag is sealed. The dialysis bag is introduced into a well-stoppered glass tube (ground glass-joint stoppered measuring flask) of suitable size, containing the solvent mixture. The whole is mounted at a suitable angle (about 45° from the vertical) on the shaft of a slowly rotating speed-reducer stirring motor. The speed of the motor is one revolution every few seconds, so that the air bubble inside the dialysis bag has ample time to travel up and down the bag, and thereby mix the solution.* Dialysis equilibrium under these conditions is rather rapid and probably complete within a few hours. A solution may be ready for refractive index measurements within 24-48 hours, with a few changes of solvent. Prolonged dialysis is not required and may lead to contamination of the solutions by microorganisms or degradation of the nucleic acids. The usual precautions should be observed: sterilized, well-rinsed dialysis bags (with water and dialysis solvent), polyethylene gloves, rapid transfer of the dialyzed solutions into well-stoppered flasks prior to refractive index analysis. Any amount of evaporation or condensation in this step may lead to disaster (compare the scatter of experimental values of $\partial n/\partial c$ compiled in Refs. *95*, *96* for support of this point). The concentration determination is not as critical and is usually performed on an aliquot after the refractive index difference between solute and solvent has been determined. Errors should be less than 1% in c and less than 1% in n. Solutions used are filtered through 0.45μ-pore Millipore filters (Millipore Filter Corp., Bedford, Mass.) prior to dialysis.

Equilibrium dialysis is not vitally important for the light-scattering measurements themselves. At low concentrations $m_u \ll m_3$ (m_u is the molality of DNA in equivalents of nucleotides, or phosphate ions, per

* Similar dialyzing devices in a commerical version are now available from BioCal instruments, ISCO, Inc., Richmond, California.

kilogram of principal solvent-water), m_3 in the DNA solution is almost identical with $m_3{}'$ in the outside, DNA-free, solution: $(m_3{}' - m_3) << m_3$. LS properties are very little affected by a small shift in salt concentration from $m_3{}'$ to m_3.* The concentration fluctuations, which lead to light being scattered, are determined by the requirements of the multicomponent theory (*18, 22*); the refractive index fluctuation associated with the concentration fluctuation is properly determined by $(\partial n/\partial c_2)_{T,\mu}$.

In summary, for $(\partial n/\partial c_2)_{T,\mu}$ good dialysis is absolutely essential; LS dialysis is recommended (to obtain the required solvent system and eliminate unwanted ions) but prolonged equilibration may be dispensed with. Dialysis may be performed at one concentration only and, for the preparation of more dilute solutions, the resulting stock solution is mixed with the dialysis equilibrium salt solution in suitable proportions. In the case of DNA samples in very short supply, solutions for LS may be prepared with sufficient care and perception without prior dialysis. (Ion-exchange or gel columns are often useful.) Initial steps of the dialysis may be undertaken in a cold room, but the final stages of the dialysis for $(\partial n/\partial c_2)_{T,\mu}$ must be executed at a temperature close to that of the LS experiment.

Refractive Index Determination

Consult previous two sections for preparation of solutions and Bauer et al. (*90*) for problems relating to the refractometry. We use the Hilger Rayleigh refractometer. The first preliminary determination is made with a 0.3-cm pathlength cell, however, a more precise determination is made with the 1-cm cell, and 3-cm pathlength cell in the case of the most dilute solutions. The 10-cm pathlength cell is not convenient because of evaporation problems (with the $\sqcup\!\sqcup$- shaped cells supplied), temperature control, and relatively large solution volume (cf. Reisler and Eisenberg (*97*) for a refractive index study of pure liquids in which some technical problems pertaining to this instrument are discussed). We use either blue light (λ = 4358 Å) or green light (λ = 5461 Å) isolated from a mercury AH4 lamp by Corning glass filters. Interference filters reduce the light intensity too much and are not recommended. In the 1-cm cell a DNA solution at a concentration of about 1 mg/ml shows a displacement of about 3 fringes. The integral fringe number is determined with white light.† With monochromatic light about 0.05 of a fringe can be estimated; $\Delta n = \delta\lambda_0/l$, where δ is the fringe shift, λ_0 is the wavelength *in vacuo* in centimeters and l is the pathlength in the same units.

* This would not necessarily be so in the vicinity of a critical point.

† Fringe shifts (*90*) may occur above 4–5 fringes and are taken into account by knowing the approximate result from measurements in the 0.3 cm cell.

Preparation of Solutions for LS

A successful LS experiment hinges on well-clarified solutions. DNA solutions for LS should not be prepared unless symmetric readings (approx. $\theta = 90°$) over a wide angular range are obtained with simple salt solutions (cf. Cohen and Eisenberg (*98*) for experimental details on LS studies of aqueous solutions). Absence of fluorescent components in the scattered beam and the ability to determine, within about 5%, the correct Rayleigh ratio and depolarization ratio of dilute aqueous solutions should be established before proceeding to the study of nucleic acid solutions.

Glassware (cells, beakers, volumetric flasks) are cleaned with hot dichromate cleaning solution and rinsed well with distilled water. We never use deionized water (which may contain microorganisms or organic debris and surface active agents) but freshly distilled water (from deionized water) in an all-glass, semiautomatic, commercially available, simple still (Fysons, Loughborough, Lincolnshire, England). The flow of deionized water to the still is regulated by an inexpensive plastic tank with a float controlled valve. Sometimes glassware is rinsed with redistilled synthetic methanol prior to drying. Glassware is usually heat sterilized (loosely covered with aluminum foil) in a laboratory oven.

For clarification of solutions we use either centrifugation at up to 20,000 rpm in a Sorvall (Ivan Sorvall, Norwalk, Connecticut) refrigerated centrifuge, or filtration through a 0.45μ-pore or 0.22μ-pore Millipore filter, or both. When using the refrigerated centrifuge beware of changes in nucleic acid concentration and composition as a result of sedimentation. We use a filter bell with gasket (manufactured by Scientific Glass Company, Bloomfield, New Jersey) as support for the 25-mm diameter all glass Millipore filter holder (Fig. 3). No suction is used and filtration proceeds by gravity only. Sometimes we use two glass flanges, without fritted-glass support for the Millipore filter, to avoid contamination and adsorption losses in very dilute solutions. A plastic bell jar (microscope cover) protects the filtering solutions from dust and other airborn material. For further discussion of methods of clarification by filtration compare Krasna and Harpst (*99*), Bernardi (*100*), as well as Refs. *83-85*.

Clarification by centrifugation is achieved in ½ to 2 hr at 20,000 rpm in the Sorvall centrifuge with about 30 ml liquid in each tube. The angle rotor head carries a number of DNA solutions and at least two solvent tubes. Solutions are removed from the centrifuge tubes by a clean 25-ml pipette equipped with a double acting valve controlled rubber bulb for suction and pressure. After the centrifuge has gently stopped, the rotor cover is removed, and the pipette tip is submerged in one of the freshly centrifuged solvents and placed against the inner wall of the centrifuge flask (closer to the center axis of the rotor) without removing the flask which is inclined toward the rotor axis, from the centrifuge rotor.* The centrifuged solvent is slowly sucked up. The first solvent is discarded, and the rinsed pipette is introduced into the second solvent flask.† This solvent is carefully run (with pipette tip touching inner side of cell wall) into a clean cell. With the same pipette and by the same procedure we then proceed to the various solutions in increasing order of concentration. Only negligible changes in concentration are produced in this way, and the concentration should at any rate be determined upon completion of the LS run. Care is taken not to wet the glass stoppers of the cells, if possible, but no tissue paper should be used to wipe off drops. Excess liquid can be carefully removed with a capillary tube and slight suction before the liquid evaporates to leave a salt layer behind.

Dilutions, or rather concentrations, are sometimes produced in different ways. It is possible to mix clarified solvent and stock solution in specific proportions. It is also possible to clarify solvent, and add successive portions of concentrated stock solution dropwise through a syringe fitted with an all plastic Millipore filter holder, and gently mix the stoppered cell. In our cells we can start with 15 ml solvent and add up to 5 ml concentrated DNA solution. Convenience usually dictates which technique should be used.

Light-Scattering Experiment

For the LS experiments we have modified the Fica instrument by (1) constructing special filter and polarizer holders for the incident and scattered beams and (2) enlarging the magnetically stirred thermostat bath to provide closer temperature control, better insulation and defogging (this allows operation down to $-5^{\circ}C$).‡ Instead of the DC amplifier for the photomultiplier output signal of the scattered beam, we have connected a self-ranging Hewlett-Packard Company (Palo Alto, California) 3440 A digital voltmeter (input impedance 10 megohm, full scale 100 mV to 1000 V, accuracy 0.05%) directly across the output of the 1P28, specially selected, photomultiplier tube. This allows precise recordable digital output over the

* Just enough pressure is applied so that, when the pipette tip pierces the surface of the liquid, no liquid from the contaminated surface enters the pipette, but also no air bubbles are expelled. Everything is done extremely gently so not to create convection currents and disturb sediment that may have formed at the bottom of the flask. To avoid introducing the pipette too deep into the centrifuge flask a paper clip is slid over the section tube of the pipette to fix the depth of immersion.

† Pipettes must not necessarily be dried (dust may be picked up electrostatically upon drying) but may be kept wet, after rinsing, in Millipore-filtered distilled water. From a clean pipette almost all fluid should drain easily (only a thin film of liquid should stay and no drops should form) and filling once and draining with clean solvent should be satisfactory.

‡ The Fica instrument was originally built with high temperature synthetic polymer solution work in mind.

entire experimental range, however, more sophisticated chopper – AC detecting systems are now available. If an occasional speck of dust enters the beam, a high transient reading results which can be damped out by shunting a condensor across the digital voltmeter output. We do not take average readings, but rather the lowest, over a period of time.

The use of vertically polarized incident light leads to better results and we use this as a rule. We work at both the blue and the green wavelengths and check whether the curves, after proper adjustment, are superimposable. We expect to obtain the same value for M, R_g, M/L, and B from either wavelength. With green light we can go to slightly lower values of h (which is useful for the extrapolation to $h \to 0$) and with blue light we can go to slightly higher values of h, therefore we can approach the asymptotic range, from which M/L is derived, more closely.

For the evaluation of LS results the quantity $Kc/\Delta R(\theta)$ is required (Eq. 3). For unpolarized light, K, for three or more component systems, as already defined, equals

$$K = (2\pi^2/N_A)(n_0^2/\lambda_0^4)(\partial n/\partial c_2)_{T,\mu}^{\ 2} \quad (11)$$

From the scattering measurements we obtain $\Delta R(\theta)$, the reduced intensity of scattering (corrected for solvent contribution) at angle θ with the incident beam, as follows

$$\Delta R(\theta) = \{\Delta I(\theta)/I_B(90)\}\, R_B(n_0/n_B)^2 \sin\theta/(1+\cos^2\theta) \quad (12)$$

where $\Delta I(\theta) = I(\theta) - I_0(\theta)$ is the intensity of light scattered at angle θ, in excess of solvent, $I_B(90)$ is the light scattered by a standard (benzene)* at $\theta = 90°$, R_B is the absolute Rayleigh ratio of benzene (15.8×10^{-6} cm^{-1} for $\lambda = 5461$ Å and 45.6×10^{-6} cm^{-1} for $\lambda = 4358$ Å (*101*), the correction factor $(n/n_B)^2$ is due to the unlike refractive indices of medium and calibrating liquid, $\sin\theta$ is due to the change in scattering volume with angle, and $(1 + \cos^2\theta)$ is due to the use of unpolarized incident light. The final working equation is derived from Eqs. (11) and (12).

$$Kc_2/\Delta R(\theta) =$$

$$(2\pi^2 n_B^{\ 2}/\lambda_0^{\ 4} N_A R_B)(\partial n/\partial c_2)_{T,\mu}^{\ 2}\,\{c_2 I_B(90)/\Delta I(\theta)\}\sin\theta/(1+\cos^2\theta) \quad (13)$$

(nonpolarized incident light)

* Benzene which is used as a standard in the calibration is purified by passage over an aluminium and a silica gel column (to absorb fluorescent impurities) and subsequent distillation and filtration (through a Millipore 0.45μ filter) directly into a clean LS cell. The calibration sample is periodically renewed, but for routine determinations we use a glass scattering secondary standard, supplied by the manufacturer, and periodically calibrated at both blue and green wavelengths against benzene. The depolarization ratio, with incident nonpolarized light, ρ_μ of benzene should not exceed 0.42 at both values of λ, at 23°C (*98*).

Note that, because of the $(n/n_B)^2$ correction, the refractive index of the medium has been replaced by the refractive index n_B of benzene. In the case of vertically polarized incident light the final working equation is slightly different,

$$Kc_2/\Delta R(\theta) = (4\pi^2 n_B{}^2/\lambda_0{}^4 N_A R_B)(\partial n/\partial c_2)_{T,\mu}{}^2 \{c_2 I_B(90)/\Delta I(\theta)\} \sin\theta \quad (14)$$

(vertically polarized incident light)

The Rayleigh ratio of benzene with vertically polarized light is $2/(1 + \rho_u)$ or about 1.41 times the Rayleigh ratio with unpolarized incident light.

In the earlier days of the study of the angular dependence of LS it had been customary to interpret experimental results in terms of the dissymmetry, the ratio $\Delta R(45)/\Delta R(135)$. This procedure is not recommended because (1) the information obtained is rather limited and useful in a few idealized cases only, and (2) from the measurements at 45°, 90°, and 135°, only, it is not possible to judge whether the solutions have been successfully clarified. Fortunately, with present day instrumentation, a wide angular range of scattering is accessible and complete scattering curves may be obtained. Measurements at low angles are extremely important (*83–85*); from the behavior at low angles and the superposition of data obtained with green and with blue light, it is usually possible to tell whether the solutions are sufficiently pure.

For the determination of M_2, B, and R_g the experimental data are extrapolated to zero angle and concentration. The data are evaluated in terms of the two equations

$$\{Kc_2/\Delta R(\theta)\}_{\lim \theta \to 0} = M_w{}^{-1} + 2Bc_2 + \ldots \quad (15)$$

and

$$\{Kc_2/\Delta R(\theta)\}_{\lim c_2 \to 0} = M_w{}^{-1}\,[1 + (16\pi^2 n^2/3\lambda_0{}^2)\,\langle R_g{}^2\rangle_z \sin^2(\theta/2) + \ldots] \quad (16)$$

R_g is derived, independent of any assumption on the shape of the macromolecules, from the ratio of the slope to intercept of Eq. (16)

$$\frac{4\pi n_0}{\sqrt{3}\lambda} R_g = \left[\frac{\text{initial slope, at } c_2 = 0, \text{ of } Kc_2/\Delta R(\theta) \text{ vs. } \sin^2(\theta/2)}{\text{intercept, at } \theta = 0, \text{ of } Kc_2/\Delta R(\theta) \text{ at } c_2 = 0}\right]^{1/2} \quad (17)$$

It is possible to perform the two extrapolations separately but, for graphical representation, it is customary to represent them in the double extrapolation known as the Zimm plot (*21*), in which $Kc_2/\Delta R(\theta)$ is plotted vs. $\sin^2(\theta/2) + kc$, where k is a constant arbitrarily chosen for reasons of convenience. Examples of Zimm plots and other representations will be shown in the next section.

A useful and convenient tabulation of various functions, dissymmetries and particle scattering factors for different types of particles may be found in an article by Chiang (Ref. *96*), p. IV-315).

Sometimes, particularly when results which involve NaDNA and CsDNA are to be compared, it is preferable to use equivalent concentrations C_u (moles of nucleotides, or phosphate, per liter) rather than weight per unit volume (c_2). Equivalent concentration units are invariant to the nature of the counterion of the DNA and, instead of the molecular weight, the number Z of nucleotides per macromolecule, $Z = M_2/M_u$, is calculated from LS experiments. The concentration ($C_u = 10^3 c_2/M_u$) and all quantities which appear in the equations discussed above can be transformed accordingly.

SOME TYPICAL RESULTS

In conclusion we show some typical results obtained in recent years. Figure 4 is a plot (*45*) of refractive index n (at constant chemical potential μ_3) versus C_u; all solutions have been separately dialyzed, but this could have been dispensed with, in view of the linear relationship.* From the slope, $(\partial n/\partial C_u)_{T,\mu}$ or $(\partial n/\partial c_2)_{T,\mu}$ can be obtained by least square analysis (in which the refractive index point of the solvent is included); cf. Appendix B for a tabulation of the results obtained. Figure 5 is a Zimm plot constructed from measurements in the angular range 30° to 150° of calf thymus NaDNA (in 0.2 M NaCl, at 25°C) broken into almost rodlike fragments by sonic irradiation. M_w of this sample is 4.65×10^5 ($Z_w = 1400$) and $\langle R_g \rangle_z = 636$ Å (*45*). Optical anisotropy (see p. 148) has not been taken into account and the second virial coefficient has not been evaluated. In this case of a relatively low-molecular-weight DNA it is not necessary to measure at lower angles to determine M_w; the extrapolation is rather short, and the value of $K\,C_u/\Delta R(\theta)$ at $\theta = 30°$, is only about 10% above the extrapolated value at $C_u = 0$ and $\theta = 0$.

The necessity to perform measurements on high-molecular-weight DNA at very low angles is exemplified in Fig. 6 where $P^{-1}(\theta)$ has been drawn vs. $h^2 R_g{}^2$ for monodisperse spheres, Gaussian coils, and rods in the range $0 < h^2 R_g{}^2 < 3.5$. It is seen that below $h^2 R_g{}^2 = 2$ it is not possible to distinguish between the latter two shapes. To show the limiting slope in the Zimm representation, the straight line $1 + h^2 R_g{}^2/3$ has also been drawn. It diverges from the complete $P^{-1}(\theta)$ curve at about $h^2 R_g{}^2 > 0.3$. For a value of $R_g \sim 700$ Å (which corresponds to a stiff rod of contour length 2400 Å, M_w of DNA about 4.85×10^5) this means that the highest scattering angle θ (for green light) should not exceed 45° for the experimental points to be on the limiting slope! Measurements in the range $1 < h^2 R_g{}^2 < 3$, where the angular range for green light is 45° to about 110° for the rod considered lie

* Still, this procedure engenders more confidence than one dialysis (which may not correspond to the correct equilibrium state) and subsequent dilution.

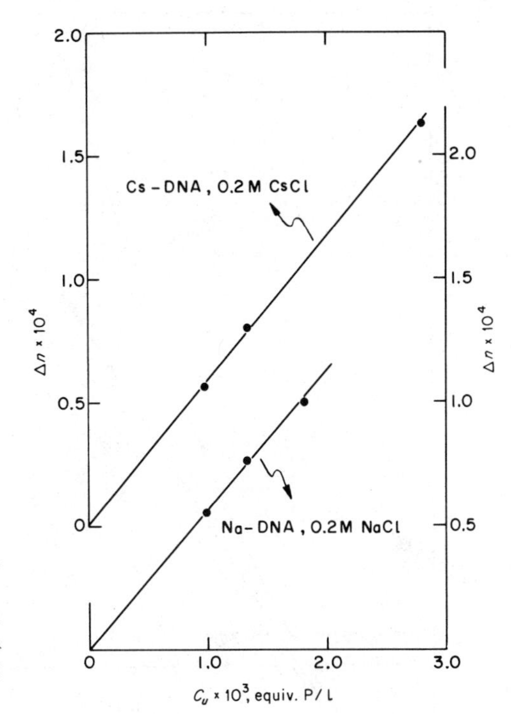

Figure 4. Refractive index differences Δn between sonicated samples of DNA and equilibrium dialyzate solutions, at $\lambda = 5461$ Å and 25°C. (Cohen and Eisenberg (*45*).)

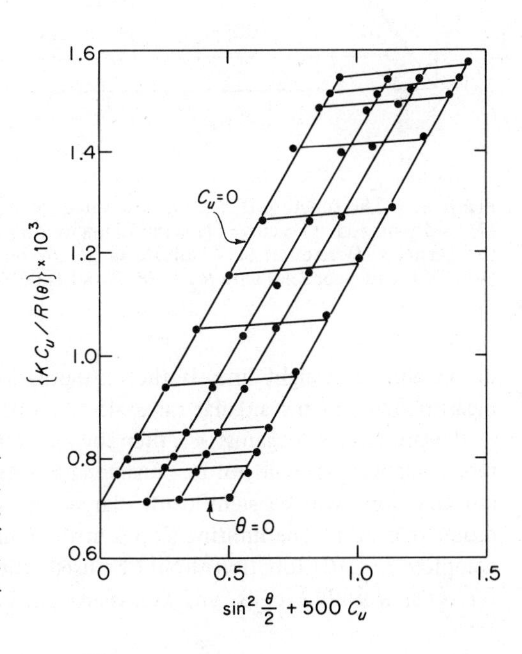

Figure 5. Zimm plot of LS data at 5461 Å at 25°C, for solutions of sonicated NaDNA in 0.2 M NaCl. The intercept equals 0.714×10^{-3} and the ration of slope (at $C_u = 0$) to intercept equals 1.28. (Cohen and Eisenberg (*45*).)

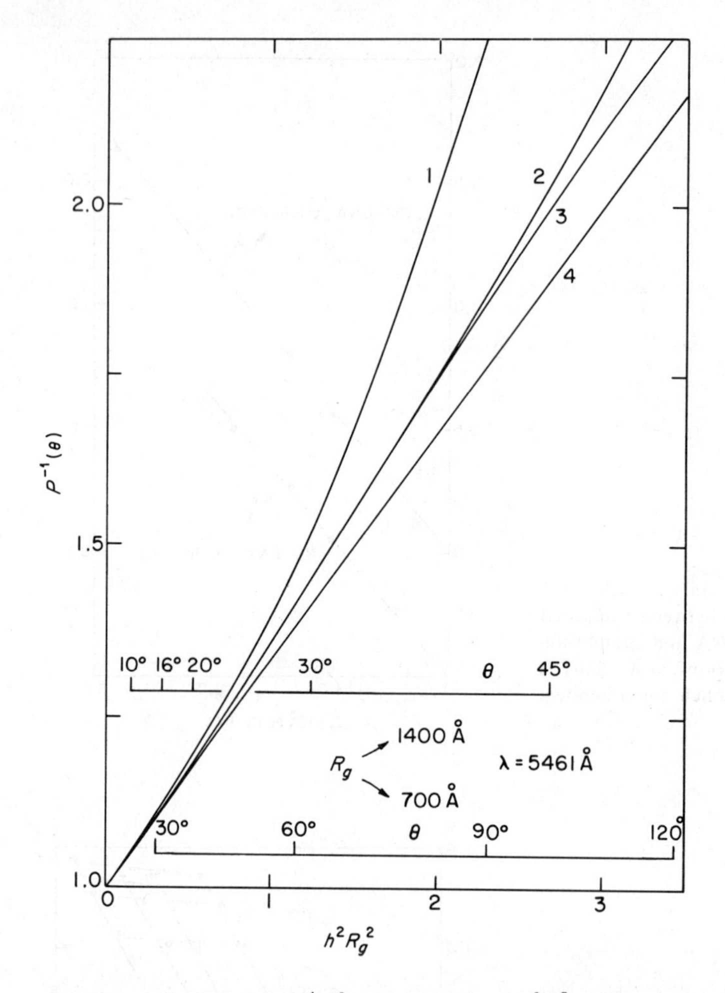

Figure 6. The function P^{-1} (θ) at low values of $h^2R_g^2$ for spheres (1), Gaussian coils (2), and rigid rods (3); curve (4) is the Zimm limiting slope, independent of the shape of the particles. Horizontal bars indicate range covered at various scattering angles θ at 5461 Å, for two particles with R_g = 700 Å and 1400 Å, respectively.

on an almost straight line. If the extrapolation to zero angle is made from measurements in the angular range 45° to 110°, by use of the limiting slope in the presentation against h^2, then the apparent intercept (and therefore the molecular weight) will not be considerably different from the true intercept but the slope will be significantly larger than the true limiting slope. Thus rather than using the limiting slopes in the Guinier or Zimm procedures, the complete P^{-1} (θ) function should be used wherever possible. Such functions for rods, wormlike coils, and Gaussian coils (without excluded volume) are

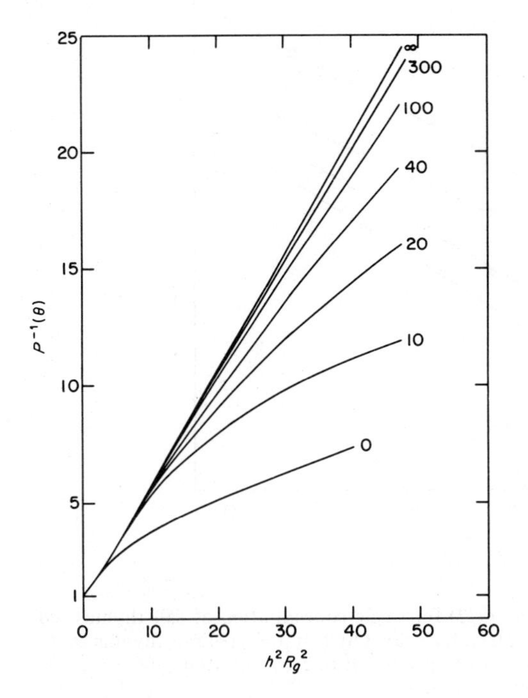

Figure 7. The function P^{-1} (θ) at larger values of $h^2R_g^2$ for wormlike chains without excluded volume, for the parameter $2L/A = \infty$ (random coil), 300, 100, 40, 20, 10, and 0 (rigid rod) (Sharp and Bloomfield (*102*)).

shown in Fig. 7. For moderate values of R_g, $P^{-1}(\theta)$ is not sensitive to the stiffness of the coils. The calculation of the detailed functions with excluded volume presents a formidable problem which has not been completely solved (*102*); Hays et al. (*25*) believe that the importance of the excluded volume in DNA studies has been overstressed.

Figure 8 represents the scattering envelope, in the angular range $\theta = 10°$ to $\theta = 60°$ for "highly polymerized" calf thymus DNA solutions (*85*), prepared for LS in various ways. For this DNA sample the angular range usually explored ($\theta > 30°$) is not sufficient, and when the measurements are carried to 10° significantly high values of M_w (about 14 x 10^6 instead of 7 x 10^6, cf. Table 1 of Ref. *85*) are obtained. In the case of phage T7 DNA, M_w about 23 x 10^6, the same phenomenon is observed. Figure 9 shows the data of Hays et al. (*85*) in the form of a Zimm plot, in the angular range between 10° and 25°. It may be noticed that the value of $\{c/R(\theta)\}_{c=0}$ at $\theta = 10°$, the lowest reading, equals 0.22 and is more than double the value (0.10) at $\theta = 0°$. Thus, although the experiments were extended with a great deal of effort to low angles, it is still necessary to perform a long and

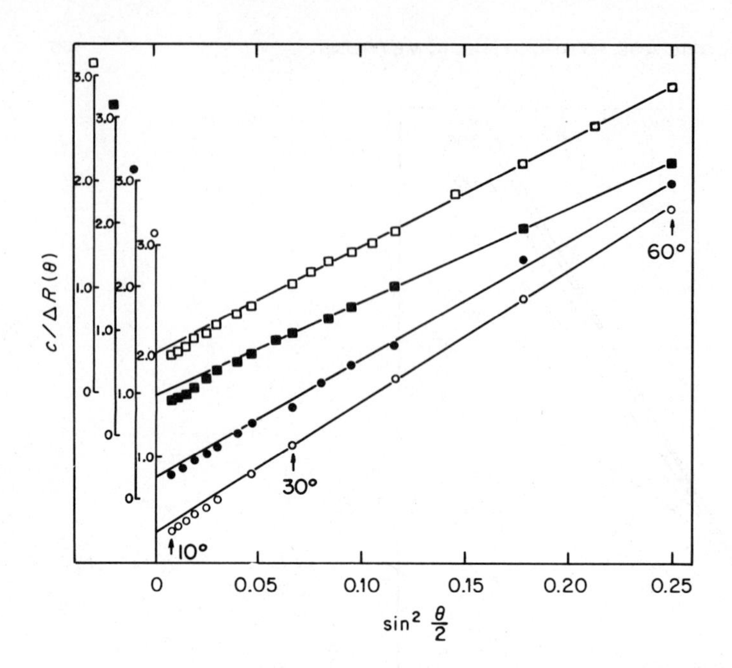

Figure 8. Plots of $c/\Delta R(\theta)$ vs. $\sin^2(\theta/2)$ for single concentration of calf thymus and T7 DNA: (□) calf thymus DNA, 41.4, μg/ml, clarified by filtration; (■) calf thymus DNA, 51.2 μg/ml, filtration; (●) T7 DNA, 45.1 μg/ml, clarified by filtration; (○) T7 DNA, 42.7 μg/ml, centrifugation (Harpst et al. (*85*)).

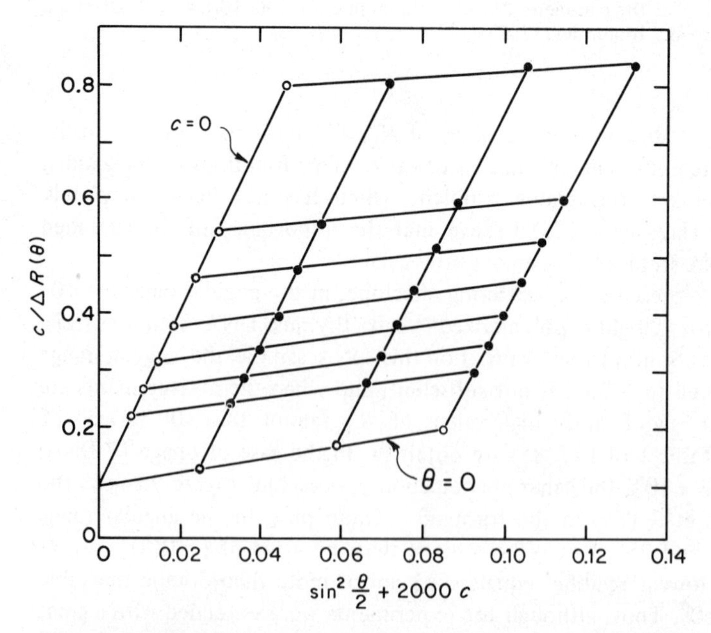

Figure 9. Reciprocal intensity plot for filtered T7 DNA, $[\eta] = 106$ dl/gm: (●) 10°–25° data at three concentrations; (○) extrapolated points at zero angle and zero concentration. Concentrations measured were: 12.6, 29.6, and 42.7 μg/ml. The molecular weight from the intercept ($c = 0, \theta = 0$) is 22.8×10^6 (Harpst et al. (*85*)).

precarious extrapolation. Note also that Figure 9 shows a positive, well-pronounced virial coefficient, which previously had not been attributed to DNA solutions under similar experimental conditions.

The dependence of R_g and of B on temperature and NaCl concentration has been studied by LS (*103*) in the case of a synthetic polynucleotide (polyriboadenylic acid, poly A) and is shown in Fig. 10. The polynucleotide undergoes, at neutral pH, a transition (with increase in temperature) from a stacked to a random coil form. There are two temperatures at which the second virial coefficient vanishes and R_g goes through a minimum over the temperature range studied.

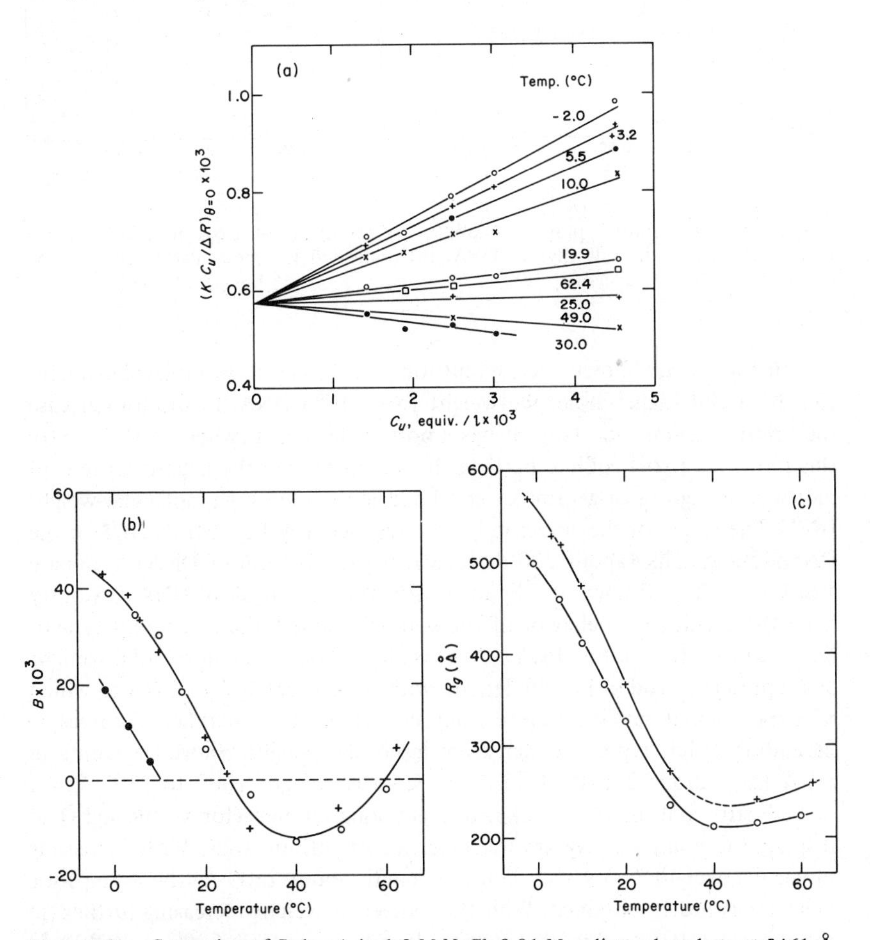

Figure 10. Scattering of Poly rA in 1.0 M NaCl, 0.01 M sodium phosphate at 5461 Å. (a) $K\ C_u/\Delta R(\theta)$, at $\theta = 0$, vs. C_u, at various temperatures; $Z_w = 1740$; (b) second virial coefficient as a function of temperature; (+) $Z = 1740$, (○) $Z_w = 1462$, (●) $Z_w = 1740$ and 1.3 M NaCl; (c) R_g versus temperature, (+) $Z_w = 1740$, (○) $Z_w = 1462$. [From H. Eisenberg and G. Felsenfeld. 1967. *J. Mol. Biol.*, **30:** 17. Reprinted with permission of the publisher, Academic Press.]

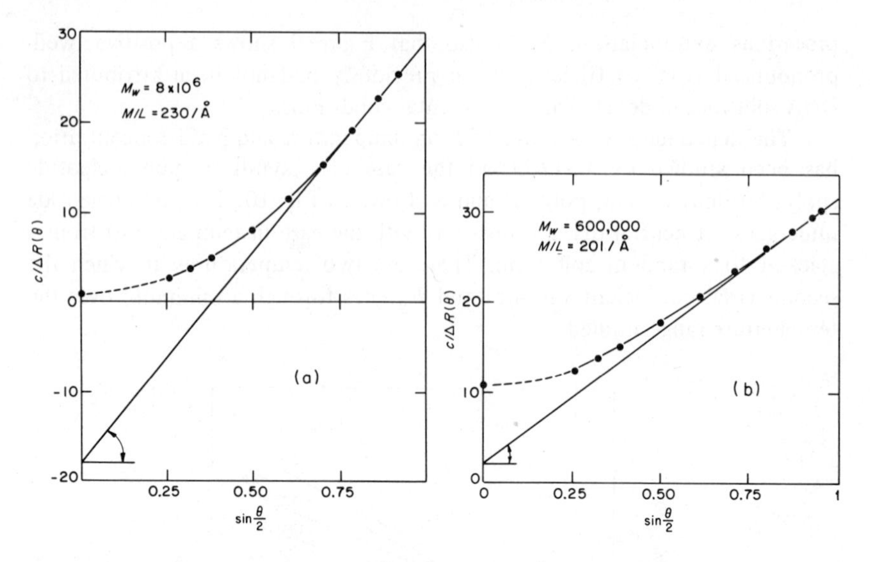

Figure 11. Asymptotic plot of $c/\Delta R(\theta)$ vs. sin $(\theta/2)$ for DNA in 1 M NaCl. (a) High-molecular-weight calf thymus DNA. (b) Sonicated low-molecular-weight chicken erythrocyte DNA (Sadron et al. (*28*)).

In Fig. 11 are shown the asymptotic plots (*28*) for low-molecular-weight (6×10^5) and higher-molecular-weight ($\sim 8 \times 10^6$) DNA. In the former case the limiting asymptotic tangent has a positive intercept (when sin $(\theta/2) \rightarrow 0$); this indicates a stiff rodlike particle. In the latter case the negative intercept indicates a zig-zag or wormlike configuration of the high-molecular-weight DNA. The slopes of the curves in both cases yield (by Eq. (6)) the M/L of the DNA double helix (about 200 daltons/Å for the B form of DNA). Compare also the work of Ptitsyn and Fedorov (*29*) for an analysis of DNA flexibility from the asymptotic behavior of the scattering curve. For a different type of representation (*31*) in Fig. 12, $h\,\Delta R(h)$ is plotted vs. h. The horizontal straight line represents a rod of infinite length, with the correct M/L of DNA to which all experimental curves asymptotically converge. The experimental curves, in ascending order, represent increasing molecular weight. Short fragments of DNA (M_W about 2×10^5 to 3×10^5, contour length 1000 to 1500 Å) are indeed stiff and almost rodlike, but not long enough (for visible light) to display the complete asymptotic behavior of infinite rods. With increasing molecular weight ($\sim 5 \times 10^5$, contour length about 2500 Å) the asymptotic behavior is nearly obtained. With the molecular weight increasing further (in the range 1×10^6 to 8×10^6, contour lengths range from 5000 to 40,000 Å) the asymptotic value is approached from above the ideal asymptotic curve. This occurs because DNA, as has already been mentioned above, is not absolutely stiff and, with increase in contour length, coiled conformations

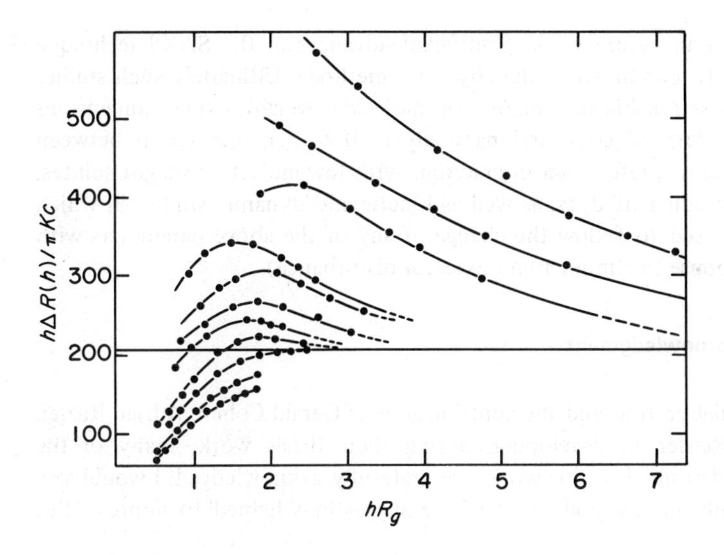

Figure 12. Asymptotic plot of $h\Delta R/\pi Kc$ vs. hR_g, for sonicated DNA; molecular weights increase in ascending order. Data of Litzler and Sadron (Litzler, Ph.D. Thesis, Strasbourg, 1967) reported by Benoit (*31*).

become increasingly significant. In SAXS of DNA solutions the asymptotic range for infinite rods may be conveniently explored (*104–107*). Because of the much lower value of λ of x-rays, the investigation may be carried to much higher values of h.*

CONCLUSION

In conclusion, for the analysis of the scattering behavior of nucleic acid and polynucleotide solutions, both the behavior of the origin peak as $h \rightarrow 0$ as well as the asymptotic behavior, at higher values of h, should be explored. LS measurements should be undertaken in classical experiments at both the blue and the green mercury lines, but the usefulness of new techniques (such as the use of coherent laser light sources and spectral analysis of the scattered beam) and developments, briefly mentioned in this work, should be explored. Corrections due to optical anisotropy should be more carefully taken into

* The correspondence of the parameters used by Luzzati et al. (*104, 105*) and by Bram and Beeman (*106, 107*) has been discussed (*19*). In the erratum to Ref. *19* a typographical error in Eq. (12) (*loc. cit.*) has been corrected, also λ in the constant K (bottom, p. 356 *loc. cit.*) refers to λ_0, the value of the wavelength *in vacuo* (see p. 140, this volume) and the correct value for the Guinier slope in the form given in the last paragraph on p. 357 (Ref. *12*) should read $-(4/3)\pi^2 R_g{}^2$.

consideration in future work. Significant advances in the SAXS technique should lead to combined studies by both methods. Ultimately such studies should disclose reliable information on molecular weight, overall dimensions (R_g), more detailed structural parameters (M/L, r_g), interaction between DNA molecules, preferential interactions with low-molecular-weight solutes, hydration, binding of dyes, as well as kinetic and dynamic studies in which LS may be used to follow the change of any of the above parameters with time (in response to a momentary external disturbance).

Acknowledgment

The collaboration and the contributions of Gerald Cohen, Adrian Raziel, and Emil Reisler in developing, during their thesis work, many of the procedures described in this work, are gratefully acknowledged. I would also like to thank all my colleagues whose suggestions helped to improve this article.

NOTE ADDED IN PROOF

Since this article was written two reviews on the theory and practice of elastic (*108*) and inelastic (*109*) scattering have appeared. Theoretical aspects and various applications of laser light scattering have also been reviewed (*110*). The diffusion coefficients of coliphages and coliphage DNA were determined by the use of optical mixing spectroscopy (*111*) from the spectrum of the scattered light, in a series of papers devoted to the molecular-weight determination of high-molecular-weight DNA. Compare also another recent review (*112*) of LS of protein solutions.

APPENDIX A

List of Symbols

a	$(R_g{}^2)_i$ of particle i is proportional to $M_i{}^{2a}$
c	concentration, gram per milliliter
d	distance from scattering element to detector
h	$4\pi \sin \theta/2$ or $4\pi \sin (\theta/2)/\lambda$
$i_n'(h)$	dependence of scattering upon angle ($i_n'(0) = 1$)
$i_n(h)$	particle scattering factor in SAXS ($i_n(0) = 1$)

m	molality, moles per kilogram of the principal solvent, component 1
m_u	molality in moles of nucleotides (or phosphate) per kilogram of component 1
n, n_0	refractive index of solution and of solvent
r	radius of rodlike particle or coil
r_g	root-mean-square radius of cross section of circular rod, or coil
x	$= L/2A$
A	Kratky-Porod persistence length
B	second virial coefficient
C_u	concentration, moles of nucleotides (or phosphate) per liter of solution
H_v, H_h, H_u	horizontal component of scattered light, with vertically, horizontally polarized and unpolarized incident light
I_0	intensity of incident beam
$I(h), I(\theta)$	intensity of scattered radiation
$\Delta I(h), \Delta I(\theta)$	excess (over solvent) intensity of scattered radiation
I_{el}	scattering of an electron
K	$= (2\pi^2/N_A)(n^2/\lambda_0^4)(\partial n/\partial c_2)_{P,T}^2$, LS constant for unpolarized incident light, for two-component system; substitute $(\partial n/\partial c_2)_{T,\mu}$ for multicomponent system
K'	$= (\partial\rho/\partial c_2)_{P,T}^2 I_{el}/N_A$, SAXS constant for two-component system; substitute $(\partial\rho/\partial c_2)_{T,\mu}$ for multicomponent system
L	length of rodlike particle, contour length of coil
LS	light scattering
M	molecular weight
M_n, M_w, M_z	number, weight, and z-average molecular weight
M_u	molecular weight of nucleotide
M/L	mass per unit length
N_A	Avogadro's number
$P(\theta)$	particle scattering factor in LS
R_g	root-mean-square radius
$\langle R_g^2 \rangle_z$	LS (z) average mean-square radius
$\Delta R(h), \Delta R(\theta)$	Rayleigh ratio (corrected for solvent contribution)
SAXS	small angle X-ray scattering

V_v, V_u	vertical component of scattered light, with vertically polarized or unpolarized incident light.
Z	number of nucleotides (or phosphates) per DNA macromolecule
α	polarizability of isotropic particle
α_{el}	polarizability of an electron
α, β	polarizability in direction of and perpendicular to axis of anisotropic particle
δ	$= (\alpha - \beta)(\alpha + 2\beta)$, optical anisotropy factor
λ, λ_0	wavelength in medium and *in vacuo*
$\theta, 2\theta$	scattering angle in LS, SAXS
μ	chemical potential
ρ	density (electrons per milliliter)
ρ_u	$= H_u/V_u$, polarization ratio with unpolarized incident light
ρ_v	$= H_v/V_v$, polarization ratio with vertically polarized incident light

APPENDIX B

Refractive Index Increments for DNA Systems at 25°C Measurements (*92*) on Sonicated Calf Thymus DNA

C_3 (mole/liter)	NaDNA–NaCl $\partial n/\partial c_2$ (ml/gm)		CsDNA–CsCl $\partial n/\partial c_2$ (ml/gm)	
	5461 Å	4358 Å	5461 Å	4358 Å
0	0.179[a]	0.186[a]		
0.200	0.168[b]	0.175[b]	0.131[b]	0.136[b]
0.200	0.176[c]	0.182[c]		

[a] dn/dc_2 in water
[b] $(\partial n/\partial c_2)_{T,\mu}$
[c] $(\partial n/\partial c_2)_{P,T,m_3}$

NOTE. For electron density increments $(\partial\rho/\partial c_2)_{T,\mu}$ (in NaCl and CsCl solutions), required in the evaluation of SAXS experiments, cf. Fig. 2 of Ref. *19*. The electron density increments have been calculated from the mass density increments given in Tables I and II of Ref. *92*.

References

1. Timasheff, S. 1963. In M. Kerker (Ed.), Electromagnetic scattering. Pergamon Press, Oxford, p. 337.
2. Kerker, M. 1968. *Ind. Eng. Chem.*, **60**: 31.
3. Kerker, M. 1969. The scattering of light and other electromagnetic radiation. Academic Press, New York.
4. Finch, J. T. and K. C. Holmes, 1967. In K. Maramorosch and H. Koprowski (Eds.), Methods in virology. Academic Press, New York, Vol. 3, p. 351.
5. Tanford, C. 1961. Physical chemistry of macromolecules. Wiley, New York.
6. Geiduschek, E. P. and A. Holtzer. 1958. *Advan. Biol. Med. Phys.*, **6**: 431.
7. McIntyre, D. and F. Gornick (Eds.). 1964. Light scattering from dilute polymer solutions. Gordon and Breach. New York.
8. Kratohvil, J. P. 1964. *Anal. Chem.*, **36**: 458 R; 1966. *ibid.* **38**: 517 R.
9. Kratohvil, J. P. 1968. In Characterization of macromolecular structure. National Academy of Sciences, Washington, D.C., p. 59.
10. Peterlin, A. 1959. *Prog. Nucleic Acid Res.*, **9**: 175.
11. Oster, G. 1960. In A. Weissberger (Ed.), Physical methods of organic chemistry. Wiley (Interscience), New York, Vol. I, Part 3, p. 2107.
12. Stacey, K. A. 1956. Light scattering in physical chemistry. Butterworths, London.
13. Kratky, O. 1963. *Progr. Biophys.*, **13**: 105.
14. Brumberger, H. (Ed.). 1967. Small angle X-ray scattering. Gordon and Breach, New York.
15. Guinier, A., and G. Fournet. 1955. Small angle scattering of X-rays. Wiley, New York.
16. Beeman, W. W., P. Kaesberg, J. W. Anderegg, and M. B. Webb. 1957. In S. Flugge (Ed.), Handbuch der Physik. Springer, Berlin, Vol. 32, p. 321.
17. Auer, H. E. and Z. Alexandrowicz. 1969. *Biopolymers*, **8**: 1.
18. Casassa, E. F. and H. Eisenberg. 1964. *Advan. Protein Chem.*, **19**: 287.
19. Eisenberg, H. and G. Cohen. 1968. *J. Mol. Biol.*, **37**: 355: *erratum*: 1969. *ibid.* **42**: 607.
20. Dubin, S. B., J. H. Lunacek, and G. B. Benedek. 1967. *Proc. Natl. Acad. Sci. U.S.*, **57**: 1164.
21. Zimm, B. H. 1948. *J. Chem. Phys.*, **16**: 1093, 1099.
22. Vrij, A. and J. Th. G. Overbeek. 1962. *J. Coll. Sci.*, **17**: 570.
23. Bünemann, H. and W. Müller. 1970. *Hoppe-Seyler's Z. Physiol. Chem.*, **351**: 107.
24. Guinier, A. 1939. *Ann. Phys.*, **12**: 161.
25. Hays, J. B., M. E. Magar, and B. H. Zimm. 1969. *Biopolymers*, **8**: 531.
26. Debye, P. 1915. *Ann. Physik.*, [4] **46**: 809.
27. Neugebauer, J. 1943. *Ann. Physik.*, [5] **42**: 509.
28. Sadron, C., J. Pouyet, A. M. Freund, and M. Champagne. 1965. *J. Chim. Phys.*, **62**: 1187.
29. Ptitsyn, O. B. and B. A. Fedorov. 1963. *Biofizika*, **8**: 659.
30. Sadron, C. 1968. *Bull. Soc. Chim. Biol.*, **50**: 233.
31. Benoit, H. 1968. *J. Chim. Phys.*, **65**: 23.
32. Krigbaum, W. R. and R. W. Godwin. 1968. *Macromolecules*, **1**: 375.
33. Kratky, O., I. Pilz, F. Cramer, F. von der Haar, and E. Schlimme. 1969. *Monatsh. Chem.*, **100**: 748.
34. Hill, W. E., J. D. Thompson, and J. W. Anderegg. 1969. *J. Mol. Biol.*, **44**: 89.
35. Sund, H., I. Pilz, and M. Herbst. 1969. *European J. Biochem.*, **7**: 517.
36. Benoit, H., A. M. Holtzer, and P. Doty. 1954. *J. Phys. Chem.*, **58**: 635.

37. Shultz, A. R. and W. Stockmayer. 1969. *Macromolecules,* **2**: 178.
38. Kratky, O. and G. Porod. 1949. *Rec. Trav. Chim.* **68**: 1106.
39. Hearst, J. E. and W. H. Stockmayer. 1962. *J. Chem. Phys.,* **37**: 1425.
40. Eisenberg, H. 1969. *Biopolymers,* **8**: 545.
41. Benoit, H. and P. Doty. 1953. *J. Phys. Chem.,* **57**: 958.
42. Landau, L. and E. Lifshitz. 1958. Statistical physics. Pergamon Press, London, p. 478.
43. Gray, Jr. H. B. and J. E. Hearst. 1968. *J. Mol. Biol.,* **35**: 111.
44. Luzzati, V. and H. Benoit. 1961. *Acta Cryst.,* **14**: 297.
45. Cohen, G. and H. Eisenberg. 1966. *Biopolymers,* **4**: 429.
46. Dev, S. B., R. Y. Lochhead and A. M. North. 1970. *Discussions Faraday Soc.,* 49: 244.
47. Hearst, J. E. and R. A. Harris. 1967. *J. Chem. Phys.,* **46**: 398.
48. Tagami, Y. 1969. *Macromolecules,* **2**: 8.
49. Bugl, P. and S. Fujita. 1969. *J. Chem. Phys.,* **50**: 3137.
50. Sanchez, I. C. and C. von Frankenberg. 1969. *Macromolecules,* **2**: 666.
51. Mauss, Y., J. Chambron, M. Daune, and H. Benoit. 1967. *J. Mol. Biol.,* **27**: 579.
52. Horn, P. and H. Benoit. 1953. *J. Polymer Sci.,* **10**: 29.
53. Horn, P. 1955. *Ann. Phys. (Paris),* [12] **10**: 386.
54. Horn, P. and H. Benoit. 1960. *Compt. Rend. Acad. Sci. (Paris),* **251**: 222.
55. Benoit, H. and G. Weill. 1957. *Coll. Czech. Commun.,* **22**: 35.
56. Weill, G., C. Hornick and S. Stoylov. 1968. *J. Chim. Phys.,* **65**: 182.
57. Utiyama, H. and M. Kurata. 1964. *Bull. Inst. Chem. Res. (Kyoto),* **42**: 128.
58. Ford, Jr., N. C., W. Lee, and F. E. Karasz. 1969. *J. Chem. Phys.,* **50**: 3098.
59. Ford, Jr., N. C., F. E. Karasz, and J. E. M. Owen. 1970. *Discussions Faraday Soc.,* 49: 228.
60. Pecora, R. 1968. *J. Chem. Phys.* **49**: 1032; 1970. *Discussions Faraday Soc.,* 49: 222.
61. Chu, B. and F. J. Schoenes. 1968. *J. Colloid Interface Sci.,* **27**: 424.
62. Foord, R., E. Jakeman, C. J. Oliver, E. R. Pike, R. J. Blagrove, E. Wood, and A. R. Peacocke. 1970. *Nature,* **227**: 242.
63. Rimai, L., J. T. Hickmott, Jr., T. Cole, and E. B. Carew. 1970. *Biophys. J.,* **10**: 20.
64. Wada, A., N. Suda, T. Tsuda, and K. Soda. 1969. *J. Chem. Phys.,* **50**: 31.
65. Yeh, Y. and R. N. Keeler. 1969. *Quart. Rev. Biophys.,* **2**: 315.
66. Miller, G. A. and C. S. Lee. 1968. *J. Phys. Chem.,* **72**: 4644.
67. Miller, G. A., F. I. San Filippo, and D. K. Carpenter. 1970. *Macromolecules,* **3**: 125.
68. Tagami, Y. and R. Pecora. 1969. *J. Chem. Phys.,* **51**: 3293.
69. Pecora, R. and Y. Tagami. 1969. *J. Chem. Phys.,* **51**: 3298.
70. Pecora, R. 1969. *Macromolecules,* **2**: 31.
71. Scheludko, A. and S. Stoylov. 1967. *Biopolymers,* **5**: 723.
72. Stoylov, S. P. and S. Sokerov. 1967. *J. Colloid Interface Sci.,* **24**: 235.
73. Brice, B. A., M. Halwer, and R. Speiser. 1950. *J. Opt. Soc. Am.,* **40**: 768.
74. Carr, Jr. C. I. and B. H. Zimm. 1950. *J. Chem. Phys.* **18**: 1616.
75. Wippler, C. and G. Scheibling. 1954. *J. Chim. Phys.,* **51**: 201.
76. Tomimatsu, Y., L. Vitello, and K. Fong. 1968. *J. Colloid Interface Sci.,* **27**: 573.
77. Utiyama, H., N. Sugi, M. Kurata, and M. Tamura. 1968. *Bull. Inst. Chem. Res. (Kyoto),* **46**: 77.
78. Utiyama, H., N. Sugi, M. Kurata, and M. Tamura. 1968. *Bull. Inst. Chem. Res. (Kyoto),* **46**: 198.
79. Schurz, J., E. Gruber, G. Warnecke and H. Pippan. 1967. *Acta Phys. Austriaca,* **26**: 211.
80. Evans, J. M., M. B. Huglin, and J. Lindley. 1967. *J. Appl. Polymer Sci.,* **11**: 2159.
81. Brice, B. A., G. C. Nutting, and M. Halwer. 1953. *J. Am. Chem. Soc.,* **75**: 824.

82. Lalanne, J. R. and P. Bothorel. 1966. *J. Chim. Phys.*, **63**: 1538.
83. Froelich, D., C. Strazielle, G. Bernardi, and H. Benoit. 1963. *Biophys. J.*, **3**: 115.
84. Harpst, J. A., A. I. Krasna, and B. H. Zimm. 1968. *Biopolymers*, **6**: 585.
85. Harpst, J. A., A. I. Krasna, and B. H. Zimm. 1968. *Biopolymers*, **6**: 595.
86. Meyerhoff, G., V. Moritz, and R. L. Darskus. 1968. *Polymer Letters*, **6**: 207.
87. Livesey, P. J. and F. W. Billmeyer, Jr. 1969. *J. Colloid Interface Sci.*, **30**: 447.
88. Keijzers, A. E. M., J. J. Van Aartsen, and W. Prins. 1965. *J. Appl. Phys.*, **36**: 2874.
89. Keijzers, A. E. M., J. J. van Aartsen, and W. Prins. 1968. *J. Am. Chem. Soc.*, **90**: 3107.
90. Bauer, N., K. Fajans, and S. Z. Lewin. 1960. In A. Weissberger (Ed.). Physical methods of organic chemistry. Wiley (Interscience), New York, Vol. I, Part 2, Chap. 18, p. 1139.
91. Brice, B. A. and M. Halwer. 1951. *J. Opt. Soc. Am.*, **41**: 1033.
92. Cohen, G. and H. Eisenberg. 1968. *Biopolymers*, **6**: 1077.
93. Stabinger, H., H. Leopold, and O. Kratky. 1967. *Monatsh. Chem.*, **98**: 436.
94. Stabinger, H., H. Leopold, and O. Kratky. 1969. *Z. Angew. Phys.*, **27**: 273.
95. Huglin, M. B. 1965. *J. Appl. Polymer Sci.*, **9**: 3963, 4003.
96. Chiang, R. 1966. In J. Brandrup and E. Immergut, (Eds.), Polymer handbook. Wiley (Interscience), New York, Part IV, p. 279.
97. Reisler, E. and H. Eisenberg. 1965. *J. Chem. Phys.*, **43**: 3875.
98. Cohen, G. and H. Eisenberg. 1965. *J. Chem. Phys.*, **43**: 3881.
99. Krasna, A. I. and J. A. Harpst. 1964. *Proc. Natl. Acad. Sci. U.S.*, **51**: 36.
100. Bernardi, G. 1964. *Makromol. Chem.*, **72**: 205.
101. Coumou, D. J. 1960. *J. Colloid Sci.*, **15**: 408.
102. Sharp, P. and V. A. Bloomfield. 1968. *Biopolymers*, **6**: 1201.
103. Eisenberg, H. and G. Felsenfeld. 1967. *J. Mol. Biol.*, **30**: 17.
104. Luzzati, V., F. Masson, A. Mathis, and P. Saludjian. 1967. *Biopolymers*, **5**: 49.
105. Luzzati, V., A. Nicolaieff, and F. Masson. 1961. *J. Mol. Biol.*, **3**: 185.
106. Bram, S. 1968. Ph.D. Thesis, University of Wisconsin.
107. Bram, S. and W. W. Beeman. 1971. *J. Mol. Biol.*, **55**: 311.
108. Lundberg, J. L., I. R. Hardin, and M. K. Tomioka. 1970. In Encyclopedia of polymer science and technology. Wiley (Interscience), New York, Vol. 12, p. 355.
109. Schaufele, R. F. 1970. In Encyclopedia of polymer science and technology. Wiley (Interscience), New York, Vol. 12, p. 397.
110. Chu, B. 1970. *Ann. Rev. Phys. Chem.*, **21**: 145.
111. Dubin, S. B., G. B. Benedek, F. C. Bancroft, and D. Freifelder. 1970. *J. Mol. Biol.*, **54**: 547.
112. Timasheff, S. N. and R. Townend. 1970. In S. J. Leach (Ed.), Physical principles and techniques of protein chemistry, Academic Press, New York, Part B Chap. 3, p. 147.

LINEAR DICHROISM WITH SPECIAL EMPHASIS ON ELECTRIC FIELD INDUCED LINEAR DICHROISM

ELLIOT CHARNEY
NATIONAL INSTITUTE OF ARTHRITIS AND METABOLIC DISEASES
NATIONAL INSTITUTES OF HEALTH
BETHESDA, MARYLAND

INTRODUCTION

The relationship between the absorption of polarized electromagnetic radiation and the structural and electronic parameters of an oriented system is probably the most direct measure of the structure and conformation of the molecules in the system aside from x-ray or neutron diffraction.* It is somewhat surprising therefore that no extensive literature exists on linear dichroism measurements,† that is measurements of the absorption of linearly polarized radiation by partially or completely oriented molecules (*1*). The surprise is somewhat tempered by the knowledge of the difficulties of achieving substantial orientation in solutions and of obtaining samples of physical dimensions in crystals appropriate to the experimental requirements.

* For very small molecules, of course, other techniques such as microwave spectroscopy and rovibrational analyses may be used, but with molecules even of the size of the nucleosides, to say nothing of polynucleotides and biological polymers, these techniques are no longer feasible because of the density of rovibrational states or because of the physical states of the molecules under reasonable experimental conditions. The largest molecule for which this type of analysis has been made are disubstituted benzenes and benzofuran (Lombardi, J. R. 1970. *J. Am. Chem. Soc.*, 92: 1831).

† Throughout this chapter the term dichroism used without a modifier refers to linear dichroism only.

Of the methods available for orientation of molecules in solution, primarily useful are hydrodynamic flow fields and electrostatic fields if a quantitative knowledge of the degree of orientation is desired. Some use has been made of oriented films produced by stroking drying polymer solutions, but at best this is a rather crude and uncertain method. Orientation in liquid crystals, while reasonably easy to achieve is also not especially amenable to quantitative measurements of the type desired.*

Magnetic and acoustical fields can also exert orienting torques on molecular systems but have been little used because the degree of orientation achievable under ordinary conditions is very small.

In this article we will discuss four principal types of dichroism measurements, but will limit the more extensive discussion to electric dichroism which has not yet received an extended review or discussion in a monograph or review paper (*2*). Birefringence measurements on optically anisotropic systems are related to dichroism measurements in such a way that it is, in principle, possible to analytically transform the information from one to that of the other. Starting with isotropic liquids or solutions, the preferred orientations required to produce the optical anisotropy can be achieved by the same techniques of electric and flow fields used for dichroism; the corresponding field dependent orientation functions which appear in the expressions for dichroism and birefringence ought to be the same for each type of field. A fairly extensive literature on both electric and flow birefringence exists and excellent reviews (*3–6*)† have appeared.

The utility of dichroism (and birefringence) methods arises from the fact that the absorption and scattering of electromagnetic radiation is governed by a quantum-mechanical parameter termed the transition moment, μ_{ab}. The integrated intensity or extinction for an absorption band corresponding to the transition between the states a and b is proportional to the square of μ_{ab}. *Most importantly this also has vector properties*. Thus it has a direction fixed in the molecule and interacts only with components of the radiation field which are polarized parallel to it. In a collection of molecules, this phenomenon can be observed only if the collection is not isotropic, so that there is at least a partially parallel alignment of transition moments. The extreme case, of course, is a single crystal in which all the molecules, and therefore their transition moments associated with optical transitions between different energy states, are rigidly arranged in a repeating, although

* The use of strong magnetic fields ($>$20 kG), strong electric fields ($>3 \times 10^4$ V/cm) or flow to further orient molecules partially oriented in liquid crystals offers no special advantages over the use of electric and flow fields in solutions, *except* for small or highly symmetric molecules oriented in host crystals, where the technique may be the only available quasi-solution method. In addition, problems associated with anisotropic light scattering, form dichroism, and induced optical activity in cholesteric liquid crystals, complicate attempts to analyze the data quantitatively.

† See also the earlier discussions and review cited in Ref. *3*.

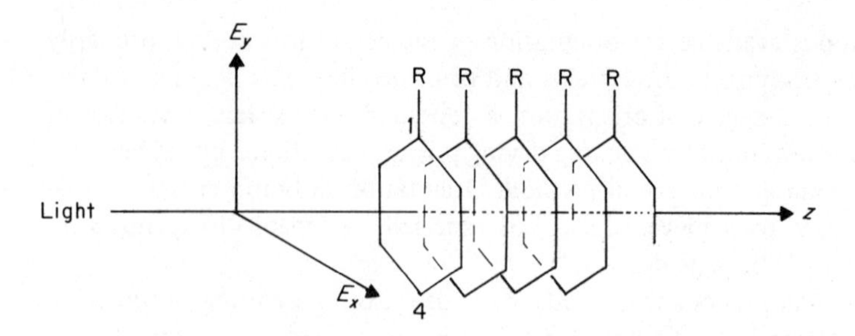

Figure 1. Schematic diagram of parallel alignment of monosubstituted benzene molecules.

not necessarily identical fashion. We shall use this extreme case to illustrate the dichroism phenomenon. Consider the chromophore in a monosubstituted benzene molecule which gives rise to the absorption band near 2600 Å. The measured extinction of a random spatial distribution of these molecules is independent of the orientation of the polarized light used to make the measurement. On the other hand a parallel alignment of the molecules such as illustrated in Fig. 1 produces linear dichroism. With the plane of polarization (E_x) perpendicular to the C_1-C_4 axis the absorbance is a maximum, while parallel (E_y) to this axis, it is zero. The transition moment μ_{2600} is therefore perpendicular to the C_1-C_4 axis and could be shown to be in the plane of the ring by another dichroism measurement in which the polarized light is propagated along the x or y direction rather than along z. In a real crystal the situation will be complicated by the presence of more complex symmetry or by intermolecular interactions. When the latter are negligible, the crystal spectra are said to be like that of an "oriented gas." Crystal dichroism was recognized very early, but its relationship to molecular parameters is a more recent discovery and the modern renewal of this field started with measurements of the polarized absorption of near infrared radiation of sucrose crystals (*7*). Just a year later, the first report of dichroism (*8*) in a crystalline biological macromolecule, hemoglobin, appeared. The heme-proteins, in fact, provide the first definitive example of the ability of crystal dichroism measurements to determine the orientation of a chromophoric group (the porphyrin) in a large macromolecule.* The theory of crystal spectroscopy including such effects as exciton (*9*) and site or crystal field splitting (*10, 11*) has been extensively developed and new impetus given to experimental work in this field. Few such investigations have been reported on nucleic acid bases (*12–17*) but much more work is now in progress.

* The orientation of the porphyrin plane in ferricytochrome *c* from crystal dichroism measurements (Eaton, W. A. and R. Hochstrasser. 1967. *J. Chem. Phys.*, **46:** 2533) agrees to within 0.5 degrees with that determined by x-ray diffraction (R. E. Dickerson et al., 1971. *J. Biol. Chem.*, **246**: 1511).

The polarized spectra of partially oriented molecules immobilized in solid films has received scant attention since the introduction (*18*) of the technique in 1947. Nevertheless, much useful information has come from it (*19*), and measurements made by this method have been central to the discussion of the existence of $\pi^* \leftarrow n$ transitions in nucleic acids and polynucleotides. Rich and Kasha found (*20*) a small band near 280 nm in the polarized spectrum of an oriented film of polycytidylic acid perpendicular to the axis of maximum birefringence and interpreted its origin as an $\pi^* \leftarrow n$ transition, perpendicular to the plane of the nucleic acid bases. Gellert (*21*) similarly measured the dichroic spectrum of an oriented film of poly dAT but found no evidence for the existence of such a band in that compound. The existence and polarization of these transitions is critical to any discussion of the optical properties which are so useful in the study of the structure and conformation of macromolecules. For example, the optical rotatory dispersion of polyriboadenylic acid at low pH has either one or two inflection points in the vicinity of 280 nm depending on the pH and ionic strength (*22*). The interpretation of these inflections as evidence for conformational changes will be quite different if the transitions which give rise to them are transitions from nonbonding or from bonding orbitals. The designation of electronic transitions in terms of the one-electron transitions between molecular orbitals such as n and π^* is, of course, an approximation, albeit a very useful approximation to the description of the change in electronic state accompanying the interaction with the radiation. One of the most useful aspects lies in the symmetry considerations which may be applied to such orbitals and states, and which permit the prediction or calculation of the direction of the transition in the molecule or chromophore. Thus the $\pi^* \leftarrow \pi$ absorption band of the monosubstituted benzene ring which occurs at about 258 nm is known to be in the plane of the ring and perpendicular to the two-fold rotation axis defined by the substituent and the carbon atom para to it (*23*). Similarly the major transition moments of the 250–280 nm absorption bands of the purine and pyrimidine bases are confined to the plane of the bases (*24*); the orientation in the plane is also defined by symmetry for pyrimidines with *axes* of symmetry in the plane but otherwise the determination of the in-plane orientation requires more sophisticated quantum mechanical calculations. Dichroism measurements may be used to obtain this information, or having it, be used in combination with the experimental results to define the relative orientation of the molecules or of the chromophore in a system of oriented molecules. As an example of the former, the measured crystal dichroism of 1-methyl thymine was used to obtain the orientation of the $\pi^* \leftarrow \pi$ transition moment in the plane of that base (*13*); the latter is exemplified by electric dichroism measurements of poly-γ-benzyl-L-glutamate (PBLG) which have been used to determine the orientation of the plane of the phenyl ring in the α-helical polypeptide in solution (*25*).

For studies in solution, the principal method used until now has been flow dichroism (*26*). The initial experimental observation of flow dichroism (*27*) led to the first studies on DNA (*28*) in the ultraviolet region and to the measurement of the infrared dichroism of synthetic polypeptides (*29*) in which the dichroism of the 3300 cm^{-1} N–H stretching vibration, the 1650 cm^{-1} amide I and the 1500 cm^{-1} amide II bands was observed. The availability of at least one commercial instrument* has spurred research using this technique. Flow dichroism measurements have been used to obtain structural information from the dichroism under constant flow fields and also to obtain hydrodynamic relaxation times from dynamic measurements (*30,31*). The interpretation of the latter requires the acceptance of various models for the shear forces acting on long rigid or flexible rods in which such complicating parameters as excluded volumes of solvents must be accounted for. Because of this and because turbulence phenomena interfere with the ability to obtain high enough shear stresses to completely orient the macromolecules, some of the interpretation of flow dichroism measurements is risky. Several different types of flow dichroism instruments have been constructed and these will be discussed briefly.

Electric dichroism is the newest of the techniques. The use of an electric field as an orienting mechanism is, of course, very old and the corresponding birefringence measurements using electric fields have been associated with the name of the discoverer, J. Kerr (*32*). The extension to dichroism measurements was proposed in 1939 by Kuhn, Dührkop, and Martin (*33*) and it was the earliest measurement on a biological macromolecule. PBLG, was reported by G. Spach (*34*). Most of the subsequent measurements have been made on nucleic acid systems, but the entire literature to date consists of approximately thirty papers.†

METHODS

We are concerned in this chapter primarily with electric dichroism. The discussion of crystal, film, and flow dichroism is therefore cursory but reference to adequate presentations will be given. Dichroism is defined in a number of ways in the literature but all of these involve some measure of the relative absorption for polarized radiation in two space- and/or molecular-fixed directions. For a completely oriented system the space- and molecule-

* The Shimadzu Flow Dichroism Apparatus, manufactured by Shimadzu Seisakusho Ltd. of Tokyo, Japan, is based on the design by A. Wada described in Ref. *26*.

† While this article is not intended as a review, because of the relative novelty of the technique and the paucity of the literature, an attempt has been made to make reference to most of these papers. We apologize in advance for the fact that our search has undoubtedly failed to locate every reference to electric-field induced dichroism.

fixed directions can be set to coincide and the maximum information is available from the minimum number of experiments. We define $\epsilon_{\parallel}$ and $\epsilon_{\perp}$, respectively, as the extinction coefficients for linearly polarized radiation parallel to the laboratory-fixed orthogonal x and y directions, z being the direction of propagation of the radiation perpendicular to the xy plane, and ϵ_x' and ϵ_y' as the extinction coefficients for plane polarized light oriented, respectively, parallel and perpendicular to an axis fixed in the molecule. If all the molecules in the system are oriented such that the direction of the molecular axis for which the extinction is ϵ_x' coincides with the x axis and the molecules are individually symmetric about this axis or their statistical distribution about the x direction is cylindrically symmetric, then $\epsilon_{\parallel} = \epsilon_x'$ and $\epsilon_{\perp} = \frac{1}{2}\epsilon_y'$; the factor of ½ enters the latter because with cylindrical symmetry only ½ the components of the perpendicular extinction lie in the xy plane. The other half are directed in the z direction and thus do not interact with the radiation field whose electric vectors may have components only along x and y. In a few cases, dichroism has been defined with respect to changes in extinction when the axis of cylindrical symmetry is oriented in the direction of propagation of the light. Thus in the flow dichroism apparatus of Callis and Davidson (*31, 35*) the direction of flow and therefore the unique axis is in the direction of propagation of the radiation. Under these conditions, the use of polarized radiation is unnecessary since the absorbance does not vary in the xy plane, i.e., $A_x = A_y \neq A_z$. With perfect orientation, the result is the same as would be observed with the radiation propagating parallel to the unique axis of a uniaxial crystal. Under conditions of imperfect orientation, the experimental "dichroism" in this case is measured by ΔA, the difference between the absorbance measured in the absence and presence of the flow gradient. This quantity is related to the molecular parameters through the expression

$$\frac{\Delta A}{A} = \frac{\epsilon_{\parallel} - \epsilon_{\perp}}{2\bar{\epsilon}} \left(\langle \cos^2 \theta_z \rangle - \tfrac{1}{3}\right)$$

with the assumption that $A = \frac{1}{3}(A_x' + 2A_y')$, where A is the isotropic absorbance for the same pathlength in the absence of flow, A_x' and A_y' are the absorbances for light propagating parallel and perpendicular to the molecular axis, $\bar{\epsilon}$ is one-third the isotropic extinction coefficient ϵ, and θ_z is the polar angle of orientation of the molecular axis with respect to the axis of propagation of the radiation and flow*; $\langle \cos^2 \theta_z \rangle$ is the space averaged value of $\cos^2 \theta_z$ for all the molecules in the system. When the orientation is perfect

* In Ref. *31*, the subscript x is used rather than z, but for consistency with the axial designations in this article, a different set of axes is chosen here. Also θ_z is defined by Callis and Davidson as the angle of orientation of a chain segment, leaving open the possibility of application to flexible polymers. For simplicity here we assume the molecule is rigid and rodlike, i.e., with an axis of cylindrical symmetry.

TABLE 1. Values of Dichroism Parameters for Fully Oriented Systems with Cylindrical Symmetry about the Axis of Orientation

	$\Delta\epsilon/\epsilon$	$R(\Vert/\perp)$	P	q (χ=0°)
Transition moment parallel to axis	3.0	∞	1.0	−0.5
Transition moment perpendicular to axis	−1.5	0	−1.0	0.5

$\theta_z = 0°$ and $\Delta A/A = -(\epsilon_{\Vert} - \epsilon_{\perp})/3\bar{\epsilon}$. Other representations of dichroism which have been used are the dichroic ratio, $R = \epsilon_y'/\epsilon_x'$, the polarization ratio, $PR = OD_A/OD_B$ where A and B refer to particular space-fixed directions and it is usual to take A and B such that PR>1, and the polarization, $P = (\epsilon_{\Vert} - \epsilon_{\perp})/(\epsilon_{\Vert} + \epsilon_{\perp})$. For flow and electric dichroism, it has been usual to use an expression of the type $\Delta\epsilon/\epsilon = (\epsilon_{\Vert} - \epsilon_{\perp})/\epsilon$ since this "reduced dichroism" can, in general, be related to an important quantity, α, defining the angle between the direction of the transition moment for the isolated absorption band under study and a fixed axis in the molecule, as well to the field parameters. Another such expression (*36, 37*) is the dichroic coefficient, $L\chi$, defined as

$$L\chi = -\frac{q_\chi}{2.303\, DE^2}$$

where $q_\chi = (I_\chi - I_{\chi(E=0)})/I_{\chi(E=0)}$, χ is the angle which the electric vector of the polarized radiation makes with the field direction in the plane perpendicular to the direction of the light propagation, D is the isotropic optical density for the solution of concentration C_0, pathlength 1 cm, and extinction coefficient ϵ, and E is the applied electric field strength in any defined units. L_χ is independent of field strength if q_χ is quadratic in field strength,* while the dichroic and polarization ratios, polarization, and reduced dichroism are field strength dependent except when orientation is complete, at which point L_χ becomes inversely proportional to E^2. The values of these experimentally observable quantities for two special cases of interest are given in Table 1.

Film Dichroism

The dichroism of films is, of course, spectacularly illustrated by the existence of polaroid sheet polarizers which depend for their properties on

* Which is, however, true only for special conditions (see below).

dichroic extinction. There are a number of types of such polarizers. In general they consist of sheets of long polymeric molecules which have been oriented by stretching or compressing and contain a dye which becomes preferentially oriented on or between the polymer chains. Measurements in both the ultraviolet and infrared regions have already been referred to (*18–21*). Most of the procedures used to date have been straightforward measurements of the intensity of polarized radiation transmitted by the film samples using monochromatic radiation. Polarizers utilized range from piles of dielectric plates inclined at Brewsters angle (*38*) to calcite prisms of Glan or Rochon type (*39*). When the source of monochromatic radiation is a prism or grating monochromator very large errors in the dichroic ratio can occur unless the polarizers pass less than about 0.01 of the unwanted component (*40*). This is owing to the fact that prisms and gratings are themselves inefficient polarizers whose polarization properties vary considerably with the wavelength of the light and the optical system in which they are used. Usually this is of little concern in the visible and near ultraviolet regions where very efficient polarizers are commercially available, but infrared polarizers tend to have either limited spectral range or poor polarization characteristics. Shurcliff (39) and others have discussed polarization phenomena at some length and these discussions apply in general to dichroism and birefringence measurements.

Another relatively sensitive method of measuring linear dichroism in the ultraviolet region has been described (*41*). In this method a quartz quarter wave plate is placed between the polarizer and the sample. At integral multiples of the parameter $\Delta n \cdot \tau / \lambda$ the emergent radiation incident on the sample becomes linearly polarized, alternately horizontally and vertically as the integer goes through successive odd and even values as λ is varied; Δn is the birefringence of the quarter wave plate at the wavelength λ and τ is the thickness of the plates. The spectrum shows a continuous series of maxima and minima if the sample is dichroic in the wavelength region studied and is smooth if there is no dichroism. The difference between the maxima and minima is a measure of the dichroism. This method can, of course, be applied to the measurement of the dichroism of any oriented sample if the orientation is static over the time required to scan the spectrum. Errors due to circular dichroism can enter if the substance is also optically active because circularly polarized light is produced at half integral values of $\Delta n \cdot \tau / \lambda$ and in fact the apparatus can be used to measure circular dichroism.

Still another method uses a technique in which the modulation is provided by rotating the polarizer at 15 cps, thus providing a 30 cycle signal which is a maximum when the polarized light is parallel with the direction of minimum absorption of the sample and vice versa (*42*). This method has the advantage that signal-to-noise improvement can be obtained by suitable use of synchronous detection techniques.

The oriented films have been produced by a variety of methods, including casting from a flowing fluid, stroking, and stretching.* The individual papers and a review by Liang (*19*) should be consulted. Recently the method has been extended to examine the effect of the aqueous ionic environment on DNA films (*43*). The observation that the dichroic ratio at 260 nm increases from 0.5 to 0.8 as the relative humidity of DNA films containing 2 to 5% salt decreases is difficult to interpret in view of the unknown degree of orientation obtained by stroking the solutions on quartz slides. The problem of interpreting data from partially oriented films and fibres has been attacked several times (*44–46*). If quantitative data are to be obtained by this method, close attention will have to be paid to determining the degree of orientation achieved.

Crystal Dichroism

Optical crystallography was one of the early techniques applied to problems in nucleic acid structure. It is only quite recently, however, that dichroism measurements have begun to yield significant information from crystalline materials of interest in nucleic acid structure. The problems of obtaining suitable crystals for this first dichroic spectrum of a crystal of a nucleic acid base are illustrative. Stewart and Davidson found it necessary to polish a small single crystal of 1-methyl thymine by a slow tedious process in order to obtain a crystal thin enough to measure its spectrum through the 260 nm absorption band (*12*). The application of reflectance techniques (*15*) to some extent alleviates this problem but introduces other experimental and theoretical ones. Most recently Eaton and Lewis (*17*) following a procedure developed by Ward (*47*) have found that by growing crystals between highly polished optical quartz flats from melts or solutions, ultrathin single crystals of nucleic acid bases could be obtained so that even absorption bands with molar extinction coefficients of about 10^4 can be measured. These crystals, are, however, very small, 100 μ or less in diameter, necessitating the use of microspectrophotometric techniques. A complete description of the microspectrophotometer is found in their paper, but the essential features are a double monochromator, a good calcite Glan polarizing prism, a polarizing microscope with strain-free Zeiss Ultrafluar objectives chromatically corrected from 2300 Å to 7000 Å and most importantly an adjustable crossed-slit field diaphragm optically imaged on the crystal. The use of the field diaphragm in this way serves to optically mask the periphery of the crystal and ensure that no light which does not penetrate the crystal enters the optical path of the detector. The use of this field diaphragm and a pinhole

* Very recently, two papers have appeared in which the latter of these two techniques has been exploited to obtain polarized absorption spectra of smaller molecules (Thulstrup, E. W., J. Michl, and J. H. Eggers. 1970. *J. Phys. Chem.*, 74, 3868, 3878).

aperture at the image plane of the objective reduces the geometric stray light to less than 1 part in 5000. Optical densities up to 2.5 can be measured with a reproducibility of 2% or better. Using this apparatus, the dichroism was measured in two different faces of a 1-methyl uracil crystal encompassing the three axes of the orthorhombic space group to which the unit cell of this crystal belongs.

The full interpretation of the data obtained from crystal dichroism measurements requires that certain crystallographic information be available; the crystallographic space group, the number of molecules per unit cell and morphological information are required. These together with geometric factors relating the electric vectors of the incident polarized radiation to the unit cell directions are sufficent to elucidate the symmetry and the orientation of the transition moments in the crystal, two parameters of considerable use in analyzing dichroic spectra. When a complete x-ray crystal structure is known, the orientation of the transition moments *in the molecules,* and the intermolecular interactions such as exciton and site splitting may also be interpreted. These factors provide a basis on which an analysis of the electronic structure or the molecular dynamics can be interpreted from the electronic (UV-visible) and the vibrational (infrared) dichroic spectral data. Classical descriptions of crystal absorptions date from Drude (*48*) while more modern descriptions are given by a number of other authors (*49, 50*). References have already been made to descriptions of exciton and site splitting, but additional useful analyses may be found (*51-53*).

Flow Dichroism

The four principal methods of measuring flow dichroism are illustrated diagrammatically in Fig. 2. In one of these (illustrated by Fig. 2a) the radiation is propagated in the direction of flow and consequently if the orientation is independent of the geometry of the channel in which the flow takes place, there is no difference in extinction for light polarized in any direction perpendicular to the direction of propagation. We have already pointed out that polarized light need not be used to make these measurements. The data must be interpreted in terms of the change in extinction as the degree of orientation of the molecules changes with flow or "shear rate." One of the earliest methods (Fig. 2c) made use of the hydrodynamic field produced between concentric cylinders; the inner cylinder is rotated with respect to the outer (*54*). In this as well as the other two methods the direction of propagation is perpendicular to the direction of flow and the radiation must be polarized to measure the *dichroism.* The flow channel is very narrow, so that the orientation tends to be dependent on the distance

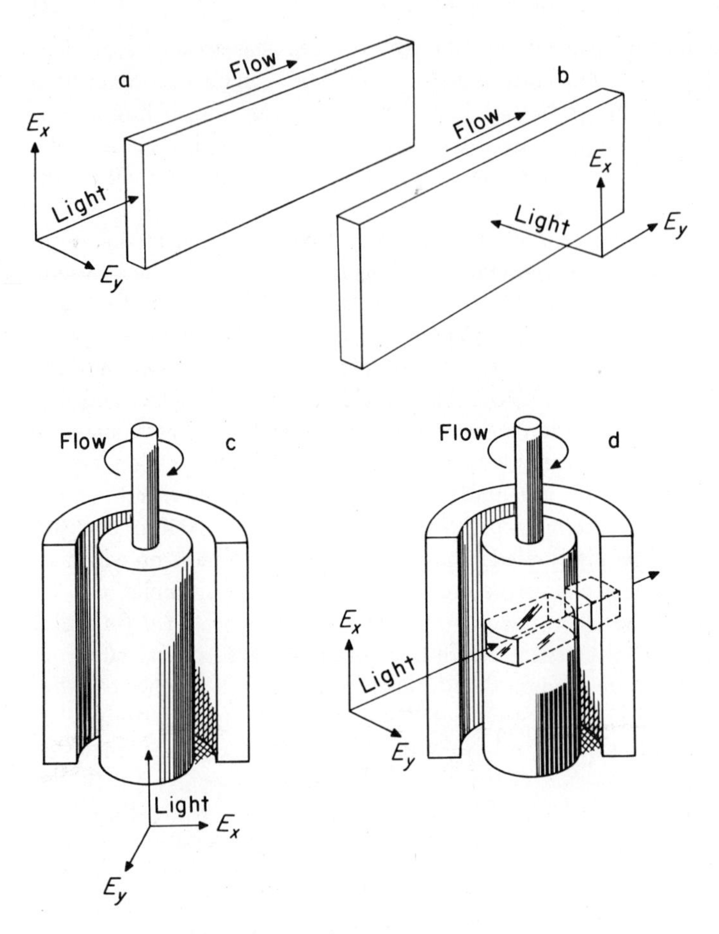

Figure 2. Flow dichroism methods.

from the walls of the apparatus in instruments of types *b*, *c*, and *d*, i.e., along the direction of propagation of the radiation. However, by making the channel high enough the orientation is made uniform in the *x* direction and, depending on the length of the flat channel in *c*, also independent of the *y* direction over the illuminated region. The theory which connects the observables with the molecular and field parameters has been discussed by several authors (*55–57*). Wada (*56*) in particular has developed the theory for the relationship of the reduced dichroism measured transversely to the direction of flow (apparatus types *b* and *d* of Fig. 1 *) to the orientation

* With some modifications this is also applicable to apparatus of type *c*.

factor $A(s)$ and the factor $B(\alpha)$ relating the orientation of the chromophoric transition moment to the molecule-fixed axes. For a rigid macromolecule whose nuclear and electronic charge distribution can be approximated as an ellipsoid of revolution, this relation is

$$\Delta\epsilon/\epsilon = [\tfrac{1}{2}(3K_1 - K_2 - 2)] \cdot [\tfrac{3}{2}(3\cos^2\alpha - 1)] = A(s) \cdot B(\alpha)$$

where α is the angle between the transition moment direction and the macromolecular axis. K_1 and K_2, the orientation distribution functions, are given by

$$K_1 = \langle \sin^2\theta \rangle = \int \sin^2\theta\, f(\theta, \phi) d\Omega$$

$$K_2 = \langle \sin^2\theta \cos 2\phi \rangle = \int \sin^2\theta \cos 2\phi\, f(\theta, \phi) d\Omega$$

which reduce, respectively, for no orientation and perfect orientation to $K_1 = 2/3, K_2 = 0$ and $K_1 = 1, K_2 = -1$ resulting in values of $\Delta\epsilon/\epsilon = 0$ and $\Delta\epsilon/\epsilon = 3/2$ $(3\cos^2\alpha - 1)$. These expressions are not limited to flow dichroism except insofar as the orientation function $f(\theta,\phi)$ is specified in terms of hydrodynamic parameters. When $f(\theta, \phi)$ is specified in electric field parameters, the same expressions will hold for electric dichroism. A viable theory which will allow the calculation of α from the measured values of $\Delta\epsilon/\epsilon$ at any degree of orientation other than perfect orientation, must therefore correctly specify the relationship between the force field and the geometrical parameters.* For the flow dichroism apparatus, θ is the angle between the rotation axis of the concentric cylinder and the macromolecular axis and ϕ is the angle between the propagation axis of the light and the projection of the macromolecular axis in the plane perpendicular to the light propagation direction. For rigid molecules, a number of theoretical models have been developed to express the field-geometry relationships. These seem to work reasonably well; a complete discussion is given by Peterlin and Stuart (*58*). Unfortunately, for flexible macromolecules the situation is complex and still somewhat ambiguous. Wada (*56*) has pointed out that the conformation and chromophore orientation in flexible molecules changes with the velocity gradient of flow making it impossible to separate the internal and external orientation factors and Tsvetkov (*5*), in discussing flow birefringence measurement, is careful to point out that the physical behavior of flexible molecules in flow fields "cannot be explained unequivocally in all details." Most of the interpretation of existing flow dichroism data of flexible macromolecules is based on the same models that are used to interpret the orientation obtained

* However, in a recent paper Mayfield and Bendet (1970. *Biopolymers,* 9: 655), by relating dichroic ratios obtained from flow dichroism of partially oriented systems to light scattering ratios from the same systems, claim to circumvent this problem.

in flow birefringence measurements.* Despite the ambiguities, considerable progress has been made in the interpretation of the birefringence measurements, based in large part on models proposed by Kuhn (*59*), Rouse (*60*), Zimm (*61*), and Cerf (*62, 63*).

In a flow dichroism experiment the dichroism reaches an equilibrium value at any given velocity gradient in a time which is dependent on the velocity gradient itself as well as on the shape and size of the macromolecule. In general as the gradient increases, the time to reach equilibrium decreases and the magnitude of the equilibrium value of the dichroism increases. Saturation values may be reached, i.e., the dichroism may no longer increase with increasing velocity gradient, well before the macromolecules in solution are completely oriented. As a consequence, the simplifying assumptions that permit the calculation of the direction of transition moments from the saturation value of the dichroism for a fully oriented system cannot be used. Flow or velocity gradients required to reach saturation of the flow dichroism signals for many macromolecules appear to be in the range of 2000 to 4000 sec^{-1}. The Shimadzu apparatus can achieve velocity gradients upwards of 3000 sec^{-1} and the colinear light and flow propagation apparatus of Callis and Davidson (*35*) has been used at gradients as low as about 200 sec^{-1} and as high as 21,250 sec^{-1}.

An interesting variant of the flow dichroism procedure has recently been described by Cram and Deering (*64*). In this procedure, ultraviolet inactivation using polarized 265 nm radiation on a flow oriented system has been used to obtain the dichroic ratios of *fd* bacteriophage. Flow gradients as high as 39,000 sec^{-1} were used in this work. Interpreting the hydrodynamic information using the method described by Scheraga, Edsall, and Gadd (*65*) to estimate the degree of orientation of the virus in the flow field, Cram and Deering conclude from the inactivation dichroic ratios that the planes of the DNA bases are tilted at 58° to 61° from the virus axis.

Electric Dichroism

The use of electric fields to induce orientation in macromolecular systems in solution offers certain advantages over flow methods, but is not without its own difficulties. These latter, however, tend to be in the nature of technical problems, which the application of ingenuity and advanced

* At this point it is worthwhile to note that while in general the orientation functions in birefringence and dichroism are the same for the macromolecule as a whole or for rigid repeating segments of it, in the vicinity of an absorption band of a chromophore which has one or more degrees of freedom with respect to the macromolecule or its rodlike segments, it is quite possible to observe a difference in the behavior of the data from these two types of measurements. At least one such observation has been made. (Milstien, J. B., and E. Charney. 1970. *Biopolymers*, 9: 991).

techniques should be largely capable of overcoming. Not the least of these difficulties is the necessity for constructing the apparatus since no commercial device for making electric dichroism measurements exists. Most of the electronic and optical parts are however commercially available and may be assembled without undue trauma if machining and electronic capabilities are at hand to help construct the electrooptical cells, and to assemble, test, and calibrate the electronic and optical equipment. Except for the elimination of one of the polarizers and in some cases a birefringent device, the equipment required is identical to that used for making electric birefringence measurements. A number of such devices have been described* and descriptions of at least two additional designs more especially directed at making dichroism measurements are currently being prepared for publication (*66, 67*). The original publications or the authors are best consulted for the details. In Fig. 3 we illustrate in block diagram form the apparatus currently in use in our laboratory for both electric birefringence and dichroism measurements. The electrical components are very similar to those described by Yamaoka (*68*). The current apparatus (*67*) provides an external square electric pulse up to 4 kilovolts from 0.05 msec to approximately 10 msec duration when the electrooptical cell impedance is sufficiently high. Since the distance between the electrodes of the cell (Fig. 4) may be as small as approximately 1 mm, field strengths up to 35 to 40 kV/cm may be achieved when the electrical conductance of the solution is low. Conductance of the solutions presents one of the major problems in these measurements. Many of the more interesting substances for which dichroism measurements can provide useful information are polyelectrolytes and some of these are not particularly stable at low ionic activity. This is one of the primary (but not the only) reasons for using short pulsed electric fields so as to prevent overheating of conducting solutions. Direct-current electric fields have been used to make electric birefringence and electric dichroism (*69*) measurements, but these are principally useful only with nonconducting solutions and do not afford the opportunity to obtain structural information related to rotational or diffusional lifetimes from the rise and decay curves. Very little advantage has been taken thus far of the opportunity to use the information available from the rise and decay of the dichroism which are schematically illustrated in Fig. 5. However considerable use of these curves from electric birefringence measurements has been made (*4, 6*). This is, in part, due to the relative novelty of the dichroism measurements. One may expect an expansion of the use of this information obtained from pulsed-field measurements, especially in relation to the use of electric dichroism measurements to examine

* Since a fairly large number of electric birefringence devices have been constructed, the following list is not intended to be exhaustive: two reviews which may be consulted are by H. Labhart. 1963. *Tetrahedron,* 19: *Suppl. 2,* 223 and the review of C. T. O'Konski cited in Ref. *4*.

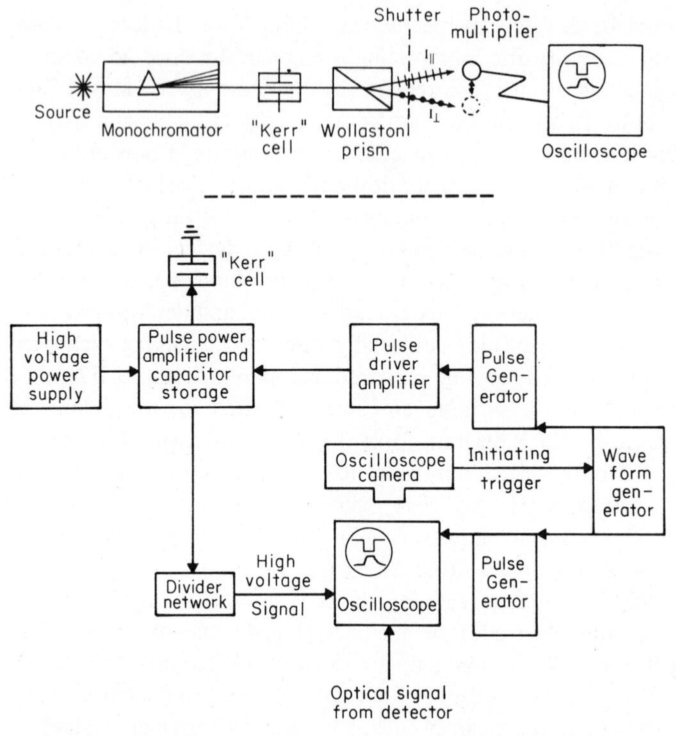

Figure 3. Block diagram of the apparatus. Upper part illustrates optical arrangement. For birefringence measurements a Glan-Taylor prism is inserted between the monochromator and the "Kerr Cell."

structural and conformational changes in reacting species. However, because birefringence measurements may be made outside of strongly absorbing regions where signal-to-noise ratios are more favorable, we may expect the rise and decay transients of electric birefringence to continue to provide the major source of information on relaxation times for the macromolecule as a whole.

Let us return briefly to Figs. 3, 4, and 5 for a description of the method. A trigger pulse manually or electronically generated is used to pulse a Tetronix pulse generator which provides a simultaneous trigger for an oscilloscope sweep and an appropriate gating signal to the high voltage supply which has stored charge in a bank of high-voltage capacitors. The stored charge of desired magnitude is released to the high-voltage electrode of the thermostatted electrooptical cell containing the solution of macromolecules. Under the influence of the resulting field between this electrode and the opposing parallel electrode maintained at ground potential, a torque is exerted on the polar or polarizable molecules which are randomly oriented in

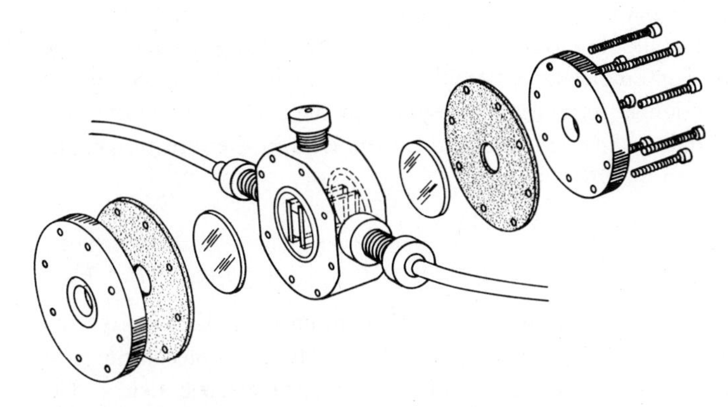

Figure 4. Electric dichroism and birefringence cell. The electrodes are polished gold plated brass or stainless steel. The body is Kel-F. Windows are Suprasil quartz. This exploded view shows end pieces, polyethylene foam pressure pads, windows, Teflon O-ring gaskets, electrodes, cell body, and electric leads.

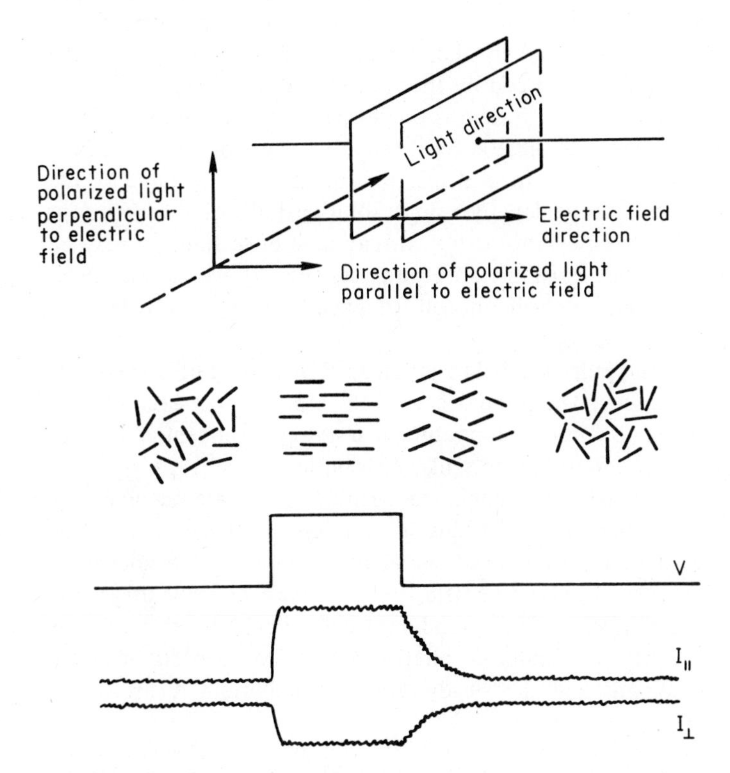

Figure 5. Schematic diagram of the electric dichroism measurement.

the absence of the field. The torque results from the action of the electric field on the electric dipole moment of the anisotropic distribution of charge in the molecule. For a molecule with cylindrical symmetry, the total moment in the field direction is to a first approximation the sum of the permanent dipole moment and the moment induced by the field in the direction (y) of the field and is given by (*70*):

$$M_y = \mu_{\|} \cos\theta + [\alpha_{\|}\cos^2\theta + \alpha_{\perp}\sin^2\theta]E_y + \tfrac{1}{2}[\beta_{\|}\cos^3\theta + 3\beta_{\perp}\sin^2\theta\cos\theta]E_y{}^2 + \ldots$$

where μ, α, and β are, respectively, the dipole moment, polarizability, and hyperpolarizability, parallel ($\|$) or perpendicular ($\perp$) to the molecular axis of symmetry, and E_y is the magnitude of the effective electric field.* The hyperpolarizability is generally very small and may be omitted from consideration except where $\mu_{\|}$ is very small or zero or at *very* high values of the field strength. Since the redistribution of charge arising from the polarizability under the influence of the field is generally extremely rapid compared even to the rise time of the pulse, the molecular orientation exhibits a rise time determined by the rotational diffusion constant of the macromolecule in the solution and by the viscosity of the solvent in which it is dissolved. Of course, if the system is polydisperse or if the molecules are not rigid or do not have a symmetry axis, there may be more than one such rotational diffusion constant. In principle this should be detectable by the inability to fit the rise curve data by a single rotational diffusion constant; in practice, the signal-to-noise ratio at the current state of technology may be such as to preclude this. For illustrative purposes (Fig. 5), we have assumed that the field strength is high enough to exert a torque sufficient to completely orient the molecules.

At the end of the pulse the field is discharged by cutting off the current to the storage capacitors. The pulse decays in the order of microseconds, but diffusional resistance to reorientation results in a slower (exponential for rigid molecules) randomization of the molecular distribution.

The field strength at which complete orientation is achieved of course depends on the moment, M_y, and this in turn on molecular parameters. Several extensive theoretical investigations of the orientation dependence on the molecular parameters and on the field strength have been made. In general the orientation dependence for a rigid molecule should be the same as expected for electric-field induced birefringence. Two aspects may be separated, the transient and the steady-state or equilibrium dichroism or

* Buckingham (Ref. *70*) has pointed out that the effective field in liquids of high dielectric constant tends to be larger than the applied field. Using the Lorentz approximation, which may not be a particularly good one but is sufficient for qualitative consideration, the effective field E_y is given by $E_y = (1/3)(D + 2)E'_y$ where E'_y is the applied external field strength and D is the dielectric constant of the solvent.

birefringence. For the first, the relevant papers on electric birefringence should be consulted (*71–75*). The magnitude of the static dichroism achieved at the height of the pulse or under a static field has been treated several times (*33, 36, 37, 76–78*) and will be discussed below.

The electrooptical system by means of which the field induced orientation is measured and from which the dichroism is calculated consists of a light source and monochromator which provide monochromatic radiation at the desired wavelength, the electrooptic cell containing the solution, a Wollaston prism which separates the horizontally and vertically polarized components of the radiation emerging from the cell, and a photomultiplier detector. The output of the detector is measured using one channel of an oscilloscope amplifier. Simultaneously the high-voltage pulse suitably attenuated through a calibrated voltage divider is fed to the other channel of the amplifier. Both signals are simultaneously displayed on the oscilloscope screen and photographed or alternatively fed to a signal averaging device. From these data and the sensitivity factors for signal amplification, the absorbance for each polarization is calculated directly (*67*) or using a suitable arrangement of two detectors, the ratio or difference between the two polarizations may be calculated (*66*).

With our instrument it is possible to make birefringence as well as dichroism measurements without disturbing the sample by inserting a Glan polarizer, which is oriented at 45° to the electric field direction, into the optical system in front of the sample cell. The Wollaston prism is then rotated in its mount to provide a polarization direction which is also 45° or 135° to the electric field direction (*67*). To determine the sign, as distinguished from the magnitude, of the birefringence an additional birefringent device must be inserted in the optical path.* In making birefringence measurements it is usual for maximum sensitivity to use the direction of polarization of the Wollaston analyzer perpendicular to that of the Glan polarizer, but it is, in principle, possible to make the birefringence measurements with the analyzer parallel to the polarizer, both of course at 45° to the electric field.

Electric Dichroism Measurement

The measurement procedure will depend on the details of the apparatus utilized. Sample preparation is standard and makes use of techniques well described elsewhere. It should be noted that the low ionic strength requirements sometimes necessitate additional dialysis or dilution. For DNA, for example, the techniques described by Eisenberg in the chapter on light

* There are many such devices which may be used; they range from a mechanically distorted and therefore birefringent dielectric plate to Soleil-Babinet and electro-optic compensators. For a description of the method see Ref. *79*.

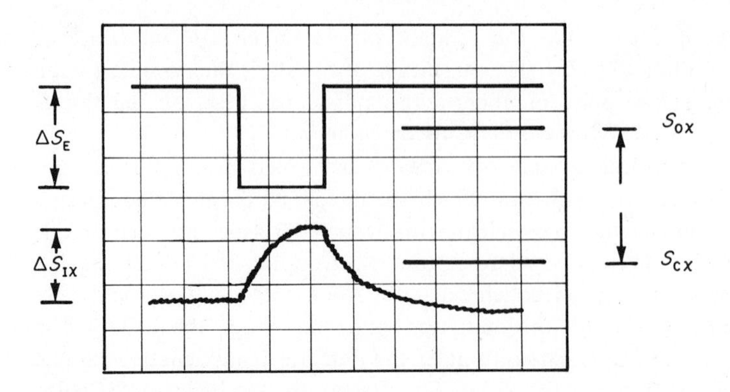

Figure 6. Schematic oscilloscope photograph. χ refers to the angle which the plane of polarization makes with the direction of the electric field. For the two extreme positions, 0° and 90°, χ is represented by $\|$ and $\perp$.

scattering and small angle x-ray scattering in this volume may be followed. To avoid shearing forces, the sample is introduced carefully into the cell with a syringe or pipette. In our apparatus (*67*), the cell is placed in a thermostattable jacket in the light beam where it is permitted to come to temperature equilibrium. Refer now to Fig. 6. (1) The sensitivity of the oscilloscope amplifier is adjusted to produce a readable difference (on 6 x 10 cm oscilloscope graticule) between the steady-state light signals generated with the photomultiplier shutter open and closed in the absence of the electric field. These signals $S_o\chi$ and $S_c\chi$ are photographed on one-third of the film of a Polaroid camera just prior to applying the electric field pulse. (2) Using the camera shutter mechanism to trigger simultaneously the camera shutter, the oscilloscope, and the high-voltage pulser, the dichroism signal, $\Delta S_I\chi$, produced by the absorbance changes resulting from the orientation of the dichroic molecules in the presence of the pulsed field, and ΔS_E, the attenuated high-voltage signal are photographed on the remainder of the film. Multiplication of $S_o\chi - S_c\chi$ and $\Delta S_I\chi$ by the sensitivity of the oscilloscope amplifier used in their measurement, and of ΔS_E by the amplifier sensitivity and the attenuation factor of the calibrated high-voltage divider network in parallel with the cell yield, respectively, $I_{0\|}$ or $I_{0\perp}$ and $\Delta I_\|$ or $\Delta I_\perp$, depending on the polarizer orientation ($\chi = \|$ or $\perp$), and the voltage, V, across the cell. The reduced dichroism at the voltage, V, and wavelength λ is calculated by means of the expression:

$$\frac{\Delta\epsilon}{\epsilon_0} = \frac{1}{A} \log \frac{1 + (\Delta I_\perp / I_{0\perp})}{1 + (\Delta I_\| / I_{0\|})}$$

where A is the isotropic absorbance determined for the same pathlength as used in the dichroism experiment.

Electric Dichroism Theory

Electric-field induced orientation is uniquely different from the orientation arising from crystal or hydrodynamic forces in that it depends on the interaction between the applied electric field and the electric dipole moment, permanent, or induced in the orienting molecules by the field. The relation between this molecular property, that is, the magnitude and direction of the dipole moment, and the dichroism was first recognized by Kuhn, Dührkop, and Martin (*33*) while the corresponding dependence for electric birefringence was first discussed by Charney and Halford (*80*). Between these two, the relation is very similar to that between optical rotary dispersion and circular dichroism and, indeed, the correlative relationship results from the fact that the optical rotatory dispersion and electric birefringence depend on the real part of the complex refractive indices while the circular and electric (linear) dichroism depends on the absorption or imaginary part of the refractive index. All are signed quantities since the contributions to the refractive indices along molecule-fixed axes depend on the direction of the electric and/or magnetic transition moments in the molecules. The two dichroism phenomena manifest themselves optically somewhat differently, however, since linear dichroism depends on the squares of the electric dipole transition moments while circular dichroism depends on the product of the electric and magnetic transition moments. As such, the latter is signed regardless of the orientation and, in fact, is usually measured on an isotropic system, while linear dichroism can only be measured on an oriented system.

The observable dichroism has a very complex dependence on the transition moments, the permanent and induced dipole moments, and the electric field strength. An extensive treatment in the language of classical physics has been given by Liptay and Czekalla (*36, 37*) and a semiclassical treatment which includes a perturbation theory examination of the effect of the electric field on the transition energy (peak position) has also been given (*76*). More recently, the theory has been further expanded (*78*). General expressions were developed for changes in intensity under the influence of the field with detailed expressions for the dependence on the molecular dipole moment components and on the polarizability tensors in the ground and excited electronic states. While the complete expressions are very complex, the limiting equations, from which structural and spectroscopic information on rigid rodlike molecules may be obtained, reduce to simpler forms at absorption band maxima or when the changes in dipole moment or polarizability induced by the external electric field is small. Although these changes may be substantial in small molecules, they are unlikely to be large in high polymers but no experimental observations or theoretical calculations have yet demonstrated this. The reduced dichroism is proportional to the difference between the electronic transition probabilities parallel and perpendicular, $\pi\|$ and $\pi\perp$, to the applied field at the frequency $(\tilde{\nu})$ in cm^{-1}:

$$\Delta\epsilon/\epsilon \sim [\pi_{\|}(\tilde{\nu}) - \pi_{\perp}(\tilde{\nu})]$$

Each of these probabilities which derive from the orientation functions and the molecular parameters has identical terms which are independent of the field strength and which cancel when the difference is taken, and large noncancelling terms which depend on the square of the field strength. The limiting form at low fields (more properly at low values of $\beta = \mu_3 E/kT$ where μ_3 is the magnitude of the total dipole moment parallel to the rod axis, E is the effective applied field strength, and kT is the Boltzmann thermal energy which tends to disorient the molecules) is*

$$\Delta\epsilon/\epsilon = \frac{\beta^2}{5}(3\cos^2\alpha(\tilde{\nu}) - 1)$$

At high field where the orientation is substantially complete and the dichroism is thus independent of field strength, the reduced dichroism reduces to

$$\Delta\epsilon/\epsilon = \tfrac{3}{2}(3\cos^2\alpha(\tilde{\nu}) - 1)$$

$\alpha(\tilde{\nu})$ is the azimuthal angle which the transition moment for the band at $\tilde{\nu}$ makes with the rod axis. This behavior is illustrated in Fig. 7 where the dashed line indicates the behavior at intermediate fields.

Two implications are at once apparent: (1) At low fields, a quadratic dependence on field strength is expected and (2) the reduced dichroism will reflect the value of α at any frequency and will be independent of the wavelength in an absorption band if the transition moment direction is independent of the wavelength.

(1) Field Strength Dependence. At intermediate values of β nonquadratic dependence on E is expected. A quantitative prediction of this behavior even for a rigid rodlike molecule requires an intimate knowledge of the dielectric properties of the molecule. These are not frequently known with much accuracy. It is likely, moreover, that deviation from the predicted E^2 dependence is also indicative of a lack of rigidity of the molecule.† Measurements on polybutylisocyanates of varying molecular weight have demonstrated (*81*) conclusively that at low fields the dichroism of smaller rigid polymers is quadratic in field strength but that a linear dependence becomes important in the same molecular-weight range in which hydrody-

* If the anisotropic polarizability of the molecule gives rise to induced moments which are not small compared to the permanent dipole moment, the low-field limiting expression takes a somewhat different form (*67*).

† Dows has examined the Kerr effect in flexible polymers (1964. *J. Chem. Phys.*, **41**: 2656) but did not explicitly comment on the effect of flexibility on the field strength dependence.

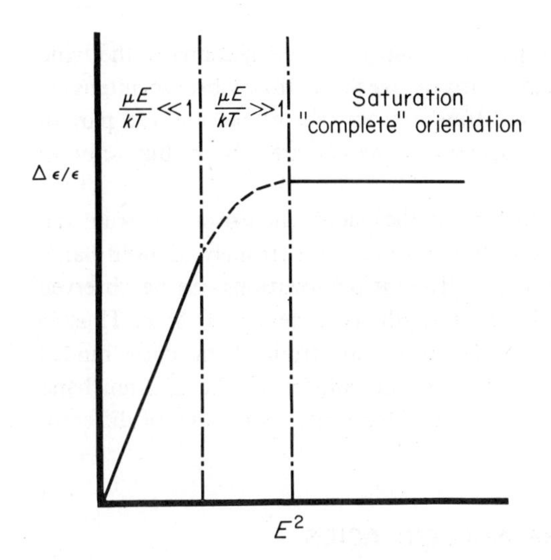

Figure 7. Electric field strength dependence of the reduced dichroism. μ in this figure refers to the total effective dipole moment, permanent and/or induced. The demarcation between $(\mu E/kT) \gg 1$ and $(\mu E/kT) \ll 1$ is not intended to be sharp since the ratio of permanent to induced moment and the mechanism of the induced moment are extremely important in determining the field strength behavior.

namic measurements indicate that the polymer is larger than its hydrodynamic persistence length. Side by side dimers of poly-γ-benzyl-L-glutamate in dioxane which may be expected to be much more rigid than the isolated monomers in ethylene dichloride have been observed *(25)* to exhibit dichroism which is quadratic in E as compared to a linear dependence on E in the latter solvent.

(2) The Wavelength Dependence. In an allowed electronic transition, which gives rise to a strong, isolated absorption band, virtually all the intensity comes from totally symmetric vibrational subbands, so that the polarization in the molecule is constant over the absorption band. In this case the reduced dichroism should be independent of wavelength over the entire width of the band. On the other hand if the transition is forbidden or only weakly allowed, the vibronic bands of symmetry other than the symmetry of the electronic part of the transition moment may appear. A graphic demonstration of this is observed *(25)* in the electric dichroism spectrum of poly-γ-benzyl-L-glutamate when the wavelength dependence of $\Delta\epsilon/\epsilon$ in the 258-nm band goes through a series of maxima and minima which correlate with known vibronic structure in the vapor spectrum of toluene. This band corresponds to the forbidden band in benzene which appears very weakly owing to vibronic interaction. In the reduced symmetry of a monosubstituted

benzene such as toluene or the phenyl group in benzyl glutamate, the band becomes allowed, but it is still relatively weak so that vibronic intensity, which gives rise to a polarization different from that of the allowed part of the band, makes important contributions which are observable only in polarized spectra.

While the high resolution achievable in the electric dichroism spectrum of a weak band may not be possible in the presence of a strongly allowed band, overlapping *electronic* transitions of different polarizations can be observed by more subtle variations in the wavelength dependence of $\Delta\epsilon/\epsilon$. Thus in polyriboadenylic acid, the reduced dichroism spectrum of the two-stranded acid form exhibits a change in slope in the middle of the 252-nm band indicative of the presence of at least two electronic transitions of different polarization in that band (*82*).

DICHROISM OF THE NUCLEIC ACIDS

Film and Crystal Dichroism

In discussing the film dichroism experiments of Rich and Kasha and of Gellert, we noted that the interpretation of the dichroism measurements of polynucleotides and nucleic acids is dependent on the knowledge of the type and orientation of the transition moments. To obtain these, measurements on the isolated bases are of utmost importance. Only a few such measurements have been reported, notably those on crystalline 1-methyl uracil and 1-methyl thymine discussed earlier and a recent study of cytosine monohydrate (*83*).* Stewart and Davidson also measured the dichroism of a crystal of an adenine–thymine dimer and inferred the direction of the transition moments in 9-methyl adenine from these data and that of the 1-methyl thymine results. In 9-methyl adenine, the transition moment is found to be at $-3°$ to the C_5-C_4 axis for the 275-nm band. However, there is some ambiguity and the alternative possibility is that the moment lies at $+45°$ to the same axis.† In the case of 1-methyl uracil, the apparently smooth band in the isotropic spectrum in aqueous solution at 2670 Å was found by Eaton and Lewis to consist of at least two components, a strong in-plane component at about

* In this work the main transition moment in the 260–270 nm region is found to be in the plane of the molecule $+12°$ to the N_1-C_4 axis. Eaton and Lewis have also examined cytosine and found the transition here to be $+14°$ to the same axis. (W. A. Eaton, 1971. *J. Am. Chem. Soc.*, 93: 2054).

† In adenine hydrochloride, this band has been analyzed as consisting of two transitions at 2800 Å and 2500 Å, inclined at $-24°$ and $+82°$, respectively from the C_4-C_5 axis (personal communication, Leigh B. Clark based on observations by Clark and Holly Ho Chen).

2700 Å interpreted as the theoretically expected $\pi^* \leftarrow \pi$ transition, and a very weak out-of-plane band at about 2640 Å which is interpreted as an $\pi^* \leftarrow n$ transition. The direction of the $\pi^* \leftarrow \pi$ transition moment was found to be almost parallel to the N_1-C_4 molecular direction. This is quite compatible with the direction found (*12*) by Stewart and Davidson for 1-methyl thymine, which lies some 19° from the same axis. These three results have important consequences in terms of the dichroism of DNA and RNA. If, as is usually assumed, the strong $\pi^* \leftarrow \pi$ transitions of uracil and thymine remain in the plane of the base in the DNA molecule then after accounting for the very weak ($\epsilon \approx 200$ compared to $\epsilon \approx 25{,}000$) $\pi^* \leftarrow n$ component, any dichroic component parallel to the helix axis arising from the thymine moiety in DNA implies that the base is tilted with respect to the helix axis. A similar comment holds true for the adenine moiety, but in addition the two possible transition moments will give different parallel components for any given base tilt. The analysis of the dichroism of DNA is thus not simple, but as more data accumulate on the direction of the transition moments in the nucleotide bases it can be expected that dichroism measurements on DNA and RNA will provide fairly detailed information on the structure of these macromolecules, in some respects perhaps comparable to x-ray data. Such data as exist come mainly from flow and electric dichroism measurements.

Flow Dichroism

The application of the hydrodynamic models (*59-63*) to flow dichroism measurements of DNA (*26, 28, 31, 35, 55, 84-86*) has not yet led to an unequivocal analysis of the structure or spectra of DNA in solution. There is considerable variation in the reported data, some of which has been attributed to differences in structure arising from the different sources of DNA. Values of the reduced dichroism of the 2600 Å band of native calf thymus DNA range from about -0.13 (*55*) to -0.65 (*26*) and still more negative values have been measured on DNA derived from other sources such as T_2 phage (*25, 86*). The theoretical value of $\Delta\epsilon/\epsilon = -1.5$ has been predicted on the basis of the assumption which is universally made that the transition moments lie in the planes of the bases. The failure to achieve this value has been attributed either to a lack of complete helicity in the DNA chain, or to kinks in the chain which do not straighten out even at high flows where saturation of the dichroism occurs. The effect of base tilt on the dichroism has been discussed, but usually in the context of partial disorder of the molecule since it is always assumed that the bases are in the Watson-Crick configuration B which occurs in the high-humidity fibers, in which the bases are found to be perpendicular to the helix axis.

Electric Dichroism

The first attempt to measure the dichroism of DNA by electric-field induced orientation was made by Dvorkin (*87, 88*). While the earliest results appear to have been spurious, Golub and Dvorkin (*89*) subsequently repeated the experiments and did observe the dichroism of DNA. A number of other measurements of the electric dichroism of DNA or DNA complexes with nucleohistones or dyes have been reported (*90–95*). The results of the four investigations (*67, 90, 92, 94*) of the electric-field induced dichroism of native calf thymus DNA are given in Table 2. Of the reported values only the data obtained from calf thymus DNA carefully sonicated to a molecular weight of 240,000 daltons gives a clear indication from the quadratic field strength dependence up to fields of about 3.5×10^4 kV/cm that the measurements were performed on rigid molecules (*78*). By comparison the dichroism of native calf thymus DNA is (*78*) almost exactly linear in field strength. The recent determination (*96*) from measurements of light scattering that the persistence length of sonicated DNA may be as large as 1500–1900 Å fits in very well with these data.

In addition to measurements on DNA, there have been several other electric dichroism investigations relative to nucleic acids. Shirai has reported a value of the dichroic ratio corresponding to $\Delta\epsilon/\epsilon = -0.10$ at 252 nm for poly A at pH = 5 in his 1965 paper (*91*) and $\Delta\epsilon/\epsilon = -0.61$ in his 1967 review article (*2*). Milstien and Charney (*82*) have examined poly A under a variety of conditions and find $\Delta\epsilon/\epsilon = -0.87 \pm 0.05$ for the two stranded acid forms. Despite the variation in the reported values of the dichroism of poly A and DNA, it appears that much useful information will be forthcoming from the use of the

TABLE 2. Electric Dichroism of Calf Thymus DNA

Reference	$-\Delta\epsilon/\epsilon$
67	0.38[a]
	0.55[b]
90	0.41[c]
92	0.50[d]
94	0.10[e]

[a] Value at maximum field strength measured, 7.7 kV/cm. Saturation was not reached at this field strength.
[b] Sample of sonicated DNA, mol. wt. = 240,000, at saturation values of field strength.
[c] Saturation value reached at 7 kV/cm.
[d] Maximum value reached at field strength approaching saturation.
[e] Field strength not specified.

electric dichroism technique. The observed values all lead to the conclusion that in DNA and poly A, there is considerable intensity parallel to the field direction. If these macromolecules behave as rigid rods, or at least attain a rodlike orientation parallel to the applied electric field, then one would have to conclude that the bases are tilted with the respect to the helix axis, or, that in the helix, unlike in the crystalline bases, there is a substantial out-of-plane component to these transitions. Hydrodynamically these molecules do not behave as straight rods at comparable molecular weights; the interpretation of the dichroism data may have to take this into account. The variation in the early results appears to be largely owing to variations in the conditions and samples examined. Measurement of the electric dichroism of RNA in ribosomes (*97*), TMV (*88*), and quantasomes from spinach chloroplasts (*98*) have also been reported.

CONCLUSIONS

Dichroism studies are becoming increasingly important sources of information on the structure and electronic properties of biological molecules. For small molecules, most of the quantitative information is coming from linear dichroism measurements on single crystals. For macromolecules, flow- and electric-field induced dichroism are both providing information, but it is likely that in many cases the complications of understanding the effects of hydrodynamic forces, especially on flexible molecules, may make electric dichroism the technique of choice. Flow dichroism measurements will continue to be useful, especially where the conducting nature of the solvent makes it difficult to obtain strong electric fields. The extension of electric dichroism studies into the areas of conformational studies and the study of moderately fast reactions is likely. One of the most important and potential uses of this technique arises out of the fact that with molecules having cylindrical symmetry orientation produces a uniaxial system, making possible the collection of all the available optical absorption information from measurements parallel and perpendicular to the electric field. The implications include the possibility of using electric dichroism to resolve spectral components which cannot otherwise be resolved even at very low temperatures.

Acknowledgments

I wish to thank my colleagues Dr. Kiwamu Yamaoka and Dr. Julie Milstien who have played a major role in the development of the electric dichroism technique, Dr. William Eaton for many discussions of dichroism methods and of the dichroism of crystalline biological molecules in particular, Dr. Martin Gellert for his helpful criticism of the first draft of this article, and Ursula L. Walz of the National Institutes of Health library for help with the bibliographic research.

References

1. Dörr, F. 1966. *Angew. Chem. Intern. Ed.*, **5**: 478.
2. A short review by M. Shirai (1970) is available in *Tampanushitsu Kakusan Koso, Bessatsu (Japan)* **9**, 46.
3. Harrington, R. E. 1967. In Encyclopedia of polymer science and technology. Wiley, New York, Vol. 7, p. 100.
4. O'Konski, C. T. 1968. In Encyclopedia of polymer science and technology. Wiley, New York, Vol. 9, p. 551.
5. Tsvetkov, V. N. 1964. In Bacon, K. (Ed.), Newer methods of polymer characterization. Wiley (Interscience), New York, Vol. 6, 563.
6. Yoshioka, K., and H. Watanabe. 1969. In Physical principles and techniques of protein chemistry, Academic Press, New York.
7. Ellis, J. W., and J. Bath. 1938. *J. Chem. Phys.*, **6**: 221.
8. Perutz, M. F. 1939. *Nature* **143**: 731.
9. Davydov, A. S. 1962. Theory of molecular excitons. McGraw-Hill, New York.
10. Hornig, D. F. 1968. *J. Chem. Phys.*, **16**: 1063.
11. Winston, H., and R. S. Halford. 1949. *J. Chem. Phys.*, **17**: 607.
12. Stewart, R. F., and N. Davidson. 1963. *J. Chem. Phys.*, 39: 255.
13. Stewart, R. F., and L. H. Jensen 1964. *J. Chem. Phys.*, **40**: 2071.
14. Rosa, E. J. 1964. Ph.D. Thesis, University of Washington, Seattle.
15. Chen, H. H., and L. B. Clark. 1969. *J. Chem. Phys.*, **51**: 1862.
16. Callis, P. R. and W. T. Simpson. 1970. *J. Am. Chem. Soc.*, **92**: 3593.
17. Eaton, W. A., and T. P. Lewis. 1970. *J. Chem. Phys.*, 53: 2164.
18. Elliott, A., and E. J. Ambrose. 1947. *Nature*, **159**: 641.
19. See, for example, the review by C. Y. Liang. 1964. In Bacon, K. (Ed.), Newer methods of polymer characterization. Wiley (Interscience), New York, Vol. 6, p. 33.
20. Rich, A., and M. Kasha. 1960. *J. Am. Chem. Soc.* **82**: 6197.
21. Gellert, M. 1961. *J. Am. Chem. Soc.* **83**: 4664.
22. Adler, A. J., L. Grossman, and G. D. Fasman. 1969. *Biochemistry*, **8**: 3846.
23. Albrecht, A. and W. T. Simpson. 1955. *J. Chem. Phys.*, **23**: 1480.
24. Mason, S. F. 1959. *J. Chem. Soc.* p. 1247.
25. Charney, E., J. B. Milstien, and K. Yamaoka, 1970. *J. Am. Chem. Soc.* **92**: 2657.
26. Wada, A. 1964. *Biopolymers*, **2**: 361.
27. Butenandt, A., H. Friedrich-Freksa, S. Hartwig, and G. Schiebe. 1942. *Hoppe-Seyler's Z. Physiol. Chem.*, **274**: 276.
28. Cavalieri, L. F., B. H. Rosenberg, and M. Rosoff. 1956. *J. Am. Chem. Soc.*, **78**: 5235.
29. Bird, G. R., and E. R. Blout. 1958. *J. Am. Chem. Soc.*, **81**: 2499.
30. Hollingsworth, C. A., and W. Poppe. 1967. *J. Appl. Phys.* **38**: 2178.
31. Callis, P. R., and N. Davidson. 1969. *Biopolymers*, **8**: 379.
32. Kerr, J. 1875. *Phil. Mag.*, **50**: 337, 446.
33. Kuhn, W., H. Dührkop, and H. Martin. 1939. *Z. Physik. Chem. (Leipzig)*, **B45**: 121.

34. Spach, G. 1959. *Compt. Rend.* **249**: 667.
35. Callis, P. R., and N. Davidson. 1969. *Biopolymers,* 7: 335.
36. Liptay, W., and J. Czekalla 1960. *Z. Naturforsch.,* **15a**: 1072.
37. Liptay, W., and J. Czekalla. 1961. *Z. Elektrochem.* **65**: 72.
38. Newman, R., and R. S. Halford. 1948. *Rev. Sci. Instr.,* **19**: 270.
39. Shurcliff, W. A. 1962. Polarized light. Harvard University Press, Cambridge, Mass.
40. Charney, E. 1955. *J. Opt. Soc. Am.,* **45**: 980.
41. Jaffe, J. H., H. Jaffe, and K. Rosenheck. 1967. *Rev. Sci. Instr.* **38**: 935.
42. Brahms, J., J. Pilet, H. Damany, and V. Chandresekharan. 1968. *Proc. Natl. Acad. Sci. U.S.* **60**: 1130.
43. Wetzel, R., D. Zirwin, and M. Becker. 1969. *Biopolymers,* **8**: 391.
44. Fraser, R. D. B. 1953. *J. Chem. Phys.,* **9**: 1151.
45. Beer, M. 1956. *Proc. Roy. Soc. (London), Ser. A,* **236**: 136.
46. Yogev, A., L. Margolies, D. Amor, and V. Mazur. 1969. *J. Am. Chem. Soc.,* **91**: 4558.
47. Ward, J. C. 1955. *Proc. Roy. Soc. (London) Ser. A.* **228**: 205.
48. Drude, P. 1904. In A. Winklemann's Handbuch der Physik. 2nd Ed., Vol. 6. Barth, Leipzig.
49. Berek, M. 1937. *Fortschr. Mineral.,* **22**: 1.
50. Ramachandran, G. N., and S. R. Ramachandran. 1961. In S. Flugge (Ed.), Encyclopedia of physics, Springer-Verlag, Berlin, Vol. XXV, p. 1; Born, M., and E. Wolf. 1969. Principles of optics, 4th Ed. Pergamon Press, London.
51. Craig, D. P., and S. H. Walmsley. 1968. Excitons in molecular crystals. Benjamin, New York.
52. McClure, D. S. 1959. Electronic spectra of molecules and ions in crystals. In F. Seitz and D. Turbull (Eds.), Solid state physics. Academic Press, New York, Parts I and II, Vols. 9 and 10.
53. Fox, D., and O. Schnepp. 1955. *J. Chem. Phys.,* **23**: 767; Simpson, W. T., and D. L. Petersen. 1957. *J. Chem. Phys.,* **26**: 588.
54. Rundle, R. E., and R. R. Baldwin. 1943. *J. Am. Chem. Soc.,* **65**: 554.
55. Wada, A., and S. Kozawa. 1964. *J. Polymer Sci. Pt. A,* **2**: 853.
56. Wada, A. 1964. *Biopolymers,* **2**: 361.
57. Hollingsworth, C. A., and W. Poppe. 1967. *J. Appl. Phys.,* **38**: 2178.
58. Peterlin, A., and H. A. Stuart. 1939. *Z. Physik,* **112**: 1, 129; *ibid.,* **113**: 663.
59. Kuhn, W., and H. Kuhn. 1943. *Helv. Chim. Acta,* **26**: 1394; 1945. *ibid.,* **28**: 1533; 1946. *ibid.,* **29**: 39, 71, 830.
60. Rouse, P. E. 1953. *J. Chem. Phys.,* **21**: 1272.
61. Zimm, B. H. 1956. *J. Chem. Phys.,* **24**: 269.
62. Cerf, R. 1955. *Compt. Rend.,* **240**: 531; *ibid.,* **241**: 496.
63. Cerf, R. 1956. *J. Polymer Sci.,* **20**: 216; 1957. *ibid,* **23**, 125; **25**, 247.
64. Cram, L. S., and R. A. Deering. 1970. *Biophys. J.,* **10**: 413.
65. Scheraga, H. A., J. T. Edsall, and H. D. Gadd, 1951. *J. Chem. Phys.,* **19**: 110.
66. Allen, F. S., and K. E. Van Holde. 1971. *Rev. Sci. Instn.,* in press.
67. Charney, E., and K. Yamaoka, in preparation.
68. Yamaoka, K. 1964. Ph.D. Thesis, University of California, Berkeley.
69. Troxell, T. C., and H. A. Scheraga. *Biochem. Biophys. Res. Commun.* 1969. **35**: 913.
70. Buckingham, A. D. 1965. *Chem. Britain,* **1**: 54.
71. Benoit, H. 1951. *Ann. Phys.,* **6**: 561.
72. Tinoco, Jr., I. 1955. *J. Am. Chem. Soc.,* **77**: 4486.
73. O'Konski, C. T., K. Yoshioka, and W. H. Orttung. 1959. *J. Phys. Chem.,* **63**: 1558.
74. Nishinari, K., and K. Yoshioka. 1969. *Kolloid-Z.,* **235**: 1189.
75. Matsumoto, M., H. Watanabe, and K. Yoshioka. 1970. *J. Phys. Chem.,* **74**: 2182.
76. Labhart, H. 1961. *Helv. Chim. Acta,* **44**: 447.

77. Liptay, W. 1965. *Z. Naturforsch.*, **20a**, 272.
78. Yamaoka, K., and E. Charney, in preparation.
79. Rabinowitch, J., and M. Rouzere. 1958. *Compt. Rend.*, **247**: 441.
80. Charney, E., and R. S. Halford. 1958. *J. Chem. Phys.*, **29**: 221.
81. Milstien, J. B., and E. Charney. 1969. *Macromolecules,* **2**: 678.
82. Milstien, J. B., and E. Charney, unpublished observations.
83. Callis, P. R., and W. T. Simpson. 1970. *J. Am. Chem. Soc.*, **92**: 3593.
84. Lee, C. S., and N. Davidson. 1968. *Biopolymers,* **6**: 531.
85. Wilhelm. F-X., and M. Daune. 1968. *Compt. Rend.*, **266**: 932.
86. Gray, D. M., and I. Rubenstein. 1968. *Biopolymers,* **6**: 1605.
87. Dvorkin, G. A. 1960. *Dokl. Akad. Nauk SSSR,* **135**: 739.
88. Dvorkin, G. A., and V. I. Krinskii. 1961. *Dokl. Akad. Nauk SSSR,* **140**: 942.
89. Golub, Ye. I., and G. A. Dvorkin. 1964. *Biofizika,* **9**: 545.
90. Houssier, C., and E. Fredericq. 1965. *Biochem. Biophys. Acta,* **120**: 113.
91. Shirai, M. 1965. *Nippon Kagaku Zasshi,* **86**: 1019.
92. Soda, T., and K. Yoshioka. 1965. *Nippon Kagaku Zasshi,* **86**: 1019.
93. Soda, T., and K. Yoshioka. 1965. *Nippon Kagaku Zasshi,* **87**: 1326.
94. Houssier, C., and E. Fredericq. 1967. *Biochim. Biophys. Acta,* **138**: 424.
95. Houssier, C. 1968. *J. Chim. Phys.,* **65**: 36.
96. Eisenberg, H. 1969. *Biopolymers,* **8**: 545.
97. Morgan, R. S. 1963. *Biophys. J.,* **3**: 253.
98. Sauer, K., and M. Calvin. 1962. *J. Mol. Biol.,* **4**: 451.

INFRARED SPECTROSCOPY OF POLYNUCLEOTIDES IN AQUEOUS SOLUTION

H. TODD MILES

LABORATORY OF MOLECULAR BIOLOGY

NATIONAL INSTITUTE OF ARTHRITIS AND METABOLIC DISEASES

NATIONAL INSTITUTES OF HEALTH

BETHESDA, MARYLAND

Molecules in solution absorb infrared light at frequencies corresponding to vibrations of the atoms which compose them. The theory of vibrational spectroscopy has been presented in the classic work of Herzberg (*1*), and excellent reviews of its application to complex organic molecules have been published by Jones and Sandorfy (*2*) and by Bellamy (*3*), among others.

Our attention in this article will be directed primarily to the infrared spectroscopy of nucleotides and polynucleotides in aqueous solution, and particularly to deriving structural and chemical information from the spectra. Polarized infrared studies of oriented nucleic acid films (*4*) have provided valuable geometrical information but will not be discussed here.

Infrared spectroscopy has several major advantages for studying the nucleic acid bases and their interactions. The spectrum of each nucleotide or polynucleotide is highly characteristic, possessing one or more strong bands at frequencies at which absorption by the other bases is negligible or weak (Fig. 1). When the polynucleotides interact to form helical complexes, moreover, their infrared spectra show large and characteristic changes in frequency and intensity of the absorption bands (*5–10*). The combination of these characteristics makes it possible to monitor simultaneously and independently different interacting bases, to recognize qualitatively which interaction products are present, and to analyze for them quantitatively (*5–10*).

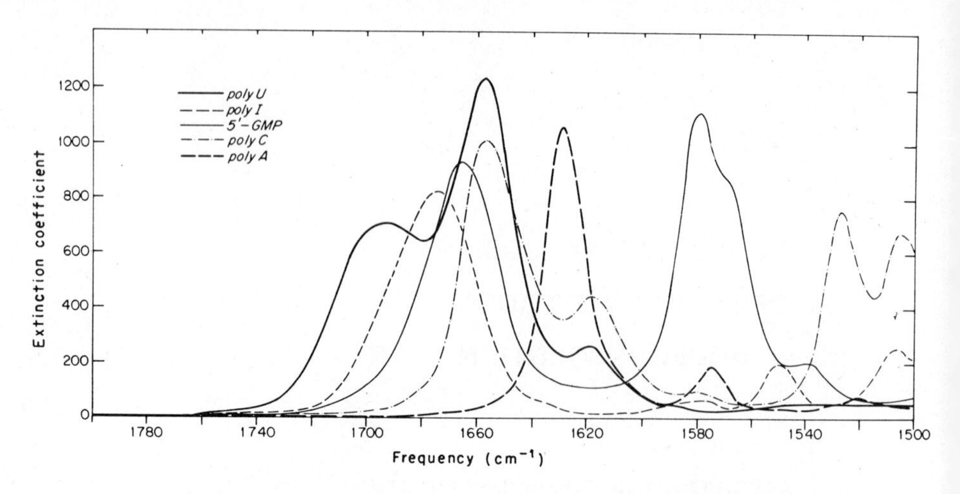

Figure 1. Infrared spectra of four polynucleotides and of 5′-GMP at pD 7. The spectra are normalized on the basis of molar absorptivity (extinction coefficient). [From H. T. Miles. 1970. In L. Grossman and K. Moldave (Eds.), Methods in enzymology; nucleic acids, Vol. XIIB, p. 256. Reproduced with permission of the copyright owner, Academic Press.]

Solvent and Frequency Range

D_2O was introduced as an infrared solvent by Gore, Barnes, and Petersen (*11*) because of its valuable windows in regions where water is almost opaque. They also showed that by using both D_2O and H_2O it is possible to observe a large part of the usual IR region in aqueous solution. D_2O has been used as a solvent by Blout and Lenormant (*12*) to make qualitative observations of DNA in the 1650 cm^{-1} region, and other applications to studies of biopolymers have been reviewed (*9,12a,13*).

In order to obtain information with a modern double monochrometer spectrophotometer on the range of solvent transparency we have observed D_2O and H_2O in a standard BaF_2 cell (Limit FH-01) of 23.8μ pathlength (the empty cell alone transmitted 12% of the light at 800 cm^{-1} and ~2% at 750 cm^{-1}). D_2O has an absorbance less than 0.8 (16% T) from 1900 cm^{-1} to 1250 cm^{-1} (less than 0.31 from 1900 cm^{-1} to 1300 cm^{-1}) and from 1190 cm^{-1} to 840 cm^{-1}, but is opaque from 1195 cm^{-1} to 1225 cm^{-1}. About 2% of the light is transmitted at 775 cm^{-1}. H_2O in the same cell is opaque above 1615 cm^{-1} but transmits ~2% of the light at 1600 cm^{-1}. The absorbance is less than 0.8 between 1505 cm^{-1} and 1000 cm^{-1}. The percent transmission is 10 at 950 cm^{-1}, ~2 at 895 cm^{-1} and 0 at 875 cm^{-1}. These data were obtained with an IR-7 spectrometer operated in the double beam mode but uncompensated.

The use of both H_2O and D_2O thus makes the region from 1990 cm^{-1}

to ~840 cm^{-1} accessible under conditions which allow enough energy for reliable routine measurements. By use of still shorter pathlengths and more than routine attention to experimental technique it should be possible to extend the observable region to frequencies somewhat below 800 cm^{-1}. When absorption by the solvent is high it is desirable, at least in the initial experiments, to observe separately the spectra of solvent and of solution in the double beam mode without compensation, and to make subsequent correction by subtraction. It is possible to encounter artifacts due to stray light or unequal compensation, and the higher the solvent absorption the greater the care necessary to avoid artifacts and obtain reliable spectra. A spectrophotometer with a double monochrometer should be used. It is worth noting here that even in the favorable region around 1650 cm^{-1} D_2O is not very transparent (absorbance ~0.25; % T ~65% at 24μ pathlength) in comparison with the usual organic infrared solvents, and one is therefore limited to unusually short pathlengths (< ~ 100μ). It is primarily for this reason that the sensitivity of detection is lower than in the ultraviolet and that higher concentrations must be used.

For general use in the double bond region D_2O is thus the aqueous infrared solvent of necessity. For a number of purposes it appears to be also the solvent of choice. NH_2 deformations, which occur in the carbonyl region, are eliminated, considerably simplifying the spectra. Comparison of spectra of nucleosides with those of alkylated model compounds in tautomeric studies (*5, 6*) are greatly aided by the use of D_2O since NH_2 deformations and coupling of N–H bending with double-bond stretching modes are eliminated. As a result the spectra of nucleosides and appropriate model compounds show a simple resemblance in D_2O that would not be observed in H_2O.

It is frequently valuable to make direct comparison of infrared and Raman spectra measured under identical conditions, and D_2O is often preferable for this purpose. Where it is necessary or desirable to make observations in H_2O, however, quite satisfactory infrared spectra can be obtained between ~1550 cm^{-1} and 950 cm^{-1} at 25μ pathlength.

The region from ~1500 to 1800 cm^{-1} has been selected for the most detailed investigation of polynucleotides not only for reasons of accessibility discussed above but also because it is most informative in chemical and structural terms. This is the region of double bond stretching vibrations (C=O, C=C, C=N) of the carbonyl groups and heterocylic rings of which the bases are composed. It is these portions of the polynucleotides which are directly involved in base pairing and stacking interactions and which thus undergo the most profound environmental changes when helical complexes are formed. The frequencies and intensities of the bands corresponding to vibrations of these chemical groupings are most sensitive to these environmental variations and undergo marked and characteristic changes as a result of them, as discussed below.

Procedure

We have usually employed fixed thickness cells manufactured by Research and Industrial Instrument Co. of London (No. FH01) with Teflon spacers, CaF_2 or BaF_2 windows, and screw seals with Teflon gaskets. The pathlength is determined by interference fringes measured with the spectrometer on the empty cell, and compensation is achieved with the same solvent in a matched cell. The cells are dimensionally stable and show little change in pathlength even after numerous cycles of heating and cooling. The cells are usually cleaned by flushing with water and methanol, but if this is not adequate the cell must be disassembled and the windows polished. With care the cells may give long service, but the windows are fairly sensitive to both mechanical and thermal shock. This is especially true of BaF_2. Freezing of solution in the cell usually causes cleavage of the windows. In filling the cell it is essential to avoid air bubbles in the light path. It is this necessity for visual inspection to assure proper filling that makes window materials which are not transparent to visible light unsuitable for many of the applications described here.

Stock solutions of polynucleotides are prepared in D_2O, lyophilized, redissolved in D_2O and the concentrations determined by ultraviolet measurement of dilutions of aliquots. Volumes are measured either with micropipettes (calibrated to be rinsed rather than to deliver) or with Micrometric syringe displacement burettes. Concentrations of the solutions for spectroscopy can be from 0.001 M to 0.1 M, but the most convenient range is ~0.01 to 0.04 M.

Temperature control is achieved by pumping fluid from a thermoregulated bath through jackets surrounding both sample and reference cells. The jackets were fabricated of brass by the NIH instrument shop and fit into the usual cell mounts of the spectrometer. All tubing inside the cell compartment is copper, and quick disconnect attachments are fitted to the copper tubing outside the cell compartment and below the level of the instrument. An alternative provision for heating was made by placing electrical heating wires coaxially with the copper tubing around the cylindrical openings into which the cells are fitted. The sample and reference cells are maintained at the same temperature.

Temperature measurements are made with Yellow Springs Instrument thermistor probes (402) inserted into the thermocouple wells of both sample and reference cells (the ceramic inserts are first removed from the wells) and observations made with either a Digitec digital thermometer or a Yellow Springs Instrument model 425C Tele-thermometer. A probable improvement in temperature measurement is the provision which has been made in the new digital conversion system for direct transmission of temperature data with absorbance and frequency for subsequent computer plotting of melting curves, but this part of the installation has not yet been reduced to practice.

Digital frequency data are collected with a Perkin Elmer shaft encoder mounted on the frequency drive and absorbance data by a strip encoder with a brush which is mounted on the pen of the strip chart.

We have used a spectrophotometer (Beckman IR-7) with a double monochrometer and provision for continuous linear scale expansion in the absorbance mode. Expansion is essential for measurements of dilute solutions and has proved to be quantitatively reliable even at tenfold expansion. A further important advantage is that weak bands of more concentrated solutions may be greatly expanded, as shown in Fig. 2. This capability is advantageous in this example in providing an additional adenine band (1575 cm^{-1}) further resolved from the uracil bands and in permitting accurate

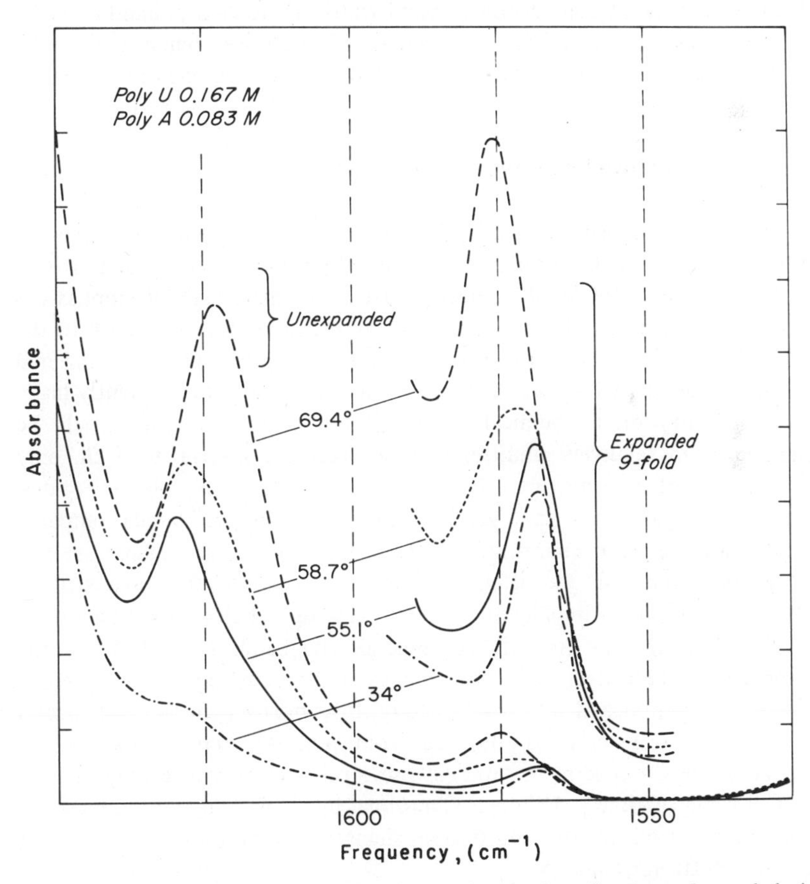

Figure 2. Temperature dependence of two adenine ring vibrations observed during melting of poly rA·2 poly rU. The weak band at ~1575 cm^{-1} can be observed clearly by ordinate scale expansion.

observation of a band which is strong in the Raman but otherwise too weak for satisfactory measurement in the infrared.

Efficient purging of the spectrometer and sample chamber is essential to obtain good spectra at high scale expansion. The air dryers supplied with the instrument (alumina towers with alternating heating cycles) are satisfactory in dry weather but completely inadequate at high relative humidity. Under the latter condition valves to a 100-liter liquid nitrogen tank are opened, and a strong auxiliary purge of dry nitrogen is maintained during spectroscopic measurements.

Frequency calibration of the spectrometer is checked with water vapor bands.

The polynucleotide absorption bands are broad in aqueous solution (e.g., $\Delta\nu_{1/2}$ 15–30 cm^{-1}), and the spectrometer has adequate energy with slits which are narrow in comparison to band width. Distortion of band shape by finite spectral slit width (*14*) observed with narrower bands and earlier designs of spectrometers has not been a problem with these measurements.

Quantitative Intensity Measurements

The great majority of infrared studies have been primarily concerned with frequencies of the bands, and have usually presented only qualitative or relative indications of band intensities. This customary lack of emphasis as compared with ultraviolet spectroscopy has probably been due both to the greater technical difficulty of making intensity measurements in the infrared and to the fact that the greater detail of vibrational spectra frequently made it possible to obtain chemical and structural information that could be obtained from relatively undifferentiated electronic spectra, if at all, only with the aid of accurate intensity measurements. With the polynucleotides, despite the larger number and greater resolution of infrared bands compared to ultraviolet spectra, quantitative infrared intensity measurements are not merely desirable but are often essential, since one of the most striking features of the spectroscopic changes occurring upon helix formation are the very large changes in intensity of some of the bands (*5–10*). When the intensities are known on an absolute basis, it becomes possible to prepare reliable summation and difference spectra, to compare experimental spectra with hypothetical ones synthesized from possible components, and to analyze unknown spectra in terms of previously determined catalog spectra (cf. following sections). Although vibrational band frequency is independent of intensity, in regions of band overlap changes of intensities alone can cause apparent shifts in frequency.

The problems encountered in making absolute infrared intensity measurements in organic solvents have been discussed in a review by Jones et al. (*14*). The same problems exist for measurements in aqueous solution, with the added difficulties of a more strongly absorbing solvent and of finding window materials which are both insoluble in water and transparent to infrared light. The short light path employed makes it particularly important that the cells be dimensionally stable and the window material inert. Accurate compensation with a reference cell is essential and requires that the cell be equal in pathlength, in solvent composition, and in temperature. We have found cells of CaF_2 and BaF_2 to be satisfactory for most purposes. The windows are hard, insoluble, and inert. They are, however, also opaque in some regions of interest (especially CaF_2), rather expensive, and somewhat brittle.

The spectra are measured on an absorbance basis (this is not essential if digital scaling is to be done later but is more convenient than percent transmission for routine use) and can either be used in this way or scaled to a molar absorptivity (or extinction coefficient) basis:

$$\epsilon_\nu = \frac{\log (I_0/I)_\nu}{ClE}$$

where $\log (I_0/I)$ is the net absorbance after subtraction of base line absorbance, C is the concentration in moles per liter, l the pathlength in centimeters, and E is the scale expansion factor. The normalizing is best carried out by a computer. The digital conversion can be done by an X-Y reader from the original chart or, preferably, simultaneously with the measurement by means of an analog to digital converter. We have previously used a converter producing punched paper tape but have recently changed to a system which transmits data directly to a central computer designed to serve a number of instruments in the same building. Normalized spectra expressed as ϵ vs. ν are shown in Figs. 1, 3, and 4.

Interpretation of Spectra

We shall consider in this and the following sections how infrared spectra may be used to determine qualitatively which materials exist in a solution and quantitatively how much of each is present. One of the great advantages of infrared spectroscopy is that it is often able to answer both questions from a single experiment. In addition structural information not available from other methods can often be obtained by the methods discussed here.

Before examining the spectrum of an actual or presumed polynucleotide complex it is important to be familiar with spectra of the individual components and if possible with their temperature dependences.

The spectra of the different polynucleotides (cf. Fig. 1) are highly characteristic, and it is often possible to tell by inspection which components are present and whether or not they are present as complexes. The bands above ~1650 cm^{-1} are predominantly carbonyl stretching vibrations and upon interaction usually undergo large changes of frequency and moderate changes of intensity. In some cases it may be difficult or impossible (see below) to assign absorption in this region to individual bases merely by inspection. In the region below ~1640 cm^{-1} we see the ring vibrations of the different bases with strong absorptions of A, G, and C well-resolved from each other (Fig. 1). These bands undergo some frequency shifts upon interaction, generally smaller than those of the carbonyl bands, but very large changes of intensity (cf. Figs. 2, 5, and 6). More detailed observations of the weaker ring vibrations can be accomplished with same solution by means of ordinate scale expansion, as illustrated in Fig. 2.

A spectrum of a solution of unknown composition should be examined for resemblances to spectra of known helical complexes. Fig. 3 is a spectrum of

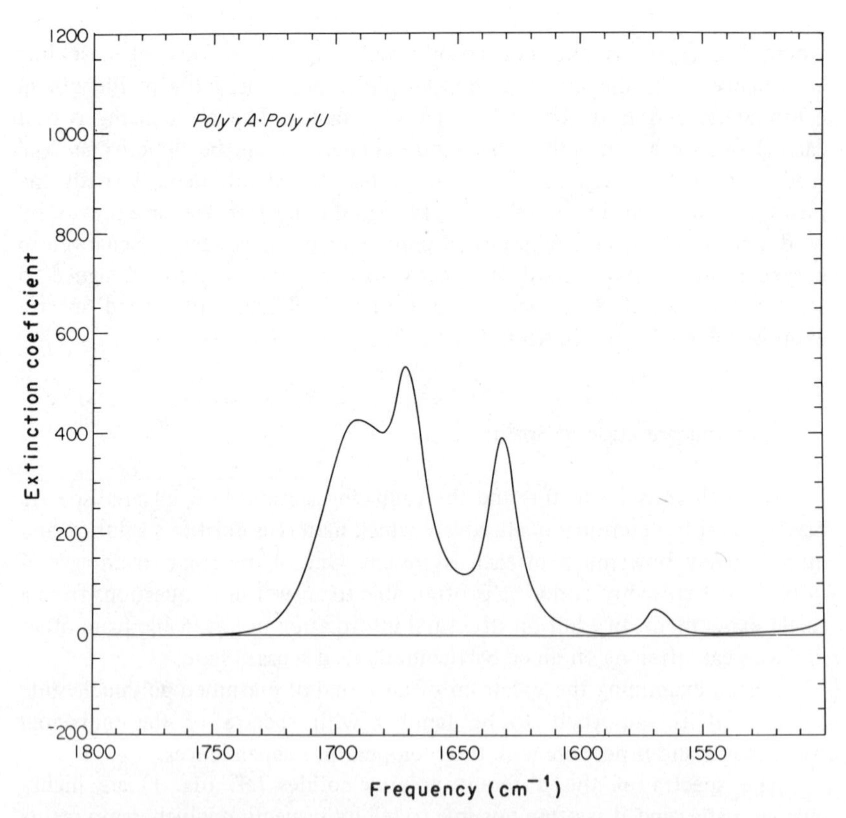

Figure 3. Standard catalog spectrum of poly rA•rU. Molar concentration is expressed in terms of total concentration of repeating units (based on data of Ref. 7).

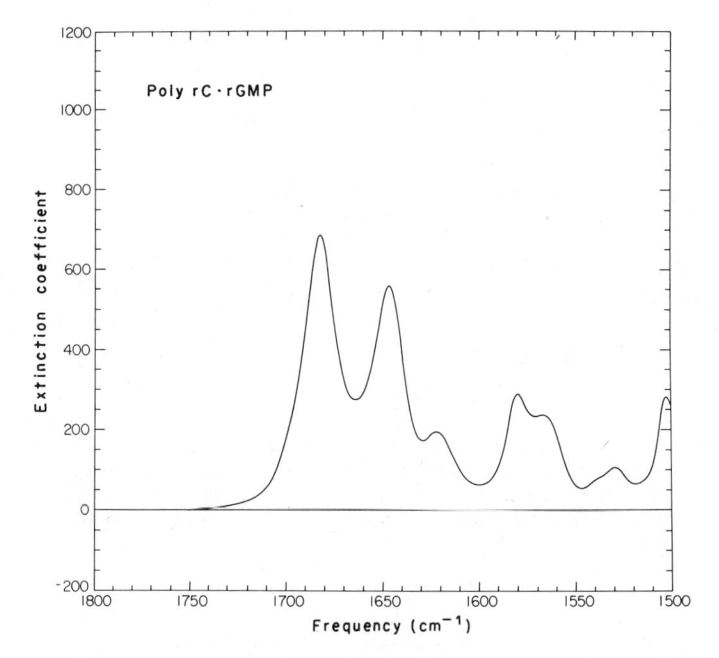

Figure 4. Catalog spectrum of two stranded helix formed between poly rC and 5′-GMP. Molar concentration is on the same basis as Fig. 3 (based on data of Ref. *10*).

poly rA·poly rU (*7*) and Fig. 4 that of a complex formed between poly C and 5′-GMP (*10*). These spectra have been normalized to a molar absorptivity basis by computer as described below. Close resemblance of the spectrum of a hypothetical complex to that of a known helix can provide strong evidence of close similarity of structure. Thus the monomer–polymer G·C spectrum (Fig. 5) shows essentially the same frequencies, absolute intensities, relative intensities, frequency shifts and intensity changes in seven vibrational bands as are observed for the polymer–polymer complex, poly rG·poly rC (*10*). We conclude that guanylic acid and poly C have formed a helical structure closely similar in the disposition of the bases to that of poly rG·poly rC. The possibility of accidental coincidence of spectral behavior is obviously remote when as many as seven bands are observed, but even with simpler spectra there are independent tests which can be applied to confirm an initial conclusion based on similarity of spectra (Ref. *15* and following sections).

Temperature Dependence of Unassociated Polymers

The spectra of most of the random coil polynucleotides change relatively little between 0° and 100°C. Where there are changes and where analyses are to be done at different temperatures it is desirable to measure several catalog

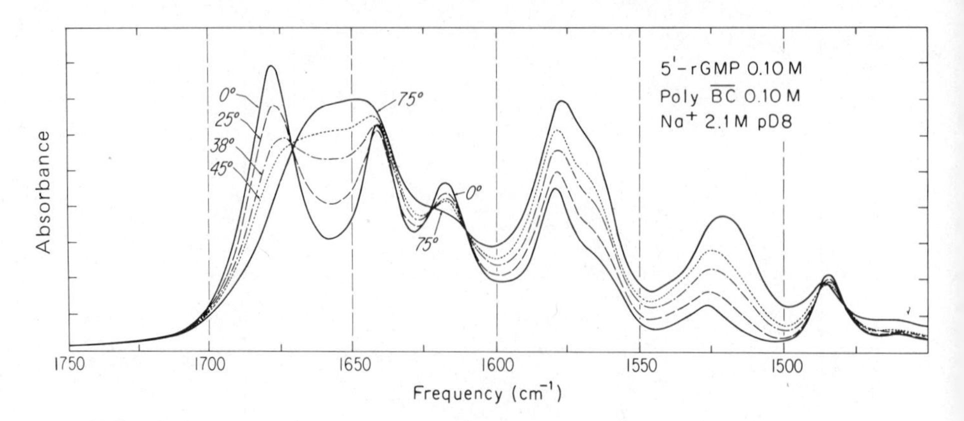

Figure 5. Infrared spectra observed during melting of the two-stranded helix formed between poly BrC and 5′-GMP. The isosbestic points indicate that only two species contribute significant absorption at these points and imply that all base residues are present only in a single helical structure or in nonordered structures over the melting range (based on data of Ref. *10*).

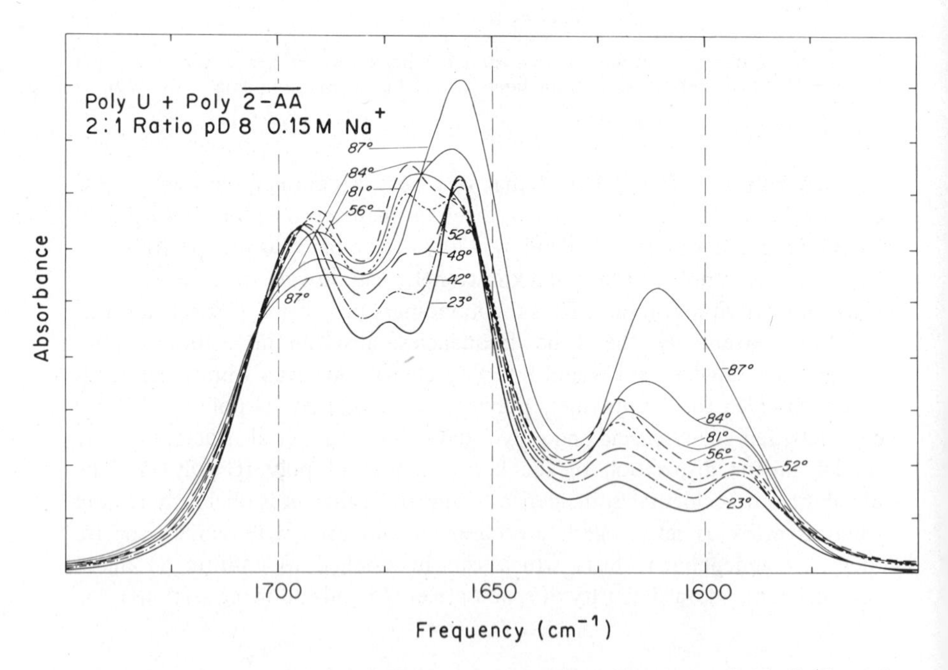

Figure 6. Spectra observed during two-step melting of three-stranded helix formed between poly U and poly 2-aminoadenylic acid. The spectra have an isosbestic point in the higher temperature range at 1669 cm^{-1}, reflecting the conversion of a two stranded helix to random coils. The apparent isosbestic at ~1659 cm^{-1} results from a fortuitous near-coincidence of extinction coefficients. (See text. Based on data from Ref. *18*).

spectra at appropriate temperatures for the same polymer. We have, for example, observed the following values (*16*) of ν_{max} (ϵ_{max}) as a function of temperature for poly U: 34°, 1692.5 (710), 1656.5 (1230), 1618.5 (310); 52°, 1691 (725), 1657.5 (1223), 1619.5 (300); 62°, 1690.5 (710), 1657.5 (1195), 1620 (285); 79°, 1690 (724), 1658 (1185), 1620.5 (285). The most intense band of neutral poly A shows a significant change of frequency but not of intensity as the temperature increases: 15°, 1631 (1085); 24°, 1629.5 (1078); 34°, 1628 (1060); 50°, 1626.5 (1050); 58°, 1626 (1060); 70°, 1624.5 (1070); 78°, 1624 (1090); 90°, 1622.5 (1090). The poly A frequency decrease presumably results from unstacking of the adenine residues and is sufficiently large that it should be taken into account in selecting poly A catalog spectra to be used in analyses of unknown spectra. The temperature dependence of each polymer may be different and should be determined in each case. Poly BrC has one band which shows a linear frequency decrease with increasing temperature, (*17*) and poly $2NH_2A$ (*18*) and poly 2 $NH_2 6NMeA$ (*15*) both show large temperature dependences of their spectra.

Thermal Dissociation – Infrared Melting Curves

In contrast to the nonspecific and generally small spectral changes discussed above are the large changes resulting from dissociation of polynucleotide helices (*7–10*). The usual pattern of change on dissociation is one of relatively large frequency shifts in the carbonyl bands above ~1640 cm^{-1} and quite large intensity increases in the ring vibrations between ~1640 and 1500 cm^{-1}, with smaller frequency shifts in the latter bands. (See Figs. 1–6).

A plot of temperature vs. infrared absorbance of a polynucleotide helix results, as do a number of other physical measurements, in a sigmoid curve reflecting a cooperative transition. Infrared spectroscopy has the major advantage, however, of permitting the selective monitoring at different frequencies of the effect of the structural transition upon the individual bases. Infrared melting curves can be used for physical studies such as dependence of Tm on cation concentration (*8*), on nucleoside concentration (*8*), or on chain length (see below). Although ultraviolet measurements are often more rapid and convenient, especially for multiple observations of a transition whose qualitative nature is already known, a probable exception is the observation of concentration-dependent phenomena such as monomer–polymer or oligomer–oligomer helix formation. For these the necessary high concentrations fall in the most convenient range for infrared measurement. We see in Fig. 7, for example, infrared melting curves of mixtures of oligo U and oligo A of total concentration 0.12 M, followed at 1625 cm^{-1} (adenine vibration). The increase in Tm with chain length is readily observed, and the same values are obtained with measurements at 1657 cm^{-1} (uridine vibration). The qualitative nature of these complexes can be shown to be

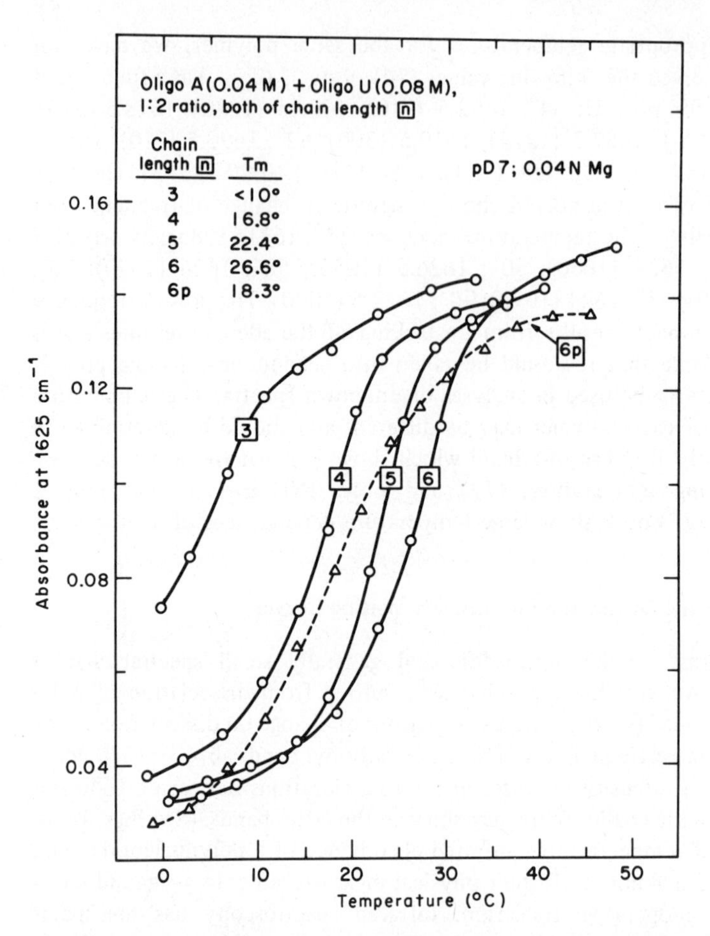

Figure 7. Infrared melting curves of complexes formed between oligo A and oligo U of varying chain length. A melting curve parallel to each of these is obtained by plotting the uridine band at 1657 cm^{-1} during the same run. The interaction of the two trimers is incomplete. The spectra of all other mixtures at low temperature are identical with that shown at 2° in Fig. 8.

three-stranded helices of 2U to 1A stoichiometry from their infrared spectra, all of which (with the exception of the $(A)_3$, $(U)_3$, which is incomplete) are identical with that of $(A)_8 \cdot 2\,(U)_8$ complex shown in Fig. 8. The observed spectra are essentially the same as that of poly rA · 2 poly rU and quite different from that of poly rA · poly rU, the 1 : 1 complex (7). From the pattern of spectral changes during thermal dissociation it is also possible to conclude that no significant amount of two-stranded helix is formed during the dissociation (see below).

Monitoring a thermal transition simultaneously by means of a very weak infrared band and a moderately strong one is shown in Figs. 2 and 9. Both the ~1625 cm^{-1} band and the ~1575 cm^{-1} bands are adenine ring vibrations, the former probably with major C=N and the latter with major C=C stretching character. Although the ~1575 cm^{-1} band is very weak, use of ordinate scale expansion provides a clear indication of its temperature dependence (Fig. 2) and a melting curve of excellent quality (Fig. 9). The two bands shift in opposite directions on dissociation, and melting curves can be obtained by plotting frequency as well as absorbance vs. temperature (Fig. 9). Both bands appear broader at mid transition (58°) because of approximately equal contributions from vibrations of helical and dissociated adenine residues, differing in frequency by about 7 cm^{-1}.

A different method of obtaining melting curves is based upon computer analysis (see below) of the complete spectra measured at different temperatures during the thermal transition. In this way the melting curve obtained is based upon actual composition rather than upon fractional change in a measured property (*7*). A further advantage of the method is that it can be applied in cases in which extensive overlap of bands makes interpretation of single frequency melting curves more difficult. Limitations of the method are the necessity of having catalog spectra for all components present during the transition (see below), and the greater amount of time required to obtain melting curves in this way.

Demonstration of Base Pairing Specificity

The spectrum of a helical complex is essentially independent of temperature in the region of helix stability, and the large changes that occur in the transition region are due to the primary effect of temperature upon secondary structure. This fact combined with the resolution of bands of the different bases provides us with the means of monitoring simultaneously and independently the participation of the different bases in an ordered structure. This possibility in turn permits an unambiguous demonstration of base pairing specificity. This method has been used with spectra which were "normal" in their resemblance to appropriate reference spectra (*7, 8, 17*) but its value is perhaps more clearly illustrated with a spectrum (Fig. 10) which does not resemble any known reference spectrum.

The infrared spectrum of a 1 : 1 mixture of poly U and the nucleoside 2-amino-6-*N*-methyladenosine (*15*) has marked temperature dependence (Fig. 10), strongly suggesting the presence of one or more ordered structures. Poly U is known to form an ordered helix (*19*), but the Tm is well below the range in which the uridine bands are still undergoing changes, and the spectrum

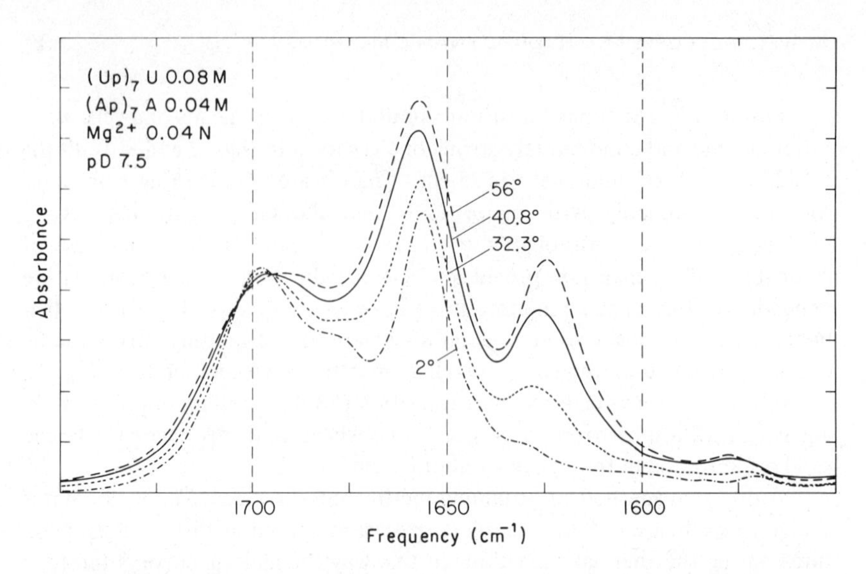

Figure 8. Infrared spectrum observed at 2° for the complex formed between $(Up)_7U$ and $(Ap)_7A$ is the same as that of poly A·2 poly U (Ref. 7) and indicates that the complex has AU_2 stoichiometry. The isosbestic points at 1703 cm^{-1} and 1690 cm^{-1} and the contrast of these spectra with those in Fig. 6 indicate that the uridine residues are present only as components of three-stranded helices or as unassociated oligomers. No significant amount of two-stranded material is formed during melting. The same spectroscopic evidence leads to the same conclusion about stoichiometry for all of the complexes whose melting curves are shown in Fig. 7.

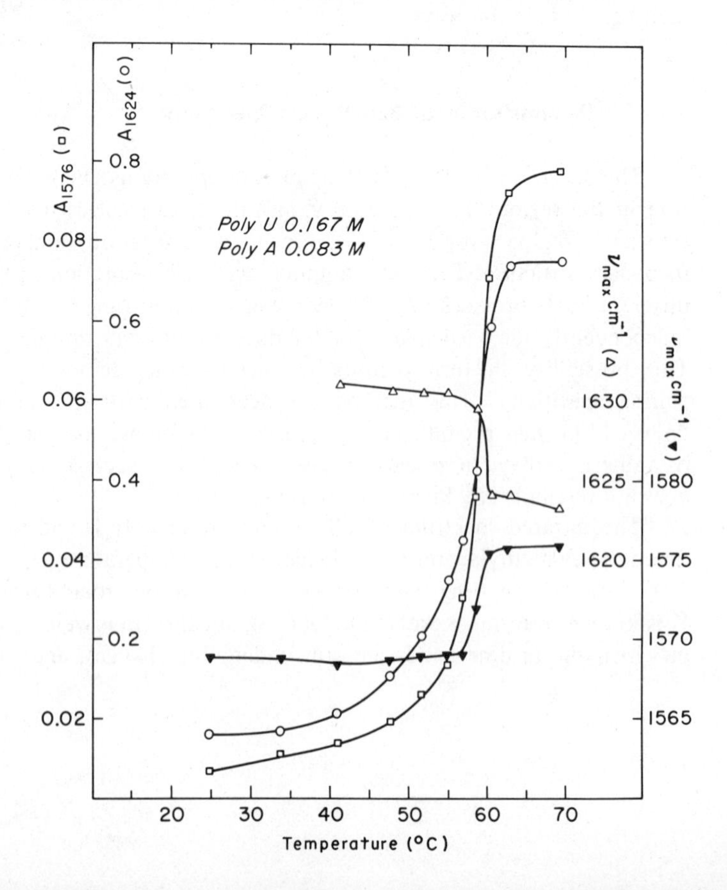

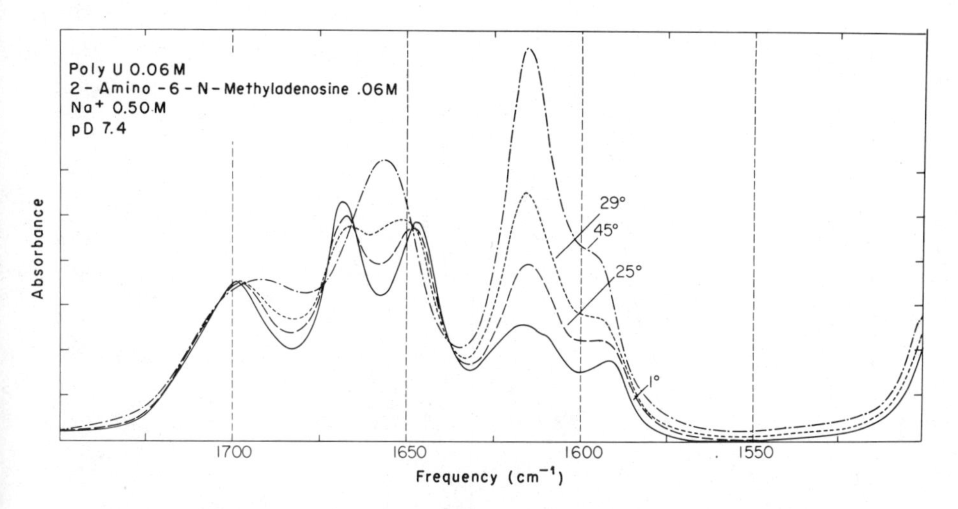

Figure 10. Infrared spectrum of 1:1 complex formed between 2-amino-6-*N*-methyladenosine and poly U. The spectrum differs from those of other two-stranded helices containing poly U in having a third strong band at 1648 cm^{-1} which the others lack. A possible explanation is given in the text. [From K. Ikeda, J. Frazier, and H. T. Miles. 1970. *J. Mol. Biol.*, **54**: 59, with permission of the copyright owner, Academic Press.]

above 1650 cm^{-1} does not resemble that of the ordered form of poly U (*8, 16*). The spectrum of the nucleoside (ν_{max} 1618.5, 1600, 1500, no bands above 1620 cm^{-1}) does not undergo intensity changes with temperature, although there are small frequency shifts (*15*). By selecting a frequency at which only the pyrimidine absorbs (1657 cm^{-1}) and one at which only the purine absorbs (1615 cm^{-1}) we can monitor the presence of both in a hypothetical ordered structure. The infrared temperature profiles (Fig. 11) are indeed sharp sigmoid curves, characteristic of cooperative transitions, but the point of particular interest here is that the purine and pyrimidine curves are completely parallel. In other experiments the counterion concentration and the reactant concentrations were varied, with resulting Tm changes, but the two melting curves remained congruent. The observed spectral changes must therefore result from the specific interaction of the two bases with each other. This conclusion does not explain the anomalous appearance of the spectrum but provides a necessary foundation for possible explanations considered in a later section.

←

Figure 9. Infrared melting curves of poly A•2 poly U taken from the spectra in Fig. 2. The intense (~1629 cm^{-1}) and weak (1576 cm^{-1}) bands (adenine ring vibrations) give parallel melting curves. The frequencies of the band maxima also reflect the melting process. There is a further but much slower change in ν_{max} of the 1625 cm^{-1} band as the temperature is increased above 60°.

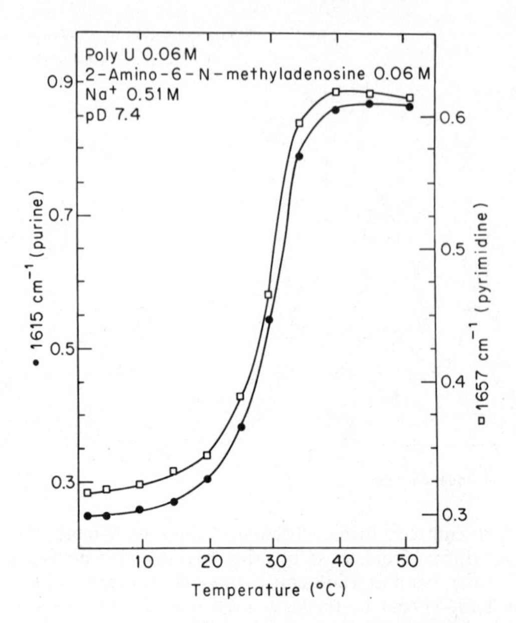

Figure 11. Infrared melting curves from the spectra in Fig. 10. The curves measured at frequencies at which only the pyrimidine (1657 cm^{-1}) or only the purine (1615 cm^{-1}) absorb significantly are parallel, indicating that the spectral changes shown in Fig. 10 result from a specific interaction of the two bases. Similar pairs of curves having different values of Tm are obtained at different concentrations of reactants (Ref. *15*). [From K. Ikeda, J. Frazier, and H. T. Miles. 1970. *J. Mol. Biol.*, 54: 59, with permission of the copyright owner, Academic Press.]

Isosbestic Points

When the spectra of a random coil polynucleotide and a helical complex with which it is in equilibrium intersect, all spectra generated by melting of the complex should pass through the intersection point (isosbestic), provided certain conditions are satisfied (*17, 20*). The presence of isosbestic points can provide valuable evidence on the number of structures in which a polynucleotide exists during thermal dissociation of a helix, as well as providing checks on self-consistency and interpretation of data. The relevant conditions for existence of isosbestic points are that the total concentration of reactants should be constant and that the spectra of the random coil polymers should be essentially independent of temperature in the region being studied. The first condition is usually satisfied, although the latter frequently is not. The existence of an isosbestic point can usually be taken as evidence that only two species have significant absorption at that frequency. In some circumstances, however, more than two may be present. The absence of an isosbestic point, on the other hand, may indicate the need for further investigation. It often indicates that a dissociation process is more complex than anticipated or that one has not employed proper conditions for production of a single complex.

These principles are illustrated by considering a G·BrC helix and two similar AU_2 helices whose spectra show sharply contrasting patterns on thermal dissociation.

Figure 5 shows the spectrum of a two-stranded helix formed by interaction of 5′-GMP with poly BrC (*10*). There is an isosbestic point at 1672 cm^{-1}, where the helix and GMP have equal absorbance and poly C makes a negligible contribution. If more than one ordered structure were present (for example, G_2C, C_2G, or GG) we would not expect to observe this isosbestic point. Similarly from the isosbestic points at 1622 cm^{-1} and 1610 cm^{-1} we conclude that the BrC residues exist in either random coil or a single regular ordered structure.

The spectrum of a three-stranded helix formed between $(Ap)_7A$ and $(Up)_7U$ is characteristic of most AU_2 helices (Fig. 8). All spectra generated by thermal dissociation pass through isosbestic points at 1703 cm^{-1} and 1690 cm^{-1}, a region in which only the uracil residues absorb. We conclude that these residues exist in only two environments during dissociation: AU_2 complex and nonordered oligo U. This conclusion is strongly reinforced by the following contrasting result.

The three-stranded complex formed between poly U and poly $2NH_2A$ (*18*) has a spectrum (Fig. 6) very similar to that of poly A·2 poly U (*7*) except that the band at ~1675 cm^{-1} has a slightly lower frequency and higher relative intensity. The 2-NH_2 group forms a third hydrogen bond to poly U and leads to a wider separation in stability of the two- and the three-stranded helices than exists in the poly A, poly U system. In the temperature range above 40° the spectrum changes until at ~56° it becomes identical with the summation of the spectra of poly $2NH_2A$·poly U and random coil poly U, each measured separately. The process occurring in this temperature range can thus be represented as

$$\text{poly } 2NH_2A \cdot 2 \text{ poly U} = \text{poly } 2NH_2A \cdot \text{poly U} + \text{poly U}$$

All spectra pass through or very close to the same point at 1659 cm^{-1} since the starting material ($\epsilon_{1659} = 930$) leads to equivalent amounts of two products having an average value of ϵ_{1659} (~820) approximately equal to that of poly $2NH_2A$·2 poly U. If there were an exact equivalence of ϵ then all curves would pass through precisely the same point, although three components absorb at that frequency. In a higher temperature range (~80°–90°) the same system shows a further series of changes and has a true isosbestic point at 1669 cm^{-1}, reflecting in this case the presence of uracil residues in only two environments, the two-stranded helix and a random coil. The spectrum of poly $2NH_2A$ has a marked temperature dependence (*18*), and there are no isosbestic points in the region of predominant purine absorption.

Tautomeric Structures of Bases

The dissociable hydrogen atoms of the bases G, U, T, and I may, in principle, be attached either to nitrogen (the keto form with double-bonded exocyclic oxygen) or to oxygen (the enol form), with quite different consequences for base pairing specificity. Similarly A and C may exist in either amino (exocyclic NH_2) or imino form (exocyclic >C=NH). Infrared spectroscopy of alkylated model compounds has been employed to establish the tautomeric forms of the monomeric bases in aqueous solution (*5, 6, 21*). Each tautomeric structure should be represented by an appropriate model compound which is isoelectronic with it. Although the comparison of spectra is empirical, it is generally unambiguous. The degree of empiricism can often be greatly reduced by obtaining independent evidence for band assignments from isotopic substitution experiments. The nucleotides for which keto structures can be written have high frequency ($<\sim$1650 cm^{-1}) bands corresponding to those of the appropriate keto models. Specific isotopic labeling with ^{18}O (*22*) has confirmed both the keto structures and the assignment of these bands to carbonyl stretching vibrations for G and I.

The question of tautomeric structure is of the greatest importance for helical complexes formed by polynucleotides, since base pairing specificity and the possibility of forming one structure rather than another depend upon this chemical property of the bases. It does necessarily follow, however, that the tautomeric form present in a monomer or random coil polymer will be the one present in a helix. The possibility exists that there may be a sufficiently large energetic advantage in the formation of a helix to permit conversion of the bases to tautomeric forms which would be less stable in the absence of base pairing. Infrared spectroscopy is the only method which has so far been able to provide convincing demonstrations of tautomeric structures of the bases in polynucleotide helical configurations.

The demonstration of the tautomeric forms present in A•U and in I•C helices have been reported previously (*5, 6*). Evidence for keto and amino structures in G•C helices is presented below.

Guanosine and GMP have strong infrared at bands at 1665 cm^{-1}, 1578 cm^{-1}, and ~1568 cm^{-1}(sh) (*21, 22*). Specific 6-^{18}O substitution (*22*) shows the 1665 cm^{-1} band to have predominant carbonyl stretching character, a conclusion consistent only with keto tautomeric structure. Further evidence for both G and C is available from studies of appropriate model compounds. Thus the keto models for G, 1-methylguanosine and 1,9-dimethylguanine have carbonyl bands at 1674 cm^{-1} (*16*) and 1671 cm^{-1} (*21*), respectively, and ring vibrations at 1591 cm^{-1}, 1550 cm^{-1} and 1590 cm^{-1}, 1548, respectively. The enol model, however, completely lacks the carbonyl absorption but has strong bands at 1617 cm^{-1} and 1594 cm^{-1} (*21*), and the enolate anion has bands at 1591 cm^{-1} and 1576 cm^{-1} (*22*). The 1576 cm^{-1}

band of the anion, like that at ~1568 cm^{-1} in the neutral molecule shows a small frequency decrease on 6-^{18}O substitution and is thus due to a normal mode involving some motion of the carbonyl oxygen.

Cytidine and poly C have strong carbonyl bands at 1650 cm^{-1} and 1656 cm^{-1} and weaker ring vibrations at 1618 cm^{-1} and 1617 cm^{-1}, closely similar in frequency, band shape, and relative intensity to those of a dimethylamino model compound at 1649 cm^{-1} and 1625 cm^{-1} (*6*). An imino model, however, has strong bands at 1671 cm^{-1}, 1657 cm^{-1}, and 1579 cm^{-1}, indicating that the cytosine residues exist in the amino form in aqueous solution in these unassociated molecules.

In a helical G•C complex the tautomeric forms of G and C residues can also be deduced from the infrared spectrum (Fig. 4). Interbase vibrational coupling and other perturbations resulting from secondary structure formation occur in these complexes (*10*), but their effects are not large enough to obscure the essential conclusion.

The highest frequency band at 1682 cm^{-1} (Fig. 4) is predominantly a G carbonyl vibration and would be entirely absent if the G residues had an enolic structure. This band has significant C character (about 25%) as a result of coupling (*10*), but the frequency of a pure G band in the absence of G•C coupling can be calculated to be 1674 cm^{-1} and would lead to the same conclusion. Confirmatory evidence is provided by the absence of strong absorption at frequencies at which the enolate anion (1591 cm^{-1}) or an enol model compound (1617 cm^{-1} and 1594 cm^{-1}) have strong bands (*21, 22*). The anion has a band at 1576 cm^{-1}, but this is not resolved from that of neutral guanosine.

If both G and C had undergone tautomeric changes upon forming a complex the imino form of C could make three hydrogen bonds to the enolic form of G in a bonding scheme having the general geometry of a Watson-Crick pair but a different fine structure. Since the G is keto, however, it follows that the C cannot be imino if it is to form three hydrogen bonds or a Watson-Crick pair. An alternative bonding is possible with the imino form of C bonded to N_7 and C_6=O of G (*23*). A bonding scheme having similar geometry has been reported (*24*), not as a two-stranded complex but as part of a three-stranded complex. In this case, however, the proton on N_3 which the C residues must have in order to make a second hydrogen bond was achieved not by tautomeric change but by protonation of the amino form of C at an unusually high pH (*24*). Since the imino structure of C does not form in this case even when it would facilitate formation of a helical strand, it appears quite unlikely to do so in a two stranded structure. Further confirmation of this point is obtainable from Fig. 4. The G•C complex has a (predominantly) C band at 1646 cm^{-1}, near that of the amino model compound and lacks strong absorption characteristic of the imino model at 1679 cm^{-1} (see above). Although a G ring vibration has about the same frequency as a C imino model compound (1578 cm^{-1} and 1579 cm^{-1}, respectively), the presence of the latter could probably be detected by a higher absorbance at that frequency than is observed in Fig. 4.

Catalog Spectra – Summations and Analyses

Measurement of infrared spectra on an absolute intensity basis and normalizing in terms of molar absorptivity permits the preparation of a library of standard spectra which can then be used in calculations to help interpret spectral changes resulting from polynucleotide interactions (*7, 8*).

The catalog spectra of unassociated polymers (Fig. 1) and of A•U and G•C helices (Figs. 3 and 4) have been used to generate synthetic spectra in Fig. 12 and to demonstrate some characteristics of an analytical system based upon them. Each of the synthetic spectra (heavier line) was formed by summation of catalog spectra, as indicated in the legend, and was then analyzed by computer to give a least squares best fit in terms of designated catalog spectra. The lighter line (when two are visible) is the spectrum resulting from the analysis, and the error spectrum gives the difference between the original spectrum and the solution of the analysis.

Individual spectra are discussed below, but several general points may be made here. When the components in terms of which the analysis is made included all those (and only those) spectra used to generate the initial spectrum, the analysis and synthesis spectra are, as they should be, identical. This conclusion merely tests the method in a system free of experimental error. When the components for the analysis include all those used in the synthesis and in addition others not present in the synthesis, the analytical program has in all cases (several dozen tests similar to those in Fig. 12) produced the correct coefficients for all components, including the value zero for those not present in the original synthesis.

When a component present in the original synthesis is omitted from the list considered for analysis, however, a very poor fit and a large error spectrum are obtained (cf. Fig. 12 D–F). The last two conclusions indicate that for combinations relevant to experimental situations the catalog spectra are sufficiently different from each other to avoid degenerate solutions. A good fit of catalog spectra to an unknown spectrum is therefore significant, and can provide a reliable quantitative measure of the composition of complex mixtures. Applications of the method to demonstrate previously unknown interactions have been published (*7, 8*). A further important application of this kind of analysis is use of infrared mixing curves to demonstrate stoichiometry of interaction (*7, 8*).

In contrast to the foregoing result, a poor fit of an analysis to an experimental spectrum suggests that a necessary component has been omitted from the analysis or that one (or more) of the catalog spectra employed does not accurately represent the intended material in the particular structure under investigation. The latter possibility might occur with spectra of paired bases, for example, if slight differences in secondary structure resulted in significantly different frequency and intensity changes from those in the complex upon which the catalog spectrum is based (cf. discussion of interbase vibrational coupling and Fermi resonance). As noted above, errors in analyses arising from nonspecific temperature dependence of nonhelical polynucleotides are minimized by measuring a number of catalog spectra at different temperatures and selecting the appropriate one for an analysis.

In addition to providing tests of the analytical system the synthetic spectra in Fig. 12 can be used to interpret changes in band contours of the natural nucleic acids. The summation spectrum Fig. 12 (B) may be taken to represent that of a hypothetical nucleic acid with equimolar amounts of the bases, fully paired. Fig. 12 (A) represents the same components, completely unassociated. The former has band maxima at 1682 cm^{-1}, ~1672 cm^{-1} (sh.), 1646 cm^{-1}, 1632 cm^{-1}, 1578 cm^{-1}, 1568 cm^{-1}, 1528 cm^{-1}, and 1502 cm^{-1}. The 1682 cm^{-1}, band is due to the ν_1 band of G·C (predominantly $G_{C=O}$ with about 25% $C_{C=O}$ character,) which overlaps at this frequency the high minimum between the 1692 cm^{-1} and 1672 cm^{-1} bands of A·U (Figs. 3 and 4). The shoulder at ~1672 is due to an interacted uridine carbonyl vibration (*7*). The band at 1646 cm^{-1} is ν_2 of G·C (*10*), and that at 1632 cm^{-1} is an interacted A ring vibration (*7*). The barely perceptible shoulder at ~1620 is due to a C ring vibration. At 1578 cm^{-1} and 1568 cm^{-1} are G ring vibrations and at 1528 cm^{-1} and 1502 cm^{-1} C ring vibrations, all of which have undergone large intensity decreases as a result of interaction.

The summation spectrum of the nonassociated polymers (Fig. 12 (A)) has an intense peak at 1658 cm^{-1} (primarily due to uridine and cytidine residues with some contribution from the G carbonyl band of ν_{max} 1665 cm^{-1}) with a somewhat less intense one at 1628 cm^{-1} (primarily the A ring vibration with small contribution from the C ring vibration at 1620 cm^{-1}). The (predominantly) C carbonyl vibration ν_2 of G·C, has disappeared as a resolved band from the band envelope because of shift to higher frequency (1656 cm^{-1}) upon loss of vibrational coupling.

It is significant that the foregoing interpretation of the synthetic spectra can account for the gross features of the spectral changes observed in DNA by Blout and Lenormant (*12, 12a*), despite differences in chemical composition and in spectroscopic procedure. They reported that denaturation of DNA resulted in shifts of absorption maxima of the band envelope from 1680 cm^{-1} to 1660 cm^{-1} and of the 1645 cm^{-1} band to 1625 cm^{-1}. We can

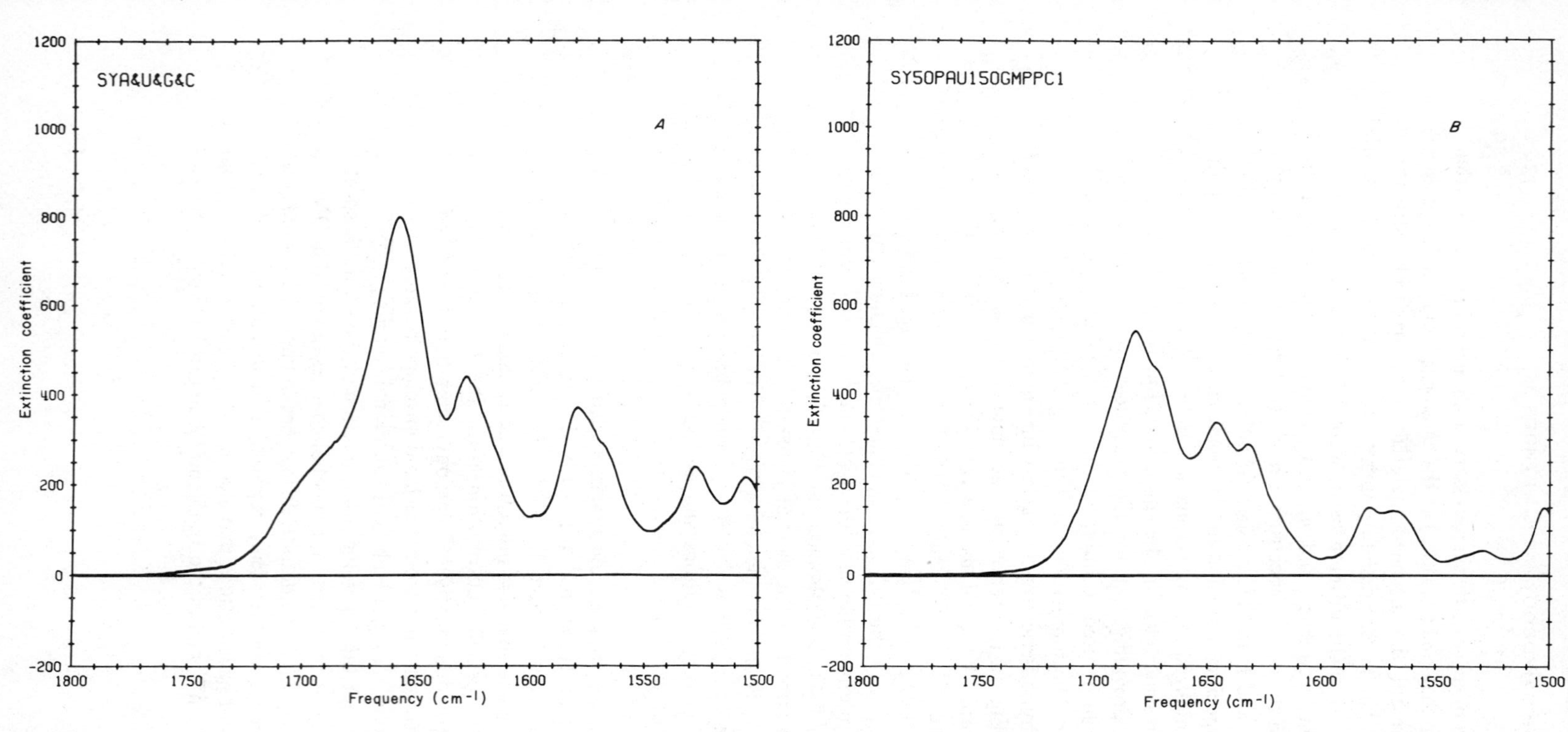

Figure 12 (*A*). Summation of spectra of poly A, poly U, poly C, and 5′-GMP (Fig. 1), each with the weight 0.25. The analysis was in terms of these components and in addition poly A•poly U and GMP•poly C. The analysis yielded coefficients of zero for the latter two and 0.25 for A, U, G, and C. The analysis curve is coincident with the synthetic spectrum. In other tests fits to the same summation curve were made in terms of the original components plus (a) poly A•poly U and (b) GMP•poly C. The latter components were found to be absent, and in both cases the analysis spectra were identical with the one in the figure. The method is thus unlikely to indicate the presence of paired bases when none exist.

Figure 12 (*B*). Summation of catalog spectra A•U (Fig. 3) and G•C (Fig. 4), each with the weight 0.5. The analysis curve in terms of these two components is coincident with the synthetic spectrum (cf. text). Fits to the same curve were also made in terms of the same components plus A, U, G, and C (as in Fig. 1). The random coil components were found to be absent, and the analysis spectrum was identical with the one in the figure.

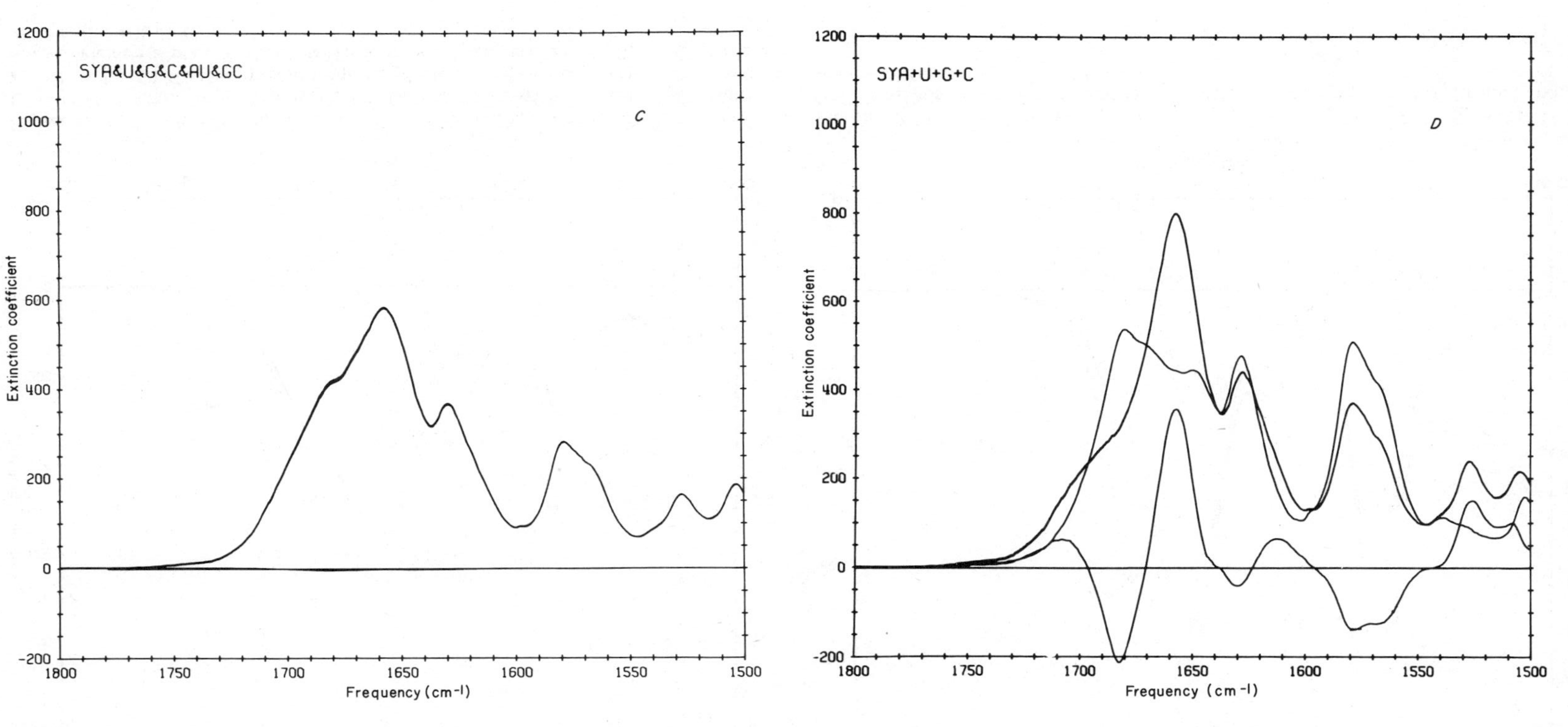

Figure 12 (*C*). Summation curve of the following spectra and coefficients: 0.15 poly A, 0.15 poly U, 0.15 poly C, 0.15 5′-GMP, 0.20 poly A•poly U, 0.20 GMP•poly C. An analysis in terms of these six components yielded the original coefficients and a curve coincident with the synthetic spectrum.

Figure 12 (*D*). Summation curve (heavy line) of poly A, poly U, poly C, and 5′-GMP, each of weight 0.25. The analysis was made in terms of poly A, 5′-GMP, poly A•poly U, and GMP•poly C, with poly U and poly C omitted. The analysis curve (lighter line) and error curve are shown, and the coefficients 0.29 poly A, 0.27 GMP, 0.17 poly A•poly U, and 0.47 GMP•poly C were obtained.

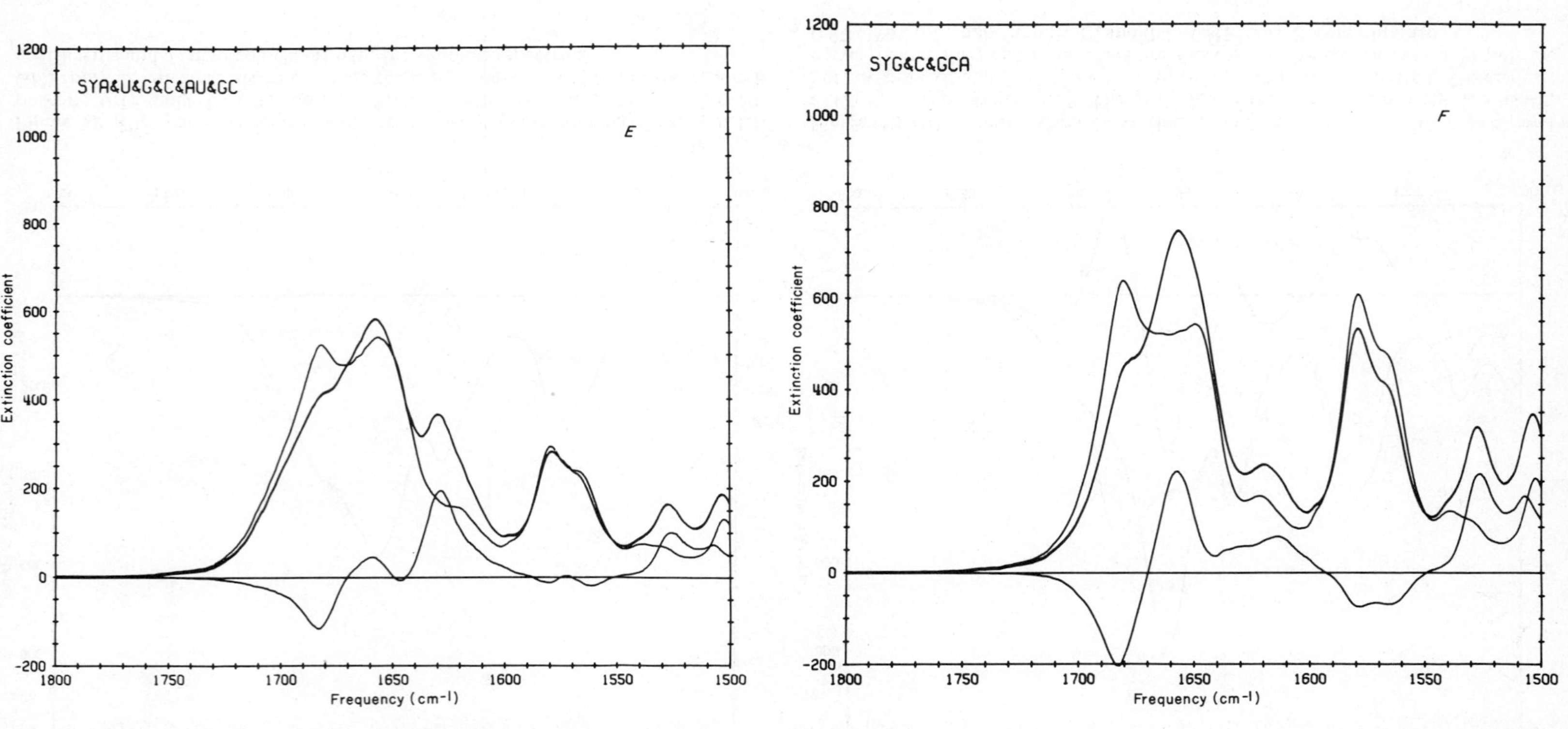

Figure 12 (*E*). Summation spectrum (heavy line) of the spectra of poly A, poly U, poly C, 5′-GMP, poly A•poly U, and GMP•poly C, with the same coefficients as in Fig. 12 (C). The analysis in terms of all these components except poly A and poly C yielded a poor fit (light line) and the coefficients 0.215 U, 0.135 G, 0.105 A•U, and 0.400 G•C.

Figure 12 (*F*). Summation spectrum (heavy line) of 0.33 G, 0.33 C, and 0.33 G•C. Analysis in terms of G and G•C, omitting C, gave a poor fit (light line) and the coefficients 0.33 G and 0.67 G•C.

now recognize from the foregoing discussion of the synthetic spectra that the latter change does not result from the shift of a single band but from the varying behavior of two different bands. The 1645 cm^{-1} band (ν_2 of G•C) has increased rather than decreased in frequency on heating and contributes to the intense absorption of the band envelope with ν_{max} ~1660 cm^{-1}, in the melted material. The shift of this strongly overlapping C band to higher frequency permits much better resolution of the adenine ring vibration at ~1625 cm^{-1}, which also becomes more prominent because of an increase in intensity on melting. The 1680 cm^{-1} →1660 cm^{-1} change in DNA can be correlated with the base pairing dependence of the carbonyl vibrations in the individual bases, as discussed above.

Interbase Vibrational Coupling

The carbonyl groups of G and of C residues in a helix are indirectly linked by the relatively weak hydrogen bonds of the base pairs. Specific ^{18}O labeling of G residues in three different G•C helices has demonstrated the existence of vibrational coupling of the G and C carbonyl vibrations (*10*). Spectra of the same G•C complex formed with normal and with 5′–GMP–6–^{18}O are shown in Fig. 13. The observation of isotopic frequency shifts in both carbonyl bands demonstrates the interstrand coupling, and the absence of isotopic shift in the lower frequency bands indicates that these vibrations are not coupled. It is possible to treat the interaction of the two highest frequency vibrations with an equation from standard first order

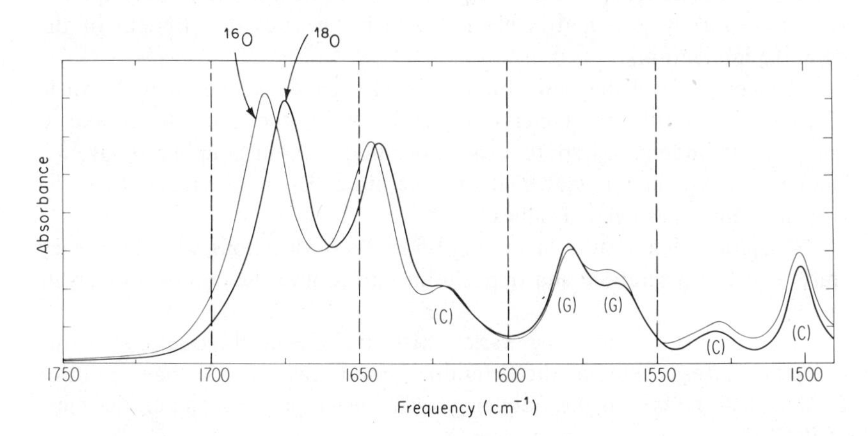

Figure 13. Infrared spectra of two-stranded helices formed between poly C and either 5′-GMP (light line) or 5′-GMP-6-^{18}O (heavy line). Isotopic frequency shifts occur in the carbonyl vibration of both G and C, though only G is isotopically labeled. The lower frequency ring vibrations do not show detectable isotopic shifts [From F. B. Howard, J. Frazier, and H. T. Miles. 1969. *Proc. Natl. Acad. Sci.,* **64**: 451, with permission of the publisher.]

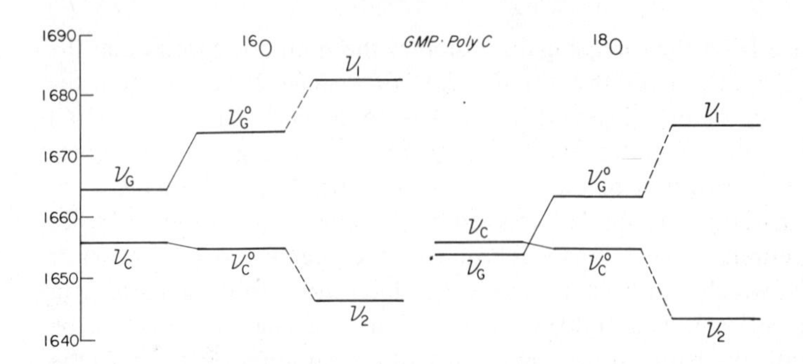

Figure 14. Energy level diagram calculated for spectra in Fig. 13. The initial G and C frequencies are considered to undergo hypothetical association shifts to the values ν_{G^0} and ν_{C^0}, which as a result of intermolecular vibrational coupling undergo shifts of equal magnitude and opposite sign to give the observed values ν_1 and ν_2.

perturbation theory. (See, e.g., Ref. *1*, Eq. II, p. 216. The same basic equation is also used to describe a different perturbation in Fermi resonance (1).) The basic equation may be reexpressed in terms of the G·C coupling as follows (*10*):

$$^k\nu_i = \tfrac{1}{2}\,(^k\nu_{G^0} + \nu_{C^0}) \pm \tfrac{1}{2}\sqrt{4W^2 + (\nu_{G^0} - \nu_{C^0})^2}$$

where ν_i is the frequency of the higher frequency coupled band when $i = 1$ and of the lower when $i = 2$; $^k\nu_{G^0}$ is the frequency of the hypothetical associated but uncoupled G vibration, and k is 16 or 18 to denote isotopic species of G residue; ν_{C^0} is the frequency of the associated but uncoupled C vibration, and W is a matrix element which measures the strength of the coupling interaction.

An energy level diagram obtained from such a calculation is shown in Fig. 14. The initial frequencies ν_G and ν_C of the separate G and C components undergo a hypothetical association without coupling to give the frequencies ν_{G^0} and ν_{C^0}, which may be either larger or smaller than the corresponding unassociated values.

Coupling then results in mixing of the two vibrations with frequency shifts equal in magnitude and opposite in sign to give the observed values ν_1 and ν_2.

The extent of coupling varies with the chemical structure of the components (e.g., whether the C residues are unsubstituted or have 5–Br or 5–Me substituents), on the frequency difference $\nu_{G^0} - \nu_{C^0}$ and on the value of W.

The elements which contribute to the hypothetical association shifts (i.e., the changes from ν_G or ν_C to ν_{G^0} or ν_{C^0} include hydrogen bonding (to solvent and to complementary base), change of dielectric constant in the helix, dipole interactions, specific and nonspecific solvent effects, and other factors (*10*).

Fermi Resonance

The anomalous A · U spectrum shown in Fig. 10 leads us to consider an intramolecular vibrational interaction in which a first overtone or combination band perturbs a fundamental having nearly the same frequency. The interaction, Fermi resonance, results in shifts in opposite directions of the two levels and in mixing of eigenfunctions of the two states (*1*). We are not concerned here with the occurrence of resonance in the unassociated molecules but rather with its occurrence as a secondary consequence of helix formation.

The spectrum in question (Fig. 10) is anomalous principally in the presence of a third band (1648 cm^{-1}) in the carbonyl region, where other two-stranded A·U helices have only two (cf. Fig. 3). The anomalous band is not merely due to the unusual purine base since the identical base, when present in a polynucleotide chain, forms a complex with poly U which has an infrared spectrum (ν_{max} 1689.5 cm^{-1}, 1668 cm^{-1}, 1616.5 cm^{-1}, and 1605 cm^{-1}, the first two being pyrimidine bands and the latter two purine bands) very similar to that of poly A·poly U (*15*).

Explanations based on a "two environment" structure for the uracil residues and on interbase vibrational coupling were considered but appeared to be unlikely (*15*). The dependence of Fermi resonance upon polynucleotide secondary structure can be rationalized in terms of spectroscopic behavior observed in certain small molecules. A number of lactones and ketones exhibit anomalous carbonyl splittings, attributed to Fermi resonance, which show marked solvent dependence (*25*). The effect of solvent in the latter cases and of secondary structure in the former would both occur through the influence of these variables on the value of δ, the frequency separation of the unperturbed vibrational levels. The functional dependence of the perturbed frequency, E, upon δ and W_{ni} (a matrix element expressing the magnitude of the perturbation) is given by the equation (Ref. *1*, p. 216):

$$E = \bar{E}_{ni} \pm \tfrac{1}{2}\sqrt{4|W_{ni}|^2 + \delta^2}$$

where $\bar{E}_{ni}$ is the mean of the unperturbed values.

The essential point of the present proposal is that small changes in detailed molecular environment may reduce separation of two frequencies enough to bring them into resonance and thus give rise to an "anomalous" spectrum as in Fig. 6. Unlike the intermolecular coupling discussed above, however, a dependence of Fermi resonance upon polynucleotide secondary structure has not been demonstrated.

References

1. Herzberg, G. Infrared and Raman spectra. 1945. Van Nostrand, Princeton, N.J.
2. Jones, R. N. and C. Sandorfy. 1956. In W. West (Ed.), Chemical applications of spectroscopy. W. Tey, New York.
3. Bellamy, L. J. 1958. The infrared spectra of complex molecules, 2nd ed. Methuen, London.
4. Sutherland, G. B. B. M. and M. Tsuboi. 1957. *Proc. Roy. Soc., Ser. A.*, **239**: 446; Bradbury, E. M., W. C. Price, and G. R. Wilkinson. 1961. *J. Mol. Biol.*, **3**: 301; Higuchi, S., M. Tsuboi, and Y. Iitaka. 1969. *Biopolymers*, **7**: 909.
5. Miles, H. T. 1958. *Biochim. Biophys. Acta*, **30**: 324.
6. Miles, H. T. 1961. Proc. Natl. Acad. Sci., *U.S.*, **47**: 791.
7. Miles, H. T. and J. Frazier. 1964. *Biochem. Biophys. Res. Commun.*, **14**: 21, 129.
8. Howard, F. B., J. Frazier, M. F. Singer, and H. T. Miles. 1966. *J. Mol. Biol.*, **16**: 415.
9. Miles, H. T. 1968. In S. Colowick and N. Kaplan (Eds.)., Methods in enzymology; Nucleic acids L. Grossman and K. Moldave, (Eds.). Academic Press, New York, Vol XIIB, p. 256.
10. Howard, F. B., J. Frazier, and H. T. Miles. 1969. *Proc. Natl. Acad. Sci., U.S.*, **64**: 451.
11. Gore, R. C., R. B. Barnes, and E. Petersen. 1949. *Anal. Chem.*, **21**: 382.
12. Blout, E. R. and H. Lenormant, Biochim. Biophys. Acta, **17**, 325 (1955).

12a. Blout, E. R., 1957. *Ann. N. Y. Acad. Sci.*, **69**: 84.

13. Jencks, W. P. 1963. In S. P. Colowick and N. O. Kaplan (Eds.), Methods in enzymology.) Academic Press, New York, Vol. XI, p. 914.
14. Jones, R. N., E. Augdahl, A. Nickon, G. Roberts, and D. J. Whittingham. 1957. *Ann. N. Y. Acad. Sci.*, **69**: 38.
15. Ikeda, K., J. Frazier, and H. T. Miles. 1970. *J. Mol. Biol.*, **54**: 59.
16. Miles, H. T. and J. Frazier, unpublished experiments.
17. Howard, F. B., J. Frazier, and H. T. Miles. 1969. *J. Biol Chem.*, **244**: 1291.
18. Howard, F. B., J. Frazier, and H. T. Miles. 1965. *J. Biol Chem.*, **240**: 801; *ibid.*, 1967. *Proc. Intern. Congr. Biochem., 7th, Tokyo.*
19. Lipsett, M. 1960. *Proc. Natl. Acad. Sci., U.S.*, **46**; 445.
20. For general discussion of isosbestic points, see, for example, H. H. Jaffé and M. Orchin, 1962. Theory and applications of ultraviolet spectroscopy. Wiley, New York, p. 561; M. D. Cohen and E. Fischer. 1962. *J. Chem. Soc.*, p. 3044.
21. Miles, H. T., F. B. Howard, and J. Frazier. 1963. *Science*, **142**, 1458.
22. Howard, F. B. and H. T. Miles. 1965. *J. Biol. Chem.*, **240**: 801.
23. Langridge, R. and A. Rich. 1960. *Acta Cryst.*, **13**: 1052.
24. Howard, F. B., J. Frazier, M. N. Lipsett, and H. T. Miles. 1964. *Biochem. Biophys. Res. Commun.*, **17**: 93.
25. Angell, C. L. 1956. *Trans. Faraday Soc.*, **52**: 1178; Bellamy, L. J. and R. L. Williams. 1959. *Trans. Faraday Soc.*, **55**: 14; Angell, C. L., P. J. Krueger, R. Lauzon, L. C. Leitch, K. Noack, R. J. D. Smith, and R. N. Jones. 1959. *Spectrochim. Acta*, **11**: 926.

ANALYSIS OF TEMPERATURE-DEPENDENT ABSORPTION SPECTRA OF NUCLEIC ACIDS

GARY FELSENFELD

LABORATORY OF MOLECULAR BIOLOGY

NATIONAL INSTITUTE OF ARTHRITIS AND METABOLIC DISEASES

NATIONAL INSTITUTES OF HEALTH

BETHESDA, MARYLAND

The ultraviolet absorption spectrum of nucleic acids in the range 2200 to 3000 Å is due principally to the absorption of the constituent bases. The spectrum is modified from that of the individual bases because of interactions between them which inevitably occur in the polymer, whether it is a single strand, largely disordered structure, or a multistrand structure with long-range order such as DNA. Modifications in the spectrum provide a useful tool for estimating the nature and extent of ordering in a polynucleotide.

The absorption bands (in the neighborhood of 2600 Å) of an ordered polynucleotide such as DNA are markedly hypochromic relative to the corresponding bands of its constituent nucleotides. (One may speak equally well of the hyperchromism of nucleotides relative to DNA.) Hypochromism is the decrease in the integrated intensity of a band, which is in turn proportional to the oscillator strength of the transition from which it arises. A detailed quantum-theoretical explanation of the origin of hypochromism has been given (*1–3*), but DeVoe (*4*) has shown that an equivalent and quite simple analysis can be given in terms of classical electromagnetic theory. From either point of view, hypochromism is a "local field" phenomenon; the field experienced by a central chromophore being irradiated at one of its absorbing frequencies is modified by the presence of other neighboring chromophores, and this in turn affects the amount of energy absorbed by the central chromophore.

The theory of hypochromism is not sufficiently complete to permit the precise deduction of secondary structure from a polynucleotide absorption spectrum, although some success has been achieved in such simple cases (*5*) as that of the dimer ApA. On the other hand, the phenomenon of hypochromism and the shape of the absorption spectrum have been exploited in a number of ways as analytical tools to give the concentration and base composition of unknown samples of native DNA, and estimates of the amounts of internal base pairing and base stacking in samples of denatured DNA and single-stranded RNA. Used simply as a measure of the extent of ordering of a polynucleotide structure, spectral properties have also proven valuable in studies of structural stability, and of the forces which contribute to ordered structure formation.

Thermal Denaturation Studies

Hyperchromism found its first and still most widespread use as a measure of the denaturation process in nucleic acids. The term "hyperchromism" is often used to describe the increase in absorbance at a single wavelength that accompanies denaturation. Although rigorous treatment requires the use of integrated intensities, the changes in absorbance near the band maximum (about 2600 Å in DNA) are roughly proportional to the band area changes, and therefore serve as a convenient measure of the extent of denaturation.

A well-ordered and cooperatively formed polynucleotide structure such as DNA or poly rA · rU undergoes a rather sharp thermal transition in which the pairing and stacking of bases is disrupted as the temperature is raised. The absorbance at 2600 Å increases by about 40% in a DNA containing half adenine–thymine (A · T) and half guanine–cytosine (G · C) pairs. The temperature at which half of the total absorbance increase has occurred is often called the "melting temperature" (T_m) by analogy with the infinitely sharp temperature-dependent phase transition that occurs in phenomena such as the melting of ice. Marmur and Doty (*6*) first observed that there was a linear correlation between T_m and the base composition of DNA under fixed solvent conditions. This observation gave rise to the most widely used method for determining the base composition of an unknown DNA. The solvent most often employed for these studies is "standard saline citrate" (SSC), which is 0.15 M NaCl and 0.015 M sodium citrate. The empirical equation relating base composition to T_m (in °C) under these conditions is

$$(\%G \cdot C) = 2.44\,(T_m - 69.3) \qquad (1)$$

This equation is the best fit to data obtained by Marmur (*6*) for DNA samples from a large number of sources, varying in base composition from about 30 to 74% in G · C content. There is considerable deviation of some experimental points from the best fit straight line; in some cases the base composition predicted by Eq. (1) differs from chemical analysis by 4% in G · C content. In part this may arise from experimental errors either in the determination of T_m or in the base composition assigned to the DNA as the result of chemical analysis. But it is likely that some of the deviation is the result of dependence of T_m on sequence as well as overall base composition. It is known (*7*) that the synthetic polymer poly dAT · dAT, with an alternating sequence of A and T in each strand, has an anomalously low T_m compared with the value predicted by Eq. (1), while the nonalternating polymer poly dA · dT has a value close to that expected. Other synthetic DNAs with repeating sequences also behave anomalously (*8*). The use of Eq. (1) to determine the base composition of an unknown DNA therefore carries with it the assumption that the base sequence is approximately random.

Although SSC is the solvent in which denaturation behavior has been most thoroughly studied, other salt concentrations or different solvent systems may be used if properly calibrated. Reducing the NaCl concentration shifts the values of T_m for DNA downward (about 17° for each factor of ten in concentration) by an amount independent of the base composition, so that the slope of the line in Eq. (1) is unchanged. Using high concentrations of $NaClO_4$ also results (*9*) in low values of T_m, but in this case the slope of the dependence of T_m on mole fraction G · C is changed. Mixed water–methanol solvents have also been used (*9, 10*). The practical advantage of solvents that lower T_m lies in the relative difficulty of achieving temperatures close to 100°C in the spectrophotometer; in the case of DNAs rich in G · C base pairs, the use of SSC as solvent makes it difficult to see the entire denaturation at temperatures below the boiling point.

Methods

The methods outlined here are applicable to thermal denaturation studies of all kinds. The equipment required in addition to a spectrophotometer is quite simple: a circulating thermostatic bath, a thermostattable cell holder, a device for measuring temperature in a cuvette, and a technique for eliminating evaporation of liquid from cuvettes over the long periods that may be necessary for measurement. A wide choice of circulating thermostatic baths is available capable of regulation to ±0.1°C. The bath need not be of large capacity. Most manufacturers of spectrophotometers now supply suitable thermostattable cuvette holders. It is useful to have one which can hold four cuvettes. If temperatures close to 100°C are to be used, it may also be necessary to cool the spectrophotometer housing with tap water in order to maintain stability of the photomultiplier response. Accessory cooling devices are available for most spectrophotometers that require cooling. The temperature of samples should be measured in a cuvette reserved for the purpose placed in one of the positions of the cuvette holder. It is much less satisfactory to rely on the temperature of the circulating bath, which at high temperature is often several degrees higher than that of the cells even after thermal equilibrium is achieved. An accuracy of ±0.1°C can be achieved easily with a calibrated thermistor as a sensing element. The thermistor chosen should be designed for immersion in water. It can be inserted into a cuvette filled with water through a hole drilled in a Teflon or polyethylene stopper. The thermistor resistance can be measured with any bridge circuit that does not pass enough current through the thermistor to heat it significantly. A suitable design has been described (*11*). Commercial thermistor bridges with digital display are also available.

The problem of sample evaporation can be dealt with in several ways. Stoppered cuvettes may be used, and sealed with silicone rubber compounds (for example, "RTV" rubber, General Electric Company). The cuvette must be nearly full to avoid evaporation of solvent from the solution and condensation on the upper walls. Many investigators use a layer of mineral oil on the surface of the solution to retard evaporation. The oil should be washed thoroughly before use to remove water soluble impurities. When oil is used it is not necessary either to fill the cuvettes completely or to stopper them tightly. Some investigators using oil obtain satisfactory results with unstoppered cuvettes. Occasional difficulty arises at higher temperatures from the deposition of bubbles of dissolved gases on the optical surfaces of the cuvette. This can be avoided by bubbling the solutions briefly with helium gas before transfer to the cuvettes. Dissolved nitrogen and oxygen are displaced by helium, which has a positive temperature coefficient of solubility.

The best procedure for determining T_m is to raise the temperature by steps and to wait in each case for thermal equilibrium to be achieved before making the absorbance measurement. In the case of DNA in SSC, the rate of denaturation is rapid and therefore the rate-limiting step is the achievement of uniform temperature throughout the cell, which may take five or ten minutes after the measured temperature reaches a stable value. The exact amount of waiting time will be determined by the geometry of cuvettes and cell holder. The achievement of a stable value of the *absorbance* is the best indication that equilibrium has been reached.

In split beam spectrophotometers with a separate reference cell compartment, where it is inconvenient to heat the reference cell, one of the positions of the thermostatted cell holder should be used for a cuvette filled with solvent. This serves to guard against unexpected shifts in the baseline. In the case of solvents which absorb light, it also provides information on the temperature dependence of solvent absorbance necessary to make corrections. All measurements of temperature-dependent optical properties must be corrected for expansion of the solvent. The thermal expansion coefficient of water may be used for aqueous solutions; the correction is about 4% for an increase in temperature from 20° to 90°C. Standard tables of the expansion coefficient exist. The coefficient is approximately given by the expression

$$V = 0.99829 + 1.045 \times 10^{-4} T + 3.5 \times 10^{-6} T^2 \quad (2)$$

where T is given in degrees centigrade. Absorbances are corrected by multiplication by V.

Further Applications of Thermal Denaturation

The use of T_m to determine composition is based upon an empirical correlation between T_m and composition. The exact source of the greater energy of stabilization of G · C pairs is not known; it may be related to the third hydrogen bond which this pair can form, but it is equally possible that other electrostatic interactions between the bases or adjacent base pairs are responsible. Further complications arise when there are long sequences of base pairs within the DNA molecule which differ significantly in base composition from neighboring sequences. If they are sufficiently long, such regions are capable of denaturing more or less independently of neighboring regions, and a detailed analysis of the distribution of T_m values will reveal their presence, although it may not be entirely justified to expect the T_m of such regions to be related to their base composition exactly in the way described for independently denaturing molecules. A number of techniques for analyzing the shapes of denaturation curves have been described. If the derivative (d absorbance/dT) is plotted vs. T, a homogeneous sample will appear as an approximately Gaussian curve, whereas in a sample with a series of discretely denaturing species or regions, a series of peaks, or peaks and shoulders are seen (*12, 13*). This method has been applied to the study of internal heterogeneity of base sequence in the DNA of certain bacterial viruses (*14*). An equivalent method of analyzing the data is to plot the denaturation curve itself on Probit paper which yields a single straight line if a single Gaussian distribution of T_m values is present, and a broken line in the presence of more than one such distribution.

Spectral Methods

The problem of base composition and compositional heterogeneity can also be approached by an examination of the nucleic acid absorption spectrum at many wavelengths. The G · C and A · T base pairs contribute differently to the DNA absorption spectrum. The spectrum of native DNA contains sufficient information to permit determination of its base composition. If the temperature is raised, the change in the spectrum (the "hyperchromic spectrum") can be used to determine the concentration and base composition of the regions that have denatured. At temperatures high enough to denature all of the DNA, the hyperchromic spectrum reflects the base composition of all the DNA that has denatured, i.e., the total sample if it was all originally native. Finally, the absorption spectrum at high temperature itself contains information that reveals the base composition of the sample. In order to extract such information, the spectra of DNA samples of known base composition must be used for calibration. As indicated above, three kinds of spectra are useful: that of native DNA, that of totally denatured DNA at high temperature, and the difference or hyperchromic spectrum between denatured and native DNA. Fortunately, the variation of all three kinds of spectra with base composition can be represented by linear equations of the form

$$\text{Absorbance (or change in absorbance)} = C[\alpha\phi(\phi + \delta) + \beta\phi + \gamma] \qquad (3)$$

where C is the DNA concentration expressed in moles of nucleotide per liter, ϕ is the mole fraction of A · T pairs, δ is a parameter related to the nonrandomness of distribution of base sequences, and α, β, and γ are functions of the wavelength only, and not the base composition (*15*). This equation arises from a consideration of the effects upon DNA spectra of interactions between stacked base pairs. If these can all be expressed as the sum of interactions between pairs of base pairs along the chain, and if the sequence along the chain is random, then Eq. (3) with δ set equal to zero is rigorously correct (*15*). If the average contribution to the spectrum of the interaction of A · T pairs with G · C pairs is the mean of the contributions of (A · T)-(A · T) and (G · C)-(G · C) interactions, then the coefficient α vanishes and a simple two-term expression remains. Examination of a large number of DNA spectra shows that α does vanish in the region 250–280 nm in the case of high-temperature ("denatured DNA") spectra and hyperchromic spectra, so that the simple result is applicable. The parameter α is not negligible in the case of native spectra, and must be included in the calculation. The methods described below provide simple techniques for solving equation set (3), given the appropriate spectra. They are easily adapted to small electronic calculators or computers.

Methods

Data are collected exactly as in the case of T_m determinations described above, except that values of the absorbance at several wavelengths are required. Determination of overall composition by methods (a) and (b) below require measurement at only one temperature, while method (c) requires both low- and high-temperature measurements. More elaborate applications (see Other Applications below) may require additional temperature points. Any buffer in the neighborhood of pH 7 and any ionic strength may be used, provided that the solvent does not absorb at the wavelength of interest. A solvent blank should be included in any case. It should be noted that some salts have absorption bands that may shift upward in wavelength with increasing temperature into the spectral region being studied. In thermal denaturation studies, the temperature should be raised until an absorbance change of less than 1 percent for a 6° temperature increase is observed. Spectra are corrected for thermal expansion to 4°C using Eq. (2). Analysis of spectra of native DNA is quite sensitive to protein contamination. DNA to be studied by this method should be extracted with phenol as well as chloroform–octanol during purification (*15*).

Calculations (*16*)

a. Native DNA

Data must be collected at 5-nm intervals from 235 to 290 nm (twelve points). If A_i is the absorbance at wavelength i, and α_i, β_i, γ_i, are the corresponding parameters in Table 1, form the three sums over all twelve points,

$$\mu_1 = \Sigma A_i \alpha_i = A_{235} \cdot \alpha_{235} + A_{240} \cdot \alpha_{240} + \ldots A_{290} \cdot \alpha_{290}$$

$$\mu_2 = \Sigma A_i \beta_i$$

$$\mu_3 = \Sigma A_i \gamma_i$$

Using the values of the constant S_i in Table 1,

$$C = \mu_1 S_4 + \mu_2 S_3 + \mu_3 S_5$$

$$C\phi = \mu_1 S_1 + \mu_2 S_2 + \mu_3 S_3$$

C and ϕ are the concentration and A · T mole fraction of the DNA.

b. Denatured DNA

Collect data at high temperature only and at four wavelengths: 250, 260, 270, and 280 nm. Correct for thermal expansion. Construct the sums

$$\mu_2 = \Sigma A_i \beta_i = A_{250} \cdot \beta_{250} + \ldots A_{280} \cdot \beta_{280}$$
$$\mu_3 = \Sigma A_i \gamma_i \tag{4}$$

using the parameters β_i and γ_i in Table 2. Here, A_i are the appropriate absorbance values at high temperature.

TABLE 1. Parameters for Three-Term Analysis of Native DNA Spectra

λ_i (nm)	α_i	β_i	γ_i
235	−2026	−656	3952
240	−1889	−1251	5031
245	−1390	−1917	6338
250	43	−2830	7480
255	−319	−1807	7616
260	−608	−1141	7307
265	2515	−3379	7052
270	871	−1409	5740
275	−386	−154	4587
280	1159	−1558	3938
285	1797	−2424	3164
290	1187	−2099	2188

Multiply each term by 10^{-7}

$S1$	0·9329	$S4$	0·2792
$S2$	2·0631	$S5$	0·2124
$S3$	0·6198	$S6$	0·8513

TABLE 2. Parameters for Two-Term Analysis of Denatured DNA Spectra (250, 260, 270, 280 nm)

λ_i (nm)	β_i	γ_i
250	−2669	9703
260	−282	9397
270	−1066	8387
280	−2887	6615

Multiply each term by 10^{-8}

L_1	3·2466
L_2	0·9568
L_3	17·0129

Solve the equations

$$C = \mu_2 L_1 + \mu_3 L_2$$
$$C\phi = \mu_2 L_3 + \mu_3 L_1 \tag{5}$$

using values of L_1, L_2, and L_3 from Table 2.

c. Hyperchromic Spectra

Calculate the difference between denatured DNA absorbance and native DNA absorbance at 250, 260, 270, and 280 nm. The absorbance values must be corrected for thermal expansion before the differences are calculated. If A_i are the absorbance *differences* then Eqs. (4) and (5) are applicable, but the parameters β_i, γ_i, and L_i must be taken from Table 3. In this case, C and ϕ are the concentration and base composition of the regions that have undergone denaturation.

TABLE 3. Parameters for Two-Term Analysis of Hyperchromic Spectra (250, 260, 270, 280 nm)

λ_i (nm)	β_i	γ_i
250	43	2264
260	1388	1981
270	−629	2892
280	−2422	2920

Multiply each term by 10^{-8}

L_1	3·4359
L_2	4·6556
L_3	14·7455

Other Applications

Spectral methods are capable of providing information about the compositional heterogeneity of a DNA sample. If the temperature is raised in small increments (as in the case of T_m determinations) and the absorbance in the range 250 to 280 nm is recorded, method (c) above may be applied to determine the base composition and concentration of each successively denaturing region of the DNA. In this way it is possible to detect separate molecules of unusual base composition, or independently denaturing regions of unusual base composition within a single molecule. The method has the advantage over determination of T_m distributions (see above) that it is much less dependent upon chain length. A DNA double helix 50 to 100 base pairs long might have an aberrant T_m compared to a long polymer of the same base composition, but its hyperchromic spectrum would differ by only 1 or 2 percent from that of the infinite chain.

This technique has been used (*17*) to detect internal heterogeneity of base composition in λ phage DNA. This DNA is not homogeneous in base sequence along the chain; one half is richer in A · T pairs than the other (*18, 19*). If this DNA is subjected to stepwise denaturation, a graph such as that shown in Fig. 1 can be constructed, in which the fraction of all A · T pairs denatured is plotted vs. the fraction of all G · C pairs denatured. If all of these identical DNA molecules denatured in an all-or-none manner, Fig. 1 would be a straight line. The break in the curve shows that there are at least two independently denaturing regions. One region comprises 58% of the DNA, and has 56% A · T pairs. The other region is 42% of the DNA and has 42% A · T pairs. By comparison, detailed shearing experiments (*20*) using base compositions determined by CsCl density gradient centrifugation, show that the "left end" of λ DNA is a region of homogeneous base composition, with 42% of the DNA and a base composition of 44% A · T. The remaining 58% of the DNA has a mean base composition of 54% A · T.

Denatured DNA

All of the spectral methods give a qualitative measure of the intactness of a sample of supposedly native DNA. Since the concentration and base composition are both determined, the molar extinction at some wavelength (usually 260 nm) of the starting material may be calculated and compared with that of authentic samples of the same base composition. A table of molar extinctions of native DNAs of various base compositions has been published (*15*). The molar extinction of the unknown DNA should be within 4% of the tabulated value. If it is higher, then some of the DNA is probably denatured. The sensitivity of this test increases if the absorbance is measured in a solvent of low ionic strength, because the denatured regions are then unable to reform short regions of accidentally matching base pairs. If the DNA is dissolved in 0.001 M NaCl and 0.001 M phosphate buffer, pH 7, any denatured regions present will contribute little to hypochromism. If the DNA is then completely heat denatured, the total DNA concentration can be determined by method (b) above, while the concentration of regions denatured by the heating process is given by method (c). If the DNA was originally completely native, the two calculations give the same concentration within about 4%.

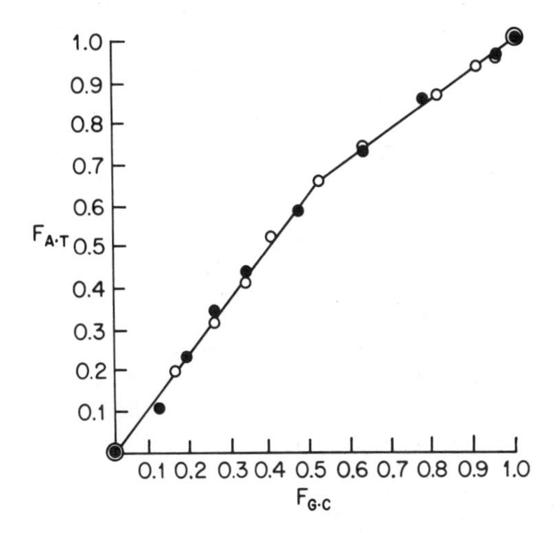

Figure 1. Fraction of all A•T pairs denatured vs. fraction of all G•C pairs denatured in λcI-857 phage. [From S. Z. Hirschman et al. (*17*).]

Methods similar to the spectral analysis given above have been developed by Fresco et al. (*21*) and Mahler et al. (*10*). The latter authors have obtained standard spectral parameters in methanol-water solvents. All of these methods must be used with caution when extensive nonrandom sequences of bases are being analyzed, since the calibrations are based upon DNAs of essentially random sequence, and spectral properties are to a certain extent sensitive to the local environment of the individual base pairs. The presence of bases other than G, C, A, and T will obviously invalidate all procedures which depend upon a calibration with DNAs not containing the unusual bases. Both the spectral analyses and T_m methods described above fall in this category, but when appropriate polynucleotides of known composition are available, a new calibration can be devised.

References

1. Tinoco, I., Jr. 1960. *J. Am. Chem. Soc.* **82:** 4785.
2. Rhodes, W. J. 1961. *Am. Chem. Soc.* **83:** 3609.
3. DeVoe, H. and I. Tinoco, Jr. 1962. *J. Mol. Biol.* **4:** 518.
4. DeVoe, H. 1963. *Nature* **197:** 1295.
5. Van Holde, K. E., J. Brahms, and A. M. Michelson. 1965. *J. Mol. Biol.* **12:** 726.
6. Marmur, J. and P. Doty. 1962. *J. Mol. Biol.* **5:** 109.
7. Bollum, F. J. 1966. In G. L. Cantoni and D. R. Davies (Eds.), Procedures in nucleic acid research, Harper & Row, New York, Vol. I, p. 581.
8. Wells, R. D., E. Ohtsuka, and H. G. Khorana. 1965. *J. Mol. Biol.* **14:** 221.
9. Geiduschek, E. P. 1962. *J. Mol. Biol.* **4:** 467.
10. Mahler, H. R., B. Kline, and B. D. Mehrotra. 1964. *J. Mol. Biol.* **9:** 801.

11. Felsenfeld, G. and G. Sandeen. 1962. *J. Mol. Biol.* **5:** 587.
12. DeLey, J. and J. Van Muylem. 1963. *Antonie van Leeuwenhoek J. Microbiol. Serol.* **29:** 344.
13. DeLey, J. 1969. *J. Theoret. Biol.* **22:** 89.
14. Falkow, S. and D. B. Cowie. 1968. *J. Bacteriol.* **96:** 777.
15. Felsenfeld, G. and S. Z. Hirschman. 1965. *J. Mol. Biol.* **13:** 407.
16. Hirschman, S. Z. and G. Felsenfeld. 1966. *J. Mol. Biol.* **16:** 347.
17. Hirschman, S. Z., M. Gellert, S. Falkow, and G. Felsenfeld. 1967. *J. Mol. Biol.* **28:** 469.
18. Hogness, D. S. and J. R. Simmons. 1964. *J. Mol. Biol.* **9:** 411.
19. Hershey, A. D. and E. Burgi. 1965. *Proc. Natl. Acad. Sci. U.S.,* **53:** 325.
20. Hershey, A. D. and E. Burgi. 1966. In Hershey A. D., Annual Report of the Director, Genetics Research Unit, *Cold Spring Harbor Lab. Biol.,* p. 559.
21. Fresco, J. R., G. C. Klotz, and E. G. Richards. 1963. *Cold Spring Harbor Symp. Quant. Biol.* **28:** 83.

MEASUREMENT
OF VISCOSITY
OF NUCLEIC ACID SOLUTIONS

B. H. ZIMM
DEPARTMENT OF CHEMISTRY
REVELLE COLLEGE
UNIVERSITY OF CALIFORNIA (SAN DIEGO)
LA JOLLA, CALIFORNIA

Measurement of the viscosity of solutions of macromolecules is a long-established technique for the determination of the molecular size of the macromolecules. General discussions are given in a number of references (see, for example, Refs. *1-3*).

The viscosity of a solution of macromolecules is an increasing function of the concentration. In the dilute range this function is approximately linear, but some curvature is usually present also. For this reason it is customary in rigorous work to extrapolate the viscosity increment to infinite dilution. The "intrinsic viscosity," $[\eta]$, or "limiting viscosity number" (the latter name recommended by the International Union of Pure and Applied Chemistry (*4*)) is the important quantity:

$$[\eta] = \lim_{c \to 0} \frac{\eta - \eta_0}{\eta_0 c} \tag{1}$$

where η is the viscosity of the solution, η_0 that of the solvent, and c is the concentration of solute in terms of weight per unit volume. If we define the *relative viscosity*, η_{rel}, as

$$\eta_{rel} = \eta/\eta_0 \tag{2}$$

and the *specific viscosity*, η_{sp}, as

$$\eta_{sp} = \eta_{rel} - 1 = (\eta - \eta_0)/\eta_0 \qquad (3)$$

then we may extrapolate plots of either η_{sp}/c or $(\ln \eta_{rel})/c$ to infinite dilution to obtain $[\eta]$. Both expressions should theoretically yield the same intercept, which should be $[\eta]$, since the first term in the expansion in Taylor series of the logarithm of η_{rel} is in fact η_{sp}. In practice, however, the plots are more or less curved, and uncertainties may arise in the extrapolation. The logarithmic plot is usually flatter and is therefore preferable, in the author's opinion. (The contrary view is expressed by Ross and Scruggs (*5*). Onyon (*3*) recommends making both plots simultaneously and averaging the intercepts.) In solutions of chain molecules at finite concentrations the value of $(\ln \eta_{rel})/c$ is almost always closer than that of η_{sp}/c to $[\eta]$; in fact, a value of the former determined at only one concentration is frequently a sufficient substitute for $[\eta]$ for practical purposes, and its reproducibility of determination is better.

In order to avoid, on the one hand, the problems of a long extrapolation and, on the other, the difficulties of measuring too small a viscosity increment, concentrations are usually selected such that η_{rel} lies between about 1.2 and 3.

We must also mention the confusion that exists with regard to units. The dimensions of $[\eta]$ are those of reciprocal concentration. Most commonly the concentration is expressed in grams of solute per deciliter, colloquially called "percent," and the limiting viscosity number then appears in terms of deciliters per gram. Not infrequently, however, cgs units, grams per milliliter, are used, with the result that the numerical value of $[\eta]$ appears to be 100 times larger. In the older German literature the standard unit was moles of monomer per liter (*Grundmolarität*). Suspicion is the only safe attitude to take regarding a literature value unless the units are clearly defined.

The measurement of viscosity for solutions of nucleic acids under 10^6 daltons in molecular size presents no unfamiliar problems. By far the overwhelming choice of instrument is some form of the capillary glass pipette. A brief discussion of procedure is given later.

With large DNA molecules the situation is quite different. As the shear stress increases, the viscosity of the solution decreases as the flow changes the shape and the orientational distribution of the macromolecules. The viscosity of such a solution is called *non-Newtonian*. With most materials the natural thermal motions of the molecules are so fast that the flow in a conventional viscometer is only a minor disturbance, but in DNA solutions the molecules are both unusually large and unusually stiff, with the result that the thermal motions are slow compared to those induced by the flow in the conventional viscometer. It is not surprising then to find that the non-Newtonian effect is serious; in fact, the values of the viscosity of a DNA solution measured in viscometers with different shear stresses may differ by as much as a factor of 2. For this reason special viscometers are preferred for large DNA, as discussed later.

TECHNIQUE OF CAPILLARY VISCOMETRY

Since capillary viscometers are familiar instruments, the following discussion is very brief and is intended to emphasize only those points of special interest to the scientist faced with the determination of the limiting viscosity number of solutions of the smaller nucleic acids. General discussions of viscometry are to be found in a number of places. The author has found the article by Swindells, Ullman, and Mark (*6*) to be especially pertinent. Onyon (*3*) writes with particular attention to the determination of intrinsic viscosity. An article by Schachman (*7*) on determining the the viscosity of protein solutions contains many points relevant also to the study of solutions of very small nucleic acids such as tRNA.

In the usual instrument a measured volume of liquid is allowed to flow under the force of gravity through the capillary and the time of flow is measured with a stopwatch. Since only the ratio of the viscosities of solution and solvent enters the viscosity number, it is not necessary to employ a viscometer with a calibration in terms of absolute viscosity. It is common practice, in fact, to record only the time of flow of the solution, t, and of the solvent, t_0, and to use these in place of η and η_0, respectively, in Eqs. (1-3).

The common type of capillary viscometer, usually called the Ostwald type, is available from many apparatus firms and can be easily constructed by any glassblower. Such an instrument, however, cannot he highly recommended for the kind of work discussed here. The usual type is likely to require corrections for kinetic energy and drainage (*6-8*) which are uncertain at best and inconvenient to apply. Since the quantity of interest is the difference between the viscosities of the solvent and of the solution, the effect of small uncertainties is magnified. It is much better, therefore, to have the viscometer constructed in such a way that the corrections are negligibly small. Such viscometers are described by Schachman (*7*) and by Cannon and Fenske (*8*). The Cannon-Fenske instrument is available commercially (*9*); it has been widely used and gives satisfactory results. A semimicro type for use with small amounts of solution is also available (*9*).

For careful work where solutions of several concentrations are to be measured for extrapolation to infinite dilution, a dilution viscometer that utilizes the suspended-level design of Ubbelohde is especially suitable. A commercially available form of the instrument (*9*), designed to minimize corrections, is shown in Fig. 1. The viscosity of the solvent, η_0, and the

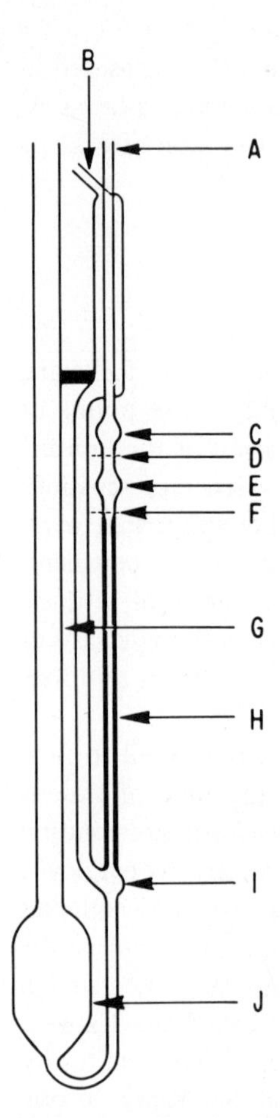

Figure 1. Cannon-Ubbelohde Dilution Viscometer (*9*). With opening B closed, the measuring bulb E is filled with liquid from the reservoir J by suction at A. The suction is removed and B opened, allowing excess liquid to drain from I. The rate of flow under gravity through the capillary H is then measured by timing the meniscus as it passes the marks D and F. [Reproduced with permission of the Cannon Instrument Company.]

viscosities of solutions of various concentrations can all be obtained in one series of measurements without emptying the viscometer. The large bulb serves as a reservior in which the mixtures can be formed and stored, but it is not involved directly in the viscosity measurement. For an example of the use of this instrument see Cohen and Eisenberg (*10*).

Precautions

We have mentioned the fact that precision is lost because of the subtraction of the viscosities of solution and solvent. Errors are magnified further in the extrapolation to infinite dilution. This is worth remembering because the technique is deceptively simple. In the author's experience the first attempt by a student to determine an intrinsic viscosity has usually been a disaster. Even experienced operators can obtain variable results because of factors such as those to be mentioned below.

Timing

With a flow time in the neighborhood of 100 seconds, the error of manual timing, about 0.1 second, amounts to one part in a thousand. A good stopwatch or clock in good condition should be capable of better precision; however, two watches may differ by more than this amount. If photoelectric or other automatic timing is used to obtain greater precision, the question of the reproducibility of the clock becomes more critical.

Temperature

Since viscosity of common solvents changes about 2% per degree, the temperature must be held within 0.05° or better to keep the temperature error less than the timing error.

Concentration Measurement

Since the limiting viscosity number has the dimensions of reciprocal concentration, its precision is no better than that of the latter, in fact, usually worse because of the extrapolation. Errors in concentration can accumulate to an alarming degree in a dilution series unless good technique is used. Serological pipettes, so common in biochemistry laboratories, are not suitable for this purpose. Pipettes should be used in the way for which they were calibrated, e.g., pipettes calibrated to deliver should not be used to measure contained amounts. With organic solvents pipettes must be recalibrated because of the different drainage corrections. For very volatile or very viscous solvents, weighing is an attractive alternative to the use of pipettes.

Dirt and Surface Films

The effect of thread of lint in a narrow capillary can be appreciable. For careful work samples should be centrifuged or filtered. For DNA, loss of solute in the filter is a danger that must be considered.

Surface films can affect the measurements in at least two ways. By changing the surface tension a surface-active agent can change the drainage and surface-tension corrections of the viscometer, but in a well-designed instrument these effects should be negligible. Much more serious is the effect of solid surface films that lead to sticking of the meniscus (*11a, 11b, 12*). Meniscus sticking is ubiquitous with solutions containing even small amounts of protein. If the protein cannot be eliminated from the solutions, as by extracting with phenol, the only recourse with conventional instrumentation is to minimize the error by working with solutions of high relative viscosity. Jacobs (*11a, 11b*) has described a capillary instrument in which the meniscus is nearly stationary and which is therefore less sensitive to surface films, but it does not seem to have been tested for the kind of determinations discussed here. Joly (*12*) describes a rotating-cylinder viscometer with a guard ring that eliminates the effect of films. The Gill-Thompson rotating-cylinder design discussed below also is insensitive to films.

Adsorption of Solute

Many solutes adsorb to glass surfaces; the author and others have encountered this with DNA. With chain molecules the adsorbed layer may extend away from the surface a distance comparable to the r.m.s. end-to-distance of the chain. Hence, adsorption can not only change the concentration of a solution, but can also narrow the capillary of the viscometer (*13a, 13b*). The effects, although small, especially in a well-designed instrument, are nevertheless hard to predict and hard to control. For this reason, as well as to minimize the effects of other errors, it is not advisable to attempt to determine the limiting viscosity number from measurements on solutions with relative viscosities less than about 1.2. Attempts to do so may only exaggerate artifacts, especially with random-chain molecules, as has been demonstrated particularly by Öhrn (*13a, 13b*) and by Tuijnman and Hermans (*14*).

Non-Newtonian Solutions

Although solutions of large DNA molecules require one of the rotating-cylinder instruments discussed below, solutions of small or denatured DNA may show mild non-Newtonian effects which can be successfully dealt with in a modified capillary viscometer. The object here is to eliminate the shear stress as an additional variable by reducing its value until the non-Newtonian effect is unimportant. A capillary viscometer especially designed for DNA work must have a very long capillary to spread out the stress and to decrease the rate of flow. There is a limit, however, to how far the flow can be reduced conveniently, since the flow must be fast enough to allow a sizable volume of liquid to pass through the capillary in a reasonable time. The volume cannot be reduced too far or meniscus effects, discussed above, have adverse effects on the precision.

The basic equations of the gravity-flow capillary viscometer show how this works out. If the density of the liquid is ρ, the acceleration of gravity g, the height of upper bulb above the lower h, the radius of the capillary r and its length l, and a volume V flows in time t, then the viscosity is calculated from the formula:

$$\eta = \pi \rho g h r^4 t / 8 V l \quad (4)$$

The magnitude of the non-Newtonian effect increases with the value of the shear stress, S, for a given liquid. The average value of S in the capillary (assuming that the non-Newtonian effects are not too great) is given by:

$$S = 8 \eta V / 3 \pi r^3 t \quad (5)$$

If we eliminate the radius between these two formulas, we can get a general expression for the shear stress in terms of the viscosity of the liquid, the volume flow rate, and the dimensions of the viscometer:

$$S = (\tfrac{1}{3})(8 V \eta / \pi t)^{1/4} (\rho g h / l)^{3/4} \quad (6)$$

Since reducing V or h makes the measurements liable to meniscus artifacts, and since it is impracticable to accomplish much by increasing t because of the one-fourth exponent, increasing the length of the capillary, l, by coiling or folding while keeping h constant is the only effective procedure for reducing the shear stress, S. The radius of the capillary is then selected to give a satisfactory flow time. The usual capillary instrument, in which the volume is about 1 ml, h and l are both about 10 cm, and the time of flow is about 100 sec for water (viscosity 0.01 poise), has an average shear stress of about 10 dyne/cm^2. With the use of a coiled capillary of about 2 meters in length and with the other dimensions kept the same the average shear stress can be reduced to 1 dyne/cm^2.

What such numbers mean quantitatively will be discussed later. At the present time no one can predict exactly how important non-Newtonian behavior may be with a novel large nucleic acid; it is therefore good practice to measure the effect in doubtful cases. For this purpose it is convenient to have two or more bulbs at different heights, h, on the same capillary. The relative viscosities and the shear stresses may then be calculated from the flow times of the several bulbs; if the viscosities are not independent of S they may be plotted against it and extrapolated to zero stress. A viscometer of this type is available commercially (*9*). Descriptions are given by a number of authors (*15–18*). If the variation with shear stress is severe, however, the best practice is to use a rotating-cylinder viscometer.

ROTATING-CYLINDER VISCOMETERS

It was found very early that solutions of large DNA molecules showed anomalous viscosities. While many of the anomalies in the early work were probably caused by association with partial denaturation and the presence of protein contamination, even highly purified preparations show large non-Newtonian effects, as indeed should be expected theoretically from the large size of the molecules, as will be discussed later. We have already seen that it is not convenient to extend the range of shear stress much below 1 dyne/cm^2 with a capillary viscometer. However, typical measurements, such as those of Eisenberg (*19*) on fish-sperm DNA, require shear stresses of 0.1 dyne/cm^2 and less to achieve meaningful results. It is possible at the present time to make such measurements only with a Couette, or rotating-cylinder, viscometer.

The Couette viscometer consists of two concentric cylinders, one inside the other, with the liquid in the annular space between them. One of the cylinders is rotated and the torque is measured. The torque developed at low shear rates in aqueous solutions is very small, and its measurement presents difficulties. Torsion wires may be used, but are fragile and need frequent calibration. Eisenberg and Frei (*20*) obtained good results with electrostatic restoring torque and a floating cylinder held in place by a jeweled pivot. This instrument has been used successfully with DNA solutions (*19, 21*). It seems to have no advantages, however, over the simpler device described below, except that the Eisenberg and Frei instrument can go to higher shear stresses before turbulent flow sets in.

Floating-Rotor Viscometer

This device (*22, 23*), shown in Fig. 2, uses a freely floating inner cylinder, supported by its own buoyancy and held in place by surface forces, and to which a constant torque is applied by the interaction of a metal pellet in the bottom with a rotating applied magnetic field. The time per revolution of the inner cylinder relative to the static outer cylinder is measured, the time being proportional to the viscosity of the liquid between the two if the torque is constant. There are no mechanical devices attached to the moving cylinder, so that all frictional dissipation of energy occurs in the liquid itself. The inner cylinder is held in place by the surface tension of the meniscus. The shape of the meniscus is crucial in establishing a centering effect; surface forces act to center an inner tube if the liquid surface rises up from its rim to the wall of the outer vessel. (In the more common case where the liquid is below the rim of both cylinders the surface tension pulls the inner vessel to one side until the two are in contact.)

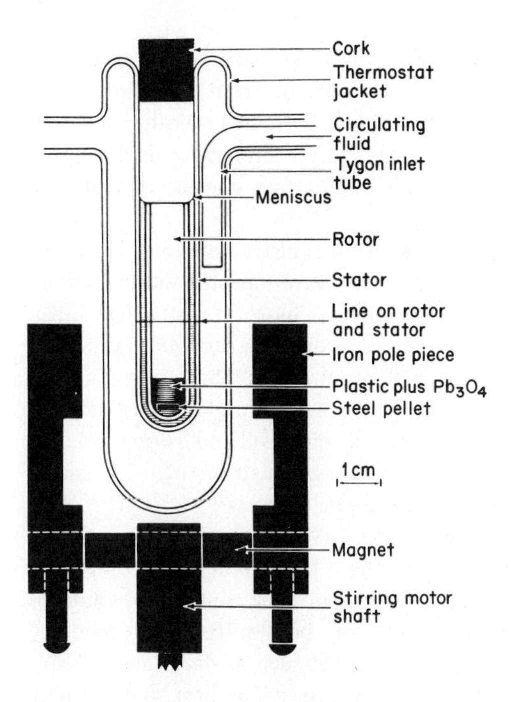

Figure 2. Floating-Rotor Viscometer of Zimm and Crothers (*22*). The time per rotation of the rotor, driven by the field from the rotating magnet, is proportional to the viscosity of the liquid.

To apply a constant torque to the rotor the interaction between a pellet of metal and a rotating magnetic field is used. If the pellet is iron, the coupling is between the field and the magnetic moment induced in the iron, which lags in direction behind the inducing field because of hysteresis. The torque thus produced is remarkably constant regardless of the speed of rotation of the field. Different shear stresses may be achieved by the use of several rotors with pellets of different sizes. If the pellet is made of an electrically conducting but nonmagnetic material such as aluminum, the torque is then produced by induced eddy currents, as in an induction motor. The torque in this latter case is proportional to the difference between the rates of rotation of the magnetic field and of the rotor; various shear stresses may thus be obtained reproducibly by using a synchronous motor with an adjustable gear box, but the calculation of the viscosity is complicated because it depends on the rotation rates of both the field and the rotor (*24, 25*). Frisman et al. (*26*) suggest changing the strength of the magnetic field by moving the magnet as a convenient means of varying the shear stress, but the author has found irreproducibility when this method is used with steel pellets. (Apparently Frisman et al. used soft iron pellets. Irreproducibility of unknown origin is also encountered sometimes with stainless-steel pellets even with constant magnetic field.) Another method, very convenient in use but requiring more elaborate equipment, is to employ an electromagnet excited with split-phase alternating current of regulated voltage as the source of field (*27*).

Precision to 0.1% has been obtained with this instrument over a range of shear stresses of 0.001–0.3 dyne/cm^2 (*22*). Details of its construction and use are given in the original references. The polytrichlorofluoroethylene rotor described in reference *25* has proved very convenient. A semimicro version is also in use in the author's laboratory which requires less than one milliliter of solution; the rotor is constructed from a precision-bore glass NMR sample tube of 5 mm outside diameter, as first suggested by Gill and Thompson (*24*) for a different type of apparatus. The basic instrument is available commercially from at least two sources (*28, 29*).

The main drawback of the floating-rotor viscometer with a free surface is its great sensitivity to solid surface films; these films sometimes completely freeze the rotor. As discussed above, such films are very common in solutions containing proteins. Sometimes the film can be liquified and rendered harmless by the addition of an antifoaming agent such as decyl alcohol (but not silicone because of lasting effects on the glass!). The best solution is to use a viscometer that is not sensitive to such films.

Gill-Thompson Cartesian-Diver Viscometer

This device, invented by Gill and Thompson (*24*), is a modification of the floating-rotor viscometer in which the rotor is completely submerged and hence is not affected at all by surface films (see Fig. 3). The submerged rotor is maintained in vertical position as a "Cartesian diver," that is, its buoyancy is a function of the expansion of the air bubble inside, the latter being controlled by the pressure applied to the chamber through the cover plate. This pressure is adjusted automatically by a servo mechanism (a recorder pen drive) which senses the rotor height by the lamp-and-photocell combination. The rotor is centered initially on the conical bottom of the chamber and a weak hydrodynamic centering effect tends to keep it in position while it is rotating. While it seems to be more difficult to obtain high precision with this design than with the floating-rotor type, the absence of sensitivity to surface films is a great asset. A successful application to lysates of *E. coli*, where the viscosity is due principally to the chromosomal DNA, has been described (*30*).

A modification of this design in which the driving torque is supplied by a split-phase electromagnet and in which recorder read-out has been incor-

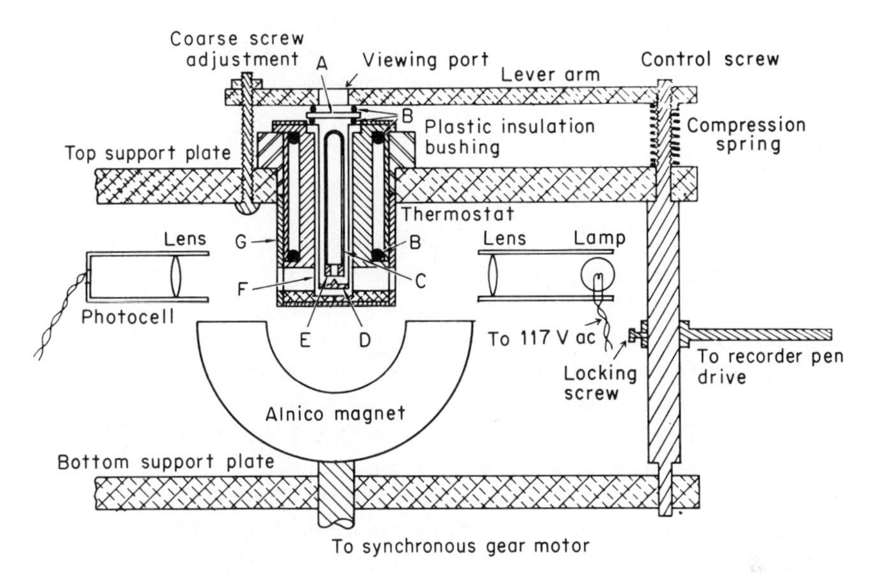

Figure 3. Submerged-Rotor Viscometer of Gill and Thompson (*24*). (*A*) glass end window, (*B*) rubber "O"-rings, (*C*) Cartesian-diver rotor with trapped air bubble, (*D*) centering cone, (*E*) metal driving ring, (*F*) outer cylinder (stator), (*G*) thermal insulation.

porated allows the measurement of the elastic as well as the viscous properties of DNA solutions (*27*).

Joly's Couette viscometer (*12*) with a guard ring also defeats the disturbing effects of surface films, but examples of its use with nucleic acids are lacking.

INTERPRETATION

Viscosity and Molecular Size

The relation of viscosity to molecular size is discussed at length in many places (*1–3*). We have already seen that the limiting viscosity number, or intrinsic viscosity, has the dimensions of volume per unit mass. If the mass is taken as that of one molecule, the corresponding volume is a measure of the molecule's influence on the hydrodynamic flow field in the solution. Einstein showed that in the case of solid spheres the intrinsic viscosity is just 5/2 over the density of the sphere, d,

$$100[\eta] = 5/2d \qquad (7)$$

(The factor 100 is needed when $[\eta]$ is given in deciliters per gram.) In other words the effective volume of the sphere is 5/2 times its actual volume. The expression of Simha gives $[\eta]$ for prolate ellipsoids of revolution with axial ratio J:

$$100[\eta]d = \frac{J^2}{15(\ln 2J - 3/2)} + \frac{J^2}{5(\ln 2J - 1/2)} + \frac{14}{15} \tag{8}$$

At large J the effective volume becomes much larger than the real volume.

The Simha formula is useful for short DNA double-stranded helices, which are stiff enough to approximate a rigid, long, thin ellipsoid. For example, Doty, McGill, and Rice (*31*) measured a series of DNA fractions of which the smallest had a molecular weight of 300,000 as determined by light scattering. With the Watson-Crick value of 3.4 Å for the thickness of each base pair, the long axis is computed to be 1650 Å. From the partial specific volume and the molecular weight, we can compute the volume of the molecule and hence the minor axis of the equivalent ellipsoid. The axial ratio is thus calculated to be 93, and $[\eta]$ from Eq. (5) is 2.9 dl/gm. The experimental value is 1.8. The discrepancy probably reflects the significant degree of flexibility that is present in a DNA molecule of this size; presumably the ellipsoid model would be more accurate for lower molecular weights.

For DNA molecules in general the best model is that of the worm-like chain of Kratky and Porod (*32*). While no closed formula of the type of Eq. (8) exists, Ullman (*33*) has made extensive numerical calculations. The author and his associates (*34, 35*) have found that both Ullman's calculations and the experimental data on DNA are well described by an empirical relation of the form

$$[\eta] + b = KM^a \tag{9}$$

where M is the molecular weight and a, b, and K are empirical constants. Such an expression gives a convenient straight line relation on a double logarithmic plot; the slope, a, is typically between 0.5 and 0.8 for chain molecules.

Nucleic acids are polyelectrolytes, and as such their dimensions are subject to change when the concentration of counterions is changed. It used to be thought that the double helix of DNA was too rigid to respond to the electrolyte changes, but Rosenberg and Studier (*36*) have shown that this is a misconception resulting from a fortuitous cancellation of specific volume and polyelectrolyte effects on the sedimentation coefficient. In fact, according to Ross and Scruggs (*5, 37*) and Rosenberg and Studier (*36*), the intrinsic viscosity of large DNA molecules changes by more than a factor of two when the salt concentration is decreased from 1M to 0.001M. The change is much greater with single-stranded DNA (*36*). Changes in the secondary structure, as with tRNA (*38*), also manifest themselves in changes in the intrinsic viscosity.

Both polyelectrolyte and structure-induced changes are readily interpreted as swelling or shrinking of the molecular coil with the use of the Flory-Fox formula,

$$[\eta] = \phi R^3/M \tag{10}$$

where R is a characteristic linear dimension of the molecule and Φ is assumed to be roughly constant. If R is taken to be the r.m.s. end-to-end length in centimeters of a linear polymer chain, Φ is then approximately 2.5×10^{21} when $[\eta]$ is given in deciliters per gram. The formula, while dimensionally correct, is not quantitatively exact, since Φ is found to be quite dependent on the structural details of the molecule, such as branching and rigidity (*39, 40*).

Viscosity and Concentration

The viscosity is found to vary as a power series in the concentration of the macromolecules. This series can be put into either of the equivalent forms:

$$\eta_{sp}/c = [\eta] + k'c \ldots \tag{11a}$$

$$\ln \eta_{rel}/c = [\eta] + k''c \ldots \tag{11b}$$

Here the constants k' and k'' are related by

$$k'' = k' - 1/2 \tag{11c}$$

where k' is called the Huggins constant, and for chain molecule solutions its value is usually near 0.4 dl^2/gm^2, which is in accord with theoretical expectations (*41*). The exact value is sensitive to intermolecular interactions, and high values are thought to be indicative of intermolecular association. For an example with DNA samples prepared by the early crude methods, see reference (*42*).

Non-Newtonian Flow

We have already mentioned the prevalence of non-Newtonian flow in solutions of large DNA molecules. Typical data are shown in Fig. 4. Since it is obviously important to measure at shear stresses where such effects are small, it would be useful to be able to predict curves such as that shown in Fig. 4. Pure theory is not yet competent to do this in detail for any model more realistic than ellipsoids (for which see Scheraga (*43*)) because of the complexity of the many interrelated hydrodynamic processes involved;

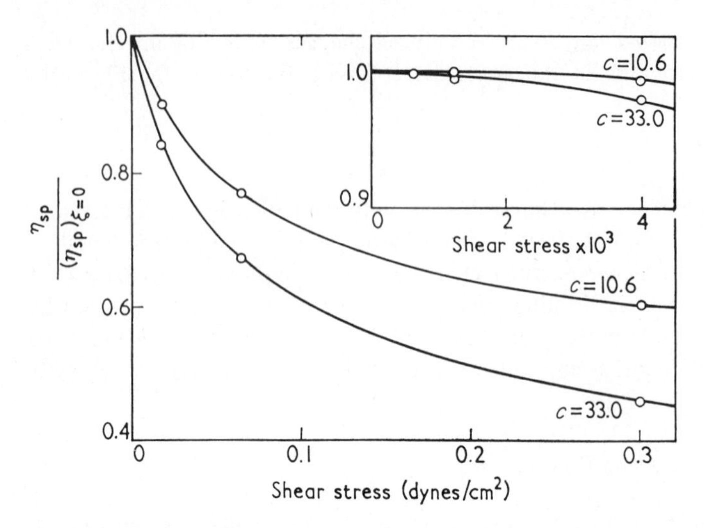

Figure 4. Dependence on shear of the specific viscosity of T2 DNA, expressed as the fraction of the value at zero shear stress (*22*). Curves are shown for two concentrations (μg/ml) of the sodium salt.

however, the author believes that a useful semiempirical generalization is possible.

The idea here is that each molecule has a characteristic relaxation time, τ, with which it responds to applied stresses. This relaxation time is related to the intrinsic viscosity, the solvent viscosity, the molecular weight, and the temperature by the relation

$$\tau = \alpha M \eta_0 [\eta]/RT \tag{12}$$

where R is here the gas constant and α is a constant that depends on the nature of the molecule. For spheres α is 400/9; for linear flexible chain molecules α is predicted to be about 40 with $[\eta]$ in dl/gm (*40*). Since we postulate that the non-Newtonian effects are caused by the distortion of the molecular distributions by the shearing solvent, we might expect that the extent of this distortion would be determined by the relative rates of shear and of molecular relaxation, which latter rate is measured by the reciprocal of τ. The rate of shear, κ, is the shearing stress, S, divided by the viscosity, η. The ratio of the two rates is then measured by the dimensionless number, $\kappa\tau$. If we assume that the solution is dilute so that $\eta = \eta_0$, introduce a value of T of 298°, and assume that α is 40, we find,

$$\kappa\tau = S\tau/\eta_0 = 1.6 \times 10^{-9} SM[\eta] \tag{13}$$

For example, we have T2 DNA of molecular weight 10^8 daltons and $[\eta]$ equal to about 300 dl/gm. The relaxation time has been measured by a dynamic technique; the value (*27*) at low concentrations is 0.5 sec, in good accord with prediction by Eq. (12). We can see in Fig. 4 that at low concentrations the specific viscosity is within 5% of its zero-shear value at S equal to 0.004, corresponding to a value of $\kappa\tau$ of 0.2, which same value is calculated from Eq. (13).

We are therefore led to conclude that non-Newtonian effects are small with chain molecules as long as the dimensionless parameter $\kappa\tau$ expressed by Eq. (13) has a small value, which value we take tentatively to be 0.2 or less on the basis of experience with T2 DNA.

As an example of the application of this reasoning, we may take the DNA sample measured by Cohen and Eisenberg with a Cannon-Ubbelohde viscometer (*10*). The shear stress may be estimated from Eq. (6). In this viscometer h and l were practically the same, V was about 1 ml, and t was 250 sec in the instrument used (*10*). The value of S was then 6 dyne/cm^2. Cohen and Eisenberg's DNA sample had a molecular weight of 465,000 and a limiting viscosity number of 3.85 dl/gm; when we substitute these values into Eq. (13) we find $\kappa\tau$ equal to 0.03. We should therefore expect no non-Newtonian effects to be manifest in the Cannon-Ubbelohde viscometer with this sample; in fact, Cohen and Eisenberg looked for such effects with the aid of another viscometer and found none, in contrast to experience with large DNA.

The pronounced non-Newtonian behavior of large DNA molecules is thus found to be a natural consequence of their unusually large molecular weights and viscosity numbers, and its presence should be expected whenever the parameter $\kappa\tau$ exceeds 0.2.

Viscosity and Sedimentation

Both the intrinsic viscosity and the sedimentation coefficient are indicators of molecular weight and size; the former is the more widely used with synthetic polymers, the latter with biopolymers. Sedimentation has the great advantage of resolving the components of mixtures, but the sedimentation coefficient is somewhat less sensitive to molecular size than is the intrinsic viscosity, and its measurement is more expensive, laborious, and time-consuming. On the other hand, concentration must be measured along with viscosity.

The combination of the two measurements is much more informative than either alone. As in Eq. (10), the sedimentation coefficient is related to molecular weight and size through the approximate formula (*1, 2*):

$$s = \psi M(1 - v\rho)/N\eta_0 R \qquad (14)$$

where s is the sedimentation coefficient, Ψ an empirical constant, v the partial specific volume of the solute, ρ the density of the solvent, N Avogadro's number, η_0 the solvent viscosity, and M and R molecular weight and linear dimension, as before. As with Eq. (10), the "constant," Ψ, is actually structure-dependent to a marked degree. Nevertheless, it was shown by Mandelkern, Flory, and Scheraga (*44, 45*) that the two equations can be combined with the elimination of R to yield an equation which, while still approximate, is of remarkably wide applicability:

$$M^{2/3} = \frac{s[\eta]^{1/3}\eta_0 N}{10^{19}\beta(1 - \nu\rho)} \tag{15}$$

Here s is in Svedberg units (10^{-13} cgs) and $[\eta]$ is in dl/gm. In this equation β is a quantity that is nearly constant over wide ranges of variation in molecular shape and structure; for example, whether a chain is linear or closed into a ring the value of β is practically independent (*40*). For DNA the best empirical value of β is discussed by Eigner and Doty (*46*); they favor a value of 2.4 for large DNA.

References

1. Flory, P. J. 1953. Principles of polymer chemistry, Cornell University Press, Ithaca, N.Y. Chapter XIV.
2. Tanford, C. 1961. Physical chemistry of macromolecules, Wiley, New York, Chapter 6.
3. Onyon, P. F. 1959. *In* P. W. Allen (ed.), Techniques of polymer characterization, Butterworths, London. Chapter 6.
4. International Union of Pure and Applied Chemistry, 1952. *J. Polymer Sci.,* **8:** 257.
5. Ross, P. D., and R. L. Scruggs. 1968. *Biopolymers,* **6:** 1005.
6. Swindells, J. F., R. Ullman, and H. Mark. 1959. *In* A. Weissberger (ed.), Physical methods of organic chemistry, 3rd edition. Interscience, New York, Chapter XII.
7. Schachman, H. K. 1957. *In* S. P. Colowick and N. O. Kaplan (eds.), Methods in enzymology. Academic Press, New York, Vol. IV, pp. 95-102.
8. Cannon, M. R., and M. R. Fenske. 1938. *Ind. Eng. Chem., Anal. Ed.,* **10:** 297.
9. Cannon Instrument Company, P.O. Box 16, State College, Pa. 16801.
10. Cohen, G., and H. Eisenberg. 1966. *Biopolymers,* **4:** 429.
11a. Jacobs, H. R. 1963. *Biorheology,* **1:** 129, 225.
11b. Jacobs, H. R. 1969. *Biorheology,* **6:** 121.
12. Joly, M. 1962. *Biorheolgy,* **1:** 15.
13a. Öhrn, O. E. 1955. *J. Polymer Sci.,* **17:** 137.
13b. Öhrn, O. E. 1956. *J. Polymer Sci.,* **19:** 199.
14. Tuijnman, C. A. F., and J. J. Hermans. 1957. *J. Polymer Sci.,* **25:** 385.
15. Krigbaum, W. R., and P. J. Flory. 1953. *J. Polymer Sci.,* **11:** 40.
16. Reichman, M. E., S. A. Rice, C. A. Thomas, and P. Doty. 1954. *J. Am. Chem. Soc.,* **76:** 3047.
17. Hermans, Jr., J., and J. J. Hermans. 1958. *Koninkl. Ned. Akad. Wetenschap. Proc. Ser.* **B, 61:** 324.
18. Collins, D. J., and H. Wayland. 1963. *Trans. Soc. Rheol.,* **7:** 275.

19. Eisenberg, H. 1957. *J. Polymer Sci.*, **25**: 257.
20. Eisenberg, H., and E. H. Frei. 1954. *J. Polymer Sci.*, **14**: 417.
21. Schumaker, V. N., and C. Bennett. 1962. *J. Mol. Biol.* **5**: 384.
22. Zimm, B. H., and D. Crothers. 1962. *Proc. Natl. Acad. Sci., U.S.*, **48**: 905.
23. Zimm, B. H. 1965. *Fractions*, 3: 2.
24. Gill, S. J., and D. S. Thompson. 1967. *Proc. Natl. Acad. Sci., U.S.*, **57**: 562.
25. Hays, J. B., and B. H. Zimm. 1970. *J. Mol. Biol.*, 48: 297.
26. Frisman, E. V., L. V. Shchagina, and V. I. Vorob'ev. 1965. *Biorheology*, **2**: 189.
27. Chapman, Jr., R. E., L. C. Klotz, D. S. Thompson and B. H. Zimm. 1969. *Macromolecules*, **2**: 637.
28. Spinco Division of Beckman Instruments, Inc., Palo Alto, California 94304.
29. W. Krannich, 3400 Göttingen, Elliehäuser Weg 17, West Germany.
30. Thompson, D. S., J. B. Hays, and S. J. Gill. 1969. *Biopolymers*, **7**: 571.
31. Doty, P., B. McGill, and S. Rice. 1958. *Proc. Natl. Acad. Sci., U.S.*, **44**: 432.
32. Kratky, O., and G. Porod. 1949. *Rec. Trav. Chim.*, **68**: 1106.
33. Ullman, R. 1968. *J. Chem. Phys.*, 49: 5486.
34. Crothers, D. M., and B. H. Zimm. 1965. *J. Mol. Biol.*, **12**: 525.
35. Hays, J. B., M. E. Magar, and B. H. Zimm. 1969. *Biopolymers*, **8**: 531.
36. Rosenberg, A. H., and F. W. Studier. 1969. *Biopolymers*, **7**: 765.
37. Scruggs, R. L., and P. D. Ross. 1964. *Biopolymers*, **2**: 593.
38. Henley, D. D., T. Lindahl, and J. R. Fresco. 1966. *Proc. Natl. Acad. Sci., U.S.*, **55**: 191.
39. Kilb, R. W., and B. H. Zimm. 1959. *J. Polymer Sci.*, **37**: 19.
40. Bloomfield, B., and B. H. Zimm. 1966. *J. Chem. Phys.*, **44**: 315.
41. Petersen, J. M., and M. Fixman. 1963. *J. Chem. Phys.*, **39**: 2516.
42. Conway, B. E., and J. A. V. Butler. 1954. *J. Polymer Sci.*, **12**: 199.
43. Scheraga, H. A. 1955. *J. Chem. Phys.*, **23**: 1526.
44. Mandelkern, L., and P. J. Flory. 1952. *J. Chem. Phys.*, **20**: 212.
45. Scheraga, H. A., and L. Mandelkern. 1953. *J. Am. Chem. Soc.*, **75**: 179.
46. Eigner, J., and P. Doty. 1965. *J. Mol. Biol.*, **12**: 549.

NMR STUDIES OF NUCLEIC ACIDS

T. YAMANE
BELL TELEPHONE LABORATORIES
MURRAY HILL, NEW JERSEY

INTRODUCTION

Nuclear magnetic resonance (NMR) spectroscopy has been used extensively on structural studies of nucleic acids and their components. In particular, with recent developments in NMR technology, considerable progress has been made in the elucidation of monomer-monomer and monomer-polymer interactions and in studying the binding of ions and small molecules to nucleic acids.

Valuable developments which have led to a sudden upsurge of biologically-inclined studies are: (1) field-frequency locked spectrometers which provide increased sensitivity and resolution by eliminating drift and consequent broadening of spectral lines, (2) time-averaging computers or computers of average transients (CAT) which make use of this stability to accumulate many traces of weak spectra, enhancing the signal-to-noise ratio in proportion to the square root of the number of scans accumulated, and most recently, (3) a spectrometer operating at 220 MHz became available which has improved sensitivity and increased peak separations. The Varian 220 MHz spectrometer, which has a helium-cooled super-conducting magnet, has sufficient magnetic field stability to obviate the need for a lock system.

The present chapter tries to demonstrate the various type of approach and degrees of success of the NMR method at this stage in its development. The theory and experimental details of the NMR method are adequately covered in reference works (*1–8*). Several review articles serve to cover early applications (*9–11*) and more recent reviews are listed (*12–15*).

BASIC CONCEPTS

The theory of NMR can be described qualitatively as follows: (1) The position of a peak in the spectrum, its chemical shift, is determined by the electronic or chemical environment of the nucleus. Nuclei having identical resonance frequencies are said to be magnetically equivalent. (2) The area under a resonance signal can give a quantitative measure of the number of resonating nuclei. For a molecule containing magnetically nonequivalent nuclei, the ratio of areas in a spectrum can thus be used to determine the ratio of the number of nuclei in each chemical group. (3) The coupling constant, J_{AB}, is a measure of the strength of the coupling between two nuclei A and B. It depends on the nature of the bond between them, and if they are separated by more than two bonds it depends on the relative orientations of the chemical bonds. It is independent of the external magnetic field. In macromolecules, these splittings are usually smaller than the intrinsic line widths. (4) Nuclear spins exchange among their energy levels at a rate determined by the rapidity of movement of the molecule or molecular group containing them. The rate is characterized by so-called "spin relaxation times" T_1 and T_2. T_1 (longitudinal or spin-lattice relaxation time) and T_2 (transverse or spin-spin relaxation time) decrease as molecular motion becomes slower. The time scale for molecular tumbling is the so-called correlation time, τ_c. (5) In liquids, generally $T_1 = T_2$, but at high viscosity or in macromolecules it is often observed that $T_1 > T_2$. When $T_1 \sim T_2$ one can often write $1/T_1 \propto \tau_c \propto \eta$, where η is the viscosity. T_2 is inversely related to the peak width, i.e., $T_2 = 1/\pi\Delta\gamma_{1/2}$, where $\Delta\gamma_{1/2}$ is the full width at half height. (6) When a small molecule binds to a macromolecule, two types of change are commonly observed in its NMR spectrum: (a) a change in chemical shift of one or more resonances, reflecting a change in the magnetic environment of one or more of the groups on binding; (b) a broadening of one or more peaks owing to decreased motional freedom of a particular part of the small molecule, or of the molecule as a whole, with a corresponding increase in correlation time and hence of relaxation rate. A working assumption has often been that those groups which interact directly with groups on the macromolecule will show both the largest change in magnetic environment and the largest decrease in motional freedom.

Depending on the lifetime of the complex, the appearance of the spectrum falls into one of three categories, referred to as the fast, intermediate, and slow-exchange cases. A thorough mathematical treatment has been reviewed (*16*).

Consider the case that involves exchange of magnetic nuclei between two equally populated sites A and B.
If

$$\frac{\tau_A + \tau_B}{2} = \tau$$

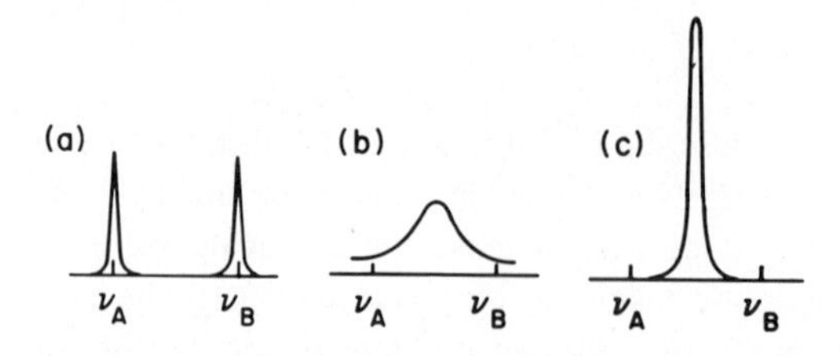

Figure 1. Change of line shape as a result of chemical exchange between two sites A and B at successively increasing exchange rates. (a) Slow exchange, $\tau >> 1/2\pi\Delta\delta$; (b) intermediate exchange, $\tau = 1/2\pi\Delta\delta$; (c) fast exchange, $\tau << 1/2\pi\Delta\delta$.

where τ is the mean lifetime, the shape of the resonance signals observed will depend upon the ratio of τ to $1/\Delta\gamma$, where $\Delta\gamma = (\gamma_A - \gamma_B)$ measured in Hz. Three conditions may be delineated: Slow exchange, $\tau >> 1/2\pi\Delta\gamma$. In this case there are separate resonance signals for the nuclei at each site (Fig. 1a). Intermediate exchange, $\tau = 1/2\pi\Delta\gamma$. If the exchange rate becomes faster (and the lifetime, τ, shorter), the two signals broaden and coalesce, giving finally a single broad resonance centered between γ_A and γ_B (Fig. 1b). Fast exchange $\tau << 1/2\pi\Delta\gamma$. A single sharp peak is observed whose chemical shift is the weighted average ($\gamma_{mean} = P_A\gamma_A + P_B\gamma_B$ where P_A and P_B are the fractional populations of A and B sites) of the shifts in the two states (Fig. 1c). Rates of the order of 5 sec^{-1} up to 10^7 sec^{-1} are measurable by this method (*17*).
(7) When metal ion binds to a molecule, two quite distinct effects are observed depending on the nature of the metal ion: Diamagnetic metal ions, such as Mg^{2+}, Zn^{2+}, Ca^{2+}, cause small chemical shifts which can be explained on the basis of charge effects. Paramagnetic ions, such as Mn^{2+}, Co^{2+}, Cu^{2+}, cause much greater chemical shift (paramagnetic shift) and line broadening of NMR spectra owing to the interaction of the nucleus with unpaired electrons. If the broadening effect is observed on more than one nucleus, it is possible to define the position of the metal ion with respect to the ligand because the degree of broadening is proportional to $1/r^6$, r being the distance between the nucleus and the unpaired electron.

EXPERIMENTAL CONSIDERATIONS

NMR measurements are carried out under biologically reasonable conditions and are nondestructive.

The volume of sample required for NMR measurements is generally 0.5 ml. A lower concentration limit to observe a proton resonance on a single sweep is about 1×10^{-3} in any given proton, depending on the spectrometer sensitivity. However, this concentration limit may be reduced to about

1×10^{-4} M by the application of Fourier transform spectroscopy (*18*). Microquantities of solution may be used in specially designed microcells (*19, 20*) with 0.03-0.05 ml, although loss of resolution usually results.

The sample is usually in D_2O solution in order to suppress the strong H_2O resonance that can overwhelm the central and aromatic regions of a polynucleotide. The exchangeable N*H*, NH_2, and O*H* protons of polynucleotides are usually replaced by exchange preliminary to running in D_2O to decrease further the intensity of the residual HDO resonance and its side bands. While the replacement of N*H* protons by deuterium results in great simplifications of the PMR spectra of polynucleotides, much potentially valuable information is lost because N*H* and NH_2 protons are intimately involved in the secondary and tertiary structures of polynucleotides (*21-23*). For work in D_2O solution (*24*), a pD is defined such that

$$\text{pD} = \text{pH} + 0.4$$

where the symbol "pH" refers to the reading obtained with a conventional "pH meter."

REFERENCE STANDARDS

The difficulties inherent in obtaining accurate measurements of field strength make an absolute scale for chemical shift impractical. Therefore line positions are measured relative to a given standard. Thus in frequency units, $\Delta\gamma = \gamma_s - \gamma_r$ (in Hz; i.e., cycles per second, cps) where $\Delta\gamma$ is the peak separation, γ_s and γ_r are the resonance frequencies of the sample and of the reference. Since $\Delta\gamma$ is proportional to the strength of the applied field, it is often expressed as a dimensionless parameter δ, where

$$\delta = \frac{\Delta\gamma}{\gamma_{osc}} \times 10^6 \text{ (in ppm)}$$

The most commonly used reference standard is tetramethyl silane (TMS), $(CH_3)_4Si$, owing to its chemical inertness and the fact that its resonance position is at one extreme of the usual range of absorption. Most other resonances occur to lower field, usually within a range of 10 ppm. The less volatile hexamethyldisoloxane (HMS) or water soluble 2,2-dimethylsila-pentane-5-sulfonate (DSS), $(CH_3)_3Si{-}(CH_2)_3{-}SO_3^- Na^+$, are also used. Water is not a useful standard since its chemical shift is sensitive to temperature and to the nature of the solution. The reference compound may be within the sample as an *internal reference*, or contained within a capillary as an *external reference*. These will usually give slightly different results depending on the difference of bulk diamagnetic susceptibility between the pure reference compound (external) and the sample solution (internal) (*25*). Internal referencing is preferably avoided since biological macromolecules have a tendency to bind small molecules. Micelle formation has been described for DSS in D_2O (*26*). The use of external reference is becoming increasingly widespread (*27–30*). The chemical shifts so measured should, however, be corrected for bulk magnetic susceptibility effects if the data are to be meaningful compared with those of other workers (Fig. 2). If variable

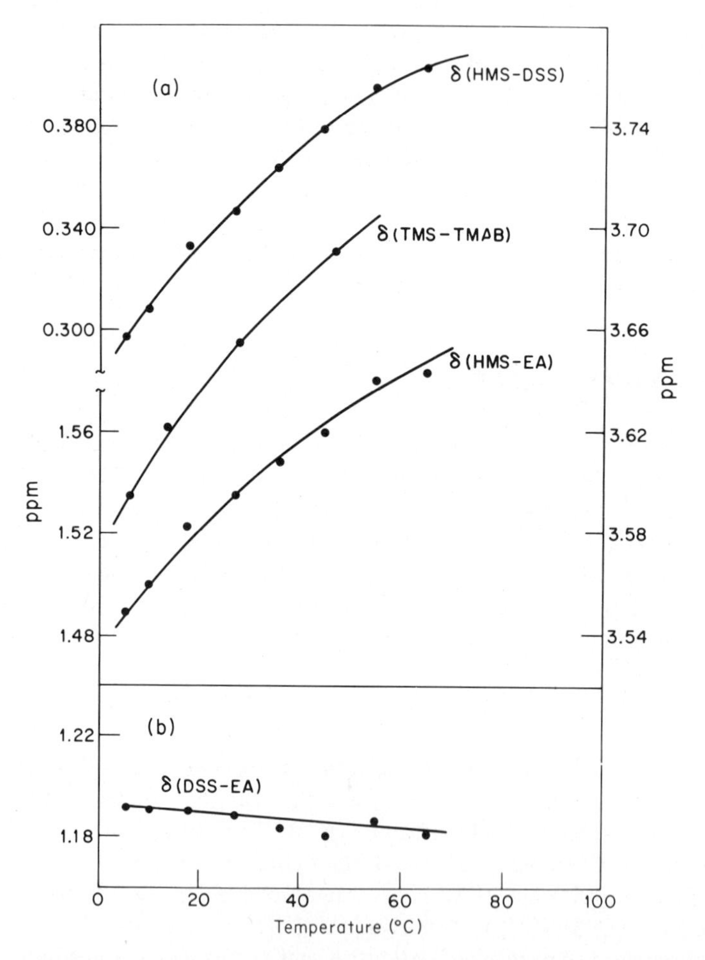

Figure 2. (a) Shift of external HMS relative to internal aqueous DSS[δ(HMS-DSS), upper left-hand scale], shift of alcohol [δ(HMS-EA), lower left-hand scale] and the shift of external TMS relative to aqueous tetramethylammonium bromide [δ(TMS-TMAB), right-hand scale], all as a function of temperature. (b) Shift of internal DSS relative to methyl resonance of ethyl alcohol as a function of temperature in aqueous solution. A positive shift indicates a resonance at higher field. All solutions were in D_2O, pD 7.0 ± 0.2. Concentrations were: DSS, 0.15 M; EA, 0.05 M; TMAB, 0.01 M (Ref. *31*).

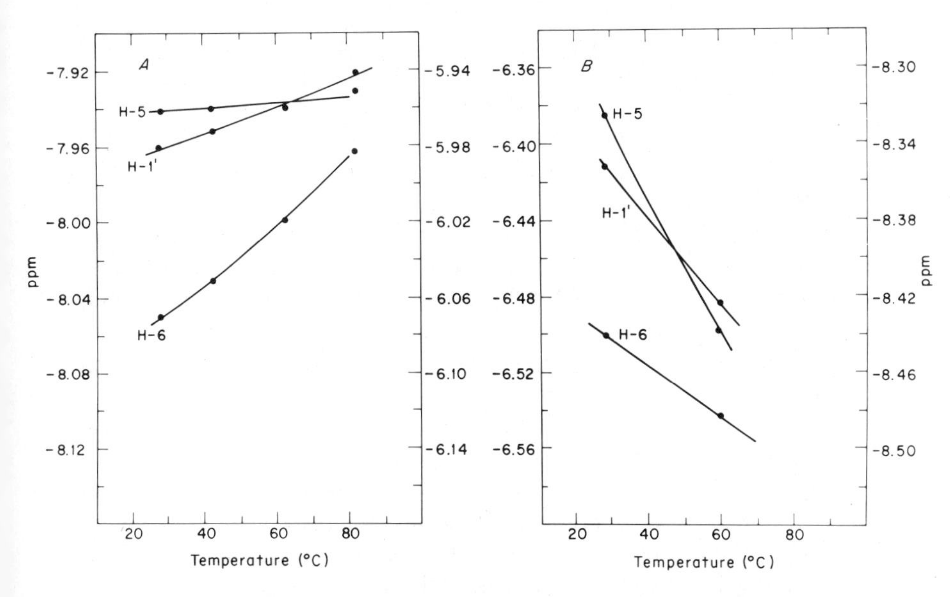

Figure 3. The chemical shift of the H-5, H-6, and H-1′ protons of uridine-5′-monophosphate as a function of temperature: (A) relative to internal DSS; (B) relative to external TMS. The shift of the H-6 proton is given by the extreme scales; the shifts of the H-5 and H-1′ protons are given by central scales. In (A) the uridine-5′-monophosphate was 0.1 M in D_2O, pD 7.0; the DSS was 0.15 M. In (B) the uridine-5′-monophosphate was 0.045 M in D_2O, pD 6.3 (Ref. *31*).

temperature measurements are made, *changes* in the correction for a given solvent-external reference pair *must* be considered (*31*) (Fig. 3). The difficulties inherent in external referencing, the instability and large spinning side-bands associated with spinning capillaries, may be minimized by the use of precision coaxial inserts (*32*). Quite accurate values for chemical shifts may be obtained using the signal of the desired reference compound as the lock signal and reading directly the difference between the fixed (lock) and variable (sweep) frequencies.

CONFORMATIONAL ANALYSIS

A detailed study of the NMR spectra of the building blocks of RNA and DNA provides a wealth of information about the structure and behavior of the monomers in solution, essential for the interpretation of polynucleotide spectra. The work to date has provided a new approach to a variety of problems. It has yielded much important information that as yet no other single technique can provide so directly.

The structural problems of purines and pyrimidines, their nucleosides and nucleotides have been approached by (1) considering the chemical shifts of base protons as a function of pH (*33, 34*), substituents (*35, 36*) and temperature (*30, 37*) (Information on the sugar–base torsion angle has been obtained (*38*).) (2) studying the dependence of H1′-H2′ coupling constants under various conditions and analyzing H2′-H3′ and H3′-H4′ coupling constants (*39, 40*), which are related to the ribose-ring puckering angles; (3) analyzing ^{31}P-H5′ coupling constants, which give information on the stereochemistry about the C5′-O5′ bond (*41–44*), and (4) analyzing the coupling constants between the proton at the 4′ position in the ribose ring and those on the exocyclic 5′-CH_2 group, leading to the dihedral angles between the relevant H–C–C planes (*43, 44*).

The first approach gives information useful for predicting base–base interactions owing to base stacking or hydrogen bonding, whereas the latter three approaches are relevant to the structure of the ribophosphate backbone. The first and the second approaches have been quite successful, the third one has shown some promise but very little has been done with the fourth method owing to difficulties in spectral analysis.

ANTI AND *SYN* CONFORMATIONS

The indication that 5′-nucleotides exist in the *anti* conformation (the base rotated about the glycosyl bond so that the 5, 6 side of the pyrimidine ring or the imidazole side of the purine ring is directed toward the 5′-position) in aqueous solution comes from the specific pH dependent deshielding effect (down-field shift) of the phosphate group on the H-8 proton, and not H-2, of purine and the H-6 proton of pyrimidine nucleotides (*34, 34, 37, 45, 46*) (Fig. 4). In this conformation the H-6 proton of the pyrimidine is much nearer the 5′-position than is the H-5 proton. This deshielding effect is not observed with the 2′- or 3′-nucleotides (*34*) (Fig. 5), and is largest when the phosphate is in the dianion form, less when in the monoanion form, and least in the monomethyl ester form. A simple calculation of the deshielding expected from electric field effects owing to the phosphate group (*47*) indicates a shift of -20 Hz for H-8 and less than -2 Hz for H-2 in the 5′-isomer. For the 2′- and 3′-nucleotides the corresponding deshieldings are estimated to be less than 1 Hz. These values are in the range of observed deshieldings. The 5′-hydroxyl group also has a specific deshielding effect on the H-6 protons considerably weaker than that from a dianionic 5′-phosphate but comparable to that from a monoanionic form (*37*). If the temperature is increased, resonances of the H-6 protons of uridine and cytidine move to a higher field (lower chemical shift), whereas the resonance of the H-5 proton does not vary appreciably

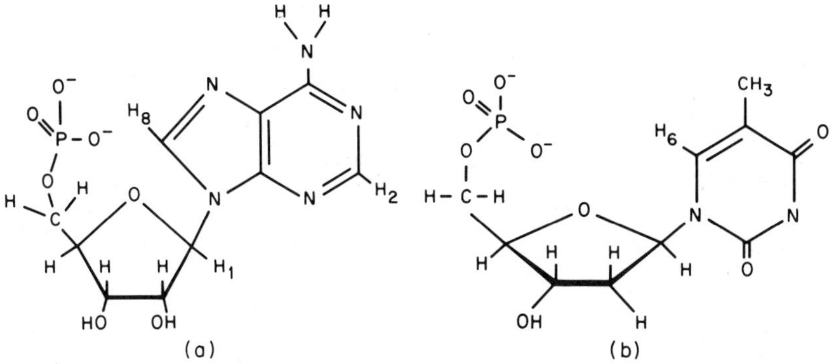

Figure 4. The *anti* configurations of (a) 5′-AMP and (b) 5′-TMP.

with temperature. This can be interpreted as due to a specific deshielding effect by a conversion of the conformation from *anti* to *syn* with the increase in temperature. A similar effect was observed with 5′-UMP and 5′-CMP. The *anti* conformation is known to be favored in the crystalline state of normal nucleosides and nucleotides (*48, 49*).

Conformation of Furanose Ring

In a very interesting theoretical study, Karplus (*50*) calculated the coupling constant among protons on the adjacent carbons of ethane as a function of the dihedral angle formed by the H–C1–C2 and C1–C2–H bond pairs. His treatment is based on the assumptions that the carbon hybrid orbitals are tetrahedral (sp^3), that the carbon–carbon bond is 1.543 Å, and that the magnitude of the coupling constant is determined mainly by the electron-spin interactions. Since the dihedral angles between C–H bonds on adjacent carbons in a furanose ring are determined by the mode and extent of buckling of the ring, a knowledge of the hydrogen coupling constants permits the determination of the stereochemistry of the ribose moiety. Approximate $J_{1'2'}$ values are readily obtained by a simple first-order treatment but the remaining couplings are less easily extracted because of considerable overlap in the 2′ to 5′ proton spectral region. Consequently, deductions regarding furanose ring conformation in earlier studies have been based mainly on the approximate value of a single coupling constant from which a single dihedral angle, $\phi_{1'2'}$ may be estimated. Suggested conformations based on a single coupling constant should be accepted with some reservation and must be confirmed by measurement of the remaining coupling constants. Recently,

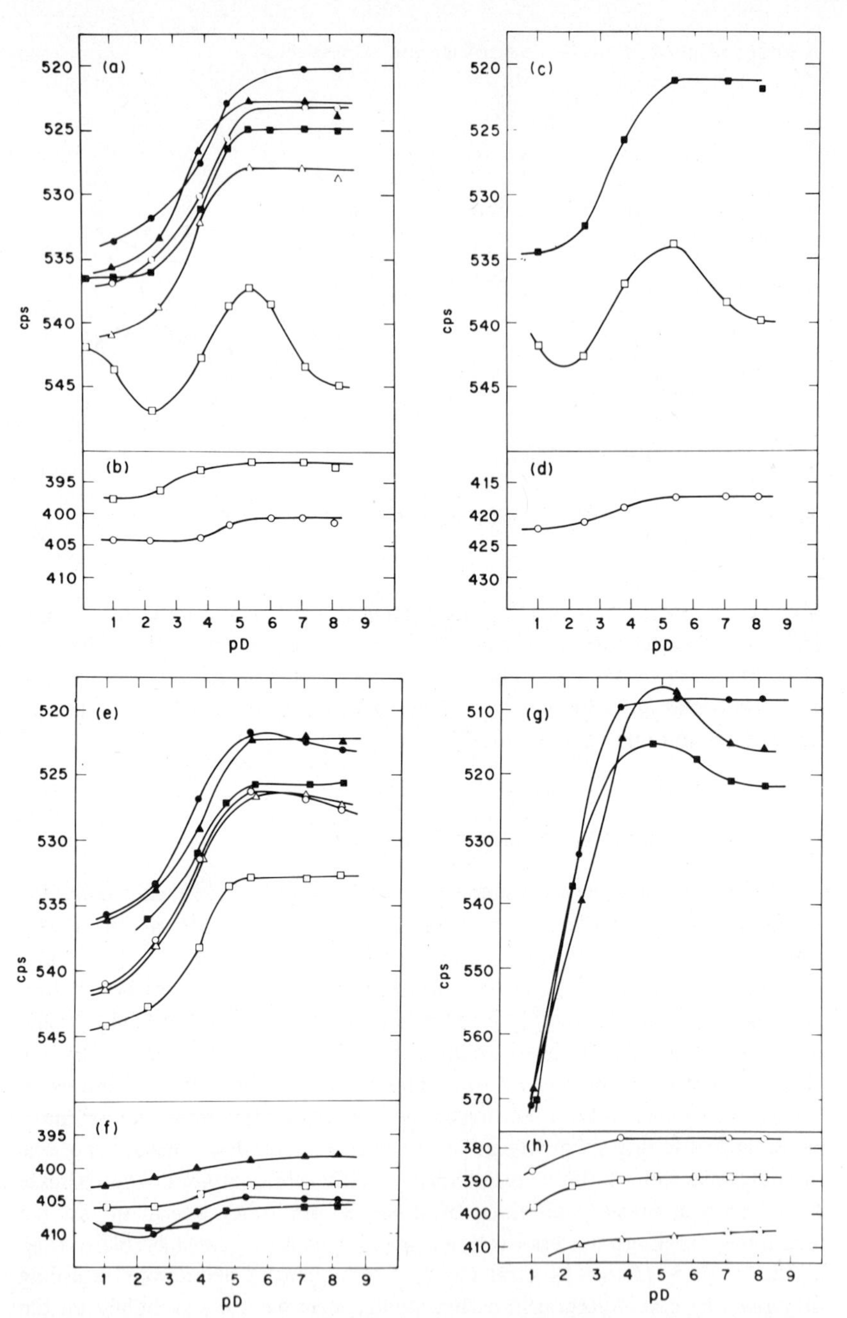

Figure 5. Proton chemical shifts (cps relative to external TMS) vs. pD of the solution. (a) Adenine H-2 (●) and H-8 (○), adenosine H-2 (▲) and H-8 (△), and adenosine 5′-monophosphate H-2 (■) and H-8 (□). (b) H-1′ of adenosine (□) and of adenosine 5′-monophosphate (○). (c) 2′-Deoxyadenosine 5′-monophosphate H-2 (■), H-8 (□), and (d) H-1′ (○). (e), (f) Adenosine 2′,3′-cyclic monophosphate H-2 (●) and H-8 (○), adenosine 3′,5′-cyclic monophosphate H-2 (■) and H-8 (△), and adenosine 2′- and 3′-monophosphate (mixture) H-2 (■) and H-8 (□). (b) H-1′ (●), (▲), (■), and (□) of compounds in a, respectively. (g), (h) Guanosine H-8 (●) and H-1′ (○), guanosine 5′monophosphate H-8 (■) and H-1′(□), and 2′-deoxyguanosine 5′-monophosphate H-8 (▲) and H-1′ (△) (Ref. *34*).

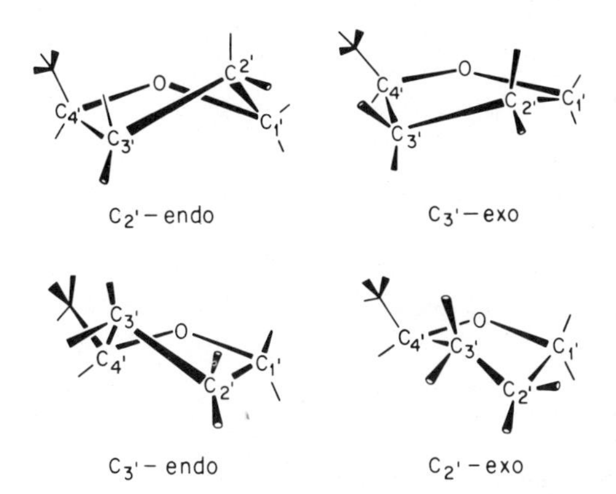

Figure 6. Possible buckled conformations for the ribose ring.

well-resolved ribose spectra of uridine (*39*) and nucleoside 3′-5′ cyclic phosphates (*40*) were obtained and the consideration of all the ribose-ring coupling constants indicates C2′ *endo* conformation (Fig. 6). (*Endo* means that the atom is located on the same side of the plane defined by C1′O1′ and C4′ as the C4′–C5′. *Exo* means that it is found on the opposite side.) Similar studies on α- and β-pseudo-uridines (*43, 44*) have revealed that the conformation of the furanose ring in these compounds can be described by an equilibrium between various puckered ring conformations.

CONFORMATION OF THE EXOCYCLIC CH_2OH GROUP

Although base-stacking interactions appear to be of primary importance in maintaining the purine and pyrimidine bases in planar stacked conformation, these forces alone cannot account for the conformational rigidity of the ribose–phosphate backbone. The conformational preference of the backbone, furanoside rings linked via C4′–CH_2–O–P–O–C3′, should be due in part to rotational barriers in the five exocyclic bonds (C4′–C5′, C5′–O5′, O5′–P, P–O3′, and O3′–C3′). NMR can be used in determining the preferred backbone conformations because of the sensitivity of proton–proton and proton–phosphorous coupling constants to the dihedral angle in H–C4′–C5′–H and H5′–C5′–O–P segments. Some attempts have been made to determine conformations of nucleotides in solution by measuring H5′–P coupling constants (*42, 43*).

The orientation of the 5′–OH relative to the furanose ring is a conformational parameter of interest in nucleoside structure. However, partly due to the difficulty in obtaining an accurate vicinal coupling constants between H4′ and H5′ protons, $J_{4'-5'}$, no detailed investigation of the

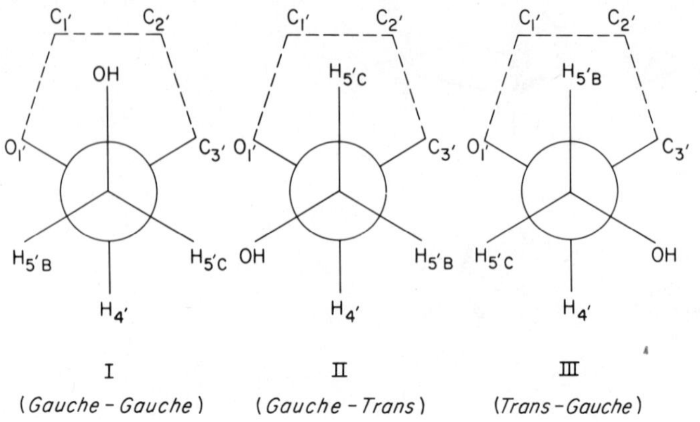

I (*Gauche - Gauche*) II (*Gauche - Trans*) III (*Trans - Gauche*)

Figure 7. Rotational isomers around the C4′–C5′ bond.

stereochemistry of the C4′–C5′ bond has been attempted until recently. A treatment of the vicinal $J_{4'-5'}$ coupling constants (*50*) can theoretically yield information about the relative rotamer populations in aqueous solution. There are three possible conformations determined by rotation about the exocyclic C4′–C5′ bond: *gauche–gauche*, the most common form found in crystalline nucleosides and nucleotides (*48*), *gauche–trans*, and *trans–gauche* (Fig. 7). A slight preference for the *gauche–gauche* form was observed for α-pseudo-uridine (*43, 44*) and β-pseudo-uridine, although molecules rotate freely about the exocyclic C4′–C5′ bond.

BASE–BASE STACKING INTERACTION; RING-CURRENT SHIFT

A particular useful NMR phenomenon in the studies of elucidation of inter- and intramolecular interactions between base monomers and oligomers is the shift of proton resonances resulting from the magnetic fields generated by ring currents in aromatic bases such as adenine, guanosine, and purine. When aromatic rings are placed in a magnetic field, circulating currents of delocalized π-electrons are set up which generate a subsidiary magnetic field opposing the main field (Fig. 8). Protons above or below the plane of the ring are partially shielded from the applied magnetic field and higher fields H_0 have to be applied to achieve the resonance condition. For instance, increasing the concentration of benzene in a nonaromatic solvent results in an up-field shift (0.5–1.0 ppm) owing to an association by hydrophobic stacking (*51*). Protons on the periphery of the ring are deshielded and their resonance peaks appear at lower fields than usual (*52*). Neither magnitudes nor

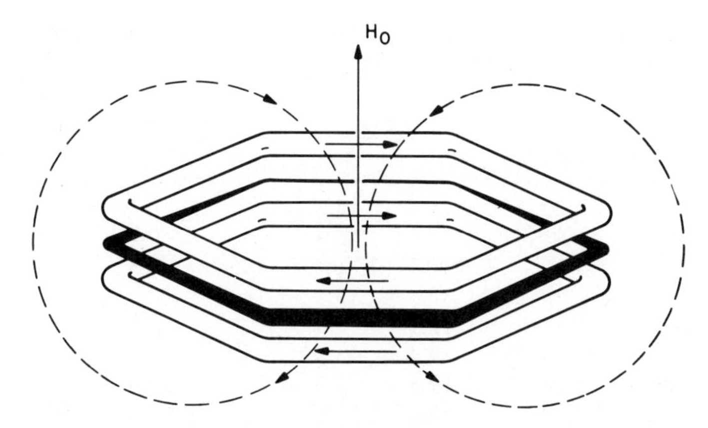

Figure 8. If the external magnetic field H_0 has a component perpendicular to the plane of an aromatic ring, ring currents are induced in the π-orbitals and subsidiary magnetic fields are generated. Nuclei in groups lying above or below an aromatic ring experience up-field resonance shifts, and nuclei in groups lying to one side of an aromatic ring experience down-field shifts.

geometrical distributions of ring-current fields of aromatic structures have been well established by theoretical calculations or experimental measurements (*6*). The ring-current field of benzene calculated by Johnson and Bovey (*53*) is generally used for semiquantitative description of ring-current fields for phenyl groups and with appropriate modifications for other aromatic structures.

The above considerations have enabled the modes of interaction of nucleic acid derivatives to be differentiated: an up-field shift is attributed to stacking of purine or pyrimidine bases and a down-field shift due to a juxtaposition of rings by specific hydrogen bonding.

The resonances of purines, their nucleosides and nucleotides shift to higher field with increasing concentration in aqueous solution (*34, 35, 54–57*), providing some detailed information about the nature and the orientation of the bases. The shift for H-2 is substantially larger than that for the H-8, H-1′, and CH_3 protons (Table 1). This can be explained by assuming a stacking model in which H-2 proton is shielded strongly by the six-membered ring of the neighboring base, and H-8 and H-1′ protons being shielded mainly by the five-membered ring.

Additionally, the latter protons will spend less time shielded by the five-membered ring relative to the time spent by the H-2 proton in the proximity of the six-membered ring. Given that the thermal energy kT is of the same order as the standard free-energy of association, the breaking up and the reforming of molecule stacks must occur rapidly, since no significant line broadening or splitting was observed. These bases associate in stacks, with the

TABLE 1. Concentration Dependence of Chemical Shifts for 11 Purine Nucleosides (0.0– 0.2 M) in D_2O[a]

Compound	Temp (°C)	Δδ, cps				
		H-2	H-8	H-6	H-1′	CH_3
Inosine	32	6.4	5.3		7.1	
1-Methylinosine[b]	33	8.9	6.4		6.8	5.3
Ribosylpurine	30	10.7	6.4	13.1	8.8	
Purine	25–27	12.6	9.6	14.2		
2′-*O*-Methyladenosine	31	13.7	7.5		8.8	
6-Methylpurine	25–27	19.4	13.3			17.0
2′-Deoxyadenosine	30	19.8	13.0		13.6	
N-6-Methyl-2′-deoxy-adenosine	32	26.0	15.8		14.0	15.2
N-6-Dimethyladenosine	28	27.2	14.5		14.4	25.5
N-6-Methyladenosine	26	32.6	17.5		12.6	18.1
2′-Deoxyadenosine[c]	30	14.8	10.0		9.8	
Adenosine[c]	32	14.8	8.3		6.9	
3′-Deoxyadenosine[c]	25	15.8	9.0		9.6	

[a]From Broom et al. (*59*).

[b]Peak positions of H-8 and H-2 are reversed with respect to the other 6-substituted nucleosides studied.

[c]Differences measured over the concentration range 0–0.10 M because of solubility limitation.

aromatic planes parallel, overlapping partially, and a ring separation of 3 to 4 Å.

Similar studies showed that purine interacts with pyrimidine nucleosides by a vertical stacking mechanism, resulting in increasing up-field shifts of pyrimidine ring protons with increasing purine concentration (*57*). Association occurs even when the potential sites for hydrogen bonding have been methylated, e.g., 1-methylinosine and *N*-6-dimethyladenosine.

It was not possible to follow the self-association of pyrimidine bases, cytidine, thymidine, and uridine (*57*), which are nonaromatic due to the lack of ring-current effects in these compounds. The magnitude of the observed shifts is of roughly the same order as might be expected from the diamagnetic susceptibility of the bulk solution. Thus in comparison to similar studies on purine bases, the concentration shifts for the pyrimidine nucleosides studies are negligible. This is in contrast to the report which showed by osmotic-pressure lowering that cytidine and uridine associate in aqueous solution (*58*), probably because their interactions cannot be monitored by PMR via the effect of the ring-current magnetic anisotropy. However, it has been shown that NMR spectra of the UpUpU and UpUpC depend critically upon

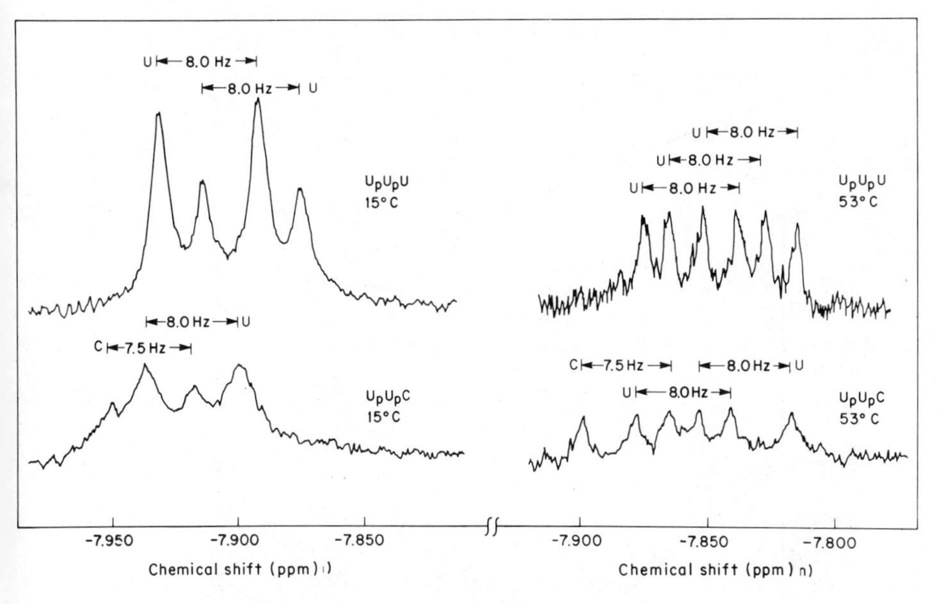

Figure 9. NMR spectra at 220 MHz of 0.02 M solutions of UpUpU and UpUpC in D_2O, pD 7.0, 0.1 M NaCl, 0.04 M $MgCl_2$. Chemical shifts are measured relative to internal 3-(trimethylsilyl) propane–sulfonic acid (Ref. *37*).

temperature and below 40°C there is a high degree of order among the pyrimidine bases (Fig. 9) (*37*).

Up-field shifts of the pyrimidine protons are observed by the addition of purine, indicating purine–pyrimidine stacking (*54*). A comparison of the purine-induced shifts of the pyrimidine base protons with the concentration shifts of purine base protons suggests that the magnitude of purine–pyrimidine interactions is approximately 50 to 60% of the purine–purine interactions.

COMPLEMENTARY BASE-PAIRING

The main phenomenon observed in a purine–aqueous system is a stacking interaction between bases, shown by a concentration-dependent up-field shift in the nonexchangeable ring proton resonance (*54–56, 59*). For instance, the C-8 ring proton and the ribose C-1′ proton of guanosine in aqueous solution show this shift. However, in DMSO (dimethyl sulfoxide) (*21*) the same protons show no change with concentration, and accordingly stacking interactions can be considered negligible in this solvent, in concentration up to 0.5 M. A slight shift of the N-1 and of the amino proton resonance

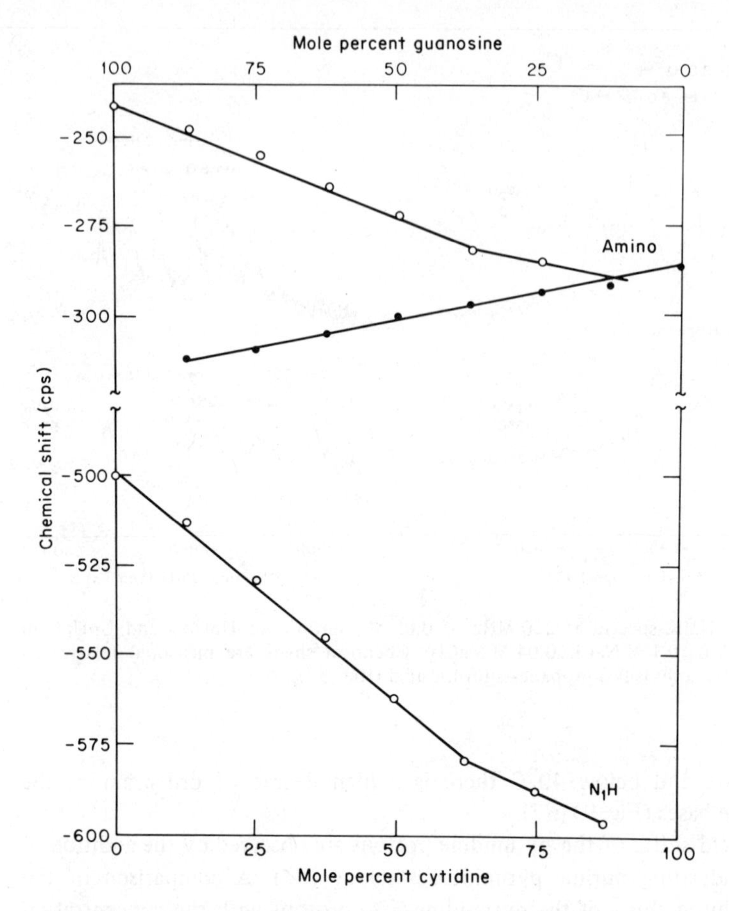

Figure 10. Chemical shift of the hydrogen-bonding protons of guanosine (○) and cytidine (●) in mixtures of the two species in DMSO. The measurements were made at 16° on solutions containing a total nucleoside concentration of 0.5 M (Ref. *21*).

is considered indicative of small degree of self-association. However, an addition of pyrimidine base causes a large down-field shift of NH protons indicating the formation of strong-hydrogen bonds (*21, 22*). In Fig. 10 are shown the large down-field shifts of the NH protons on the formation of the G-C pair in DMSO. Such studies have also been extended to association between less common bases (*60*).

The discrepancy that the complementary base-pairing of nucleosides and nucleotides derivatives was observed only in nonaqueous solution (*21, 22*) and not in aqueous solution was explained by the fact that the large up-field shifts resulting from stacking can mask smaller down-field shifts owing to hydrogen-bonded interactions. However, in nonaqueous media the tendency for hydrophobic interactions is reduced, and that of the weaker hydrogen-bonding is increased (*61*). It also became possible in nonaqueous solvents to observe the purine and pyrimidine amino protons which are directly involved in the formation of hydrogen-bonds, since they are not rapidly exchanging with solvent.

However, since NH and NH_2 proton resonances have been observed to good advantage in proteins in aqueous solution (*62*), there is a priori no reason not being possible to detect it in polynucleotides in aqueous solution. Recently, resonances of exchangeable protons in the low field of the 220 MHz spectra of RNA and in the mixtures of adenine oligomers and poly U have been observed (*23*).

BINDING OF SMALL MOLECULES

As an aid to understanding the interactions between dye molecules and nucleic acids, NMR studies of solutions of bases, nucleosides, and nucleotides in the presence of acridine orange have been carried out (*63*). Up-field shifts of the NMR resonances of the nucleosides and nucleotides demonstrated that the dye interactions occurred via a mechanism of base-stacking rather than coplanar ionic attraction. Comparisons of the relative magnitudes of the up-field shifts induced in ribo- and deoxyribonucleotides showed that the interaction was stronger for the latter.

The three resonance peaks of purine (0.1 M) are shifted down-field in the presence of Zn^{2+} (0.01 M) by 2.0, 2.7, and 3.7 Hz for the H-2, H-6, and H-8 proton resonances, respectively (*64*). The addition of Cu^{2+} (0.1 M) to purine (0.1 M) in DMSO, causes line broadening in the order H-6 > H-8 > H-2, indicating that both Zn^{2+} and Cu^{2+} bind to purine around the N-7 position.

The addition of Zn^{2+} ion to cytosine (0.1 M) solution in DMSO causes the H-5 and the H-6 signals to be shifted down-field to an equal extent, and Cu^{2+} ion (10^{-3} M) causes a slightly greater line broadening for the H-5 than for the H-6 resonance. Thus the N-3 is preferred for both Zn^{2+} and Cu^{2+} ions binding in cytosine (*65*).

NH protons are not affected by Cu^{2+} ions (*65*), indicating that Cu^{2+} does not bind to amino groups, confirming the earlier proposals (*66, 67*).

The ^{31}P spectra of dAMP and dTMP showed considerable broadening of the phosphate resonances in the presence of Cu^{2+}, indicating that Cu^{2+} binds to the phosphate moieties (*64–66*) and it also caused line broadening of H-2 and H-8, the latter to a considerably greater extent. The ribose proton lines were little influenced by Cu^{2+} (*65*). Therefore, Cu^{2+} ion is bound to the phosphate and to the adenine moiety around N-7 in dAMP.

Recent results seem to indicate that the position of the phosphate affects profoundly the binding of Cu^{2+} to bases. For instance, H-8 and H-2 are broadened simultaneously in adenosine, 2′-AMP and the cyclic AMP. However, H-8 is preferentially broadened in 3′- and 5′-AMP (*68*).

A thorough and a detailed NMR study of the binding sites and conformations of Mn^{2+}, Co^{2+}, and Ni^{2+} complexes with ATP and nucleic acids have revealed the following (*69–74*): (1) Mn^{2+} and Co^{2+} ions bind to the phosphates of AMP and RNA. (2) In Mn^{2+}-ATP and Co^{2+}-ATP complexes, the metal ion binds to the phosphates approximately 100% of the time that it is bound. The evidence strongly favors simultaneous binding to all three phosphates. (3) In concentrated solutions of Mn^{2+}-ATP, the metal is simultaneously bound to the phosphate moiety of one nucleotide and to the adenine ring N-7 of the second nucleotide, forming a 1 : 2 metal–ligand complex. At low concentration, 1 : 1 complex is formed and the results are the metal ion being located ca. 3.8 Å from the H-8 proton, separated from the adenine ring by a coordination shell water molecule.

The relaxation enhancement technique (*75, 76*) is very useful for the study of a wide range of chemical and biological problems involving magnetic ions.

In solutions of paramagnetic ions such as Mn^{2+}, spin relaxation of the water protons is dominated by the metal ions. The effective correlation time τ_c is thought to be determined by Brownian rotation of the hydrated Mn^{2+} complex. In the presence of a macromolecule, τ_c may be increased by hindered rotation in the bound state, in which case spin relaxation of the solvent water protons becomes even more efficient. A proton relaxation enhancement parameter, ϵ, may be defined as follows:

$$\epsilon = R^*/R$$

where R^* and R are the proton relaxation rates for a certain ion concentration in the presence of the macromolecules under study and for the aqueous solution, respectively. R is obtained from the experimentally observed relaxation rate $1/T_1$ by subtracting the contribution to the relaxation rate which has nothing to do with the paramagnetic ions (generally $1/T_1$ for pure water). R^* is obtained in a corresponding manner. The three limiting cases of no ion binding, ion binding in accessible sites, and exterior ion binding may be characterized by $\epsilon = 1$, $\epsilon \ll 1$, and $\epsilon \gg 1$, respectively. By suitable titrations it is possible to determine the characteristic value for each complex at the tight binding sites as well as calculate the number of binding sites and association constants (*77*). It was shown (*77*) that the behavior patterns of ribosomal and transfer RNAs have two features in common: (1) there are at least two classes of binding sites, a smaller number of strong than weak binding sites and (2) the binding within the class of strong sites is cooperative.

This relaxation enhancement technique has been extensively used for the studies of paramagnetic ions, mainly Mn^{2+}, to ribosomes and tRNAs (*77–80*).

DINUCLEOTIDES

The relative conformation of the two bases in a dinucleotide can be studied conveniently if one of the bases is a purine or a purine derivative, because the ring-current magnetic anisotropy can influence the magnetic environment of the ring protons of the opposing base. For instance, information about the intramolecular base-stacking interaction can be obtained from the observed chemical shifts of the base protons as a function of temperature and pH (*81*), or by investigating the mode of interaction between a "probe" molecule such as purine and the dinucleotide (*81, 82*). For instance, the PMR spectrum of ApA in the region of the aromatic protons consists of four distinct resonance peaks corresponding to two adenine H-2 and two H-3 protons.

The two H-2 resonances in ApA both appear at significantly higher fields than the spectral position of this proton in ApU, UpA, and the other adenine mononucleotides, indicating that the two adenine bases in ApA are intramolecularly stacked (Table 2). If the temperature is increased, the two adenine bases destack and H-8(5′) and the up-field H-2(3′) resonances exhibit significant down-field shifts. This indicates that both adenine bases are preferentially oriented in the *anti* conformation, as they are in a helical DNA segment of sequence dApdA.

The temperature dependence of the H1′–H2′ coupling constants of the ribose moieties (*82, 83*) indicates that at low temperatures (below 30°C) the $J_{H1', H2'}$ is noticeably smaller for ApA than for its monomer (2.3 Hz instead of 5.3) and approaches that of the monomer as the temperature increases. This provides strong evidence that at low temperature the intramolecular stacking interaction of the two adenine bases in ApA changes the ribose conformation of both nucleosides from 2′-*endo* toward 3′-*endo*, whereas at high temperatures they resemble the conformation of the monomer.

Since purine bases are known to associate in stacks (*35, 36, 54–57*), ApA is also expected to do so. In fact, a pronounced concentration effect on the PMR spectrum of the adenosine protons of ApA has been observed (*81*). The up-field shifts observed for all the proton resonances with increasing concentration can be attributed to an extensive self-association, via external stacking without base intercalation.

Addition of purine to ApA results in the formation of the purine-intercalated dinucleotide complex (*81, 92*). Purine intercalation was monitored by the purine induced chemical shifts observed for the adenine H-2 and H-8 proton resonances and the H1′ ribose resonances, the variation of the coupling constants between the H1′ and H2′ ribose protons ($J_{H1' H2'}$) with purine concentration, and the line broadening of the purine resonances. The variation of the H1′–H2′ coupling constants suggests that the ribose conform-

TABLE 2. Summary of Chemical Shifts of the Adenine Protons in ApA, ApU, UpA, and Related Mononucleosides and Nucleotides at 31°C[a]

Molecule	Concn, M	Solution pD	Chemical shifts,[b] ppm			
			H8(5′)	H8(3′)	H2(5′)	H2(3′)
ApA (sodium salt)	0.01	7–8	8.69	8.65	8.57	8.43
	0.003	7–8	8.70	8.67	8.60	8.47
	infinite dilution[c]	7–8	8.71	8.68	8.61	8.48
ApU (sodium salt)	0.01	7.7	—	8.77	—	8.63
	0.003	7–8	—	8.80	—	8.66
	0.001	7–8	—	8.80	—	8.67
UpA (sodium salt)	0.01	7–8	8.84	—	8.66	—
	0.003	7–8	8.85	—	8.67	—
3′-AMP	0.01	5.8	—	8.79	—	8.68
	0.01	11	—	8.70	—	8.68
5′-AMP	0.009	6.2	8.92	—	8.68	—
	0.01	10	9.06	—	8.70	—
2′-AMP	0.01	5	8.79		8.68	
Adenosine 3′, 5′-cyclic phosphate	0.01	—	8.67[d]		8.67[d]	
Adenosine	0.003	—	8.76		8.68	

[a] From Chan and Nelson (*82*).
[b] Downfield relative to external TMS; not corrected for bulk susceptibility.
[c] Extrapolated.
[d] More exact positions of resonances: 8.674, 8.665.

ation of the two furanose rings of the ApA dinucleotide changes when purine intercalation occurs, from 2′-*endo* toward 2′-*endo*, because two adenine bases have to move apart to incorporate a purine base, restoring the ribose conformation of that of the mononucleotides.

There are reports on the variation of line widths from different commercial preparations of several dimers owing to paramagnetic metal ion contaminations (*30*). Therefore the broadening of the guanine H-8 observed depending on the neighboring base in dimers reported previously (*84*) should be repeated. In all dimers reported the nucleosidyl units all have the *anti* conformation, and the turn of the (3′-5′) screw axis of the stack is right handed.

Addition of purine to pyrimidine dinucleotides resulted not only in up-field shifts, but also a splitting in certain of the pyrimidine absorptions into two sets with equal intensities indicating purine insertion between the pyrimidine bases (*81*).

Thymine Resonance

Two separate resonances have been observed for thymine methyl protons in a series of trinucleotides (*86*) and dinucleotides (*87*), and the chemical shift depends upon whether the 5′-neighbor is a purine or a pyrimidine. The T–CH_3 resonance of TMP is a single line located at 1.89 ppm, independent of the temperature. For dimers in which thymine is in the 5′-neighbor position regardless of the nature of the 3′-neighbor, e.g., TpT, dTpC, dTpA, and dTpG, the T–CH_3 resonance occurs from 1.81 to 1.84 ppm at 20°C and from 1.86 to 1.89 ppm at 90°C. The resonance position is similar when thymine has a pyrimidine neighbor on the 5′ position. However, the T–CH_3 proton resonances of dimers in which thymine has a purine neighbor on the 5′ side (dApT 1.61 ppm and dGpT 1.68 ppm) are shifted considerably to higher field values at 20°C. This shift decreases as temperature is increased. The high-field shifts of T–CH_3 proton resonances of dApT and dGpT at 20°C is due to intramolecular complex formation stabilized by stacking of the purine and thymine bases in which the favored configuration is one in which the thymine–methyl group lies well within the magnetic field of the purine ring-current. As the temperature is increased, the equilibrium between these stacked forms and forms in which there is a random relationship between the two bases of the dimer is shifted toward the randomized configurations which should provide a T–CH_3 proton resonance at about 1.89 ppm. Since the resonance signal observed is narrow, fast exchange between the stacked and randomized configurations seems to be likely.

The spectrum of denatured (single-stranded) calf thymus DNA at 90°C shows 5 distinct spectral regions (Fig. 11) (*87*). The region 7.3 to 8.3 ppm

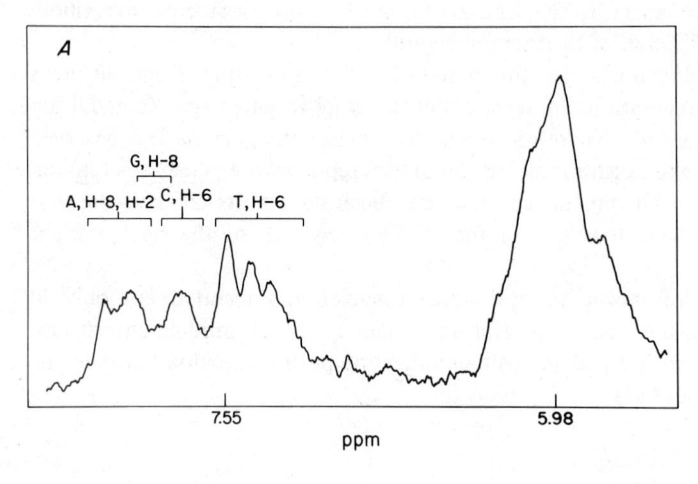

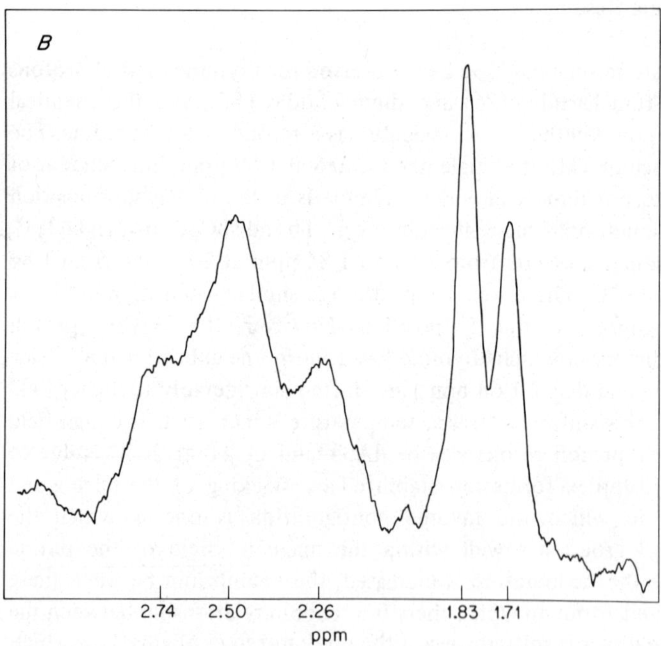

Figure 11. (A) Low field regions of PMR spectrum of calf thymus DNA at 220 Mc/s. Concentration, 30 mg/ml. D_2O, pD 7, 93°C, 50 spectra averaged.
(B) High field regions of PMR spectrum of calf thymus DNA at 220 Mc/s. Concentration 20 mg/ml. in D_2O, pD 7.0, 93°C, 20 spectra averaged in computer of average transients (Ref. *87*).

comprises the H-2 and H-8 protons of adenine, the H-8 protons of guanine, and the H-6 protons of cytosine and thymine; those from 5.8 to 6.4 ppm come from the H-5 protons of cytosine and the H-1′ deoxyribose residue. The H-3′ signals occur in the region from 2.1 to 2.8 ppm, and the T–CH_3 proton resonances cover the range 1.6 to 1.9 ppm. The HDO resonance obscures the resonance region of the H-4′ and H-5′ protons. Two T–CH_3 signals at 1.71 and 1.83 ppm are completely resolved with a frequency separation of ca. 28 cps. The relative intensities provide a measure of the frequency of occurrence of purine and pyrimidines at this position; the ratio was found to vary considerably for DNA from different species in a manner which paralleled differences in their purine/pyrimidine contents as determined by other methods (*85, 88*). The observation of large nearest-neighbor effects in single strand DNA (denatured) at 90°C indicated that even under these drastic conditions a considerable amount of base stacking still exists.

Polynucleotides

Direct NMR studies of polynucleotides have been less frequent and less successful than those on polypeptides. This has been partly due to the low resolving power of earlier instruments and complications which are not present for instance in the polypeptide systems. No NMR spectrum is observable for aqueous solution of DNA at temperatures below the thermal transition point (*85, 89*). This is due to the rigid structure of DNA which would result in very slow molecular motions that were ineffective in averaging out anisotropic dipole–dipole interactions. Thus below the thermal transition the NMR spectrum of DNA contains lines that are too broad to detect. The spectrum of DNA in Fig. 11 was obtained at 90°C and arises, therefore, from single-stranded, random-coil DNA.

The PMR of poly A shifts progressively to lower fields as the temperature is increased from 25° to 60°C, as a result of the conversion of a partially ordered configuration at room temperature to a random-coil configuration (*87, 90*). The spectral line widths decreased at the same time. The resonance frequencies of the PMR spectra of poly I or poly C did not shift by more than a few hertz as the temperature was increased from 25° to 90°C although some narrowing of the resonances was observed. In particular, the resonances of poly I were broad at 25°C but narrowed appreciably by 60°C indicating increased motion of the polymer molecules in solution at higher temperatures (Fig. 12).

When poly A and poly U form the well-known helical poly (A·U) and poly (A·2U) complexes (*81, 91, 92*), it results in broad spectra which are not detected by NMR spectrometer. In contrast, the PMR spectra of single-stranded poly A and poly U are rather sharp because of rapid local motion of

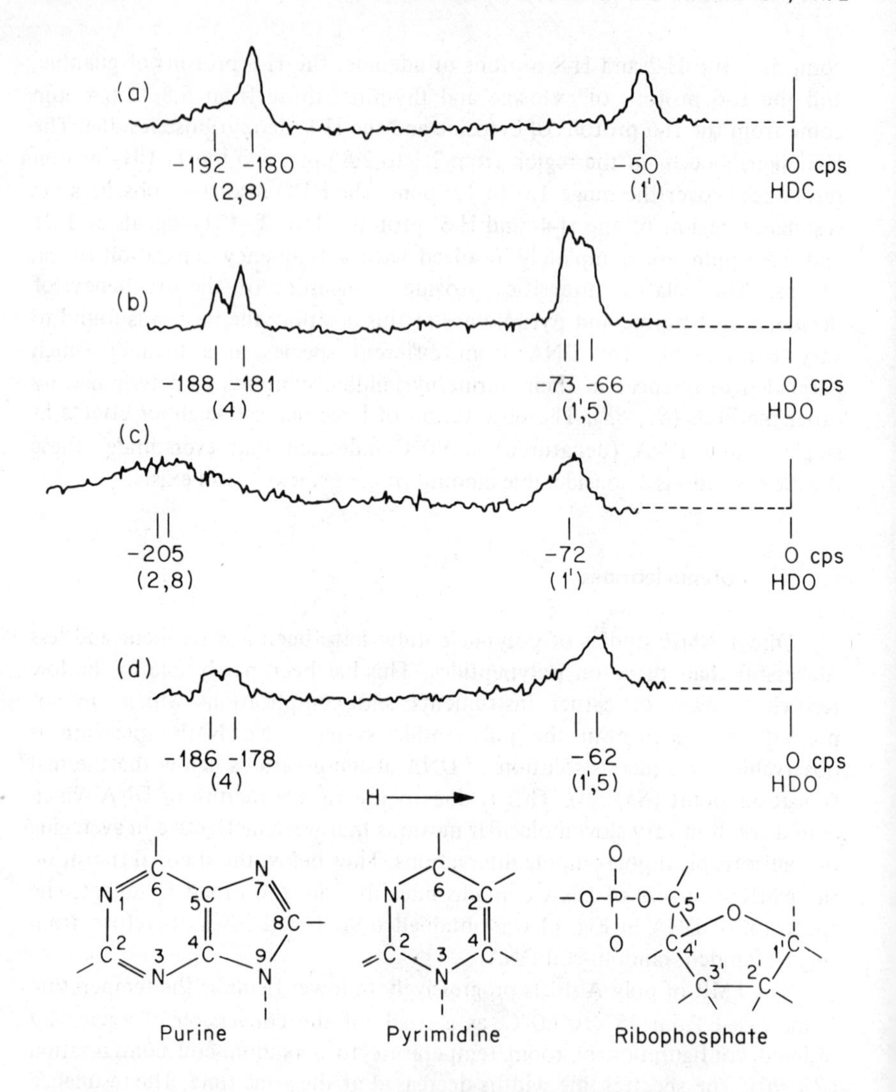

Figure 12. (a) Poly A, 32.0 mg/ml in D_2O, 35.5°C, 8 traces stored; (b) poly U, 31.8 mg/ml in D_2O, 35°C, 8 traces stored; (c) poly I 20.2 mg/ml in D_2O, 75°C, 10 traces stored; (d) poly C, 21.0 mg/ml in D_2O, 60°C, 10 traces stored. (Ref. *89*).

the polymer segments, although the line widths of the uridine proton resonances in poly U are 1.5–2.0 Hz, about twice as great as those for the uridine monomer. The loss of motional freedom usually takes place when an ordered, rigid complex is formed, resulting in an increase in line widths. Thus the formation of poly (A·U) and poly (A·2U) can be followed quantitatively from the intensities of the PMR spectra of solutions containing these polynucleotides. "Melting" curves of a solution of poly A and poly U can be obtained by plotting the averaged observed spectral intensities as a fraction of the expected intensities of poly A and poly U assuming random-coil configuration.

PMR studies on the interactions of poly U and poly C with nucleotides and nucleosides triphosphates have been reported (*93, 94*).

The addition of purine to poly U caused the up-field shift of the three monitored resonances (H-6, H-5, and H-1′) with the magnitudes in the order H-5 > H-6 > H-1′. Since the addition of purine to the uridine solution causes similar shifts for the uridine resonances, these up-field shifts show that purine interacts with both monomer and polymer by stacking with the uracil bases.

While $J_{H5\text{-}H6}$ remained unchanged, $J_{H1'\text{-}H2'}$ increased slightly as purine was added to poly U, indicating a slight change in ribose conformation. At low purine concentrations the H-2 resonance occurs at lower field in the presence of poly U than in a solution of purine alone. This suggests that poly U has the effect of disrupting the purine self-association. No broadening of the poly U proton resonances is observed, indicating that the intercalation of purine between adjacent uracil bases of poly U does not result in an appreciably rigid or ordered structure for the complex. The rate of exchange of purine molecules between poly U binding and the free purine is rapid on the NMR time scale (10^3 sec^{-1}), since the purine resonances represent an average of bound and free purine, rather than a superposition of broad resonances from bound purine and narrow resonances from free purine. Similarly, the exchange of uracil bases between free and purine-bound environments is rapid as indicated by the lack of separate uridine resonances appearing as a consequence of purine binding.

The mode of interaction of adenosine with poly U depends on temperature. In the range 37° to 26°C, the line widths of monitored resonances (H-8, H-6, and H-2) are constant and slight chemical shifts are observed (Figs. 13, 14, 15). Below 26°C, however, all resonances broaden markedly with decreasing temperature. Paralleling the line width behavior, the adenosine resonances shift abruptly to higher fields with decreasing temperature. The narrow temperature range over which the PMR spectral behavior of this system changes and the concomitant line broadening suggests the formation of a rigid, ordered complex. The up-field shifts of the adenine proton resonances indicate adenine–adenine base-stacking in the complex. Adenine–uracil base-stacking is absent because no appreciable shifts in the uracil base proton are observed. From the variation of the A:U base ratio and the comparison of the intensities of the adenosine resonances at high and low temperatures indicated that the exchange of adenosine between the free and

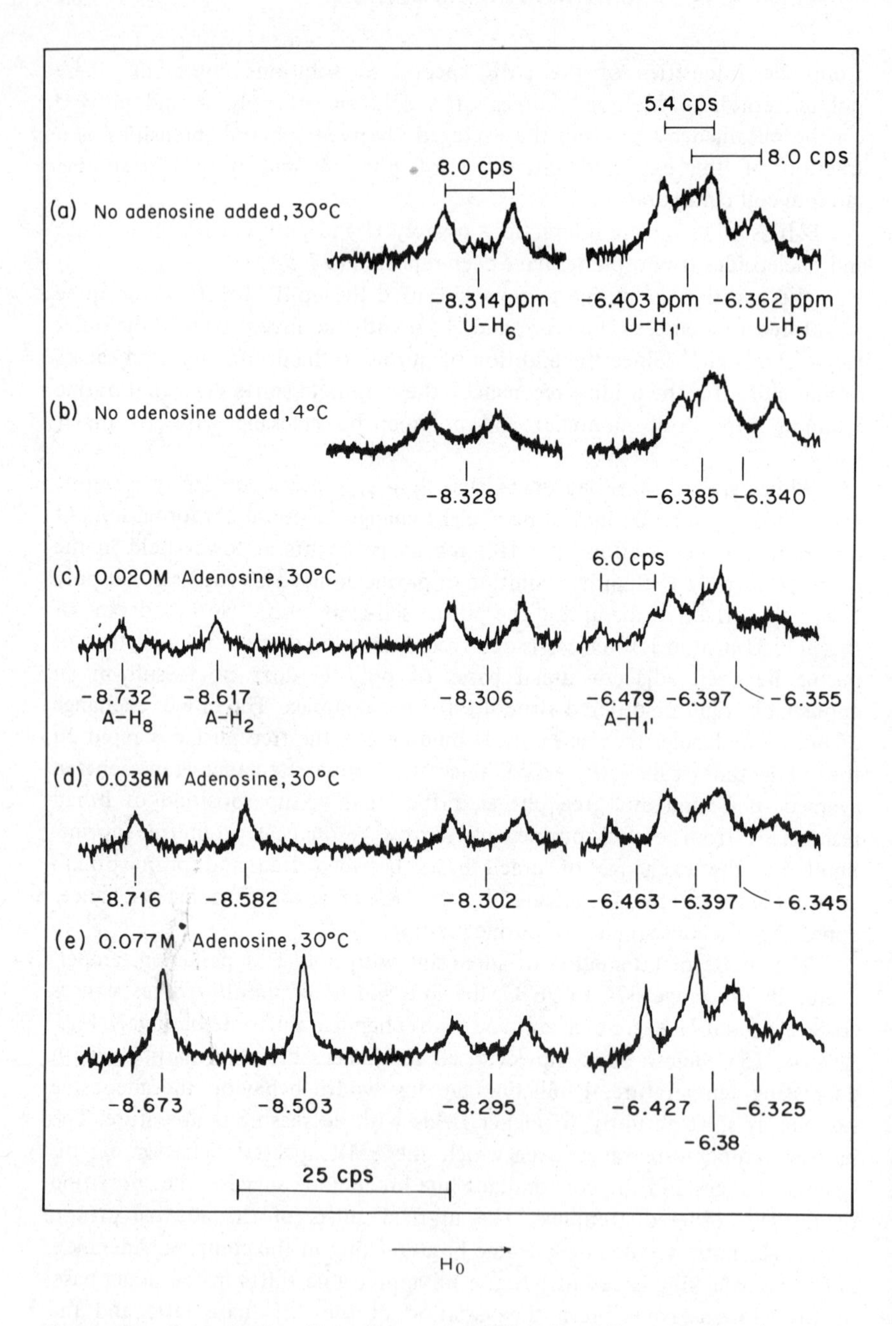

Figure 13. Polyuridylic acid (sodium salt), 0.078 M in uridine, and adenosine proton resonances at 100 Mc/sec; (a) no adenosine added, 30°C; (b) no adenosine added, 4°C; (c) 0.020 M adenosine, 30°C; (d) 0.038 M adenosine, 30°C; (e) 0.077 M adenosine, 30°C (Ref. *93*).

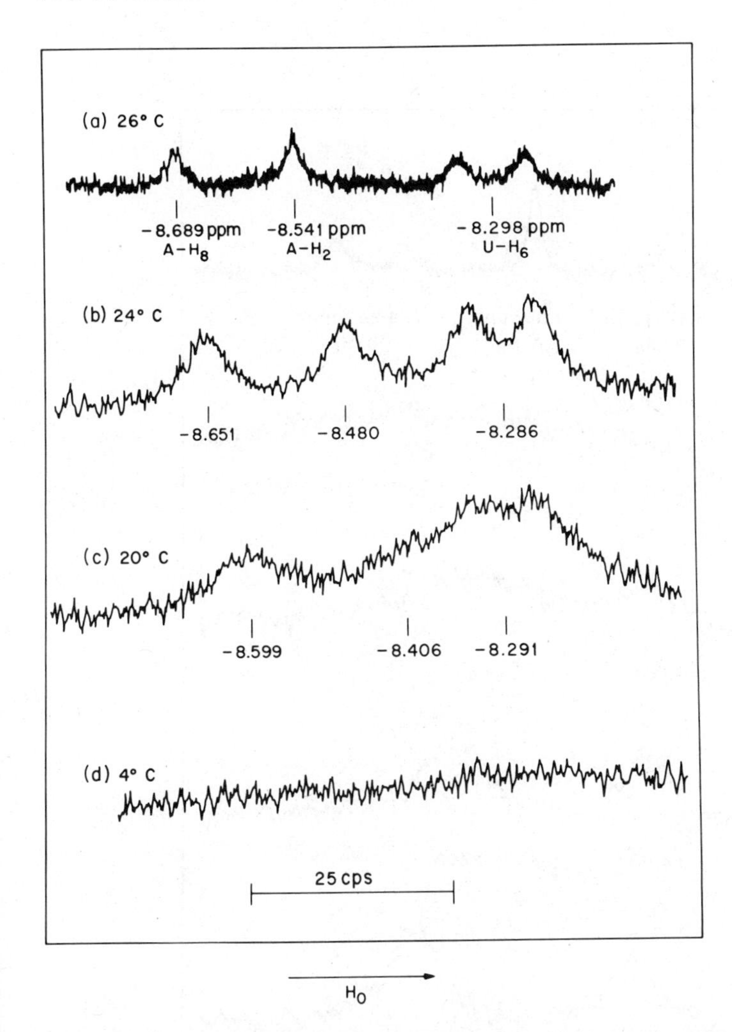

Figure 14. Poly U H6, 0.078 M uridine, and 0.038 M adenosine H8 and H2 resonances at 100 Mc/s. Base ratio A : U = 1 : 2. (a) 26°C, single scan; (b) 24°C, 17 scans; (c) 20°C, 26 scans; (d) 4°C, 20 scans (Ref. *93*).

bound forms is slow on the NMR scale below 20°C and the stoichiometry of the complex involves 2U per A. Thus, below 26°C, a rigid triple-stranded 1 A · 2U complex is formed.

The interaction of ApA with poly U has been reported (*95*). Since ApA is itself extensively stacked (*78, 79*), it would be sterically impossible to intercalate between adjacent uracil bases of poly U. No interaction was observed above 32°C, but below this temperature, a rigid triple-stranded complex involving 1A · 2U is formed, as has previously been reported (*96*).

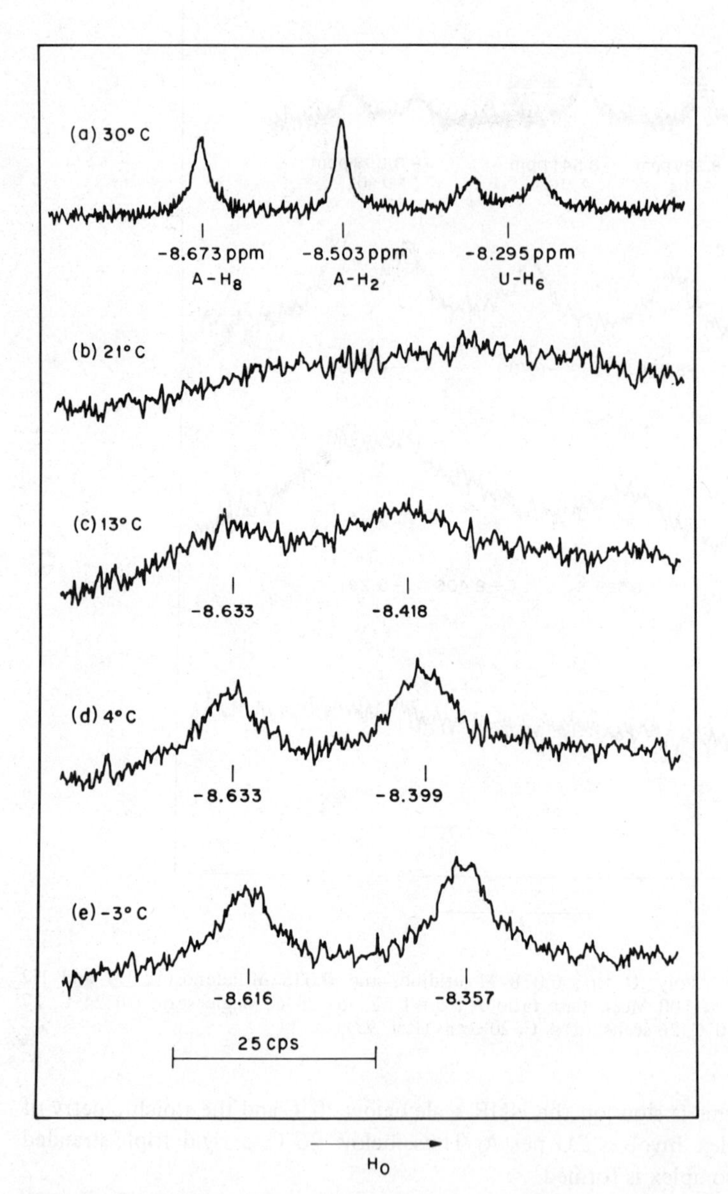

Figure 15. Polu U H6, 0.078 M in uridine, and 0.077 M adenosine H8 and H2 resonances at 100 Mc/s. Base ratio A : U = 1 : 1. (a) 30°C, single scan; (b) 21°C, 20 scans; (c) 13°C, 20 scans; (d) 4°C, 15 scans; (e) −3°C, 15 scans (Ref. *93*).

Several NMR studies on transfer RNA have been reported (*27, 90, 97–99*). The resolution of the peaks has been disappointing. There is almost no better resolution in the 220 MHz spectrum of purified alanine tRNA than there is in the 60 MHz spectrum of the unfractionated tRNA. Apparently the chemical shift heterogeneity of the bases, and the intrinsic widths of the NMR lines, are still sufficiently large to obviate resolution of individual base resonances. It has been suggested (*27*) that earlier NMR evidence for the involvement of the ribose moieties of tRNA in bonds stronger than those experienced by the bases (*90, 99*) was an artifact of experimental conditions. The growth of two broad peaks, representing roughly the aromatic and sugar protons, have been attributed to the thermal disaggregation of salt-induced tRNA aggregates. In view of the extreme broadness of the peaks of fractionated tRNA species,

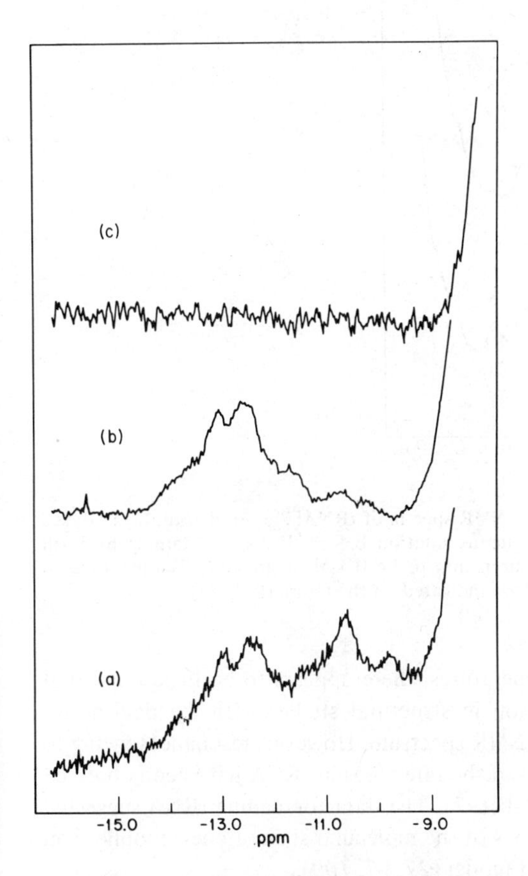

Figure 16. The low-field 220-MHz NMR spectra of unfractionated tRNA from yeast, 80 mg/ml: (a) sample dialyzed against EDTA and then distilled water, (b) sample in H_2O before dialysis containing high salt (estimated from conductivity measurements to be greater than 0.4 M, (c) sample dissolved in D_2O instead of H_2O (Ref. *23*).

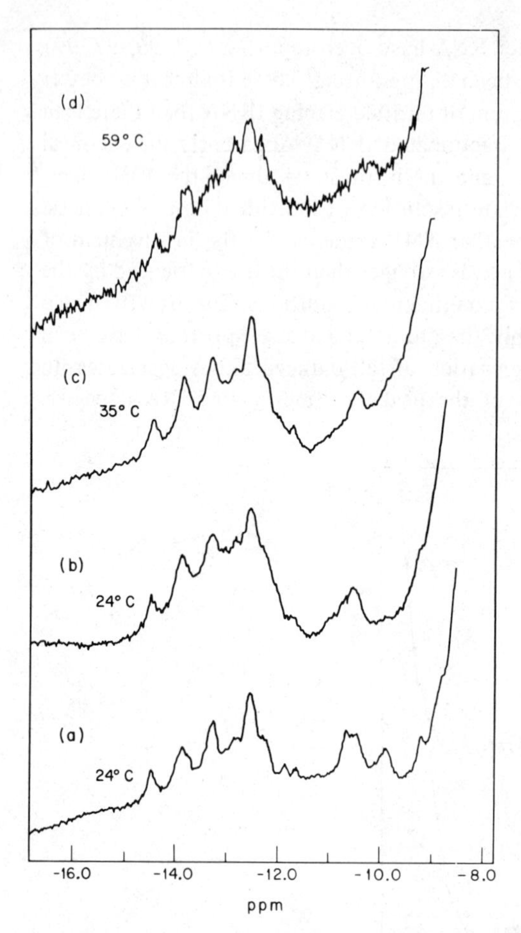

Figure 17. The low-field 220-MHz NMR spectra of $tRNA^{Phe}_{yeast}$ 80 mg/ml. (a) Dialyzed sample, 24°C. (b)–(d) Sample in aqueous solution before dialysis containing high salt (estimated from conductivity measurements to be 0.4 M or greater). Temperatures at which each spectrum were recorded are indicated on the figure (Ref. *23*).

even at 220 MHz at high temperatures, there appears to be little hope that NMR will find useful application in structural studies with the rigid native conformation in this region of NMR spectrum. However, resonances owing to the methyl and dihydro groups of the rare bases in tRNA have been observed at 100 MHz (*27*) and at 220 MHz (*97*). The data for alanine tRNA suggested that the rare bases were in regions of the molecular structure less mobile than suggested by an open clover-leaf model (*27, 97, 100*).

Although NMR studies of nonexchangeable protons of tRNA in D_2O found to be not particularly well resolved, NMR studies of low-field resonances from the relatively few hydrogen bonded protons of tRNA in aqueous solution show a great promise. Figure 16 shows the 220 MHz spectrum of unfractionated tRNA in H_2O and D_2O (*23*). The peaks observed between -10 and -15 ppm are more than twenty times weaker in the D_2O sample indicating that these resonances come from exchangeable protons. The same spectral region for purified $tRNA^{Phe}_{yeast}$ shows 23 ± 3 exchangeable protons (Fig. 17) (*23*). On raising the temperature, certain of these low-field resonances are differentially broadened.

It is evident that the 220-MHz NMR spectra of hydrogen bonded protons contains much useful information about the secondary and tertiary structure of tRNA molecules, although more experimental data is required to utilize this information.

NUCLEI OTHER THAN PROTONS

The main reason for less use of nuclei other than protons in NMR studies on polynucleotides is due to (1) its low natural abundance of the isotope (e.g., ^{13}C, ^{17}O), (2) sensitivity tends to be low, and (3) nuclei with spin quantum numbers $1 > ½$ (e.g., ^{14}N) are limited in their usefulness because they tend to give broad peaks.

^{31}P nuclei have only one disadvantage, a low sensitivity, but ^{31}P resonance lies well within the range of present techniques. ^{31}P magnetic resonance has been studied on studies of the binding of Mn^{2+} and Co^{2+} ions to RNA (*69–72*). The scalar coupling constant A was evaluated for the Co-^{31}P interaction from the magnitude of T_1 in the presence of Co^{2+} ions. ^{13}C and ^{15}N have not been widely used in biopolymers because of their low abundance and inherent lack of sensitivity, though they are both of spin $I = ½$ and potentially they are very valuable for NMR work. ^{15}N and proton magnetic resonance spectra of several derivatives of pyrimidine have been reported (*101*). 1-Methylcytosine $-^{15}N_3$ was shown to protonate at N_3 in agreement with previously reported work on 1-methylcytosine labeled with ^{15}N only in the amino group (*102, 103*). Recently ^{13}C magnetic resonance spectra of nucleotides and their derivatives have been reported (*104, 105*).

The limitations of the PMR are dependent on the need for all sites of the molecular framework to be proton bearing. For instance, in GMP there is only one proton (H-8) on the purine ring, and thus the effect of inter- and intramolecular interactions at other sites of the guanine ring cannot be observed. In this sense, ^{13}C magnetic resonance complements the PMR and in combination with Fourier transform, this field will undoubtedly be of great importance in the near future. Further, the greater ^{13}C chemical shift range enables relatively small differences in structure to be detected.

References

1. Pople, J. A., W. G. Schneider, and H. J. Bernstein. 1959. High resolution nuclear magnetic resonance. McGraw-Hill, New York.
2. Roberts, J. D., 1959. Nuclear magnetic resonance. Benjamin, New York.
3. Abragam, A. 1961. The principles of nuclear magnetism. Oxford University Press, London.
4. Roberts, J. D. 1962. An introduction to the analysis of spin–spin splitting in high resolution nuclear nagnetic resonance spectra. Benjamin, New York.
5. Slichter, C. P., 1963. Principles of magnetic resonance. Harper & Row, New York.
6. Emsley, J. W., J. Feeney, and L. H. Sutcliffe. 1965. High resolution nuclear magnetic resonance spectroscopy. Pergamon Press, New York, Vols. I and II.
7. Carrington, A., and A. D. McLachlan. 1967. Introduction to magnetic resonance. Harper & Row, New York.
8. Becker, E. D. 1969. High resolution NMR. Academic Press. New York.
9. Ehrenberg, A., B. G. Malmström, and T. Vänngård (Eds.). 1967. Magnetic resonance in biological systems. Pergamon Press, New York.
10. Kowalsky, A., and M. Cohn. 1964. *Ann. Rev. Biochem.*, **33**: 481.
11. Jardetzky, O., and C. D. Jardetzky. 1962. In D. Glick (Ed.), Methods of biochemical analysis. Wiley (Interscience), New York, Vol. 9, p. 236.
12. Dwek, R. A., and R. E. Richards. 1967. *Ann. Rev. Phys. Chem.*, **18**: 99.
13. Cohen. J. S. 1969. In C. Nicolau (Ed.), Experimental methods in molecular biology. Wiley, New York.
14. Sheard, B., and E. M. Bradbury. 1970. In J. A. V. Butler and D. Noble (Eds.), Progress in biophysics and molecular biology. Pergamon Press, New York, Vol 20, p. 187.
15. Rowe, J. J. M., J. Hinton and K. L. Rowe, 1970. *Chem. Rev.*, **70**: 1.
16. Johnson, Jr., C. S. 1965. In J. S. Waugh (Ed.), Advances in magnetic resonance. Academic Press, New York, Vol. 1, p. 33.
17. Saunders, M. 1967. In A. Ehrenberg, B. G. Malmström, and T. Vänngård (Eds.), Magnetic resonance in biological systems. Pergamon Press, London, p. 85.
18. Ernst, R. R., and W. A. Anderson. 1966. *Rev. Sci. Instr.*, **37**: 93; Ernst, R. R. 1966. In J. S. Waugh (Ed.), Advances in magnetic resonance. Academic Press, New York, Vol. 2, p. 1.
19. Shoolery, J. N. 1962. *Discussions Faraday Soc.*, **34**: 104.
20. Hall, G. E. 1968. In E. F. Mooney (Ed.), Annual review of NMR spectroscopy. Academic Press, New York, p. 227.
21. Katz, L., and S. Penman. 1966. *J. Mol. Biol.*, **15**; 220.
22. Shoup, R. R., H. T. Miles, and E. D. Becker. 1966. *Biochem. Biophys. Res. Commun.* **23**: 194.
23. Kearns, D. R., D. J. Patel, and R. G. Shulman, 1971 *Nature*, **229**: 338.
24. Glascol, P. K., and F. A. Long. 1960. *J. Phys. Chem.*, **64**: 188.
25. Tiers, G. V. D. 1958. *J. Phys. Chem.*, **62**: 1151.
26. Donaldson, B. R., and J. C. P. Schwarz. 1968. *J. Chem. Soc. (B)* p. 395.
27. Smith, I. C. P., T. Yamane, and R. G. Shulman. 1968. *Science*, **159**: 1360.
28. Hruska, F. H., and S. S. Danyluk. 1968. *Biochim. Biophys. Acta*, **157**: 238.
29. Kopple, K. D., and M. Ohnishi. 1969. *J. Am. Chem. Soc.*, **91**: 962; K. D. Kopple and D. H. Marr. 1967. *ibid.*, **89**: 6193.
30. Ts'o, P. O. P., N. S. Kondo, M. P. Schweizer, and D. P. Hollis. 1969. *Biochemistry*, **8**: 997.
31. Blackburn, B. J., F. E. Hruska, and I. C. P. Smith. 1969. *Can. J. Chem.*, **47**: 4491.

32. Wilmad Glass Co., Buena, New Jersey, Cat. No. 520.
33. Prestegard, J. H., and S. I. Chan. 1968. *J. Am. Chem. Soc.,* **91**: 2843.
34. Danyluk, S. S., and F. E. Hruska. 1968. *Biochemistry,* **7**: 1038.
35. Schweizer, M. P., A. D. Broom, P. O. P. Ts'o, and D. P. Hollis. 1968. *J. Am. Chem. Soc.,* **90**: 1042.
36. Gronowitz, S., B. Norman, B. Gestblom, B. Mathiasson, and R. A. Hoffmann. 1964. *Arkiv Kemi,* **22**: 65.
37. Smith, I. C. P., B. J. Blackburn, and T. Yamane. 1969. *Can. J. Chem.,* **47**: 513.
38. Donohue, J., and K. N. Trueblood. 1960. *J. Mol. Biol.,* **2**: 363.
39. Fujiwara, S., and M. Uetsuki. 1968. In S. Fujiwara and L. H. Piette (Eds.), Recent developments in magnetic resonance in biological systems. Hirokawa, Tokyo, p. 1.
40. Smith, M., and C. D. Jardetzky. 1968. *J. Mol. Spectr.,* **28**: 70.
41. Tsuboi, M., M. Kainosho, and A. Nakamura. 1968. In S. Fujiwara and L. H. Piette (Eds.), Recent developments in magnetic resonance in biological systems. Hirokawa, Tokyo, p. 43.
42. Tsuboi, M., F. Kuriyagawa, K. Matsuo, and Y. Kyogoku. 1967. *Bull. Chem. Soc., Japan,* **40**: 1813.
43. Hruska, F. E., A. A. Grey, and I. C. P. Smith. 1970. *J. Am. Chem. Soc.,* **92**: 214; *ibid.,* **92**, 4088.
44. Grey, A. A., I. C. P. Smith, and F. E. Hruska, *J. Am. Chem. Soc.,* in press.
45. Blackburn, R. J., A. A. Grey, F. E. Hruska, and I. C. P. Smith. 1970. *Can. J. Chem.,* **48**, 2866.
46. Emerson, T. R., R. J. Swan, and T. L. V. Ulbricht. 1967. *Biochemistry,* **6**: 843.
47. Buckingham, A. D., 1960. *Can. J. Chem.* **38**: 300.
48. Sundaralingan, M., and L. H. Jensen. 1965. *J. Mol. Biol.,* **13**: 930.
49. Haschmeyer, A. V. E., and A. Rich. 1967. *J. Mol. Biol.,* **27**: 69.
50. Karplus, M. 1959. *J. Chem. Phys.,* **30**: 11; 1963. *J. Am Chem. Soc.,* **85**: 2870.
51. Zimmerman, J. R., and M. R. Foster. 1957. *J. Phys. Chem.,* **61**: 282.
52. Pople, J. A. 1956. *J. Chem. Phys.,* **24**: 1111.
53. Johnson, C. E., and F. A. Bovey. 1958. *J. Chem. Phys.,* **29**: 1012.
54. Chan, S. I., M. P. Schweizer, P. O. P. Ts'o, and G. K. Helmkamp. 1964. *J. Am. Chem. Soc.,* **86**: 4182.
55. Jardetzky, O. 1964. *Biopolymers Symp. IPP 501; ibid., Biopolymers,* **1**: 501.
56. Ts'o, P. O. P., G. K. Helmkamp, and C. Sander. 1962. *Proc. Natl. Acad. Sci. U.S.* **48**: 686.
57. Schweizer, M. P., S. I. Chan, and P. O. P. Ts'o. 1965. *J. Am. Chem. Soc.,* **87**: 5241.
58. Ts'o, P. O. P., and S. I. Chan. 1964. *J. Am. Chem. Soc.* **86**: 4176.
59. Broom, A. D., M. P. Schweizer, and P. O. P. Ts'o. 1967. *J. Am. Chem. Soc.,* **89**: 3612.
60. Scheit, K. H. 1967. *Angew. Chem.,* **6**: 179.
61. DeVoe, H., and I. Tinoco. 1962. *J. Mol. Biol.,* **4**: 500.
62. McDonald, C. C., and W. D. Phillips. 1970. In G. Fasman (Ed.) Biological macromolecules, Vol. II. Dekker, New York.
63. Hruska, F. E., and S. S. Danyluk. 1968. *Biochim. Biophys. Acta,* **161**: 250.
64. Wang, S. M., and C. N. Li. 1966. *J. Am. Chem. Soc.,* **88**: 4592.
65. Eichhorn, G. L., P. Clark, and E. D. Becker. 1966. *Biochemistry,* **5**: 245.
66. Schneider, P. W., and H. Brintzinger. 1964. *Helv. Chim. Acta,* **47**: 1717.
67. Moll, H., P. W. Schneider, and H. Brintzinger. 1964. *Helv. Chim. Acta,* **47**: 1837.
68. Berger, N. A., and G. L. Eichhorn, *Abstr. Am. Chem. Soc. 160th Natl. Mtg,* 1970.
69. Shulman, R. G., H. Sternlicht, and B. J. Wyluda. 1965. *J. Chem. Phys;* **43**: 3116.
70. Sternlicht, H., R. G. Shulman, and E. W. Anderson, 1965. *J. Chem. Phys.,* **43**: 3123.

71. Sternlicht, H., R. G. Shulman, and E. W. Anderson. 1965. *J. Chem. Phys.*, **43:** 3133.
72. Shulman, R. G., and H. Sternlicht. 1965. *J. Mol. Biol.*, **13:** 952.
73. Sternlicht, H., D. E. Jones, and K. Kustin. 1968. *J. Am. Chem. Soc.*, **90:** 7110.
74. Hammes, G. G., and S. A. Levison. 1964. *Biochemistry,* **3:** 1504; G. G. Hammes and D. L. Miller. 1967. *J. Chem. Phys.*, **46:** 1533.
75. Eisinger, J., R. G. Shulman, and W. E. Blumberg. 1961. *Nature,* **192:** 963.
76. Eisinger, J., R. G. Shulman, and B. M. Szymanski. 1962. *J. Chem. Phys.*, **36:** 1721.
77. Eisinger, J., F. Fawaz-Estrup, and R. G. Shulman. 1965. *J. Chem. Phys.*, **42:** 43.
78. Sheard, B., S. H. Miall, A. R. Peacocke, I. O. Walker, and R. E. Richards. 1967. *J. Mol. Biol.*, **28:** 389.
79. Cohn, M., A. Danchin, and M. Grunberg-Manago. 1969. *J. Mol. Biol.*, **39:** 199.
80. Danchin, A., and M. Guéron. 1970. *J. Chem. Phys.*, **53:** 3599.
81. Chan, S. I., B. W. Gangerter, and H. H. Peter. 1966. *Proc. Natl. Acad. Sci. U.S.*, **55:** 720.
82. Chan, S. I., and J. H. Nelson. 1969. *J. Am. Chem. Soc.*, **91:** 168.
83. Hruska, F. H., and S. S. Danyluk. 1968. *J. Am. Chem. Soc.*, **90:** 3266.
84. Inoue, Y., and S. Aoyagi. 1967. *Biochem. Biophys. Res. Commun.*, **28:** 973.
85. Chan, S. I., and J. H. Nelson. 1969. *J. Am. Chem. Soc.*, **91:** 168.
86. Scheit, K. H., F. Cramer, and A. Franke. 1967. *Biochim. Biophys. Acta,* **145:** 21.
87. McDonald, C. C., W. D. Phillips, and J. Lazar. 1967. *J. Am. Chem. Soc.*, **89:** 4166.
88. Josse, J., A. D. Kaiser, and A. Kornberg. 1961. *J. Biol. Chem.*, **236:** 864.
89. McDonald, C. C., W. D. Phillips, and S. Penman. 1964. *Science,* **144:** 1234.
90. McDonald, C. C., W. D. Phillips, and J. Penswick. 1965. *Biopolymers,* **3:** 609.
91. Rich, A., and D. R. Davies. 1956. *J. Am. Chem. Soc.*, **78:** 3548.
92. Warner, R. C. 1956. *Federation Proc.*, **15:** 379.
93. Bangerter, B. W., and S. I. Chan. 1968. *Biopolymers,* **6:** 983; *ibid. Proc. Natl. Acad. Sci., U.S.*, **60:** 1144.
94. Ts'o, P. O. P., and M. P. Schweizer. 1968. *Biochemistry,* **7:** 2963.
95. Kreishman, G. P., and S. I. Chan, *Biopolymers,* in press.
96. Cantor, C. R., and W. W. Chin. 1968. *Biopolymers,* **6:** 1745.
97. Smith, I. C. P., T. Yamane, and R. G. Shulman. 1968. *Proc. IUPAC Symp. Macromolecular Chem.*, Toronto.
98. McDonald, C. C., and W. D. Phillips, in ref. 9, p. 3.
99. McTague, J. P., V. Ross, and J. H. Gibbs. 1964. *Biopolymers,* **2:** 163.
100. Holley, R. W., J. Apgar, G. A. Everett, J. T. Madison, M. Marquisee, S. H. Merrill, J. R. Penswick, and A. Zamir. 1965. *Science,* **147:** 1462.
101. Roberts, B. W., J. B. Lambert, and J. D. Roberts. 1965. *J. Am. Chem. Soc.*, **87:** 5439.
102. Miles, H. T., R. B. Bradley, and E. D. Becker. 1963. *Science,* **142:** 1569.
103. Becker, E. D., H. T. Miles, and R. B. Bradley. 1965. *J. Am. Chem. Soc.*, **87:** 5575.
104. Dorman, D. E., and J. D. Roberts. 1970. *Proc. Natl. Acad. Sci. U.S.*, **65:** 19.
105. Jones, A. J., M. W. Winkley, D. M. Grant, and R. K. Robins. 1970. *Proc. Natl. Acad. Sci., U.S.*, **65:** 27.

SECTION

B

CENTRIFUGATION TECHNIQUES

SEDIMENTATION VELOCITY EXPERIMENTS IN THE ANALYTICAL ULTRACENTRIFUGE

WILLIAM BAUER
DEPARTMENT OF CHEMISTRY
UNIVERSITY OF COLORADO
BOULDER, COLORADO

JEROME VINOGRAD
DIVISION OF CHEMISTRY AND CHEMICAL ENGINEERING
AND DIVISION OF BIOLOGY
CALIFORNIA INSTITUTE OF TECHNOLOGY
PASADENA, CALIFORNIA

INTRODUCTION

Two methods are available for the sedimentation velocity analysis of nucleic acids in the analytical ultracentrifuge. These methods, which differ in the concentration distribution of the macrospecies at the beginning of the experiment, are illustrated diagrammatically in Fig. 1. In band sedimentation (*1*) the nucleic acid is layered onto the sedimentation solvent in the spinning rotor, and the species move as bands with their characteristic sedimentation velocities. In boundary sedimentation (*2, 3*) the nucleic acid is initially uniformly distributed throughout the sedimentation solvent, and the species form trailing boundaries as they move down the cell.

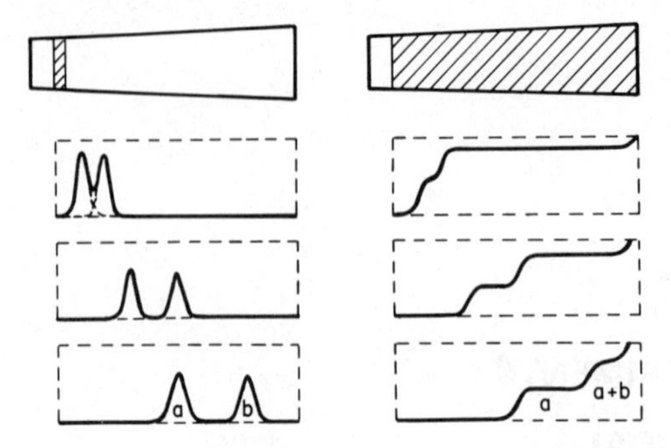

Figure 1. Two methods for performing sedimentation velocity experiments in the analytical ultracentrifuge. Left, the band velocity method; right, the boundary velocity method. Reproduced from *Engineering and Science*, Nov. 1968. Published at the California Institute of Technology.)

Figure 1 illustrates the two sedimentation procedures for a sample containing two components with sedimentation coefficients differing by about 50%. In order for either of these methods to be free from convective disturbances, it is necessary that the density of the solution (including the macrospecies) increase continuously in the direction of increasing field.* In principle, this stabilization is achieved in the region of a boundary owing to the positive density gradient associated with increasing polymer concentration. The front edge of a moving band is, however, intrinsically unstable, because the polymer concentration in this region decreases in the direction of the field. The associated negative density gradients are compensated by small positive density gradients which are generated automatically upon layering a light sample solution on a denser sedimentation solvent. The principal stabilizing gradient, formed by the diffusion of low-molecular-weight species between the two solutions, is normally adequate for stabilization. It is significantly supplemented in the latter stages of an experiment by the density gradients that arise from the sedimentation of the denser species in the binary sedimentation solvent. The perturbation of the density distribution in the solvent, however, is sufficiently small to have only an insignificant effect on the velocity of the sedimenting macromolecules.

Several of the features of each method of sedimentation velocity analysis are listed in Table 1. Band sedimentation is often the method of choice for studies of homogeneous or paucidisperse preparations of nucleic acids. As indicated in the table, the method is distinguished by the requirement for much smaller amounts of material and the possibility of obtaining standard sedimentation coefficients at infinite dilution in a single experiment with physically separated components. Boundary sedimentation is often the method of choice when polydisperse nucleic acids are to be analyzed.

* Density changes due solely to compression are ineffective in this respect.

TABLE 1. **Band and Boundary Sedimentation of Nucleic Acids**

	Band sedimentation	Boundary sedimentation
1. Amount of nucleic acid[a]	0.2 μg	15 μg
2. Optical systems	Absorption system only	Absorption, Rayleigh, and Schlieren[b]
3. Dialysis	Generally unnecessary	Usually necessary
4. Choice of solvent	Self-generating density gradient required[c]	No special requirements, except for light transmission
5. Slowly sedimenting absorbing impurities	Generally do not interfere	Opacity precludes experiments; lower optical densities interfere with macromolecular analysis
6. Dilution of sedimenting species during the experiment	The concentration maximum is reduced and the bands are spread by heterogeneity in the sedimentation coefficient, by diffusion, and by radial dilution.	The plateau concentration is reduced by radial dilution only.
7. Detection of concentration-dependent sedimentation	Back sharpening and forward spreading are sensitive indications of concentration-dependent sedimentation.[d] Symmetrical bands usually indicate that the measured quantity corresponds to s^0	Measurements of the sedimentation coefficients as a function of concentration are usually necessary.
8. Separation of components	Resolved components are physically separated.	Resolved components are never completely separated.
9. Hydrodynamic interactions between components: effect on sedimentation coefficients.	Mutual concentration dependence does not affect resolved components	Slow components reduce the sedimentation coefficients of all faster species.
10. Hydrodynamic interactions between components: effect on macromolecular analysis.	Resolved components are unaffected.	Johnston-Ogston effects (*4*) may cause serious errors. A series of experiments as a function of concentration is necessary.
11. The use of solvents in which the macrospecies is marginally stable.	Observations may be made within 5 min of contact between macrospecies and sedimentation solvent.	Generally 30 min or more elapse before the first record is made.
12. Nature of the initial distribution.	An approximate step function which is expected to vary somewhat in quality with the details of the layering process.	Homogeneous except for the redistributions that occur during acceleration.

[a] For operation with UV scanner or film. The examples were calculated for a slowly diffusing, single component in a 4° centerpiece.

[b] The refractometric systems normally can be used only with small nucleic acids, such as tRNA. The high concentrations required lead to significant hydrodynamic interactions for the larger nucleic acids.

[c] Normally the sedimentation solvent, usually a salt solution, should have a density of at least 0.02 gm/ml greater than that of the sample solvent in order to prevent convection. This corresponds to 0.5 M in the case of NaCl; at lower ionic strengths, D_2O is normally added.

[d] The presence of skewing does not necessarily mean that concentration dependence is significant except in the early stages of the experiment. The initial skewing process augments the dilution of the band. The band may thereafter remain skewed if the diffusion coefficient is low.

BAND SEDIMENTATION

Choice of Centerpiece

Three types of band-forming centerpieces (*5*), fabricated from charcoal-filled Epon, are available commercially in both single- and double-sector versions (Fig. 2).* The sample solution transfers through the channel in the Type I centerpiece face to the top of the sedimentation solvent at speeds below 1000 rev/min. The top of the liquid column must lie below the channel port for transfer to occur. This requirement results in a sedimentation path reduced by about 1 mm, relative to the other two types. Transfer in the Type II centerpiece occurs through the gap between the window and the centerpiece face, separated by a raised bead in Kel-F centerpieces and/or by a polyethylene or Kel-F gasket. Standard centerpieces may be readily modified to form Type II centerpieces. Transfer of sample in Type III centerpieces is brought about by displacement of the sample solution by the denser sedimentation solvent via the lower channel. This centerpiece must be filled with an adequate volume of sedimentation solvent for transfer to occur but, as for Type II, there is no special upper limit on the column height. The Type III centerpiece may also be used for flotation experiments in which the sample is initially delivered to the bottom of the centerpiece.

Cell Assembly

1. Position the lower window holder, window, and centerpiece in the housing. Before insertion of the centerpiece, it is advisable to apply a very

* Type II is presently not available in double-sector. The single-sector versions of Types I and II are currently available as modified Kel-F centerpieces.

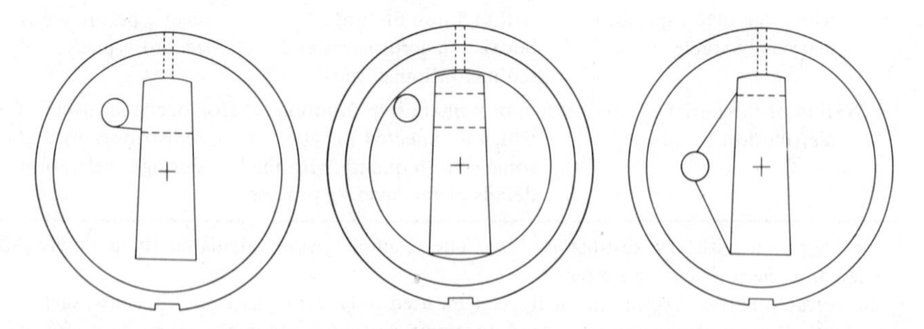

Figure 2. Band-forming centerpieces. The horizontal dashed lines represent the level of the sedimentation solvent. (Reproduced from *Biopolymers*, Ref. 5.)

light layer of Kel-F or silicone grease to the centerpiece faces to prevent intersector leakage.

2. Inject a measured volume of sample solution into the sample well with a Hamilton 10-μl or 50-μl syringe. These syringes may conveniently be used, but with less accuracy, as air pistons by fitting a disposable length of Intramedic tubing onto the needle (PE20 tubing for 10-μl, PE50 tubing for 100-μl syringes). Care should be taken to avoid formation of an air bubble underneath the liquid in the sample well. An appropriate volume in a double-sector (2.5°), 12-mm centerpiece is 10–15 μl, but as little as 2 μl may be used.

3. Complete the cell assembly and insert the sedimentation solvent through the housing fill-holes.

4. Adjust the cell and rotor assembly to approximately the desired run temperature, commonly 20°C.

Selection of Experimental Conditions

Choice of Rotor Speed

The radial position for minimum band stability lies near the center of the sedimentation path in solvents containing more than 0.5 M salt. It is advisable for this reason to select a speed which will place the fastest sedimenting band at cell-center after approximately 40 min. As a guide to the selection of rotor speed, the appropriate choice is 44,000 rev/min for a DNA with a standard sedimentation coefficient of 20 S. The experimental conditions selected for this example are: 2.5° sector angle, 1.2-cm centerpiece thickness, 10-μl sample volume, and sample solution density approximately 1.0 gm/ml. For a 4° sector angle the corresponding sample volume would be increased by a factor of 1.6, and by an additional factor of 2.5 if a centerpiece of 3.0-cm thickness is used. The actual uncorrected sedimentation coefficient for such a DNA is approximately 16 S in a CsCl solution, 1.35 gm/ml (2.85 M), as well as in 1.0 M NaCl and 0.1 M NaCl in 50% D_2O. The approximate speed for any other nucleic acid may be estimated from the relationship

$$\text{rev/min} = 44{,}000(20/s_{20,w})^{1/2} \quad (1)$$

In solvents of high viscosity, such as dimethylsulfoxide, speeds up to 56,000 rev/min have been used (*6*).

Elimination of Optical Artifacts

Large refractive index gradients may be encountered at the interface between sample solution and sedimentation solvent immediately after layering. These gradients result in failure of the light passing through this region of the cell to reach the camera lens. The effect persists for approximately one minute per microliter of sample solution, when the sample is layered onto a 1.35 gm/ml CsCl solution in a Beckman Model E ultracentrifuge fitted with a 2-inch camera lens. The loss of early information which results from the artifact may be avoided by waiting for the appropriate time at a lower speed (5,000–10,000 rev/min) before accelerating to the final rotor speed, or by adding CsCl to the sample solvent.

Excessive refractive index gradients are also formed by sedimentation of CsCl at the top and bottom of a 3.0-cm cell. These effects are compensated with a 2° negative wedge window in the cell assembly.

Choice of Sedimentation Solvent

Table 2 presents a summary of many of the pertinent properties of the commonly used sedimentation solvents. The correction factors listed represent approximate values of the ratio $s_{20,w}/s_{obs}$ at 20°C. For experiments performed at other temperatures, the temperature dependence of the viscosity and the density of the solution should also be taken into account. These factors will be discussed at greater length in the section on conversion of observed to standard sedimentation coefficients.

Choice of Amount of DNA in Sample Solvent

Single Homogeneous Components. Hydrodynamic interactions increase with the effective hydrodynamic volume and with the concentration of a nucleic acid. It is therefore necessary to reduce the concentration as the effective hydrodynamic volume increases. A small, compact molecule, such as nicked SV40 DNA, 3.0×10^6 daltons, forms symmetrical bands in 1.35 gm/ml CsCl with up to 1.0 μg in a 2.5° sector, 1.2-cm cell. A larger, less compact molecule, such as linear λ DNA, 30×10^6 daltons, forms approximately symmetrical bands in a 3.0-cm centerpiece with up to 0.2 μg. The lower limit of detection of a moving band with the photoelectric scanner has been reported to be approximately 0.01 μg. The effective hydrodynamic volume decreases at high pH and larger amounts of DNA may then be used.

Polydisperse Samples. Heterogeneity of sedimentation coefficients results in increased band spreading and in rapid diminution of the band height during an experiment. The greater loads required in these experiments dictate the use of a 3.0-cm centerpiece to minimize the effects of concentration-dependent sedimentation in the early stages of the experiment. In cases of extreme heterogeneity, boundary sedimentation is the method of choice.

Paucidisperse Samples. In the event that the fastest component is present in high concentration relative to the slower components, resolution

TABLE 2. Commonly Used Sedimentation Solvents

Solvent	$s_{20,w}/s_{obs}$	Comments
A. CsCl, 1.35 gm/ml; 0.01 M Tris	~1.5[a]	Greatly reduces concentration dependence of s and increases symmetry of bands; large density stabilization; low viscosity; requires special procedure for correction to $s_{20,w}$.
B. CsCl, 1.35 gm/ml; 0.1 M KOH	(~1.3)[b]	Denaturing solvent; correction to obtain $s_{20,w}$ of sodium form is approximate, adequate concentration of chelating agents in sample solvent is required to complex divalent metals[c]; other comments as in A.
C. NaCl, 1.0 M; 0.05 M Na_2HPO_4	1.17	Svedberg correction procedure to obtain $s_{20,w}$ is applicable.[d]
D. NaCl, 0.9 M; 0.1 M NaOH	1.16	Denaturing solvent; Svenberg correction to obtain $s_{20,w}$ is applicable except for uncertainty in $\bar{v}$; chelating agent required as in B.
E. D_2O, 50%; 0.1 M NaCl	1.20	Low ionic strength sedimentation solvent; requires correction of 0.5% for isotope exchange; solutions are hygroscopic.
F. DMSO, 100%		Denaturing solvent used for RNA; high viscosity; opaque at wavelengths $\leqslant$270 nm.

[a] This factor is approximate and depends upon the buoyant density of the DNA, as explained in the section on conversion of observed to standard sedimentation coefficients.
[b] The uncertainties in the correction procedure for alkaline CsCl are greater than in neutral CsCl. It is customary to report uncorrected sedimentation coefficients obtained at 20°C.
[c] Failure to add chelating agent results in the failure to form bands in alkaline systems.
[d] The value $\bar{v}$ = 0.556 ml/gm, customarily used in the Svedberg correction procedure, may be high by 8% (7).

between slow components is improved. If the slowest component is present in high concentration relative to the fast components, the latter bands are spread as they move out of the domain of the slow component. Again, 3.0-cm centerpieces are indicated to minimize this effect.

Recording of Results

Figure 3 presents results obtained with the two available absorption optical recording systems. The film method (Fig. 3a) is used to best advantage with the addition of the monochromator and high intensity light source accessories, which reduce the exposure times and improve optical quality. Single-sector, charcoal-filled Epon or Kel-F centerpieces may be used. Dural and aluminium-filled Epon should be avoided because of the possibility of

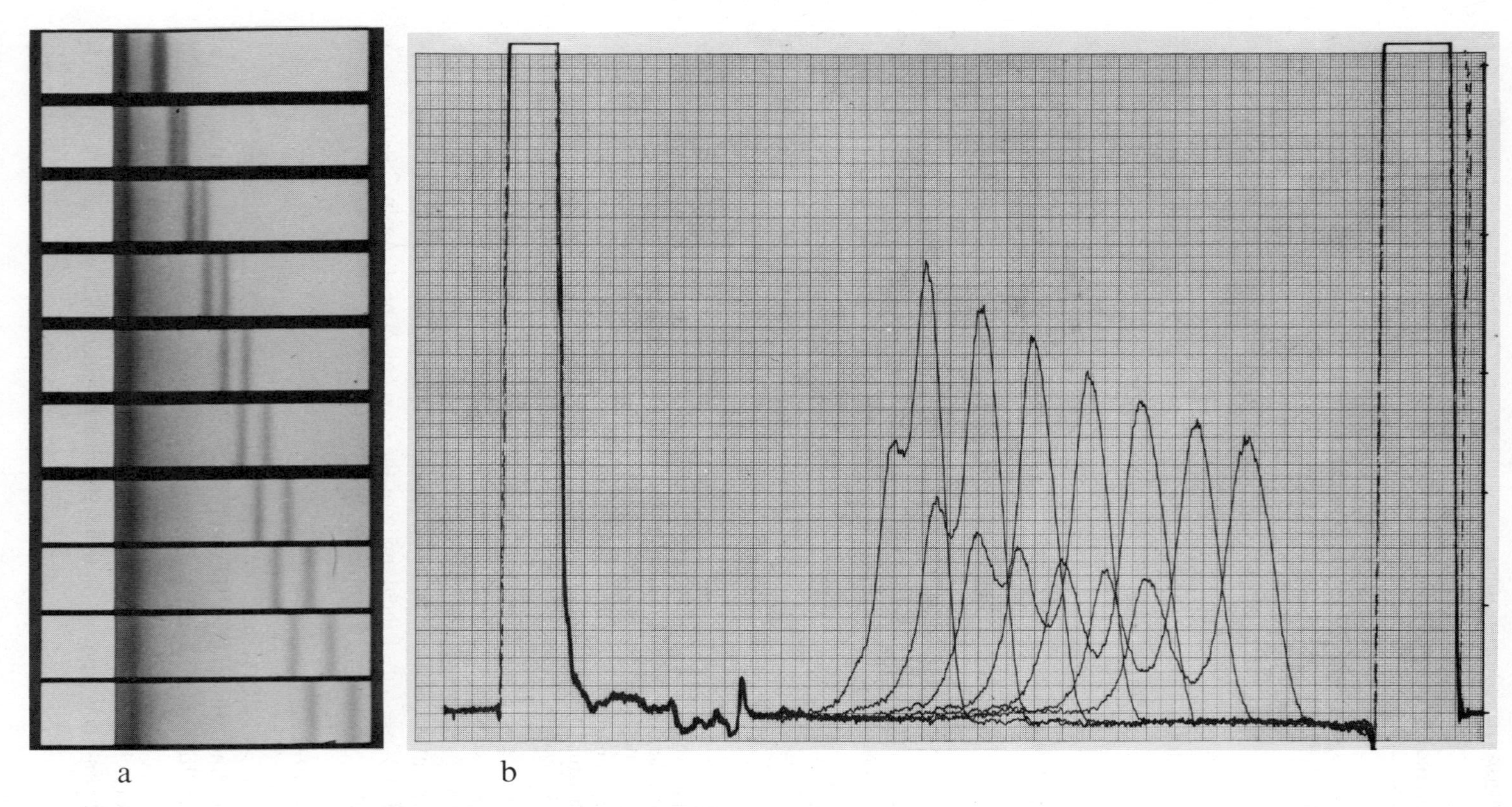

Figure 3. (a) A photographic record of a band sedimentation velocity experiment. (Reproduced from *Proc. Natl. Acad. Sci., U.S.*, Ref. *1*.) (b) Photoelectric scans of moving bands in a sedimentation velocity experiment. A mixture of closed and nicked SV40 DNA. The bands are moving to the right in 1.35 gm/ml CsCl, 44,000 rev/min at 20°C. Successive scans at regular intervals were recorded on a 7001 AM Moseley X-Y recorder, attached in parallel to an Offner recorder provided with the Beckman photoelectric scanner.

undesirable chemical reactions. Up to three cells may be run, with difficulty, with the use of an An-F rotor, side wedge windows, and a mechanical beam interrupter (alternator). Densitometry of the films is performed with a Joyce-Loebl microdensitometer or, with reduced sensitivity, with a Beckman Analytrol. A linear relation between densitometer response and concentration is not required for locating the position of band maximum and, in fact, this coordinate may often be located visually on the screen of a projection comparator (Nikon 6F or Gartner M2001-RS-P).

The photoelectric scanner obviates the need for a densitometer (Fig. 3b). The double-beam feature allows for automatic subtraction of solvent absorbance, and both double-sector and single-sector cells may be used. Up to five cells may be used in a single experiment with the An-G rotor and the multiplex accessory, provided that experimental conditions are so chosen as to allow for the collection of sufficient data.

Calculation of Sedimentation Coefficients

If the band is approximately symmetrical, the location of the band maximum is measured at several equal time intervals and the logarithm of this coordinate plotted as a function of time. The observed sedimentation coefficient, s, is calculated from the slope of this plot with the relationship (*8*)

$$s = \frac{3.500 \times 10^{13}}{(\text{rev/min})^2} \frac{d \log r_m}{dt} \tag{2}$$

where the logarithm is taken to the base 10; r_m is the distance from the center of rotation to the band maximum; t is in minutes; and s is expressed in Svedberg units. The error introduced by the use of band maximum is in most cases no greater than 0.1%.

If the band is highly skewed, it may be necessary to use the exact relationship (*9*)

$$s = \frac{3.500 \times 10^{13}}{(\text{rev/min})^2} \frac{d < \log r >}{dt} \tag{3}$$

where the mass average logarithm is

$$< \log r > = \frac{\int_{-\infty}^{\infty} (\log r) c dr^2}{\int_{-\infty}^{\infty} c dr^2} \tag{4}$$

or, in numerical form,

$$< \log r > = \frac{\sum_{\text{band}} c_i r_i \log r_i}{\sum_{\text{band}} c_i r_i} \tag{5}$$

This method should be attempted only with data obtained with the photoelectric scanner, because of the necessity for the measurement of precise concentration.

Conversion of Observed to Standard Sedimentation Coefficients

The Svedberg Method (*2*). The sedimentation coefficient measured as described above depends not only on the molecular weight and frictional coefficient of the macrospecies, but also on the viscosity and buoyant effect of the solvent. Sedimentation coefficients obtained with nucleic acids are generally converted to "standard" conditions with an equation analogous to the one suggested by Svedberg for the calculation of the standard sedimentation coefficient of proteins. The latter quantity, $s_{20,w}$, for the sodium form of nucleic acids, is used to facilitate the comparison of the results of experiments performed in different solvents at different temperatures, and is given by

$$s_{20,w} = s\left(\frac{\eta_{T,w}}{\eta_{20,w}}\right)\left(\frac{\eta_{T,\text{solv}}}{\eta_{T,w}}\right)\frac{(1 - \bar{v}_{\text{ref}}\rho_{20,w})}{(1 - \bar{v}_{\text{ref}}\rho_{T,\text{soln}})} \tag{6}$$

where η is the viscosity; T is the temperature of the experiment; ρ is the density; and $\bar{v}_{\text{ref}}$ is the estimated value of the DNA partial specific volume, taken equal to 0.556 ml/gm for Na DNA (*10*). A more rational correction equation would replace $\bar{v}_{\text{ref}}$ in the numerator by $\bar{v}_{20,w}$, and in the denominator by $\bar{v}_{T,\text{soln}}$. Here, $\bar{v}_{20,w}$ refers to the value which $\bar{v}_{20}$ would assume in the limit as the ionic strength approaches zero under the condition that the duplex structure remains unchanged. Unfortunately, there has been no careful study of the problems involved in making this extrapolation. Cohen and Eisenberg have obtained some indication that $\bar{v}$ increases with ionic strength between 0.1 M and 4 M NaCl (*7*). In addition, the data of Chapman, cited in Gray and Hearst (*11*), indicate that $\bar{v}$ in 0.015 M NaCl increases approximately 0.1% per degree. In view of the small magnitude of this temperature effect and the present uncertainties in the ionic strength dependence of $\bar{v}$, present practice is to use Eq. (6) with $\bar{v}_{\text{ref}} = 0.556$ ml/gm.

The Bruner-Vinograd Method (*12*). In concentrated binary solvents, such as 1.35 gm/ml CsCl, the Svedberg correction has been shown to be inapplicable. The relationship between the product $s\eta_r$ and the density of the solution, ρ, is linear for both T7 DNA and MS-2 RNA in CsCl solutions.

Since the observed sedimentation coefficient vanishes at a solution density equal to the buoyant density at 20°C, θ_{20}, the following method of calculation may be used to obtain

$$(s_{20,w})_{Cs} = s\,\frac{\eta_T\,(\theta_{20} - 0.998)}{\eta_{20}\ \ (\theta_{20} - \rho_{20})} \tag{7}$$

The buoyant density increases with temperature by 4.5 x 10^{-5} gm/ml per degree and $\theta_{20} = \theta_{25} - 0.002$. The relative viscosity η_r at 20°C is 0.952 at a solution density of 1.38 gm/ml measured at 25°C. Data at other CsCl salt concentrations, in the density range 1.1 to 1.88, is given by Bruner and Vinograd (*12*).

The ratio of $(s_{20,w})_{Cs}$, calculated with Eq. (7), to $(s_{20,w})_{Na}$ calculated with Eq. (6), has been determined to be 1.31 with T7 DNA and 1.37 with MS-2 RNA.

Macromolecular Analysis

Sedimentation velocity experiments are more often performed for the purpose of macromolecular analysis than for the determination of a sedimentation coefficient. These experiments are used to determine the absolute or, more commonly, the relative amount of any component.

Absolute Mass Determination. This procedure requires a prior calibration to determine the relationship between corrected pen excursion and optical density of the nucleic acid at any given wavelength. This plot is generally a straight line, with a slope, k cm OD^{-1}. The total mass in the band is then given by

$$m = \frac{\phi(\Delta r)\epsilon'}{k} \sum_{\text{band}} r_i p_{c,i} \tag{8}$$

where m is in μg if ϵ' is in μg/ml/OD ($\epsilon' = 50$ μg/ml/OD for a native DNA at 42% GC and 257.9 nm (*13*)), ϕ is the sector angle in radians, Δr is the uniform difference in centimeters between ordinate measurements taken in the summation, and $p_{c,i}$ is the pen excursion corrected with the attenuator for system nonlinearity. With only a small error, the summation in Eq. (8) may be approximated by

$$\sum_{\text{band}} r_i p_{c,i} \simeq r_m \sum_{\text{band}} p_{c,i} \tag{9}$$

Relative Mass Determinations within the Same Cell. For components having the same extinction coefficients, the parameters ϕ, ϵ', k, and Δr are unnecessary. The fractional mass in the jth band is given by

$$f_{m,j} = \frac{r_{m,j} \sum_{\text{band } j} p_{ci,j}}{\sum_{\text{bands}} r_{m,j} \sum_{\text{band } j} p_{ci,j}} \tag{10}$$

BOUNDARY SEDIMENTATION

The problems encountered in boundary sedimentation velocity have been treated in the monographs by Schachman (*3*) and by Svedberg (*2*). We, therefore, limit this discussion to a few selected aspects of the technique.

Calculation of the Sedimentation Coefficient

Equation (2) may be used to calculate s by the boundary method in experiments with absorption optics, and with r_m taken as the inflection point of the moving boundary. The inflection point is satisfactorily approximated in most cases by the location of the midpoint in the concentration-distance distribution. This procedure is applicable only to the case of a free boundary, one contained between two regions of constant concentration.

Concentration Dependence

Concentration-dependent sedimentation sharpens the boundaries and tends to cancel the broadening effects of diffusion and of sedimentation coefficient heterogeneity. It is, therefore, generally necessary to perform a set of experiments at different initial concentrations in order to determine s^0. This is generally done by extrapolating a plot of $1/s$ vs. c to infinite dilution. Neither the shape of the boundary nor the absence of curvature in the log r_m vs. t plot is a reliable criterion for the absence of a significant lowering in s due to concentration dependence.

Macromolecular Analysis of Paucidisperse and Polydisperse Systems

The boundary method is suitable for determining the relative amounts of the nucleic acid components in the absence of hydrodynamic interaction between components. The mass fraction of component j is given by

$$f_j = \frac{\Delta p_{c,j} r_j^2}{\sum_{\text{components}} \Delta p_{c,j} r_j^2} \tag{11}$$

where $\Delta p_{c,j}$ is the corrected pen excursion over boundary j, and corresponding r_j is the boundary position. The presence of hydrodynamic interactions must be assumed, however, and errors caused by the Johnston-Ogston effect (*4*) avoided by extrapolating the analytical results for f_j to infinite dilution.

In the case of a polydisperse system, in which separate components do not resolve, the weight-average sedimentation coefficient may be obtained by following the location of the midpoint of the boundary. If a centripetal plateau region does not form, the weight-average sedimentation coefficient may be calculated with the use of the exact equation

$$\bar{s}_w = \frac{1.75 \times 10^{13}}{(\text{rev/min})^2} \frac{d}{dt} \log \left\{ 1 - \frac{2(\Delta r)}{p_{c,R} R^2} \sum_{r_a}^{R} r_i p_{c,i} \right\} \tag{12}$$

where r_a is the meniscus and R is any arbitrary coordinate in the centrifugal plateau region.

Boundary Stability

Redistributions of the macrospecies in boundary sedimentation give rise to positive density gradients, and there is no intrinsic density instability. At the low concentrations required to avoid hydrodynamic interactions, this intrinsic stabilization is often inadequate to protect the boundary from the convective effects caused by small thermal gradients. The thermal gradient may be materially reduced by preequilibrating the rotor in the chamber before the run, by switching off the heating element in the chamber with the refrigeration set at a pressure corresponding to 50–55°F, and by employing a clear anodized rotor.

The Effect of Angular Velocity on the Sedimentation of Large DNAs

Very large molecular-weight DNAs, with $M > 30 \times 10^6$, generally exhibit complex and nonideal sedimentation behavior which becomes more pronounced as rotor speed and concentration are increased (*14, 15*). Characteristically, the plateau ahead of the boundary tilts upward and the boundary velocity estimated from the movement of the inflection point is noticeably high. These effects may be minimized by performing the experiments at low speeds, coupled with a series of determinations as a function of concentration (*14*). No dependence upon angular velocity has been reported in band sedimentation experiments.

Acknowledgments

This work was supported by U.S. Public Health Service Grant No. GM15327, from the National Institute of General Medical Sciences, and No. CA08014, from the National Cancer Institute. Dr. Bauer is the recipient of Grant No. GB19190 from the National Science Foundation. This is contribution no. 4175 from the Division of Chemistry and Chemical Engineering, California Institute of Technology.

References

1. Vinograd, J., R. Bruner, R. Kent, and J. Weigle, 1963. *Proc. Natl. Acad. Sci., U.S.,* **49**: 902.
2. Svedberg, T., and K. O. Pedersen. 1940. The ultracentrifuge. Oxford University Press, New York.
3. Schachman, H. K. 1959. Ultracentrifugation in biochemistry. Academic Press, New York.
4. Johnston, J. P., and A. G. Ogston. 1946. *Trans. Faraday Soc.,* **42**: 789.
5. Vinograd, J., R. Radloff, and R. Bruner. 1965. *Biopolymers,* **3**: 481.
6. Strauss, J. H., Jr., R. B. Kelly, and R. L. Sinsheimer. 1968. *Biopolymers,* **6**: 793.
7. Cohen, G., and H. Eisenberg. 1968. *Biopolymers,* **6**: 1077.
8. Vinograd, J., and R. Bruner. 1966. *Biopolymers,* **4**: 131.
9. Schumaker, V. N., and J. Rosenbloom. 1965. *Biochemistry.* **4**: 1005.
10. Hearst, J. E. 1962. *J. Mol. Biol.,* **4**: 415.
11. Chapman, R. 1968. Ph.D. Thesis, Harvard University, Cambridge, Massachusetts. Cited in H. B. Gray, Jr. and J. E. Hearst. 1968. *J. Mol. Biol.,* **35**: 111.
12. Bruner, R., and J. Vinograd. 1965. *Biochim. Biophys. Acta,* **108**: 18.
13. Felsenfeld, G., and S. Z. Hirschman. 1965. *J. Mol. Biol.,* **13**: 407.
14. Hearst, J. E., and J. Vinograd. 1961. *Arch. Biochem. Biophys.,* **92**: 206.
15. Aten, J. B. T., and J. A. Cohen. 1965. *J. Mol. Biol.,* **12**: 537.
16. Vinograd, J. 1968. *Eng. Sci. (Pasadena),* **32**: 12.

EQUILIBRIUM DENSITY GRADIENT CENTRIFUGATION *

WACLAW SZYBALSKI AND ELIZABETH H. SZYBALSKI
McARDLE LABORATORY
UNIVERSITY OF WISCONSIN
MADISON, WISCONSIN

INTRODUCTION

Equilibrium density gradient centrifugation, introduced by Meselson, Stahl, and Vinograd (*1*), has proved to be a highly versatile tool for the separation, purification, and characterization of a variety of macromolecules, including nucleic acids, proteins, and nucleoproteins, e.g., viruses and cell particulates. The theory of the technique has been discussed in depth (*1–13*).

Equilibrium density gradient centrifugation should not be confused with equilibrium sedimentation, which does not employ any supporting gradient (*14, 14a*), or with other forms of density gradient centrifugation that were designed to measure the sedimentation velocity of macromolecules, i.e., zonal (or band) centrifugation in sucrose, deuterium oxide, or other gradients (*14, 15*). Although all use the same tool, the analytical or preparative ultracentrifuge, they are in principle quite different methods.

Equilibrium density gradient centrifugation is an *equilibrium* (*static*) method, whereby the separation or characterization of the macromolecular species is based solely on its *buoyant density*. The gradient is *self-generating* in the centrifugal field and must encompass the *entire* density range of the macromolecules. It is usually formed in concentrated aqueous solutions of

* Support for writing this chapter was provided by grants from the National Science Foundation (GB-2096), the National Cancer Institute (CA-07175), and the Alexander and Margaret Stewart Trust Fund.

heavy metal salts, such as CsCl or Cs_2SO_4, which are highly water-soluble, preferably do not absorb UV light, and create steep gradients in the density range of materials such as nucleic acids. Centrifugation must be continued until equilibrium is achieved, i.e., until the macromolecules reach the *isodensity* position, at which their buoyant density corresponds to the density of the solvent, at a defined level in the gradient column. Equilibrium density gradient centrifugation belongs to the class of *isopycnic* gradient centrifugation techniques (*15a*), in which the buoyant density of the solvent at the bottom of the gradient column is higher than that of the macromolecules. The innovation that differentiates it from earlier isopycnic methods, including the historical opus of Galilei (*15b*) and the techniques of Linderstrøm-Lang and Lanz (*16*) and Behrens (*17*), is that the density gradient is perfectly stable and reproducible, since it is generated and maintained during centrifugation rather than preformed by layering.

The second density-gradient technique, denoted *zonal* (or *band*) *sedimentation*, is a *kinetic* method used to determine the *sedimentation coefficient* of the macromolecular species. In this case the macromolecules move continually through the gradient and eventually settle at the bottom of the tube if centrifugation is not interrupted. Usually, the gradient is artificially *preformed* from solutions substantially less dense than the macromolecular species; its only purpose is to stabilize the sedimenting zone of macromolecules and thus prevent convection. Various compounds have been used to prepare preformed gradients, including sucrose, glycerol, D_2O, various salts, and even CsCl. Band sedimentation in diffusion-generated CsCl density gradients, introduced by Vinograd and Bruner (*18, 19*), also belongs to the class of kinetic methods, as well as the temperature perturbation-controlled method of Schumaker and Wagnild (*20*)

Only equilibrium density gradient centrifugation will be discussed here. To summarize the procedure, at the beginning of the run a homogeneous solution containing the cesium salt and the macromolecules to be analyzed is prepared. The density of the salt solution is adjusted to match approximately the buoyant density of the macromolecules. If the latter density is not known, several concentrations of salt need to be tested. During centrifugation, the cesium salt sediments in the direction of the centrifugal field and eventually establishes a sedimentation-diffusion equilibrium gradient. During this period the macromolecular species seeks out a unique position in the gradient column (the isodensity region), corresponding to its own buoyant density, and forms a band. The width and the more or less Gaussian profile of the band depend on the molecular weight and density homogeneity of the macromolecules. Skewness of the band is an indication of density heterogeneity. If two or more macromolecular species of sufficiently different densities are present in the tube, two or more bands will appear at equilibrium (bimodal or polymodal distribution).

By this method the buoyant density of macromolecules can be determined with a great degree of precision and reproducibility when the original density of the cesium salt solution and the physical parameters of the centrifugation are known. Also of paramount importance, the samples are preserved intact during the course of the measurements, and thus can be examined by other assay procedures.

Buoyant density measurements on viruses, nucleic acids, and proteins provide important clues as to their chemical composition. For instance, the base composition of DNA can be inferred from its density, and the presence of rare bases can be detected. An estimate of the molecular weight and heterogeneity of macromolecules can be derived from the shape of the band. By preparative equilibrium density gradient centrifugation, macromolecules differing sufficiently in buoyant density in the native state, or after labeling with isotopes or base analogs, can be separated and purified. The buoyant density of nucleic acids is also sensitive to other changes in structure, e.g., denaturation and binding of ligands such as Ag^{+} or Hg^{2+} salts, of polynucleotides, and of intercalating dyes or antibiotics. In certain cases these chemical reactions can be utilized for preparative fractionation of nucleic acids or for separation of the complementary DNA or RNA strands. These and other applications of buoyant density measurements in the equilibrium density gradient will be reviewed, following a technical description of the method.

INSTRUMENTATION AND TECHNIQUES

Laboratory Space Requirements

One or two analytical ultracentrifuges, together with the supporting equipment, can be placed in one 10 x 20 ft laboratory, which should be well ventilated and lighted but readily convertible into a dark room. The latter feature facilitates critical alignment of the optical system, simplifies the photography, and permits loading and developing the films in the same room. An adequate supply of electricity, water, and filtered compressed air are of paramount importance, together with desk-height benches permitting convenient assembling of the cells.

Analytical Ultracentrifuge

The Model E analytical ultracentrifuge with ultraviolet absorption optics (2-inch camera lens), An-F rotor, and two or three four-cell sets, as described in the section on four-cell operation, can be purchased from Beckman Instruments, Spinco Division, Palo Alto, California (abbreviated as Spinco Co. throughout this text). Several accessories, including a monochromatic light source, scanning equipment, and multicell operation devices, will be discussed in the appropriate sections. The Schlieren and interference optics can be deleted from centrifuges used only for density gradient centrifugation of nucleic acids and nucleoproteins, especially when a monochromator is furnished. However, it is sometimes convenient to observe the formation of the gradient with the Schlieren system, and solvents opaque to ultraviolet light or solutes with low extinction coefficients require its use.

In the authors' laboratory ultracentrifuge drives require very infrequent replacement (5000–10,000 hr of operation) when used for long equilibrium density gradient runs, especially if the brushes are replaced in the drive motor after each 1000–2000 hr of operation and the carbon deposit is removed at the same time.

Operation of the Ultracentrifuge

The operation of the Model E ultracentrifuge is described in detail in Instruction Manual E-1M-3 supplied by the Spinco Co. (1964 or later editions) and is outlined in Ref. *21*.

Alignment of the Ultraviolet Optics

The UV optics should be meticulously aligned, following the procedure outlined by Schachman et al. (*22*) and in Spinco Technical Bulletin 6124 (1962) and Instructions for Model E Engineers, "UV Alignment Procedure" (1963), using a cell half-filled with water. The alignment should then be repeated under actual conditions of density-gradient centrifugation after equilibrium is reached (24–28 hr). Adjust the position of the UV light source, moving it along the front to rear axis, to center the images of all cells in the middle of the vertical tube outlet (Fig. 1,G) just under the mirror (Fig. 1,H) which must be removed during this aligning operation and replaced by ground glass. Check that the images of all the cells are centered on the camera lens (Fig. 1,M). Adjust the light source to assure even illumination of the cell; a tracing of the photograph obtained with the filled cell should produce a horizontal line (line B–M, Fig. 2). Adjust the masks of the upper and lower collimating lenses (inside the rotor chamber) to eliminate any light that might pass outside the rotor (necessary only for the four-place rotor F). If necessary, shorten the slit in these two mask apertures by blocking off the front-facing end of the slit with black tape or some special mask. Any stray light in the UV system causes reflections and false images.

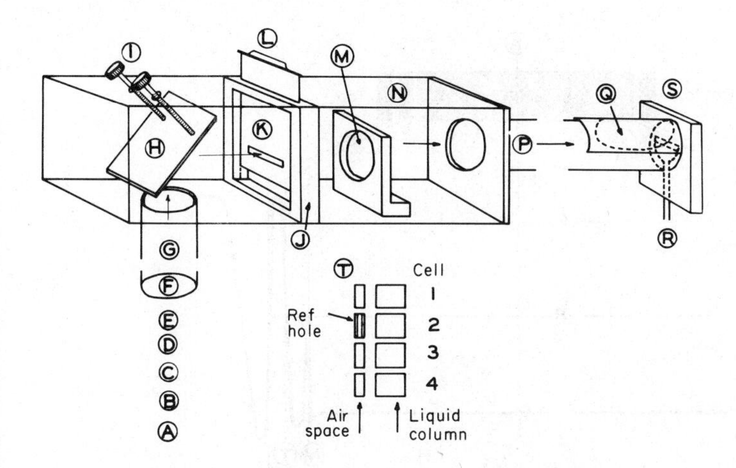

Figure 1. A schematic diagram of the optical system for four-cell operation. A, light source; B, mask over light source (3 x 3 mm); C, chlorine–bromine filter; D, lower collimating lens; E, cell and rotor; F, upper collimating lens; G, vertical tube; H mirror; I, thumb screws for mirror adjustment, projecting through the top of the Lucite housing; J, screen-mask assembly made of a 2¼ x 3¼-in. wooden film holder with its back removed and with a 5 x 40 mm slit (K) cut into the movable film protector plate (L). The latter is painted white or covered with white tape to serve also as a projection screen for the images of the four cells. The wooden frame of the screen-mask assembly is mounted in the Lucite housing (N); K, slit (5 x 40 mm); L, vertically movable plate; M, camera lens (2 inch); N, Lucite housing replacing standard metal cover and permitting observation of the cell images and alignment of mirror H with the cover in place; P, horizontal tube; Q, cover (easily removable and light-tight) over opening in horizontal tube next to shutter (R); R, shutter, S, camera; T, image of the four cells projected on screen L located in the focal plane of the upper collimating lens, as viewed from the left end of the centrifuge.

The optimally focused position of the camera lens should be marked, since any repositioning of this lens would change the magnification factor and might cause the image to become slightly out of focus. Refocusing is necessary when damaged lenses are replaced or when a different wavelength is selected in centrifuges equipped with a Spinco monochromator. A monochromator that does not require this wavelength-dependent adjustment has been described (*23*). A similar reflectance optical system is now available from the Spinco Co. and consists of a concave mirror replacing mirror H and camera lens M (see Fig. 1).

Cleanliness of the Optical System

The rotor chamber should be maintained clean and free of oil deposits. Both collimating lenses, especially the upper lens, should be removed and cleaned with ethanol before each run. Blow off with filtered compressed air any dust specs remaining on both sides of the upper collimating lens, since

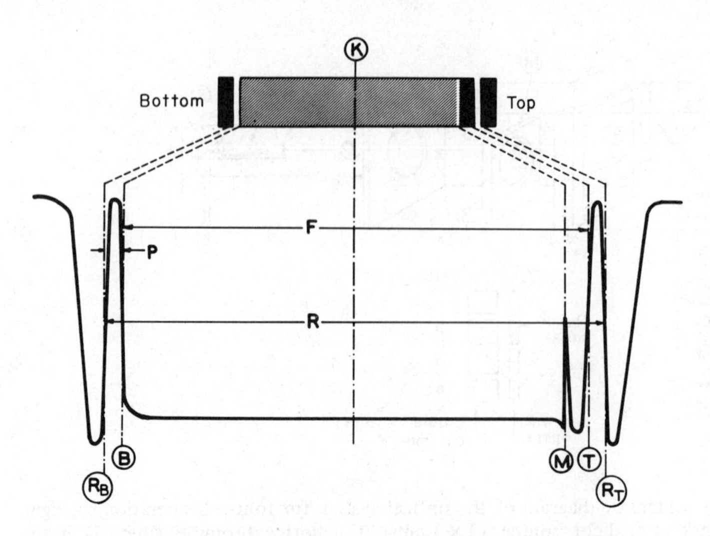

Figure 2. Schematic drawing representing a photograph of the solvent-filled cell (upper shaded drawing) and its microdensitometric tracing. T, top edge of cell cavity; B, bottom edge of cell cavity; K, center of cell cavity; F, distance from T to B; R_T, inside edge of top reference hole (proximal to the center of rotation); R_B, inside edge of bottom reference hole (distal to the center of rotation), R, distance between inside edges (R_T and R_B) of reference holes; P, distance between bottom edge of cell cavity (B) and inside edge of bottom reference hole (R_B); M, meniscus.

otherwise these will appear on the photograph. Seal the whole optical system with adhesive tape (Lucite housing N, and the ends of vertical tube G (Fig. 1) to prevent dust and oil from entering the opening under the mirror and being deposited on the upper collimating lens. Lining tube G (Fig. 1) above the collimating lens with black, lint-free absorbent paper prevents oil deposition on the lens (A. D. Hershey, personal communication).

Four-Cell Operation

Since a density gradient run takes 20–40 hr, it is usually wasteful to perform the centrifugation on only one sample. For this reason four-cell operation was developed in this laboratory (*24–28*), and a description of the procedure is included here.

Special Accessories. (Spinco catalog number is included in parentheses):

1. Analytical rotor F (307382)
2. Four cells with 12 mm, 2° Kel-F centerpieces (301193) and with the following windows (see Spinco price list PL-246E):
   (a) all upper windows: minus 1° wedge window assemblies (307071) (green)

(b) lower windows:
cell 1: 1° (up) side-wedge window assembly (307487) (blue)
cell 2: flat window assembly (301171 and 301730) (black)
cell 3: 1° (down) side-wedge window assembly (326893) (brown)
cell 4: 2° (down) side-wedge window assembly (326897 or 350073) (brown)

2° Kel-F centerpieces are preferred, since compared with 4° centerpieces they require half as much DNA and salt solution (0.3–0.4 ml), are more rigid, leak less often, and have a longer cavity (1.45 vs. 1.40 cm); they require, however, twice the time for a photographic exposure. Charcoal-filled epoxy centerpieces are also recommended, especially since they are stronger than Kel-F cells.

3. Mask with a 3 x 3 mm opening (Fig. 1,B) to be placed over the light source.
4. Screen-mask assembly (Fig. 1,J) with a 5 x 40 mm vertically movable slit to be placed in the proximity of the camera lens.
5. Lucite cover (Fig. 1,N) over the mirror-camera lens assembly and long adjusting screws (Fig. 1,I) for the mirror.
6. Opening in the horizontal tube next to the camera, with light-tight, easily removable cover (Fig. 1,Q).

Operation. The operation is analogous to one-cell operation with the exception of the cell assemblage and photography. Proper assembling of the wedge windows in all four cells can be checked by observing the direction and degree of deviation of a vertical line drawn on the wall observed through the hand-held cell at a distance of 1–2 ft. The cells should be tightened with a cell torque wrench (327119) at 20–30 sec of 135 lb torque.

After the gradient is established, the UV lamp is turned on, the shutter is opened, and the room should be darkened. Yellow-green dim images (Fig. 1,T) of the four cells should appear on the white screen (accessory No. 4) in front of the camera lens. It is possible to make the images brighter by temporary removal of the chlorine–bromine filter (Fig. 1,C). If the ultracentrifuge is equipped with a monochromator, it should be shifted to the visible-light range.

The final (equilibrium) photographs can usually be taken after 20 hr at 44,770 rpm, or 40 hr at 31,410 rpm, for DNA or RNA of molecular weight above 10 million; longer centrifugation is required for smaller molecules. Equilibrium is reached when the position and shape of the band does not change in two photographs taken on two consecutive days. It is helpful sometimes to take photographs early during the establishment of the equilibrium, to check for leaks, to note the kinetics of the band formation, or to distinguish spurious "bands" (crystals of salt or dirt particles on the outside of the cell windows, or buoyant particles and precipitates) from true bands.

The principle of four-cell photography depends on photographing the image of one cell at a time by permitting it to pass through the 5 x 40 mm slit (accessory No. 3, Fig. 1,K) while blocking off the other three cell images. The slit must be near the camera lens, since all four images of the cells merge again in the focal plane.

1. The 5 x 40-mm slit should be moved to permit only the image of cell No. 1 to pass.

2. The cover (Fig. 1,Q) next to the camera should be opened and the mirror adjusted in such a way to center the cell image over the open shutter (the edges of the shutter slit (Fig. 1,R) must be covered with white tape to permit better visualization of the cell image).

3. After the shutter is closed, the chlorine-bromine filter is replaced, or the monochromator is reset to 265.4 mμl, and photographs can be taken, usually at exposures of 1 and 2 min (Kodak Commercial film).

4. These operations are repeated for the other three cells.

A complete system for four-cell operation with an automatic four-cell rotating mask can also be purchased from the Spinco Co. Its operation is outlined in Ref. *21* and, in more detail, in Instruction Manual E-TB-019A of the Spinco Co. Other systems for multicell operation in conjunction with the scanning procedure have also been described (*29–31*).

Photography and Microdensitometry

Kodak Commercial film, 2 1/4 x 3 1/4 in. (standard size available through wholesale photographic suppliers and to be used with modified film holder) or 2 1/2 x 3 1/2 in. (available from the Spinco Co., or on special order from the Kodak Co., Rochester, New York) is used in conjunction with Kodak D-11 developer and acid fixer. Several other companies supply similar blue and UV light-sensitive, fine-grain film and corresponding developers. Small volumes of developer are required when employing a 10 x 15-cm developing tray in a little dish placed on a gently rotating platform. The developer should be changed each day. Water or dust spots on the film should be stringently avoided; the film should be dusted off just before it is inserted in the film holder, and after it has been developed, fixed, and thoroughly rinsed it should be dried in dust-free air.

In the experience of this laboratory, only the Joyce-Loebl double-beam recording microdensitometer (Mark IIIC), especially equipped with a cylindrical condensor lens and 7.5 x arm ratio (on special order from the supplier), has proved uniformly reliable. Other microdensitometers with separate film and recorder drives often give distorted nonreproducible tracing distances. The dimensions of the scanning beam focused in the plane of the film should be approximately 4 x 0.05 mm. A beam height of 4 mm, obtainable only with a cylindrical condensor lens, compensates for the film graininess and results in rather smooth tracings. A proper arm ratio (7.5 x) permits fitting all the tracings across the standard 8 1/2 x 11 in. tracing sheets.

The distances on the tracings should be measured immediately since the dimensions of the tracing paper change under conditions of varying humidity. The 25-cm Lucite Farol cursors with metric scale and a vernier readout available from the Martin Sweets Co. (Louisville, Kentucky) (about $30) were found well suited for measuring distances on microdensitometric tracings. When two or more peaks overlap, the duPont 310 Curve Resolver (Instrument Products Division, E. I. duPont de Nemours and Co., Wilmington, Delaware) is very useful, since it permits locating the centers of the overlapping peaks and integrating the area under them.

Photography and microdensitometry are now being replaced by photoelectric scanning (*29–31*). A commercially available scanner (Spinco Bulletin SB-264) which is mounted directly in the optical path of the centrifuge, permits independent scanning of up to six cells without the necessity of using wedge windows in conjunction with a Multiplex Accessory (Spinco Catalog No. 350121). In order to distinguish between the solute and precipitate bands, and between the true bands and dirt spots in the optical system, scans should also be performed before equilibrium is established, since spurious bands are usually formed before true bands. Differential scans with double-sector cells compensate for the optical imperfections common to both sectors but require more complex cells, sometimes prone to intersector leakage during long centrifugation runs. It is suggested that the provisions for regular UV photography be retained when converting the centrifuge for scanning operation.

Reagents

The cesium chloride or cesium sulfate should be free of any UV-absorbing materials and divalent metal ions, especially calcium and magnesium. Optical grade CsCl and Cs_2SO_4 can be purchased from a number of companies (including Harshaw Chemical Co., Cleveland, Ohio). Technical CsCl or Cs_2SO_4, purchased from American Potash and Chemical Corporation at approximately $40 per pound proved to be satisfactory in most cases after subjection to a simple purification procedure. This includes filtration of the solution (100 gm CsCl or Cs_2SO_4 per 100 ml glass-distilled water) through exhaustively washed Whatman No. 1 filter paper, addition of an equal volume of a slurry of activated charcoal (Norit A previously boiled in 0.1 N HCl and thoroughly washed with distilled water), boiling for 10 min, filtering off the charcoal through a very dense acid-washed filter, evaporation of excess water, and two recrystallizations of the salt from boiling water. Care should be exercised during all these operations not to contaminate the CsCl or Cs_2SO_4 with any UV-absorbing materials (or traces of nucleases).

Two other purification steps are useful with some batches of cesium salts: (1) firing the salt at a temperature of 600–700°C overnight in an electric muffle furnace, a procedure which removes most of the organic contaminants; and (2) precipitation from saturated water solution by the addition of an equal volume of 100% ethanol. Simple purification procedures for cesium salts have been described (*32*).

The absorbance of a saturated solution of the purified CsCl or Cs_2SO_4 should be less than 0.05 OD units (10-mm cell) at 260 nm. Usually the salt solutions are buffered with 10^{-3} to 10^{-2} M Tris, borate, Na_2HPO_4-citric acid, or HCl-cacodylate at pH 8.5 (for native DNA) or 5.5–7 (for RNA). To band DNA in the denatured state, 10^{-2} to 10^{-1} M NaOH or carbonate buffer was employed (*33, 34*).

Other cesium salts, rubidium and lithium salts, or mixed gradients have also been employed, including CsBr and CsI (*3, 6*), Cs formate, Cs oxalate, and Cs acetate (*3, 6, 10, 35–37a*), RbCl and RbBr (*5, 10, 37a*), LiCl and LiBr (*5, 37a*), CsCl–LiCl (*38*), CsBr–LiBr (*3*), CsCl–$(NH_4)_2SO_4$ (*158*), NaI (*39*), KBr (*5*), and NaBr–sucrose (*39a*).

Preparation of Samples

To fill the 12-mm, 2° sector analytical cell, approximately 0.4–0.5 ml of solution is sufficient. Twice as much is required for the 12-mm 4° sector cell. When CsCl or Cs_2SO_4 is used for preparative runs in the SW39 swinging bucket rotor, 2.5 ml of solution per tube is sufficient, providing the tube is filled to the top with paraffin oil. When angle rotors 40, 50, or 50Ti are employed, at least 4–5 ml of solution per tube are required, and the tube must be filled to the rim with paraffin oil.

To prepare 0.5 ml of CsCl–DNA solution for analytical gradient centrifugation, one requires 0.4 ml of a saturated solution of CsCl (at 23–24° C), 0.1 ml of 10^{-3} M Tris buffer (pH 8.5), and 0.25–0.5 μg of high-molecular-weight DNA per band. The final density of the solution should be adjusted to permit banding of the DNA sample near the center of the cell, usually 1.69–1.75 gm/cm^3, depending on the type of DNA and the density label.

To prepare 0.5 ml of Cs_2SO_4–DNA solution, one requires 0.2 ml of a saturated solution of Cs_2SO_4 (at 23–24° C), 0.3 ml water or 10^{-3} M Tris buffer, and 0.25–0.5 μg DNA per band. The final density should be adjusted to 1.41–1.45 gm/cm^3.

To prepare an RNA sample, a saturated Cs_2SO_4 solution is mixed with 0.1 M phosphate–citrate buffer containing 0.5–1.0 μg RNA, and usually

1–2% formaldehyde to prevent aggregation and precipitation of high-molecular-weight samples (*40*). The density should be adjusted to approximately 1.6 gm/cm^3 (n_D = 1.381 or 8°10′). Mixed Cs_2SO_4–CsCl solutions are described below.

Preformed or partially preformed ("step") density gradients for equilibrium centrifugation of labile macrospecies, e.g., viruses, cells, cell particulates, and enzymes, have been prepared from CsCl (*39b, 39c*), RbCl, potassium tartrate or citrate (*37a, 39d*), and polymers such as polyglucose (*39e*), Ficoll (*39e*), and Urografin (*39f, 39g*). Most of these gradients are thermodynamically unstable. In preformed CsCl gradients, the sample is introduced at a level close to the expected isodensity position (*39b*), or mixed with the upper, more dilute portion of the step gradient (*39c*). The centrifugation time required to reach equilibrium is considerably shortened, and the sample is exposed to relatively lower concentrations of salts or to less toxic buoyant media.

Measurement of the Density of the Solution

The density of the CsCl or Cs_2SO_4 solution is measured either pycnometrically or refractometrically. For routine adjustment of the density refractometric measurements are quite satisfactory.

A 15–20 μl drop is sufficient to fill the Bausch and Lomb Abbe-3L refractometer with horizontal prisms, or any similar instrument that has a special narrow gap between the prisms, which should be specified at the time of ordering. The drop of solution is delivered onto the center of the prism; care should be taken not to touch the glass surface, which can be easily scratched. The prisms are then immediately closed without splattering. The temperature of the prisms should be maintained at 25°C with a circulating thermostatic bath. The relationship between the refractive index and the density of CsCl and Cs_2SO_4 solutions is presented in Table 1.

Pycnometric measurements can be performed either with a conventional pycnometer, preferably equipped with a thermometer and vacuum jacket, or with a capillary pycnometer when only small volumes of solution are available. A capillary pycnometer of approximately 0.2-ml volume can easily be prepared from a broken fragment of a 1-ml serological glass pipette with the two ends drawn into fine capillaries. Such a pycnometer, which weighs 2–3 gm, is filled or emptied by attaching fine polyethylene tubing to one end and controlling air pressure by mouth while holding the pycnometer in the Walton cross-action sterilizer forceps (Cat. No. 7L240; Lawton Co., New York).

The density of the solution at 25°C is determined by consecutive weighings of the capillary pycnometer, which rests on a simple handmade

TABLE 1. Density (gm/cm^3) and Refractive Indices ($n_D^{25°}$, or expressed in degrees and minutes) of CsCl and Cs_2SO_4 Solutions at 25°C[a]

	CsCl		Cs_2SO_4	
Density (gm/cm^3)	$n_D^{25°}$	degrees, minutes	$n_D^{25°}$	degrees, minutes
1.4000	1.3722	7°16′	1.3666	6°42′
1.5000	1.3815	8°12′	1.3740	7°27′
1.6000	1.3905	9°8′	1.3814	8°12′
1.7000	1.3996	10°4′	1.3885	8°56′
1.8000	1.4086	11°0′	1.3957	9°40′

[a]Note that a diagram can be drawn from these data for graphic interpolation of intermediate figures.

wire support attached to the pan of the balance. The density, ρ_0 (gm/cm^3), is calculated as

$$\rho_0 = \frac{\text{net weight of salt solution}}{\text{net weight of water}} \times 0.99704$$

and is then corrected, if necessary, for temperature, which should be measured during the weighings inside the semimicro- or microanalytical balance. When the weighing is performed at 25°C, no correction is needed. For each degree below 25°C, *subtract* from the density the value 0.0003 gm/cm^3 (for Cs_2SO_4 solutions at 1.4-1.5 gm/cm^3), 0.0004-0.0005 (for Cs_2SO_4 solutions at 1.6-1.8 gm/cm^3), or 0.0005 (for CsCl solutions at 1.6-1.8 gm/cm^3). The same values should be *added* to the density for each degree above 25°C.

The buoyant density can be measured directly in the gradient by adding as density markers small glass floats (*40a, 40b*), immiscible liquids (*45*), or pieces of plastic wrapping of known density, e.g., polyvinyl chloride (Saran), which has a density of about 1.7 gm/cm^3 (Szybalski, unpublished data).

Filling the Cells

To avoid shearing, the solutions should be transferred slowly to the cell, using 1- or 2-ml syringes and blunt 22-gauge needles. The filling hole in the centerpiece can be enlarged with a precision drill, if 20- or 21-gauge needles or Pasteur pipettes are used. The cells should always be filled to the same height (to avoid variation in the hydrostatic pressure), and with the air space occupying only 5% of the cell height. With practice, it is possible to fill the cells without spilling any liquid outside the centerpiece. It should be remembered that CsCl and Cs_2SO_4 solutions are very corrosive to aluminum and easily damage the cell housing, especially the threads of the housing plug and the key strip. The metal parts of the cell should be washed well and frequently treated with silicone lubricant.

When banding DNA at very high rotor speeds (59,780 rpm), Wake and Baldwin (*41*) and Inman and Baldwin (*42*) used only half-filled cells to attain rapid equilibrium, to compensate for the high density of cesium salt solutions, and to avoid overloading the rotor, although the latter precaution is not considered necessary (L. Gropper, Spinco Co., personal communication).

Calibration of the Density Gradient and Calculation of the Buoyant Density

Methods for calculating the equilibrium density distribution and the buoyant density of the macromolecular species are given by Ifft et al. (*5*), Ludlum and Warner (*43*), and Erikson and Szybalski (*27*). These methods, however, disregard the perturbing effects on the density gradient of pressure and preferential interaction of the macromolecular species with the solute. A three-component theory of density gradient equilibrium which takes into account these parameters is presented by Hearst and Vinograd (*4*). The application of computer programming to the calculation of the density distribution has been described (*5, 13, 44*).

In this section a practical and simple semiempirical method will be outlined for determining (1) the *absolute* buoyant density of the DNA or RNA based on its position in the cell and on the average density of the CsCl or Cs_2SO_4 solution, and (2) the *relative* buoyant density of the macromolecules based on their position in relation to another DNA or RNA sample of known buoyant density, i.e., a density marker. The effects of hydrostatic pressure and other factors are considered in the graphic solutions. All routine calculations are simple.

Determination of the Magnification Factor and Dimensions of the Centerpiece

To determine the magnification factor, install the cell containing the CsCl or Cs_2SO_4 solution in the rotor opposite the counterbalance, spin overnight at standard speed (31,410 or 44,770 rpm), and take photographs. The magnification factor (M_f) is the ratio of the traced distance R (cm) between the images of the inner edges of the reference holes in the counterbalance (Fig. 2) to the actual distance (1.60 cm, as specified by the manufacturer):

$$M_f = \frac{R}{1.60}$$

The operation should be repeated with each centerpiece. Then the relative length F/M_f of the centerpiece and the distance from the cell bottom to the center of rotation $(7.30 - P/M_f)$ should be determined (see Fig. 2). It is advantageous if a matched set of four centerpieces with identical F and P dimensions can be used.

When Kel-F centerpieces are used, these dimensions depend somewhat on the rotor speed and on the height and density of the gradient column because of deformation of the plastic. Charcoal-filled Epon centerpieces are more rigid.

Effects of Distance from the Rotation Center and of Hydrostatic Pressure on the Distance between Two Bands

If identical pairs of DNA samples are banded at different positions in the cell by varying the initial density of the CsCl or Cs_2SO_4 solution, the distance D between the bands (Fig. 3) changes somewhat: the bands come closer together when located nearer the bottom of the cell (farther from the center of rotation). This result is to be expected, since the density gradient, being a function of the radius, becomes steeper near the bottom of the cell.

Under ideal conditions, the relative increment in the distance between two bands should be inversely proportional to the distance from the center of rotation. This is not always the case, however, since two DNAs banding at different depths are under unequal hydrostatic pressures and since the compressibility coefficients of DNAs differ from those of CsCl or Cs_2SO_4 solutions.

To determine the relationship of the distance between the bands to the positions of the bands in the cell, it is necessary to band two DNA samples at four or more positions and to plot the experimental data thus obtained. Figure 4 represents such a plot. It shows the relationship of the value E (E_1 = distance between the cell center and the midpoint A between the two bands) to the correction factor $k = (D_0 - D)/D$ where D_0 represents the distance between the two bands when spaced symmetrically in the center of the cell at $E = 0$, and D is the band distance in individual experiments with positive or negative E values (Fig. 3). By using the diagram in Fig. 3 and the factor k estimated from Fig. 4, it is possible to correct the distance D between the bands to the "true" distance $D_0 = D + kD$.

This semiempirical computation of D_0 permits determination of the *absolute* buoyant density of any reference DNA or other macromolecular

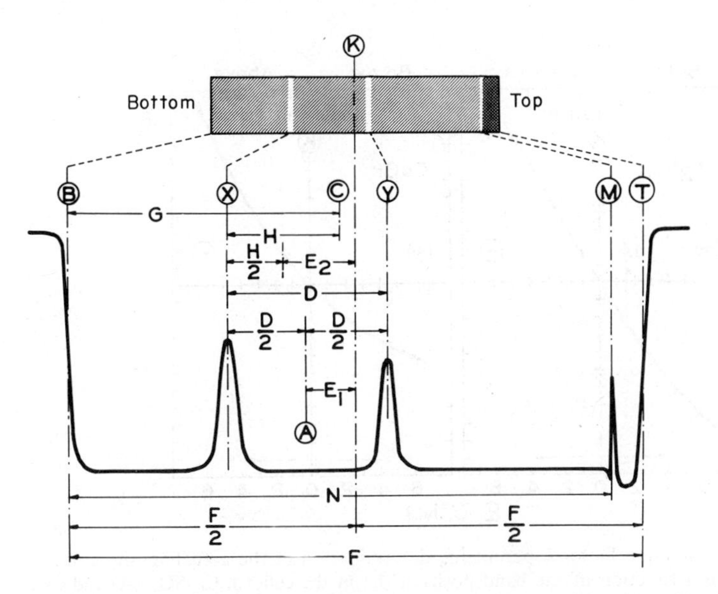

Figure 3. Schematic drawing representing a photograph of the cell filled with CsCl or Cs_2SO_4 solution (shaded area), with two DNA bands (*X* and *Y*), and a tracing of such a photograph. *N*, distance between bottom edge (B) of cell cavity and meniscus (M); *D*, distance between DNA bands (*X* and *Y*); *A*, midpoint between bands X and Y; E_1, distance between *A* and center of cell cavity (K); *C*, point of isoconcentration, usually corresponding to geometric mean between distances from center of rotation to bottom of cell cavity and to meniscus, respectively; E_2, distance between center of cell cavity (K) and midpoint between *X* and *C*; *G*, distance between bottom edge of cell cavity (B) and point *C*.

$$G = 7.3\,M_f - P - \sqrt{\frac{(7.3\,M_f - P)^2 + (7.3\,M_f - P - N)^2}{2}}$$

where M_f = [R(cm)]/1.60 is the magnification factor. Other designations are identical to those in Fig. 2.

species – and determination of the *actual* density gradient, as will be shown below. When these two values are known, the buoyant density of any DNA or RNA in relation to the known density of the marker can be readily assessed.

Determination of the Actual Density Gradient Coefficient and the Absolute Buoyant Density of the Macromolecular Species

This determination will be illustrated for *E. coli* DNA banded for 44 hr in a CsCl or Cs_2SO_4 gradient at 31,410 rpm and 25°C. For highest accuracy and reproducibility the calibration should be repeated when the centrifugation is performed for different periods of time, at different speeds, temperatures, or density ranges, or with any batch of CsCl or Cs_2SO_4 containing different adjuncts or impurities. The procedure is as follows.

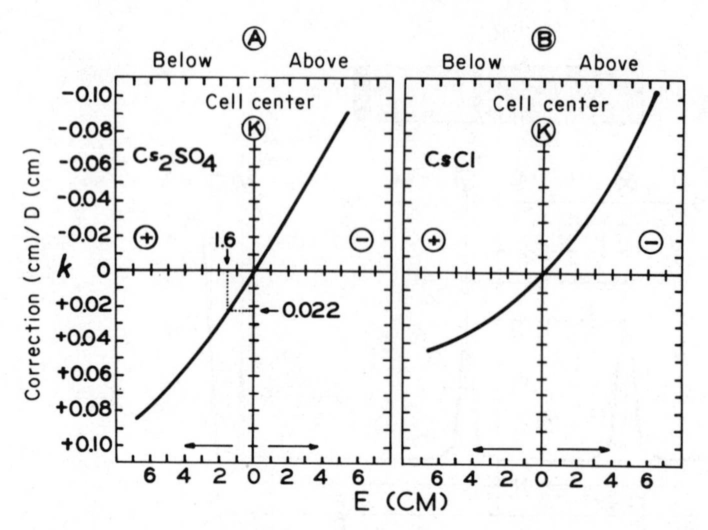

Figure 4. Correction factor k permitting determination of the actual density gradient distribution as a function of the band position (E) in the cell, for Cs_2SO_4 (A) and CsCl (B) equilibrium density gradients. A numerical example of the use of these diagrams is described in the text. The distance (D) between the X and Y bands or the distance (H) between band X and isoconcentration point C are to be corrected in the same manner. The height of the liquid column should be kept reasonably constant ($N/F = 0.90–0.95$) and the density of the CsCl or Cs_2SO_4 solution so adjusted as to keep point A within the central 1/3 of the cell. Symbols A, C, D, E, (E_1, E_2), F, H, K, N, X, and Y are explained in Figs. 2 and 3.

1. The DNA is banded in two cells at different positions by adjusting the initial density of the CsCl or Cs_2SO_4 solution; then it is photographed and traced.

2. The densities, ρ_0' and ρ_0'', of homogeneous mixtures of the contents of each cell at 25°C are determined pycnometrically as described above.

3. Line C (Fig. 3) is drawn on the two tracings. C corresponds to the center between the squares of the distances (cm) from the center of rotation to the cell bottom, r_B, and to the meniscus, r_M:

$$\sqrt{\frac{r_B{}^2 + r_M{}^2}{2}} \tag{1}$$

Equation (2) below gives the distance G (cm) for the measured distance N (cm) (Fig. 2 and 3). Instead of calculating G each time, it is convenient to prepare a graph or table, which gives the distance G corresponding to each measured distance N.

$$G = 7.3M_f - P - \sqrt{\frac{(7.3M_f - P)^2 + (7.3M_f - P - N)^2}{2}} \tag{2}$$

4. Assuming that band X in Fig. 3 represents *E. coli* DNA, the distances H′ and H″ (cm) for the two cells are measured.

5. The distances H′ and H″ are corrected in the same manner as has been described for distance D, measuring E_2 and using factor k from Fig. 4b. (For further discussion of D see the next subsection.) The corrected distances will now be designated H_0' and H_0''.

6. Calculate the actual density-gradient coefficient α, and the buoyant density ρ (gm/cm^3) of the DNA (band X) from the two equations:

$$\begin{aligned} \rho - \rho_0' &= \alpha H_0' \text{ (for cell 1)} \\ \rho - \rho_0'' &= \alpha H_0'' \text{ (for cell 2)} \end{aligned} \tag{3}$$

that is,

$$\alpha = \frac{\rho_0'' - \rho_0'}{H_0' - H_0''} \text{ and } \rho = \rho_0' + \alpha H_0' \tag{4}$$

This calculation is based on the assumption that the buoyant density at point C (point of isoconcentration) corresponds to the pycnometrically determined density of the total CsCl or Cs_2SO_4 solution. If this assumption is correct, banding of *E. coli* DNA at different positions in the cell should always lead to identical computed values of α and ρ.

The point of isoconcentration can be independently determined by running three cells with different initial densities of CsCl or Cs_2SO_4 and determining the three unknowns – α, ρ, and "true G" – from the three equations, one equation for each cell. (In each case substitute $H_0 - x$ for the H_0 in Eq. (3), solve for α, ρ, and x, and calculate the "true G" as equal to $G + x$, with G being derived as previously from Eq. (2).)

The position of point C (distance G) can also be determined by an iteration procedure. The determination of ρ and α should be repeated several times, and it is preferable to band more than one DNA in each cell and to average the values for α and ρ. The most reliable buoyant-density value is obtained when the DNA band happens to coincide with the true point of isoconcentration (C, Fig. 3).

When this method was used with highly purified grades of CsCl and Cs_2SO_4 at 25°C, the buoyant density of *E. coli* DNA was equal to 1.7035 gm/cm^3 in CsCl (44,770 rpm) and 1.4260 gm/cm^3 in Cs_2SO_4 (31,410 rpm) (*27, 41*). The values of α are listed in Table 2.

Determination of the Buoyant Density of a DNA Sample in Relation to a Reference DNA of Known Buoyant Density

Let us assume that Fig. 3 represents the banding pattern of T6 coliphage DNA (band X) and *E. coli* DNA (band Y) after 44 hr centrifugation in a Cs_2SO_4 gradient at 31,410 rpm and 25°C.

TABLE 2. Conversion Coefficient (α) of Distances (cm) in the Cell (1X) or on the Tracing (14X) to Density Increments (gm/cm^3) for Cs_2SO_4 and CsCl Gradients[a]

	α		
	Cs_2SO_4		CsCl
Magnification factor (M_f)	31,410 rpm	44,770 rpm	44,770 rpm
14x	0.00750	0.01500	0.00848
1x	0.105	0.210	0.1187

[a]Example: For the corrected distance between the traced bands X and Y, $D_0 = 3.34$ cm (Fig. 3). The density differential equals 3.34 x 0.00750 = 0.0251 gm/cm^3. The conversion coefficient 0.00750 is shown for Cs_2SO_4 at 31,410 rpm; $M_f = 14$ x.

On the tracing, the measured distance D between the bands equals 3.27 cm, and the center A between the bands is shifted from the cell center K by the distance $E_1 = 1.6$ cm (Fig. 3). From Fig. 4b the correction factor k equals +0.022 for $E = 1.6$ cm (below cell center).

The abbreviated correction for the band distance would be kD = +0.07 cm (= 0.022 x 3.27). Thus, the corrected distance between the bands would correspond to $D_0 = D + kD = 3.34$ cm (= 3.27 + 0.07). Since the tracing was made at magnification factor 14X, factor α equals 0.00750, as found in Table 2.

The density of the lighter band (Y) of *E. coli* DNA was calculated from the equation:

$$\rho = \rho_0 - \alpha D_0 \tag{5}$$

where ρ_0, the density of the reference DNA (in this case T6 coliphage DNA), equals 1.4510 gm/cm^3. Thus, the *E. coli* DNA density ρ equals 1.4510 – (0.00750 x 3.34) = 1.4510 – 0.0250 = 1.4260 gm/cm^3.

This simple procedure for the determination of buoyant density can be summarized as follows:

1. The distance in millimeters between the "reference" and "unknown" peaks is measured.
2. This value is corrected, using the diagram in Fig. 4 and the distance between the center of the cell and the center between the peaks (E_1 in Fig. 3)
3. This corrected value is multiplied by coefficient α from Table 2 to obtain the difference in buoyant density between the peaks in gm/cm^3.
4. The latter figure is added to (or subtracted) the known buoyant density of the "reference" DNA. The whole procedure takes approximately 1 or 2 min per tracing.

The partial specific volume, buoyant density, refractive index increment, and preferential interaction for various DNAs in the CsCl density gradients are given by Cohen and Eisenberg (*12a*).

Effect of Temperature

Equilibrium density gradient centrifugation is usually carried out at 25°C. However, lower temperatures may offer certain advantages: (1) degradation of the macromolecules, especially DNA and RNA, or inactivation of viruses is reduced; (2) renaturation of nucleic acids is retarded, when separation of the complementary strands, or of molecules with cohesive ends (e.g., sheared halves of coliphage λ DNA) is the aim; (3) resolution of the bands increases; (4) somewhat lower concentrations of cesium salts can be employed. The obvious disadvantage is the decreased solubility of cesium salts at low temperatures, with possible crystallization at the bottom of the tube.

The buoyant density of DNA increases with the temperature, due to loss of water from the hydrated CsDNA complex. The temperature coefficient of the density was determined, and may be used to normalize results between 5 and 60°C to 25°C (*11a*).

Preparative Equilibrium Density Gradient Centrifugation

The foregoing discussion was devoted primarily to *analytical* equilibrium density gradient centrifugation. This technique is quite precise, but permits localization of the bands only by their optical properties, mainly OD measurements. When the biological activity or radioactivity of macromolecules must be determined, *preparative* density gradient centrifugation is required for fractionation or purification of the samples.

The centrifugation can be performed either in swinging bucket rotors (*46, 47*) or in fixed angle rotors (*48–50*). The latter method gives somewhat higher resolution, but the density adjustment before the run is quite critical, since the gradient column in the angle rotor is quite short and the macromolecules should band above the hemispherical part of the tube. A special fixed-angle rotor with horizontal tubes was described by Kozinski and Shahn (*51*).

The recommended volumes of solvent were specified in the preceding section on sample preparation. To resolve two bands of native DNA less than 20 μg of DNA per band should be used in the ½ x 2-in. tube of the SW 39L rotor. Two to four times more of denatured DNA can be employed. The sample capacity of the tubes in angle rotor 40 or 50 is at least five times higher.

Centrifugation is usually carried out for 2–3 days at 30,000–39,000 rpm for DNA, and one day at 20,000–30,000 rpm for viruses, in both cases preferably at low temperature (10°C). At still lower temperatures the CsCl may crystallize at the bottom of the tube and form a plug that interferes with the collection of fractions. In that event, by careful manipulation the plug can be dislodged slightly with the tip of the needle used to pierce the bottom of the tube, permitting flow around the plug as it slowly dissolves.

A multispeed centrifugation procedure, denoted the "density gradient relaxation method," was recently devised for the more efficient separation of milligram quantities of a mixture of DNAs (*52*).

DNA, especially denatured DNA, adsorbs tenaciously to the surfaces of glass vials and centrifuge tubes, particularly nitrocellulose tubes. The loss of DNA caused by nonspecific adsorption can often be lessened by adding such detergents as Sarkosyl NL30 (sodium lauroyl sarcosinate) or Dupanol (sodium dodecyl sulfate -- SDS), by boiling the tubes in 0.1 M EDTA or by coating the glass surfaces with denatured DNA or a 1:1:1 mixture of albumin (*52a*), Ficoll, and polyvinylpyrrolidone (Szybalski, unpublished data).

Many types of fractionation devices have been described (*53-66*). With the fractionator used in our laboratory since 1958 (*47, 53*) the bottom of the tube is gently pierced and small drops of the solution are delivered through a short syringe needle (gauge 19) under pressure controlled by the height of a water column. The fractions, as small as 1 or 2 drops (approximately 10 μl per drop), are collected in 9 x 30-mm vials, which are subsequently sealed with plastic mending tape and stored in the refrigerator or freezer.

The absorbance of as small as two-drop fractions can be assayed without dilution by using a 10-μl microcuvette with a 2-mm light path, 2-mm width, 2-3 mm liquid column height, the sample being transferred with a Clinac micropipetter (Lapine Scientific Co., Chicago) equipped with a polyethylene tip. Before determining radioactivity or biological activity (infectivity of viruses or nucleic acids, transforming or template activity), the concentration of the cesium salt usually must be drastically reduced by either dilution, dialysis (*67*), or other techniques.

DNA bands can be visualized in the tube, before collecting fractions, by adding ethidium bromide or comparable fluorescent dye to the salt-DNA solution and illuminating the tube with long-wave UV light (*10, 68*). However, the buoyant density of the DNA-dye complex may differ from that of the pure DNA and resolution of the bands may be impaired. Recently it was reported that in NaI gradients, in contrast to CsCl gradients the separation of two DNA bands is relatively unaffected by the addition of ethidium bromide, and sufficient dye can be used to render the bands visible by ordinary light, simplifying collection procedures (*39*). The banded DNA can then be recovered by piercing the wall of the tube just below the boundary with a tuberculin syringe. Incidentally, this is the collection method of choice for preparative centrifugation of virus particles and proteins, which generally form visibly turbid bands.

The determination of buoyant densities in preparative density gradient centrifugation by calculation of the density distribution or by direct measurement, and the sources of error in the fraction collection procedure, have been discussed by Ifft et al. (*5*) and Rüst (*13*).

The differential penetration of animal virus particles and protein contaminants through an immiscible layer of organic solvents into a CsCl gradient has been applied in the preparative centrifugation of viruses (*68a*).

APPLICATIONS

Selected applications of CsCl and Cs_2SO_4 equilibrium density gradient centrifugation will be presented here, and the properties of the two gradients will be compared. The buoyant densities of selected cellular and viral DNAs and RNAs, and of synthetic polynucleotides, are presented in Tables 3-7.

Comparison of CsCl and Cs_2SO_4 Density Gradients

The properties of density gradients prepared with CsCl and Cs_2SO_4 are rather different. At equivalent rotor speeds Cs_2SO_4 forms approximately twice as steep a gradient as CsCl. Consequently, CsCl gradients provide better resolution of DNA bands, but Cs_2SO_4 gradients are preferable for fractionation of macromolecules with widely different buoyant densities, e.g., normal vs. bromouracil-labeled DNA (*24-26, 41, 42*).

The CsCl gradient is better suited for routine determination of the guanine + cytosine (GC) content of DNA, since in this medium there is an almost linear relationship between the percent GC (20-80%), the buoyant density, and the melting temperature of the DNA (*21, 69-71*). In Cs_2SO_4 gradients there is a lesser and nonlinear dependence of density on GC content (*27*). However, it is useful to determine the percent GC of a given DNA sample from its densities in both gradients, for if different values are obtained some unusual bases or abnormal structural features can be expected. For example, glucosylated T4 coliphage DNA is denser than *Escherichia coli* DNA in the Cs_2SO_4 gradient (1.443 vs. 1.426 gm/cm^3) but less dense in the CsCl gradient (1.700 vs. 1.710 gm/cm^3) (*27*). These values reflect another important difference between the two salts. DNA is much more heavily hydrated in Cs_2SO_4 gradients, with the density averaging 1.4 gm/cm^3 as compared with 1.7 gm/cm^3 in CsCl gradients (*3, 27, 41*).

The lower buoyant density of DNA in Cs_2SO_4 gradients permits the use of lower concentrations of this salt than of CsCl. Moreover, RNA can be banded in Cs_2SO_4 gradients formed from solutions of about 1.6 gm salt per milliliter, well below the solubility limits (*3, 40*). Other specific advantages of

TABLE 3. Buoyant Densities of Selected Cellular DNAs[a]

Source of DNA	Buoyant density (gm/cm^3) Cs_2SO_4 [b]	$CsCl$ [c]	$CsCl$ [d]	G + C (mol %)
Mycoplasma mycoides var. *capri*	1.4194	1.6856	1.6791	25
Spirillum linum	1.4207	1.6911	1.6846	29
Clostridium perfringens	1.4212	1.6915	1.6850	31
Staphylococcus aureus	1.422	1.694	1.687_5	33
Cytophaga johnsonii	1.4216	1.6945	1.6880	34.5
Rhinchosciara angelae (Diptera)	1.422	1.695	1.688_5	35
Bacillus cereus	1.422	1.696	1.689_5	36
Vicia faba	1.422	1.696	1.689_5	36
Proteus vulgaris	1.422	1.698	1.691_5	38
Proteus mirabilis	1.422	1.6986	1.6921	38.5
Calf thymus	1.422	1.699	1.692_5	39
Rabbit kidney cells	1.422	1.699	1.692_5	39
Human cell line D98	1.422	1.6994	1.6929	39.5
Acinetobacter anitratus	1.423	1.699_5	1.693	39.5
Ehrlich ascites	1.423	1.700	1.693_5	40
Mouse liver	1.423	1.700	1.693_5	40
Chicken erythrocytes	1.423	1.7008	1.6943	41
Neisseria catarrhalis	–	1.701	1.694_5	41
Acinetobacter haemolysans	–	1.701	1.694_5	41
Moraxella nonliquefaciens	–	1.701	1.694_5	41
Moraxella bovis	–	1.702_5	1.696	42.5
Physarum polycephalum	1.424	1.702_5 [e]	1.696[e]	42.5
Moraxella liquefaciens	1.424	1.703	1.696_5	43
Moraxella osloensis	–	1.703	1.696_5	43
Moraxella phenylpyrouvica	–	1.703	1.696_5	43
Acinetobacter lwoffi	–	1.703	1.696_5	43
Herring sperm	1.424	1.703	1.696_5	43
Bacillus subtilis	1.4240	1.7034	1.6969	43.5
Moraxella kingii	–	7.704_5	1.698	44.5
Neisseria caviae	–	1.704_5	1.698	44.5
Neisseria ovis	–	1.704_5	1.698	44.5
Neisseria flavescens	–	1.707_5	1.701	47.5
Neisseria cinerea	–	1.709	1.702_5	49
Escherichia coli	1.4260	1.7100	1.7035	50
Brucella abortus	1.428	1.717	1.710	56
Xanthomonas oryzae	–	1.723	1.716_5	63
Xanthomonas pruni	1.432	1.725	1.718_5	65
Xanthomonas translucens	–	1.7277	1.7212	67.5
Streptomyces chrysomalus	1.4345	1.7305	1.7240	70
Streptomyces griseus	1.435	1.731	1.724_5	71
Micrococcus lysodeikticus	1.435	1.731	1.724_5	71
Micrococcus luteus	1.4352	1.7311	1.7246	72

[a] Buoyant densities determined at 25°C, 44,770 rpm (CsCl) or 31,410 rpm (Cs_2SO_4). Densities abbreviated to nearest 0.001 (or 0.0005) gm/cm^3 for 2–3 determinations, or measured with an accuracy of $\pm 0.0002^3$. *Acinetobacter, Moraxella,* and *Neisseria* data taken from Bøvre, Fiandt, and Szybalski (77).

[b] Measured vs. *Escherichia coli* DNA, density 1.4260 gm/cm^3.

[c] Measured vs. *Escherichia coli* DNA, density 1.7100 gm/cm^3.

[d] Measured vs. *Escherichia coli* DNA, density 1.7035 gm/cm^3.

[e] Second peak at 1.713 gm/cm^3, or 1.706_5, respectively.

Cs_2SO_4 over CsCl include: (1) the magnified density difference between native and denatured DNA (*24, 27*); (2) the approximately linear dependence of density on the extent of glucosylation of T-even coliphage DNA (*27*); and (3) the possibility of studying complexes between DNA and Hg^+ or Ag^+ ions (which precipitate as the chlorides) (*23, 72–75*).

Base Composition and Melting Temperature of Nucleic Acids

Equilibrium density gradient centrifugation is widely used to assess the base composition of DNA. As already stated, CsCl is the salt of choice for forming the density gradient because of the largely linear relationship between the buoyant density and the GC content in this gradient (*21, 69–71*) (Fig. 5 and Tables 3 and 4). However, density determinations in both CsCl and Cs_2SO_4 gradients are recommended, since discrepancies between the GC contents derived by the two methods may indicate the presence of rare bases, as in the case of T-even coliphage DNA, containing glucosylated hydroxymethylcytosine (hmC) (Table 5) (*27*), and *Bacillus subtilis* phages PBS2 and SP8 DNAs, containing uracil or hydroxymethyluracil (hm^5U) (Table 4).

Differences in base composition between DNAs, if sufficiently large, can be utilized for their separation by preparative density gradient centrifugation. For example, phage DNA was separated from the host bacterial DNA; transforming DNA was separated from the viral DNA of a transducing phage (*78, 79*); and transforming DNA was fractionated into subcomponents (*80*). In like manner, animal and plant cell DNA of nuclear or cytoplasmic origin was fractionated into major and minor (satellite) components (*70, 75a, 80a–80h*), leading to studies on the localization, site of origin, and function of satellite DNA. Bacterial plasmid DNA (extrachromosomal DNA, including the F and R factors) was separated from the host chromosomal DNA as satellite DNA bands in the CsCl gradient (*80i–80k*).

Systematic and taxonomic purposes are also served, since phylogenetic relationships are reflected in the GC content of DNA, a chemical constant, and for a given species the distribution is generally quite characteristic (*71, 75a–77*).

The base composition (mol % GC) of bihelical DNA can be calculated from the formula of Schildkraut et al. (*21, 70*),

$$\%\,GC = (\rho - 1.660)/0.00098$$

or of de Ley (*71*),

$$\%\,GC = 1020.6\,(\rho - 1.6606)$$

or the simplified formula of the authors,

$$\%\,GC = 1000\,(\rho - 1.660)$$

TABLE 4. Buoyant Densities of Selected Viral DNAs[a]

Source of DNA (host/phage)	Buoyant density (gm/cm^3) Cs_2SO_4[b]	$CsCl$[c]	$CsCl$[d]	G + C (mol %)	Comments
Staphylococcus aureus phages					
44A	1.419	1.687	1.680_5	27	–
81	1.421	1.695	1.688_5	35	–
80	1.422	1.697	1.690_5	37	–
Fowl pox virus	1.4213	1.6945	1.6880	35	–
Rabbit pox virus	1.4217	1.6959	1.6894	36	–
Bacillus anthracis phage γ	1.422	1.696	1.689_5	36	–
Bacillus tiberius phage α	1.424	1.704	1.697_5	44	–
Bacillus stearothermophilus phage $\phi\mu4$	1.425	1.706	1.699_5	46	–
Brucella abortus phage (Tbilisi)	1.4251	1.7089	1.7024	49	–
Salmonella phage χ	1.428	1.715	1.708_5	55	–
Actinophage ϕ17	1.429	1.719	1.712_5	59	–
Xanthomonas pruni phage XP5	1.430	1.722	1.715_5	62	–
Escherichia coli phages					
T5	–	1.7016	1.6951	42	–
T1	–	1.7057	1.6992	46	–
$\lambda dgal_{A\text{-}J\text{-}b2}$	–	1.7083	1.7018	48	mol. wt. 23×10^6
$\lambda dgal_{L\text{-}J\text{-}b2}$	–	1.7090	1.7025	49	mol. wt. 34×10^6
λ^+	1.426	1.7093	1.7028	49	mol. wt. 31×10^6
T7	1.426	1.710	1.703_5	50	–
21	–	1.7100	1.7035	50	–
λb2b5	–	1.7108	1.7043	51	mol. wt. 24×10^6
λb2cI	–	1.7110	1.7045	51	mol. wt. 26×10^6
T3	1.426	1.7105	1.7040	50	–
ϕ80	–	1.7121	1.0560	52	–
M13	1.4465	1.7223	1.7158	41	Single-stranded
f1	–	1.7226	1.7161	41	Single-stranded
fr	–	1.7223	1.7158	41	Single-stranded
ϕX174	–	1.725	1.718_5	42	Single-stranded
ϕR	–	1.7243	1.7178	42	Single-stranded
Pseudorabies virus	1.436	1.732	1.725_5	73	–
Bacillus subtilis phages					
ϕ15	1.420	1.694	1.687_5	34	–
ϕ29	1.420	1.694	1.687_5	34	–
SP3	1.421	1.695	1.688_5	35	–
pKc	1.421	1.695	1.688_5	35	–
SPα	1.424	1.703	1.696_5	43	Defective phage
SPX	1.424	1.703	1.696_5	43	Defective phage

TABLE 4 (continued)

Source of DNA (host/phage)	Buoyant density (gm/cm^3) Cs_2SO_4[b]	$CsCl$[c]	$CsCl$[d]	G + C (mol %)	Comments
SP50	1.424	1.704	1.697$_5$	44	–
ϕ1	1.424	1.704	1.697$_5$	44	–
ϕ2	1.424	1.704	1.697$_5$	44	–
ϕ14	1.424	1.704	1.697$_5$	44	–
F	–	1.709	1.702$_5$	49	–
SP10	1.439	1.715–1.723	1.708$_5$–1.716$_5$	44	Density decreases upon storage
PBS1	1.433	1.722	1.715$_5$	28	Uracil replaces thymine
PBS2	1.433	1.722	1.715$_5$	28	Uracil replaces thymine
SP70	1.433	1.722	1.715$_5$	–	Possibly related to PBS1 and 2
SP80	1.433	1.722	1.715$_5$	–	Possibly related to PBS1 and 2
SP90	1.433	1.722	1.715$_5$	–	Possibly related to PBS1 and 2
SP100	1.433	1.722	1.715$_5$	–	Possibly related to PBS1 and 2
SP8	1.455	1.742	1.735$_5$	43	hm^5U[e] replaces thymine
SP82	1.455	1.742	1.735$_5$	43	hm^5U replaces thymine
SPO–1	1.455	1.742	1.735$_5$	43	hm^5U replaces thymine
ϕe	1.455	1.743	1.736$_5$	43	hm^5U replaces thymine
2C	1.455	1.742	1.735$_5$	43	hm^5U replaces thymine
ϕ25	1.455	1.742	1.735$_5$	43	hm^5U replaces thymine
SP60	1.455	1.742	1.735$_5$	43	hm^5U replaces thymine

[a]Buoyant densities determined at 25°C, 44,770 rpm (CsCl) or 31,410 rpm (Cs_2SO_4). Densities abbreviated to nearest 0.001 (or 0.0005) gm/cm^3 for 2–3 determinations, or measured with an accuracy of ± 0.0002 gm/cm^3.

[b]Measured vs. *Escherichia coli* DNA, density 1.4260 gm/cm^3.

[c]Measured vs. *Escherichia coli* DNA, density 1.7100 gm/cm^3.

[d]Measured vs. *Escherichia coli* DNA, density 1.7035 gm/cm^3.

[e]hm^5U = 5-hydroxymethyluracil.

[f]Density and mol. wt. decrease in alkali. Probably contains ribotide linkages.

TABLE 5. Buoyant Densities of T-Even Coliphage DNAs[a]

	Buoyant density (gm/cm^3)			G = hm^5C	Glucose hm^5C	Biose hm^5C
DNA[b]	Cs_2SO_4[c]	CsCl[d]	CsCl[e]	(mol %)	(mol %)	(mol %)
T2(o)s[f]	1.4314	1.7060	1.6995	34	(0)	–
T2gt[g]	–	1.7057	1.6992	34	(0)	–
T6(o)[h]	1.4298	1.7052	1.6987	34	14	7
T2	1.4392	1.7020	1.6955	34	80	5
T2 X $\overline{T2}$	1.4403	1.7018	1.6953	34	85	1.5
$\overline{T2}$	1.4430	1.7005	1.6940	34	100	0
T4	1.4430	1.7005	1.6940	34	100	0
C16	1.4478	1.7103	1.7038	34	130	65
T6	1.4510	1.7105	1.7040	34	148	72

[a] Taken from Erikson and Szybalski (*27*).
[b] Glucosylated hm^5C (5-hydroxymethylcytosine) replaces cytosine.
[c] Measured vs. *Escherichia coli* DNA, density 1.4260 gm/cm^3, and poly d(A-T), density 1.4240 gm/cm^3.
[d] Measured vs. *Escherichia coli* DNA, density 1.7100 gm/cm^3.
[e] Measured vs. *Escherichia coli* DNA, density 1.7035 gm/cm^3.
[f] Enzymatically synthesized; glucose-free.
[g] Glucose transferase-deficient mutant (Sheldrick, unpublished data).
[h] Grown in $UDPG^-$ host.

where ρ is the buoyant density in gm/cm^3, and the density of the *E. coli* reference DNA is taken to be 1.710 gm/cm^3 and its GC content 50%.

The melting temperature (T_m) of DNA is also a linear function of the buoyant density, ρ, as well as of the GC content within the range of 30–70% GC (*71, 81, 82*). The T_m in SSC, pH 7, can be calculated from the formula of de Ley (*71*),

$$T_m(\text{SSC}) = 429.76\,(\rho - 1.5002)$$

or in SSC/10, pH 7, from the formula of Mandel et al. (*82*),

$$T_m\,(\text{SSC}/10) = 489.4\,(\rho - 1.556)$$

The T_m in SSC/10 is increased by $16.3 \pm 0.5°C$ when the buffer is SSC. Significant departure from the linear relationships given above may indicate the presence of rare bases or sugar moieties (*82*).

The buoyant density of single DNA strands of known base composition can be calculated from the formula of Riva et al. (*83*),

$$\rho = 1.6267\text{A} + 1.7580\text{G} + 1.7681\text{C} + 1.7424\text{T}$$

where ρ is the density in the neutral CsCl gradient and A, G, C, and T are the molar fractions of adenine, guanine, cytosine, and thymine, respectively.

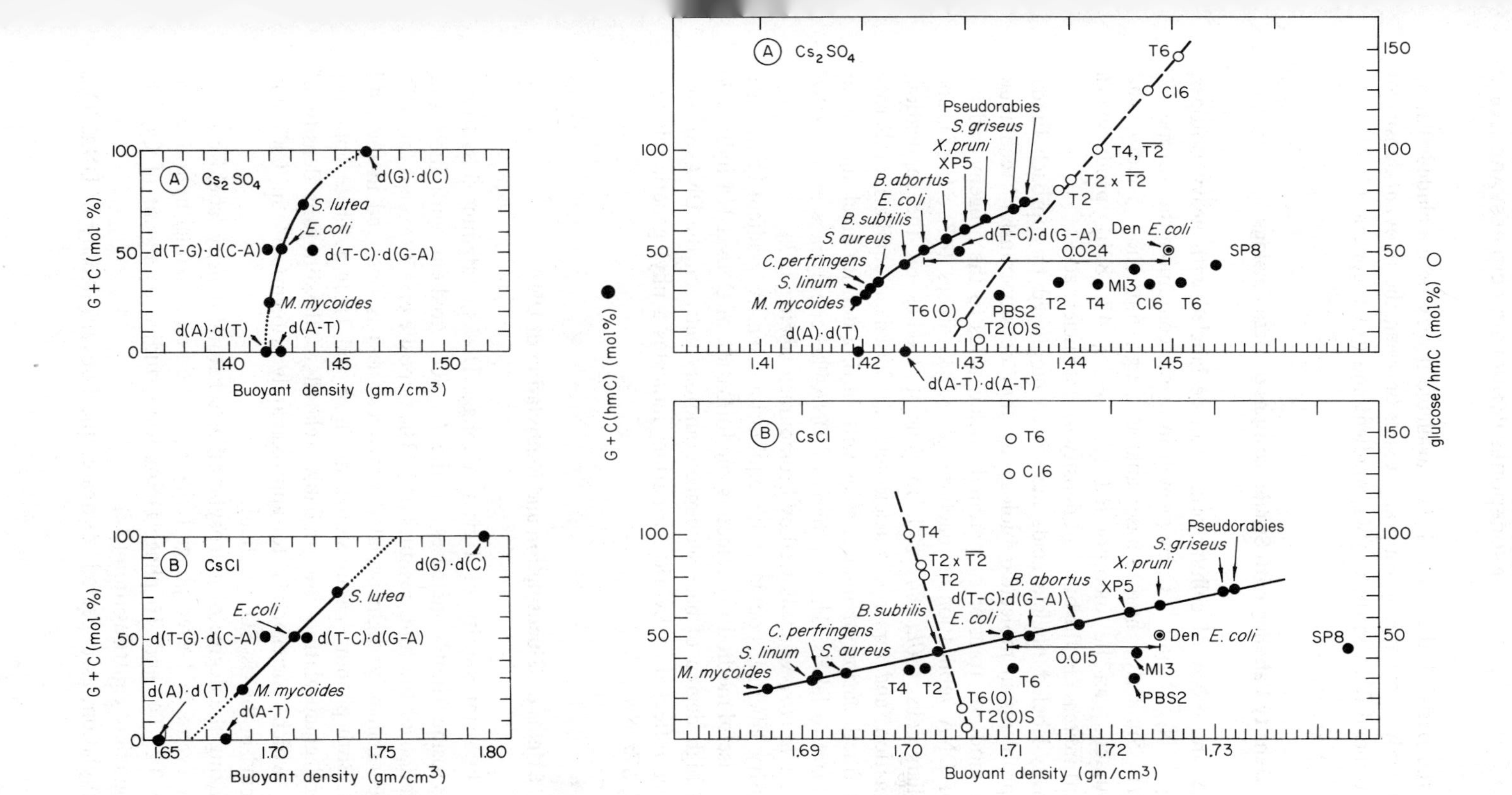

Figure 5. Buoyant densities (25°C) of native DNAs in Cs_2SO_4 (A) and CsCl (B) gradients, with contracted (left) and expanded (right) density scales, as a function of base composition (•, solid lines), expressed as mol % G + C or G + hmC (hydroxymethyl cytosine, hmC), or as a function of the glucose to hmC ratio (glucose/hmC; mol %), the latter for T-even phage DNA (⊙, dashed lines). For numerical data and complete names of bacteria, viruses, and phages see Tables 3–7. Denatured *E. coli* DNA (⊙, denatured *E. coli*) was prepared by heating the DNA (10 μg/ml) for 10 min at 96–100°C in 0.002 M sodium citrate (pH 7.8) and rapid chilling at 0°C.

On the basis of the limited data available it may be concluded that an approximately linear relationship also exists between the buoyant density of double-stranded RNA in the Cs_2SO_4 gradient and its GC content (*172*).

Density Labeling with Stable Isotopes or Base Analogs

DNA and RNA of cellular origin can be labeled with heavy isotopes, including ^{15}N, ^{2}H, and ^{13}C, by growth in media containing the isotopes (*2, 35, 83a*). Similarly, heavy base analogs, e.g., 5-bromo-, 5-iodo-, and 5-fluorouracil, can be incorporated (*1, 24*–26, 47, 83b), producing a significant increase in the buoyant density of the nucleic acid.

Density-labeled nucleic acids have been employed to approach fundamental problems in molecular biology by the "density-transfer" technique, e.g., the mode of transfer of parental nucleic acids to the progeny and the mode of DNA replication in bacteria (*2, 83b*), viruses (*47, 84, 85*), and mammalian cells (*80h, 86, 87*). In general, studies on the replication, recombination, and repair of nucleic acids are based on equilibrium density gradient fractionation of nucleic acids with both a density label and one or two radioactive labels. The synthesis of DNA-like copolymers by the DNA polymerase *in vitro* was followed by bromouracil labeling (*41*).

Density labeling, coupled with equilibrium density gradient centrifugation, is a useful method for detecting the formation of DNA–DNA hybrids *in vitro*. A high degree of base sequence homology between the DNAs of two species, as reflected in hybrid formation, indicates a phylogenetic relationship (*27, 87a–87d*).

Effects of Glucosylation and Methylation of DNA

The buoyant density of T-even coliphage DNAs containing 5-hydroxymethylcytosine (hmC) increases in the Cs_2SO_4 gradient with increasing glucosylation of the hmC residues (*27*). The opposite effect (except for DNA with predominantly diglucosylated hmC residues) is observed in the CsCl gradient. Both phenomena are shown in Fig. 5 and Table 5. These opposing effects of glucosylation are probably related to the buoyant density of glucose, which corresponds to approximately 1.6 gm/cm^3 in CsCl and 1.5 gm/cm^3 in the Cs_2SO_4 gradient.

Utilizing the degree of glucosylation as a natural density label for DNA, the early events in T-even phage DNA replication and thermal hybridization between T2, T4, and T6 phage DNAs were studied by means of Cs_2SO_4 density gradient centrifugation (*27*).

Methylation, as expected, decreases the buoyant density of DNA. This can be inferred from determinations on DNAs of the same base composition (28% GC) but containing either uracil in place of thymine (phage PBS2, Table 4) or 5-methyluracil (thymine), the normal DNA component. In the former case the experimentally determined density is 1.722 gm/cm^3 in CsCl and 1.433 gm/cm^3 in Cs_2SO_4. In the latter it should be 1.688 gm/cm^3 in CsCl and 1.420 gm/cm^3 in Cs_2SO_4 (Fig. 5). The increment amounts to approximately 1 mg/cm^3 in CsCl and 0.4 mg/cm^3 in Cs_2SO_4 gradients per 1% methylated base (*88*).

Effects of Denaturation and Formaldehyde on Nucleic Acids

The buoyant density of denatured DNA is higher than that of the corresponding native bihelical DNA (*2, 69b, 70*). This density bias increases with decreasing GC content and is a function of the residual secondary structure of the denatured DNA and interaction between the complementary strands. For example, the density of the replicative form of ϕX174 DNA is 1.707 gm/cm^3 in the bihelical state and 1.719 gm/cm^3 in the denatured form, whereas the density of the single viral strand is 1.725 gm/cm^3 (*88*). The density bias between bihelical and single-stranded DNA is greater in Cs_2SO_4 than in CsCl gradients (*27*).

Formaldehyde reacts with the amino groups of denatured DNA, decreasing the amount of secondary structure. The presence of 1% formaldehyde during denaturation (followed by dilution or dialysis) increases the density of the DNA strands, providing greater resolution of the native and denatured DNA bands. However, its addition after denaturation results in a progressive decrease in the density of the denatured DNA (*24*) or RNA (*40*). At high concentrations formaldehyde cross-links bihelical DNA (*89*).

Separation of the Complementary DNA Strands

Some denatured DNAs give rise to two bands in equilibrium density gradients, which correspond to the separated complementary strands. The buoyant density bias can be further accentuated by the addition of guanine-rich ribopolymers, which preferentially interact with one strand (*90–100*). Complexing with methylmercuric hydroxide has also been proposed (*101*). At present poly (U, G) is the most effective known agent for the preparative separation of the complementary DNA strands. Centrifugation in alkaline density gradients can be employed in certain cases, as described in the following section. Isolated DNA strands have many applications, including the study of transcription (*90, 95, 102–106*), DNA replication (*100*), and electronmicroscopic cytogenetics of DNA heteroduplexes (*107, 108*).

TABLE 6. Buoyant Densities of Synthetic Polynucleotides

Polynucleotide[a]	Buoyant density (gm/cm^3)							
	Cs_2SO_4	Ref.	Cs_2SO_4 alkaline	Ref.	CsCl	Ref.	CsCl alkaline	Ref.
rA	1.570[b]	*(178)*						
rG	1.693[c]	*(88)*						
	1.75	*(179)*						
rU	1.650[b]	*(178)*						
rC	1.583[c]	*(88)*			1.873_5[d]	*(180)*		
	1.59	*(179)*						
	1.63[b]	*(134)*						
dA	1.379[b]	*(178)*	1.379	*(178)*	1.617[e]	*(181)*	1.622	*(182)*
					1.678_5[d]	*(88)*		
dI			1.45–1.46	*(34)*				
dG	1.539[c]	*(88)*	1.54	*(34)*	1.760[f]	*(190)*	1.791	*(188)*
					1.763[e]	*(188)*		
dT	1.424[b]	*(178)*	1.456	*(178)*	1.739[e]	*(181)*	1.771	*(182)*
dC	1.418[b]	*(190)*	1.40–1.41	*(34)*	1.685[e]	*(188)*	1.722	*(188)*
	1.42[b]	*(134)*						
d(br^{5}C)			1.73	*(34)*				
rA • rU	1.660[b]	*(178)*						
rA • 2rU	1.702[b]	*(178)*						
rI • rC	1.62[b]	*(134)*						
rG • rC	1.66[b]	*(134)*						
	1.685	*(179)*						
dA • dT	1.417[c]	*(88)*			1.637_5[e]	*(181, 182, 191)*		
	1.419[g]	*(183, 191)*			1.640_5[d]	*(88)*		
	1.432[b]	*(178)*						
dA • 2dT	1.492[b]	*(178)*						
dI • dC	1.48[b]	*(34)*			1.754[e]	*(181, 191)*		

	1.467[c]	*(27)*			1.787[d]	*(27)*		
	1.49[b]	*(134)*			1.789_5[d]	*(70, 180)*		
rG • dG					1.797[f]	*(190)*		
rA • dT	1.433[b]	*(178)*						
rA • 2dT	1.519[b]	*(178)*						
(rA • dT) • rU	1.582[b]	*(178)*						
(rA • rU) • dT	1.584[b]	*(178)*						
(dA • dT) • rU	1.525[g]	*(183)*						
	1.536[b]	*(178)*						
dA • 2rU	1.620[b]	*(178)*						
rI • dC	1.54[b]	*(134)*						
2dI • rC	1.68[b]	*(134)*						
rG • dC	1.552[g]	*(190)*						
	1.58[b]	*(134)*						
dG • rC	1.57[b]	*(134)*			1.853_5[d]	*(180)*		
r(A–U) • r(A–U)	1.614[b]	*(184)*						
r(A–br^{5}U) • r(A–br^{5}U)	1.695[b]	*(184)*						
d(A–U) • d(A–U)	1.439[c]	*(189)*			1.711_5[d]	*(189)*		
d(A–hm^{5}U) • d(A–hm^{5}U)	1.473[c]	*(189)*			1.727_5[d]	*(189)*		
d(A–br^{5}U) • d(A–br^{5}U)	1.540[b]	*(34, 136)*	1.54–1.56	*(34)*				
	1.550[g]	*(188)*						
	1.557[b]	*(178)*						
d(A–T) • d(A–T)	1.424	*(27, 183)*	1.40	*(34)*	1.671_5	*(27, 189, 192)*	1.722	*(182, 185)*
	1.425	*(191)*	1.416	*(185)*	1.672	*(181–183, 191)*		
	1.426	*(34, 134, 178)*			1.672_5	*(70)*		
d(A–s^{2}T) • d(A–s^{2}T)	1.4646[c]	*(192)*						
d(A–s^{4}T) • d(A–s^{4}T)[h]	1.4373	*(193)*			1.720[e]	*(193)*		
	1.5269	*(193)*						
d(n^{2}A–T) • d(n^{2}A–T)					1.718	*(186)*		
d(I–C) • d(I–C)	1.453[g]	*(185, 191)*	1.436	*(185)*	1.735[e]	*(181, 185, 191)*	1.766	*(185)*

TABLE 6 (continued)

Polynucleotide[a]	Buoyant density (gm/cm^3)							
	Cs_2SO_4	Ref.	Cs_2SO_4 alkaline	Ref.	CsCl	Ref.	CsCl alkaline	Ref.
d(G–C) • d(G–C)	1.448[g]	*(187, 191)*	1.465	*(191)*	1.741[e]	*(191)*	1.793	*(191)*
d(A–C)					1.689[e]	*(182)*	1.684	*(182)*
d(T–C)	1.460[g]	*(183)*						
d(T–G)					1.777[e]	*(182)*	1.828	*(182)*
d(T–C) • d(G–A)[h]	1.439[c]	*(88)*						
	1.428[g]	*(187, 191)*			1.708_5[d]	*(88)*		
	1.466[g]	*(183, 191)*			1.711[e]	*(182, 191)*		
d(T–G) • d(C–A)	1.420[c]	*(88)*			1.690_5[d]	*(88)*		
	1.423[g]	*(183, 187, 191)*			1.690_5[e]	*(182, 191)*		
					1.702[e]	*(193)*		
d(s^4T–G) • d(C–A)					1.714[e]	*(193)*		
d(T–I) • d(C–A)					1.691[e]	*(193)*		
d(T–T–C) • d(G–A–A)	1.427[g]	*(187, 191)*			1.685_5[e]	*(182, 191)*		
d(T–T–G) • d(C–A–A)	1.420[g]	*(187, 191)*			1.683[e]	*(182, 191)*		
d(T–A–C) • d(G–T–A)	1.422[g]	*(187, 191)*			1.713[e]	*(182, 191)*		
d(A–T–C) • d(G–A–T)	1.418[g]	*(187, 191)*			1.687[e]	*(182, 191)*		
(rA–dU) • (rA–dU)	1.500[b]	*(184)*						
d(T–C) • d(G–A) • r(U–CH$^+$)	1.520[g]	*(183)*						

[a] The abbreviations for the polynucleotides conform to those proposed by the IUPAC-IUB Commission on Biochemical Nomenclature.
[b] Measured vs. d(A–T) • d(A–T), density 1.426 gm/cm^3 or d(A–br^5U) • d(A–br^5U), density 1.540 gm/cm^3.
[c] Measured vs. *Escherichia coli* DNA, density 1.4260 gm/cm^3 or d(A–T) • d(A–T), density 1.4240 gm/cm^3 (27).
[d] Measured vs. *Escherichia coli* DNA, corrected to density 1.7035 gm/cm^3 (27) or d(A–T) • d(A–T), density 1.6715 gm/cm^3.
[e] Measured vs. d(A–T) • d(A–T), density 1.672 gm/cm^3 or phage T4 DNA, density 1.693 gm/cm^3.
[f] Measured vs. phage λ DNA, corrected to density 1.705 gm/cm^3.
[g] Measured vs. phage T4 DNA, density 1.444 gm/cm^3 or d(A–T) • d(A–T), density 1.424 gm/cm^3.
[h] Probably exists in two different molecular configurations in concentrated Cs_2SO_4 solutions.

Alkaline Equilibrium Density Gradients

At alkaline pHs exceeding a critical level (>11), DNA is denatured and increases in buoyant density (*33, 83b*). At the same time, the thymine and guanine residues are deprotonated and are neutralized by cesium ions. The total density increment is independent of the GC content of the DNA and amounts to approximately 0.06 gm/cm^3 in CsCl gradients – 0.015 gm/cm^3 due to denaturation plus 0.045 gm/cm^3 due to deprotonation and neutralization for DNA containing 50% GC (*33*).

Alkaline density gradients have been applied to study the denaturation and strand separation of spontaneously renaturing DNAs, such as circular polyoma DNA (*109*) and synthetic deoxyribopolymers (*34, 110*) (Table 6), and to distinguish between native and denatured states of normal vs. bromouracil-labeled DNA (*83b*).

Some natural DNAs exhibit a small density bias between the complementary strands in alkaline density gradients (*93, 111*), and a few, e.g., certain mitochondrial DNAs (*112*) and other satellite DNAs (*80d–80f*) show a large density bias. In the latter case, preparative separation of the complementary DNA strands by alkaline density gradient centrifugation is feasible.

Effects of Irradiation on DNA

The buoyant density of native DNA in CsCl gradients increases upon ultraviolet irradiation. This was originally construed to indicate some denaturation-like process (*113*). That simple interpretation is not tenable, however, since the density increments for DNA irradiated with various doses of UV are two to four times lower in Cs_2SO_4 than in CsCl, whereas denaturation should have the reverse effect (*24*).

The crosslinking of DNA, and the effects of halogenated uracil analogs on the UV and x-ray sensitivity of normal and crosslinked DNA from bacteria and human cell lines have been studied by means of equilibrium density gradient centrifugation (*24–26, 113–115*).

Complexing of Silver and Mercury Ions by DNA

DNA binds the heavy metal ions, Ag^+ and Hg^{2+}, with a resulting large increase in buoyant density (*72–75*). The binding is reversible, and the DNA can be recovered in its original form by removal of the ligand with complexing agents, such as Cl^-, CN^-, or EDTA.

The reaction is selective, Hg^+ preferentially binding to AT-rich DNAs and Ag^+ to GC-rich DNAs. Therefore, large buoyant density differences can be created between DNAs of different base compositions. For example, the d(A-T)-like component of cancer crab DNA was isolated and purified by preparative centrifugation in a Hg^{2+}–Cs_2SO_4 gradient, mammalian DNA was fractionated (*80e*), and coliphage T4 DNA was separated from *E. coli* DNA (*73*).

CsCl gradients are not applicable because of the insolubility of the chlorides, and the Cs_2SO_4 must be free of complexing agents that bind Ag^+ or Hg^{2+}. The amount of ligand bound at equilibrium can be manipulated by adjusting the pH, since Ag^+ and Hg^{2+} ions displace protons in the base moieties of DNA.

By the same principle, selective binding of Hg^{2+} ions has been utilized for a very efficient separation of half molecules of coliphage λ and λb_2 DNA, which differ in GC content by 6–9% (*75, 116,-118*) and also of DNA fragments of the λ and φ80 phages (*118a, b*).

Denatured or single-stranded DNA binds Ag^+ and Hg^{2+} ions much more strongly than does native DNA, and the augmented density bias can be utilized for their preparative separation (*73*).

The use of methylmercuric hydroxide, which denatures DNA and preferentially binds to thymine-rich strands, has been proposed as a means for preparative separation of complementary DNA strands differing in base composition (*101*).

Interaction of DNA with Antibiotics and Dyes

Three groups of antibiotics that inhibit RNA synthesis, the anthracyclines, actinomycins, and chromomycins, form complexes with DNA, which are stable at high ionic strength (*119*). These agents cause a progressive decrease in the buoyant density of the DNA, in both CsCl and Cs_2SO_4 gradients, with increasing antibiotic concentration and DNA GC content. As a result, it is possible to measure the binding of the antibiotics and to study their mode of interaction with DNA. *Sarcina lutea* DNA was separated into two bands when centrifuged in the presence of olivomycin (*119*).

Covalent crosslinking of the complementary DNA strands by the mitomycins and porfiromycins (*120*), as well as naturally existing crosslinks (*32, 121, 122*), have been studied with equilibrium density gradients.

It has been shown that closed circular DNAs, when compared with linear or nicked circular molecules, cannot bind as much of intercalating dyes, e.g., ethidium bromide or its analogs, and consequently exhibit higher buoyant densities at saturating dye concentrations (*10, 68, 123, 124*). The DNA-dye bands can be visualized by virtue of their color and fluorescence. Moreover, the dye can be subsequently removed quantitatively by passage of the DNA sample through a 0.8 x 4.5-cm column of analytical grade Dowex-50 resin.

Equilibrium density gradient centrifugation in the presence of intercalating dyes is now the method of choice for the detection and preparative isolation of closed circular DNAs of viral, mitochondrial, chloroplast or other plasmidic origin.

Equilibrium Density Gradient Centrifugation of RNA, Ribosomes, and RNA-DNA Hybrids

CsCl density gradient centrifugation is not directly applicable to RNA preparations, because the buoyant density of RNA is quite high and CsCl is not sufficiently soluble to provide a suitable gradient under the usual conditions. This problem can be partially circumvented by centrifugation at 40–50°C; in this temperature range the solubility of CsCl is high enough to establish a gradient in which RNA will not sediment to the bottom of the cell (*125–126b*). Technical difficulties, which include the inherent thermal lability of RNA as well as oil fogging of the optical components of the standard centrifuge, limit the usefulness of this procedure. Recently it was shown that RNA can be banded in the CsCl gradient at 25°C if D_2O is substituted for H_2O as the solvent (*126a*).

The use of cesium formate (*35*) and cesium acetate (*36, 37*) solutions is hampered by their high viscosity, and the precipitation of high-molecular-weight, single-stranded RNA in Cs_2SO_4 solution limits the application of this salt. However, some single-stranded RNAs, e.g., R17 phage RNA and *B. subtilis* ribosomal 18s RNA, do not form precipitates in Cs_2SO_4 gradients (*40*). Similarly, all double-stranded RNAs and DNA-RNA hybrid molecules form true Gaussian bands in this gradient.

Three mixed solvents applicable to determination of the buoyant densities of all RNAs tested have been described by Lozeron and Szybalski (*40*). *Solvent 1* consists of 0.2 ml of 0.1 M phosphate-citrate buffer, pH 6.5, and 0.3 ml saturated Cs_2SO_4 solution containing 1–2% formaldehyde. In this solvent most RNAs (2–3 μg/band) can be banded in the lower half of the cell without formation of precipitates or excessive loss of material from the bands.

If banding in the center of the cell is desired, a 1:1 mixture of saturated solutions of CsCl and Cs_2SO_4 (0.4 ml) is combined with 0.1 ml of formaldehyde-RNA solution in buffer, bringing the final concentration of formaldehyde to 1% (*Solvent 2*).

When the presence of formaldehyde cannot be tolerated, as in the case of infectivity assays for RNA, *Solvent 3* can be employed. This consists of 0.05 ml buffer, 0.4 ml saturated CsCl solution, and 0.05 ml saturated Cs_2SO_4 solution. In this solvent, however, most RNAs can be banded only in the lower one-third of the cell. Centrifugation is at 44,770 rpm and 25°C for 20–44 hr, with a longer period for lower molecular weight RNAs.

Both Solvents 2 and 3 completely eliminate the interfering phenomena observed when RNA is centrifuged in pure Cs_2SO_4 gradients, i.e., the formation of a precipitate band of one RNA or of coprecipitate bands when two or more RNAs are present, and the partial or complete disappearance of RNA from the UV-absorbing band (*40*).

Recently, it was shown that the tendency of single-stranded RNAs to aggregate in Cs_2SO_4 gradients can be reduced by the addition of the denaturing solvent, dimethylsulfoxide (DMSO), in low concentrations at neutral or acid pH. The Cs_2SO_4-DMSO system was used to separate duplex DNA and single-stranded and double-stranded RNA, and a method for calculating the buoyant density of the macromolecule was given (*126b*).

The Cs_2SO_4 density gradient was employed for the characterization of the single and double-stranded RNAs of a number of viruses and bacteriophages. The buoyant density data were summarized by Erikson and Franklin (*127*) and are presented in Table 7.

The mechanism of action of the DNA-dependent RNA polymerase was studied with the Cs_2SO_4 gradient (*128–135*), as well as the formation of natural DNA–RNA complexes during *in vitro* and *in vivo* synthesis of RNA (*135–138*). Similarly, artificial DNA–RNA complexes produced by the DNA–RNA hybridization procedure were detected and isolated (*145a, 156*).

When banded in a CsCl gradient containing Mg^{2+} in high concentrations, some ribosomes lose a part of their protein and separate into two peaks (*139, 140*), unless fixed previously with formaldehyde (*141, 141a*). On the other hand, bentonite-washed ribosomes are preserved in Cs_2SO_4 gradients (14.3 gm/cm^3), even in the absence of magnesium salts (*142*). Normal ribosomal and transfer RNA (*35, 40, 88*) and ribosomal RNA synthesized in the presence of chloramphenicol (*143*) were also banded in Cs_2SO_4 gradients. Ribosomal subunits were split off and fractionated by CsCl density gradient centrifugation (*144*), leading to extensive studies on ribosomal assemblage (*145*).

Determination of Molecular Weight of Macromolecules

In equilibrium density gradient centrifugation, there is an inverse relationship between the width of the band and the molecular weight of the macromolecules. The width (and the skewness) of the band is also determined by the density heterogeneity of the sample. In their original paper Meselson, Stahl, and Vinograd (*1*) derived equations for computing the molecular weights of macromolecules from the shape of the bands. In the past, however, this method proved to be rather unreliable and generally resulted in too low values (*1–4, 146–151*), which was attributed to density heterogeneity (*147*), optical distortions (*149*), and other ill-defined causes (*148*). Only recently,

TABLE 7. Buoyant Densities of Viral and Cellular RNAs in the Cs_2SO_4 Gradient

Source of RNA	Buoyant density (gm/cm^3) Single-stranded —	+ 1% HCHO	Double-stranded (RF)	Reference
Calf thymus synthetic	1.55[a]	–	–	*129*
φX174 coliphage synthetic	1.590–1.601[b]	–	–	*130, 131*
MS2 coliphage	1.607[g]	–	–	*162*
	1.626	–	1.609[c]	*163, 164*
	1.630	–	–	*165*
R17 coliphage	1.621	1.616	1.607[d]	*40, 127*
	1.630	–	1.606[e]	*166*
fr coliphage	1.634	–	1.609	*167*
M12 coliphage	1.634	–	1.614	*168*
Poliovirus	1.632	–	–	*37, 169*
	1.64	–	1.58	*171*
	1.65	–	1.60	*170*
Wound tumor virus	–	–	1.599	*88*
Reovirus	–	–	1.610	*126b, 133*
EMC (encephalomyocarditis virus)	1.63	–	1.57	*173*
	1.69	–	1.635	*174*
Neurospora crassa ribosomal	1.637(ppt)[f]	1.625	–	*88*
Bromegrass mosaic virus	1.631(ppt)	1.616	–	*88*
TMV (tobacco mosaic virus)	1.635(ppt)[h]	1.627	–	*40*
	1.640(ppt)	–	1.601	*163*
	1.675(ppt)	–	1.620–1.628	*175*
TMV (fl^5U–labeled)	–	1.652	–	*40*
TYMV (turnip yellow mosaic	1.642–1.65 (ppt)	–	1.635–1.643	*175, 176*
virus)	1.642 (ppt)	–	1.617	*172*
Bacillus subtilus 5s	1.643	1.636	–	*40*
Bacillus subtilis 18s	1.649	1.634	–	*40*
Bacillus subtilis 23s	1.653	1.637	–	*40*
Escherichia coli ribosomal	1.663(ppt)	–	–	*166*
Newcastle disease virus	1.68	–	–	*177*

[a] Calf thymus native DNA, 1.425; denatured DNA, 1.451; DNA·RNA synthetic hybrid 1.490 gm/cm^3 (*129*).

[b] φX174 DNA, 1.452 gm/cm^3; DNA·RNA synthetic hybrid, 1.491–1.510 gm/cm^3 (*130, 131*).

[c] Density in CsCl, 1.868 gm/cm^3 (51.9% GC) (*125*).

[d] + 1% HCHO, 1.606 gm/cm^3 (*88*).

[e] R. I. (replicative intermediate), 1.616 gm/cm^3 (*127, 166*).

[f] (ppt) indicates formation of an RNA precipitate (*40*).

[g] Density in CsCl, 1.895 gm/cm^3 (*126, 126a*).

[h] Density at pH 4.2, 1.649 gm/cm^3 (ppt), and in presence of 10 vol % dimethylsulfoxide, 1.573 (pH 4.3) and 1.548 (pH 7) gm/cm^3 (*126b*). Density in CsCl at 40°, 1.913 $gm/cm^{\circ 3}$ (*126a*).

the effects of thermodynamic nonideality were evaluated, and the corrections were presented in the form of a virial expansion and applied for measuring the molecular weights of coliphage DNAs (*151*). Other methods for calculating molecular weights also were recently described (*12, 151a, 157*), and the relationship between DNA heterogeneity and its number-average molecular weight was evaluated (*151b*), and the application of density gradient centrifugation to the problem of compositional distribution in polymers was discussed (*157a*).

In the special case of the lambdoid coliphages the *relative* molecular weights of the DNAs can be accurately assessed by measuring the buoyant densities either of the whole phages (*46, 152–154*) or of the cohesive complexes between normal and density-labeled phage DNAs (*155*).

Equilibrium Density Gradient Centrifugation of Proteins, Nucleoproteins and Polysaccharides

Plasma proteins, various enzymes, and other proteins have been banded in equilibrium density gradients (*37a, 39a, 158–161*). For proteins, the Schlieren optical system was most frequently employed to record the equilibrium distribution, although absorbance monochromatic optics at 280 nm should be preferable. The method and theory, as applied to proteins, have been discussed (*12, 158, 159, 160b*).

Ribonucleoproteins, e.g., ribosomes (*141, 141a*), and deoxyribonucleoproteins, e.g. chromatin (*161a*), can also be banded in equilibrium density gradients after fixation with formaldehyde. The differing protein/DNA ratios (and consequently buoyant densities) of nucleoproteins can be utilized for their fractionation (*161b*). Equilibrium density gradient centrifugation has also been widely applied for the isolation and purification of DNA and RNA viruses, although inactivation of some viruses in concentrated cesium salt solutions has necessitated the use of other, usually preformed gradients (*39d, 39e*).

Densities of polysaccharides in CsCl and Cs_2SO_4 gradients were measured (*27*), and HeLa cell polyglucose was preparatively separated from the nucleic acids and characterized by centrifugation in a CsCl density gradient (*161c*).

References

1. Meselson, M., F. W. Stahl, and J. Vinograd. 1957. *Proc. Natl. Acad. Sci., U.S.*, **43:** 581.
2. Meselson, M., and F. W. Stahl. 1958. *Proc. Natl. Acad. Sci., U.S.*, **44:** 671.
3. Hearst, J. E., and J. Vinograd. 1961. *Proc. Natl. Acad. Sci., U.S.*, **47:** 825, 999, 1005.

4. Hearst, J. E., J. B. Ifft, and J. Vinograd. 1961. *Proc. Natl. Acad. Sci. U.S.*, **47:** 1015.
5. Ifft, J. B., D. H. Voet, and J. Vinograd. 1961. *J. Phys. Chem.*, **65:** 1138.
6. Vinograd, J., and J. E. Hearst. 1962. *Progr. Chem. Org. Natl. Prod.*, **20:** 372.
6a. Hearst, J. E. 1962. *J. Mol. Biol.*, **4:** 415.
7. Vinograd, J. 1963. In S. P. Colowick and N. O. Kaplan (Eds.), Methods in enzymology. Academic Press, New York. Vol. VI, p. 854.
8. Meselson, M., and G. M. Nazarian. 1963. In J. W. Williams (Ed.), Ultracentrifugal analysis in theory and experiment. Academic Press, New York, p. 131.
9. Hearst, J. E. 1965. *Biopolymers,* **3:** 1, 57.
10. Bauer, W., and J. Vinograd. 1969. *Ann. N. Y. Acad. Sci.*, **164:** 192.
11. Hearst, J. E., and J. Vinograd. 1961. *J. Phys. Chem.* **65**: 1069.
11a. Vinograd, J., R. Greenwald, and J. E. Hearst. 1965. *Biopolymers* **3:** 109.
12. Ifft, J. B. 1969. In P. Alexander and H. P. Lundgren (Eds.), A laboratory manual of analytical methods of protein chemistry. Pergamon Press, New York, Vol. 5, p. 151.
12a. Cohen, G. and H. Eisenberg. 1968. *Biopolymers.* **6**: 1077.
13. Rüst P. 1968. Biopolymers **6**: 1077.
14. Schachman, H. K. 1959. Ultracentrifugation in biochemistry. Academic Press, New York.
14a. Yphantis, D. A. 1964. *Biochemistry,* **3:** 297.
15. Anderson, N. G. (Ed.). 1966. *Natl. Cancer Inst. Monogr.*, **21:** 241.
15a. Anderson, N. G. 1955. *Exptl. Cell Res.*, **9:** 446.
15b. Galilei, G. 1638. Discorsi e dimostrazioni matematiche intorno à due nuove scienze. Elsevier, Leyden, p. 114.
16. Linderstrøm-Lang, K., and H. Lanz, Jr. 1938. *Compt. Rend. Trav. Lab. Carlsberg (Chim),* **21:** 315.
17. Behrens, M. 1938. *Z. Physiol. Chem.*, **253:** 185.
18. Vinograd, J., R. Bruner, R. Kent, and J. Weigle. 1963. *Proc. Natl. Acad. Sci., U.S.*, **49:** 902.
19. Vinograd, J., and R. Bruner. 1966. *Biopolymers,* **4:** 131, 157.
20. Schumaker, V. N., and J. Wagnild. 1965. *Biophys. J.*, **5:** 947.
21. Mandel, M., C. L. Schildkraut, and J. Marmur. 1968. In L. Grossman and K. Moldave (Eds.), Methods in enzymology. Academic Press, New York. Vol. XIIB, p. 184.
22. Schachman, H. K., L. Gropper, S. Hanlon, and F. Putney. 1962. *Arch. Biochem. Biophys.*, **99:** 175.
23. Cummings, D. J., and L. Mondale. 1966. *Biochim. Biophys. Acta,* **120:** 448.
24. Opara-Kubinska, Z., Z. Kurylo-Borowska, and W. Szybalski. 1963. *Biochim. Biophys. Acta,* **72:** 298.
25. Erikson, R. L., and W. Szybalski. 1963. *Cancer Res.*, **23:** 122.
26. Erikson, R. L., and W. Szybalski. 1963. *Radiation Res.*, **20:** 252.
27. Erikson, R. L., and W. Szybalski. 1964. *Virology,* **22:** 111.
28. Szybalski, W. 1962. In The molecular basis of neoplasia. Univ. of Texas Press, Austin, p. 147.
29. Lamers, K., F. Putney, I. Steinberg, and H. K. Schachman. 1963. *Arch. Biochem. Biophys.*, **103:** 379.
30. Spragg, S. P., S. Travers, and T. Saxton. 1965. *Anal. Biochem.*, **12:** 259.
31. Cheng, P.–Y., and J. L. Littlepage. 1966. *Anal. Biochem.*, **15:** 211.
32. Wright, R. R., W. S. Pappas, and J. A. Carter. 1966. *Natl. Cancer Inst. Monogr.* **21:** 241.
33. Vinograd, J., J. Morris, N. Davidson, and W. F. Dove, Jr. 1963. *Proc. Natl. Acad. Sci., U.S.*, **49:** 12.

34. Inman, R. B., and R. L. Baldwin. 1964. *J. Mol. Biol.,* **8:** 452.
35. Davern, C. I., and M. Meselson. 1960. *J. Mol. Biol.,* **2:** 153.
36. Stanley, Jr., W. M. 1963. Ph.D. Thesis, Univ. of Wisconsin, Madison.
37. Zolotor, L., and R. Engler. 1967. *Biochim. Biophys. Acta,* **145:** 52.
37a. Hu, A. S. L., R. M. Bock, and H. O. Halvorson. 1962. *Anal. Biochem.,* **4:** 489.
38. Kohn, K. W., and C. L. Spears. 1967. *Biochim. Biophys. Acta,* **145:** 720.
39. Anet, R., and D. R. Strayer. 1969. *Biochem. Biophys. Res. Commun.,* **37:** 52.
39a. Adams, G. H., and V. N. Schumaker. 1969. *Ann. N. Y. Acad. Sci.,* **164:** 130.
39b. Polson, A., and J. Levitt. 1963. *Biochim. Biophys. Acta,* **75:** 88.
39c. Brunk, C. F., and V. Leick. 1969. *Biochim. Biophys. Acta,* **179:** 136.
39d. McCrea, J. F., R. S. Epstein, and W. H. Barry. 1961. *Nature,* **189:** 220.
39e. Oroszlan, S., L. W. Johns, Jr., and M. A. Rich. 1965. *Virology,* **26:** 638.
39f. Schatz, G., E. Haslbrunner, and H. Tuppy. 1964. *Biochem. Biophys. Res. Commun.,* **15:** 27.
39g. Tamir, H., and C. Gilvarg. 1966. *J. Biol. Chem.,* **241:** 1085.
40. Lozeron, H. A., and W. Szybalski. 1966. *Biochem. Biophys. Res. Commun.,* **23:** 612.
40a. Møller, K. M., and P. Ottolenghi. 1964. *Compt. Rend. Lab. Carlsberg.,* **34:** 169.
40b. Griffith, O. M., M. J. Gordon, and J. Patterson. 1967. *Anal. Biochem.,* **21:** 329.
41. Wake, R. G., and R. L. Baldwin. 1962. *J. Mol. Biol.,* **5:** 201.
42. Inman, R. B., and R. L. Baldwin. 1962. *J. Mol. Biol.,* **5:** 185.
43. Ludlum, D. B., and R. C. Warner. 1965. *J. Biol. Chem.,* **240:** 2961.
44. Quétier, F., E. Guillé, and L. Lejus. 1969. *Arch. Biochem. Biophys.,* **130:** 685.
45. Richard, A. J., J. Glick, and R. Burkat. 1970. *Anal. Biochem.* **37:** 378.
46. Weigle, J., M. Meselson, and K. Paigen. 1959. *J. Mol. Biol.,* **1:** 379.
47. Kozinski, A. W., and W. Szybalski. 1959. *Virology.,* 9: 260.
48. Hershey, A. D., E. Burgi, and C. I. Davern. 1965. *Biochem. Biophys. Res. Commun.,* **18:** 675.
49. Fisher, W. D., G. B. Cline, and N. G. Anderson. 1964. *Anal. Biochem.,* **9:** 477.
50. Flamm, W. G., H. E. Bond, and H. E. Burr. 1966. *Biochem. Biophys. Acta.,* **129:** 310.
51. Kozinski. A. W., and E. Shahn. 1966. *Virology.,* **28:** 346.
52. Anet, R., and D. R. Strayer. 1969. *Biochem. Biophys. Res. Commun.* **34:** 328.
52a. Billen, D., and R. Hewitt. 1966. *Anal. Biochem.,* **15:** 177.
53. Szybalski, W. 1960. *Experientia.,* **16:** 164.
54. Britten, R. J., and R. B., Roberts. 1960. *Science.,* **131:** 32.
55. Heckly, R. J. 1960. *Anal. Biochem.,* **1:** 97.
56. Martin, F. G., and B. N. Ames. 1961. *J. Biol. Chem.,* **236:** 1372.
57. Brakke, M. K. 1963. *Anal. Biochem.,* **5:** 271.
57a. Ellison, S. A., and H. S. Rosenkranz. 1963. *Anal. Biochem.,* **5:** 263
58. Dresden, M., and M. B. Hoagland. 1965. *Science.,* **149:** 647.
59. Salo, T. 1965. *Anal. Biochem.,* **10:** 344.
60. Newton, N. 1965. *Anal. Biochem.,* **10:** 361.
60a. Berg, R., and D. P. Durand. 1966. *Appl. Microbiol.,* **14:** 687.
61. Ron, E. Z., R. E. Kohler, and B. D. Davis. 1966. *Science.,* **153:** 1119.
62. Watkins, J., and D. E. H. Tee. 1966. *Sci. Tools.,* **13:** 7.
63. Coleman, G., and J. Sykes. 1966. *Sci. Tools.,* **15:** 43.
63a. Oumi, T., and S. Osawa. 1966. *Anal. Biochem.,* **15:** 539.
64. Romani, R. J., and L. K. Fisher. 1967. *Anal. Biochem.,* **21:** 333.
65. Oppenheim, J., J. Scheinbuks, and L. Marcus. 1968. *Anal. Biochem.* **25:** 252.
66. Pattee, P. A., D. L. Berryhill, and P. A. Hartman. 1968. *Applied Microbiol.,* **16:** 958.

67. Rüst, P. 1966. *Anal. Biochem.*, **17:** 316.
68. Radloff, R., W. Bauer, and J. Vinograd. 1967. *Proc. Natl. Acad. Sci., U.S.*, **57:** 1514.
68a. Trautman, R., S. S. Breese, Jr., and H. L. Bachrach. 1962. *J. Phys. Chem.*, **66:** 1976.
69. Rolfe, R., and M. Meselson. 1959. *Proc. Natl. Acad. Sci. U.S.*, **45:** 1039.
69a. Sueoka, N., J. Marmur, and P. Doty. 1959. *Nature,* **183:** 1429.
69b. Doty, P., J. Marmur, J. Eigner, and C. Schildkraut. 1960. *Proc. Natl. Acad. Sci., U.S.*, **46:** 461.
70. Schildkraut, C. L., J. Marmur, and P. Doty. 1962. *J. Mol. Biol.*, **4:** 430.
71. de Ley, J. 1970. *J. Bacteriol.*, **101:** 738.
72. Davidson, N., J. Widholm, U.S. Nandi, R. Jensen, B. M. Olivera, and J. C. Wang. 1965. *Proc. Natl. Acad. Sci., U.S.*, **53:** 111.
73. Nandi, U. S., J. C. Wang, and N. Davidson. 1965. *Biochemistry,* **4:** 1687.
74. Jensen, R. H. and N. Davidson. 1966. *Biopolymers* **4:** 17.
75. Davidson, N. and J. C. Wang, 1968. In A. Rich and N. Davidson, (Eds.), Structural chemistry and molecular biology. Freeman, San Francisco, p. 430.
75a. Sueoka, N. 1961. *J. Mol. Biol.*, **3:** 239.
76. Mandel, M. 1969. *Ann. Rev. Microbiol.*, **23:** 239.
77. Bøvre, K., M. Fiandt, and W. Szybalski. 1969. *Can. J. Microbiol.*, **15:** 335.
78. Okubo, S., M. Stodolsky, K. Bott, and B. Strauss. 1963. *Proc. Natl. Acad. Sci., U.S.*, **50:** 679.
79. Bott, K., and B. Strauss. 1965. *Virology.*, **25:** 212.
80. Szybalski, W., and Z. Opara-Kubinska. 1965. *Symp. Biol. Hungarica.*, **6:** 43.
80a. Kit, S. 1961. *J. Mol. Biol.*, **3:** 711.
80b. Borst, P., A. M. Kroom, and G. J. M. Ruttenberg. 1967. In D. Shugar (Ed.), Genetic elements. Properties and function. Academic Press, New York, p. 81.
80c. Schildkraut, C. L., and J. J. Maio. 1968. *Biochim. Biophys. Acta,* **161:** 76.
80d. Schildkraut, C. L., and J. J. Maio. 1969. *J. Mol. Biol.*, **46:** 305.
80e. Corneo, G., E. Ginelli, and E. Polli. 1970. *J. Mol. Biol.* **48:** 319.
80f. Flamm, W. G., P. M. B. Walker, and M. McCallum. 1969. *J. Mol. Biol.*, **40:** 423; **42:** 441.
80g. Granick, S., and A. Gibor. 1967. *Progr. Nucleic Acid Research,* **6:** 143.
80h. Rabinowitz, M. 1968. *Bull. Soc. Chim. Biol.*, **50:** 311.
80i. Rownd, R., R. Nakaya, and A. Nakamura. 1966. *J. Mol. Biol.*, **17:** 376.
80j. Marmur, J., R. Rownd, S. Falkow, L. S. Baron, C. L. Schildkraut, and P. Doty. 1961. *Proc. Nat. Acad. Sci., U.S.*, **47:** 972.
80k. Falkow, S., D. K. Haapala, and R. P. Silver. 1969. In G. E. W. Wolstenholme and M. O'Connor (Eds.), Ciba foundation symposium on bacterial episomes and plasmids, Churchill, London, p. 136.
81. Marmur, J., and P. Doty. 1962. *J. Mol. Biol.*, **5:** 109.
82. Mandel, M., L. Igambi, J. Bergendahl, M. L. Dodson, Jr., and E. Scheltgen. 1970. *J. Bacteriol.*, **101:** 333.
83. Riva, S., I. Barrai, L. Cavalli-Sforza, and A. Falaschi. 1969. *J. Mol. Biol.*, **45:** 367.
83a. Marmur, J., and C. L. Schildkraut. 1961. *Nature*, **189:** 636.
83b. Baldwin, R. L., and E. M. Shooter. 1963. *J. Mol. Biol.*, **7:** 511.
84. Meselson, M., and J. J. Weigle. 1961. *Proc. Natl. Acad. Sci., U.S.*, **47:** 857.
85. Volkin, E. 1965. *Progr. Nucleic Acid Res. Mol. Biol.*, **4:** 51.
86. Djordjevic, B., and W. Szybalski. 1960. *J. Exptl. Med.*, **112:** 509.
87. Simon, E. H. 1961. *J. Mol. Biol.*, **3:** 101.
87a. Schildkraut, C. L., J. Marmur, and P. Doty. 1961. *J. Mol. Biol.*, **3:** 595.

87b. Schildkraut, C. L., K. L. Wierzchowski, J. Marmur, D. M. Green, and P. Doty. 1962. *Virology,* **18:** 43.
87c. DeLey, J., and S. Friedman. 1964. *J. Bacteriol.,* **88:** 937.
87d. Lozeron, H. A., and W. Szybalski. 1969. *Virology,* **39:** 373.
88. Szybalski, W. 1968. In L. Grossman and K. Moldave (Eds.) Methods in enzymology. Acadamic Press, New York. Vol. XIIB, p. 330.
89. Freifelder, D., and P. F. Davison. 1963. Biophys. J., **3:** 49.
90. Taylor, K., Z. Hradecna, and W. Szybalski. 1967. *Proc. Natl. Acad. Sci., U.S.,* **57:** 1618.
91. Kubinski, H., Z. Opara-Kubinska, and W. Szybalski. 1966. *J. Mol. Biol.,* **20**: 313.
92. Sheldrick, P., and W. Szybalski. 1967. *J. Mol. Biol.,* **29:** 217.
93. Hradecna, Z., and W. Szybalski. 1967. *Virology,* **32:** 633.
94. Murata, N., and W. Szybalski. 1968. *J. Gen. Appl. Microbiol,* **14:** 57.
95. Summers, W. C., and W. Szybalski. 1968. *Virology,* **34:** 9.
96. Guha, A., and W. Szybalski. 1968. *Virology,* **34:** 608.
97. Summers, W. C., and W. Szybalski. 1968. *Biochim. Biophys. Acta,* **166:** 371.
98. Summers, W. C. 1969. *Biochim. Biophys. Acta,* **182:** 269.
99. Lozeron, H. A., and W. Szybalski. 1969. *Bacteriol. Proc.,* p. 192.
100. Ihler, G., and W. D. Rupp. 1969. *Proc. Natl. Acad. Sci., U.S.,* **63:** 138.
101. Gruenwedel, D. W., and N. Davidson. 1967. *Biopolymers,* **5:** 847.
102. Szybalski, W., K. Bøvre, M. Fiandt, A. Guha, Z. Hradecna, S. Kumar, H. A. Lozeron, V. M. Maher, H. J. J. Nijkamp. W. C. Summers, and K. Taylor. 1969. *J. Cell. Physiol.,* **74.** Supplement 1, 33.
103. Kumar, S., and W. Szybalski, 1969. *J. Mol. Biol.,* **40:** 145
104. Grau, O., A. Guha, E. P. Geiduschek, and W. Szybalski. 1969. *Nature,* **224:** 1105.
105. Lozeron, H. A., W. Szybalski, A. Landy, J. Abelson, and J. D. Smith. 1969. *J. Mol. Biol.,* **39:** 239.
106. Bøvre, K., and W. Szybalski. 1969. *Virology,* **38:** 614.
107. Westmoreland, B. C., W. Szybalski, and H. Ris. 1969. *Science.,* **163:** 1343.
108. Hradecna, Z., and W. Szybalski. 1969. *Virology,* **38:** 473; 1970. *ibid.***40:** 178.
109. Vinograd, J., and J. Lebowitz. 1966. *J. Gen. Physiol.,* **49:** No. 6, Part 2: 103.
110. Chamberlin, M. J., and D. I. Patterson. 1965. *J. Mol. Biol,* **12:** 410.
111. Hogness, D. S. 1966. *J. Gen. Physiol,* **49,** No. 6, Part 2: 29.
112. Borst, P., and C. Aaij. 1969. *Biochem. Biophys. Res. Commun.,* **34:** 358.
113. Marmur, J., W. E. Anderson, L. Matthews, K. Berns, E. Gajewska, D. Lane, and P. Doty. 1961. *J. Cell. Comp. Physiol.,* **48:** Supplement 1, 33.
114. Summers, W. C., and W. Szybalski. 1967. *J. Mol. Biol.,* **26:** 107, 227.
115. Cremonese, M., C. Giampaoli, M. Matzeu, and G. Onori, 1969. *Biophys. J.* **9:** 1451.
116. Wang, J. C., U. S. Nandi, D. S. Hogness, and N. Davidson. 1965. *Biochemistry,* **4:** 1697.
117. Hershey, A. D., and E. Burgi. 1965. *Proc. Natl. Acad. Sci., U.S.,* **53:** 325.
118. Cohen, S. N., U. Maitra, and J. Hurwitz. 1967. *J. Mol. Biol.,* **26:** 19.
118a. Skalka, A., E. Burgi, and A. D. Hershey. 1968. *J. Mol. Biol,* **34:** 1.
118b. Skalka, A. 1969. *J. Virol.,* **3:** 150.
119. Kersten, W., H. Kersten, and W. Szybalski. 1966. *Biochemistry,* **5:** 326.
120. Szybalski, W., and V. N. Iyer, 1964. *Federation Proc.,* **23:** 946.
121. Szybalski, W. 1964. *Abhandl. Deut. Akad. Wiss. Berlin Kl. Med. No.* **4:** 1.
122. Marmur, J., R. Rownd, and C. L. Schildkraut, 1963. *Progr. Nucleic Acid Res.,* **1:** 231.
123. Bauer, W., and J. Vinograd. 1968. *J. Mol. Biol.,* **33:** 141; 1970. *ibid.,* **54**: 281.
124. Hudson, B., W. B. Upholt, J. Devinny, and J. Vinograd. 1969. *Proc. Natl. Acad. Sci., U.S.,* **62:** 813.

125. Kelly, R. B., J. L. Gould, and R. L. Sinsheimer. 1965. *J. Mol. Biol.,* **11:** 562.
126. Bruner, R., and J. Vinograd. 1965. *Biochim. Biophys. Acta,* **108:** 18.
126a. Daniel, E. and D. Banin. 1970. *Biochim. Biophys. Acta,* **224**: 311.
126b. Williams, A. E., and J. Vinograd. 1971. *Biochim. Biophys. Acta,* **228**: 423.
127. Erikson, R. L., and R. M. Franklin. 1966. *Bacteriol. Rev.,* **30:** 267.
128. Chamberlin, M., and P. Berg. 1964. *J. Mol. Biol.* **8:** 297; see also 1963. *Cold Spring Harbor Symp. Quant. Biol.,* **28:** 67.
129. Warner, R. C., H. H. Samuels, M. T. Abbott, and J. S. Krakow. 1963. *Proc. Natl. Acad. Sci., U.S.,* **49:** 533.
130. Bassel, A., M. Hayashi, and S. Spiegelman. 1964. *Proc. Natl. Acad. Sci., U.S.,* **52:** 796.
131. Sinsheimer, R. L., and M. Lawrence. 1964. *J. Mol. Biol.,* **8:** 289.
132. Robinson, N. S., W.-T. Hsu, C. F. Fox, and S. B. Weiss. 1964. *J. Biol. Chem.,* **239:** 2944.
133. Shatkin, A. J. 1965. *Proc. Natl. Acad. Sci., U.S.,* **54:** 1721.
134. Chamberlin, M. J. 1965. *Federation Proc.,* **24:** 1446.
135. Hayashi, M. 1965. *Proc. Natl. Acad. Sci., U.S.,* **54:** 1736.
136. Hayashi, M. N., and M. Hayashi. 1966. *Proc. Natl. Acad. Sci., U.S.,* **55:** 635.
137. Konrad, M. W., and G. S. Stent. 1964. *Proc. Natl. Acad. Sci., U.S.,* **51**: 647.
138. Paoletti, C., N. Dutheillet-Lamonthezia, A. Obrenovitch, D. Aubin, and P. Jeanteur. 1965. *Compt. Rend. Acad. Sci.,* **261:** 1775.
139. Brenner, S., F. Jacob, and M. Meselson, 1961. *Nature.,* **190:** 576.
140. Meselson, M., M. Nomura, S. Brenner, C. Davern, and D. Schlessinger. 1964. *J. Mol. Biol.,* **9:** 696.
141. Spirin, A. S., N. V. Belitsina, and M. I. Lerman. 1965. *J. Mol. Biol.,* **14:** 611.
141a. Perry, R. P., and D. E. Kelley. 1966. *J. Mol. Biol.,* **16:** 255.
142. DeFilippes, F. M. 1965. *Science,* **150:** 610.
143. Dubin, D. T., and A. T. Elkort. 1964. *J. Mol. Biol.,* **10:** 508.
144. Traub, P., and M. Nomura. 1968. *J. Mol. Biol.,* **34:** 575.
145. Nomura, M. 1969. *Sci. Amer.,* **221:** No. 4. 28.
145a. Becker, W. M., A. Hell, J. Paul, and R. Williamson. 1970. *Biochim. Biophys. Acta.,* **199:** 348.
146. Baldwin, R. L. 1959. *Proc. Natl. Acad. Sci., U.S.,* **45:** 939.
147. Sueoka, N. 1959. *Proc. Natl. Acad. Sci., U.S.,* **45:** 1480.
148. Thomas, C. A., and T. Pinkerton. 1962. *J. Mol. Biol.,* **5:** 356.
149. Cummings, D. J. 1963. *Biochim. Biophys. Acta.,* **72:** 475.
150. Eisenberg, H. 1967. *Biopolymers.,* **5:** 681.
151. Schmid, C. W., and J. E. Hearst. 1969. *J. Mol. Biol.,* **44:** 143.
151a. Daniel, E. 1969. *Biopolymers,* **7:** 359; **8:** 303, 553.
151b. Jacob, M., and G. Pouyet. 1967. *J. Mol. Biol.,* **24:** 355.
152. Kellenberger, G., M. L. Zichichi, and J. Weigle. 1960. *Nature.,* **187:** 161.
153. Kellenberger, G., M. L. Zichichi, and J. Weigle. 1961. *J. Mol. Biol.,* **3:** 399.
154. Weigle, J. 1961. *J. Mol. Biol.,* **3:** 393.
155. Baldwin, R. L., P. Barrand, A. Fritsch, S. A. Goldthwait, and F. Jacob. 1966. *J. Mol. Biol.,* **17:** 343.
156. Spiegelman, S., and S. A. Yankofsky. 1965. In V. Bryson and H. J. Vogel (Eds.), Evolving genes and proteins. Academic Press, New York, p. 537, and 1962. *Proc. Natl. Acad. Sci., U.S.,* **48:** 106.
157. Kotaka, T., and R. L. Baldwin. 1969. *Biopolymers,* **7:** 87.
157a. Hermans, J. J. 1969. *Ann. N. Y. Acad. Sci.,* **164**: 122.
158. Cox, D. J., and V. N. Shumaker. 1961. *J. Amer. Chem. Soc.,* **83:** 2439, 2445.
159. Ifft, J. B., and J. Vinograd. 1962. *J. Phys. Chem.* **66:** 1990.
160. Fessler, J. H., and A. J. Hodge. 1962. *J. Mol. Biol.,* **5:** 446.

160a. Griffith, O. M., M. J. Gordon, and J. Patterson. 1967. *Anal. Biochem.*, **21**: 329.
160b. Creeth, J. M., and M. A. Denborough. 1970. *Biochem. J.*, **117**: 879.
161. Pace, N. R., and S. Spiegelman. 1966. *Proc. Natl. Acad. Sci., U.S.*, **55**: 1608.
161a. Hancock, R. 1970. *J. Mol. Biol.*, **48**: 357.
161b. Lehnert, S., and S. Okada. 1965. *Biochim. Biophys. Acta.*, **109**: 557.
161c. Graves, I. L. 1970. *Biopolymers*, **9**: 11.
162. Shimura, Y., R. E. Moses, and D. Nathans. 1965. *J. Mol. Biol.*, **12**: 266.
163. Burdon, R. H., M. A. Billeter, C. Weissmann, R. C. Warner, S. Ochoa, and C. A. Knight. 1964. *Proc. Natl. Acad. Sci., U.S.*, **52**: 768.
164. Billeter, M. A., C. Weissmann, and R. C. Warner. 1966. *J. Mol. Biol.*, **17**: 145.
165. Doi, R. H., and S. Spiegelman. 1963. *Proc. Natl. Acad. Sci., U.S.*, **49**: 353.
166. Erikson, R. L. 1966. *J. Mol. Biol.*, **18**: 372.
167. Kaerner, H. C., and H. Hoffmann-Berling. 1964. *Z. Naturforsch.*, **19b**: 593.
168. Ammann, J., H. Delius, and P. H. Hofschneider. 1964. *J. Mol. Biol.*, **10**: 557.
169. Engler, R., and O. Tolbert. 1965. *Virology*, **26**: 246.
170. Bishop, J. M., D. F. Summers, and L. Levintow. 1965. *Proc. Natl. Acad. Sci., U.S.*, **54**: 1273.
171. Pons, M. 1964. *Virology*, **24**: 467.
172. Bockstahler, L. E. 1967. *Mol. Gen. Genetics*, **100**: 337.
173. Montagnier, L., and F. K. Sanders. 1963. *Nature*, **199**: 664.
174. Dalgarno, L., E. M. Martin, S. L. Liu, and T. S. Work. 1966. *J. Mol. Biol.*, **15**: 77.
175. Ralph, R. K., R. E. F. Matthews, A. I. Matus, and H. G. Mandel. 1965. *J. Mol. Biol.*, **11**: 202.
176. Mandel, H. G., R. E. F. Matthews, A. T. Matus, and R. K. Ralph. 1964. *Biochem. Biophys. Res. Commun.*, **16**: 604.
177. Kingsbury, D. W. 1966. *J. Mol. Biol.*, **18**: 195.
178. Riley, M., B. Maling, and M. J. Chamberlin. 1966. *J. Mol. Biol.*, **20**: 359.
179. Haselkorn, R., and C. F. Fox. 1965. *J. Mol. Biol.*, **13**: 780.
180. Schildkraut, C. L., J. Marmur, J. R. Fresco, and P. Doty. 1961. *J. Biol. Chem.*, **236**: PC3.
181. Burd, J. F., and R. D. Wells. 1970. *J. Mol. Biol.*, **53**: 435.
182. Wells, R. D., and J. E. Blair. 1967. *J. Mol. Biol.*, **27**: 273.
183. Morgan, A. R., and R. D. Wells. 1968. *J. Mol. Biol.* **37**: 63.
184. Chamberlin, M. J. 1966. In G. L. Cantoni and D. R. Davies (Eds.), Procedures in nucleic acid research. Harper & Row, New York, p. 513.
185. Grant, R. C., S. J. Harwood, and R. D. Wells. 1968. *J. Am. Chem. Soc.* **90**: 4474.
186. Cerami, A., D. C. Ward, E. Reich, and I. H. Goldberg. 1966. *Abstr. Second Intern. Biophys. Congr. Vienna*, No. 150.
187. Wells, R. D., and J. E. Larson. 1970. *J. Mol. Biol.*, **49**: 319.
188. Wells, R. D., personal communication.
189. Cassidy, P., F. Kahan, and P. Doty, Manuscript in preparation.
190. Champoux, J. J. 1969. Ph.D. thesis, Stanford University, p. 77.
191. Wells, R. D., J. E. Larson, R. C. Grant, B. E. Shortle, and C. R. Cantor. 1970. *J. Mol. Biol.* **54**: 465.
192. Lezius, A. G. 1970. *European J. Biochem.* **14**: 154.
193. Lezius A. G., and E. M. Gottschalk. 1970. *Hoppe-Seylers Z. Physiol. Chem.* **351**: 119, 413.

SPECIAL METHODS FOR MEASURING LENGTHS OF LAMBDA DNAs*

ROBERT L. BALDWIN
DEPARTMENT OF BIOCHEMISTRY
STANFORD UNIVERSITY SCHOOL OF MEDICINE
STANFORD, CALIFORNIA

Both methods of measuring DNA lengths are limited to phages derived from or related to λ. The first method, which is a calibration of DNA length vs. phage buoyant density, has the advantages of simplicity, high precision, and requiring only small amounts of phage; however it is an indirect method. The second method, which compares the buoyant density of a mixed DNA dimer with those of the two input DNAs – one being density labeled – is a direct method and also quite simple, but it requires high-titer phage stocks and is less precise. To workers in other fields it may seem incongruous that two special methods of measuring DNA molecular weights should be developed for just one group of phages. In part this reflects the large volume of research now being done on λ derivatives of different lengths: these include deletion mutants, duplication mutants, specialized transducing phages containing genes for bacterial functions, and hybrid phages resulting from crosses between λ and related phages, such as phage 80. In part, also, this chapter illustrates methods used successfully with λ that may be applied to other groups of phages. It is likely that a calibration of phage buoyant density against DNA length will be possible for many phages. Mixed DNA dimers can be formed by other groups of phages, unrelated to λ, since several other phage DNAs are known to have specific cohesive ends (*1*, *2*).

* Supported by research grants from the U.S. National Science Foundation (GB 8016) and the U.S. National Institutes of Health (AM04763).

CALIBRATION OF DNA LENGTH AGAINST PHAGE BUOYANT DENSITY

Because of its simplicity and high precision, this is one of the most useful methods for determining λ DNA lengths. It requires only small amounts of phage: 10^{10} phage for the analytical centrifuge, or about 10^5 phage for the preparative centrifuge. The method is limited to phages which have the normal complement of λ structural proteins (phages that are deletion and substitution mutants of λ or of λ hybrids), since the phage buoyant density depends also on the nature of the phage proteins. Whenever a mutant is suspected of having an altered structural protein, its DNA length should be checked by a more direct method.

The buoyant density of a phage is given by

$$\theta(\text{phage}) = \sum_i \alpha_i \theta_i \quad (1)$$

where α_i is the volume fraction and θ_i is the buoyant density of the *i*th constituent. Suppose that there are n constituents of the phage (DNA, proteins, H_2O, CsCl, . . .) which affect its buoyant density in CsCl:

$$\theta(\text{phage}) = \alpha_1\theta_1 + \alpha_2\theta_2 + \ldots \alpha_n\theta_n \quad (1a)$$

with λDNA being labeled as 1.

Computation of DNA lengths from phage buoyant densities was first suggested by Weigle, Meselson, and Paigen (*2a*), who discovered that transducing λ phages could be separated preparatively from the wild type by banding them in an equilibrium CsCl gradient. Starting from an equation equivalent to Eq. (1), they derived a relation (the W.M.P. equation) showing how the phage buoyant density might be expected to vary with DNA length, but they warned that it contained several untested assumptions. It was assumed that θ(phage) depends only on two phage constituents, DNA (1) and protein (2), that the protein content is invariant, and that θ_1, θ_2, are constants equal to the buoyant densities in CsCl of free λ DNA (taken as 1.71) and of phage ghosts (assumed to be about 1.30).

Soon afterwards Hearst and Vinograd (*3*) showed that the buoyant density of T4 DNA is not a constant, but varies in different Cs salts according to the water activity at the DNA banding density, changing from an extrapolated value of 2.12 at zero water activity to 1.44 in Cs_2SO_4 and 1.23 in Cs silicotungstate. Their findings, which were confirmed by Cohen and Eisenberg (*4*) for calf thymus DNA using a type of measurement that can be made solely in CsCl solutions, cast doubt on three assumptions in the W.M.P. equation. First, that θ_1 can be treated as a constant; second, that an appropriate value for θ_1 is that of free λ DNA (the DNA bands at a considerably lower water activity than does the phage); and third, that neglect of the solvent (CsCl and H_2O) as a phage constituent can be justified.

This problem was studied by Costello (*5*) who measured θ for λ ghosts and for three phages with (1) no deletion, (2) the b_2 deletion, and (3) the b_{221} deletion, in a series of different Cs salts. He found that θ(phage) does depend strongly on the water activity in a manner determined chiefly by the phage DNA. He also found an apparent dependence of the net DNA hydration on DNA length, which probably is a result of an increase in free volume inside the phage head when there is a DNA deletion. This is a large effect in Cs_2SO_4 where the net DNA hydration is large. Finally he found that if ambient values for θ_1 and θ_2 are used in the W.M.P. equation (i.e., values taken from graphs of θ vs. water activity, evaluated at the λ phage banding density), then the W.M.P. equation gives wrong DNA lengths. Although θ for λ could be predicted with fair accuracy from Eq. (1) as a function of water activity, it was not found possible to predict $\Delta\theta$ satisfactorily in this way: $\Delta\theta$ is the difference between the buoyant densities of phage λ and a deletion mutant. The reason appears to be that $\Delta\theta$ depends sensitively on the relationship between DNA length and net DNA hydration in the phage, and this relation is not known.

It becomes necessary to inquire then whether $\Delta\theta$ is indeed a unique function of DNA length and, if so, how this function might be determined empirically. According to Eq. (1a) $\Delta\theta$ will be a function only of DNA length as long as all the quantities on the right-hand side of Eq. (1a) are either constants or functions only of DNA length. Specifically, the volume made available by different DNA deletions must always be filled by the same constituents, in a ratio that depends only on DNA length. In the simplest case, these constituents will be the solvent constituents (CsCl and H_2O) and their ratio will be determined by the net hydration of the DNA.

Recently duplication mutants of λ have become available and have been used to study this question (*6*). These mutants have the property that, once the primary duplication event has occurred yielding a phage *add* $X_{(1)}$, then secondary duplication mutants (*add* $X_{(2)}$, *add* $X_{(3)}$, . .) that carry additional copies of this duplication can be generated by recombination. Thus the DNA of each secondary duplication mutant has a length greater than its predecessor by an amount that is constant, precise to the nucleotide level. The most useful set of these mutants spans the range of DNA lengths from 77% to 106% of λ DNA with 6 members, the size of the duplication being 5.7% of λ DNA. The results show that $\Delta\theta$ is a unique, almost linear, function of DNA length in this range. To check this conclusion, $\Delta\theta$ was measured for the b_2 and b_{221} deletions and plotted against their electron microscope (or E.M.) lengths (*7, 8*) on the same graph. Then the E.M. lengths were used to determine the constant relating absolute DNA length to relative DNA lengths of the duplication mutants. Since current E.M. measurements of λ DNA lengths have a precision of 0.1%, the calibration could be done now using only $\Delta\theta$ and E.M. measurements. The most precise current value for the b_2

deletion ($f = 0.121 \pm 0.001$) (*9*) happens to agree with the average of the two published values ($f = 0.112$ and $f = 0.130$) used in fixing the calibration equation.

The empirical relation between λ DNA length and phage buoyant density suggested by Bellett et al. (*6*) is

$$f = 8.64\Delta\theta\,(1 - 2.4\Delta\theta) \tag{2}$$

where f is the DNA deletion relative to λ DNA = 1.00 and $\Delta\theta$ is $(\theta_\lambda - \theta)$ in gm/ml, CsCl, 25°. In using the calibration equation it is not necessary to take account of pressure dependence, which contributes to $\Delta\theta$ but is included in the calibration equation. The temperature dependence of $\Delta\theta$ is small, and $\Delta\theta$ is insensitive both to pH in the region 7–8 and to the presence or omission of 0.01 m Mg^{2+}. No account has been taken of the dependence of θ(phage) on the AT content of the added or deleted DNA segment. Both f and $\Delta\theta$ have positive signs in Eq. (2) if the mutant phage contains a DNA deletion.

Table 1 shows that λ DNA lengths can be determined with high precision from measurements of $\Delta\theta$. The best way of doing this is to use two phage density markers (λ and λb_{221}) and to determine $\Delta\theta$ for an unknown phage relative to an assigned value of 0.0271 gm/ml for $\Delta\theta(\lambda b_{221})$ – the value used in the calibration equation. In addition to standardizing $\Delta\theta$ in this way, many possible sources of error cancel, such as an error in rotor speed or magnification factor. In Table 1 the results of four new experiments with a mixture of $\lambda c\mathrm{I}_{857}$, $\lambda b_2 c$, and $\lambda b_{221} c_{26}$ are shown. In (A) the absolute values found for $\Delta\theta$ (using the procedure described by Bellett et al. (*6*)) are shown, and the corresponding values of f are given from the calibration equation. The resulting precision in DNA lengths is ±0.2% and the values agree with standard values (ones used in the calibration equation) within ±0.5%. This may seem surprisingly good, but it can be improved by the simple device of using two density markers. In (B) the same data are used to compute relative values of $\Delta\theta$ for λb_2 from a standard value of 0.0271 for $\Delta\theta(\lambda b_{221})$; $\Delta\theta_{rel} = (0.0271)(\Delta\theta)/\Delta\theta(\lambda b_{221})$. Now the DNA lengths computed for λb_2 have an internal precision of ±0.1% and agree within ±0.1% with the λb_2 length used in the calibration equation. No special effort was made to achieve high precision in these experiments, for which the rotor speed was 44,000 rpm. At a rotor speed of 30,000 rpm (which is convenient for banding phages) the bands would be twice as far apart, and there should be a corresponding gain in precision. The band positions in these experiments were measured directly from the films by use of a microcomparator. This procedure is both rapid and accurate when the phage bands are symmetrical.

TABLE 1. Precision of the Method When (A) Absolute and (B) Relative Values of $\Delta\theta$ Are Used

	Calibration equation		A					B		
Phage	$\Delta\theta$[a]	f[d]	Expt. no.	$\Delta\theta(\lambda b_2 c)$[b]	f[d]	$\Delta\theta(\lambda b_{221} c_{26})$[b]	f[d]	Expt. no.	$\Delta\theta_{rel}(\lambda b_2 c)$[c]	f[d]
$\lambda c I_{857}$	0	0	1	0.0148	0.123	0.0279	0.225	1	0.0144	0.120
$\lambda b_2 c$	0.0145	0.121	2	0.0148	0.123	0.0278	0.224	2	0.0144	0.120
$\lambda b_{221} c_{26}$	0.0271	0.219	3	0.0149	0.124	0.0274	0.221	3	0.0147	0.123
			4	0.0147	0.123	0.0273	0.221	4	0.0146	0.122

[a] Values of $\Delta\theta$ ($= (\theta_\lambda - \theta)$, CsCl, 25°) reported by Bellett et al. (*6*)
[b] New values of $\Delta\theta$ determined in four experiments, for this article.
[c] Values of $\Delta\theta_{rel}$ for $\lambda b_2 c$ were computed by use of two density markers, λ and λb_{221}: $\Delta\theta_{rel}(\lambda b_2 c) = (0.0271)(\Delta\theta \lambda b_2 c)/\Delta\theta \lambda b_{221} c_{26}$.
[d] $f = 8.64\Delta\theta(1 - 2.4\Delta\theta)$.
[e] Conditions: analytical centrifugation at 44,000 rpm, 25°, pH 7.4, in a Spinco Model E ultracentrifuge with UV optics; CsCl, density 1.48, in 0.01 M $MgCl_2$, 0.01 M tris Cl.

DNA LENGTHS BY THE MIXED DIMER METHOD

The principle of this method is that a DNA dimer, containing one density-labeled DNA molecule (H) joined via cohesive ends to a second unlabeled molecule (L), will band in an equilibrium CsCl gradient at a position determined by the relative amounts (by volume) of the L and H DNAs. Since the volume of a DNA molecule is closely proportional to its length, this provides a direct method of determining the ratio of the lengths

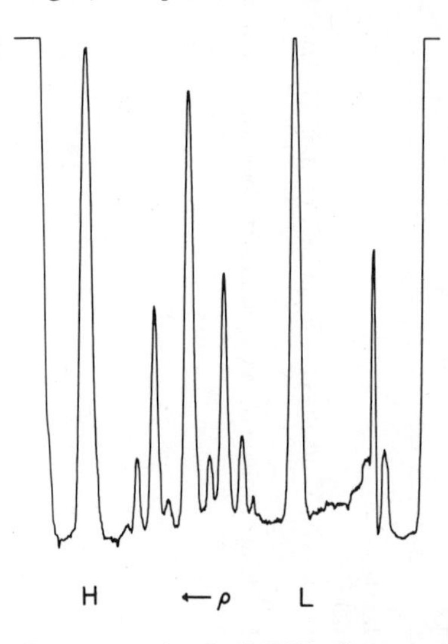

Figure 1. A mixed DNA dimer experiment with unlabeled (L) and BU-labeled (H) DNAs from $\lambda cI_{857}susS_7$. Note the sharpness of the H peak, and the way in which the mixed oligomer peaks decrease in height going from dimer to trimer to tetramer. (a) Conditions of centrifugation: 44,000 rpm, 25°, 0.60ml CsCl, initial density 1.75, 0.01 M tris Cl, pH 8; the filling hole of the 12 mm Kel-F centerpiece has been drilled out to admit a wide-bore Pasteur pipette. (b) Conditions for forming mixed dimers: 2.5×10^{11} of each phage (titer at least 10^{13}/ml; stored in 0.01 M $MgCl_2$, 0.01 M tris Cl, pH 7.4) were mixed together and lysed with ½ volume CsCl–sarkosyl–EDTA (CsCl, density 1.75; 0.2% sarkosyl (Geigy NL 30); 0.02 M EDTA, pH 8 and the mixture was incubated with gentle shaking at 50° for 1 hour. A capillary pipette was used to transfer 50 μl to the CsCl solution used for centrifugation. (c) Production of BU-labeled phage (a gift from Dr. M. T. Record, Jr.): $\lambda cI_{857}susS_7$ was used to lysogenize *E. coli* 159T⁻ (Meselson; given to us by Dr. A. D. Kaiser); the lysogen was grown at 30° to 5×10^8/ml in a synthetic medium (3 X D) and 20 μg thymidine/ml; the culture was chilled rapidly, the cells were centrifuged and resuspended at the same concentration in fresh 3 X D medium with 20 μg 5-bromodeoxyuridine/ml in place of thymidine; after 10 min at 45°, the temperature was reduced to 41° for 1 hour., and then to 38° for two more hours; the cells were then resuspended in 0.01 volumes 0.01 M $MgCl_2$, 0.01 M tris Cl, pH 7.5, lysed with $CHCl_3$, and treated with a minimal amount of pancreatic DNase to reduce the viscosity before pelleting and purifying in a CsCl gradient. (Composition of 3 X D medium: per liter of H_2O: 4.5 gm KH_2PO_4, 10.5 gm Na_2HPO_4, 10.5 gm Na_2HPO_4, 3.0 gm NH_4Cl, 15.0 gm vitamin-free casamino acids (nontechnical grade, filtered through Norit), 25.0 ml glycerol; after autoclaving, 0.5 ml 10% $MgSO_4 \cdot 7H_2O$ and 0.5 ml 0.01 M $CaCl_2$ are added.)

of the L and H DNAs. Only the band positions of the L and H DNAs and of the L : H dimer are needed, although the positions of the trimers L_2 : H and L : H_2 may also be used. Two technical requirements are: (1) the unknown DNA must form a mixed dimer with a standard DNA, and (2) the density labeling of the H DNA must be uniform enough to give a sharp DNA band and should provide a sizable increase in buoyant density over the L DNA. Uniformly BU-labeled λ DNA can be prepared which satisfies the latter criteria (see Fig. 1). All the phages related to or derived from λ that have been tested from mixed DNA dimers with λ: phages 424, 434, and 21 (*1*), and phage 80 (*10*).

The ratio of the volumes (V_L/V_H) of the L and H DNAs is found from

$$(n_H/n_L)(1-x) = (V_L/V_H)x \tag{3}$$

where n_H and n_L are the numbers of H and L molecules in a DNA oligomer. The quantity x is computed from the position of the oligomer band, $r_K{}^0$, relative to those of the L and H DNAs , $r_L{}^0$ and $r_K{}^0$,

$$x = \frac{r_K{}^0 - r_L{}^0}{r_H{}^0 - r_L{}^0} \tag{4}$$

The dimer band can be identified as the major mixed oligomer band, the trimer bands are the next most prominent ones, and so forth (see Fig. 1). When only the position of the dimer band is used, (V_L/V_H) is computed directly from Eq. (3) with $(n_H/n_L) = 1$. When the trimer band positions are also used, $(n_H/n_L)(1-x)$ is plotted against x and (V_L/V_H) is found from the slope.

The theory of the method is reviewed in the Appendix; Eq. (3) is exact, whereas Eq. (4) is an approximation.

Since a density-labeled DNA can have a different volume than the corresponding unlabeled DNA, it is necessary first to determine (V_L/V_H) for λ DNA. We will denote this ratio as q (BU) when the H DNA is BU-labeled.

$$q(BU) = V_L(\lambda)/V_H(\lambda) \tag{5}$$

In earlier experiments (*1*) $q(BU)$ was found to be 1.00, within the experimental error. Using more uniformly labeled BU DNA (see Fig. 1), we now find $q(BU)$ to be 0.98 (Table 1).

Let l denote the DNA length. To determine $l(\phi)/l(\lambda)$ from a mixed DNA dimer experiment in which the λ DNA is BU-labeled and the test DNA(ϕ) is unlabeled, one first determines $V_L(\phi)/V_H(\lambda)$ by Eq. (3) and then computes

$$\frac{l(\phi)}{l(\lambda)} = \frac{V_L(\phi)}{V_H(\lambda)} \Big/ q(BU) \tag{6}$$

(No account is taken of a possible dependence of l/V on the % GC of the

TABLE 2. Some Tests of the Mixed Dimer Method

A				B				
Phage	Expt. no.	V_L/V_H(dimer)[a]	V_L/V_H(slope)[b]	Phage	Expt. no.	f(dimer)[c]	f(slope)[d]	f(E.M.)
$\lambda cI_{857}susS_7$	1	0.98_0	0.987 ± 0.005(s.d.)					
	2	0.97_6	0.983 ± 0.006					
λb_2c	3	0.85_5	0.872 ± 0.013	λb_2c	3	0.13	0.11	0.130[e], 0.112[f], 0.121[g]
	4	0.85_2	0.871 ± 0.015		4	0.13	0.11	
$\lambda b_{221}c_{26}$	5	0.76_1	0.770 ± 0.010	$\lambda b_{221}c_{26}$	5	0.22	0.21	0.223[f]
	6	0.76_7	0.775 ± 0.007		6	0.22	0.21	

[a] $(1-x)/x$ for the dimer band. The H DNA is BU-labeled DNA from λcI_{857} $susS_7$ in all experiments.
[b] computed by least squares from the slope of $(n_H/n_L)(1-x)$ against x: four points were used in each case, corresponding to $(n_H/n_L) = 0, 0.5, 1$, and 2.
[c] computed as $1-(V_L/V_H)/0.98$ (see Eqs: (6) and (7) using results for the dimer band only.
[d] Computed as in footnote c, but using V_L/V_H from the slope.
[e] Ref. *8*.
[f] Ref. *7*.
[g] Ref. *9*.

unlabeled DNA.) The extent of a deletion, relative to λDNA = 1.00, is defined by

$$f = [l(\lambda) - l(\phi)]/l(\lambda) \qquad (7)$$

Some new measurements of the lengths of the b_2 and b_{221} deletions, using this method and the λ(BU) DNA shown in Fig. 1, are given in Table 2: f is found to be 0.11–0.13 for the b_2 deletion and 0.21–0.22 for the b_{221}, deletion. These agree (Table 2) with accepted values from electron microscopy (*7, 8*). Since the approximation used for x [Eq. (4)] might introduce curvature in a plot of $(n_H/n_L)(1 - x)$ against x, the values found by least squares from the slope of this plot are compared with the values found from the dimer band alone. There are differences, but these are of the same order as the experimental error.

APPENDIX

In a typical density gradient experiment, the buoyant density (θ) of a DNA is a directly measured quantity, obtained from the density of the solution at the center of the DNA band, when that solution is brought to 1 atm pressure (*11*). In this article we use the term buoyant density in a more general sense, and we refer to the buoyant density found from the density of a DNA band as the *banding density*, θ^0, of that DNA. Mixed DNA oligomers, containing both density-labeled (H) and unlabeled (L) DNAs, form a series of bands, each at a different CsCl concentration and pressure. It is necessary to define and evaluate the buoyant densities θ_H and θ_L in each of these mixed oligomer bands.

In general terms, the buoyant density is an intensive thermodynamic quantity, which is a function of temperature, pressure, and the composition of the solution. In specific terms the buoyant density of a DNA is known to depend on the water activity and hence on the CsCl concentration (*3*), and less strongly on pressure (*12*). A straightforward definition of the buoyant density θ_i is the weight per unit volume of the hydrated species i.

$$\theta_i \equiv \frac{(1 + \Gamma_i')}{(\bar{v}_i + \Gamma_i')} \qquad \text{(A-1)}$$

where $\bar{v}_i$ is the partial specific volume of unhydrated i and Γ_i' is its net hydration (gm H_2O bound per gm of i) The partial specific volume of the bound water is taken to be 1, and species i is required to be an electrically neutral molecule (e.g., CsDNA). Eq. (A-1) may be used to determine Γ_i' at the banding density (*3, 13*). However Γ_i' may be determined at any solution composition by use of other methods (cf. Cohen and Eisenberg, Ref. *4*), so that Eq. (A-1) is an operational definition.

To relate the banding density of a mixed DNA oligomer (species K) to the buoyant densities of its constituents, it is necessary only to note that the buoyant density of a complex molecule or particle is the volume-fraction average of the buoyant densities of its constituents (*1*). Thus the banding density of oligomer species K is given by

$$\bar{\theta}_K{}^0 = \alpha_L \theta_L + (1 - \alpha_L)\theta_H \tag{A-2}$$

where the bar over $\bar{\theta}_K^0$ denotes only that it is a complex molecule. The volume fraction α_L of LDNA is given by

$$\alpha_L = (n_L V_L)/(n_L V_L + n_H V_H) \tag{A-3}$$

where n_L and n_H are the numbers of L and H DNA molecules in the oligomers, and V_i is the partial molar volume of hydrated i.

$$V_i \equiv M_i(\bar{v}_i + \Gamma_i') \tag{A-4}$$

In Eq. (A-4), M_i is the molecular weight of anhydrous i, and $\bar{v}$ for the bound water is again taken to be 1.

The problem in determining (V_L/V_H) from Eqs. (A-2) and (A-3) is that θ_L, θ_H, V_L, and V_H vary with r, the position in the ultracentrifuge cell, through the dependence of CsCl concentration and also of pressure on r. In the treatment of Baldwin et al. (*1*), this problem was bypassed by an approximation. We show below how this approximation is derived, and how it could be refined when additional data are available. The quantity which has been approximated is denoted by x, and it is defined by

$$x \equiv \frac{\theta_K{}^0 - \theta_L}{\theta_H - \theta_L} \tag{A-5}$$

where θ_H and θ_L are the buoyant densities of L and H DNA in the oligomer band at r_K^0. The problem is to evaluate x in terms of the band positions r_K^0, r_L^0, and r_H^0, and the approximate expression which has been used is

$$x \simeq \frac{(r_K{}^0 - r_L{}^0)}{(r_H{}^0 - r_L{}^0)} \tag{A-6}$$

To derive (A-6) we first expand θ and ρ as Taylor's series in r and omit terms higher than first-order.

$$\theta_i(r) = \theta_i(r_i^0) + \left(\frac{d\theta_i}{dr}\right)_{r_i^0}(r - r_i^0) + \ldots \tag{A-7}$$

$$\rho(r) = \rho(r_{HL}{}^0) + \left(\frac{d\rho}{dr}\right)_{r_{HL}^0}(r - r_{HL}{}^0) + \ldots \tag{A-8}$$

In Eq. (A-7) the buoyant density of i has been expanded about its banding density θ_i^0; in the following equation we treat $d\theta_i/dr$ as a constant and omit the subscript r_i^0.

$$\theta_i(r) = \theta_i + \frac{d\theta_i}{dr}(r - r_i^0) + \ldots \tag{A-7a}$$

In Eq. (A-8) ρ is the compositional density, or the density of the solution at r when the pressure is reduced to 1 atm; r_{HL}^0 is the band position of the mixed DNA dimer. Next the numerator and denominator of (A-5) are expressed by (A-9a):

$$\theta_{\mathrm{H}} - \theta_{\mathrm{L}} = (\theta_{\mathrm{H}}^0 - \theta_{\mathrm{L}}^0) + \frac{d\theta_{\mathrm{H}}}{dr}(r_{\mathrm{K}}^0 - r_{\mathrm{H}}^0) - \frac{d\theta_{\mathrm{L}}}{dr}(r_{\mathrm{K}}^0 - r_{\mathrm{L}}^0) \tag{A-9a}$$

$$\bar{\theta}_{\mathrm{K}}^0 - \theta_{\mathrm{L}} = (\bar{\theta}_{\mathrm{K}}^0 - \theta_{\mathrm{L}}^0) - \frac{d\theta_{\mathrm{L}}}{dr}(r_{\mathrm{K}}^0 - r_{\mathrm{L}}^0) \tag{A-9b}$$

Since θ_i^0 by definition equals ρ at r_i^0, we may use Eq. (A-8) to write

$$(\theta_{\mathrm{H}}^0 - \theta_{\mathrm{L}}^0) = \left(\frac{d\rho}{dr}\right)_{r_{\mathrm{HL}}^0}(r_{\mathrm{H}}^0 - r_{\mathrm{L}}^0) \tag{A-10a}$$

$$(\bar{\theta}_{\mathrm{K}}^0 - \theta_{\mathrm{L}}^0) = \left(\frac{d\rho}{dr}\right)_{r_{\mathrm{HL}}^0}(r_{\mathrm{K}}^0 - r_{\mathrm{L}}^0) \tag{A-10b}$$

The resulting expression for x (Eq. A-5) is

$$x = \frac{(r_{\mathrm{K}}^0 - r_{\mathrm{L}}^0)}{(r_{\mathrm{H}}^0 - r_{\mathrm{L}}^0)} \frac{\left[\left(\frac{d\rho}{dr}\right)_{r_{\mathrm{HL}}^0} - \left(\frac{d\theta_{\mathrm{L}}}{dr}\right)\right]}{\left(\frac{d\rho}{dr}\right)_{r_{\mathrm{HL}}^0} - \left(\frac{d\theta_{\mathrm{H}}}{dr}\right)(r_{\mathrm{H}}^0 - r_{\mathrm{K}}^0) - \left(\frac{d\theta_{\mathrm{L}}}{dr}\right)(r_{\mathrm{K}}^0 - r_{\mathrm{L}}^0)} \tag{A-11}$$

It reduces to Eq. (A-6) when

$$\frac{d\theta_{\mathrm{H}}}{dr} \simeq \frac{d\theta_{\mathrm{L}}}{dr} \tag{A-12}$$

To obtain a more accurate expression for x, one could follow the route outlined above and retain one higher term both in Eqs. (A-7) and (A-8), as well as using known values for $d\theta_{\mathrm{H}}/dr$ and $d\theta_{\mathrm{L}}/dr$. Data for θ_{H} as a function of water activity and pressure would be needed. It is not certain whether the experimental precision of the method (see Table 1) would warrant this. In the present procedure, which uses the simple approximation (Eq. A-6) and a calibration of the method using L and H λ DNAs (see Table 1), errors in Eq. (A-6) probably are compensated in part by this calibration.

References

1. Baldwin, R. L., P. Barrand, A. Fritsch, D. A. Goldthwait, and F. Jacob. 1966. *J. Mol. Biol.*, **17**: 343
2. Wetmur, J. G., N. Davidson, and J. V. Scaletti. 1968. *Biochem. Biophys. Res. Commun.*, **25**: 684
2a. Weigle, J., M. Meselson, and K. Paigen. 1959. *J. Mol. Biol.*, **1**: 379.
3. Hearst, J. E., and J. Vinograd. 1961. *Proc. Natl. Acad. Sci., U.S.*, **47**: 825.
4. Cohen, G., and H. Eisenberg. 1968. *Biopolymers*, **6**: 1077.
5. Costello, R. C. 1970. Ph.D. Thesis, Stanford University, Stanford, California.
6. Bellett, A. J. D., H. G. Busse, and R. L. Baldwin. 1970. In A. D. Hershey (Ed.), The bacteriophage lambda. Cold Spring Harbor Press.
7. Davis, R. W., and N. Davidson. 1968. *Proc. Natl. Acad. Sci., U.S.*, **60**: 243.
8. Westmoreland, B. C., W. Szybalski, and H. Ris. 1969. *Science*, **163**: 1343.
9. Davis, R. W. 1970. Private communication.
10. Yamagishi, H., K. Nakamura, and H. Ozeki. 1965. *Biochem. Biophys. Res. Commun.*, **20**: 727.
11. Vinograd, J., and J. E. Hearst. 1962. *Fortschr. Chem. Org. Naturstoffe.* **20**: 372.
12. Hearst, J. E., J. B. Ifft, and J. Vinograd. 1961. *Proc. Natl. Acad. Sci., U.S.*, **47**: 1015.
13. Williams, J. W., K. E. Van Holde, R. L. Baldwin, and H. Fujita. 1958. *Chem. Rev.*, **58**: 715.

SECTION C

MISCELLANEOUS TECHNIQUES

TEMPERATURE-JUMP METHODS

D. M. CROTHERS

DEPARTMENT OF CHEMISTRY

YALE UNIVERSITY

NEW HAVEN, CONNECTICUT

INTRODUCTION

Many reactions of nucleic acids produce a change in their optical properties. Important examples are the helix-coil or melting transition and the binding of substances such as dyes, drugs, and antibiotics. The nature of the optical change, be it in absorbance, fluorescence, or rotatory properties, provides useful diagnostic information for understanding the reaction. Even more information can be obtained when spectroscopic studies are coupled with kinetic techniques. The principle is to study the time response of the optical change, and the new source of data is the resolution of separate effects on the time axis.

As an illustration of this principle, consider the example of proflavine binding to DNA. This reaction can be formulated according to the mechanism

$$P + D \underset{k_{21}}{\overset{k_{12}}{\rightleftharpoons}} (PD)_{out} \underset{k_{32}}{\overset{k_{23}}{\rightleftharpoons}} (PD)_{in}$$

where P is proflavine, D a DNA binding site, and $(PD)_{out}$ and $(PD)_{in}$ are outside-bound and intercalated complexes, respectively (*1*). At equilibrium, as much as 30% (depending on conditions) of the bound dye can be in the outside-bound form, and when the static optical properties are examined, one measures an average of the characteristics of the two complex forms, and may even be unaware of the existence of two forms. However, conditions can be found under which the outside-bound form equilibrates much more rapidly with free dye than with the intercalated form. Thus, a rapid perturbation of the complex produces a fast spectral change due to the first equilibrium followed by a slower change corresponding essentially to the intercalation reaction. The time axis provides a separation of effects due to the two complex forms.

Aside from the important function of demonstrating and characterizing multiple molecular forms, kinetic measurements provide valuable information in the magnitude of the time constants. The kinetic constants and their variation with conditions helps one distinguish between plausible reaction mechanisms. Furthermore, information on the time scale of particular processes can be very useful in understanding their biological or biochemical function.

Relaxation methods are particularly appropriate to the study of nucleic acids and the weak (noncovalent) complexes they form. The principle of the technique is to bring a reversible reaction system to equilibrium, perturb the state by altering one of the external variables such as temperature or pressure, and observe the rate at which the system "relaxes" to its new equilibrium. The perturbation can be periodic, as in sound absorption, or a step-function change, as in a temperature jump. Eigen and DeMaeyer (*2*) examining the mathematics of relaxation under various conditions such as constant entropy or constant pressure, show that in dilute solutions one expects essentially the same relaxation time for a given chemical reaction.

A further useful generalization is that a system of coupled chemical equilibria gives rise to a set of relaxation times. The reason for this is basically the same one that produces a set of vibrational frequencies for coupled oscillators, and the mathematics used to describe the two situations are closely related. If a given system can exist in N different states, a small, rapid perturbation that produces an initial displacement x_j^0 of the concentration of species j from its equilibrium value will result in relaxation that can be represented by the sum of exponential decay terms,

$$x_j = x_j^0 \sum_{i=1}^{N-1} \beta_i \exp(-t/\tau_i) \tag{1}$$

where τ_i are the relaxation times of the system. These, in general, have a definite relation to the chemical reaction rate constants and to concentrations of the various species present.

For aqueous solutions, the most widely applied relaxation technique has been the temperature-jump method, originally developed by Eigen and DeMaeyer (*2*). The common method of producing the temperature jump is to discharge a high-voltage capacitor through the sample; a feasible procedure only with ionic solutions. Very large rates of instantaneous energy dissipation in the sample can be achieved in this way. For example, if a capacitor (C) of 0.02 μF is charged to a voltage (V) of 50 kV and discharged across a cell with resistance (R) of 100 Ω, the initial current is 500 A, and the initial power is

25 MW. The exponential decay time for the discharge current is RC, or 2 μsec, while the decay time for the heating power is $RC/2$, or 1 μsec. The total energy stored is $CV^2/2$, or 25 J, which, if applied to a 1 ml sample, would produce a temperature rise of about 6°C. It is also possible to produce a temperature jump by a laser or microwave pulse, but to achieve the power rating of the simple discharge system requires considerable engineering in these alternative methods.

One possible objection to the use of electrical heating is the accumulation of electrode products. Fortunately, the amount of such material is small. Using the figures of the above example, the charge passed is $Q = CV$, or 10^{-3} coulombs, about 10^{-8} moles of electrons. It is, however, advisable to work with buffered solutions, and, if the biological material studied is particularly sensitive to metal ions, to provide appropriate chelating agents for any ions that may be produced from the electrodes, usually gold or platinum.

The use of relaxation methods means, by definition, that the magnitude of concentration changes will be small following the perturbation. Consequently, the change in optical properties will also be small, and accurate measurement requires an instrument with a much lower noise level than conventional spectrophotometers. There are two main sources of noise in a photometric detection system: fluctuations in the average source intensity, and random fluctuations in the photon count that corresponds to a given average intensity. An example of a source of the former kind is wander of the arc in an arc lamp or variations in the power supplied to the lamp. The noise due to these can be greatly reduced by operating the lamp from a highly regulated supply, and by using a split-beam comparison detector to compensate fluctuations.

Noise from random photon fluctuations which cannot, of course, be reduced by a double-beam operation becomes more serious the shorter the time resolution desired. At a given intensity I, the number of photons reaching the detector in time interval Δt is proportional to $I\Delta t$. The expected random fluctuations in this quantity increase with its square root, or $(I\Delta t)^{1/2}$. The ratio of signal- $(I\Delta t)$ -to-noise $(I\Delta t)^{1/2}$ is thus proportional to $(I\Delta t)^{1/2}$. Hence, in order to increase signal-to-noise at a given time resolution Δt, one must increase I as far as is technically feasible. Consequently (in the absence of suitable laser sources) temperature-jump systems usually incorporate very fast optics, meaning poorly collimated beams of relatively large cross section and poor spectral resolution, operating with intense arc sources. At present, it is possible to produce an instrument with peak-to-peak noise that is less than 1 part in 10^4 of the visible light signal (the difference in detector output with the light on and off) in time ranges from 1 μsec resolution to several seconds. This corresponds to a noise level of about 4×10^{-5} OD units. Because source intensities are weaker in the ultraviolet, the noise level with fast time resolution (1 μsec or less) is somewhat larger than this, perhaps by a factor of 3 at 266 nm. Such an instrument permits accurate kinetic measurement of absorbance changes less than 0.01 OD unit for a few degrees temperature increase.

INSTRUMENTATION

One of the main reasons temperature-jump spectroscopy has not been more widely practiced has been the lack of completely satisfactory commercial instruments. Perhaps the best version that can be purchased ready-made is that manufactured by Messanlagenstudiengesellschaft, W. Germany. However, since this organization does not have extensive service facilities, the user will have to expect to make most repairs himself. An instrument is also marketed by American Instrument Company, Silver Spring, Maryland. It remains true, however, that the performance of the commercial equipment can be bettered by one built in the laboratory. In general, this can be done using available electronic components and some machine shop work. Temperature-jump design features have recently been discussed in some detail by French and Hammes (*3*); some alternatives to their suggestions are provided here.

The Cell

A cell that is supposed to sustain 50,000 V across an aqueous solution has certain obvious design requirements. First, it must be sturdy enough to withstand the forces developed, and it must be thick enough to provide electrical insulation for the high-voltage electrode. It is usual to keep one electrode at ground potential; thermostatting of the contents occurs by heat flow through that electrode. Figure 1 shows a cross section drawing of the standard cell developed by Eigen, DeMaeyer, and their collaborators. The electrodes are located above and below the beam through the cell, whose pathlength is about 1 cm, or less in semimicro cells. The cross-sectional area of the solution heated is thus about 1 cm^2, and the electrodes are roughly 2.5 cm apart. Current cell designs have conical windows to accommodate the poorly collimated beams that accompany high light intensities. The inner surface of the conical window is about 3 mm in diameter, and the beam itself is only slightly smaller than this.

The surface area of the electrodes is larger than the cross section of fluid heated, which is the volume in the narrow central region of the cell. By this design, one is ensured of uniform heating of the solution, since the electric field will be homogeneous through the whole cross section. Nonuniform

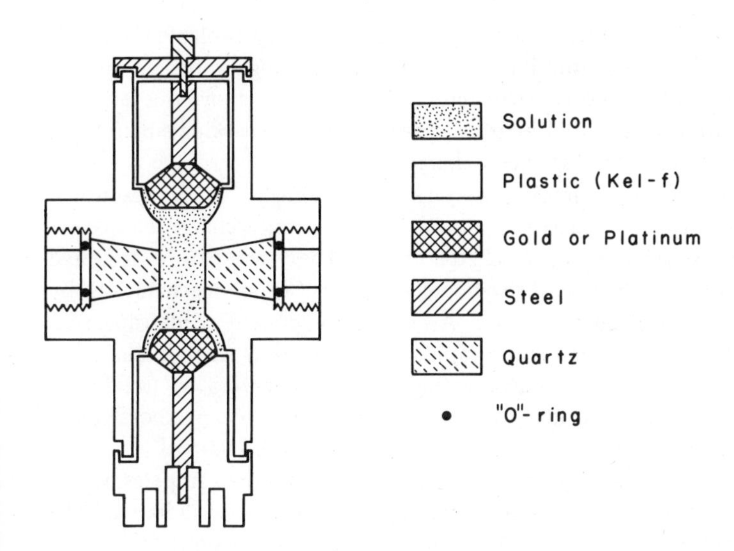

Figure 1. Cross section of a standard temperature-jump cell.

heating produces serious optical disturbances from Schlieren effects and must be avoided. This cell is moderately subject to convective disturbances in which the unheated solution around the upper electrode comes pouring down into the beam after the temperature jump. At or below room temperature this disturbance appears after roughly two seconds, but becomes faster at higher temperature.

The "standard cell" shown contains about 6 ml of solution, still rather large for scarce biological materials. We have reduced this volume to about 1 ml, still keeping the basic features of the standard cell design (*4*). In particular, the volume around the electrode is greatly reduced by providing only a small gap between it and the plastic wall of the cell. This, incidentally, reduces the problem of convection following the jump by reducing the volume of nonheated solution, and making it more difficult for it to flow down into the beam. We have constructed cells of Kel F and of polycarbonate plastics.

High-Voltage Discharge System

Careful design of the high-voltage discharge unit is essential if one wishes to make measurements in the time range of 20 μsec and below. As mentioned in the introduction the heating time is $RC/2$, so for a fixed resistance R (governed by the salt concentration used) it is desirable to use small capacitors C to obtain rapid heating. However, the energy available is $CV^2/2$, and a reasonable temperature jump will result only if high voltage is used. A typical upper limit on most instruments is about 50 kV; above this level it is increasingly difficult to avoid dielectric breakdown and sparking either in the cell or at some other point in the discharge circuit.

The basic discharge circuit is very simple. The high-voltage capacitor is connected to the cell via a spark gap. After the capacitor is charged, a spark is triggered across the gap and discharge ensues. The spark can be triggered electrically, as described by French and Hammes (*3*), or mechanically by forcing the two electrodes together. A critical component is the capacitor, which should be constructed for rapid discharge, with small inductance. Suitable capacitors are, for example, the series E marketed by Condensor Products, Brooksville, Florida. The instrument in use in our laboratory is designed to accept a range of capacitor sizes, from 0.005 to 0.05 μF.

The maximum discharge current may be several hundred amperes, which clearly provides a large potential source of electromagnetic radiation. The feature of central importance in avoiding this is cylindrical symmetry of the discharge circuit, so that the ground level return current to the capacitor occurs in a cylinder around the high-voltage current to the cell. Detailed instructions for building a discharge system are provided by French and Hammes (*3*).

It has been our experience that, once cylindrical symmetry of the discharge system is established, the major source of disturbance signals associated with the discharge is coupling with the power supply used to charge the capacitor. Such disturbance effects can be very large, and considerably longer in time scale than the discharge itself. These phenomena can be avoided by providing means for disconnecting the power supply from the capacitor before discharge. We use the supply KV-60S from Kilo Volt Corporation, Hackensack, New Jersey with several special features. The most important is a solenoid switch at the high-voltage output that serves to connect and disconnect the load. Also built in is a 50-MΩ load resistor to limit the charging rate, and provision for "surge on." In addition to isolating the power supply, steps must be taken to prevent the cable that connects it to the discharge capacitor from acting as an antenna for the discharge. We accomplished this by placing a 10-MΩ resistor at the end of the cable that is inserted into the shielded discharge system.

Optical System

A well designed optical system for a temperature-jump apparatus is optimized for two factors: maximum stability of the light source and maximum light flux. This generally requires a sacrifice of spectral purity,

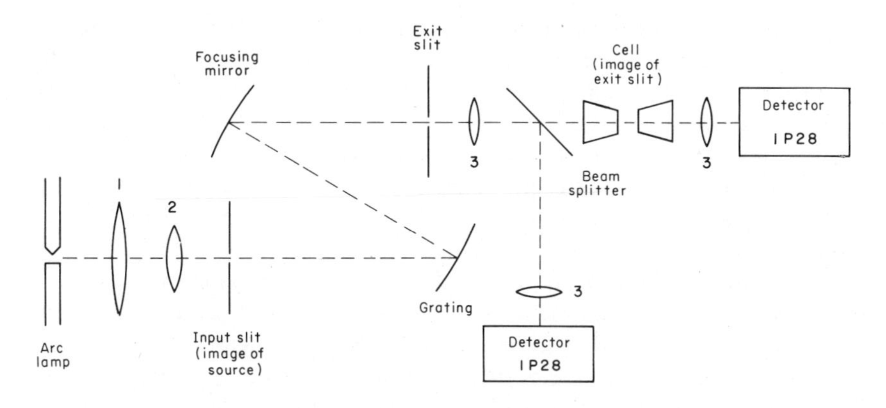

Figure 2. Schematic diagram of a temperature-jump optical system. Lenses (quartz): (1) 62-mm diameter, 42-mm focal length; (2) 1.5-in. diameter, 100-mm focal length; (3) 1-in. diameter, 50-mm focal length.

except with laser sources. An outline of suitable optics is shown in Fig. 2. The light source, usually an arc lamp, should be mounted with provision for external adjustment of the lamp position. We have used the housing Model 366Q made by Electro Powerpacs, Cambridge, Massachusetts, modified by addition of a water-cooled sleeve around the lens holder, and a larger heat-dissipating aluminum mount for the upper electrode. In lamp selection, we have had best results with a 200-W xenon–mercury arc (Hanovia). This lamp is considerably more stable than the smaller varieties, although there is great variation from one sample to another in all arc lamps. Sometimes these will show a rather slow (~1 sec) periodic oscillation in arc position, which appears as a variation in intensity at the detector. This motion can often be entirely prevented by placing a magnet (such as a large stirring bar) on the outside of the lamp housing. The best position of the magnet should be adjusted by trial.*

A highly regulated power supply is essential for good lamp stability (*5*). Lamps may be operated directly with regulation for constant current or constant power; the latter is more convenient, but supplies of the former type are more generally available. If a constant voltage supply is used, a series resistor is required, since the voltage drop across the lamp is determined by the arc characteristics. We have used the constant current supply QRC 40-15A manufactured by Sorenson, Norwalk, Connecticut, which has proved quite satisfactory except for one trivial modification.†

* I thank Mr. C. R. Rabl for pointing this out to me.

† The current sensing transistors Q18 and Q19 were found to be subject to thermal-drift effects. A small cardboard screen was folded around these two adjacent transistors, shielding them from air currents and eliminating the short-term drift.

The 40-V level provided by this supply is not sufficient to start an arc lamp. We therefore use another, much cheaper, Sorenson supply DCR 80-10, which provides 80 V and can be used in current limiting mode to start the lamps. The two supplies must be protected from each other by series diodes, with provision to short around the diode on the main supply after the auxiliary supply is switched out of the circuit. Once the lamp is ignited, the current load is shifted manually, one ampere at a time, from auxiliary to main supply.

To protect the lamp from excessive current at the moment of ignition (the current comes from discharge of the output capacitance of the supply) we use a 2-Ω, 200-W resistor in series with the auxiliary supply. The necessary high-voltage, radio frequency starting pulse is provided by an ignition device (for example, #357 Igniter from Electro Powerpacs, Cambridge, Massachusetts).

The Electro Powerpac Lamp housing that we use provides a large quartz collecting lens. We added a second lens, of focal length 100 mm and 1½ in. diameter to project and focus an image of the source at the entrance slit of a Bausch and Lomb high-intensity grating monochromator, number 33–86* using entrance slit of 2.5 mm and exit slit 1.5 mm. It is important that the geometry be adjusted so that light is not lost at any optical element; for example, the illuminated area at the grating should be just sufficient to fill the grating. The image of the exit slit is then focused at the center of the cell by a 50-mm, 1-in. diameter quartz lens. A 50-mm, 1-in. lens in the photomultiplier housing (for example, D-21-20 from Oriel Optics, Stamford, Connecticut) collimates the beam, which then illuminates the photosensitive area of the photomultiplier tube (typically a 1P28). A beam splitting mirror in front of the cell reflects about 25% of the light to a reference detector identical to the main detector. This mirror is designed to be easily removed to allow single beam operation for very fast relaxation times.

Modern operational amplifiers provide simple, high-speed, and very stable means for amplifying the photo current, converting it to a voltage level detected by an oscilloscope. Our instrument provides signal-to-noise ratios in excess of 10^4 under good conditions. Noise pick-up in leads to the oscilloscope can easily reach a tenth of a millivolt unless special precautions are taken, and we have therefore found it convenient to amplify the signal to about 5V within the shielded amplifier housing. Figure 3 shows a circuit that will produce 5V from a 50 μA photocurrent, with a minimum rise time of about 0.1 μsec. The operational amplifiers were purchased from Analog Devices, Cambridge, Massachusetts. A two-stage amplifier was used to provide the desired time response.

* This monochromator passes appreciable amounts of stray light, expecially in the ultraviolet.

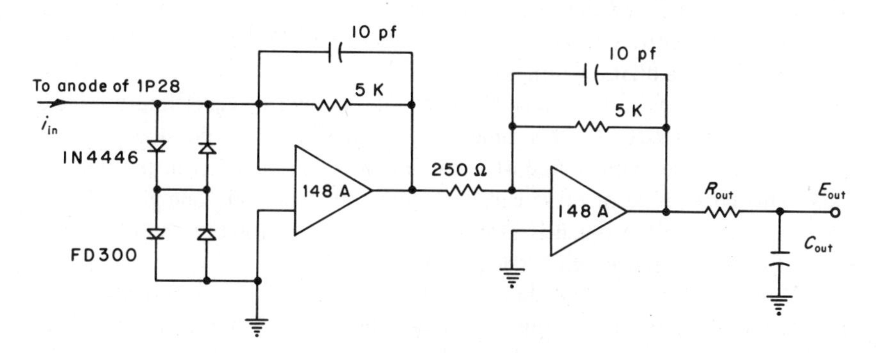

Figure 3. Amplifier to convert the photocurrent of a photomultiplier tube to a low-impedance output voltage. The purpose of the diodes at the input to the first operational amplifier is protection against accidental high voltage.

$E_{out} = i_{in} \times 10^5$ V/A
$t_{const} = R_{out}C_{out}$

For minimum $t_{const} = 0.1\ \mu$sec
$R_{out} = 200\Omega$
$C_{out} = 0$

The current produced by a photomultiplier tube is linearly related to light intensity only for small values of the photocurrent. Thus, in order to utilize all the light that falls on the photocathode, it is necessary to provide a means for varying the gain within the photomultiplier tube. This is commonly done by using a restricted number of the dynodes or amplification stages. Unwanted dynodes should be connected to the anode, from which the photocurrent is drawn. Over the UV and visible wavelength ranges the instrument must detect widely varying light intensities. It is therefore essential to have a means for switching externally the number of dynodes connected to the anode, keeping constant at the same time the volts per dynode stage actually in use. At several of the peak emission wavelengths of the xenon-mercury arc lamp, we are easily able to draw 50 μA of photocurrent in using the 1P28 as a photodiode, that is, with no additional dynode stages.

In double beam operation, the output from the two detectors is compared at the differential oscilloscope inputs (for example, 3A3 input amplifier in a Tektronix 564 storage oscilloscope with enhanced writing rate). Several means are possible for balancing the two channels against each other. In our instrument the two photomultiplier tubes are powered by separate, continuously variable high-voltage supplies (6102A from Hewlett Packard, Berkeley Heights, New Jersey). In this way both detectors can be operated at identical photocurrents. The output of the main detector is continuously monitored with a digital voltmeter.

We use the instrument in single beam mode only for relaxation times less than about 100 μsec. In this time range lamp flicker is too slow to be a problem, and the main source of noise is random fluctuation in the photon flux. This latter type of noise can be reduced by using only one photodetector. In this mode the signal is examined with the oscilloscope inputs set for ac, so that no reference signal is required.

Double-beam operation should reduce the noise due to lamp flicker by at least a factor of 10. This improvement will be achieved, however, only if the position of each component is carefully adjusted to give maximum light transmission. If this is not done, slight vibrations will appear as intensity variations which may well not compensate from one channel to the other. Our instrument is therefore built in modular form, so that components can be moved to allow for small changes in cell and lamp geometry.

Sources of Noise

In addition to noise problems associated with the optical system, arising from lamp flicker, component vibration and limited photon flux, there are two disturbances resulting from the temperature rise whose nature should be recognized. One of these is convective mixing of the cell contents, which limits the upper time range of the usual temperature-jump instrument to roughly two seconds, below room temperature. As the cell temperature increases the convective disturbance appears at shorter times. The major source for convection is unheated liquid above the volume element heated by the electric discharge. The heavier, cooler liquid flows down through the cell and across the optical axis. The result is a large, and irreproducible, change in light transmission, which can usually be easily recognized. To minimize this phenomenon, the cell design should reduce as far as possible the unheated volume of liquid in the cell above the observation windows.

Cooling also results from heat transport out of the cell, mainly through the metal electrodes. Transport in the absence of convection leads usually to an initially linear rate of optical density change, generally quite small, and producing a slightly sloping base line in slower relaxation phenomena. In addition, the rapidly cooled liquid adjacent to the upper electrode will tend to initiate convection, with its associated irreproducible consequences. There seems to be no way of avoiding or correcting for this latter process, except to place the electrodes farther away from the observation path.

Another serious disturbance that sometimes limits temperature-jump measurements is cavitation in the liquid as a result of the discharge. At temperatures above the density maximum of water (4°C), the volume of an aqueous solution must increase following a temperature jump if the pressure is held constant. However, pressure equalization is generally slower than the electrical heating, since the former involves processes at the speed of sound. Thus, at the instant of heating the liquid volume is constant and, because of the small compressibility of water, there is large transient pressure increase. A rough value for several degrees temperature rise at room temperature is 20 atm. When this pressure shock is reflected from a free surface, it returns as a negative pressure wave. This produces microscopic bubbles in the solution, or cavitation. The optical result is a large decrease in light transmission, appearing a few microseconds after the temperature jump and extending out to variable times at which it disappears as suddenly as it appeared. Careful degassing of solutions and use of relatively small temperature jumps make it possible to avoid serious consequences of this phenomenon up to about 50°C. Beyond this temperature, the cavitation disturbance extends over longer and longer time ranges, so that the time from 5 to 50 μsec may be blocked out. We are currently building a high-pressure compartment for the temperature-jump cell in an attempt to overcome this limitation.

Slower Temperature-Jump Instruments

Several cells have been described which allow measurement of slower phenomena than accessible with the high-voltage discharge unit. These cells are designed to fit into conventional spectrophotometers, and are thermostatted by circulating baths to hold temperature constant after the jump. Two baths are provided, at different temperatures, and circulation is switched from one to the other at the moment of the temperature change. The cell described by Pohl (*6*) relies on rapid equilibration with the circulating baths for the temperature change, and incorporates thin-walled cells and a large surface-to-volume ratio for rapid heat flux. Spatz and Crothers (*7*) described a cell in which a low-voltage, high-frequency heating current is used to raise the contents of the cell from the temperature of the lower to that of the higher bath in less than 1 sec. Thus a combination of these relatively simple cells with fast temperature-jump equipment permits measurement of kinetic phenomena from 10^{-6} seconds to many thousands of seconds.

Future Developments

One obvious development of temperature-jump equipment currently underway is incorporation of laser light sources. The excellent collimation of these beams permits design of truly micro cells and the high intensity should further reduce noise levels in the fast time range. Performance in the fast time range should also be improved by use of repeated small temperature pulses with signal averaging. There seems to be no reason why accurate measurement of relaxation times as short as 100 nsec cannot be made with improved temperature-jump techniques.

Another important development will be to include automatic data analysis for resolution of multiple relaxation times. Instruments described up to the present provide only for analog recording of data in the form of oscilloscope traces. Present rapid analog to digital converters should permit storage of data in digital form, including simultaneous recording in several time ranges. A computer can then be used to calculate the relaxation spectrum.

INTERPRETATION OF DATA

The output of a temperature-jump spectrophotometer is customarily the variation with time of the light transmitted by the solution. Let I_0 be the intensity before the perturbation; I_0 is proportional to the signal at the output of the detection amplifier, Fig. 3. If I is the light intensity at any time, the change in solution absorbance A from the beginning value is given by

$$\begin{aligned} \Delta A &= \log_{10}(I/I_0) \\ &= \log_{10}(1 + \delta I/I_0) \end{aligned} \tag{2}$$

where δI is the change in light intensity. In most temperature jump experiments, $\delta I/I_0$ is smaller than 0.05, so that an expansion of the logarithm in Eq. (2) may be used, resulting in

$$\Delta A = \delta I/2.3I_0 \tag{3}$$

As a consequence, the variation in signal intensity is proportional to the change in absorbance of the solution, and is therefore also linear in concentration.

There are two main kinds of information available from a temperature jump experiment. One is the time constant or constants with which the system responds to the rapid temperature perturbation, the so-called relaxation times, whose mathematical form is discussed below. Less use has been made of the other observable parameter, the amplitude of the absorbance change that is associated with each relaxation time, although Eigen (*8*) has recently emphasized the utility of such measurements.

Relaxation Times

The relation between kinetic constants and relaxation times for simple one step reactions can be easily derived. When the reaction system includes more than one step, the analysis is a little more complicated, involving a normal coordinate analysis closely analogous to that used for coupled oscillators. Consider a simple first-order reaction.

$$\mathrm{A} \underset{k_{21}}{\overset{k_{12}}{\rightleftharpoons}} \mathrm{B} \tag{4}$$

The rate of change of the concentration [A] of component A is

$$\frac{d[\mathrm{A}]}{dt} = -k_{12}[\mathrm{A}] + k_{21}[\mathrm{B}] \tag{5}$$

The concentrations [A] and [B] can be expressed in terms of the equilibrium value following the perturbation, $\bar{\mathrm{A}}$ and $\bar{\mathrm{B}}$, respectively, and the time-dependent displacement from that equilibrium:

$$\begin{aligned} [\mathrm{A}] &= \bar{\mathrm{A}} + x \\ [\mathrm{B}] &= \bar{\mathrm{B}} - x \end{aligned} \tag{6}$$

(Because of conservation of mass and the stoichiometry of reaction (4), the displacement of B from equilibrium must be the negative of the displacement of A from equilibrium.) Thus we can write

$$\frac{d(\bar{\mathrm{A}} + x)}{dt} = -k_{12}(\bar{\mathrm{A}} + x) + k_{21}(\bar{\mathrm{B}} - x) \tag{7}$$

Since the perturbation is a step function, $\bar{\mathrm{A}}$ is independent of time after the temperature jump, and $d\bar{\mathrm{A}}/dt$ is zero. Furthermore, by the condition for chemical equilibrium,

$$k_{12}\bar{\mathrm{A}} = k_{21}\bar{\mathrm{B}} \tag{8}$$

Therefore, Eq. (7) can be simplified to

$$\frac{dx}{dt} = -(k_{12} + k_{21})x \tag{9}$$

which is a simple linear first-order differential equation, with solution

$$x = x_0 e^{-t/\tau} \tag{10}$$

where

$$1/\tau = k_{12} + k_{21} \tag{11}$$

and x_0 is the initial value of the displacement from equilibrium at $t = 0$ (see below). Equation (10) predicts an exponential decay of x toward zero.

Unlike a first order reaction, the relaxation time of a second order reaction depends on concentration. Consider the equilibrium

$$\mathrm{A + B} \underset{k_{21}}{\overset{k_{12}}{\rightleftharpoons}} \mathrm{C} \tag{12}$$

with rate equation

$$\frac{d[\mathrm{A}]}{dt} = -k_{12}\,[\mathrm{A}]\,[\mathrm{B}] + k_{21}\,[\mathrm{C}] \tag{13}$$

Proceeding as before, with $[\mathrm{A}] = \bar{\mathrm{A}} + x$, and neglecting second-order terms in the (small) displacement x, we arrive at the equation

$$\frac{dx}{dt} = -\,[k_{12}(\bar{\mathrm{A}} + \bar{\mathrm{B}}) + k_{21}]x \tag{14}$$

which is again a linear, first order differential equation, with solution

$$x = x_0 e^{-t/\tau} \tag{15}$$

where the reciprocal of the relaxation time is given by

$$1/\tau = k_{12}(\bar{\mathrm{A}} + \bar{\mathrm{B}}) + k_{21} \tag{16}$$

If $1/\tau$ is measured and plotted against $(\bar{\mathrm{A}} + \bar{\mathrm{B}})$, the slope gives k_{12} and the intercept k_{21}.

A system of coupled chemical reactions generally gives more than one relaxation time. There is a close mathematical analogy with a system of coupled oscillators, as, for example, in the internal vibrational modes of a molecule. In that case the vibrational frequencies result from the harmonic oscillation of a set of normal coordinates, which do not necessarily correspond to any one particular bond. A similar normal coordinate transformation is applied to a set of coupled chemical reactions in order to get the relaxation times. These have a meaning analogous to the reciprocal frequencies of a vibrating system, except that in the case of chemical reactions one predicts only exponential decay terms under relaxation conditions, with no oscillatory behavior.

The mathematical operations are most easily understood in connection with a specific example. Consider the reaction mechanism

$$\mathrm{A + B} \underset{k_{21}}{\overset{k_{12}}{\rightleftharpoons}} \mathrm{C} \underset{k_{32}}{\overset{k_{23}}{\rightleftharpoons}} \mathrm{D} \tag{17}$$

which has been proposed for the binding of proflavine by DNA (*1*), where A is a free binding site, B free proflavine, C an externally bound dye molecule, and D the intercalated complex. One must allow for two independent terms

x_1 and x_2, describing the displacement from equilibrium. We let

$$[\mathrm{A}] = \bar{\mathrm{A}} + x_1; \qquad [\mathrm{B}] = \bar{\mathrm{B}} + x_1$$
$$[\mathrm{D}] = \bar{\mathrm{D}} + x_2; \qquad [\mathrm{C}] = \bar{\mathrm{C}} - x_1 - x_2 \tag{18}$$

Kinetic equations for the disappearance of A and the appearance of D

$$\frac{d[\mathrm{A}]}{dt} = -k_{12}[\mathrm{A}][\mathrm{B}] + k_{21}[\mathrm{C}]$$
$$\frac{d[\mathrm{D}]}{dt} = k_{23}[\mathrm{C}] - k_{32}[\mathrm{D}] \tag{19}$$

are then readily transformed into

$$\frac{dx_1}{dt} = -M_{11}x_1 - M_{12}x_2$$
$$\frac{dx_2}{dt} = -M_{21}x_1 - M_{22}x_2 \tag{20}$$

where

$$M_{11} = k_{12}(\bar{\mathrm{A}} + \bar{\mathrm{B}}) + k_{21}; \qquad M_{12} = k_{21}$$
$$M_{21} = k_{23}; \qquad M_{22} = k_{23} + k_{32} \tag{21}$$

Equation (20) can be written compactly in vector notation as

$$\dot{\mathbf{x}} = -\mathbf{M}\mathbf{x} \tag{22}$$

where $\dot{\mathbf{x}}$ is the vector of time derivatives dx_i/dt, $\mathbf{x}$ the vector of displacements x_i and $\mathbf{M}$ the matrix array of coefficients M_{ij}.

All physically realistic matrix operators $\mathbf{M}$ can be diagonalized by a similarity transformation $\mathbf{T}$:

$$\mathbf{T}^{-1}\mathbf{M}\mathbf{T} = \Lambda \tag{23}$$

$$\mathbf{T}^{-1}\mathbf{T} = \mathbf{T}\mathbf{T}^{-1} = \mathbf{E} \tag{24}$$

where Λ is a diagonal matrix, and $\mathbf{E}$ the unit matrix. Let

$$\mathbf{T}^{-1}\mathbf{x} = \mathbf{y} \tag{25}$$

Multiplying Eq. (22) by $\mathbf{T}^{-1}$ and inserting the unit matrix between $\mathbf{M}$ and $\mathbf{x}$ yields

$$\mathbf{T}^{-1}\dot{\mathbf{x}} = -\mathbf{T}^{-1}\mathbf{M}\mathbf{T}\mathbf{T}^{-1}\mathbf{x} \tag{26}$$

or, using Eqs. (23) and (25),

$$\dot{\mathbf{y}} = -\Lambda\mathbf{y} \tag{27}$$

Since Λ is a diagonal matrix, (27) is equivalent to a set of linear first order differential equations

$$\frac{dy_i}{dt} = -\lambda_i y_i \tag{28}$$

where λ_i are the diagonal elements of Λ. The solution to Eq. (28) is

$$y_i = y_i^0 e^{-t/\tau_i} \tag{29}$$

where $1/\tau_i = \lambda_i$. The quantities y_i, which are linear combinations of the x_i, are thus the normal coordinates and the relaxation times τ_i are the reciprocal eigenvalues of the matrix **M**. Thus, in the case of the specific mechanism (17), the eigenvalues are determined by solving the determinantal equation

$$\begin{vmatrix} M_{11} - \lambda & M_{12} \\ M_{21} & M_{22} - \lambda \end{vmatrix} = 0 \tag{30}$$

which has solutions

$$\lambda = \frac{M_{11} + M_{22}}{2} \pm \tfrac{1}{2}\sqrt{(M_{11} + M_{22})^2 + 4M_{12}M_{21} - 4M_{11}M_{22}} \tag{31}$$

Further simplification of the expression for the separate eigenvalues depends on the values of the elements M_{ij}. However, if both eigenvalues can be measured with good experimental accuracy, it is convenient to use the relations

$$\lambda_1 + \lambda_2 = M_{11} + M_{12} \tag{32}$$

and

$$\lambda_1 \lambda_2 = M_{11}M_{22} - M_{12}M_{21} \tag{33}$$

which are useful because they give a linear variation in the concentration $(\bar{A} + \bar{B})$.

According to Eq. (29), each normal coordinate decays exponentially, with relaxation time τ_i. Since the normal coordinates are combinations of concentrations, which are linearly related to absorbance, the difference ΔA between initial and final absorbance will decay toward zero with a spectrum of relaxation times τ_i:

$$\Delta A = \sum_i A_i^0 e^{-t/\tau_i} \tag{34}$$

where A_i^0 is the amplitude of the absorbance change associated with relaxation time τ_i. (The exact relation between A_i and the concentration displacements x_i^0 at time zero will be discussed in the following section.) When the reaction is complicated, and several relaxation times are present, extensive analysis may be necessary to resolve the component decay times. A useful paper relating to this problem is that of Laiken and Printz (*9*).

Relaxation Amplitudes

The amplitude of the total absorbance change following a temperature jump depends on extinction coefficients, reaction enthalpies, and equilibrium constants. Consider the first order reaction (4). The equilibrium constant

$$K = \bar{B}/\bar{A} \tag{35}$$

$$\bar{A} + \bar{B} = C_T \tag{36}$$

varies with temperature according to the van't Hoff relation

$$\frac{d \ln K}{dT} = \frac{\Delta H}{RT^2} \tag{37}$$

Since, by conservation of mass, $d\bar{A} = -d\bar{B}$, differentiation of (35) yields

$$\frac{1}{\bar{B}} + \frac{1}{\bar{A}} \quad \frac{d\bar{B}}{dT} = \frac{\Delta H}{RT^2} \tag{38}$$

The absorbance change dA is related to concentration changes by

$$dA = d\bar{B}\Delta\epsilon \tag{39}$$

where $\Delta\epsilon$ is the difference in extinction coefficients of B and A, $\Delta\epsilon = \epsilon_B - \epsilon_A$. Thus we convert (38) to

$$\frac{dA}{dT} = \frac{K}{(1+K)^2} \frac{\Delta H}{RT^2} \Delta\epsilon C_T \tag{40}$$

and if two of the three parameters K, $\Delta\epsilon$, and ΔH are known, the third can be determined. If none of these is known, one can still use measurements at varying wavelength to determine a relative difference spectrum since, assuming ΔT and K constant,

$$dA(\lambda_1)/dA(\lambda_2) = \Delta\epsilon(\lambda_1)/\Delta\epsilon(\lambda_2) \tag{41}$$

The real utility of this latter procedure comes in coupled reaction systems with which rapid perturbation techniques permit measurement of difference spectra for a single step, information not accessible to static techniques.

In a second-order reaction, measurement of relaxation amplitudes at varying total concentration permits determination of two of the three parameters K, ΔH, and $\Delta\epsilon$ if one of them is known. For illustration, consider a dimerization reaction

$$A + A \rightleftharpoons B \tag{42}$$

with equilibrium constant K,

$$K = \bar{B}/\bar{A}^2 \tag{43}$$

and total concentration C_T

$$C_T = 2\,\overline{\mathrm{B}} + \overline{\mathrm{A}} \tag{44}$$

Differentiation of (43) with respect to temperature, and use of the van't Hoff relation yields

$$\frac{d\overline{\mathrm{B}}}{dT} = \frac{\Delta H}{fRT^2} \tag{45}$$

where $f(K, C_T)$ is given by

$$f = 1/\overline{\mathrm{B}} + 4/\overline{\mathrm{A}} \tag{46}$$

and can readily be calculated in terms of K and C_T from Eqs. (43) and (44). The change in absorbance dA is given by

$$dA = \Delta\epsilon d\overline{\mathrm{B}} \tag{47}$$

where

$$\Delta\epsilon = \epsilon_{\mathrm{T}} - 2\epsilon_{\mathrm{A}} \tag{48}$$

and the ϵ are molar extinction coefficients. Thus, Eq. (45) becomes

$$\frac{dA}{dT} = \frac{\Delta H \Delta\epsilon}{fRT^2} \tag{49}$$

The use of this equation involves measurement of the absorbance change dA for fixed dT at different values of C_T. A trial choice of K permits calculation of $1/f$ for each C_T. The measured dA/dT should vary linearly with $1/f$; if K is improperly chosen, the variation will not be linear. When linearity is obtained the slope is $\Delta H\Delta\epsilon/RT^2$. Hence, K and ΔH can be determined if $\Delta\epsilon$ is known.

For coupled reactions, we will consider only the problem of how to calculate the amplitude of each exponential effect in the relaxation spectrum. Suppose that an optically absorbing substance can exist in $N + 1$ states. An example might be a dye that can form N different complexes with DNA, plus the state of free dye. The total absorbance change ΔA^0 following the perturbation is given by

$$\Delta A^0 = \sum_{i=1}^{N+1} \epsilon_i x_i^0 \tag{50}$$

where x_i^0 is the difference between equilibrium concentrations of form i before and after the perturbation. (Calculation of x_i^0 requires some knowledge of, or assumption about, the heat and equilibrium constants of each reaction step.) Since the total concentration of the dye is subject to conservation, there are only N independent variables, with

$$x_{N+1}{}^0 = -\sum_{i=1}^{N} x_i{}^0 \tag{51}$$

Thus

$$\Delta A^0 = \sum_{i=1}^{N+1} \Delta\epsilon_i x_i{}^0 \tag{52}$$

where

$$\Delta\epsilon_i = \epsilon_i - \epsilon_{N+1} \tag{53}$$

Experimental observations should lead to a spectrum of relaxation effects, with the absorbance varying with time according to

$$\Delta A = \sum_{i=1}^{N} A_i^0 e^{-t/\tau_i} \tag{54}$$

where $A_i{}^0$ is the amplitude of the absorbance change associated with τ_i. Our problem is to find the relation between $x_i{}^0$ and $A_i{}^0$.

From Eqs. (25) and (29) we can write the time variation of the normal coordinate y_i in terms of the $x_i{}^0$ as

$$y_i = e^{-t/\tau_i} \sum_j T_{ij}^{-1} x_j{}^0 \tag{55}$$

where T_{ij}^{-1} is the element ij of the transformation $\mathbf{T}^{-1}$. Now, since

$$\mathbf{x} = \mathbf{T}\,\mathbf{y} \tag{56}$$

it follows that

$$x_k = \sum_i e^{-t/\tau_i} T_{ki} \sum_j T_{ij}^{-1} x_j{}^0 \tag{57}$$

Multiplying by $\Delta\epsilon_k$ and summing over k to obtain the absorbance yields

$$\begin{aligned} \Delta A &= \sum_k \Delta\epsilon_k x_k \\ &= \sum_k \Delta\epsilon_k \sum_i e^{-t/\tau_i} T_{ki} \sum_j T_{ij}^{-1} x_j^0 \end{aligned} \tag{58}$$

The coefficient of e^{-t/τ_i}, defined by Eq. (54) as $A_i{}^0$, is thus

$$A_i{}^0 = \sum_k \Delta\epsilon_k T_{ki} \sum_j T_{ij}^{-1} x_j{}^0 \tag{59}$$

which is the desired relation.

In some circumstances, all of the amplitude of an observed effect can, to a good approximation, be ascribed to a single relaxation step. For example, in the binding of proflavine by DNA, high concentrations make the equilibration of the outside bound form with free dye very rapid compared to the intercalation reaction (1). In this case, the amplitude of the fast effect can be treated like a simple second-order reaction, and, for example, a difference spectrum for the external complex can be measured. Since only a small fraction of the dye is usually in the outside complex, its spectral properties are inaccessible by static measurements.

FIELDS OF APPLICATION

Two general areas of nucleic acid chemistry have been extensively investigated using temperature-jump methods. One is the denaturation process, usually involving melting of the double helix. Both DNA (*7*) and oligonucleotides (*10, 11*) have been examined. The results are a clearer understanding of the mechanism of these processes, including the nature of nucleation events and the rate of the elementary steps of base pair formation.

The other subject studied has been complex formation by nucleic acids with biologically active dyes (*1, 12*). In this case also, kinetic studies have helped clarify the mechanism of the binding process, and have demonstrated the existence of multiple complex forms.

References

1. Li, H. J. and D. M. Crothers. 1969. *J. Mol. Biol.*, 39: 461.
2. Eigen, M. and L. DeMaeyer. 1963. In S. L. Freiss, E. S. Lewis, and A. Weissberger (Eds.), Technique of organic chemistry, Vol. 8, Part 2. Wiley (Interscience), New York.
3. French, T. C. and G. G. Hammes. 1969. In K. Kustin (Ed.), Methods in enzymology, Vol. XVI. Academic Press, New York.
4. Cole, P. E. 1971. Ph.D. Thesis, Yale University, New Haven.
5. Green, M., R. H. Breeze, and B. Ke. 1968. *Rev. Sci. Instr.*, **39:** 1968.
6. Pohl, F. 1968. *European J. Biochem.* **4:** 373.
7. Spatz, H.C.H. and D. M. Crothers. 1969. *J. Mol. Biol.*, **42:** 191.
8. Eigen, M. 1970. Address to 14th Annual Meeting of the Biophysical Society.
9. Laiken, S. L. and M. P. Printz. 1970. *Biochemistry*, **9:** 1547.
10. Eigen, M. 1967. In S. Claesson (Ed.), *Nobel Symposium No. 5.* Almquist and Wiskell, Stockholm.
11. Craig, M., D. M. Crothers, and P. Doty, in preparation.
12. Schmeckel, D. E. V. and D. M. Crothers, 1970. *Biopolymers*, in press.

HYDROGEN-EXCHANGE MEASUREMENTS ON NUCLEIC ACIDS*

BRUCE McCONNELL
DEPARTMENT OF BIOCHEMISTRY AND BIOPHYSICS
UNIVERSITY OF HAWAII
HONOLULU, HAWAII

PETER H. VON HIPPEL
INSTITUTE OF MOLECULAR BIOLOGY AND
DEPARTMENT OF CHEMISTRY
UNIVERSITY OF OREGON
EUGENE, OREGON

INTRODUCTION

Hydrogen exchange as a potential method of macromolecular conformation analysis was introduced and first developed by Linderstrøm-Lang and his colleagues in the 1950s. Since that time conceptual and methodological refinements have improved the technique to the point that several variants of it are now being used extensively in protein structure studies (for reviews, see Refs. *1-3*).

More recently, this approach has also been extended to an analysis of DNA structure (*4–8*) using the tritium-gel filtration method introduced by Englander (*9*). It has been shown that the potentially exchangeable hydrogens

* The research referred to which was carried out in the laboratories of the authors was supported by U.S. P.H.S. Research Grants AM-03412, AM-12215 and GM-15792.

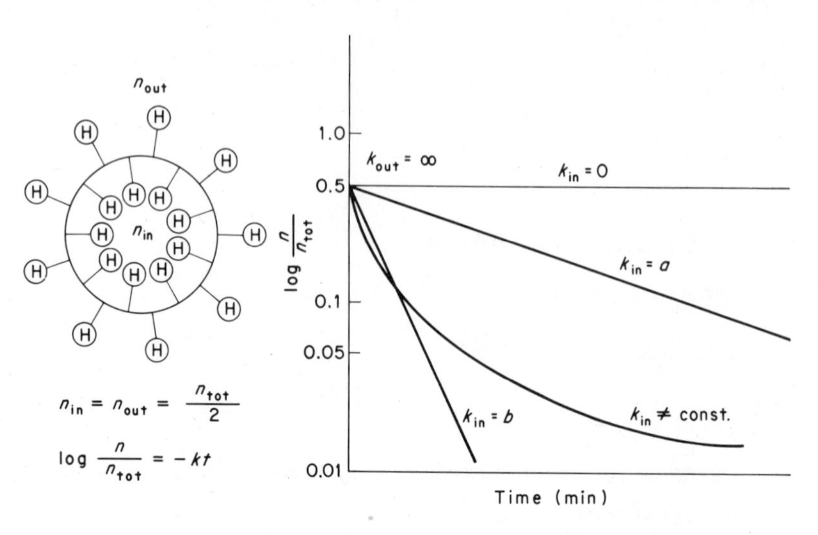

Figure 1. Schematic presentation of an idealized macromolecular hydrogen-exchange system.

of DNA can be fully labeled by brief incubation in tritiated water, and that after passage of the sample through a Sephadex column to remove most of the free tritium, the exchange-out of the hydrogens involved in interchain hydrogen bonding can be followed (after another gel filtration step) on a time scale calibrated in seconds. Similar findings have been made with tRNA (*10, 11*) and various polynucleotides (*5, 12, 13*). Some aspects of hydrogen-exchange methodology as applied to nucleic acids have recently been reviewed (*14*).

The situation illustrated in Fig. 1 provides an over-simplified and schematic view of the types of information which, in principle, may be derived from hydrogen-exchange data. The spherical structure on the left represents an idealized macromolecule, containing n potentially exchangeable hydrogens. (Exchangeable hydrogens are defined as those attached to nitrogen, oxygen, and sulfur atoms; hydrogens attached directly to carbon are generally considered nonexchangeable.) One-half of these hydrogens (n_{out}) are in direct contact with solvent; the rest (n_{in}) are located in the "interior" of the macromolecule and as a consequence are partially or completely shielded from the solvent environment. Possible patterns for the kinetics of exchange of this idealized macromolecule are shown on the right in Fig. 1.

For the ideal case, the experimental time scale is arranged so that all the exchangeable hydrogens which are fully exposed to solvent (n_{out}) exchange with solvent hydrogens immeasurably fast (k_{out} = "∞"). The shielded exchangeable hydrogens (n_{in}) exchange more slowly and at a rate which depends on solvent access and on the rate and/or extent to which these hydrogens are exposed to the solvent by structural fluctuations. The kinetics of exchange of these hydrogens can assume a variety of forms. One extreme possibility is that these hydrogens are buried within the macromolecular structure in a way which removes them completely from contact with water. If in addition the macromolecule is totally devoid of conformational motility, the rate constant for the exchange of the n_{in} hydrogens will be zero. If these hydrogens are all partially accessible to the solvent (and accessible to the same extent) or if periodic fluctuations in the structure of the macromolecule expose all the n_{in} hydrogens to the solvent at the same rate or to the same extent, then this group will exchange with solvent as a single exchange class characterized by a single rate constant (e.g., $k_{in} = a$ or b; see Fig. 1). If various of the n_{in} hydrogens differ in their accessibility to solvent or in the rate or extent to which they are exposed by structural fluctuations, these hydrogens will not exchange as a single kinetic class. In this case the measured kinetics will consist of the sum of up to n_{in} first-order reactions (Fig. 1; $k_{in} \neq$ constant).

Both equilibrium and kinetic parameters characteristic of the macromolecule can in principle be derived from data such as those schematized in Fig. 1. The intercept with the ordinate provides a measure of the total number of slowly exchanging hydrogens per macromolecule (although the extrapolation to zero time may be difficult if exchange is rapid and the exchange curve is not linear), while the slope(s) of the curve provides a measure of the inaccessibility to solvent of the interior hydrogens and/or of the conformational motility of the structure.

In the discussion that follows we will restrict ourselves primarily to considering DNA, both because it has a well-defined secondary structure and because most of the presently available hydrogen exchange studies of nucleic acids have been made on it. In the last section of this article extension to other nucleic acid and polynucleotide containing systems will be considered.

Theory

Equilibrium Measurements

If hydrogen exchange is to serve as a direct measure of the hydrogens involved in interchain hydrogen bonding in nucleic acids, kinetic data obtained with native DNA which has been shown to be fully helical by standard criteria should extrapolate to a value representing 2 hydrogens per dA·T nucleotide pair (np) and 3 per dG·dC nucleotide pair; i.e., to 2.5 H/np for a DNA containing 50 mol % dA·T pairs. This appears to be the case under slow-exchange conditions for calf thymus and T4 DNA and for a variety of bacterial DNAs (*4, 5, 7, 8*), however, see Refs. *15, 26–28*, Note Added in Proof, and below.

Kinetic Measurements

The rate at which hydrogens located in any particular environment exchange with solvent can be written:

$$dH_i/dt = k_i H_i \tag{1}$$

where H_i is the number of hydrogens remaining in all sites of type i at time t, and k_i is the characteristic first-order rate constant for exchange of these hydrogens under a particular set of environmental conditions. For a macromolecule containing a number, n, of kinetically distinct types of sites which exchange independently, the integrated rate equation becomes

$$H_t = \sum_{i=0}^{n} H_{o,i} e^{-k_i t} \tag{2}$$

where H_i = total number of hydrogens per molecule *not* exchanged at time t, and $H_{o,i}$ and k_i are, respectively, the number of hydrogens initially present and the observed exchange rate for each kinetic class.

The various rate constants, k_i, reflect the intrinsic chemical rate of exchange of the groups involved, as well as the additional limitations to exchange imposed by local steric and other environmental conditions. Exchange of protons is subject to both general and specific acid and basic catalysis, as well as possible non-pH-dependent catalysis by water itself (*7*). For an "open" exchangeable hydrogen, i.e., one fully exposed to solvent, the observed exchange rate can therefore be divided into pH-dependent and pH-independent processes:

$$k_i = k_0 + k_{H_3O^+}\ [H_3O^+] + k_{OH^-}\ [OH^-] + \sum_i k_{HA_i{}^{\overset{+}{0}}} [HA^{\overset{+}{0}}] + \sum_i k_{A_i^{\overset{0}{-}}} [A_i^{\overset{0}{-}}] \tag{3}$$

where k_i is a pseudo first-order rate constant in units of hydrogens per mole nucleotide pair per second, k_0 represents the rate constant for proton transfer to and from DNA sites by pH-independent processes (e.g., concerted water catalysis mechanisms), $k_{H_3O^+}$ and k_{OH^-} are the second-order rate constants (in 1./mol-sec) for proton transfer to and from DNA sites by H_3O^+ and OH^-, respectively, and $k_{HA_i{}^{\overset{+}{0}}}$ and $k_{A_i^{\overset{0}{-}}}$ are the second-order rate constants for proton transfer to and from DNA sites by other proton donors (positively charged or neutral) and acceptors (neutral or negatively charged) added to the system at concentrations $[HA_i{}^{\overset{+}{0}}]$ and $[A_i^{\overset{0}{-}}]$, respectively. (For further discussion of the rates of proton transfer reactions in simple systems, see Ref. *16*).

For the amino and imino hydrogens of DNA (as well as the 2′OH ribose hydrogens of RNA), the specific acid and base catalyzed exchange of "fully open" sites (e.g., in mononucleotides) proceeds at diffusion-controlled rates. This means that the rate of the exchange reactions catalyzed by H_3O^+ and

OH^- does not depend on the specific pK values of the various exchangeable DNA hydrogens. Furthermore, the minimum rate of exchange for open sites should fall near neutrality and the pseudo first-order rate constant should be approximately 10^3 sec^{-1} (i.e., $t_{1/2} \simeq 10^{-3}$ sec) at ~ pH 7 (using 10^{10} l./mol-sec as the second-order rate constant for a diffusion-controlled reaction). For other (general) acid–base catalysts the rate of exchange will depend on the pK of both the catalyst and the various DNA sites. In small molecule model compounds k_0 has been shown to be negligible. For a detailed discussion of these matters, see Ref. *7*.

Since, as indicated above, it is observed that the interstrand hydrogens of the DNA double helix exchange 10^5 to 10^7 times more slowly than the corresponding hydrogens of free nucleotides under the same conditions, this slowing must reflect some aspects of the secondary structure and conformational motility of the native DNA helix. Detailed analysis of the exchange behavior of DNA under a variety of solvent conditions gives some insight into the nature of the conformational motility of the native DNA reflected in these results. These data have recently been summarized, and models of the dynamic structure of DNA based on these data have been proposed (*7, 8*).

METHODS

Gel Filtration Chromatography

In its general aspects the application of the gel filtration technique to hydrogen–tritium exchange in DNA has been described by Englander (*14*). Various modifications designed to speed up the exchange procedure and make it possible to achieve very early time points have been made by Hanson (*15*) and by Englander and Printz and their colleagues (manuscripts in preparation). In the following section we will outline the bases of the gel filtration technique, and then describe its application (together with necessary experimental and computational modifications) to an especially difficult exchange situation. Procedures to be used in more straightforward situations can easily be inferred by the reader.

The gel filtration procedure is designed to separate free tritiated water (THO) from DNA and involves two distinct phases: (1) An initial separation is carried out to separate the DNA from the highly tritiated solvent used to incorporate the isotope into the exchangeable sites of the macromolecule. This starts the exchange-out process. (2) A subsequent separation is then made of the additional free tritium which exchanges out of the macromolecule into the solvent on further incubation. Both of these phases of the procedure can be achieved in either a single (interrupted) chromatographic step (one-column technique), or in a two-step (two-column) system. In the latter technique several columns are used at different times for the second phase, i.e., to separate the exchanged-out tritium from aliquots of the DNA eluate of an initial column used for the first phase.

Since the column effluents in the one- and two-column procedures are analyzed for their tritium content (scintillation counting) and DNA content (absorbance at 260 nm), the column buffers should not contain components that interfere with these measurements. When such components are not present, both the one- and two-column procedures are used routinely in a single exchange experiment: The two-column procedure is used to obtain kinetic points from 150 to 2500 sec (or longer), and the one-column technique is used in the same experiment to obtain kinetic points at exchange times less than 150 sec. This routine use of both procedures compensates for the fact that the half-time for the rate of tritium loss from DNA is frequently less than the time required to obtain the first kinetic point by the two-column procedure.

It is desirable in many instances to investigate the effect on the exchange kinetics of variable concentrations of a solvent component that seriously interferes with the subsequent analysis by: (1) absorbing at 260 nm; (2) changing the extinction coefficient of the nucleic acid; or (3) acting as a quenching agent in scintillation counting. In this situation the one-column method (see Refs. *7, 8*) cannot be used, since in this procedure the solvent of interest (the contact solvent) must be the same as the column buffer. On the other hand, the versatility of the two-column technique here becomes apparent, since only the column used for the first separation (the "first" column) need be equilibrated with the contact solvent containing an interfering substance. The "second" column, from which the fractions that are actually to be analyzed for DNA and tritium are eluted, can be equilibrated with a noninterfering solvent in which exchange proceeds at minimal rates. A typical operation is described below for the situation in which the contact solvent of interest represents the most rigorous experimental condition likely to be encountered: i.e., a high salt content resulting in chromatographic problems as well as in serious quenching of counts, UV interference, and destabilization of the nucleic acid.

Exchange Procedures

Tritiated water (50 mCi) is added to a 2.5 ml solution containing 5 mg of DNA previously dialyzed at 0° against the contact solvent. If the solvent contains a component that destabilizes DNA, the solution is kept at ice-water temperature and allowed to equilibrate for enough time to ensure that the tritium/hydrogen ratio of the DNA reaches that of the solvent. One and a half hours is sufficient for DNA at 0°C.

During this period the "first" column (15 x 2 cm) previously prepared in water and cooled to 0° (as are the "second" columns) is equilibrated against the contact solvent. If the contact solvent contains large concentrations of components representing the environmental variable in the experiment (for example, 4 M sodium trichloroacetate, Ref. *8*) equilibration may have to be accomplished in stages (e.g., via an intermediate column wash with 1 or 2 M solute). Otherwise the layering of a limited amount of high density solution over a G-25 Sephadex column made up in water may lead to instability and floating of the top part of the column bed into the high density layer. Also, if the contact solvent contains large amounts of salt, the height of the first column should be about 20 rather than the usual 15 cm. Such salt solutions are often viscous, and pressure applied by a manual pressure bulb and monitored by a low-pressure gauge may be necessary to force the solvent through the column in a reasonable amount of time. With properly siliconized columns, pressures up to 4.5 psi can be tolerated without loss in separation efficiency.

Since the flow of contact solvent through the first column usually results in a dimensional change of the column bed, it is necessary to determine the void volume of the column with Blue Dextran 2000 (Pharmacia) dissolved in 2 ml of the contact solvent. To this end the liquid level is allowed to drain to the top of the column bed, flow is stopped with a pinch clamp, and a small graduated cylinder (25 ml) is placed under the column. Two milliliters of the Dextran solution is layered over the top of the resin and allowed to sink onto the column. Flow is stopped once more (under these conditions capillarity will not stop flow automatically) and contact solvent sufficient to wash the Dextran through is layered onto the bed. Flow is then restarted and pressure is applied if the solution is highly viscous. It is advantageous at this point not only to note the column pressure but also to start a clock to ensure that the Dextran (and later the DNA) will be eluted in a reasonable amount of time.

When the Blue Dextran just begins to emerge from the column, the graduated cylinder is removed and further collection is made in 1- to 1.2-ml aliquots into a series of small tubes. In high salt considerable diffusion of the Dextran band will occur and the elution of the Dextran will occupy six or more tubes, of which the ones containing the maximum Dextran concentration can be determined easily by visual inspection. The operational void volume is taken as the volume collected in the graduated cylinder. This is the volume which will be discarded in an actual run prior to beginning collection of the pooled fractions of the eluted DNA. The volume of the pool should be no more than one-half the total volume of the eluted Dextran fraction; i.e., it should include only those tubes collected prior to (plus one-half the volume of) the tube containing the maximum concentration of Dextran. If the maximum concentration is distributed into two tubes the first, plus those eluted previously, represents the desired pool volume.

It is advantageous to determine the void and pool volumes in this manner before each experiment to ensure that only half of the DNA of the total band is collected in the exchange pool. The reason is that under more rigorous column conditions of high salt viscosity, only the *first* half of the DNA emerging from the first column is separated efficiently from bulk tritium. At some convenient time later in the experiment a routine check of pool radioactivity and DNA absorbance provides a useful preview of the exchange rate and of the success of the seperation from free tritium.

Before beginning the run a number of "second" columns (12 x 2 cm) are washed with two bed volumes of a buffer which contains no components that interfere with DNA or radioactivity determinations and which serves to establish a minimal exchange rate environment for the macromolecule. For the "first" column, the level of the contact solvent is brought to the surface of the Sephadex bed and the graduated cylinder marked with the void volume is set underneath.

The experiment is started by transferring 2 ml of the tritiated DNA solution to the top of the column with a *previously chilled pipette.* Column flow is started and timing is begun as soon as the DNA has sunk into the column. Contact solvent is layered over the column bed and sufficient pressure is applied to give a reasonable flow rate.

After elution of the void volume, the pooled front one-half of the DNA peak is collected in a calibrated tube surrounded by ice water. After collection of the pool the column is clamped off, the pool tube is inverted two to three times and a 1-ml aliquot is transferred immediately to the first of the "second" columns. The additional manipulation required has been described by Englander (*14*).

In this type of experiment the second columns not only remove exchanged tritium from the DNA, but also remove the interfering components of the contact solvent. The latter separation can be checked by analyzing (by UV spectrophotometry) the collected fractions for the interfering solute, as well as for DNA and tritium.

Calculation of Exchange Data

This two-solvent modification for the two-column procedure differs from the one-solvent technique in three respects important in calculation and interpretation of the kinetic data.

1. Calculations of the Molarity of Water Hydrogens. The first consequence is in the conversion of the data (counts per minute, cpm, and A_{260}) to hydrogens per nucleotide pair (H/np). Normally, this calculation is made on the basis that the water concentration in the solution is 55.5 M, and thus that the gram-atom concentration of water hydrogens is 111 M. Thus:

$$\text{H/np} = 111\,\bar{R}\left(\frac{\epsilon}{C_0}\right) \tag{4}$$

In this expression $\bar{R}$ is the average of three or more values of $(\text{cpm})_{\text{corr}}/A_{260}$ obtained from fractions of the eluted tritium-DNA, $(\text{cpm})_{\text{corr}}$ is the measured counts per minute minus the background, and ϵ is often multiplied by 2 to express the data in terms of nucleotide pairs, since the hydrogens on hydrogen-bonded base pairs appear to be the only ones whose exchange rates are slow enough to measure by this gel filtration technique. For DNA, these extinction values are obtained by interpolation of the data provided by Felsenfeld and Hirchman (*17*). C_O is the counting rate (in counts per minute) of the original DNA–THO equilibrium mixture. In the absence of equilibrium isotope effects the equilibrium tritium/hydrogen ratio in DNA is the same as that of the solvent and is defined by $C_O/111$. However, if the contact solvent contains more than 1 M solute the data must be corrected since the water concentration is no longer 55.5 M (i.e., the solvent isotope ratio is greater than $C_O/111$).

The water concentration of the sample (and thus the molarity of solvent hydrogens) can be determined most conveniently by weighing successive 10 ml aliquots of the contact solvent transferred to a tared flask placed upon a pan balance. The average weight minus the weight of salt in the volume added to the solution provides the weight and therefore the concentration of water. Alternatively, the DNA can be tritiated in the presence of low-salt and equilibrated with the high-salt contact solvent when separating the bulk tritium in the "first" column. In this case, the tritium/hydrogen ratio of the tritiated DNA will correspond to the isotope ratio in 55.5 M water and no correction is needed.

2. "Second"-Column Rate Corrections. Since the exchange-out rate of tritium from DNA in the "second" column (minimal exchange rate) buffer has been established previously, the kinetic points can be corrected for the time of travel of the DNA through the "second" column as follows:

$$(\text{H/np})_t = k_1(t - t_2) + k_2(t_2) \tag{5}$$

where $(\text{H/np})_t$ are the hydrogens per nucleotide pair exchanged out in time t, k_1 and k_2 are the exchange rates (in units of hydrogens per nucleotide pair per second) for the contact solvent and "second" column buffer, respectively, and t_2 is the time spent by the DNA in the second column buffer. If $k_1/k_2 \leqslant 2$, the corrected and uncorrected kinetic curves are essentially superimposable since t_2 is usually about 60 sec and t for the first kinetic point in such experiments is often as much as 350 to 400 sec. For this particular situation the uncorrected curve will differ from the corrected one by less than 10% at the *first* kinetic point, which difference is just within the experimental error of the method. It is obvious that as t increases the correction becomes negligible. In general, this correction becomes important when k_1 is much larger than k_2 and t is *not* much greater than t_2. The uncorrected curves at early points will then tend to flatten out and appear somewhat more linear than the corrected curve. The making of such corrections for the slower exchange occurring in the second column requires recognition of the underlying assumption that the effect on the DNA of the contact solvent is removed instantaneously when the pool aliquot is applied to the second column. Two factors determine whether this assumption is reasonable. The first is the time actually required to exchange the contact solvent for the minimal buffer in the second column. Estimations of this time involve the same considerations as those used for calculating the zero time of exchange (the time after starting the clock when exchange-out actually begins at the start of the experiment, see Ref. *14*). Usually this time is less than 10 sec and does not amount to a serious correction at the longer exchange times associated with the two-column procedure. The second factor here is the rate at which the equilibrium is established in the interaction between the solute of the contact solvent and DNA. Since the separation time is less than 10 sec (above) this is not a serious consideration unless the forward and reverse rates of the interaction are so slow that equilibrium is not established during much of the time the DNA is in the second column. However, even this extreme situation actually represents a *decrease* in the magnitude of the second-column correction, which in any case is small.

3. Loss of Early Time Points. A third consideration in applying this two-solvent modification to DNA-hydrogen exchange is the loss of the early time points on the exchange-out curve. This comes about not only because the usual rapid one-column technique cannot be used, but also because the high viscosity of the contact solvent, coupled with the intrinsic slowness of the two-column method, results in a first kinetic point which falls at times as long as 400 sec after the initiation of exchange-out. Since the solvent of interest usually increases exchange rates, the kinetic curve obtained in such experiments may represent only a small fraction of the total exchangeable hydrogens of DNA. Therefore, curve shape analysis as described below is important in deciding whether to carry the interpretation of the observed data to the hydrogens which exchanged out at rates too fast to measure. Such analysis requires accurate data, and thus the most overriding experimental consideration is good temperature control, since the temperature coefficient of DNA-hydrogen exchange is large (*6-8*).

DNA Preparation

The two-column technique requires up to 4 mg of DNA in a 2 ml volume to ensure reliable results, although less than one-third of this amount has been successfully used. At such concentration levels the DNA is too viscous for rapid and successful gel filtration and must be sonicated to ensure proper chromatographic behavior. This treatment does not appear to affect the hydrogen exchange properties; sonication for a 5 to 6 min under low power at 20 kc/sec is sufficient to reduce the viscosity of the DNA (*4*). The sonication is best performed on 15 ml quantities of DNA (2 mg/ml in either standard or dilute saline-citrate, pH = 7) in a small water-jacketed glass vessel under an atmosphere of scrubbed nitrogen. Before sonication the nitrogen is bubbled under the surface of the solution for 5 min. The titanium sonication horn tip is placed ⅛ to ¼ in. under the solution surface. After sonication, the DNA solution is filtered (Millipore), transferred in 3-ml aliquots to dialysis bags and dialyzed against several changes each of 0.01 M NaCl, 0.01 M Na_2EDTA, pH 7 and then 0.01 M NaCl, 0.001 M Na_2EDTA, pH 7, at 5°. This treatment removes the metal contaminants injected during sonication and results in the reestablishment of normal melting behavior for the DNA. The bags can be stored indefinitely in the latter buffer and transferred to the contact solvent for dialysis when needed. Dialysis against a high-salt contact solvent generally leads to a drastic reduction of the volume of the DNA solution within the dialysis membrane. Equilibration is usually sufficient at this stage to allow quantitative transfer of the bag contents to a graduated cylinder or volumetric flask and readjustment to the original volume with the dialyzate. The dialyzate is also used as the contact solvent to equilibrate the first column (see above).

Other Methods

Nucleic acid hydrogen exchange can be studied by a variety of methods which have been devised to fit special circumstances. Some of them will be mentioned very briefly here. These methods can be divided broadly into indirect techniques (isotope exchange) and direct techniques involving spectroscopic observation of the exchanging proton. The isotope-exchange methods always involve the two separation phases, i.e., a preliminary separation of bulk tritium (or deuterium) from the macromolecule into the tritium-free solvent. With high-molecular-weight nucleic acids and ribosomes the separation is based on the enormous difference in size between the exchanging isotope and the macromolecule. This size differential is particularly important in the "fast" gel filtration methods, which utilize high-pressure flow through small column beds made of finely divided Sephadex beads of uniform particle size (e.g., see Hanson (*15, 26*: Englander, et al., *28*) and in the dialysis technique described by Englander and Crowe (*14, 18*). The slower dialysis technique can be applied to DNA if the first separation phase is performed by gel filtration chromatography, rather than by dialysis. The chief advantage of the combined gel filtration–dialysis technique is the smaller amount of nucleic acid required (about one-tenth that required for the two-column technique). A typical experiment would involve the application of 0.25 ml of the original DNA–THO equilibrium mixture (see above) to a 1 x 6 cm column of fine-grade G-25 Sephadex. The DNA eluted in a volume of 1–1.5 ml is transferred to a dialysis membrane prepared and mounted on the apparatus described by Englander and Crowe (*18*).

A novel technique that actually amplifies the size differential is the filter-trapping method, in which the tritiated nucleic acid is allowed to form a nucleic acid–polypeptide complex that can be washed free of tritium on a filter membrane (*19*). The advantages of this technique are the small amount of macromolecule required, the fact that the exchange of either macromolecule in the complex can be measured separately, and the rapidity (in principle) with which the free tritium can be separated from the complex. Sonication of the DNA to reduce viscosity is not necessary, which might be advantageous in studying the interaction of a protein with intact viral DNA. Lower-molecular-weight nucleic acids such as tRNA, which are not always completely excluded by filtration gels and dialysis membranes, might be examined by the filter technique if a nonfilterable complex can be formed with a larger component, e.g., ribosomes. The chief limiting factor of the filter technique is the rate with which the complex dissociates, relative to the time it takes to wash the complex free of tritium. Another problem which must always be considered is whether the formation of the complex itself alters the kinetic or equilibrium hydrogen-exchange properties of the macromolecules involved (see Ref. *19*).

For direct observation of the exchanging proton in nucleic acids, nuclear magnetic resonance spectrometry can provide broadened signals for the amino protons of AMP, CMP, GMP, and polyadenylate in aqueous solution (*20*). Although quantitation of signal parameters is complicated by the proximity of the water proton signal and by relaxation of the nitrogen quadrapole, these effects can be separated to allow estimates of proton lifetimes by studying signal width variations as a function of the concentration of various solvent components and environmental factors.

Among the spectroscopic techniques, Raman spectra are of some interest since aqueous solutions of nucleotides can be investigated. Although changes in line intensity and frequency accompany the titration of the endocyclic nitrogens of adenine, cytosine, guanine, and thymine, Raman spectroscopy does not provide rate data for proton exchange. However the method is useful for identifying the exchange site and for studying the extent of intermolecular hydrogen bonding of nucleotides in aqueous solution. For details see Peticolas (*21*). Infrared spectrometry can also be used for these purposes by carrying out studies in D_2O solutions.

LIMITATIONS, RESERVATIONS, AND CONCLUSIONS

The Rate of Exchange

Nucleic acid hydrogen exchange is hampered by the fact that under the slowest of exchange conditions the resolution time of the techniques used is of the same order of magnitude as the half-time for exchange. This is in marked contrast to protein hydrogen exchange, where exchange rates can be slowed to achieve half-times of hours or even days. The available evidence suggests that the rate of exchange of hydrogens involved in secondary structure in both proteins and nucleic acids is slowed to comparable extents (by factors of 10^4 to 10^7) compared to appropriate "open" small molecule model compounds. Thus it appears that the differences between the exchange rates of the two types of macromolecules reflect primarily the fact that the specific acid or base catalyzed exchange of the free nucleotide protons is diffusion-controlled, while that of peptide hydrogens is not.

Equilibrium Consequences

As indicated in the Introduction, one use to which nucleic acid hydrogen exchange measurements can be put is to determine the number of hydrogens per molecule which exchange more slowly than "free" hydrogens because of some type of involvement in secondary structure. Under ideal conditions (see Fig. 1) the free hydrogens exchange "instantaneously" (on the time scale of the method) while the structurally involved hydrogens exchange sufficiently slowly to permit reliable extrapolation of exchange-out plots to zero time. In principle, then, this method provides a measure (in nucleic acid and polynucleotide structures) of the number of hydrogens per nucleotide (or per nucleotide pair, or per molecule) involved in inter- and intrachain hydrogen bonding, and offers a useful adjunct to direct structure determination by x-ray diffraction and other means. However, problems exist:

1. What Hydrogen Bonds are Actually Measured? In DNA and polydeoxyribonucleotides the only exchangeable hydrogens are those attached to amino and imino nitrogens, and both types exchange at comparable rates in aqueous solutions. It appears that under some circumstances these two chemically distinct types might be discriminated by selective general acid-base catalysis of exchange (see Ref. 7). In RNA and polyribonucleotides, the possibility of slow exchange of the 2′-hydroxyl of the ribose ring also has to be considered. This has been a problem in transfer RNA studies, since attempts to use the number of "slow" hydrogens as a quantitative measure of double-helical structure have been hampered by the uncertainty as to whether some of the slow hydrogens might not originate from the 2′OH position of ribose. Resolution of this question is important to

the evaluation of proposed models of tRNA and polyribonucleotide secondary structure. Chemical approaches to settle this matter are currently being evolved (e.g., see Refs. *13, 22* and Note Added in Proof).

The question of the chemical types of hydrogen bonds which are actually being measured also arises in studies of natural or artificial nucleic acid–protein complexes; e.g., ribosomes (*23, 24*) or DNA–polypeptide complexes (*19*). However in some situations a clear and useful discrimination can be made on the basis of the pH dependence of the exchange rates. Indeed, under favorable circumstances it has been shown that the exchange of the protein partner in the complex can be studied near pH 4, where DNA hydrogens exchange "instantaneously," and the exchange of the nucleic acid hydrogens can be studied near pH 7, where the peptide hydrogens exchange "instantaneously" (see Ref. *19*). Thus the structural consequences of complex formation on both partners can be examined and compared with controls containing each component separately.

2. Do the "Free" (Non-Intra- or Interchain Hydrogen Bonded) Hydrogens Actually Exchange "Instantaneously"? It has been shown that the relevant hydrogens of mononucleotide model compounds exchange with diffusion-controlled rates; e.g., in aqueous solution the half-times for exchange of these protons are equal to or smaller than 10^{-3} sec (*16, 20, 25*). Furthermore it has been shown that the exchangeable hydrogens of single-stranded (stacked or unstacked) polynucleotides such as polyadenylic or polyuridylic acid at neutral pH exchange "instantaneously" on the time scale of the gel filtration hydrogen exchange method (see Refs. *6, 12, 15*). Also, as mentioned above, experiments with ordered DNA samples in which the first kinetic point is taken at 90 to 100 sec after the initiation of exchange-out show that the expected number of hydrogens are obtained when the data are extrapolated to zero time (*4, 5, 8*). On the other hand, recently data have been obtained with DNA with methods which provide kinetic points at times as early as 10 sec after the initiation of the exchange-out process (*15*). These studies seem to reveal the presence of additional hydrogens, and if these turn out to be some of the nonbonded amino hydrogens of DNA this suggests that the latter hydrogens exchange much more slowly in the environment of the grooves of native DNA than in free solution. Englander and coworkers (private communication) also find slowly exchanging hydrogens in dG-dC polymers in excess of the number expected on the basis of a simple DNA-like, ordered, double-helical structure. The situation is not yet clear, but these studies suggest that the assumption that "free" exchangeable hydrogens in macromolecules actually exchange at the model compound rate must be checked in each case with appropriate control experiments before using the number of extrapolated "slow" hydrogens per molecule as a basis for counting inter- or intramolecular hydrogen bonds and thus supporting or rejecting proposed structural models.

Kinetic Consequences

The fast rate of nucleic acid hydrogen exchange also poses problems in another sense, since the most common type of information obtained from the tritium-gel filtration procedures often is derived by a semiquantitative or qualitative comparison of kinetic curves by inspection. The chief question to be examined in this connection is whether experimental data representing as little as 10% of the total exchangeable hydrogens can be assumed to apply to the rest of the hydrogens which had exchanged out before the first kinetic point could be obtained. This problem arises even with gel filtration techniques, since the interesting experiments are often those in which increases in exchange rate are investigated which result from the addition to the DNA solution of a particular exchange catalyst, salt, or physically interacting substance. However, these problems can often be handled since DNA provides a relatively simple macromolecular model for hydrogen exchange (as compared to proteins) and because the data can be accurate enough to place curve inspection upon a quantitative basis.

To assess the effect of a change in solvent environment on the exchange kinetics, a comparative inspection of kinetic curves involves first a determination of whether the environmental change affects the observed kinetic classes differently, or whether it affects all classes equally. This requires an analysis of the curve shapes, which can fall into two extreme groups:

1. A series of kinetic curves plotted on the same graph will appear to be displaced in a parallel fashion to indicate that the solvent variable has induced an abrupt shift of slow hydrogens into an "instantaneous" class, leaving the remainder of the hydrogens unaltered in exchange-out behavior. This can be tested experimentally by demonstrating that the *difference* (in units of H/np) between two kinetic curves (measured along a series of vertical lines drawn between the curves on a plot with hydrogens per nucleotide pair remaining at time t plotted on the ordinate and exchange-out times plotted on the abscissa) per unit H/np remaining in either curve at all points is a constant. Thus the fraction of total hydrogens lost to an "instantaneous" class, as in the partial denaturation of DNA by destabilizing salts (*8*), can be assessed without requiring extrapolation of either kinetic curve to zero time.

2. A series of kinetic curves compared on the same graph appear to "fan out," indicating that the solvent variable has affected all exchange classes equally. Here the ratio of exchange-out times corresponding to any particular value of H/np for each curve will remain constant at all points on the curve. In other words, a series of *horizontal* lines drawn through two exchange-out curves intersects each curve at times on the abscissa which show a constant ratio for all values of H/np (e.g., see Table I, Ref. *7*).

This particular analysis, which tests whether or not all measured kinetic classes are affected equally, is important in two respects. First, a positive test means that the kinetic curves can be compared on a rate basis, since a corollary to underlying assumptions that permit such an analysis is that in comparing the slopes of tangents drawn at the same value of H/np for any two curves, one is observing the exchange of the same protons. Secondly, a similarity in curve shape is suggestive (but certainly not conclusive) evidence for the fact that the kinetic curves originated at the same zero-time point. A positive test, therefore, is the first requirement in deciding whether to extend interpretation of the experimental part of the curve to the majority of hydrogens which have exchange rates too fast to measure. The seriousness of the assumptions involved in such extension is decreased if, as is often the case, experimental points for most of the exchangeable hydrogens can be measured on the curve in the series which represents the least added catalyst or other environmental perturbant (*7*).

Interpretations of Exchange Kinetics

The ultimate aim of kinetic studies of macromolecular hydrogen exchange is to provide insight into the structure and conformational dynamics of macromolecules under a variety of conditions. The various reservations listed above suggest that this method, like other physicochemical techniques of macromolecular structure analysis, is most powerful when used in conjunction with other approaches, after careful calibration with molecules of known structure and predictable dynamic modes (see Refs. *7* and *8* as examples of the types of considerations which must be entertained).

NOTE ADDED IN PROOF

Since this paper went to press additional studies have appeared which make it possible to provide more definitive statements about certain points raised above. Hanson (*26*) has extended his earlier observations with rapid hydrogen-exchange techniques and shown definitively that various DNAs and synthetic polynucleotides contain more slowly exchanging hydrogens than can be accounted for solely by those involved in interchain hydrogen bonding. In all cases he finds that the total number of hydrogens measured at short exchange-out times is close to the total present on the nucleotides, i.e., 3 per dA·T base pair and 5 per dG·dC base pair. Englander and von Hippel (*27*) have extended these findings by carrying out a very careful hydrogen-exchange analysis of native calf thymus DNA down to exchange times as short as 7 sec, and have confirmed Hanson's conclusion that on this time scale 5 hydrogens are measured per dG·dC pair and 3 per dA·T pair. Furthermore, Englander and von Hippel have shown that under "minimal exchange rate conditions" the native DNA exchange-out curve can be broken down into two exchange classes: one corresponding numerically to the "outside" hydrogens attached to the amino groups of adenine, cytosine, and guanine; the other corresponding to the "interior" internucleotide hydrogen bonded hydrogens. The group which corresponds numerically to the "exterior" hydrogens is characterized by a half-time of exchange about 15-fold greater than that corresponding to the "interior" hydrogens. According to this analysis all the "exterior" hydrogens have completed exchange within about 100 sec, in keeping with the earlier findings of Printz and von Hippel (*4, 6*) that exchange-out data obtained at 100 sec or more after exchange is initiated extrapolate very close to the number expected for internucleotide hydrogen-bonded hydrogens. However this latter finding (i.e., that the exchange-out curves for DNA – and perhaps for the DNA-like synthetic polynucleotide complex poly dA·T (*26*)) which can be broken down into two kinetic classes of hydrogens, one corresponding roughly to the "exterior" hydrogens and the other the "interior" hydrogens, is *not* general for nucleic acids of different conformation. Data obtained with poly dG·dC, poly rG·rC, and transfer RNA (*13, 28*) show that the exchange-out curves for these materials extrapolate in essentially one continuous curve up to values which correspond to 5 slowly exchanging hydrogens per G·C pair and 3 slowly exchanging hydrogens per A·T (or A·U) pair. These studies of Englander et al. on tRNA and on poly rG·rC also prove that the 2′–OH groups of polyribonucleic acids do not exchange slowly in any of these structures.

The above data thus demonstrate that in all double helical polynucleotide structures *all* the hydrogens attached to the purine and pyrimidine bases exchange sufficiently slowly to be measured by gel filtration experiments if experimental points can be obtained at times down to about 10 sec, thus indicating that the exchange of *all* these hydrogens is slowed as a consequence of involvement in double helical, hydrogen-bonded structure. Possible explanations for this slowing of the "exterior" as well as the "interior" hydrogens are considered elsewhere (*26–28*). The practical general consequences of this appears to be that one can most reliably count the number of *base pairs* per molecule involved in the double-helical structure in any particular nucleic acid or synthetic polynucleotide complex by obtaining exchange-out data at times as short as 10 sec after exchange out has been initiated, and then attributing 5 hydrogens to each G-C pair and 3 hydrogens to each A·T or A·U pair.

Acknowledgments

The authors are grateful to Dr. S. Walter Englander and Dr. Carl Hanson for letting us read the manuscripts cited above prior to publication.

References

1. Hvidt, A. and S. O. Nielsen. 1966. *Advan. Protein Chem.*, **21**: 288.
2. Harrington, W. F., R. Josephs, and D. M. Segal. 1966. *Ann. Rev. Biochem.*, **35**: 599.

3. Englander, S. W. 1967. In G. Fasman (Ed.), Biological macromolecules. Dekker, New York. Vol I, p. 339.
4. Printz, M. P. and P. H. von Hippel. 1965. *Proc. Natl. Acad. Sci., U.S.*, **53**: 363.
5. von Hippel, P. H. and M. P. Printz. 1965. *Federation Proc.*, **24**: 1458.
6. Printz, M. P. and P. H. von Hippel. 1968. *Biochemistry*, **7**: 3194.
7. McConnell, B. and P. H. von Hippel. 1970. *J. Mol. Biol.*, **50**, 297.
8. McConnell, B. and P. H. von Hippel. 1970. *J. Mol. Biol.*, **50**, 317.
9. Englander, S. W. 1963. *Biochemistry*, **2**: 798.
10. Englander, S. W. and J. J. Englander. 1965. *Proc. Natl. Acad. Sci., U.S.*, **53**: 370.
11. Gantt, R. R., S. W. Englander, and M. V. Simpson. 1969. *Biochemistry*, **8**: 475.
12. Printz, M. P. 1970. *Biochemistry*, 9: 3077.
13. Englander, J. J., N. Kallenbach, and S. W. Englander. 1970. *Federation Proc.* **29**, Abstr. 2400.
14. Englander, S. W. 1968. In L. Grossman and K. Moldave (Eds.), Methods in enzymology, Vol. XII. Academic Press, New York, p. 379.
15. Hanson, C. V. 1969. *Anal. Biochem.*, **32**: 303.
16. Eigen, M. 1964. *Angew. Chem.*, **3**, 1.
17. Felsenfeld, G. and S. Z. Hirschman. 1965. *J. Mol. Biol.*, **13**, 407.
18. Englander, S. W. and D. Crowe. 1965. *Anal. Biochem.*, **12**: 579.
19. Lees, C. W. and P. H. von Hippel. 1968. *Biochemistry*, **7**: 2480.
20. McConnell, B., M. Raszka, and M. Mandel. 1971. *Biophys. Soc. Abstr.*, WPM-M9.
21. Peticolas, W., this volume, p. 94.
22. Rosenfeld, A., C. L. Stevens, and M. P. Printz. 1970. *Biophys. Soc. Abstr.* WPM-D6.
23. Page, L. A., S. W. Englander, and M. V. Simpson. 1967. *Biochemistry*, **6**: 968.
24. Sherman, M. I. and M. V. Simpson. 1969. *Proc. Natl. Acad. Sci, U.S.*, **64**: 1388.
25. Marshall, T. H. and E. Grunwald. 1969. *J. Am. Chem. Soc.*, **91**: 4541.
26. Hanson, C. V. 1971. *J. Mol. Biol.*, in press.
27. Englander, J. J., and P. H. von Hippel. 1971. *J. Mol Biol.*, in press.
28. Englander, J. J., N. Kallenbach and S. W. Englander. *J. Mol. Biol.*, in press.

USE OF INTERCALATING DYES IN THE STUDY OF SUPERHELICAL DNAs

JAMES C. WANG
CHEMISTRY DEPARTMENT
UNIVERSITY OF CALIFORNIA
BERKELEY, CALIFORNIA

Dyes which bind to DNA by intercalation produce a change in the helical rotation of the DNA as a result of the insertion of the dye between two base pairs (*4, 16, 17*). In the case of covalently bonded circular DNA, the binding of a significant amount of dye will result in changes in the superhelicity of the DNA that can be readily detected by hydrodynamic methods (*1, 3, 8*). In this chapter we describe the use of these dye-DNA complexes to investigate the properties of superhelical DNA.

PREPARATION OF SUPERHELICAL DNAs CONTAINING VARIABLE NUMBER OF SUPERHELICAL TURNS (*1*)

A number of phage DNAs with cohesive ends can be conveniently used as the starting material. We have used 186 DNA (19 megadalton), λb2b5c DNA (24 megadalton), λ DNA (31 megadalton), and N 1 DNA (31 megadalton). In the procedures described below, λb2b5c DNA is chosen as an example.

Phage λb2b5c grown on *E. coli* W 3110, W 3550, or 1100 is purified by two cycles of differential centrifugation and then banded in CsCl (density 1.45 gm/cm^3 containing 0.01 M tris buffer, pH 7.5, and 0.01 M $MgSO_4$) at 35,000 rpm and 20°C for at least 24 hr (while layering of the phage suspension on a preformed CsCl gradient is time saving, we found that sometimes the phage band collected in this way contained appreciable amounts of UV absorbing impurities). The phage band is collected through a hole punctured on the side of the centrifuge tube right under the band. After dialyzing against 0.01 M tris, pH 7.5, 0.01 M $MgSO_4$, the phage is extracted three times with phenol (freshly saturated with 0.01 M tris, pH 7.5). The phenol used should have been distilled under nitrogen or argon and stored at −20°C. The aqueous layer is dialyzed exhaustively against 2 M NaCl, 0.01 M Na_3 EDTA, pH 7.6. The DNA so prepared contains no detectable single-chain scissions and can be stored at 4°C for a few years.

To convert the DNA to the hydrogen bonded cyclic form, the DNA solution is diluted to an A_{260} of 0.1–0.3 with 2 M NaCl, 0.01 M Na_3 EDTA, disaggregated by heating at 75°C for 5 min, cooled in an ice bath, and then annealed at 50°C for 2 hr. After these treatments the DNA contains mostly cyclic monomers, but a significant amount of cyclic dimer (~10% by mass if the solution is annealed at an A_{260} of 0.1) is also present. If the dimeric species is objectionable, the annealing should be done at a lower concentration. The DNA is concentrated (by dialysis against dry sucrose, for example) to an A_{260} of ~1, and then dialyzed exhaustively against a medium containing 0.01 M tris, pH 8, 0.002 M $MgCl_2$ and 0.001 M Na_3 EDTA. [Hydrogen bonded cyclic λ DNA without Mg^{2+} (~10^{-3} *M*) or a moderate concentration of Na^+ (~ 0.1 *M*) is very sensitive to hydrodynamic shear. Dialysis into media lacking counterions at appropriate concentrations should be avoided.]

A calculated amount of an ethidium bromide stock solution is then added and the solution is gently mixed. In the enzyme reaction medium, the binding of ethidium by DNA is strong (*1, 2*). Therefore, to a good approximation the bound dye concentration equals the total dye concentration. Let ν be the number of dye molecules bound per DNA phosphate, N the number of nucleotide per DNA molecule, and ϕ the winding angle in radians of the DNA helix per dye molecule bound. If the DNA is converted to the covalently closed form by ligase in the presence of this amount of dye, and the dye molecules are subsequently removed, the number of superhelical turns τ introduced owing to the removal of dye is

$$\tau = \frac{N\nu\phi}{2\pi}$$

The superhelical density σ, defined as the number of superhelical turns per ten base pairs (*3*) is

$$\sigma = \frac{\tau}{N/20} = \frac{10\nu\phi}{\pi}$$

Using the model building value of $\phi = -\pi/15$ (*4*), $\sigma = -\nu/1.5$. Therefore if σ is the desired superhelical density (in the enzyme reaction medium), the number of moles of ethidium needed, E, is

$$E \approx \nu P = 1.5 \mid \sigma \mid P$$

where P is the number of moles of DNA in nucleotides. This is equivalent to ~90|σ|μg of the dye per OD unit of DNA (1 ml of a solution with an A_{260} of 1). Exposure to light should be minimized after the addition of dye. Stock solutions of DPN and bovine plasma albumin are then added successively to final concentrations of 3 μg/ml and 50 μg/ml, respectively.

An appropriate amount of *E. coli* polynucleotide ligase (*5*) predetermined by trials on small samples, is then added. Band sedimentation in an alkaline medium is a convenient method to determine the percent of covalently closed DNA (*6*). Usually ca. 50% by mass of the DNA can be converted to the covalently closed form if $-\sigma$ is less than 0.06. The yield decreases beyond $-\sigma \approx 0.06$. The DEAE fraction of ligase (*5*) is quite satisfactory for this preparation procedure, and is stable when stored in 60% v/v glycerol at −20°C. After the addition of the enzyme and gentle mixing, the solution is *immediately* incubated at 30°C for ~30 min. The reaction is stopped by the addition of 0.2 M Na_3 EDTA to a final concentration of 0.01 M in EDTA.

The solution is then weighed. To each gram of the solution, 1.04 gm of solid CsCl and 13 μl of a 10 mg/ml ethidium bromide solution are added. The solution is then spun at 35,000 rpm and 20°C for 48-60 hr in a SW 39 or SW 50 rotor. At the end of the banding two bands should be visible (*7*). The denser band, which is the superhelical DNA, is collected and immediately extracted 4 times with *n*-butanol at 0°C, each time with approximately equal volumes of the two solvents. The aqueous phase is then dialyzed exhaustively against the desired medium. If the sample is to be stored, we recommend 2 M NaCl, 0.01 M Na_3EDTA, pH 7.6, as the storage medium.

The procedures described above can also be used when a covalently closed DNA duplex, such as ϕX 174 RF I, is used as the starting material. In this case the DNA is first treated briefly with pancreatic DNase to introduce a few single-chain scissions into each molecule. The DNase is then heat-inactivated, and the DNA sample processed as described.

DETERMINATION OF THE SUPERHELICAL DENSITY OF A SUPERHELICAL DNA

Hydrodynamic Measurements

Band Sedimentation (*9*)

A convenient bulk sedimentation medium is 3 M CsCl, 0.01 M Na_3EDTA. Varying amounts of an ethidium bromide stock in 3 M CsCl, 0.01

M Na_3EDTA are added to the medium for different runs (*1*). Thirty-millimeter type III centerpieces are recommended. If a photoelectric scanner is used, double sector cells are advantageous. The sedimentation coefficients are then plotted as a function of the total dye concentration in the bulk sedimentation medium. Since the amount of DNA used in each run is small, the total dye concentration is also the free dye concentration in equilibrium with the DNA-dye complex. At the point where the DNA-dye complex contains no superhelical turn, the sedimentation coefficient is a minimum and should be very close to that of the same DNA with at least one single-chain scission, measured at the same free dye concentration. The free dye concentration c_f at this point is designated c_f^0. The superhelical density is calculated from $-\nu(c_f^0)/1.5$. The functional dependence of ν on c_f, $\nu(c_f)$, is termed "the binding curve." $\nu(c_f^0)$ is the value of ν at $c_f = c_f^0$. Since at $c_f = c_f^0$ the affinity for the dye is the same for the covalently closed DNA (which contains no superhelical turn at this dye concentration) and the same DNA with at least one single-chain scission, it is sufficient to determine the binding curve for the latter DNA, which is usually more available. For the dye ethidium in 3 M CsCl at 20°C, the binding curve in the range of interest here is satisfactorily represented by the equation

$$\frac{\nu}{c_f} = 2.66 \times 10^4 \ (0.23 - \nu)$$

The band sedimentation method has the advantage that very little material is needed (a few μg), and that the method is insensitive to nicked DNA which might be present. With a four-cell rotor, the superhelical density of a DNA can be determined in about a day. It has the further advantage that if a sample contains species different in superhelical densities, some quantitative information such as the relative quantity of each species and its superhelical density, can usually be obtained by analyzing the sedimentation patterns at different dye concentrations.

Boundary Sedimentation (*3, 8*)

If it is desirable to measure the superhelical density in a low-salt medium, the band sedimentation method is inadequate. While replacing the high-salt sedimentation medium by a low-salt medium with D_2O as the solvent is plausible for the band sedimentation method, the free dye concentration in a low-salt medium is quite low and therefore difficult to be maintained at a constant level. Thus boundary sedimentation runs are more convenient for low-salt media.

Usually DNA solutions containing ~ 15 μg/ml of DNA and varying amounts of ethidium in the desired salt medium are prepared and the sedimentation coefficients of the DNA-dye complexes are measured as a function of the total dye concentration c_T. At the minimum of this curve, the sedimentation coefficient should be very close to that of the same DNA with at least one single-chain scission. The total dye concentration at this point is designated $c_T{}^0$. The number of dye molecules bound per nucleotide, ν^0, is calculated by solving the simultaneous equations $c_T{}^0 = c_f{}^0 + \nu^0 [\mathrm{P}]$ and $\nu^0 = \nu(c_f{}^0)$, where [P] is the concentration of DNA nucleotides. If the binding is strong, $c_f{}^0$ is negligible comparing with $\nu^0 [\mathrm{P}]$, and $\nu^0 \approx c_T{}^0/[\mathrm{P}]$.

The boundary sedimentation method has the disadvantage that considerably more material is needed. This is partly alleviated by keeping the cells at 4°C after each run for several hours, and resuspending the DNA-dye complex by occasional gentle shaking. A known amount of dye solution can be added to each solution in the centrifuge cell via a microsyringe. The sedimentation coefficient of this sample can then be measured again.

At low-DNA concentration and low-salt concentration, boundary instability could be a serious problem. With an unblackened rotor, it is helpful to leave the temperature regulating unit of the Model E ultracentrifuge on **Indicate** during the run, so that convection currents owing to heating are avoided (*10*). The temperature drift during a run is small. Since in a low ionic strength medium the binding curve for ethidium is insensitive to temperature (*2*), no significant error is introduced owing to a small drift in temperature.

It should be pointed out that in the calculation of σ from $\sigma = 10\phi\nu/\pi$, ϕ is the *average* winding angle per bound dye. It is unknown at present whether ϕ is a function of the salt concentration if the only mode of interaction is intercalation. If an appreciable amount of "outside" binding is occurring at low ionic strength, it is plausible that ϕ is itself salt dependent (*19*).

Viscometry

Corresponding to the minimum of the sedimentation coefficient of a covalently closed double-stranded DNA at zero superhelical density, the relative viscosity is a maximum. Therefore if an analytical ultracentrifuge is not available, a viscometer such as the Cartesian-diver viscometer can be used (*11*). A capillary viscometer especially designed for the measurement of superhelical density has been reported recently (*18*). Viscometry has the disadvantage, however, that if the solution contains several DNA species, only the average property is measured. This is a severe drawback if molecules of different superhelical densities are present, or if a significant number of single-chain scissions are introduced during a series of measurements.

Binding Curve Measurements

By determining the value of ν^0 at which the free dye concentration $c_f{}^0$ in equilibrium with the DNA-dye complex is the same for a superhelical DNA and its nicked form, the degree of superhelicity is readily calculated (*1, 3*). The methods commonly used for the determination of the dye-binding curves include absorption spectroscopy, fluorescence spectroscopy, equlibrium dialysis, and density-gradient centrifugation. If the DNA sample contains species of different superhelical densities, then perhaps with the exception of the last method, these methods give a certain complicated average superhelical density for the sample. Therefore whenever heterogeneity in superhelical density is suspected, the band or boundary sedimentation methods are the methods of choice. In principle at least, the density gradient centrifugation method is capable of yielding information about the dispersion in superhelical density for such a sample from an analysis of the shape of the buoyant band in a dye–CsCl gradient. However, this has yet to be done.

If it is known that the superhelical DNA has a sharp distribution in superhelical density, then a comparison of the dye-binding curves of the superhelical DNA and the same DNA with at least one single-chain scission per molecule not only gives the superhelical density of the former, but also yields useful information such as the free energy of superhelicity (*12*).

In the following discussion, only the spectroscopic methods are described. Equilibrium dialysis procedures used for studying dye-binding by nucleic acids do not differ from those used for studying other small molecule–macromolecule interactions and therefore will not be described here. The density gradient centrifugation procedures have already been described in detail in the original works (*3, 12*).

UV Absorption (*2*)

While absorption measurements in the visible region are commonly used in the studying of dye binding, it is frequently impractical for superhelical DNAs owing to the limited amount of DNA available. Therefore absorption measurements in the UV become advantageous because of the higher molar absorptivity of the dye in the UV. A high-salt medium is recommended. In a low-salt medium, the free dye concentration is low and accurate results are difficult to achieve. There is one difficulty for a high-salt medium, however. The molar absorptivity of the bound dye, ϵ_b, cannot be measured directly. This is because both the DNA and the dye absorb in the UV. In order to approach quantitative binding of the dye by DNA, a high concentration of DNA is required and therefore measurements in the UV for such a sample cannot be performed. To evaluate ϵ_b indirectly, it is assumed that the binding curve for the nicked DNA is known. Even if the nicked DNA is also only available in small quantity, a readily available DNA of a base composition close to this DNA can be used in obtaining the binding curve by absorption spectroscopy in the visible for example (*13*).

A carefully measured quantity (1–2 ml) of the nicked DNA with an

A_{260} of ~1 is placed in a UV cuvette with a stopper. It is usually more accurate if the weight of the solution is measured and its volume is then calculated from the weight and the density of the solution. A Teflon coated "flea" is placed in the cuvette so that the solution can be stirred on a magnetic stirrer. The spectrum of the solution in the UV is taken in a spectrophotometer with a thermostatted cell holder. Aliquots of an ethidium (or other dye) solution in the same salt medium is added to the DNA solution. Five microliters of a 5×10^{-4} M solution is adequate each time. An automatic burette with a push-button control, such as the Radiometer ABU-1, or a microburette of the Sanz type (Misco) is convenient for the dye titration. After each addition of dye, the solution is stirred for several minutes and then kept in the cell holder. When thermal equilibrium has been reached, the absorbance at 287 nm (or the complete spectrum if desired) is measured.

The binding curve $\nu = \nu(c_f)$ is obtained from the equations

$$\nu = \frac{A[1 + (v/V^0)] - A^0_{\mathrm{DNA}} - \epsilon_f c^0_T v/V^0}{(\epsilon_b - \epsilon_f)[\mathrm{P}]^0}$$

and

$$c_f = \frac{c^0_T v - V^0 [\mathrm{P}]^0 \nu}{(V^0 + v)}$$

where A is the absorbance observed, $A^0{}_{\mathrm{DNA}}$ is the absorbance of the original DNA solution before any dye is added, $[\mathrm{P}]^0$ is the concentration of the original DNA solution in nucleotides, $c^0{}_T$ is the concentration of ethidium in the ethidium bromide titrant v is the volume of the ethidium bromide titrant added, V^0 is the volume of the original DNA solution, ϵ_f is the molar absorptivity of the free dye, and ϵ_b is the molar absorptivity of the bound dye.

Since the binding curve for the nicked DNA is known, we can treat ϵ_b as an adjustable parameter until the best fit is obtained.

The procedure is then repeated for the superhelical DNA, using the ϵ_b value obtained in the previous titration. During the binding measurements for the superhelical DNA, unnecessary exposure of the DNA solution to light, either inside or outside the spectrophotometer, should be avoided. Otherwise an appreciable fraction of the DNA may be converted to the nicked form by photochemical reactions.

The crossover of the binding curves of the nicked and the superhelical DNA gives ν^0 and therefore σ for the superhelical DNA. The free energy of superhelicity can be obtained by analyzing the binding curves (*12*).

If the superhelical density is the only quantity desired, it is not necessary to obtain the binding curves. Identical volumes of the superhelical and the nicked DNA of the same concentration are placed in two matched cuvettes and the difference in absorbance at 287 nm is taken after each addition of *equal* quantities of ethidium solution to the cuvettes. The difference in absorbance is small (~0.01 OD) and accurate measurements are important. The difference in absorbance, ΔA, is plotted against ν, the total volume of the ethidium stock solution added to each cuvette. The point $\nu = \nu^0$ at which $\Delta A = 0$ is located from such a plot. The free dye concentration c_f^0, and the number of dye bound per nucleotide ν^0 can be calculated from the equations

$$c_f^0 = \frac{1}{V^0 + \nu^0} (\nu^0 c_T^0 - \nu^0 V^0 [P]^0)$$

and

$$\nu^0 = \nu(c_f^0)$$

where V^0, c_T, $[P]^0$ are defined as before, and $\nu(c_f)$ is the binding curve for the nicked DNA.

Fluorescence (*2*)

When ethidium is bound to a double-stranded DNA helix, there is a large increase in the fluorescence intensity. Therefore binding measurements can be carried out in a fluorimeter in the same way as described for absorption measurements. An emission wavelength of 590 nm and an excitation wavelength of 546 or 365 nm are used. The fluorimetric method has the advantage of a higher sensitivity compared with UV absorption. A DNA solution of a concentration as low as 5 μg/ml may be used. Furthermore, contrary to the UV absorption method, the fluorescence intensity of bound ethidium can be measured directly, since DNA itself does not fluoresce at the excitation wavelength.

If the superhelical density is the only quantity desired, measurements are carried out in a manner identical to the procedure described in the UV absorption method. The quantity ν^0 is readily obtained when both the nicked and the superhelical samples have equal fluorescence intensities (measured at the same time so that any change in the intensity of the excitation light will not affect the result).

If the binding curves are desired, it is necessary to make corrections for the absorption of the excitation light by the solution, to monitor the constancy of the intensity of the excitation light, and to check the linearity of the dependence of the fluorescence intensity on the concentration of both free and bound dye. These are described in the work by LePecq and Paoletti (*2*).

Determination of the Maximal ν

In the presence of excess ethidium, the number of ethidium molecules bound per nucleotide approaches a limiting value ν_{max}. ν_{max} for a covalently closed DNA duplex is in general lower than ν_{max} of the same DNA with at least one single-chain scission. Furthermore, for a closed DNA ν_{max} is related to the superhelical density of the DNA in the absence of the dye. Therefore, while in the previous section the superhelical density σ is calculated from ν^0, it is also possible to obtain σ from a measurement of ν_{max}. This was done by Gray et al. in their buoyant-density method (*14*). Paoletti et al. also measured ν_{max} by fluorimetry for a number of superhelical DNAs (*15*). The procedures for such measurements are described in detail in the original works.

DETERMINATION OF THE RELATIVE WINDING ANGLES OF INTERCALATING DYES

For a given superhelical DNA, if measurements are made of the ν^0 values for different intercalating dyes (or more generally, substances which change the winding angle of the DNA helix), the relative winding angles of these substances can be calculated readily (*19*). The procedure described below is convenient if the binding of a certain substance is reversible and does not inhibit the ligase reaction. Actinomycin D is used as an example (*20*).

Cyclized λb2b5c DNA (~40 μg/ml) and a calculated amount of actinomycin D (to give 0.01–0.03 actinomycin molecules per DNA nucleotide) are mixed in the ligase reaction medium without ligase, and equilibrated at 25°C for 1 hr in the dark. ν_{AD}, the number of actinomycin bound per nucleotide, is calculated from the known binding parameters. The binding is fairly strong in the ligase reaction medium, and ~90% of actinomycin is bound. Ligase is then added and the mixture is incubated at 25°C for ~1 hr. At the end of the incubation, EDTA is added to stop the reaction as described previously. The mixture is phenol-extracted and then ether-extracted to remove all the actinomycin present. The covalently closed form of the DNA is isolated as described previously, and its superhelical density is determined with ethidium as described. Let ν^0 be the number of ethidium bound per nucleotide in 3 M CsCl at 20°C so that the DNA contains no superhelical turns. If ϕ_{AD} and ϕ are the winding angles (in radians) of actinomycin and ethidium, respectively, it is easy to show that

$$\phi_{AD}\nu_{AD} = \phi\nu^0 + 0.0016$$

The last term in the equation is due to a change of medium from the ligase reaction medium to 3 *M* CsCl, and a change in temperature from 25°C to 20°C (*1*). For three samples with $\nu_{AD} = 0.01$, 0.02, and 0.04, we obtained $\phi_{AD}/\phi = 1.0$ in each case. This procedure has the advantage that the substance under investigation is present only during the ligase reaction and introduces no complication during subsequent measurements of the superhelical density of the DNA.

References

1. Wang, J. C. 1969. *J. Mol. Biol.*, **43**: 25, 263.
2. LePecq, J-B. and C. Paoletti 1967. *J. Mol. Biol.*, **27**: 87.
3. Bauer, W. and J. Vinograd. 1968. *J. Mol. Biol.*, **33**: 141.
4. Fuller, W. and M. J. Waring. 1964. *Ber. Bunsen*, **68**: 805.
5. Olivera, B. M. and I. R. Lehman. 1967. *Proc. Natl. Acad. Sci., U.S.*, **57**: 1426.
6. Vinograd, J. and J. Lebowitz. 1966. *J. Gen. Physiol.*, **49**: 103.
7. Radloff, R., W. Bauer, and J. Vinograd. 1967. *Proc. Natl. Acad. Sci., U.S.*, **57**: 1514.
8. Crawford, L. V. and M. J. Waring. 1967. *J. Mol. Biol.*, **25**: 23.
9. Vinograd, J., R. Bruner, R. Kent, and J. Weigle. 1963. *Proc. Natl. Acad. Sci., U.S.*, **49**: 902.
10. Studier, F. W. 1965. *J. Mol. Biol.*, **11**: 373.
11. Gill, S. J. and D. S. Thompson. 1967. *Proc. Natl. Acad. Sci., U.S.*, **57**: 562.
12. Bauer, W. and J. Vinograd. 1970. *J. Mol. Biol.*, **47**: 419; **54**: 281.
13. Waring, M. J. 1965. *J. Mol. Biol.*, **13**: 269.
14. Gray, H. B., W. B. Upholt, and J. Vinograd. 1971. *J. Mol. Biol.*, in press.
15. Paoletti, C., J. B. LePecq, and I. R. Lehman. 1971. *J. Mol. Biol.*, **55**: 75.
16. Lerman, L. S. 1961. *J. Mol. Biol.*, **3**: 18.
17. Neville, D. M., Jr. and D. R. Davies. 1966. *J. Mol. Biol.*, **17**: 57.
18. Révet, B. M. J., M. Schmir and J. Vinograd. 1971. *Nature*, **229**: 10.
19. Waring, M. 1970. *J. Mol. Biol.*, **54**: 247.
20. Wang, J. C. 1971. *Biochim. Biophys. Acta*, **232**: 246.

CALORIMETRY OF NUCLEIC ACIDS

PHILIP D. ROSS

LABORATORY OF MOLECULAR BIOLOGY

NATIONAL INSTITUTE OF ARTHRITIS AND METABOLIC DISEASES

NATIONAL INSTITUTES OF HEALTH

BETHESDA, MARYLAND

INTRODUCTION

Calorimetric studies have given a general idea of the magnitude of the energies involved in maintaining the helical two- and three-stranded structures occurring in natural and synthetic polynucleotides. From consideration of the results of these energy measurements, certain important conclusions concerning the nature of the forces stabilizing ordered nucleic acid structures have been drawn. Additionally, a knowledge of the thermodynamic quantities characterizing polynucleotides is necessary for the application of various statistical mechanical theories of the melting of the polynucleotide helix.

CALORIMETRIC METHOD

Calorimetry is the art and science of the measurement of energy change. Since heat, by its nature, is not easily confined within the boundaries of the vessel in which a thermal phenomenon is taking place, its measurement presents some difficulties. Solutions containing nucleic acids are generally quite dilute in the macromolecular component, and as a consequence small quantities of heat are associated with transformations involving such minute amounts of material. Very sensitive calorimetric equipment is needed for the measurement of these small quantities of energy.

Two main types of calorimetric techniques have been employed in the investigations of nucleic acids. In one kind of measurement the heat of reaction accompanying mixing is studied. In the other method the heat of transition necessary for the melting of an ordered structure is determined directly.

Heat of Reaction Calorimetry

In this type of experiment the change of enthalpy accompanying the mixing of two reacting components is measured. For example, the thermal effects accompanying the acid denaturation of DNA may be observed by titration of DNA with acid in the calorimetric vessel (Fig. 1). Alternatively, the formation of the helical poly (rA·rU) complex may be studied by measurement of the heat liberated upon mixing poly rA and poly rU. The calorimetric method is extremely general and nonspecific in that heat from all of the thermal processes accompanying the mixing of the components contributes to the resultant heat which is observed. The appropriate heat of reaction can be deduced from measurements of this type only after proper attention has been given to the individual contribution to the measured heat from each likely source. These may include the heats of dilution of both the salt and polymer components and the heat of ionization or of protonation of

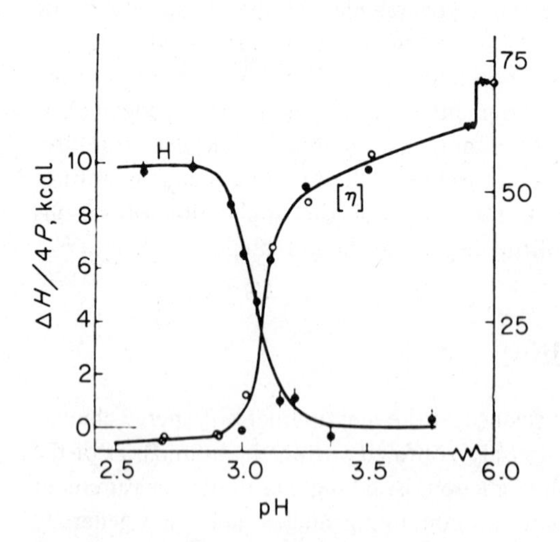

Figure 1. Enthalpy and viscosity changes accompanying the acidification of salmon sperm DNA at 25°C and 0.1 M ionic strength (*11*).

either the substrate or buffer components. The magnitude of these thermal effects invariably must be determined through separate experiments. The importance of such corrections has been emphasized in a review by Sturtevant (*1*). The measurement of the heat of reaction is essentially carried out at a constant temperature. The method is capable of the highest accuracy of better than 1%. High precision is an essential precondition for obtaining significant differences between similar energies characteristic of a general type of reaction, as for example, in the determination of the effect of the variation of chemical structure of one component on the energetics of formation of a helical complex.

Thermal Transition Calorimetry

The interest in making a direct determination of the energy absorbed in the melting of an ordered polynucleotide structure has led to methods based on the measurement of heat capacity as the second major type of calorimetric technique employed in the study of nucleic acids. The helix-to-coil transformation is associated with an abrupt change in the heat capacity taking place within the narrow temperature range over which the transition occurs (Fig. 2a). The change in the heat capacity arising from the disordering of the macromolecular component is evaluated as a function of temperature and the energy absorbed in the transition is obtained from integration of the curve of the heat capacity anomaly vs. temperature.

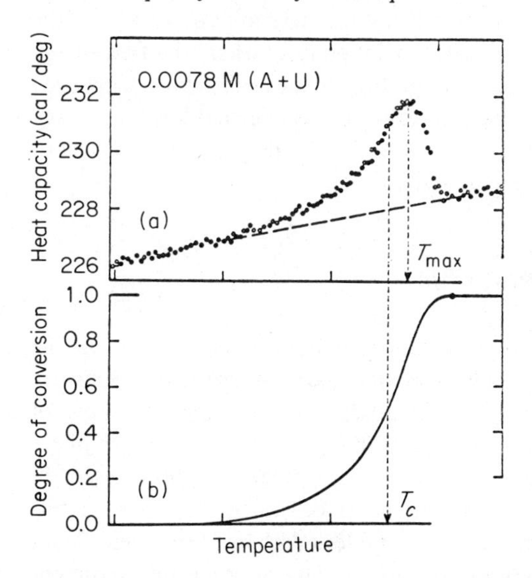

Figure 2. Course of the helix-coil transition of poly (rA•rU) as a function of temperature (*24*). (a) Heat capacity anomaly. (b) Degree of conversion which is directly proportional to the energy absorbed in the transition.

The direct measurement of the heat of transition has been accomplished in two ways. In the classical heat capacity technique, the temperature rise of a known mass of solution resulting from the input of an accurately known quantity of electrical energy is recorded at various ambient temperatures. This yields a point by point determination of the heat capacity at individual temperatures. In the case of a macromolecule undergoing an order–disorder transformation this method traces out the shape of the anomaly in the heat capacity vs. temperature curve (Fig. 2a). In order to determine the effects arising from the macromolecule alone, the contribution to the heat capacity by the solvent as a function of temperature must be determined by separate experiment and subtracted.

In order to accomplish all of this more directly, differential heat capacity instruments employing the twin calorimeter principle have been constructed in which identical amounts of solution, one containing nucleic acid plus solvent and the other containing solvent alone are simultaneously and continuously heated and the differential absorption of energy due to the macromolecular component is recorded as a function of temperature. In these instruments the data are presented in two ways. Either the heat capacity anomaly (the instantaneous temperature difference between the two cups) is delineated as a function of temperature (Fig. 2a) or the total integrated power needed to compensate the transition enthalpy is displayed directly (Fig. 2b). This gives a curve in units of energy of the familiar sigmoid shape which also corresponds to the fractional course of the transition.

These studies yield extremely valuable information – delineating the shape of the transition curve and permitting the determination of both the enthalpy and the entropy at the transition temperature, where the free energy change is essentially zero. On account of the technical difficulties involved these heat capacity methods are perhaps capable of somewhat lower accuracy (5%) than the heat of reaction studies. Nevertheless, they give a great deal of information from a single experiment and are elegant in concept.

CALORIMETRIC APPARATUS

In order to conduct the type of experiments described above on the thermal properties of nucleic acids it is necessary to use highly sensitive calorimetric apparatus. This requirement arises from the limited supply of some of the materials available for study and the preference to work in dilute solutions of nucleic acids at concentrations similar to those used in spectroscopic and hydrodynamic measurements described elsewhere in this book (Section A, chapters by Gratzer, Felsenfeld, Crothers, Zimm, and Bauer and Vinograd). A detailed description of appropriate calorimetric equipment is beyond the scope of this chapter. The methodology, principles, and instrumentation involved for the determination of the heat of reaction accompanying mixing has been well described in a number of review articles (*2-5*).

Several calorimeters employing the classical heat capacity technique have been discussed in considerable detail by Ackermann (*6*). For descriptions of the various differential heat capacity calorimeters the interested reader is referred to some of the original articles (*7-10*) which also contain additional references. Several calorimeters of each of the major types described above are commercially available.

PROCEDURE FOR A CALORIMETRIC EXPERIMENT

Prerequisite to a meaningful calorimetric measurement, one must know the amount of materials used, the stoichiometry, and the extent of reaction. Typical studies of the poly (rC·rI) complex are described below.

Heat of Reaction Measurement

Solutions: Poly rC (0.67 mg) dissolved in 3 ml of 0.1 M NaCl, pH 8.0 and immersed in warm bath (70°C) for 15 min. Poly rI (0.67 mg) dissolved in 5 ml of 0.1 M NaCl, pH 8.0, and immersed in warm bath (70°C) for 15 min.

Concentration determination: 0.5 ml of each solution diluted to 5 ml and optical density in the ultraviolet determined spectrophotometrically at 268 nm and 248 nm, respectively. Using values for the molar extinction coefficient, of $\epsilon_{268} = 6200$ liters/mol-cm for poly rC and $\epsilon_{248} = 10{,}000$ liters/mol-cm for poly rI the concentrations were found to be 7.19×10^{-4} M poly rC and 3.76×10^{-4} M poly rI.

Loading calorimeter: 2.000 gm of poly rC solution (1.437×10^{-6} moles) and 4.100 gm of poly rI solution (1.542×10^{-6} moles) were introduced into the two sections of the calorimeter cell by means of a syringe fitted with narrow gauge polyethylene tubing. Samples containing 2.0 gm and 4.1 gm of 0.1 M NaCl were similarly loaded into the two sections of the tare calorimeter vessel. The calorimeter temperature was 25.0°C.

Calorimetric procedure: The calorimeter used in the author's laboratory for these type of studies is a calibrated heat leak type of instrument. After an equilibration time of about 1 hr, the steady state differential thermopile output which may be about one microvolt is balanced out by the amplifier offset and the true zero of the instrument is set and checked for stability on the pen recorder. The reactants are mixed by rotation of the calorimeter vessels. The formation of the poly (rC·rl) complex produces a rapid rise of temperature in the reaction cell, causing deflection of the recorder pen. This temperature rise then decays exponentially to the original base line, a process that takes about 16 min. The area under this curve of differential thermopile voltage vs. time is determined by a ball and disc type integrator attached to the recorder pen. The energy equivalent of unit area has been previously determined by electrical or chemical calibration. Thus, from the total number of integrator counts and the known calibration constant the number of millicalories liberated in the reaction is determined. Dividing the number of calories by the number of moles reacted gives ΔH in units of calories/mol. In the specific example chosen, the experiment was run on the 30-μV scale of the amplifier, 803.8 integrator units were obtained, corresponding to 8.038 mcal thus $\Delta H = -5594$ calories/mol of base pairs.

Heat of Transition Measurement

Approximately 1.6 mg each of poly rC and poly rI are dissolved in 1.5 ml of 0.05 M NaCl. The concentrations are determined spectrophotometrically as above by taking 15 λ aliquots dissolved in 3 ml of solvent. The concentrations are then adjusted to make the solutions equimolar. The complex is formed by combining 1.4 ml of each solution. Then 2.6 ml of the mixed solution which is now poly (rC•rI) is pipetted into a calorimeter cup and the exact amount is recorded by weight. An identical amount of 0.05 M NaCl is introduced into the tare calorimeter vessel. The cups are fastened to the thermopile-heater assembly which is placed in the calorimeter block and the whole system closed up for an equilibration period of an hour. The thermal transition run is started by passing current through the identical heaters located in each calorimeter vessel and activating the adiabatic environmental control system which maintains the calorimeter block within 0.01°C of the calorimeter vessels. At a heating rate of 15°/hr a run spanning the temperature range from 25 to 70° may take 3 hr. One observes on the recorder initially a steady base line followed by a sigmoid transition curve as in Fig. 2b. This curve on the author's equipment gives a direct reading of the differential energy required to maintain the two vessels at the same temperature which in fact corresponds to the energy absorbed in the transition. The energy absorbed divided by the number of moles used gives the heat of the transition in units of calories per mole.

RESULTS OF CALORIMETRIC STUDIES

In the following section the results of calorimetric investigations of nucleic acids are reviewed. No attempt at an encyclopedic survey of the literature has been made. Rather, it is hoped that the work discussed will indicate the current state of knowledge with respect to some of the leading problems in this field. The emphasis has been exclusively placed upon studies on the associative interaction between nucleic acid bases. Other important topics, such as the heats of hydrolysis of various phosphate ester bonds, are not considered in this review.

STUDIES ON DNA

Acid Denaturation

Measurements of the enthalpy change accompanying the acid denaturation of DNA by Sturtevant, Geiduschek, and their associates (*11, 12*) were the first calorimetric studies on DNA to appear since the renewed modern interest in nucleic acids generated by the structural proposals of Watson and Crick in 1953. This work on the acid denaturation of DNA serves as an excellent example of the study of a cooperative transition induced by alteration of the solvent conditions in a calorimeter designed to measure the heat of reaction.

The experiments consisted of measuring the enthalpy change resulting from mixing DNA solutions in 0.1 M NaCl with varying amounts of HCl in 0.1 M NaCl. Starting with the DNA solution at pH 6 at 25°C, no heat effect was observed until the final pH exceeded pH 3.3. Between pH 3.3 and pH 2.9 increasing amounts of heat were absorbed with a saturation at about 5 kcal/base pair below pH 2.9. These results are illustrated in Fig. 1. The marked change in the value of the observed ΔH took place over the same narrow pH range in which a sharp decrease in the intrinsic viscosity was noted. This is also shown in Fig. 1. These results served to identify the sudden enthalpy change with the configurational collapse of the DNA molecule. Observed values of ΔH for the acid denaturation of DNA were 3,5, and close to 6 kcal/base pair at 5°, 25°, and 40°C. The important conclusion was drawn from these early studies that the behavior of DNA upon acid denaturation in no way lent support for the existence of hydrogen bonds between base pairs in the native molecule and that the stability of DNA mainly arose from other sources.

In a subsequent investigation by Bunville et al. (*13*) the observed heats of acid denaturation of DNA were corrected for the heat effects arising from the protonation of the bases, which had originally (*11, 12*) been believed to be negligible. This was accomplished according to the formula

$$\Delta H_{\text{trans}} = \Delta H_{\text{obs}} + \Delta H_{\text{ion}} \tag{1}$$

The values for the heats of ionization of the bases in the helical and coiled forms of DNA were taken to be those of the ionization heats of the isolated deoxynucleotides. The latter were determined in a separate investigation (*14*) described below. Upon making this correction, a value of $\Delta H = 8300$ cal/mol of base pairs was obtained at 25°C for the hypothetical process.

$$\text{DNA (native, pH 6)} \rightarrow \text{DNA (denatured, pH 6)} \qquad (2)$$

No effect of base composition on ΔH of transition was found for DNA samples with guanine-cytosine contents ranging from 37 to 64%. The magnitude of these results were interpreted to be consistent with the idea that hydrophobic interactions are important in the stabilization of the DNA helix.

Urea Denaturation

The absorption of heat accompanying the denaturation of DNA by going from 0.8 to 2.2 M urea at pH 3.25 and 27°C has been studied by Rialdi and Profumo (*15*). After correction of the measured energies for the heat of dilution of urea, a value of $\Delta H = 9.5 \pm 1.5$ kcal/mol of base pairs was reported for the helix to coil transition. This quantity is obtained as a small difference between two fairly large observed heats. It may be estimated from titration curves that the DNA is approximately one-third protonated under the conditions chosen for this work.

These experiments add evidence by means of an independent route that the value of ΔH of transition for DNA is between 8 to 10 kcal/mol of base pairs.

Thermal Denaturation

The energy required to melt native DNA under a wide variety of conditions of salt concentration and pH has been the subject of an extensive study by Privalov, Ptitsyn, and Birshtein (*16*) following earlier work of Privalov and his associates (*17*). This investigation was carried out using a differential heat capacity type of calorimeter to observe the thermal transition. The kind of information obtained with this instrument is shown schematically in Fig. 2a where the instantaneous temperature difference or the difference in heat capacity between the sample and the solvent, resulting from the energy absorption required for melting the DNA sample, is displayed as a function of the rising temperature. If, as in some instruments, the cumulative energy supplied to nullify the temperature difference is displayed as a function of temperature, one obtains the energy involved in the transition directly in the form of familiar shape shown in Fig. 2b. The ΔH

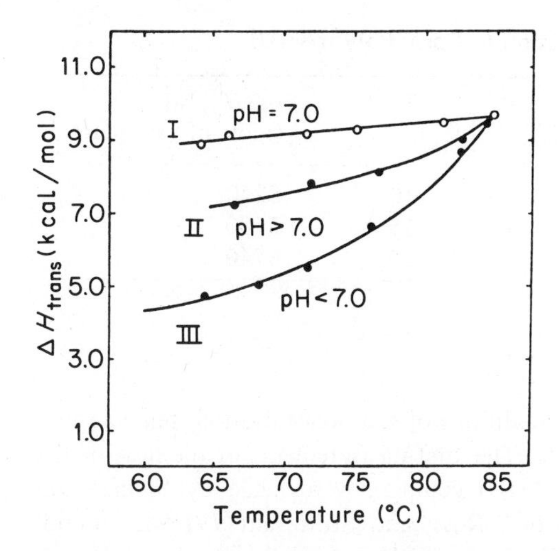

Figure 3. Heat of denaturation of DNA, ΔH_{trans} as a function of transition temperature (*16*). (I) Variable ionic strength, $I = 0.01$ to 0.2 M, pH 7.0. (II) Increasing pH, $I = 0.2$. (III) Decreasing pH, $I = 0.2$.

results of Privalov et al. are summarized as a function of T_m in Fig. 3. It was found that the measured ΔH of transition at neutral pH was close to 9 kcal/base pair and that this value was quite insensitive to changes in the simple salt concentration from 0.01 to 0.2 M. Considerably greater variation in the observed value of ΔH was found as the pH was altered, especially under acidic conditions. From consideration of both the pH dependence and the ionic strength dependence of the measurements of ΔH of transition, it was concluded that the difference in free energy between the native and denatured states of DNA was $\Delta G = 1.2$ kcal/mol of base pairs under approximate physiological conditions of neutral pH, ionic strength 0.2 M, 37°. This value of the free energy represents the maximum stability of the native structure. The magnitude of the stabilization free energy was found to decrease either as the pH was changed from neutrality or as the ionic strength was reduced.

STUDIES ON SYNTHETIC POLYNUCLEOTIDES

Formation of Poly (rA · rU)

The synthetic polyribonucleotides, polyriboadenylic acid (poly rA) and polyribouridylic acid (poly rU) can react to form either two- or three-stranded helical complexes poly (rA·rU) and poly (rA·2rU), respectively,

TABLE 1. Heat of the Reaction to Form Poly (rA·rU)

M KCl	$-\Delta H^{a}$, cal (mole of base pairs)$^{-1}$	T, °C	$-\Delta H^{b}$, cal (mole of base pair)$^{-1}$
0.10	5900	10	5220
0.50	5250	25	5780
1.00	4750	40	6740

[a] All work at 25° *(18)*.
[b] All work, 0.1 M KCl, pH 7.0*(19)*.

depending upon experimental conditions of salt concentration, temperature, and the mole ratio of reactants. The first measurements of the heat of the reaction to form the poly (rA·rU) complex were made by Steiner and Kitzinger (*18*) and by Rawitscher, Ross, and Sturtevant (*19*) who found, respectively, that the measured value of ΔH depended upon both the salt concentration (*18*) and the temperature (*19*). These results are summarized in Table 1.

The result that the measured exothermic heats decrease with increasing salt concentration and decreasing temperature is probably mainly due to the degree of internal order or self-structure of poly rA itself. In order to gain an idea of the intrinsic enthalpy of the reaction between hypothetically random poly rA and poly rU, it is necessary to extrapolate calorimetric data obtained with this system to conditions where the poly rA is essentially completely disordered. An attempt to perform this extrapolation was made by Rawitscher, Ross, and Sturtevant (*19*) who by combining data on ultraviolet hypochromism and their own calorimetric data on the enthalpy of formation of the ordered acid form of poly rA (which may not be wholly appropriate) obtained a value of $\Delta H = -8700$ cal/mol of base pairs for the heat of interaction for the hypothetical reaction at the transition temperature between "random coil" poly rU with "random coil" poly rA to form the poly (rA·rU) complex at pH 7.

The heat of the addition of the second strand of poly rU to poly (rA·rU) to form the three-stranded poly (rA·2rU) complex has been determined by Ross and Scruggs (*20*). A value of $\Delta H = -3800$ cal/mol of poly (rA·2rU) formed was found to be fairly insensitive to the experimental conditions of salt concentration and temperature employed. This result lends support to the idea that the major portion of the variation of the observed ΔH of the poly rA and poly rU reaction with experimental conditions is attributable to differences in the conformation and internal order of poly rA. From these results on the formation of both the poly (rA·rU) and the poly (rA·2rU) complexes, the entropy change accompanying the incorporation of a single strand of either poly rA or poly rU into an ordered helical structure was estimated to be $\Delta S = 11.5$ cal/mol-°C at the melting temperature. In this study (*20*) it was found that the enthalpy change associated with the two- to three-strand conversion reaction Eq. (3).

$$2 \text{ poly (rA}\cdot\text{rU)} \rightleftharpoons \text{poly (rA}\cdot 2\text{rU)} + \text{poly rA} \tag{3}$$

induced by raising the temperature was unfavorable. It follows that this interesting reaction (3) is entropically driven.

Thermal Denaturation; Poly rA, Poly rU System

The heats of the thermally induced conformational transition of poly (rA·rU) and poly (rA·r2U) have been measured by Krakauer and Sturtevant (*21*) and by Ackermann and his coworkers (*22–24*) (Fig. 2). Within the combined experimental uncertainties, the results of both laboratories appear to be in agreement. Krakauer and Sturtevant (*21*) made an extensive investigation of the effect of ionic strength and counterion on the transition enthalpies. They reported an interesting and unexplained difference in the measured heats which were found to be somewhat lower in the presence of K^+ than in the presence of Na^+. Hinz, Schmitz, and Ackermann (*23*) established that the transition enthalpy was concentration independent over the range 1 to 4×10^{-3} M poly (rA·rU), justifying the significance of calorimetric data obtained at finite concentrations. In order to eliminate the temperature dependent intramolecular secondary structure of poly rA, Neumann and Ackermann (*24*) extrapolated their measured transition enthalpies to 95°C where poly rA is essentially devoid of intramolecular order. This extrapolation yielded a value of $\Delta H^0 = 9.3$ kcal/mol of base pairs for the thermal dissociation of poly (rA·rU) and $\Delta H^0 = 13.5$ kcal/mol of poly (rA·2rU) for the dissociation of the three-stranded poly (rA·2rU) complex to the randomly coiled single-strand polymers. One may calculate, using Hess' law of the additivity of thermochemical equations, that the ΔH for adding the third strand of poly rU to poly (rA·2rU) to form poly (rA·2rU) is −4.2 kcal/mol of poly (rA·2rU). All of these results are in essential agreement with those found previously (*18–20*) by heat of reaction calorimetry. They possess the virtue, however, of having the thermodynamic data determined at the respective transition temperatures.

Neumann and Ackermann (*24*) analyzed their results in terms of the statistical mechanical theory of cooperative transformations and deduced a value for the molar stacking free energy of poly (rA·rU) $\epsilon(298°\text{K}) = -3$ kcal/mol of poly (rA·rU) and a value for the mean nucleation parameter, τ, of 400. The reader should consult the original reference (*24*) for details of the treatment.

Poly rA

There are two main areas of thermodynamic investigation concerning the properties of poly rA. The first is concerned with defining the extent of intramolecular interaction in neutral solutions of poly rA and determining the amount of energy necessary to disrupt that partial order, and the second is concerned with measurement of the energetics of the two-stranded poly rA structure formed at acidic pH.

Poly rA in Neutral Solution

Epand and Scheraga (*25*) determined the heat of solution of solid poly rA at 25° and 45° and performed heat capacity measurements on the solid and the solutions. Calculations utilizing the appropriate thermochemical cycle yielded a value of $\Delta H = -9$ kcal/mol of adenine for the enthalpy of formation of the totally stacked structure. This result depends upon an estimate of the change in the fraction of ordered bases upon going from 25° to 45° which, in turn, is strongly dependent upon the values assigned to the optical parameters for the completely ordered and disordered forms.

In contrast to their results, Neumann and Ackermann (*24*) estimated an approximate value of $\Delta H = 4.5 \pm 2$ kcal/mol of poly rA for the melting of the neutral poly rA structure. This latter value is in reasonable agreement with the values deduced from other calorimetric measurements of the variation of the measured ΔH for the formation of poly (rA·rU) as a function of temperature (*19–21*). The thermal properties of neutral poly rA require further careful study. Thermochemical information on this topic is necessary for the understanding of both the poly rA, poly rU system and the acid ordered form of poly rA.

Poly rA; Acid Structure

The enthalpy change associated with the two-stranded ordered structure of poly rA that forms at acidic pH has been examined both by heat of reaction (*19*) and heat capacity (*26*) techniques. Rawitscher, Ross, and Sturtevant (*19*) titrated poly A from neutral to acid pH in the reaction calorimeter and found a marked change in apparent heat content over a narrow range of pH, which was coincident with the conformational change as observed spectrophotometrically and titrimetrically. At lower pHs a continuing evolution of heat was attributed to the binding of protons by the ordered poly rA helix. Subtracting the protonation heat from the observed transition heat yielded a value of $\Delta H = -3100$ cal/mol of base pairs at 25° and a value of about 1 kcal less at 10° for the hypothetical reaction

$$rA\ (coil,\ pH\ 7) + \alpha H^+ = rA\ (helix,\ pH\ 5, \alpha = 0) \qquad (4)$$

where α is fraction of adenine groups protonated.

Klump et al. (*26*) studied the melting of acid poly rA over the pH range 5.7 to 4.5 by determining the heat capacity of the solution through the transition region. The enthalpy of transition, ΔH was found to vary from 3.4 kcal/base pair at pH 5.5 to 5.6 kcal/base pair at pH 4.7. These data were extrapolated to 100°C, presumably to eliminate intramolecular interaction in poly rA, and a value of ΔH of 5.6 kcal/mol of base pairs was obtained. This result correctly represents the change in enthalpy under the conditions of the experiment, which are complicated. These complexities include varying degrees of protonation of the polymer at different temperatures and absence of knowledge concerning the heat of ionization of both the helical and coil forms and the dependence of these ionization heats upon temperature. The simplicity suggested by this system containing only one polymeric component is clearly deceptive.

Poly dAT Copolymer

Few calorimetric studies have been carried out with synthetic polydeoxyribonucleotides on account of the unavailability of sufficient quantities of these materials. The only investigation reported is that on the poly dAT copolymer carried out by Scheffler and Sturtevant (*27*). They employed the differential heat capacity method and found a ΔH of 7.9 kcal/mol of base pairs under conditions of salt concentration such that the helical copolymer melted at 40°C. This result is about 1 kcal greater than that found for poly (rA + rU) at the same transition temperature, although such a comparison is complicated by what are undoubtedly different extents of intramolecular interactions of the melted components in the two cases.

Poly (rC·rI)

The heat of the reaction between poly rC and poly rI to form the two-stranded poly (rI·rC) complex has been determined by Ross and Scruggs (*28*) who, employing an isothermal reaction calorimeter, found a value of $\Delta H = -5.6$ kcal/base pair over the range of Na^+ concentration of 0.02 to 0.4 M at both 20 and 37°C. If it is possible to extrapolate these results to the transition temperature, which is not unreasonable given the insensitivity of the measured heats to varying salt concentration and temperature, then it is clear that the poly (rC·rI) complex is characterized by values of ΔH and

ΔS at the transition temperature markedly different from those found for poly (rA·rU). Consequently, it appears that different nucleic acid base pairs may have different heats and entropies of transition.

The results obtained with the poly (rC·rI) complex represent a very interesting case, since the measured ΔH is independent of salt concentration and temperature. This is in marked contrast to the results obtained with the formation of poly (rA·rU) (*18, 19*). Either this is an intrinsic thermodynamic difference resulting from the differing chemical compositions of the two sets of base pairs which affects the strength of the base pair in the complex, or the energy difference may arise from differing thermodynamic states of the coiled forms in solution. Recently Hinz, Haar, and Ackermann (*29*) have reported transition enthalpies of poly (rC·rI) with somewhat different results.

Substituent Effects in Poly (rC·rI) Complexes

Ross et al. (*30*) have recently determined the heats of interaction between two poly rC derivatives substituted in the 5-position of the cytosine ring and poly rI. Although both poly 5-MeC and poly 5-BrC both raise the melting temperatures of their respective 1 : 1 complexes with poly rI by about the same amount, the observed enthalpies of reaction were widely different. The poly (5-BrC·rI) complex exhibited very large heats compared to poly (rC·rI). The exact values of these energies depended upon conditions of salt concentration and temperature. The 5-methyl group contributed only about 500 cal/mol of base pairs to the enthalpy of stabilization. These results provide additional examples of differing values of thermodynamic parameters associated with the different base pairings.

MONOMER–POLYMER INTERACTIONS

Interactions Between Poly rU and Adenine Derivatives

In order to examine the effect of chemical structure upon the thermodynamics of nucleic acid complex formation, the reaction between a series of adenine derivatives with a common substrate, poly rU has been studied by Scruggs and Ross (*30*). A heat of reaction, $\Delta H = -12.8$ kcal/mol of A·2poly rU complex was found for the interaction between poly rU and either adenine, adenosine, or deoxyadenosine in 0.6 M NaCl at 20°. This result indicates that the presence or absence of the sugar group or of the 2′ OH group contributes little to the ΔH of these monomer–polymer complexes. Complexes of poly rU with 2-aminoadenosine and 2,6-diaminopurine, which can form three hydrogen bonds with the first strand of poly rU, were found to be 3 kcal more stable, both by reaction calorimetry and by differential heat capacity measurement of the energy absorbed in melting these complexes. This work again illustrates that different base pairings can have different thermodynamic properties. The presence of the third hydrogen bond produces a particularly striking effect increasing ΔH by 3 kcal, a result that suggests that the adenine–uracil and guanine–cytosine pairs in ribonucleic acids may be characterized by considerably different values of ΔH.

STUDIES ON NUCLEIC ACID MONOMERS

Heats of Ionization

The heats of ionization of a number of deoxynucleotides and related compounds were determined at 25°C at an ionic strength close to 0.1 M by Rawitscher and Sturtevant (*14*). The motivation for this work was to obtain the heats of ionization of deoxynucleotides in order to correct the heats observed for the acid denaturation of DNA (*11, 12*) discussed above for the thermal effect of the uptake of protons by DNA upon acidification. Values of $\Delta H = 2640$, 4280, and 140 cal/mol were found for the ionization of 5′-deoxyadenylic acid, 5′-deoxycytidylic acid, and 5′-deoxyguanylic acid, respectively. No calorimetric data are available on the temperature dependence of the heats of ionization for these nucleic acid derivatives. There are indications that large and complex effects may exist for the temperature dependence of the ionization heats of compounds of this type.

Monomer Interactions in Aqueous Solutions

The thermodynamics of the self-association of a number of monomeric purine (*32, 33*) and pyrimidine (*34*) derivatives in water has been investigated by Gill and coworkers. Under these conditions it is possible for both stacking and hydrogen bonding interactions to take place. The calorimetric work consisted of measuring the heats of dilution of the purine or pyrimidine in water, diluting to as low a final concentration as possible. From these data the relative partial molal enthalpies were evaluated. The method assumes that all heats of dilution effects arise from the dissociation of complex species. The nature and kind of associated species present can be represented by various sets of chemical equations. In this case, two model schemes have been applied, that of dimerization and that of multistep equilibria of the form

$$A + A_{n-1} = A_n \qquad (5)$$

with each step having equal enthalpy and free energy. If either equilibrium constant or osmotic coefficient data are available, as they were in this case, it is possible to combine that information with the relative molal enthalpies to obtain ΔH^0 and K characterizing the self-association of these purine or pyrimidine derivatives according to one of the model schemes. The standard free energy, ΔG^0 and the entropy, ΔS^0 follow from these determinations. The results of Gill and his associates are summarized in Table 2 (*33, 34*). It may be seen that the equilibrium constants for self-association vary from 0.6 to 13 M^{-1} and the enthalpies from −2 to −6 kcal/mol. Large variations in the entropies, ΔS^0, are noted. It is difficult to interpret these results in terms of chemical structure. This work clearly demonstrates, however, the importance and variability of entropy effects accompanying the self-association of purines and pyrimidines. Presumably these entropy effects arise from changes in the water solvent structure about the solute molecules.

TABLE 2. Thermodynamics of Self-Association of Purine and Pyrimidine Derivatives[a]

Compound	ΔH^0 (kcal/mol)	ΔG^0 (kcal/mol)	ΔS^0 (cal/deg-mol)	K M^{-1}
Purine	−4.2	−0.44	−13	2.1
6-Methylpurine	−6.0	−1.12	−16	6.7
Purine riboside	−2.5	−0.38	−7	1.9
Deoxyadenosine	−3.7	−1.5	−7	12
Caffeine	−3.4	−1.5	−6	13
Cytidine	−2.8	0.08	−10	0.87
Uridine	−2.7	0.29	−10	0.61

[a]Water, 25° (*33, 34*).

Monomer Interactions in Nonaqueous Solutions

Adenine and uracil derivatives are known to associate in $CHCl_3$ and CCl_4 under conditions where stacking interactions are greatly reduced. Binford and Holloway (*35*) utilizing heat of dilution data, obtained a value of $K = 100\ M^{-1}$ and $\Delta H = -6.2$ kcal/mol for the 1 : 1 base pairing interaction between 9-ethyladenine and 1-cyclohexyluracil in $CHCl_3$. Further calorimetric studies of interactions in nonaqueous media would be of considerable interest.

SUMMARY

Calorimetry has proven to be a useful and valuable tool for the investigation of nucleic acids. It has been unequivocally demonstrated in a number of papers reviewed above that a value of ΔH of about 9 kcal/base pair stabilizes the DNA (*11–17*) and the poly (rA·rU) (*18–24*) duplex helices. It appears that different types of nucleic acid base pairs may have different thermodynamic parameters characterizing their association (*28, 30, 31*). This field will have to be described calorimetrically before the effect of variation of chemical structure can be understood in further detail.

There are many aspects of nucleic acid chemistry open for calorimetric investigation. For instance, heats of polymerization, interactions between nucleic acids and small molecules (binding), and the hydrolysis of various internucleotide and nucleotide–peptide bonds might all be profitably studied calorimetrically. The calorimetric method possesses considerable potential as a tool for kinetic studies and for the determination of equilibrium constants. Finally, the extreme generality of calorimetric technique makes it applicable to situations where lack of transparency precludes the use of optical methods. This condition prevails in many particulate systems of interest; for example, chromosomes, bacteria, yeasts, or tissue slices might readily be examined. In principle, calorimetry may be used for the study of virtually any system both *in vivo* and *in vitro*. In practice, all that is required is a suitable modification of equipment and some ingenuity in the design of experiments.

References

1. Sturtevant, J. M. 1962. In H. A. Skinner (Ed.), Experimental thermochemistry. Wiley (Interscience), New York. Vol. 2, p. 427.
2. Wadsö, I. 1970. *Quart. Rev. Biophysics,* **3**: 383.
3. Skinner, H. A., J. M. Sturtevant, and S. Sunner. 1962. In H. A. Skinner (Ed.), Experimental thermochemistry. Wiley (Interscience), New York, Vol. 2, p. 157.
4. Sturtevant, J. M. 1949. In A. Weissberger (Ed.), "Calorimetry," technique of organic chemistry, 2nd edition, Wiley (Interscience), New York, Vol. 1.
5. Skinner, H. A. 1969. In H. D. Brown (Ed.), Biochemical microcalorimetry. Academic Press, New York, p. 1.
6. Ackermann, T. In H. D. Brown (Ed.), Biochemical microcalorimetry, Academic Press, New York, p. 235.
7. Privalov, P. L., D. R. Monaselidze, G. M. Mrevlishvili, and V. A. Magaldadze. 1964. *J. Exptl. Theoret. Phys. (USSR)*, **47**: 2073.
8. Gill, S. J., and K. Beck. 1965. *Rev. Sci. Instr.,* **36**: 274.
9. Danforth, R., H. Krakauer, and J. M. Sturtevant. 1967. *Rev. Sci. Instr.,* **38**: 484.
10. Clem, T. R., R. L. Berger, and P. D. Ross. 1969. *Rev. Sci. Instr.,* **40**: 1273.
11. Sturtevant, J. M., and E. P. Geiduschek. 1958. *Am. Chem. Soc.,* **80**: 2911.

12. Sturtevant, J. M., S. A. Rice, and E. P. Geiduschek. 1958. *Discussions Faraday Soc.*, **25**: 138.
13. Bunville, L. G., E. P. Geiduschek, M. A. Rawitscher, and J. M. Sturtevant. 1965. *Biopolymers,* **3**: 213.
14. Rawitscher, M. A., and J. M. Sturtevant. 1960. *J. Am. Chem. Soc.,* **82**: 3739.
15. Rialdi, G., and P. Profumo. 1968. *Biopolymers,* **6**: 899.
16. Privalov, P. L., O. B. Ptitsyn, and T. M. Birshtein. 1969. *Biopolymers,* **8**: 559.
17. Privalov, P. L., K. A. Kafiani, and D. P. Monaselidze. 1965. *Biofizika,* **10**: 393.
18. Steiner, R. F., and C. Kitzinger. 1962. *Nature,* **194**: 1172.
19. Rawitscher, M. A., P. D. Ross, and J. M. Sturtevant. 1963. *J. Am. Chem. Soc.,* **85**: 1915.
20. Ross, P. D., and R. L. Scruggs. 1965. *Biopolymers,* **3**: 491.
21. Krakauer, H., and J. M. Sturtevant. 1968. *Biopolymers,* **6**: 491
22. Neumann, E., and T. Ackermann. 1967. *J. Phys. Chem.,* **71**: 2377.
23. Hinz, H. J., O. J. Schmitz, and T. Ackermann. 1969. *Biopolymers,* **7**: 611.
24. Neumann, E., and T. Ackermann. 1969. *J. Phys. Chem.,* **73**: 2170.
25. Epand, R. M., and H. A. Scheraga. 1967. *J. Am. Chem. Soc.,* **89**: 3888.
26. Klump, H., T. Ackermann, and E. Neumann. 1969. *Biopolymers,* **7**: 423.
27. Scheffler, I. E., and J. M. Sturtevant. 1969. *J. Mol Biol.,* **42**: 577.
28. Ross, P. D., and R. L. Scruggs. 1969. *J. Mol. Biol.,* **45**: 567
29. Hinz, H. J., W. Haar, and T. Ackermann. 1970. *Biopolymers,* 9: 923.
30. Ross, P. D., R. L. Scruggs, F. B. Howard, and H. T. Miles, in press.
31. Scruggs, R. L., and P. D. Ross. 1970. *J. Mol. Biol.,* **47**: 29.
32. Stoesser, P. R., and S. J. Gill. 1967. *J. Phys. Chem.,* **71**: 564.
33. Gill, S. J., M. Downing, and G. F. Sheats. 1967. *Biochemistry,* **6**: 272.
34. Farquhar, E. L., M. Downing, and S. J. Gill. 1968. *Biochemistry,* **7**: 1224.
35. Binford, J. S., Jr., and D. M. Holloway. 1968. *J. Mol. Bio.,* **31**: 91.

tRNA CRYSTALLIZATION

DAVID R. DAVIES
NATIONAL INSTITUTE OF ARTHRITIS AND METABOLIC DISEASES
NATIONAL INSTITUTES OF HEALTH
BETHESDA, MARYLAND

B. P. DOCTOR
DIVISION OF BIOCHEMISTRY
WALTER REED ARMY INSTITUTE OF RESEARCH
WALTER REED ARMY MEDICAL CENTER
WASHINGTON, D.C.

The crystallization of tRNA was a landmark in the physical chemistry of nucleic acids (*1*). The subsequent production of large single crystals has made it likely that a detailed description of the tertiary structure of a nucleic acid will be determined by x-ray diffraction in the near future (*2–7*). In this chapter we describe some methods currently used for obtaining tRNA crystals.

There are basically two methods. One involves salting out of the tRNA with concentrated solutions of salts such as ammonium sulfate. The other involves precipitation by organic solvents. The concentration of the precipitating agent in both cases has generally been controlled by vapor diffusion.

Both methods have been influenced by the very small quantities of pure nucleic acid available to the investigator. This has necessitated the use of microtechniques involving small volumes (~25 μl).

MATERIALS AND METHODS

tRNA

Highly purified unfractionated tRNA as well as purified single species of tRNA have successfully been crystallized. The unfractionated tRNA is deproteinized by phenol extraction and further purified by gel filtration in 1 M NaCl on Sephadex G-100. The preparation is exhaustively dialyzed against a dilute buffer containing $MgCl_2$.

The purified single species of tRNA is also similarly dialyzed exhaustively prior to use. In some instances 5×10^{-4} M EDTA (pH 7.0) is used in the first dialyzate. The concentration of tRNA is usually 10 mg/ml and the final dialysis is carried out against distilled water. The samples are then either lyophilized or used as such.

Organic Solvents

(a) Ethanol: Commercially available 95% ethanol is used with desired dilution. (b) 1-4 Dioxane: The commercial preparation is distilled prior to use. (c) 2-Methyl-2, 4-pentanediol is also distilled prior to use.

Inorganic Salts and Buffers

$MgCl_2$, $MnCl_2$, KCl, spermine hydrochloride, spermidine hydrochloride, sodium cacodylate buffer, and Tris HCl buffer are prepared in distilled water as described or required.

Equipment and Glassware

Quartz capillaries (0.7 or 1.0-mm diameter) are acid washed, siliconized, and dried prior to use. The single cavity microculture *slides* (0.5 cm thick) with 1-cm diameter and 0.2–0.3-cm depth indentation are siliconized and dried. Small circular dishes 10-cm diameter and 5-cm deep, with a ground glass lip and a plate glass lid have been used as equilibration chambers for the samples and solvent and permit easy examination with a microscope. In some cases a Lucite lid has been used with stoppered ports that can be opened for easy addition of sample and solvents (*9*).

CRYSTALLIZATION PROCEDURE: THE VAPOR DIFFUSION METHOD

The vapor diffusion method offers a slow approach to the precipitating conditions by permitting a very gradual increase in the concentration of the precipitating agent. Its advantages over other methods such as dialysis lie in one's ability to work with very small quantities of material and in the ease with which it may be set up.

The precipitant used may be either ammonium sulfate or an organic solvent (ethanol, dioxane, or 2-methyl-2,4-pentanediol). In both cases the method is essentially the same and involves concentrating the precipitant by placing the tRNA solution in an enclosed chamber with a reservoir containing a higher concentration of precipitant.

Since the volume of the reservoir is considerably larger than that of the tRNA sample, the final concentration of precipitant in the sample will be approximately the same as the concentration of precipitant in the reservoir. Correspondingly there will be some increase in the tRNA and salt concentration in the sample.

In Table 1 are shown some representative conditions which have resulted in large crystals of tRNA. The solution is initially prepared with the appropriate concentration of tRNA and various salts. It should be noted that Mg^{2+} ion is universally used. The concentration of precipitant in this initial solution is kept well below that which will precipitate the tRNA. In the case of volatile precipitants such as ethanol it is not really necessary to add any directly to the initial tRNA solution.

A small volume (5-20 μl) of this solution is then placed in the well of a microculture slide or in a thin-walled glass capillary which in turn is placed inside the equilibration chamber. Along with the samples is placed a reservoir containing a buffered solution of precipitant at that concentration estimated most likely to lead to the formation of large crystals. The chamber is sealed and allowed to equilibrate at either room temperature or at a lower temperature. As shown in Table 1, 8°C has been a popular temperature for crystallization; we find however that the more readily attainable 4°C is a satisfactory alternative.

After 2 or 3 days the samples may be examined with a microscope for the formation of crystals. Depending on the purity of tRNA, the particular species of tRNA, or upon other factors not fully understood, crystals may appear within 2 days or may take 2–3 weeks. If no crystals appear after 2 weeks, the concentration of precipitant in the reservoir may be increased by a few percent. If an amorphous precipitate appears in these samples, the contents of the reservior may be replaced with water and the solution equilibrated against water until a clear solution is obtained. The equilibration may then be repeated, but against a lower concentration of precipitant.

TABLE 1. Some Representative Crystallization Conditions

tRNA	Method	% tRNA	Equilibration solution	Salts (Initial concentrations)	pH	Temp.	Ref.
Yeast Phe	Vapor diffusion	1.5	Dioxane–water 26%	$<$1 mM Mg^{2+} $<$10 mM K^+ $<$10 mM EDTA 17% dioxane	a	Room temp.	4
Yeast Val	Precipitation from solution	0.75		10 mM $MgCl_2$ 60 mM KCl 5 mM $MnCl_2$ Dioxane + 1% tertiary butyl alc. in H_2O 20%	a	8°	7
Yeast fMet	Vapor diffusion	0.4–1.0	Ammonium sulfate 50% sat. at 8° 65% sat. at room temp.	5 mM $MgCl_2$ 50 mM NH_4Cl 5 mM Cac 1-2 mM of one of Hg^{2+}, Co^{2+}, Mn^{2+}, spermine, spermidine, 30% sat'd ammonium sulfate	6.0	8° and room temp.	8
Yeast fMet	Vapor diffusion	0.3	EtOH/H_2O	5 mM Mg^{2+} 50 mM KCl 5 mM Tris 1 mM Co^{2+}	7.4	8–10°	9
Yeast fMet	Vapor diffusion	0.34	EtOH/H_2O 7%	5 mM Mg^{2+} 5mM KCl 1 mM Mn^{2+} 5 mM Tris	7	4°	10

Yeast Phe	Vapor diffusion	4–5	Methylpentane diol. 40%	5 mM $MgCl_2$ 5 mM Cac or Tris 15% MPD	6.5 or 7.5	Room temp.	11
E. coli Phe	Vapor diffusion	0.45–0.5	$EtOH/H_2O$	5 mM $MgCl_2$ 50 mM KCl 5 mM Tris 1 mM $CoCl_2$	7.4	8–10°	9
E. coli Leu	Vapor diffusion	0.4–1.0	Ammonium sulfate 35–40% sat'd	5 mM $McCl_2$ 50 mM NH_4Cl 5 mM Cac plus 1–2 mM of one of Hg^{2+}, Co^{2+}, Mn^{2+}, spermidine, 30% sat'd ammonium sulfate	6.0	8°	8
E. coli fMet	Vapor diffusion	1	Methylpentane diol. 40%	5 mM $MgCl_2$ 5 mM Cac 15% MPD	6.5	Room temp.	11
Yeast unfractionated	Vapor diffusion	5–30	$Dioxane/H_2O$ 40–50% at 32° C 30–40% at 23° C	10 mM $MgCl_2$ 150 mM KCl 10 mM Cac 0.1 mM EDTA	7	Room temp. and 32°	5,6

[a] pH uncertain owing to prior dialysis with distilled water.

The interesting observation has been made that in the case of yeast fMet-tRNA the type of crystals obtained (orthorhombic, hexagonal, or tetragonal) was to a certain extent determined by the temperature, the final concentration of precipitant, and the speed with which it was attained. Thus the tetragonal and orthorhombic forms were produced preferentially when the $(NH_4)_2SO_4$ concentration in the reservoir was rapidly raised to 2.7 M at 4°C. The hexagonal form was favored at room temperature when the $(NH_4)_2SO_4$ concentration was gradually raised to 2.3 M over 10–14 days. The production of hexagonal crystals could be guaranteed by "seeding." The seeding solution was prepared by crushing a hexagonal crystal in a stabilizing solvent containing 2.6 M $(NH_4)_2SO_4$ 0.005 M $MgCl_2$, 0.002 M spermine, and 0.005 M cacodylate pH 6.0; the solution was centrifuged to remove large fragments and diluted in the stabilizing solvent to minimize the number of nucleation sites. The sample to be crystallized was first equilibrated against 2.26 M $(NH_4)_2SO_4$, then touched with a glass fiber which had been wetted with the seeding solution, and finally equilibrated against 2.3 M $(NH_4)_2SO_4$. When crystals of suitable size have been grown they may be removed from the mother liquor with the aid of a fine capillary constructed by drawing out the tip of a Pasteur pipette. They can then be transferred to a suitable container for microscopic or x-ray examination. However, it should be noted that it is the general experience of most investigators that the crystals are rather soft and deteriorate rapidly with manipulation. For x-ray diffraction work it is therefore probably desirable to grow them in the thin-walled glass capillaries in which they will ultimately be used.

Frequently it is extremely difficult to remove crystals from the tiny drops in which they have been grown. In such cases it is sometimes possible to increase the volume of the drop by the addition of a solution containing precipitant and salts only. Providing that the concentration of precipitant and salts in this solution is made high enough it is frequently possible to keep the crystals for long periods of time without their redissolving. The optimum concentration of precipitant and salts should be determined empirically but should in any case be higher than the final concentration required to produce crystals.

If an attempt is to be made to crystallize a hitherto uncrystallized tRNA then a broad range of conditions should be investigated; we suggest that a matrix of conditions be set up involving a large number of small variations on the general procedures described above. Parameters that might be varied include the choice of precipitant, concentration of precipitant, concentration of tRNA, and temperature of crystallization. However, it should be borne in mind that the above methods have not been universally applicable to all species of pure tRNA and radically different procedures involving quite different concentrations or precipitants may be required with some tRNAs.

Crystallization Procedure Not Involving Vapor Diffusion

Paradies and Sjöquist (*7*) have used a method in which precipitant (dioxane plus 1% tertiary butyl alchohol) is added directly to the tRNA solution (Table 1) to produce a concentration just below the precipitating concentration. The solution is then placed in sealed containers and allowed to stand. In this way large crystals of *E. coli* $tRNA_{Val}$ have been obtained which gave x-ray diffraction patterns of remarkable resolution and clarity.

Crystallization Procedure for Unfractionated tRNA

For unfractionated tRNA Blake et al. (*6*) have found that single crystals can be obtained under the conditions of Table 1. Note the very high concentration of tRNA used and the high temperatures of crystallization. In general it would appear that the solubility of tRNA in the presence of organic solvents increases very rapidly with temperature.

CONCLUSION

In the past two years since the first crystals of tRNA were obtained, striking progress has been made in improving the quality of the crystals. This is reflected in the resolution of the x-ray diffraction pattern obtained from the crystals, which has gone from approximately 15 Å for the earliest crystals to as much as 3 Å for many species at the present. With this resolution, together with the known nucleotide sequences of the tRNA it should be possible to determine the three-dimensional structure with some precision.

In addition, as more tRNA species are purified, it should become possible to obtain these also in crystalline form for x-ray diffraction studies. This will enable comparative studies to be made of the three-dimensional structures of tRNA.

References

1. Clark, B. F. C., B. P. Doctor, K. C. Holmes, A. Klug, K. A. Marcker, S. J. Morris, and H. H. Paradies. 1968. *Nature*, **219**: 1222.
2. Kim, S. H., and A. Rich. 1968. *Science*, **162**: 1381.
3. Hampel, A., M. Labanauskas, P. G. Connors, L. Kirkegard, U. L. Rajbhandary, P. B. Sigler, and R. M. Bock. 1968. *Science*, **162**: 1384.
4. Cramer, F., F. Van Der Haar, W. Saenger, and E. Schlimme. 1968. *Angew. Chem.*, 80: 969.

5. Fresco, J. R., R. D. Blake, and R. Langridge. 1968. *Nature,* **220**: 5174.
6. Blake, R. D., J. R. Fresco, and R. Langridge. 1970. *Nature,* **225**: 32.
7. Paradies, H. H. and J. Sjöquist. 1970. *Nature,* **226**: 159.
8. Young, J. D., R. M. Bock, S. Nishimura, H. Ishikura, Y. Yamada, U. L. Rajbhandary, M. Labanauskas, and P. G. Connor. 1969. *Science,* **166**: 1527.
9. Hampel, A., and R. Bock. 1970. *Biochemistry,* 9: 1873.
10. Kim, S. H., and A. Rich. 1969. *Science,* **166**: 1621.
11. Doctor, B. P., and D. R. Davies, unpublished information.
12. Johnson, C. D., K. Adolph, J. J. Rosa, M. D. Hall, and P. B. Sigler. 1970. *Nature,* **226**: 1246.

SPECIFIC LABELING IN DNA FOR AN ELECTRON MICROSCOPIC STUDY OF BASE SEQUENCE

MICHAEL BEER
THOMAS C. JENKINS DEPARTMENT OF BIOPHYSICS
JOHNS HOPKINS UNIVERSITY
BALTIMORE, MARYLAND

Some time ago a method was suggested for the electron microscopic determination of the sequence of bases in nucleic acids (*1*). The procedures visualized require that base selective reagents can be found which are: (1) sufficiently specific in their attachment to bases; (2) detectable in the electron microscope; and (3) sufficiently small that the labeled site can be identified with adequate resolution. After selective labeling of bases, the nucleic acid molecules must be deposited in an untangled form on support films suitable for high-resolution electron microscope. With contemporary technology, the macromolecules and the attached selective markers can not be clearly detected in the electron microscope. For sufficient visibility the anionic groups of the nucleic acid phosphates and the base specific reagents must be "stained" with massive cations to render them detectable in the electron microscope.

In this chapter we shall review procedures for base selective attachment of identifying groups and describe briefly the deposition of the macromolecules on the supporting films. In addition, we shall outline a problem still not completely solved. This is the "staining" or ion exchange with the massive cations and the subsequent electron microscopy.

Preparation of Nucleic Acids in which the Guanine Bases are Selectively Marked with Benzenedisulfonic Acid Groups (*2–4*)

2-Amino-*p*-benzenedisulfonic acid is purified by repeated precipitation with concentrated HCl from an alkaline solution. The diazonium salt is made by dissolving 126 gm of the purified amine in 400 ml of water and 22 ml of 50% sodium hydroxide. Forty grams of $NaNO_2$ is added in the cold, the solution is filtered, and 100 ml of 12 N HCl is added. The diazonium salt precipitates as a white slurry. It is filtered in the cold and stored at a concentration of 0.8 M in 0.05 N HCl at –70°C. Under these conditions, the diazonium activity decreases by less than 50% during 1 year.

The nucleic acid to be marked is dissolved in 1 ml of water or 0.1 M $NaHCO_3$ buffer at pH 9. Then 1 ml of diazonium slurry at 0°C is brought to pH 8 to 9 by the addition of about 0.03 ml of 50% NaOH. This is added to the nucleic acid solution; thereafter the pH is maintained at 9.0 with a saturated solution of K_2CO_3 adjusted to pH 11.5 with HCl. The reaction is rapid at first, but slows down as the diazonium decays. After 10 or 15 min more diazonium is added and the procedure repeated three times. Finally, the reacted nucleic acid is purified on Sephadex G-75.

The macromolecular component obtained has benzenedisulfonic groups on the guanine residues and the others. To convert the product into one where primarily guanines are labeled, the nucleic acid must be incubated at pH 4. To do this, the nucleic acid is precipitated by ethanol and redissolved in 0.1 M sodium acetate at pH 4 and incubated for 1 hr at 37°C. The product obtained is a nucleic acid in which primarily the guanine residues carry the benzenedisulfonic groups.

The spectra of samples of nucleic acid can be used to estimate the extent of reaction of the guanine base. Thus Moudrianakis and Beer (*3*) found for salmon sperm DNA the following results:

OD_{386} nm/OD_{260}	% Guanine bases reacted with diazonium
0	0
0.08	31
0.15	61
0.23	80

Should the extent of the reaction be insufficient, diazotization of the nucleic acid can be repeated.

Often the reaction with the diazonium salt leads to some fragmentation of the nucleic acid molecules, but the unbroken components can be recovered by sucrose gradient centrifugation or appropriate Sephadex chromatography. This has been most clearly examined by Erickson and Beer (*4*), who showed that after 30% of the guanine residues of MS-2 RNA had been diazotized about 5% of the RNA showed an undiminished sedimentation velocity. Electron microscopy on this fraction showed that the RNA molecules were essentially unbroken. Similar data on the fragmentation of DNA during diazotization are not available at present.

Preparation of Deoxyribonucleic Acid in Which Thymine Bases are Selectively Marked (*5–7*)

It is possible to produce deoxyribonucleic acid (DNA) in which approximately 90% of the thymines are converted to an addition product of the thymine base with 1 mole of osmium tetroxide and 2 moles of cyanide ion. The reaction, which requires single strands, is gentle and leaves molecules as long as ϕX 174 DNA essentially unbroken. As the reaction proceeds the buoyant density of the macromolecule increases. This can be shown by cesium chloride equilibrium density gradient centrifugation. A convenient method of preparation is the following.

A reagent solution is made using OsO_4 and KCN neutralized with HCl and dissolved in water from a quartz still to give a solution 0.2 M OsO_4 and 0.2 M KCN at pH 7.0. To the solution an equal volume of denatured DNA is added in concentration up to 100 μg/ml in 0.01 M NaCl. When double stranded DNA is used it is advisable to denature it before adding the OsO_4 and KCN. The resulting solution is sealed in a glass tube and incubated in a water bath at 55°C for 4 hr. After the reaction the solution is passed through a column containing Sephadex G-75 (particle size, 40–120 μ), and eluted with 0.01 M NaCl. The fractions are pooled. It appears that the reaction product is not completely stable and should be used for electron microscopy as soon as possible after preparation. The decay appears to have a half-life of perhaps 100 hr at 25°C.

Deposition of Labeled Nucleic Acid Molecules on Thin Carbon Films

After labeling the nucleic acid must be deposited on supporting films. For high-resolution microscopy today, it is still best to use exceedingly thin carbon support films, perhaps only 20 to 30 Å thick. These must be supported on thicker perforated films which can be produced as reviewed by Bradley (*8*). The transfer of the single stranded nucleic acid molecules to the carbon films is best accomplished by the method of Koller et al. (*9*). The grids with the carbon side down are floated on a very dilute aqueous solution (10^{-5}%) of alkyldimethylbenzyl ammonium chloride (BAC). This material is commercially available as the disinfectant Zephiran (Winthrop Laboratories,

New York, N.Y.). After floating overnight the grids are removed, blotted dry on filter paper and touched to the surface of a solution containing 0.1 to 1.0 μg/ml DNA and 0.001 to 0.1 M ammonium acetate or potassium chloride at pH 5 to 7. The excess liquid is left on the grid for about 15 sec and then removed with absorbent filter paper. The grid is finally washed on distilled water. Under these conditions, the distribution of small DNA molecules, such as φX 174 DNA, over the whole grid is regular. The adsorption can be carried out with the nucleic acid solution at elevated temperature and so one can gain some insight into the configurational changes accompanying warming. For example, φX 174 DNA at 20°C appears as a compact structure whereas at 55°C it occurs as a larger and thinner ring. Similarly the gradual melting of double-stranded nucleic acid can be followed.

Staining of Nucleic Acid Molecules with Electron Scattering Cations

To render visible the nucleic acid molecules, as well as the selectively bound organic reagents, appropriate cations must be bound to the various anionic groups. This "staining" is generally carried out by floating the grid on suitable solutions to bring about the ion exchange. Solutions used with some success contain 10^{-3} M Ta_6Br_{14} at pH 5-6 or 10^{-3} M UO_2 $(COO-)_2$ at pH 4. After staining for some minutes the grids are washed for several seconds on salt-free water and then dried on filter paper. The best micrographs, obtained by Bartl, Erickson, and Beer (*10*), using these procedures give strong evidence for the visibility of the base specific groups but not for the phosphate groups. The recent results obtained by Crewe, Wall, and Langmore (*11*) suggest strongly that with the energy selecting scanning microscope single heavy atoms can be detected. Thus results from that instrument would be expected to reveal the positions of a phosphates as well as the base specific reagents. To date, however, no such micrographs are available.

References

1. Beer, M., and E. N. Moudrianakis. 1962. *Proc. Natl. Acad. Sci., U.S.,* **48:** 409.
2. Moudrianakis, E. N., and M. Beer. 1965. *Biochem. Biophys. Acta,* **95:** 23.
3. Moudrianakis, E. N., and M. Beer. 1965. *Proc. Natl. Acad. Sci., U.S.,* **53:** 564.
4. Erickson, H. P., and M. Beer. 1967. *Biochemistry,* **6:** 2694.
5. Beer, M., S. Stern, D. Carmalt, and K. H. Mohlhenrich. 1966. *Biochemistry,* **5:** 2283.
6. Highton, P. J., B. L. Murr, F. Shafa, and M. Beer. 1968. *Biochemistry,* 7: 825.
7. DiGiamberardino, L., T. Koller, and M. Beer. 1969. *Biochem. Biophys. Acta,* **182:** 523.

8. Bradley, D. E. 1965. In D. H. Kay (Ed.), Techniques for electron microscopy, 2nd edition. F. A. Davids, Philadelphia.
9. Koller, T., A. G. Harford, Y. K. Lee, and M. Beer. 1969. *Micron,* **1:** 110.
10. Bartl P., H. P. Erickson, and M. Beer. 1970. *Micron,* **1**: 374.
11. Crewe, A. V., J. Wall, and J. Langmore. 1970. *Science,* **168:** 1338.

A SPECIFIC METHOD FOR THE DETERMINATION OF THE ORIGIN AND HOMOLOGY OF tRNA

V. DANIEL AND U. Z. LITTAUER*

DEPARTMENT OF BIOCHEMISTRY

WEIZMANN INSTITUTE OF SCIENCE

REHOVOT, ISRAEL

Principle

Labeled *N*-acetylaminoacyl-tRNA is hybridized with DNA. The method involves the enzymatic charging of tRNA with a labeled amino acid. The labeled aminoacyl-tRNA is then acetylated chemically to yield *N*-acetylaminoacyl-tRNA. This is done because N-substituted aminoacyl-tRNA derivatives are more resistant to hydrolysis than the corresponding unblocked aminoacyl-tRNA derivatives (*1-4*). The *N*-acetylaminoacyl-tRNA derivatives are then hybridized with DNA under conditions that preserve the *N*-acetylaminoacyl-tRNA ester bond. Since the radioactive label is present only in the aminoacyl moiety of the *N*-acetylaminoacyl-tRNA, one is able to monitor the hybridization of the tRNA specific for the given amino acid.

Materials

E. coli tRNA is prepared as described by Littauer et al. (*5*), followed by digestion with deoxyribonuclease to remove any DNA that still might be

* Fogarty Scholar in residence 1969–1970, U.S., National Institutes of Health.

present. The tRNA solution is incubated for 60 min at 37° in a solution containing 10 mM Tris-HCl buffer (pH 7.4), 10 mM $MgCl_2$ and 30 μg/ml electrophoretically purified DNase. Preincubated pronase (200 μg/ml) is added and the tRNA solution is further incubated for 4 hr at 37°. The RNA solution is mixed with 0.2 vol of 5 M $NaClO_4$ and is then extracted twice with an equal volume of chloroform and half a volume of phenol. The tRNA is precipitated by the addition of 2 vol of cold ethanol and the precipitate washed three times with ethanol and dissolved in water. If the pronase treatment of tRNA is omitted, the tRNA should be stripped of amino acids by incubation in 100 mM Tris-HCl buffer (pH 8.8) for 3 hr at 37°. All the tRNA preparation should be tested for ribonuclease contamination by incubating an aliquot for 20 hr at 37° in 10 mM Tris buffer (pH 7.4) and measuring its amino acid acceptor activity as compared with nonincubated samples (*5*).

Pronase B grade (Calbiochem) is preincubated for 5 hr at 37° in 10 mM Tris-HCl buffer (pH 7.5) to reduce endogenous nuclease contamination. *E. coli* aminoacyl-tRNA synthetases are prepared as described by Muench and Berg (*6*) up to the hydroxyapatite step. The aminoacylation of tRNA is performed under the conditions optimal for the formation of the given aminoacyl-tRNA. In addition to the respective (^3H)-labeled amino acid, the reaction mixture contains the other 19 nonlabeled amino acids.

(^3H)-Labeled amino acids with high specific activities (1-50 Ci/mmole) are obtained from Schwarz, Bioresearch, Orangeburg.

N-Hydroxysuccinimide acetyl ester is prepared as described by de Groot et al. (*7*); it may also be purchased from Miles-Yeda Co., Rehovot, Israel.

tRNAs from T4 infected cells and T4 DNA are prepared as described previously (*8*).

E. coli DNA is prepared by a modification of Marmur's method (*9*). The denaturation of DNA is carried out by alkali (*10*). To a DNA solution (100 μg/ml) 1 N NaOH is added to bring the pH to 13. After standing for 10 min, the solution is neutralized with 1 N HCl. The denaturation reaction is followed by the increase in A_{260} of the solution.

Nitrocellulose membrane filters (Type B6-coarse 27 mm) are purchased from Schleicher and Schuell.

Dimethylsulfoxide is distilled before use and kept dry.

T_1 ribonuclease (Calbiochem) 5000 units, is dissolved in 1.0 ml 1 x SSC (standard saline citrate buffer) pH 5.0, heated for 10 min at 80° and then cooled in ice.

SSC contains 0.15 M NaCl and 0.015 M sodium acetate. The pH is adjusted to 5 by the addition of 1 M citric acid.

All glassware are sterilized and all buffers and solutions are autoclaved before use to ensure destruction of any ribonuclease contamination.

PROCEDURE

Preparation of *N*-Acetylaminoacyl-tRNA

(^{3}H)-Aminoacyl-tRNA (300 μg) dissolved in 0.1 ml of 0.01 M sodium acetate buffer (pH 5.0) is reacted with a solution of 0.2 ml of dimethylsulfoxide containing 20 mg of *N*-hydroxysuccinimide acetyl ester. The mixture is shaken briefly and kept at room temperature for about 15 hr. The *N*-acetyl-(^{3}H)-aminoacyl-tRNA is precipitated with 0.2 vol of 5 M NaCl and 3 vol of ethanol, centrifuged, and the precipitate washed twice with ethanol and dissolved in 1.0 ml of 0.01 M sodium acetate buffer, pH 5.0.

Hybridization Procedure

N-Acetyl-(^{3}H)-aminoacyl-tRNA (130 μg) is mixed with a 25 ml solution of 1 mg denatured DNA in 4 x SSC, pH 5.0. The mixture is incubated for 45 min at 70°. Aliquots containing 50–100 μg DNA are loaded onto 27 mm nitrocellulose filters previously washed with 2 x SSC. The loading of the hybrid is carried out under slow suction. (The amount of DNA to be loaded on the filter is estimated by calculating the expected percentage of the genome specifying the given tRNA species so that the amount of labeled aminoacyl-tRNA hybridized will be enough to permit the monitoring of radioactivity.) The actual amount of DNA bound on the filter is calculated from the difference between A_{260} readings of the hybridization mixture before and after passing it through the nitrocellulose filters. The filters are then washed with 50 ml of 2 x SSC on each side and incubated with T_1 ribonuclease (12 units/ml) in 2 x SSC (pH 6.0) for 60 min at 23°. The filters are removed and washed with 50 ml of 2 x SSC on each side, dried, and counted in a scintillation spectrometer.

Blank values are obtained by heating a solution of *N*-acetyl-(^{3}H)-aminoacyl-tRNA for 45 min at 70° in the absence of DNA and passing it through nitrocellulose filters followed by washings with 2 x SSC and T_1 ribonuclease treatment as described above. In order to eliminate the possibility of nonspecific *N*-acetyl-(^{3}H)-aminoacyl-tRNA binding to the filters, a control reaction is prepared in which the *N*-acetyl-(^{3}H)-aminoacyl-tRNA is hybridized with an unrelated denatured DNA (e.g., in an assay of homology in a mammalian DNA-tRNA system use bacterial DNA).

In certain instances the system can be made even more sensitive and specific. Increased resolution would be needed, for example, when the homology of a minor tRNA species is determined in the presence of a more abundant heterologous tRNA specifying the same amino acid. In a case such as this a 10- to 20-fold excess of the uncharged heterologous tRNA is

TABLE 1. The Hybridization of T4 *N*-Acetyl-(^{3}H)-aminoacyl-tRNA with T4 DNA [a]

T4 *N*-acetyl-(^{3}H)-aminoacyl-tRNA	DNA hybridized (cpm/mg DNA)	
	T4	Calf thymus
N-acetylleucyl-	2600	460
N-acetylisoleucyl-	1700	122
N-acetylarginyl-	1900	96
Control *E. coli* tRNA		
N-acetylleucyl-	0	–

[a]T4 *N*-acetyl-(^{3}H)-aminoacyl-tRNAs (65 μg) are mixed with denatured T4 DNA (500 μg) or calf thymus DNA (500 μg) in 12 ml of 5 x SSC, pH 5.0, and incubated for 45 min at 70°. The hybridization is carried out in the presence of a 13-fold excess uncharged *E. coli* tRNA (850 μg). Aliquots containing 80 to 160 μg of DNA are removed and loaded onto 27-mm nitrocellulose filters. The filters are washed with 2 x SSC, pH 5.0 (50 ml on each side), incubated in 4.0 ml of 2 x SSC, pH 6, with T_1 ribonuclease (12 units/ml) for 1 hr at room temperature, then washed again with 2 x SSC, pH 5.0. The specific activities of the various (^{3}H)-labeled amino acid varied between 0.8–1.4 x 10^7 cpm/mμmole. All values are corrected for nonspecific binding of *N*-acetyl-(^{3}H)-aminoacyl-tRNA to filters. These blank values are obtained by heating a solution of *N*-acetyl-(^{3}H)-aminoacyl-tRNA to 70° in the absence of DNA, passing the solution through filters, washing the filters with 2 x SSC, and then incubating with T_1 RNase. The blank values were 1000 cpm, 230 cpm, and 300 cpm, for *N*-acetyl derivatives of (^{3}H) leucyl-, (^{3}H) isoleucyl-, and (^{3}H) arginyl-tRNA, respectively.

added to the hybridization mixture. Such a procedure was recently employed to identify T4 bacteriophage-induced tRNA molecules (*11–13*). A major difficulty in this particular system arises from the fact that the T4-specific tRNA represents only a small fraction of the total amino acid-specific accepting tRNA found in T4-infected *E. coli* cells. Thus, when a given *N*-acetyl-(^{3}H)-aminoacyl-tRNA is prepared from tRNA isolated from T4 phage-infected cells, the T4 specific *N*-acetyl-(^{3}H)-aminoacyl-tRNA will be mixed together with a large excess of host *E. coli* *N*-acetyl-(^{3}H)-aminoacyl-tRNA molecules. In order to overcome any possible nonspecific hybridization of the *E. coli* *N*-acetyl-(^{3}H)-aminoacetyl-tRNA with the T4 DNA a 15-fold excess of unchanged *E. coli* tRNA was added to the annealing mixture. This enabled the identification of a number of T4 specific tRNA molecules (Table 1).

The stability of the ester bond in *N*-acetyl-(^{3}H)-aminoacyl-tRNA during the hybridization procedure was examined by determining the TCA precipitable counts of aliquots removed from the hybridization mixture before and after the heating period. Most of the *N*-acetyl-(^{3}H)-aminoacyl-tRNA deriva-

TABLE 2. Stability of *N*-Acetyl-(^{3}H)-aminoacyl-tRNA to Heating at 70°[a]

	Percentage of TCA remaining insoluble	
N-Acetyl-(^{3}H)-aminoacyl-tRNA	45 min	90 min
Valine	100	100
Tyrosine	100	97
Leucine	93	87
Arginine	97	94
Histidine	67	52
Methionine	86	78
Phenylalanine	90	81
Alanine	100	81
Isoleucine	100	93
Lysine	100	100
(^{3}H)Leucyl-tRNA	30	–

[a]The various *N*-acetyl-(^{3}H)-aminoacyl-tRNA samples were heated at 70° in 2 x SSC, pH 5.0, for the indicated periods (*12, 13*).

tives remained stable (80–100%) even after heating for 90 min at 70° (Table 2). Owing to their heat stability, the *N*-acetyl-(^{3}H)-aminoacyl-tRNA solutions may be recovered from filtrates of the hybridization reaction and then reused in a second hybridization experiment.

General Remarks

The acylation of tRNA with radioactive aminoacids and the subsequent hybridization of the aminoacyl-tRNA to DNA is a very useful procedure for studying both the structural homology and the origin of the tRNAs. Since the radioactive label is found only in the aminoacyl moiety, it allows the hybridization of amino acid-specific tRNA to be followed. The conditions for tRNA hybridization to homologous DNA, as determined for the *E. coli* system (*14, 15*) include incubation of the hybridization mixture at 65–70° in order to disrupt the tRNA secondary structure. When the hybridization of (^{3}H)-aminoacyl-tRNA with the respective DNA is carried out, it is necessary to preserve the aminoacyl-tRNA ester bond during the annealing reaction. To avoid hydrolysis of the aminoacyl-tRNA, the latter is acetylated with *N*-hydroxysuccinimide acetyl ester to yield *N*-acetyl-(^{3}H)-aminoacyl-tRNA, and the pH of the annealing mixture is lowered to 5.0. Using the *N*-acetyl-(^{3}H)-aminoacyl-tRNA–DNA hybridization method we have recently been able to demonstrate the presence of T4 bacteriophage specific tRNA molecules in *E. coli* infected cells (*11–13*). Different *N*-acetyl-(^{3}H)-aminoacyl-tRNA molecules prepared from tRNA isolated from T4 bacteriophage infected *E. coli* cells were found to hybridize with T4 DNA without being competed for *E. coli* tRNA. At the same time no hybridization was observed between the N-acetyl-(^{3}H)-aminoacyl-tRNA prepared from tRNA from uninfected cells and T4 DNA (Table 1).

Another way to preserve the aminoacyl-tRNA ester bond during the annealing reactions is to lower the temperature of the hybridization. Organic solvents such as formamide or dimethylsulfoxide were shown to reduce the mean thermal denaturation temperature (T_m) of nucleic acids (*16*). Bonner et al. (*17*) hybridized pea DNA and RNA in 30% formamide at 0° and 24°. However, these conditions are not suitable for the hybridization of rRNA and tRNA from *E. coli* (*18*). Formamide solutions of higher concentrations (50%), pH 5.0 and incubation at 30–33° were necessary for the effective hybridization of the tRNA molecules. Using these conditions Weiss et al. (*19*) hybridized leucyl-tRNA from T4 phage infected *E. coli* cells to T4 DNA to show the formation of T4 phage leucyl-tRNA. In addition, Nass and Buck (*20*) have been able to demonstrate by this method that rat liver mitochondrial leucyl-tRNA differs from cytoplasmic leucyl-tRNA in its primary base sequence and is transcribed on the mitrochondrial DNA. However, not all aminoacyl-tRNA ester bonds for the different aminoacids have the same stability under the conditions of hybridization. The preparation of N-substituted derivatives of the aminoacyl-tRNAs increases the stability of these compounds and enables the hybridization of the aminoacyl-tRNA to be carried out under the same conditions as for the uncharged tRNA (e.g., at 70°).

Acknowledgments

These studies were supported, in part, by USPHS Research Grant No. R05-TW-00260 and by agreement 455114.

References

1. Simon, S., U. Z. Littauer, and E. Katchalski. 1964. *Biochim. Biophys. Acta,* **80:** 169.
2. Littauer, U. Z., S. Simon, and E. Katchalski. 1964. *Symp. Nucleic Acids, Hyberabad,* Saraswaty Press, Calcutta, p. 246.
3. Katchalski, E., S. Yankofsky, A. Novogrodsky, Y. Galenter, and U. Z. Littauer. 1966. *Biochim. Biophys. Acta,* **123**: 641.

4. Yankofsky, S. A., S. Yankofsky, E. Katchalski, and U. Z. Littauer. 1970. *Biochim. Biophys. Acta,* **199:** 56.
5. Littauer, U. Z., S. A. Yankofsky, A. Novogrodski, H. Bursztyn, Y. Galenter, and E. Katchalski. 1969. *Biochim. Biophys. Acta,* **195:** 29.
6. Muench, K. H., and P. Berg. 1966. In G. L. Cantoni and D. R. Davies (Eds.). Procedures in nucleic acid research. Harper & Row, New York, p. 375.
7. de Groot, N., Y. Lapidot, A. Panet, and Y. Wolman. 1966. *Biochim. Biophys. Res. Commun.* **25:** 17.
8. Daniel V., S. Sarid, and U. Z. Littauer. 1968. *FEBS Letters,* **2:** 39.
9. Daniel, V., S. Lavi, and U. Z. Littauer. 1969. *Biochim. Biophys. Acta,* **182:** 76.
10. Gillespie, D., and S. Spiegelman. 1965. *J. Mol. Biol.,* **12**: 829.
11. Daniel, V., S. Sarid, and U. Z. Littauer. 1969. *Abstr. 6th FEBS Mtg., Madrid,* p. 200.
12. Littauer, U. Z., and V. Daniel. 1969. *J. Cell. Physiol.,* **74**: suppl. 1, 71.
13. Daniel, V., S. Sarid, and U. Z. Littauer. 1970. *Science,* **167:** 1682.
14. Giacomoni, D., and S. Spiegelman. 1962. *Science,* **138**: 1328.
15. Goodman, H. M., and A. Rich. 1962. *Proc. Natl. Acad. Sci. U.S.,* **48**: 2101.
16. McConaughy, B. L., C. D. Laird, and B. J. McCarthy. 1969. *Biochemistry,* **8:** 3289.
17. Bonner, J., G. Kung, and I. Bekhor. 1967. *Biochemistry,* **6:** 3650.
18. Daniel, V. and U. Z. Littauer. 1967. Personal observations.
19. Weiss, S. B., W. T. Hsu, J. W. Foft, and N. H. Scherberg. 1968. *Proc. Natl. Acad. Sci., U.S.,* **61:** 114.
20. Nass, M. M. K., and C. A. Buck. 1969. *Proc. Natl. Acad. Sci., U.S.,* **62:** 506.

CHROMATOGRAPHY OF NUCLEIC ACIDS ON HYDROXYAPATITE COLUMNS*

GIORGIO BERNARDI

LABORATOIRE DE GÉNÉTIQUE MOLECULAIRE

INSTITUT DE BIOLOGIE MOLECULAIRE

FACULTÉ DES SCIENCES

PARIS, FRANCE

I. INTRODUCTION

Hydroxyapatite† (HA)‡ columns, originally developed by Tiselius, Hjertén, and Levin (*1-3*) for protein chromatography, were first used with nucleic acids by Semenza (*4*), in Tiselius' laboratory, and by Main et al. (*5-7*). In this early work it was seen that native DNA could be adsorbed on and eluted from the columns, that DNA degraded by pancreatic DNase or by acid had a lower affinity for HA than undegraded DNA, and that protein, TMV RNA, and poly A could be partially separated from DNA.

Work begun in 1959 in the author's laboratory (*8, 9*) as a development of previous investigations on the chromatography of phosphoproteins on HA

* This article also appears in *Methods in Enzymology*, Volume XXI, edited by L. Grossman and K. Mololave, Academic Press, New York, 1971.

† Hydroxyapatite, not hydroxylapatite, is the name recommended by Wyckoff (*106*), since "hydroxyl" implies the derivatives being named after the substituted ion, a usage which is not observed in the corresponding fluorine and chlorine derivatives (e.g. "fluorapatite," "chlorapatite," not "fluoridapatite," "chloridapatite").

‡ Abbreviations: HA, hydroxyapatite; NaP, KP, equimolar mixtures of NaH_2PO_4, Na_2HPO_4 and KH_2PO_4, K_2HPO_4, respectively (pH is close to 6.8; ionic strength is equal to about twice the molarity). The abbreviation PB (phosphate buffer) used by some authors does not indicate the cation; since the eluting power of phosphate is quite different for different salts, it is advisable not to use the abbreviation PB.

columns (*10*) led to the recognition (*11*) that HA could discriminate nucleic acids endowed with different secondary structures, rigid, ordered structures having more affinity for HA than flexible, disordered ones (*12–17*), a general rule also holding for proteins (*18*). Although most fractionations of nucleic acids reported so far are based on this property of HA, subtler differences in secondary and tertiary structures can also be discriminated by HA columns (*12, 17, 19, 20*) since they originate different distributions of groups available for interaction with the adsorbing sites of HA (see Section IX).

The aim of this chapter is to review the known facts and the current ideas on the chromatography of nucleic acids on HA. This is a rather difficult task for two main reasons: (1) chromatography of biopolymers on HA is at present in a stage of fast development, both experimentally and theoretically; (2) most of the experiments which will be discussed here were performed using conditions chosen in an empirical way; a systematic exploration of the parameters involved in the chromatography of nucleic acids (like that done by Kawasaki and Bernardi (*21, 22*) for proteins) is not yet available. It should be mentioned that a theory of the chromatography of rigid macromolecules has been developed (*23, 24*) based on experimental results and general conclusions arrived at in Tiselius' as well as in the author's laboratories. It is likely that this theory is valid for nucleic acids having a rigid structure.

II. MATERIALS AND METHODS

A. Preparation of Hydroxyapatite

1. Preparation Procedure of Tiselius et al. (*1*)

The following is a description of this procedure as used in our laboratory, taking into account minor modifications suggested by Levin (*3*), Miyazawa and Thomas (*25*), and Bernardi (*15*).

Materials. In the author's laboratory the following analytical grade reagents (Merck, Darmstadt, Germany) are routinely used: $CaCl_2 \cdot 2H_2O$ (Merck catalog no. 2382), $Na_2HPO_4 \cdot 2H_2O$ (no. 6580), $NaH_2PO_4 \cdot H_2O$ (no. 6346), $K_2HPO_4 \cdot 3H_2O$ (no. 5099), KH_2PO_4 (no. 4873).

Preparation of Brushite, $CaHPO_4 \cdot 2H_2O$. Two liters each of 0.5 M $CaCl_2$ and 0.5 M Na_2HPO_4 are fed at a flow rate of 250 ml/hr (using a multichannel peristaltic pump; two separatory funnels with Pasteur pipettes as outlets may also be used) into a 5-liter beaker containing 200 ml of M NaCl. The addition is done under stirring just sufficiently strong to avoid sedimentation of the brushite precipitate. At the end of this step, brushite is allowed to settle; the supernatant is decanted and the precipitate is washed with two 4-liter volumes of distilled water.

Conversion of Brushite into Hydroxyapatite, $Ca_{10}(PO_4)_6(OH)_2$. Brushite is suspended in 4 liters of distilled water and stirred, 100 ml of 40% (w/w) NaOH is added, the mixture is then heated in 40-50 minutes to boiling, and boiled for 1 hour with simultaneous stirring. The precipitate is then allowed to settle completely and the supernatant is siphoned off. The precipitate is then washed with 4 liters of water. The supernatant is siphoned off when a 2-cm layer of precipitate is formed on the bottom of the beaker. This is the only time during the whole procedure when a complete settling of the precipitate is not allowed in order to eliminate the "fines." The precipitate is then washed twice more, allowing a complete settling. At this point, the precipitates from two preparations are pooled together and suspended in 4 liters of 0.01 M sodium phosphate buffer, pH 6.8 (NaP) and just brought to boiling; boiling at this point should be avoided. The precipitate is then suspended in 4 liters of 0.01 M NaP and boiled for 5 min. This operation is repeated once more using 0.01 M NaP and then again using 0.001 M NaP; in both cases boiling is done for 15 min. About 400-500 ml of packed precipitate are obtained from two pooled preparations.

Storage of Hydroxyapatite. The final precipitate, formed by blade-like crystals, can be stored in 0.001 M NaP for several months at 4° without any change to its chromatographic properties. The addition of chloroform as a preservative is not necessary. When resuspending HA crystals strong agitation should be avoided, since this breaks down the crystals and their aggregates, thus rendering the material unsuitable for column chromatography.

2. Alternative Preparation Procedures

Other methods for preparing hydroxyapatite have been described by Main, Wilkins, and Cole (*7*), Anacker and Stoy (*26*), Jenkins (*27*), and Siegelman et al. (*28, 29*). Not many results have been reported with these preparations so that it is therefore difficult to judge their relative merits.

3. Commercial Hydroxyapatite Preparations

A preparation obtained according to the procedure of Tiselius et al. (*1*) is sold by Bio-Rad Laboratories (Richmond, Calif.) either as a suspension in 0.001 M NaP, or as a dry powder. Another preparation is sold by Clarkson Chemical Co. (Williamsport, Pa.). Comments on the preparations sold during the past two years have been generally favorable.

B. Experimental Techniques with Columns

For general instructions on column chromatography the reader is referred to Determann (*30*) and Fischer (*31*). Some features which apply more specifically to HA columns, will be briefly discussed here.

1. Packing of the Columns

Packing the columns is accomplished by adding a suspension of HA crystals in Na or K phosphate buffers, pH 6.8 (NaP or KP), to columns partially filled with the same buffer; the column outlet is progressively open only after a 1-cm layer of HA is settled. Further additions of the HA suspensions are then made to fill the column. The filling operation may be facilitated by the extension of the column with a glass tube of the same diameter. Alternatively, columns may be prepared by adding the HA suspension to a funnel mounted on the top of the column, the whole system being full of starting solvent. The HA suspension in the funnel is stirred during the preparation of the column. This procedure, suggested by Flodin (*32*) for Sephadex, allows for very homogeneous packing.

2. Adsorption and Elution

As a rule, the sample is loaded in the solvent with which the column was previously equilibrated, generally a low-molarity NaP or KP.

As a rule, NaP or KP of increasing molarities are used to elute nucleic acids. NaP cannot be used at 4° at molarities higher than 0.5 M because of the limited solubility of Na_2HPO_4 at this temperature. Columns are normally operated under a slight pressure (30–50 cm of water). If controlled by a pump, the flow rate should not be kept higher than that of a column flowing under a slight hydrostatic pressure. The phosphate molarity in the column effluent may be checked by refractive index measurement, phosphorus analysis, conductimetry.

3. Column Regeneration

If elution of adsorbed material is complete, the column may be simply reequilibrated with the starting buffer and reused, preferably after removal of the top layer. The same column can be reused 3–4 times.

4. Recovery of Irreversibly Adsorbed Materials

The HA bed may be extruded from the column and treated in one of the following ways: (1) put in dialysis bags and dissolved by dialysis vs. 1 M EDTA, pH 8.0; (2) eluted with 0.1 M NaOH; (3) dissolved in 1 N HCl.

C. The Adsorption–Elution Process

As already mentioned, a systematic exploration of the parameters involved in the chromatography of nucleic acids on HA columns is yet to be done. Under these circumstances, it seems useful at least to review briefly the basic features of the adsorption–elution process and to present the limited information obtained so far on this subject in our laboratory.

1. Adsorption

This may be done in a batch or on a column. Five sets of parameters should be considered: (1) the HA bed, (2) the material to be adsorbed, (3) the solvent, (4) the temperature at which adsorption takes place, and (5) the time of contact of the nucleic acid solution with HA, respectively.

HA Bed. The total volume of packed HA crystals, V_t (total volume), is equal to the sum of three terms: the volume of the dry crystals, V_c (crystal volume), the volume of the solvent bound to the HA crystals and inaccessible to the material to be adsorbed, V_i (inner volume), and the volume of the solvent between the HA crystals and accessible to the material to be adsorbed, V_o (outer volume):

$$V_t = V_c + V_i + V_o \tag{1}$$

The total volume of the packed HA bed, V_t, can be easily determined by measuring its dimensions. The outer volume, V_o, can be determined by measuring the elution volume of a nonadsorbed substance, such as methyl orange, eosin, fuchsin, methyl red (*3*), i.e., the volume of the solvent which leaves a HA column between loading and appearance of this substance in the effluent. The inner volume, V_i, can be calculated from the difference $(V_o + V_i) - V_o$, the term $(V_o + V_i)$ being determined by measuring the loss in weight, at 110°, of a known amount of packed HA crystals. The crystal volume, V_c, may be calculated from the difference $V_t - (V_o + V_i)$.

HA preparations obtained as described in Section II.A.1, packed under stirring and equilibrated with 0.001 M KP, exhibit linear flow rate vs. pressure drop diagrams, a pressure drop (hydrostatic pressure divided by the length of the column) of 10 causing a flow rate of ~100 ml/cm²/hr. For these preparations, $V_o = 0.82$, $V_i = 0.10$, and $V_c = 0.08$ ml/ml HA bed. The density of the packed HA crystals (wet) is equal to 1.17 gm/ml. The value found for V_o is quite reproducible for preparations obtained according to the method described above and definitely higher than that (0.60–0.75) reported by Levin (*3*). Obviously, HA preparations obtained according to different procedures or preparations in which crystals were broken down may have different properties. Since HA crystals are in the form of lamellae, it is likely that mechanical breakdown does not cause a very large increase in the surface available for adsorption.

The Material to be Adsorbed. Two parameters are of interest. The first is the amount of material to be adsorbed, which should be established knowing the capacity of HA. As an indication of this point it can be mentioned that the amount of native DNA which can be adsorbed per milliliter of packed HA crystals equilibrated with 0.001 M KP is about 10 A_{260} units. The second parameter is the concentration of the material. The coexistence of different materials to be adsorbed should also be considered, since this will lead to competition for the adsorbing sites and cause displacement effects.

The Solvent. The concentration of eluting ions (phosphate ions, as a rule) at the adsorption step is obviously a critical parameter in determining the capacity of HA for a given material to be adsorbed. The presence in the solvent of substances having a stronger affinity for calcium than phosphate, such as EDTA, citrate, may decrease the capacity of HA to zero. On the other hand, an increase in ionic strength of the solvent due to ions having an affinity for calcium lower than phosphate, e.g., chlorides, decreases the eluting power of phosphate (see also Section III.A.4 and Section IX). KP is remarkably more effective than NaP as an eluting agent.

Temperature. Temperature has an effect on the adsorption phenomenon itself (adsorption isotherm), the ionization of phosphate ions, and the secondary structure of the nucleic acids to be adsorbed. The effect of temperature on adsorption and on phosphate ionization is not important, yet deserves to be investigated in detail; the effect on the nucleic acid structure may cause serious changes in their affinity for HA (see Section V.F. and VI.D).

Time of Contact. The time of contact between nucleic acids and HA necessary to reach adsorption equilibrium is of the order of half an hour if a solution of native DNA ($A_{260} = 0.4$) in 0.001 M KP is put in contact with a large excess of HA. If adsorption is done on a column rather than in batch one should consider the flow rate while loading the nucleic acid solution.

2. Elution

Elution may be performed by increasing the concentration of eluting ions (usually phosphate), either stepwise or continuously. Stepwise elution may be used with both batches and columns; molarity gradient elution can be used with columns only. In both cases, the flow rate of the eluent should be kept within certain limits to avoid a deformation of the chromatographic peaks; flow rates comprising between 5 and 50 ml/hr/cm^2 were used in most experiments presented here. A third way of eluting nucleic acids which has been applied to rigid, ordered structures is to increase the temperature of the column to the point where a helix-coil transition occurs with concomitant decrease in affinity for HA. If the phosphate molarity at which melting takes place is high enough to elute disordered structures, elution occurs (*25*). This procedure has, however, nothing to do with a chromatographic elution and will be discussed later (Section V.F).

Stepwise Elution. This procedure is very useful when separating two (or more) adsorbed substances which have known different elution molarities. Its two main disadvantages, when used with columns, are: (1) tailing of the peaks and (2) "false peaks." In the first case substances with strongly curved adsorption isotherms and therefore extended elution ranges cannot be eluted by a solvent of constant composition without tailing, unless elution is so strong that R_f is close to 1.0 (*1*). In the second case single substances with strongly curved isotherms may give rise to several zones with each new molarity step of the eluent releasing an additional amount of substance (*1*).

Gradient Elution. Two parameters are very important in determining the resolving power of the columns: the length of the column and the slope of the gradient in the column, *grad*. In the usual case of linear molarity gradients *grad* may be calculated as follows:

$$grad = \frac{\Delta M}{V} \frac{S}{V_o/V_t} \tag{2}$$

where ΔM is the difference in phosphate molarity between the initial and the final eluent, V the total volume of the eluent, S the cross-sectional area of the column, V_o and V_t the outer volume and the total volume of the column already defined. Using the above definition *grad* represents the increase in phosphate molarity per centimeter of column, if S and V are expressed in cm^2 and cm^3, respectively.

It is important to stress that most experiments described in the following pages were done under conditions of low resolution, namely using columns shorter than 10 cm and *grad* values of the order of 1 mM/cm.

When elution is done by a linear phosphate molarity gradient, the chromatographic behavior of a nucleic acid is characterized by two parameters (*21*). The first is the elution molarity, m_{elu}, which is defined as the phosphate molarity at which the center of gravity of the nucleic acid peak is eluted where the center of gravity of the peak is given by

$$\bar{V} = \int V f \, dV \Big/ \int f \, dV \tag{3}$$

In Eq. (3) f is the distribution function of the peak and V the volume of the solvent. The second parameter is the width of the peak which can be calculated as its standard deviation and should be normalized by dividing it by S:

$$\sigma = \left[\int (V - \overline{V})^2 f\, dV / \int f\, dV \right]^{1/2} \frac{1}{S}$$

Both chromatographic parameters, m_{elu} and σ, depend upon several factors, such as column length, slope of the gradient, presence of other chromatographic components.

III. CHROMATOGRAPHY OF NATIVE DNA

A. Chromatography of Native DNA

1. Properties of Native DNA after Chromatography

No significant changes in the physical, chemical, and biological properties of native DNA take place in the adsorption-elution process, as indicated by the following results (*15*).

DNA samples from calf thymus, chicken erythrocytes, and *E. coli*, displaying molecular weights (as determined by light-scattering) in the 4-6 x 10^6 molecular-weight range, did not show, after the adsorption-elution process, any significant difference with respect to the original samples in any of the following properties: light-scattering envelope (and therefore weight-average molecular weight and radius of gyration), sedimentation coefficient, ultraviolet spectrum, and ultraviolet melting curve.

DNA samples from the same sources, but showing a molecular weight higher than 6 x 10^6, often had a lower molecular weight after the adsorption-elution process. This phenomenon was apparently due, in some cases, to the removal of a small amount of large aggregates from DNA sample. In other cases, in which aggregates samples had been prepared by using steps leading to an aggregation of DNA (e.g., alcohol precipitation at an early stage of the deproteinization procedure), the adsorption-elution process caused a dis-aggregation of the intermolecular complexes possibly linked through protein material.

Results obtained with DNA samples from phages T1, T2, T4, T5, and λ showed that the sedimentation constants of the loaded and the eluted samples were identical. In the case of T2 DNA, preparations of "whole" molecules (mol wt = 1.3 x 10^8), as obtained by chromatography on methy-lated serum albumin-kieselguhr (MAK) columns, were used. It appears, therefore, that breakage by shearing does not occur during the elution of very high molecular weight DNA from HA.

Transforming *Haemophilus influenzae* DNA was adsorbed on and eluted

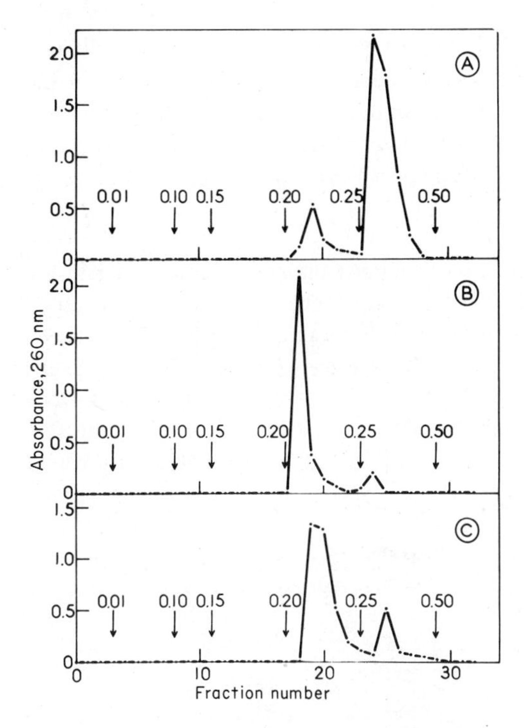

Figure 1. Chromatography of native calf thymus DNA (preparation A 1). (A) Chromatography of 1.28 mg of DNA on a 1.3 x 5-cm column. (B) Rechromatography of the 0.25 M fraction; 0.78 mg of DNA were loaded on a 1.3 x 3-cm column. In all cases 3-ml fractions were collected. Rechromatography experiments were done on pooled 0.20 and 0.25 M fractions from two chromatographic experiments. DNA loading took place at fraction 0; the stepwise increase in KP molarity is indicated by the vertical arrows. Reproduced from G. Bernardi. 1961. *Biochem. Biophys. Res. Commun.*, **6**: 54.

from HA without any modification in the biological activity of 3 different genetic markers (the ultraviolet monitoring system was not used in these experiments).

2. Recovery of Native DNA from the Columns

In the large majority of cases a complete recovery was obtained, as judged from A_{260} measurements. Incomplete recoveries from HA may be obtained with DNA preparations containing aggregated material (see above).

3. Stepwise Elution

When elution was carried out stepwise according to the scheme shown in Fig. 1, DNA samples from calf thymus and chicken erythrocytes (mol wt = 4–6 x 10^6) were eluted at 0.20 M and 0.25 M KP (Fig. 1A).

Occasionally, minor additional fractions were eluted when the KP molarity was further raised to 0.30 M and 0.50 M. On rechromatography each one of the two fractions (as well as the occasional minor peaks) was eluted again in two peaks, at 0.20 M and 0.25 M KP, respectively (Figs. 1B, 1C), indicating that these peaks may be considered as "false peaks" (1, 2, and Section II.C.2.).

4. Gradient Elution

When elution was performed with a linear molarity gradient of KP, the chromatogram obtained with DNAs from higher organisms or bacteria (mol wt = 4-6 x 10^6) showed only one peak centered at 0.20-0.22 M (Figs. 2A, 2B) whereas samples from T5 and T2 phages were eluted in single peaks centered at 0.27 M (Figs. 2C, 2D). It has been reported (*33*) that DNA from mouse lymphoma was only partially eluted by 0.26 M NaP, the rest being removed from the column by 1.0 M NaP under conditions where

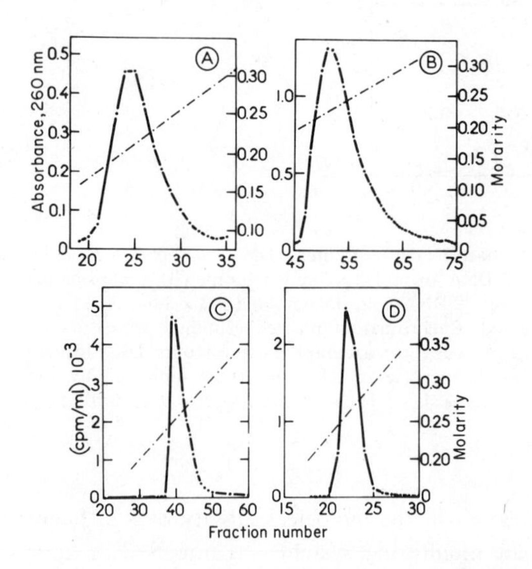

Figure 2. (A) Chromatography of native calf thymus DNA (preparation B 15). A 2-ml sample of a solution having an $A_{260\ nm}$ = 5.0 was loaded on a 1 x 4.5-cm column. This experiment was carried out at 4°. Fractions of about 3.8 ml were collected. Elution was carried out with 100 + 100 ml of 0.001 and 0.5 M KP. (B) Chromatography of *H. influenzae* DNA. A solution (4 ml) having an $A_{260\ nm}$ = 5.01 was loaded on a 1 x 10-cm column. Fractions of 3 ml were collected. Elution was carried out with a molarity gradient (150 + 150 ml) of KP (0.001-0.5 M). Flow rate, 36 ml/hr. Recovery was 92%. (C) Chromatography of ^{32}P-labeled T5 DNA on a 1 x 6-cm column. Elution was carried out with a molarity gradient of KP. Fractions of 2.6 ml were collected. Then 1.2 x 10^5 counts/min were loaded and recovered form the column. (D) Chromatography of ^{32}P-labeled T2 DNA on a 1 x 10 cm column. Elution was carried out with a molarity gradient of KP containing 1% formaldehyde. Fractions of 5.2 ml were collected. Recovery was 92%. Flow rate, 40 ml/hr. Reproduced from G. Bernardi. 1969. *Biochim Biophys. Acta,* 74: 423.

mouse thymus DNA was eluted in 94% yield by 0.26 M NaP. The meaning of this finding is not clear, however, since the starting DNA and the fractions were not characterized.

Interestingly enough, gradient elution can also be carried out at a practically constant ionic strength, using gradients formed by 0.001 M KP + M KC1 as the starting buffer and 0.5 M KP as the final buffer. In both these solvents the ionic strength is equal to 1.0, since the ionic strength of NaP and KP is equal to about twice their molarities (see footnote p. 455). In this case, DNA is eluted at the same phosphate molarity as in the absence of KCl. If a still higher ionic strength is used, the two buffers being, for instance, 0.001 M KP *plus* 2 M KCl and 0.5 M KP *plus* 1 M KCl, the phosphate eluting molarity was higher and equal to 0.27 M. This increase in phosphate eluting molarity is already observed at a 1 M level when using a Na^+ system. For example, using 0.001 M NaP + M NaCl and 0.5 M NaP as the limiting buffers, DNA was eluted at a NaP of 0.31 (*15*).

B. Fractionation of Native DNA

1. Fractionation According to Molecular Weight

HA columns have a very low degree of discrimination toward molecular weight. For instance, under experimental conditions similar to those of Fig. 2, calf thymus DNA samples ranging in molecular weight from 6×10^6 to 1×10^5 obtained by limited degradation by spleen acid DNase (an enzyme known to break both DNA strands at the same level and to cause no significant changes in the secondary structure of DNA in the molecular weight range under consideration, *34, 35*) are eluted at the same KP molarity (0.20–0.22). Similar results were obtained with DNA samples sheared in a high-speed Vir-Tis homogenizer in the presence of chloroform-isoamyl alcohol (E. G. Richards, 1962, unpublished experiments). In contrast to this behavior, DNA samples from bacteria and higher animals displaying molecular weights in excess of 10^7 were eluted at a slightly higher KP molarity (0.22–0.25) and the large DNAs from T2 and T5 phages were eluted at about 0.27 M KP, as already mentioned.

Moderate degrees of fractionation were obtained when running artificial mixtures of degraded and undegraded DNA. For instance T2 phage DNA "whole" molecules (mol wt = 1.3×10^8) could be separated to a fair extent from sonicated T2 DNA (mol wt $\cong 5.10^5$), but not all from T2 DNA "half" molecules (*15*). Similarly, artificial mixtures of intact DNA and of DNA partially degraded by spleen acid DNase (see above) could be fractionated to some extent by stepwise elution (*15*). In these cases it is possible that fractionation depends on the displacement of the shorter molecules by the large ones, and on physical fractionation according to the hydrodynamic volume of DNA molecules during the flow through the HA column.

2. Fractionation According to Secondary or Tertiary Structures, Glucosylation, and Base Composition

When DNA samples from calf thymus and chicken erythrocytes or *H. influenzae* are chromatographed under experimental conditions of low resolution, no fractionation with respect to base composition or genetic markers can be detected.

In a few cases, native DNAs have a chromatographic behavior different from that just described because of their particular secondary or tertiary structures, glucosylation, and base composition. These DNAs can therefore be separated from those exhibiting the "usual" behavior.

The Single-Stranded DNA from ϕX 174 Phage. This DNA is eluted using the stepwise elution at 0.10 M and 0.15 M KP (Fig. 3A), therefore at molarities much lower than double-stranded DNA molecules (see also Sections V and IX).

The Twisted Circular DNA from Polyoma Virus. This DNA is eluted at a lower molarity than the open linear and circular forms of the same DNA (Fig. 3B); the latter forms being eluted at the same molarity. This interesting observation (*36, 37*), so far the only one concerning a tertiary structure of DNA, will be discussed in Section IX.

Glucosylated DNA from T-Even Phages. This DNA is eluted at a higher molarity than nonglucosylated *E. coli* or T5 (*15, 38*: see Fig. 3C). DNA from a nonglucosylated mutant strain or T4 bacteriophage (which, however, contains 5-hydroxymethylcytosine, like T4 DNA) is also separated, although to a lesser extent, from *E. coli* DNA (Fig. 3D). It is also interesting that the separation of T4 DNA and *E. coli* DNA is not observed anymore when these DNAs are chromatographed in a denatured state (*38*).

Mitochondrial Yeast DNAs. These DNAs are eluted at a higher molarity than nuclear yeast DNAs (*19, 20*; Fig. 4). This separation is not related to a difference in molecular weight, since both DNAs had approximately the same molecular weight, nor to a difference in the tertiary structure, since both DNAs were formed by open, linear molecules. (A very small percentage of open circular molecules present in mitochondrial DNA showed the same chromatographic behavior as the open, linear ones (*19*), a finding in agreement with the similar observation on the polyoma DNA (*36, 37*).) Yeast mitochondrial DNAs are rather exceptional in their base composition. In fact, DNAs from wild-type cells have AT contents of 83%, and DNAs from different cytoplasmic "petite" mutants have AT contents ranging from 85%

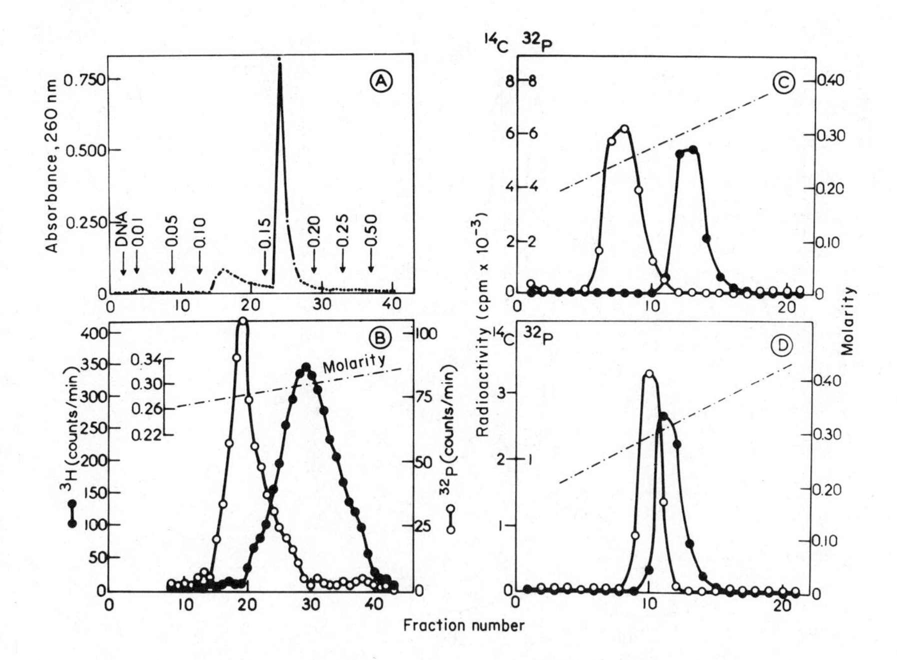

Figure 3. (A) Chromatography of ϕX 174 DNA on a 1.3 x 3 cm; 5 ml of DNA solution having an A_{260nm} = 1.68 was loaded; 3.3-ml fractions were collected. Recovery was 96%. Reproduced from G. Bernardi. 1969. *Biochim. Biophys. Acta,* **174**: 423. (B) Chromatography of native polyoma virus DNA. A sample of tritium-labeled component II (16 S, untwisted, circular) mixed with component I (20 S, twisted, circular) labeled with ^{32}P was loaded on a column and eluted with a linear concentration gradient (0.23–0.32 M) of NaP; 0.5 ml-fractions were collected. Reproduced from P. Bourgaux and D. Bourgaux-Ramoisy. 1967. *J. Gen. Virol.,* **1**: 323. (C) Chromatography of a mixture (1.0 ml) of ^{32}P-labeled T4 DNA (0.3 μg) and ^{14}C-labeled *E. coli* DNA (1.6 μg). The sample in 0.05 M phosphate, pH 7.0, was applied to a 1 x 3-cm column and was eluted by a linear molarity gradient of phosphate, pH 7.0, (0.18–0.40). Fractions of 2 ml were collected. Recovery of the DNA was 85% for T4 DNA and 87% for *E. coli* DNA. Reproduced from M. Oishi. 1969. *J. Bacteriol.,* **98**: 104. (D) Chromatography of a mixture (1.0 ml) of ^{32}P-labeled T4 gt DNA (0.4 μg) and ^{14}C-labeled *E. coli* DNA (1.6 μg). Experimental conditions as in part C. Recovery of the DNA was 99% for T4 gt DNA and 96% for *E. coli* DNA. Reproduced from M. Oishi. 1969. *J. Bacteriol,* **98**: 104.

to 96% according to the mutants (*20*). Furthermore, yeast DNAs contain not only alternating dAT : dAT stretches, but also nonalternating dA : dT ones (*19, 20, 39*). It is probable that the presence of the latter causes mitochondrial DNA to be eluted at a higher molarity. In fact biosynthetic nonalternating poly (dA : dT) has a high elution molariy, about 0.49 M NaP, whereas the alternating poly (dAT : dAT) has an elution molarity close to that of nuclear DNA. These results are very interesting in that they show that HA can discriminate slightly different native DNA structures.

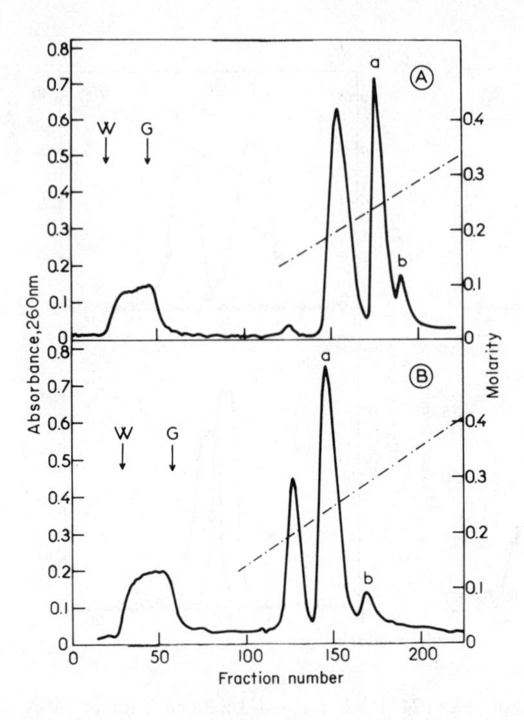

Figure 4. (A) Chromatography of a DNA preparation from a wild-type yeast. A 100-ml sample of DNA solution in 0.1 M NaP, A_{260} = 0.820, was loaded on a 2 x 40-cm HA column. The column was then washed with 100 ml of 0.1 M NaP and elution was carried out with a linear gradient (450 ml + 450 ml) of NaP (0.1–0.5 M). Loading was started at fraction 0, washing at fraction marked by arrow W, gradient at fraction marked by arrow G. Fractions of 3.8 ml were collected. Flow rate was close to 55 ml/hr. A_{260} recovery was 98%. (B) Chromatography of a DNA preparation from a "petite" cytoplasmic mutant. A 100-ml sample of DNA solution in 0.1 M NaP, A_{260} = 0.800, was loaded on a 2 x 34-cm HA column. Then 3.5 ml fractions were collected. Flow rate was about 50 ml/hr. A_{260} recovery was 100%. All other indications as above. Reproduced from G. Bernardi, F. Carnevali, A. Nicolaieff, G. Piperno, and G. Tecce. 1968. *J. Mol. Biol.*, **37**: 493.

IV. CHROMATOGRAPHY OF NUCLEOHISTONES

As described in Section III.A, native DNA can be eluted from HA columns by phosphate buffers in the presence of chlorides without any major change in its chromatographic behavior. Since nucleohistones can be progressively dissociated into their DNA and histone components by exposure to increasing salt concentrations and since histones are less strongly adsorbed by HA than DNA in the presence of salt (*40*), it should be feasible to separate on HA histone fractions, released by salt, from the residual partial nucleoproteins. This is indeed what happens, as shown in Fig. 5 (*41*). In these

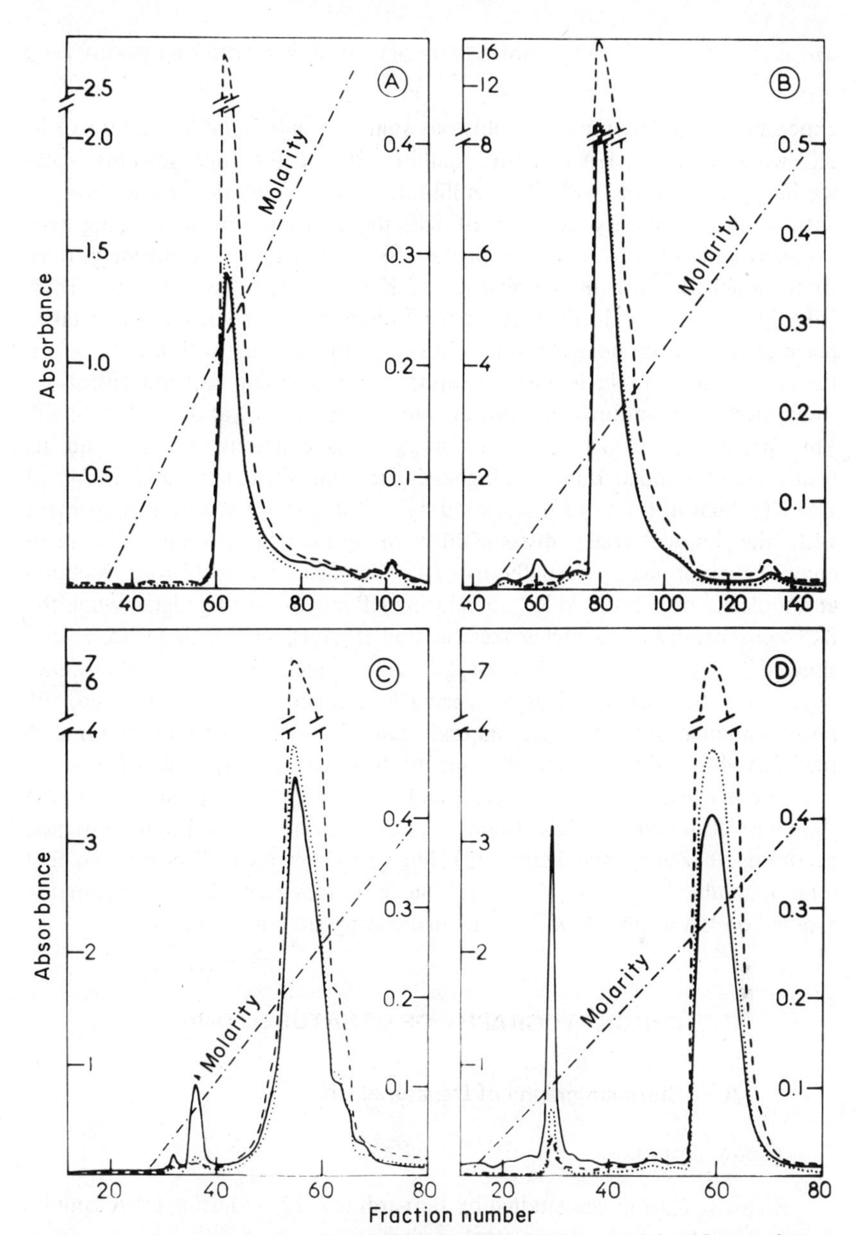

Figure 5. Chromatography of calf thymus nucleohistone on 2.3 x 19-cm columns. Elution was performed with a linear KP molarity gradient (450 ml; 0.001–0.5 M). The concentration of KCl in the eluting buffer varied from 0 to 3 M in different experiments. Curves indicated absorbances at 260 nm (broken line), at 280 nm (dotted line) and at 235nm (solid line). (A) Chromatography in the absence of KCl. Total load was 160 A_{260} units; fraction volume 6.7 ml; recovery 93%. Absorbance ratios of eluted material were $A_{235}/A_{260} = 0.532$ and $A_{280}/A_{260} = 0.553$. (B) Chromatography in the presence of 0.75 M KCl. Total load was 900 A_{260} units; fraction volume 4.4 ml; recovery 97%. Absorbance ratios of the main peak were $A_{235}/A_{260} = 0.525$ and $A_{280}/A_{260} = 0.555$. (C) Chromatography in the presence of 1 M KCl. Total load was 450 A_{260} units; fraction volume 6.8 ml; recovery 96%. Absorbance ratios of the main peak were $A_{235}/A_{260} = 0.500$ and $A_{280}/A_{260} = 0.548$. (D) Chromatography in the presence of 3 M KCl. Total load was 450 A_{260} units; fraction volume 6.8 ml; recovery 95%. Absorbance ratios of the main peaks were $A_{235}/A_{260} = 0.455$ and $A_{280}/A_{260} = 0.532$. Reproduced from I. Faulhaber and G. Bernardi. 1967. *Biochim Biophys. Acta,* **140**: 561.

experiments, calf thymus nucleohistone solutions in 0.7 mM KP and 0 to 3 M KCl were loaded on HA columns equilibrated with the same solvents. After washing the columns with the equilibration solvent, elution was performed with a linear molarity gradient of KP, the KCl concentration being kept constant and equal to that used in the dissociation step. The elution pattern of nucleohistone run in the absence of KCl resembles that of native DNA (Fig. 5A). It is very likely that a partial dissociation of nucleohistone takes place at its elution molarity, since the A_{260} tracing trails both the A_{235} and the A_{280} patterns. Nucleohistone solutions in increasing KCl concentrations show increasing amounts of a protein component eluting at 0.07–0.10 M KP. This protein peak had an A_{280}/A_{260} ratio equal to 1.2–1.3 and its lysine–arginine molar ratio was higher in fractions dissociated at 0.75 M and 1 M KCl than in fractions dissociated by 2.0 M and 3.0 M KCl, in agreement with the known easier dissociability of lysine-rich histones. The main component, formed by partially or totally dissociated nucleohistone, is eluted at a molarity of 0.2–0.3 M KP, the eluting KP molarity being higher when the KCl concentration was higher (see Section III.A.4). The nucleohistone peak showed A_{280}/A_{260} and A_{235}/A_{260} ratios which approached the values obtained with pure DNA as increasing KCl concentrations were used, the ratios obtained with the material dissociated by 3 M KCl being the same as for DNA obtained from nucleohistone by the detergent procedure (*34*).

The original work of Faulhaber and Bernardi (*41*) was performed on the "soluble" fraction of calf thymus nucleohistone preparations obtained according to Zubay and Doty (*42*). More recent work (*43*) has shown that total ("soluble" + "gel" fractions) nucleohistone can also be chromatographed on HA using 3 M KCl as the dissociating salt concentration.

V. CHROMATOGRAPHY OF DENATURED DNA

A. Chromatography of Denatured DNA

1. Stepwise Elution

Stepwise elution was studied by Bernardi (*11, 12, 16*) using DNA samples from calf thymus and chicken erythrocytes.

DNA Partially or Totally Denatured by Heat. DNA denatured by heating for 15 min at 100° at a concentration of 50–100 μg/ml in 0.13 M NaCl–0.01 M KP, and fast cooled, showed a chromatographic behavior quite different from that of native DNA (shown in Fig. 6A), since it was eluted in three fractions at 0.15 M, 0.20 M, and 0.25 M KP (Fig. 6C). Minor fractions were occasionally eluted at 0.10 and 0.50 M KP. When DNA was heated to temperatures between 85° and 100°, the elution patterns were intermediate between those of native and fully denatured (100°) DNA. As denaturing temperatures were increased, increasing amounts of material were eluted at lower molarities; in other words, a gradual shift to the left of the elution pattern was obtained when running DNA samples which had been heated up to increasing temperatures in the range 85° to 100°. As an example, Fig. 6B shows a chromatogram obtained with a DNA sample heated up to 90°.

Heat-Denatured DNA Reacted with Formaldehyde. The behavior of heat-denatured, fast cooled DNA, reacted for 24 hours at 25° with 1% (final concentration) neutralized formaldehyde, was studied using KP containing 1% formaldehyde as the eluent. The chromatographic pattern was slightly different from that just described for heat-denatured DNA, since most of the material was eluted by 0.15 M KP and smaller fractions were eluted at 0.10 M, 0.20 M, and 0.25 M KP; occasionally, a minor fraction was eluted by 0.50 M KP (Fig. 6E). The elution profile appeared therefore shifted to the left when compared with that obtained when using heat-denatured DNA which had not been reacted with formaldehyde. In contrast, native DNA treated with formaldehyde and eluted with formaldehyde-containing KP showed the same elution pattern as native DNA run in the usual conditions (Fig. 6D; compare this figure with Fig. 6A).

Rechromatography experiments performed on the fractions obtained from heat-denatured, fast-cooled, formaldehyde-reacted DNA showed the following results: (1) The 0.15 M and 0.10 M fractions contained, respectively, very little and no material eluting at molarities higher than 0.15 M (Figs. 7B, 7A). (2) The 0.20 M fraction showed two main fractions eluting at 0.15 M and 0.20 M, and a minor one eluting at 0.10 M (Fig. 7C). (3) The 0.25 M and 0.50 M fractions contained, respectively, very little and no material eluting at molarities lower than 0.25 M (Figs. 7D, 7F). On a third chromatography, the 0.25 M fraction did not show any material eluting at a lower molarity (Fig. 7E).

These rechromatography experiments suggest the existence of two distinct fractions in heat-denatured DNA: a large one eluting at molarities lower than 0.20 M, and a small one eluting at molarities higher than 0.20 M.

DNA Denatured by Heat in the Presence of Formaldehyde. DNA which had been heated up to 100° in the presence of 1% formaldehyde (under which conditions the melting temperature is lowered by 10–15°; see Ref. *44*) showed, on chromatography with formaldehyde-containing KP, an elution pattern further shifted to the left as compared with that of heat-denatured, formaldehyde-treated DNA (Fig. 6F; compare this figure with Fig. 6B). An important feature of the elution profile obtained under these conditions is the absence of fractions eluting at molarities higher than 0.20 M KP.

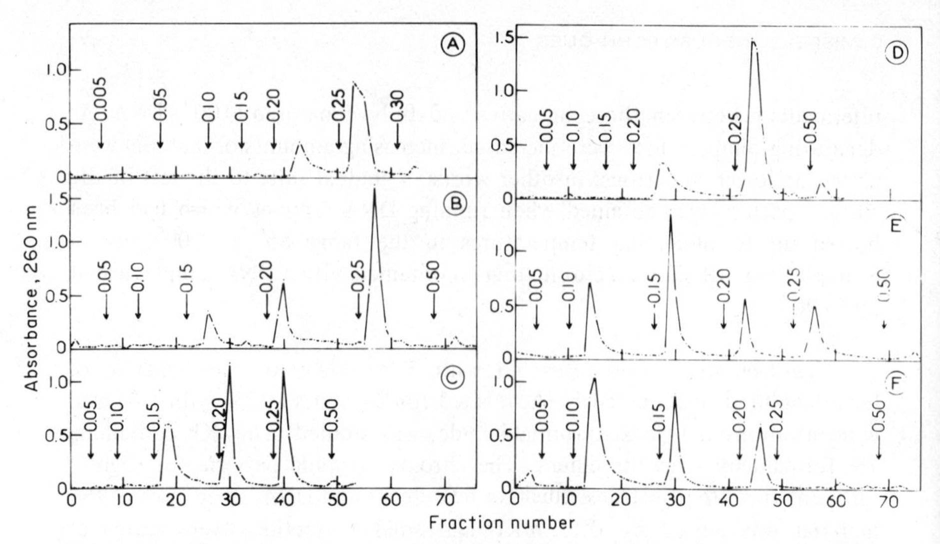

Figure 6. Chromatography of calf-thymus DNA (preparation B3) on 1.3 x 7 cm hydroxyapatite columns. DNA solutions (10–20 ml), having an $A_{260\,nm}$ in the 1–2.5 range, were loaded at fraction number zero. Then 3.8-ml fractions were collected. Recoveries were 100%, except where otherwise stated. Stepwise elution of: (A) native DNA; (B) DNA heated to 90° and then fast-cooled; (C) DNA heated to 100° and then fast-cooled; in this case the recovery was 95%. Stepwise elution in the presence of 1% formaldehyde, of: (D) native DNA; (E) DNA heated to 100°, fast-cooled, and then reacted with formaldehyde; (F) DNA heated to 100° in the presence of formaldehyde; in this case the recovery was 93%. Reproduced from G. Bernardi. 1965. *Nature,* **206**: 779.

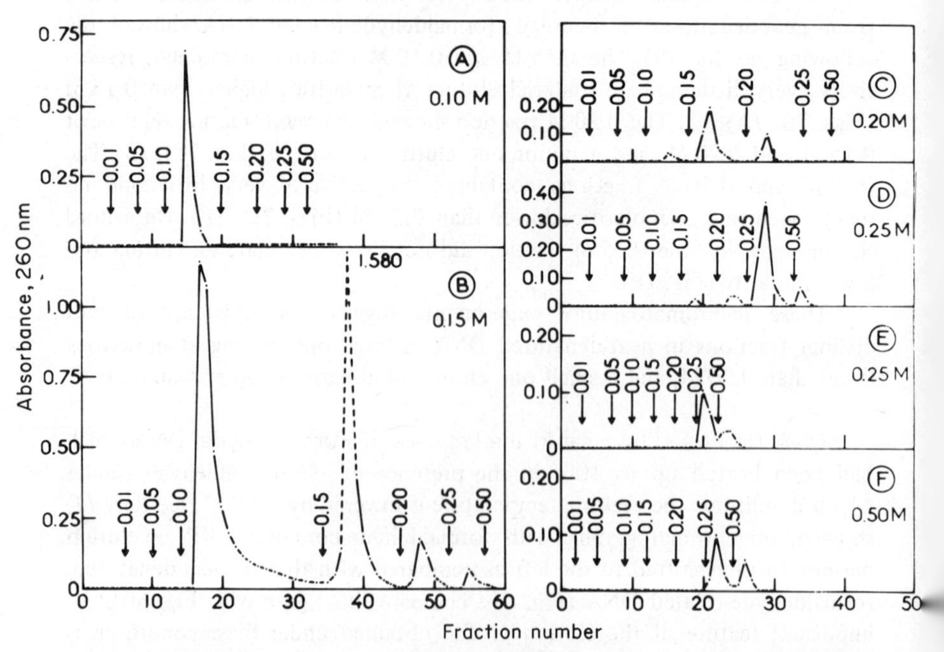

Figure 7. Rechromatography experiments performed on the fractions obtained from heat-denatured, fast-cooled, and formaldehyde-reacted DNA. Stepwise elution of: (A) 0.10 M fraction; (B) 0.15 M fraction; (C) 0.20 M fraction; (D) 0.25 M fraction; (E) 0.25 M fraction from B (third chromatography); (F) 0.50 M fraction. Reproduced from G. Bernardi. 1969. *Biochim. Biophys. Acta,* **174**: 423.

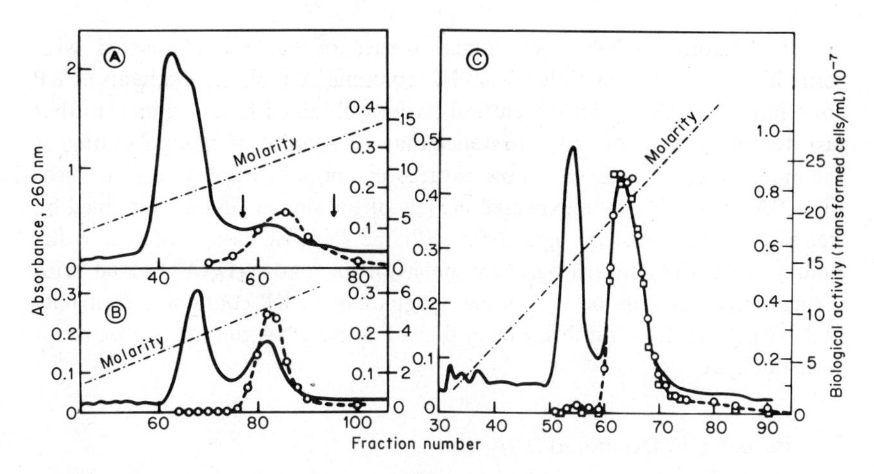

Figure 8. (A) Chromatography of alkali-denatured *H. influenzae* DNA. A DNA solution of 200 ml (sample N_2A; 37 μg/ml in 0.15 M NaCl–0.01 M phosphate, pH 7.0) was adjusted to pH 12.8 with 5 N NaOH at room temperature. After about 10 min at this pH, the solution was neutralized with 2 M KH_2PO_4, diluted with 0.15 M NaCl 0.01 M phosphate, pH 7.0 to 200 ml and loaded on a 1 x 20-cm column. Elution was carried out with a linear molarity gradient (150 + 150 ml) of KP (0.001–0.5 M, inner scale); 2.4-ml fractions were collected. Recovery of both A_{260} and biological activity (cathomycin marker, shown as circles) was 51%. (B) Rechromatography of fractions 57 to 75 from previous chromatogram (pooled and brought to 400 ml with 0.15 M NaCl–0.01 M phosphate pH 7.0) on a 1 x 10-cm column. Elution was carried out with a linear molarity gradient (100 + 100 ml, inner scale) of KP; 2.4-ml fractions were collected. Recovery of A_{260} was 60%, of biological activity 49%. (C) Chromatography of a mixture of native (streptomycin marker) and heat-denatured (cathomycin marker) *H. influenzae* DNA. A DNA solution of 31 ml (sample N_2; 25 μg/ml in 0.15 M NaCl–0.01 M phosphate, pH 7.0) was heat-denatured, added to 5 ml of a native DNA solution (sample S; 75 μg/ml in 0.15 M NaCl–0.01 M phosphate, pH 7.0), and loaded on a 2 x 15-cm column. Elution was carried out with a linear molarity gradient (100 + 100 ml) of KP (0.001–0.5 M); 2.7-ml fractions were collected. Circles indicate the cathomycin activity (right-hand, inner scale); squares indicate the streptomycin activity (right-hand, outer scale). The elution molarity of the first peak was 0.14 M; that of the second peak, 0.21 M. Recovery of A_{260} was 76%; recovery of streptomycin activity was 76%; recovery of cathomycin activity was 62%. Biological activity was tested at a DNA concentration of 0.05 μg/ml. Reproduced from M. R. Chevallier and G. Bernardi. 1968. *J. Mol. Biol.*, **32**: 437.

2. Gradient Elution

Using the gradient elution procedure, denatured DNA from animal tissues or bacteria is eluted in one main fraction at 0.12–14 M KP, followed by a smaller fraction at about 0.20–0.22 M (Figs. 8A and 10A). The chromatographic validity of these fractions can be demonstrated by rechromatography experiments (see, for example, Fig. 8B). The two fractions shown by gradient elution are equivalent to those eluting below and above 0.20 M, respectively, in the stepwise chromatography.

If denatured DNA from animal tissues or bacteria is reacted with formaldehyde and then eluted from HA columns by molarity gradients of KP containing formaldehyde, the elution pattern obtained is very similar to that just described with the only difference that the amount of material eluting at the molarity of native DNA is now reduced in amount (Fig. 9B; compare also, Fig. 10A), a result not unexpected in view of the similar findings obtained by stepwise elution. Also in agreement with the stepwise elution results, is the finding that the chromatographic behavior of native DNA treated with formaldehyde and eluted by a molarity gradient of KP containing formaldehyde (Fig. 8A) does not differ from that of native DNA run in the absence of formaldehyde.

3. Recovery of Denatured DNA

Recovery of denatured DNA from the columns was often found to be incomplete and yields of only 50–80% were not rare. Several findings suggest

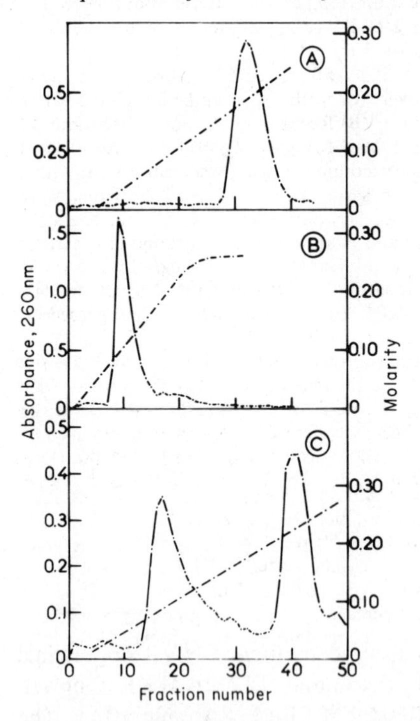

Figure 9. Chromatography of calf thymus DNA (preparation B3) on 1.3 x 7-cm columns. A DNA solution of 10–20 ml (A_{260} in the 1–2.5 range) was loaded at fraction number zero. Then 3.8 ml fractions were collected. Recoveries were 100%, except where otherwise stated. Gradient elution in the presence of 1% formaldehyde of (A) native DNA; (B) heat-denatured DNA; recovery 95%; (C) a 1 : 1 mixture of native and heat-denatured DNA. Reproduced from G. Bernardi. 1965. *Nature,* **206**: 779.

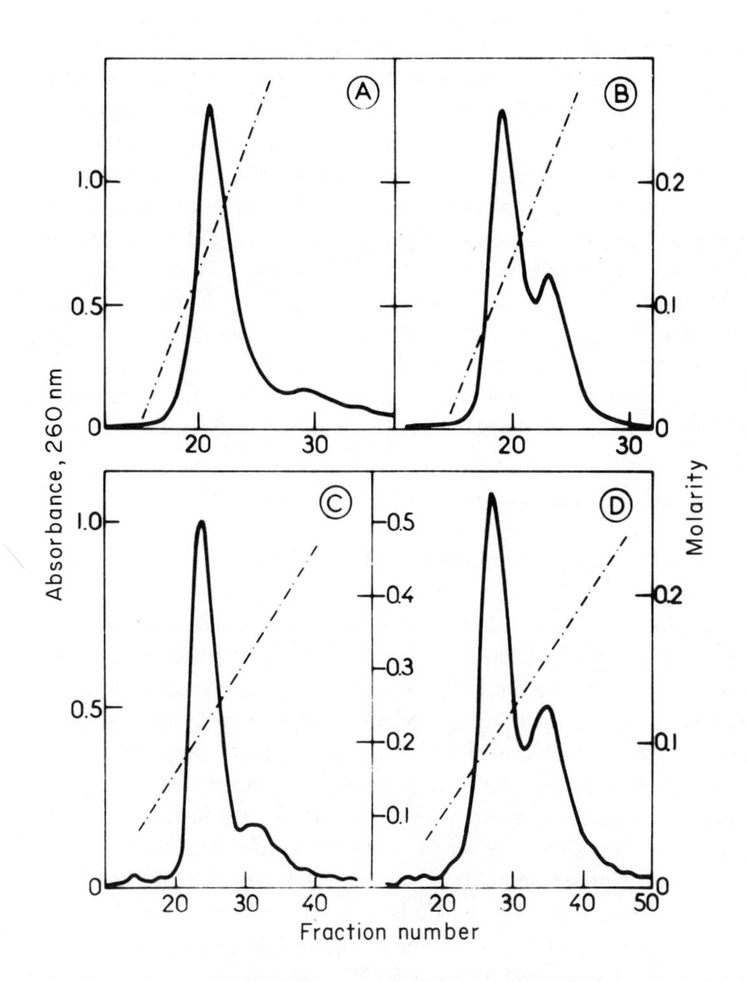

Figure 10. Chromatography of denatured salmon sperm DNA. (A) Intact DNA; (B) DNA treated with mustard gas (230 moles/20,000 nucleotides). Both samples were dissolved in 5×10^{-4} KP at a concentration of 2 mg/ml; after 48 hr at room temperature, the samples were diluted to 50 ml with double-distilled water and kept 30 min at 37°. After cooling to room temperature, samples were made 0.15 M in NaCl by addition of 1 M NaCl. The addition of 1 M NaCl caused a decrease in the $A_{260\ nm}$ of the samples of 19%. Both samples were then treated with 1% formaldehyde and chromatographed in the presence of formaldehyde. In both runs, 27-ml solutions of DNA ($A_{260\ nm}$ = 0.910 for A; $A_{260\ nm}$ 0.725 for B) were loaded on 1 x 10-cm columns. Then 100 ml of each 0.001 M KP were used for the gradient elution. The experiments shown in (C) and (D) also concern intact and alkylated salmon sperm DNA, respectively; the only difference with those reported in (A) and (B) is that formaldehyde treatment was omitted. Reproduced from G. Bernardi. 1969. *Biochim. Biophys. Acta,* **174**: 435.

that low recoveries may be due to aggregations of denatured DNA molecules, mediated by residual protein and/or intermolecular base pairing. (1) Repeated deproteinization treatments of native DNA samples from bacteria or higher organisms with chloroform–isoamylalcohol improves the recovery from the columns of these samples after they have been denatured. Since this treatment shears at the same time as it deproteinizes DNA, it is impossible to decide whether a decrease in molecular weight or deproteinization, or both, are responsible for the better yields obtained in this case. (2) Treatment of denatured DNA with formaldehyde improves the recovery. (3) Heat denaturation of DNA in the presence of formaldehyde gives, as a rule, complete recoveries. (4) Raising the ionic strength of the DNA solutions up to 2–3 just before cooling, or neutralizing if alkali-denaturation was used, raises the recovery of denatured DNA to over 90% (*43*). (5) Low recoveries predominantly affect the first large fraction of single-stranded molecules (see below).

B. The Native-like Fraction of Denatured DNA

Both stepwise and gradient elution results show that the bulk of denatured DNA from either bacteria or animal tissues is eluted at a lower phosphate molarity than a native DNA. The properties of the main fraction of denatured DNA are those of single-stranded DNA: (1) its melting curve shows a slow continuous increase of A_{260} and a low hyperchromicity, 10–15% (*16*); (2) its reaction with formaldehyde at room temperature is complete and no further increase is obtained upon heating (*16*); (3) its buoyant density is 15 mg/cm^3 higher than that of native DNA (*43*); (4) its electron-microscopic appearance is that of single-stranded DNA (*14*); (5) its chromatographic behavior on HA is similar to that of single-stranded DNA from ϕX 174 phage; (6) in the case of transforming *Haemophilus influenzae* DNA, its biological activity is extremely low (*13, 14*).

In contrast, a small fraction of denatured DNA is eluted at the same phosphate molarity as native DNA. This fraction, first recognized several years ago (*11*), has been called "native-like" since its properties are similar to those of native DNA (*12–14, 16*). The existence of a native-like fraction in denatured DNAs has been confirmed by independent work from Doty's laboratory (*45–48*), where it was isolated from both bacterial and animal tissues using the aqueous dextran–polyethylene glycol two-phase system of Albertsson (*49*), and also by Walker and McLaren (*50*) who prepared it according to the procedure described from sonicated mouse DNA. As just mentioned, the properties of the native-like DNA fraction are similar to those of native DNA: (1) its melting curve shows a sharp increase in A_{260} in the 80°–100° range; yet, some increase already takes place in the 50°–80° range and the total hyperchromicity is only 25% (*16, 14*); (2) its reaction with formaldehyde at room temperature is very scarce, whereas heating the DNA solution in the presence of formaldehyde elicits a hyperchromic shift of about 25% (*16*); (3) its chromatographic behavior on HA is that of double-stranded DNA, from which it cannot be separated, at least at the level of resolution used (Fig. 8C); (4) in the case of *H. influenzae* DNA, its microscopic appearance is that of double-stranded DNA, with, however, frequent single-stranded ends (*14*); (5) in the case of *H. influenzae* DNA, the native-like fraction is the carrier of the residual transforming activity, surviving denaturation, exhibited by all tested genetic markers.

The different properties of the native-like fraction compared to native DNA are due, in part, to the presence of contaminating single-stranded molecules from the main fraction, and in part to real structural differences (*16, 14*). It should be mentioned here that, if DNA is denatured in the presence of formaldehyde, no native-like fraction appears in the chromatogram.

Even if the properties of the native-like fraction from bacterial and animal sources are similar enough to be described together, their origin is different. In the case of bacterial DNAs, the available evidence (*13, 14, 45, 48*) suggests that the native-like fraction is formed by fragments of the bacterial genome whose strands never came apart completely because of interstrand cross links of unknown origin. Two explanations have been suggested for the origin of the cross links (*47*), namely that of a biological origin within the cell and that of an induction by shearing during the preparation of DNA. These explanations are not mutually exclusive. The cross links present in bacterial DNAs might have indeed a biological origin and exist prior to the DNA extraction (see Section V.E for one possible explanation). In addition, some cross links might be the result of chemical reactions caused by shearing occurring during DNA preparation (*47*). This latter possibility is supported by the fact that this type of artifact has been also found in mammalian DNAs (*47*; see also below) and in sheared DNA from SP82 (*47*). Regarding the chromatographic behavior of DNA which has been cross-linked *in vitro* and then denatured, see Section V.C.

In the case of DNAs from animal cells, the nature and the origin of the native-like fraction certainly is much more complex. In calf thymus DNA, for instance, the fast-renaturing satellite DNAs are present in the native-like fraction (*16, 43*); these molecules undergo strand separation during the renaturation process, but this is followed by a rapid reassociation, very probably due to the repetitive nucleotide sequences (*51*) they contain. Similarly, the presence of satellite DNA has been reported in the native-like fraction of denatured mouse DNA (*52*). Other DNA molecules which renature rapidly like mitochondrial DNAs, might be present as well in the native-like fraction. Two other species of DNA molecules appear to be present in the native-like fraction, namely, DNA molecules which have been cross-linked *in vivo* or during the extraction procedure (*43, 46, 53*) and single-stranded molecules having a base composition, and consequently a secondary structure, such that their elution molarity is particularly high (*43;* see also Section V.D.).

C. Chromatography of Denatured, Cross-Linked DNA and Renatured DNA

1. Denatured, Cross-Linked DNA

The results obtained by running salmon sperm DNA, cross-linked by mustard gas and then denatured are shown in Fig. 10. A much larger amount of DNA was eluted at the molarity of native DNA in the cross-linked samples compared to the untreated samples. In both cases, reaction with formaldehyde reduced the amount of material eluting like native DNA.

In another series of experiments, *Micrococcus lysodeikticus* DNA was crosslinked by treatment with nitrous acid at pH 4.2 for various lengths of time and then heat denatured (*16*). In this case the amount of material eluted at the molarity of native DNA increases with increasing contact time with the cross-linking agent.

2. Renatured DNA

As expected, bacterial renatured DNA showed the chromatographic behavior of native DNA, except for the presence of some residual denatured material eluting at a lower molarity (*16*).

D. Fractionation of Denatured DNA

Denatured DNA molecules are fractionated by HA according to their average base composition (*12, 16*). Experiments carried out on formaldehyde-reacted, denatured calf thymus and chicken erythrocytes DNAs by stepwise elution have shown higher levels of A and T in the low-eluting fractions and lower levels of A and T in the high-eluting fractions. A and T, and G and C, respectively, are not present in equimolar amounts in the fractions eluting below 0.20 M. The results obtained with formaldehyde-reacted polyribonucleotides (see Section VI.D) suggest that this fractionation of single-stranded DNA molecules is due to the fact that they have slightly different structures as a consequence of their different base sequences (see also Section IX). It should be remarked that the native-like fraction of calf thymus DNA fits by chance the trend of increasing GC contents with increasing elution molarity shown by single-stranded DNA, since it is mainly formed by satellite DNAs having a higher GC contents than the main DNA. This coincidence and the fact that our early analytical results on the fractions indicated equimolar amounts of A and T, and G and C, respectively, led us to suggest that the fractionation of denatured DNA was a fractionation of double-stranded molecules containing different amounts of disordered A–T rich regions (*12*).

E. Separation of Native and Denatured DNA

1. Separation at Room Temperature

Because of their widely different elution molarities, native and denatured DNA can be easily separated by either stepwise or gradient elution, in the absence or in the presence of formaldehyde. Examples of separations of artificial mixtures of native and denatured DNAs are shown in Figs. 8C and 9C. Obviously, in these cases, the native-like fractions of denatured DNA are eluted together with native DNA, at least under the low-resolution conditions used in the experiments shown.

Separations of native and denatured DNA at room temperature, using the gradient elution technique, have been performed in several laboratories, particularly in connection with problems related to DNA replication. Okazaki et al. (*54–56*) and Oishi (*57–59*) have reported that newly synthesized DNA from *Escherichia coli* and *Bacillus subtilis* is in the form of small pieces about 1,000 nucleotides long and that at least some of this DNA is single-stranded rather than double-stranded. Similar results have been reported by Painter and Schaefer (*60*) and Habener, Bynum, and Shack (*61*) for newly replicated HeLa cell DNA. Pauling and Hamm (*62*) have provided evidence that newly-replicated *E. coli* DNA exists temporarily in a form that rapidly renatures following heat-denaturation and perhaps represents the forked molecule postulated as an intermediate in DNA replication. It would be interesting to know whether any relationship exists between this fast-renaturing material and the native-like fraction of bacterial DNA (see Section V.B). HA chromatography has also been used by Lark (*63*) and Schandl and Taylor (*64*) to investigate the chromosome replication in *E. coli* and Chinese hamster cells, respectively. Other examples of separations of native from denatured DNA can be found in investigations on phage DNAs by Thomas et al. (*65–67*). Along a different line of research, Zimmerman et al. (*68*) have developed an assay method for DNA ligase based on the separation of denatured λDNA from denatured λDNA which has become covalently bound to cross-linked λDNA.

2. Separation at High Temperatures (50°–70°)

As can be expected from the results just mentioned, the separation of denatured and native DNA on HA columns can take place at any temperature below the melting range of DNA (*25*). Fractionation at 50°–60° has been intensively used (*69–71*) to investigate the reassociation kinetics of DNAs from different sources. In a typical experiment (Fig. 11A) DNA is sheared (and simultaneously denatured) by passing it twice through a needle with a pressure drop of 3.4 kilobars, and incubated at concentrations ranging from 2 to 8600 μg/ml at 60° in 0.12 M phosphate; at various times, samples are passed over a HA column kept at 60° and equilibrated with 0.12 M NaP to separate nonretained single-stranded fragments from adsorbed fragments which contain double-stranded regions. The reassociation kinetics measured in the way just described parallels the optical reassociation kinetics (Fig. 11B), yet appears to progress more rapidly, the half-reaction time being higher by a factor of 2 to 3. This discrepancy has been explained as owing to the fact that the optical method measures the fraction of the length of the DNA that is actually paired, whereas HA measures the fraction of the total number of DNA fragments that have some portion of their length reassociated. Reaction with formaldehyde at room temperature (*16*) of the reassociated fraction would probably permit a check of whether the rate over-

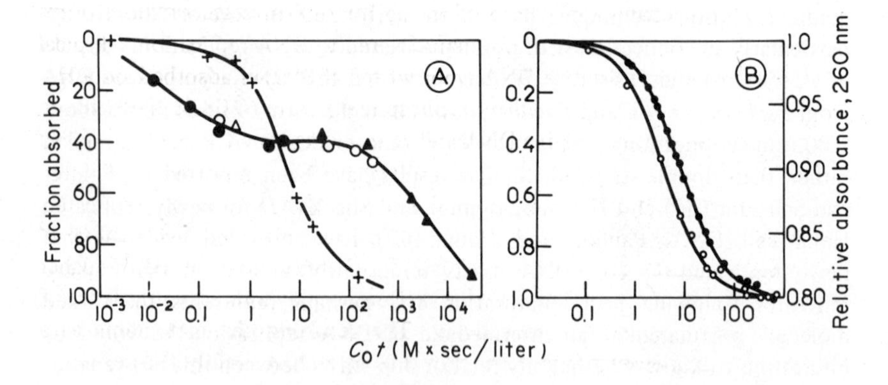

Figure 11. (A) The kinetics of reassociation of calf thymus DNA measured with hydroxyapatite. The DNA was sheared at 50,000 psi and incubated at 60° in 0.12 M PB. At various times samples were diluted, if necessary (in 0.12 M PB at 60°), and passed over a hydroxyapatite column at 60°. The DNA concentrations during the reaction were: (△) 2; (●) 10; (○) 600; (▲) 8600 micrograms per milliter; (+) radioactively labeled *E. coli* DNA at 43 μg/ml present in the reaction containing calf thymus DNA at 8600 μg/ml. Reproduced from R. J. Britten and D. E. Kohne. 1967. *Carnegie Inst. Washington Yr. Bk*, **66**: 73. (B) The time course of reassociation of sheared (50,000 psi) *E. coli* DNA. Solid circles and right scale decrease in optical density at 260 nm. Open circles and left scale binding to hydroxyapatite. Reproduced from R. J. Britten and D. E. Kohne. 1966. *Carnegie Inst. Washington Yr. Bk.*, **65**: 73.

estimation obtained from the HA values is only due to this factor. A batch procedure for the thermal elution of DNA from HA has been described by Brenner et al. (*72*).

Contrary to the opinion expressed by McCallum and Walker (*73*), there is no essential difference between the results obtained at 50°-70° and those obtained at room temperature, particularly if chromatography is performed after reaction with formaldehyde using formaldehyde containing buffers.

Szala and Chorazy (*74*) have fractionated calf thymus DNA treated at different temperatures within the melting range on HA columns thermostatted at 3°-5° lower than the temperature used in the partial denaturation. Except for the fact that columns were thermostatted at relatively high temperatures, the approach is that used in previous experiments by Bernardi (*12*; see Fig. 6). Accordingly, the results are the same, namely, as the denaturation temperature is increased, the amount of DNA eluting at the position of denatured DNA increases. As expected, the fraction eluted as native DNA, being enriched in the higher melting molecules, shows an increase in its T_m as the fractionation is performed at increasingly higher temperatures.

F. Thermal Chromatography

A very interesting application of the ability of HA to discriminate native and denatured DNA is the "thermal chromatography" (*25*). In a typical experiment, sonicated native DNA (mol wt = 200,000) is adsorbed on a HA column. This is then percolated by a phosphate buffer (0.08 M NaP) able to elute denatured, but not native DNA and at the same time the temperature of the column is increased in steps to slightly over 100°. At each temperature within the melting range a certain fraction of the double-stranded segments is denatured and washed through the column. As expected, fractions eluted at lower temperatures are richer in A and T, whereas those eluted at higher temperatures are richer in G and C; all fractions display a molar equivalence of A and T, and G and C, respectively. The resulting chromatograms as a function of temperature (Fig. 12A) proves to be almost exactly the derivative of the optical melting curve (Fig. 12B). The slightly higher T_m shown by the thermal chromatogram has been ascribed to the higher effective ionic strength prevailing in the neighborhood of the highly charged HA crystal. Figure 12C shows that the mole fraction G + C of the fractions is generally proportional to temperature, indicating a variation of 2.0% per degree – a value which is to be compared with 2.4% per degree obtained by optical melting of a large number of DNAs from different species (*75*). In view of the significant differences in solvent and molecular weight between the two experiments, the apparent agreement between these two numbers may be fortuitous. Several

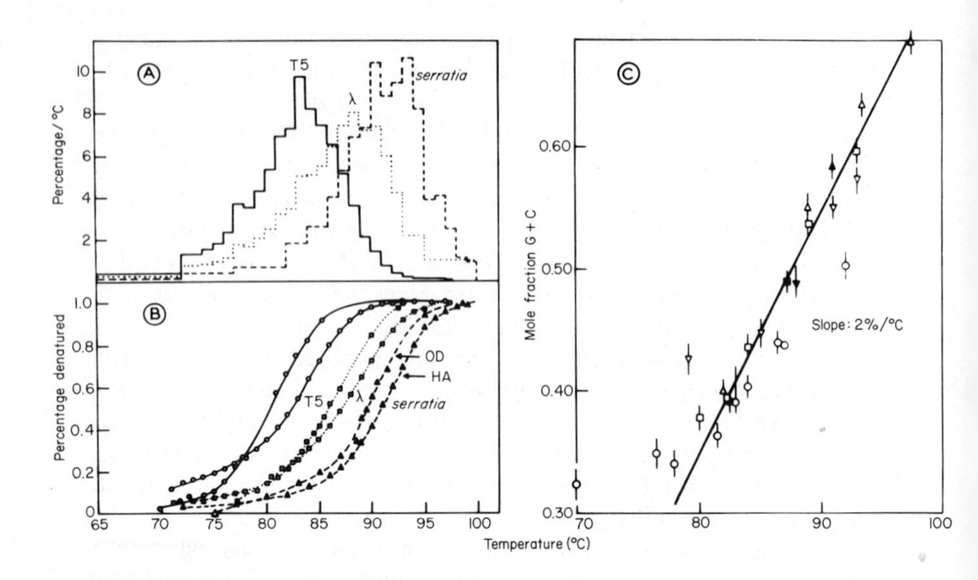

Figure 12. (A) Thermal chromatograms of three different DNAs. The results with T2 DNA are superimposable with those of T5. Those with *Haemofilus* DNA are very similar to T5 and T2 DNA. (B) Optical melting curves plotted with integral form of thermal chromatograms. Notice that the optical melting curves and the integral form of the thermal chromatograms shown in (A) are very similar. (C) Results of nucleotide analysis. The observed mole fraction of G + C is plotted against the mean value of the temperature over which they were collected. The solid points represent results obtained on sonic fragments which were not subjected to thermal chromatography. Nucleotide analysis made of fractions coming from the low-temperature edges of the chromatogram which contain less than 1%/°C, give G + C values which fall above the line. (○) T5 fractions; (●) T5 unfractionated; (□) λ fractions; (■) unfractionated; (△) *Serratia* fractions; (▲) *Serratia* unfractionated; (▽) *E. coli* fractions; (▼) *E. coli* unfractionated. Reproduced from Y. Miyazawa and C. A. Thomas, Jr. 1965. *J. Mol. Biol.*, **11**: 223.

reservations have been made by Miyazawa and Thomas (*25*) to the general conclusion that thermal chromatograms are direct reflections of the heterogeneity of composition existing among the various segments. (1) The form of the thermal chromatograms depends on the length of the duplex segments. This is not surprising since this is in the range of strong dependence of T_m on fragment size. (2) When nucleotide analysis is made on the fractions in the leading edge of the thermal chromatogram which contain less than 1% per degree, the mole fraction G + C departs from the line drawn in Fig. 12C. This probably represents a nonspecific leakage of material which was damaged by sonic treatment, or exceptionally short fragments which would melt at a lower temperature. (3) Repeated thermal chromatography and nucleotide analysis of T5 DNA segments generate points which fall on a line of lower slope. These observations are statistically significant and may indicate that there are still other effects not known which affect the temperature-composition relation as seen in these experiments. In spite of these reservations, Fig. 12C shows that it is possible to fractionate and identify segments having compositions which are quite far removed from the mean value for a given DNA; for instance, T5 DNA produces segments which have G + C contents ranging from 32 to 50%, λDNA from 38 to 60%, *Serratia* DNA from 40 to 69% and *E. coli* from 42 to 57%. T2 DNA and *H. influenzae* DNA show thermal chromatograms identical to that of T5 DNA.

The most interesting result obtained by thermal chromatography obviously is the demonstration of the compositional heterogeneity of relatively long DNA segments. The technique has been also used, however, to obtain other information.

The specific association of ribooligonucleotides of known chain length synthesized by RNA polymerase on T4 or T7 DNA with denatured T4 or T7 DNAs has been investigated by thermal chromatography (*76*). The results obtained indicate that in experiments testing interspecies homology by RNA-DNA hybridization about 12 perfectly complementary nucleotides must be involved in the complex. Thermal chromatography has also been used to investigate the stability of oligoadenylate-polyuridylate complexes (*77*). The temperature of release from HA was found to be a function of the chain length of the oligonucleotides, the presence of Mg^{2+} and the presence or absence of terminal phosphates on the oligonucleotides; unfortunately it was impossible to decide whether 1A : 1U or 1A : 2U structures predominated at the surface of the HA crystals. Other recent applications of thermal chromatography have concerned the preparation of poly (dAT : dAT) from crab DNA (*78*), the thermal stability of reassociated DNA (*71, 79*), and the characterization of poliovirus-specific double-stranded RNA (*80*).

VI. CHROMATOGRAPHY OF RNAs AND SYNTHETIC POLYRIBONUCLEOTIDES

Investigations on the chromatography of RNAs and synthetic polyribonucleotides have been much less extensive than those on DNAs. This drawback is particularly serious if one considers the multiplicity of secondary structures which exist among natural and synthetic polyribonucleotides and the potential usefulness of HA columns in investigating those structures.

A. Chromatography of Ribosomal and Viral RNAs

Using the stepwise elution technique, the ribosomal RNAs from yeast and from Ehrlich ascites tumor cells and the RNAs from tobacco mosaic

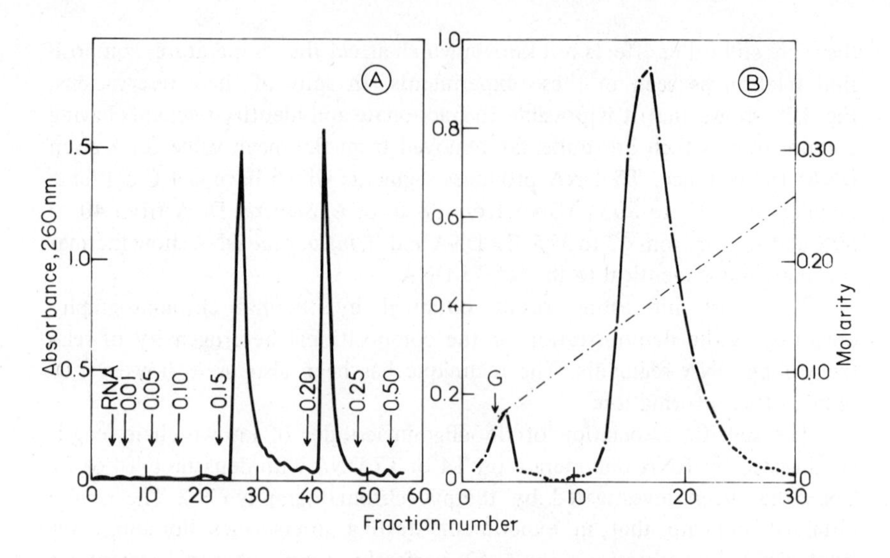

Figure 13. (A) Chromatography of ribosomal RNA from Ehrlich ascites tumor on a 1.3 x 6-cm column. An RNA solution of 8 ml having an $A_{260\ nm}$ = 3.12 was loaded on the column at fraction number four. Then 3.2 ml-fractions were collected, using the stepwise elution technique. Recovery was 100%. (B) Chromatography of 20.8 $A_{260\ nm}$ units of ribosomal RNA from yeast to a 1 x 10-cm column. Loading took place at fraction number zero. A molarity gradient of KP was started at the fraction marked with G. Fractions of 2.9 ml were collected. Recovery was 91%. This experiment was performed at 4°. Reproduced from G. Bernardi. 1969. *Biochim. Biophys. Acta,* 174: 449.

virus, turnip yellow mosaic virus, and alfalfa mosaic virus were all eluted in two peaks at 0.15 M and 0.20 M KP (Fig. 13A). When elution was carried out with a molarity gradient, high-molecular-weight RNAs were all eluted as single peaks (Fig. 13B) and at a molarity close to 0.15 M KP. These results suggest that the two peaks obtained with high-molecular-weight RNAs are "false" peaks.

No detailed investigation has been done so far on the dependence of the elution molarity of high-molecular-weight RNAs on temperature or molecular weight. No supporting evidence was obtained in our laboratory in favor of the claim (*81, 82*) that HA columns fractionate high-molecular-weight RNAs on the basis of molecular weight.

It should be pointed out that in several cases the sedimentation coefficient of high-molecular-weight RNAs eluted from HA were found to be lower than those of the starting materials. This decrease in sedimentation coefficients seems to be due to the fact that HA "disaggregates" the large polynucleotide fragments forming RNA "molecules" which contain "hidden breaks." That HA columns do not cause by themselves any breakage of large RNA molecules is shown by the good recovery of infectivity obtained when viral RNAs are chromatographed on HA (*83, 84*).

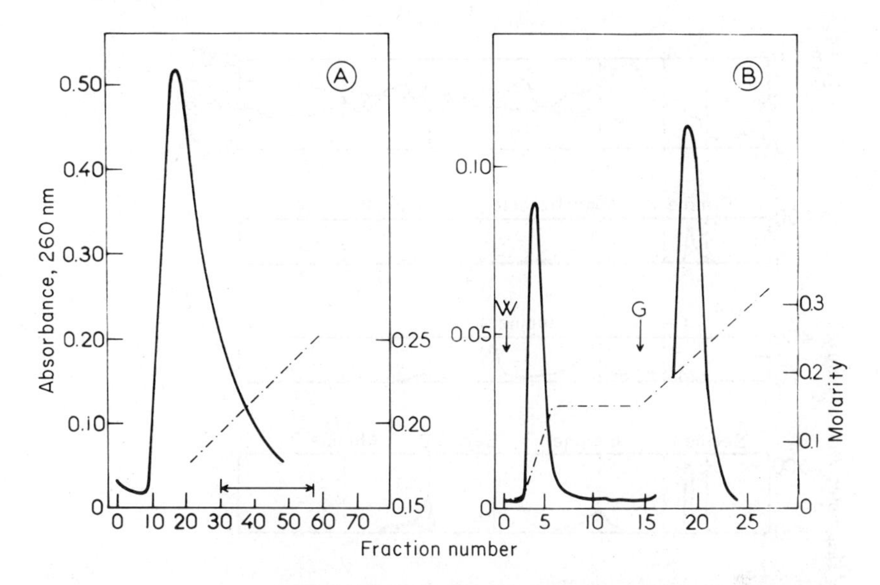

Figure 14. (A) Chromatography of total RNA from tobacco leaves harvested 2 days after infection with alfalfa mosaic virus. A 20-mg sample was loaded on a 2 x 15-cm column equilibrated with 0.001 M KP. The column was then washed with 0.15 M KP until the ultraviolet absorption of the eluent was negligible. A molarity gradient (0.15–0.30 M KP) was applied to the column at fraction zero. Then 5-ml fractions were collected. Fractions indicated by the arrow were rechromatographed. (B) Rechromatography of fractions eluted by 0.20–0.25 M KP of total RNA from tobacco leaves harvested 8 days after infection. Then 1.4 A_{260} units were loaded on a 1 x 10-cm column, equilibrated with 0.001 M KP. The column was then washed with 0.15 M KP (arrow W) until the ultraviolet absorption of the effluent was negligible. A molarity gradient (0.15–0.50 M KP) was then applied to the column (arrow G); 2-ml fractions were collected. Reproduced from L. Pinck, L. Hirth, and G. Bernardi. 1968. *Biochem. Biophys. Res. Commun.*, 31: 481.

B. Chromatography of Double-Stranded RNA

In contrast with the claim that the replicative intermediate of viral RNAs are not separated from ribosomal RNA on HA columns (*85*), both the replicative form of turnip yellow mosaic virus RNA (*86*) and the replicative intermediate of alfalfa mosaic virus RNA (*87*) can be easily separated from single-stranded RNA on HA columns, since these double-stranded RNAs are eluted only by 0.20–0.22 M KP, like native DNA (Fig. 14). This separation is therefore analogous to that of denatured and native DNA (see Section V.E).

C. Chromatography of Transfer RNAs

Transfer RNAs were first fractionated on HA columns using the stepwise elution procedure by Hartmann and Coy (*88*). Subsequent work by several

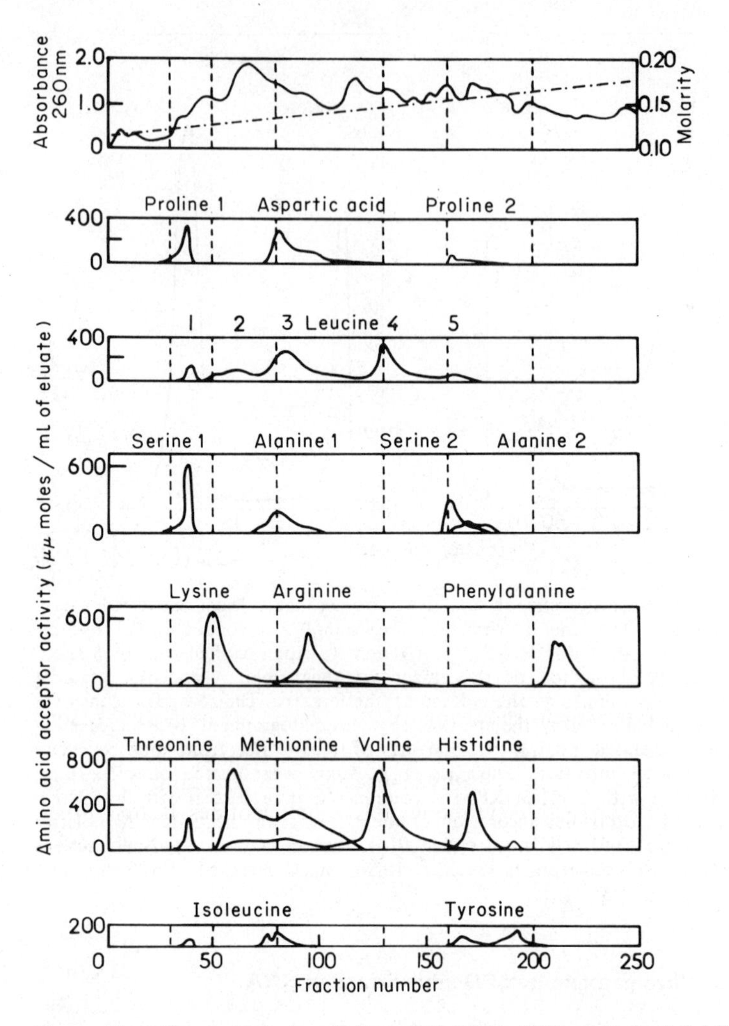

Figure 15. Chromatography of *E. coli* B tRNA (100 mg in 2 ml 0.1 M NaP) on a 2 x 68-cm column at room temperature. A NaP molarity gradient of 2 liters (0.1–0.2) was used at a flow rate of 12 ml/hr. Then 4-ml fractions were collected. Reproduced from R. L. Pearson and A. D. Kelmers. 1966. *J. Biol. Chem.*, **241**: 767.

groups of investigators (*89–94*) led to remarkable separations (Fig. 15) which represent one of the best examples of the discriminating power of HA with respect to the secondary and tertiary structures of rather similar molecules. It may be interesting in this connection to recall that under conditions of low resolution (similar to those of Fig. 2) tRNA is eluted as a single, sharp peak (*12*). Muench and Berg (*91*) and Dirheimer (*92*) used as the eluent phosphate buffer having a pH of 5.8 and 5.4, respectively; this allows isotopically labeled aminoacyl tRNAs to be chromatographed as such. It is evident that, in combination with other available methods, HA chromatography is very useful in the separation and purification of transfer RNAs.

D. Chromatography of Synthetic Polyribonucleotides

Poly U is eluted at 0.10 M KP at 25° (Fig. 16A) under conditions where it is completely devoid of any secondary structure (*95*); at 4°, where it has some sort of helical secondary structure (as shown by its hypochromism and increase in positive optical rotation, *95, 96*), poly U acid is only eluted by 0.15 M KP (Fig. 16B). Likewise, poly C is eluted at room temperature by 0.12 M KP (Fig. 16D) and by a slightly higher molarity at 4°. Recovery of poly U and poly C from the columns has always been complete.

In contrast with the two polypyrimidinic acids just mentioned, poly I gave very irregular results and poor recoveries and poly A was eluted in a broad peak around 0.2 M KP with recoveries of only 60–85% (Fig. 16C). It is known that, at room temperature and neutrality, poly A has a structure which is random with respect to total conformation, but ordered in terms of short-range interactions, the ordered regions having a single-stranded, stacked, helical structure (*97–100*).

Experiments intended to show how HA columns can discriminate among different single-stranded structures and also to help understand the fractionation of heat-denatured and formaldehyde-reacted DNA according to its base composition were done with synthetic polyribonucleotides heated up to 100° for 5 min in 0.01 M KP containing 1% formaldehyde. Recoveries were complete in all cases and the elution molarities were 0.05 M KP for poly U, 0.12 M for poly C, 0.16 M for poly I and 0.17 M for poly A (Fig. 17).

The artificial complexes of poly U and poly A were also examined under conditions where double-stranded poly A–poly U or triple-stranded 2 poly U–poly A were formed. The double-stranded complex was eluted in the molarity region where the native DNA is eluted, whereas the triple-stranded complex was eluted only by 0.45–0.50 M KP (Fig. 18). The double-stranded poly I–poly C complex was also eluted at a molarity close to 0.20 M KP, like poly A–poly U (*17*).

VII. CHROMATOGRAPHY OF OLIGONUCLEOTIDES

Only a very limited number of reports have dealt so far with the chromatography of oligonucleotides. Main and Cole (*5*) were the first to report that the pancreatic DNase digest is eluted at very low phosphate molarities, most of the material not being retained by the column equilibra-

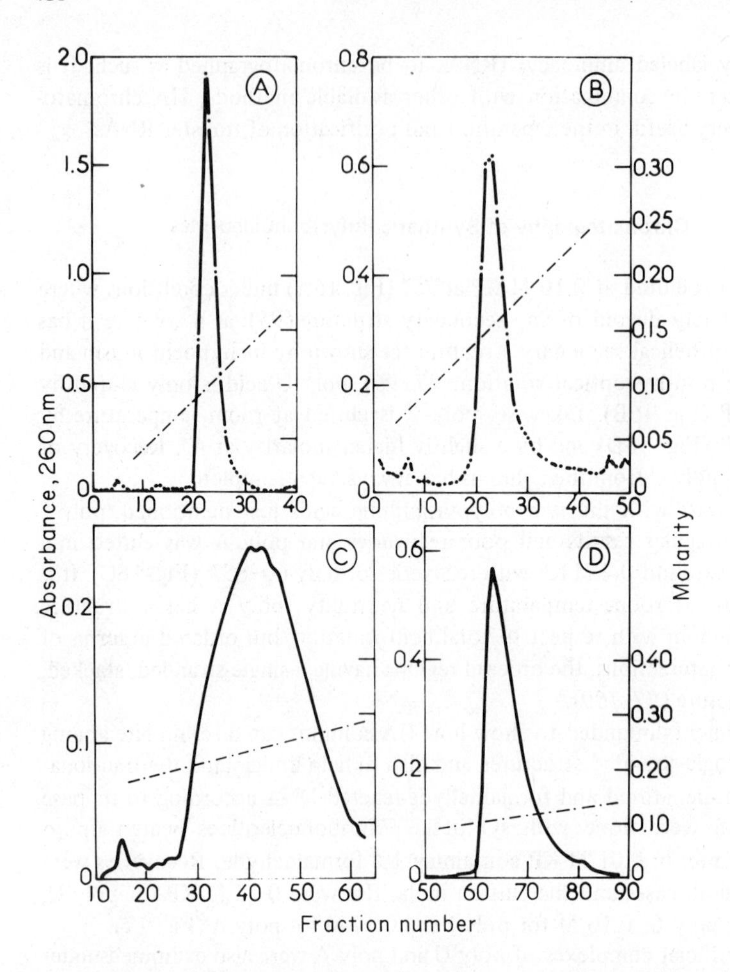

Figure 16. (A) Chromatography of polyuridylic acid. This experiment was performed at 25° and 20 A_{260} units were loaded. Recovery was 98% and 2.9-ml fractions were collected. (B) Chromatography of polyuridylic acid. This experiment was performed at 4° and 6.5 A_{260} units were loaded and recovered. (C) Chromatography of polyadenylic acid. Experiment carried out at room temperature and 20.9 A_{260} units were loaded. Recovery was 81% and 3-ml fractions were collected. (D) Chromatography of polycytidilic acid and 9.9 A_{260} units were loaded and recovered; 2.9-ml fractions were collected. In all cases 1 x 10-cm columns were used and loading took place at fraction zero. Reproduced from G. Bernardi. 1969. *Biochim. Biophys. Acta,* 174: 449.

ted with 0.005 M NaP; however, a minor peak was eluted at 0.045 M NaP. RNA "core," the water-undialyzable fraction of the oligonucleotides formed by pancreatic RNase digestion, was mostly eluted at 0.015 M KP, with a minor peak eluting about at 0.10 M KP (*12*). The higher eluting molarity of the RNA "core" compared to the pancreatic DNase digest might be due to

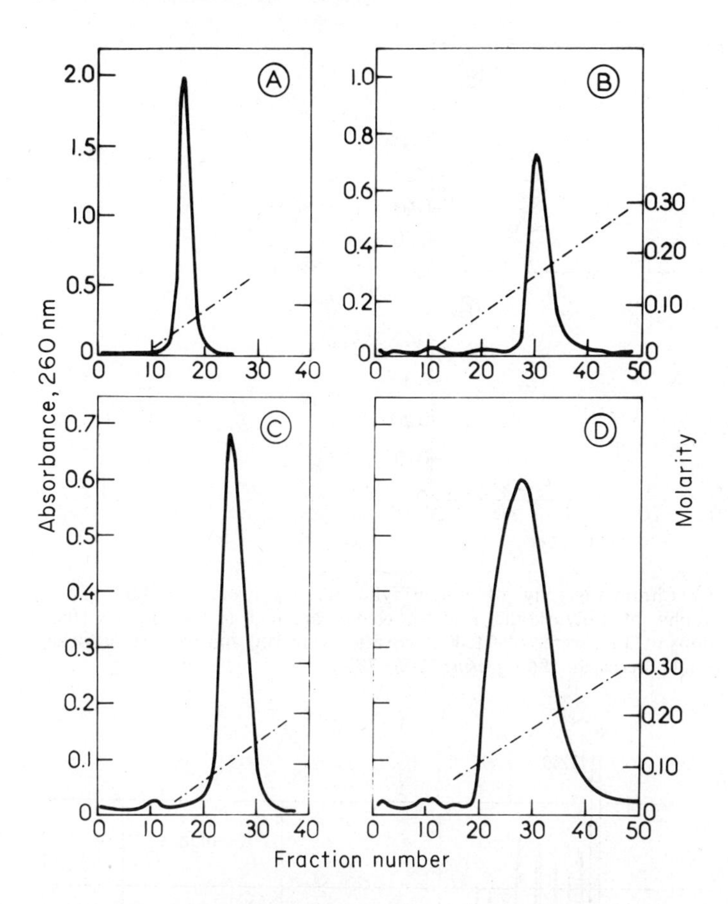

Figure 17. (A) Chromatography of 11 $A_{260\ nm}$ units of polycytidilic acid. (B) Chromatography of 23.3 $A_{260\ nm}$ units of polyadenylic acid. (C) Chromatography of $A_{260\ nm}$ units of polyurydilic acid. (D) Chromatography of 11.7 units of polyinosinic acid. Polyribonucleotides were heated to 100° for 5 min and fast-cooled in 0.01 M KP containing 1% neutralized formaldehyde which was also present in the eluting buffers. Columns of 1 x 20 cm were used and 3-ml fractions were collected. Recovery was 100% in all cases. From G. Bernardi. 1969. *Biochim. Biophys. Acta,* **174:** 449

the larger size of the fragments and/or to their secondary structure (RNA "core" has a high purine contents).

The fractionation of large oligonucleotides on HA columns, both in the presence and in the absence of urea or formaldehyde, seems to be very promising: the results of Mundry (*101*) and Bernardi et al. (*20*) indicate an excellent reproducibility of the complex patterns which are obtained (Fig. 19). It is evident that large oligonucleotides are eluted in a relatively high molarity range.

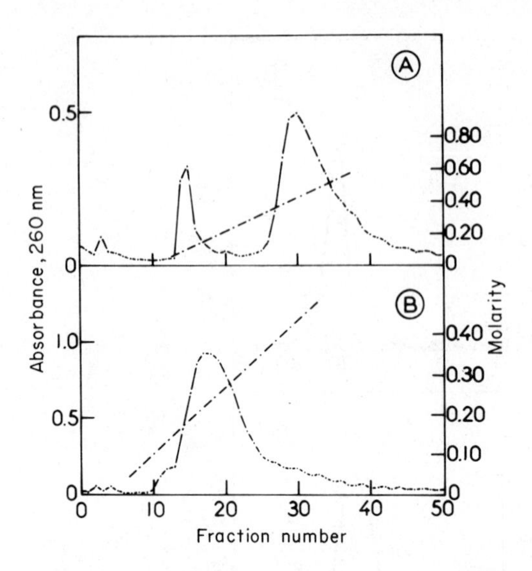

Figure 18. (A) Chromatography of polyuridylic acid + polyadenylic acid (3 : 1); (B) Chromatography of polyuridylic acid + polyadenylic acid (1 : 3) on 1 x 10-cm columns. Fractions of 3 ml were collected. Recoveries were 100 and 95%, respectively. Reproduced from G. Bernardi. 1965. *Nature,* **206**: 779.

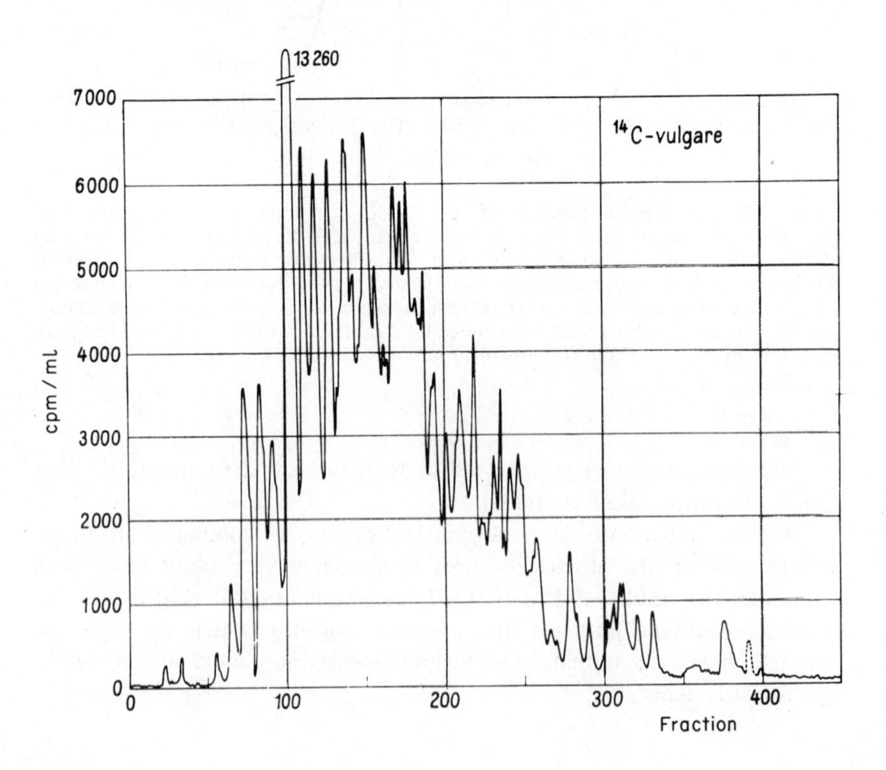

VIII. CHROMATOGRAPHY OF BASES, NUCLEOSIDES, AND NUCLEOSIDE MONO- AND POLYPHOSPHATES (*103*)

Purine and pyrimidine bases (adenine, thymine, cytosine, methylcytosine, and uracil), ribo- and deoxyribonucleosides (adenosine, cytidine, guanosine, thymidine, and uridine), and coenzyme derivatives like thiamine and riboflavin, were not retained by HA columns equilibrated with 0.001 M KP.

Nucleoside monophosphates (2′, 3′ mixed isomers and 5′ isomers of the ribo series, 3′ and 5′ isomers of the deoxyribo series), thiamine monophosphate and riboflavin 5′ phosphates were also eluted by 0.001 M KP, but they were slightly retarded. Employing long enough columns they could be separated from nucleosides.

Nucleoside polyphosphates were strongly adsorbed and needed rather high KP molarities to be eluted. As an example, Fig. 20 shows a chromatogram obtained with a mixture of AMP, ADP, ATP, and adenosine tetraphosphate. The complete separation obtained was controlled by rechromatography experiments. The peaks were identified on the basis of their typical elution molarity by running the single compounds separately.

Similar experiments showed that other nucleoside 5′-diphosphates (CDP, UDP) and thiamine pyrophosphate were eluted at the same molarity as UDP and that other nucleoside 5′-triphosphates (deATP, CTP, UTP) were eluted like ATP. An interesting exception is that GDP and GTP and, at a smaller extent, IDP and ITP are eluted at higher phosphate molarities than the other di- and triphosphates, respectively. This finding is probably related to phenomena of intermolecular association of the type described by Gellert et al. for GMP (*104*).

Compounds in which the pyrophosphate has no free secondary acid group, such as ADP ribose, NAD, FAD, UDP glucose, were not retained by the columns equilibrated with 0.001 M KP. NADP and coenzyme A which, besides a pyrophosphate with no free secondary acid group, have a terminal phosphate were retained slightly more than nucleoside monophosphates. Finally phosphoribose pyrophosphate was eluted at a higher molarity than triphosphates.

These observations, while of great interest for our understanding of the mechanism of adsorption on HA, also have practical usefulness since they indicate the potentialities of HA columns for the separation of nucleic acid derivatives from nucleic acids and proteins.

Figure 19. Elution profile of Vulgare-TMV-RNA, hydrolyzed with T1 ribonuclease, precipitated with 1/9 volume of 20% trichloroacetic acid in absolute ethanol, and dissolved in 0.001 M NaP–7 M urea. A 2.2 x 45-cm column was used. Elution was done with a linear molarity gradient (2 + 2 liters) of 0.001 M NaP–0.2 M NaP, both buffers containing 7 M urea. Fractions of 5 ml were collected. Flow rate was 30 ml/hr. The dotted peak and the obvious asymmetry of the "last peak" are artifacts due to interruption of the elution by an accident. Reproduced from K. W. Mundry. 1965. *Z. Vererbungslehre*, 97: 281.

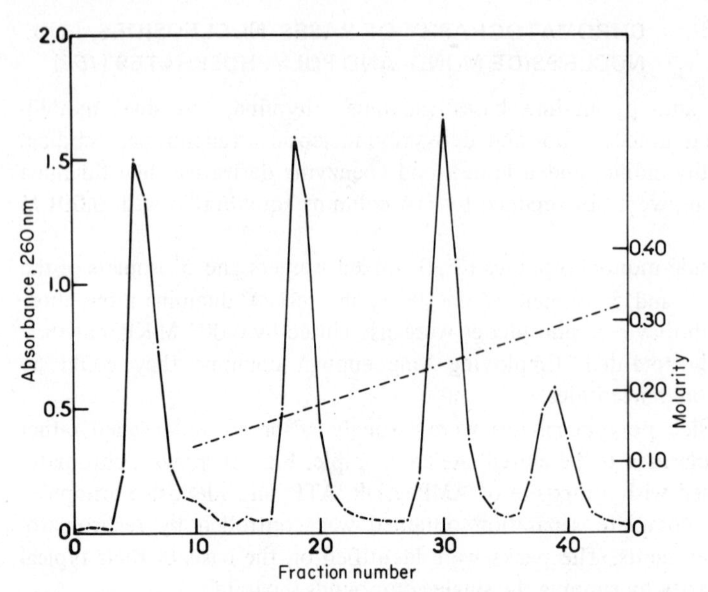

Figure 20. Chromatography of a mixture of AMP, ADP, ATP, and adenosine tetraphosphate. Recoveries were 133, 113, and 50%, respectively, owing to the fact that ADP, ATP, and, to a much larger extent, adenosine tetraphosphate, were contaminated by the lower phosphates, as indicated by separate experiments. The overall recovery was 101%. Fractions of 3 ml were collected. Elution was carried out with a KP molarity gradient, 0.001 M–0.5 M (100 + 100 ml) and a 1 x 10 cm column. Reproduced from G. Bernardi. 1964. *Biochim. Biophys. Acta,* 91: 686.

IX. MECHANISM OF ADSORPTION AND ELUTION OF NUCLEIC ACIDS – RELATIONSHIPS BETWEEN ELUTION MOLARITY AND SECONDARY STRUCTURE OF NUCLEIC ACIDS

The main factor involved in the adsorption of nucleic acids on HA seems to be the interaction between the phosphate groups of nucleic acids and calcium ions on the surface of HA crystals. This is suggested by several facts.

1. Treatment of HA with compounds having a very strong affinity for calcium, i.e., EDTA and citrate, decreases its adsorption capacity for nucleic acids. Compounds having a lower degree of affinity for calcium, i.e., phosphates and several carboxylic compounds, may be used as eluents; ions having a very low affinity for calcium, i.e., chlorides, practically do not interfere with the adsorption of nucleic acids.

2. Electrophoresis of HA prepared according to Tiselius et al. (*1*) shows that HA crystals have a net positive charge in 0.001 M KP (*17*). The isoelectric point was not determined; it is known from the work of Mattson et al. (*105*) that HA crystals are amphoteric and that the isoelectric point of

HA varied from 6.5 to 10.2 for different preparations examined by those authors.

3. The sedimentation rate of HA crystals in 0.001 M KP is greatly increased by the addition of DNA; this may be due to a decrease in electrostatic repulsion among the positively charged crystals.

4. Phosphoproteins have a much higher affinity for HA than nonphosphorylated proteins.

5. Chromatography of nucleosides, some coenzymes, and their mono- and polyphosphate compounds shows that the elution molarity of these materials is only dependent upon their phosphate group(s). The nonphosphorylated compounds are not retained by HA equilibrated with 0.001 M KP; monophosphates are eluted by 0.001 M KP, but they are retarded, and di-, tri-, and tetraphosphates elute at increasingly higher, characteristic phosphate molarities, independently of the organic molecules to which they are bound.

It should be pointed out, however, that, if the interaction between phosphates and calcium certainly plays a prominent role in the mechanism of adsorption of nucleic acids, amino groups of nucleotides, when exposed, might also become involved in the interaction with HA. In this connection, it should be mentioned that mononucleotides run in 0.001 M NaH_2PO_4, pH = 5.2, are fractionated on HA columns (*40*), a finding implying that bases may be involved in the interaction with HA when their amino groups are ionized and available. It has also been shown (*40*) that basic polyamino acids show a strong interaction with HA.

Elution of polynucleotides from HA by phosphate appears to be due to a specific competition between phosphates of the buffer and the phosphates of polynucleotides for adsorbing sites on HA and not simply to an increase in ionic strength. In fact, as mentioned in Section III.A, elution can be performed at a practically constant ionic strength with native DNA being eluted at the same phosphate molarity independently of the ionic strength of the eluting buffer. In some cases, mentioned in Section III.A, addition of chlorides to the eluting buffer causes an increase in the phosphate molarity eluting native DNA. It is likely that this effect is due to the repression of phosphate dissociation rather than to a slight change in the structure of native DNA at high ionic strength; the latter phenomenon cannot be ruled out altogether, however.

Concerning the relationships between the elution molarities of polynucleotides and their secondary (or tertiary) structures, the results discussed so far lead to the following conclusions:

1. In the case of polynucleotides endowed with a rigid structure, it is evident that all those possessing a double-stranded structure (native and renatured DNA, replicative RNA, poly A-poly U, and poly I-poly C) are eluted in the same molarity range. Differences in the elution molarities exist, however (see, for example, Section III.B), and suggest that conditions of

higher resolution may lead to interesting separations. The only triple-stranded polynucleotide examined, 2 poly U–1 poly A, is eluted at a much higher phosphate molarity (about 0.45 M KP). The higher elution molarity of this complex may be possibly explained by the fact that a triple-stranded polynucleotide has a higher linear charge density than double-stranded polynucleotides, and therefore a larger number of phosphate groups per unit length is available for interaction with HA, and/or by the fact that the distribution of these groups in 2 poly U–1 poly A matches the distribution of adsorbing sites on HA better than that of phosphates of double-stranded polynucleotides.

2. As far as polynucleotides endowed with a "random-coiled" structure are concerned, their chromatographic behavior needs to be investigated under a much wider set of experimental conditions than has been done so far, since their structure is strongly dependent on ionic strength, temperature, and presence of compounds like formaldehyde, urea, etc. In all cases, however, randomly-coiled polynucleotides are eluted from HA columns by lower phosphate molarities than the corresponding rigid, helical polynucleotides. The property of HA of discriminating these two types of structures seems to be of general validity since in the case of proteins, too, the disruption of their secondary and tertiary structures causes a drastic decrease in their affinity for HA. An explanation for this phenomenon is that the groups which were available for the interaction with the adsorbing sites on HA in the rigid, ordered structures, greatly decrease in number on the "outer surface" of the randomly coiled, denatured nucleic acids or proteins. A similar explanation may hold for the decreased affinity for HA observed for the twisted circular form compared to both the linear open and the linear circular forms of polyoma virus DNA (see Section III.B). Furthermore, local concentrations of interacting sites due to the existence of rigid secondary and tertiary structures disappear upon denaturation. An example of the importance of charge distribution is given by the chromatographic behavior of nucleoside triphosphates and trinucleotides. In spite of the fact that these compounds have the same charge, nucleoside triphosphates are eluted by 0.2 M KP, whereas trinucleotides are eluted by 0.001 M KP. Finally. it should be mentioned that, as in the case of rigid polynucleotides, HA can discriminate among different structures existing in flexible polynucleotides; the different chromatographic behavior of formaldehyde-reacted polyribonucleotides and the fractionation of single-stranded, denatured DNA being typical examples.

In conclusion, HA chromatography is a new, powerful technique in the field of nucleic acids. Among its advantages over other techniques, three are of special importance. (1) Since the chromatographic material is an inorganic, crystalline, insoluble salt, it is possible to work over a wide range of temperatures, as well as in the presence salts and of organic reagents, such as formaldehyde, phenol, chloroform, urea. (2) The adsorption–elution phenomenon is basically associated with the density and distribution of interacting groups at the surface of polynucleotides; this makes the process understandable in physicochemical terms and permits a theoretical approach (*23, 24*). (3) The technique is simple, reproducible, and does not require any expensive equipment or reagent.

NOTE ADDED IN PROOF

Chromatography of nucleic acids on HA columns is enjoying a rapidly increasing popularity, as witnessed by many investigations published after this chapter was written. Some of these recent papers will be mentioned here since they provide examples of difference applications of the technique.

The most widespread use of HA columns remains the separation of denatured from native, or renatured DNA (*107–117*); of RNA from DNA (*118*); and of single-stranded from double-stranded RNA (*119*). Investigations on naturally (*120*) or artificially (*121*) cross-linked DNAs have also been reported. Thermal chromatography has been used for the partial purification of sea urchin satellite DNA (*122*), although a fractionation involving a polyethylene glycol–dextran system seems to be preferable in this case (*123*). It has also been used for determining the genetic relatedness between *Leptospira* strains (*124*), for separating polyribo (GC) from template polyribo (IC) (*125*), and for preparing high GC segments from *Mycoplasma* DNA (*126*) and high GC satellite DNAs (*127*). In this latter case 7.2 M sodium perchlorate was used in the eluting buffer to decrease the melting temperature. HA columns have also been used for the separation of DNA–RNA hybrids (*128–133*). The separation of yeast mitochondrial DNA from nuclear DNA reported above has been used by Fukuhara (*134, 135*).

Several investigations in the area of polyribonucleotides have been presented in a symposium on hydroxyapatite chromatography held in Strasbourg. The symposium dealt with the separation of ribooligonucleotides and the relationship between elution molarity and chain length (*136*), the fractionation of tRNA (*137*), the chromatography of nucleic acids from uninfected and virus-infected plants (*138*), the separation of three 5-S RNA species from wheat germs (these seem to differ only in their terminal phosphorylation (*139*) and the isolation of ribooligonucleotides obtained from yeast RNA by pancreatic RNase digestion (these contain 72% G and are eluted at 0.45 M NaP (*140*)).

References

1. Tiselius, A., S. Hjertén, and O. Levin. 1956. *Arch. Biochem. Biophys.*, **65**: 132.
2. Hjertén, S. 1959. *Biochim. Biophys. Acta*, **31**: 216.
3. Levin, Ö. 1962. *Methods Enzymol.*, **5**: 27.
4. Semenza, G. 1957. *Arkiv Kemi*, **11**: 89.

5. Main R. K., and L. Cole. 1957. *Arch. Biochem. Biophys.*, **68**: 186.
6. Main, R. K., M. J. Wilkins, and L. Cole. 1959. *Science*, **129**: 331.
7. Main, R. K., M. J. Wilkins, and L. Cole. *J. Am. Chem. Soc.*, **81**: 6490.
8. Bernardi, G. 1961. *Biochem. Biophys. Res. Commun.*, **6**: 54.
9. Bernardi, G., and S. N. Timasheff. 1961. *Biochem. Biophys. Res. Commun.*, **6**: 58.
10. Bernardi, G., and W. H. Cook. 1960. *Biochim. Biophys. Acta*, **44**: 96.
11. Bernardi, G. 1962. *Biochem. J.*, **83**: 32 P.
12. Bernardi, G. 1965. *Nature*, **206**: 779.
13. Chevallier, M. R., and G. Bernardi. 1965. *J. Mol. Biol.*, **11**: 658.
14. Chevallier, M. R., and G. Bernardi. 1968. *J. Mol. Biol.*, **32**: 437.
15. Bernardi, G. 1969. *Biochim. Biophys. Acta*, **174**: 423.
16. Bernardi, G. 1969. *Biochim. Biophys. Acta*, **174**: 435.
17. Bernardi, G. 1969. *Biochim. Biophys. Acta*, **174**: 449.
18. Bernardi, G., and T. Kawasaki. 1968. *Biochim. Biophys. Acta*, **160**: 301.
19. Bernardi, G., F. Carnevali, A. Nicolaieff, G. Piperno, and G. Tecce. 1968. *J. Mol. Biol.* **37**: 493.
20. Bernardi, G., M. Faures, G. Piperno, and P. Slonimski. 1970. *J. Mol. Biol.*, **48**: 23
21. Kawasaki, T., and G. Bernardi. 1970. *Biopolymers*, **9**: 257.
22. Kawasaki, T., and G. Bernardi. 1970. *Biopolymers*, **9**: 269.
23. Kawasaki, T. 1970. *Biopolymers*, **9**: 277.
24. Kawasaki, T. 1970. *Biopolymers*, **9**: 291.
25. Miyazawa, Y., and C. A. Thomas, Jr. 1965. *J. Mol. Biol.*, **11**: 223.
26. Anacker, W. F., and V. Stoy. 1958. *Biochem. Z.*, **33**: 141.
27. Jenkins, W. T. 1962. *Biochem. Prepn.*, **9**: 83.
28. Siegelman, H. W., G. A. Wieczorek, and B. C. Turner. 1965. *Anal. Biochem.*, **13**: 402.
29. Siegelman, H. W., and E. F. Firer. 1964. *Biochemistry*, **3**: 418.
30. Determann, H. 1968. Gel chromatography. Springer Verlag, New York.
31. Fischer, L. 1969. In T. S. Work and E. Work (Eds.), Laboratory techniques in biochemistry and molecular biology. North Holland Pub. Co., Amsterdam, Vol. I, p. 151.
32. Flodin, P. J. 1961. *J. Chromatog.*, **5**: 103.
33. Patel, A. S., and K. S. Korgaonkar. 1969. *Experientia*, **25**: 25.
34. Bernardi, G., and C. Sadron. 1964. *Biochemistry*, **3**: 1411.
35. Young, E. T., II, and R. L. Sinsheimer. 1965. *J. Biol. Chem.*, **240**: 1274.
36. Bourgaux, P., and D. Bourgaux-Ramoisy. 1967. *J. Gen. Virol.*, **1**: 323.
37. Bourgaux-Ramoisy, D., N. Van Tieghem, and P. Bourgaux. 1967. *J. Gen. Virol.*, **1**: 589.
38. Oishi, M. 1969. *J. Bacteriol*, **98**: 104.
39. Bernardi, G., and S. N. Timasheff. 1970. *J. Mol. Biol.*, **48**: 43.
40. Bernardi, G., experiments to be published.
41. Faulhaber, I. and G. Bernardi. 1967. *Biochim. Biophys. Acta*, **140**: 561.
42. Zubay, G., and P. Doty. 1959. *J. Mol. Biol.*, **I**: 1.
43. André, M., and G. Bernardi, in preparation.
44. Grossman, L., S. S. Levine, and W. S. Allison. 1961. *J. Mol. Biol.*, **3**: 47.
45. Rownd, R., D. M. Green, R. Sterglanz, and P. Doty. 1968. *J. Mol. Biol.*, **32**: 369.
46. Alberts, B. M., and P. Doty. 1968. *J. Mol. Biol.*, **32**: 379.
47. Alberts, B. M. 1968. *J. Mol. Biol.*, **32**: 405.
48. Mulder, C., and P. Doty. 1958. *J. Mol. Biol.*, **32**: 423.
49. Albertsson, P. A. 1962. *Arch. Biochem. Biophys. Suppl.*, **1**: 264.
50. Walker, P. M. B., and A. McLaren. 1965. *Nature*, **208**: 1175.

51. Waring, M., and R. J. Britten. 1966. *Science,* **154**: 791.
52. Walker, P. M. B., and A. McLaren. 1969. *Nature,* **221**: 771.
53. McLaren, A. 1969. *Mol. Gen. Genetics,* **104**: 104.
54. Okazaki, R., T. Okazaki, K. Sakabe, K. Sugimoto, and A. Sugino, 1968. *Proc. Natl. Acad. Sci., U.S.,* **59**: 598.
55. Sugimoto, K., T. Okazaki and R. Okazaki. 1968. *Proc. Natl. Acad. Sci., U.S.,* **60**: 1356.
56. Sugimoto, K., T. Okazaki, Y. Imae, and R. Okazaki. 1969. *Proc. Natl. Acad. Sci.* **63**: 1343.
57. Oishi, M. 1968. *Proc. Natl. Acad. Sci., U.S.,* **60**: 329.
58. Oishi, M. 1968. *Proc. Natl. Acad. Sci., U.S.,* **60**: 691.
59. Oishi, M. 1968. *Proc. Natl. Acad. Sci., U.S.,* **60**: 1000.
60. Painter, R. B., and A. Schaefer. 1969. *Nature,* **221**: 1215.
61. Habener, J. F., B. S. Bynum, and J. Shack. 1969. *Biochem. Biophys. Acta,* **186**: 412.
62. Pauling, C., and L. Hamm. 1969. *Biochem. Biophys. Res. Commun.,* **37**: 1015.
63. Lark, K. G. 1969. *J. Mol. Biol.,* **44**: 217.
64. Schandl, E. K., and J. H. Taylor. 1969. *Biochem. Biophys. Res. Commun.,* **34**: 291.
65. Rhoades, M., L. A. MacHattie, and C. A. Thomas, Jr. 1968. *J. Mol. Biol.,* **37**: 21.
66. Rhoades, M., and C. A. Thomas, Jr. 1968. *J. Mol. Biol.,* **37**: 41.
67. Reznikoff, M. S., and C. A. Thomas, Jr. 1969. *Biology,* **37**: 309.
68. Zimmerman, S. B., J. W. Little, C. K. Oshinsky, and M. Gellert. 1967. *Proc. Natl. Acad. Sci., U.S.,* **57**: 1841.
69. Britten, R. J., and D. E. Kohne. 1966. *Carnegie Inst. Washington. Yr. Bk.,* **65**: 73.
70. Britten, R. J., and D. E. Kohne. 1967. *Carnegie Inst. Washington Yr. Bk.,* **66**: 73.
71. Britten, R. J., and D. E. Kohne. 1968. *Science,* **161**: 529.
72. Brenner, D. J., C. R. Fanning, A. V. Rake, and K. E. Johnson. 1969. *Anal. Biochem.,* **28**: 447.
73. McCallum, M., and P. M. B. Walker. 1967. *Biochem. J.,* **105**: 163.
74. Szala, S., and M. Chorazy. 1969. *Bull. Acad. Polon. Sci.,* **17**: 277.
75. Marmur, J., and P. Doty. 1962. *J. Mol. Biol.,* **5**: 109.
76. Niyogi, S. N., and C. A. Thomas, Jr. 1967. *Biochem. Biophys. Res. Commun.,* **26**: 51.
77. Niyogi, S. N., and C. A. Thomas, Jr. 1968. *J. Biol. Chem.,* **243**: 1220.
78. Brzezinski, A., P. Szafranski, P. H. Johnson, and M. Laskowski, Sr. 1969. *Biochemistry,* **8**:1228.
79. McConaughy, B. L., C. D. Laird, and B. J. McCarthy. 1969. *Biochemistry,* **8**: 3289.
80. Bishop, J. M., G. Koch, and L. Levintow. 1967. In J. S. Colter and W. Paranchych (Eds.), The molecular biology of viruses. Academic Press, New York, p. 355.
81. Vizoso, A. D., and A. T. H. Burness. 1960. *Biochem. Biophys. Res. Commun.,* **2**: 102.
82. Burness, A. T. H., and A. D. Vizoso. 1961. *Biochim. Biophys. Acta,* **49**: 225.
83. Pinck, L. 1969. Ph.D. Thesis, University of Strasbourg.
84. Brown F., J. F. E. Newmann, and D. L. Stewarts. 1963. *Nature,* **197**: 590.
85. Franklin, R. M. *Proc. Natl. Acad. Sci. U.S.,* **55**: 1504.
86. Bokstahler, L. E. 1967. *Mol. Gen. Genetics,* **100**: 337.
87. Pinck, L., L. Hirth, and G. Bernardi. 1968. *Biochim. Biophys. Res. Commun.,* **31**: 481.
88. Hartmann, G., and U. Coy. 1961. *Biochim. Biophys. Acta,* **47**: 612.
89. Pearson, R. L., and A. D. Kelmers. 1966. *J. Biol. Chem.,* **241**: 767.

90. Harding, U., H. Schauer, and G. Hartmann. 1966. *Biochem. Z.*, **346**: 212.
91. Muench, K. H., and P. Berg. 1966. *Biochemistry*, **5**: 982.
92. Dirheimer, G. 1968. *Bull. Soc. Chim. Biol.*, **50**: 7.
93. Legocki, A. B., and J. Pawelkiewcz. 1967. *Bull. Acad. Polon. Sci.*, **15**: 517.
94. Giege, R., J. Heinrich, J. H. Weil, and J. P. Ebel. 1967. *Biochim. Biophys. Acta*, **174**: 43.
95. Richards, E. G., C. P. Flessel, and J. R. Fresco. 1965. *Biopolymers*, **1**: 431.
96. Lipsett, M. N. *Proc. Natl. Acad. Sci., U.S.*, **46**: 445.
97. Luzzati, V., A. Mathis, F. Masson, and J. Witz. 1964. *J. Mol. Biol.*, **10**: 28.
98. Van Holde, K. E., J. Brahms, and A. M. Michelson. 1965. *J. Mol. Biol.*, **12**: 726.
99. Leng, M., and G. Felsenfeld. 1966. *J. Mol. Biol.*, **15**: 455.
100. Michelson, A. M., J. Massoulie, and W. Guschlbauer. 1967. *Progr. Nucleic Acid Res.*, **6**: 83.
101. Mundry, K. W. 1965. *Z. Vererbungslehre*, **97**: 281.
102. Mundry, K. W. 1969. *Mol. Gen. Genet.*, **105**: 361.
103. Bernardi, G. 1964. *Biochim. Biophys. Acta*, **91**: 686.
104. Gellert, M., M. N. Lipsett, and D. R. Davies. 1962. *Proc. Natl. Acad. Sci., U.S.*, **48**: 2013.
105. Mattson, S., E. Kontler-Anderson, R. B. Miller, and K. Vantras. 1951. *Kgl. Lantbruks-Hogskol. Ann.*, **18**: 493 quoted by S. Larsen. 1966. *Nature*, **212**: 212.
106. Wyckoff, R. W. G. 1951. Crystal structures. Wiley (Interscience), New York.
107. Barzilai, R. and C. A. Thomas. 1970. *J. Mol. Biol.*, **51**: 145.
108. Berger, H. and J. L. Irvin. 1970. *Proc. Natl. Acad. Sci., U.S.*, **65**: 152.
109. Dasgupta, R. and S. Mitra. 1970. *Biochem. Biophys. Res. Commun.*, **40**: 793.
110. Habener, J. F., B. S. Bynum, and J. Shack. 1970. *J. Mol. Biol.*, **49**: 157.
111. Laszlo, J., D. S. Miller, O. Brown, and E. Baril. 1970. *Biochem. Biophys. Res. Commun.*, **38**: 112.
112. Leffler, A. T., E. Creskoff, S. W. Luborsky, V. McFarlan, and P. T. Mora. 1970. *J. Mol. Biol.*, **48**: 455.
113. Lusby, E. W. and S. R. Dekloet. 1970. *Biochem. Biophys. Acta*, **209**: 263.
114. Nozawa, R. and D. Mizuno. 1970. *Proc. Natl. Acad. Sci., U.S.*, **63**: 904.
115. Sato, S., M. Tanaka, and T. Sugimura. 1970. *Biochem. Biophys. Acta*, **209**: 43.
116. Votavova, H., J. Sponar, and Z. Sormova. 1970. *European. J. Biochem.*, **12**: 208.
117. Doenecke, D. and C. E. Sekeris. 1970. *FEBS Letters*, **8**: 61.
118. Jannsen, S., E. R. Lochmann, and R. Megnet. 1970. *FEBS Letters*, **8**: 113.
119. Black, D. R. and C. A. Knight. 1970. *J. Virol.*, **6**: 194.
120. Chevallier, M. R. and M. L. Greth. 1969. *Mol. Gen. Genet.*, **105**: 344.
121. Berns, K. I. and C. Silverman. 1970. *J. Virol.*, **5**: 299.
122. Stafford, D. W. and W. R. Guild. 1969. *Exptl. Cell. Res.*, **55**: 347.
123. Patterson, J. B. and D. W. Stafford. 1970. *Biochemistry*, **9**: 1278.
124. Haapala, D. K., M. Rogul, L. B. Evans, and A. D. Alexander. 1960. *J. Bacteriol.*, **98**: 421.
125. Karstadt, M. and J. S. Krakow. 1970. *J. Biol. Chem.*, **245**: 752.
126. Ryan, J. L. and H. J. Morowitz. 1969. *Proc. Natl. Acad. Sci., U.S.*, **63**: 1282.
127. Graham, D. E. 1970. *Anal. Biochem.*, **36**: 315.
128. Walker, P. M. B. 1969. In J. N. Davidson and W. E. Cohn (eds.), Progress in nucleic acid research and molecular biology, Academic Press, London, vol. 9, p. 301.
129. Siebke, J. C. and T. Ekren. 1970. *European. J. Biochem.*, **12**: 380.
130. Davidson, E. H. and B. R. Hough. *Proc. Natl. Acad. Sci., U.S.*, **63**: 342.
131. Becker, W. M., A. Hell, J. Paul, and R. Williams. 1970. *Biochem. Biophys., Acta*, **199**: 348.

132. Colli, W. and M. Oishi. 1969. *Proc. Natl. Acad. Sci., U.S.*, **64**: 642.
133. Bekhor, I., J. Bonner, and G. K. Dahmus. 1969. *Proc. Natl. Acad. Sci., U.S.*, **62**: 271.
134. Fukuhara, H. 1969. *European J. Biochem.*, **11**: 135.
135. Fukuhara, H. 1970. *Mol. Gen. Genet.*, **107**: 58.
136. Mundry, K. W. 1970. *Bull. Soc. Chim. Biol.*, **52**: 873.
137. Zimmer, A. and G. Hartmann. 1970. *Bull. Soc. Chim. Biol.*, **52**: 867.
138. Pinck, L. 1970. *Bull. Soc. Chim. Biol.*, **52**: 843.
139. Soave, C., E. Galante, and G. Torti. 1970. *Bull. Soc. Chim. Biol.*, **52**: 857.
140. Piperno, G. and G. Bernardi. 1970. *Bull. Soc. Chim. Biol.*, **52**: 885.

HYDROXYAPATITE TECHNIQUES FOR NUCLEIC ACID REASSOCIATION

D. E. KOHNE AND R. J. BRITTEN

CARNEGIE INSTITUTION OF WASHINGTON

DEPARTMENT OF TERRESTRIAL MAGNETISM

WASHINGTON, D.C.

INTRODUCTION

Main et al. (*1, 2*) reported the use of hydroxyapatite for fractionation of DNA. Bernardi (*3*) and Miyazawa and Thomas (*4*) showed convincingly that double- and single-stranded DNA could be separated on hydroxyapatite. Miyazawa and Thomas further demonstrated that DNA could be fractionated on the basis of its base composition by thermal chromatography on a hydroxyapatite column.

The hydroxyapatite technique described here provides a powerful tool with which to attack problems of nucleic acid interactions and has been utilized for the characterization and fractionation of DNAs from many sources. It has been used to: (a) measure the kinetics of reassociation of many different DNAs (*5–8*); (b) isolate and purify repeated DNA sequences from many different DNAs (*8*); (c) determine the extent of similarity between the DNA sequences of different species (*7, 9, 10*); (d) isolate and purify the specific DNA cistrons which code for ribosomal RNA in bacteria (*11*); (e) measure the hybridization of RNA (messenger RNA and ribosomal RNA) with DNA in both animals and bacteria (*12, 11,* Kohne unpublished data); (f) rapidly purify DNA from a number of tissues (*13*).

The demonstrated advantages of using hydroxyapatite are: (a) an experimental system where virtually complete reaction of the DNA is routine and easily detectable; (b) the ability to easily separate and recover both the reassociated fraction and the nonreassociated fraction; (c) the easy characterization and fractionation of the DNA by thermal chromatography; (d) the capability of including "internal standards" in the nucleic acid solutions; (e) the potential for easily handling widely varying amounts of DNA in a routine manner (0.0001 μg to 2000 μg); (f) the ability to easily maintain specific conditions of salt concentration and temperature during the fractionation of the DNA. The major disadvantages of the technique as described here are the inability to do nucleic acid competition experiments and the inability to do a large number of analyses at one time. Recently, however, a hydroxyapatite technique which enables the handling of many different samples simultaneously has been developed (*14*).

MATERIALS

1. Hydroxyapatite. This can be purchased from Bio-Rad, Richmond, California, the Clarkson Chemical Co., Williamsport, Pa., or made in the laboratory (*4*).
2. Water jacketed columns. These columns must withstand temperatures of 100°C.
3. A combination water heater and pump with good temperature control.
4. Phosphate Buffer (PB). Made by mixing equimolar quantities of Na_2HPO_4 and NaH_2PO_4.
5. Log C_0t plot (*8*). A method for presentation of reassociation kinetics in which the fraction of DNA reassociated is plotted against log C_0t. C_0t is the product of DNA concentration (C_0) and the time (t) of reassociation. C_0t is most easily calculated as the optical density at 260 nm of dissociated DNA/2, times the hours of incubation (^{OD} x hours = C_0t). This approximates the usual units (seconds x mol/liter) since the extinction coefficient of DNA under these conditions is about 7200.

GENERAL COMMENTS

Single- and double-stranded DNAs have different affinities for hydroxyapatite (*3, 4, 8*). Adsorption of DNA to hydroxyapatite apparently is controlled by the phosphate ion concentration (*3*). At low phosphate ion concentrations, both double- and single-stranded DNA adsorb to hydroxyapatite. At intermediate phosphate ion concentrations only double-stranded DNA remains adsorbed while single-stranded DNA can be washed away. At a high phosphate ion concentration double-stranded DNA is released from hydroxyapatite. Double-stranded DNA also can be released from hydroxy-

TABLE 1. Effect of Phosphate Ion Concentration on the Adsorption of Single-Stranded DNA to Hydroxyapatite[a]

Phosphate concentration (M)	Percent binding of single-stranded DNA to hydroxyapatite
0.08	51.0
0.10	8.0
0.11	2.7
0.12	0.7
0.13	0.5
0.14	0.4

[a] Radioactive, sheared *E. coli* (50,000 psi) DNA (51% G + C) at a very low concentration was heat denatured in one of the phosphate ion concentrations specified above and immediately passed over hydroxyapatite equilibrated to 60°C.

apatite at an intermediate phosphate ion concentration by raising the temperature of the hydroxyapatite above the dissociation temperature of the double-stranded DNA. At this elevated temperature the DNA becomes single-stranded, no longer adsorbs to hydroxyapatite, and can be eluted from the column (*4*).

The hydroxyapatite technique used by the early investigators has been modified in order to simplify the separation of single- and double-stranded DNA (*7, 8*). A number of salt and temperature combinations were explored to find a condition at which double-stranded DNA adsorbs to hydroxyapatite, while single-stranded DNA does not adsorb and, therefore, passes through the hydroxyapatite column. The salt and temperature condition of 0.12 M PB (phosphate buffer, pH = 6.8, prepared by mixing equimolar amounts of the NaH_2PO_4 and Na_2HPO_4) and 60°C permits adsorption of double-stranded DNA to hydroxyapatite but does not permit adsorption of single-stranded DNA. Table 1 presents data used in selecting this condition. Under these conditions, radioactive bacterial DNA which is almost wholly reassociated (as judged by optical criteria) is adsorbed almost completely (98 to 99%) to hydroxyapatite at 60°C in 0.12 M PB or 0.14 M PB. At 50°C a higher PB concentration (0.14 M) is needed to prevent extensive adsorption of single-stranded *E. coli* DNA. DNAs with significantly higher guanine-cytosine contents than *E. coli* may necessitate higher PB concentrations and/or higher temperatures of column operation to effect a condition where single-stranded DNA does not adsorb.

Different batches of hydroxyapatite vary widely in their affinity for single-stranded DNA. Each batch should be checked by passing a small amount (about 0.1 μg) of freshly denatured sheared radioactive DNA over the hydroxyapatite in 0.12 M PB at 60°C. Less than 2% of the radioactivity should be adsorbed. It is important that the DNA used for this test be completely single-stranded. Thus, a DNA which reassociates very rapidly or contains fractions which reassociate rapidly should not be used. Bacterial DNA serves nicely for this purpose. Hydroxyapatite which binds a significant portion of the single-stranded DNA can be treated to reduce this binding to acceptable levels. Boiling hydroxyapatite in 0.14 M PB for 15–30 minutes prior to using is sometimes sufficient to reduce the single-strand binding to below 2% of the input DNA. The presence of 0.4% SLS in the DNA solution and washing buffer usually reduces the extent of single-stranded binding to below 1%. It is not necessary to boil the hydroxyapatite when SLS is used. The SLS has little or no effect on the reassociation kinetics or thermal stability of the DNA.

Under the fractionation conditions described above the *useful capacity* of the hydroxyapatite is generally about 200 μg of double-stranded DNA per milliter of wet packed hydroxyapatite. Various batches of hydroxyapatite differ in capacity and each new batch should be checked. The total binding capacity is much higher than the *useful capacity*. Hydroxyapatite will initially bind *all* of the double-stranded DNA passed through it. However, as more DNA is passed through the column the hydroxyapatite adsorbs only a part of it until eventually it is completely saturated and can bind no more DNA. For measurement of reassociation the useful capacity is the maximum amount of double-stranded DNA a given quantity of hydroxyapatite can totally adsorb.

Hydroxyapatite has several practical operational advantages. Air can be blown through the hydroxyapatite bed during the analysis without affecting the results. A very rapid column flow rate (5–20 ml per minute) can be used and the flow rate from a column can be increased by utilizing positive air pressure. The column bed can also be stirred frequently during the analysis without affecting the result.

MEASURING REASSOCIATION WITH HYDROXYAPATITE (*7, 8*)

The hydroxyapatite procedure used to measure the extent of reassociation of DNA is as follows:

1. Dissociate sheared DNA in 0.12 M PB (or 0.14 M PB) by heating at 100°C for 2–3 min.

2. Incubate the dissociated DNA at 60°C for specified times.

3. Pass the incubation mixture through a water jacketed hydroxyapatite column which has been equilibrated at 60°C and 0.12 M PB (or 0.14 M PB). If necessary the column should be equilibrated with PB to which SLS has been added to a final concentration of 0.4%. In this case SLS should also be added to the incubation mixture (final concentration 0.4%) before passing it through the column.

4. Wash single-stranded DNA from the column with 0.12 or 0.14 M PB. If SLS is being used it should also be added to the wash buffer at a final concentration of 0.4%.

5. (a) If the thermal stability of the reassociated DNA is not going to be measured, remove the adsorbed DNA by high salt (0.4 M PB) or high temperature elution (100°C in 0.12 or 0.14 M PB). (b) To measure the thermal stability of the DNA adsorbed to the hydroxyapatite the temperature of the column is raised in discrete steps. After each temperature rise the DNA which has been rendered single strand is washed off the column in 0.12 or 0.14 M PB. The volume of the wash is usually 2-4 times the bed volume of the column. After the wash at 100°C the column is washed with 0.4 M PB to elute any remaining DNA.

6. It is possible to analyze the DNA by eluting it from the hydroxyapatite with a PB gradient (*3*).

KINETICS OF DNA REASSOCIATION

Bacterial DNA

The time course of DNA reassociation can be followed by measuring the fraction of DNA which adsorbs to hydroxyapatite at specified times after the initiation of the reassociation reaction. In order to achieve reproducible rates or reassociation the temperature of incubation, cation concentration, and DNA fragment size must be carefully controlled (*8, 15*).

Another method for measuring reassociation kinetics depends on the fact that double-stranded DNA adsorbs less ultraviolet light at 260 nm than does single-stranded DNA. Thus, reassociation kinetics can also be monitored by following the decrease with time of the ultraviolet absorbancy of a DNA solution (15-18).

Figure 1 shows reassociation kinetic data of *E. coli* DNA measured by the hydroxyapatite technique. The C_0t method (*8*) has been used to plot this data. The curves represent ideal second-order reactions. The bimolecular nature of the reassociation reaction has been shown by the concentration dependence of the rate of reaction (*8*).

With the hydroxyapatite method the half-time of reassociation is easily calculated. Ideally, the beginning (zero adsorption) and end (100% adsorption) points of the reaction are known; practically, the beginning and end points are determined experimentally and the data points corrected as designated in Fig. 1. The half-time of reaction is more difficult to calculate with the ultraviolet absorbancy technique. The reassociation reaction generally is initiated by heating the DNA solution to 100°C and then quickly cooling the solution to 60°. The absorbancy of the DNA at the beginning of

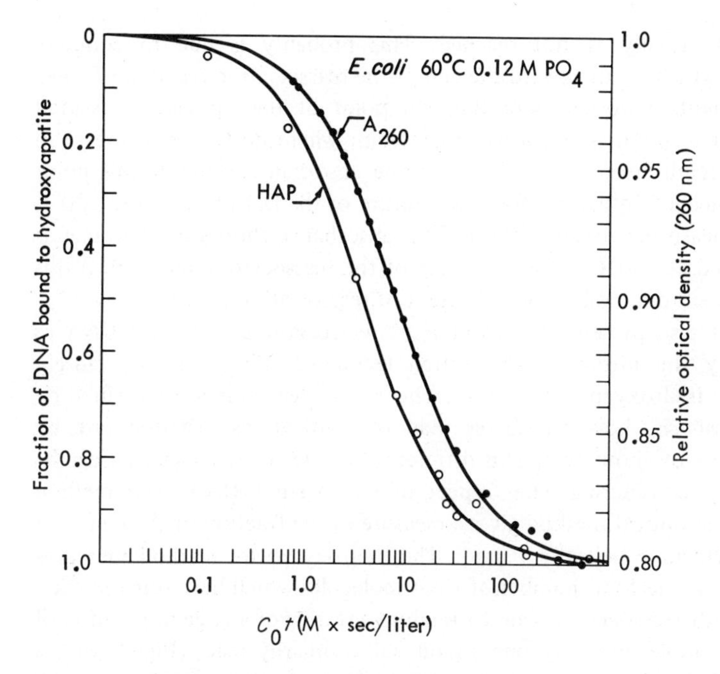

Figure 1. Reassociation kinetics of sheared (50,000 psi) *E. coli* DNA as measured by two different methods; optical hypochromicity at 260 nm and hydroxyapatite. Open circles represent hydroxyapatite data while closed circles record optical data. The curves drawn are theoretical second-order reaction curves. Each hydroxyapatite data point represents an aliquot taken from an incubation mixture containing 504 μg/ml of DNA. The DNA was incubated at 60°C in 0.12 M PB and aliquots taken out at specified times and passed over a hydroxyapatite column equilibrated to 60°C and 0.12 M PB. The DNA in the various fractions was measured by optical density at 260 nm. The maximum observed hydroxyapatite reassociation of the DNA was 91.5% of the optical density present. The data points have been corrected for nonreassociating material. The optical reassociation kinetic experiment was performed using a Zeiss spectrophotometer possessing an automatic recorder and sample changer. The reaction mixture was identical to that used for the hydroxyapatite experiment. The DNA solution (504 μg DNA/ml in 0.12 M PB) in a 1-mm cell was placed in a heated cell block in the spectrophotometer and denatured at 100°C. The temperature of the cell was then rapidly dropped to and held constant at 60°C. Reassociation was monitored by the decrease in absorbance of the DNA solution.

the reaction must be estimated since a drop in absorbancy which is not due to reassociation takes place when the temperature of the DNA solution is changed from a dissociating temperature of 100°C to the incubation temperature of 60°. This decrease in absorbancy is caused by formation of intrastrand secondary structure induced by the temperature drop. Under the conditions used here the drop in absorbancy is 2–3% of the absorbancy at 100°C. The true end point of the optically measured reassociation reaction also is difficult to determine since the ultraviolet absorbancy expected for native DNA usually is not reached. This probably is due to complex interactions which occur toward the end of the reassociation reaction (*7*). For the experiments reported here the end point of the optically measured reassociation reaction is taken to be the point where no further decrease in absorbancy is observed. The validity of the absorbance value at this point should be checked by raising the temperature of the DNA solution to 100°C and again measuring its absorbance. This absorbance should be the same as that measured at 100°C before the start of the reassociation reaction. If it is not, the hypochromic value for the extent of reassociation is suspect.

Figure 1 also presents the kinetics of reassociation of *E. coli* DNA as measured by the ultraviolet absorption technique. These kinetics closely resemble the hydroxyapatite kinetics. The rate of reassociation measured by hydroxyapatite is, however, faster than the optical reassociation rate by about a factor of about two. This difference in rates of reassociation can be explained by considering what aspect of reassociated DNA each method measures. The optical method gives a measure of the fraction of the length of the DNA which is helically paired. The hydroxyapatite method measures the fraction of the total number of DNA molecules which have some portion of their length reassociated. Due to random shearing of a large population of similar DNA molecules, any one strand will ordinarily react (Fig. 2) with a strand which has been sheared at a different point. Thus, DNA molecules which are only partially complementary collide and reassociate, with the result that double-stranded molecules are formed which possess both reassociated and single-strand regions (Fig. 2F). The first step in reassociation is the formation of double-stranded molecules which have on the average about 50% of their length as helical structure. Since these molecules will adsorb to hydroxyapatite a reassociation reaction will appear to progress more rapidly when measured by hydroxyapatite than when measured optically. The observed factor of about two between the half reaction times observed by the two methods is consistent with that predicted by approximate mathematical analysis (*7*).

Characterization of Reassociated DNA.

Comparison of the thermal stability of reassociated DNA to that of native DNA from the same source provides an indication of the precision of nucleotide pairing between the component strands of the reassociated DNA. Figure 3 presents the thermal stability profile of reassociated *E. coli* DNA as measured by two different methods. Both methods, the hydroxyapatite thermal elution chromatography and the optical hyperchromicity technique, give almost identical results. Miyazawa and Thomas (*4*) also have compared the optical and hydroxyapatite measured thermal stabilities of several

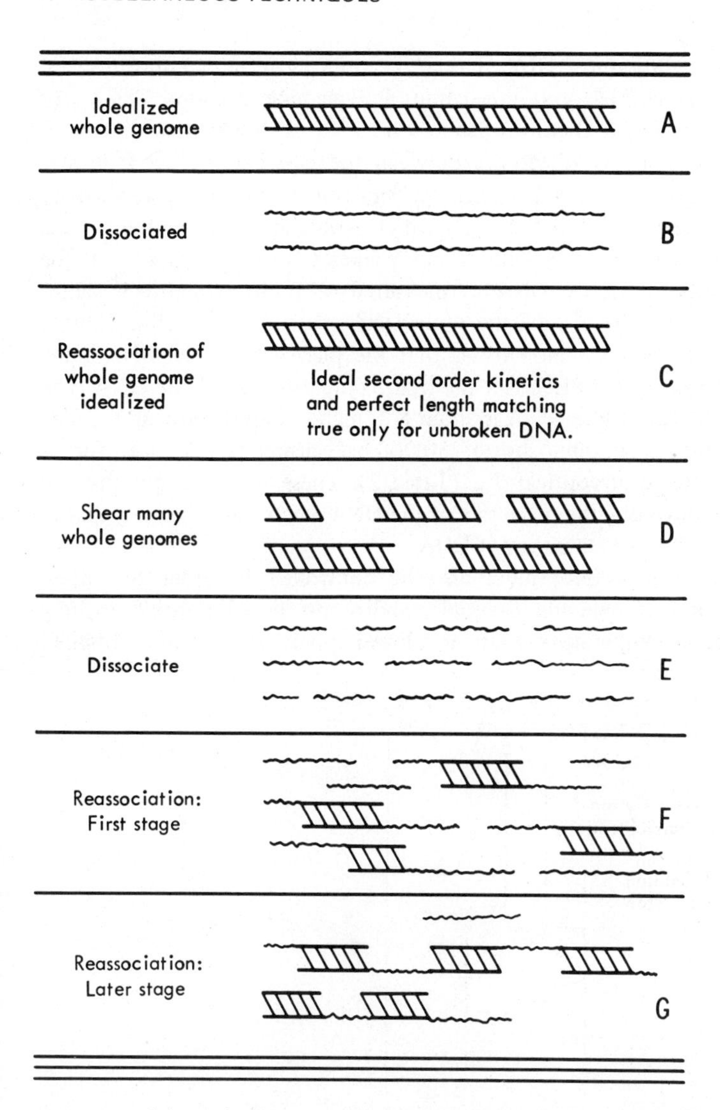

Figure 2. Schematic diagram to illustrate the reassociation of sheared DNA.

different native DNAs and found little difference in the results obtained by the different methods. Figure 4 presents data which further demonstrates the ability of hydroxyapatite to characterize DNA on the basis of its thermal stability. For this experiment completely reassociated unlabeled *E. coli* DNA was mixed together with completely reassociated ^{32}P-labeled *Proteus mirabilis* DNA and the DNA adsorbed to hydroxyapatite. The DNAs then were thermally eluted from the hydroxyapatite in 0.14 M PB (as described in

Fig. 4). *Proteus mirabilis* DNA has a guanine-cytosine composition of 38-40% while *E. coli* DNA contains about 52% guanine-cytosine. The T_m of *Proteus mirabilis* DNA, therefore, should be 4° to 5° lower than that of *E. coli* DNA (*16*). The T_m difference between the two DNAs (Fig. 4) is very close to that predicted. Further, the T_m's measured by the hydroxyapatite technique are very close to the T_m's which would be seen by the optical hyperchromicity method. McCallum and Walker (*9*) have reported that the thermal stability of native DNA as measured on hydroxyapatite is significantly higher than the T_m of the same DNA measured optically. Table 2 summarizes data which demonstrates that the piece size of the DNA being characterized has a great effect on the hydroxyapatite T_m of the DNA. The hydroxyapatite T_m of long DNA is about 8°C higher than its optical T_m. The hydroxyapatite T_m of the shorter 50,000 psi sheared DNA, however, is almost the same as its optical T_m (Table 2). These results emphasize the necessity of controlling the piece size of the DNA when using hydroxyapatite to measure the thermal stabilities of DNA.

The elution procedure must also be controlled in order to obtain reproducible and comparable thermal stability profiles. Extensive washing after each rise in temperature results in a lower apparent T_m than is obtained

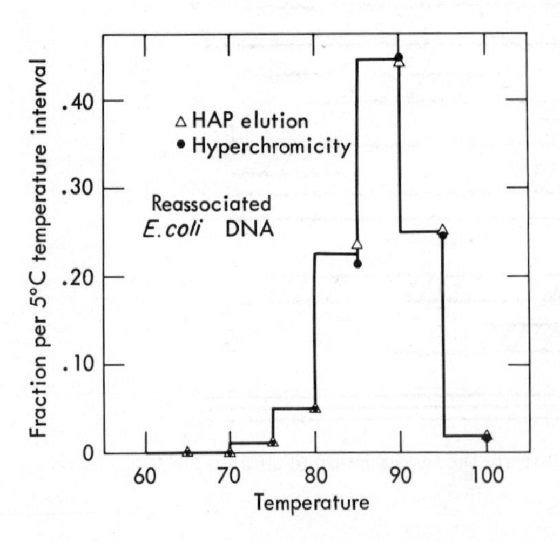

Figure 3. The correspondence of optical (closed circles) hydroxyapatite (triangles) thermal stability curves is shown in the following manner: (1) Bind the DNA to hydroxyapatite column by passing it through the column in 0.12 M PB at 60°C. (2) Wash the column with 0.12 M PB to remove single-stranded DNA. (3) Raise the temperature of the column in 5°C steps and wash any dissociated DNA from the column at each step. (4) Determine the absorbancy (at 260 nm) of the eluate. The optical hyperchromicity melting profile (in 0.12 M PB) was determined in a Zeiss spectrophotometer fitted with a heated cell block. Increase in absorbancy at 260 nm is plotted as the fraction of the total increase between 60°C and 100°C.

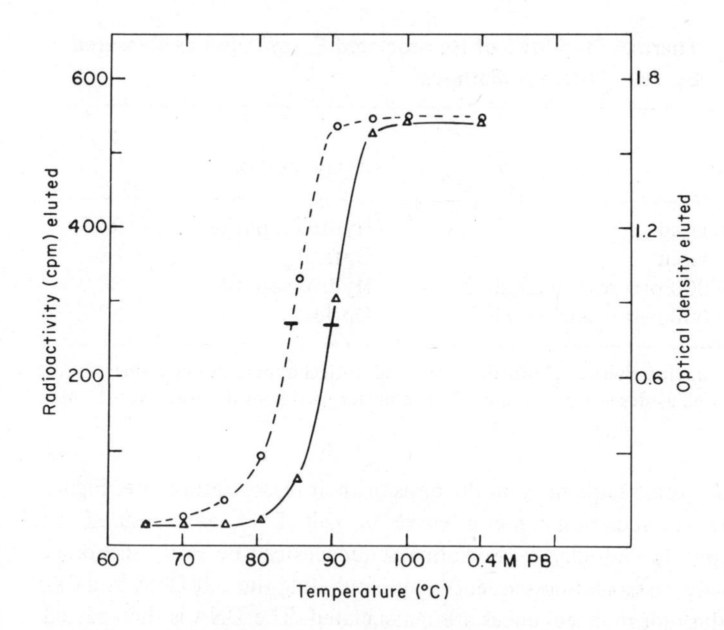

Figure 4. Thermal stability profile of a mixture of reassociated *Proteus mirabilis* DNA (circles) and reassociated *E. coli* DNA (triangles). Sheared (50,000 psi) ^{32}P-labeled *Proteus mirabilis* and unlabeled *E. coli* DNAs were denatured separately and completely reassociated in 0.14 M PB. A small amount of the labeled Proteus DNA (8 μg) and the mixture passed over hydroxyapatite equilibrated to 0.14 M PB and 60°C. Ninety-eight percent of the optical density and 95% of the radioactivity adsorbed to hydroxyapatite. The thermal chromatogram was obtained as described in Fig. 3, with the exception that 0.14 M PB was used. Radioactivity was monitored by assaying for Cerenkov radiation. The unlabeled DNA was measured by its absorbance at 260 nm.

with less washing. However, if the washing procedure is standardized T_m's obtained by hydroxyapatite thermal chromatography are quite reproducible (Table 2). A very good way to obtain comparable hydroxyapatite T_m's is to include an "internal standard" in each thermal chromatography run. This internal standard can be labeled or unlabeled DNA of a known base composition and the T_m's of other DNAs can be accurately measured relative to the standard DNA.

Measurement of Reassociation of Higher-Organism DNA

Higher-organism DNAs contain some nucleotide sequences which are present many times per haploid cell and others which are present only once (*8*). The hydroxyapatite technique described here is quite useful for measuring the kinetics of reassociation of higher-organism DNA and also for

TABLE 2. Thermal Stabilities of Reassociated *E. coli* DNA as Measured by Two Different Methods[a]

State of DNA	Assay method	T_m (°C)
Long reassociated	Hydroxyapatite	94.5
Long reassociated	Optical	88
Sheared (50,000 psi) reassociated	Hydroxyapatite	88.7
Sheared (50,000 psi) reassociated	Optical	87

[a] The hydroxyapatite thermal chromatography and optical hyperchromicity measurements were done as described in Figure 5. The buffer used in both cases was 0.12 M PB.

separating nucleotide sequences on the basis of their reassociation rate. Figure 5 shows the reassociation kinetic curve of calf DNA as measured by hydroxyapatite. The rapidly reassociating sequences can be easily separated from the slowly reassociating sequences by incubating the calf DNA to a C_0t where only the repeated sequences are reassociated. The DNA is then passed over hydroxyapatite where the repeated DNA adsorbs while the nonrepeated single-stranded DNA passes through the column.

Reassociated repeated DNA from most organisms is primarily composed of double-stranded DNA which contains an appreciable fraction of unpaired base in the reassociated regions. The capacity of hydroxyapatite for this type of double-stranded DNA is somewhat lower than that for DNA with perfectly matched base pairs. Allowance must be made for this. When working with repeated DNA it is also important that the hydroxyapatite column criteria and the criterion of DNA incubation be the same since changes in conditions can alter the amount of repeated DNA detected. The piece size of the DNA is of particular importance when working with higher-organism DNA. Interspersion of repeated and nonrepeated DNA sequences often leads to the formation of large aggregates by the DNA fragments and these cannot always be eluted from hydroxyapatite. Reduction of the DNA fragment size by shearing at 50,000 psi avoids this irreversible binding.

Hydroxyapatite Measurement of RNA : DNA Hybridization

RNA : DNA hybrids can also be detected using exactly the same hydroxyapatite conditions already described. While single-stranded DNA does not bind to hydroxyapatite single-stranded RNA does to a large extent. For this reason the indicator used to monitor the RNA : DNA reaction is *radioactive DNA*. The basic technique of reacting labeled DNA with

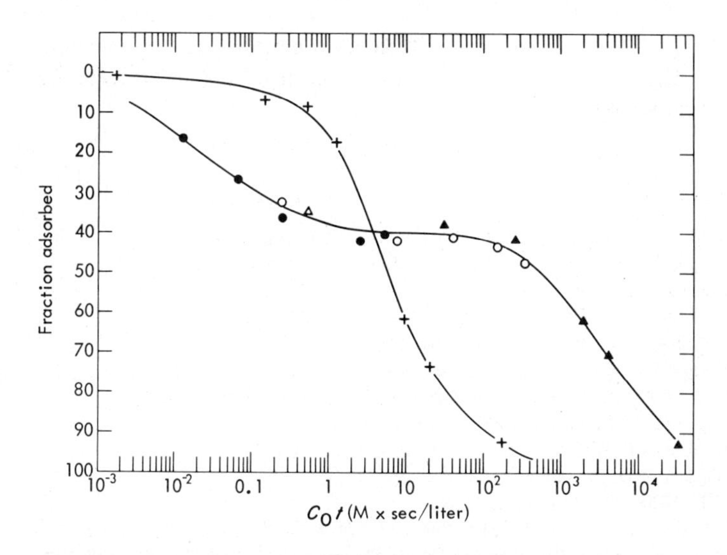

Figure 5. The kinetics of reassociation of calf-thymus DNA measured with hydroxyapatite. The DNA was sheared at 3.4 kilobars and incubated at 60°C in 0.12 M phosphate buffer. At various times, samples were diluted, if necessary (in 0.12 M phosphate buffer at 60°C), and passed over a hydroxyapatite column at 60°C. The DNA concentrations during the reaction were (μg/ml): open triangles, 2; closed circles, 8600. Crosses are radioactively labeled *E. coli* DNA at 43 μg/ml present in the reaction containing calf thymus DNA at 8600 μg/ml.

unlabeled RNA has broad application for RNA : DNA hybrid studies. It can be used to purify any specific DNA sequence for which sufficient product RNA can be obtained. This technique is described in detail in Ref. *11*.

The capacity of hydroxyapatite for RNA is less than that for DNA. Excess RNA can prevent the binding of RNA : DNA hybrids to hydroxyapatite. Care must be taken to use an amount of RNA which will not prevent binding of RNA : DNA hybrids. Different batches of hydroxyapatite have different capacities. Dry hydroxyapatite (Bio-Rad HTP) has a very low capacity for RNA. The best results have been obtained with Bio-Rad HT where the *useful* RNA capacity is about 40 μg of RNA per milliliter of packed hydroxyapatite. It is also possible to measure RNA/DNA hybrid formation with hydroxyapatite after treatment with RNase (*12, 19*) under conditions which depolymerize the free RNA and preserve the hybrids. In this method Bio-Rad HTP is satisfactory since a capacity to bind RNA is not required. It may be useful to remove the RNase and digestion products by passage over Sephadex G-200. Another technique which shows promise is the use of hydroxyapatite in the presence of 8 M urea and in the DNA purification method. Little or no binding of RNA is observed under these conditions and the hybrids will bind. The optimum PB concentration has not yet been determined for this purpose. These latter methods can be used with either labeled RNA or DNA and with either one in excess and driving the hybrid formation reaction.

References

1. Main, R. and L. J. Cole. 1957. *Arch. Biochem. Biophys.*, **68**: 181.
2. Main, R., M. J. Wilkins, and L. J. Cole. 1959. *J. Am. Chem. Soc.*, **81**: 6490.
3. Bernardi, G. 1965. *Nature,* **206**: 779.
4. Miyazawa, Y and C. D. Thomas, Jr. 1965. *J. Mol. Biol.,* **11**: 223.
5. Walker, P. and A. McClaren. 1965. *Nature*, **208**: 1175.
6. Walker, P. and M. McCallum. 1966. *J. Mol. Biol.,* **18**: 215.
7. Britten, R. J. and D. E. Kohne. 1966. *Carnegie Inst. Wash. Yr. Bk.,* **65**: 73.
8. Britten, R. J. and D. E. Kohne. 1968. *Science,* **161**: 529.
9. McCallum, M. and P. Walker, 1967. *Biochem. J.,* **105**: 163.
10. Brenner, D. J. and D. B. Cowie. 1968. *J. Bacteriol.* **95**: 2258.
11. Kohne, D. E. 1968. *Biophys. J.,* **8**: 1104.
12. Rake, A., A. Gelderman, and R. J. Britten. 1971. *Proc. Natl. Acad. Sci., U.S.* **68**: 172.
13. Britten, R. J., M. Pavich, and J. Smith. 1970. *Carnegie Inst. Wash. Yr. Bk.,* **68**: 401.
14. Brenner, D. J., G. R. Fanning, A. Rake, and C. Johnson. 1969. *Anal. Biochem.,* **28**: 447-459.
15. Wetmur, J. G. and N. Davidson. 1968. *J. Mol. Biol.,* **31**: 349.
16. Marmur, J. and P. Doty. 1961. *J. Mol. Biol.,* **3**: 585.
17. Subirana, J. A. and P. Doty. 1966. *Biopolymers,* **4**: 171.
18. Thrower, K. J. and D. R. Peacocke. 1966. *Biochim. Biophys. Acta,* **119**: 652.
19. Davidson, E. H. and B. R. Hough, 1969. *Proc. Natl. Acad. Sci., U.S.,* **63**: 342.

SECTION

D

PREPARATION AND PURIFICATION OF NUCLEIC ACIDS

FRACTIONATION OF TRANSFER RIBONUCLEIC ACID ON COLUMNS OF HYDROXYLAPATITE

KARL H. MUENCH

DEPARTMENTS OF MEDICINE AND BIOCHEMISTRY

UNIVERSITY OF MIAMI SCHOOL OF MEDICINE

MIAMI, FLORIDA

Hydroxylapatite* was proposed for protein chromatography in 1956 (*1*) and thereafter was used for chromatography of nucleic acids (*2-4*). Since that time it has been used successfully for the resolution of tRNA (*5-9*). The advantages of hydroxylapatite chromatography of tRNA may be summarized as follows: high capacity, high resolution, resolution on a basis different from that of other methods, no inactivation or alteration of tRNA, application on either the analytical or preparative scale, application to either aminoacyl-tRNA or to uncharged tRNA, and usefulness as a probe of secondary and tertiary structure. Two major obstacles have prevented wider use of hydroxylapatite chromatography: The difficulty in repeatedly preparing hydroxylapatite with similar chromatographic characteristics, and low column flow rates of some batches of hydroxylapatite. Both of these disadvantages have been overcome in the following procedures.

* I shall use *hydroxylapatite* to mean a modified calcium phosphate, which has been called semihydroxylapatite, and which approaches but does not reach the composition of true *hydroxylapatite*, $Ca_{10}(PO_4)_6(OH)_2$. *Hydroxyapatite* and *hydroxylapatite* are synonymous.

Preparation

Formation of Calcium Phosphate

The two steps to be used in the formation of hydroxylapatite are taken directly from Main, Wilkins, and Cole (*4*) and have the advantages of reproducibility, high yield, simplicity of procedure, and formation of a product with a high flow rate. The first step differs from the procedure usually used (*1, 10*) in several respects. Sodium phosphate buffer, pH 6.7,* is used instead of Na_2HPO_4. The sodium phosphate is held in constant excess to the other reactant, $CaCl_2$; therefore, the precise rate of addition of reactants is not a critical factor in crystal formation. Because of the lower pH, the product formed is not brushite ($CaHPO_4 \cdot 2H_2O$) but anhydrous secondary calcium orthophosphate ($CaHPO_4$) (*4*). The latter has a larger crystal size and provides a flow rate approximately four times that of brushite when packed in a chromatographic column. $CaHPO_4$ does not adsorb nucleic acid well (*4*).

Procedure

Place 1800 ml of 0.5 M sodium phosphate buffer, pH 6.7,* in a 4-liter beaker equipped with a magnetic stirring bar at 23°. With gentle stirring of the buffer add 1500 ml of 0.5 M $CaCl_2$ dropwise at such a rate that discrete drops do not fuse into a stream. The addition of $CaCl_2$ may be accelerated by the use of more than one dropping vessel. After all $CaCl_2$ has been added continue gentle magnetic stirring for 1 hr, and then allow the $CaHPO_4$ to settle. Decant the supernatant solution, which will have a pH of 3.3. Resuspend the $CaHPO_4$ in 3 liters of water, allow the $CaHPO_4$ to settle, and decant the supernatant solution.

Formation of Hydroxylapatite

Transfer the washed $CaHPO_4$ to a 6-liter Erlenmeyer flask with 3 liters of H_2O and add 1 ml of 1% phenolphthalein. Place the flask on a heated magnetic stirrer, add enough concentrated ammonia to maintain the red color of phenolphthalein, and heat the mixture to boiling while gently stirring. Continue to add concentrated ammonia as needed throughout the boiling step, and boil the suspension for 30 min. Allow the hydroxylapatite to settle and decant the supernatant suspension while it is still hot. The demarcation between the desired hydroxylapatite and the unwanted, fine material in the supernatant will be quite clear. Wash the hydroxylapatite by repeated gentle

* Because of the variation of pH with phosphate concentration, pH 6.7 will designate a buffer composed of equimolar Na_2HPO_4 and NaH_2PO_4.

suspension, sedimentation, and decantation with the use of seven 3-liter portions of 0.005 M sodium phosphate buffer, pH 6.7.

The total yield should be 350 ml as packed for chromatography. Procedures employing NaOH for the conversion of calcium phosphate into hydroxylapatite (*1, 10*) provide lower yields (*11*). The hydroxylapatite may be stored in suspension at 0–2°. I have seen no evidence for change in properties during storage for at least one year, nor have such changes been reported by others. The hydroxylapatite is unstable at pH values less than 5 and to chelating agents such as EDTA and citrate (*10*). All batches of hydroxylapatite made in my laboratory by this procedure have been alike in chromatographic properties and flow rate.

The hydroxylapatite appears as large single crystals and crystal clusters under the microscope. The x-ray diffraction pattern is identical to that of hydroxylapatite prepared by the method of Tiselius. The two products are also identical in the calcium to phosphorus ratio of 1.53, a value approaching that of 1.67 for pure hydroxylapatite, $Ca_{10}(PO_4)_6(OH)_2$ (*4*). Main et al. (*4*) found that flow rates for the $CaHPO_4$ and the hydroxylapatite made from it were about the same, whereas brushite had a lower flow rate, which improved after the formation of hydroxylapatite by the method of Tiselius. I have regularly and easily achieved a flow rate of 30 ml/hr with the use of moderate pressure on hydroxylapatite packed into 1 x 100-cm chromatographic columns. In a direct comparison of five hydroxylapatite preparations, the first made by the procedure described, the second by the method of Tiselius (*1*), and the other three from two different commercial sources, all had identical chromatographic properties in terms of adsorption and desorption of tRNA. However, on identical 1 x 4-cm columns under a pressure head of 30 cm of water the respective flow rates were 21, 9, 10, 6, and 5 ml/hr.

In addition to the methods already mentioned, other procedures for preparation of hydroxylapatite have been published (*12, 13*). A disadvantage of commercial preparations is batch variability which may result from mechanical damage to the delicate crystal structure during shipment. Hydroxylapatite–cellulose, a commercial preparation, gave little or no resolution of tRNA (*6*).

Column Preparation

Gently suspend the stored hydroxylapatite in about three times its packed volume and pour the suspension into a chromatographic column containing a few centimeters of buffer. When the hydroxylapatite has packed by gravity alone to a level of 3 to 5 cm, open the stopcock and continue pouring the suspension until the column has reached the desired length. In an alternative procedure the column may be formed by pouring the suspension of hydroxylapatite and allowing packing to occur by gravity alone over a 12-hr period. When the latter procedure is followed, some additional packing will occur when buffer flow is started under pressure. In early experiments I found that long columns provided superior resolution, and since then I have regularly used columns 90 x 1 cm. I have not done extensive comparisons between long and short columns, and shorter columns may very well suffice, particularly for certain applications. Others (*6, 8, 9*) have apparently found long columns advantageous. Once the column has been formed passage of a methyl orange marker (*10*) is a good way to detect defects. Equilibrate the column at pH 5.8 by passage of 0.05 M potassium phosphate buffer, composed of 1 mole of K_2HPO_4 for every 10 moles of KH_2PO_4. After resolution of tRNA and reequilibration with buffer of low concentration each column may be reused at least twice without repouring, although removal of particulate matter from the top of the column may be necessary.

Preparation of tRNA and Aminoacyl-tRNA

The preparation of tRNA from a variety of sources has been described by several laboratories (*14–16*). For *E. coli* we follow the procedure of Zubay (*14*), but to break aminoacyl ester bonds we substitute for the incubation at high pH, an incubation in neutral partition solvent containing triethylammonium phosphate (*17*). We isolate the tRNA from the partition solvent with detergent (*18*). The preparation from *E. coli* of 20 partially purified and stabilized aminoacyl-tRNA synthetases, each free of interfering enzymes, has been described (*18, 19*) with a general procedure for the formation of aminoacyl-tRNA. The aminoacyl-tRNA is isolated from the reaction mixture by precipitation with potassium acetate buffer, pH 5, and ethanol, then dissolved in 1 or 2 ml of 0.01 M potassium phosphate buffer, pH 5.8, and applied to the hydroxylapatite column. When desired, the aminoacyl-tRNA can be entirely freed of reactants by passage over a Sephadex G-25 column prior to hydroxylapatite chromatography (*7*).

Chromatography

Procedure

After application of up to 500 A_{260} units* of tRNA to the 90 x 1-cm column, wash the column with dilute buffer. Develop the column by a linear gradient from dilute to concentrated potassium phosphate buffer, pH 5.8, in a total volume of 2000 ml. The dilute buffer may range from 0.10 to 0.15 M,

* One A_{260} unit of tRNA has an absorbance at 260 nm of 1.0 when dissolved in 1.0 ml of 5 mM KH_2PO_4-5 mM K_2HOP_4 buffer and in a 1.0-cm optical path.

and the concentrated buffer from 0.20 to 0.50 M. Because different batches of hydroxylapatite may require slightly different concentrations of buffer to elute tRNA, try the steeper gradients initially. Maintain a flow rate of 30 ml/hr or higher with a suitable pump. Stepwise elution of the columns is unsatisfactory except for initial screening experiments, because it produces false multiple peaks (*7, 11*). Column capacity for tRNA probably exceeds 600 A_{260} units per cm^2 of column cross section. At pH 5.4 excellent resolution was obtained for 3600 A_{260} units/cm^2 (*9*), and at pH 6.8, where tRNA is less tightly bound to the column, 1100 A_{260} units/cm^2 was resolved (*8*).

For analytical work on aminoacyl-tRNAs carrying a ^3H or ^{14}C label in the aminoacyl residue mix a suitable aliquot of each fraction with carrier RNA, add excess 2 M HCl, collect the precipitate on a glass fiber disc, dry, immerse in a liquid scintillator, and count (*7*). Another aliquot may be removed for determination of A_{260}, or alternatively, the column effluent may be monitored by a suitable UV-recording device.

For preparative work in which the tRNA must be isolated for further studies, first determine the conductivity of each fraction and then bring each fraction to the conductivity of 0.40 M potassium phosphate buffer, pH 5.8,

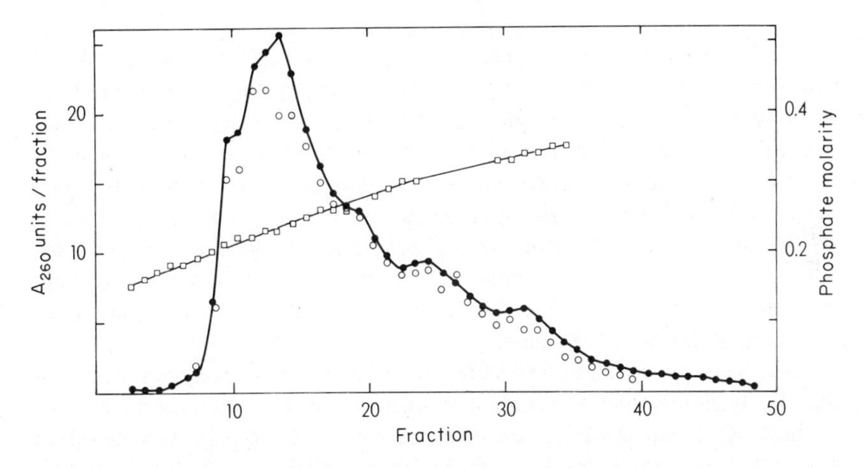

Figure 1. Recovery of tRNA with hexadecyltrimethylammonium chloride. A 90 x 0.9-cm column of hydroxylapatite was charged with 320 A_{260} units of *E. coli* tRNA in 1.0 ml and developed at a rate of 30 ml/hr by a linear gradient from 0.15 to 0.50 M potassium phosphate buffer, pH 5.8, in a total volume of 2000 ml with collection of 40-ml fractions. The phosphate concentration (□) was determined by conductivity, and each fraction was brought to 0.4 M potassium phosphate buffer by addition of 1 M buffer. Then, tRNA was isolated by use of hexadecyltrimethylammonium chloride and ethyl ether. The tRNA present in each fraction was determined spectrophotometrically before (●) and after (○) isolation. Recovery of A_{260} units in fractions 8 through 40 was 89%; 107% of applied A_{260} units appeared in the column effluent.

by addition of water or 1 M buffer. Add 0.05 volume of 0.10 M hexadecyltrimethylammonium chloride and 0.5 volume of diethyl ether containing 0.01% 2-mercaptoethanol. Mix vigorously and hold for 10 min at 37°; then allow the two-phase mixture to cool to 23°. The tRNA collects as a solid phase at the interface of the two liquid phases and is easily recovered by solution in 1 M NaCl and precipitation with two volumes of ethanol. The procedure is modified from the work of Mirzabekov (*20*). As shown in Fig. 1 the method efficiently recovers tRNA even at concentrations lower than 0.05 A_{260} unit/ml^3 and does not discriminate between the tRNAs present in different parts of the chromatogram.

Results

Typical chromatographic resolutions for leucyl- and alanyl-tRNAs and for glycyl-tRNA are shown in Figs. 2 and 3, respectively. So far, 35 isoacceptors have been found in *E. coli* B for the 17 amino acids examined (*6–8, 21*). In brewers' yeast 25 isoacceptors for 14 amino acids have been described (*9*).

Discussion

As has been the rule for virtually every method so far described for the resolution of tRNA, hydroxylapatite chromatography may resolve a given set of tRNA isoacceptors either better than other methods or not as well. Thus, alanyl-tRNA is better resolved on hydroxylapatite than on gradient partition columns (*17*), whereas the reverse situation applies to glycyl-tRNA and leucyl-tRNA. Others (*6*) have obtained better resolution of leucyl-tRNA at pH 6.8 and 23° than I have obtained at pH 5.8 and 0°. The reverse is true for other amino acids. I have recommended the procedure at pH 5.8, particularly for aminoacyl-tRNAs, because of the increased stability of the aminoacyl-ester bond at the lower pH value.

The basis for resolution of tRNA on columns of hydroxylapatite is not completely understood. Certainly, secondary structure is an important factor for both RNA and DNA; nucleic acids lacking secondary structure desorb at distinctly lower concentrations of phosphate buffer (*11, 22–24*). Molecular size is apparently not an important factor in fractionation of nucleic acids on hydroxylapatite (*22, 25*). Bernardi has advanced the theory that Ca^{2+} ions on the crystal surface of hydroxylapatite specifically bind the negatively charged phosphate residues in nucleic acids and that binding is therefore dependent on the distribution of negative charges on nucleic acids. The ionic interaction is specific, since presence of 1 M KCl has no effect on the concentration of phosphate buffer needed to elute DNA (*22*). However, other factors undoubtedly are important; the glucosylation of T4 bacteriophage DNA

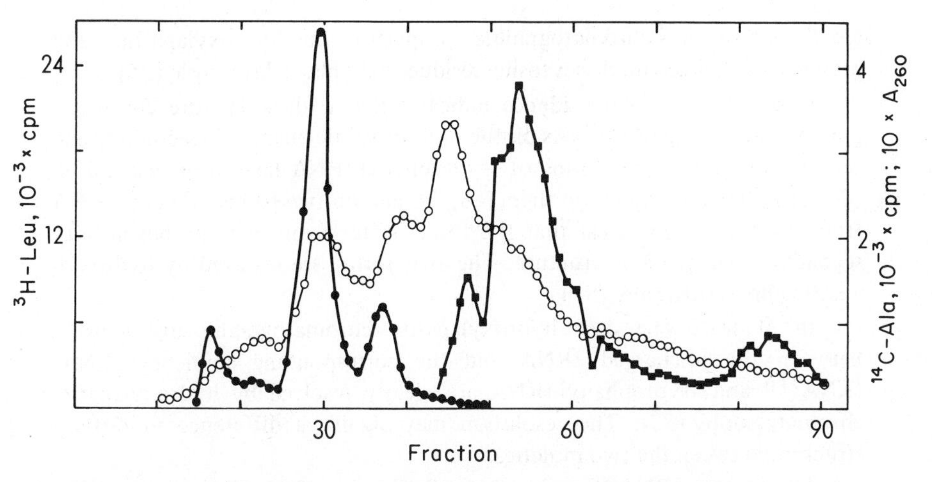

Figure 2. Hydroxylapatite chromatography of alanyl- and leucyl-tRNAs. A sample containing 12 nmoles of L-^{14}C-alanyl-tRNA (17×10^3 cpm/nmole) (■) and 29 nmoles of L-^{3}H-leucyl-tRNA (13×10^3 cpm/nmole) (●) in 212 A_{260} units was placed on a 106 x 0.9-cm hydroxylapatite column. The column was washed with 0.2 M potassium phosphate buffer, pH 5.8, then, beginning with fraction 1, was developed by a linear potassium phosphate buffer gradient, first from 0.2 M to 0.3 M in a total volume of 1000 ml, then from 0.3 M to 0.5 M in a total volume of 2000 ml. The flow rate was 30 ml/hr. Fraction volume was 22 ml. A_{260} (○) revealed 97% of applied A_{260} units in the effluent. The positions of alanyl- and leucyl-tRNAs were determined by acid precipitation of fraction aliquots, collection on glass fiber discs, and differential scintillation counting.

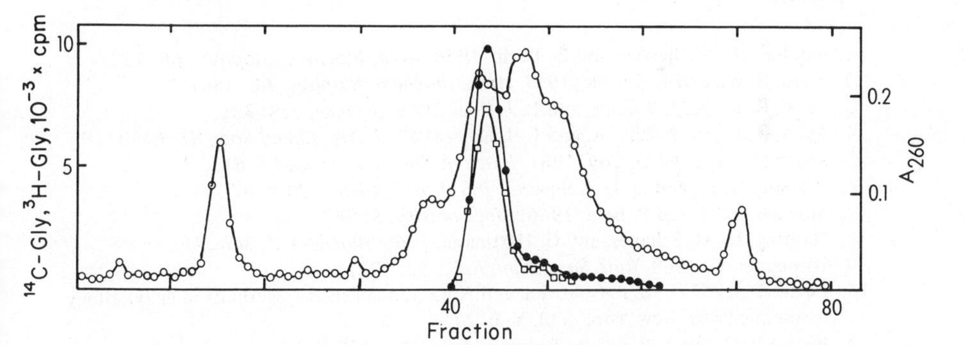

Figure 3. Hydroxylapatite chromatography of glycyl-tRNA. A sample containing 2.0 nmoles of ^{14}C-glycyl-tRNA (73×10^3 cpm/nmole) (●) from wild-type *E. coli* Y_{mel} and 4.0 nmole ^{3}H-glycyl-tRNA (45×10^3 cpm/nmole) (□) from mutant *E. coli* A58 Su1 in a total of 100 A_{260} units was placed over a 96 x 0.9 cm hydroxylapatite column. The column was washed with 0.15 M sodium phosphate buffer, pH 5.8, at 30 ml/hr with collection of 22-ml fractions. At the start of fraction 28 a linear gradient was started from 0.15 to 0.40 M sodium phosphate buffer, pH 5.8, in a total volume of 2000 ml. Aliquots of each fraction were removed for determination of A_{260} (○) and acid-precipitable radioactivity as already described. Of A_{260} units applied 104% appeared in the effluent.

greatly alters its chromatographic properties on hydroxylapatite, and presence of hydroxymethylcytosine residues may play a lesser role (*25*).

Strong but indirect evidence indicates that hydroxylapatite chromatography may distinguish tRNAs on the basis of subtle changes in secondary and tertiary structure. Two forms of tryptophanyl-tRNA have been recognized (*26, 27*). Physical studies on them (*28, 29*) and on two forms of leucyl-tRNA from yeast (*30, 31*) reveal that the forms differ from one another in both secondary and tertiary structure. The two forms are resolved by hydroxylapatite chromatography (*32*).

In at least one case hydroxylapatite chromatography distinguishes between the uncharged tRNA and the corresponding aminoacyl-tRNA: $tRNA^{Trp}$ and tryptophanyl-tRNA are clearly resolved by hydroxylapatite chromatography (*32*). The resolution may signify a difference in tertiary structure between the two moieties.

The various tRNAs may be arranged in order of lipophilic character by their order of elution from various chromatographic columns (*17, 33–35*). Since there is no correlation between lipophilic character and behavior of tRNA in hydroxylapatite chromatography, this is a logical method to be used in conjunction with those methods resolving tRNAs on the basis of lipophilic qualities. With that strategy methionyl-$tRNA^{F}$ (*36*) and tryptophanyl-tRNA (*37*) of *E. coli* were purified to homogeneity in three steps.

References

1. Tiselius, A., S. Hjertén, and Ö. Levin. 1956. *Arch. Biochem. Biophys.*, **65**: 132.
2. Main, R. K., and L. J. Cole. 1957. *Arch. Biochem. Biophys.*, **68**: 186.
3. Main, R. K., M. J. Wilkins, and L. J. Cole. 1959. *Science,* **129**: 331.
4. Main, R. K., M. J. Wilkins, and L. J. Cole. 1959. *J. Am. Chem. Soc.*, **81**: 6490.
5. Hartmann, G., and U. Coy. 1961. *Biochim. Biophys. Acta.*, **47**: 612.
6. Pearson, R. L., and A. D. Kelmers. 1966. *J. Biol. Chem.*, **241**: 767.
7. Muench, K. H. and P. Berg. 1966. *Biochemistry,* **5**: 982.
8. Harding, U., H. Schauer, and G. Hartmann. 1966. *Biochem. Z.* **346**: 212.
9. Dirheimer, G. 1968. *Bull. Soc. Chim. Biol.*, **50**: 1221.
10. Levin, Ö. 1962. In S. P. Colowick and N. O. Kaplan (Eds.), Methods in enzymology. Academic Press, New York, Vol. V, p. 27.
11. Bernardi, G. 1969. *Biochem. Biophys. Acta,* **174**: 423.
12. Jenkins, W. T. 1962. *Biochem. Prepn.* **9**: 83.
13. Siegelman, H. W., G. A. Wieczorek, and B. C. Turner. 1965. *Anal. Biochem.* **13**: 402.
14. Zubay, G. 1966. In G. L. Cantoni and D. R. Davies (Eds.), Procedures in nucleic acid research. Harper & Row, New York, Vol. I, p. 455.
15. Brunngraber, E. F. 1962. *Biochem. Biophys. Res. Commun.*, **8**: 1.
16. Holley, R. W., J. Apgar, B. P. Doctor, J. Farrow, M. A. Marini, and S. H. Merrill. 1961. *J. Biol. Chem.*, **236**: 200.
17. Muench, K. H., and P. Berg. 1966. *Biochemistry,* **5**: 970.
18. Muench, K. H., and P. A. Safille. 1968. *Biochemistry*, **7**: 2799.

19. Muench, K. H., and P. Berg. 1966. In G. L. Cantoni and D. R. Davies (Eds.), Procedures in nucleic acid research. Harper & Row, New York, Vol. I, p. 375.
20. Mirzabekov, A. D., A. I. Krutilina, V. I. Gorshkova, and A. A. Baev. 1964. *Biokhimiya,* **29**: 1158.
21. Muench, K. H. 1969. *Biochemistry,* **8**: 4872.
22. Bernardi, G. 1965. *Nature,* **206**: 779.
23. Bernardi, G. 1969. *Biochim. Biophys. Acta,* **174**: 435.
24. Bernardi, G. 1969. *Biochim. Biophys. Acta,* **174**: 449.
25. Oishi, M. 1969. *J. Bacteriol.,* **98**: 104.
26. Gartland, W. J., and N. Sueoka. 1966. *Proc. Natl. Acad. Sci., U.S.,* **55**: 948.
27. Muench, K. H. 1966. *Cold Spring Harbor Symp. Quant. Biol.,* **31**: 539.
28. Ishida, T., and N. Sueoka. 1967. *Proc. Natl. Acad. Sci., U.S.,* **58**: 1080.
29. Ishida, T., and N. Sueoka. 1968. *J. Biol. Chem.,* **243**: 5329.
30. Adams, A., T. Lindahl, and J. R. Fresco. 1967. *Proc. Natl. Acad. Sci., U.S.,* **57**: 1684.
31. Lindahl, T., A. Adams, and J. R. Fresco. 1967. *J. Biol. Chem.,* **242**: 3129.
32. Muench, K. H. 1969. *Biochemistry,* **8**: 4880.
33. Weiss, J. F., and A. D. Kelmers. 1967. *Biochemistry,* **6**: 2507.
34. Gillam, I., S. Millward, D. Blew, M. von Tigerstrom, E. Wimmer, and G. M. Tener. 1967. *Biochemistry* **6**: 3043.
35. Gillam. I., D. Blew, R. C. Warrington, M. von Tigerstrom, and G. M. Tener. 1968. *Biochemistry,* **7**: 3459.
36. Marcker, K. A., S. K. Dube, and B. F. C. Clark. 1968. In L. O. Fröholm and S. G. Laland (Eds.), Structure and function of transfer RNA and 5S-RNA. Academic Press, New York, p. 53.
37. Joseph, D. R., and K. H. Muench. 1970. *Federation Proc.,* **29**: 467 (Abstr.).

tRNA SEPARATIONS USING BENZOYLATED DEAE–CELLULOSE*

K. L. ROY,† A. BLOOM, and D. SÖLL
DEPARTMENT OF MOLECULAR BIOPHYSICS AND BIOCHEMISTRY
YALE UNIVERSITY
NEW HAVEN, CONNECTICUT

Introduction

tRNA extracted from cells is a complex mixture of probably more than 60 structurally similar nucleic acid molecules. Therefore, the separation and purification of individual tRNA species requires techniques capable of detecting rather subtle differences between macromolecules of similar charge, molecular weight, and structure. The first successful approach to this problem was the countercurrent distribution method (*1*). Although it is a powerful procedure, the practical execution is quite laborious. In addition, it is primarily a preparative tool and cannot be readily scaled down to the analytical level. Many column chromatographic procedures have been described which allow rapid, high-resolution separation of tRNAs. They include chromatography on DEAE-cellulose (*2*), DEAE-Sephadex (*3*), hydroxylapatite (*4*), or methylated-albumin-kieselguhr (*5*). Two very successful techniques have been developed which were designed specifically for tRNA fractionation – the reversed phase chromatographic systems (*6*), and the benzoylated DEAE-cellulose exchanger developed by Tener's group (*7*).

* Research conducted in this laboratory was supported by grants from the National Institutes of Health, the National Science Foundation, and the American Cancer Society.
† Current address: Department of Microbiology, University of Alberta, Edmonton, Canada.

BD–Cellulose* is an anion exchanger with the high capacity and technical convenience of DEAE-cellulose, yet with the greater selectivity of the more hydrophobic, but low-capacity methylated-albumin-Kieselguhr. BD-cellulose can be used in two ways: (1) direct chromatographic fractionation of tRNA and (2) the isolation of specific tRNA by derivatization.

This article is based on our experience primarily with *E. coli* and yeast tRNA. Fractionation on BD-cellulose of tRNA from tobacco leaves (*8*), *L. acidophilus* (*9*), and other sources has also been reported.

Materials

BD-Cellulose

This may be prepared in the laboratory or purchased commercially. The Regis Chemical Co., Chicago, Illinois; Schwarz Bio-Research, Orangeburg, N.Y.; P. L. Biochemicals, Milwaukee, Wisconsin; Boehringer Mannheim Corp., New York; and Serva Entwicklungslabor, Heidelberg, Germany all sell BD-cellulose. We have not tested any of these commercial preparations.

The procedure of Gillam et al. (*7*) is used for the preparation of BD-cellulose. Reagent grade chemicals are used throughout. DEAE-Cellulose powder (not microgranular) of 0.9–1.0 meq/gm dry weight is dried thoroughly before use by heating to 80° in a vacuum oven for 12–18 hours. The dried DEAE-cellulose (100 gm) is added slowly to 2.5 liters of dry pyridine in a 5-liter round-bottom flask. The flask must be swirled vigorously during *all* phases of the preparation to ensure a homogeneous and complete reaction. Benzoyl chloride (345 gm, 4 mole/mole of anhydroglucose) is added slowly with continuous swirling, after which the flask is transferred to a heating mantle. A reflux condenser and drying tube are connected to the flask and controlled heat is applied. It should take about 45 min heating to reach a boil, and during this time the contents of the flask will darken. Boiling should continue for 45 min during which the mixture will become a clear, dark brown, viscous solution. Very few cellulose fibers should be seen at this stage. The flask is allowed to cool slowly and the contents become more viscous. After 30 min the solution of BD-cellulose in pyridine is slowly poured into a large volume (20–40 liters) of cold water (a plastic garbage can or large chromatography tank is a suitable container) with continuous stirring. Because of the strong odor of pyridine this should be done in, or in front of, a fume hood. As the pyridine solution of BD-cellulose enters the water it forms a long rope from which the pyridine must be extracted. The long strands of BD-cellulose are left in the water for several hours to harden and to allow pyridine to diffuse out. The aqueous solution is decanted and

* Abbreviations: BD–cellulose, benzoylated DEAE–cellulose; UV, ultraviolet; one A_{260} unit is the amount of material per milliliter of solution which produces an absorbance of 1 in a 1-cm light path at 260 nm.

replaced at least 3–4 times with vigorous stirring each time. The brittle rope of BD-cellulose is broken into much smaller pieces during this procedure. After nearly all the pyridine has been removed, the product is washed repeatedly with ethanol to remove benzoic acid and other UV-absorbing impurities. It is finally washed with 2 M NaCl + 25% ethanol. To obtain a material suitable for chromatography the particle size must be greatly reduced. BD-Cellulose is suspended in water and subjected to very brief grinding in a blender. Great care should be taken to avoid prolonged grinding since it will produce BD-cellulose too fine to allow rapid chromatography. The size of the particles should be small enough to pass through a 30-mesh standard sieve, but not through a 200-mesh sieve. Very fine particles can be removed by repeated suspension of the BD-cellulose in water, followed by decantation after the desired material has settled. We generally check the quality of every new batch by examining the resolution of a standard tRNA sample upon chromatography.

BD-Cellulose is stored at 5° in a sodium chloride solution (2 M or greater) containing 0.5% sodium azide. After BD-cellulose has been used for chromatography it may be cleaned efficiently by washing with 6 M guanidine–HCl.

N-Hydroxysuccinimide Esters

Schwarz BioResearch and Boehringer Mannheim Corp., New York, sell the phenoxyacetyl and 2-naphthoxyacetyl esters of *N*-hydroxysuccinimide.

The preparation of the phenoxyacetyl and 2-naphthoxyacetyl esters of *N*-hydroxysuccinimide is as described by Gillam et al. (*10*). Phenoxyacetic acid or 2-naphthoxyacetic acid is dissolved in dry dioxan (4 ml/mole). An equimolar amount of *N*-hydroxysuccinimide is added, followed by an equimolar amount of dicyclohexylcarbodiimide.* The flask containing the reactants is stoppered and shaken to dissolve the dicyclohexylcarbodiimide, then allowed to stand for 2 hr. Then the crystals of dicyclohexylurea are filtered off, the filtrate is concentrated to a thick syrup on a rotary evaporator and the product crystallized from 2-propanol. Yields of 60 to 80% are obtained. The melting points for the naphthoxyacetyl and phenoxyacetyl ester are 147° and 103°, respectively.

Chromatographic Procedures

Column Size

The column size chosen for chromatographic separation may be varied within rather wide limits. We have obtained good results with columns having

* Dicyclohexylcarbodiimide is a skin irritant and should be used with appropriate precautions.

length to diameter ratios from 5 : 1 to 150 : 1. The amount of tRNA loaded has varied between 5 mg and 0.05 mg per milliliter of packed column volume. For unfractionated tRNA, we would normally recommend a length to diameter ratio from 10 : 1 to 25 : 1 and a load of about 1–2 mg/ml. An example of the high capacity of BD-cellulose for tRNA is shown in Fig. 1.

Packing

Columns are routinely packed by partially filling with 2 M NaCl, then pouring in a thin slurry of BD-cellulose in 2 M NaCl. The slurry is added continually as the salt solution runs out of the column. When the packed resin reaches the desired height it is washed with 2 M NaCl under a hydrostatic head of 0.5–1.5 m. After a thorough wash with 2 M NaCl the column should be washed with the equilibrating solution. This is normally the solution in which the tRNA is to be applied and with which elution is to be started. Several column volumes should be used to equilibrate the adsorbent, as determined by monitoring absorbance and conductivity of the effluent.

Loading

Despite the high capacity of BD-cellulose for tRNA it should be remembered that tRNA is itself a polyanion and can therefore act as an eluent. For this reason tRNA should be dissolved in the equilibrating buffer and loaded at a concentration of not more than 1–2 mg/ml/cm^2 of column cross-sectional area. Because the earliest tRNAs to elute from BD-cellulose emerge at NaCl concentrations of 0.5 M or greater, the equilibration solution should be 0.3 to 0.4 M in NaCl.

Elution

Usually NaCl is the eluting salt and linear gradients give very satisfactory results. Above 1.2 M NaCl very little material is eluted, and ethanol is used to elute any remaining material. A gradient of ethanol from 0 to 15% in 1.2 M NaCl is capable of removing all nonderivatized tRNAs. The total volume of eluent should be at least 8 x the column volume, and may be as much as 25 x column volume.

Assay of Fractions for Acceptor Activity

The eluted fractions can be assayed by several procedures for the ability to accept a specific amino acid, with or without prior dialysis (see Ref. *11*). Dialysis may not be necessary in fractions containing large amounts of tRNA. However, the possible inhibitory effect of high salt concentrations should be checked.

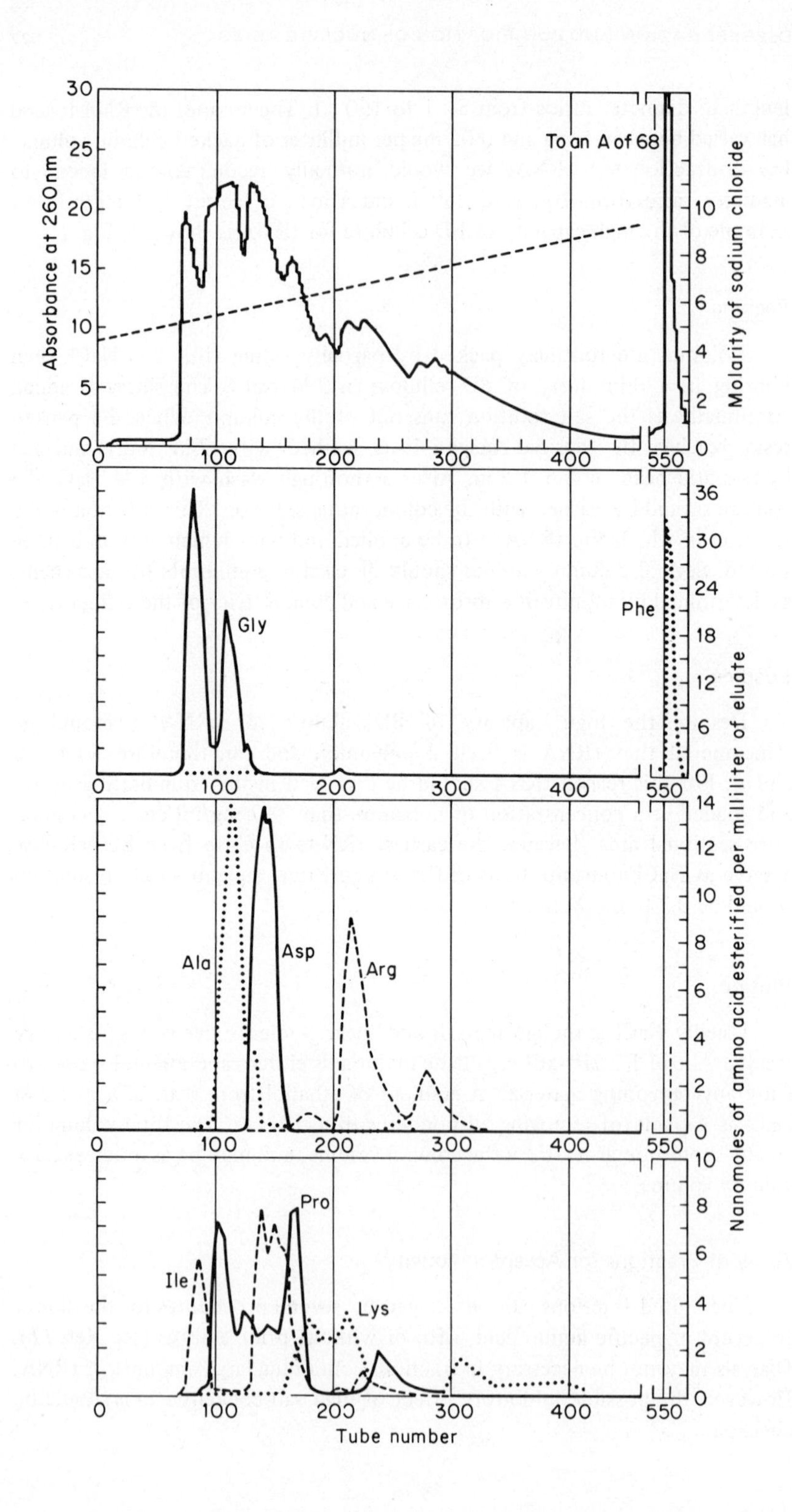

Absorbance at 260 nm
To an A of 68
Molarity of sodium chloride
Gly
Phe
Ala
Asp
Arg
Pro
Ile
Lys
Nanomoles of amino acid esterified per milliliter of eluate
Tube number

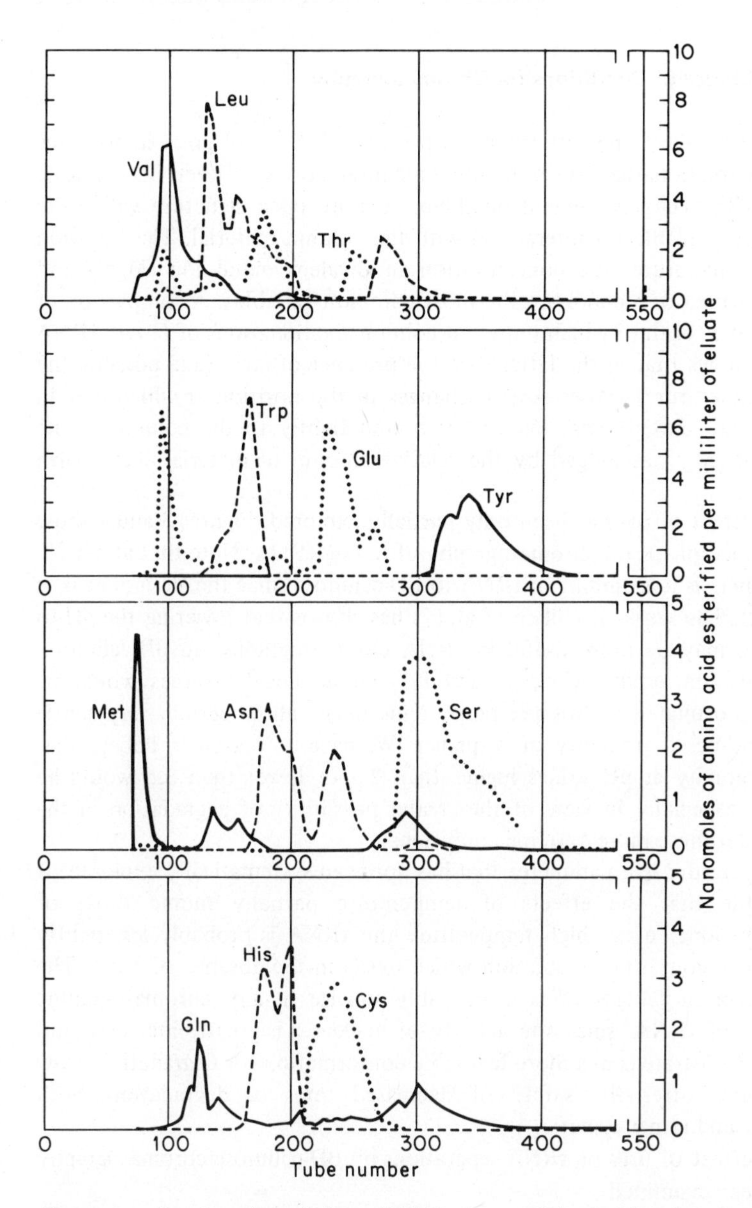

Figure 1. Fractionation of 5 gm (70,000 A_{260} units) of tRNA from brewers' yeast (Boehringer) on a column (3.2 x 110 cm) of BD-cellulose. The sample was applied in 500 ml of 0.45 M sodium chloride and 0.01 M magnesium sulfate and eluted with the indicated (dashed line) gradient of sodium chloride solution (a total of 10 liters) containing 0.01 M magnesium sulfate. Fractions were 20 ml/15 min. At the end of the gradient (tube 497) elution was continued with 1.0 M sodium chloride–0.01 M magnesium sulfate containing 10% (v/v) 2-methoxyethanol. Total recovery, about 66,000 A_{260} units. Chromatography was performed at room temperature (22°). (Figure from Ref. 7, reprinted with permission of the American Chemical Society).

Choice of Conditions for Chromatography

The secondary and tertiary structures of tRNA are of great importance for tRNA fractionation, since the nonionic interactions between BD-cellulose and the tRNA largely depend on them. A more open structure will make more bases available for interaction with the column material. The variation of pH, temperature, the concentration of divalent metal ion (Mg^{2+}) and agents like urea or alcohol are the most common variables. A comparison of Figs. 2 and 3, both of which represent column fractionations of *E. coli* tRNA at pH 5, shows that in the latter case the presence of Mg^{2+} (and possibly the higher temperature) causes several changes in the position at which certain acceptors are eluted. Less material is bound tightly to the column in the presence of Mg^{2+} as judged by the relative amount of material eluted with ethanol.

The effect of pH has been only partially explored. Figures 3 and 4 show the influence of pH on chromatography of *E. coli* tRNA. Note that at pH 7.5 (Fig. 4) there is less binding of tRNA to BD-cellulose since the alcohol peak is very small. The work of Gillam et al. (*7*) has shown that lowering the pH to 3.5 or 4.0 may be quite useful in tRNA chromatography on BD-cellulose. Presumably the amino group of cytidyl and adenosyl residues would be partially protonated in this pH range. This might allow certain very subtle separations based primarily on sequence. We have no reason to believe that chromatography at pH values higher than 7.5 or lower than 3.5 would be advisable, especially in view of the greater possibility of degradation of the tRNA under these more extreme conditions.

Temperature is a parameter that has not been systematically explored. It is probable that the effects of temperature partially mimic those of magnesium ion; i.e., at high temperature the tRNA is probably less tightly folded, analogous to the situation which exists in the absence of Mg^{2+}. The use of elevated temperatures is probably unwise in any chromatographic separation of tRNA, since the activity of nucleases is greatly increased and the tRNA substrate is in a more favorable conformation for degradation. Low temperatures offer the safety of decreased rates of degradation, both enzymatic and nonenzymatic.

The effect of urea on tRNA separations by BD-cellulose chromatography has not been examined.

Chromatography of Aminoacyl-tRNA

The chromatographic separation of aminoacyl-tRNA (usually radioactively labeled in the amino acid) is a very useful analytical tool. The pH of choice for these separations is 4.5, a pH at which the stability of the aminoacyl linkage is greatest. These separations are very fast and easy to assay. However, consideration should be given to the fact that there will be some deacylation of the aminoacyl-tRNA during chromatography and therefore column fractions should be counted for radioactivity after TCA precipitation. Examples of this type of separation are shown in Figs. 5 and 6.

The elution position of an aminoacyl-tRNA may vary from the position of the same tRNA species in the uncharged form. This has been exploited with BD-cellulose to give excellent purifications of tyrosine, tryptophan, and phenylalanine tRNA (*12–14*). The aromatic amino acids esterified to tRNA will increase greatly the nonionic interactions of these species with BD-cellulose and therefore be very much retarded and eluted with ethanol. This is also obvious in comparing the phenylalanine and tyrosine profiles in Figs. 3, 5, and 6.

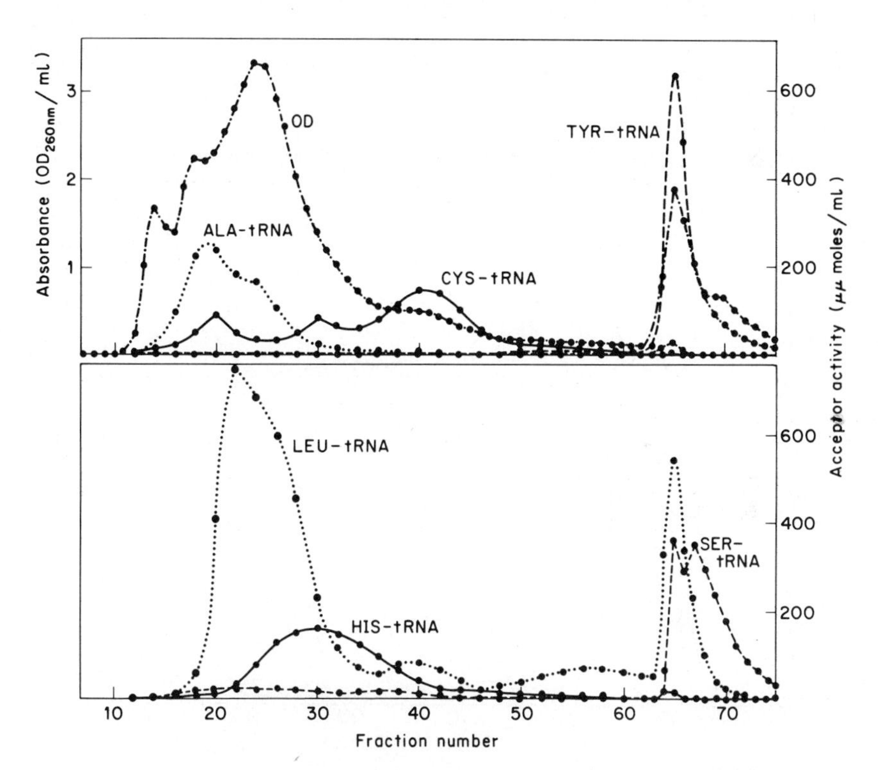

Figure 2. Chromatography of *E. coli* tRNA on BD-cellulose at pH 5.0. tRNA (400 A_{260} units) was chromatographed on a column (2.5 x 15 cm) of BD-cellulose at 2°. Elution was carried out with a linear gradient. The mixing vessel contained 500 ml of 0.4 M NaCl–0.05 M sodium acetate (pH 5.0) and the reservoir contained an equal volume of 1.1 M NaCl–0.05 M sodium acetate (pH 5.0). After completion of the gradient, the tRNAs which were still adsorbed were eluted with a solution containing 1.5 M NaCl–0.05 M sodium acetate (pH 5.0) and 15% ethanol. Fractions of 15.5 ml were collected every 17 min. (Figure from Ref. 15, reprinted with permission of Elsevier, Amsterdam).

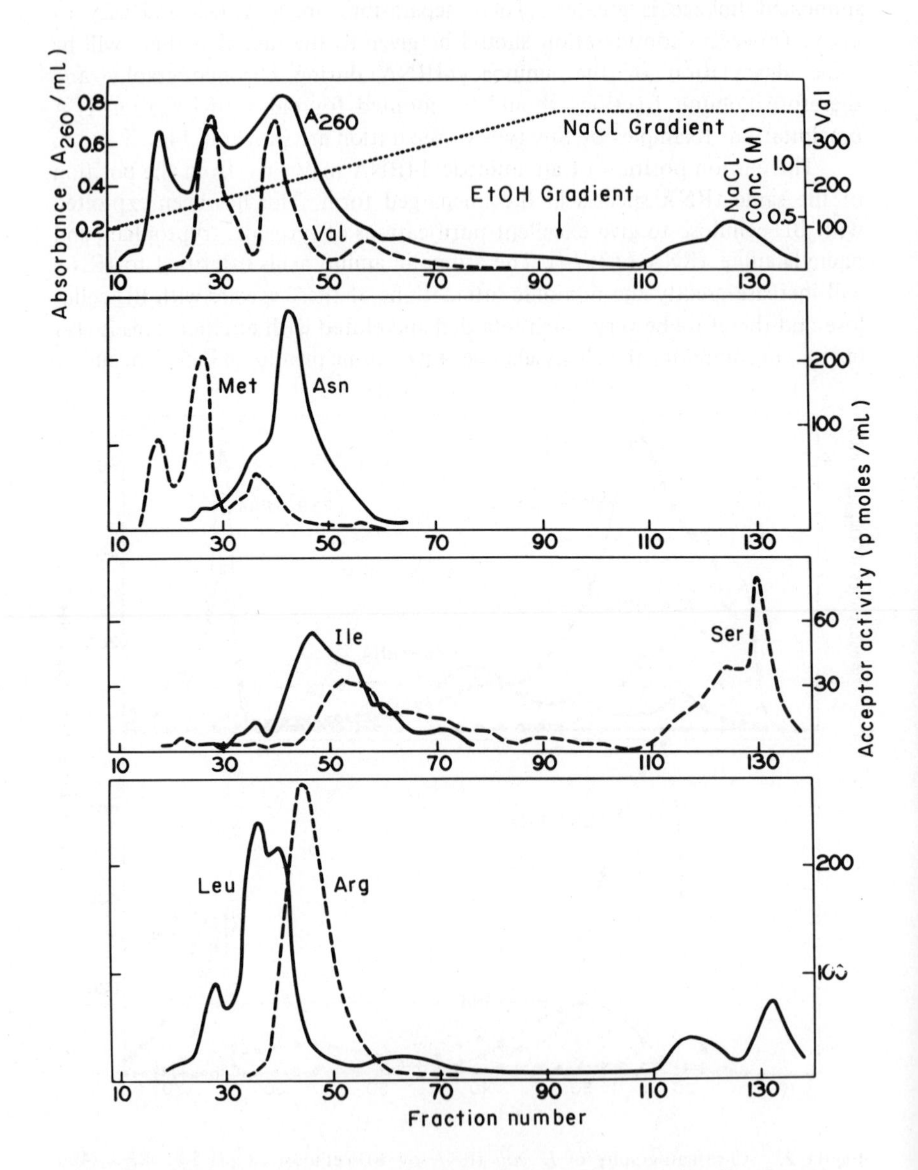

Figure 3. Chromatography of *E. coli* tRNA on BD-cellulose at pH 5.0 in the presence of Mg^{2+}. tRNA (200 A_{260} units) was applied in 0.2 M NaCl, 0.01 M $MgCl_2$, and 0.05 M sodium acetate (pH 5.0) to a column (1 x 75 cm) of BD-cellulose and eluted with a linear gradient (dotted line) of NaCl from 0.4 to 1.5 M (total volume 700 ml) followed by a linear gradient of ethanol (total volume 400 ml) from 0 to 30% in 1.5 M NaCl. Buffer and Mg^{2+} remained constant throughout. This experiment was conducted at 2°.

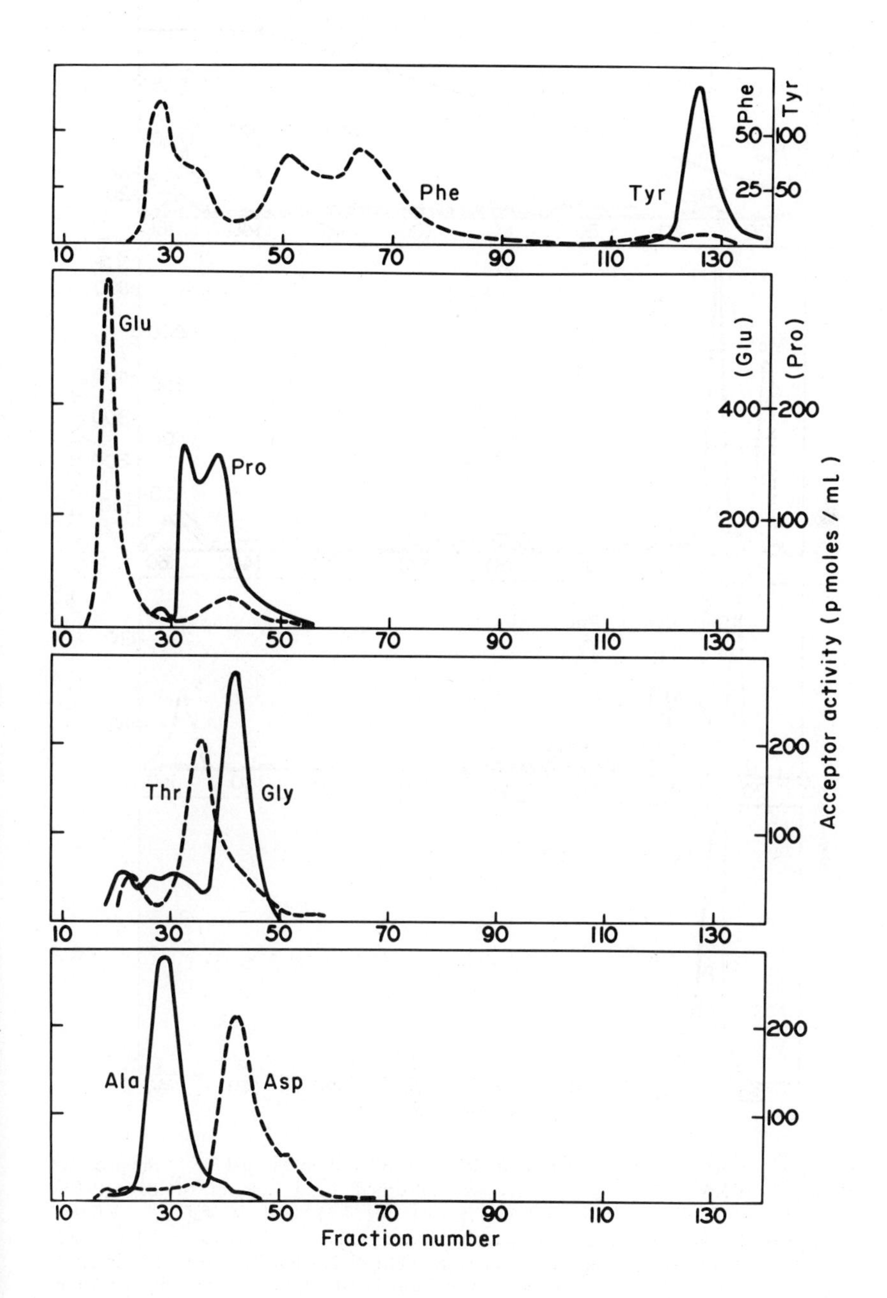
Phe
Tyr
Phe
Tyr
50
100
25
50
Glu
Pro
(Glu)
(Pro)
400
200
200
100
Thr
Gly
200
100
Ala
Asp
200
100
10
30
50
70
90
110
130
Fraction number
Acceptor activity (p moles / mL)

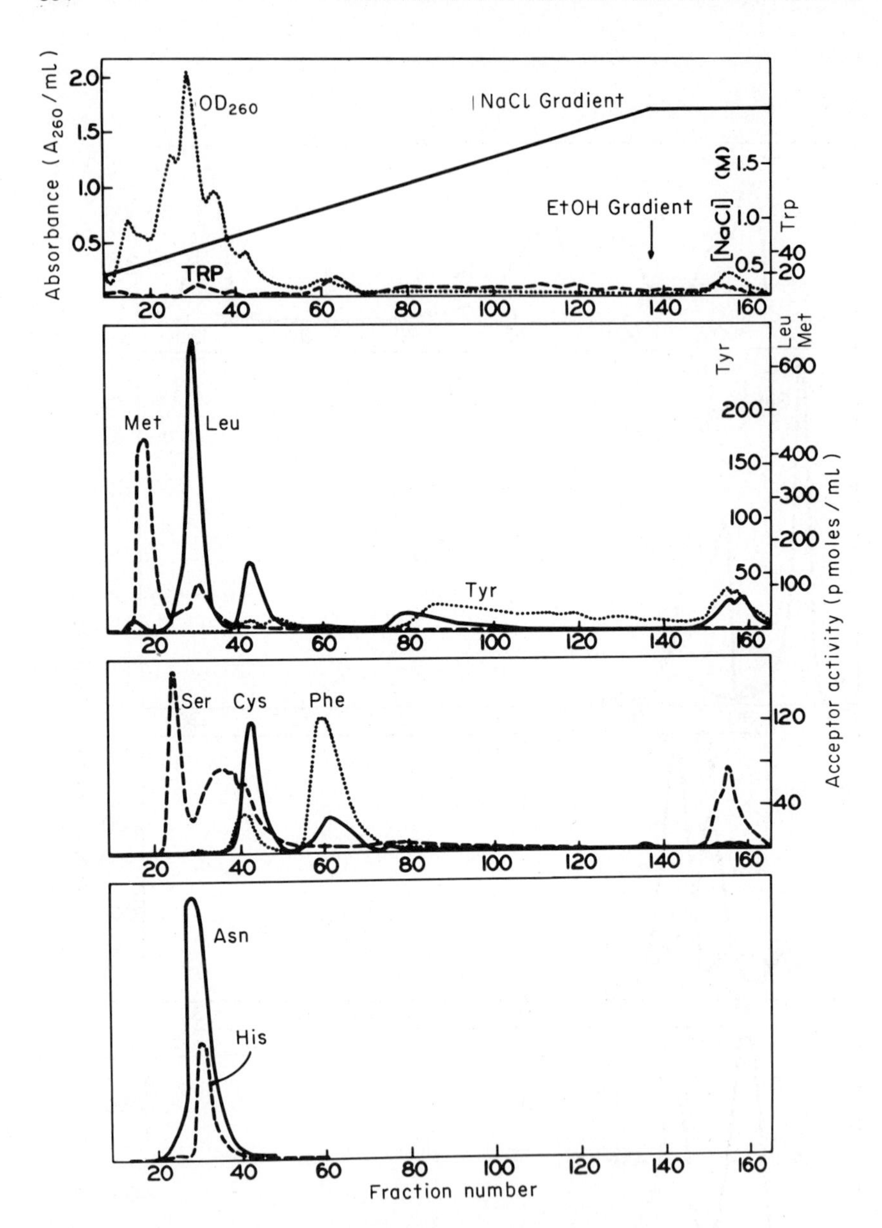

Figure 4. Chromatography of *E. coli* tRNA on BD-cellulose at pH 7.5 in the presence of Mg^{2+}. tRNA (195 A_{260} units) was adsorbed to a column of BD-cellulose (0.7 x 94 cm) in 0.2 M NaCl, 0.01 M Tris-HCl (pH 7.5), 0.01 M $MgCl_2$, and 0.04% sodium azide. Elution was performed with a linear gradient of NaCl from 0.4 to 2.0 M (700 ml total volume) followed by a linear gradient of ethanol from 0 to 20% in 2.0 M NaCl. Tris, $MgCl_2$, and sodium azide were present in all solutions at the concentrations described above. A flow rate of 15 ml/hr was maintained and 5 ml fractions were collected (conducted at room temperature, 22–24°).

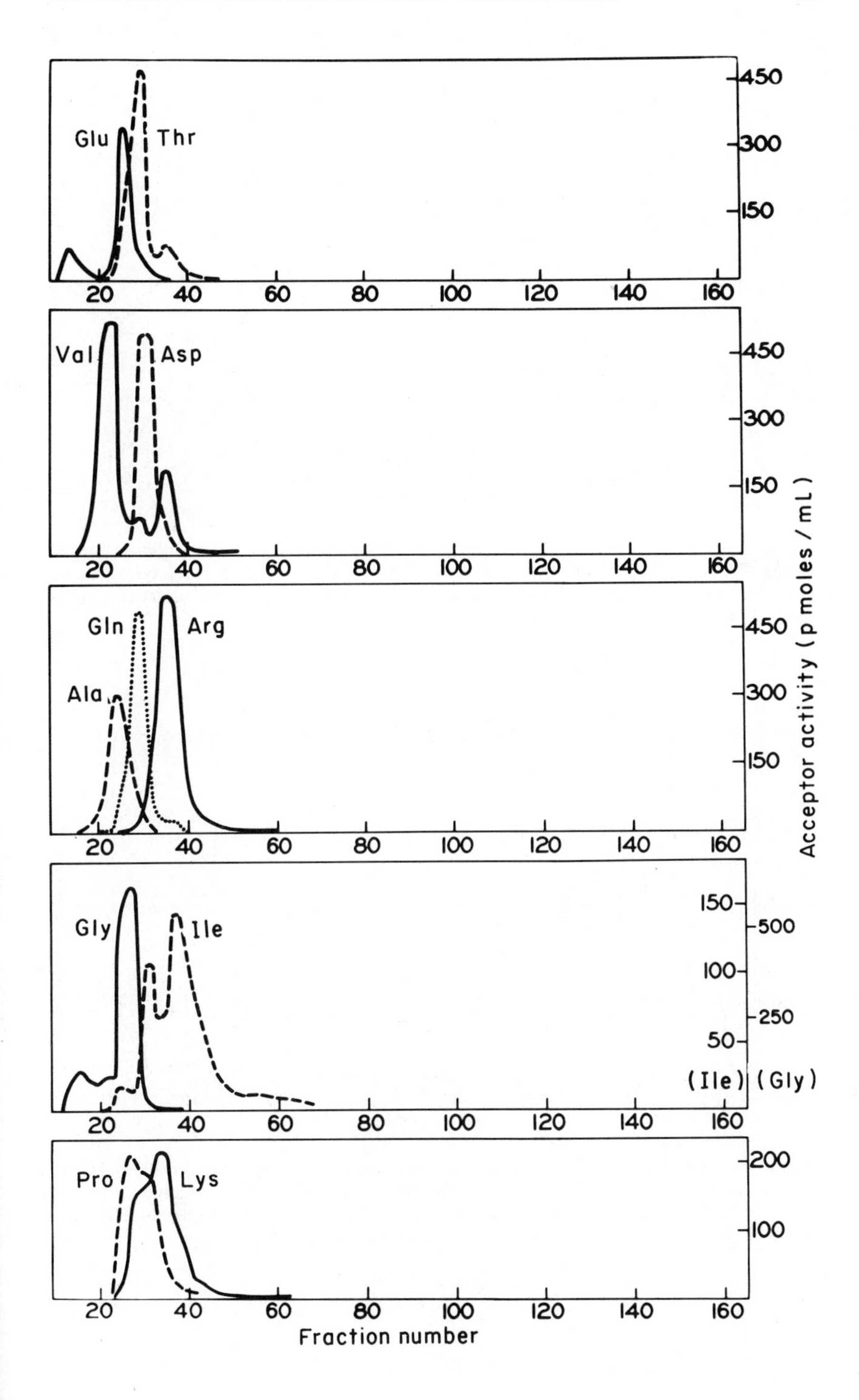

Glu
Thr
450
300
150
20
40
60
80
100
120
140
160
Val
Asp
Gln
Arg
Ala
Acceptor activity (p moles / mL)
Gly
Ile
150
100
50
500
250
(Ile)
(Gly)
Pro
Lys
200
100
Fraction number

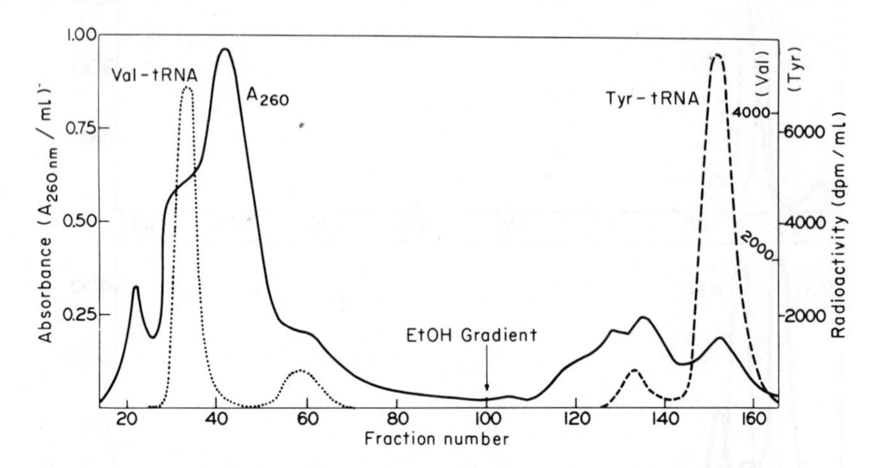

Figure 5. Chromatography of valyl- and tyrosyl-tRNAs on BD-cellulose. Unfractionated *E. coli* tRNA (200 A_{260} units) to which [^{14}C]-valine and [^{3}H]-tyrosine were esterified, was chromatographed on a BD-cellulose column (1 x 80 cm) under conditions identical to those described in Fig. 3.

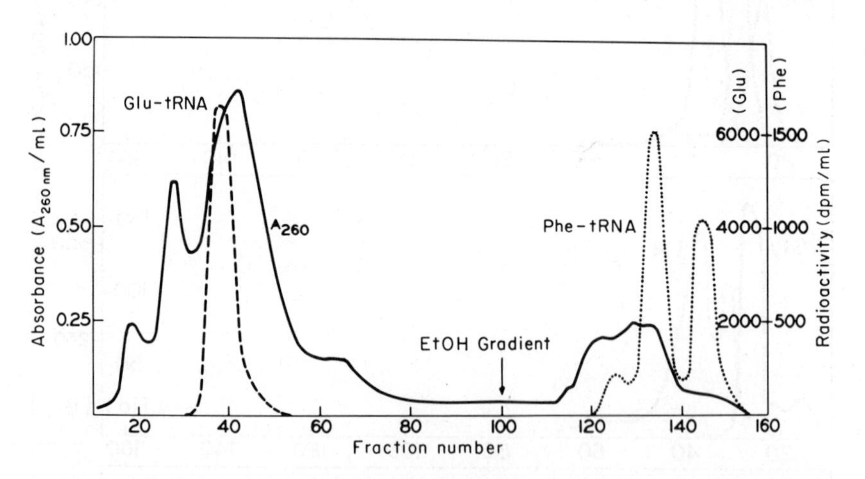

Figure 6. Chromatography of glutamyl- and phenylalanyl-tRNAs on BD-cellulose. Unfractionated *E. coli* tRNA (200 A_{260} units), to which [^{3}H]-glutamic acid and [^{14}C]-phenylalanine were esterified, was chromatographed on a BD-cellulose column (1 x 80 cm) under conditions identical to those described in Fig. 3.

Isolation of Individual tRNAs by Derivatization

The principle underlying the derivatization procedure is that the addition of a highly hydrophobic group to a tRNA molecule will increase the nonionic binding of that tRNA to BD-cellulose and thus retard the species on chromatography. Such a hydrophobic group is introduced by acylation of the amino group of enzymatically formed aminoacyl-tRNA. All nonderivatized tRNAs will elute with high salt or in the presence of a very small amount (about 3%) of ethanol. The derivatized tRNA will be held very firmly and only released with high alcohol concentration (up to 40%). After deacylation a family of purified isoaccepting tRNAs is obtained if crude tRNA was the starting material.

Aminoacylation of tRNA

Aminoacylation must be very nearly complete if good yields are to be obtained. A highly purified aminoacyl-tRNA synthetase is not necessary, but the preparation used should be known to have good activity for the amino acid of interest, and to be free of nucleases and amino acids. It is advisable that the tRNA be completely discharged since any aminoacyl-tRNA already present in the tRNA preparation will also be derivatized and may become an impurity in the derivatized fraction. To avoid this possibility a "sham"-acylation has been used (*10*). Optimum conditions (ATP, Mg^{2+}, amino acid concentration) for aminoacylation of the tRNA sample should be determined for each case by a small scale experiment to ensure that the large scale reaction reaches completion rapidly (5–10 min). It is advantageous if the amino acid used in the large scale aminoacylation contains a small quantity of $[^{14}C]$-amino acid (giving a specific activity of 1.0–5.0 mCi/mmole). This allows the purification to be readily quantitated. It is not necessary to extract the reaction mixture with phenol since a small amount of protein will not interfere with subsequent steps.

Derivatization of tRNA

The derivatization reagent of greatest use is the 2-naphthoxyacetyl ester, except in those cases where a tRNA accepting an aromatic amino acid is desired. Acceptors of aromatic amino acids may be selectively adsorbed more firmly to BD-cellulose by derivatization with the phenoxyacetyl ester or by aminoacylation alone (*12–14*).

Gillam et al. (*10*) have found that derivatization of threonyl-tRNA was incomplete under all conditions tested. We have had similar difficulty in attempts to derivatize seryl-tRNA. Possibly, the β-hydroxyl group common to both amino acids may interfere with this reaction.

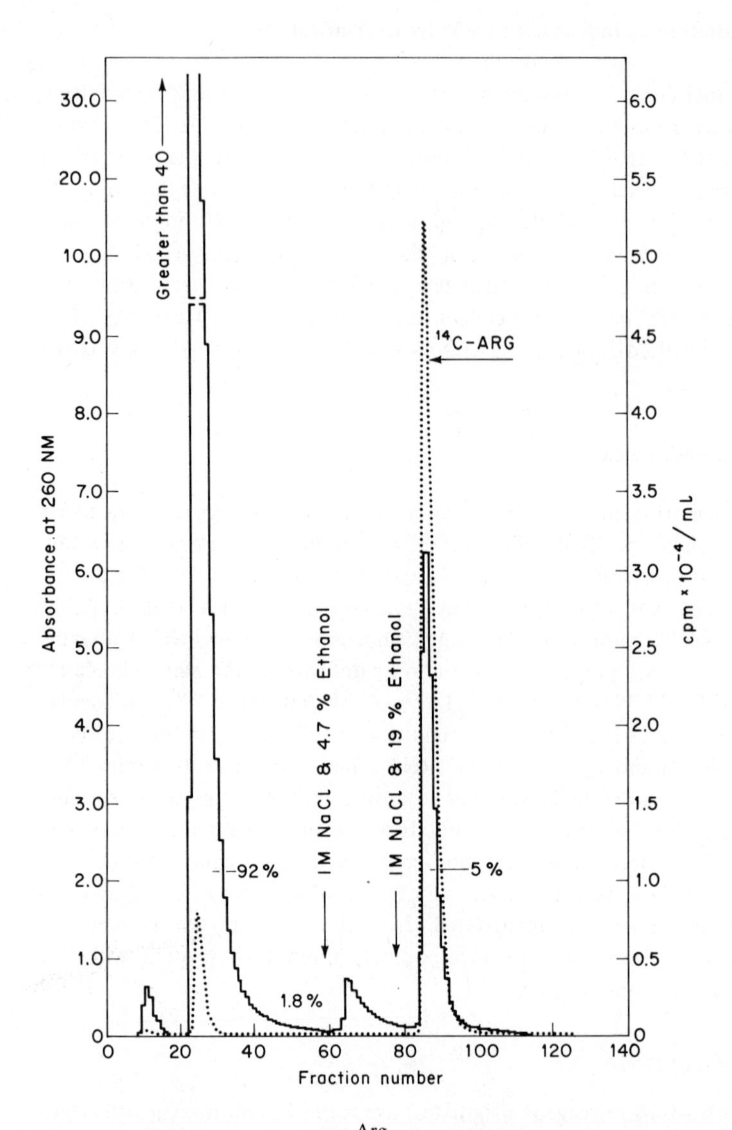

Figure 7. Isolation of $tRNA^{Arg}$ by naphthoxyacetylation. *N*-2-Naphthoxyacetylarginyl-tRNA from 2700 A_{260} units of brewers' yeast tRNA was applied to a column (3.5 x 12 cm) of BD-cellulose in a solution of 0.01 M $MgCl_2$, 0.01 M sodium acetate (pH 4.5), 0.3 M NaCl, and washed with the same. At fraction 16 elution with a solution containing 0.01 M $MgCl_2$, 0.01 M sodium acetate (pH 4.5), and 1.0 M NaCl was started. At the indicated points (arrows) elution with the indicated solutions was begun. Both of these contained 0.01 M $MgCl_2$ and 0.01 M sodium acetate (pH 4.5). (Figure from Ref. *10*, reprinted with the permission of the American Chemical Society).

General Procedure for Derivatization of Aminoacyl-tRNA

The aminoacyl-tRNA is dissolved in 0.05 M sodium acetate (pH 4.5). This and all subsequent steps must be done at 0°. Triethanolamine-HCl or Tris-HCl (2 M, pH 8.0) and dioxan are added to give a final buffer concentration of 0.2 M and a dioxan content of 25% (i.e., 0.4 ml dioxan and 0.16 ml of 2 M buffer per milliliter of tRNA solution). The final tRNA concentration should be 20 mg/ml. The phenoxyacetyl or naphthoxyacetyl ester of *N*-hydroxysuccinimide, dissolved in dioxan (50 mg/ml), is added in two portions to the stirred tRNA solution. The final amount used should be 0.1 ml per milliliter of reaction mixture. This represents at least a 100-fold excess of the derivatizing reagent over the aminoacyl group. The results of the aqueous dioxan solution cause little, if any, precipitation of the reagent, unlike the water-tetrahydrofuran solution used by Gillam et al. (7). After 10 min at pH 8.0 glacial acetic acid is used to lower the pH to 4.0-4.5. Then the tRNA is precipitated by the addition of two volumes of ethanol. The precipitate is centrifuged, washed twice with cold ethanol, and finally dissolved in buffer (pH 4.5) for chromatography.

Chromatography of Derivatized Aminoacyl-tRNA

The derivatized aminoacyl-tRNA is chromatographed on BD-cellulose at pH 4.5. The low pH buffer is needed to stabilize the aminoacyl ester linkage. The bulk of the tRNA is eluted with 1 M NaCl and the remaining tRNA, including the derivatized species, is then eluted with a step of ethanol (Fig. 7), or a gradient of ethanol (0 to 40%) (Fig. 8) containing the other ingredients of the buffer. The derivatized species will, in most cases, be the last UV-absorbing peak eluted, especially when the naphthoxyacetyl reagent is used.

Removal of the naphthoxyacetyl or phenoxyacetyl amino acid from the tRNA is accomplished by incubation of the derivatized tRNA in Tris-HCl buffer at pH 8 and room temperature or 37°. The concentration of Tris-HCl should be at least 1 M and incubation should proceed for 2 hr. The extent of removal of the derivatized amino acid should be checked during this incubation by the filter paper assay method.

The resulting purified isoaccepting tRNAs may be separated by chromatography on BD-cellulose or by other methods.

General Considerations

The strategy to be used in purification of any tRNA cannot be unequivocally defined, but certain points should be emphasized. Good purification of tRNAs can be achieved without the use of the derivatization

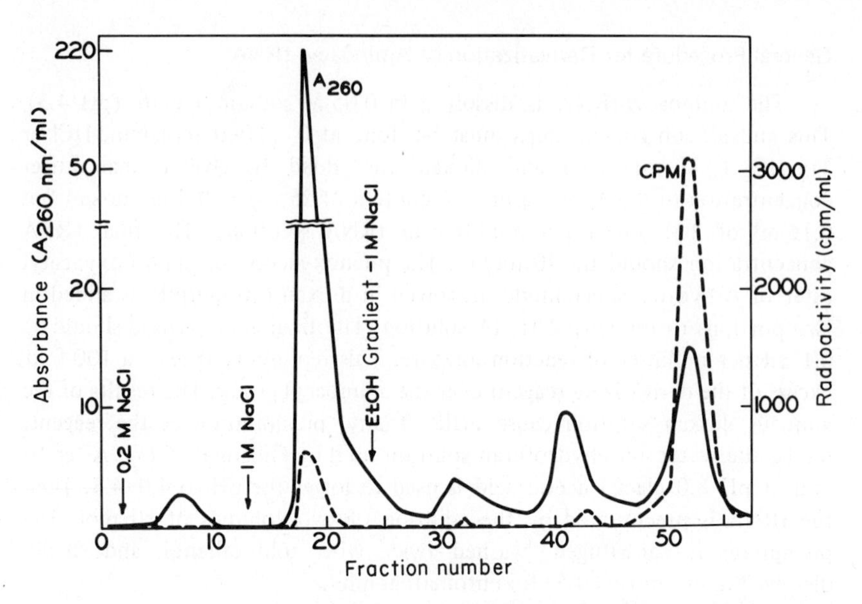

Figure 8. Isolation of $tRNA^{Ile}$ by naphthoxyacetylation. *N*-2-Naphthoxyacetylisoleucyl-tRNA from 8000 A_{260} units of *Mycoplasma* tRNA was applied to a column (2 x 30 cm) of BD-cellulose in a solution of 0.2 M NaCl, 0.01 M $MgCl_2$, 0.05 M sodium acetate (pH 5.0), and 0.04% sodium azide. At the indicated points (arrows) elution was started with 1 M NaCl solution and with a linear gradient (0 to 40%) of ethanol (total volume of gradient 400 ml). Fraction size was 16 ml. All solutions contained 0.01 M $MgCl_2$, 0.05 M sodium acetate (pH 5.0), and 0.04% sodium azide.

procedure, but, where suitable, the derivatization method is the obvious choice for the first step. This is particularly true when one is interested in a single acceptor or family of isoaccepting tRNAs.

The final yield of purified tRNA depends on several factors. Yields after derivatization are sometimes reduced by incomplete aminoacylation, or by hydrolysis of the aminoacyl ester linkage during the derivatization reaction.

Finally, in all purifications the need for very clean glassware cannot be overemphasized, since the introduction of trace quantities of any nuclease can significantly reduce acceptor activity.

References

1. Doctor, B. P., this volume, p. 558.
2. Cherayil, J. D. and R. M. Bock. 1965. *Biochemistry,* 4: 1174.
3. Nishimura, S., this volume, p. 542.
4. Muench, H. K., this volume, p. 515.
5. Sueoka, T., this volume, p. 608.
6. Novelli, G. D. 1969. *J. Cell. Physiol.,* 74: Suppl. 1, 121.

7. Gillam, I. C., S. Millward, D. Blew, M. von Tigerstrom, E. Wimmer, and G. M. Tener. 1967. *Biochemistry,* **6**: 3043.
8. Johnson, M. S. and R. J. Young. 1969. *Virology,* **38**: 607.
9. Peterkofsky, A., and C. Jesensky. 1969. *Biochemistry,* **8**: 3798.
10. Gillam, I. C., D. Blew, R. C. Warrington, M. von Tigerstrom, and G. M. Tener. 1968. *Biochemistry,* **7**: 3459.
11. Bollum, F. J. 1966. In G. L. Cantoni and D. R. Davies (eds.), *Procedures in nucleic acid research*. Harper & Row, New York, Vol. I, p. 296.
12. Wimmer, E., I. H. Maxwell, and G. M. Tener. 1969. *Biochemistry,* **7**: 2623.
13. Maxwell, I. H., E. Wimmer, and G. M. Tener. 1968. *Biochemistry,* **7**: 2629.
14. Litt, M. 1968. *Biochem. Biophys. Res. Commun.,* **32**: 507.
15. Roy, K. L. and D. Söll. 1968. *Biochim. Biophys. Acta,* **161**: 572.

FRACTIONATION OF TRANSFER RNA BY DEAE-SEPHADEX A-50 COLUMN CHROMATOGRAPHY

SUSUMU NISHIMURA
BIOLOGY DIVISION
NATIONAL CANCER CENTER
RESEARCH INSTITUTE
TOKYO, JAPAN

Introduction

The application of DEAE-Sephadex A-50 column chromatography for the fractionation of amino acid specific tRNAs was first described by Kawade et al. (*1*) and later by Cherayil and Bock (*2*). DEAE-Sephadex chromatography in the presence of dimethylformamide or of borate was used to purify tRNAs from yeast as was recently reported by Miyazaki et al. (*3*) and Miyazaki and Takemura (*4*). In our laboratory, extensive studies on the fractionation of tRNA from *E. coli* (*5–7*), yeast (*8, 9*), and rat liver (*10*) using Sephadex were performed. The major modification adopted in our work is the use of lower NaCl concentrations than previously used to elute tRNA. This gives much better resolution of the different tRNAs. We performed the chromatography more than 20 times using different batches of the Sephadex and tRNA over the last three years. We have observed that the procedure is reliable and reproducible, since almost identical chromatographic profiles were obtained in each case.

DEAE-Sephadex column chromatography has the following advantages. First, the column has a relatively large capacity, and tRNA can be eluted from the column at the concentration of 10 to 80 OD units* per milliliter. Furthermore, the amount of tRNA loaded on the column does not affect the chromatographic profile. Consequently, small amounts of tRNA such as ^{32}P-labeled tRNA can be fractionated in the same fashion (*10, 11*). Second, the fractions eluted from the column can be assayed directly for amino acid acceptor activity without any further isolation procedure, and the tRNA eluted can easily be precipitated by the addition of 2.5 volumes of cold ethanol. Third, the column is operated under very mild conditions. The elution of tRNA is performed at pH 7.5 in the presence of magnesium ion, and tRNA is never exposed to drastic conditions. Therefore, it may be expected that the isolated tRNA would not be modified at all with respect to the intactness of its molecular conformation or the structure of labile minor constituents.

Combination of DEAE-Sephadex column chromatography with other column chromatographic procedures such as reverse phase partition chromatography (*12*) and benzoylated DEAE-cellulose (BD-cellulose) chromatography (*13*) seems to be the most promising procedure so far devised to obtain purified amino acid specific tRNAs. By use of these techniques, $tRNA^{Met}_{f}$ (*7*), $tRNA^{Met}_{m1}$ (*7*), $tRNA^{Met}_{m2}$ (*7*), $tRNA^{Tyr}_{1}$ (*5*), $tRNA^{Tyr}_{2}$ (*5, 14*), $tRNA^{Ser}_{1}$ (*6, 15*), $tRNA^{Ser}_{3}$ (*6, 15*), $tRNA^{Val}_{1}$ (*5, 16*), $tRNA^{Phe}$ (*5, 16*), $tRNA^{Cys}$ (*16*), $tRNA^{His}$ (*16*), $tRNA^{Asp}$ (*16*), and $tRNA^{Leu}_{2}$ (*15*) from *E. coli*; $tRNA^{Tyr}$ (*10*), $tRNA^{Val}$ (*10*), $tRNA^{Phe}$ (*10*), $tRNA^{Lys}$ (*17*), and $tRNA^{Glu}$ (*17*) from rat liver; and $tRNA^{His}$ (*8*), $tRNA^{Glu}$ (*18*), $tRNA^{Phe}$ (*18*), and $tRNA^{Met}_{f}$ (*19*) from yeast were isolated. In addition, three other $tRNA^{Leu}$'s (*20*) – $tRNA^{Arg}$ (*21*), $tRNA^{Lys}$ (*21*), and $tRNA^{Try}$ (*21*) – from *E. coli* were considerably purified. These purified tRNAs were essential for the studies on tRNA problems at the molecular level, such as sequence determination, characterization of minor components, and crystallization. Furthermore, availability of numerous tRNAs in relatively large amounts enabled us to show the regular presence of particular minor components in relation with codon recognition of tRNA (*22, 23*).

This chapter deals primarily with the details of large scale fractionation. However, it should be emphasized that a small size (0.5 x 150 cm) column can be used for fractionation of a small amount of tRNA, provided that the slope of the gradient and the flow rate of the buffer are maintained constant with respect to the column area.

Materials

DEAE-Sephadex A-50 (particle size, 40–120 μ) was obtained from Pharmacia Fine Chemicals.

* OD units refers to the absorbance measurement in 1 ml volume at 260 nm and at neutral pH using a quartz cuvette with 1-cm light path.

Unfractionated tRNA from *E. coli* B

Unfractionated tRNA was prepared from cells of *E. coli* B harvested late in the log phase as described by Zubay (*24*) except that the alkaline treatment to strip acylated amino acids was omitted. The final precipitate of tRNA from isopropanol was dissolved in 0.006 M β-mercaptoethanol at a concentration of 2000 OD_{260} units/ml.

Unfractionated tRNA from yeast

Yeast tRNA was prepared according to Holley's procedure (*25*) with a modification described previously (*8*).

Unfractionated tRNA from Rat Liver

Rat liver tRNA was prepared by the combined use of the methods described by Brunngraber (*26*) and Zubay (*24*). Four kilograms of rat liver, obtained from 600 Wistar male rats, were homogenized with 6 liters of 88% phenol and 6 liters of 1 M NaCl, 0.0005 M EDTA, 0.02 M Tris-HCl, pH 7.5, and 0.01 M $MgCl_2$. The homogenate was shaken for 1.5 hr at a room temperature and centrifuged for 20 min. To the supernatant, 2 vol of cold ethanol was added. The precipitate collected by centrifugation was homogenized with 4 liters of cold 1 M NaCl using a mixer. The suspension was centrifuged at 9000 rpm for 30 min. The precipitate was resuspended in 2 liters of 1 M NaCl and centrifuged. To the combined supernatant, 2 vol of cold ethanol was added to precipitate tRNA and DNA. The precipitate was suspended in 1.75 liters of 0.3 M sodium acetate. Then 700 ml of isopropanol was added to the suspension to precipitate DNA. It was found that the amount of isopropanol required to precipitate DNA, but not to precipitate tRNA was less than that required for the fractionation of *E. coli* tRNA. The exact amount of isopropanol was easily determined by adding 80 ml of isopropanol after the viscosity of the suspension dropped sharply. The precipitate containing DNA was removed by centrifugation at 9000 rpm for 5 min, and resuspended in 1 liter of 0.3 M sodium acetate. A 500-ml sample of isopropanol was added to the suspension which was then centrifuged. tRNA was finally precipitated from the combined supernatants by adding 1.6 liters of isopropanol. Yield: 42,000 OD units.

Methods

Pretreatment of DEAE-Sephadex A-50

The Sephadex was pretreated with both alkali and acid before use. Then 800 gm of DEAE-Sephadex A-50 (an amount sufficient for 2 columns of

6 x 150 cm) were suspended in 20 liters of 0.1 M Na_2HPO_4 and 1 M NaCl. The supernatant was removed by decantation. The Sephadex was washed in a Büchner funnel (diameter, 40 cm; height 20 cm) with distilled water until it started to become swollen. Then, the Sephadex was suspended in 20 liters of 0.1 M NaH_2PO_4 and washed with distilled water until it was swollen up to the top of the funnel. The Sephadex was suspended in 10 liters of a buffer containing 0.02 M Tris-HCl (pH 7.5), 0.008 M $MgCl_2$, and 0.375 M NaCl (buffer A). The pH of the suspension was adjusted to 7.5 by the addition of NaOH. The Sephadex was again washed with 20 liters of buffer A without aid of suction, and finally suspended in 10 liters of buffer A.

The Sephadex can be used repeatedly. Used DEAE-Sephadex was poured out from the column, washed in a Büchner funnel with 10 liters of 1 M NaCl, and then distilled water until the Sephadex was swollen to the top of the funnel. Swollen Sephadex was treated with buffer A as described in the pretreatment of the Sephadex.

Packing of the Sephadex into a Column

The Sephadex suspension was poured into a column (diameter, 6 cm; height, 200 cm) having a sealed-in coarse sintered glass disk. After allowing it to stand overnight, the Sephadex was packed to give a column 120–150 cm in length. The supernatant was removed by gentle suction before use.

Operation of a Column

An *E. coli* tRNA solution (100 ml) (total 200,000 OD units) was mixed with 400 ml of buffer A, and then loaded onto the column. After the tRNA solution was completely absorbed into the Sephadex, elution was started at room temperature using a linear gradient with 12 liters of buffer containing 0.02 M Tris-HCl (pH 7.5), 0.016 M $MgCl_2$, and 0.525 M NaCl (buffer B) in the reservoir and 12 liters of buffer A in the mixing chamber. The flow rate was approximately 100–200 ml/hr. It should be mentioned that better resolution was obtained with slower flow rates, especially in the case of the tRNAs eluted in the middle fractions. When the flow rate was maintained at less than 200 ml/hr most of the tRNAs, including $tRNA^{Met}$, $tRNA^{Tyr}$, $tRNA^{His}$, $tRNA^{Asp}$, $tRNA^{Glu}$, $tRNA^{Lys}$, $tRNA^{Val}$, $tRNA^{Ser}$, and $tRNA^{Phe}$ were quite well separated.

BD-Cellulose Column Chromatography

BD-Cellulose was prepared according to the procedure described by Gillam et al. (*13*). BD-Cellulose column chromatography was performed at 4°C.

Reverse-Phase Partition Column Chromatography

The reverse-phase column chromatography was carried out as described by Kelmers et al. (*12*) with a slight modification. Chromosorb-W was coated with maximal amount of the organic solvent to increase loading capacity. Chromosorb-W (AW, 100-200 mesh, from Sugimoto Technical Co., Tokyo, or DMCS treated, AW, 100-120 mesh, from Johns-Manville Products Co.) was suspended in excess amount of isoamyl acetate containing 4% dimethyl-dilaurylammonium chloride (Aliquot 204). Excess isoamyl acetate was removed from the Chromosorb by means of a Büchner funnel. The Chromosorb was then washed with the buffer containing 0.02 M Tris-HCl (pH 7.5) and 0.4 M NaCl, until no more organic solvent was released. The washed Chromosorb was suspended in the same buffer, and poured into a column (diameter 1.5 cm; length, 150 cm) which was previously filled with the above buffer.

Aminoacyl-tRNA Synthetase

Crude aminoacyl-tRNA synthetases from *E. coli*, yeast, and rat liver were prepared, respectively, as described previously (*5, 8, 10*).

Assay of Amino Acid Acceptor Activity of tRNA Fractions Eluted from the Column

The reaction mixture contained 0.005–0.05* ml of a column fraction, 10 μmoles of Tris-HCl (pH 7.5), 1 μmole of KCl, 0.2 μmole of ATP, 0.01–0.02 μCi of ^{14}C-labeled amino acid (specific activity†, 40–350 μCi/μmole) and appropriate amounts of aminoacyl-tRNA synthetase (0.001–0.005 ml, 60–300 μg, depending on the batch of the enzyme preparation) in a total volume of 0.1 ml. The reaction mixture was incubated at 37°C for 10 min. Aliquots (0.05–0.08 ml) of the reaction mixture were applied to a Whatman 3MM filter paper disk (diameter, 24 mm). The filter paper disks were immersed in cold 5% trichloroacetic acid, washed twice with fresh portions of the acid solution at intervals of 10 to 15 min, then they were washed with ethanol–ether (1 : 1, v/v) mixture in the cold, and finally with ether. Dried filter disks were assayed for radioactivity in a liquid scintillation counter. In the assay of cysteine acceptor activity, the reaction mixture contained 6 μmoles of β-mercaptoethanol in addition to other components; in the washing procedure, the 5% trichloroacetic acid contained 0.002 M cold cysteine and 0.006 M β-mercaptoethanol. For the

* Usually 5 μl of the fraction is enough to give a count more than 1000 cpm in the maximal peak.

† In case of assay for the acceptor activity of methionine, glutamine, and asparigine, 0.02 μCi of ^{14}C-amino acid with specific activity of 40 μCi/μmole was used.

assay of acceptor activity of tRNAs from yeast and rat liver, similar procedures were used except for modifications in the reaction mixtures described previously (*8, 10*).

Assay of Formate Acceptor Activity of tRNA

The reaction mixture and the assay procedure of transformylation of methionyl-tRNA was prepared as described previously (*9*).

Results

Fractionation of *E. coli* tRNA by DEAE-Sephadex A-50 Column Chromatography

Figures 1a and 1b* represent two typical experiments when 200,000 OD units of unfractionated *E. coli* tRNA were fractionated by the DEAE-Sephadex column chromatography. More than 95% of ultraviolet-absorbing material was recovered. Assay of the fractions for acceptance of all 20 amino acids indicated that individual tRNAs were eluted as rather sharp peaks, and were reasonably well separated from each other.

The best separation was obtained in the case of methionine tRNA. Methionine tRNA (mixture of $tRNA_f^{Met}$ and $tRNA_m^{Met}$) was eluted at the beginning of the chromatography and the material in the first half of the methionine-tRNA peak was purified more than 20 times. The methionine tRNA thus obtained is more than 80% pure, and accepts none of the other amino acids except slight amounts of isoleucine and lysine. It should be pointed out that a partial resolution between $tRNA_f^{Met}$ and $tRNA_m^{Met}$ was not obtained by this procedure (*5, 7*).

In addition to methionine tRNA, $tRNA_2^{Tyr}$, $tRNA_2^{Leu}$, and $tRNA_3^{Ser}$ which were eluted at the end of the chromatography were also considerably purified. Other tRNAs eluted in the middle region were usually purified about 5-fold.

The presence of isoaccepting species for certain amino acids is readily evident. The number of multiple components for a given amino acid evidenced by the Sephadex procedure is 2 for tyrosine; 2 for valine; 4 for serine; 2 for phenylalanine; 3 for leucine; 2 for glutamic acid; 2 for glutamine; 2 for glycine; and 2 for tryptophan. Chromatographic profiles for $tRNA^{Arg}$, $tRNA^{Ile}$, $tRNA^{Thr}$, $tRNA^{Lys}$, and $tRNA^{Ala}$ were rather complex suggesting the presence of multiple components. On the other hand, histidine, aspartic acid, proline, cysteine, and asparagine gave single tRNA peaks. It is clear from

* *Abbreviations*: Phe, Leu, Ser, Tyr, Cys, Try, Pro, His, Gln, Arg, Ile, Met, Thr, Asn, Lys, Val, Ala, Asp, Glu and Gly listed in the figures indicate transfer RNA for a given amino acid.

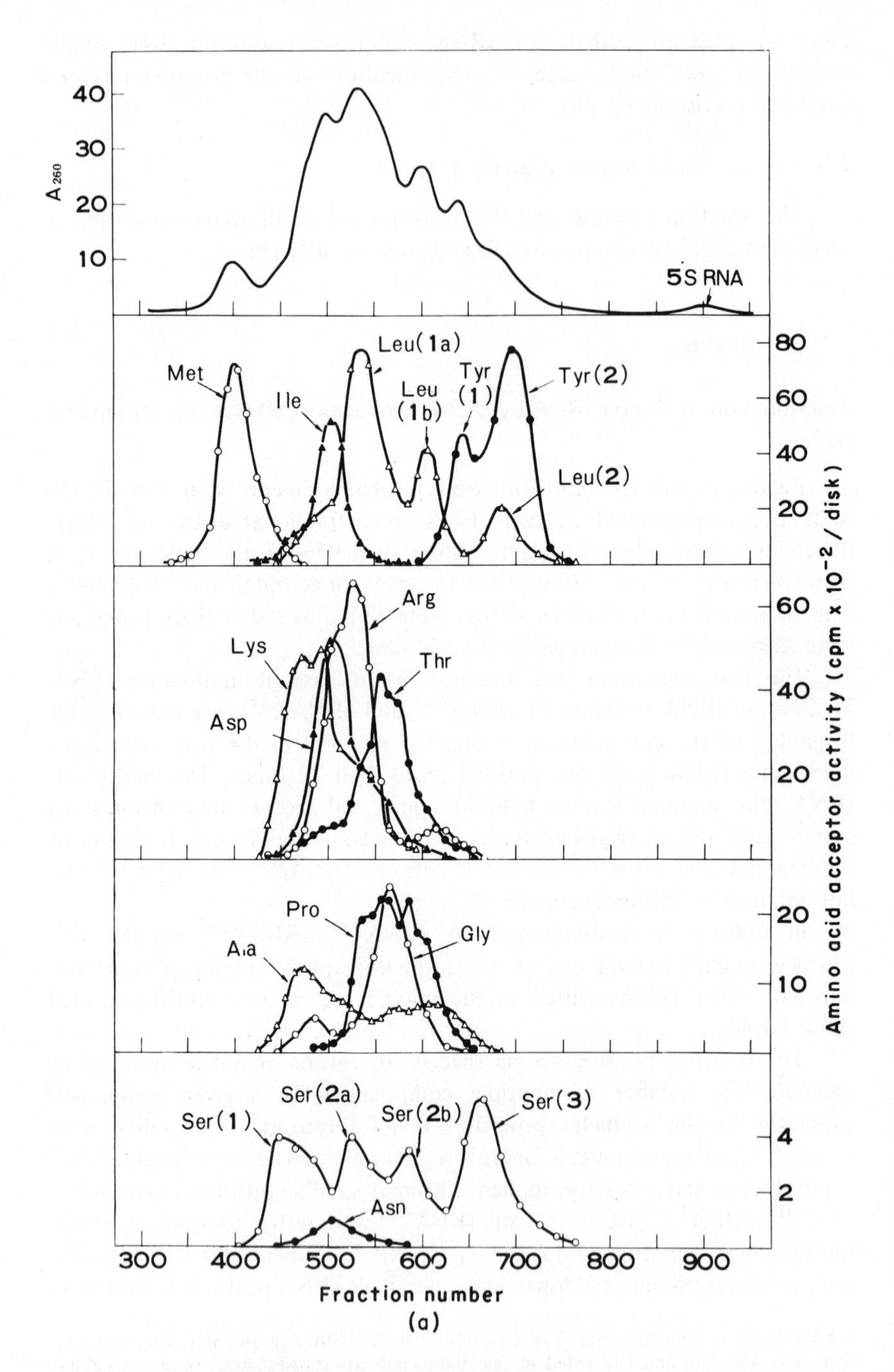

Figure 1. Fractionation of *E. coli* tRNA by DEAE-Sephadex A-50 column chromatography. Each fraction contained 20 ml of the effluent.

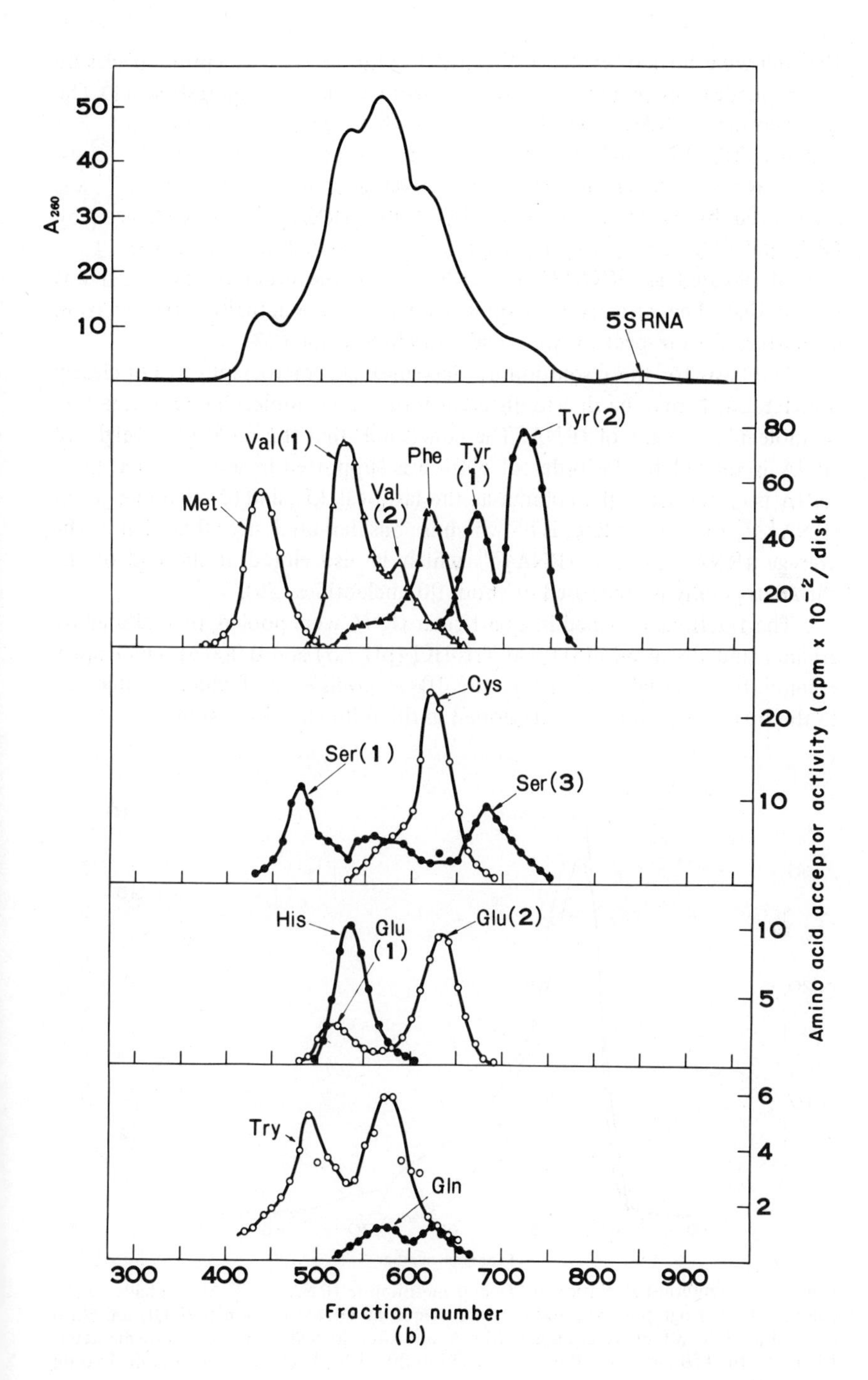
A260
50
40
30
20
10
5S RNA
Met
Val(1)
Val (2)
Phe
Tyr (1)
Tyr(2)
80
60
40
20
Amino acid acceptor activity (cpm x 10^{-2}/ disk)
Ser(1)
Cys
Ser(3)
20
10
His
Glu (1)
Glu(2)
10
5
Try
Gln
6
4
2
300
400
500
600
700
800
900
Fraction number

(b)

the following lines of evidence that the resolution of isoaccepting species by the Sephadex procedure is not due to artifacts such as aggregation. (1) The isoaccepting tRNAs showed different codon recognition properties. For instance, $tRNA_1^{Ser}$ and $tRNA_2^{Ser}$ were recognized by codons of UC series, while $tRNA_3^{Ser}$ were recognized by AGU and AGC (*6*). $tRNA_1^{Val}$ was recognized by GUU, GUA, and GUG, while $tRNA_2^{Val}$ was recognized by GUU and GUC (*21*). (2) Tyrosine tRNA was separated into two species. They were designated as $tRNA_1^{Tyr}$ and $tRNA_2^{Tyr}$ in the order of elution. It was shown that their primary nucleotide sequences were actually different from each other with respect to two nucleotides in S-region (*27*).

The basis for the resolution by Sephadex chromatography is not clearly understood. It may be due to differences of charge, molecular conformation, or molecular weight of tRNA. The conclusion that the molecular weight of tRNA is related to the order of elution is supported by the fact that (1) 5 sRNA is eluted from the column as the last peak (*5*), and (2) chain length of $tRNA^{Tyr}$, that elutes late, is 85 which is considerably longer than that of the average tRNAs (*27, 28*). $tRNA_2^{Leu}$ which is also eluted at the end of the chromatography is composed of about 90 nucleotides (*29*).

The fractions enriched in a particular tRNA were pooled, precipitated by ethanol and dissolved in 0.01 M Tris-HCl (pH 7.5) and 0.006 M β-mercaptoethanol. In the usual case, one obtains 10–15 pools each of which is subjected to the further fractionations described in the following discussion.

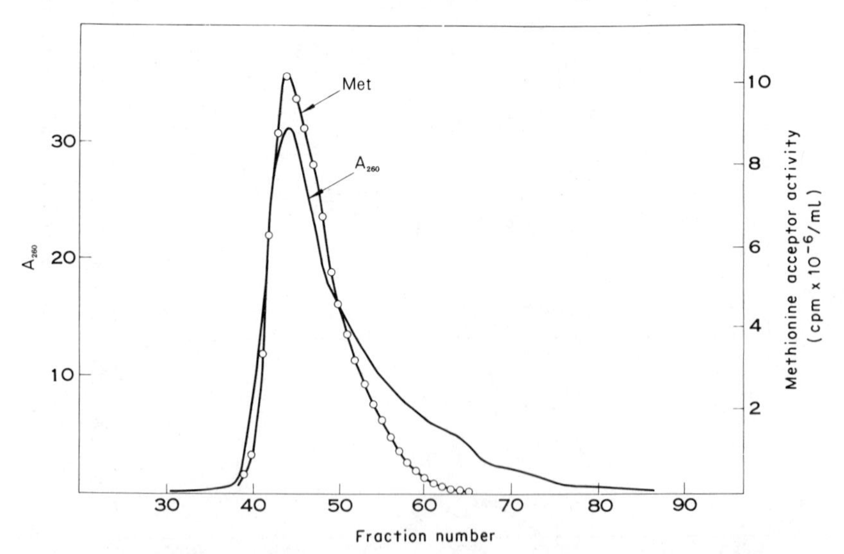

Figure 2. Preferential elution of *E. coli* methionine tRNA by DEAE-Sephadex A-50 column chromatography. Six milliters of methionine tRNA solution (630 OD units/ml) were diluted to 30 ml by adding buffer A, and then loaded on the column (diameter, 1 cm; length, 150 cm). The flow rate was about 20 ml/hr. Each fraction contained 10 ml of the effluent.

Rechromatography of Methionine tRNAs on DEAE-Sephadex A-50

The methionine tRNA fractions eluted from a column of DEAE-Sephadex and pooled (for instance, the fractions from 376 to 420 of Fig. 1a) were loaded again onto the DEAE-Sephadex A-50 column, and elution was carried out simply with buffer A. Methionine tRNA was further purified as a result of its preferential elution by this buffer as shown in Fig. 2 (*7*). Alternatively, if one wants to obtain only methionine tRNA, unfractionated *E. coli* tRNA can be directly applied on a column of the DEAE-Sephadex and eluted with buffer A. In any case, this salt concentration of 0.375 M was found to be critical for the preferential elution of methionine tRNA.

Fractionation of tRNAs from Rat Liver and Yeast by DEAE-Sephadex A-50 Column Chromatography

DEAE-Sephadex column chromatography can be also used for the separation of tRNAs from sources other than *E. coli*. Figures 3 and 4 show chromatographic profiles of the fractionation of tRNA from rat liver and yeast, respectively. As in the case of *E. coli* tRNA, tRNAs corresponding to each amino acid were well separated from each other.

It should be emphasized again that isoaccepting tRNAs having different codon recognition properties could be separated (valine and serine from rat liver, see Ref. *10*; methionine and glutamic acid, see Refs. *9* and *30*). A comparison of chromatographic profiles of tRNAs from *E. coli*, rat liver, and yeast, makes it clear that there is no relationship between the elution position of tRNA for a given amino acid from different species. For instance, *E. coli* $tRNA^{Tyr}$ was eluted at the end of the chromatography, while rat liver $tRNA^{Tyr}$ was eluted at the beginning of the chromatography. It is noteworthy that the DEAE-Sephadex procedure resulted in a 10- to 25-fold purification of $tRNA^{His}$ from yeast, $tRNA^{Tyr}$, $tRNA^{Val}$, $tRNA^{Phe}$, and $tRNA^{Lys}$ from rat liver. Further purification of these tRNA's could be achieved by the successive application of BD-cellulose and/or the reverse-phase partition column chromatography.

Isolation of *E. coli* $tRNA_f^{Met}$, $tRNA_{m1}^{Met}$ and $tRNA_{m2}^{Met}$

Purified *E. coli* methionine tRNA obtained by DEAE-Sephadex column chromatography is a mixture of $tRNA_f^{Met}$ and $tRNA_m^{Met}$'s. Separation of methionine tRNA into three isoaccepting species (i.e., $tRNA_f^{Met}$, $tRNA_{m1}^{Met}$, and $tRNA_{m2}^{Met}$) can be easily achieved by its further fractionation with BD-cellulose column chromatography (*7*). Figure 5 shows a typical chromatographic pattern of methionine tRNA (fractions from 40 to 50 of Fig. 2) on a column of BD-cellulose. The elution profile resulted in three peaks with respect to the absorbance at 260 nm. $tRNA_f^{Met}$ was eluted as the first peak

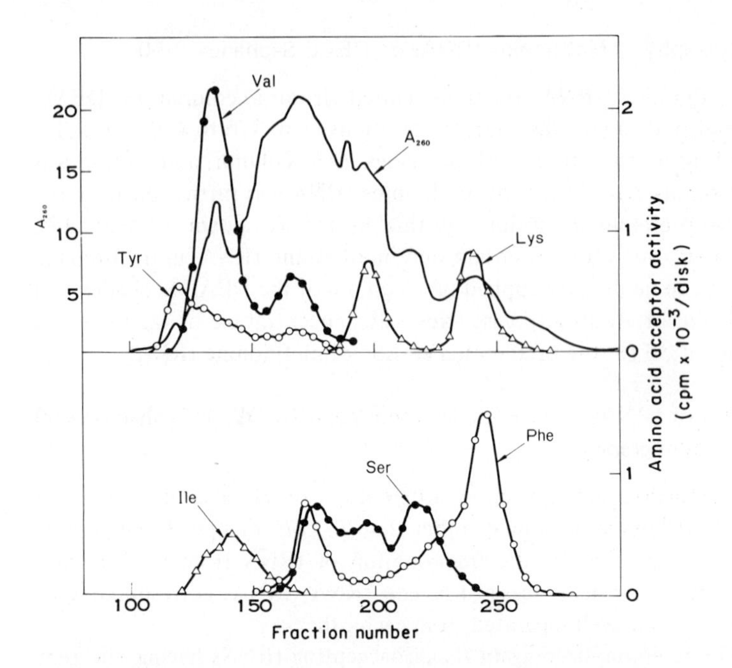

Figure 3. Fractionation of rat liver tRNA by DEAE-Sephadex A-50 column chromatography. Rat liver tRNA (20,000 OD units) was loaded on a column (diameter, 2 cm; length, 150 cm) of DEAE-Sephadex A-50. The linear gradient elution was carried out with 3 liters of buffer A and 3 liters of buffer B. Each fraction contained 10 ml of the effluent.

and was clearly separated from $tRNA_m^{Met}$'s and other contaminating tRNAs such as $tRNA^{Lys}$ and $tRNA^{Ile}$. $tRNA_f^{Met}$ thus obtained seems to be more than 95% pure, as judged by the characterization elution profile of the RNase T_1 digest after column chromatography on DEAE-Sephadex A-25 (7). Another fact to support this conclusion is that 90% of $tRNA_f^{Met}$ in this fraction can be charged with methionine and then formylated, if one assumes that 1 OD_{260} unit of tRNA is equal to 1.66 mμmoles. It should be mentioned that if one wishes to isolate only $tRNA_f^{Met}$, the methionine tRNA fraction can be directly applied on a column of BD-cellulose without rechromatography on DEAE-Sephadex. The yield of $tRNA_f^{Met}$ from 200,000 OD units of unfractionated *E. coli* tRNA is approximately 2,500 OD units.

It is interesting to note that BD-cellulose chromatography gave two methionine $tRNA_m$'s, neither of which was involved in transformylation. Both methionine $tRNA_m$'s were not separated from contaminating $tRNA^{Lys}$ and $tRNA^{Ile}$. However, the acceptor activities for these two amino acids were negligible when calculated on a molar basis. Judged from the ratio of the incorporated radioactivity to absorbance at 260 nm, both methionine

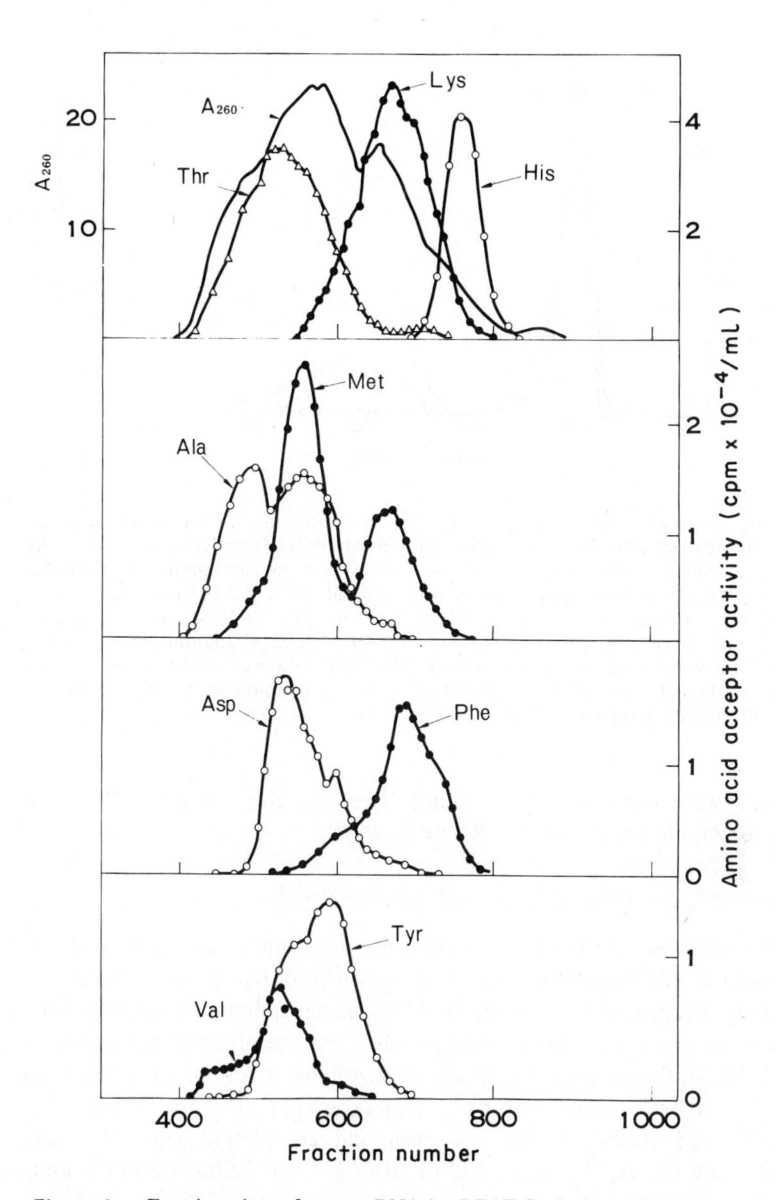

Figure 4. Fractionation of yeast tRNA by DEAE-Sephadex A-50 column chromatography. Yeast tRNA (100,000 OD units) was loaded on a column (diameter, 6 cm; length, 150 cm) of DEAE-Sephadex A-50. The conditions for chromatography are the same as described in the legend of Fig. 1.

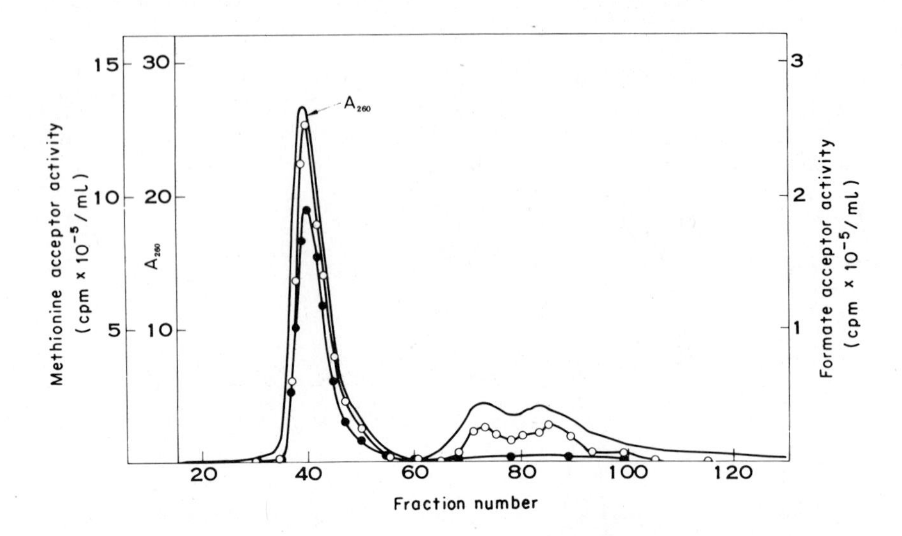

Figure 5. Separation of isoaccepting species of methionine tRNA by BD-cellulose column chromatography. Purified methionine tRNA (1000 OD units) was loaded on the column (diameter, 1 cm; length, 100 cm) which was preequilibrated with buffer containing 0.02 M sodium acetate buffer (pH 6.0) and 0.5 M NaCl. The linear gradient elution was carried out using 400 ml of 0.02 M sodium acetate buffer (pH 6.0)–1.5 M NaCl in the reservoir and 400 ml of 0.02 M sodium acetate buffer (pH 6.0)–0.5 M NaCl in the mixing vessel. The flow rate was about 5 ml/hr. Each fraction contained 3 ml of the effluent: (○——○) methionine acceptor activity: (●——●) formate acceptor activity.

$tRNA_m$'s were more than 90% pure.* This was also confirmed from the chromatographic patterns of the RNase T_1 digest.

Isolation of *E. coli* $tRNA_2^{Tyr}$, $tRNA_3^{Ser}$, and $tRNA^{Leu}$

Reverse-phase partition column chromatography as developed by Kelmers et al. (*12*) is another useful procedure to obtain purified tRNAs by combining this procedure with the DEAE-Sephadex column chromatography. Figure 6 shows the chromatographic profile of reverse-phase chromatography of the DEAE-Sephadex column chromatography fractions enriched with $tRNA_2^{Tyr}$, $tRNA_3^{Ser}$, and $tRNA_2^{Leu}$ (fractions 711 to 750 of Fig. 1b). $tRNA_2^{Tyr}$ was eluted as the last peak and completely separated from $tRNA_3^{Ser}$ and $tRNA_2^{Leu}$. Final yield of $tRNA_2^{Tyr}$ was 1.200–1.500 OD units

* High salt concentration inhibits somewhat the charging of methionine tRNA. Specific activity of methionine acceptance of $tRNA_m^{Met}$ fraction was less than that of $tRNA_f^{Met}$ as seen in Fig. 5. However this is due to the effect of NaCl contained in the effluent. $tRNA_m^{Met}$s isolated by ethanol precipitation gave almost the same methionine acceptor activity as compared with $tRNA_f^{Met}$.

from 200,000 OD units of unfractionated tRNA. This $tRNA_2^{Tyr}$ accepts at least 1.33 mμmoles of ^{14}C-tyrosine per 1 OD unit of the tRNA, and is found to be more than 95% pure as judged from chromatographic profiles of RNase T_1 and pancreatic RNase digests. It was used for the determination of the total primary sequence (*28*).

The fractions containing $tRNA_3^{Ser}$ and $tRNA_2^{Leu}$ separated from $tRNA_2^{Tyr}$ (fractions from 51 to 76 of Fig. 6) were pooled and further purified by BD-cellulose column chromatography as shown in Fig. 7 (*15*). $tRNA_3^{Ser}$ and $tRNA_2^{Leu}$ were completely separated from each other and obtained as single homogeneous peaks. Both tRNAs seem to be completely pure as judged from their amino acid acceptances and from the chromatographic profiles of the RNase digest. The yield of these tRNAs is 500–700 OD units, respectively, per original 200,000 OD units of the unfractionated tRNA. This $tRNA_2^{Leu}$ was recognized by codons of CU series. It was crystallized and used for x-ray crystallography (*31*). Although not shown in the figure, $tRNA_1^{Tyr}$ could be purified similarly to $tRNA_2^{Tyr}$ by reverse-phase partition chromatography of $tRNA_1^{Tyr}$ fraction of the Sephadex procedure.

Isolation of *E. coli* $tRNA_1^{Ser}$

E. coli $tRNA_1^{Ser}$ was purified by applying the $tRNA_1^{Ser}$ fraction (fractions from 421 to 476 of Fig. 1a) to BD-cellulose (*15*). As seen in Fig. 8, serine tRNA was eluted as two peaks in the last portions of the NaCl-ethanol gradient elution; both species of $tRNA^{Ser}$ accept 1.6 mμmoles of ^{14}C-serine per 1 OD unit of the tRNA. The major portion of serine tRNA eluted as the second peak was designated as $tRNA_1^{Ser}$ here. A small peak of $tRNA^{Ser}$ eluted before $tRNA_1^{Ser}$ might be $tRNA_{2a}^{Ser}$. Its homology with respect to nucleotide sequence was confirmed by the analysis of oligonucleotides derived by RNase T_1 and pancreatic RNase digestion. Figure 8 also showed that lysine tRNA was well-separated from leucine and serine tRNAs, and consisted of two isoaccepting species. These lysine tRNA fractions could be further purified by the reverse-phase partition column chromatography at pH 7.5.

Isolation of *E. coli* $tRNA^{Phe}$, $tRNA^{Cys}$, and $tRNA_2^{Glu}$

The fraction containing mainly $tRNA^{Phe}$, $tRNA^{Cys}$, and $tRNA_2^{Glu}$ obtained by the DEAE-Sephadex procedure (fractions from 601 to 651 of Fig. 1b) was further purified on a BD-cellulose column as shown in Fig. 9*(*61*). This resulted in complete separation of $tRNA_2^{Glu}$ from $tRNA^{Phe}$

* The BD-cellulose used in this experiment was different from that used for other experiments. Since this BD-cellulose was not fully benzoylated, tRNAs were eluted with low salt concentration.

and $tRNA^{Cys}$. However, $tRNA^{Phe}$ and $tRNA^{Cys}$ were not separated from each other by this procedure.

Final purification of $tRNA_2^{Glu}$ was achieved by the fractionation of the $tRNA_2^{Glu}$ fraction (fractions from 100 to 120 of Fig. 9) on a reverse phase column. As shown in Fig. 10, $tRNA_2^{Glu}$ was further purified and separated from contaminating other species of tRNA such as proline- and threonine-tRNAs. Elution pattern of 260 nm absorbance and glutamic acid-acceptor activity coincide in the major fraction of $tRNA_2^{Glu}$, suggesting that $tRNA_2^{Glu}$ thus obtained was substantially pure. This $tRNA_2^{Glu}$ accepts 1.1 mμmoles of ^{14}C-glutamic acid per 1 OD unit of the tRNA,* and does not accept other amino acids except a slight amount of leucine. High purity of this $tRNA_2^{Glu}$ was proved from a chromatographic profile of RNase T_1 and pancreatic RNase digests.

The fraction containing $tRNA^{Cys}$ and $tRNA^{Phe}$ (fraction from 151 to

* The low acceptance of glutamic acid in the tRNA may not be due to the impurity, since charging of glutamic acid is very much affected by salt concentration.

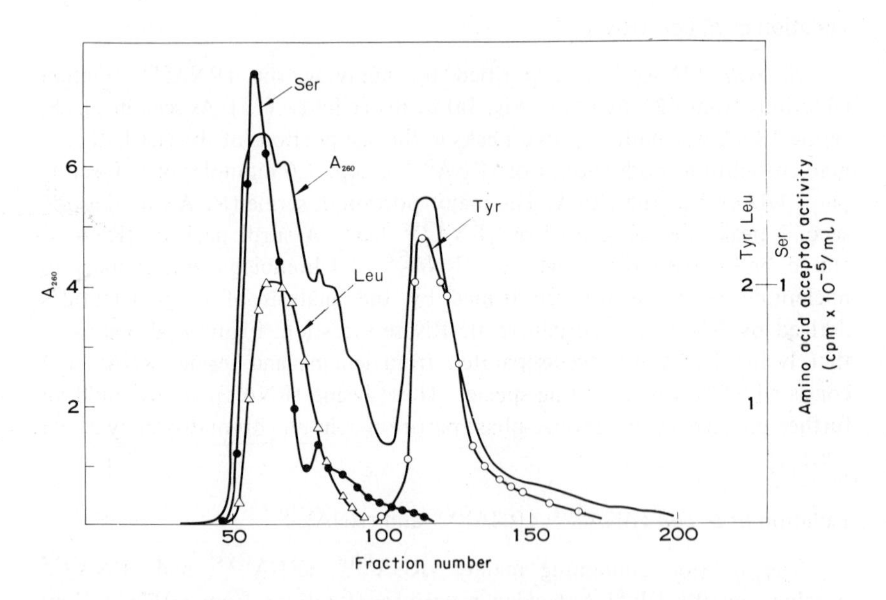

Figure 6. Isolation of *E. coli* $tRNA_2^{Tyr}$ by reverse-phase column chromatography. The tRNA fractions (6800 OD units) dissolved in 30 ml of water was diluted to 60 ml by adding 0.02 M Tris-HCl (pH 7.5)–0.4 M NaCl, and loaded on the column (diameter, 1.5 cm; length, 150 cm). The linear gradient elution was performed using 3 liters of 0.02 M Tris-HCl (pH 7.5)–0.012 M $MgCl_2$–0.7 M NaCl saturated with isoamyl acetate in the reservoir, and 3 liters of 0.02 M Tris-HCl (pH 7.5)–0.4 M NaCl saturated with isoamyl acetate in the mixing chamber. The flow rate was 55 ml/hr. Each fraction contained 20 ml of the effluent.

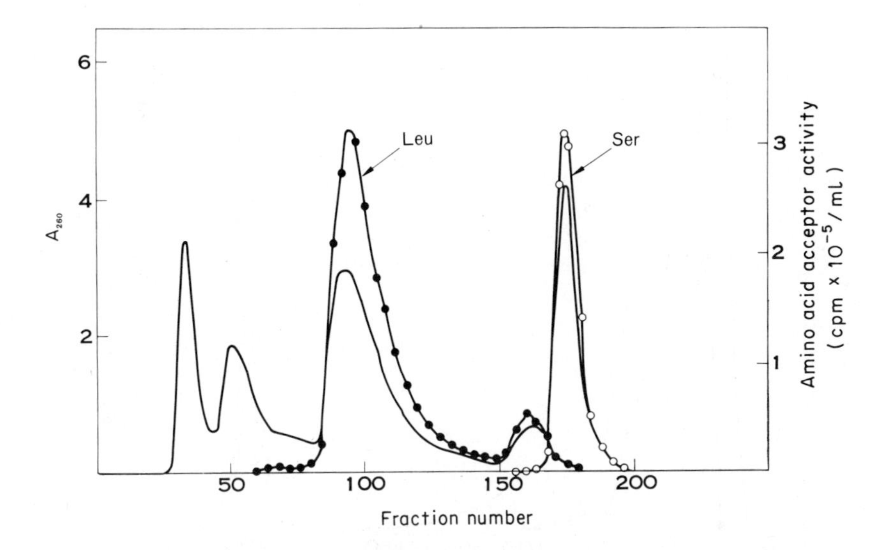

Figure 7. Isolation of *E. coli* $tRNA_3^{Ser}$ and $tRNA_2^{Leu}$ by BD-cellulose column chromatography. The tRNA fractions (2030 OD units) dissolved in 5 ml of water were mixed with 20 ml of sodium acetate buffer (pH 6.0) containing 0.5 M NaCl, and loaded on a column (diameter, 1.5 cm; length, 130 cm) of BD-cellulose. The first linear gradient elution was performed using 1 liter of 0.02 M sodium acetate buffer (pH 6.0)–1.5 M NaCl in the reservoir, and 1 liter of 0.02 M sodium acetate buffer (pH 6.0)–0.5 M NaCl in the mixing chamber. After 1.25 liters of the effluent were collected, the elution was switched to use the second linear gradient with 500 ml of 0.02 M sodium acetate buffer (pH 6.0)–2.5 M NaCl–20% ethanol in the reservoir, and 500 ml of 0.02 M sodium acetate buffer (pH 6.0)–1.15 M NaCl in the mixing chamber. Each fraction contained 10 ml of the effluent.

185 of Fig. 9) was also further subjected to the reverse-phase column chromatography. As shown in Fig. 11, $tRNA^{Cys}$ and $tRNA^{Phe}$ were eluted in the last peak, and completely separated from the bulk of the other tRNA fraction. Partial resolution between $tRNA^{Cys}$ and $tRNA^{Phe}$ was finally obtained by this step. Thus, one-third of $tRNA^{Cys}$ eluted at the beginning and one-third of $tRNA^{Phe}$ eluted at the end were obtained in a purity of more than 70%.

Isolation of $tRNA_I^{Val}$, $tRNA^{His}$, and $tRNA^{Asp}$

A reverse-phase column chromatography performed at pH 4.3 was another useful procedure to use in combination with DEAE-Sephadex column chromatography for the isolation of amino acid specific tRNAs (*5*). The conformation or net charge of tRNA may be changed under the conditions of low pH, thus facilitating the separation of a particular type of tRNA. This procedure was used to separate $tRNA_I^{Val}$, $tRNA^{His}$, and

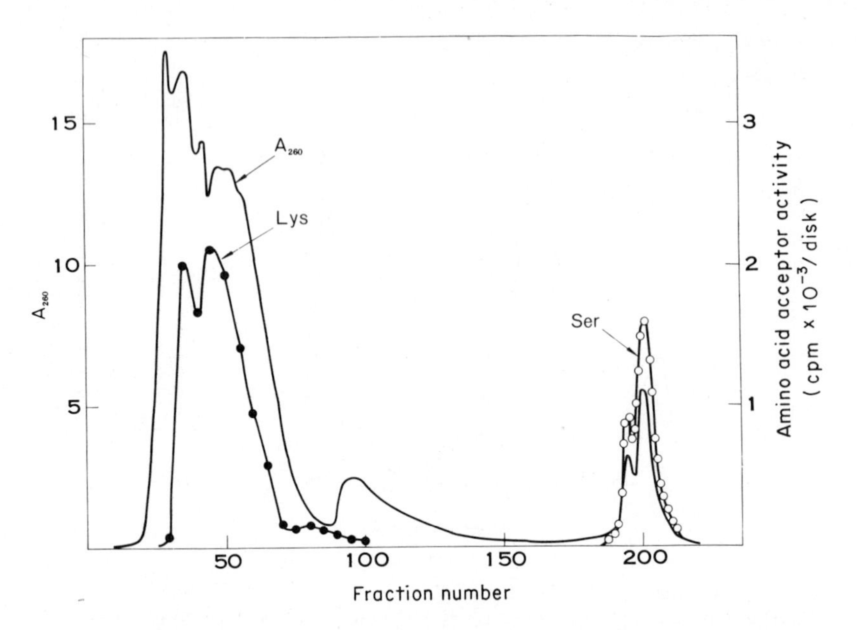

Figure 8. Isolation of *E. coli* $tRNA_I^{Ser}$ by BD-cellulose column chromatography. The $tRNA_2^{Ser}$ fraction (12,250 OD units) dissolved in 12.5 ml of water was mixed with 40 ml of sodium acetate buffer (pH 6.0)–0.5 M NaCl, and loaded on a column (diameter, 1.5 cm; length 80cm) of BD-cellulose. The first linear gradient elution was performed using 1 liter of 0.02 M sodium acetate buffer (pH 6.0)–1.5 M NaCl in the reservoir, and 1 liter of 0.02 M sodium acetate buffer (pH 6.0)–0.5 M in the mixing chamber. After 1.5 liters of the effluent were collected, a second linear gradient elution was started with 500 ml of 0.02 M sodium acetate buffer (pH 6.0)–2.5 M NaCl–20% ethanol in the reservoir, and 500 ml of 0.02 M sodium acetate buffer (pH 6.0)–1.25 M NaCl in the mixing chamber. Each fraction contained 10 ml of the effluent.

$tRNA^{Asp}$ (*16*). These tRNAs were not resolved at all with a usual reverse-phase chromatography at neutral pH. The fraction containing $tRNA_I^{Val}$, $tRNA^{His}$, and $tRNA^{Asp}$ (fractions from 511 to 545 of Fig. 1b) was subjected to the acidic reverse phase chromatography as shown in Fig. 12. Mixture of $tRNA^{Asp}$, $tRNA^{Lys}$, and $tRNA^{Ile}$ was passed through the column, while $tRNA_I^{Val}$ and $tRNA^{His}$ were eluted later and considerably purified.

The $tRNA_I^{Val}$ region (fractions from 75 to 90 of Fig. 12) was further purified by BD-cellulose column chromatography as seen in Fig. 13. The valine tRNA was eluted at the beginning from the column, and was separated from contaminants. The final preparation of this $tRNA_I^{Val}$ accepted 1.6 mμmoles of ^{14}C-valine per 1 OD unit of the tRNA and failed to accept all other 19 amino acids. Analysis of oligonucleotides derived by the RNases digestion also revealed that the $tRNA_I^{Val}$ thus obtained was more than 95% pure. Yield of $tRNA_I^{Val}$ was approximately 1300 OD units from 200,000 OD

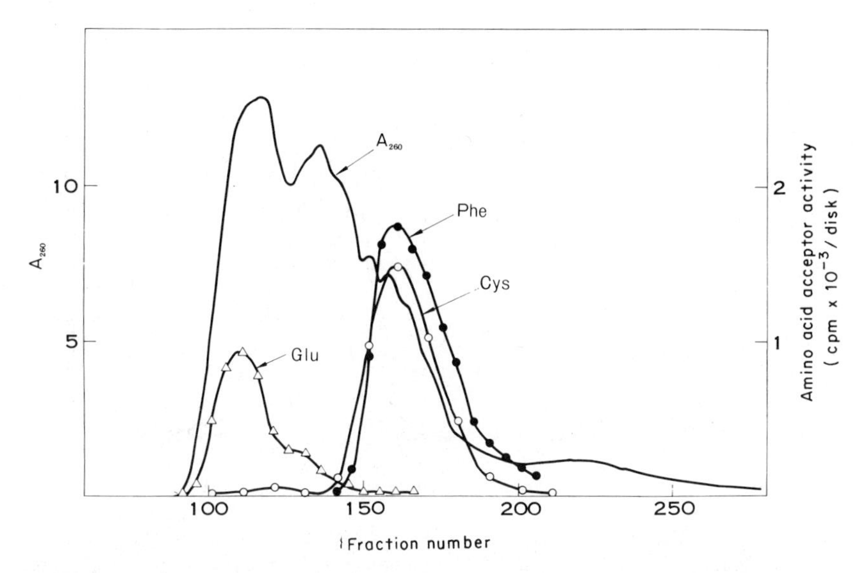

Figure 9. Fractionation of *E. coli* tRNA^{Phe}, tRNA^{Cys}, and tRNA$_2^{Glu}$ by BD-cellulose column chromatography. The tRNAs fractions (13,250 OD units) dissolved in 80 ml of 0.02 M sodium acetate buffer (pH 6.0) were loaded on a column (diameter, 3 cm, length 80 cm) of BD-cellulose. The linear gradient elution was performed using 3 liters of 0.02 M sodium acetate buffer (pH 6.0)–1.4 M NaCl in the reservoir, and 3 liters of 0.02 M sodium acetate buffer (pH 6.0)–0.4 M NaCl in the mixing chamber. Each fraction contained 20 ml of the effluent.

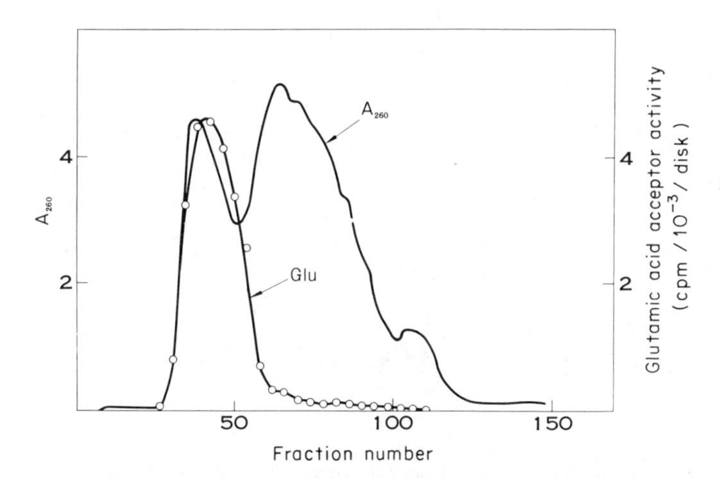

Figure 10. Final purification of *E. coli* tRNA$_2^{Glu}$ by reverse-phase column chromatography. The tRNA$_2^{Glu}$ fraction (4,600 OD units) dissolved in 20 ml of water was diluted to 40 ml by adding the buffer containing 0.02 M Tris-HCl (pH 7.5)–0.4 M NaCl, and loaded on the column (diameter, 1.5 cm; length, 150 cm). The conditions for the chromatography were the same as described in the legend of Fig. 6. Each fraction contained 15 ml of the effluent.

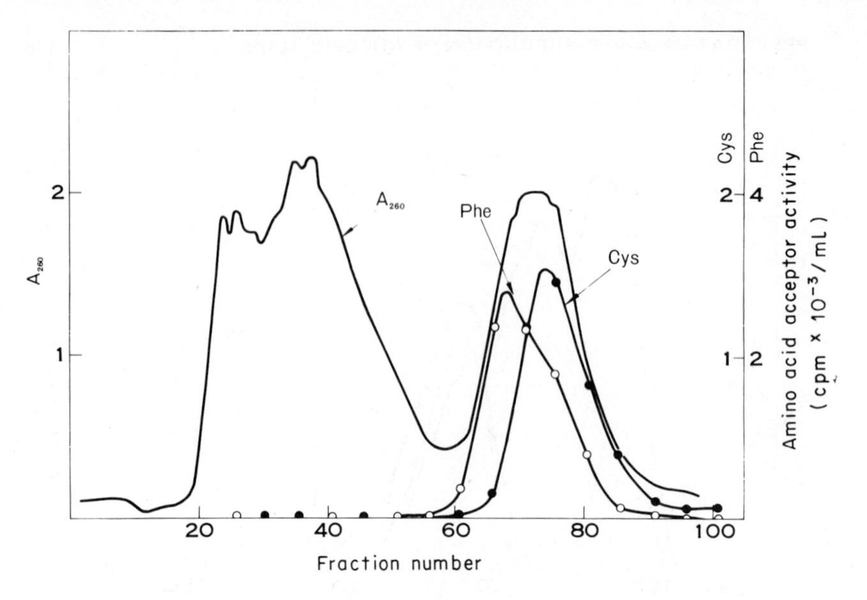

Figure 11. Isolation of *E. coli* $tRNA^{Phe}$ and $tRNA^{Cys}$ by reverse-phase column chromatography. The $tRNA^{Phe}$–$tRNA^{Cys}$ fraction (1760 OD units) dissolved in 7 ml of water were mixed with equal volume of 0.02 M Tris-HCl (pH 7.5)–0.4 M NaCl, and loaded on the column (diameter, 1.5 cm; length, 150 cm). The linear gradient elution was carried out using 2 liters of 0.02 M Tris-HCl (pH 7.5)–0.4 M NaCl saturated with isoamyl acetate in the mixing chamber, and 2 liters of 0.02 M Tris-HCl (pH 7.5)–0.47 M NaCl–0.012 M $MgCl_2$ in the reservoir. Each fraction contained 17 ml of the effluent.

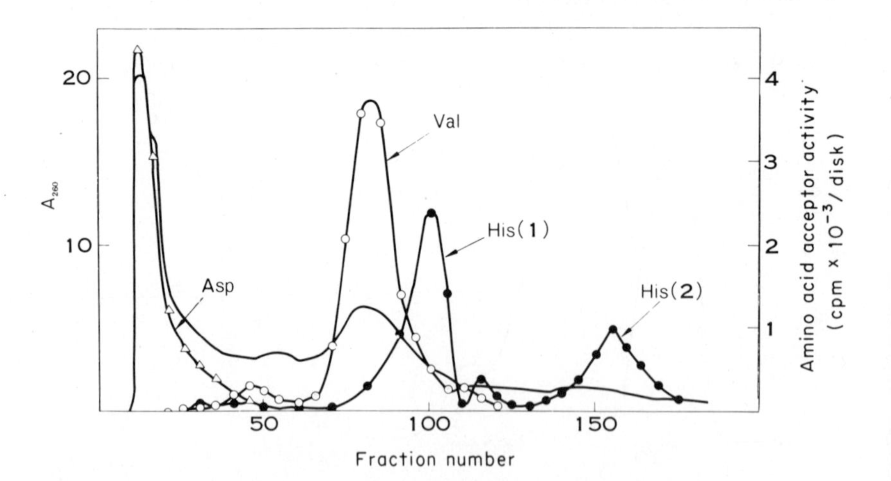

Figure 12. Fractionation of *E. coli* $tRNA_1^{Val}$, $tRNA^{His}$, and $tRNA^{Asp}$ by the reverse-phase column chromatography at pH 4.3. The tRNA fractions (15,000 OD units) dissolved in 15 ml of water were mixed with an equal volume of the buffer containing 0.05 M sodium acetate buffer (pH 4.3) and 0.45 M NaCl. The pH of the tRNA solution was adjusted to 4.3 by adding 2 M acetic acid, and loaded on the column (diameter, 1.5 cm; length, 150 cm). The linear gradient elution was carried out using 3 liters of 0.05 M sodium acetate buffer (pH 4.3)–0.45 M NaCl in the mixing chamber, and 3 liters of 0.05 M sodium acetate buffer (pH 4.3)–0.7 M NaCl in the reservoir. Each fraction contained 20 ml of the effluent. Flow rate was 65 ml/hr.

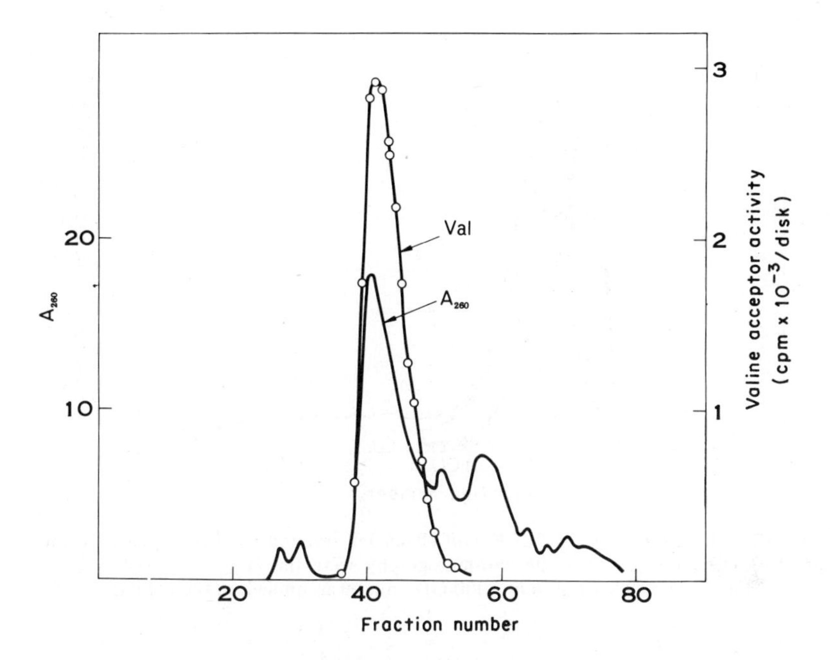

Figure 13. Isolation of *E. coli* $tRNA^{Val}$ by BD-cellulose column chromatography. The $tRNA^{Val}$ fraction (3060 OD units) was applied on the column (diameter, 1.5 cm; length, 100 cm). The linear gradient elution was carried out using 900 ml of 0.02 M sodium acetate buffer (pH 6.0)–1.5 M NaCl in the reservoir, and 900 ml of 0.02 M sodium acetate buffer (pH 6.0)–0.5 M NaCl in the mixing chamber. Flow rate was 12 ml/hr. Each fraction contained 9 ml of the effluent.

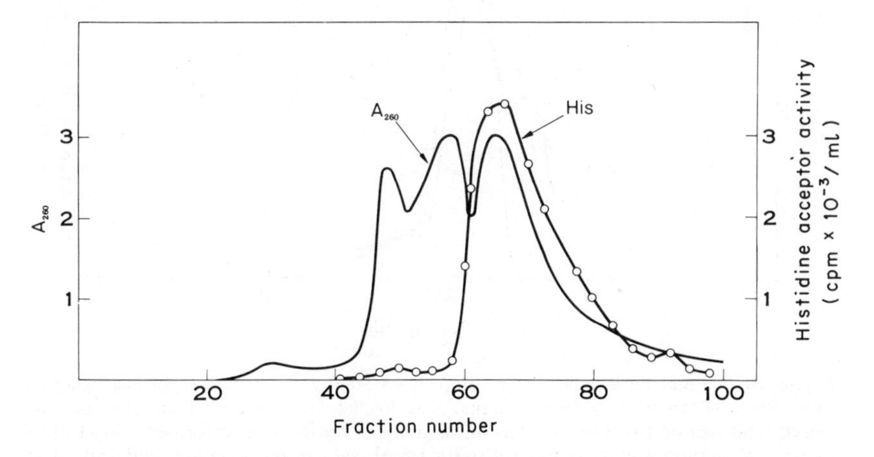

Figure 14. Isolation of *E. coli* $tRNA_1^{His}$ by BD-cellulose column chromatography. The $tRNA_1^{His}$ fraction (450 OD units) was chromatographed as described in the legend of Fig. 5. Each fraction contained 4.7 ml of the effluent.

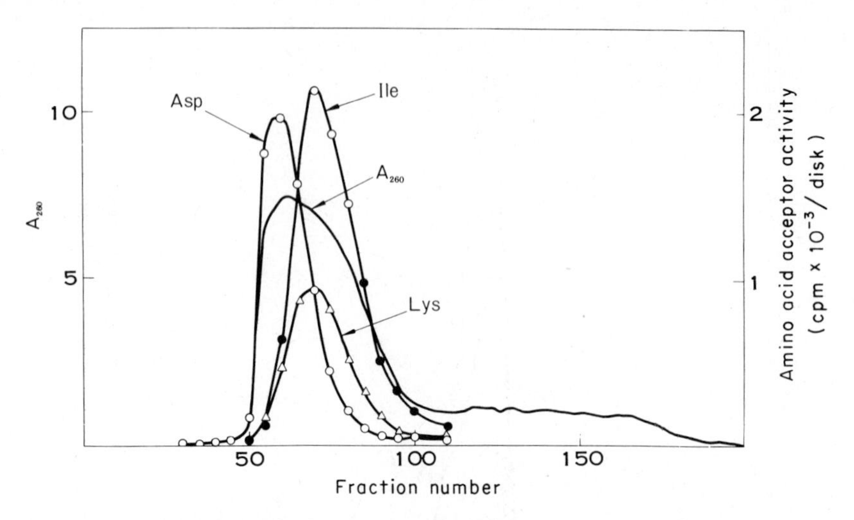

Figure 15. Rechromatography of *E. coli* tRNA^{Asp} fraction by reverse-phase chromatography. The conditions for the chromatography were the same as described in the legend of Fig. 6. The tRNA fraction (6000 OD units) was applied on the column.

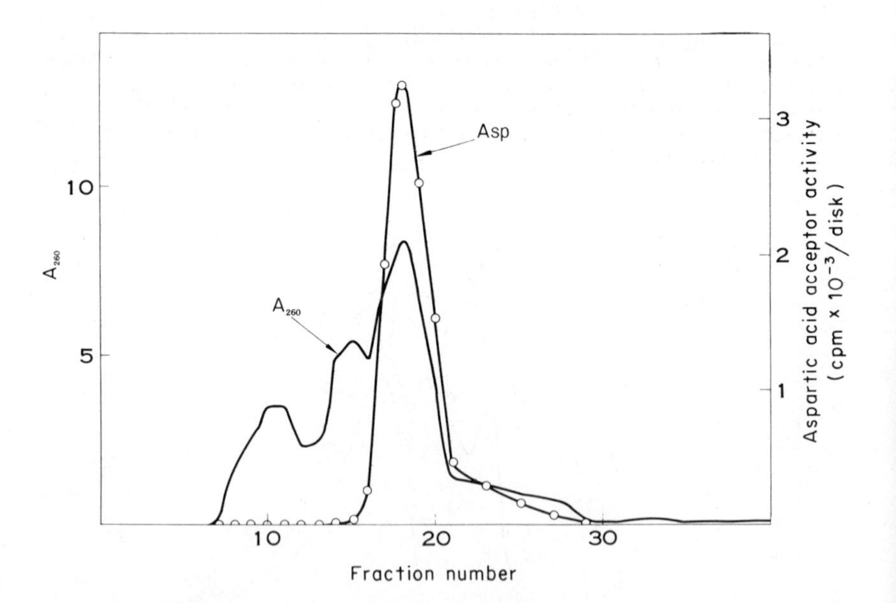

Figure 16. Final purification of *E. coli* tRNA^{Asp} by BD-cellulose chromatography. The tRNA^{Asp} fraction (450 OD units) was loaded on a column (diameter, 0.5 cm; length, 100 cm) of BD-cellulose. The linear gradient elution was performed using 150 ml of 0.02 M sodium acetate buffer (pH 6.0)–1.5 M NaCl in the reservoir, and 150 ml of 0.02 M sodium acetate buffer (pH 6.0)–0.5 M NaCl in the mixing chamber. Each fraction contained 5 ml of the effluent.

units of unfractionated tRNA. This $tRNA_1^{Val}$ was used for the determination of its primary sequence (*32*). $tRNA_2^{Val}$ which was only partially separated from $tRNA_1^{Val}$ by the first DEAE-Sephadex column chromatography was found to emerge very late from a column of BD-cellulose. Therefore, if any $tRNA_2^{Val}$ is contained in the $tRNA_1^{Val}$ fraction in the step of reverse-phase column chromatography, it is completely removed by the final BD-cellulose chromatography.

In a similar fashion, the $tRNA_1^{His}$-rich fraction (fraction from 91 to 105 of Fig. 12) was further purified by BD-cellulose chromatography as shown in Fig. 14. $tRNA_2^{His}$ was also purified by the same procedure although not shown in the figure. These $tRNA^{His}$'s accept 1.6 mμmoles of ^{14}C-histidine per 1 OD unit of the tRNA, indicating that they are substantially pure.

The fraction enriched with $tRNA^{Asp}$ (fractions from 11 to 25 of Fig. 12) was again subjected to the reverse-phase column chromatography at neutral pH. As shown in Fig. 15, this chromatography resulted in the partial resolution of $tRNA^{Asp}$ from $tRNA^{Ile}$ and $tRNA^{Lys}$. The $tRNA^{Asp}$ fraction (fractions from 52 to 60 of Fig. 15) was then chromatographed on a column of BD-cellulose as shown in Fig. 16. The final preparation of $tRNA^{Asp}$ thus obtained accepted 1.6 mμmoles of ^{14}C-aspartic acid per 1 OD unit of the tRNA.

References

1. Kawade, Y., T. Okamoto, and Y. Yamamoto. 1963. *Biochem. Biophys. Res. Commun.,* **10**: 200.
2. Cherayil, J. D., and R. M. Bock. 1965. *Biochemistry*, **4**: 1174.
3. Miyazaki, M., M. Kawada, and S. Takemura. 1966. *J. Biochem. (Tokyo),* **60**: 519.
4. Miyazaki, M., and S. Takemura. 1966. *J. Biochem. (Tokyo),* **60**: 526.
5. Nishimura, S., F. Harada, U. Narushima, and T. Seno. 1967. *Biochim. Biophys. Acta,* **142**: 133.
6. Ishikura, H., and S. Nishimura. 1968. *Biochim. Biophys. Acta,* **155**: 72.
7. Seno, T., M. Kobayashi, and S. Nishimura. 1968. *Biochim. Biophys. Acta,* **169**: 80.
8. Takeishi, K., S. Nishimura, and T. Ukita. 1967. *Biochim. Biophys. Acta,* **145**: 605.
9. Takeishi, K., T. Ukita, and S. Nishimura. 1968. *J. Biol. Chem.,* **243**: 5761.
10. Nishimura, S., and I. B. Weinstein. 1969. *Biochemistry,* **8**: 832.
11. Seno, T., and S. Nishimura. 1968. *Biochim. Biophys. Acta,* **157**: 97.
12. Kelmers, A. D., G. D. Novelli, and M. P. Stulberg. 1965. *J. Biol. Chem.,* **240**: 3970.
13. Gillam, I., S. Millward, D. Blew, M. von Tigerstrom, E. Wimmer, and G. M. Tener. 1967. *Biochemistry*, **6**: 3043.
14. Nishimura, S., F. Harada, and M. Hirabayashi. 1969. *J. Mol. Biol.,* **40**: 173.
15. Ishikura, H., Y. Yamada, and S. Nishimura. 1971. *Biochim. Biophys. Acta,* **228**: 471.
16. Nishimura, S., F. Ishii, H. Hara, and K. Oda, manuscript in preparation.
17. Nishimura, S., and F. Ishii, manuscript in preparation.
18. Yoshida, M., K. Takeishi, and T. Ukita. 1971. *Biochim. Biophys. Acta,* **228**: 153.
19. RajBhandary, U. L., and H. P. Ghosh. 1969. *J. Biol. Chem.,* **244**: 1104.
20. Nishimura, S., K. Ishii, Y. Yamada, and H. Ishikura, manuscript in preparation.

21. Nishimura, S., unpublished data.
22. Nishimura, S., Y. Yamada, and H. Ishikura. 1969. *Biochim. Biophys. Acta,* **179**: 517.
23. Ishikura, H., Y. Yamada, K. Murao, M. Saneyoshi, and S. Nishimura. 1969. *Biochem. Biophys. Res. Commun.,* **37**: 990.
24. Zubay, F. 1962. *J. Mol. Biol.,* **4**: 347.
25. Holley, R. W. 1963. *Biochem. Biophys. Res. Commun.,* **10**: 186.
26. Brunngraber, E. F. 1962. *Biochem. Biophys. Res. Commun.,* **8**: 1.
27. Goodman, H. M., J. Abelson, A. Landy, S. Brenner, and J. D. Smith. 1968. *Nature,* **217**: 1019.
28. RajBhandary, U. L., S. H. Chang, H. J. Gross, F. Harada, F. Kimura, and S. Nishimura. 1969. *Federation Proc.,* **28**: 409.
29. Ishikura, H., Y. Yamada, K. Ishii, and S. Nishimura, unpublished data.
30. Sekiya, T., K. Takeishi, and T. Ukita. 1969. *Biochim. Biophys. Acta,* **182**: 411.
31. Young, J. D., R. M. Bock, S. Nishimura, H. Ishikura, Y. Yamada, U. L. RajBhandary, M. Labanauskas, and P. C. Connors. 1969. *Science,* **166**: 1527.
32. Harada, F., F. Kimura, and S. Nishimura. 1969. *Biochim. Biophys. Acta,* **195**: 590.

FRACTIONATION AND PURIFICATION OF tRNA ON COLUMNS OF METHYLATED ALBUMIN SILICIC ACID

R. STERN
NATIONAL INSTITUTE OF DENTAL RESEARCH
NATIONAL INSTITUTES OF HEALTH
BETHESDA, MARYLAND

U. Z. LITTAUER *
DEPARTMENT OF BIOCHEMISTRY
WEIZMANN INSTITUTE OF SCIENCE
REHOVOT, ISRAEL

RESOLUTION OF tRNAs

Introduction

Columns of methylated serum albumin absorbed on kieselguhr (MAK) were first introduced for the chromatography of nucleic acids by Lerman (*1*) and Mandell and Hershey (*2*). The column has been used extensively for the chromatography of tRNA (*3*). A limitation of the MAK column has been its small capacity. A similar column of increased capacity was needed, a requirement fulfilled by methylated albumin absorbed on a support of silicic acid (MASA), which was first suggested by Okamoto and Kawade (*4*) and later adapted specifically for the resolution of tRNAs (*5, 6*). This column has

* Fogarty Scholar in residence 1969–1970, U.S., National Institutes of Health.

a 50- to 100-fold greater adsorptive capacity and in some instances has a better ability to resolve the various components of a given aminoacyl-tRNA than the MAK column.

Reagents for Preparation of the Column

1. Jacketed column (2 cm i.d. x 20 cm).
2. Silicic acid (325 mesh), 20 gm, Fisher Scientific Co.
3. Methylated albumin prepared according to the method of Mandell and Hershey (*2*).
4. Buffered solutions
   a. 0.05 M sodium phosphate (pH 6.8)
   b. 0.05 M sodium acetate (pH 5.5) containing 0.8 M NaCl (initial buffer)
   c. 0.05 M sodium acetate (pH 5.5) containing 1.15 M NaCl (terminal buffer)
5. Acid-washed glass beads (200 μ average diameter), Minnesota Mining and Manufacturing Co.
6. Kieselguhr, 2 gm (British Drug House)

Preparation

The silicic acid is suspended and decanted 20 times with tap water and twice with distilled water. The suspension is treated with 1 N HCl, washed with water until the pH reaches that of water, and then dried in an oven. The powder is stored at room temperature. Stock solutions of 1.0 M buffer without NaCl are adjusted so as to give the desired pH on dilution to 0.05 M at 23°. The diluted buffers are not readjusted following the addition of NaCl.

To prepare a column (2 x 12 cm), 20 gm of silicic acid are suspended in 100 ml of 0.05 M sodium phosphate buffer (pH 6.8), boiled for 2 min, and cooled. The methylated albumin (1 gm) is dissolved in 100 ml of water. The methylated albumin solution is added slowly to the silicic acid solution, with stirring, at room temperature. Stirring is continued for 2 hr. The material is washed several times by decanting and resuspending the protein–silicic acid mixture in terminal buffer. These washings permit a more rapid flow rate. Acid-washed glass beads are placed at the bottom of the column to a height of 2 cm. A 10-ml aliquot of the methylated albumin–silicic acid mixture (100 ml) is pipetted onto this bed. The column outlet is opened and the excess buffer forced out under an air pressure of 2–3 psi, just down to the level of the packed material, and the addition of 10-ml aliquots is repeated as above. When all of the methylated albumin–silicic acid mixture has been applied, additional buffer is added to wash excess protein from the walls of the column. Kieselguhr (2 gm) is suspended in 10 ml of the initial buffer, boiled for 2 min, cooled, and added to the top of the column bed as a protective covering. The column is then washed with terminal buffer until the optical density at 260 nm falls below 0.02 (about 1 liter). The column is then washed with 100 ml of initial buffer.

Column Chromatography

Labeled aminoacyl-tRNAs can be prepared as described below. The labeled aminoacyl-tRNA (3–80 mg) is dissolved in 10 ml of initial buffer, mixed with 2 ml of water, and applied to the column. The column is washed with 50 ml of initial buffer and the tRNA eluted with a linear gradient containing 200 ml of initial buffer and 200 ml of terminal buffer. Fractions (about 120) of 3.3 ml each are collected at a flow rate of 1 ml/min using a hydrostatic pressure of 1 m between gradient reservoir and column outlet. The column is run at 16° by circulating water from a thermostatically controlled cooling bath. Each receiving tube contains 0.2 ml of 1 M sodium acetate buffer (pH 4.0) to minimize aminoacyl-tRNA hydrolysis before the tubes can be assayed. Optical density at 260 nm is determined, aliquots of the fractions are then removed, 1 drop of carrier RNA (5 mg/ml) is added, and the RNA is precipitated with equal volumes of 10% cold trichloroacetic acid and filtered onto membrane filters (Sartorius Membranfilter, Göttingen, Germany, 0.6 μ). The filters are washed three times with 10 ml of 2.5% trichloroacetic acid and dried. The membrane filters are placed in 10 ml of toluene scintillation fluid and radioactivity counted.

The optical density at 260 nm usually emerged between fractions 15 and 100. In most RNA preparations, a variable narrow peak of absorbance appears preceding the main peak. In some cases it is entirely absent.

To reuse, the column is washed with 0.05 M sodium acetate buffer (pH 5.5) containing 2.0 M NaCl until the $A_{260\ \mathrm{nm}}$ falls below 0.02 and then washed with 100 ml of initial buffer. After repeated use, the optical density profile becomes increasingly constricted, with more material being eluted at lower salt concentrations. Also the protective kieselguhr layer at the top of the column becomes tarnished, but can easily be replaced. In addition, the flow rate decreases as the uppermost part of the kieselguhr layer becomes matted with fibrous DNA which contaminates the tRNA preparations. The flow rate of the column is restored by resuspending a portion of the kieselguhr layer.

Recovery of Biological Activity

It has been reported that recovery of amino acid acceptor activity from methylated albumin columns is unsatisfactory (*4*). However, this difficulty, due to methylated albumin eluted from the column, can be circumvented by pronase treatment of the column fractions (*7*). Proteolysis of the methylated albumin is accompanied by hydrolysis of aminoacyl-tRNA bonds. On reacylation of the recovered material all of the original amino acid acceptor activity is obtained. However, in the case of Phe-tRNA only 60–70% of the transfer ability to polypeptide is regained after this treatment.

Extensive dialysis against water followed by lyophilization causes inactivation of tRNA. Consequently tRNA fractions should be dialyzed against 10^{-3} M NaCl. Fractions are then lyophilized to a small volume and precipitated with 0.2 volume of 5 M NaCl and 2.5 volumes of 96% ethanol at 0°. The material is allowed to precipitate for several hours at −15° and centrifuged, and the pellet dissolved in 1.0 ml of 0.01 M Tris-HCl (pH 7.4) incubated at 37° for 5 hr with 0.7 ml of pronase solution. To reduce possible traces of nucleases, the pronase solution is preincubated for 90 min at 37° (2 mg/ml in 0.01 M Tris-HCl buffer, pH 7.4) before use. Some preparations of pronase required up to 48 hr of preincubation (*8*).

At the end of the incubation the solution is cooled to 4°. Phenol (0.7 volume of 70%) and chloroform (0.07 volume) are added and shaken for 30 min at 4°. The mixture is centrifuged and the phenol phase is saved. To the aqueous phase, 0.4 volume of 70% phenol and 0.04 volume of chloroform are added, shaken for 30 min at 4°, and centrifuged. The two phenol phases are combined, shaken briefly with 0.5 volume of 0.01 M Tris-HCl buffer (pH 7.4), and centrifuged, and the aqueous phase combined with the other aqueous fraction. NaCl (0.2 volume of 5 M) and ethanol (2.2 volumes of 96%) are added and stored at −15° for several hours. The suspension is centrifuged and the pellet dissolved in a small volume of water. The material is then ready for reacylation. Recovery is approximately 70%.

Resolution of tRNA

The MASA column has been used extensively for the chromatography of aminoacyl-tRNAs and in some instances provides a better resolution than MAK column chromatography. This is particularly pronounced with the tRNA extracted from a relaxed methionine-requiring mutant of *E. coli* grown in a medium deficient in methionine (*9*). MASA column chromatography of such a methionine-starved tRNA preparation (Fig. 1) resolves four peaks of the Phe-tRNA (*6, 11*). The first two peaks correspond to methyl-deficient tRNA, while the last peak corresponds to a fully methylated normal tRNA.

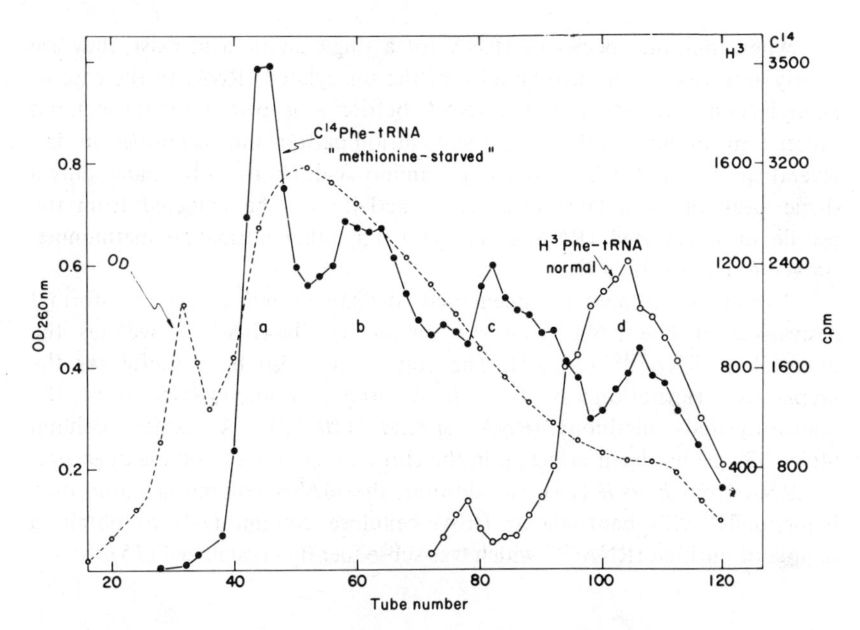

Figure 1. MASA column chromatography of normal and methionine-starved Phe-tRNA from *E. coli* A-19. Normal tRNA (0.2 mg) was charged with [^{3}H]-phenylalanine and methionine-starved tRNA (0.2 mg) was acylated with [^{14}C] phenylalanine. The two phenylalanyl-tRNAs were mixed together with 4.5 mg of uncharged methionine-starved *E. coli* A-19 tRNA in 10 ml of 0.05 M sodium acetate buffer (pH 5.5) containing 0.4 M NaCl. This mixture was applied to a MASA column and chromatographed as described in the text (*6*).

Thus, MASA columns provide a means of separating methyl-deficient from normal Phe-tRNA. On the other hand, only three partially separated components are obtained on chromatography of methionine-starved Phe-tRNA on a MAK column (*7*).

The MASA column resolves RNA on the basis of charge, size, G + C content, and conformation. An additional basis of separation of tRNA on the methylated albumin columns can be inferred from an empirical observation made to ensure successful chromatography. It was found that a lower limit exists for the amount of tRNA which can be placed on the column (*6*). This suggests a displacement phenomenon, since minimal amounts of tRNA are required to saturate sites on the column bed.

It has also been shown that extent of ordered structure plays an important role in RNA resolution (*10, 11*). The column has been used to distinguish between uncharged, aminoacylated and *N*-formylaminoacyl-tRNAs, probably on the basis of differences in conformation. The unacylated species of tRNA have similar profiles on the MASA column and generally elute with the major peak of $A_{260\ \mathrm{nm}}$.

When multiple species of tRNA for a single amino acid exist, they are poorly resolved on chromatography of the unacylated tRNA. In the case of phenylalanine, acylation of the tRNA before it is placed on the column causes a major shift in the subsequent elution pattern and also brings out the several species of tRNA for the single amino acid. On the other hand, only a single peak of phenylalanine acceptor activity can be detected from the profile of unacylated tRNA extracted from either normal or methionine-starved *E. coli* (*5, 10, 11*).

The MASA column has been used to distinguish *N*-acetyl or *N*-formyl derivatives of Phe-tRNA from the unblocked Phe-tRNA as well as the unacylated $tRNA^{Phe}$ (*10, 11*). The column has also been useful for the preparative separation of *E. coli* *N*-formylmethionyl-tRNA from the nonformylated methionyl-tRNA species (*10–12*). A larger column (4.4 x 33 cm) has been effective in the chromatography of 500 mg quantities of tRNA from *E. coli* (*13*). In addition, the MASA column has been used sequentially with benzoylated DEAE-cellulose column (*14*) to obtain a sample of purified $tRNA^{Phe}$ which was subsequently crystallized (*15*).

Resolution of Other RNAs

The profile of the total extractable RNA from *E. coli* has peaks of 4S, 5S, and higher-molecular-weight RNAs (Fig. 2). Higher-molecular-weight RNAs are more tightly bound to the column and the partially resolved peaks of high-molecular-weight RNA are suggestive of the two species of ribosomal RNA. However, high-molecular-weight RNAs tend to form complex aggregates at high salt concentrations and it has been observed in the case of the MAK column fractions that these RNAs are not resolved from each other when reexamined by polyacrylamide gel electrophoresis (*16*). Therefore methylated albumin columns appear to be limited to the resolution of low-molecular-weight RNAs.

However, the MASA column has been used in the separation of 5S RNA from tRNA. Two configurational states of 5S RNA have also been resolved on a preparative level using the MASA column (*17*).

Properties of the Column

Several mesh sizes of silicic acid had been evaluated early in the adaptation of the MASA column. Materials of smaller mesh size have too slow a flow rate. Larger particles have a greatly diminished capacity for methylated albumin absorption and for tRNA. The source of silicic acid is also important. Several commercial preparations of silicic acid were tested and, of these, the

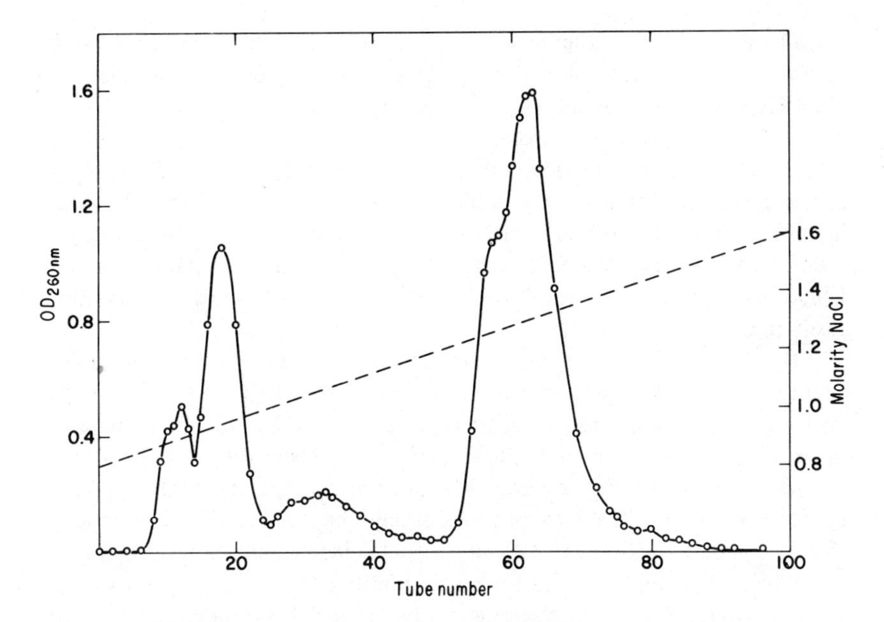

Figure 2. MASA column chromatography of total *E. coli* RNA. *E. coli* MRE 600 cells grown to the middle of the logarithmic growth phase and stored as a frozen cell paste, were the source of total cell RNA which was extracted with phenol and SDS. One hundred A_{260} units were applied to the column which was then washed with 50 ml of 0.05 M sodium acetate buffer (pH 5.5), 0.8 M in NaCl. The RNA was eluted with a gradient formed from 200 ml each of 0.8 M and 1.6 M NaCl in 0.05 M sodium acetate buffer (pH 5.5). Fractions (4 ml) were collected. The column was run at room temperature (*6*).

Fisher product was superior for the resolution of tRNA. It should be noted that commercially prepared methylated bovine serum albumin is not effective in resolution of tRNA.

A number of parameters must be controlled to obtain maximum resolution and to ensure reproducibility of tRNA profiles on the MASA column. The pH and salt concentration of the initial solution containing the RNA sample are critical. Apparently the configurational state of the tRNA remains fixed after being applied to the column and the subsequent chromatographic behavior of the tRNA is determined by this initial configurational state. This is the single most critical variable. Chromatography is also pH dependent. For maximum resolution an optimal pH exists for each aminoacyl-tRNA species. *E. coli* Phe-tRNA is best resolved at pH 5.5 and inferior patterns are obtained at pH 5.0. Temperature is also a variable. The chromatographic profiles are optimal at 16° and resolution diminishes with increasing temperature, particularly above 23°.

Both a lower as well as an upper limit exist for the tRNA which can be placed on the MASA column (*6*). For optimal resolution, 5–80 mg can be placed on the MASA column. When less than 2 mg of methionine-starved Phe-tRNA is placed on the column, the four components of Phe-tRNA are not resolved. Whenever chromatography of a small amount of labeled aminoacyl-tRNA is required, nonlabeled carrier tRNA should be added to enhance the resolution on the column. It was also observed that the upper limit for optimal resolution of tRNAs on the columns is below its capacity to retain tRNA. Similar observations have been made with the MAK column (*5*). Unlike the MAK column the MASA column can be allowed to go dry and is easily rewetted.

Despite the previously mentioned precautions the resolution of aminoacyl-tRNAs is still subject to some variation. Some of this variation depends on the growth conditions of the bacteria as well as the particular preparation of methylated albumin and silicic acid used. Therefore, it is unwise to compare the relationship between tRNAs from two separate chromatographic profiles. When possible two isotopes should be employed and aminoacyl-tRNAs compared within a single run by double-label counting techniques.

When the column is used for resolution of tRNA on a preparative scale some methylated albumin continues to be leached from the column despite prolonged use and protracted prewashing. Methylated albumin has been shown to complex with tRNA (*18*) with a loss of biological activity. Treatment of the complex with pronase and reextraction of the tRNA with phenol restores amino acid acceptor activity, while phenol extraction alone does not (*6, 7, 18*). Similarly tRNA from methylated albumin columns gives spurious results when tested in ribosome-dependent binding studies for its coding properties. This miscoding effect is eliminated by pronase treatment followed by phenol extraction, but not by phenol alone (*19*). However, in more sensitive tests of biological function of tRNA, as in studies of the transfer of phenylalanine from Phe-tRNA into polyphenylalanine, or into phenylalanine containing peptides of hemoglobin in an *in vitro* protein synthesizing system, only 70% of the transfer function is restored following both pronase and phenol treatments (*19*).

PURIFICATION OF tRNA BY CHROMATOGRAPHY OF AROMATIC N-BLOCKED DERIVATIVES OF AMINOACYL-tRNA

Introduction

While chromatography on the MASA column does not afford appreciable purification of tRNA, it is possible to achieve substantial purification through the use of aromatic *N*-blocked derivatives of aminoacyl-tRNA. Attachment of an aromatic substituent onto the amino group of aminoacyl-tRNA increases its affinity for methylated albumin-silicic acid, thus allowing it to be separated from the unsubstituted tRNA (*6*). The method is described in detail below.

Preparation of Unfractionated tRNA

The isolation and purification of amino acid-specific tRNA species by this method demands the availability of tRNA preparations with a high degree of purity. In addition, RNA preparations may contain trace amounts of bound nucleases that survive the exposure to phenol (*20*). Unless even trace quantities of nucleolytic activity are eliminiated from the starting materials, there is little prospect of ultimately recovering biologically active preparations of purified, amino acid-specific tRNA species. The procedure described below gives tRNA preparations having the requisite purity.

Escherichia coli K_{12} $W6RC^{rel}Met^-$ is grown at 37° with vigorous aeration in a medium consisting of 1% Difco yeast extract, 2.18% K_2HPO_4, 1.7% KH_2PO_4, and 1% glucose (*21*). Cells are harvested near the end of their logarithmic growth phase and washed with 0.9% NaCl. A modification of the Zubay (*22*) procedure is employed for tRNA isolation (*25*). The essential operations involve: (1) leaching the tRNA out of whole cells with phenol; (2) extraction of the leached material with 1 M NaCl; (3) differential precipitation with isopropanol of non-tRNA macromolecules from the NaCl-soluble fraction; (4) extraction of the tRNA solution with a mixture of aqueous phenol, $NaClO_4$ and chloroform; and (5) exposure of the tRNA solution to incubation at pH 8.8.

Throughout the purification of tRNA care should be taken not to introduce nuclease by using *sterile* buffer solutions and glassware. Unless otherwise noted, all operations are carried out at 4°. Freshly harvested cells (500 gm wet wt) are suspended in 1 liter of cold water, 660 ml of 70% phenol is added, and the mixture is shaken vigorously for 1 hr. Phenol (Mallinckrodt, 88% analytical grade reagent) is redistilled prior to use. The resulting emulsion is centrifuged at 15000 *g* for 30 min and the turbid upper aqueous phase carefully decanted and saved. The lower phenolic phase is discarded. The RNA is precipitated from the aqueous extract by the addition of 0.2 vol of 5 M NaCl and 2.5 vol of 96% ethanol. The precipitate is allowed to settle overnight at −15° after which as much as possible of the alcoholic supernatant is siphoned off and the remaining slurry centrifuged at 15000 *g* for 10 min. Excess supernatant is drained off by inverting the tubes and the pellet fraction is suspended in a total of 50 ml of ice-cold 1 M NaCl. The suspension is centrifuged and the sediment discarded (in a few cases, this initial NaCl-insoluble sediment is extracted once more with 25 ml of cold 1 M

NaCl). The RNA is precipitated by the addition of 2 vol of ethanol and the resulting sediment is washed with 75% ethanol. This washed pellet is then dissolved in 180 ml of 0.3 M sodium acetate (pH 7.0) and adjusted to a nucleic acid concentration of between 3.0 and 3.5 mg/ml. After warming to 20°, the solution is stirred and 0.54 vol of 2-propanol slowly added over a 15-min period at 20°. Precipitated material is collected by centrifugation at 20° and the supernatant fraction is saved. The pellet is taken up in one-half the original volume (90 ml) of acetate buffer, then again precipitated at 20° with 0.54 vol of 2-propanol. This second 2-propanol pellet is discarded. Both 2-propanol supernatants are combined, chilled to 4°, and mixed with a further 0.34 vol of chilled 2-propanol. After 10 min at 4°, precipitated tRNA is collected by centrifugation in the cold, washed with 75% ethanol, and dissolved in 10 ml of water. At this stage, solutions are mixed with 0.2 vol of 5 M $NaClO_4$, 0.5 vol of 88% phenol, and 0.1 vol of chloroform, and shaken for 20 min at 20°. The suspension is cooled to 4°, then centrifuged at 10,000 *g* for 5 min at 4°. The upper aqueous layer is removed and extracted twice more for 10 min at 20° with 0.5 vol of 88% phenol and 0.1 vol of chloroform. The RNA is precipitated from the final aqueous phase by adding 2.5 vol of cold 96% ethanol; the pellets so obtained are washed with 75% ethanol, dissolved in 20 ml of water, and dialyzed overnight against 0.001 M NaCl.

Certain precautions are advisable in preparing dialysis tubing for use. The tubings (Visking) are freed from heavy metal and ultraviolet-absorbing impurities (*23*) by boiling twice in 5% Na_2CO_3. The dialysis bags are then rinsed with water, boiled in 0.05 M EDTA (pH 8.0), and stored in 0.1 mM EDTA. Before use the dialysis bags are boiled and rinsed in distilled water. Contamination of the bags by nuclease from fingers is avoided by wearing gloves (*24*).

The tRNA solution is diluted with 1 vol of autoclaved 0.2 M Tris-HCl buffer (pH 8.8), then held at 37° over chloroform for 5 hr. Solutions are chilled to 4°, after which 0.4 vol of 1 M sodium acetate buffer (pH 5.0) and 2.5 vol of cold 96% ethanol are added. The precipitated material is recovered by centrifugation, washed with 75% ethanol dissolved in water and stored at −15°. Yields are approximately 400 mg of tRNA.

The use of frozen cell batches is avoided. Freshly harvested cells permit higher yields of RNA and contain lower levels of nucleases. The best preparations result from cultures harvested during exponential growth and washed with 0.9% NaCl prior to the phenol treatment. Another point is that the phenol phase is not reextracted with buffer as is done in many procedures. This step is omitted despite a lower tRNA recovery since its inclusion yields higher protein and ribonuclease contaminations. The NaCl extraction step is best carried out using the smallest possible volume of solvent. More numerous extractions and/or larger volumes than noted above lead to higher levels of non-tRNA polynucleotide in the final product. (It is best to assay the amino acid acceptor capacity of each of the NaCl-extracts. If the second NaCl extract shows a lower value than the first NaCl-extract it should be discarded). Finally, successful isopropanol fractionation is achieved only when crude starting material is dissolved in acetate buffer at concentrations between 3.0 and 3.5 mg/ml.

It is also imperative that each preparation of tRNA be assayed for nuclease content as described below and in Ref. *25*. If traces of nuclease activity are found, the preparation should be discarded. The limit of acceptable ribonuclease activity corresponds to a loss of 15% of amino acid acceptor activity of the tRNA after 24 hr of incubation in pH 8.8 buffer at 37°.

tRNA preparations contain traces of nonamino acid accepting RNA which are strongly bound to MASA and are eluted from the column at salt concentrations higher than that of the bulk of the tRNA, but similar to that of the aminoacyl-tRNA derivatives. To increase the purification and to obtain higher specific activity of the final product this material must be removed from the tRNA preparation. This is achieved by a preliminary passage of the unfractionated tRNA through the MASA column before aminoacylation and subsequent preparation of the aminoacyl-tRNA derivative is begun.

E. coli tRNA (200 mg) is dissolved in 6 ml of water and mixed with 10 ml of sodium acetate buffer 0.05 M (pH 5.5) containing 0.78 M NaCl and applied onto a MASA column (2 x 12 cm). The tRNA is eluted from the column with 0.05 M sodium acetate buffer (pH 5.5) containing 1.0 M NaCl and isolated by precipitation with 2.1 vol of ethanol. Following removal of this contaminant insignificant amounts of nonspecific binding are observed on the MASA column, as measured with (^{32}P) labeled tRNA (*9*). This is in contrast with other techniques of RNA column chromatography in which constant high levels of nonspecific binding are observed (*15, 26*) which persist after recycling of the preparation through the column.

Assay for Nuclease Contamination in tRNA

All the tRNA preparations are routinely assayed for the presence of nuclease contamination. Aliquots of the tRNA solution are diluted with one volume of autoclaved 0.2 M Tris-HCl buffer (pH 8.8), and held at 37° over chloroform for 25 hr. The tRNA is then isolated by ethanol precipitation and assayed for its capacity to accept valine or phenylalanine. tRNA preparations showing more than 15% decrease in their acceptor activity should be discarded.

Aminoacyl-tRNA Synthetase Preparation

The objective of this purification procedure is esterification of tRNA with only one type of amino acid. It is therefore desirable to work with purified aminoacyl-tRNA synthetase preparations. Furthermore, the enzyme preparation should not contain free amino acids. It is freed of amino acids by being dialyzed or passed through a Sephadex G-50 column *immediately before use*. It is also important that the preparation be freed of proteases that might release free amino acids during the incubation period. Finally, the preparation should be nuclease-free.

In addition the amino acid used for the esterification of tRNA must be devoid of traces of other amino acids. It is also possible to work with partially purified aminoacyl-tRNA synthetase preparations provided they are free of nucleases. However, most efficient purification is attained with more purified enzyme preparations. The assay for nuclease content of the enzyme preparation is performed by charging a nuclease-free tRNA sample with an amino acid (e.g., valine) discharging the resulting aminoacyl-tRNA (pH 8.8 for 5 hr) and then recharging the tRNA with the same amino acid. No loss of amino acid acceptance should be observed.

The following procedure leads to partially purified Val-, Phe-, and Met-tRNA synthetase preparations devoid of significant nuclease contamination. *E. coli* K_{12} W6RC^{rel} Met$^-$ cells are grown in a glycerol-lactate minimal medium (*27*) containing 0.01 M potassium phosphate buffer (pH 7.4) supplemented with 50 μg/ml of L-methionine at 37° with forced aeration. Cells are harvested in the middle of the logarithmic growth phase, washed once with 0.9% NaCl, and frozen. Two hundred grams of frozen cells are suspended in 600 ml of Tris-HCl buffer, (pH 7.8) containing 0.01 M $MgCl_2$ and 1 μg/ml of DNase at 4°. The suspension is homogenized in a Mantan-Gaulin homogenizer (Everett, Massachusetts) by recycling it three times at 4000 psi. The lyzate is centrifuged for 10 min at 15,000 *g* and the precipitate discarded. The supernatant fluid is centrifuged at 105,000 *g* for 2 hr. The 105,000 *g* supernatant is stirred and 0.1 vol of 10% streptomycin sulfate solution (freshly prepared and neutralized to pH 7.0) is added. After 30 min of stirring, precipitated material is removed by a 10-min centrifugation at 15,000 *g*, and the supernatant fraction adjusted to pH 8.0 with 0.2 N NaOH (this solution should not be frozen and the subsequent steps should be carried out immediately). To each 100 ml of the streptomycin supernatant 21 gm of ammonium sulfate is slowly added while mixing; after 20 min the mixture is centrifuged at 20,000 *g* for 30 min and the precipitate discarded. To the supernatant fluid additional ammonium sulfate is added (6.3 gr/100 ml of streptomycin supernatant) and the resulting precipitate is dissolved in 60 ml of 0.02 M Tris- HCl buffer, pH 7.4 dialyzed overnight against the same buffer, and saved (ammonium sulfate fraction *b*, serves as a source of Phe- and Met-tRNA synthetase). To the supernatant fluid additional ammonium sulfate is added (12.5 gr/100 ml of streptomycin supernatant) and the resulting precipitate is dissolved in 25 ml of 0.02 M potassium phosphate buffer, pH 7.5. The solution is then dialyzed overnight against the same buffer (ammonium sulfate fraction *c*, serves as a source of Val-tRNA synthetase).

Phe-tRNA Synthetase

To the ammonium sulfate fraction *b* is added: 0.2 vol of 0.1 M neutralized ATP solution, 0.2 vol of 0.1 M L-phenylalanine, 0.04 vol of 0.5 M $MgCl_2$, 0.14 vol of 0.05 M KF, and 0.06 vol of 0.5 M of K_2HPO_4. This mixture is incubated at 55° for 45 min, and denatured protein is removed by chilling and centrifugation. Saturated ammonium sulfate solution (1.6 vol) is now added to the supernatant fraction. The precipitate is collected by centrifugation, dissolved in 12 ml of 0.02 M Tris-HCl buffer, pH 7.4, and dialyzed overnight against the same buffer. Traces of residual nuclease are removed by chromatography through a 2 x 20-cm DEAE-cellulose column (*28*). The column is equilibrated with initial buffer consisting of 0.02 M potassium phosphate buffer, (pH 7.5), 0.02 M 2-mercaptoethanol and 0.001 M $MgCl_2$. The extract is applied, and the column is washed with 500 ml of the initial buffer. The enzyme is eluted with a buffer containing 0.25 M potassium phosphate (pH 6.5), 0.02 M 2-mercaptoethanol, and 0.001 M $MgCl_2$. Peak activity fractions are pooled, dialyzed against 0.01 M Tris-HCl buffer (pH 7.4), and dithiothreitol is then added to a final concentration of 0.25 mg/ml. The enzyme solution is divided into small aliquots and stored frozen at −15°.

Met-tRNA Synthetase

Ammonium sulfate fraction *b* is processed exactly as for Phe-tRNA synthetase except that the enzyme is incubated at 55° in the presence of L-methionine, and 2-mercaptoethanol is omitted from the buffer used for the DEAE-cellulose chromatography. In addition, dithiothreitol is not added to the final enzyme solution.

Val-tRNA Synthetase

For each 10 ml of the ammonium sulfate fraction *c* solution 0.8 ml of 1 M potassium phosphate buffer, (pH 7.5) is added and the mixture is held at 37° for 90 min and then chilled. The enzyme solution is then incubated at 55° for 45 min in the presence of L-valine, ATP, $MgCl_2$, KF, and K_2HPO_4 as already described. After being heated, undenatured protein is concentrated by $(NH_4)_2SO_4$ precipitation, dissolved in phosphate buffer, dialyzed, and passed through a DEAE-cellulose column as described above.

Preparation and Isolation of (^{14}C) Aminoacyl-tRNA

The reaction mixture for the preparative aminoacylation of tRNA with (^{14}C) valine contains 2 mM ATP, 5 mM KCl, 30 mM Tris-HCl buffer (pH 7.5), 2 mM $MgCl_2$, 5 mM 2-mercaptoethanol, 25 nmoles/ml of [^{14}C] valine (1.4×10^4 cpm/nmole), 2 mg/ml of unfractionated tRNA, and saturating amounts of enzyme. Again it should be stressed that just before use the enzyme preparation must be dialyzed for 2 hr at 4° against 0.01 M Tris-HCl buffer, pH 7.4, containing 0.01 M in 2-mercaptoethanol or passed through a Sephadex G-50 column. The reaction is carried out at 37° for 30 min, then stopped by chilling at 4°. For each 10 ml of reaction mixture, 1.2 ml of 1.0 M sodium acetate buffer (pH 5.5), 0.04 ml of glacial acetic acid, 1.0 ml of 5 M $NaClO_4$, 1.0 ml of chloroform, and 6.0 ml of 70% phenol is added with stirring. The mixture is shaken in the cold for 30 min and centrifuged. The aqueous upper phase is saved and RNA is precipitated with ethanol. The resulting pellets are washed with 96% ethanol and dissolved in 0.8 ml of 0.01 M acetate buffer (pH 5.5). Free valine and other low-molecular-weight substances are removed by gel filtration through a 1 x 25 cm Sephadex G-50 column using 0.005 M NaCl as eluent. Peak fractions are pooled, lyophilized to dryness in small aliquots of known RNA content, and stored at –15°. Other aminoacyl-tRNA compounds are similarly prepared.

Amino Acid Acceptance Assay

Unfractionated tRNA or purified $tRNA^{Val}$ preparations are assayed for valine acceptance activity according to a slight modification of the procedure of Berg et al. (*29*). The reaction mixture (0.1 ml) contains 30 μmole Tris-HCl buffer (pH 7.4), 0.2 μmole $MgCl_2$, 0.5 μmole KCl, 0.5 μmole 2-mercaptoethanol, 2.5 nmole of [^{14}C] valine (3.1×10^5 cmp/nmole), 2.0 nmole of each of the 19 other (^{12}C) amino acids, either 100 μg unfractionated tRNA or 2 μg purified $tRNA^{Val}$, and enough enzyme to obtain maximal charging. After a 15-min incubation at 37°, the assay tubes are chilled to 4°, mixed with 0.05 ml of carrier serum albumin solution (4 mg/ml), and 3 ml of ice-cold 5% trichloroacetic acid is added. After 10 min at 4° the precipitates are collected onto membrane filters, washed twice with 5-ml portions of 2.5% trichloroacetic acid, dried, placed in 10 ml of toluene scintillation fluid, and radioactivity counted in a scintillation spectrophotometer.

For methionine acceptance assay, the mixture (0.1 ml) contains 3.0 μmole of Tris-HCl buffer (pH 7.5), 0.2 μmole ATP, 1.0 μmole $MgCl_2$, 1.1 nmole (^{14}C) methionine (5.9×10^5 cpm/nmole), 2.0 nmole of each of the 19 other (^{12}C) amino acids, 100 μg unfractionated tRNA or 7 μg purified $tRNA^{Met}$, and enzyme.

Phenylalanine acceptance assay mixtures (0.05 ml) contain 2.7 μmole of Tris-HCl buffer, pH 7.8; 0.1 μmole of ATP; 0.1 μmole of GSH (neutralized); 1.0 μmole of $MgCl_2$; 0.15 nmole of (^{14}C) phenylalanine (4×10^5 cpm/nmole); 0.2 nmole of each of the 19 other (^{12}C) amino acids; 4–8 μg of unfractionated tRNA or about 0.4 μg of purified $tRNA^{Phe}$, and enzyme.

Just before use, the frozen concentrated Val- and Phe-tRNA synthetases are thawed and diluted in 0.02 M potassium phosphate buffer (pH 7.5), made 0.05 M in 2-mercaptoethanol, and the Met-tRNA synthetase diluted in 0.01 M Tris-HCl buffer (pH 7.4).

Nucleic Acid Determination

The concentration of tRNA is estimated by measuring the absorption at 260 nm. One milligram tRNA contains an A_{260} of 30 units in 0.01 N NaOH (*29*) or, alternatively, an A_{260} of 25 units in 0.5 M NaCl. The values measured in NaOH are more accurate, and less variable than those obtained in NaCl.

Preparation of Aromatic N-Blocked Aminoacyl-tRNAs

For the purification of Phe-tRNA and Tyr-tRNA good results are obtained with *N*-carbobenzyloxy-[^{14}C]Phe-tRNA (*6*). For other aminoacyl-tRNAs, *N*-diphenylacetylaminoacyl-tRNA provides superior purification (*8*).

Preparation of *N*-Carbobenzyloxy-[^{14}C]Phe-tRNA

[^{14}C]Phe-tRNA (200 mg) is dissolved in 40 ml of 0.5 M KPO_4 buffer (pH 7.0) in a 250-ml glass-stoppered Erlenmeyer flask and stirred on a shaker at 0° for 30 min. At 0, 10, and 20 min, 2.0-ml aliquots of 100% carbobenzyloxy chloride solution (Miles-Yeda, Rehovot, Israel) are added. The residual carbobenzyloxy chloride is removed by adding 80 ml of ethyl ether and shaken briefly in a separatory funnel. The aqueous layer is removed, leaving behind the ethyl ether layer and the interphase. The aqueous layer is then reextracted with ethyl ether until all traces of carbobenzyloxy chloride have been removed. Ethanol (2.1 volume) and 5 M NaCl (0.2 volume) are added and the precipitate is centrifuged and dissolved in 0.05 M sodium acetate buffer (pH 5.5) containing 0.78 M NaCl. Under these conditions, the amino acid acceptance of tRNA is not impaired. However, if the reaction is run at pH 8.0, over 40% of the amino acid acceptance capacity of the tRNA is destroyed.

Preparation of *N*-Diphenylacetyl Aminoacyl-tRNA

The use of diphenylacetyl chloride for the preparation of *N*-diphenylacetyl aminoacyl-tRNAs is preferred for the purification of the tRNAs specific for nonaromatic amino acids.

Preparation of Diphenylacetyl Chloride. This compound is synthesized according to Staudinger (*30*). Diphenylacetic acid (8.5 gm) and phosphopentachloride (8.3 gm) are mixed together in a round-bottom flask and 25 ml of benzene is added. The mixture is stirred until most of the solid material is in solution. The solvent is then removed by distillation at 60° under reduced pressure. The residue is recrystallized from hot petroleum ether and dried *in vacuo* over NaOH.

Preparation of *N*-Diphenylacetyl-[^{14}C] Val-tRNA. To 77 ml of [^{14}C] Val-tRNA (200 mg), 11.7 ml of 1.0 M K-phosphate buffer (pH 6.0) and 1.17 ml of 1.0 M $MgCl_2$ are added. The solution is stirred on a wrist shaker at 0° for 30 min. At 0, 10, and 20 min 9.0-ml aliquots of the diphenylacetyl chloride solution are added (1.0 gm reagent dissolved in 27 ml of dioxane).

At the end of the reaction the tRNA is precipitated with 3 vol of ethanol and the resulting pellet dissolved in 0.05 M Na-acetate buffer, pH 5.5. The final molarity of the solution is adjusted to 0.5 M by the addition of water and checked on a refractometer. The extent of esterification is assayed by hydrolysis with purified *N*-acetylaminoacyl-tRNA hydrolase (*31, 32*) or by alkaline hydrolysis and paper chromatography of the products (butanol : acetic acid : H_2O, 78 : 5 : 17). The reaction goes to over 90% of completion.

Column Chromatography

It must be stressed that when attempting to recover biologically active tRNA it is necessary to prepare the column using a purified source of albumin. The removal of nucleases from the albumin preparation increased the yield of biologically active tRNA from the column. A considerable amount of nuclease activity (75%) can be removed from the albumin prior to its methylation by passing the protein through a DEAE-cellulose column in the presence of Mg^{2+} (*28*). When pancreatic ribonuclease-like activity is assayed in the DEAE column fractions, most of the activity is found in the initial wash (*33*) (0.025 M Tris-HCl buffer, pH 7.8). Additional smaller peaks of nuclease-like activity are found using highly [^{3}H]-labeled poly C as a stubstrate (*33, 34*). A broad low peak of activity is also found corresponding to the $A_{280\ nm}$ profile of the serum albumin itself. This may represent the nonspecific esterase activity which has been reported for albumin (*35*).

Following methylation, most of the ribonuclease-like activity disappears when [^{3}H]-labeled poly C is used as the substrate in the assay.

Chromatography of *N*-Carbobenzyloxy-[^{14}C] Phe-tRNA

A new MASA column is used for these experiments. The column is washed with 2 M NaCl containing 0.05 M sodium acetate buffer (pH 5.5) and then equilibrated with 0.78 M NaCl containing 0.05 M sodium acetate buffer (pH 5.5).

The *N*-carbobenzyloxy-[^{14}C]Phe-tRNA (91 mg in 13 ml of 0.78 M NaCl containing 0.05 M sodium acetate buffer, pH 5.5) is applied onto a MASA column (2 x 12 cm). The bulk of the RNA is eluted with a steep linear sodium chloride gradient formed from 60 ml each of 0.78 M and 1.78 M NaCl in 0.05 M sodium acetate buffer (pH 5.5). At the end of this gradient the *N*-carbobenzyloxy-[^{14}C]phenylalanyl-tRNA is still bound to the column. The column is further eluted with 50 ml of 1.78 M NaCl-sodium acetate, a small peak of $A_{260\ nm}$ material appears together with most of the ^{14}C-labeled derivative. The labeled material coincides precisely with the optical density profile. The small residual amount of *N*-carbobenzyloxy [^{14}C]Phe-tRNA is eluted with a linear gradient formed from 60 ml each of 1.78 M and 4.0 M NaCl in sodium acetate buffer (Fig. 3). The column is run at 16° and 2.5-ml fractions are collected. Peak fractions are combined which represent about 70% of the initial amino acid specific tRNA.

Chromatography of *N*-Diphenylacetyl-Val-tRNA and *N*-Diphenylacetyl-Met-tRNA

The procedure is similar to that used with the *N*-carbobenzyloxy derivatives. The *N*-diphenylacetyl aminoacyl-tRNA solution is placed on a new MASA column and eluted with a gradient formed from 100 ml each of 0.78 M and 1.1 M NaCl in 0.05 M sodium acetate buffer (pH 5.5). At the end of this gradient the column is further washed with 1.1 M NaCl-acetate until the $A_{260\ nm}$ is below 0.2 (too extensive washing initiate elution of the *N*-diphenylacetyl aminoacyl-tRNA). A linear gradient formed from 60 ml each of 1.1 M NaCl and 2.0 M NaCl in sodium acetate buffer is then applied. The *N*-diphenylacetyl aminoacyl-tRNA is eluted between 1.3 to 1.5 M NaCl. It is useful to rechromatograph this material on a second MASA column. This provides a 5–25% increase in specific activity of the labeled derivative tRNA by the criterion of cpm/$A_{260\ nm}$.

tRNA^{Phe} can not be purified by this procedure since *N*-diphenylacetyl-Phe-tRNA is strongly bound to the MASA column and is only partially eluted at high NaCl concentration. For this purpose the *N*-carbobenzyloxy derivative is used.

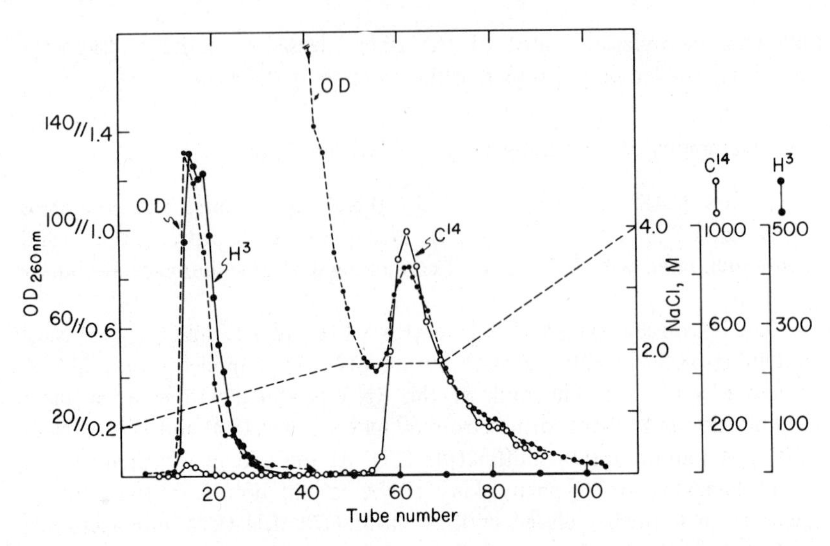

Figure 3. MASA chromatography of *N*-carbobenzyloxy-[^{14}C]phenylalanyl-tRNA and [^{3}H]phenylalanyl-tRNA. A new MASA column was used for these experiments. The column was washed with 4 M NaCl containing 0.05 M sodium acetate buffer (pH 5.5) and then equilibrated with 0.78 M NaCl containing 0.05 M sodium acetate buffer (pH 5.5). *E. coli* A-19 tRNA was first purified by passing it through a MASA column (2 x 12 cm) to remove traces of high-molecular-weight RNA that are strongly held to the column. *E. coli* A-19 tRNA (200 mg) was dissolved in 6 ml of water and mixed with 10 ml of sodium acetate buffer (0.05 M, pH 5.5) containing 0.78 M NaCl and applied on a MASA column. The tRNA was eluted from the column with 0.05 M sodium acetate buffer (pH 5.5) containing 1.8 M NaCl and isolated by ethanol precipitation. The isolated tRNA was then charged with [^{14}C]phenylalanine. Before use, the phenylalanine tRNA synthetase preparation was freed of any amino acids by Sephadex G-50 gel filtration. The [^{14}C]Phe-tRNA was precipitated with ethanol, dissolved in 40 ml of 0.5 M KPO_4 buffer (pH 7.0), and treated with carbobenzyloxy chloride. *N*-Carbobenzyloxy-[^{14}C]phenylalanyl-tRNA (84,000 cpm) (91 mg) was mixed with 2 mg of [^{3}H]phenylalanyl-tRNA (27,000 cpm) in 13 ml of 0.78 M NaCl containing 0.05 M sodium acetate buffer (pH 5.5). This solution was placed on a new MASA column (2 x 12 cm). The RNA was eluted with a linear gradient formed from 60 ml each of 0.78 M and 1.78 M NaCl solutions in 0.05 M sodium acetate buffer (pH 5.5). The column was further eluted with 50 ml of 1.78 M NaCl-sodium acetate and then a linear gradient of 60 ml each of 1.78 and 4.0 M NaCl in sodium acetate buffer. Fractions (2.5 ml) were collected at 16° (*6*).

Liberation of Free tRNA from N-Blocked Aminoacyl-tRNA

It had been observed that the ester link between *N*-carbobenzyloxy phenylalanine and tRNA is more resistant to alkaline hydrolysis than the ester bond in Phe-tRNA. In this respect the blocking of the amino group of phenylalanine with a carbobenzyloxy group has the same effect on the stability of the ester linkage as the introduction of a polypeptide group (*36-39*). The same effect is operative with an N-blocking group on any aminoacyl-tRNA. Free tRNA can be liberated from the stable N-blocked aminoacyl group by digestion with pronase or by incubation in alkaline buffer.

Pronase Digestion

In our initial work (*6*) pronase digestion was used to liberate tRNA from *N*-carbobenzyloxy-Phe-tRNA (*25*). A major difficulty in the pronase digestion method is avoiding RNase contamination during the incubation with the protease. Trace amounts of RNase bound to tRNA or introduced with the pronase digestion cause significant inactivation of the purified $tRNA^{Phe}$

Incubation with Tris buffer at pH 8.8

An alternative method to the pronase digestion is incubation with Tris buffer at an alkaline pH. This procedure is simpler and for recent work has been used routinely. This is now the preferred procedure. The N-blocked aminoacyl-tRNA derivative is incubated with 0.1 M Tris-HCl (pH 8.8), 0.01 M $MgCl_2$, and 0.5 M KCl at 37°. The addition of KCl and $MgCl_2$ enhances the rate of hydrolysis (*40*). Before hydrolysis the N-blocked aminoacyl-tRNA solution should be extracted with phenol. If this phenol extraction is omitted, the biological activity of the tRNA decreases.

In a typical experiment, the pooled fractions of N-blocked aminoacyl-tRNA are concentrated by ultrafiltration through a UM-10 membrane (Diaflow ultrafiltration cell, Amicon Corp., Lexington, Mass.) to a volume of 4 ml. The concentrated solution is diluted with 40 ml of water and then concentrated again to about 2.0 ml. The solution is further concentrated to 0.5 ml by flash evaporation and precipitated with 2.1 volume of ethanol at −15°. The precipitate is taken up in 0.5 ml of H_2O and mixed together with 0.1 ml of 5 M $NaClO_4$; 0.1 ml chloroform and 0.25 ml of 80% phenol. The mixture is shaken at 20° for 60 min, centrifuged, the aqueous phase removed, and the N-blocked aminoacyl-tRNA derivative is precipitated by the addition of 2.1 vol of ethanol at −15°. The N-blocked aminoacyl-tRNA derivative is then taken up in 0.5 ml of water and discharged by incubation with 0.1 M Tris-HCl buffer (pH 8.8) 0.01 M $MgCl_2$, and 0.5 M KCl at 37°. The length of incubation depends on the type of aminoacyl-tRNA used and must be determined in pilot experiments. Thus, *N*-diphenylacetyl-Val-tRNA is incubated for 6 hr while 1 hr suffices to cleave the ester bond of *N*-diphenylacetyl-Met-tRNA. Following incubation the solution is neutralized by the addition of 0.1 ml of 1M sodium acetate buffer (pH 4.5) and the RNA precipitated by the addition of 0.2 vol of 5M NaCl and 2.1 vol of ethanol. The RNA precipitate is then taken up in water and assayed for amino acid acceptance activity.

Criteria for Purity of the tRNA Product

The purity of the N-blocked aminoacyl-tRNA derivatives eluted from the MASA column can be estimated from the specific activity, cpm/mg RNA. Table 1 demonstrates that the values, expressed as nmoles/mg RNA, are 34, 31, and 28 for the N-blocked derivatives of Phe-, Val-, and Met-tRNA, respectively.

However to liberate free tRNA from the stable N-blocked aminoacyl-tRNA it is necessary to use relatively drastic procedures, as outlined in the previous section. Both pronase digestion and incubation in alkaline buffer cause some loss of amino acid acceptor activity of the liberated tRNA. Table 1 shows that pronase digestion of *N*-carbobenzyloxy Phe-tRNA causes a decrease in biological activity from 34 to 18 nmoles/mg RNA. Incubation of *N*-diphenylacetyl-Val-tRNA in Tris buffer at pH 8.8 causes a decrease of from

TABLE 1. Amino Acid Acceptance of Purified tRNAs

Sample	Specific activity[a] (nmoles/mg RNA)	Amino acid acceptor activity (nmoles/mg RNA)					
		Phenylalanine	Valine	Methionine	Lysine	Isoleucine	Mixture of 15 [^{14}C]-labeled amino acids[b] cpm/μg RNA
Initial tRNA preparation	–	1.10	2.90	0.96	0.55	–	3900
Purified tRNA^{Phe}[c]	34	18	0.04	0.03	0.11	–	300
Initial tRNA preparation	–	1.13	2.60	1.94	0.65	1.50	4050
Purified tRNA^{Val}[d]	31	0.11	20	0.06	0.12	0.14	498
Purified tRNA^{Met}[e]	28	0.09	0.17	28	0.06	0.15	518

[a] Specific activity of the N-blocked aminoacyl-tRNA derivatives prior to discharge.
[b] Assay with crude aminoacyl-tRNA synthetase and a mixture of 15-L-[^{14}C] amino acids together with unlabeled phenylalanine, valine, or methionine.
[c] Purified via *N*-carbobenzyloxy-Phe-tRNA and discharged by pronase digestion (*6*).
[d] Purified via *N*-diphenylacetyl-Val-tRNA and discharged by incubation in alkaline buffer (*8*).
[e] Purified via *N*-diphenylacetyl-Met-tRNA and discharged by incubation in alkaline buffer (*8*).

31 to 20 nmoles/mg RNA. No loss in acceptor activity of $tRNA^{Met}$ is observed. In all probability the initial higher acceptor activity of the tRNA preparations represented the more accurate values for the purity of the tRNA.

Sequential Purification of Additional tRNA Species

For the sequential purification of additional amino acid specific tRNAs from the same batch of tRNA, the peak A_{260} tubes eluted from the MASA column are combined. The RNA is precipitated by adding 2.1 vol of ethanol. The precipitate is taken up in H_2O and extracted from 60 min at room temperature with 0.5 vol of 80% phenol, 0.2 vol of 5 M sodium perchlorate, and 0.1 vol of chloroform. The suspension is spun and the aqueous phase is removed and the RNA precipitated with 2.1 vol of ethanol. The precipitate is dissolved in H_2O. The tRNA is then discharged by incubation with 0.5 M KCl, 0.01 M $MgCl_2$ 0.1 M Tris (pH 8.8) at 37°C for 3 hr. The solution is brought to neutral pH by addition of 1 M sodium acetate buffer pH 4.5 and dialyzed for 18 hr at 4°C against 50 H_2O. The tRNA is then precipitated with 2.1 vol of ethanol and 0.2 vol 5M NaCl, and the pellet taken up in 5.0 ml water. The dialysis step is then repeated for 60 min at 4°C. This dialysis step removes phosphate ions which interfere with the aminoacylation reaction. The tRNA is now ready for charging with a second amino acid and the subsequent purification of the corresponding tRNA (*8*).

Concluding Remarks

The use of N-blocked aminoacyl-tRNA derivatives for the purification of amino acid specific tRNA is a relatively easy and rapid procedure. The yield is high and with some modifications it can be adapted as a general procedure for any tRNA species. However, special precaution will be required for highly labile tRNA ester bonds such as found with glycyl and arginyl-tRNAs (*41*) and cysteinyl-tRNAs (*42*).

The procedure in many cases yields a product with a specific activity sufficient for oligonucleotide fingerprint and sequence work. The insignificant level of nonspecific binding of RNA to the MASA column affords a distinct advantage for these kinds of investigations. The limitations of the system are its dependence on the availability of a purified aminoacyl-tRNA synthetase preparation. It should also be recalled that traces of methylated albumin bind tenaciously to tRNA (*19*) and may modify some of its biological properties. There is a persistent loss of specific activity concomitant with the liberation of free tRNA from the N-blocked derivative

(2–30%). However, such loss of activity is of no significance for fingerprint and sequence work.

Another limitation of the system is that it does not separate isoaccepting tRNAs. By combining this procedure with another separatory technique such as column chromatography or countercurrent distribution, one can obtain a preparation with a purity of over 90%, as shown by amino acid acceptor activity. Such efforts will be necessary in any case if different isoaccepting tRNAs are to be separated.

Acknowledgment

These studies were performed at the Weizmann Institute of Science, Rehovot, Israel and supported, in part, by grants from NIH. During the course of the studies Robert Stern was a special post-doctoral fellow of the National Institutes of Health, U.S. Public Health Service (1-F3-CA-28, 111-01).

References

1. Lerman, L. S. 1955. *Biochim. Biophys. Acta,* **18**: 132.
2. Mandell, J. D., and A. D. Hershey. 1960. *Anal. Biochem.,* **1**: 66.
3. Sueoka, N., and T. Yamane. 1962. *Proc. Natl. Acad. Sci., U.S.,* **48**: 1454.
4. Okamoto, T., and Y. Kawade. 1963. *Biochem. Biophys. Res. Commun.,* **13**: 324.
5. Littauer, U. Z., M. Revel, and R. Stern. 1966. *Cold Spring Harbor Symp. Quant. Biol.,* **31**: 501.
6. Stern, R., and U. Z. Littauer. 1968. *Biochemistry,* **7**: 3469.
7. Revel, M., and U. Z. Littauer. 1965. *Biochem. Biophys. Res. Commun.,* **20**: 187.
8. Ellman-Zutra, L., A. Patchornik, and U. Z. Littauer, manuscript in preparation.
9. Methionin-starved tRNA refers to the RNA extracted from a relaxed methionine requiring mutant of *E. coli* (K_{12}W6, A-19, or G-15) grown in a medium deficient in methionine. Methyl-deficient tRNA refers only to those species which contain few or no methyl groups. Methionine-starved tRNA is an approximately equal mixture of methyl-deficient tRNA and fully methylated normal tRNA.
10. Stern, R., L. Ellman-Zutra, and U.Z. Littauer. 1969. *Biochemistry,* **8**: 313.
11. Littauer, U. Z., and R. Stern. 1968. In L. O. Fröholm and S. G. Laland (Eds.), *Proc. Fourth. Mtg. Federation European Biochem. Soc., Oslo,* Symposium on Structure and Function of Transfer RNA and 5S RNA, Universitets-forlaget, Oslo, p. 93.
12. Leder, P., and H. Bursztyn. 1966. *Proc. Natl. Acad. Sci., U.S.,* **56**: 1579.
13. Carbon, J., and J. B. Curry. 1968. *Proc. Natl. Acad. Sci., U.S.,* **59**: 467.
14. Gillam, I., S. Millward, D. Blew, M. von Tigerstrom, E. Wimmer, and G. M. Tener. 1967. *Biochemistry,* **6**: 3043.
15. Vold, B. S. 1969. *Biochem. Biophys. Res. Commun.,* **35**: 222.
16. Ingle, J., and J. L. Key. 1968. *Biochem. Biophys. Res. Commun.,* **30**: 711.
17. Aubert, M., J. F. Scott, M. Reynier, and R. Monier. 1968. *Proc. Natl. Acad. Sci., U.S.,* **61**: 292.
18. Goldin, H., and I. I. Kaiser. 1969. *Biochem. Biophys. Res. Commun.,* **36**: 1013.

19. Stern, R., F. Gonano, E. Fleissner, and U. Z. Littauer. 1970. *Biochemistry,* 9: 10
20. Littauer, U.Z., and M. Sela. 1962. *Biochim. Biophys. Acta,* **61**: 609.
21. Littauer, U. Z., and A. Kornberg. 1957. *J. Biol. Chem.,* **226**: 1077.
22. Zubay, G. 1962. *J. Mol. Biol.,* **4**: 347.
23. Littauer, U. Z., and H. Eisenberg. 1959. *Biochim. Biophys. Acta,* **32**: 320.
24. Holley, R. W., J. Apgar, and S. H. Merrill. 1961. *J. Biol. Chem.,* **236**: PC 42.
25. Littauer, U. Z., S. A. Yankofsky, A. Novogrodsky, H. Bursztyn, Y. Galenter, and E. Katchalski. 1969. *Biochim. Biophys. Acta,* **195**: 29.
26. Roy, K. L., and D. Söll. 1968. *Biochim. Biophys Acta,* **161**: 572.
27. Hershey, A. D., and M. Chase. 1952. *J. Gen. Physiol.,* **36**: 39.
28. Muench, K. H., and P. Berg. 1966. In G. L. Cantoni and D. R. Davies (Eds), *Procedures in nucleic acid research,* Harper & Row, New York, Vol. I, p. 375.
29. Berg, P., F. H. Bergmann, E. J. Ofengand, and M. Dieckmann. 1961. *J. Biol. Chem.,* **236**: 1726.
30. Staudinger, H. 1911. *Ber. Deut. Chem. Ges.,* **44**: 1619.
31. Cuzin, F., N. Kretchmer, R. E. Greenberg, R. Hurwitz, and F. Chapeville. 1967. *Proc. Natl. Acad. Sci., U.S.,* **58**:2079.
32. Fogel, Z., A. Zamir, and D. Elson. 1968. *Proc. Natl. Acad. Sci., U.S.* **61**: 701.
33. Stern, R. 1969. unpublished observations.
34. Zimmerman, S. B., and G. Sandeen. 1965. *Anal. Biochem.,* **10**: 444.
35. Peterson, E. A. 1969. personal communication.
36. Simon, S., U. Z. Littauer, and E. Katchalski. 1964. *Biochim. Biophys. Acta,* **80**: 169.
37. Littauer, U. Z., S. Simon, and E. Katchalski. 1964. *Symp. Nucl. Acids. Hyderabad,* Saraswaty Press, Calcutta, p. 246.
38. Katchalski, E., S. Yankofsky, A. Novogrodsky, Y. Galenter, and U. Z. Littauer. 1966. *Biochim. Biophys. Acta,* **123**: 641.
39. Yankofsky, S. A., S. Yankofsky, E. Katchalski, and U. Z. Littauer. 1970. *Biochim. Biophys. Acta.* **199**: 56.
40. Novogrodsky, A. 1971. *Biochim. Biophys. Acta,* **228**: 688.
41. Coles, N., M. W. Bukenberger, and A. Meister. 1962. *Biochemistry* **1**: 317.
42. Peterkofsky, A., and C. Jesensky. 1969. *Biochemistry.* 8: 3798.

COUNTERCURRENT DISTRIBUTION OF TRANSFER RIBONUCLEIC ACID*

B. P. DOCTOR
DIVISION OF BIOCHEMISTRY
WALTER REED ARMY INSTITUTE OF RESEARCH
WALTER REED ARMY MEDICAL CENTER
WASHINGTON, D. C.

Since most of the transfer ribonucleic acids from a given species have similar molecular weight and similar net charge, in the early stages, it was thought to be difficult if not impossible to separate these molecules effectively. The use of countercurrent distribution procedures furnished the first practical method for the fractionation of these molecules. These procedures were successfully employed by several investigators for the separation and purification of tRNAs isolated from yeast, *E. coli,* and rat liver (*1-25*). The aim of this chapter is to fully describe these procedures including the modifications which were introduced in recent investigations. Using these procedures by themselves or in combination with other available chromatographic procedures it is possible to fractionate grams of tRNA and obtain up to and more than 100 mg of purified specific tRNA.

* A portion of this chapter is taken from a previously published article by the author in "Methods in Enzymology," Vol. XII A (L. Grossman and K. Moldave, Eds.). Copyright ©1967 by Academic Press, New York. Reprinted with permission.

FRACTIONATION OF tRNAs BY COUNTERCURRENT DISTRIBUTION

Countercurrent distribution procedures have been employed successfully by several investigators for the separation and purification of tRNAs isolated from yeast, *E. coli,* and rat liver (*1–25*). It is often possible to separate several tRNAs specific for a single amino acid in a given species (isoaccepting tRNAs). One advantage of countercurrent distribution techniques is that by the use of theoretical calculations, the extent of purity of a single homogenous tRNA can be determined. Finally, once the procedure is worked out, the method is quite simple to use. There are three solvent systems that are commonly used, which are presented here. In order to obtain separated or purified fractions of specific tRNA by these procedures it is essential that no amino acid acceptor activity be lost during the process of separation. The major cause of loss of activity appears to be due to the presence of "nucleases" as contaminants in crude tRNA preparations (*26*). It is thus essential that if a nuclease is found to be present, the tRNA preparation to be used be freed of the contaminant. This can be done by an additional phenol extraction.

Phosphate Buffer Solvent System (*27*)

All the procedures presented here are formulated for a 200-tube apparatus, each tube having the capacity to hold 10 ml of each phase. If an apparatus having different dimensions is employed, corresponding modifications according to its dimensions should be adopted. Increase or decrease of sample quantity relative to the volume of the solvent system does not appear to have an appreciable effect on the pattern of distribution. However, this solvent system is quite sensitive to changes in temperature, apparently due to the change in composition of the solvent system. It is thus important to perform the experiment in a constant temperature area (23° ± 1°).

Dipotassium hydrogen phosphate (550 gm) and sodium dihydrogen phosphate monohydrate (850 gm) are dissolved by addition of 4000 ml glass distilled water (final volume 4470 ml). To this buffer solution 300 ml formamide and 1300 ml isopropanol are added and the contents are shaken vigorously to equilibrate the two phases. The two-phase solvent system thus formed is allowed to equilibrate overnight at 23° ± 1°. The ratio of volumes of upper phase to lower phase is approximately 1 : 1.

The 200-tube countercurrent fractionator apparatus is filled with 10 ml each of the previously separated upper and lower phases. Tube no. 200 is connected to a tube no 1. with a connecting tube which is filled with 10 ml of each phase. Approximately 2 gm of yeast or rat liver or 1 gm of *E. coli* tRNA is dissolved in 252 ml of the phosphate buffer described above. When all the tRNA is dissolved, 17 ml formamide, 73.5 ml isopropanol, and 18 ml of lower phase are added very slowly (dropwise) and the contents are shaken to equilibrate the sample. The phases are separated and a small amount of insoluble material at the interface is discarded. The contents of the first 20 tubes are emptied and replaced with 10 ml of each phase containing the tRNA sample (approximately 18 tubes). The flask and separatory funnel used in the preparation of the sample are rinsed with enough upper and lower phase to fill tubes 19–20 with 10 ml of each phase. The machine is set for 200 transfers (2 min or 15 shakes and 4 min settling time) and the distribution is performed.

Upon completion of 200 transfers, the machine is emptied and the contents of every 5 tubes across the distribution train are pooled so as to obtain 40 fractions. To determine the RNA content of each fraction a small aliquot of the upper and lower phases are withdrawn from each fraction and diluted to measure the absorbance at 260 nm. Corresponding amounts of each phase are used as blanks.

The material is then transferred to a 500-ml separatory funnel and extracted with 100 ml of anhydrous peroxide-free ether. The lower phase is transferred directly to another separatory funnel. To the lower phase is added 150 ml of water (thus adjusting the concentration of phosphate to approximately 0.4 M) and an aqueous solution of cetyltributyl ammonium bromide (Cetavlon) (*28, 29*) and the contents are shaken gently. The amount of cetavlon added should be 5–6 times the amount of tRNA on w/w basis. After allowing the mixture to stand at room temperature for 20–30 minutes, 50 ml of ether is added and the contents are shaken. After 10 min, the quaternary complex of tRNA floats at the interface and is recovered by carefully draining the lower phase and collecting the interface material in a small test tube. The remainder of the liquid from the interface material is removed by aspiration or centrifugation.

The tRNA is then converted to the sodium salt by adding 1–2 ml of 2 M NaCl. To precipitate tRNA, 2.5 vol of cold 95% ethanol is added. After the samples are allowed to stand in the cold to complete the precipitation of tRNA, they are centrifuged, washed once with cold 80% ethanol and twice with cold 95% ethanol, and dried under vacuum. The recovery of A_{260} nm absorbing material and amino acid acceptor activity from the samples containing 0.05 mg/ml or more of tRNA in 0.4 M phosphate is better than 95%. At the lower concentrations of tRNA the recovery is slightly less. The tRNA fractions thus obtained are dissolved in distilled water and used for amino acid incorporation assay using one or more of the several available procedures.

Many species of tRNA isolated from *E. coli* contain 4-thiouridylic acid.

Recently, it has been shown that cetavlon reacts with this thio nucleotide in the presence of ultraviolet light (*30*). In our attempt to isolate *E. coli* tRNA fractions from the countercurrent distribution solvent systems, we have experienced the loss of 335 nm absorbance of tRNA when the fractions were exposed to cetavlon for long periods of time in laboratories fitted with flourescent lights. Whether the loss of 335 nm absorbance is accompanied by the loss of amino acid acceptor activity of any tRNA species or not, is being investigated at the present time. Preliminary results indicate that both these phenomena are not affected in most cases. However, there is some element of doubt. Therefore, an alternative procedure for the isolation of tRNA from the fractions is presented here.

The contents of every 5 tubes across the train are pooled, thus obtaining 40 fractions. Each fraction is extracted with equal volume of ether. The lower phase is transferred to another 500-ml separatory funnel and 35 ml of 2-methoxyethanol is added and the contents are shaken. The lower phase is drained off and discarded. To the upper phase, 100 ml of *n*-butanol and 150 ml of ether is added and the contents shaken. A small volume of lower phase containing tRNA samples are dialyzed in the cold against two or more changes of water. The samples are concentrated to a small desired volume by evaporation at 37° under vacuum using rotary evaporator. The concentrated samples are stored frozen at −20°.

For the presentation of results, a plot is made showing the fraction number or transfer number on the ordinate and milligrams or A_{260} nm units per fraction of tRNA on the left abscissa. Then from the amino acid incorporation assay results, amino acid acceptor activity equivalent to starting tRNA is calculated and expressed on the right abscissa. Such a plot clearly shows the extent of purification achieved by the countercurrent distribution for each tRNA fraction. A representative result obtained for yeast tRNA by Apgar et al., (*8*), is shown in Fig. 1a and for *E. coli* tRNA, is shown in Fig. 1b.

The solvent system described here contains 1.25 M phosphate buffer. Solvent systems containing 1.90 M and 1.7 M phosphate buffer are also employed (*4, 14*). These two solvent systems give slightly better resolutions but on a long run phosphate crystallizes in the apparatus, thus changing the partition coefficient. It is still useful to use the different molarity of phosphate buffer in conjunction with other solvent systems, if one wishes to purify a given amino acid specific tRNA. It should also be mentioned that increasing the formamide concentration in this solvent system increases the partition coefficient and the reverse is true for isopropanol concentration. Thus one can develop a wide range of solvent systems (as may be required for any specific purpose) by simply varying either the phosphate buffer, or formamide, or isopropanol concentration.

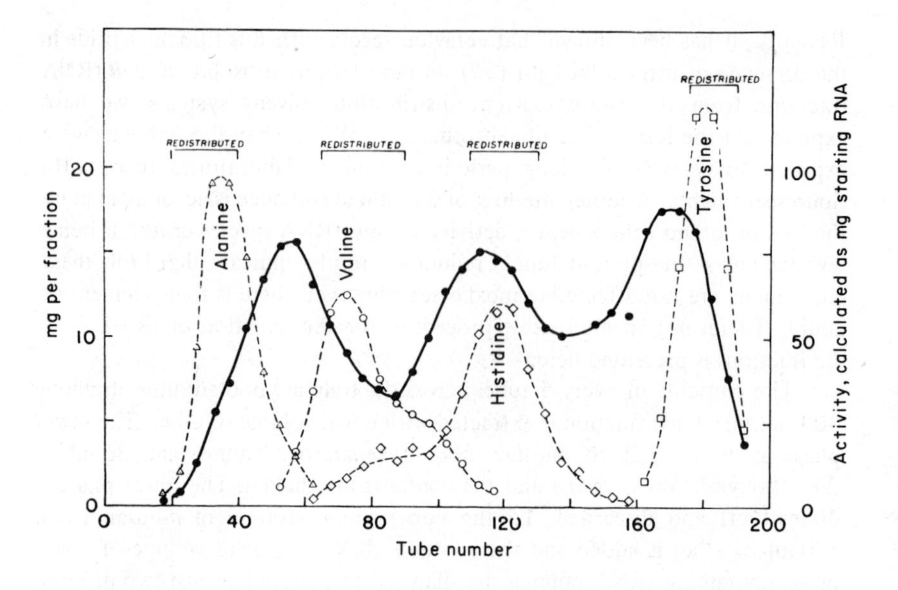

Figure 1A. Two hundred-transfer countercurrent distribution of 500 mg of yeast tRNA in a solvent system similar to the one described under procedure 1.——, Milligrams of RNA; △-△ alanine acceptor activity; ○-○, valine acceptor activity; □-□, histidine acceptor activity; ◇-◇, tyrosine acceptor activity. Reproduced from J. Apgar, R. W. Holley, and S. H. Merrill, 1962. *J. Biol. Chem.* **237**: 796 with permission.

Ammonium Sulfate Solvent System

This solvent system is a modification of the solvent system described by Kirby (*31*). The pattern of distribution is essentially the same as the one obtained with the solvent system described in the first procedure. However, the solubility of tRNA is greater in this solvent system than in the phosphate system. Also, this system is less sensitive to variations in temperature.

Twelve hundred grams of ammonium sulfate is dissolved in approximately 3 liters of distilled water. Next 40 ml of glacial acetic acid is added. After stirring throughly, 8 ml of concentrated ammonium hydroxide and enough distilled water are added to make up the final volume to 4 liters (pH 4.0). To this, 150 ml of formamide and 1600 ml 2-ethoxyethanol are added and the contents are shaken vigorously. The two phase solvent system thus formed is allowed to equilibrate overnight at constant temperature (23° ± 1°). The ratio of volumes of upper phase to lower phase is approximately 1 : 1.

Three grams of yeast or rat liver or 2 g of *E. coli* tRNA is dissolved in approximately 180 ml distilled water; 82.5 gm ammonium sulfate, 2.75 ml glacial acetic acid, and 0.55 ml concentrated ammonium hydroxide are added

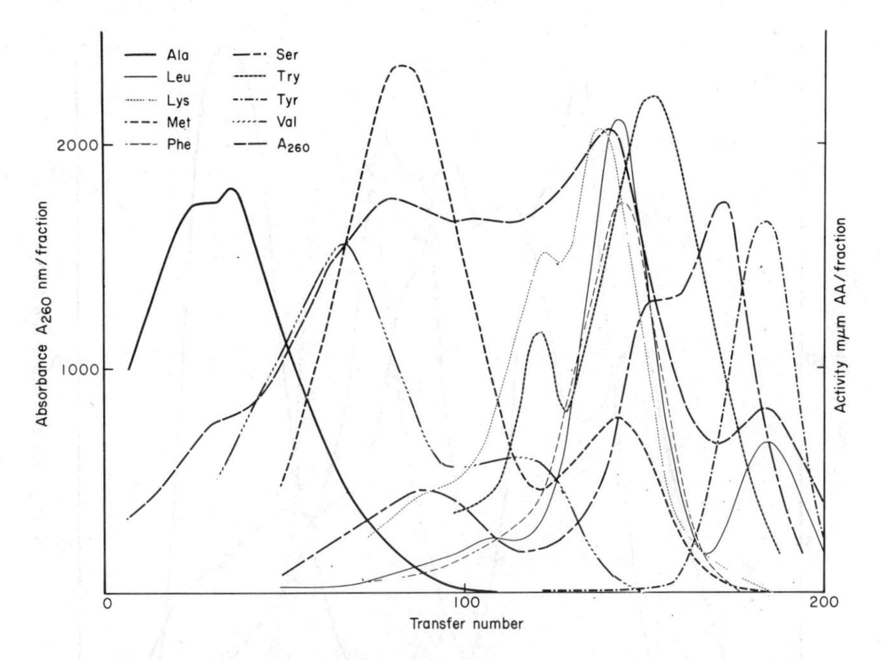

Figure 1B. Two hundred and twenty-five transfer countercurrent distributions of 1.8 gm of *E. coli* B tRNA according to the procedure 1. The capacity of tubes were 40 ml each phase per tube and each fraction consists of contents of six tubes. The legends for the identification of nine tRNAs are shown in the figure.

slowly to the tRNA sample with constant stirring. The solution was made up to a final volume of 275 ml and 11 ml of formamide is added. To this, 110 ml 2-ethoxyethanol is added dropwise with constant shaking. This is essential in order to dissolve desired amounts of tRNA in given volume. The procedures for the charging of the machine, introduction of sample, and performing the countercurrent distribution are the same as described for the phosphate buffer solvent system. Also the procedure for the recovery of tRNA fractions after the distribution is the same except for the following modification. After extraction of the sample with equal volume of anhydrous ether, the lower phase is diluted with 350 ml water and the cetavlon solution added. After allowing to stand for 20–30 minutes, 80 ml of anhydrous ether is added and the contents shaken. The recovery of tRNA is essentially the same as the phosphate solvent system. A representative result obtained for *E. coli* tRNA by such a procedure is shown in Fig. 2.

The alternate procedure of the isolation of tRNA from the phosphate buffer solvent system as described previously can also be employed for this solvent system, but with the following modification.

After extraction of sample with equal volume of ether, the lower phase is

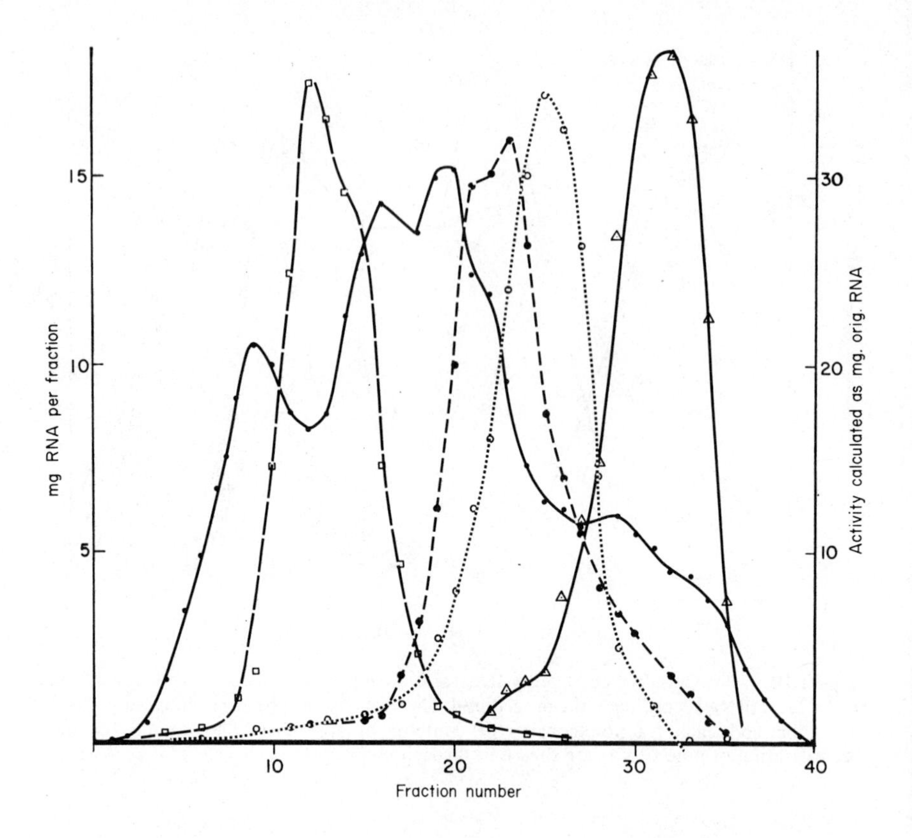

Figure 2. Two hundred-transfer countercurrent distribution of 300 mg of *E. coli* tRNA in a solvent system described under procedure 2. .______. mg RNA per fraction; □-□ proline acceptor activity; .___ __. lysine acceptor activity; ○-○ phenyl alanine acceptor activity; △-△ tyrosine acceptor activity.

transferred to 250-ml Erlenmeyer flask. Thirty-five ml 2-methoxyethanol is added and the contents shaken and allowed to stand for a few minutes. If an ammonium sulfate precipitate appears in this solution it is removed by filtration through filter paper or centrifugation at room temperature. To the filtrate 125 ml *n*-butanol and 150 ml ether are added and the contents shaken. The lower phase containing tRNA is collected and dialyzed. The rest of the procedure is the same as previously described.

Variations in partition coefficient of tRNA in this solvent system can also be accomplished by varying the pH, formamide, and 2-ethoxyethanol concentration. Increasing any of the three factors mentioned above increases the partition coefficient of tRNA. The partition coefficient of tRNA can be lowered by the addition of 2-butoxyethanol.

Inorganic Salt-Free Solvent System

The solvent system described in this procedure was developed by Zachau et al., (*6, 9*). Some of the features of this system are listed below: (1) the tRNA is converted to the tri-*n*-butylammonium salt prior to countercurrent distribution; (2) the solvent system does not contain inorganic salts and this facilitates the recovery of the tRNA after the distribution; (3) tRNA is very soluble in this solvent system (up to 200 mg yeast tRNA per milliliter of lower phase (*18*); (4) the solvent system is less sensitive to temperature, but is dependent on tRNA concentration (partition coefficient increases slightly at higher concentration; (5) for several tRNAs the distribution pattern is different from the patterns obtained with the systems described in the first two procedures.

Two liters of *n*-butanol, 2600 ml distilled water, 200 ml tri-*n*-butylamine, 60 ml glacial acetic acid, and 440 ml peroxide-free *n*-butyl ether were mixed together and shaken. The ratio of the volumes of upper phase to lower phase is approximately 1 : 1. Commercially available technical grade tri-*n*-butylamine should be purified prior to use in the preparation of the tRNA sample and the solvent system (*32*). The amine is heated with 15-20% (w/v) of phthalic acid anhydride until homogeneous mixture is obtained. After cooling, the amine is filtered and decanted, stirred for several hours with an equal volume of 2-3 N NaOH. The phases are allowed to separate and the tri-*n*-butylamine phase is washed four times with distilled water and vacuum distilled. With purified amine, a partition coefficient of 5 is obtained using 0.2 mg yeast tRNA tri-*n*-butylamine salt per milliliter in a solvent system prepared with 20 parts *n*-butyl ether per 100 parts *n*-butanol. Using technical grade amine, a lower partition coefficient is obtained. However, this can be compensated by lowering the *n*-butyl ether content of the solvent system.

Two grams of tRNA is dissolved in water, precipitated at 0° by addition of enough HCl to make the final concentration 0.1 N, centrifuged in the cold for 3 min at 5000 rpm, washed twice with 0.07 N HCl, and immediately dissolved in 30 ml water and 2.3 ml tri-*n*-butylamine (with stirring). The solution is concentrated to 10-15 ml by flash evaporation and introduced into the first two tubes of the previously charged apparatus after equilibration with equal volume of upper phase.

The desired number of transfers are performed with 2 min of shaking and 10-15 min settling time. The samples are withdrawn from the machine and pooled as described. The tRNA from the fractions is recovered by precipitation with 2% potassium acetate and 1 vol of ethanol, washed once with ethanol and dried under vacuum.

The recovery is about 95%. A representative distribution pattern of yeast tRNA is shown in Fig 3.

Because of good solubility to tri-*n*-butylamine salt of tRNA in this

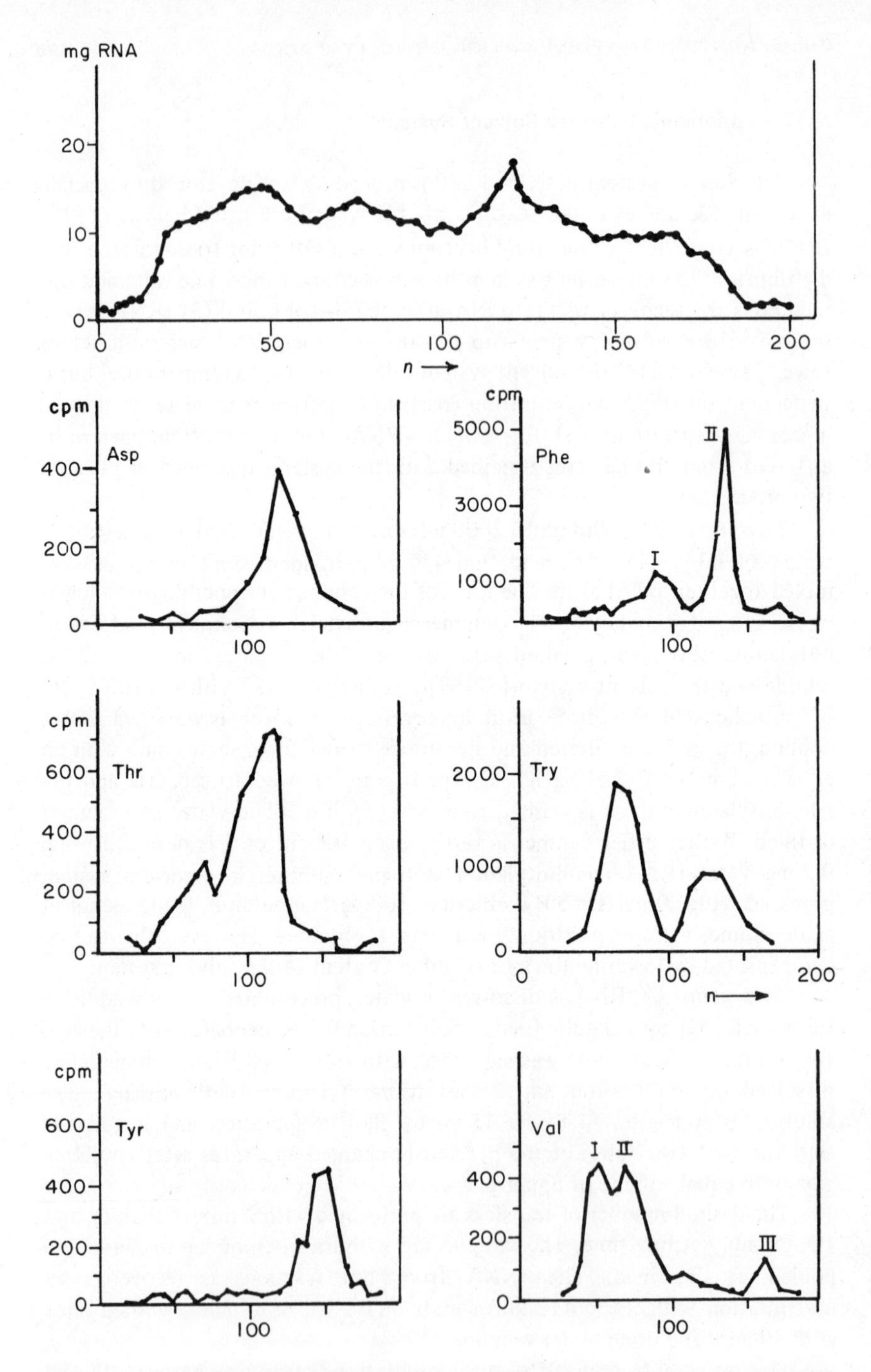

Figure 3. Two hundred-transfer countercurrent distribution of 1.5 gm of yeast tRNA in a solvent system described under procedure 3. Reproduced from R. Thiebe and H. G. Zachau. 1965. *Biochem. Biophys. Acta,* **103:** 568 with permission.

solvent system, it is an excellent system to employ for the fractionation of tRNA on preparative scale. Up to 40 gm of tRNA have been reported to be used on a batchwise scale in this solvent system.

These three solvent systems and the procedures that employ them are the basic methods for the separation and purification of tRNA by countercurrent distribution. In addition steady-state countercurrent distribution apparatus have also been employed for the separation and purification of tRNA (*33*). In this apparatus both upper and lower phases are transferred. Thus by proper programming of the apparatus, one can keep any given tRNA in the machine and remove the rest of the material from the machine. This can be done two ways: (1) By altering the concentrations of the contents of the solvent system in such a fashion the tRNA to be purified has a partition coefficient of 1.0 in this modified solvent system. The mixed tRNA is then introduced in the center tube of the steady-state distribution train. The machine is then set to perform one upper phase and one lower phase transfer alternatively. In this way material having the partition coefficient of 1.0 will stay in the center of the train where as all other material (having either $K > 1$ or $K < 1$) will migrate away from the center of the distribution train. (2) Instead of adjusting the contents of solvent system so that a given tRNA has a $K = 1.0$, the number of upper phase and lower phase transfers are programmed in such a way that the materials having the desired K value either do not appreciably migrate or migrate at the slowest rate. The purification of any tRNA is thus theoretically possible, although this has not yet been tried in practice.

Normally there are approximately 45–55 tRNA molecules present in a mixed tRNA of any one given species. One would like to obtain either the separation of all the isoaccepting tRNAs of a given species or the purification of one or more of these isoaccepting tRNAs. Thirdly, one would wish to accomplish the separation and purification of individual tRNAs. It is possible to accomplish all these goals by countercurrent distribution. So far as separation of isoaccepting tRNA or individual tRNAs from each other is concerned, this can be accomplished by extending the countercurrent distribution beyond the capacity of the apparatus if it cannot be accomplished within the normal 200 transfer distributions. This is done as follows: at the end of 200 transfers performed with either solvent system described in the first two procedures, the contents of the tubes 1 to 50 are left in the distribution train and the rest are carefully emptied. The emptied tubes are refilled with fresh upper and lower phases and the apparatus is set to recycle for 600 more transfers on the material contained in the first 50 tubes. Next, a second 200 transfer countercurrent distribution, with exactly the same amount of RNA under the identical conditions, is performed. This time, the contents of the tubes 1 to 50 and 101 to 200 are emptied and replenished with fresh solvent. Additional 600 transfers are performed by recycling. Similarly, by performing two more countercurrent distributions with exactly the same amounts of RNA and recycling the material in tubes 101 to 150 and 151 to 200, respectively, for 600 additional transfers each, one can obtain a composite pattern of a total of 800 transfer countercurrent distributions. An example of such a pattern for 400 transfer countercurrent distributions of yeast tRNA has been described (*4*).

Most of the isoaccepting tRNAs of any given amino acid are well separated by 800 transfer countercurrent distributions. At the end of 200 transfers, the upper phase leaving tube 200 is collected by means of a fraction collector, whereas fresh upper phase is continuously fed into the apparatus through tube no. 1. This process is continued until all the tRNAs have been moved out of the distribution train. By this method, the tRNAs having greater partition coefficients are relatively less purified and less well separated (of course, the purification and separation is better than the 200 transfer distribution) than the tRNAs having lower partition coefficients. For example, by performing approximately 1000 to 1200 transfer distributions of yeast tRNA in phosphate buffer solvent system, the only tRNA remaining in the distribution train is $tRNA^{Ala}$. This material is essentially pure and is obtained in very high yield. Also, by performing over 1000 transfer distributions of *E. coli* tRNA in ammonium sulfate system (procedure 2), one can obtain $tRNA^{Met}$ in essentially pure form and in high yield. This leads us into the use of countercurrent distribution in purification of tRNA.

Purification

Since the initial successful application of countercurrent distribution techniques for the purification of tRNA, several other techniques have been developed. Among them (1) benzoylated DEAE-cellulose column chromatography (*34*), (2) DEAE-Sephadex column chromatography (*35*), and (3) reverse-phase column chromatography (*36*) have proved to be of greatest value. The purification of tRNA therefore can be accomplished by three ways: (1) using countercurrent distribution techniques only, (2) using a combination of countercurrent distribution techniques along with one or more column chromatographic techniques, and (3) by employing one or more of the column chromatographic techniques. The present status of the field is at a stage where one wishes to obtain a fairly large amount of purified tRNA in good yield employing the least amount of effort. Most of the tRNAs thus far purified are the ones which can be easily separated from the bulk of the tRNAs by one of the above mentioned methods. Theoretically, one can purify any tRNA by any of these methods. However, the yield and quality of purified tRNA will be a limiting factor in attempting to purify tRNA by a single method. Thus, it is advisable to employ the combination of the above methods for this purpose. Because of the capacity of the countercurrent distribution procedure to process large quantities of tRNA, on one hand, and at the same time to be able to accomplish excellent resolution of various tRNAs from each other, this procedure can be the method of choice for the initial steps in tRNA purification. Also in many individual cases (particularly in the case of tRNAs which have either very low partition coefficients (Ala, Val, Pro, Met) or very high partition coefficients (Tyr, Phe, Lys)), one can obtain the purification of these tRNAs by countercurrent distribution only in either one or two steps.

There are at present tRNAs isolated from two species that are used generally, namely, yeast and *E. coli.* In the case of yeast tRNA, $tRNA^{Ala}$, $tRNA^{Val}$, $tRNA^{Met}$, $tRNA^{Phe}$, $tRNA^{Tyr}$, $tRNA^{Lys}$, $tRNA^{His}$, $tRNA^{Ile}$, and $tRNA^{Ser}$ have been purified. In the case of *E. coli* the following tRNAs are purified, $tRNA^{Ala}$, $tRNA^{Val}$, $tRNA^{Gly}$, $tRNA^{Met}$, $tRNA^{Glu}$, $tRNA^{Arg}$, $tRNA^{Phe}$, $tRNA^{Tyr}$ and $tRNA^{Leu}$. The following is a brief account regarding the procedures that one can adopt for the purification of these and other tRNAs. Some of these procedures were taken from the literature, others have been tried out in the author's laboratories and still others are possible suggestions.

Purification of Yeast tRNAs

$tRNA^{Ala}$

This tRNA can be purified by performing 1000–1200 transfer countercurrent distributions according to procedure 1. If the material thus obtained is still not of desirable purity, a low salt gradient column chromatography on BD cellulose has proved to be adequate for the removal of small amounts of impurities (*37*).

$tRNA^{Val}$

$tRNA^{Val}$ can be purified by countercurrent distribution according to either procedure 1 or 2. The $tRNA^{Val}$-rich fraction can be redistributed for 400 and 850 transfers according to Apgar et al. (*8*), to obtain it in purified form. Alternatively, the $tRNA^{Val}$-rich fraction from the initial countercurrent distribution is chromatographed on BD-cellulose with 0.3–0.7 M NaCl gradient containing 0.001 M $MgCl_2$ and 0.01 M KOAC, pH 5.5 (approximately 3 to 4-fold purification), and finally chromatographed on DEAE-Sephadex A-50 column, using 0.375 M NaCl, 0.02 M Tris, pH 7.5, and 0.008 M $MgCl_2$ without any gradient.

$tRNA^{Met}$

Using countercurrent distribution and BD-cellulose and DEAE-Sephadex A-50 column chromatography, Rajbhandary et al. (*24*), have purified this tRNA in good yield.

$tRNA^{Phe}$ and $tRNA^{Tyr}$

These two tRNAs have essentially close partition coefficients in the solvent systems employed either in procedure 1 or 2 described earlier. The fraction enriched in these tRNAs is chromatographed on BD-cellulose 0.01 M acetate, pH 5.5, and 0.01 M $MgCl_2$. Using a 0.3 to 1.0 M NaCl gradient $tRNA^{Tyr}$ elutes between 0.7 and 0.9 M NaCl whereas $tRNA^{Phe}$ is stripped from the column with 1.0 M NaCl buffer containing 10% ethanol. The separated $tRNA^{Tyr}$ and $tRNA^{Phe}$ can then be purified further using reverse-phase column chromatography (*36*) either at pH 4.5 and/or at pH 7.5.

$tRNA^{Lys}$

After performing 200-transfer countercurrent distribution of yeast tRNA according to procedure 2, the $tRNA^{Lys}$-rich fractions are distributed in essentially the same fractions as $tRNA^{Tyr}$- and $tRNA^{Phe}$-rich fractions. Further purification of $tRNA^{Lys}$ can be accomplished by reverse-phase column chromatography.

$tRNA^{Ser}$

Using combinations of procedures 1 and 2, Connelly and Doctor have described the purification and separation of two isoaccepting $tRNA^{Ser}$ (*21*). Karau and Zachau (*17*) used the combination of all three procedures to obtain these two isoaccepting tRNAs in pure form in large quantities. However, 4 to 5 redistributions of $tRNA^{Ser}$-rich fractions were required, a procedure which is time consuming and results in mechanical loss of material. By introducing BD-cellulose column chromatography steps after the initial 200-transfer countercurrent distributions of crude yeast tRNA, 2 to 3 redistribution steps can be eliminated. After BD-cellulose column chromatography the essentially pure $tRNA^{Ser}$ fractions are distributed for 600–800 transfers according to procedure 2. This procedure completely separates the two isoaccepting species of $tRNA^{Ser}$, and the $tRNA^{Ser}$ thus obtained is essentially pure.

$tRNA^{His}$

This tRNA has been purified by using repeated countercurrent distribution (*8*). The tRNAs mentioned above are purified either by countercurrent distribution alone or in combination with other chromatographic methods. In addition several other tRNAs have been purified by column chromatography only.

Purification of *E. coli* tRNAs

$tRNA^{Ala}$

The methods of purification of $tRNA^{Ala}$ from either *E. coli* or yeast is essentially the same.

$tRNA^{Val}$

After 200-transfer countercurrent distributions of *E. coli* tRNA, $tRNA^{Val}$-rich fraction is subjected to BD-cellulose column chromatography (2 x 100 cm) using 0.3–0.7 M NaCl gradient in 0.01 M $MgCl_2$ and 0.01 M KOAC, pH 5.5, buffer. The tRNA fractions eluted early from the column are the $tRNA^{Val}$-rich fractions. The $tRNA^{Val}$-rich fraction thus obtained is rechromatographed on (0.8 x 160 cm) BD-cellulose column using the same gradient to obtain $tRNA^{Val}$ in purified form.

$tRNA^{Met}$

A simple method for the separation of $tRNA_f^{Met}$ and $tRNA^{Met}$ using combinations of countercurrent distribution and DEAE-Sephadex A-50 column chromatography have been described by Doctor et al. (*38*). Since this method offers a good example of obtaining a large quantity of purified tRNA in excellent yield by combinations of countercurrent distribution and column chromatography, it is presented here in detail.

Approximately 15 gm of *E. coli* tRNA is dissolved in 1.5 liters of 30% ammonium sulfate, pH 4.0, according to the procedure 2 described before. Sixty milliliters of formamide and 600 ml of 2-ethoxyethanol are added to it and shaken. After separation of two phase, the upper and lower phases containing the tRNA are introduced in the first three tubes of 30-tube countercurrent distribution apparatus (with a capacity of 350 ml of each phase per tube, or one can use 1-liter separatory funnels) and 40 transfers are performed, the upper phases leaving the 30th tube are collected. At the end of the 40 transfers, 0.1 ml of each phase from every second tube was diluted to 3 ml and absorbances at 260 nm were determined to measure the amount of tRNA in each fraction. Ten microliters of these diluted samples were used directly in an assay for the methionine acceptor activity of the RNA fractions. The tRNA from the desired fractions was isolated batchwise by the procedure previously described. The $tRNA_f^{Met}$ and $tRNA^{Met}$ (in Fig. 4 referred to as Met_f and Met_m) fractions thus obtained were deproteinized

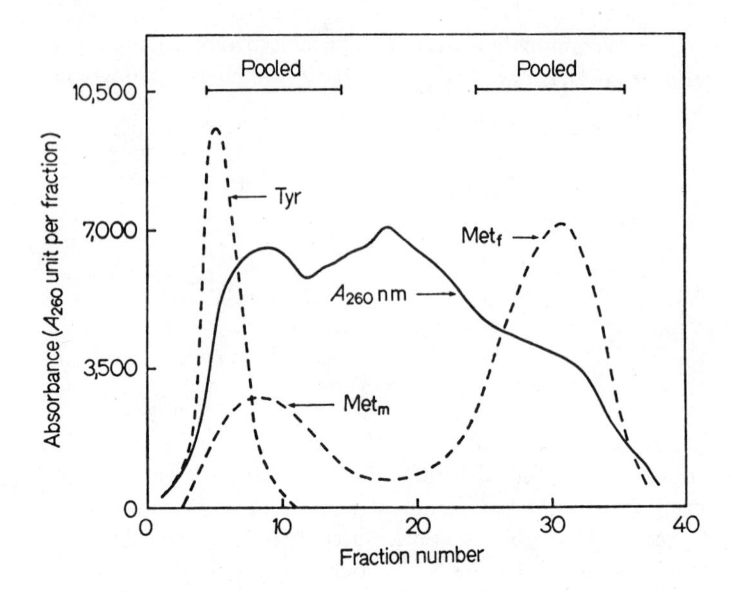

Figure 4. Forty-transfer countercurrent distribution of 10 gm *E. coli* tRNA in the ammonium sulfate solvent system. The solid line represents the absorbance per fraction, measured at 260 nm and the dotted lines as indicated show methionine and tyrosine acceptor activities in arbitrary units as a reference. The fractions marked "pooled" were recovered and chromatographed on DEAE-Sephadex A-50 columns as described in Figs. 5 and 6. The faster running material is found in the lower numbered fractions.

once more by phenol extraction prior to application to the column for further procedure.

Figure 4 shows the results of a 40-transfer countercurrent distribution run of 10 gm of *E. coli* tRNA. The methionine acceptor activities of the $tRNA_f^{Met}$ and $tRNA^{Met}$ fractions (pooled as described in Fig. 4) were approximately 200 pmoles and 60 pmoles per A_{260} unit, respectively. These fractions contained approximately 3.5×10^4 and 6.0×10^4 A_{260} units, respectively, per 10 gm of starting tRNA.

For further purification of these two tRNAs and the separation of their subspecies, the pooled fractions were subjected to DEAE-Sephadex A-50 column chromatography according to Nishimura et al. (*35*).

DEAE-Sephadex A-50, bead form, was prepared according to this procedure and a column of 4 x 100 cm was packed. The column was washed with 1 liter of initial buffer consisting of 0.375 M NaCl in 0.02 M Tris, pH 7.5, and 0.008 M $MgCl_2$. The sample was dissolved in 150 to 200 ml of initial buffer and applied to the column at the rate of 30 ml/hr. A gradient consisting of 3 liters of initial buffer and 3 liters of 0.390 M NaCl in 0.02 M Tris pH 7.5 and 0.016 M $MgCl_2$ was used and 10-ml fractions were collected.

The NaCl $MgCl_2$ concentrations mentioned here are different from the ones used in experiments described in Fig. 5 and 6, since these concentrations were later found to be more efficient. The tRNAs from the desired fractions were recovered by precipitation with two volumes of ethanol stored at $-20°$ overnight and recovered by centrifugation. The fractions were dried under vacuum and stored in dried form at $-20°$.

Figure 5 describes the results of DEAE-Sephadex column chromatography of a $tRNA_f^{Met}$ fraction obtained from a countercurrent distribution indicated in Fig. 4. Transfer RNA fractions for which absorbance and methionine acceptor activity peaks are coincident accepted approximately 1.6 nmoles amino acid per A_{260} unit. This value is taken for designating specific activity of 1 in Fig. 5. From the results presented here, it is quite evident that methionine acceptor activity is resolved into two peaks. Approximately 2,200 A_{260} units were obtained as pure $tRNA_f^{Met}$ which constitutes approximately 3.5 μmoles of methionine acceptor activity. By rechromatography of the tRNA fractions containing methionine acceptor activity, one can obtain additional 750 to 1,000 A_{260} units of pure $tRNA_f^{Met}$. This constitutes approximately 50-60% recovery of $tRNA_f^{Met}$ as purified tRNA.

Figure 6 shows the results of DEAE-Sephadex column chromatography of a $tRNA^{Met}$ fraction separated as indicated in Fig. 4. Because of the limiting capacity of a column of these dimensions only half of the fraction obtained from countercurrent distribution was applied to the column at one time. Approximately 350 A_{260} units tRNA were obtained in purified form (by the criteria mentioned for $tRNA_f^{Met}$). The presence of two methionine acceptor activity peaks is obvious for $tRNA^{Met}$ also.

It is quite evident from these results that the combination of countercurrent distribution and column chromatography is a very efficient way to purify and separate tRNA.

$tRNA^{Tyr}$ and $tRNA^{Phe}$

As mentioned for yeast tRNA, these two tRNAs isolated from *E. coli* also have close partition coefficients in solvent systems described under procedures 1 and 2. However, they are sufficiently separated so that they can be used for further purification either by redistribution in countercurrent distributions or by column chromatographic procedures. $tRNA^{Phe}$-enriched fraction obtained from 200-transfer countercurrent distributions of *E. coli* tRNA according to either procedure 1 or 2 is redistributed for 400-500 transfers in 1.9 M phosphate buffer solvent system (*4*) containing 42 ml isopropanol (100 ml phosphate). The $tRNA^{Phe}$-enriched fraction obtained from this redistribution is redistributed in the same solvent system for 800-1000 transfers to obtain it in purified form. Alternatively, $tRNA^{Phe}$-

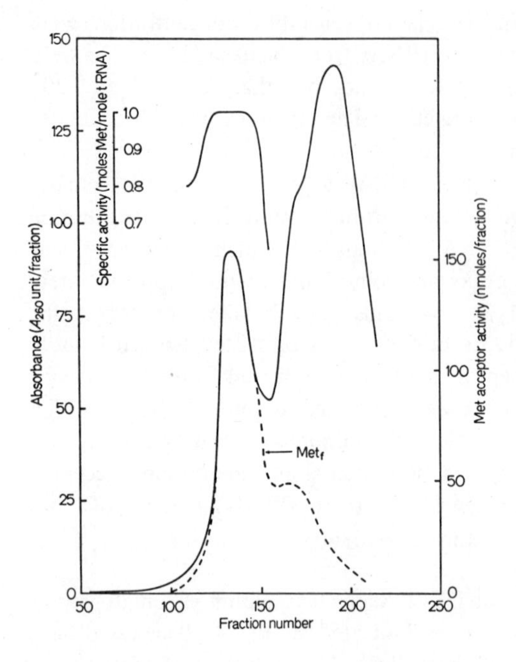

Figure 5. DEAE-Sephadex A-50 column chromatography of $tRNA_f^{Met}$ fraction obtained from a 40-transfer countercurrent distribution of *E. coli* tRNA as shown in Fig. 4 was chromatographed on a DEAE-Sephadex A-50 column (4 x 100 cm). The solid line signifies the absorbance (at 260 nm) per fraction and the dotted line methionine acceptor activity in nanomoles per fraction. A solid line also signifies the specific activity of the fractions (i.e., moles of methionine incorporated per mole of tRNA). The values obtained were calculated using the assumption that when 1.6 nmoles of methionine are incorporated into 1 A_{260} unit tRNA, 1 mole of methionine equals 1 mole of tRNA. The values are plotted in such a manner that when 1 mole of methionine is incorporated into 1 mole of tRNA the A_{260} curve will coincide with methionine acceptor activity curve. Approximately 110 mg of purified $tRNA_f^{Met}$ was obtained.

enriched fraction from 200-transfer countercurrent distributions is subjected to column chromatography on DEAE-Sephadex A-50 (same dimensions as for $tRNA^{Met}$) using 0.4 M NaCl in 0.01 M Tris, pH 7.5, and 0.010 M $MgCl_2$ as mixing gradient and 0.45 M NaCl in 0.02 M Tris, pH 7.5, and 0.025 M $MgCl_2$ as limit gradient. The phenylalanine tRNA peak is essentially pure but has some overlap. The initial 200-transfer countercurrent distribution step can be replaced by passage of mixed tRNA through BD-cellulose (packed with 1 M NaCl, 0.01 M KOAC, pH 5.5, and 0.01 M $MgCl_2$). The tRNA is dissolved in the buffer and the column washed with the buffer until A_{260} nm of effluent is below 0.2 A_{260} nm. The tightly bound tRNA containing $tRNA^{Phe}$ is eluted with the buffer containing 10% ethanol. The fraction thus obtained is subjected to DEAE-Sephadex A-50 column chromatography as described to accomplish final purification of $tRNA^{Phe}$. Tyrosine tRNA (containing two isoaccepting species) can be obtained by 200-transfer countercurrent distribution of tRNA by either procedure 1 or 2. The $tRNA^{Tyr}$-enriched fraction (approximately 12-fold purified) is redistributed for 400 transfers according to procedure 1, with the following two modifications: (1) the addition of a 1.9 M phosphate buffer system (*4*) containing 46 ml isopropanol (100 ml

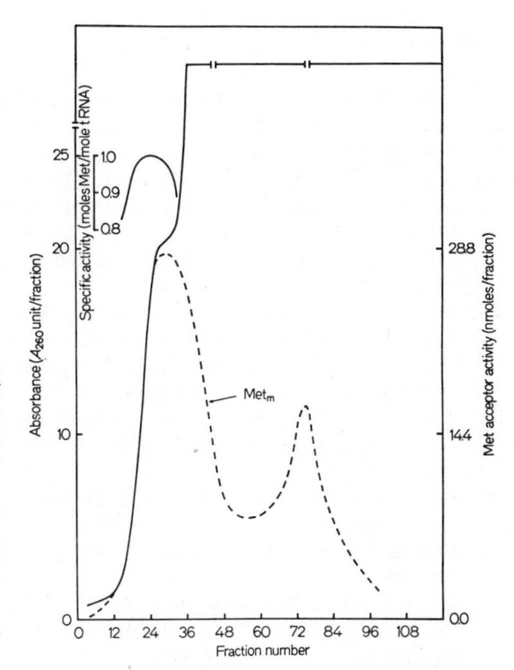

Figure 6. DEAE-Sephadex A-50 column chromatography of $tRNA^{Met}$ fraction obtained (approximately half of the total material) from 40-transfer countercurrent distribution of *E. coli* tRNA, as shown in Fig. 4 was chromatographed on a DEAE-Sephadex A-50 column (4 x 100 cm). The curves signify quantities as in Fig. 5. Approximately 17 mg of purified $tRNA^{Met}$ were obtained.

phosphate), and (2) the distribution is carried out at 26–27° instead of 24°C (room temperature). Two isoaccepting $tRNA^{Tyr}$'s can be separated in this manner. The separated $tRNA^{Tyr}$ fractions are redistributed separately for additional 800 transfers each to obtain them in purified form. Alternatively the $tRNA^{Tyr}$-enriched fraction from either 200-transfer initial countercurrent distribution or ethanol (NaCl buffer (1 M) containing 10% ethanol washed from BD-cellulose column as described for $tRNA^{Phe}$ which also contains $tRNA^{Tyr}$) is subjected to DEAE-Sephadex A-50 column chromatography as described for $tRNA^{Phe}$.

As mentioned at the start of this portion of the chapter, in addition to the tRNA for which the purification procedures are described, $tRNA^{Gly}$, $tRNA^{Leu}$, $tRNA^{Glu}$, $tRNA^{Arg}$, and several other tRNAs from *E. coli* have been purified using the column chromatographic procedures only. The readers may find the procedures for their purification elsewhere in this volume.

Acknowledgments

The author wishes to thank Mrs. Benita L. Murphy and Miss Mary Ann Sodd for their efforts in the preparation of this manuscript.

References

1. Holley, R. W., and S. H. Merrill. 1959. *J. Am. Chem. Soc.*, **81**: 753.
2. Holley, R. W., B. P. Doctor, S. H. Merrill, and F. M. Saad. 1959. *Biochim. Biophys. Acta,* **35**: 272.
3. Holley, R. W., J. Apgar, and B. P. Doctor. 1960. *Ann. N Y. Acad. Sci.*, **88**: 745.
4. Doctor, B. P., J. Apgar, and R. W. Holley. 1961. *J. Biol. Chem.*, **256**: 1117.
5. Doctor, B. P., and C. M. Connelly. 1961. *Biochim. Biophys. Res. Commun.*, **6**: 201.
6. Zachau, H. A., M. Tada, W. B. Lawson, and M. Schweiger. 1961. *Biochim. Biophys. Acta,* **53**: 221.
7. Apgar, J., R. W. Holley, and S. H. Merrill. 1961. *Biochim. Biophys. Acta,* **53**: 220.
8. Apgar, J., R. W. Holley, and S. H. Merrill. 1962. *J. Biol. Chem.*, **237**: 796.
9. Tada, M., M. Schweiger, and H. A. Zachau. 1962. *Z. Physiol. Chem.*, **328**: 85.
10. Wiesmeyer, H., K. Kjellin, and H. Boman. 1962. *Biochim. Biophys. Acta,* **61**: 625.
11. Apgar, J., and R. W. Holley. 1962. *Biochem. Biophys. Res. Commun.*, **8**: 391.
12. Weisblum, B., S. Benzer, and R. W. Holley. 1962. *Proc. Natl. Acad. Sci., U. S.,* **48**: 1449.
13. Doctor, B. P., C. M. Connelly, G. W. Rushizky, and H. A. Sober. 1963. *J. Biol. Chem.,* **238**: 3985.
14. Goldstein, J., T. P. Bennett, and L. C. Craig. 1964. *Proc. Natl. Acad. Sci., U. S.,* **51**: 119.
15. Apgar, J., and R. W. Holley. 1964. *Biochem. Biophys. Res. Commun.*, **16**: 121.
16. Rushizky, G. W., H. A. Sober, C. M. Connelly, and B. P. Doctor. 1965. *Biochem. Biophys. Res. Commun.,* **18**: 489.
17. Karau, W., and H. A. Zachau. 1964. *Biochim. Biophys. Acta.,* **91**: 549.
18. Hoskinson, R. M., and H. A. Khorana. 1965. *J. Biol. Chem.*, **240**: 2129.
19. Ingram, V. M., and J. A. Sjoquist. 1963. *Cold Spring Harbor Symp. Quant. Biol.,* **28**: 133.
20. Von Ehrenstein, G., and D. Dais. 1963. *Proc. Natl Acad. Sci., U.S.,* **50**: 81.
21. Connelly, C. M., and B. P. Doctor. 1966. *J. Biol. Chem.*, **241**: 715.
22. Kellogg, D. A., B. P. Doctor, J. E. Loebel, and M. W. Nirenberg. 1966. *Proc. Natl Acad. Sci., U.S.,* **55**: 912.
23. Madison, J. H., G. A. Everett, H. K. Kung. 1966. *Cold Spring Harbor Symp. Quant. Biol.* **31**: 409.
24. Rajbhandary, V. L. and H. A. Ghosh. *J. Biol. Chem.*, **244**: 1104.
25. Doctor, B. P. 1967. In L. Grossman and K. Moldave (Eds.), Methods in enzymology. Academic Press, New York, Vol. XIIA, p. 634.
26. Holley, R. W., J. Apgar, and S. H. Merrill. 1961. *J. Biol. Chem.*, **236**: PC42.
27. Holley, R. W., J. Apgar, G. A. Everett, J. T. Madison, S. H. Merrill, and A. Zamir. 1963. *Cold Spring Harbor Symp. Quant. Biol.,* **28**: 117.
28. Mirzabekov, A. D., A. T. Krutilina, V. I. Gorshkova, and A. A. Boev. 1964. *Biokhimiya,* **29**: 1158.
29. Lipsitt, M. N., and B. P. Doctor. 1967. *J. Biol. Chem.*, **242**: 4072.
30. Pless, M., H. Ochiai, and P. A. Cerutti. 1969. *Biochem. Biophys. Res. Commun.,* **34**: 70.
31. Kirby, K. S. 1960. *Biochim. Biophys. Acta.,* **41**: 338.
32. Zachau, H. G. 1965. *Z. Physiol. Chem.*, **342**: 98.
33. Reeves, R. H., N. Imura, H. Schwan, and G. B. Weiss, 1968. *Proc Natl. Acad. Sci. U.S.,* **60**: 1450.
34. Gillam, I., S. Millward, D. Blew, M. von Tigerstrom, E. Wimmer, and G. M. Tener. 1967. *Biochemistry,* **6**: 3943.

35. Nishimura, S., R. Harada, V. Narushima, and T. Seno. 1967. *Biochim. Biophys. Acta,* **142**: 133.
36. Weiss, J. F., and A. D. Kelmers. 1967. *Biochemistry,* **6**: 2507.
37. Loehr, J. S. and E. B. Keller. 1968. *Proc. Natl. Acad. Sci., U.S.,* **61**: 1115.
38. Doctor, B. P., B. J. Wayman, S. Cory, P. S. Rudland, and B. F. C. Clark. 1969. *European J. Biochem.,* **8**: 93.

FRACTIONATION OF NUCLEIC ACIDS BY METHYLATED ALBUMIN-KIESELGUHR COLUMN CHROMATOGRAPHY

T. ISHIDA
ASAHI CHEMICAL INDUSTRY CO., LTD.
TOKYO, JAPAN

J. KAN
CITY OF HOPE MEDICAL CENTER
DUARTE, CALIFORNIA

T. KANO-SUEOKA
DEPARTMENTS OF BIOLOGY AND
BIOCHEMICAL SCIENCES
PRINCETON UNIVERSITY
PRINCETON, NEW JERSEY

The methylated albumin-kieselguhr (MAK) column was first used by Lerman (*1*) to fractionate pneumococcal DNA. He demonstrated that DNA for the streptomycin resistant marker was fractionated at a unique position in the salt gradient elution profile. Subsequently properties of the column for fractionation of nucleic acids have been elucidated by several investigators (*2–5*) and MAK column chromatography has been successfully used to fractionate various DNA and RNA molecules.

PROPERTIES OF METHYLATED ALBUMIN-KIESELGUHR COLUMN

DNA is fractionated on the MAK column according to size, extent of hydrogen bonds, and base composition (*2, 3*). Generally, the larger the molecular weight of DNA, the higher the salt concentration required for elution. Denatured DNA is eluted at a higher salt concentration than native DNA. The column also recognizes the base composition of both native and denatured DNA, i.e., the higher the GC content of DNA, the lower the salt concentration required for elution. The MAK column is also useful for the fractionation of various species of RNA (*3–5*). Thus, the column separates 4s (transfer RNA) and 16s and 23s (ribosomal RNA) RNA of *E. coli*, which are eluted in that order; that is, in the order of increasing molecular weight of the RNA. tRNA specific for the different amino acids and isoaccepting tRNAs for particular amino acid are resolved by MAK column chromatography, presumably by virtue of differences in the primary and secondary structures of the tRNA (*5*). Tryptophan tRNA in *E. coli* can assume two conformations – the native and the denatured form – which are interconvertible (*6*). This difference in the secondary structure of the same primary sequence is sufficient to enable the two forms to be resolved by the MAK column. In this instance, the native form is eluted at a lower salt concentration than the denatured form.

We shall consider the basis for the fractionation of nucleic acids on the MAK column. The albumin, which coats the diatomaceous kieselguhr, has been esterified with methanol, thus permitting the basic residues of the protein molecule to interact with reactive groups on the nucleic acid. Other kinds of interaction with the protein might also be involved. In essence, this is similar to the histone columns previously described (*7, 8*). One can visualize the following ionic interaction as the major factor for fractionation. Positively charged groups in the basic amino acid residues of the albumin (the imidazole group in histidine, the guanido group in arginine, and the ϵ-amino group in lysine) can form ionic bonds with negatively charged phosphate groups of the nucleic acid. Hydrogen bonding, however, may also be involved. The carbonyl and imide groups of the peptide backbone of the methylated albumin may be able to form hydrogen bonds with the amino and keto groups of the base residues in the nucleic acid.

PROCEDURE

Reagents

Methylated albumin
Kieselguhr

Sodium phosphate buffer, 1 M (pH 6.3 or 6.7 at 0.05 M)
For fractionation of aminoacyl-tRNA pH 6.3 is used to minimize deacylation. For the rest of the nucleic acids, pH 6.7 can be used.
NaCl, 4 M

Preparation of Methylated Albumin

The preparation of methylated albumin was originally described by Fraenkel-Conrat and Olcott (*9*). The procedure of Mandell and Hershey (*2*) is described here. Five grams of bovine serum albumin (from Fraction V bovine serum albumin powder obtainable either from Armour Laboratories, Chicago or from Nutritional Biochemicals Corp., Cleveland) is esterified by the addition of 4.2 ml of 12 M HCl and 500 ml of absolute methyl alcohol. The methyl alcohol should be freshly opened. The mixture is stored in the dark for 3 days or more and shaken several times each day. This can be done in a one-liter Erlenmeyer flask covered completely with aluminum foil. During this time the protein dissolves and reprecipitates. The precipitate is collected by centrifugation and washed twice with absolute methyl alcohol and twice with anhydrous ether. When most of the ether has evaporated, the precipitate is dried *in vacuo* over potassium hydroxide pellets. The dried esterified protein is reduced to powder which can be stored over potassium hydroxide for at least two years. For use, a 1% solution of methylated albumin in water can be made by allowing the protein to dissolve with occasional gentle shaking in the cold. The solution is stable for several months when kept refrigerated.

Preparation of Buffered Saline Solutions

Buffered saline solutions are composed of 0.05 M sodium phosphate buffer plus varying amounts of NaCl. Solutions are prepared by diluting stock solutions of 1 M sodium phosphate buffer (107.8 gm of $NaH_2PO_4 \cdot H_2O$ plus 31.9 gm of Na_2HPO_4 for 1 liter, pH 5.6), and 4 M NaCl solution. The pH of the buffer is about 6.3 at 0.05 M. Although the pH of the buffer becomes lower with the addition of salt, the solutions are not adjusted to 6.3, but used as such. For pH 6.7, use 78 gm of $NaH_2PO_4 \cdot H_2O$ and 61.8 gm Na_2HPO_4 for 1 liter to make 1M buffer.

Preparation of Methylated Albumin-Kieselguhr Column

Any good chromatographic column can be used. A simple column having an inner joint with a coarse sealed-in fritted glass disc (e.g., Kontes Glass Co.)

is conveniently cleaned after each use. The size of the column is largely determined by the amount of nucleic acid to be fractionated. Since a packed volume of excessive height causes channeling, especially in the case of DNA, the scale should be changed by varying the diameter of the column rather than the height. In general, the height should be between 1 and 5 cm for the fractionation of DNA and up to 7 cm for RNA. We routinely use a column with a 7-cm height for the fractionation of tRNA.

The column originally reported by Mandell and Hershey (*2*) had three layers with two layers of MAK using different ratios of methylated albumin to kieselguhr. A simplified MAK column was later introduced (*3*) consisting of two layers with one layer of MAK. No significant difference between the two was observed for fractionation of RNA and most of the DNA preparations having a molecular weight of 10 to 30 x 10^6. However, using high-molecular-weight DNA, e.g., intact phage T2 or T4 DNA molecules (130 x 10^6 mol wt), on a simplified column is likely to create a channeling problem (*2*). The method for the simplified MAK column is described below.

Preparation of a column with 30 ml of MAK suspension will be used as an example. The 30-ml MAK suspension is prepared by gently stirring 1.5 ml of 1% methylated albumin solution into a boiled and cooled kieselguhr suspension (6 gm of kieselguhr and 30 ml of 0.1 M buffered saline: 0.1 M NaCl in 0.05 M sodium phosphate buffer, pH 6.3 or 6.7). The kieselguhr can be obtained from Amend Drug & Chemical Co., Inc., New York, as "infusorial Earth." A chromatographic column of 18 mm i.d. for the fractionation of RNA, or 30 mm i.d. for the fractionation of DNA, is first packed with paper powder. Dry paper powder is put into the column which contains 20 ml of 0.1 M buffered saline, and the suspension is stirred to eliminate air bubbles. The paper powder is allowed to settle by gravity to form a flat layer of 3-5 mm. The 30-ml MAK suspension is then poured slowly on top of the paper powder layer with the bottom outlet of the column closed. After the entire 30 ml suspension is placed in the column, the outlet is opened and air pressure (ca. 2 psi) is applied to the top of the column so that a uniform layer of MAK will be formed. During the entire packing and loading operation the column should not be permitted to dry. The column is washed with about 20 ml of 0.1 M buffered saline. A previously boiled and cooled kieselguhr suspension (2 gm of kieselguhr in 15 ml of 0.1 M buffered saline) is now poured slowly on top of the MAK layer. Sometimes the kieselguhr is replaced by silicic acid for the top layer which removes contaminating proteins (*10*). Air pressure is similarly applied to form this top layer containing only kieselguhr or silicic acid. The top layer serves as protection for the MAK layer on which the nucleic acid is fractionated. The column is then washed with about 100 ml of the starting buffer. The washing operation can be monitored by measuring the absorbancy at 260 nm of the eluent during the wash. Although the column is generally used the same day, it may be left at room temperature for use the next day. The capacity of this column is 3 mg of nucleic acids.

Preparation of Samples

Transfer RNA

For radioactive aminoacyl-tRNA, 5000 cpm or more is suspended in 40 ml of 0.3 M buffered saline. The total amount of tRNA is adjusted to 2-3 mg by adding nonradioactive tRNA as the carrier. The resolution is better when the column is loaded to near maximal capacity.

High-Molecular-Weight RNAs

It is desirable to remove DNA since the inclusion of appreciable amounts of DNA in the sample to be chromatographed can lead to slow flow rates and to channeling which will adversely affect the resolution of the RNA. The purified RNA (up to 3 mg) is suspended in 40 ml of 0.4 M buffered saline.

DNA

The concentration of DNA, either native or denatured, is adjusted to about 20-100 μg/ml in standard saline citrate (0.15 M NaCl and 0.015 M sodium citrate). Up to 3 mg of DNA is applied to the column. For separation of complementary strands of DNA, DNA denatured by heat or by alkali are subjected to dialysis in the cold for 18 hr against 1 or 2 liters of 0.6 M buffered saline. The final concentration of the DNA solution should not exceed 25 μg/ml. Denatured DNA (1.5-2.5 mg) is applied to the column. The recovery of denatured DNA is usually less than 70% of the input DNA. Shearing the DNA sonically may improve recovery.

Adsorption to the Column

Diluted nucleic acid samples are applied to the column with as little disturbance as possible to the upper surface. After loading, the column is washed with 40 ml of buffered saline at the concentration of the starting buffer. The effluent of the sample solution may be kept to measure absorbancy at 260 nm to check for overloading.

Elution of the Column

Gradient Elution

The charged column is connected to a mixing chamber of a gradient apparatus containing the starting buffer and the final buffer. A gradient which gives a linear increase in salt concentration (*11, 12*) is satisfactory. The salt gradient is established with 120 ml of the starting buffer and 120 ml of the final buffer. The flow rate is set to about 1 ml/min. Once the elution is started, readjustment of the flow rate should be avoided as this produces an artifact in the elution profile. During the elution, the head of the column (the buffer above the surface of the kieselguhr layer) should remain low (0.5 cm) and the column should never become dry. The salt concentration of the starting buffer is generally 0.3 M for the fractionation of tRNA, 0.4 M for the higher-molecular-weight RNA and DNA, and 0.6 M for the strand separation of DNA. Concentration of the final buffer is generally 0.8-1.0 M for tRNA, 1.2 M for higher-molecular-weight RNA and DNA, and 1.4 M for denatured DNA. The elution is carried out at room temperature. Salt concentrations of fractions can be measured with a refractometer. After completion of the chromatography, the inner joint with the sealed-in glass disc should be immersed in a dichromate-sulfuric cleaning solution overnight to remove trapped material.

Stepwise Elution

Elution of the nucleic acid can be accomplished by passing stepwise through the column a series of buffered saline solutions with increased salt concentration, starting with the same concentration as used for equilibrating the column. Each time the buffered saline is driven down to the level of the top layer. The amount of eluting solution per step, as well as the concentration increment of salt for each step, should be determined empirically. An example of such an elution procedure has been reported (*3*). In general, recovery of denatured DNA is better obtained by stepwise elution.

Intermittent Gradient Elution

This procedure was first used by Hogness and Simmons (*13*) to separate two halves of phage λ DNA. Recently, Rudner, Karkas, and Chargaff (*14*) were able to separate opposite strands of *B. subtilis* DNA. Denatured DNA was eluted by means of a linear salt gradient using either 0.6 M to 1.2 M or 0.7 M to 1.4 M NaCl. The absorbance of the eluates was recorded by a continuous-flow ultraviolet monitor. As soon as the first peak began to emerge (at approximately 0.8-0.9 M NaCl) the flow of the saline of higher molarity into the mixing chamber was stopped. When the absorbance peak reached its maximum, the gradient was restarted and maintained until the end of the experiment.

Analysis of Fractions

Fractions eluted from the MAK column can be analyzed by various physical, chemical, and biological methods, some examples of which will be described here.

Ultraviolet Absorption

The amount of nucleic acid in each fraction can be estimated from absorbance at 260 nm.

Elution Profile of Aminoacyl-tRNA

Radioactivity from (^{14}C) and (^{3}H) aminoacyl-tRNA can be determined for each fraction by precipitating the tRNA and collecting it on an appropriate filter. The aminoacyl-tRNA is precipitated with trichloroacetic acid (final concentration of 10%) with a carrier such as crude salmon-sperm DNA (100 μg/ml). The precipitate is collected by filtering on, for example, a glass filter disc, washed with additional 10% TCA, dried under a heat lamp and placed in scintillation fluid (0.4% Omnifluor in toluene, New England Nuclear Corp., Boston). Radioactivity is measured in a liquid scintillation spectrometer.

Elution Profile of Nonaminoacylated tRNA

An aliquot from each fraction is used to determine the amino acid acceptor activity of a particular amino acid, as described by Berg et al. (*15*). Since a microgram quantity of tRNA is sufficient in most cases for obtaining an elution profile, one MAK column fraction can be used for esterifying various amino acids. The profiles of aminoacylated and unacylated tRNA are quite similar except that there is a slight shift in peak positions.

Hybridization

Transfer and ribosomal RNAs from each fraction may be annealed to DNA, or DNA fractions may be hybridized with DNA or RNA. Weiss et al. (*16*) have been able to identify a leucine tRNA fraction which is synthesized after bacteriophage T4 infection and coded by T4 genome in *E. coli.* Likewise, Oishi (*17*) reports that by using the MAK column only one of the two complementary strands of DNA (the H strand) in *B. subtilis* is used for the transcription of transfer and ribosomal RNA.

Codon Assignment of tRNA Fractions

MAK column fractions can be tested for their codon response by using either the ribosome binding technique (*18–20*) or amino acid incorporation (*18, 21, 22*). Fractions are extensively dialyzed against H_2O and lyophilized. When aminoacyl-tRNA is fractionated on the column, the lyophilized material is directly used, while, when nonacylated tRNA is used to fractionate, tRNA is charged and reisolated before use.

Transforming Activity

DNA fractionated on the MAK column can be used for bacterial transformation. Lerman (*1*), who introduced the MAK column separation of nucleic acids, assayed the transforming activity of streptomycin resistance of various fractions from the MAK column. Roger, Beckmann, and Hotchkiss (*23*) used the MAK column to separate the two complementary strands of pneumococcal DNA and assayed for transforming activity of two genetic markers after renaturing the separated strands.

APPLICATION

Fractionation of DNA

Mandell and Hershey (*2*) developed MAK column chromatography to the point of permitting separation of DNA and RNA by their molecular size. Using the column, Hershey and Burgi (*24*) showed that phage T2 DNA can be isolated as rather uniform large molecules. Furthermore, by applying controlled shearing to the DNA they were able to identify fragments one-half and one-quarter the size of the original molecules on the column.

Sueoka and Cheng (*3*) showed that the base composition of DNA, both native and denatured, is recognized by the MAK column. DNA of various bacteria with different GC content were separated by the column (*3*). Fractionation of calf thymus DNA, mouse DNA, and *Bacillus subtilis* DNA, has shown the clear effects of base composition within each DNA sample (*25*). This effect has been used for separating two halves of native phage λ DNA (*13*).

The separation of two complementary strands of DNA on the MAK column was first demonstrated by Roger, Beckmann, and Hotchkiss on pneumococcal DNA (*23*). Since then it has also been shown in other systems (*14, 17*). The fact indicates that the fractionation of single-stranded DNA is affected not only by the GC content as previously shown (*3*), but also by the difference between A and T and/or between G and C. The column also distinguishes native and denatured forms of DNA. This property of the column has been used to isolate poly dAT of the marine crab by denaturing main DNA by the "heat-and-quickly-cool" method (*3, 26*). The poly dAT resumed a double-helical structure upon rapid cooling, thus eluting from the column at a lower concentration of salt than that for denatured main DNA.

Fractionation of RNA

The MAK column can fractionate cellular RNA into soluble RNA, 16S and 23S ribosomal RNAs in the order of increasing salt concentration (*3-5,*

27). Messenger RNA distributes in a wide range of salt concentration (*28*) and a particular mRNA can be identified by the column (*29*). All the amino acid transfer RNA comes to the soluble RNA fraction (*5*). Individual tRNAs either aminoacylated or nonacylated show unique profiles, which provide the means of identifying and fractionating tRNA.

Advantageous features of the MAK column for the fractionation of tRNA are: (1) ability to fractionate on a small scale; (2) ability to compare two samples directly for the same or different amino acid by double radioactive labeling; (3) simple and quick procedure. Its disadvantage lies in its relatively poor resolution. The rather small capacity of the column can be improved by using silicic acid instead of kieselguhr (see next section). Thus the method is particularly suitable for exploratory studies on tRNA, particularly when a number of samples are to be compared.

Ribosomal 16S and 23S RNA can be clearly separated on the MAK column. 5S RNA needs refractionation (*30, 31*).

MAS Column

Okamoto and Kawade (*32*) have reported that silicic acid (S) can hold more methylated albumin than kieselguhr, thus increasing the capacity of the column. Preparation of the MAS (methylated albumin–silicic acid) column is similar to that of the MAK column, except that 30 mg MA is added for each gram of silicic acid. The MAS column has a capacity for 5 mg tRNA per milliliter of packed column volume. To avoid a slow flow rate, silicic acid is suspended in distilled water, and the supernatant, after settling, is decanted. This washing should be repeated at least three times. The fractionation patterns of aminoacyl-tRNAs are almost identical with the MAK column. The MAS column is advantageous for large-scale fractionation.

References

1. Lerman, L. S. 1955. *Biochim. Biophys. Acta,* **18**: 132.
2. Mandell, J. D., and A. D. Hershey. 1960. *Anal. Biochem.,* **1**: 66.
3. Sueoka, N., and T. Y. Cheng. 1962. *J. Mol. Biol.,* **4**: 161.
4. Philipson, L. 1961. *J. Gen. Physiol.,* **44**: 899.
5. Sueoka, N., and T. Yamane. 1962. *Proc. Natl. Acad. Sci., U.S.,* **48**: 1454.
6. Gartland, W. J., and N. Sueoka. 1966. *Proc. Natl. Acad. Sci., U.S.,* **55**: 148.
7. Brown, G. L., and M. Watson. 1953. *Nature,* **172**: 339.
8. Brown, G. L., and A. V. Martin. 1955. *Nature,* **176**: 971.
9. Fraenkel-Conrat, H., and H. S. Olcott. 1945. *J. Biol. Chem.,* **161**: 259.
10. Sueoka, N., and J. Hardy. 1968. *Arch. Biochem. Biophys.,* **125**: 558.
11. Parr, C. W. 1954. *Biochem. J.,* **56**: xxvii.
12. Britten, R. J., and R. B. Roberts. 1960. *Science,* **131**: 32.

13. Hogness, D. S., and J. R. Simmons. 1964. *J. Mol. Biol.,* 9: 411.
14. Rudner, R., J. D. Karkas, and E. Chargaff. 1968. *Proc. Natl. Acad. Sci., U.S.,* **60**: 630.
15. Berg, P., F. H. Bergmann, E. J. Ofengand, and M. Dieckmann. 1961. *J. Biol. Chem.,* **236**: 1726.
16. Weiss, S. B., W. T. Hsu, J. W. Foft, and N. H. Scherberg. 1968. *Proc. Natl. Acad. Sci., U. S.,* **61**: 114.
17. Oishi, M. 1969. *Proc. Natl. Acad. Sci., U.S.,* **62**: 256.
18. Littauer, U. Z., M. Revel, and R. Stern. 1966. *Cold Spring Harbor Symp. Quant. Biol.,* **31**: 501.
19. Peterkofsky, A., C. Jesensky, and J. D. Capra. 1966. *Cold Spring Harbor Symp. Quant. Biol.,* **31**: 515.
20. Kano-Sueoka, T., M. Nirenberg, and N. Sueoka. 1968. *J. Mol. Biol.,* **35**: 1.
21. Yamane, T., T. Y. Cheng, and N. Sueoka. 1963. *Cold Spring Harbor Symp. Quant. Biol.,* 28: 569.
22. Kan, J., M. Nirenberg, and N. Sueoka. 1970. *J. Mol. Biol.,* **52**: 179.
23. Roger, M., C. O. Beckmann, and R. D. Hotchkiss. 1966. *J. Mol. Biol.,* **18**: 174.
24. Hershey, A. D., and E. Burgi. 1960. *J. Mol. Biol.,* **2**: 143.
25. Cheng, T. Y., and N. Sueoka. 1963. *Science,* **141**: 1194.
26. Sueoka, N., and T. Y. Cheng. 1962. *Proc. Natl. Acad. Sci., U.S.,* **48**: 1851.
27. Otaka, E., H. Mitsui, and S. Osawa. 1962. *Proc. Natl. Acad. Sci., U.S.,* **48**: 425.
28. Kano-Sueoka, T., and S. Spiegelman. 1962. *Proc. Natl. Acad. Sci., U.S.,* **48**: 1942.
29. Hayashi, M., S. Spiegelman, N. C. Franklin, and S. E. Luria. 1963. *Proc. Natl. Acad. Sci., U.S.,* 49: 729.
30. Galibert, F. 1965. *Nature,* **207**: 1039.
31. Morell, P., I. Smith, D. Dubnau, and J. Marmur. 1967. *Biochemistry,* **6**: 258.
32. Okamoto, T., and Y. Kawade. 1963. *Biochem. Biophys. Res. Commun.* **13**: 324.

PREPARATION AND PROPERTIES OF *Escherichia coli* 5S RNA

ROGER MONIER

CENTRE DE BIOCHIMIE ET DE BIOLOGIE

MOLÉCULAIRE CENTRE NATIONAL DE LA

RECHERCHE SCIENTIFIQUE

MARSEILLES, FRANCE

The larger ribosomal subunit from all organisms studied so far has been found to contain one molecule of a polyribonucleotide with a molecular weight of 40,000, the so-called "5S RNA" (*1*). The complete primary structure of several 5S RNAs is now known (*2–4*). This RNA is remarkable, particularly for the complete absence of minor bases and methylated ribose. Apart from its intrinsic interest as a universal ribosomal constituent, 5S RNA is a good model for various studies on polynucleotides, since it can be prepared in a pure form with relatively little work. As a matter of fact, it has been used by Sanger and his group (*2*) to establish the validity of their methods for sequence determination.

The procedure described below was specifically devised with reference to *Escherichia coli*, but can be applied to many microorganisms with only minor modifications.

Bacterial Strain and Growth Conditions

Although all *Escherichia coli* strains, when used with proper care, will give similar results, it is advisable to start the preparation from RNase I^-

mutants, such as the RNase I_{10}^{-} strain of Gesteland (*5*). *Escherichia coli* MRE 600 (*6*) is also frequently employed with success.

Since 5S RNA is an exclusively ribosomal component, the cells should be harvested after growth on a rich medium, to ensure a high ratio of ribosomes to total cellular weight. The following composition, in grams per liter, has been found convenient:

K_2HPO_4	21.25 gm
KH_2PO_4	10.62 gm
Difco Yeast Extract	10 gm
Glucose	10 gm

After growth at 37°C on this medium, the cells are harvested in mid log phase in a refrigerated continuous centrifuge and can be frozen and stored at –20°C for several months before extraction.

Preparation of Ribosomes

Although several methods of ribosome preparation can be used, 5S RNA of the highest purity will only be obtained when purified ribosomes serve as a starting material for RNA extraction. The technique described by Kurland (*7*) has been found particularly efficient in this respect and can be followed without modification.

Extraction of RNA

All manipulations are performed at temperatures between 0° and 5°C. The quantities indicated are those needed when 100 gm of bacteria (wet weight) are used for the preparation.

RNA is extracted from the ribosomes by means of the phenol procedure. The pellets are first frozen either in liquid nitrogen or in an acetone–Dry Ice bath, removed from the centrifuge tube without complete thawing, and homogenized in a blender with 200 ml of TM buffer (0.05 M Tris-HCl, pH 7.6; 0.01 M $MgCl_2$), 40 ml of a Macaloid* suspension (12.5 mg/ml in TM buffer), and 290 ml of phenol–water (90 : 10) (w/w). After 2 min blenderizing, the emulsion is stirred in an Erlenmeyer flask with a magnetic stirrer for 60 min. The phases are separated by low-speed centrifugation at 5°C and the aqueous (upper) layer is carefully removed with an automatic pipetting device. The phenolic phase and the denatured protein layers are washed twice by shaking with 20 ml of TM buffer and centrifugation. The three aqueous

* Macaloid, a product from the American Tansul Co., Baroid Division, National Lead Co., Houston, Texas.

layers are collected and the extracted RNA is precipitated by the addition of 0.1 volume of 20% sodium acetate, pH 5.0, and of 2.5 vol of 95% ethanol, precooled at $-10^\circ C$. After low-speed centrifugation and careful decantation, the pelleted RNA is dissolved in the minimal volume of 0.05 M NaCl and reprecipitated with 2.5 vol of cold 95% ethanol. After a third ethanolic precipitation, the RNA is dissolved in 100 ml of 0.05 M NaCl. Usually, the absorbance at 260 nm of this solution is about 30,000 to 32,000 A_{260} units.

Chromatography on DEAE-Cellulose

The RNA solution is filtered through a DEAE-cellulose column equilibrated with 0.02 M potassium phosphate buffer, pH 7.7. Whatman DE-11 DEAE-cellulose,* precycled and equilibrated according to the directions of the Whatman Advanced Ion-Exchange Celluloses Laboratory Manual, can be used for this purpose; the amount of dry DE-11 powder necessary for complete adsorption is usually 140 gm. After application of the sample, the column is washed with a 0.35 M NaCl solution until the absorbance at 260 nm in the effluent becomes negligible. 5S RNA, together with tRNA and some 16S and 23S RNAs, is then eluted with a 1 M NaCl solution. The number of A_{260} units recovered in this fraction should be about 3 000 to 3,300. The eluate from the DEAE-cellulose column is too dilute for further processing and must be concentrated. This can be achieved by filtration in an Amicon filtration unit,† equipped with a Diaflo UM 20 filter until the inside concentration reaches about 50 A_{260} units. Ethanolic precipitation is then possible. After centrifugation the RNA precipitate is dissolved in 25 ml of 0.01 M acetate buffer, pH 5.0, containing 0.75 M NaCl and 1% (v/v) methanol.

Molecular Sieving on Dextran Gel

The final separation of 5S RNA from the tRNAs and high-molecular-weight ribosomal RNAs still contained in the sample is finally achieved by molecular sieving on G-100 Sephadex,‡ equilibrated with 0.75 M NaCl, 0.01 M sodium acetate buffer, pH 5.0, and 1% (v/v) methanol. A column of 5 x 100 cm, equipped for upward circulation, can be conveniently used. A circulating pump, with a flow rate of 60 ml/hr, a UV monitoring device and a

* H. Reeve Angel and Co. Ltd, 14 New Bridge Street, London EC 4, England.
† Amicon Corporation, 25 Hartwell Ave., Lexington, Mass. 02173.
‡ Pharmacia Fine Chemicals, Upsala, Sweden.

recycling valve* are necessary to obtain the best resolution between 5S RNA and tRNA, after one recycling of the 5S RNA fractions through the column.

The 5S RNA-containing effluent is finally concentrated by filtration as before and recovered by ethanolic precipitation. The final yield is about 500 A_{260} units (ca. 20 mg).

Control of 5S RNA purity

In the absence of a biological assay, analysis by polyacrylamide gel electrophoresis is the most reliable test of purity for 5S RNA preparations of bacterial origin. Hindley (*8*) has actually shown that gel electrophoresis permits the complete separation of 5S RNA from all other metabolically stable RNA species in crude extracts from *Escherichia coli*. The procedure of Richards, Coll, and Gratzer (*9*), using a 10% gel and a discontinuous buffer system at pH 8.9, can be followed. Urea (7 M) can be incorporated in the gel and buffers, in order to permit the detection of eventual hidden breaks. Under these conditions, a pure sample of 5S RNA will give two incompletely separated bands. The partial splitting in two bands is not an indication of chemical heterogeneity, but originates from conformation modifications (*8*).

Comments on the Procedure

The procedure described above is the most convenient for the large scale preparation of electrophoretically homogeneous samples. Simpler procedures, starting from direct phenol extraction of unbroken cells or from crude ribosome fractions, lead to samples which always contain a few percent of a polyribonucleotide contaminant, which runs on polyacrylamide gels in an intermediate position between 5S RNA and tRNAs and is unrelated to 5S RNA. Therefore these simpler procedures are not to be recommended, when a final purification by gel electrophoresis is not possible. On the contrary, they are very useful for the preparation of radioactively labeled samples, which can be purified by the electrophoresis technique of Adams et al. (*10*).

Although the DEAE-cellulose step is not essential for the isolation of 5S RNA from tRNAs and high molecular weight ribosomal RNAs, it efficiently eliminates traces of proteins and polysaccharides, which usually contaminate the phenolic extracts.

* Re Cy Chrom Micro Valve type 4911 A, from L.K.B.-Produkter AB, P.O. Box 76, Stockholm – Bromma 1, Sweden.

Since it has been shown that 5S RNA can exist under several stable conformations, only one of which is recognized by the ribosomes (*11*), it is important to avoid denaturing conditions throughout the preparation.

Some Properties of *Escherichia coli* 5S RNA

5S RNA has a molecular weight of 40,000. $s_{20,w}$ values between 4.5 and 4.8 have been reported (*12*). The $\epsilon_{(P)260\ nm}$ is $6.90 \pm 0.05 \times 10^3$, when measured at 22°C in 0.15 M NaCl, 0.015 M Na citrate, pH 7.0 (*13*). As judged from studies of the optical properties of its solutions (*12-14*), 5S RNA has a high degree of secondary structure with a double helical content of about 60%. The mole fractions (±0.02) of nucleotide constituents are: AMP 0.19, UMP 0.17, GMP 0.34, CMP 0.30. The total length of the molecule is 120 nucleotides. The complete nucleotide sequence of 5S RNA from two *Escherichia coli* strains (MRE 600 and K_{12} CA 265) has been determined (*2*). In both strains, two molecular species, which differ from each other by only one nucleotide substitution, have been characterized. This multiplicity is probably related to the redundancy of 5S RNA cistrons in bacterial genomes (*15*).

References

1. Rosset, R., R. Monier, and J. Julien. 1964. *Bull. Soc. Chim. Biol.*, **46**: 87.
2. Sanger, F., G. G. Brownlee, and B. G. Barrell. 1968. *J. Mol. Biol.*, **34**: 379.
3. Forget, B. G. and S. M. Weissman. 1969. *J. Biol. Chem.*, **244**: 3148.
4. Williamson, R. and G. G. Brownlee. 1969. *FEBS Letters*, **3**: 306.
5. Gesteland, R. F. 1966. *J. Mol. Biol.*, **16**: 67.
6. Cammack, K. A. and H. E. Wade. 1965. *Biochem. J.*, **96**: 676.
7. Kurland, C. G. 1966. *J. Mol. Biol.*, **18**: 90.
8. Hindley, J. 1967. *J. Mol. Biol.*, **30**: 125.
9. Richards, E. G., J. A. Coll, and W. B. Gratzer. 1965. *Anal. Biochem.* **12**: 452.
10. Adams, J. D., P. G. N. Jeppesen, F. Sanger, and B. G. Barrell. 1969. *Nature*, **223**: 1009.
11. Aubert, M., J. F. Scott, M. Reynier, and R. Monier. 1968. *Proc. Natl. Acad. Sci., U.S.* **61**:292.
12. Boedtker, H. and D. G. Kelling, 1967. *Biochem. Biophys. Res. Commun.*, **29**: 758.
13. Scott, J. F., R. Monier, M. Aubert, and M. Reynier. 1968. *Biochem. Biophys. Res. Commun.* **33**: 794.
14. Cantor, C. R. 1968. *Proc. Natl. Acad. Sci., U.S.*, **59**: 478.
15. Smith, I., D. Dubnau, P. Morell, and J. Marmur. 1968. *J. Mol. Biol.*, **33**: 123.

POLYACRYLAMIDE GEL ELECTROPHORESIS OF RNA

C. WESLEY DINGMAN AND ANDREW C. PEACOCK
CHEMISTRY BRANCH
NATIONAL CANCER INSTITUTE
NATIONAL INSTITUTES OF HEALTH
BETHESDA, MARYLAND

Analytical fractionation of RNA by polyacrylamide gel electrophoresis has been a recent addition to the biochemist's armamentarium of techniques for studying nucleic acid structure and metabolism (e.g., *1-4*). It is probable that the full usefulness of this high-resolution technique has yet to be realized. In this chapter we will describe the relevant procedures currently in use in our laboratory.

Isolation of RNA

In most cases, the RNA is obtained as rapidly as possible by treating tissues or subcellular fractions briefly with 0.2 to 0.5% sodium dodecyl sulfate followed by shaking or stirring with an equal volume of phenol and subsequent separation of the phases by centrifugation. Following these steps, one has three options: (1) direct application to the gel of an aliquot of the aqueous phase; (2) repeated phenol treatments of the aqueous phase followed by ethanol precipitation of the RNA and solution of the washed precipitate in a suitable salt solution (e.g., 0.05 M NaCl, 2 mM Na_3EDTA); or (3) concentration of the RNA in the aqueous phase by treating it with up to 4 volumes of 90% phenol (w/w) followed by direct application to the gel of an aliquot of the aqueous phase. This latter procedure, useful in dealing with small tissue samples, does not significantly alter the amount or the relative proportions of the RNA species present in the aqueous phase, but will lead to preferential loss of DNA from the aqueous phase (*5*). Because other solutes such as salts and sucrose are also concentrated by this technique, it is important to anticipate this in the solute make up of the initial aqueous phase, since high salt concentrations (i.e., greater than about 0.15 M) interfere with the resolution of RNA species during electrophoresis. The presence of phenol does not interfere with the electrophoretic separation, and low-molecular-weight oligonucleotides present in the aqueous phase migrate with sufficient velocity to separate them from all RNA species of 24,000 molecular weight and higher. The phenol present in the aqueous phase does prevent the treatment of the RNA sample with ribonuclease or deoxyribonuclease (*2*), and estimation of the amount of RNA present by absorption at 260 nm.

The optimum concentration of RNA in the sample to be analyzed depends on its molecular weight and the monomer concentration of the gel to be used. In general, concentrations in the range of 200 to 1000 μg/ml are favorable.

The biggest obstacle to success is the lability of RNA to very small amounts of ribonuclease. Cleanliness in the laboratory, as well as in the glassware, has an importance difficult to overestimate. Whenever enzymatic digestions are performed, one should use disposable equipment, or else wash the equipment with cleaning solution followed by copious rinsing. Ribonuclease may contaminate enzymes and reagents unexpectedly. Some estimate of the extent of accidental degradation may be made from inspection of stained gels. Lines should be sharp with a clear background. Degradation products often migrate somewhat faster than the transfer RNAs, and their presence is an indicator of the extent of degradation.

The Electrophoresis Cell

RNA electrophoresis is customarily performed either with flat gel slabs permitting the simultaneous electrophoresis of several samples or in cylindrical tubes, each one of which is usually used as a separate analysis. Suitable equipment has been described by Raymond (*6*) for the slab technique (E-C Apparatus Company, Philadelphia) and by Ornstein (*7*) and Davis (*8*) for the cylindrical tube technique (Canalco Company, Rockville, Maryland).

The advantages of the slab technique are (1) all samples (including marker RNA species for reference) are analyzed under identical conditions on the same gel; (2) one can compare samples having RNAs of similar mobility; (3) there is good temperature control; (4) it is readily adapted to a two-dimensional gel (*9*). The advantages of the cylindrical technique are (1) the ease of investigation of the effects of gel composition, buffer strength, etc., when studying a single kind of sample; (2) the length of gel may be varied over wide limits with almost no restriction; (3) sample application is somewhat easier; (4) the cost of equipment is lower.

Reagents

Acrylamide and agarose must be of good quality. Not all items of commerce are equally suitable. Purification by recrystallization of acrylamide has been reported by Loening (*3*), but we have been more successful with the standard grades of acrylamide (Eastman Kodak 5521) and agarose (SeaKem, manufactured by Marine Colloid, Inc., distributed by Bausch and Lomb).

Acrylamide gels are prepared with or without the incorporation of 0.5% agarose, depending upon the acrylamide concentration. Gels 3.5% or lower in acrylamide generally require the use of agarose to provide the necessary strength and flexibility.

The following stock solutions are used:

1. Acrylamide : *N,N′*-methylenebisacrylamide ("bis"), 19 : 1, 20%. One liter contains 190 gm of acrylamide and 10 gm of methylenebisacrylamide.

2. Ammonium persulfate, 1.6%, in water, stored in a dark bottle, freshly prepared every 2 weeks.

3. Buffers (stock solutions, prepared 10 times more concentrated than the final solution): (a) Borate buffer (pH 8.3). One liter contains 106 gm Tris, 55 gm boric acid, and 9.3 gm disodium EDTA. (b) Tris-HCl buffer, 250 mM in Tris-HCl, pH 8.0. Magnesium ion, EDTA, or other ions may be added to this buffer as required.

4. *N,N,N′,N′*-Tetramethylenediamine (TEMED), 0.5% (v/v) in water. Yellow TEMED solutions should not be used.

5. Indicator solution, sucrose-EDTA-bromphenol blue. This solution is 40% in sucrose, 0.001 M in Na_2EDTA, and 0.1% in bromphenol blue, pH 6.2.

Preparation of Gel

Raymond's cell requires 160 ml of gel mixture. To make a 2.0% acrylamide-0.5% agarose gel for this cell, 0.8 gm agarose, 113 ml H_2O, 16 ml of the 20% acrylamide-bis solution, 10 ml of TEMED, 16 ml of stock buffer, and 5 ml 1.6% ammonium persulfate are used. Different acrylamide concentrations require the addition of different amounts of water to bring the volume to 160 ml. The agarose is dissolved first in the indicated amount of water by heating and stirring until the solution is brought to a boil, preferably under reflux. After 15 min at reflux temperature, the agarose solution is cooled to approximately 40° by holding the flask under running tap water at about 35°. The concentrated 20% acrylamide-bis solution, the 10x stock buffer, and the TEMED are mixed, warmed to approximately 35°, and added to the still warm (40°) agarose solution. The ammonium persulfate is added, and after rapid mixing the solution is poured into the electrophoresis cell, previously equilibrated at 20°.

The slot formation is better if the slot former is chilled in the freezer (–20°) before putting it in place. In addition, we place a small test tube, about half-full of water frozen in the –20° freezer, in front of the slot former to promote gelation of the agarose in this part of the cell which is not otherwise cooled.

Gels without agarose are prepared in the same general way, except that solutions are brought directly to 20° before mixing.

Electrophoresis

Stock buffers are diluted tenfold, placed in the electrode vessels, and the gel equilibrated by a prerun electrophoresis of approximately 1 hr. During this time the thermostat control is adjusted to 0° or any other desired temperature.

The sample to be analyzed is diluted 4 : 5 in the sucrose-bromphenol blue solution and the desired quantity (5–50μl) applied to the gel. The dye permits easier visualization of the sample, and its migration affords an indication of the progress of the electrophoresis.

The power supply is connected so that the RNA will run toward the positive pole and the voltage set to approximately 10 V/cm. The duration of the run depends upon the type of RNA under study. The 4S RNA migrates approximately 2.5 cm/hr in a 10% gel and approximately 6.5 cm/hr in a 2% gel at 0°C.

Analysis of Gels

Several means of analyzing the gels are available. Gel strips may be scanned directly in the ultraviolet. For this purpose, gels with an intrinsically low UV absorbance are necessary. This requirement is best met by using TEMED as the catalyst and low concentration gels (see above). Alternatively, gels may be stained with either methylene blue (*2*) or with "stains-all" (*9*). They may then be photographed (using a green filter) (Fig. 1) and/or scanned in the visible at the appropriate wavelength (i.e., 570 nm for "stains-all"). We have found the Gilford Recording Spectrophotometer, Model 240, with a

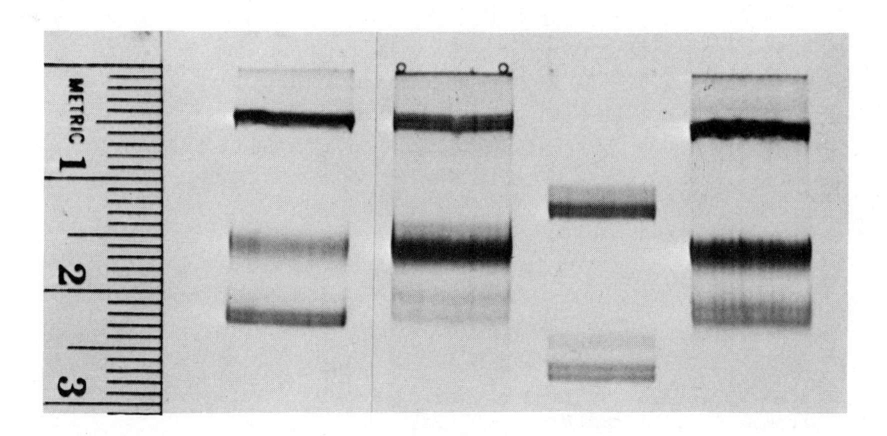

Figure 1. Electrophoresis of ribosomal RNAs; (left to right) hamster embryo from all culture, mouse mammary gland, *Escherichia coli*, and hamster embryo (repeat). The aqueous phase from the phenol treatment was applied directly to the gel as described in the text. The gel is 2.5% acrylamide–0.5% agarose composite gel. The more slowly migrating bands (5–14 mm) are the larger ribosomal RNA, and the more rapidly migrating (20–30 mm) are the smaller RNA. The band at 16 mm is DNA. Stained with "stains-all."

Gelscan attachment particularly suitable for scanning at a variety of wavelengths.

Radioactivity, present in the various RNA species, may be estimated by autoradiography after drying the gels (*10*), by planchet counting (*3*), or liquid scintillation counting (*2, 11*) of individual gel slices. Gels may be sliced in a number of ways depending upon their shape and the resolution desired. Automatic devices for slicing cylindrical gels are available (*12*). We have usually used a hand-operated slicer containing multiple stainless steel blades set 1 mm apart. Some gels may be sliced directly (e.g., composite gels containing 1.9% acrylamide and 0.5% agarose), but others slice better after being frozen on a slab of Dry Ice and then thawed just to the point where they may be sliced without undue pressure.

The determination of radioactivity in gel slices by liquid scintillation spectrometry may be accomplished in a number of ways (*2, 11, 13*). Recent studies in this laboratory indicate that a satisfactory procedure for quantitative recovery and reproducibility is to treat the gel with 0.2 ml of 0.5 N NaOH overnight at room temperature, followed by 1 ml of NCS reagent (Amersham/Searle) and 10 ml of toluene scintillator containing PPO and POPOP. Comparable results were achieved for tritium-labeled RNA when gel slices were oxidized in an automatic sample oxidizer (*14*).

The staining process results in some loss of low-molecular-weight (i.e., less than 20,000 mol wt) RNA species and nucleotides. To assay for radioactivity present in these very low-molecular-weight species, gels should be sliced and counted without staining and destaining.

Finally, the molecular weight of the various RNA species present may be estimated if suitable markers of assumed molecular weight are also present in the same gel, because in most cases the electrophoretic mobility is inversely proportional to the logarithm of the molecular weight (*4, 15, 16*).

References

1. Richards, E. G., and W. B. Gratzer. 1964. *Nature,* **204**: 878.
2. Peacock, A. C., and C. W. Dingman. 1967. *Biochemistry,* **6**: 1818.
3. Loening, U. E. 1967. *Biochem. J.,* **102**: 251.
4. Bishop, D. H. L., J. R. Claybrook, and S. Spiegelman. 1967. *J. Mol. Biol.,* **26**: 373.
5. Dingman, C. W., unpublished observations.
6. Raymond, S. 1962. *Clin. Chem.,* **8**: 455.
7. Ornstein, L. 1964. *Ann. N. Y. Acad. Sci.,* **121**: 321.
8. Davis, B. J. 1964. *Ann. N. Y. Acad. Sci.,* **121**: 404.
9. Dahlberg, A. E., C. W. Dingman, and A. C. Peacock. 1969. *J. Mol. Biol.,* **41**: 139.
10. Lim, R., J. J. Huang, and G. A. Davis. 1969. *Anal. Biochem.,* **29**: 48.
11. Dingman, C. W., and A. C. Peacock. 1968. *Biochemistry,* **7**: 659.
12. Maizel, J. V., Jr. 1966. *Science,* **151**: 988.
13. Tishler, P. V., and C. J. Epstein. 1968. *Anal. Biochem.,* **22**: 89.
14. Peacock, A. C., unpublished observations.
15. Peacock, A. C., and C. W. Dingman. 1968. *Biochemistry,* **7**: 668.
16. Leoning, U. E. 1969. *Biochem. J.,* **113**: 131.

ISOLATION OF NATURAL HOMOPOLYNUCLEOTIDES ON POLYNUCLEOTIDE CELLULOSES

MARY EDMONDS
DEPARTMENT OF BIOCHEMISTRY
FACULTY OF ARTS AND SCIENCES
UNIVERSITY OF PITTSBURGH
PITTSBURGH, PENNSYLVANIA

Polynucleotides fixed to solids are reagents uniquely suited for fractionations which can exploit the highly specific and easily controlled interactions between complementary bases of nucleic acids (*1-5*). This is most strikingly observed in the recent achievements in the recognition of gene-specific complexes by hybridization of DNA or RNA to homologous DNA immobilized either on cellulose nitrate membrane filters (*3, 4*) or on hydroxylapatite columns (*5*). Recognition of the hybrid has generally depended on radioactive labeling of one of the nucleic acids, although in a few cases the technique has been scaled up to allow isolation of chemically significant quantities of bacterial messenger RNA (*6*).

In addition to methods for the isolation of gene-specific complexes, several techniques have been described for attaching polynucleotides to solids (*7-9*). Among these the most frequently used have been the polynucleotide celluloses developed by Gilham (*9*), in which chemically synthesized oligonucleotides are covalently bonded to cellulose. Deoxythymidylate oligomers fixed in this way are effective templates for RNA polymerase and serve as primers for DNA polymerase and the terminal deoxynucleotide transferase (*10*). Columns of polythymidylate (poly T) cellulose have been used to fractionate oligomers from ribonuclease digests on the basis of the length of their contiguous adenosine sequences (*11*). They have also been particularly effective for the detection and isolation of the polyadenylate-rich (poly A) species which are found in many RNA preparations (*12, 13*). This report will focus primarily on the use of polynucleotide celluloses for the analysis and isolation of these relatively large naturally occurring homopolynucleotides. It is apparent that if the complementary polynucleotide celluloses were available, the technique could be extended readily to the isolation of other homopolynucleotides, as well as the purine and pyrimidine isostichs present in many nucleic acids. The tedium of the chemical synthesis of these celluloses has somewhat limited their use, but this situation may have been considerably improved by a recent modification in the synthesis described by Gilham (*14*). Use of the water-soluble carbodiimide derivative, *N*-cyclohexyl-N′-β-(4-methyl morpholinium) ethyl carbodiimide *p*-toluenesulfonate, permits any nucleotide or polynucleotide derivative containing a phosphomonester group to be esterified directly to cellulose.

GENERAL APPROACH

The isolations of AMP-rich oligonucleotides and polynucleotides reported thus far have been on temperature-controlled columns of poly T cellulose (*11–13*). Although these columns are highly effective in resolving short adenosine oligomers by progressive thermal denaturation (*11*), resolution of oligomers larger than ten or twelve monomers becomes difficult as thermal denaturation temperatures approach a limit. Although a lack of oligomers in this size range has generally limited investigation of this point, it is supported by theoretical considerations (*15*), as well as by extrapolation of melting-transition measurements obtained by thermal chromatography of short adenosine oligomer–poly U duplexes on hydroxylapatite columns (*16*).

It was therefore concluded that except for very large-scale operations, columns offered no advantage for the isolation of poly A which was not more than offset by the multiple analyses which could be done rapidly with minimal dilution on poly T cellulose filters. Conditions of binding and elution, in the filter assay to be described, are such that all oligomers larger than 10 or 12 monomers would be bound and released without fractionation. Other techniques, such as polyacrylamide gel electrophoresis can reveal heterogeneity within the poly A fraction.

PREPARATION AND PROPERTIES OF POLY T CELLULOSE

Materials

1. Cellulose

Selection of a commercial cellulose for polynucleotide attachment will depend on the problem under investigation. The rapid flow rates attainable with Muncktell's cellulose powder No. 400 (Grycksbro Papperbruk AB, Grycksbro, Sweden) make it suitable for column use. A more finely divided cellulose such as Whatman CF-11 is more suited to filter or batch-type operations. A comparison of these two celluloses with respect to thymidylate residues bound and poly A binding capacity is shown in Table 1.

2. Chemicals

The 5′-deoxymononucleotides obtained as ammonium salts from Schwarz Bio Research are prepared for polymerization by conversion to pyridinium salts by passing through a column containing a large excess of Dowex-50 in the pyridinium form. Pyridine used in the polymerization reaction is dried over calcium hydride before distillation *in vacuo*. Other reagents are used as obtained from commercial sources.

Chemical Synthesis of Poly T Cellulose

The synthesis is carried out in two stages essentially as described by Gilham (*9*). In the first stage, the 5′-deoxythymidylate is polymerized with dicyclohexyl carbodiimide in anhydrous pyridine. After 5 days, cellulose powder dried *in vacuo* over phosphorus pentoxide is added to the reaction along with fresh carbodiimide and the reaction is continued for another 5 days. After the appropriate washings, the cellulose is stored either as a dry powder or in aqueous suspension.

Yield of Polythymidylate Residues

A phosphate analysis on an ashed sample of the dried material is the most convenient and reliable method for estimating the number of TMP residues bound. The Whatman CF-11 preparation of Table 1 contained more than 30% of the TMP added to the reaction. Examination of the poly A binding capacity of both preparations of Table 1 indicates that only a fraction of this TMP (about 15%) functions in binding poly A in the assay

TABLE 1. Properties of Two Polythymidylate Celluloses

Source of cellulose	μmoles phosphate per mg cellulose	μmoles AMP bound per mg cellulose
1. Muncktell	0.03	0.005
2. Whatman CF-11	0.18	0.025

used here. Much of the TMP must either be at sites inaccessible to poly A or in oligomers too short to form a stable complex under the conditions of this assay. Gilham has estimated the length of the oligomers to average 5 to 10 residues (*9*) which is considerably below the number required to give the most stable duplex (*17*). Jovin and Kornberg (*10*) found the stability of a poly dA (200 residues)-poly dT cellulose duplex could be greatly increased

* Absorbance units are defined throughout as the product of the absorbance at 260 nm (1-cm light path) and the volume in milliliters.

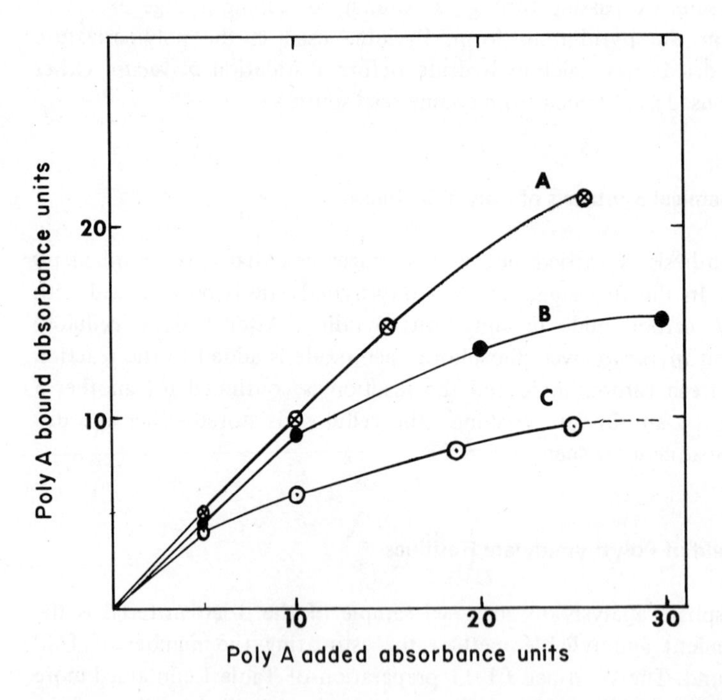

Figure 1. Binding capacity of poly T cellulose. Increased amounts of ^{3}H-poly A were hybridized to 125 mg of poly T cellulose as described. See footnote for definition of absorbance units.* Curves *A* and *B* were obtained with two different poly T celluloses both synthesized on Whatman CF-11 cellulose. Curve *C* is preparation B after three cycles of poly A binding and release.

by reacting it with thymidine triphosphate and DNA polymerase. The melting temperature after repair with DNA polymerase was 26° higher. The existence of such short poly dT sequences in poly T cellulose may be particularly advantageous for the isolation of poly A from natural nucleic acid mixtures, since an enhanced discrimination against weaker interactions involving mismatched or looped-out sequences might be expected.

Binding Capacity of Poly T Cellulose

A hybridization-competition assay has been used to determine the binding capacity of poly T cellulose. A series of reactions containing identical tracer amounts of tritium-labeled poly A but increased quantities of unlabeled poly A are reacted with an identical weight of poly T cellulose in the assay to be described shortly. Once poly T sites are saturated, the competition for binding sites in those reactions containing excess poly A will be manifested as a diminished radioactivity released in the poly A fraction. Saturation curves constructed from such data can be used to measure poly T cellulose binding capacity. In Fig. 1 the amount of poly A bound has been plotted against the total poly A present in the reaction. The amount of poly A bound, in this case, is the product of that fraction of the total radioactivity bound and the total poly A present in each assay.

Poly A assays can be based on absorbance rather than radioactivity measurements, but these can be complicated by the release of ultraviolet absorbing impurities from poly T cellulose at the low ionic strengths used to recover poly A in these assays.

Storage and Reutilization

The simplicity of regeneration of poly T cellulose compensates for the inconveniences involved in the chemical synthesis necessitated by the current lack of commercial sources. Preparations stored in 1 M sodium chloride are stable for years. Not unexpectedly, however, repeated use results in gradual losses of binding capacity. This is seen in curve *C* of Fig. 1 where binding capacity declined about 40% after three successive cycles of poly A binding and elution. Most of this loss is incurred during the washing of the cellulose at low ionic strength where it becomes partially colloidal and thus difficult to recover by centrifugal force. This wash is necessary, however, to remove radioactivity which accumulates on cellulose treated with highly radioactive nucleic acid preparations.

Enzymatic degradation is another potential source of loss of binding capacity in experiments where RNA has been treated with DNase before

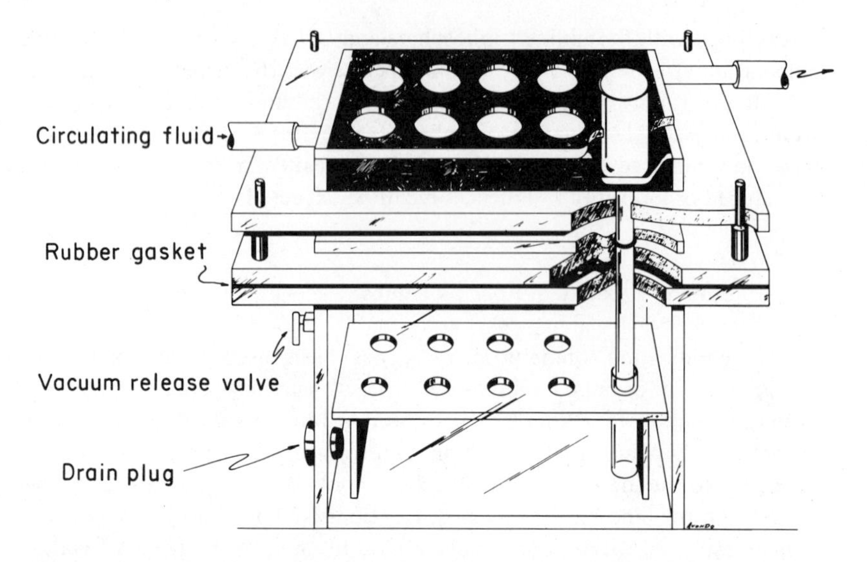

Figure 2. Temperature-controlled vacuum filtration apparatus. The upper box containing the circulating fluid is easily lifted from the four brass pegs on which it rests. The top layer of the vacuum box contains a series of holes ringed with Neoprene gaskets to seal the stems of the Hirsh funnels mounted in the upper box. A series of hand-tightened screws (not shown) seal the upper layer of the vacuum box through a thin pad of rubber attached to the edge of the lower layer. Vacuum is applied at an outlet (not shown) on the upper left side of the box.

applying to poly T cellulose. An excess of ethylenediaminetetraacetic acid (EDTA) routinely included in the binding step and in the poly T cellulose during storage should remove this hazard.

ISOLATION OF AMP-RICH POLYNUCLEOTIDES ON POLY T CELLULOSE

Apparatus

The procedure was developed for multiple isolations of small quantities of labeled poly A from radiolabeled heterogeneous RNA preparations. However, with minor modifications it should be suitable for detection of a variety of homopolynucleotide interactions. Each of the three stages of the assay, i.e., binding, washing, and elution, are carried out at a characteristic temperature and ionic strength. To facilitate multiple analyses, a vacuum filtration box was designed to permit fine control of the temperature of each stage. The apparatus shown in Fig. 2 consists essentially of two boxes, one for vacuum filtration and another on top of it containing a set of wells surrounded by circulating fluid for temperature control. The two boxes are connected through the stems of Hirsh funnels (15 ml; medium porosity) in each well, which pass through air tight "O"-rings sealed into the top of the vacuum filtration box. Poly A is released from the poly T cellulose layered in the funnel into a test tube mounted below each funnel in a removable rack. The rack is replaced by a set of tall narrow beakers during preliminary washing cycles to prevent contamination with the high levels of radioactivity normally washed out of the cellulose.

Isolation of Nucleic Acids

The method selected for the isolation of poly A from cells should minimize the extraction of protein and DNA which can be trapped in cellulose. The hot phenol detergent method of Warner et al. (*18*) gives high yields of poly A from HeLa and Ehrlich ascites cells in RNA preparations relatively free of protein and DNA. In many RNA preparations from mammalian cells and nuclei, however, low levels of polynucleotides accompany poly A through the hybridization assay. Since poly A levels never exceed 1% of the total RNA of such preparations, even very low levels of contaminants can obscure the presence of an AMP-rich fraction. Such contaminants could arise either through nonspecific binding to poly T cellulose or through hydrogen bonding of adenine isostichs contained within larger polynucleotide chains. Treatment with appropriate nucleases before analysis can remove either form of contamination, but it should be noted in the case of the latter possibility, that release of a large adenine isostich would be indistinguishable from a free poly A of the same size. The question of the origin of a poly A species derived from a ribonuclease treated preparation is beyond the scope of this discussion. It has been considered in some detail in reports of the characterization of poly A in Ehrlich ascites (*13*) and HeLa cells (*19*).

The nuclease treatment used here removes RNA contaminants of the type just described from either the total or nuclear RNA of HeLa and ascites cells. RNA from other sources may require modified treatments or none at all. Poly A essentially free of other nucleotides can be recovered from such preparations with minimal losses and no degradation. These criteria have been established through routine monitoring of the assay with an added labeled poly A marker.

Preparation of the RNA Sample

The RNA (250–500 μg) pellet recovered from 70% ethanol is dissolved in 1.5 ml of 0.033 M Tris pH 7.5. If necessary, the sample should be

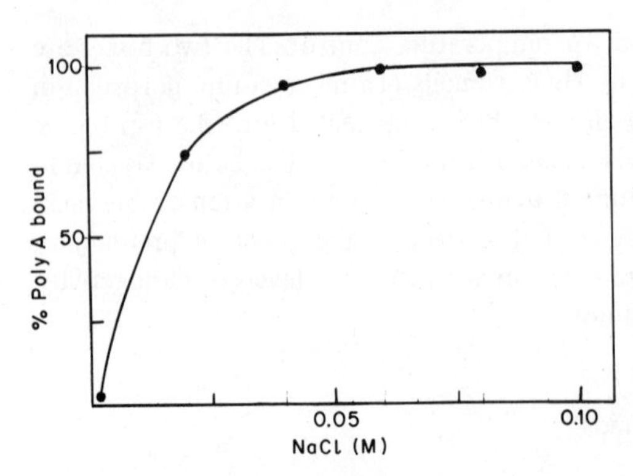

Figure 3. Effect of salt concentration on poly A binding to poly T cellulose. ^{3}H-poly A (250 μg 0.05 μCi) was bound to a series of poly T cellulose filters. Each filter was washed with NT buffer at 2°C as described. The temperature was raised to 23° and each funnel was washed with 0.01 M Tris pH 7.5 containing sodium chloride at one of the concentrations shown.

supplemented with an RNase-free carrier RNA to attain these levels. Although normally unnecessary, in some cases it may be useful to suppress low levels of endogenous nuclease activity in the sample by adding sodium dodecyl sulfate (SDS). Levels up to 0.20% in the assay are without effect on the poly A recovery. It is convenient to take aliquots of the RNA sample for the determination of acid-insoluble radioactivity at this stage of the assay.

Nuclease Treatment of the Polynucleotide Sample

The following components are added to the cold RNA sample:

$MgCl_2$	4	μmoles
DNase I*	10	μg
T_1 RNase†	100	units (about 5 μg)
RNase A‡	0.10	μg
^{3}H-Poly A§	0.025	μCi(-50 μg)
H_2O to final volume	2.00	ml

After 15 min at 37°, the sample is made 0.05 M with EDTA to stop DNase activity. NaCl is added to give a final concentration of 0.10 M in a total volume of 3.0 ml before chilling to 0°.

Hybridization to Poly T Cellulose

The vacuum filtration apparatus is precooled to maintain a constant temperature of 2°C within each funnel. A slurry containing 125 mg (dry

* Preparation DPFF from Worthington Biochemicals.

† Takadiastase ribonuclease T_1 is the Sankyo preparation purchased from Calbiochem. Corp.

‡ RNase A was preparation XII from Sigma Biochemicals.

§ ^{3}H-poly A and unlabeled poly A were from Miles Laboratories.

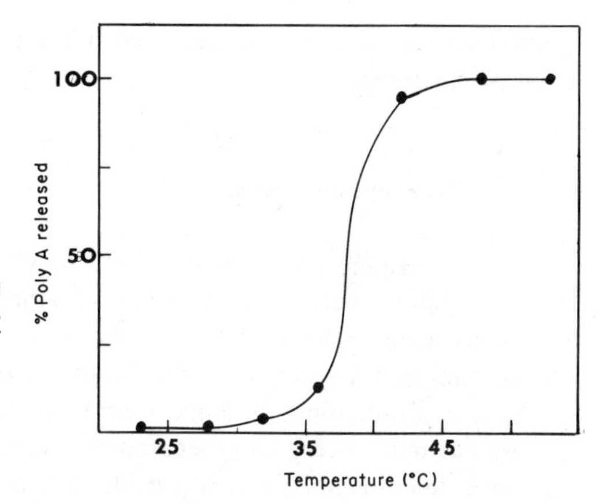

Figure 4. Effect of temperature on poly A binding. Conditions were those of Fig. 3. Washing of filters was as described in methods. ^{3}H-poly A was eluted with 0.1 M sodium chloride; 0.01 M Tris pH 7.5 at the temperatures shown.

weight) of poly T cellulose in 0.1 M NaCl: 0.01 M Tris pH 7.5 (NT) is allowed to settle on each funnel before it is tightly packed by evacuating the lower chamber. An identical aliquot of the cellulose slurry is added to the cold RNA solution. After 15 min at 0°, during which time it is occasionally resuspended by stirring, the slurry is carefully poured onto the poly T cellulose layer on the evacuated funnel and the liquid is rapidly removed.

Washing of Poly T Cellulose

Each funnel is rapidly washed with four 3-ml portions of NT solution at 2°C. The temperature of the filtration unit is raised to 23° and an identical washing with NT solution is carried out at 23°. Examination of Fig. 3 shows no poly A is released at this salt concentration and temperature.

Elution of Poly A from Poly T Cellulose

Poly A can be released either by reducing the salt concentration (Fig. 3) or by raising the temperature (Fig. 4). The former approach was originally selected because ionic strength is more easily varied over a broad range than is temperature. The ready availability of the temperatures selected here, i.e., 2° and 23°C, eliminates the necessity for temperature-controlled funnels.

A series of seven 1-ml aliquots of 0.01 M Tris pH 7.5 at 23° are collected in sequence in test tubes placed below each funnel. The process is carried out rapidly to minimize cooling of the filter which could occur during a prolonged period of solvent evaporation from the funnel. Radioactivity measurements can be made directly on aliquots of the total 7-ml eluate, although the poly A content of the RNA is most conveniently and accurately determined on aliquots acid precipitated with carrier and assayed simultaneously with acid-precipitated aliquots of the RNA sampled before nuclease treatment.

Recovery of Poly A

The remaining eluate is made 0.1 M with NaCl and 200 μg of RNA carrier is added before 2½ volumes of ethanol. The AMP-rich fraction is recovered from ethanol by centrifugation for 30–45 min. at 15,000 *g*. Nucleotide analyses can be carried out on an alkaline hydrolyzate of the precipitate. Contamination is most easily detected with ^{32}P-labeled RNA. A convenient method for rapid separation and analysis of ^{32}P in AMP and CMP has been described (*13*) which provides a measure of RNA contamination of the AMP-rich fraction. If poly A is to be recovered for further analysis, the

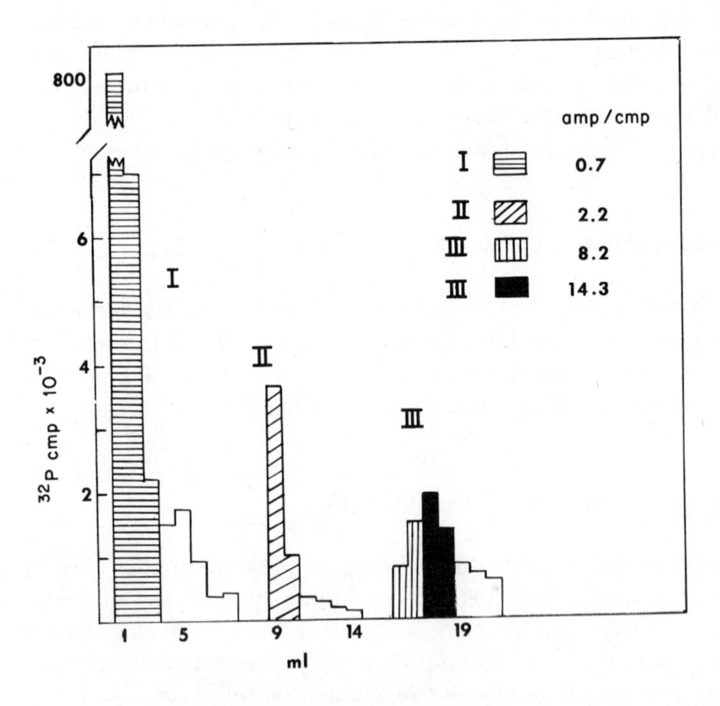

Figure 5. Fractionation of the total RNA of Ehrlich ascites cells on poly T cellulose filters. RNA from cells labeled with ^{32}P for 3 hr *in vitro* in a medium described for HeLa cells (*19*). The isolated RNA (4 absorbance units) was preincubated for 15 min at 37°C with 100 units of T_1RNase before treating with poly T cellulose. Fractions from region I were collected at 2°C in 0.1 M NaCl; 0.01 M Tris pH 7.5. Fractions of II with same solution at 23°C. Fractions of III were collected in 0.01 M Tris pH 7.5 at 23°C. The ^{32}P content of AMP and CMP in alkaline hydrolyzates of the indicated fractions was done by anion-exchange chromatography (*13*).

eluate is taken up into a 10-ml Luer-Lok syringe to which a Swinnex-13 filter unit has been attached. It is slowly forced through a 13-mm, 0.22-μ Type G-S Millipore filter in the filter unit. This removes fine particles of poly T cellulose which may be released into the eluate at low ionic strength.

A complete analysis carried out on ^{32}P-labeled RNA from the total RNA of Ehrlich ascites cells is shown in Fig. 5. Although this particular sample received less than the usual nuclease treatment, nonspecific binding even at low temperatures accounted for less than 1% of the total radioactivity. This material was readily washed out by raising the temperature to 23° before the AMP-rich fraction was eluted by lowering the ionic strength. A poly A fraction almost free of other nucleotides can be recovered if the sample is given the more extensive nuclease treatment described above.

Isolation of AMP-Rich Polynucleotides from Other Sources

This procedure has been used to detect AMP-rich polynucleotides in both yeast and *E. coli* (*20*). The isolation and preparation of polynucleotides from such sources for poly T cellulose hybridization may require modification. Methods for the isolation of undegraded RNA preparations from these sources are available in the literature. It should be added that the requirement for nucleases to remove polynucleotide contaminants may differ from those described here for mammalian RNA preparations.

References

1. Bautz, E. F. K., and B. D. Hall. 1962. *Proc. Natl. Acad. Sci., U.S.*, **48**: 400.
2. Bolton, E. T., and B. J. McCarthy. 1962. *Proc. Natl. Acad. Sci., U.S.*, **48**: 1390.
3. Nygaard, A. P., and B. D. Hall. 1963. *Biochem. Biophys. Res. Commun.*, **12**: 98.
4. Gillespie, D., and S. Spiegelman. 1965. *J. Mol. Biol.*, **12**: 829.
5. Miyazawa, Y., and C. A. Thomas, Jr. 1965. *J. Mol. Biol.*, **11**: 223.
6. Riggsby, W. S. 1969. *Biochemistry*, **8**: 222.
7. Adler, A. J., and A. Rich. 1962. *J. Am. Chem. Soc.*, **84**: 3977.
8. Britten, R. J. 1963. *Science*, **142**: 963.
9. Gilham, P. T. 1964. *J. Am. Chem. Soc.*, **86**: 4982.
10. Jovin, T., and A. Kornberg. 1968. *J. Biol. Chem.*, **243**: 250.
11. Gilham, P. T., and W. E. Robinson. 1964. *J. Am. Chem. Soc.*, **86**: 4985.
12. Edmonds, M., and R. Abrams. 1963. *J. Biol. Chem.*, **238**: PC 1186.
13. Edmonds, M., and M. G. Caramela. 1969. *J. Biol. Chem.*, **244**: 1314.
14. Gilham, P. T. 1968. *Biochemistry*, **7**: 2809.
15. Magee, W. S., Jr., J. S. Gibbs, and B. H. Zimm. 1963. *Biopolymers*, **1**: 133.
16. Niyogi, S. K., and C. A. Thomas, Jr. 1968. *J. Biol. Chem.*, **243**: 1220.
17. Walker, P. M. B., *Progr. Nucl. Acid Res. Mol. Biol.*, 1969. **9**: 305.

18. Warner, J. R., R. Soeiro, H. C. Birnboim, M. Girard, and J. E. Darnell. 1966. *J. Mol. Biol.*, **19**: 349.
19. Edmonds, M., M. H. Vaughan, Jr., and H. Nakazato. 1971. *Proc. Natl. Acad. Sci., U.S.*, **68** (in press).
20. Edmonds, M., and D. Kopp. 1970. *Biochem. Biophys. Res. Commun.*, **41**: 1531.

SECTION

E

CHEMICAL MODIFICATIONS

CHEMICAL MODIFICATIONS OF tRNA AND rRNA

F. VON DER HAAR, E. SCHLIMME, AND D. H. GAUSS

MAX-PLANCK-INSTITUT FUR EXPERIMENTELLE MEDIZIN

GÖTTINGEN, GERMANY

INTRODUCTION

Chemical modifications of nucleic acids are undertaken to obtain an insight into their three-dimensional structure as well as their function by discriminating between reactive and nonreactive nucleotides. A chemical reaction suitable for this purpose should fulfill certain conditions: (1) The reagent has to have a well-defined specificity toward certain elements of the nucleic acid. (2) The overall three-dimensional structure of the nucleic acid must be preserved during the reaction, neither chain must break nor should unzipping of base-paired regions occur. (3) The reaction product should be stable during analysis and easily detectable, e.g., by radioactive labeling or spectroscopic measurements. (4) If side reactions occur they should not disturb the analytical evaluation of the main reaction.

The chemical modifications of RNA can be divided grossly into two groups. Those concerned with the minor nucleosides are rather specific and selective since the molecules they attack are chemically quite distinct from the common nucleosides. The work concerned with the common nucleosides is much more complicated. Even a reaction studied extensively in model systems is often difficult to understand when carried out with RNA. tRNA and rRNA are believed to be globular nucleic acid molecules with well-defined three-dimensional structures. Different tRNA species can behave quite differently in the same chemical reaction (which was the authors' experience in a comparative study of $tRNA^{Phe}_{yeast}$ with $tRNA^{Ser}_{yeast}$ as well as $tRNA^{Ser}_{yeast}$ monomer and $tRNA^{Ser}_{yeast}$ dimer).

The analytical procedure after a modification can cover (1) investigations on changes in chemical and physical properties of the nucleic acid, (2) analysis of nucleoside composition yielding the type and percentage of nucleosides accessible for the reagent, and (3) sequence analysis and location of the modification. Only in the last case a maximum of unambiguous information is obtained.

Many chemical modifications of tRNA are done with the intention of obtaining insight into enzymatic mechanisms. Also with this type of work a well-defined chemical reaction and location of the modification is necessary prior to the enzymatic investigation.

This chapter deals with chemical modifications of tRNA (*1-5*) and rRNA (*4-8*) carried out on the polymer level and published in the years 1966 to 1969. Only in certain cases work with monomers or DNA or earlier publications are mentioned. With regard to series of papers by the same authors on the same procedure, only the latest publications are cited. Enzymatic methods, physical modifications, mode of action of antibiotics, radiation-induced reactions, methods for preparing radioactive labeled compounds, reactions with protons, hydroxyl ions or metals, and modified polymers generated by random *in vivo* incorporation of modified monomers are not within the scope of this chapter. Biological tests are also not mentioned here. The earlier literature is covered by a number of reviews and books (*1-11*); furthermore the references cited with respect to the special reactions may be taken as guide references.

CHEMICAL MODIFICATIONS LOCALIZED IN THE RNA SEQUENCE

Cyanoethylation with Acrylonitrile

RNAs can be cyanoethylated with acrylonitrile. Studies with nucleic acid components showed that the reagent (I)

$$CH_2{=}CH{-}CN \; (I) \; + \; \text{uracil (II)} \longrightarrow \text{N-3-}(CH_2{-}CH_2{-}CN)\text{-uracil (III)}$$

I II III

reacts preferentially with pseudouridine (II), inosine, and 4-thiouridine yielding N1-substituted (e.g., III) or S-substituted nucleosides (*12–14*). Reactions with other nucleosides, e.g., uridine, are slower. Various investigations with unfractionated tRNA from yeast or *E coli* showed different reactivities among the pseudouridines and inosines and a change of three-dimensional structure together with loss of biological activity (*15–21*). The common sequence T–Ψ–C–G of tRNA bulk is not cyanoethylated (*19, 21*). The effect of cyanoethylation of inosine in the anticodon in $tRNA^{Ala}_{yeast}$ and $tRNA^{Val}_{yeast}$ on the coding properties of tRNA was checked (*22, 23*). Primarily in $tRNA^{Ala}_{yeast}$ mainly one of the two pseudouridines and the inosine are cyanoethylated; as shown by an enzymatic digest the pseudouridine in the T–Ψ–C–G sequence is resistant against the reagent (*24*) indicating an exposed anticodon and a buried T–Ψ–C–G loop.

Reaction with Water-Soluble Carbodiimide

The water soluble carbodiimide, *N*-cyclohexyl-*N'*-β-(4-methylmorpholinium)-ethylcarbodiimide cation (IV) reacts preferentially with the N3 of uridine (V, VI) and the N1 of guanosine; it also attacks pseudouridine and thymidine.

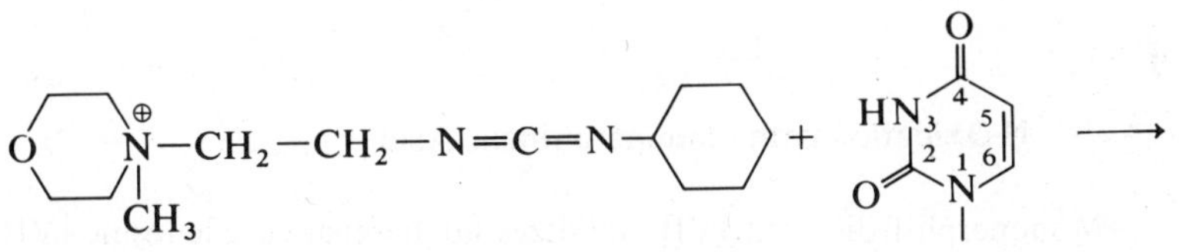

$$\mathrm{O(C_4H_8)\overset{\oplus}{N}(CH_3)-CH_2-CH_2-N{=}C{=}N-C_6H_{11}} \quad + \quad \text{V} \longrightarrow$$

IV V

$$\mathrm{O(C_4H_8)\overset{\oplus}{N}(CH_3)-CH_2-CH_2-NH-C({=}N-C_6H_{11})-N3(\text{uracil})}$$

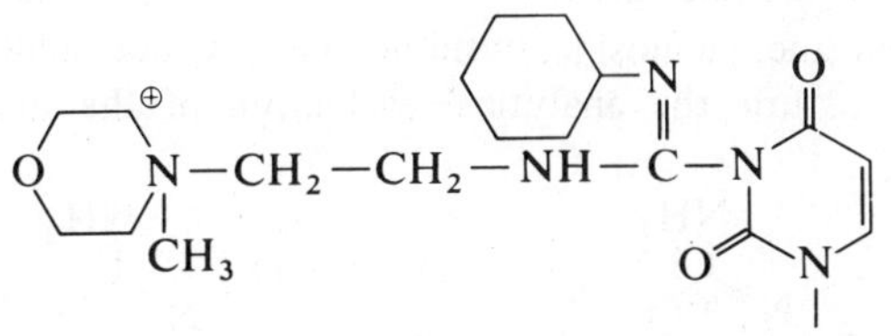

VI

In RNA the reaction occurs preferentially with nucleosides that are not base paired (*10, 25–30*). The reagent was applied to ribosomes (*31*) and unfractionated tRNA (*32–35*). Other RNase splitting patterns (*25, 36*) than usual, and lowered or destroyed activities in protein biosynthesis, were found. Under physiological conditions the tRNA seems to be very compact but with an exposed anticodon. Reaction of $tRNA^{Val\,1}_{yeast}$ with water soluble carbodiimide was investigated under conditions stabilizing the structure (*37*). Brostoff and Ingram (*38*) reacted the water soluble carbodiimide with $tRNA^{Ala}_{yeast}$. A sequence analysis clearly proved the high reactivity of the nucleosides of the anticodon loop whereas no reactivity of the T–Ψ–C–G loop was found indicating that this loop is somehow involved in the tertiary structure.

Nitrous Acid Treatment

Reaction of nucleic acids with nitrous acid results mainly in deamination, especially of cytidine. The reaction was carried out frequently (*39*) in DNA mutation research, and was also applied to tRNA from *E. coli* (*40*). Nelson, Ristow, and Holley (*41*) reacted $tRNA^{Ala}_{yeast}$ with nitrous acid and with the brominating agent *N*-bromosuccinimide. The locations of the most sensitive nucleosides were then investigated by enzymatic digest and chromatographic analysis. It was found that the T–Ψ–C–G loop is less reactive than expected from the cloverleaf arrangement and presumably is protected in some way by the tertiary structure. Carbon and Curry (*42, 43*) treated tRNA from *E. coli* with nitrous acid. From biological tests they conclude that the anticodon of one $tRNA^{Gly}$ species might be altered from UCC to UCU.

N-Oxidation with Monoperphthalic Acid

Monoperphthalic acid (VII) oxidizes nonbasepaired adenosine (VIII) to adenosine-1-*N*-oxide (IX) in RNA selectively under defined conditions (*44–46*). No oxidation of adenosine in poly A·poly U can be observed (*45, 47*). Side reactions include generation of cytidine-3-*N*-oxide and some destruction of adenosine, guanosine, cytidine, and pseudouridine. These side reactions do not disturb the analytical evaluation of the main reaction.

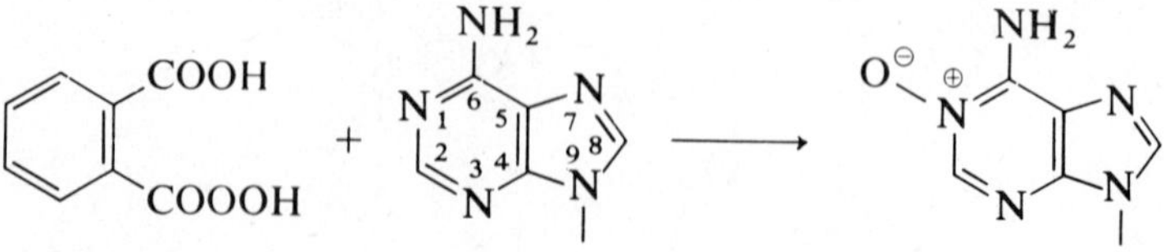

VII + VIII ⟶ IX

Cramer et al. (*44, 47*) found that in $tRNA^{Phe}_{yeast}$ at 20° only the 3′-terminal adenosine and three adenosines in the anticodon loop are oxidized, although 17 adenosines are present in the primary structure out of which 9 are unpaired in the cloverleaf. $tRNA^{Phe}_{yeast}$ oxidized only in the anticodon loop could be obtained by oxidizing a tRNA lacking the terminal adenosine followed by readdition of the adenosine with C-C-A pyrophosphorylase (EC 2.7.7.20). $tRNA^{Ser}_{yeast}$ monomer and dimer were subjected to N-oxidation. Like the oxidation of $tRNA^{Phe}_{yeast}$ oxidation of $tRNA^{Ser}_{yeast}$ indicated a rather compact molecule with an accessible anticodon loop and terminal adenosine. In the dimer only the terminal adenosines are oxidizable. From these experiments and other data Cramer et al. proposed a model for the tertiary structure of tRNA (*44, 45*). In *E. coli* 23S, 16S, and 5S rRNA the amount of oxidizable adenosine was established using the N-oxidation method; the data were compared with the proposed structures of 5S rRNA (*48*).

Reaction with Kethoxal

Litt et al. (*49, 50*) reacted $tRNA^{Phe}_{yeast}$ with tritiated kethoxal (X) (β-ethoxy-α-ketobutyraldehyde) at 37°. It reacts rapidly only with guanosine (XI) in single-stranded regions (XII); an unfolded cloverleaf exhibits seven guanosines in the single-stranded regions. After a 30-min reaction only the

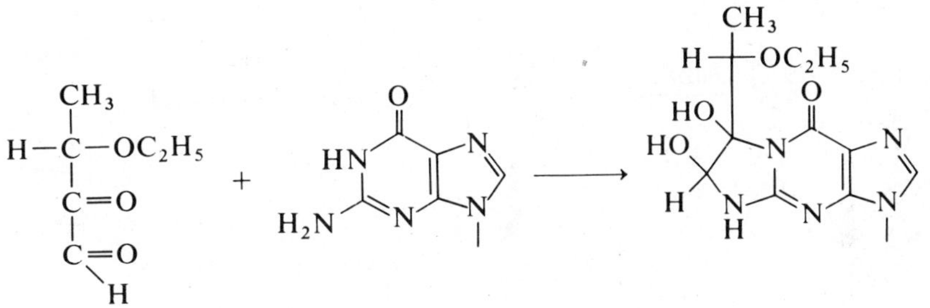

X XI XII

guanosines at position 20 (in the dihydrouridine loop of the tRNA chain) and 34 (in the anticodon) were labeled to a significant extent. This is consistent with the tertiary structure model proposed for $tRNA^{Phe}_{yeast}$ by Cramer et al. (*44, 45*).

Methylation

Methylation of tRNA with dimethyl sulfate (*51–56*) in water or dimethyl formamide (*55, 56*) as well as of rRNA (*55*) leads mainly to 7-methylguanosine, 1-methyladenosine, and 3-methylcytidine. Denaturation and chain breaks can occur (*53, 55*). The enzymatic splitting pattern of rRNA is altered after methylation (*36*). Other methylating agents were also used (*57, 58*). Bollack, Dirheimer, and Ebel showed that methylation of $tRNA^{Phe}_{yeast}$ attacks the anticodon, the 3′-end and the dihydrouridine loop but not the T-Ψ-C-G loop (*59*). Enzymatic methylation plays an important role *in vivo* and can also be carried out *in vitro* (*60*).

CHEMICAL MODIFICATIONS OF MINOR NUCLEOSIDES

The existence of several minor nucleosides with chemical characteristics differing from the common nucleosides offers the opportunity of selective chemical attack at these positions.

Borohydride Reduction of Dihydrouridine and Uridine

In model reactions it was found, that dihydrouridine (XIII) (*61*), N^4-acetylcytidine (XV) (*62*) and 4-thiouridine (XVII) (*63*) react with excess borohydride ($BH_4^{\ominus}$) at about pH 9 to give *N*-ribosyl-3-ureidopropan-l-ol

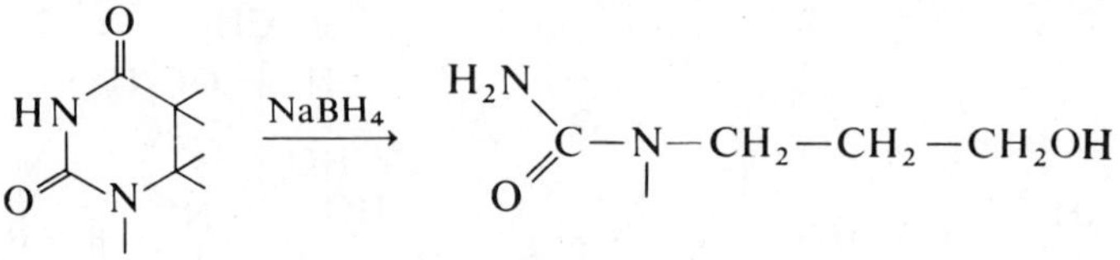

XIII $\xrightarrow{NaBH_4}$ $H_2N-CO-N(-)-CH_2-CH_2-CH_2OH$ XIV

XV $\xrightarrow{NaBH_4}$ XVI

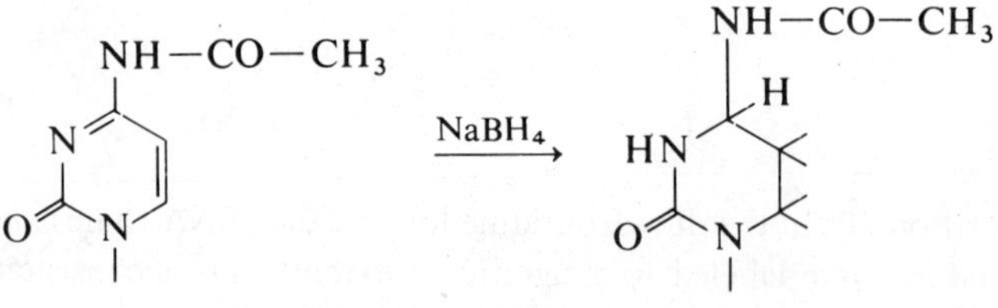

XVII $\xrightarrow{NaBH_4}$ XVIII

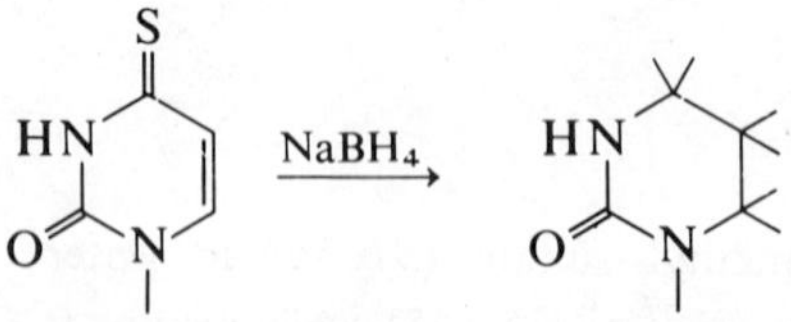

(XIV), *N*-ribosyl-N^4-acetyl tetrahydrocytosine (XVI) and *N*-ribosyl-2-oxo hexahydropyrimidine (XVIII). Reduction of various tRNAs (*64–66*) using tritiated borohydride gives products analogous to those obtained by reduction of the monomers. The N^4-acetylcytidine of $tRNA^{Ser}_{yeast}$, situated in a base-paired region in the cloverleaf structure, is not reduced by borohydride (*67*) in contrast to the three dihydrouridines. In $tRNA^{Phe}_{yeast}$ the unknown nucleoside Y in the anticodon loop and the two dihydrouridines are attacked (*67*).

Borohydride can also reduce uridine (thymidine) on the monomer and polymer level in a UV-induced reaction to dihydrouridine (*68, 69*). The dihydrouridine produced in this way is then susceptible to the dark reaction yielding finally, at least in part, the ureido compound. When about 8 uridines per mole of tRNA are reduced to dihydrouridines the biological activity is lost (*70, 71*). It is not known if in tRNA this reaction is random or there are preferred regions of attack.

Reactions of Pseudouridine

Because of the C-glycosidic linkage pseudouridine (XIX, ΨMP) can be cleaved by UV-irradiation and by periodate oxidation at ~ pH 9 at slightly elevated temperatures (*72–75*). The nature of Z is not established. In periodate oxidation 5-formyluracil (XX) and 5-carboxyluracil are formed. At

$$\text{XIX} \xrightarrow[\text{NaIO}_4]{h\nu \text{ or}} \text{XX} + \text{Z} + \text{H}_3\text{PO}_4$$

about pH 7 the reaction is slower and only 5-carboxyluracil is found. In the UV-induced cleavage there are side reactions with cytidine and uridine (*76*). There is evidence that the reaction can be used to split $tRNA^{Ala}_{yeast}$ into large segments. The same authors reporting on the biological activity of dissected $tRNA^{Ala}_{yeast}$ used the classical limited enzymatic digest instead of the UV-splitting, however, (*77*). For further reactions of pseudouridine, e.g., see Cyanoethylation (p. 644), and miscellaneous modifications (p. 656).

Reactions of Thiouridine

Thiouridine can be modified chemically in different ways (see pp. 644 and 656); the reduction with borohydride is cited under Borohydride Reduction (p. 648). 4-Thiouridine (XXI) can be oxidized by OsO_4 (*78*) or $NaIO_4$ (*79, 80*) yielding a sulfonic acid derivative (XXII). The sulfonic acid group is a very good leaving group and this derivative can be converted to uridine in the presence of OH^- as nucleophile (XXIII) or to cytidine and derivatives thereof if ammonia or primary amines are present in the reaction (XXIV).

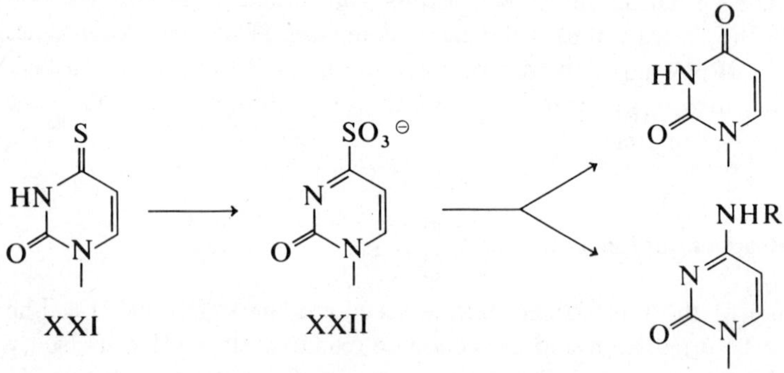

XXI → XXII → XXIII / XXIV

Since periodate is more specific, this reagent has to be preferred; only the 3′-ribose is attacked in addition to 4-thiouridine (see Modifications of the 3′-end) and even this can be prevented if the tRNA is charged with its amino acid prior to oxidation.

In a second reaction type the enol form of 4-thiouridine is fixed. Thus in $tRNA^{Tyr}_{E.\ coli}$ containing two 4-thiouridine residues on oxidation with I_2/KI an intramolecular disulfide bridge is formed (*81, 82*). If a second sulfur compound is present mixed disulfides can be produced. The iodine oxidation, which leads to partial inactivation of the acceptor function of tRNA, can be reverted by reduction with thiosulfate (*83*). 2-Thiouridine, also found in tRNA from *E. coli*, behaves similarly on I_2/KI oxidation (*84*). The thio group can be titrated with *p*-mercury benzoate ion (*85, 86*). In the same type of reaction cyanogen bromide (*87*), ethylenimine (*88*), and *N*-ethylmaleimide (*83*) can be added to the 4-thio group (XXV, XXVI).

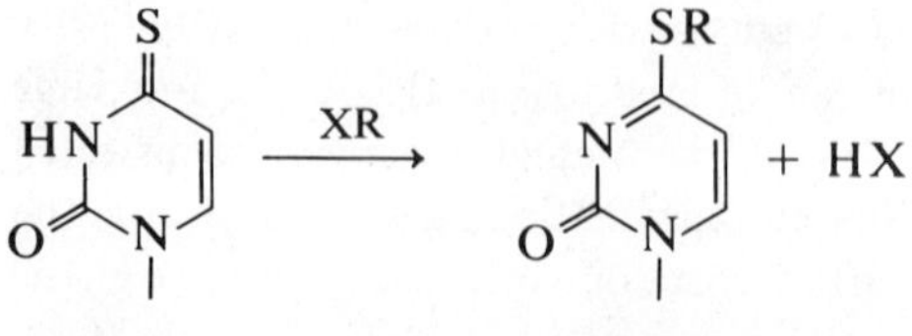

XXV —XR→ XXVI + HX

4-Thiouridine exhibits an UV-maximum at 330 nm (*89*), which lies outside the normal UV absorption of nucleic acids and thus offers the possibility of a specific photochemistry of 4-thiouridine in the RNA chain. 4-Thiouridine in the hexadecyltrimethylammonium salt of tRNA from *E. coli* was oxidized to uridine by irradiation at 330 nm in an air-saturated tertiary butanol. In the presence of primary amines in the irradiation mixture 4-thiouridine was converted partially to uridine and partially to cytidine derivatives (*89*). A very interesting photochemical conversion of 4-thiouridine could be achieved in the case of $tRNA^{Val}_{E. coli}$ (*90, 91*). In this case a covalently linked addition product between 4-thiouridine (position 8 in the tRNA chain) and a cytidine (position 13) was found. This indicates the proximity of the two nucleosides in the tRNA three-dimensional structure. The exact chemical nature of the product has not been established.

Modification of the Minor Nucleoside Next to the Anticodon

The still unknown fluorescent nucleoside Y next to the anticodon of $tRNA^{Phe}_{yeast}$ and $tRNA^{Phe}_{wheat\ germ}$ undergoes modification to a still fluorescent Y′ on slightly alkaline treatment (*92*). Under more drastic conditions it is converted to a nonfluorescent constituent susceptible to pancreatic RNase splitting (*93*). Its borohydride reduction was mentioned earlier (p. 649). On acidic treatment the glycosidic linkage of the nucleoside Y is split giving the opportunity to isolate the base for investigations of its structure (*94*). The remaining $tRNA^{Phe}_{yeast}$ is then susceptible to the Whitfeld degradation (*95, 96*) (see Modifications of the 3′-end, p. 654). In this way a $tRNA^{Phe}_{yeast}$ with a single chain break in the anticodon loop is produced by chemical attack; the two halves can be separated and studied independently (*96*).

In the position analogous to the nucleoside Y in $tRNA^{Phe}_{yeast}$, in certain other tRNAs, for instance, $tRNA^{Ser}_{yeast}$ another minor base, N^6-(Δ^2- isopentenyl)-adenosine is found. During reaction with I_2/KI this base is modified to a so far unknown product (*97*). It is known, however, that the reaction is no simple addition of iodine to the double bond of the N^6-substituent. Spectral shifts indicate that also the purine ring system is involved in this reaction. On reaction with permanganate the isopentenyl residue is removed yielding an adenosine instead of the modified adenosine (*98*).

MODIFICATIONS OF THE 3′-END OF tRNA

The 3′-end of tRNA is susceptible to specific enzymatic degradation and restoration with C-C-A pyrophosphorylase yielding a molecule with a modified 3′-end (*99*). The terminal *cis* diol group can be oxidized by periodate ($IO_4^{\ominus}$) yielding two aldehyde groups which can undergo further reactions, i.e., condensation, oxidation, and reduction, without affecting the remainder of the molecule. The 3′-end contains the amino acid acceptor function of the molecule and modification may lead to insight in the aminoacylation mechanism. Furthermore two types of models of the three-dimensional structure of tRNA have been suggested in which the 3′-end may or may not be involved. Thus chemical modification may help to indicate the correct model. The specific reactivity of the 3′-end, unique to all tRNAs, makes it attractive to look for possibilities of isomorphous substitution for X-ray analysis.

Exchange of 3′-Terminal Cytidines and Adenosine

The most clearcut modification of the 3′-end is achieved by incorporation of ATP or CTP analogs using the C-C-A pyrophosphorylase (EC 2.7.7.20). This enzyme is able to regenerate the missing 3′-terminal -C-C-A sequence of tRNA. tRNA with a missing 3′-terminal AMP (tRNA-C-C) or missing 3′-terminal CMP(s) and AMP (tRNA-C and tRNA . . .) can be obtained either by chemical degradation (see Modification of the 3′-terminal ribose, p. 654) or enzymatic digestion. The last nucleotide, AMP, is readily removed by low amounts of snake venom phosphodiesterase (EC 3.1.4.1) in a short time. It is somewhat more difficult to obtain tRNA-C or tRNA . . . Carefully controlled conditions have to be used; the general procedure given by Zubay (*99*) must be adapted for each specific tRNA (*100*). The terminal -C-C-A is regenerated with ATP and CTP or their analogs in the regeneration mixture.

Thus the nucleoside antibiotics tubercidin, toyocamycin, and sangivamycin can be incorporated in tRNA (*101*). Formycin (XXVIII, formycin triphosphate) is a fluorescent nucleoside and can be used for fluorescence studies of tRNA (*102*) (XXIX). The incorporation of the thioanalog (XXX, adenosine-5′-*O*-(1-thiotriphosphate)) yields the first modification of the phosphodiester bond (XXXI) and gives the possibility to study the influence of the terminal phosphodiester linkage on enzymatic reactions of tRNA (*103*). Incorporation of analogs containing heavy atoms (e.g., 5-iodocytidine (XXXIV), x = Br or I) is interesting for isomorphous substitution in X-ray analysis of crystalline tRNA (*104, 105*) (XXXV, XXXVI). Incorporation of AMP into tRNA-C gives tRNA-C-A and of UMP into suitable intermediates gives tRNA-C-U, tRNA-U, tRNA-U-C, and tRNA-U-A. This can be achieved by using rat liver C-C-A-pyrophosphorylase which obviously is less specific than the equivalent enzymes from microorganisms (*106*–108). Specific deamination of the 3′-terminal adenosine to an inosine can be tRNA—C—C + HO—P(=O)(OH)—O—P(=O)(OH)—O—P(=O)(OH)—O—[ribose, 2′,3′-OH; base NH$_2$, H, N]

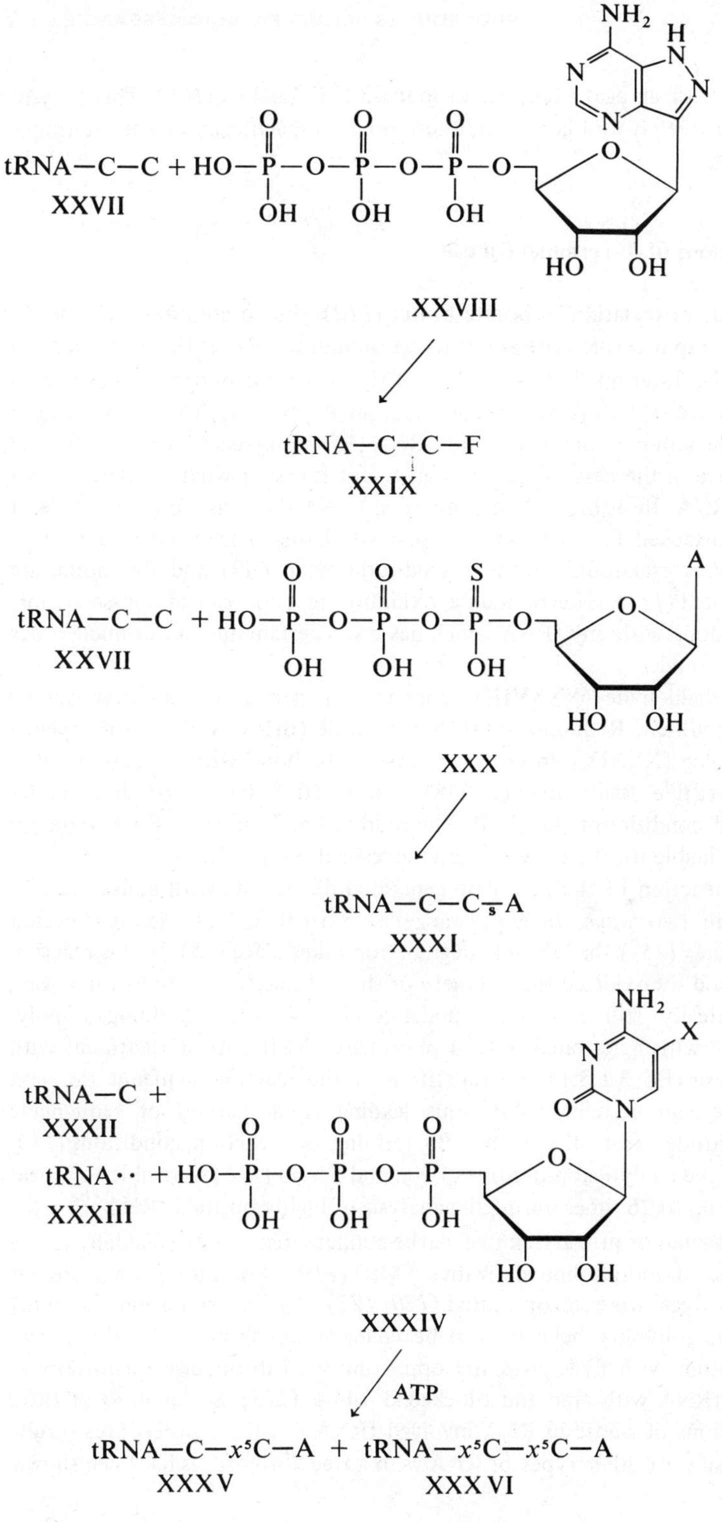

XXVII

XXVIII

↓

tRNA—C—C—F

XXIX

tRNA—C—C + HO—P(=O)(OH)—O—P(=O)(OH)—O—P(=S)(OH)—O—[ribose-A, HO, OH]

XXVII

XXX

↓

tRNA—C—C$_s$A

XXXI

tRNA—C +

XXXII

tRNA·· + HO—P(=O)(OH)—O—P(=O)(OH)—O—P(=O)(OH)—[ribose, HO, OH; cytosine base NH$_2$, X]

XXXIII

XXXIV

↓ ATP

tRNA—C—x^5C—A + tRNA—x^5C—x^5C—A

XXXV XXXVI

achieved with an agal adenylate deaminase (EC 3.5.4.6) (*109*). This enzyme requires a free *cis* diol group and therefore acts specifically on the 3′terminal adenosine.

Modifications of 3′-Terminal Ribose

On aminoacylation in borate buffer (*110*), the borate ester (XL) of the *cis* diol group prevented the aminoacylation indicating that the reaction takes place at the 3′-terminal ribose moiety of the tRNA. Oxidation with periodate splits the C2′–C3′ linkage of the 3′-terminal ribose (XXXVII) yielding a dialdehyde group at the end of the tRNA (*95*). This reaction is specific and quantitative in the case of the monomer, but it is somewhat restricted in the case of tRNA. In 4-thiouridine containing tRNA the 4-thiouridine is at least partially attacked (see Thiouridine, p. 650). Using more reagent and raised temperatures pseudouridine (see Pseudouridine, p. 649) and thymidine are also attacked (*111*). The periodate oxidation reaction can of course analogously be done with any tRNA which has lost a certain number of nucleotides from the 3′-end.

The dialdehyde (XXXVIII) generated by periodate oxidation can be further modified. Reduction with borohydride ($BH_4^{\ominus}$), yields a ring-opened ribose analog (XXXIX). In order to prevent the borohydride reactions cited in Borohydride Reduction (p. 648) the reaction has to be done under controlled conditions (*112, 113*). The oxidized and reduced tRNA seems to be very valuable for further enzymatic investigations (*112*).

The reaction of the periodate generated dialdehyde with amines can be directed in two ways. In a pH range of 5 to 9 and at slightly elevated temperatures (45°) the Whitfeld degradation takes place (*95*). In this reaction the base and the oxidized sugar moiety of the last nucleotide are lost in a way, mechanistically still not quite understood (*114, 115*), yielding a polynucleotide with a terminal 3′-(2′-) phosphate (XLII). After treatment with phosphatase (EC 3.1.3.1) and repetition of the reaction sequence the next nucleoside can be eliminated, thus leading to a method of sequencing polynucleotides from the 3′-end. By refining the reaction conditions (*116, 117*) and the analytical procedures Uziel and Khym (*118*) were able to repeat the cycle up to 26 times during the analysis of highly purified $tRNA^{Phe}_{E.\ coli\ B}$.

Hydrazines or primary amines can be condensated with the dialdehyde to a six-membered morpholine derivative (XLI) (*119*). For fluorescence studies acriflavin dyes were incorporated (*120–122*). By incorporation of bulky groups the solubility behavior can be changed, aiding in purification (*119*). Condensation with dyes gives the opportunity to distinguish quantitatively between tRNA with free and blocked 3′-ribose (*123*). So far most of these modifications of ribose in RNA involved tRNA but these procedures should apply equally to other types of RNA with a free 3′-ribose, as has been shown

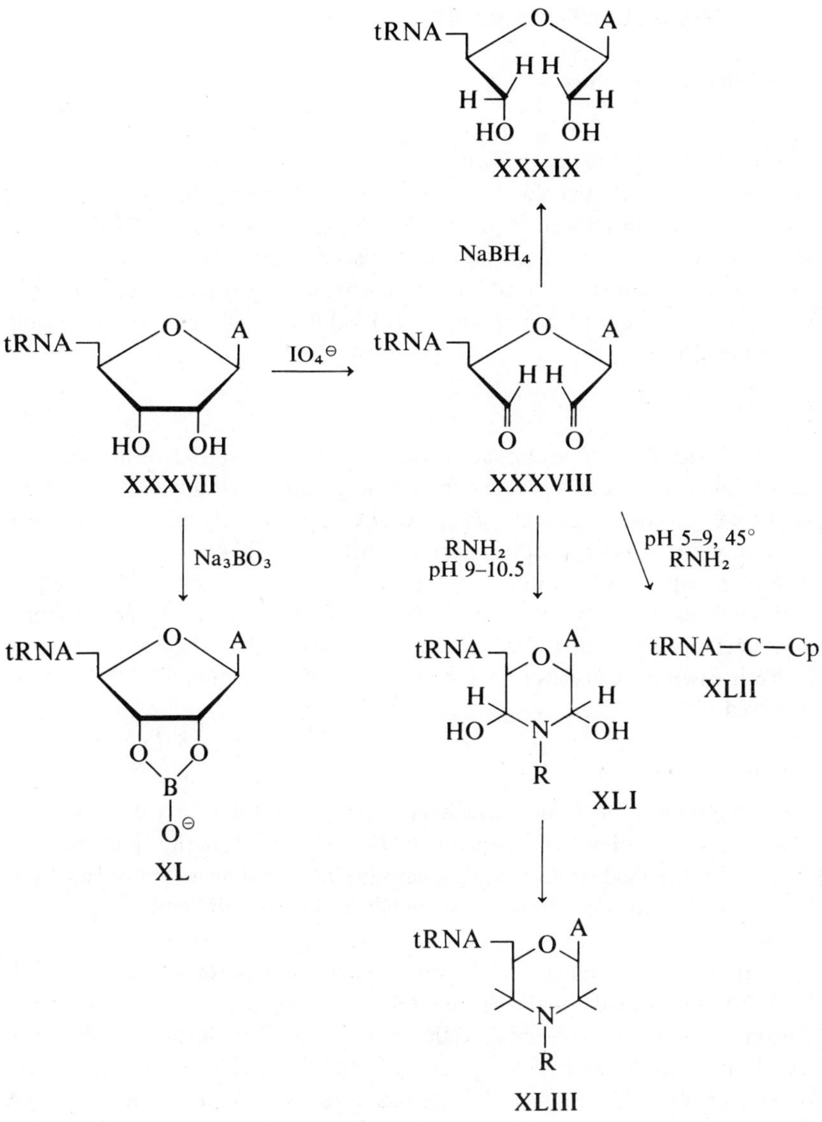

tRNA— O A
H H
H— —H
HO OH
XXXIX

$NaBH_4$

tRNA— O A
HO OH
XXXVII

$IO_4^{\ominus}$

tRNA— O A
H H
O O
XXXVIII

Na_3BO_3

RNH_2
pH 9–10.5

pH 5–9, 45°
RNH_2

tRNA— O A
O O
B
$O^{\ominus}$
XL

tRNA— O A
H H
HO N OH
R
XLI

tRNA—C—Cp
XLII

tRNA — O A
N
R
XLIII

by the incorporation of isonicotinic acid hydrazide into rRNA (*124*). The 3,5-dihydroxymorpholine derivative (XLI) generated during the reaction between amine and dialdehyde is hydrolytically unstable. It can be reduced with borohydride to a morpholine derivative free of hydroxyl groups (*125, 126*) (XLIII). In order to prevent elimination this reaction has to be done at a pH above 9 and therefore all the reactions of the borohydride with tRNA listed in Borohydride Reduction (p. 648) have to be taken into account.

MISCELLANEOUS MODIFICATIONS

Formaldehyde Treatment

Modification of RNA (e.g., *127, 10*) as well as of nucleosides or nucleotides (*128–130*) with formaldehyde has been repeatedly investigated. The point of attack and the mode of the reaction are strongly dependent on the reaction conditions and must be established in each case. rRNA (*129*) as well as tRNA from yeast and *E. coli* (*130–134*) were reacted with formaldehyde. Under defined conditions methylene bridges were formed (*129, 132*). In the presence of Mg^{2+} tRNA exhibits a tight core not accessible for the reagent (*134*).

Acetylation

Acetylation with acetic anhydride of tRNA and rRNA in dimethyl formamide preferentially attacks the amino group of cytidine (*135*). Whereas in tRNA no chain breaks occur, rRNA is degraded to a polydisperse mixture with an average sedimentation constant of 10S (*136*). Acetylation of tRNA in aqueous solution with acetic anhydride leads to preferential acetylation of the 2′-hydroxy group of the ribose moieties; the acetyl group may be removed from tRNA by treatment with hydroxylamine (*137–139*). In both cases of acetylation a change of tertiary structure of tRNA can be observed.

Halogenation

Chlorination of ribonucleic acid can be carried out with *N*-chlorosuccinimide in dimethyl formamide (*140, 141*). Primarily guanosine is involved in the modification while adenosine and cytidine are inert. In rRNA, but not in tRNA, chain breaks were detected; the three-dimensional structure was altered.

Bromination of nucleic acids was investigated extensively by Ebel, Weil, et al. They found that in dimethyl formamide 8-bromoguanosine, 5-bromouridine, and 5-bromocytidine are formed. The tertiary structure of tRNA and rRNA from yeast is altered, and in rRNA chain breaks were observed (*141, 142*). Biological functions were tested with brominated tRNA and with tRNA brominated after detachment of the 3′-terminal -C-C-A and subsequent readdition of the C-C-A (*56, 143*). Brominated tRNA recognizes other than the normal triplets (*144*). Similar studies with methylation were reported in the same papers. Nelson et al. (*41*) brominated $tRNA^{Ala}_{yeast}$ with *N*-bromosuccinimide, carried out an enzymatic digest and separated the oligonucleotides. Comparing these oligonucleotides with those from nitrous acid treatment they could draw conclusions on the tRNA tertiary structure (see Nitrous Acid Treatment, p. 646).

Iodination of tRNA with I_2/KI in aqueous solution (see Modification of Minor Nucleosides, p. 651) and with ICI in dimethyl formamide is reported. Uridine is specifically attacked under these conditions, yielding an uridine derivative different from 5-iodouridine (*145*). Another report states that 5-iodouridine, 5-iodocytidine, and 8-iodoguanosine are produced (*146*).

Reaction with Semicarbazide

Only cytidine is reported to be attacked on treatment of tRNA with semicarbazide at pH 4.2, but specificity of the reaction with tRNA has not been investigated by sequence analysis. The reaction can be monitored by UV-spectroscopy in the range of 300–340 nm (*147–150*).

Reaction with Hydroxylamine

At pH 10 hydroxylamine attacks uridine preferentially, and at pH 6 it attacks cytidine (*10*). The selectivity in both cases is rather low (*151*). The reaction is specific for cytidine if *O*-alkylhydroxylamines are used (*151*). It offers an advantage for the preparation of large segments of tRNA (*152*).

Reaction with Hydrazine

Uridine, thymidine, and cytidine are attacked by hydrazine. In this reaction on the polymer level the pyrimidines are finally excised and ribosylhydrazones are formed (*153–155*). In contrast to unsubstituted hydrazine, 2,4-dinitrophenylhydrazine reacts randomly (*154*).

Reaction with Girard-P-Reagent

Acetohydrazide pyridinium chloride, Girard-P-reagent, reacts selectively with cytidine (*156*). Optimal reaction conditions are pH 4.2, 37°C. The reaction is completed after 90 hr. In $\text{tRNA}^{\text{bulk}}_{\text{yeast}}$ (*157*) 70% of the total cytidine residues were reacted. No chain breaks could be observed but the three-dimensional structure of tRNA was remarkably altered. An increasing number of modified nucleosides corresponds to a decrease of aminoacylation.

Various Chemical Modifications

Allyl bromide in dimethyl formamide reacts with the guanosine in tRNA and rRNA yielding 7-allylguanosine (*158, 159*). Higher allylation extents lead to degradation; rRNA is degraded more rapidly than tRNA. Englander et al. investigated the hydrogen–tritium exchange with RNA using Sephadex chromatography and rapid microdialysis (*160–163*). With the powerful oxidizing agents OsO_4 (*164–166*) and MnO_4^- (*167, 168*) RNAs can be oxidized; under controlled conditions pyrimidines in single-stranded regions are preferentially attacked (*164, 167*). Other methods of modification of RNA include hydrogenation (*169*), reaction of the 5′-terminal phosphate (*170, 171*), and the use of hydrogen peroxide (*172*), succinic anhydride (*173*), and 4-(*N*-2-chlorethyl-*N*-methylamino)-benzaldehyde (*174*). Guanosine in RNA is attacked by N^2-fluorenylhydroxylamine and *N*-acetoxy-2-acetylaminofluorene (*175–177*). The enzymatic cleavage of tRNA after modification of guanosine with glyoxal was investigated (*178, 179*).

CONCLUSION

The chemical modifications of ribonucleic acids together with other methods have led to insight into the structure and function of these molecules. Even when X-ray analyses and more detailed results of other methods, e.g., spectroscopic methods, will be available, chemical modifications may be important with respect to certain structural problems, e.g., in solution.

Sequence analysis of a nucleic acid by electron microscopy will perhaps become another field for chemical modifications (*10, 180, 181*).

Finally aminoacyl-tRNAs, peptidyl-tRNAs, and their derivatives, e.g., acylated species, can be prepared by chemical reactions (*182–194*); in elucidation of the mechanism of protein biosynthesis these compounds are interesting tools.

NOTE ADDED IN PROOF

Chemical modifications of tRNA and rRNA reported since this chapter was submitted for publication are as follows: cyanoethylation of *E. coli* $tRNAF^{Met}$ (*195, 196*) and $tRNA^{Arg}$ (*197*); reaction of yeast $tRNA^{Ala}$ with water soluble carbodiimide (*198*); treatment of yeast $tRNA^{Ala}$ with nitrous acid (*199*); hypermethylation of *E. coli* tRNA (*200*); chain scission of tRNA at 7-methylguanosine (*201*); spin labeling of 4-thiouridine in *E. coli* tRNA (*202*); modification of thiopyrimidines in *E. coli* tRNA with *S*-benzylthioisothiourea (*203*); reaction of 4-thiouridine in *E. coli* tRNA bulk (*204*) and $tRNA^{Tyr}$ (*205*) with cyanogen bromide; reaction of thiolated bases in rRNA with *N*-ethylmaleimide (*206*); conversion of 4-thiouridine in *E. coli* tRNA to N^4-methylcytidine with OsO_4 (*207*); modification of isopentenyl adenosine in yeast $tRNA^{Tyr}$ with bisulfite (*208*); terminal labeling of RNA with dimedone (*209*), and of rabbit reticulocyte rRNA with ^3H-isoniazid (*210*) after periodate treatment; reaction of *E. coli* $tRNA^{Phe}$ with formaldehyde (*211*); inactivation of yeast tRNA with diketen (*212*); modification of yeast $tRNA^{Ser}$ with iodine (*213*); modification of cytidine with methoxamine in yeast $tRNA^{Val}$ (*214*) and *E. coli* $tRNA^{Tyr}$ (*215*) with location of the modifications within the sequence; interaction of tRNA with 4-(*N*-2-chloroethyl-*N*-methylamino)-benzaldehyde (*216*); modification of *E. coli* tRNA with *N*-acetoxy-2-acetylaminofluorene (*217*); preparation and further reactions of aminoacyl-tRNA, acyl-aminoacyl-tRNA and peptidyl-tRNA (*218–223*); preparation of spin-labeled *E. coli* aminoacyl-tRNA (*224*); structure elucidation of the minor nucleoside Y (*N**) in yeast $tRNA^{Phe}$ is reported (*225*).

References

1. Zachau, H. G. 1969. *Angew. Chem.*, **81**: 645. *Angew. Chem. Intern. Ed.*, **8**: 711.
2. Miura, K.-I. 1967. *Progr. Nucleic Acid Res. Mol. Biol.*, **6**: 39.
3. Ebel, J. P. 1968. *Bull. Soc. Chim. Biol. France,* **50**: 2255.
4. Cox, R. A. 1968. *Quart. Rev.*, **22**: 499.
5. Fröholm, L. O., and S. G. Laland (Eds.). 1968. Structure and function of transfer RNA and 5S-rRNA. Universitetsforlaget, Oslo and Academic Press, New York.
6. Monier, R. 1968. *Bull. Soc. Chim. Biol. France.*, **50**: 2277.
7. Osawa, S. 1968. *Ann. Rev. Biochem.*, **37**: 109.
8. Spirin, A. S., and L. P. Gavrilova. 1969. The ribosome. Springer-Verlag, New York.
9. Grossman, L., and K. Moldave (Eds.). 1967–1968. *Methods Enzymol.* **XII A, B**
10. Kochetkov, N. K., and E. I. Budowsky. 1969. *Progr. Nucleic Acid Res. Mol. Biol.*, **9**: 403.
11. Michelson, A. M. 1963. The chemistry of nucleosides and nucleotides. Academic Press, New York.
12. Ofengand, J. 1965. *Biochem. Biophys. Res. Commun.* **18**: 192.
13. Ofengand, J. 1967. *J. Biol. Chem.*, **242**: 5034.
14. Chambers, R. W. 1965. *Biochemistry.*, **4**: 219.
15. Lake, A. V., and G. M. Tener. 1966. *Biochemistry.*, **5**: 3992.
16. Millar, D. B. 1969. *Biochim. Biophys. Acta.*, **174**: 32.
17. Yoshida, M., and T. Ukita. 1965. *J. Biochem.*, **57**: 818.
18. Yoshida, M., and T. Ukita. 1965. *J. Biochem.*, **58**: 191.
19. Yoshida, M., and T. Ukita. 1966. *Biochim. Biophys. Acta.*, **123**: 214.
20. Yoshida, M., and T. Ukita. 1968. *Biochim. Biophys. Acta.*, **157**: 455.
21. Yoshida, M., and T. Ukita. 1968. *Biochim. Biophys. Acta.*, **157**: 466.
22. Yoshida, M., Y. Furuichi, T. Ukita, and Y. Kaziro. 1967. *Biochim. Biophys. Acta.*, **149**: 308
23. Yoshida, M., Y. Furuichi, Y. Kaziro, and T. Ukita. 1968. *Biochim. Biophys. Acta,* **166**: 636.
24. Yoshida, M., Y. Kaziro, and T. Ukita. 1968. *Biochim. Biophys. Acta,* **166**: 646.
25. Ivanova, O. I., D. G. Knorre and E. G. Malygin. 1967. *Mol. Biol.*, (*Russ.*) **1**: 335.
26. Gilham, P. T. 1962. *J. Am. Chem. Soc.*, **84**: 687.
27. Naylor, R., N. W. Y. Ho, and P. T. Gilham. 1965. *J. Am. Chem. Soc.*, **87**: 4209.
28. Ho, N. W. Y., and P. T. Gilham. 1967. *Biochemistry,* **6**: 3632.
29. Metz, D. H., and G. L. Brown. 1969. *Biochemistry,* **8**: 2312.
30. Augusti-Tocco, G., and G. L. Brown. 1965. *Nature,* **206**: 683.
31. Budker, V. G., A. S. Girshovich, D. G. Knorre, and L. J. Stefanovich. 1969. *Mol. Biol.* (*Russ.*), **3**: 250.

32. Bekker, J. M., Yu. N. Molin, A. S. Girshovich, and M. A. Grachev. 1969. *Mol. Biol. (Russ.)*, **3**: 366.
33. Knorre, D. G., E. G. Malygin, G. S. Mushinskaya, and V. V. Favorov. 1966. *Biokhimiya,* **31**: 334.
34. Girshovich, A. S., D. G. Knorre, O. D. Nelidova, and M. N. Ovander. 1966. *Biochim. Biophys. Acta,* **119**: 216.
35. Metz, D. H., and G. L. Brown. 1969. *Biochemistry,* **8**: 2329.
36. Brownlee, G. G., F. Sanger, and B. G. Barrell. 1968. *J. Mol. Biol.,* **34**: 379.
37. Girshovich, A. S., M. A. Grachev, and L. V. Obukhova. 1968. *Mol. Biol. (Russ.)*, **2**: 351.
38. Brostoff, S. W., and V. M. Ingram. 1967. *Science,* **158**: 666.
39. Shapiro, R., and S. H. Pohl. 1968. *Biochemistry,* **7**: 448.
40. Carbon, J. A. 1965. *Biochim. Biophys. Acta,* **95**: 550.
41. Nelson, J. H., S. C. Ristow, and R. W. Holley, 1967. *Biochim. Biophys. Acta,* **149**: 590.
42. Carbon, J. and J. B. Curry. 1968. *J. Mol. Biol.,* **38**: 201.
43. Carbon, J. and J. B. Curry. 1968. *Proc. Natl. Acad. Sci., U.S.A.,* **59**: 467.
44. Cramer, F., H. Doepner, F. von der Haar, E. Schlimme, and H. Seidel. 1968. *Proc. Natl. Acad. Sci., U.S.A.,* **61**: 1384.
45. Cramer, F., V. A. Erdmann, F. von der Haar, and E. Schlimme. 1969. *J. Cell Physiol.,* **74**: 163 (*Suppl.*).
46. Gangloff, J., and J. P. Ebel. 1969. *Bull. Soc. Chim. Biol. France,* **50**: 2335.
47. von der Haar, F., V. A. Erdmann, E. Schlimme, and F. Cramer, submitted for publication.
48. Erdmann, V. A., and F. Cramer. 1968. *Nature,* **218**: 92.
49. Litt, M., and V. Hancock. 1967. *Biochemistry,* **6**: 1848.
50. Litt, M., 1969. *Biochemistry,* **8**: 3249.
51. Pillinger, D. J., J. Hay, and E. Borek. 1969. *Biochem. J.,* **114**: 429.
52. Hay, J. 1969. *Biochem. J.,* **114**: 28P.
53. Zakharyan, R. A., T. V. Venkstern, and A. A. Baev. 1967. *Biokhimiya,* **32**: 1068.
54. Zakharyan, R. A., T. V. Venkstern, and A. A. Baev. 1968. *Biokhimiya,* **33**: 111.
55. Bollack, C., G. Keith, and J. P. Ebel. 1965. *Bull. Soc. Chim. Biol. France,* **47**: 765.
56. Weil, J. H. 1965. *Bull. Soc. Chim. Biol. France,* **47**: 1303.
57. Griffin, B. E. 1968. *Methods Enzymol.* **XIIA**: 141.
58. Matsumoto, H., and H. H. Higa. 1966. *Biochem. J.,* **98**: 20C.
59. Bollack, C., G. Dirheimer, and J. P. Ebel. 1969. *FEBS-Mtg., Madrid, Abstr.* **127**.
60. Borek, E., and P. R. Srinivasan. 1966. *Ann. Rev. Biochem.,* **35**: 275.
61. Cerutti, P., and N. Miller. 1967. *J. Mol. Biol.,* **26**: 55.
62. Miller, N., and P. Cerutti. 1967. *J. Am. Chem. Soc.,* **89**: 2767.
63. Cerutti, P., J. W. Holt, and N. Miller. 1968. *J. Mol. Biol.,* **34**: 505.
64. Cerutti, P. 1968. *Biochem. Biophys. Res. Commun.,* **30**: 434.
65. Shugart, L., and M. P. Stulberg. 1969. *Federation Proc.,* **28**: Abstr. 859.
66. Molinaro, M., L. B. Sheiner, F. A. Neelon, and G. L. Cantoni. 1968. *J. Biol. Chem.,* **243**: **1277**.
67. Igo-Kemenes, T., and H. G. Zachau. 1969. *European J. Biochem.,* **10**: 549.
68. Cerutti, P., K. Ikeda, and B. Witkop. 1965. *J. Am. Chem. Soc.,* **87**: 2505.
69. Balle, G., P. Cerutti, and B. Witkop. 1966. *J. Am. Chem. Soc.* **88**: 3946.
70. Cerutti, P. 1968. *Methods Enzymol.* **XIIB**: 461.
71. Adman, R., and P. Doty. 1967. *Biochem. Biophys. Res. Commun.,* **27**: 579.
72. Tomasz, M., and R. W. Chambers. 1964. *J. Am. Chem. Soc.,* **86**: 4216.
73. Tomasz, M., Y. Sanno, and R. W. Chambers. 1965. *Biochemistry,* **4**: 1710.

74. Chambers, R. W. 1966. *Progr. Nucleic Acid Res. Mol. Biol.*, **5**: 349.
75. Tomasz, M., and R. W. Chambers. 1966. *Biochemistry,* **5**: 773.
76. Schulman, L. H., and R. W. Chambers. 1968. *Proc. Natl. Acad. Sci., U.S.*, **61**: 308.
77. Imura, N., G. B. Weiss, and R. W. Chambers. 1969. *Nature,* **222**: 1147.
78. Burton, K. 1969. *Biochem. J.*, **114**: 30P.
79. Ziff, E. B., and J. R. Fresco. 1969. *Biochemistry,* **8**: 3242.
80. Ziff, E. B., and J. R. Fresco. 1969. *J. Am. Chem. Soc.*, **90**: 7338.
81. Lipsett, M. N. 1967. *J. Biol. Chem.*, **242**: 4067.
82. Lipsett, M. N., and B. P. Doctor. 1967. *J. Biol. Chem.* **242**: 4072.
83. Carbon, J., and H. David. 1968. *Biochemistry,* **7**: 3851.
84. Carbon, J., L. Hung, and D. Jones. 1965. *Proc. Natl. Acad. Sci., U.S.*, **53**: 979.
85. Lipsett, M. N. 1966. *Cold Spring Harbor Symp. Quant. Biol.*, **31**: 449.
86. Kaiser, I. I. 1969. *Biochim. Biophys. Acta,* **182**: 449.
87. Saneyoshi, M., and S. Nishimura. 1967. *Biochim. Biophys. Acta,* **145**: 208.
88. Reid, B. R. 1968. *Biochem. Biophys. Res. Commun.*, **33**: 627.
89. Pleiss, M., H. Ochiai, and P. A. Cerutti. 1969. *Biochem. Biophys. Res. Commun.*, **34**: 70.
90. Favre, A., M. Yaniv, and A. M. Michelson. 1969. *Biochem. Biophys. Res. Commun.*, **37**: 266.
91. Yaniv, M., A. Favre, and B. G. Barrell. 1969. *Nature,* **223**: 1331.
92. Yoshikami, D., and E. B. Keller. 1969. *Federation Proc.*, **28**: Abstr. 849.
93 RajBhandary, U. L., R. D. Faulkner, and A. Stuart. 1968. *J. Biol. Chem.*, **243**: 575.
94. Thiebe, R., and H. G. Zachau. 1968. *European J. Biochem.*, **5**: 546.
95. Whitfeld, T. R. 1954. *Biochem. J.*, **58**: 390.
96. Philippsen, P., R. Thiebe, W. Wintermeyer, and H. G. Zachau. 1968. *Biochem. Biophys. Res. Commun.*, **33**: 922.
97. Fittler, F., and R. H. Hall. 1966. *Biochem. Biophys. Res. Commun.*, **25**: 441.
98. Kline, L., F. Fittler, and R. H. Hall. 1969. *Biochemistry,* **8**: 4361.
99. Zubay, G. L. 1968. *Methods Enzymol.* **XII B**: 227.
100. von der Haar, F., E. Schlimme, M. Gomez-Guillen, and F. Cramer. 1970, in preparation.
101. Uretsky, S. C., G. Acs, E. Reich, M. Hori, and L. Altwerger. 1968. *J. Biol. Chem.*, **243**: 306.
102. Ward, D. C. et al. 1969. *J. Biol. Chem.*, **244**: 3243.
103. Schlimme, E., F. von der Haar, F. Eckstein, and F. Cramer. 1970. *European J. Biochem.*, **14**: 351.
104. von der Haar, F., E. Schlimme, and F. Cramer. 1970. *Hoppe Seyler's Z. Physiol. Chem.*, **351**: 113.
105. Soffer, R. L., S. Uretsky, L. Altwerger, and G. Acs. 1966. *Biochem. Biophys. Res. Commun.*, **24**: 376.
106. Daniel, V., and U. Z. Littauer. 1965. *J. Mol. Biol.*, **11**: 692.
107. Daniel, V., and U. Z. Littauer. 1963. *J. Biol. Chem.*, **238**: 2102.
108. Klemperer, H. G., and E. S. Canellakis. 1966. *Biochim. Biophys. Acta,* **129**: 157.
109. Li, Ch.-Ch., and J.-Ch. Su. 1967. *Biochem. Biophys. Res. Commun.*, **28**: 1068.
110. Hecht, L. I., M. L. Stephenson, and P. C. Zamecnik. 1959. *Proc. Natl. Acad. Sci., U.S.*, **45**: 505.
111. Hayward, R. S., and S. B. Weiss. 1966. *Proc. Natl. Acad. Sci., U.S.*, **55**: 1161.
112. Cramer, F., F. von der Haar, and E. Schlimme. 1968. *FEBS Letters,* **2**: 136.
113. Rajbhandary, U. L. 1968. *J. Biol. Chem.*, **243**: 556.
114. Ogur, M., and J. O. Small. 1960. *J. Biol. Chem.*, **235**: PC 60.
115. Neu, H. C., and L. A. Heppel. 1964. *J. Biol. Chem.*, **239**: 2927.

116. Khym, J. X., and W. E. Cohn. 1961. *J. Biol. Chem.*, **236**: PC 9.
117. Khym, J. X., and M. Uziel. 1968. *Biochemistry*, **7**: 422.
118. Uziel, M., and J. X. Khym. 1969. *Biochemistry*, **8**: 3254.
119. Zamecnik, P. C., M. L. Stephenson, and J. F. Scott. 1960. *Proc. Natl. Acad. Sci., U.S.*, **46**: 811.
120. Millar, D. B., and R. F. Steiner. 1965. *Biochim. Biophys. Acta*, **102**: 571.
121. Millar, D. B., and R. F. Steiner. 1966. *Biochemistry*, **5**: 2289.
122. Millar, D. B., and M. McKenzie. 1966. *Biochem. Biophys. Res. Commun.*, **23**: 724.
123. Russev, G. 1969. *Anal. Biochem.*, **27**: 244.
124. Midgeley, J. E. M., and D. J. McIlreavy. 1967. *Biochim. Biophys. Acta*, **145**: 512.
125. Khym, J. X. 1963. *Biochemistry*, **2**: 344.
126. Brown, D. M., and A. P. Read. 1965. *J. Chem. Soc.*, p. 5072.
127. Penniston, J. T., and P. Doty. 1963. *Biopolymers*, **1**: 145.
128. Eyring, E. J., and J. Ofengand. 1967. *Biochemistry*, **6**: 2500.
129. Feldman, M. Ya. 1967. *Biochim. Biophys. Acta*, **149**: 20.
130. Boedtker, H. 1967. *Biochemistry*, **6**: 2718.
131. Millar, D. B., and M. McKenzie. 1967. *Biochemistry*, **6**: 2520.
132. Axelrod, V. D., M. Ya. Feldman, I. I. Chuguev, and A. A. Bayev. 1969. Biochim. Biophys. Acta **186**: 33.
133. Verhulst, A.-M., J. Werenne, and J. Charlier. 1968. *Arch. Intern. Physiol. Biochem.*, **76**: 970.
134. Rosenfeld, A., and C. L. Stevens. 1969. *Federation Proc.*, **28**: Abstr. 3564.
135. Keith, G., and J. P. Ebel. 1968. *Compt. Rend. Acad. Sci. Paris*, **D266**: 1066.
136. Keith, G., and J. P. Ebel. 1968. *Biochim. Biophys. Acta*, **166**: 16.
137. Knorre, D. G., N. M. Pustoshilova, N. M. Teplova, and G. G. Shamovskii. 1965. *Biokhimiya*, **30**: 1218.
138. Knorre, D. G., N. M. Pustoshilova, and A. P. Sevast'yanov. 1968. *Biokhimiya*, **33**: 56.
139. Knorre, D. G., A. N. Malysheva, N. M. Pustoshilova, A. P. Sevast'yanov and G. G. Shamovskii. 1966. *Biokhimiya*, **31**: 1181.
140. Duval, J., and J. P. Ebel. 1967. *Bull. Soc. Chim. Biol. France*, **49**: 1665.
141. Duval, J., and J. P. Ebel. 1966. *Compt. Rend. Acad. Sci. Paris* **D263**: 1773.
142. Duval, J., and J. P. Ebel. 1965. *Bull. Soc. Chim. Biol. France*, **47**: 787.
143. Ebel, J. P., J. H. Weil, B. Rether, and J. Heinrich. 1965. *Bull. Soc. Chim. Biol. France*, **47**: 1599.
144. Bakes, J., C. Ehresmann, N. Befort, J. H. Weil, and J. P. Ebel. 1968. *European J. Biochem.*, **4**: 490.
145. Yoshida, H., J. Duval, and J. P. Ebel. 1966. *Compt. Rend. Acad. Sci. Paris*, **C262**: 233.
146. Ascoli, F. and F. M. Kahan. 1966. *J. Biol. Chem.*, **241**: 428.
147. Hayatsu, H., and T. Ukita. 1964. *Biochem. Biophys. Res. Commun.*, **14**: 198.
148. Muto, A., K. I. Miura, H. Hayatsu, and T. Ukita. 1965. *Biochim. Biophys. Acta*, **95**: 669.
149. Hayatsu, H., K. I. Takeishi, and T. Ukita. 1966. *Biochim. Biophys. Acta*, **123**: 445.
150. Hayatsu, H., and T. Ukita. 1966. *Biochim. Biophys. Acta*, **123**: 458.
151. Brown, D., and P. Schell. 1965. *J. Chem. Soc.*, p. 208.
152. Seifert, W., and W. Zillig. 1967. *Hoppe-Seyler's Z. Physiol. Chem.*, **348**: 1017.
153. Verwoerd, D. W., and W. Zillig. 1963. *Biochim. Biophys. Acta*, **68**: 484.
154. Patel, A. B., and H. D. Brown. 1967. *Nature*, **214**: 402.
155. Brown, D. M. 1967. *Methods Enzymol.* **XIIA**: 31.
156. Kikugawa, K., H. Hayatsu, and T. Ukita. 1967. *Biochim. Biophys. Acta*, **134**: 221.
157. Kikugawa, K., A. Muto, H. Hayatsu, K.-I. Miura, and T. Ukita. 1967. *Biochim. Biophys. Acta*, **134**: 232.

158. Bollack, C., and J. P. Ebel. 1968. *Bull. Soc. Chim. Biol. France,* **50**: 2351.
159. Bollack, C., E. Klein, and J. P. Ebel. 1968. *Bull. Soc. Chim. Biol. France,* **50**: 2363.
160. Englander, S. W. 1968. *Methods Enzymol.* **XIIB**: 379.
161. Gantt, R. R., S. W. Englander, and M. V. Simpson. 1969. *Biochemistry.* **8**: 475.
162. Page, L. A., S. W. Englander, and M. V. Simpson. 1967. *Biochemistry,* **6**: 968.
163. Hanson, C. V. 1969. *Anal. Bioch.,* **32**: 303.
164. Burton, K., N. F. Varney, and P. C. Zamecnik. 1966. *Biochem. J.,* **99**: 29C.
165. Burton, K., and W. T. Riley. 1966. *Biochem. J.,* **98**: 70.
166. Burton, K. 1967. *Biochem. J.,* **104**: 686.
167. Hayatsu, H., and T. Ukita. 1967. *Biochem. Biophys. Res. Commun.,* **29**: 556.
168. Holbrook, J. J., A. S. Jones, and M. J. Welch. 1965. *J. Chem. Soc.,* p. 3998.
169. Griffin, B. 1969. *Biochem. J.,* **114**: 31P.
170. Ralph, R. K., R. J. Young, and H. G. Khorana. 1963. *J. Am. Chem. Soc.,* **85**: 2002.
171. RajBhandary, U. L., R. J. Young, and H. G. Khorana. 1964. *J. Biol. Chem.,* **239**: 3875.
172. Priess, H., and W. Zillig. 1965. *Hoppe-Seyler's Z. Physiol. Chem.,* **342**: 73.
173. Takeishi, K., H. Hayatsu, and T. Ukita. 1969. *Biochim. Biophys. Acta,* **195**: 304.
174. Belikova, A. M., T. E. Vakhrusheva, V. V. Vlasov, N. I. Grineva, D. G. Knorre, and V. A. Kurbatov. 1969. *Mol. Biol.* (*Russ.*), **3**: 210.
175. Kriek, E. 1965. *Biochem. Biophys. Res. Commun.,* **20**: 793.
176. Miller, E. C., U. Juhl, and J. A. Miller. 1966. *Science,* **153**: 1125.
177. Kriek, E., J. A. Miller, U. Juhl, and E. C. Miller. 1967. *Biochemistry,* **6**: 177.
178. Kochetkov, N. K., E. I. Budowsky, N. E. Broude, and L. M. Klebanova. 1967. *Biochim. Biophys. Acta,* **134**: 492.
179. Broude, N. E., E. I. Budowsky, and N. K. Kochetkov. 1967. *Mol. Biol.* (*Russ.*), **1**: 214.
180. Gal-Or, L., J. E. Mellema, E. N. Moudrianakis, and M. Beer. 1967. *Biochemistry,* **6**: 1909.
181. Erickson, H., and M. Beer. 1967. *Biochemistry,* **6**: 2694.
182. De Groot, N., I. Fry-Shafrir, and Y. Lapidot. 1969. *European J. Biochem.,* **8**: 571.
183. Lapidot, Y., N. De Groot, S. Rappoport, and D. Elat. 1969. *Biochim. Biophys. Acta,* **190**: 304.
184. De Groot, N., Y. Groner, and Y. Lapidot. 1969. *Biochim. Biophys. Acta,* **186**: 286.
185. Lapidot, Y., N. De Groot, and S. Rappoport. 1969. *Biochim. Biophys. Acta,* **182**: 105.
186. Lapidot, Y., and N. De Groot. 1969. *Biochim. Biophys. Acta,* **179**: 521.
187. Lapidot, Y., N. De Groot, S. Rappoport, and A. Panet. 1968. *Biochim. Biophys. Acta,* **157**: 433.
188. Lapidot, Y., D. Inbar, N. De Groot, and H. Kössel. 1969. *FEBS Letters,* **3**: 253.
189. Stern, R., L. E. Zutra, and U. Z. Littauer. 1969. *Biochemistry,* **8**: 313.
190. Littauer, U. Z., S. A. Yankofsky, A. Novogrodsky, H. Bursztyn, Y. Galenter, and E. Katchalski. 1969. *Biochim. Biophys. Acta,* **195**: 29.
191. Zagrebel′nyi, S. N., and D. G. Knorre. 1966. *Biokhimiya,* **31**: 893.
192. Gottikh, B. P., A. A. Kraevskii, L. L. Kiselev, and L. Yu. Frolova. 1967. *Mol. Biol.* (*Russ.*), **1**: 767.
193. Zachau, H. G., and H. Feldmann, 1965. *Progr. Nucleic Acid Res. Mol. Biol.,* **4**:217.
194. Hoffman, B. M., P. Schofield, and A. Rich. 1969. *Proc. Natl. Acad. Sci., U.S.,* **62**: 1195.
195. Siddiqui, M. A. Q., M. Krauskopf, and J. Ofengand. 1970. *Biochem. Biophys. Res. Commun.,* **38**: 156.
196. Siddiqui, M. A. Q., and J. Ofengand. 1970. *J. Biol. Chem.,* **245**: 4409.

197. Wagner, L. P. and J. Ofengand. 1970. *Biochim. Biophys. Acta,* **204**: 620.
198. Brostoff, S. W. and V. M. Ingram. 1970. *Biochemistry,* **9**: 2372.
199. May, M. S., and R. W. Holley. 1970. *J. Mol. Biol.,* **52**: 19.
200. Hay, J., D. J. Pillinger, and E. Borek. 1970. *Biochem. J.* **119**: 587.
201. Wintermeyer, W. and H. G. Zachau. 1970. *FEBS Letters,* **11**: 160.
202. Hara, H., T. Horiuchi, M. Saneyoshi, and S. Nishimura. 1970. *Biochem. Biophys. Res. Commun.,* **38**: 305.
203. Saneyoshi, M. 1970. *Biochem. Biophys. Res. Commun.,* **40**: 1501.
204. Saneyoshi, M. and S. Nishimura. 1970. *Biochim. Biophys. Acta,* **204**: 389.
205. Walker, R. T. and U. L. RajBhandary. 1970. *Biochem. Biophys. Res. Commun.,* **38**: 907.
206. Cotter, R. I., and W. B. Gratzer. 1970. *Biochem. Biophys. Res. Commun.,* **39**: 766.
207. Burton, K. 1970. *FEBS Letters,* **6**: 77.
208. Furuichi, Y., Y. Wataya, H. Hayatsu and T. Ukita. 1970. *Biochem. Biophys. Res. Commun.,* **41**: 1185.
209. Glitz, D. G. and D. S. Sigman. 1970. *Biochemistry,* **9**: 3433.
210. Hunt, J. A. 1970. *Biochem. J.,* **120**: 353.
211. Rosenfeld, A., C. L. Stevens, and M. P. Printz. 1970. *Biochemistry,* **9**: 4971.
212. Erhardt, F. and H. G. Zachau. 1970. *Hoppe-Seylers Z. Phys. Chem.,* **351**: 567.
213. Hirsch, R. and H. G. Zachau. 1970. *Hoppe Seyler's Z. Phys. Chem.,* **351**: 563.
214. Jilyaeva, T. I. and L. L. Kisselev. 1970. *FEBS Letters,* **10**: 229.
215. Cashmore, A. R. 1970. *FEBS Letters,* **12**: 90.
216. Grineva, N. L., D. G. Knorre, and V. A. Kurbatov. 1970. *Mol. Biol. Russ.,* **4**: 814.
217. Fink, L. M., S. Nishimura, and I. B. Weinstein. 1970. *Biochemistry,* **9**: 496.
218. Igarashi, S. and W. Paranchych. 1970. *J. Biochem.,* **67**: 123.
219. Yankofsky, S. A., S. Yankofsky, E. Katchalski, and U. Z. Littauer. 1970. *Biochim. Biophys. Acta,* **199**: 56.
220. Hamburger, A. D., N. De Groot, and Y. Lapidot. 1970. *Biochim. Biophys. Acta,* **213**: 115.
221. Lapidot, Y., D. Elat, S. Rappoport, and N. De Groot. 1970. *Biochem. Biophys. Res. Commun.,* **38**: 559.
222. De Groot, N., A. Panet, and Y. Lapidot. 1970. *European J. Biochem.,* **15**: 215.
223. Fölsch, G. 1970. *Acta Chem. Scand.,* **224**: 1115.
224. Schofield, P., B. M. Hoffman, and A. Rich. 1970. *Biochemistry,* **9**: 2525.
225. Nakanishi, K., N. Furutachi, M. Funamizu, D. Grunberger, and I. B. Weinstein. 1970. *J. Am. Chem. Soc.,* **92**: 7617.

MODIFIED POLYNUCLEOTIDES

FRITZ ECKSTEIN AND KARL-HEINZ SCHEIT*

MAX-PLANCK-INSTITUT FÜR EXPERIMENTELLE MEDIZIN

CHEMISCHE ABTEILUNG

GÖTTINGEN, GERMANY

In addition to use in investigations of enzyme specificities, modified polynucleotides can be of interest for a number of reasons. Polymers which absorb light in the visible or UV region outside of absorbance maxima of nucleotides and proteins can be valuable for studies of nucleic acid-protein interactions. Polymers which are resistant to nucleases or which are at least degraded at slower rates than normal polynucleotides might offer advantages in studies of systems where a long-lived polymer is required.

Modified polynucleotides can be prepared either by modification of the polymer or by polymerization of modified nucleoside di- or triphosphates. This discussion will be restricted to the second approach. In a recent review the synthesis of modified single stranded polynucleotides is summarized (*1*). The preparation of fluorescent polynucleotides has been reported recently (*2-4*). The enzymatic synthesis of some modified polymers which either have special optical properties or exhibit some resistance to nucleases will be described below.

Materials

The chemical synthesis of the following substrates was published elsewhere: Uridine- and adenosine 5′-*O*-(1-thiotriphosphate), ppp_s-U† and

* Present Address: Max-Planck-Institut für Biophysikalische Chemie, Karl-Friedrich-Bonhoeffer-Institut, Göttingen, Germany.

† The symbolism for nucleosides phosphorothioates and 4-thiopyrimidine nucleotides was suggested by Dr. W. E. Cohn, Oak Ridge, Tennessee, and will be published in the *Handbook of Biochemistry* (H. A. Sober ed.), Chemical Rubber Co., Cleveland, Ohio, second edition.

ppp_s-A, (*5*), 4-thiouridine 5′-diphosphate (s^4UDP) (*11*); 4-thiouridine 5′-triphosphate (s^4UTP) (*11*); 4-thiothymidine 5′-triphosphate (s^4dTTP) (*11*); 4-methylthiopyrimidine-2-on-1-β(D-riboside-5′-diphosphate) (CH_3s^4UDP) (*9*); (^{3}H)CH_3s^4UDP (*9*); (^{35}S)s^4UDP (*12*). dATP was purchased from Schwartz Bioresearch Inc. (Orangeburg, N.Y.), ATP and UTP from Zellstoffabrik Waldhof (Mannheim). Polynucleotide phosphorylase (EC 2.7.7.8) with a specific activity of 165 μmoles UDP per hour per milligram protein was prepared from *E. coli* according to the literature (*14*). RNA polymerase (EC 2.7.7.6) was isolated from *E. coli* following the method of Burgess (*7*). It had a specific activity of 1475 μmoles ATP per 10 min per milligram of protein. RNA polymerase purified according to Zillig (*6*) had a specific activity of 1140* DNA polymerase (EC 2.7.7.7) was isolated from *E. coli* according to Richardson et al. (*15*) up to step VII. The enzyme preparation used had a specific activity of 1448 nmoles dATP per 30 min per milligram protein. DNA polymerase from *B. subtilis* was obtained as described by Okasaki and Kornberg (*16*). For the experiments reported below the phosphocellulose fraction was used.

The following enzymes were commercial preparations: Pancreatic DNase I (Grade DPFF) (EC 3.1.4.5; Worthington Biochemical Corporation, Freehold, N.Y.); snake venom phosphodiesterase (EC 3.1.4.1, C.F. Boehringer, Mannheim); micrococcal nuclease (EC 3.1.4.7; Sigma Chemical Corp., St. Louis).

POLYRIBONUCLEOTIDES CONTAINING A PHOSPHOROTHIOATE-SUGAR BACKBONE†

Synthesis

Poly r($-A_s-U_s$) (*5*)

The reaction solution contained 8 mM TrisAc, pH 7.9, 8 mM $MgAc_2$, 0.12 mM NH_4Cl, 1 mM of each [^{35}S]ppp_s-A and ppp_s-U, 0.3 mM poly d(A-T), and 420 units of RNA-polymerase (*6*) in a final volume of 3 ml. After approximately 6 hr incubation at 37°, 1 mg of DNase I (DPFF) in 600 μl of water was added. After 20 min‡ protein was removed by repeated extraction with chloroform-isoamylalcohol (5 : 2, v/v) and the aqueous layer was concentrated to approximately 0.5 ml in a rotary evaporator at 25° *in vacuo*. The aqueous solution was applied on a Sephadex G-25 column which was eluted with doubly-distilled water. The polymeric material eluted with the void volume was concentrated as described above and chromatographed

* Both RNA preparations were kindly provided by Dr. H. Sternbach, Göttingen.

† See footnote, p. 665.

‡ It is preferable to add 1.2 ml of 4 percent SDS-solution before extracting protein.

on a Sephadex G-200 column with 0.05 M Tris HCl, pH 7.0. The polymer eluted with the void volume was concentrated and desalted on a Sephadex G-25 column eluted with doubly-distilled water. The aqueous solution of the polymer was concentrated, the UV-spectrum measured, and quickly frozen in acetone-Dry Ice. The material was lyophilized and stored at –20°. Yield 6.9 OD_{260} units (20%).

Poly r($_{s}$A-U)

The reaction solution contained 8 mM TrisAc, pH 7.9, 8 mM $MgAc_2$, 0.12 mM NH_4Cl, 1 mM of each [^{14}C]UTP and ppp_S-A, 0.15 mM poly d(A-T), and 75 units of RNA-polymerase (*7*) in a final volume of 1 ml. After 22 hr at 37°, 0.3 mg of DNase I (DPFF) in 0.3 ml of water was added and after 20 min 1 ml of 4% SDS solution was added. Extraction of protein and purification of the polymer was carried out as described for poly r($_{s}$A$_{s}$U). Yield 3.1 OD_{260} units (26%).

Poly r(A$_{s}$U)

The reaction was carried out as described for poly r(-A-U) except that 1 mM each of [^{14}C]ATP and ppp_S-U was used. The yield was 2.75 OD_{260} units (23%).

Properties

Physical Properties

The introduction of phosphorothioate instead of phosphate groups in poly r(A-U) (*8*) does not influence the UV-spectrum (Fig. 1), the midpoint of the optical density transition upon heating (Fig. 2), or the increase in optical density at 260 nm upon melting. The molecular weight, however, decreases with increasing phosphorothioate content (Table 1).

Degradation by Nucleases

Poly r(A$_{s}$U), poly r($_{s}$A-U), and poly r($_{s}$A$_{s}$U) are not entirely resistant to nucleases. They are, however, degraded at a considerably slower rate than poly r(A-U) as shown in two examples (Figs. 3 and 4). The polymers become more resistant to micrococcal nuclease with increasing phosphorothioate content, whereas no such effect can be observed with snake venom phosphodiesterase. All polymers are degraded completely by concentrations of enzymes higher than those indicated in Figs. 3 and 4.

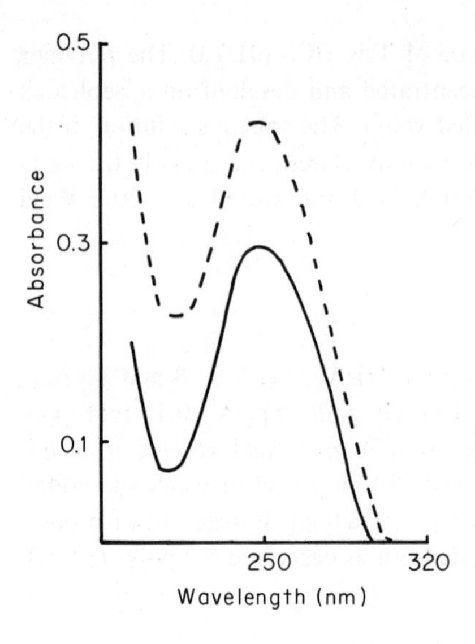

Figure 1. UV-Spectrum of poly r(A-U) (———) and poly r(${}_sA_{\bar{s}}U$) (– – –) in water, pH 6.5.

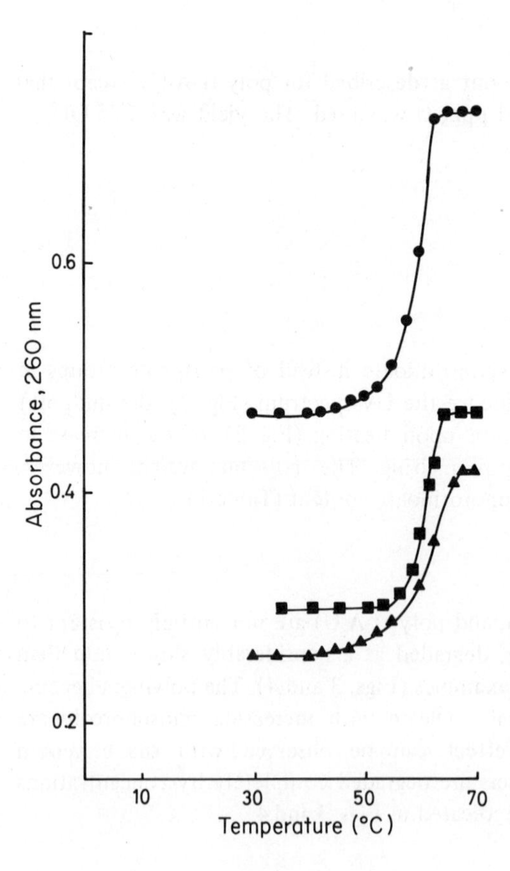

Figure 2. Temperature melting profile of poly r (A-U) (●-●), poly r($A_{\bar{s}}U$) (■-■) and poly r(-$A_{\bar{s}}U$) (▲-▲). The spectra were taken in 0.1 M sodium citrate buffer, pH 7.5. Reproduced by permission, from *European J. Biochem.* (Ref. *5*).

TABLE 1. Molecular Weight and Sedimentation Values of Polymers[a]

Polymer	$M_w^{c=o}$	$S_{20,w} \times 10^{13}$
Poly r(A-U)	128,200	7.6
Poly r(A$_s$-U)	105,000	7.4
Poly r($_s$A$_s$U)	30,000	5.0

[a] Determined by low-speed equilibrium centrifugation in a Spinco Model E analytical untracentrifuge equipped with a photoelectric scanner in 0.005 M sodium citrate buffer. Reproduced by permission, from *European J. Biochem.* (Ref. *5*).

Template Activity (*5*)

Like poly r(A-U), poly r($_s$A$_s$U) can serve as template for the polymerization of ATP and UTP as well as ppp$_S$-U and ppp$_S$-A with RNA-polymerase in the presence of Mn^{2+}. The rates of polymerization, however, are somewhat reduced.

Induction of Interferon (*9*)

Substitution of phosphorothioate for phosphate in the alternating copolymer poly r(A-U) causes a significant increase of induction of interferon and cellular resistance to virus infection (*10*). Poly r(A$_s$U) reduces vesicular stomatitis virus plaque formation in human skin fibroplasts at a 10- to

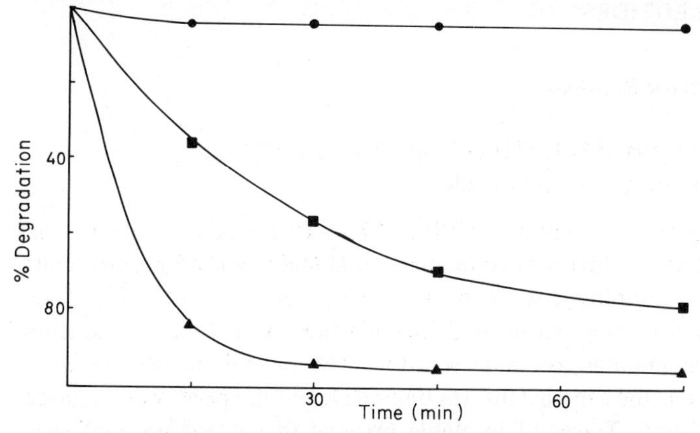

Figure 3. Degradation of poly r(A-U) (▲ ▲), poly r(A$_s$U) (■ ■) and poly r($_s$A$_s$U) (● ●) by snake venom phosphodiesterase at 37°. Experimental conditions: 0.1 mM substrate, 0.1 M Tris-HCl pH 8.0, and 3 μg of enzyme in a total volume of 400 μl.

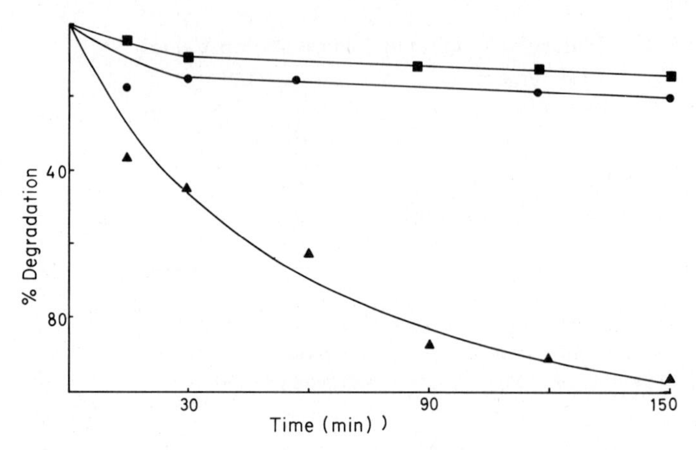

Figure 4. Degradation of poly r(A-U) (▲ ▲), poly r(A_sU) (■ ■) and poly r($_sA_sU$) (● ●) by micrococcal nuclease at 37°. Experimental conditions: 0.1 mM substrate, 0.01 M $CaCl_2$, 0.02 M Tris-HCl pH 8.8, and 0.5 μg of enzyme in a total volume of 400 μl.

100-fold lower concentration and produces 12-fold higher amounts of interferon in human skin fibroplasts than poly r(A-U). Poly r($_sA_sU$) is active at 100- to 10,000-fold lower concentrations and produces 20-fold (human skin fibroplasts) or 100-fold (rabbits) higher amounts of interferon. The higher activity of the phosphorothioate polymers can be correlated to their reduced degradation by RNase.

POLYNUCLEOTIDES CONTAINING 4-THIOPYRIMIDINE NUCLEOTIDES

Enzymatic Synthesis

Large-Scale Synthesis of Poly s^4U (17) and Poly CH_3s^4U (13) by Polynucleotide Phosphorylase from *E. coli*

Ten micromoles of either s^4UDP or CH_3s^4UDP (sodium salt) in 1 ml 0.1 M Tris-HCl buffer (pH 8.3) containing 2 mM $MgCl_2$ and 15 enzyme units polynucleotide phosphorylase were incubated for 6 hr at 37°C. The enzymatic reaction was terminated by addition of 0.1 ml 1% aqueous cetyltrimethylammonium bromide solution. The precipitate was collected by centrifugation, the supernatant was discarded, and the pellet was dissolved in 0.5 ml methanol. Traces of insoluble material (denatured protein) were removed by centrifugation. An equal volume of 1% Na ClO_4 in acetone (w/v) was added to the clear methanolic solution. The resulting precipitate was collected by centrifugation, dissolved in a minimal volume of 1% (w/v) aqueous $NaClO_4$ and the sodium salts of polynucleotides precipitated by addition of excess ethanol. For further purification the crude mixture of polynucleotides was separated on Sephadex G-200 (column size 2 x 67 cm). The buffer used for separation of poly s^4U contained 0.1 mM dithiothreitol. The polynucleotides eluted with the breakthrough volume exhibited sedimentation coefficients $s^0_{20,w}$ of 3.85 S for poly s^4U, and 5.0 S for poly $CH_3s^4U^4$. The yields were 50 OD_{330} units (30%) poly s^4U and 48 OD_{300} units (50%) poly CH_3s^4U.

The Synthesis of Poly r(A-s^4U) by RNA Polymerase from *E. coli*

The incubation mixture contained in 2.5 ml: 3 μmoles ATP, 3 μmoles s^4UTP, 3 μmoles dithiothreitol, 0.03 M $MgCl_2$, 0.6 OD_{260} units poly d(A-T), and 358 enzyme units RNA-polymerase. The mixture was incubated at 37° and 20-μl aliquots were withdrawn at intervals, diluted with 1 ml H_2O and the optical density measured at 320 nm. After 6 hr 50% of the theoretical hypochromicity was observed. The incubation mixture was diluted by 0.7 ml of 4% SDS solution and the protein removed by repeated extraction with chloroform–isoamylalcohol (5 : 2, v/v). The aqueous phase containing polynucleotide was subjected to Sephadex G-200 filtration using 0.05 M Tris-HCl (pH 7.0) and 0.1 mM dithiothreitol as buffer. Poly r(A-s^4U) was eluted with the breakthrough volume of the Sephadex column. The appropriate fractions were pooled and stored in frozen state. The yield was 6 OD_{330} units (20%).

The Synthesis of Poly d(A-s^4T) by DNA Polymerase

When s^4dTTP is substituted for dTTP in the enzymatic synthesis of poly d(A-T) by *E. coli* DNA polymerase (*19*) formation of a polymer was observed either by incorporation of radioactive labeled dAMP or by a hypochromic effect at 260 nm and at 335 nm, the latter corresponding to the ultraviolet absorption of s^4dTMP. Both the initial velocity and the extent of polymer synthesis were strictly dependent on the concentration of template poly d(A-T). The product closely resembling poly d(A-T) was found to contain equal amounts of dAMP and ds^4TMP in alternating sequence. However, the enzymatic synthesis of poly d(A-s^4T) by *E. coli* DNA polymerase has the disadvantage that stoichiometric amounts of template poly d(A-T) are required. Since the template was only partially degraded by the 5′-specific exonuclase during polymerization the isolated poly d(A-s^4T) was usually contaminated with poly d(A-T). Considerable progress has been made with the observation that DNA polymerase from *Bacillus subtilis* was able to synthesize poly d(A-s^4T) extensively and *de novo.* (*20*).

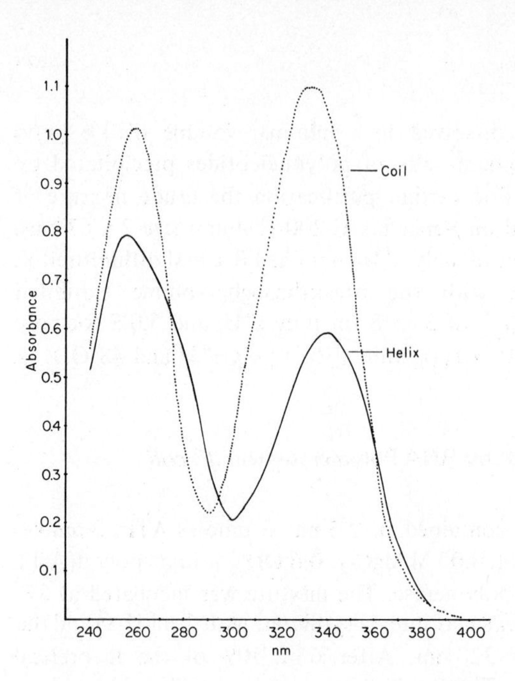

Figure 5. Absorption spectrum of poly r(A-s^4U). The absorption spectrum of poly r(A-s^4U) was measured in 0.08 M sodium cacodylate (pH 7.0) at 20° (———) and at 65° (. . . . .).

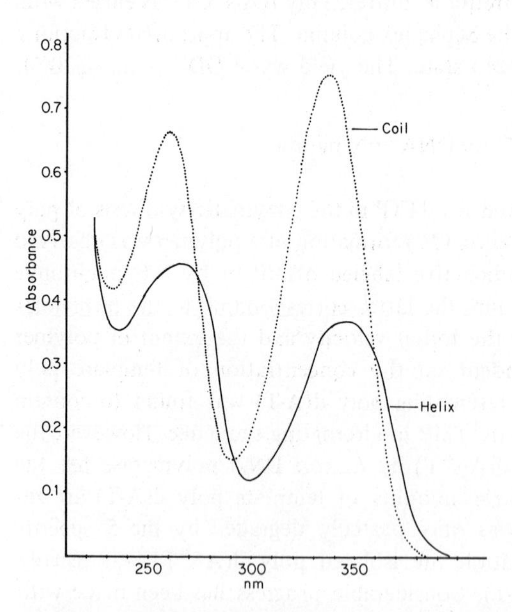

Figure 6. Absorption spectrum of poly d(A-s^4T). The absorption spectrum of poly d(A-s^4T) was measured in 0.015 M NaCl–0.0015 M sodium citrate (pH 7.0) at 20° (———) and 60° (. . . . .).

Synthesis of Poly d(A-s^4T) by *B. subtilis* DNA Polymerase

The incubation mixture contained in 2 ml: 200 μmoles Tris-succinate pH 8.2; 20 μmole $MgCl_2$: 1.55 μmoles s^4dTTP; 1.8 μmoles dATP; 0.1 OD_{260} unit poly d(A-T); 3 μmoles dithiothreitol; and 15 units *B. subtilis* DNA polymerase (phosphocellulose fraction). The mixture was incubated at 35°. An aliquot of 0.25 ml of the incubation mixture was transferred into a 0.1-cm quartz cuvette. The synthesis was controlled by measuring continuously the OD_{335} using a Zeiss PMQ II spectrophotometer with its cell compartment being thermostatted at 35°. Polymerization started after a lag period of 30–60 min. The synthesis of poly d(A-s^4T) proceeded with a constant rate of 1 OD_{335} unit/hr corresponding to 0.075 μmoles/ml/hr (for calculation see below). When, usually after 5–6 hr, the optical density had decreased by 4–5 units another 1.55 μmoles s^4dTTP and 1.8 μmoles dATP were added and the incubation continued for 10 to 12 hr. When the decrease in OD_{335} had ceased the enzyme was destroyed by addition of 0.2 ml 2 M NaCl–1 M trisodium citrate and heating at 75° for 5 min. The viscous solution was transferred into a small dialysis bag, dialyzed for 6 hr against 250 ml 0.02 M NaCl–0.01 M trisodium citrate containing 2 mM dithiothreitol and for an additional 12 hr against 0.01 M NaCl–0.01 M Tris-HCl pH 7.5 containing 1 mM EDTA and 2 mM dithiothreitol. The yield of poly d(A-s^4T) was 76% relative to s^4dTTP. The mean sedimentation coefficient of that particular poly d(A-s^4T) preparation was found to $s_{20,w}$ = 9 S.*

Physical Properties

The absorption spectra of s^4UMP and s^4dTMP measured at pH 7 exhibit maxima at 380 nm and 335 nm, respectively. This fact allows a convenient and very precise investigation of purity as well as composition of the polynucleotides poly s^4U, poly CH_3s^4U, poly r(A-s^4U), and poly d(A-s^4T) by means of their absorption spectra. Poly s^4U (*17*) and poly CH_3s^4U (*13*) were found to be more than 95% pure. The nucleotide composition of the alternating copolymers poly r(A-s^4U) and poly d(A-s^4T) calculated from the absorption spectra after total enzymatic hydrolysis corresponded to the expected values (see Table 2).

Spectrophotometrical titration of poly s^4U with poly A followed at 330 nm clearly indicated the formation of a helical complex (poly $s^4U)_2$ (poly A) (*17*). Its rather unusual properties such as increased breadth of

*Sedimentation experiments were performed in a Spinco Model E ultracentrifuge equipped with a photoelectric scanner and a multiplexer. The ultracentrifuge was run at 36,000 rpm and 20°. $s_{20,w}$-values were calculated from photoelectric scanner tracings (330 nm) by the moving-boundary method. Solvent: 0.05 m NaCl – 0.01 M Tris-HCl pH 7.0 containing 1 mM EDTA and 0.1 mM dithiothreitol.

TABLE 2. Spectral and Physical Properties of Polynucleotides containing 4-Thiopyrimidine Nucleotides

Polynucleotide	λ_{max} (nm)	Total hyperchromicity (%)[e] after enzymatic hydrolysis	Hyperchromicity of helix–coil transition (%)[e]	T_m (°C)	σ_{T_m}[k] (°C)	ϵ(P)[g]	Absorption ratios	Composition[g]
Poly s^4U[a]	338;244	20 (330 nm)	7	8[h]	8	1.68×10^4 (330 nm)	$A_{330} : A_{260}$ = 6.60	
Poly CH_3s^4U[b]	300;280 (inflection); 312 (inflection)	40 (312 nm)				9.6×10^3 (300 nm)		
Poly r(A-s^4U)[c]	255;340	31.0 (260 nm) 57.4 (330 nm) 52.0 (280 nm)[f]	24.5 (260 nm) 51.0 (330 nm)	52[i]	3	1.03×10^4 (330 nm)	$A_{340} : A_{255}$ = 0.747	AMP: s^4UMP = 1.12
Poly d(A-s^4T)[d]	344;265	63 (335 nm) 43 (260 nm)	50.4 (335 nm) 31 (260 nm)	64[i]	15	0.77×10^4 (335 nm) 0.95×10^4 (260 nm)	$A_{344} : A_{260}$ = 0.84	dAMP: s^4dTMP= 1
[Poly s^4U]$_2$ [poly A][a]	330;260	32 (380 nm) 20 (280 nm)[f]	25 (330 nm) 15 (280 nm)	62[h]	20			

[a] Simuth, J., K. H. Scheit, and E. M. Gottschalk. 1970. *Biochim. Biophys. Acta,* **204**: 371.
[b] Scheit, K. H. 1970. *Biochim Biophys. Acta,* **209**: 445.
[c] Scheit, K. H. and F. Cramer, manuscript in preparation.
[d] Lezius, A. G., manuscript in preparation; Lezius, A. G. and K. H. Scheit, 1967. *European J. Biochim.,* **3**: 85.
[e] Hyperchromicity was defined as

$$\frac{\Delta A}{A_{max}} \times 100$$

[f] Hypochromicity.
[g] $\epsilon_{(P)}$-values and nucleotide ratios were calculated using the following ϵ-coefficients (pH 7, H_2O):

$\epsilon_{330}^{s^4UMP} = 2.1 \times 10^4$; $\epsilon_{260}^{s^4UMP} = 2.85 \times 10^3$; $\epsilon_{335}^{s^4dTMP} = 2.1 \times 10^4$;

$\epsilon_{260}^{s^4dTMP} = 2.5 \times 10^3$; $\epsilon_{260}^{AMP} = \epsilon_{260}^{dAMP} = 1.54 \times 10^4$

[h] 0.05 M sodium cacodylate (pH 7), 0.1 M KCl.
[i] 0.08 M sodium cacodylate (pH 7).
[j] 0.015 M sodium citrate (pH 7)–0.150 M NaCl.
[k] δT means the temperature range in which 90% of the thermal hypochromicity appeared.

transition, little hyperchromicity at 330 nm, lack of hypochromicity at 260 nm, and no stabilization of helical structure by Mg^{2+} cations may be the result of base pairing between 4-thiouracil and adenine in a reversed Hoogsteen (*21*) manner with the 4-thioketogroup being excluded from hydrogen bonding. The alternating copolymers poly r(A-s^4U) and poly d(A-s^4T) likewise form helical structures (see Figs. 5 and 6). The characteristic data of their thermal transitions are listed in Table 2. The tremendous hypochromicities near 330 nm are striking. Poly r(A-s^4U) seems to be similar to poly r(A-U) with respect to T_m and δT_m values (*18*). There exists, however, growing evidence that poly d(A-T) and poly d(A-s^4T) (*22*) have different helical structures. Interesting physical properties are displayed by poly CH_3s^4U (*13*). The absorption spectra of poly CH_3s^4U before and after complete enzymatic hydrolysis are shown in Fig. 7. Despite the lack of secondary structure poly CH_3s^4U exhibited 40% hyperchromicity at 312 nm. Although CH_3s^4UMP shows no detectable fluorescence the polynucleotide poly CH_3s^4U (*13*) possesses an emission spectrum with λ_{max} of emission at 510 nm and λ_{max} of excitation at 300 nm. The corrected emission spectrum of poly CH_3s^4U is given in Fig. 8. The quantum yield of emission was calculated to 0.3×10^{-3}.

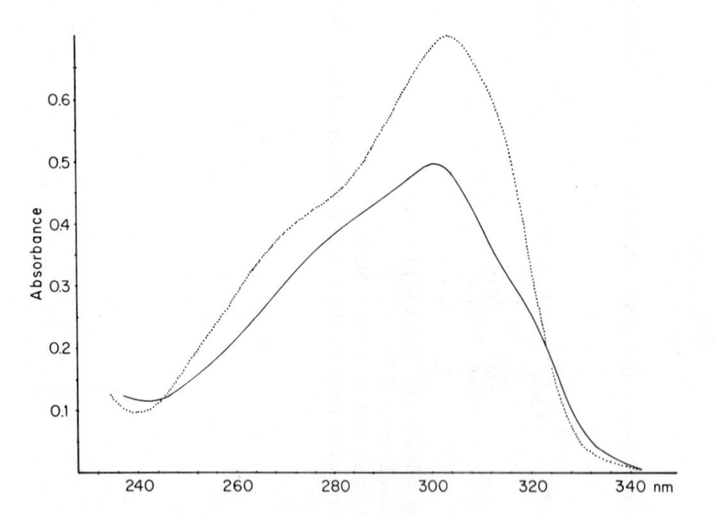

Figure 7. Absorption spectrum of poly CH_3s^4U. (——) 0.5 OD_{300} units poly CH_3s^4U in 1 ml 0.05 M Tris-HCl (pH 8.0). (. . .) 0.5 OD_{300} units poly CH_3s^4U in 1 ml 0.05 M Tris-HCl (pH 8.0) after incubation with 0.75×10^{-2} units snake venom phosphodiesterase at 37°. Reproduced by permission, from *Biochem. Biophys. Acta*, (Ref. *13*).

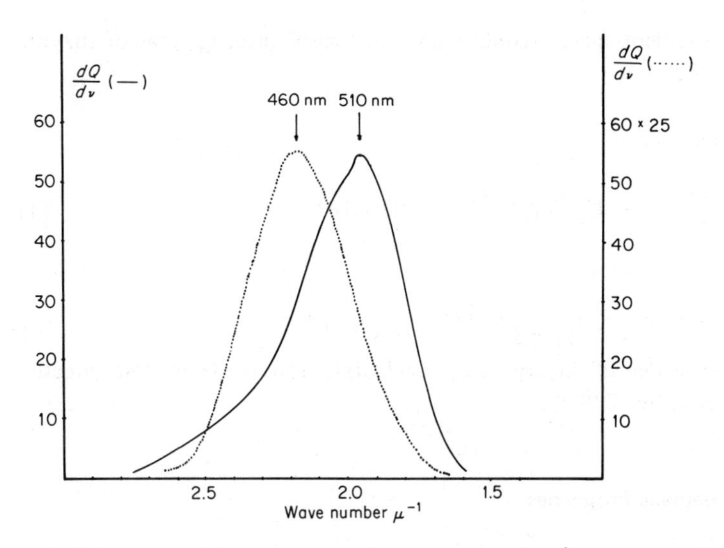

Figure 8. Corrected emission spectrum of poly CH_3s^4U. (———) poly CH_3s^4U in water (absorbance at 303 nm 0.276). (. . .) quinine sulfate in 0.1N H_2SO_4 (1 mg/1; absorbance at 303 nm 0.55×10^{-2}). The estimation of quantum yield from corrected emission spectra followed a procedure, given by C. A. Parker, and W. T. Rees, 1960. *Analyst,* **85**: 587. The two apparent emission spectra from which the corrected spectra have been derived were obtained at the same instrument settings. *dQ/dr* represents quanta per unit frequency interval given in relative energy units. Reproduced by permission, from *Biochem Biophys. Acta* (Ref. *13*).

Biological Properties and Some Applications of Polynucleotides containing 4-Thiopyrimidine Nucleotides

The polynucleotides poly s^4U, poly CH_3s^4U, and poly r(A-s^4U) are degraded by a great variety of nucleases such as snake venom phosphodiesterase, spleen phosphodiesterase, micrococcus nuclease from *Staphylococcus aureus*, and pancreatic RNase. The course of hydrolysis can easily be followed measuring the increase in optical density at the appropriate hyperchromic absorption bands. Enzymatic hydrolysis of poly d(A-s^4T) (*19, 20*) was carried out in two steps using pancreatic DNase and finally snake venom phosphodiesterase to achieve total hydrolysis to mononucleotides. Poly CH_3s^4U was successfully employed to study the kinetics of the exonucleolytic degrading enzymes venom phosphodiesterase and polynucleotide phosphorylase (*13*). Poly d(A-s^4T) was found to be a poor template for the synthesis of poly d(A-T) by DNA polymerase (*19*). The enzymatic polymerizations of either dATP and s^4dTTP by DNA-polymerase or ATP and s^4UTP by RNA polymerase on a poly d(A-T) template proved to be very convenient spectroscopic assays for kinetic investigations of enzyme action (*19, 22*).

Using the relations

$$\epsilon_{330}^{\text{polym}} = \epsilon_{330}^{\text{s}^4\text{UMP}} - \epsilon_{(\text{P})\,330}^{\text{poly r(A-s}^4\text{U)}} = 1.07 \times 10^4 \tag{1}$$

and

$$\epsilon_{335}^{\text{polym}} = \epsilon_{335}^{\text{ds}^4\text{TMP}} - \epsilon_{(\text{P})\,335}^{\text{poly d(A-s}^4\text{T)}} = 1.33 \times 10^4 \tag{2}$$

the concentration of incorporated nucleotide can be estimated directly dividing $\Delta A_{(t)}$ by ϵ^{polym}.

Chemical Properties

The 4-thioketo groups of 4-thiopyrimidine nucleotides in polynucleotides are the target of many specific chemical reactions: alkylations, oxidations, reactions with mercurials, or photochemical reactions (*23-30*). These reactions can be used either for the purpose of identification and modification or for the introduction of radioactive as well as spectroscopic labels in corresponding polynucleotides. The latter possibility looks rather promising since s^4UMP or s^4dTMP can now be incorporated in most nucleic acids and polynucleotides by means of the appropriate enzymatic system.

References

1. Michelson, A. M., I. Massoulie, and W. Guschlbauer. 1967. In I. N. Kavison and W. E. Cohn (Eds.), Progress in nucleic acid research. Academic Press, New York, Vol. 6. p. 84.
2. Ward, D. C., E. Reich, and L. Stryer. 1969. *J. Bio. Chem.*, **244**: 1228.
3. Pochon, F., M. Lang, and A. M. Michelson. 1968. *Biochim. Biophys. Acta,* **169:** 350.
4. Ward, D. C., and E. Reich. 1968. *Proc. Natl. Acad. Sci., U.S.*, **61:** 1494.
5. Eckstein, F., and H. Gindl. 1970. *European. J. Biochem.*, **13**: 558.
6. Zillig, W., E. Fuchs, and R. Milette. 1966. in G. L. Cantoni and D. R. Davies (Eds.), Procedures in nucleic acid research. Harper & Row, New York, Vol. I, p. 323.
7. Burgess, R. R. 1969. *J. Biol. Chem.*, **244:** 6160.
8. Chamberlin, M. J., 1966. in G. L. Cantoni and D. R. Davies (Eds.), Procedures in nucleic acid research. Harper & Row, New York, Vol. I, p. 513.
9. De Clercq, E., F. Eckstein, H. Sternback, and T. C. Merigan. 1970. *Vivology,* **42**: 421.
10. De Clercq, E., F. Eckstein, and T. C. Merigan. 1969. *Science,* **165**: 1137.
11. Scheit, K. H., 1968. *Chem. Ber.* **101:** 1141.
12. Scheit, K. H., and E. Gaertner. 1969. *Biochim. Biophys. Acta,* **1:** 182.
13. Scheit, K. H. 1970. *Biochim. Biophys. Acta,* **209**: 445.

14. Khimi, Y., and U. Z. Littauer. 1968. In L. Grossman and K. Moldave (Eds.), Methods in enzymology. Academic Press, New York, Vol. XII, p. 513.
15. Richardson, C. C., C. L. Schildkraut, H. V. Aposhian, and A. Kornberg. 1964. *J. Biol. Chem.,* **239**: 222.
16. Otkazaki, T., and A. Kornberg. 1964. *J. Biol. Chem.,* **239**: 259.
17. Simuth, J., K. H. Scheit and E. M. Gottschalk. 1970. *Biochim. Biophys. Acta,* **204**: 377.
18. Scheit, K. H., H. Sternbach, and F. Cramer, manuscript in preparation.
19. Lezius, A. G., and K. H. Scheit. 1967. *European J. Biochem.,* **3**: 85.
20. Lezius, A. G., manuscript in preparation.
21. Saenger, W., and D. Suck. 1970. *Nature (London),* **227**: 1046.
22. Scheit, K. H., unpublished results.
23. Reid, B. 1968. *Biocheim. Biophys. Res. Commun.,* **33**: 834.
24. Scheit, K. H., 1969. *Biochim. Biophys. Acta,* **195**: 294.
25. Scheit, K. H., 1968. *Biochim. Biophys. Acta,* **166**: 285.
26. Carbon, J., and H. David. 1968. *Biochemistry,* **7**: 3851.
27. Lipsett, M. N., and B. P. Doctor. 1967. *J. Biol. Chem.,* **242**: 4072.
28. Ziff, E., and J. R. Fresco. 1968. *J. Am. Chem. Soc.,* **90**: 7338.
29. Burton, K., 1967. *Biochem. J.,* **104**: 686.
30. Favre, A., M. Yaniv, and A. M. Michelson. 1969. *Biochem. Biophys. Res. Commun.,* **37**: 266.

THE SELECTIVE CHEMICAL DEGRADATION OF DNA: APURINIC AND APYRIMIDINIC ACIDS, PYRIMIDINE, AND PURINE ISOSTICHS

HANS TÜRLER

DEPARTMENT OF MOLECULAR BIOLOGY

UNIVERSITY OF GENEVA

GENEVA, SWITZERLAND

Since highly specific enzymes for the selective degradation of DNA are still lacking, methods based on the chemical differences between purines and pyrimidines are the only ones available to isolate specific degradation products of DNA. These methods led to studies on the nucleotide arrangement and on the complementarity and polarity of double-stranded DNA and made possible the differentiation, by analysis of the pyrimidine sequences, of DNA from different organisms with identical overall base composition. Furthermore, such degradation products are used for studies on their activity as templates, primers, or substrates for enzymes.

The principal degradation steps are shown in Fig. 1. The treatment of DNA under mild acid conditions leads to quantitative removal of the purines to give *apurinic acid*. On the other hand, the treatment of DNA with hydrazine causes destruction of the pyrimidine bases yielding as a first product a derivative of apyrimidinic acid, the hydrazone, which after reaction with benzaldehyde yields *apyrimidinic acid*; this is complementary to apurinic acid, when double-stranded DNA is the starting material. Hydrolysis of apurinic or apyrimidinic acid liberates, respectively, the pyrimidine or purine clusters, i.e., the total population of isolated nucleotides and of uninterrupted sequences of pyrimidine or purine oligonucleotides having the

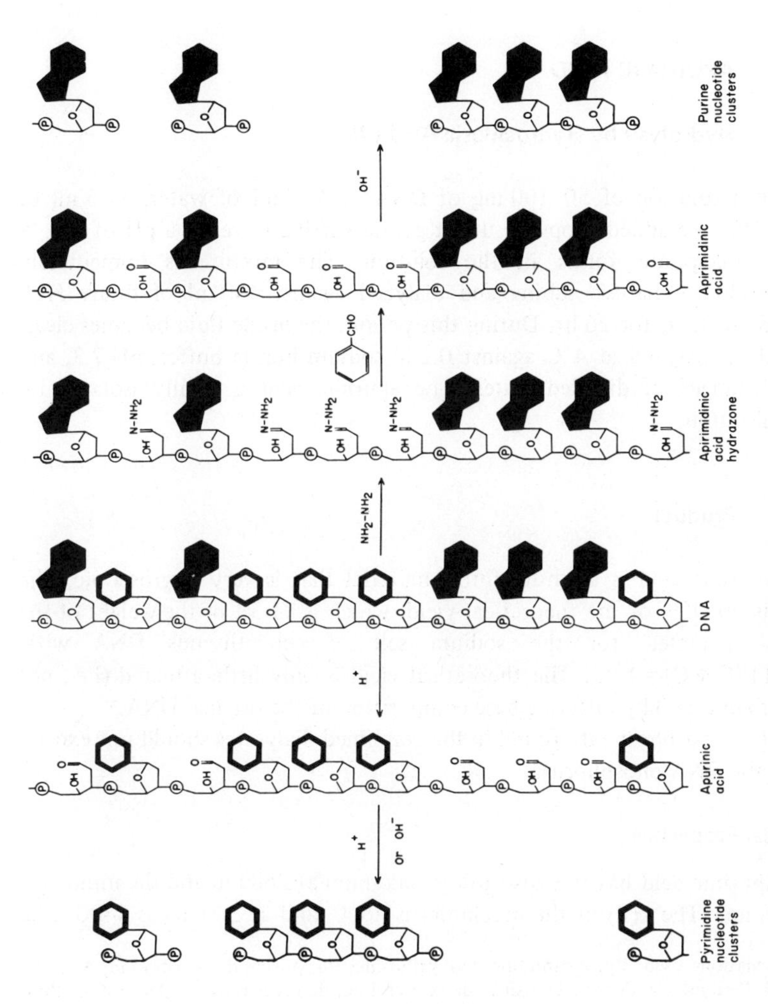

Figure 1. Schematic representation of the selective degradation of DNA. (Reproduced from E. Chargaff. Aspects of the nucleotide sequence in nucleic acids. In M. Sela, New Perspectives in Biology. B. B. A. Library, Vol. 4, Elsevier, Amsterdam, 1964).

general formula $Py_n p_{n+1}$ or $Pu_n p_{n+1}$.* These can be separated chromatographically into a series of size groups, each comprising all oligonucleotides of the same length, with $n = 1, 2, 3, \ldots$, etc. To designate such a group, the term *isostich* was proposed (*1*).† The isostichs can be further separated into their components, differing in base composition, by paper chromatography and paper electrophoresis.

APURINIC ACID

Hydrolysis by Hydrochloric Acid (*3*)

To a solution of 50–100 mg of DNA in 42.5 ml of water, 12.5 ml of 0.1 N HCl are added dropwise under gentle stirring to reach a pH of 1.6. A heavy precipitate forms in the solution. The mixture is immediately transferred to dialysis tubing and dialyzed against 440 ml of dilute HCl, pH 1.6, at 37°C for 26 hr. During this period, the inside fluid becomes clear. It is then dialyzed at 4°C against 0.2 M sodium borate buffer, pH 7.3, and three changes of distilled water. The apurinic acid is finally isolated by lyophilization.

Product

Apurinic acid is a white, fluffy material that is very hygroscopic. The yield is 90–95% of the theoretical yield, which is 81.3% of the weight of the DNA, calculated for the sodium salt of calf thymus DNA with $(A + T)/(G + C) = 1.33$. The theoretical yield is only little affected (i.e., not more than 0.5%) by differing base composition of the original DNA.

The total phosphate found in the combined dialysates should not exceed 3% of the DNA phosphorus.

Physical Properties

Apurinic acid has the absorption maximum at 268 nm and the minimum at 238 nm. The $\epsilon(P)$ at the maximum is 4600, and at 260 nm is 4400. The

*Abbreviations used: Py, pyrimidine deoxyriboside; Pu, purine deoxyriboside; A, C, G, M, and T stand for deoxyadenosine, deoxycytidine, deoxyguanosine, deoxy-5-methylcytidine, and thymidine, respectively. Esterified phosphate is indicated by p; placed at the left of the nucleoside symbol it is bound to the 5′-hydroxyl; placed at the right, to the 3′-hydroxyl.

†From the Greek "isos," equal, and "stichos," line. To give an example: in the purine isostich 2 there are four dinucleotides, pApAp, pGpGp, pApGp and pGpAp, the last two being at the same time isostichs and isomers.

Instead of isostich the term isoplith is also used (*2*).

absorbance ratios are A_{250}/A_{260} : 0.77; A_{280}/A_{260} : 0.75; A_{290}/A_{260} : 0.25 (*3, 4*). All these values were determined in 0.1 M phosphate buffer pH 7.1.

The average molecular weight was found to be about 15,000 and the sedimentation constant $s_{w,20}$, 1.55×10^{-13} (*5*).

Chemical Properties

The base analysis of apurinic acid gives thymine and cytosine in unchanged proportions, whereas adenine and guanine are completely absent or certainly below 2% of their original amount. The actual recoveries of the pyrimidines with respect to phosphorus are 47–49%.

The deoxyribose residues, determined by means of diphenylamine (*6*) using 5′-deoxyadenylic acid as the standard are the same as in DNA, 50 moles per 100 gram-atoms of phosphorus.

Apurinic acid gives all the aldehyde reactions characteristic for reducing sugars, such as color formation with reduced fuchsine and the reactions with *m*-phenylenediamine and ammoniacal silver solution. It forms derivatives with hydroxylamine, semicarbazide, and 2,4-dinitrophenylhydrazine (*5, 7*).

APYRIMIDINIC ACID

For the preparation of apyrimidinic acid one faces a double difficulty – the pyrimidine glycosidic bond is quite strong, but the purine glycosidic bond is weak. Once the pyrimidines are removed, the product is also sensitive to alkali; therefore the conditions have to be kept anhydrous or at neutral pH. Hydrazinolysis of the pyrimidines followed by reaction of the intermediate apyrimidinic acid hydrazone with benzaldehyde is so far the only method for the preparation of apyrimidinic acid (*4*).

Procedure

Anhydrous hydrazine is prepared from 95% hydrazine (Eastman) or hydrazine hydrate by refluxing and distilling over solid sodium hydroxide, if necessary several times. Hydrazine is inflammable, poisonous, and very caustic.

One hundred milligrams of DNA are placed in a 5-ml ampoule and dried in a vacuum at 60°C over phosphorus pentoxide in a drying pistol for at least 2 hr. After the addition of 1 ml freshly distilled anhydrous hydrazine the vessel is sealed and kept in a water bath for 4 hr at 60°C. Of this time 30–45 min are required for the DNA to go into a clear and colorless solution.

The ampoule is cooled in Dry Ice–alcohol until the solution is frozen, opened quickly, and placed in a vacuum desiccator containing a beaker of sulfuric acid. The hydrazine is evaporated in the frozen state in a vacuum and the dry residue dissolved in 50 ml of 0.2 M sodium borate buffer pH 6.9. Then 3 ml of benzaldehyde are added and the mixture is stirred vigorously in a stoppered tube at 4°C for 12–15 hr. The excess of benzaldehyde and the yellow benzalazine formed are extracted gently avoiding the formation of an emulsion with 15-ml portions of ether. The extractions are continued until the aqueous phase is free of benzaldehyde (6–10 extractions). The aqueous solution is subjected to dialysis in the cold, each time for 12–15 hr, twice against 500 ml of 0.02 M sodium borate buffer pH 6.9 and three times against 500 ml of distilled water. The combined dialyzates are concentrated and used for the determination of total phosphorus. The clear dialyzed solution is evaporated in the frozen state, yielding *apyrimidinic acid.*

If *apyrimidinic acid hydrazone* is to be isolated, the solution in borate buffer is dialyzed as described above and subjected to lyophilization omitting the reaction with benzaldehyde.

Product

Apyrimidinic acid is isolated as a white or slightly yellowish, fluffy material in 90–95% of the theoretical yield. This, calculated again for the sodium salt and for calf thymus DNA, is 84.6% for apyrimidinic acid and 86.8% for the hydrazone derivative. Both substances are very hygroscopic.

Physical Properties

The absorption maximum of apyrimidinic acid is at 256 nm, the minimum at 226.5 nm. The $\epsilon(P)$ was determined as 6000 at the maximum and 5950 at 260 nm. The absorbance ratios are A_{250}/A_{260} : 0.98; A_{280}/A_{260} : 0.42; A_{290}/A_{260} : 0.17. All values were determined in 0.1 M phosphate buffer pH 7.1. The optical properties of apyrimidinic acid hydrazone are very similar (*4*).

The sedimentation constant of a calf thymus apyrimidinic acid was $s_{w,20}$: 1.74×10^{-13} (*4*). This is very similar to the sedimentation constant found for apurinic acid.

Chemical Properties

The base analysis of apyrimidinic acid reveals the complete absence of the pyrimidine bases. Not in all but in many cases, the A/G ratio was slightly decreased, due to some preferential losses of adenine during the hydrazinolysis, corresponding to 2–4% of the total adenine. The actual recoveries of the purines with respect to phosphorus were 46–48%.

The deoxyribose residues determined with diphenylamine (*6*), using 5′-deoxyadenylic acid as the standard, were 90–100 moles/100 gram-atoms phosphorus. For the apyrimidinic acid hydrazone, this value was considerably decreased to 40–50 moles/100 gram-atoms phosphorus (*4*).

Apyrimidinic acid, but not the hydrazone, gives the typical aldehyde reactions with reduced fuchsine and *m*-phenylenediamine.

In solution it must be kept at neutral pH, because in alkali it is degraded and in acid it is depurinated.

Comments

The suitability of the hydrazinolysis reaction for the preparation of apyrimidinic acid was questioned recently (*8, 9*). The procedure described above permitted a number of good apyrimidinic acid preparations to be obtained from alcohol-precipitated calf thymus DNA, which were well characterized. However, difficulties were encountered when this procedure was applied to several lyophilized bacterial DNA samples. The products obtained were degraded and 5–40% of the total phosphorus became dialyzable (*4*).

Conditions of Hydrazinolysis and Losses of Adenine

The minimal time required for complete destruction of the pyrimidine bases was determined by kinetic studies using the pyrimidine mononucleotides. At 60°C this is 4 hr and at 40°C, it is 15–16 hr (*10*). At 30°C 30 hr of incubation are required (*9*). Hydrazinolysis at 37°C or 30°C led also to some losses of adenine amounting to 2–4% of the total adenine (*9, 11*).

Degradation of Apyrimidinic Acid During Hydrazinolysis

The total phosphorus found in the combined dialyzate is a direct indication of the quality of the product. In good apyrimidinic acid preparations it should not exceed 3% of the DNA phosphorus. The reasons for the degradation of the apyrimidinic acids from bacterial DNAs are not entirely clear. Preliminary results suggest that lyophilized DNA binds water irreversibly on storage, leading to an alkaline pH during the hydrazinolysis (*12*). It is recommended to use, whenever possible, alcohol-precipitated and acetone-dried DNA or freshly lyophilized DNA.

ISOLATION OF THE PYRIMIDINE AND PURINE ISOSTICHS

Liberation of Py_np_{n+1} and Pu_np_{n+1}

Direct Hydrolysis of DNA to Py_np_{n+1}

Because the acid hydrolysis conditions required to split the phosphodiester bonds between two deoxyribose residues are stronger than those needed for depurination, it is not necessary to isolate the apurinic acid first; the DNA may be directly hydrolyzed to the pyrimidine isostichs Py_np_{n+1}. Two methods are used: degradation by formic acid–diphenylamine and hydrolysis by sulfuric acid.

Hydrolysis by Formic Acid–Diphenylamine (*13*). The DNA is incubated at a concentration of 0.5–1.5 mg/ml in 66% formic acid containing 2% diphenylamine (analytical-reagent grade or recrystallized) at 30°C for 17 hr. To the reaction mixture an equal volume of water is added and the diphenylamine is removed by several extractions with ether. The solution is concentrated and freed of the remaining formic acid by repeated evaporations in a rotary evaporator. If the purines are to be separated from the reaction mixture, it is passed, after the ether extractions, through a column of Dowex-50 (H^+) (1 x 4 cm). The column is washed with water until the eluate has an absorbance of less than 0.05 (at 260 nm). The combined eluates are neutralized and concentrated.

Hydrolysis by 0.2 N Sulfuric Acid (*14*). A 2% suspension of DNA in 0.2 N sulfuric acid is heated in a sealed ampoule (or under reflux) at 100°C for 35 min. The tube is cooled in ice, opened, and the hydrolyzate is neutralized with 2 N ammonia. If necessary, the solution is clarified by centrifugation. The clear hydrolyzate is used for column or paper chromatography.

The Liberation of Inorganic Phosphate. The liberation of the pyrimidine isostichs Py_np_{n+1} from DNA by acid, and also of Py_np_{n+1} from apurinic acid or of Pu_np_{n+1} from apyrimidinic acid by alkali, are thought to occur by a series of β-elimination reactions (*15*), which release from a randomly-arranged, double-stranded, and base-paired DNA 25% of its phosphorus as inorganic phosphate. In fact, when DNA is hydrolyzed with formic acid–diphenylamine the inorganic phosphate reaches a value of 25–28% and then remains stable with prolonged hydrolysis (*13*). Under the conditions of sulfuric acid hydrolysis only 15–20% of the phosphate is released and no definite endpoint of the reaction is reached with continued hydrolysis (*16, 17*). However, the pyrimidine isostichs determined by fractionation on a DEAE-cellulose column are released in comparable amounts under both conditions (*18, 19*), and no compounds other than Py_np_{n+1} are found in the hydrolyzate (*20*). On the other hand, Jones et al. reported the detection of deoxyribose residues apparently bound to the nondialyzable pyrimidine oligonucleotides formed by sulfuric acid hydrolysis (*21*).

Incomplete liberation of inorganic phosphate is also found upon acid hydrolysis of apyrimidinic acid (*17*). The question remains open, whether some organic phosphate is retained between deoxyribose residues because of incomplete eliminations in the sugar-phosphate backbone, or is carried in some yet unidentified phosphate containing degradation product.

Comparison of the Two Hydrolysis Methods. When the two methods are compared by analyses of the pyrimidine isostichs produced, the following results are found: After fractionation on a DEAE-cellulose column, the pyrimidine isostich distribution based on phosphate analysis is quite similar with both methods (*18, 19*). On direct two-dimensional paper chromatography of the sulfuric acid hydrolyzate, the recovery of isostichs 1–4 amounted to about 65% of that obtained in a similar mapping of a formic acid–diphenylamine hydrolyzate. For isostich 5, the results were the same for both methods (*1*).

The sulfuric acid hydrolysis is an easy and fast method for the isolation of the pyrimidine isostichs; nevertheless recent publications (*18, 19, 21*) recommend the use of formic acid-diphenylamine for quantitative analysis.

Alkaline Hydrolysis of Apurinic Acid to Py_np_{n+1}

A 1–2% solution of apurinic acid in 0.2 N potassium hydroxide is heated in a sealed ampoule at 100°C for 35 min. The hydrolyzate is cooled in ice and neutralized with 2 N perchloric acid. The crystalline precipitate of potassium perchlorate is centrifuged, washed with a small amount of ice-cold water, and the two supernatants are combined (*1*).

The recovery and the frequency of the pyrimidine isostichs and of their components released by acid from DNA and by alkali from apurinic acid, estimated after two-dimensional paper chromatography of the hydrolyzates, were very similar (*1*).

The Liberation of Inorganic Phosphate. Since the same products, Py_np_{n+1}, are formed as in acid hydrolysis, the reaction mechanism probably also comprises a series of β-eliminations. However, the liberation of inorganic phosphate remains again incomplete (*22*). Under the conditions described above, the inorganic phosphate amounts to 12–15% after 35 min and to 22–25% after 2 hr (*17*). In alkaline hydrolyzates of apurinic and apyrimidinic acid, 8–10% of the phosphate was found in a degradation product which was identified as 2-oxocyclopent-1-enylphosphate (*22*).

Alkaline Hydrolysis of Apyrimidinic Acid to Pu_np_{n+1}

Various conditions have been used for the alkaline hydrolysis of apyrimidinic acid or its hydrazone derivative. When apyrimidinic acid is isolated, the same conditions used successfully for the hydrolysis of apurinic acid can be applied. The liberation of inorganic phosphate from apurinic and apyrimidinic acid under the same conditions is very similar but is slower from the hydrazone derivative (*17, 23*).

Because some depurination, especially losses of adenine, may occur during alkaline hydrolysis (*17, 24*) the mildest possible conditions should be chosen. For this reason it is recommended to hydrolyze the apyrimidinic acid rather than the hydrazone derivative (*23*). The fate of the hydrazone group during the alkaline hydrolysis and its influence on the cleavage of the phosphodiester bonds have not been studied in detail.

For apyrimidinic acid the best results were obtained with either 0.2 N KOH, 100°C, 35 min or 0.3 N KOH, 60°C, 8 hr (*17*). The hydrazone derivative was hydrolyzed with 0.3 N KOH at 100°C for 1 hr (*11*). The alkaline hydrolyzate is neutralized with perchloric acid and separated from the precipitate.

Fractionation of the Pyrimidine and Purine Isostichs on DEAE-Cellulose Columns

Fractionation of Py_np_{n+1} and Pu_np_{n+1} with a Lithium Chloride Gradient

This is the original procedure worked out by Spencer and Chargaff (*25, 26*) who adapted a procedure used for the separation of synthetic oligonucleotides. It has been used both for the pyrimidine and the purine isostichs.

DEAE-Cellulose Chromatography. From 1.0 gm of DEAE-cellulose (Cl^-), 325 mesh, suspended in 50 ml of 0.01 M lithium acetate pH 5.0* a column of about 0.8 x 12 cm is prepared and washed with the same buffer. The neutralized hydrolyzate of 20–30 mg of DNA or of apurinic or apyrimidinic acid is diluted to 100 ml with the lithium acetate buffer. Small aliquots are removed for the determination of total and inorganic phosphate. The solution is applied to the column, which is then washed with 200–300 ml of the lithium acetate buffer. The isostichs are eluted with a linear gradient of lithium chloride increasing up to 0.32 M, in 0.01 M lithium acetate, pH 5.0. The total gradient volume is 1400 ml. A constant flow throughout the separation is maintained by a pump delivering about 1 ml/min to the column. Fractions of 6 ml are collected. The ultraviolet absorption of the eluate may be monitored continuously. At the end of the gradient, which elutes isostichs 1–9, the column is washed with 100 ml of 1 M lithium chloride to elute the

*Later experiments on the fractionation of Py_np_{n+1} showed the optimal pH to be 5.5 (*27*). At this pH the partial separation of the cytosine and thymine nucleotides shown in Fig. 2 is prevented and each isostich elutes as one sharp peak (*19, 27*).

higher isostichs. These are normally pooled with isostich 9. The total phosphate recovered from the column is practically 100% of that loaded.

Separation of Py_np_{n+1}. The fractionation of the pyrimidine isostichs from a sulfuric acid hydrolyzate of *E. coli* DNA is shown in Fig. 2. The free purines are not adsorbed to the column and run through. Most of the inorganic phosphate formed during hydrolysis is found with the purines and in the washes; a small part may also be eluted with isostich 1 (*26*). Using sulfuric acid hydrolysis there is, however, a constant elution of small amounts of inorganic phosphate throughout the gradient complementing it to 25–28% of the total phosphate eluted. If formic acid–diphenylamine hydrolysis is used, all the inorganic phosphate is eluted in the washes, occasionally in fraction 1, but the other isostichs are free of it (*19*). In the elution profile shown in Fig. 2 the isostichs are partially separated. This is typical when the elution is made at pH 5.0, but already at pH 5.5 the isostichs elute as single, sharp peaks (cf., footnote p. 688).

Separation of Pu_np_{n+1}. Figure 3 shows the fractionation of the purine isostichs from an alkaline hydrolyzate of calf thymus apyrimidinic acid. Two peaks (A and B) are eluted before isostich 1. Peak A is also found in alkaline hydrolyzates of apurinic acid and contains presumably some sugar degradation products and all the inorganic phosphate present in the hydrolyzate. Peak B contains both organic and inorganic phosphate. The total phosphate

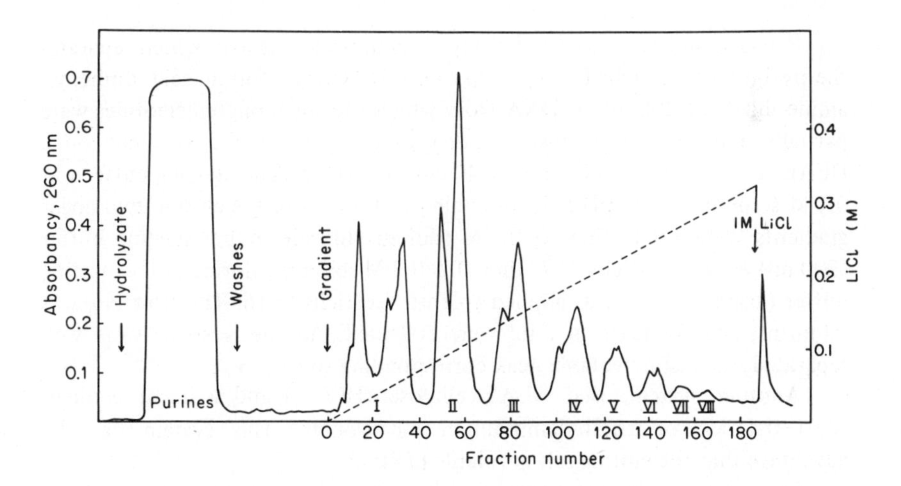

Figure 2. Chromatography of an acid hydrolyzate of *E. coli* DNA on a DEAE-cellulose column. The pyrimidine isostichs Py_np_{n+1} are numbered I–VIII. The gradient was formed by means of a 9-chamber "Varigrad"; the total gradient volume was 1350 ml. For experimental details see text. (Reproduced from *Biochim. Biophys. Acta*, Ref. *26*).

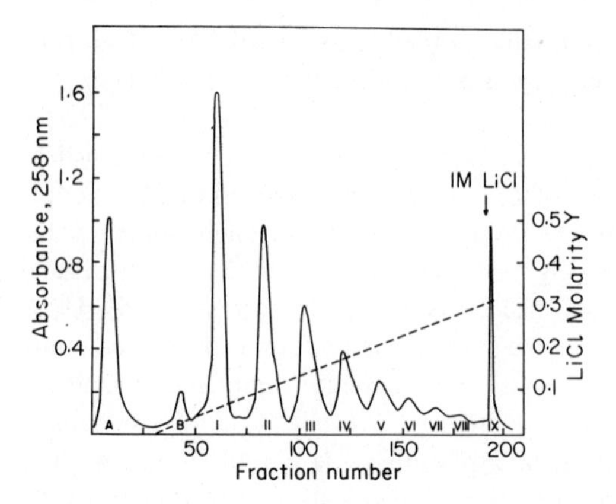

Figure 3. Chromatography of an alkaline hydrolyzate of apyrimidinic acid (calf thymus) on a DEAE-cellulose column. The purine isostichs $Pu_n p_{n+1}$ are numbered I–VIII. The gradient was formed by means of a 9-chamber "Varigrad"; the total gradient volume was 1350 ml. For experimental details see text. (Reproduced from *Biochim. Biophys. Acta*, Ref. *17*).

in peaks A and B amounts to close to 25% of that eluted. Peaks A and B were analyzed for the presence of free purines or purine nucleosides, but only traces of these could be detected (*17*).

Other Systems

Petersen and Reeves (*27*) developed an improved system which separates the pyrimidine isostichs $Py_n p_{n+1}$ up to $n = 16$. Using a formic acid–diphenylamine digest of 200 mg of DNA from which the small oligonucleotides were partially removed by dialysis (*27, 28*) they separated the isostichs on a DEAE-cellulose (Cl^-) column (1 x 40 cm) at pH 5.5. The starting buffer was 0.1 M sodium acetate pH 5.5, containing 7 M urea. For the elution two linear gradients were used: first, 0–0.2 M sodium chloride in the starting buffer (700 ml) eluted isostichs 1–7, then 0.2–0.3 M sodium chloride in the starting buffer (again 700 ml) was applied to elute isostichs 8–16. The flow rate was 12 ml/hr, and fractions of 2 ml were collected. All the isostichs were well separated, the last detectable peak corresponding to $n = 16$.

Another system used DEAE-cellulose (HCO_3^-) and a linear gradient (0.01–0.41 M) of triethylammonium bicarbonate. This system has the advantage that the eluting salt is volatile (*18*).

Some authors prefer to remove the terminal phosphate groups with a phosphomonoesterase before fractionation of the isostichs. This is done to prevent a spontaneous release of terminal phosphate that might partially displace isostichs into the lower isostich fraction. The dephosphorylation of the purine isostichs with *E. coli* alkaline phosphatase (EC 3.1.3.1) is described by Cape and Spencer (*11*). Py_np_{n-1} were separated on a DEAE-Sephadex (HCO_3^-) column (2 x 18 cm) by elution with a linear gradient of triethylammonium bicarbonate, 0.01 M pH 8.3 to 1.3 M pH 8.7, with a total gradient volume of 3 liters (*2*). Pu_np_{n-1} have also been separated on DEAE-Sephadex (Cl^-) with a linear gradient of sodium chloride, 0–0.48 M in 0.005 M Tris-HCl, pH 7.5, containing 7 M urea (volume 2.5 liters) (*29*); or on DEAE-cellulose (Cl^-) with a lithium chloride gradient, 0–0.4 M in 0.005 M Tris-HCl, pH 7.5, containing 7 M urea (1 liter) (*11*).

Calculation of the Isostich Distribution

There are three means to calculate the isostich distribution after the chromatographic fractionation: by analysis of total or organic phosphate, or by ultraviolet absorption. If the degradation reactions proceeded uniformly and quantitatively, the results should be the same for all three methods. Comparison of the results and the recoveries obtained are, in fact, a good control for the procedures used. As long as the mechanisms involved in these reactions are not completely understood, it is dangerous to base the results on only one criterion.

The results for the isostich distribution are usually expressed as moles pyrimidine (or purine) per 100 gram-atoms pyrimidine (or purine) nucleotide phosphorus, corrected to the theoretical recovery of 100. When the DNA phosphorus is used instead of the pyrimidine nucleotide phosphorus, the results are corrected to the theoretical recovery of 50. The sum of the results obtained for each isostich should be within about 3% to the theoretical value.

Calculation from Total or Organic Phosphate

Total and inorganic phosphate are usually determined colorimetrically in small aliquots of the pooled eluate fractions under each peak. To obtain the moles of pyrimidine or purine in each isostich, the presence or absence of the terminal phosphate groups have to be taken into account. For isostich n the relationship used is:

$$\frac{\mu\text{moles pyrimidine}}{100\ \mu\text{g-atoms pyrimidine P}} = \frac{\mu\text{g-atoms } P_n \times f_n \times 200}{\mu\text{g-atoms P put on column}}$$

where P_n is the phosphate in the pooled fractions of isostich n (total or organic according to the method of calculation); and f_n is the conversion factor for isostich n considering the presence or absence of terminal phosphate groups. For example, f_n is 0.5 for Py_1p_2; 0.67 for Py_2p_3; 0.75 for Py_3p_4; etc.

The calculation from phosphate values gives good results for formic acid-diphenylamine digests, because total and organic phosphate are practically the same for isostichs 2-9. For sulfuric acid and alkaline hydrolyzates, careful phosphorus balances have to be made because of the elution of inorganic phosphate with the isostichs. The calculation from phosphate values will be difficult, if the DNA contains considerable amounts of non-DNA phosphorus, as is often the case in bacterial DNA preparations.

Calculation from the Ultraviolet Absorption

For all isostichs an average extinction coefficient $\Delta\epsilon_{260-320}$ can be used, which is calculated from the $\Delta\epsilon_{260-320}$ of the mononucleotides according to the T/C or A/G ratios in the DNA, apurinic or apyrimidinic acid. The individual extinction coefficients at neutral pH are:

Purine nucleotides (*30*):	$\Delta\epsilon_{260-320}$	pA	15.3×10^3
		pG	11.8×10^3
Pyrimidine nucleotides:	$\Delta\epsilon_{260-320}$	pC	7.6×10^3 (*15*)
		pT	9.3×10^3 (*15*)
		pM	5.1×10^3 (*31*)

Since the T/C and A/G ratios may vary in the individual isostichs, some errors are introduced by this method. To avoid these, Spencer et al. used the isosbestic wavelength (i.e., the wavelength of equal molar absorption of the two nucleotides) (*11, 29*). The following values were determined for the isosbestic wavelengths and the extinctions:

pC + pT (*32*):	pH 5	249 nm	ϵ: 4.74×10^3
		270 nm	ϵ: 8.68×10^3
	pH 7.5	252 nm	ϵ: 5.76×10^3
		274 nm	ϵ: 8.41×10^3
pA + pG (*11*):	pH 7.5,7 M urea	270 nm	ϵ: 9.74×10^3

For isostich n the relationship used is

$$\frac{\mu\text{moles pyrimidine}}{100\ \mu\text{g-atoms pyrimidine P}} = \frac{[A_{nm} \times \text{volume (ml)}]_n \times 200}{\epsilon_{nm} \times 10^{-3} \times \mu\text{g-atoms P loaded}}$$

where A_{nm} is the absorption of the isostich solution at the wavelength used.

Calculation from the ultraviolet absorption with these extinction coefficients gives good results for the pyrimidine isostichs, because these show practically no hypochromic effect. In the case of the purine isostichs the calculated recovery was only 80% (*17*). This is most probably due to the retention of hypochromicity by the purine isostichs, as demonstrated already for the dinucleotides (*11*).

ANALYSIS OF THE COMPONENTS OF THE ISOSTICHS

Direct Analysis of the Pyrimidine Oligonucleotides without Previous Fractionation of the Isostichs

The paper used for paper chromatography (e.g., Whatman No. 1, 46 x 56 cm) is preferably washed chromatographically with 1 N HCl and water before use. The analyses are done by descending paper chromatography.

Analysis of $Py_n p_{n+1}$ *(20)*

The pyrimidine oligonucleotides, $Py_n p_{n+1}$, up to the hexanucleotides, were analyzed by two-dimensional paper chromatography.

Solvent 1: Isopropanol-water, 7 : 3 v/v, with 0.35 ml of concentrated ammonia ($d = 0.88$) per liter of tank volume in a beaker placed at the bottom of the tank.

Solvent 2: Isobutyric acid-0.5 N ammonia, 5 : 3 v/v, pH 3.7.

Aliquots of the neutralized hydrolyzates, corresponding to 1-2 mg of DNA, were spotted on the papers. The first irrigation in solvent 1 is done for 72 hours, the optimal temperature being 30°C. The dried paper is irrigated in the second dimension with solvent 2 for 36 hours.

A schematic representation of the spots obtained is shown in Fig. 4.

Analysis of $Py_n p_{n-1}$

This is a more elaborate method and allows the separation of pyrimidine components up to nonanucleotides (*13, 33*).

The formic acid-diphenylamine hydrolyzate, freed of the purines, was treated with phosphomonoesterase until 70-75% of the phosphate was released (*13*). Aliquots corresponding to 2 mg of DNA were applied to the paper. The first stage was two-dimensional chromatography using as solvents

1. isopropanol-5 N ammonia, 65 : 35 v/v
   Duration: 3 hours in addition to the time required for the solvent to reach the lower edge of the paper.
2. *t*-butanol-conc. HCl-water, 11 : 1 : 3 v/v
   Duration: until the solvent front is at the lower edge of the paper.

The clearly separated spots are cut out and determined spectrophotometrically. The unresolved area was eluted, evaporated to dryness in a stream of nitrogen and separated in a second stage.

First dimension: electrophoresis in formate buffer pH 2.7. Forty-five milliliters of 98% formic acid are added to 6 liters of water and adjusted with

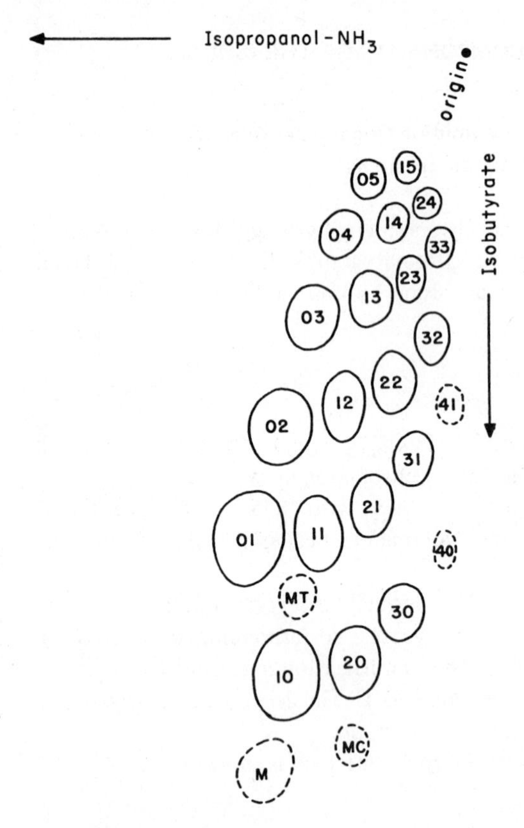

Figure 4. Schematic representation of a two-dimensional chromatogram of an acid hydrolyzate of human DNA (spleen). The compositions of the pyrimidine oligonucleotides Py_np_{n+1}' in each spot are indicated as the number of C units followed by that of T units, e.g., 10 is Cp_2, 01 is Tp_2, 23 is $C_2T_3p_6$, etc. For methylcytosine-containing compounds the composition is indicated by the nucleoside symbols. The broken lines show the positions of components found in other experiments with calf thymus DNA. (Reproduced from *Biochim. Biophys. Acta*, Ref. *20*).

concentrated ammonia to pH 2.7. Electrophoresis with approximately 9 V/cm and 10–13 mA for 15 hr (*34*).

Second dimension: paper chromatography by irrigation, twice in the same direction, with *t*-butanol–0.078 M ammonium formate pH 3.8, 1 : 1 v/v for 20–24 hr each.

Analysis of the Individual Isostichs After Fractionation

Desalting of the Isostichs

Before paper chromatography the isostichs have to be concentrated and desalted. When a volatile salt is used for the gradient elution, it is sufficient to evaporate several times in a rotary evaporator. Otherwise one of the following methods have to be used.

Adsorption to Charcoal. So far this method was used for desalting the pyrimidine isostichs only. To the concentrated isostich solution activated charcoal (Darco G 60) is added until the supernatant is free of ultraviolet absorption. The charcoal is filtered off, washed with water and the pyrimidine isostichs are eluted with ethanol–conc. ammonia ($d = 0.88$)–water, 25 : 1 : 25 v/v. It is useful to add an internal standard, e.g., thymidylic acid (200–400 μg), to determine the recovery (*25, 26*).

Desalting with DEAE-cellulose (*35*). The purine isostichs Pu_np_{n+1} were desalted by adsorption on small DEAE-cellulose columns (*17*).

The isostich fractions are diluted with water to three times their volume and passed through a small DEAE-cellulose (HCO_3^-) column prepared from 300 mg of dry material. The column is washed subsequently with 100 ml water and 100 ml 0.01 M ammonium bicarbonate. Then the isostichs are eluted with 5 ml portions of 2 M ammonium bicarbonate until the ultraviolet absorption in the eluate is less than 0.025 (4–5 times). The eluate is evaporated several times until the residue is free of salt. The recoveries in terms of phosphate and ultraviolet absorbance were 75% for purine isostich 1, 80–85% for isostich 2, and 90–100% for the higher isostichs.

Partial release of terminal phosphate groups was observed in the isostichs Py_np_{n+1}, when they were not immediately desalted (*25*) and also during desalting of the purine isostichs (*17*). Complete enzymatic dephosphorylation of the isostichs before analysis by paper chromatography will prevent the formation of double spots of the type (Py_np_{n+1} plus Py_np_n).

Paper Chromatography of the Pyrimidine Isostichs

Py_np_{n+1} or Py_np_{n-1} can be separated by two-dimensional chromatography with the same solvents used for the direct analysis of the pyrimidine oligonucleotides.

However, up to isostich 6 good results are also obtained by chromatography in one dimension with isopropanol–conc. HCl ($d = 1.19$)–water, 17 : 4 : 4 (*2, 18*).

Paper Chromatography of the Purine Isostichs

After dephosphorylation the purine isostichs 2–5 were separated in one dimension with isobutyric acid–0.5 N ammonia, 5 : 3 v/v, pH 3.7 (*17*). Usually 2–4 ultraviolet absorbance units were applied to the paper.

Separation of the Components of the Higher Pyrimidine Isostichs by Column Chromatography

A considerable improvement for the analysis of the pyrimidine clusters was achieved by the subfractionation of the pyrimidine isostichs on DEAE-cellulose columns at acid pH, which separates the isostich components in order of increasing thymidine content (*27*).

The isostich fractions are diluted to twice their volume with distilled water and loaded without desalting onto a DEAE-cellulose column (formate, 1.5 x 25 cm, prepared from 5 gm DEAE-cellulose, 325 mesh). The column is washed with 0.1 M formic acid and eluted with a linear gradient to 1.0 M ammonium formate pH 3.0 in 0.1 M formic acid (pH 2.7). The total gradient volume is 1 liter and fractions of 5 ml are collected (*40, 41*).

Similar separations were obtained with a DEAE-Sephadex (Cl^-) column eluted with a linear gradient of 0 to 0.4 M NaCl in 7 M urea-HCl pH 3.0 (*42*).

Characterization and Estimation of the Isostich Components

The components separated according to base composition by paper or column chromatography are characterized by their position (r_f-value) and spectral ratios, and estimated spectrophotometrically. If enough material can be recovered the purity of the spots should be checked by base analysis and base/phosphorus ratios.

The pyrimidine oligonucleotides are eluted from the paper with 0.1 N HCl. Data for the identification and estimation of a large number of pyrimidine oligonucleotides are listed in the literature, such as the r_f-values for $Py_n p_{n+1}$ in the two-dimensional system relative to pT (*25*) and pTp (*20*); the spectral ratios for all possible pyrimidine oligonucleotides (*25, 33*) including those containing methylcytosine (*25*); the molar extinctions at various wavelengths (*20, 25*), as well as the isosbestic wavelengths and the corresponding extinctions at pH 1 (*33*), pH 5, and pH 7.5 (*32*). For the purine oligonucleotides, eluted from the paper with 0.02 M ammonium bicarbonate, these data are not listed in the literature. The spectral ratios and the extinction coefficients for the mixed oligonucleotides were calculated from those of the mononucleotides (*17*). No allowance could be made for possible hypochromic effects, which remain to be determined for the purine oligonucleotides.

The results are usually expressed as moles pyrimidine (or purine) per 100 gram-atoms pyrimidine (or purine) nucleotide phosphorus. Some authors use the DNA phosphorus as a reference. Several ways have been chosen to present the results:

1. The uncorrected values are given, as they were obtained from the eluates of the spots. This is always done for results obtained by direct paper chromatography (*1, 20, 26*).

2. The values "found" are corrected for losses during the manipulation, especially during desalting, as determined by the recovery of an internal standard (*26, 33*) or of phosphate (*18, 36*).

3. The values are recalculated for the isostich distribution obtained from fractionation of the isostichs on DEAE-cellulose columns, corrected to a 100% recovery (*26, 37*).

4. The contribution of the components to each isostich is presented as mole percent. This allows results obtained by different procedures to be compared and avoids all corrections (*17*).

In any case the method of calculation and the corrections used should be stated clearly whenever results are presented.

THE PROBLEM OF NUCLEOTIDE SEQUENCE ANALYSIS IN DNA

The further analysis of the individual isostich components encounters many difficulties. Each spot or peak that can be isolated by the procedures described comprises, except for the homologous oligonucleotides, a number of sequence isomers; e.g., the spot CT_2p_2 contains presumably the three isomers CpTpT, TpCpT, and TpTpC. Only in the case of the shortest mixed oligonucleotides could the isomers CpT, TpC and ApG, GpA be separated and estimated by chromatography or by enzymatic procedures (*13, 15, 30*). With increasing length of the oligonucleotides the difficulties of sequence analysis rise rapidly: the number of possible sequence isomers increase greatly, and at the same time there is a decrease of the amount of material available. The selective removal of one of the pyrimidine (or purine) bases will facilitate the analysis of isomeric groups in many cases. The selective destruction of cytosine in DNA or in pyrimidine oligonucleotides offers one such possibility (*38*). Another is suggested, perhaps, by experiments with permanganate-oxidized DNA, in which all the bases are destroyed except adenine (*24, 39*). For sequence analysis of higher isomeric oligonucleotides the degradation procedures will have to be combined with chemical or enzymatic end-group analyses (*43*).

Today we are still very far from an unambiguous solution of nucleotide sequences in DNA. Investigations on the nucleotide arrangement of cellular DNAs can be interpreted only in a very general statistical manner, because of the heterogenous, fragmented starting material (*44–46*). Similar studies on separated strands of cellular DNA will provide more information, but in principle will not change the problem. Considerable progress has been made with structural studies of small single-stranded DNA molecules. The complete catalog of all nonisomeric pyrimidine and purine oligonucleotides of the bacteriophages S 13 (*41*) and φX 174 (*2, 42*) were established by analyses of the pyrimidine nucleotide clusters in the viral strand and the double-stranded replicative form (*41*) or the complementary strand made *in vitro* (*42*). With the same methods structural studies should be possible for separated strands of defined fragments, such as single genes or recognition sites for enzymes or other proteins, isolated from high-molecular-weight DNA.

Acknowledgments

The great part of the work referred to in this article was done at the Cell Chemistry Laboratory (Columbia University, New York) under the direction of Erwin Chargaff.

I am very grateful to Drs. P. Rüst and R. Hancock for their help in preparing this manuscript.

References

1. Shapiro, H. S. and E. Chargaff. 1964. *Biochim. Biophys. Acta* **91**: 262.
2. Hall, J. B. and R. L. Sinsheimer. 1963. *J. Mol. Biol.*, **6**: 115.
3. Tamm, C., M. E. Hodes, and E. Chargaff. 1952. *J. Biol. Chem.*, **195**: 49.
4. Türler, H. and E. Chargaff. 1969. *Biochim. Biophys. Acta*, **195**: 446.
5. Tamm, C. and E. Chargaff. 1953. *J. Biol Chem.*, **203**: 689.
6. Giles, K. W. and A. Myers. 1965. *Nature*, **206**: 93.
7. Coombs, M. M. and D. C. Livingston. 1969. *Biochim. Biophys. Acta*, **174**: 161.
8. Ellery, B. W. and R. H. Symons. 1966. *Nature*, **210**: 1159.
9. Cashmore, A. R. and G. B. Petersen. 1969. *Biochim. Biophys. Acta*, **174**: 591.
10. Temperli, A., H. Türler, P. Rüst, A. Danon, and E. Chargaff. 1964. *Biochim. Biophys. Acta*, **91**: 462.
11. Cape, R. E. and J. H. Spencer. 1968. *Can. J. Biochem.*, **46**: 1063.
12. Türler, H., in preparation.
13. Burton, K. and G. B. Petersen. 1960. *Biochem. J.*, **75**: 17.
14. Shapiro, H. S. and E. Chargaff. 1957. *Biochim. Biophys. Acta*, **26**: 608.
15. Shapiro, H. S. and E. Chargaff. 1957. *Biochim. Biophys. Acta*, **26**: 596.
16. Shapiro, H. S. and E. Chargaff. 1960. *Biochim. Biophys. Acta*, **39**: 68.
17. Türler, H., J. Buchowicz, and E. Chargaff. 1969. *Biochim. Biophys. Acta*, **195**: 456.
18. Neulat, M. M. 1967. *Biochim. Biophys. Acta*, **149**: 422.
19. Spencer, J. H., R. E. Cape, A. Marks, and W. E. Mushynski. 1969. *Can. J. Biochem.*, **47**: 329.
20. Shapiro, H. S. and E. Chargaff. 1963. *Biochim. Biophys. Acta*, **76**: 1.
21. Jones, A. S., J. R. Tittensor, and R. T. Walker. 1966. *Nature*, **209**:296.
22. Brammer, K. W., A. S. Jones, A. M. Mian, and R. T. Walker. 1968. *Biochim. Biophys. Acta*, **166**: 732.
23. Chargaff, E., P. Rüst, A. Temperli, S. Morisawa, and A. Danon. 1963. *Biochim. Biophys. Acta*, **76**: 149.
24. Jones, A. S. and R. T. Walker. 1964. *Nature*, **202**: 24.
25. Spencer, J. H. and E. Chargaff. 1963. *Biochim. Biophys. Acta*, **68**: 9.
26. Rudner, R., H. S. Shapiro, and E. Chargaff. 1966. *Biochim. Biophys. Acta*, **129**: 85.
27. Petersen, G. B. and J. M. Reeves. 1966. *Biochim. Biophys. Acta*, **129**: 438.
28. Sutton, W. D. and G. B. Petersen. 1969. *Biochim. Biophys. Acta*, **174**: 155.
29. Sedat, J. and R. L. Sinsheimer. 1964. *J. Mol. Biol.*, **9**: 489.
30. Chargaff, E., J. Buchowicz, H. Türler, and H. S. Shapiro. 1965. *Nature*, **206**: 145.

31. Shapiro, H. S. and E. Chargaff. 1960. *Biochim. Biophys. Acta,* **39**: 62.
32. Spencer, J. H., R. E. Cape, A. Marks, and W. E. Mushynski. 1968. *Can. J. Biochem.* **46**: 627.
33. Petersen, G. B. 1963. *Biochem. J.,* **87**: 495.
34. Petersen, G. B. 1961. *Biochim. Biophys. Acta,* **51**: 212.
35. Rushizky, G. W. and H. A. Sober. 1962. *Biochim. Biophys. Acta,* **55**: 217.
36. Ooka, T. and M. M. Neulat-Portier. 1969. *Biochim. Biophys. Acta,* **182**: 542.
37. Spencer, J. H. and E. Chargaff. 1963. *Biochim. Biophys. Acta,* **68**: 18.
38. Shapiro, H. S. and E. Chargaff. 1966. *Biochemistry,* **5**: 3012.
39. Jones, A. S. and R. T. Walker. 1964. *Nature,* **202**: 1108.
40. Černý, R., W. E. Mushynski, and J. H. Spencer. 1968. *Biochim. Biophys. Acta* **169**:439.
41. Černý, R., E. Černá, and J. H. Spencer. 1969. *J. Mol. Biol.,* **46**: 145.
42. Darby, G., L. B. Dumas, and R. L. Sinsheimer. 1970. *J. Mol. Biol.,* **52**: 227.
43. Székely, M. and F. Sanger. 1969. *J. Mol. Biol.,* **43**: 607.
44. Burton, K. In P. N. Campbell and G. D. Greville (Eds.) Essays in biochemistry, Academic Press, New York, Vol 1, 1965, p. 57.
45. Shapiro, H. S. In T. Hayashi and A. G. Szent-Györgyi (Eds.), Molecular architecture in cell physiology, Prentice-Hall, Englewood Cliffs, N.J., 1966, p. 123.
46. Chargaff, E. In J. N. Davidson and W. E. Cohn (Eds.), Progress in nucleic acid research and molecular biology, Academic Press, New York, 1968, vol 8, p. 297.

SECTION

F

NUCLEIC ACID-PROTEIN INTERACTION

THE ISOLATION OF REPRESSORS

VINCENZO PIRROTTA, MARK PTASHNE,
PAUL CHADWICK, AND ROBERT STEINBERG
THE BIOLOGICAL LABORATORIES
HARVARD UNIVERSITY
CAMBRIDGE, MASSACHUSETTS

The Repressors

Three genetically defined repressor molecules have been isolated thus far. Two of these control certain genes of the coliphages λ and 434, and the third controls the genes of the *lac* operon of *E. coli*. Each of these repressors binds specifically and with high affinity to specific regions of DNA called the operators, suggesting that, *in vivo*, these repressors function by blocking the transcription of DNA to RNA. Furthermore, it has been explicitly demonstrated that the λ repressor functions *in vitro* to block transcription of specific species of RNA by purified RNA polymerase upon a λ DNA template (*18*).

The Genetics

The isolation of a repressor has depended on a close correlation of biochemistry with a prior exhaustive genetic analysis of the systems under study. The repressor hypothesis was first elaborated in 1961 by Jacob and Monod (*1*), and other reviews have summarized the more recent genetic experiments in the systems discussed here (*2–4*). We might summarize the fundamental genetic results for the *lac* operon as follows. There is a single gene, the *i* gene, the product of which turns off the three genes of the *lac* operon that code for enzymes involved in the metabolism of lactose. Mutations which delete or truncate the *i* gene or its product result in constitutive synthesis of the *lac* enzymes, that is, continual synthesis in the absence of lactose. Ordinarily, the *lac* genes function only in the presence of lactose or of certain lactose analogs (e.g., IPTG: isopropyl-1-thio-β-*d*-galactopyranoside), because these small molecules, or their metabolic products, bind directly to the *lac* repressor and release it from its target site. The site of action of the repressor is genetically defined by operator mutations, located just before the three *lac* genes, that result in constitutive enzyme synthesis in the presence of active repressor.

The hypothesis that certain λ and 434 phage genes are controlled by repressors is based on genetic arguments analogous to those developed for the *lac* operon. In these cases, a single gene of the phage, called the C_I gene, codes for a repressor (either the λ phage repressor or the 434 phage repressor) which directly blocks two groups of early phage genes. Mutations in two operators have been characterized in phage λ (*5*), and somewhat less extensively in phage 434 (*6*).

Inducers for the phage repressors are not known. It is known, however, that various treatments of lysogenic cells, such as irradiation with low doses of ultraviolet light, abolish immunity and allow expression of phage genes. The induction process requires the synthesis of a bacterial gene product but the molecular basis of induction remains unknown.

Detection of Repressors in Cell-Free Extracts

Equilibrium Dialysis

The *lac* repressor was first detected by its IPTG-binding activity in partially fractionated extracts of *E. coli* (*7*). The experiment was predicated on the observation that IPTG is an apparently nonmetabolized inducer of the *lac* operon, suggesting that IPTG interacts directly with the repressor. The procedure entailed floating dialysis bags containing concentrated protein mixtures, in a solution containing ^{14}C-labeled IPTG and searching for a cellular fraction (for example, an ammonium sulfate fraction) which would concentrate the label in the bag. A small but reproducible effect was found and the appropriate controls showed that the activity was missing from strains bearing amber or deletion mutations in the *i* gene. The *lac* repressor was then purified by following the IPTG-binding activity.

When considering the applicability of this method to isolation of other controlling molecules, certain problems encountered in this case should be kept in mind. First, an ordinary *E. coli* cell contains only about ten *lac* repressor molecules per *i* gene copy, representing roughly 0.002% of the cell's proteins. Since the dissociation constant for the IPTG-repressor complex is about 2×10^{-6} moles/liter, and since a cell pellet is about 10^{-9} M in cells, it is clear that some 10 to 100-fold purification had to be achieved before any effect could be seen, pending the isolation of certain special strains which could produce augmented levels of the *lac* repressor (see below). Moreover, there was no way of knowing whether there might not be other IPTG-binding proteins in the extracts which would mask the activity of the repressor. All these problems must be faced in attempting to isolate other repressors using this method, even where exhaustive genetic analysis has been performed.

The specific affinity of the *lac* repressor for IPTG has also been used to follow the purification of repressor by column chromatography (*11*). This was accomplished by equilibrating and eluting the column with buffer containing ^{14}C-labeled IPTG. The repressor-containing fractions contain radioactivity peaks above background. An alternative method is to detect ^{14}C-IPTG binding activity in the fractions using the membrane filter method. Samples containing repressor are incubated with ^{14}C-IPTG and then passed through a nitrocellulose filter. Since *lac* repressor is retained by the filter, IPTG bound to the repressor is also retained while free IPTG is washed through.

These and other assay methods including the DNA-binding assay discussed below, are reviewed by Bourgeois (*19*).

Differential Synthesis

The λ and 434 phage repressors were first detected as proteins labeled with radioactive amino acids in cells in which the relative rate of repressor synthesis had been raised some 5000-fold above its ordinary value (*8, 9*). This was accomplished by irradiating UV-sensitive *E. coli* cells with massive doses of ultraviolet light, a treatment which drastically decreases host protein synthesis but leaves relatively intact the capacity to synthesize proteins when undamaged phage DNA is injected into the cells. When irradiated cells were infected with λ phages, primarily phage proteins were synthesized. If however the irradiated cells were lysogenic for phage λ, the repressor present in the cells turned off the genes of the incoming phage, with the exception of the repressor gene itself.

The λ repressor was identified by infecting aliquots of irradiated lysogenic cells on the one hand with ordinary λ phages and on the other with λ phages bearing an amber mutation in the C_I gene and labeling the two cultures separately with ^{3}H- and ^{14}C-leucine. After incubation the two cultures were mixed, sonicated, and fractionated. The repressor was isolated by DEAE-cellulose chromatography as a single protein labeled with ^{3}H but not with ^{14}C and comprising 5–10% of the total counts incorporated. The supposition that the labeled protein was indeed the product of the C_I gene was confirmed by the findings that phages bearing C_I mutations produced altered forms of the isolated protein.

The UV-irradiation method produces dilute repressor preparations (approximately 10^{-10} moles/liter), highly impure biochemically but essentially pure radiochemically. The method should be easily applicable to the isolation of any phage repressor, and in principle should work for the isolation of the product of any gene incorporated into a transducing phage chromosome. Such genes (for example, the *C* gene of the arabinose operon) would probably have to function at levels close to or above those of the phage repressors in the irradiated cells to be detectable; also they should not be subject to the high levels of catabolite repression characteristic of irradiated cells.

Making More Repressors

The phage repressors are synthesized *in vivo* at levels similar to that of the *lac* repressor, an average cell containing about 50 molecules of repressor constituting about 0.005% of the cell's proteins.

Two techniques have been employed to augment *in vivo* repressor concentrations (*10*). The first is to increase the number of repressor gene copies. This is done in the *lac* case by thermally inducing a λ-*lac* transducing prophage which produces a thermolabile λ repressor and, in addition, contains a mutation in the *S* gene which prevents cell lysis. This increases the specific activity of the repressor about 20-fold. A similar increase in the concentration of λ repressor is obtained using a doubly lysogenic strain containing a λ prophage with a wild-type repressor and a 434 prophage which produces a thermolabile repressor. In addition, the prophage attachment region is deleted in one of the two prophages, and each prophage bears mutations in the *Q* and *S* genes which prevent cell lysis. When this strain is thermally induced, the prophages detach in a tandem circle which produces increased levels of λ repressor as it replicates. This technique is easily modified to produce augmented levels of 434 repressor. In this case the double lysogen is constructed with phages which produce a thermolabile λ repressor and a wild-type 434 repressor.

The second method for increasing repressor yields is to select for mutations in the promoter of the repressor gene which increase the rate at which the gene is transcribed. Such mutations have been selected for *lac* (*10*) and, when placed on the λ-*lac* transducing phage mentioned above, strains are produced in which repressor synthesis constitutes up to 2.5% of the cell's protein.

Using these techniques it is now possible to isolate and purify on a large scale all three repressors.

Repressor Purification

Under the appropriate conditions, all three repressors can be made to bind to phosphocellulose and DEAE-cellulose columns at low ionic strength and are eluted at higher ionic strengths. In addition, at sufficiently high concentrations they sediment at 6–7 S. These properties provide a simple scheme for purification by chromatography followed by a sizing step. Such a purification procedure for the *lac* repressor is described by Riggs and Bourgeois (*11*). Using strains which overproduce repressor, pure *lac* repressor can be obtained simply by elution of an ammonium sulfate fraction from a phosphocellulose column. A procedure for this purification is given by Müller-Hill et al. (*12*). Repressor obtained this way retains four IPTG binding sites per tetramer, but gradually loses operator binding activity.

The phage repressors are, in practice, purified by a more elaborate procedure described below.

Repressor-DNA Binding

Two methods have been used to study the binding of repressors to DNA. Specific binding was first shown using the glycerol gradient method (*13, 14*). In this assay radioactive repressor is mixed with unlabeled DNA and sedimented through a glycerol gradient. DNA-bound repressor sediments with the DNA while unbound repressor sediments much more slowly. The position of the repressor in the gradient is determined by collecting fractions from the gradient and assaying for radioactivity.

A more convenient assay was developed by Riggs et al. (*15*) for the *lac* repressor and is based on the repressor's affinity for nitrocellulose filters. If the repressor is mixed with ^{32}P-labeled DNA containing the *lac* operator and then passed through a filter, the DNA bound to the repressor is retained on the filter, while the unbound DNA is washed through. This method has been adapted subsequently to assay the λ and 434 phage repressors. This is the method of choice for studying quantitative aspects of the binding reaction and for guiding the purification of unlabeled repressors. The assay is extremely sensitive because each repressor binds one molecule of DNA containing the operator. If one uses the intact chromosome of phage λ containing the appropriate operators, each repressor-DNA complex contains 3×10^7 daltons or 10^5 phosphates and one can easily measure operator concentrations in the range of 10^{-14} M. Repressor in crude extracts can be assayed provided that a large excess of unlabeled DNA not containing the operator is added to the binding mixture. Proteins present in the extract which bind DNA nonspecifically are then adsorbed to the unlabeled DNA and the repressor is detected by the amount of specific operator-DNA binding activity.

The membrane filter assay is, in principle, applicable to detecting any DNA binding protein provided that: (1) the protein binds tightly to the filter, at least when complexed with DNA; (2) there is an available supply of ^{32}P-labeled DNA containing the protein binding region, in relatively high concentration; (3) the DNA dissociates from the protein slowly compared with the time required to pass the mixture through a filter and wash it, i.e., a dissociation half-life of a few minutes or more.

PROCEDURES FOR REPRESSOR ASSAYS

Preparation of Phage DNA

Phage stocks are prepared by temperature induction of *E. coli* M65 ($\lambda\ C_{I}ts_{857}\ S_7$)/$\lambda$ or W3102 (λimm434 $C_{I}ts\ S_7$)/λ, grown with aeration to a concentration of 2×10^8 cells/ml in TB medium (10 gm Difco Bactotryptone and 5 gm NaCl per liter). The temperature is raised to 42°C for 15-20 min, then lowered to 38°C and vigorous aeration is continued for 4 hr. The S_7 mutation allows phage maturation but prevents cell lysis, thus increasing the burst size to nearly 1000 phage/cell. The cells are collected by centrifugation, resuspended and lysed by stirring at room temperature with enough chloroform to saturate the solution for 30 minutes. Pancreatic DNase (0.2 μg/ml) is added and stirring is continued overnight at 4°C. The lysate is centrifuged at low speed to remove debris, then at high speed to pellet the phage. The phage is resuspended and purified by banding three times on CsCl block gradients. DNA is extracted from the phage by rolling with neutral, water-saturated phenol. After three extractions, the DNA is dialyzed extensively against buffer containing 0.05 M KCl, 0.01 M Tris, pH 7.5, and 0.001 M EDTA.

To prepare ^{32}P-labeled phage DNA, cells are grown in 40 ml of TB broth to a concentration of 2.5×10^8 cells/ml. The temperature is raised to 43°C for 20 min. The culture is then chilled and the cells are pelleted and resuspended in 40 ml of dephosphorylated TB containing 2×10^{-4} M phosphate and a total of 1.0 mCi of carrier-free $H_3{}^{32}PO_4$, which has been prewarmed to 38°C. Incubation is continued for 4 hr at 38°C with aeration. Lysis is accomplished as for unlabeled phage. No DNase is added. The debris is removed from the lysate by a low-speed spin, and the phage are purified by banding twice on CsCl block gradients before extracting the DNA with phenol.

Glycerol Gradient Sedimentation

Reaction Mixture

Two-tenths milliliter contain: 0.01 M Tris, pH 7.5; 0.01 M KCl; 10^{-3} M EDTA; 10^{-4} M dithiothreitol; and 1-10 μg of λ or λimm434 DNA.

One-tenth milliliter of labeled repressor is added and the mixture is allowed to stand for 10 min at room temperature. The mixture is then layered on a 3.5 ml, 7–30% glycerol gradient containing 0.01 M Tris, pH 7.5; 0.01 M KCl; 10^{-3} M EDTA; and 25 μg/ml bovine serum albumin, and centrifuged at 60,000 rpm and 5°C for 1.5 hr in an International SB 405 rotor. The gradient is collected in fractions of ten drops into 20 scintillation vials and counted with Biosolve BBS-3 (Beckman) and toluene-fluor scintillation fluid. The repressor-DNA complex appears as a sharp peak of radioactivity sedimenting halfway down the gradient.

Membrane Filter Binding

Reaction Mixture

Seven-tenths milliliter contains: 0.01 M piperazine-*N, N'-bis*(2-ethane sulfonic acid) (PIPES) at pH 7.0, 0.05 M KCl, 10^{-3} M EDTA, 10^{-4} M dithiothreitol, 5% dimethyl sulfoxide, 5% glycerol, 25 μg/ml bovine serum albumin, 150 μg sheared chick blood DNA (Calbiochem), 0.05 μg ^{32}P-labeled, and 10 μg unlabeled λ or λimm434 DNA.

A few microliters of repressor sample are added to the binding mixture and incubated at 20°C for 25 min to achieve equilibrium. Three 0.2-ml aliquots are then filtered and washed with two volumes of buffer on three Schleicher and Schuell B6 membrane filters, with a flow rate of about 0.1 ml/10 sec. The filters are then dried and counted in a gas flow counter. The triplicate samples are generally within 5% of each other. At the operator concentration used here the binding of repressor is stoichiometric if the amount of repressor added is such that no more than 30% of the DNA is bound to the filter. For nonquantitative assays, e.g., to assay fractions from column chromatography, lower operator concentrations may be used and the unlabeled phage DNA may be omitted. The purpose of the chick blood DNA is to reduce the amount of background due to DNA binding impurities. It does not affect the binding of repressor to the operator and may be omitted when using highly purified repressor. To determine the amount of nonspecific binding activity in the sample, a control assay is performed, using labeled DNA of opposite specificity, i.e., λimm434 DNA in the case of λ repressor assays.

The assay is extremely sensitive: 1 μl of crude supernatant in a repressor preparation such as described below, generally suffices to saturate 1.0 μg of λ DNA, in the presence of 150 μg of chick blood DNA.

PROCEDURES FOR ISOLATION OF REPRESSORS

Isolation of Labeled Repressors

To isolate labeled λ repressor, *E. coli* 159(λ *ind*$^-$) are grown in A medium containing 7 gm Na_2HPO_4, 3 gm KH_2PO_4, 2.5 gm NaCl, 1.0 gm

NH_4Cl, 0.12 gm $MgSO_4$, 0.5 mg $FeCl_3$, and 4 gm maltose per liter. When the cells reach a concentration of 10^9 cells/ml, they are chilled and diluted to 3×10^8 cells/ml by the addition of preconditioned medium (A medium which has previously supported the growth of cells to a concentration of 3×10^8 cells/ml). The Mg^{2+} concentration is raised to 0.02 M with $MgSO_4$ and the cells are irradiated 80 ml at a time in a petri dish 5.5 cm in diameter kept in ice, with constant swirling. The ultraviolet source is two General Electric, 15 watt germicidal lamps and the cells receive a total of about 50,000 erg/mm^2 in 12 min. Irradiated cells (140 ml) are then infected at a multiplicity of infection of 20 with λ $susN_7$ $susN_{53}$ phage. After incubation for 5 min at 37°C, 1 mCi of ^{3}H-leucine (New England Nuclear) is added and incubation is continued for one hour with shaking. The cells are chilled and washed three times with buffer containing 0.01 M Tris at pH 7.5 and 0.005 M $MgSO_4$. They are then suspended in 1.5 ml of the same buffer containing 0.4 M KCl and sonicated in ice with four 15-sec pulses. The sonicate is centrifuged in an International SB 405 rotor at 60,000 rpm for 20 min. The supernatant is dialyzed against buffer containing 0.01 M Tris at pH 7.5, 0.05 M KCl, 0.005 M $MgSO_4$, 10^{-4} M EDTA, and 10^{-4} M dithiothreitol. The sample is then applied to a 1 x 15 cm column of DEAE-cellulose equilibrated with the same buffer. The repressor is eluted with a 0.05–0.2 M KCl gradient and appears in the effluent at approximately 0.15 M KCl. About 40 fractions of 4 ml each are collected in tubes containing a drop of 0.5% BSA and are assayed by counting 0.1 ml of each in scintillation vials. The fractions containing a sharp radioactivity peak are pooled and concentrated by dialysis against dry Sephadex G-200 to a volume of about 5 ml, containing a total of 100,000–200,000 cpm. If the irradiated cells are infected with λ ind^- $susN_7$ $susN_{53}$ phage, labeled ind^- repressor is obtained. The λ ind^- repressor is eluted from the DEAE-cellulose column at about 0.07 M KCl instead of 0.15.

Labeled 434 repressor can be isolated in the same way, starting with *E. coli* 159(λ *imm*434*T* ind^-) and infecting with λ *imm*434*T* $susN_7$ phage. Phosphocellulose chromatography is substituted for DEAE-cellulose. The repressor extract is applied to the column in buffer containing 0.02 M KPO_4 pH 7.5, 10^{-4} M EDTA and 10^{-4} M dithiothreitol, and is eluted with a gradient 0.02–0.15 M KPO_4. The repressor appears in the effluent at a KPO_4 concentration of about 0.07 M.

Isolation of Unlabeled λ ind^- Repressor

Growth of Cells

E. coli cells, strain MR41(λ *imm*434 C_Its $susQ_{73}$ $susQ_{501}$ S_7, λ ind^- $susQ_{73}$ $susQ_{501}$ S_7 b_{104}) are grown to a concentration of 5×10^8 cells/ml at 32°C in 2 x YT medium (16 gm Difco Tryptone, 10 gm Difco yeast extract,

TABLE 1. Purification Procedure

	Total protein (mg)	Total specific DNA binding activity[a]	λ binding activity/mg protein	Yield
Crude supernatant	210,000	28,000	0.13	100%
Ammonium sulfate fraction	61,000	10,000	0.16	36%
Pooled DEAE I peak	14,500	8,200	0.57	29%
Pooled phosphocellulose peak	1,250	7,500	6.0	27%
Pooled DEAE II peak	105	6,400	61	23%
Pooled glycerol gradient fractions	42	4,500	107	16%

[a]Activity expressed as milligrams of ^{32}P-labeled DNA that can be bound to membrane filters. Specific binding activity is the λ-DNA binding activity after subtraction of nonspecific λ *imm*434 DNA binding activity, assayed in the presence of 250 μg/ml chicken blood DNA.

5 gm NaCl, and 0.1 ml Dow Corning Antifoam A per liter). The temperature is then raised to 42°C for 15–20 min, then lowered to 38°C and aeration continued for 4–5 hr. The induced cells, which at this time become much enlarged and full of phage DNA, are collected and frozen. About 7 gm/liter of cells are obtained.

The procedure detailed here describes the isolation from 4 kg of cells (grown at the facilities of the New England Enzyme Center). However, scaled down versions have been applied successfully to as little as 50 gm of cells. A summary of the purification steps is shown in Table 1.

Lysis of Cells

Before use, 4 kg of cells are thawed and resuspended in 10 liters of buffer containing 0.2 M NaCl, 0.01 KPO_4, pH 7.2, and 10^{-3} M EDTA. Probably because of accumulation of phage lysozyme, the cells begin to lyze when the pellet is thawed. To complete the lysis process the cells are homogenized in a refrigerated Waring Blendor for 5 min. The Mg^{2+} concentration is then raised to 0.01 M and 5 mg of DNase (Worthington, electrophoretically pure) are added. The suspension is homogenized for another 10 min. When the extremely viscous solution becomes fluid (additional gentle stirring at room temperature for 10–15 min may be required), it is centrifuged in a continuous flow rotor to remove ribosomes and cell debris. All further steps are carried out in the cold room.

Ammonium Sulfate Fractionation

To the supernatant (fraction I, about 12 liters) are added 2780 gm of solid ammonium sulfate (0–40%) with stirring. The solution is allowed to stand for several hours before centrifuging in a Lourdes continuous flow centrifuge at 14,000 rpm and 250 ml/min. To the supernatant (fraction II) are added 1670 gm of ammonium sulfate (40–60%). The suspension is allowed to stand overnight before centrifuging. The precipitate is collected and resuspended in 700 ml of buffer A (0.01 M Tris, pH 7.5, 10^{-3} M EDTA, 5% glycerol, and 10^{-3} M 2-mercaptoethanol). The solution is then dialyzed against 30 liters of the same buffer containing 0.04 M KCl. Dialysis against low salt concentration may result in sizable losses of repressor. It is preferable therefore to interrupt dialysis after about 10 hr, before equilibrium is reached, and dilute the dialyzate with buffer until the conductivity corresponds to the desired ionic strength, in this case 0.03 M KCl (fraction III).

DEAE-Cellulose Chromatography

Fraction III (about 5 liters) is loaded onto a 1800 ml DEAE-cellulose column equilibrated with the same buffer at pH 7.5 and 0.03 M KCl. The flow rate is adjusted at 800 ml/hour. When all the sample is loaded, the top half of the column appears dark brown with a yellow band beginning to separate at the bottom. The top 30% of the column is stirred and allowed to settle again before washing with 6 liters of buffer at 0.05 M KCl. The repressor is then eluted with 0.15 M KCl buffer and appears in the effluent together with a bright yellow colored material. The λ DNA binding activity is found in about 5 liters of yellow colored effluent and then drops suddenly. The active fraction is precipitated with 1950 gm of ammonium sulfate (60%) and resuspended in 250 ml of buffer containing 0.02 M KPO_4, pH 7.0, 10^{-3} M EDTA, 5% glycerol, and 10^{-4} M dithiothreitol. This material (fraction IV), is dialyzed against 7 liters of the same buffer containing 0.04 M KPO_4 for 10 hr with two buffer changes and then diluted to 600 ml to give a concentration of 0.04 M KPO_4.

Phosphocellulose Chromatography

The dialyzed fraction IV is loaded onto a 700 ml phosphocellulose column equilibrated at pH 7.0 and 0.04 M KPO_4. The column is then washed with 1.5 liters of buffer containing 0.15 M KPO_4 and the repressor is eluted with 0.3 M KPO_4. The effluent is collected in 20-ml fractions. The fractions containing the repressor are pooled to give fraction V, precipitated with 39 gm of ammonium sulfate per 100 ml (60%) and resuspended in 40 ml of buffer A containing 0.04 M KCl-Fraction V is dialyzed against 4 liters of buffer A containing 0.04 M KCl for 10 hr. Some precipitate is observed in the dialyzate but there is no appreciable loss of activity in the solution. The dialyzate is diluted to give 100 ml at 0.04 M KCl.

DEAE-Cellulose Chromatography II

The dialyzed fraction V is applied to a 200 ml, 2.5 x 40-cm column of DEAE-cellulose, equilibrated with the same buffer at 0.04 M KCl. After loading, the column is washed with two volumes of buffer at 0.04 M KCl and then with 0.1 M KCl. The repressor chromatographs slowly and appears in the effluent after about 600 ml of 0.1 M KCl buffer have passed through. The active fractions are pooled, precipitated with 39 gm ammonium sulfate per 100 ml, resuspended with 5 ml of buffer A, and dialyzed against the same buffer containing 0.2 M KCl (fraction VI).

Glycerol Gradient Sedimentation

Fraction VI is layered in 1-ml portions onto 11.5 ml, 10–30% glycerol gradients containing 0.01 M Tris, pH 7.5, 0.5 M KCl, 10^{-4} M EDTA, and 10^{-4} M dithiothreitol. The gradients are centrifuged for 45 hr at 40,000 rpm and 5°C in an International SB 283 rotor. The gradient is collected from the bottom in fractions of 30 drops. The OD_{280} profile shows two peaks, one sedimenting near 2.9 S and one near 6.2 S. The λ DNA binding activity is located only under the 6.2 S peak. Fractions around that peak are pooled and stored at −10°C (fraction VII).

SDS-acrylamide gel electrophoresis of this material shows one major component which accounts for more than 90% of the total protein and migrates as a molecule of 27,000 daltons. One minor component is visible which migrates as a large protein (about 100,000 mol wt).

Isolation of the 434 Repressor

The 434 repressor can be isolated in a similar way from a doubly lysogenic bacterial strain. In this case the λ prophage carries the thermosensitive repressor mutation, while the λ *imm*434 prophage has a wild-type C_I gene. The bacterial strain used is therefore MR41(λ *imm*434*T* $susQ_{73}$ $susQ_{501}$ S_7, λ C_Its_{857} $susQ_{73}$ $susQ_{501}$ S_7 b_{104}). The basic procedures for the isolation are the same as for the λ repressor. Only the differences will, therefore, be outlined here.

Ammonium Sulfate Fractionation

Under the same conditions, the 434 repressor is found in the 30–40% ammonium sulfate fraction rather than 40–60% as λ.

Phosphocellulose Fractionation

The first chromatography step is on a phosphocellulose column rather than DEAE. The dialyzed ammonium sulfate fraction is applied to a column equilibrated with 0.03 M KPO_4, pH 7.5, and eluted with a gradient 0.03–0.2 M KPO_4. The 434 repressor elutes around 0.09 M KPO_4.

DEAE-Cellulose Chromatography

At pH 7.5 the 434 repressor binds to a DEAE-cellulose column equilibrated with 0.02 M KCl and is eluted with a 0.02–0.1 M KCl gradient. The repressor appears around 0.05 M KCl.

Ind^- variants of the 434 repressor are known which chromatograph slightly differently from the wild type. One, produced by λ *imm*434 ind_7^- phage, is eluted from a phosphocellulose column slightly later, and one, produced by λ *imm*434*T* ind_T^-, slightly earlier than the wild type.

Some Properties of the Isolated Repressors

The phage repressors exist as oligomers composed of identical subunits of molecular weight 27,000 (λ repressor) and 25,000 (434 repressor). The oligomers are in rapid, concentration dependent equilibrium with the monomers. The binding of λ ind^- repressor to its operator exhibits all the properties expected if the form which binds is a dimer. At low concentrations (10^{-9} M or less) the λ and 434 repressors sediment at 2.4 S. However, at high concentration (10^{-5} M) the λ ind^- repressor sediments in a glycerol gradient at 6.2 S, about the rate expected for a tetramer of 2.4 S subunits. The λ repressor is a weakly acidic protein with isoelectric point of 6.3, while the 434 repressor is basic, becoming neutral near pH 8.5.

In contrast the *lac* repressor is a stable tetramer with subunit molecular weight of 40,000 daltons and an isoelectric point of 5.5. Its S value is 7.2. Monomers have been detected in *lac* repressor preparations, particularly in old preparations; although the subunits bind IPTG, they do not bind to DNA.

All three repressors bind very tightly to DNA containing their respective operators. The *lac* repressor–operator complex has a dissociation constant of 10^{-13} moles/liter in 0.01 M Mg^{2+} and 0.01 M KCl (*17*). The presence of Mg^{2+} is probably required for specific binding. The phage repressors bind to DNA in the filter assay in the form of dimers. The dissociation constant for the reaction:

$$2R + O \rightleftharpoons R_2O$$

where R is the repressor monomer and O is the operator, is 2×10^{-22} $(\text{moles/liter})^2$ at 20°C and 0.05 M KCl (*18*). The binding of the phage repressors but not of the *lac* repressors becomes stronger at lower pH and lower temperature. For all three repressors the binding becomes weaker at higher ionic strength. Mg^{2+} is not required for the binding assay of phage repressor but, while 0.01 M Mg^{2+} has no detectable effect on the binding of the λ repressor, it abolishes almost completely the binding of the 434 repressor.

References

1. Jacob, F. and J. Monod. 1961. *J. Mol. Biol.,* **3**: 318.
2. Brenner, S. 1965. *Brit. Med. Bull.,* **21**: 244.
3. Beckwith, J. 1967. *Science,* **156**: 597.
4. Dove, W. F. 1968. *Ann. Rev. Genetics,* **2**: 305
5. Ptashne, M. and N. Hopkins. 1968. *Proc. Natl. Acad. Sci. U.S.,* **60**: 1282.
6. Pirrotta, V. 1969. Ph.D. thesis, Harvard University, Cambridge, Mass.
7. Gilbert, W. and B. Müller-Hill. 1966. *Proc. Natl. Acad. Sci., U.S.,* **56**: 1891.
8. Ptashne, M. 1967. *Proc. Nat. Acad. Sci., U.S.,* **57**: 306.
9. Pirrotta, V. and M. Ptashne. 1969. *Nature,* **222**: 541.
10. Müller-Hill, B., L. Crapo, and W. Gilbert. 1968. *Proc. Natl. Acad. Sci., U.S.,* **59**: 1259.
11. Riggs, A. D. and S. Bourgeois. 1968. *J. Mol. Biol.,* **34**: 361.
12. Müller-Hill, B., K. Beyrether, and W. Gilbert. In L. Grossman and K. Moldave (Eds.), Methods in enzymology, nucleic acids. Academic Press, New York, Vol. XXI, in press.
13. Ptashne, M. 1967. *Nature,* **214**: 232.
14. Gilbert, W. and B. Müller-Hill. 1967. *Proc. Natl. Acad. Sci., U.S.,* **58**: 2415.
15. Riggs, A. D., S. Bourgeois, R. F. Newby and M. Cohn. 1968. *J. Mol. Biol.,* **34**, 365.
16. Pirrotta, V., P. Chadwick, and M. Ptashne. *Nature,* in press.
17. Riggs, A. D., H. Suzuki, and S. Bourgeois. 1970. *J. Mol. Biol.* **48**: 67.
18. Chadwick, P., V. Pirrotta, R. Steinberg, N. Hopkins, and M. Ptashne. 1970. *Cold Spring Harbor Symp. Quant. Biol,* **35**, in press.
19. Bourgeois. S. In L. Grossman and K. Moldave (Eds.), Methods in enzymology, nucleic acids. Academic Press, New York, Vol. XXI, in press.

INTERACTION OF MS2 COAT PROTEIN AND MS2 RNA

TSUTOMU SUGIYAMA
CENTRAL RESEARCH DEPARTMENT
E. I. DU PONT DE NEMOURS AND COMPANY
WILMINGTON, DELAWARE

INTRODUCTION

The coat protein of the RNA phages of the f2 group (f2, MS2, R17, fr, etc.) functions as the structural unit of the virion which encoats viral RNA and also as the regulatory unit which controls the synthesis of the viral RNA replicase in the infected cells (*1*). These two functions of coat protein can be demonstrated *in vitro* by employing the viral RNA and viral coat protein isolated from purified virions. Viral RNA which has been extracted by any conventional procedure may be employed for this purpose; however, the coat protein is rather unstable and hence it is necessary to use certain precautions during its isolation.

The general procedure for the growth and assay of MS2 virus as well as comments concerning laboratory contamination by MS2 have been described in Volume I, pp. 498–512 (*2*).

MEDIA

MS Broth (*3*)

Dissolve in 1 liter of H_2O:

Bacto-tryptone (Difco)	10 gm

NaCl	8 gm
Bacto-yeast extract (Difco)	1 gm

Adjust to pH 7.4 with NaOH and autoclave at 120°C for 15 min.

After cooling, add the following:

10% glucose (autoclaved)	10.0 ml
10% $CaCl_2$ (autoclaved)	2.0 ml
1% thiamine·HCl (sterilized by filter)	1.0 ml

Modified TPA (*m*-TPA) (*4*)

Dissolve in 900 ml of H_2O:

cysteine, tyrosine, leucine	50 mg each
other 17 common L-amino acids	150 mg each
NaCl	0.5 gm
KCl	8.0 gm
NH_4Cl	1.1 gm
tris(hydroxymethyl)aminomethane	12.1 gm
Na pyruvate	0.8 gm
4.6% KH_2PO_4	1.0 ml
2.3% Na_2SO_4	1.0 ml
1% adenine (in 0.1 N HCl)	1.0 ml

Adjust to pH 7.4 with HCl and autoclave at 120°C for 15 min.

After cooling add:

0.0025% $FeCl_3 \cdot 6H_2O$	4.0 ml
10% $CaCl_2$	2.0 ml
10% $MgSO_4$	2.0 ml
1% thiamine·HCl (sterilized by filter)	1.0 ml
10% glucose	10.0 ml

GROWTH AND PURIFICATION OF MS2 VIRUS

The yield of the virus depends greatly on the extent of the aeration of the culture during its growth. A vat fermenter may be conveniently used for vigorous aeration of the large scale culture (*2*), while a shaker-type incubator must be used with vigorous shaking for medium scale cultures. In order to obtain necessary aeration in a shaking culture, a relatively small volume of the medium should be used per flask (see below). The use of a baffled flask may further increase the efficiency of aeration.

It is advisable that each experimenter empirically determine the adequate culture size and aeration rate to obtain the maximum virus yield with his own growth apparatus. The following may serve as examples.

Growth of Unlabeled MS2

Inoculate 4 liters of MS broth with 20 ml of an overnight culture of *Escherichia coli* Hfr 3000 and distribute the medium into twelve 2-liter-baffled flasks (330 ml per flask). Grow the culture at 37°C with moderate aeration (100 gyrations per minute, 2.5-cm gyration circle). Add MS2 phage (approximately 5×10^{11} plague forming units (pfu) per flask) when a cell density of about 2×10^8/ml is reached. Ten to 15 min later, increase the rate of shaking (250 gyrations per minute) to obtain maximum aeration. Three hours after the addition of the phage, harvest and pool the lyzate. The lyzate should have 5×10^{11} to 2×10^{12} pfu/ml.

Growth of [^{3}H]-Phenylalanine-Labeled MS2

Inoculate 500 ml of *m*-TPA (minus phenylalanine) with 2.5 ml of the overnight culture of *E. coli* Hfr 3000 in the same medium and divide the medium into two 2-liter-baffled flasks. Grow the culture at 37°C with moderate aeration. When a cell density of about 2×10^8/ml is reached, add MS2 phage (3×10^{11} pfu per flask) and [^{3}H]-L-phenylalanine (~ 3Ci/mM, 1 mCi per flask). Continue the culture as described for the unlabeled MS2.

The yield of virus from *m*-TPA culture may be considerably lower than in MS broth culture (~ 10^{11} pfu/ml). MS broth may be added (2%) to *m*-TPA as a supplement to increase the yield of virus, but this will reduce the specific radioactivity of the virus unless a greater quantity of the isotope is used in compensation.

Purification of MS2 Virus (*5*)

To the 4 liters of pooled lyzate (MS broth or *m*-TPA), add 100 ml of 0.1M EDTA (pH 7.0) and 1120 gm of ammonium sulfate with vigorous stirring. After 2 to 18 hr in the cold, the precipitate is spun out at 10,000 *g* for 15 min in a refrigerated centrifuge. It is then thoroughly mixed with 45 ml of a buffer containing 0.1 M NaCl, 0.05 M tris-HCl and 0.01 M EDTA (pH 7.6). One milliliter of egg white lysozyme (10 mg/ml) is then added and the mixture is incubated at 37°C for 10 min. Next, 80 ml of Freon 113 (CCl_2FCClF_2) is added, and the resulting mixture is vigorously mixed in an Omnimixer for 5 min. The emulsion is then broken by centrifugation for 10 min at 10,000 *g*. The upper aqueous layer is decanted and to the Freon layer is added 35 ml of the above buffer, and the mixture is again treated in the Omnimixer and subjected to centrifugation. The two aqueous layers are pooled and retreated with 50 ml of fresh Freon. To the final aqueous layer, solid CsCl is added (0.55 gm per ml of the aqueous layer), and the solution is divided into 12 polycarbonate tubes and spun at 45,000 rpm (140,000 *g*) for 24 hr at 4°C in Spinco Ti 50 rotor. On completion of the run, two bands are usually visible: the lower virus band at 1.47 gm/ml and the upper virus-pili complex band at 1.44 gm/ml. The virus band is collected through holes punctured in the bottom of the tubes, pooled, and subjected to rebanding. Normally, a total of three to four cycles of bandings are necessary to obtain virus band completely free from pili-virus band and other cellular materials. After the final banding, virus suspension is dialyzed against 0.1 M tris-HCl buffer (pH 7.0).

The concentration of the virus is calculated from its absorption at 260 nm in a buffer containing 0.1 M NaCl and 0.01 M tris-HCl, pH 7.6 (specific absorbancy 8.03/mg/ml). The particle weight of MS2 virus is 3.6×10^6. The yield of the purified virus ranges from 5 to 25 mg per liter of the original lysate for MS broth culture and 2 to 10 mg per liter for *m*-TPA culture supplemented with 2% MS broth.

For extraction of coat protein, a virus preparation can be stored in an ice bath for weeks. Alternatively, it can be divided into many small tubes and stored at −70°C for several months. For the extraction of RNA, a virus preparation should be used immediately after purification.

PREPARATION OF MS2 RNA AND MS2 COAT PROTEIN

MS2 RNA

To a suspension of purified MS2 at about 5 mg/ml in 0.1 M tris-HCl (pH 7.0) is added an equal volume of 80% redistilled phenol and the mixture is stirred at about 45°C for 5 min on a magnetic stirrer. The emulsion is broken by centrifugation at 6000 *g* for 5 min at about 25°C. The aqueous layer is removed and retreated with equal volume of fresh phenol. The phenol treatment is repeated (normally a total of 3 to 4 times) until no denatured protein layer is visible at the interphase. To improve the yield of MS2 RNA, a volume of 0.1 M tris-HCl equal to the initial volume of phage suspension is added to the phenol layer from the first extraction, stirred and centrifuged as before. The aqueous layer is then successively used to extract the phenol layers from the second and third (and fourth, if employed) extractions. This aqueous layer is then combined with the initial aqueous layer, and the RNA precipitated by the addition of 3 vol of cold ethanol and 1/20 vol of 3 M sodium acetate (pH 4.5). After a few hours the precipitate is sedimented by centrifugation at 6000 *g* for 10 min in the cold. The supernatant solution is discarded and the precipitate dissolved in H_2O of about one-fifth the initial volume of the virus suspension. The RNA is precipitated repeatedly in the same manner (normally a total of 4 to 5 times) until the UV absorption spectrum of the alcoholic supernatant shows no characteristic spectrum of phenol (main peak at 270 nm). The final RNA precipitate is taken up in H_2O of one-tenth the initial volume of phage suspension and centrifuged at 10,000 *g* for 20 min to remove insoluble material. The RNA solution is stored at –70°C.

The concentration of MS2 RNA is determined from its optical density at 260 nm in 0.05 M tris-HCl, pH 7.0 (specific absorbancy 25.1/mg/ml). The molecular weight of MS2 RNA is 1.1×10^6 (*5*). The recovery of extracted RNA ranges from 80 to 90% as based on an RNA content of 31.5% for MS2 virus.

MS2 Coat Protein (*6*)

A glass centrifuge tube containing glacial acetic acid (melting point 16.6°C) is gently placed in an ice bath and kept unstirred for 15 min so that it will chill without freezing. One-half volume of a cold suspension of purified MS2 at about 5 mg/ml in 0.1 M tris-HCl (pH 7.0) is added and quickly mixed with the chilled glacial acetic acid. The mixture is kept in an ice bath for about 60 min with occasional stirring and then centrifuged at 10,000 *g* for 20 min in the cold. The white precipitate of RNA is discarded and the clear protein solution in 67% acetic acid is transferred to a dialysis tubing which has been treated in 67% acetic acid for several hours. The protein solution is then dialyzed in the cold against 2 changes of 1000 vol of 0.001 M acetic acid which has been prepared using double-distilled or distilled and deionized water. The dialysis flask is tightly covered and is placed in an ice bath during dialysis. After completion of dialysis, the pH of the protein solution is normally about 3.2. If the pH is higher, especially if it is above 4.0, protein is frequently lost due to precipitation. Precipitation of protein also frequently occurs if the concentration of protein is higher than 3 mg/ml. As the last step, the protein solution is centrifuged at 10,000 *g* for 20 min to remove insoluble materials.

The concentration of protein is obtained from its optical density at 280 nm in 0.001 M acetic acid (specific absorbancy 1.1/mg/ml). The molecular weight of MS2 coat protein is 1.4×10^4 (*7, 8*). The protein solution may or may not remain clear for days in an ice bath. However, for reproducible results it is best to use freshly prepared protein solution. The protein solution prepared as above contains almost exclusively MS2 coat protein and little, if any, maturation protein, a minor component of MS2 virion, which is apparently lost during the preparation (*9*).

FORMATION OF RIBONUCLEOPROTEIN COMPLEXES AND THEIR ANALYSES (*6*)

Two types of ribonucleoprotein complexes can be formed by the interaction of MS2 RNA and MS2 coat protein *in vitro*. At the input protein-to-RNA molar ratio of 10:1 or less, complex I which has the same sedimentation coefficient as free MS2 RNA (27S) is formed predominantly. Complex I is a structure consisting of a molecule of viral RNA and not more than 7 or 8 molecules of coat protein. The exact nature of the interaction of coat protein with RNA is yet unknown. At higher input ratio of protein to RNA, complex II is also formed. This has a sedimentation coefficient of about 70 S (the sedimentation coefficient of MS2 virion is 81 S). At the input ratio of 180:1 or higher, complex II is formed exclusively. Complex II is a virus-like structure which is noninfectious, presumably due to its lack of maturation protein.

Formation of Complex I

MS2 RNA (10 mg/ml in H_2O)	0.100 ml (1.0 mg, 0.9 x 10^{-9} moles)
H_2O	0.667 ml
0.5 M tris-HCl (pH 7.8)	0.200 ml
MS2 coat protein (3 mg/ml in 0.001 M acetic acid)	0.033 ml (0.1 mg, 7 x 10^{-9} moles)
Total	1.000 ml

MS2 RNA, H_2O, and tris buffer are mixed in a small test tube in an ice bath. Tris buffer may be substituted for by other buffers such as 5X concentration of the buffer used for the *in vitro* protein-synthesizing system from *E. coli* (0.05 M tris-HCl, 0.01 M Mg acetate, 0.03 M KCl, 0.03 M NH_4Cl, and 0.005 M 2-mercaptoethanol, pH 7.5 (*10*)). Coat protein is then slowly added into the mixture with gentle stirring. If the protein is not added at the last step, precipitation of coat protein may occur. The formation of complex I is spontaneous and no incubation is necessary. No technique has been

developed to separate complex I from free MS2 RNA. Therefore, a larger amount of coat protein may be used to ensure the conversion of all the RNA into complex I.

Formation of Complex II

MS2 RNA (10 mg/ml in H_2O)	0.050 ml (0.5 mg, 5 x 10^{-10} moles)
H_2O	0.250 ml
0.5 M tris-HCl (pH 7.8)	0.200 ml
MS2 coat protein (3 mg/ml in 0.001 M acetic acid)	0.500 ml (1.5 mg, 1 x 10^{-7} moles)
Total	1.000 ml

MS2 coat protein is added slowly with gentle stirring to the mixture of MS2 RNA, H_2O, and the buffer. The mixture is incubated at 37°C for 60 min. The insoluble material formed at the time of the addition of coat protein may disappear quickly during incubation. The trace of insoluble material present at the end of the incubation is removed by centrifugation at 10,000 *g* for 10 min.

Analysis of Complex I and Complex II by Sucrose Gradient Centrifugation

Coat protein prepared from [^{3}H]-phenylalanine-labeled MS2 virus is used to facilitate the sucrose gradient analysis.

A 0.3 ml sample of complex I or complex II, as prepared above, is placed on top of a 5 to 20% linear sucrose gradient in the same buffer used for the formation of the complex. The gradients are centrifuged in a Spinco SW65 rotor at 45,000 rpm (150,000 *g*) at 4°C for 85 min or 225 min to analyze complex II (~70 S) or complex I (~27 S), respectively. Fractions are collected from the bottom of the tube, and to each are added 2.0 ml of 0.1 M tris-HCl buffer (pH 7.0) containing 0.1% sodium dodecylsulfate. After vigorous mixing to dissolve coat protein, a 0.5 ml aliquot is mixed with 10 ml Bray's solution and the radioactivity of the coat protein is determined in a scintillation counter. The rest of the sample is then used to measure the UV absorbance of the RNA.

It is important to note here that, at low concentrations, coat protein can easily dissociate from MS2 RNA and then adhere to the walls of the tubes in the presence of salts. Therefore, all measurement of the radioactivity of the coat protein should be made after thoroughly dissolving it in a buffer containing sodium dodecylsulfate.

Analysis of Complex I by its Messenger Activity in the *in vitro* Protein Synthesizing System

Coat protein of complex I inhibits the translation of viral RNA polymerase cistron without inhibiting the translation of coat protein cistron in the *in vitro* protein-synthesizing system derived from *E. coli.* The translation of these two cistrons can be monitored by measuring the incorporation into hot TCA-insoluble materials of radioactive histidine which is present in RNA polymerase but not in coat protein and any other amino acid, such as phenylalanine, which is present in both proteins. For such a study, complex I prepared as above is added into a suitable *in vitro* protein synthesizing system derived from *E. coli* (*1, 10*) and the stimulation of the incorporation of these two amino acids by complex I is compared to that by free MS2 RNA. An example of such an experiment is shown in Table 1. Complex I strongly reduces the incorporation of [^{14}C]-histidine while the incorporation of [^{14}C]-phenylalanine is only moderately reduced. Complex II does not stimulate incorporation of any amino acid.

TABLE 1. Incorporation of [^{14}C]-Phenylalanine and [^{14}C]-Histidine into Hot TCA-Insoluble Material[a]

Messenger	Input molar ratio of coat protein to MS2 RNA	[^{14}C]-Phenylalanine incorporation (cpm)		[^{14}C]-Histidine incorporation (cpm)	
MS2 RNA	0	83,200	(100%)	14,651	(100%)
complex I	2.1	78,100	(94%)	4,770	(33%)
	4.2	70,900	(85%)	5,180	(36%)
	6.0	65,800	(79%)	3,140	(21%)
	8.4	66,400	(80%)	3,590	(25%)
	10.1	62,900	(76%)	2,430	(17%)
	12.6	60,000	(72%)	2,700	(18%)
	15.1	66,200	(79%)	2,290	(16%)

[a] MS2 RNA and complex I prepared at the input coat protein-to-RNA ratio indicated were added into the *in vitro* protein synthesizing system (total volume 0.3 ml) and incubated at 35°C for 20 min (*1, 10*). The reaction was terminated by cold TCA and the radioactivity in the hot TCA-insoluble materials was determined. The *in vitro* protein synthesizing system contained per milliliter: 50 μmoles tris-HCl, pH 7.5; 10 μmoles magnesium acetate; 30 μmoles KCl; 30 μmoles NH_4Cl; 5 μmoles 2-mercaptoethanol; 1 μmole ATP; 0.05 μmoles GPT; 0.05 μmoles phosphoenolpyruvate; 20 μg pyruvate kinase (Sigma Chemical Co.); 100 μg stripped tRNA (General Biochemicals, Inc.); 0.05 μmoles each of 20 L-amino acids except the one added as the labeled amino acid; 1 μCi of [^{14}C]-phenylalanine (0.37 Ci/mmole) or [^{14}C]-histidine (0.24 Ci/mmole) (New England Nuclear); 1.3 mg (protein) of *E. coli* extract, and 500 μg of MS2 RNA.

References

1. Sugiyama, T. and D. Nakada. 1967. *Proc. Natl. Acad. Sci., U.S.*, **57**:1744.
2. Billeter, M. A. and C. Weissmann. 1966. In G. L. Cantoni and D. R. Davies (Eds.), Procedures in nucleic acid research. Harper & Row, New York, Vol. I., p. 498.
3. Davis, J. E. and R. L. Sinsheimer. 1963. *J. Mol. Biol.*, **6**: 203.
4. Kelly, R. B., J. L. Gould, and R. L. Sinsheimer. 1965. *J. Mol. Biol.*, **11**: 562.
5. Strauss, J. H. and R. L. Sinsheimer. 1963. *J. Mol. Biol.*, 7: 43.
6. Sugiyama, T., R. R. Hebert, and K. A. Hartman. 1967. *J. Mol. Biol.*, **25**: 455.
7. Lin, J., C. M. Tsung, and H. Fraenkel-Conrat. 1967. *J. Mol. Biol.*, **14**: 528.
8. Weber, K. 1967. *Biochemistry*, **6**: 3144.
9. Sugiyama, T. and D. Nakada. 1968. *J. Mol. Biol.*, **31**: 431.
10. Sugiyama, T., H. O. Stone, Jr., and D. Nakada. 1969. *J. Mol. Biol.*, **42**: 97.

ISOLATION AND CHARACTERIZATION OF A COMPLEX BETWEEN AN AMINOACYL-tRNA SYNTHETASE AND ITS tRNA SUBSTRATE

ULF LAGERKVIST
DEPARTMENT OF MEDICAL BIOCHEMISTRY
UNIVERSITY OF GOTHENBURG
GOTHENBURG, SWEDEN

The recognition by an aminoacyl-tRNA synthetase of its tRNA* substrate is an essential step in the translation of genetic information into amino acid sequences in proteins. The structural basis of this recognition has therefore become a central problem in modern biochemistry. While a full understanding of the mechanism of recognition must ultimately be based on a thorough knowledge of the structure of both tRNA and enzyme our present and rather limited information is necessarily derived from more indirect evidence. At the same time the recent reports of the crystallization of a number of tRNAs (*1–4*) as well as one of the activating enzymes (*5*) in a form that shows promise for crystallographic study are hopeful indications that we

* The abbreviations used are: $tRNA^{Val}$, transfer ribonucleic acid specific for valine; valyl-tRNA, transfer ribonucleic acid esterified with valine; $tRNA^{Lys}$, transfer ribonucleic acid specific for lysine; E^{Val}, valyl ribonucleic acid synthetase; E^{Val}–$tRNA^{Val}$, a complex involving the enzyme E^{Val} and transfer ribonucleic acid specific for valine: AMP-Val · E^{Val} · valyl-tRNA, a complex involving the enzyme E^{Val}, AMP, valine, and valyl-tRNA; $tRNA^{Val}$CCp, $tRNA_1{}^{Val}$ that has been stripped of its 3′-terminal adenosine; $tRNA^{Val}$–UCACCA, $tRNA_1{}^{Val}$ that has been stripped of a 3′-terminal sequence with the average composition UCACCA; $tRNA^{Val}$–AUCACCA, $tRNA_1{}^{Val}$ that has been stripped of a 3′-terminal sequence with the average composition AUCACCA.

may soon be able to build our theories of recognition on the firm foundation of structural studies.

In the meantime it is useful to try to summarize some of the experimental approaches to this problem and the conclusions that have been drawn from the results obtained. Attempts in several laboratories to use structurally modified tRNAs or oligonucleotides of known composition as competitive inhibitors of aminoacyl-tRNA synthetases of varying degrees of purity have given conflicting results (*6–13*). Hayashi and Miura (*6, 7*) using crude preparations of phenylalanyl-, lysyl-, and prolyl-tRNA synthetases from yeast obtained competitive inhibition of the aminoacylation reaction in the presence of oligonucleotides corresponding to the anticodon for the amino acid in question. These results, however, were not confirmed by other investigators (*11, 12*) using similar systems. The conclusion that the anticodon represents the sole region of recognition in the tRNA molecule is therefore not sufficiently supported by the experimental evidence. Interesting results regarding the role of the anticodon region have been obtained by Mirzabekov et al. using $tRNA^{Val}$ from yeast where part of this region had been excised by digestion with T_1-RNase (*14–16*). The structurally modified $tRNA^{Val}$ still completely retained its ability to be esterified with valine. Attempts by Chambers and co-workers (*17*) to demonstrate the ability of smaller fragments of $tRNA^{Ala}$ to act as substrates for the cognate aminoacyl-tRNA synthetase have given less convincing results since the acceptor activity of these fragments was only a small fraction of the theoretically expected value. A previous conclusion by Schulman and Chambers (*18*) that the first three base pairs of the cloverleaf model constitute the recognition site of all tRNA molecules appears unlikely for reasons that will be discussed more fully below.

The transfer of the aminoacyl group from the isolated enzyme-AMP-aminoacyl complex to tRNA has been studied by several groups. An interesting difference was found among the threonyl-tRNA synthetase from rat liver (*19, 20*) and the seryl-tRNA synthetase from yeast (*20a*) and the isoleucyl- and valyl-tRNA synthetases from *E. coli* and yeast (*21, 22*) in that the transfer of the aminoacyl group to tRNA required Mg^{2+} when catalyzed by the threonine enzyme or serine enzyme while the isoleucine and valine enzymes had no such requirement. This apparent contradiction can probably be resolved if one considers the effect of Mg^{2+} on the conformation of tRNA (*23*). A reasonable hypothesis would then be that the enzyme per se does not require Mg^{2+} in order to transfer the aminoacyl group from the enzyme-AMP-aminoacyl complex to tRNA but that Mg^{2+} is sometimes required to keep the tRNA in its proper conformation. This hypothesis was in fact suggested by Bluestein et al. (*20a*) on the basis of their finding that spermidine, and Ca^{2+}, can successfully replace Mg^{2+} in the transfer reaction step, while they can not replace Mg^{2+} in the amino acid activation step. A further indication that the effect of Mg^{2+} may be nonspecific and that the target may be the tRNA rather than the enzyme is given by the results of Waldenström (*24*) who was able to show a pronounced effect of monovalent cations on the rate of the transfer reaction using the lysine enzyme from *E. coli.* This is consistent with recent results obtained by Fresco and co-workers* indicating a strong influence of monovalent cations on tRNA conformation at ionic strengths comparable to those used by Waldenström.

The formation of an enzyme-AMP-aminoacyl complex has in the case of the arginyl-tRNA synthetase from *E. coli* been shown to be dependent on the presence of the cognate tRNA as indicated by the inability of the enzyme to catalyze isotope exchange between ATP and inorganic pyrophosphate $-^{32}P$ in the absence of tRNA (*25*). This should provide an interesting model system for studying the influence of structural modifications on the substrate function of the tRNA. These results indicate a change in conformation at the amino acid and/or AMP site when tRNA is bound to the enzyme. Yarus and Berg (*26*) have recently reported that the binding of isoleucine to the isoleucyl-tRNA synthetase from *E. coli* (in the absence of ATP) markedly increases the rate at which $tRNA^{Ile}$ enters and leaves its binding site. It would thus appear that at least for some of these enzymes there is an interdependence between the tRNA and amino acid sites so that binding of substrate to one of the sites influences the properties of the other site. The methionine enzyme from *E. coli* has been shown to undergo a conformational change in the presence of ATP, methionine, and Mg^{2+} that leads to a sharp drop in the number of SH-groups accessible to a thiol reagent (DTNB) as compared to the native enzyme (*27*).

The search for a stable enzyme substrate complex between an aminoacyl-tRNA synthetase and its specific tRNA was originally prompted by the possibility of using this complex as a model system for the recognition between enzyme and tRNA. Valyl-tRNA synthetase from yeast was the first enzyme where such a complex could be demonstrated (*22, 28*). Since then similar complexes have been found with valyl-tRNA synthetase from *E. coli* (*29*) and lysyl-tRNA synthetase from yeast (*5, 30*) as well as with crude preparations of leucyl-tRNA synthetase from *E. coli* (*31*) and arginyl-, glycyl-, and valyl-tRNA synthetase from *Bacillus stearothermophilus* (*32*). Yarus and Berg (*33*) have shown that isoleucyl-tRNA synthetase from *E. coli* can form a specific and stable complex with $tRNA^{Ile}$ under conditions where the enzyme is denaturated by filtration through a nitrocellulose filter. This elegant technique was used to investigate the influence of isoleucine on the tRNA binding site as discussed above.

* J. R. Fresco, personal communication.

In the following pages we describe the isolation and some properties of a complex between an enzyme in the native form (valyl-tRNA synthetase from yeast) and its tRNA substrate including $tRNA^{Val}$ structurally modified at the 3′-hydroxyl end (*34*).

EXPERIMENTAL PROCEDURE

Materials

Valyl-tRNA synthetase from yeast may be prepared according to Lagerkvist and Waldenström (*35*). The concentration of the enzyme, usually expressed on a molar basis, is calculated from determinations of protein according to Lowry et al. (*36*) or from determinations of protein nitrogen by a ninhydrin method (*37*) based on a nitrogen content of 16% in the enzyme protein. The Lowry method is standardized by using enzyme preparations in which protein nitrogen had been determined by the ninhydrin method. A molecular weight of 112,000 (*35*) for valyl-tRNA synthetase is used in the calculations.

Crude tRNA from yeast may be obtained from Boehringer, Mannheim, Germany. The concentration of tRNA is expressed in terms of its optical density. The unit used (A_{260}) is defined as the amount of material that, when dissolved in 1.0 ml, gives an absorbance of 1.0 at 260 nm with a light path of 1.0 cm. When the molar concentration of tRNA is calculated from its optical density 1 A_{260} unit is taken to equal 1.8 nmoles of tRNA. The separation of the two major components of $tRNA^{Val}$ from yeast ($tRNA_1{}^{Val}$ and $tRNA_2{}^{Val}$) is described in Ref. *28*. Unless otherwise stated these fractions have been used without further purification. Partially purified $tRNA_1{}^{Val}$ and $tRNA_2{}^{Val}$ was obtained by chromatography on DEAE-Sephadex according to Miyazaki et al. (*38, 39*). Highly purified $tRNA_1{}^{Val}$ was prepared from crude yeast tRNA by the method of Gillam et al. (*40*). The product obtained in the final purification step had a ratio of esterified valine to A_{260} that was close to the value expected for pure $tRNA^{Val}$ (1.8 nmoles per A_{260}). However, after removal of the valine, the preparation gave only 1.2 to 1.3 nmoles per A_{260} in the standard assay. Nevertheless better than 90% of the A_{260} present in the preparation could be obtained as a complex with the enzyme (see below). This would be consistent with the assumption that the purified material contained only $tRNA^{Val}$ although some of the chains had been sufficiently altered in the purification procedure to make them inactive in the standard assay. The product was free of $tRNA_2{}^{Val}$ as indicated by its inability to form a stable complex with the enzyme on Sephadex (see below). Terminally modified $tRNA_1{}^{Val}$ ($tRNA^{Val}$ CCp, $tRNA^{Val}$–UCACCA, and $tRNA^{Val}$–AUCACCA) was obtained from highly purified $tRNA_1{}^{Val}$ as described in Ref. *34*.

Sephadex G-100 was a product of Pharmacia, Uppsala, Sweden. Pevikon – C 870, a copolymer of polyvinylacetate and polyvinylchloride, was obtained from Fosfatbolaget, Stockholm, Sweden. The material was ground to a size of 200 mesh or finer and washed with concentrated acetic acid and water before use.

Methods

Assays for enzyme activity (*35*) as well as for ability of tRNA to form aminoacyl-tRNA were performed as previously described (*22*).

Formation and Isolation of Complexes Between Enzyme and $tRNA^{Val}$

All operations were carried out at a temperature of 0°–4°. Valyl-tRNA synthetase and tRNA were mixed together and cacodylate buffer (pH 7.0), was added to give a concentration of 100 mM in a final volume of 0.15 ml. The complex could then be isolated either by filtration on Sephadex G-100 (*22*), by sucrose gradient centrifugation, or by electrophoresis (*28*).

For the *Sephadex filtration* method the sample was mixed with an equal volume of 1 M ammonium sulfate and added to the top of a column of Sephadex G-100 (0.8 x 40 cm) that had previously been equilibrated with 0.5 M ammonium sulfate containing 0.01 M sodium cacodylate buffer (pH 7.0). The column was eluted with the same buffer at an approximate rate of 10 ml per hour and fractions of 0.7–1.0 ml were taken.

In the *centrifugation* procedure the sample was layered over a linear sucrose gradient of 2.5 ml of 3% sucrose containing 10 mM cacodylate buffer (pH 7.0), and 250 mM ammonium sulfate against 2.5 ml of 20% sucrose containing the same concentration of buffer and ammonium sulfate. The sample was then centrifuged at 40,000 rpm for 22 to 24 hr in a Spinco SW-39 swinging bucket rotor. The tube was punctured and fractions of approximately 0.2 ml were taken.

When the complex was isolated by *electrophoresis* the enzyme and tRNA, dissolved in 250 mM ammonium phosphate buffer (pH 6.5), final volume 0.20 ml, were added to the top of a Pevikon column, 0.95 x 30 cm, containing 250 mM ammonium phosphate buffer (pH 6.5). The column was jacketed and maintained at a temperature of 2°–3°. The electrophoresis was carried out at 7 V/cm and 27–28 mA for 10 to 12 hr. The column was then eluted with the same buffer as above. The flow rate was 7–8 ml/hr and fractions of 0.5 to 1.0 ml were taken.

RESULTS

Complex Formation with Sephadex Filtration Technique

Complex Formation with Unfractionated tRNA

Chromatography of a mixture of enzyme and unfractionated yeast tRNA gave two peaks of $tRNA^{Val}$. The first peak (I) was associated with the enzyme activity peak, while the second peak (II) appeared in its expected position, together with the bulk of the tRNA. When the chromatogram was assayed for $tRNA^{Lys}$, only one peak, coinciding with the main tRNA peak was obtained. No $tRNA^{Lys}$ peak was found associated with the enzyme (Fig. 1). The same result was obtained in similar experiments in which the chromatograms were assayed in parallel for $tRNA^{Val}$ and tRNAs specific for the amino acids alanine, aspartic acid, glycine, leucine, phenylalanine, proline, serine, threonine, and tyrosine. In the absence of either enzyme or tRNA, or if the chromatographic fractions were treated with RNase before assay, Peak I disappeared completely. Yeast tRNA that had been stripped of Mg^{2+} by treatment with EDTA gave approximately the same yield of complex as did untreated tRNA. Addition of Mg^{2+} to the incubation medium containing the stripped tRNA did not increase the yield, nor did the presence of EDTA lower it.

The isolated complex E^{Val}–$tRNA^{Val}$ was quite stable, and could be rechromatographed on Sephadex after 6 hr at 0° with a yield of about 90%. The yield after 24 hr at 0° was approximately 75%. With increasing ratios of enzyme to $tRNA^{Val}$ in the incubation mixture a maximum of complex formation corresponding to 25–30% of the $tRNA^{Val}$ incubated was obtained. Further increase of the amount of enzyme in the incubation did not increase the yield of complex formation indicating that the tRNA preparation was heterogeneous and that only part of it could form an enzyme–substrate complex stable on Sephadex.

Complex Formation with $tRNA_1{}^{Val}$ and $tRNA_2{}^{Val}$

Chromatography of $tRNA^{Val}$ from yeast on DEAE-cellulose separated it into two components, $tRNA_1{}^{Val}$ and $tRNA_2{}^{Val}$. When $tRNA_1{}^{Val}$ and $tRNA_2{}^{Val}$ were tested for their ability to form a complex with the enzyme, stable enough to be isolated by Sephadex filtration, it was found that only the minor fraction, $tRNA_2{}^{Val}$ corresponding to 25–30% of the total $tRNA^{Val}$, could form such a complex. In the presence of excess enzyme more than 90% of the $tRNA_2{}^{Val}$ incubated was obtained as a complex with the enzyme. The major fraction, $tRNA_1{}^{Val}$, on the other hand, gave no detectable formation of stable complex even when a large excess of enzyme was used in the incubation.

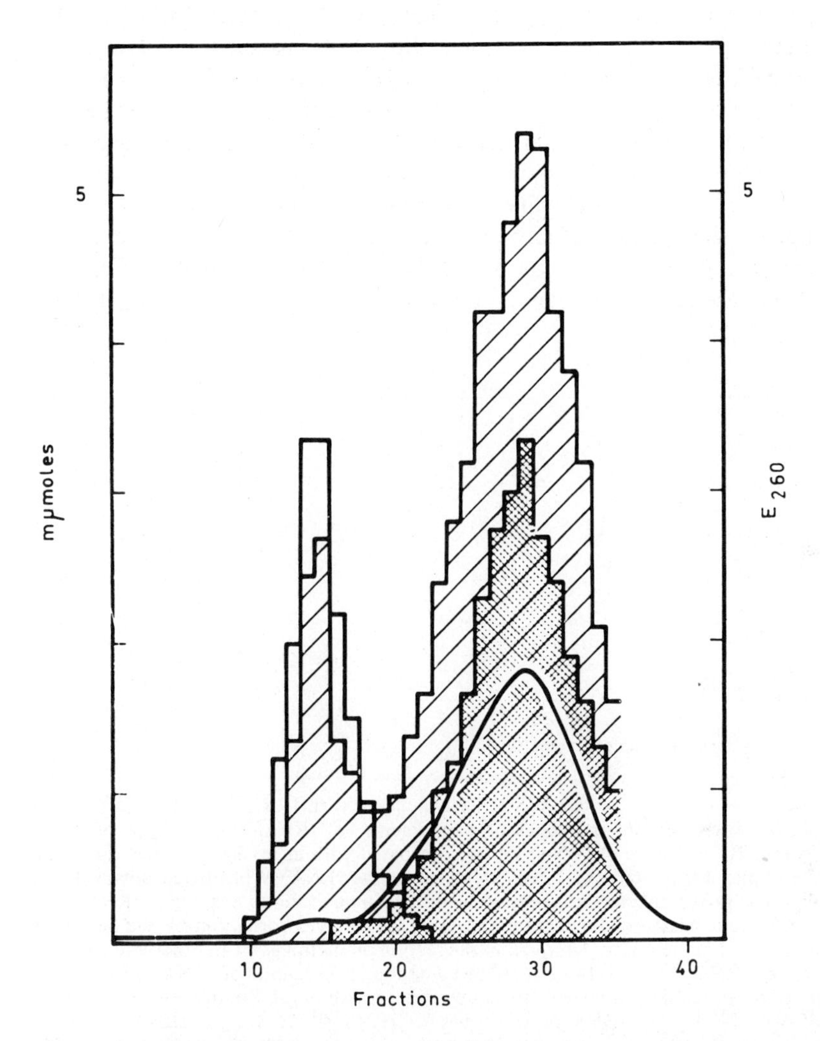

Figure 1. Separation of an enzyme–$tRNA^{Val}$ complex on Sephadex G-100. Valyl-tRNA synthetase (2.7 nmoles) and unfractionated yeast tRNA (20 A_{260} of tRNA containing 1.0 nmoles of $tRNA^{Val}$) were incubated and the resulting complex was isolated by chromatography on Sephadex G-100 as described in the section on Methods. Aliquots of the chromatographic fractions were assayed for enzyme activity and for $tRNA^{Val}$ and $tRNA^{Lys}$. Enzyme, nmoles x 10, is represented as plain area; $tRNA^{Val}$, nmoles x 10^2, as hatched area; and $tRNA^{Lys}$, nmoles x 10^2, as crosshatched area. Optical density at 260 nm is represented as a continuous line. The enzyme values were calculated from the enzyme activity actually measured in the chromatographic fractions and the known specific activity of the incubated enzyme. They have not been corrected for the yield of enzyme activity obtained in this chromatogram (60%). The recovery of protein was quantitative. Reproduced with permission from *J. Biol. Chem.* **241**: 5391 (1966).

Previous attempts to show the formation of the complex AMP-Val · E^{Val} · valyl-tRNA were hampered by the ability of the enzyme to catalyze a slow hydrolysis of the ester linkage in valyl-tRNA (*22*). This difficulty was overcome by isolating the complex on a Sephadex column that had been equilibrated with an elution buffer that contained labeled valine, ATP, and Mg^{2+} so that any valyl-tRNA hydrolyzed by the enzyme would be immediately reesterified. With this technique it was possible to show the formation of the complex although a comparatively large excess of enzyme had to be used in the incubation mixture to get a detectable amount of complex formation. Furthermore, the separation of the complex AMP-Val · E^{Val} · valyl-tRNA

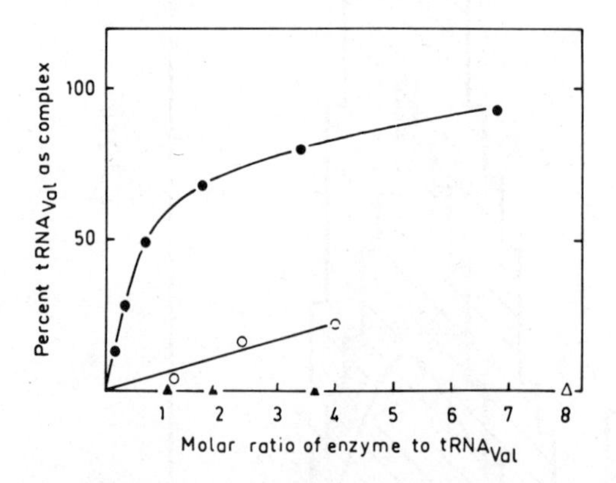

Figure 2. Formation of enzyme-tRNA complexes with $tRNA_1^{Val}$ and $tRNA_2^{Val}$ with Sephadex filtration. In one series of experiments (▲) $tRNA_1^{Val}$ (9.4 A_{260} of tRNA containing 1.35 nmoles of $tRNA_1^{Val}$) was incubated with increasing amounts of enzyme and complex formation, expressed as percent of the $tRNA^{Val}$ incubated obtained as a complex with the enzyme, was plotted against the molar ratio of enzyme to $tRNA^{Val}$ in the incubation mixture. Incubation and assay of complex formation by the Sephadex filtration technique was as described in the section on methods. In an otherwise identical series (●) $tRNA_2^{Val}$ (15.5 A_{260} of tRNA containing 1.55 nmoles of $tRNA_2^{Val}$) was used. The third series was intended to show the formation of complexes of the type AMP-Val E^{Val} • valyl-tRNA. In this series (○), $tRNA_2^{Val}$ (16 A_{260} of tRNA containing 1.6 nmoles of $tRNA^{Val}$) was incubated with 10 mM $MgCl_2$, 1.3 mM ATP, 2.5 μM L-valine-^{14}C, 13 mM GSH, 10 mM cacodylate buffer (pH 7.0), and enzyme as indicated in the figure in a final volume of 0.15 ml. Incubation was for 5 min at 37° and was terminated by chilling in ice and adding 0.15 ml of 1 M ammonium sulfate containing 10 mM cacodylate buffer (pH 7.0). The sample was then applied to a column, 0.8 cm^2 x 40 cm, of Sephadex G-100 that had previously been equilibrated with 0.5 M ammonium sulfate containing 10 mM cacodylate buffer (pH 7.0), 10 mM $MgCl_2$, 1 mM ATP, and 1.85 μM ^{14}C-L-valine. The column was eluted with the same solution at an approximate rate of 10 ml/hr and fractions of 1 ml were taken. Aliquots were assayed for ^{14}C-valyl-tRNA by precipitation with perchloric acid in the presence of bovine serum albumin as described previously. Complex formation was expressed as percent of ^{14}C-valyl-tRNA obtained as a complex with the enzyme and plotted as above. In one experiment (△) of this series $tRNA_1^{Val}$ was substituted for $tRNA_2^{Val}$. Reproduced with permission from *J. Biol. Chem.*, **244**: 2476 (1969).

from free valyl-tRNA was less satisfactory than in the case of the complex E^{Val}-$tRNA^{Val}$, indicating a greater lability of the complex with a continuous breakdown during chromatography. Attempts to form a similar complex with esterified $tRNA_1{}^{Val}$ were unsuccessful, as might be expected from the inability of unesterified $tRNA_1{}^{Val}$ to form a stable complex with the enzyme under these conditions. These findings are summarized in Fig. 2.

Complex Formation with Sucrose Gradient Centrifugation

Separation of Complex from Free Enzyme and $tRNA^{Val}$

Centrifugation through a sucrose gradient gave a good separation of the complex E^{Val}-$tRNA^{Val}$ from free $tRNA^{Val}$ and a partial separation of the complex from the free enzyme. *s*-Values calculated from nine sucrose gradient centrifugations, based on an $s^0_{20,w}$ of 5.5 S for the free enzyme and assuming a linear relationship between *s* and the distance of the substance from the top of the gradient, gave a mean value of 6.9 for the complex. The molar ratio of $tRNA^{Val}$ to enzyme protein in the fractions containing the complex was 0.7 to 0.9 except in the distal part of the complex peak where there was an admixture of free enzyme in such experiments in which only part of the enzyme protein was obtained as complex (Fig. 3). The same molar ratio of tRNA to enzyme was obtained whether $tRNA^{Val}$ was determined in the standard assay based on its ability to be esterified with valine or total tRNA was calculated from the optical density of the complex at 260 nm. Centrifugation of the enzyme through the sucrose gradient was always accompanied by a considerable decrease in specific activity regardless of whether tRNA was present or not. The specific activity after centrifugation varied with different enzyme preparations, but was generally 20 to 40% of the original specific activity. It should be pointed out that the specific activity in the fractions containing the enzyme in complex with $tRNA^{Val}$ was the same as in the fractions containing free enzyme regardless of the proportions of complex to free enzyme in the sucrose gradient. When the enzyme was incubated with $tRNA_1{}^{Val}$ at a molar ratio of $tRNA_1{}^{Val}$ to enzyme of 0.6 in the incubation, 50 to 60% of the enzyme protein was shifted over from the position of the free enzyme to that of the complex. Increasing the proportions of $tRNA_1{}^{Val}$ to enzyme in the incubation mixture did not produce any further shift of enzyme protein into the complex position (Fig. 3A). With $tRNA_2{}^{Val}$ similar results were obtained except that in this case essentially all of the enzyme protein could be shifted over into the complex position at a molar ratio of $tRNA_2{}^{Val}$ to enzyme of 1.6 in the incubation mixture. This was indicated by the appearance in the complex position of a symmetrical peak of enzyme protein without any detectable shoulder in the region corresponding to free enzyme (Fig. 3D). The maximum

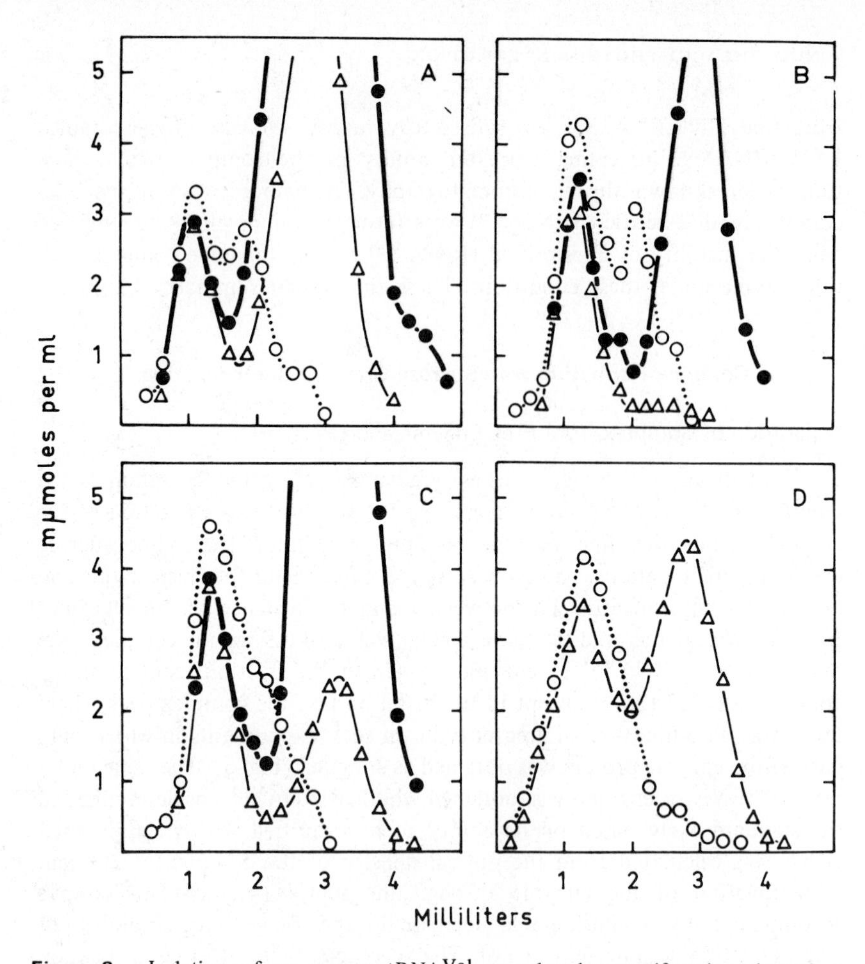

Figure 3. Isolation of an enzyme-tRNA^{Val} complex by centifugation through a sucrose gradient. The tRNA preparations used in these experiments had been fractioned into tRNA$_1^{Val}$ and tRNA$_2^{Val}$ as described in the text and in all the experiments except that described in (D) the fractions had been further purified by chromatography on DEAE-Sephadex (*38, 39*). Total tRNA calculated on the A_{260} of the fractions has not been plotted in (D) since the ultraviolet profile of the complex was not sufficiently separated from that of the main tRNA peak when this comparatively crude tRNA$_2^{Val}$ fraction was used. (A) tRNA$_1^{Val}$ (22.5 A_{260} of tRNA containing 13.3 nmoles of tRNA$_1^{Val}$) and 5.5 nmoles of valyl-tRNA synthetase were incubated and the resulting complex was isolated by sucrose gradient centrifugation. (○) Enzyme, (△) tRNA^{Val}, (●) total tRNA. The enzyme values were calculated from determinations of enzyme protein in the fractions while total tRNA was obtained from the optical density at 260 nm corrected for the absorbance of the enzyme. For further experimental details the reader is referred to the section on methods. The specific activity of the enzyme in the complex position was 2500 as compared to 2800 in the position corresponding to free enzyme. (B) This experiment was similar to (A) except that tRNA$_2^{Val}$ (4.8 A_{260} of tRNA containing 2.5 nmoles of tRNA^{Val}) was used instead of tRNA$_1^{Val}$. The specific activity of the enzyme in the complex position was 2700 and in the position corresponding to free enzyme 2650. (C) This experiment was identical with (B) except that 9.6 A_{260} of tRNA containing 5.0 nmoles of tRNA$_2^{Val}$ were used in the incubation. The specific activity of the enzyme in the complex position was 2000 and in the position corresponding to the free enzyme 2200. (D) This experiment was the same as (B) except that 82 A_{260} of tRNA containing 9.0 nmoles of tRNA$_2^{Val}$ were used in the incubation. Reproduced with permission from *J. Biol. Chem.*, **244**: 2476 (1969).

percentage of enzyme protein that can be obtained as a complex with $tRNA_1^{Val}$ appears to vary with the method of preparation of the enzyme. With enzyme that was prepared with some minor modifications of the usual procedure (*35*) about 80% of the protein could form a complex with $tRNA_1^{Val}$.

Specificity of Complex Formation

When sucrose gradients with a low ionic strength were used for the isolation of the complex, there was a tendency for the enzyme to become associated with tRNA chains other than $tRNA^{Val}$. This tendency to unspecific complex formation increased with increasing ratios of enzyme to tRNA in the incubation mixture. The unspecific association between the enzyme and nonsubstrate tRNA was completely abolished by the addition of ammonium sulfate at a concentration of 200 mM to the sucrose gradient. Under these conditions only $tRNA^{Val}$ showed a peak that appeared together with the enzyme (Fig. 4). The ratio of $tRNA^{Val}$ to A_{260} in the complex, which was

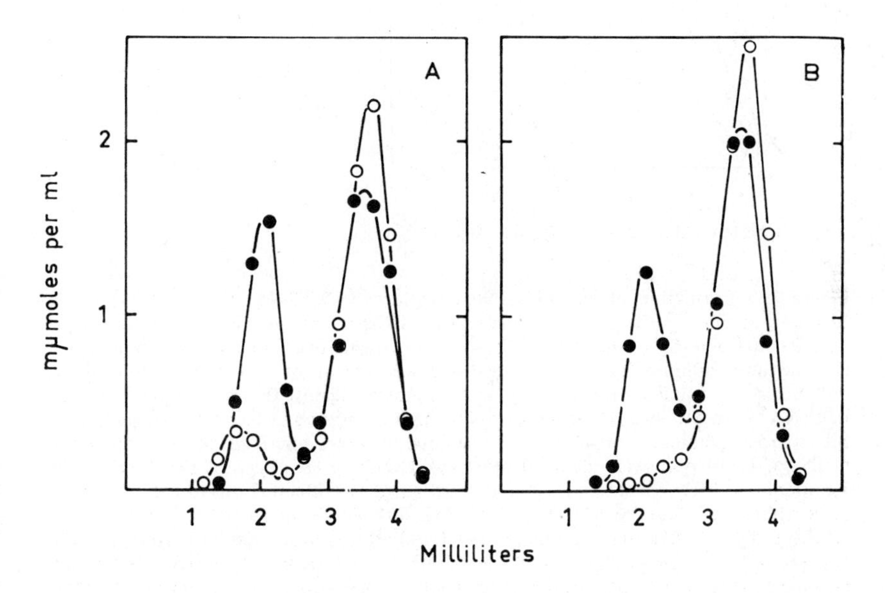

Figure 4. Specificity of complex formation with the sucrose gradient centrifugation; dependence on ionic strength in the gradient. (A) $tRNA_2^{Val}$ (26 A_{260} of tRNA containing 2.6 nmoles of $tRNA_2^{Val}$) and 1.6 nmoles of valyl-tRNA synthetase were incubated and the resulting complex was centrifuged through a sucrose gradient containing 50 mM ammonium sulfate. The procedure was otherwise the same as that described in the section on methods except that centrifugation was for 16 hr. (●) $tRNA^{Val}$, (○) $tRNA^{Lys}$. (B) This experiment was identical with (A) except that the sucrose contained 200 mM ammonium sulfate. Reproduced with permission from *J. Biol. Chem.*, **244**: 2476 (1969).

close to that expected for pure $tRNA^{Val}$, is also consistent with a completely specific complex formation (Fig. 3).

When the effect of different salts on the stability of the complex was investigated it was found that substituting ammonium phosphate (pH 6.5), for ammonium sulfate in the gradient had no effect on complex formation and stability. However, when potassium phosphate (pH 6.5), was used the yield of complex was approximately 50% of that of the control with ammonium sulfate. Under these conditions the complex appeared as a shoulder and

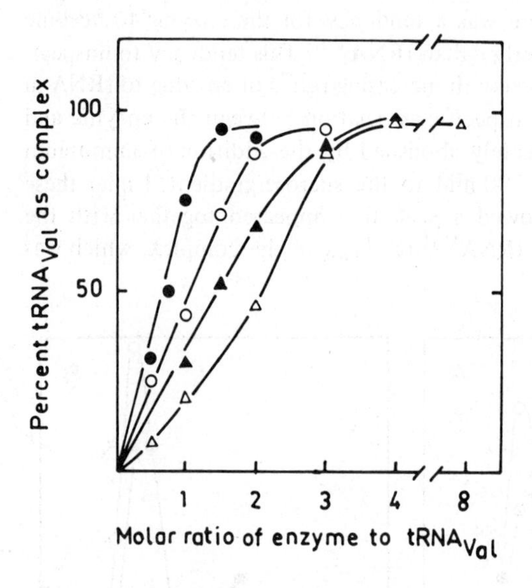

Figure 5. Formation of enzyme-tRNA complexes with $tRNA_1^{Val}$ and $tRNA_2^{Val}$ with sucrose gradient centrifugation. In one series of experiments (▲), $tRNA_1^{Val}$ (2.8 A_{260} of tRNA containing 0.81 nmoles of $tRNA_1^{Val}$) was incubated with increasing amounts of enzyme and complex formation, expressed as percent of the $tRNA^{Val}$ incubated obtained as a complex with the enzyme, was plotted against the molar ratio of enzyme to $tRNA^{Val}$ in the incubation mixture. Incubation and assay of complex formation by the sucrose gradient centrifugation technique was as described in the section on methods. In an otherwise identical series (●), $tRNA_2^{Val}$ (16 A_{260} of tRNA containing 1.6 nmoles of $tRNA_2^{Val}$) was used. The third series was intended to show the formation of complexes of the type AMP-Val • E^{Val} • valyl-tRNA. In this series (○), $tRNA_2^{Val}$ (16 A_{260} of tRNA containing 1.6 nmoles of $tRNA_2^{Val}$) was incubated with 10 mM $MgCl_2$, 1.3 mM ATP, 2.5 μM L-valine-^{14}C, 13 mM GSH, 10 mM cacodylate buffer (pH 7.0), and enzyme as indicated in the figure in a final volume of 0.15 ml. Incubation was for 5 min at 37° and was terminated by chilling in ice followed by centrifugation through a sucrose gradient as described in the section on methods with the difference that the gradient contained in addition 10 mM $MgCl_2$, 1 mM ATP, and 1.85 μM L-valine-^{14}C. The fractions obtained were assayed for ^{14}C-valyl-tRNA by precipitation with perchloric acid in the presence of bovine serum albumin as described previously (*22*). In an otherwise identical series of experiments (△), $tRNA_1^{Val}$ (2.8 A_{260} of tRNA containing 0.81 nmole of $tRNA_1^{Val}$) was substituted for $tRNA_2^{Val}$. Reproduced with permission from *J. Biol. Chem.*, **244**: 2476 (1969).

not as a distinct peak separated from the free $tRNA^{Val}$, indicating a poor stability of the complex in potassium phosphate. When Tris-maleate buffer (pH 6.5) was used instead of ammonium sulfate there was no detectable formation of a stable complex.

Complex Formation with $tRNA_1^{Val}$ and $tRNA_2^{Val}$

The sucrose gradient centrifugation technique allowed the demonstration of stable complex formation with both $tRNA_1^{Val}$ and $tRNA_2^{Val}$. When enough enzyme was used more than 90% of the $tRNA^{Val}$ present in the incubation mixture could be obtained as a complex with the enzyme. The same result was obtained when esterified $tRNA_1^{Val}$ or $tRNA_2^{Val}$ was used to form the complex AMP-Val · E^{Val} · valyl-tRNA which was isolated by centrifugation through a gradient containing labeled valine, ATP, and Mg^{2+}. With an excess of enzyme practically all of the incubated valyl-tRNA was recovered in complex form (Fig. 5). Ths separation of the complex from free valyl-tRNA, in experiments with a low ratio of enzyme to valyl-tRNA in the

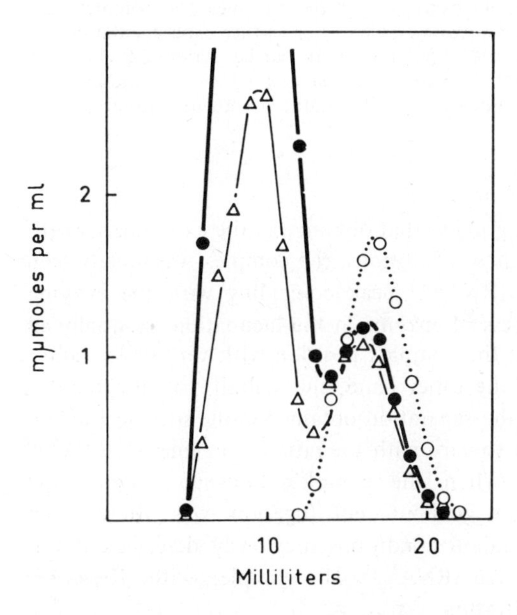

Figure 6. Isolation of an enzyme-$tRNA^{Val}$ complex by electrophoresis on a Pevikon column. $tRNA_2^{Val}$ (22 A_{260} of tRNA containing 17 nmoles of $tRNA_2^{Val}$) and 8.2 nmoles of valyl-tRNA synthetase were incubated and the resulting complex was isolated by electrophoresis. (○) Enzyme, (△) $tRNA^{Val}$, (●) total tRNA. The enzyme values were calculated from determinations of enzyme protein in the fractions while total tRNA was obtained from the optical density at 260 nm corrected for the absorbance of the enzyme. For further experimental details see section on methods. Reproduced with permission from *J. Biol. Chem.*, **244**: 2476 (1969).

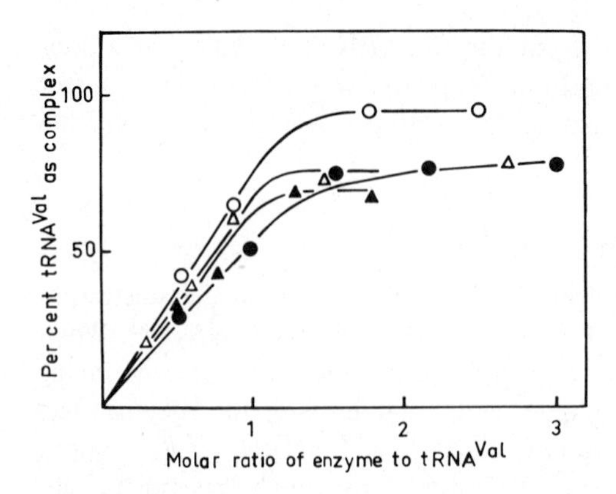

Figure 7. Formation of enzyme-tRNA complexes with tRNA$_1^{Val}$ preparations structurally modified at the 3′-hydroxyl end. In one series of experiments (○), 3.56 A_{260} of native tRNA$_1^{Val}$ was incubated with increasing amounts of enzyme and complex formation, expressed as percentage of the A_{260} incubated obtained as a complex with the enzyme, was plotted against the molar ratio of enzyme to tRNA$_1^{Val}$ in the incubation mixture. The A_{260} of the tRNA in the complex was obtained after correction for the known ultraviolet absorbance of the enzyme. The complex was isolated by centrifugation through a sucrose gradient. For further experimental details see the section on methods and related references. In similar series, 2.6 A_{260} of tRNA^{Val} CCp (●), 2.4 A_{260} of tRNA^{Val}–UCACCA (▲), and 2.1 A_{260} of tRNA^{Val}–AUCACCA (△) were incubated in the same way. Reproduced with permission from *J. Biol. Chem.*, **245**: 435 (1970).

incubation mixture was not as good as that obtained in the experiments with unesterified tRNA^{Val}. Under these conditions, the complex was mostly seen as a leading shoulder of the tRNA^{Val} peak coinciding with the enzyme activity. However, with an excess of enzyme in the incubation essentially all of the valyl-tRNA appeared in the complex position with very little trailing behind the enzyme peak. On the other hand, the stability of the complex E^{Val}–tRNA^{Val}, as judged by the separation obtained in the sucrose gradient centrifugations, does not seem to vary with the ratio of enzyme to tRNA^{Val} in the incubation mixture. When the complex between enzyme and tRNA$_1{}^{Val}$, isolated by sucrose gradient centrifugation, was filtered over Sephadex G-100 with the standard conditions previously described it was completely broken down to free tRNA^{Val}. The complex with tRNA$_2{}^{Val}$ gave a recovery of 40% after filtration.

Isolation of Complex by Electrophoresis

Electrophoresis on a Pevikon column separated the complex E^{Val}–tRNA^{Val} from free tRNA^{Val} but gave no separation of the complex and the free enzyme (Fig. 6). The recovery of protein from the electrophoresis was quantitative, but the specific activity of the enzyme recovered was only

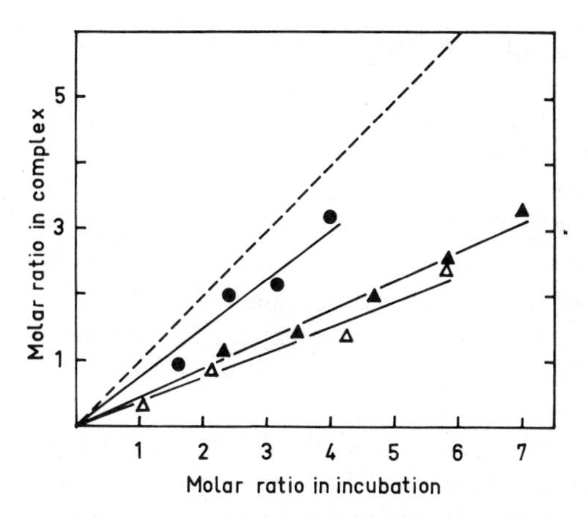

Figure 8. Competitive inhibition of complex formation between enzyme and native $tRNA_1^{Val}$ in the presence of structurally modified $tRNA_1^{Val}$ preparations. In one series of experiments (●), 1.8 A_{260} of native $tRNA_1^{Val}$ were incubated with 5.7 nmoles of enzyme in the presence of increasing amounts of $tRNA^{Val}$ CCp and the resulting complex was isolated by sucrose gradient centrifugation. The A_{260} corresponding to native $tRNA_1^{Val}$ present in the complex was determined by the usual assay procedure using a $tRNA_1^{Val}$ preparation as standard that was assumed to be 100% pure on the basis of its ability to be completely complexed by the enzyme. The amount of $tRNA^{Val}$ CCp was calculated as the difference between total tRNA in the complex, obtained from the A_{260} corrected for protein absorbance, and the native $tRNA^{Val}$. The molar ratio of the $tRNA^{Val}$ CCp to native $tRNA_1^{Val}$ in the complex was then plotted against the molar ratio of the same compounds in the incubation mixture. For further experimental details see the section on methods and related references. In otherwise identical series of experiments the same amounts of native $tRNA_1^{Val}$ and enzyme were incubated with increasing concentrations of $tRNA^{Val}$–UCACCA (▲) or $tRNA^{Val}$–AUCACCA (△). A dotted line with a slope of 1 is included for comparison. Reproduced with permission from *J. Biol Chem.*, **245**: 435 (1970).

10–15% of that observed before electrophoresis. With a low ionic strength in the electrophoresis, there was a tendency for the enzyme to become associated with tRNA chains other than $tRNA^{Val}$. This unspecific association was completely absent when the electrophoresis took place in 250 mM ammonium phosphate (pH 6.5), as indicated by the absence of tRNA specific for amino acids other than valine. The ratio of $tRNA^{Val}$ to A_{260} was close to that expected for pure $tRNA^{Val}$ which further supports the specificity of the complex formation. Complex formation could be shown both with $tRNA_1{}^{Val}$ and $tRNA_2{}^{Val}$ and with enough enzyme in the incubation all of the $tRNA^{Val}$ could be obtained as complex.

Complex Formation with Modified $tRNA_1{}^{Val}$

When preparations of highly purified $tRNA_1{}^{Val}$ modified at the 3′-hydroxyl end ($tRNA^{Val}$CCp, $tRNA^{Val}$-UCACCA, and $tRNA^{Val}$-AUCACCA) were incubated with the enzyme followed by centrifugation through a sucrose gradient the results obtained were identical with those already described for the native $tRNA_1^{Val}$.

The centrifugation thus gave a good separation of $tRNA_1^{Val}$ complexed to the enzyme from free $tRNA_1^{Val}$ and a partial separation of the complex and the free enzyme. The complex formation was completely specific as indicated by the inability of highly purified $tRNA^{Lys}$, either native or structurally modified, to form a complex with the enzyme. When enough enzyme was used, 70-80% of the A_{260} of the modified $tRNA_1^{Val}$ preparations could be obtained as a complex with the enzyme. This should be compared with a transfer of more than 90% of the A_{260} of native $tRNA_1^{Val}$ into the complex position in similar experiments (Fig. 7).

$tRNA^{Val}$CCp, $tRNA^{Val}$-UCACCA, and $tRNA^{Val}$-AUCACCA also acted as competitive inhibitors of complex formation with native $tRNA_1^{Val}$. This finding is consistent with the previous data and makes it possible to compare the affinity of the enzyme for modified and native $tRNA_1^{Val}$. In Fig. 8 the molar ratio of modified to native $tRNA_1^{Val}$ in the complex has been plotted against the same ratio in the incubation mixture. The slope of the straight line obtained then gives the relative affinity of the enzyme for the modified preparation to that for the native one. If a correction is made for the fact that only 70-80% of the modified $tRNA_1^{Val}$ preparations can form a complex with the enzyme (Fig. 7) the relative affinities for the more extensively modified preparations, $tRNA^{Val}$-UCACCA and $tRNA^{Val}$-AUACCA, can be calculated to be 0.5 to 0.6.

CONCLUDING REMARKS

Although great progress has been made in the elucidation of the primary structure of a number of tRNAs we still have no real information regarding what parts of the molecule play a role in the recognition by an aminoacyl-tRNA synthetase of its tRNA substrate.

The valyl-tRNA synthetase from yeast forms a stable and specific enzyme-substrate complex with $tRNA^{Val}$ from the same source that can be isolated by filtration on Sephadex, centrifugation through a sucrose gradient, or column electrophoresis. This provides us with a model for the study of the recognition mechanism separate from the overall catalytic function of the enzyme. Using this model we have found that extensive modification at the 3′-hydroxyl end of $tRNA_1^{Val}$ with the loss of an average of 7 nucleotide residues by digestion with snake venom phosphodiesterase did not affect the ability of the $tRNA_1^{Val}$ to form a stable complex with the enzyme that could be isolated by sucrose gradient centrifugation. If the ability to form a stable complex with the enzyme is accepted as an indication of recognition of the tRNA by the enzyme the results of this investigation would suggest that the first three base pairs that form part of the double-stranded stem in the cloverleaf model of $tRNA_1^{Val}$ cannot be the sole recognition site of this molecule.

References

1. Clark, B. F. C., B. P. Doctor, K. C. Holmes, A. Klug, K. A. Marcker, S. J. Morris, and H. H. Paradies. 1968. *Nature,* **219**: 1222.
2. Fresco, J. R., R. D. Blake, and R. Langridge. 1968. *Nature,* **220**: 1285.
3. Young, J. D., R. M. Bock, S. Nishimura, H. Ishikura, Y. Yamada, U. L. RajBhandary, M. Labanauskas, and P. G. Connors. 1969. *Science,* **166**: 1527.
4. Kim, S. -H., and A. Rich. 1969. *Science,* **166**: 1621.
5. Rymo, L., U. Lagerkvist, and A. Wonacott. 1970. *J. Biol. Chem.,* **245**: 4308.
6. Hayashi, H., and K. -I. Miura. 1964. *J. Mol. Biol.,* **10**: 345.
7. Hayashi, H., and K. -I. Miura. 1966. *Nature,* **209**: 376.
8. Deutscher, M. 1965. *Biochem. Biophys. Res. Commun.,* **19**: 283.
9. Torres-Gallardo, J., and M. Kern. 1965. *Proc. Natl. Acad. Sci., U.S.,* **53**: 91.
10. Korzhov, V. A., and L. S. Sandakhchiev. 1966. *Biokhimiya,* **31**: 71.
11. Letendre, C., A. M. Michelson, and M. Grunberg-Manago. 1966. *Biochem. Biophys. Res. Commun.,* **23**: 442.
12. Holten, V. Z., and K. B. Jacobson. 1967. *Biochemistry,* **6**: 1293.
13. Roy, K. L., and G. M. Tener. 1967. *Biochemistry,* **6**: 2847.
14. Mirzabekov, A. D., L. Y. Kazarinova, D. Lastity, and A. A. Bayev. 1969. *FEBS Letters,* **3**: 268.
15. Mirzabekov, A. D., D. Lastity, and A. A. Bayev. 1969. *FEBS Letters,* **4**: 281.
16. Mirzabekov, A. D., E. S. Levina, and A. A. Bayev. 1969. *FEBS Letters,* **5**: 218.
17. Imura, N., G. B. Weiss, and R. W. Chambers. 1969. *Nature,* **222**: 1147.
18. Schulman, L. H., and R. W. Chambers. 1968. *Proc. Natl. Acad. Sci., U.S.,* **61**: 308.
19. Allende, J. E., G. Mora, M. Gatica, and C. C. Allende. 1965. *J. Biol. Chem.,* **240**: PC 3229.
20. Allende, C. C., J. E. Allende, M. Gatica, J. Celis, G. Mora, and M. Matamala. 1966. *J. Biol. Chem.,* **241**: 2245.
20a. Bluestein, H. G., C. C. Allende, J. E. Allende, and G. L. Cantoni. 1968. *J. Biol. Chem.,* **243**: 4693.
21. Norris, A. T., and P. Berg. 1964. *Proc. Natl. Acad. Sci., U.S.,* **52**: 330.
22. Lagerkvist, U., L. Rymo, and J. Waldenström. 1966. *J. Biol. Chem.,* **241**: 5391.
23. Fresco, J. R., A. Adams, R. Ascione, D. Henlye, and T. Lindahl. 1966. *Cold Spring Harbor Sym. Quant. Biol.,* **31**: 527.
24. Waldenström, J. 1968. *European J. Biochem.,* **5**: 239.
25. Mehler, A. H., and S. K. Mitra. 1967. *J. Biol. Chem.,* **242**: 5495.
26. Yarus, M., and P. Berg. 1969. *J. Mol. Biol.,* **42**: 171.
27. Bruton, C. J. 1967. Thesis, Trinity College, Cambridge, England.
28. Lagerkvist, U., and L. Rymo. 1969. *J. Biol. Chem.,* **244**: 2476.
29. Yaniv, M., and F. Gros. 1969. *J. Mol. Biol.,* **44**: 17.
30. Letendre, C., J. M. Humphreys, and M. Grunberg-Manago. 1969. *Biochim. Biophys. Acta,* **186**: 46.
31. Seifert, W., G. Nass, and W. Zillag. 1968. *J. Mol. Biol.,* **33**: 507.
32. Okamoto, T., and Y. Kawade. 1967. *Biochim. Biophys. Acta,* **145**: 613.
33. Yarus, M., and P. Berg. 1967. *J. Mol. Biol.,* **28**: 479.

34. Lagerkvist, U., and L. Rymo. 1970. *J. Biol. Chem.*, **245**: 435.
35. Lagerkvist, U., and J. Waldenström. 1967. *J. Biol. Chem.*, **242**: 3021.
36. Lowry, O. H., N. J. Rosebrough, A. L. Farr, and R. J. Randall. 1951. *J. Biol. Chem.*, **193**: 265.
37. Strid, L. 1961. *Acta Chem. Scand.*, **15**: 1423.
38. Miyazaki, M., M. Kawata, K. Najazawa, and S. Takemura. 1967. *J. Biochem. (Tokyo)*, **62**: 161.
39. Kawata, M., M. Miyazaki, and S. Takemura. 1967. *J. Biochem. (Tokyo)*, **62**: 287.
40. Gillam, I., D. Blew, R. C. Warrington, M. von Tigerstrom, and G. M. Tener. 1968. *Biochemistry*, **7**: 3459.

DNA-CELLULOSE CHROMATOGRAPHY OF PROTEINS

E. K. F. BAUTZ AND J. J. DUNN
MOLEKULARE GENETIK DER UNIVERSITÄT
HEIDELBERG, WEST GERMANY

Introduction

Some enzymes involved in nucleic acid metabolism, notably the DNA and RNA polymerases, have relatively acidic isoelectric points but are nevertheless able to bind readily to phosphate groups. For this reason, a combination of DEAE-cellulose and phosphocellulose chromatography has been extremely useful in the purification of these enzymes (*1–3*), since there are very few proteins which adsorb to both resins at moderate ionic strengths. It has been suspected, for example, that the binding of DNA polymerase to phosphocellulose is the result of a specific affinity of some portion of the enzyme for phosphate groups and that by using DNA instead of phosphocellulose the binding might be even more specific. In order to make DNA part of a solid matrix for column chromatography, one would have to employ a resin that binds DNA tightly but which does not carry charged groups itself. Since this is obviously not feasible, attempts were made to immobilize physically the DNA on a cellulose matrix. One such attempt was to irradiate a slurry of DNA and cellulose with UV light in order to produce a crosslinked DNA trapped on the cellulose matrix. This method has been applied successfully to the purification of DNA polymerase from *Micrococcus lysodeikticus* (*4*). Alternatively, Alberts et al. (*5*) have found that just mixing a concentrated solution of native DNA with cellulose powder and bringing the resulting cake to dryness results in a nearly irreversible adsorption of the

DNA to the cellulose matrix. Under suitable conditions this material can be used for column chromatography of DNA-binding proteins. Although both methods can be used for the purification of RNA polymerase, we prefer the latter procedure for the preparation of DNA-cellulose as the DNA appears to be in a more native state and consequently the exclusion of other proteins binding to denatured but not to native DNA is more complete.

We describe here a DNA-cellulose chromatography procedure for the purification of DNA-binding proteins which is based on the procedure developed by Alberts et al. and which has been adopted in this laboratory for the isolation of DNA-dependent RNA polymerase (EC 2.7.7.6) from *E. coli* to greater than 95% purity.

Procedure

In the procedure described the DNA used is from phage T4, which in our experience is preferable to the commercially available calf thymus DNA. If only commercial DNA is available, we suggest to inquire as to which product obtainable is of the highest molecular weight. Since we suggest the use of T4 DNA whenever possible, we have included an easy method for the isolation of T4 DNA in quantities of several hundred milligrams.

Large volume lyzates of T4 phage are concentrated and purified by precipitation with polyethylene glycol (*6*). The pelleted phage are gently resuspended in 0.1 M sodium phosphate buffer (pH 7.2) at a final concentration of 20–30 OD units at 260 nm. The phage suspension is adjusted to a final concentration of 1% with respect to sodium dodecyl sulfate (SDS) by the addition of 0.11 vol of a 10% solution of SDS. The phage suspension is heated in a water bath until it reaches 65°C and is held at 65°C for 3–5 min. The suspension which has now become quite viscous is chilled in an ice bath and sufficient 2 M KCl is added with gentle swirling to yield a final concentration of 0.3 M KCl. The potassium salt of the dodecyl sulfate and some protein are removed by centrifugation at 27,000 *g* for 15 min. The DNA is decanted and extracted once with an equal volume of phenol containing 0.08% 8-hydroxyquinoline and saturated with phosphate buffer. The aqueous phase is separated by low-speed centrifugation and then centrifuged at 27,000 *g* for 15 min. The DNA is carefully decanted not to disturb any pelleted protein and dialyzed exhaustively against 0.01 M Tris-HCl (pH 7.4), 1 mM Na_3 EDTA (TE buffer).

The DNA solution having a concentration of approximately 1 mg/ml is placed in a beaker and dry cellulose (Whatman CF-11) is slowly added with stirring to achieve a thick paste, i.e., about 1 gm of cellulose per 3 ml of DNA solution. The cellulose paste is dried by blowing a stream of air (25°C) from a mechanical blower over it. The dried paste is powdered and thoroughly dried

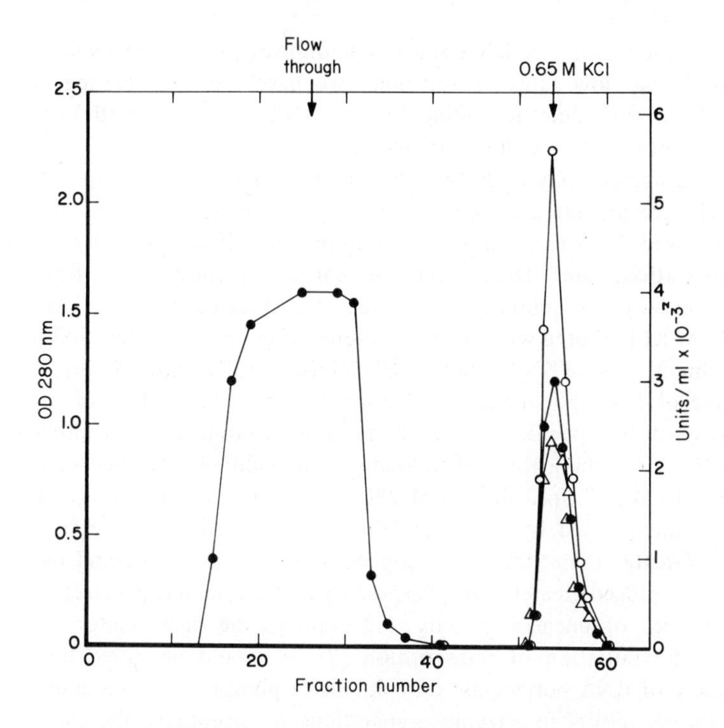

Figure 1. DNA-cellulose chromatography of RNA polymerase. Closed circles indicate protein; open circles and triangles represent the activities on T4 and calf thymus DNA, respectively. A flow rate of approximately 1 ml/min was maintained during the loading and washing steps. Column size was 4 x 15 cm and the DNA concentration was 400 μg per packed milliliter. Fractions of 15 ml were collected.

under vacuum. The dried DNA-cellulose is hydrated by adding 2 liters of TE buffer per 100 gm of dry DNA-cellulose and allowing it to remain overnight at 4°C. The supernatant is decanted and the DNA-cellulose is collected by low-speed centrifugation, washed with two 500 ml volumes of TE buffer, and stored, if desired, as a frozen slurry in TE buffer + 0.15 M KCl.

The amount of DNA immobilized on the cellulose matrix is determined by boiling a small amount of DNA-cellulose resuspended in water for 10 min and measuring the amount of 260 absorbing material released. Usually up to 50% of the input DNA remains on the cellulose, yielding columns with 400 to 500 μg of DNA per packed milliliter of cellulose.

The DNA-cellulose is thawed and resuspended in 10 vol of TE buffer + 0.15 M KCl and poured into a column 4 cm in diameter. The cellulose is allowed to settle under gravity and is equilibrated with 0.02 M Tris-HCl (pH 8.1), 1 mM Na_3EDTA, 10% glycerol, and 0.2 mM dithiothreitol (DC buffer). The amount of DNA cellulose necessary is dependent on the concentration of immobilized DNA and the amount of protein expected to be adsorbed. Using flow rates of 1 ml/min, we have found that 1 mg of immobilized DNA is capable of binding 2 mg of RNA polymerase with little or no activity appearing in the flow through.

For the chromatography of RNA polymerase on DNA-cellulose, 500 gm of *E. coli* cells are disrupted by grinding with glass beads, and the nucleic acids are removed by treatment with streptomycin sulfate, polyethylene glycol fractionation, and DEAE-cellulose chromatography. The RNA polymerase activity is eluted from the DEAE-cellulose with DC buffer + 0.23 M KCl, diluted with an equal volume of DC buffer (minus KCl), and applied directly to the DNA-cellulose. The elution profile shown in Fig. 1 is of a sample of RNA polymerase containing approximately 80 mg of enzyme (based upon the specific activity of the purified product) at an initial purity of 10%. After application of the sample, the column is washed with DC buffer + 0.15 M KCl until the OD at 280 nm is less than 0.05. The RNA polymerase is then eluted by increasing the ionic strength (DC buffer + 0.65 M KCl). Greater than 90% of the applied activity is recovered and the eluted enzyme is judged greater than 95% pure by polyacrylamide gel electrophoresis; it is free of nuclease activity and contains the sigma factor (*7*) necessary for the initiation of transcription (*8*). It should be noted that chromatography of RNA polymerase on columns of phosphocellulose under similar conditions results in enzyme preparations of essentially the same purity but lacking the sigma factor.

While the procedure given here has been used in our laboratory exclusively for the isolation of RNA polymerase, it should be applicable to the purification of any protein that binds tightly to DNA. An example is the T4 gene 32 product which has been isolated in this way by Alberts et al. (*5*). We suggest that DNA columns should be especially useful for the purification of repressor proteins, since these are specifically adsorbed to their operator sites. This specificity could be utilized by constructing a double-layer DNA column, similar to those used in the isolation of gene specific RNA (*9*), containing in the upper layer DNA missing the operator region, and in the lower layer DNA possessing the operator. This procedure should yield in the upper layer all DNA binding proteins while the lower layer should contain only the repressor.

References

1. Richardson, C. C. 1966. In G. L. Cantoni and D. R. Davies (Eds.), Procedures in nucleic acid research. Harper & Row, New York, Vol. I, p. 263.
2. Aposhian, H. V. 1966. In G. L. Cantoni and D. R. Davies (Eds.), Procedures in nucleic acid research. Harper & Row, New York, Vol. I, p. 277.
3. Burgess, R. R. 1969. *J. Biol. Chem.*, **244**: 6160.

4. Litman, R. M. 1968. *J. Biol. Chem.,* **243**: 6222.
5. Alberts, B. M., F. J. Amodio, M. Jenkins, E. D. Gutmann, and F. L. Ferris. 1968. *Cold Spring Harbor Symp. Quant. Biol.*, **33**: 289.
6. Yamamoto, K., B. Alberts, R. Benzinger, L. Lawhorne, and G. Treihen. 1970. *Virology*, **40**: 734.
7. Burgess, R. R., A. A. Travers, J. J. Dunn, and E. K. F. Bautz. 1969. *Nature,* **221**: 43.
8. Dunn, J. J., and E. K. F. Bautz. 1969. *Biochem. Biophys. Res. Commun.,* **36**: 925.
9. Bautz, E.K. F., and E. Reilly. 1966. *Science,* **151**: 328.

SECTION

G

METHODS FOR RNA SEQUENCE DETERMINATION

FRACTIONATION AND SEQUENCE ANALYSIS OF RADIOACTIVE NUCLEOTIDES

B. G. BARRELL
MEDICAL RESEARCH COUNCIL
LABORATORY OF MOLECULAR BIOLOGY
CAMBRIDGE, ENGLAND

INTRODUCTION

This chapter describes the separation of digests of radioactive RNA by a two-dimensional "fingerprinting" procedure (*1*) and the sequence analysis of the nucleotides. The first dimension employs ionophoresis on cellulose-acetate at pH 3.5, since sharper spots are obtained on this system than in paper ionophoresis. Ionophoresis on DEAE-cellulose paper at acid pH is used as the second dimension. This combines the advantages of ion exchange and ionophoresis. The positive charges on the paper cause a rapid electroendosmotic flow of buffer from the cathode to the anode, carrying the nucleotides through the paper, subjecting them to ion-exchange chromatography; superimposed on this is the fractionation due to ionophoresis. The importance of the latter is shown by the fact that better separations are achieved by this technique than chromatography on DEAE-cellulose paper. The effect is that the more acidic nucleotides move faster by ionophoresis but slower by ion exchange so that essentially the two effects are opposing each other. Figure 1 shows a ribonuclease T_1 digest of *E. coli* 5S RNA fingerprinted by this procedure and a diagram showing the structure of the nucleotides.*

The separation of larger nucleotides, e.g., those obtained by limited

* Throughout this chapter, except where stated otherwise, the letters A, C, G, and U are used for the 3′ nucleotide residues. Thus G refers to guanosine 3′-phosphate and ACG to the trinucleotide ApCpGp.

digestion with nucleases or long ribonuclease T_1 products, and the determination of their sequence is also described. This technique referred to as "homochromatography" (*2*) is used for resolving those nucleotides that cannot be fractionated by the "standard fingerprinting" procedure. Here the first dimension is the same as before but the second dimension uses ascending chromatography on DEAE-cellulose thin layer plates with a mixture of nonradioactive nucleotides to displace and thus fractionate the radioactive ones. In effect the radioactive nucleotides are displaced by a series of anions of different valences or affinity for the DEAE groups. This system is produced when a concentrated mixture of nonradioactive nucleotides is used as a developing medium for the chromatogram, saturating the DEAE groups and displacing one another, thus generating a series of fronts. Therefore the smaller ones with lower valences or affinity are displaced by the larger ones and so move faster; the radioactive nucleotides move with the different fronts and are fractionated according to their affinity for the DEAE.

Figure 2 shows a complete ribonuclease T_1 digest of bacteriophage R17

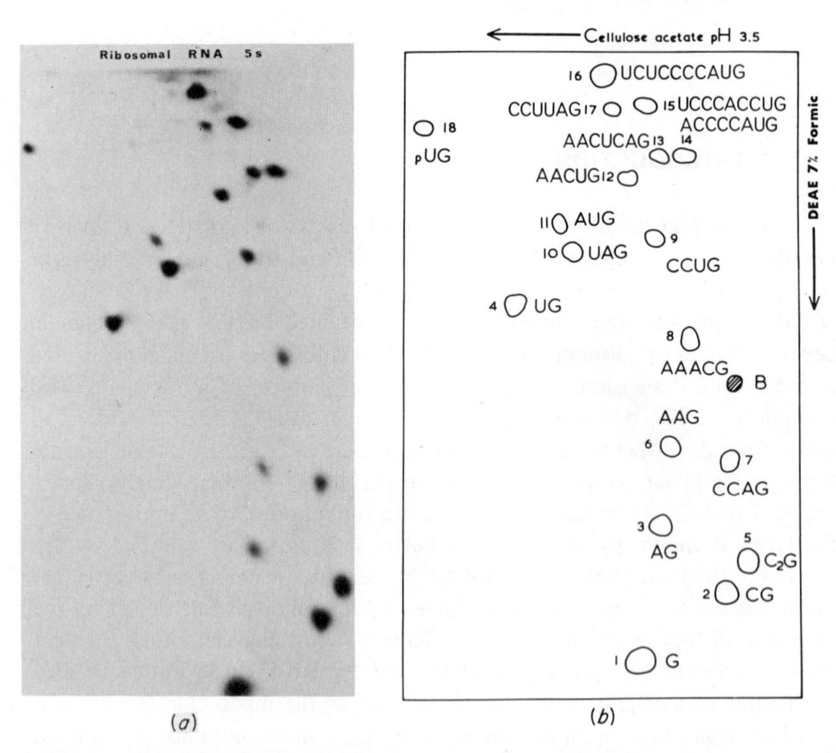

Figure 1. Ribonuclease T_1 digest of *E. coli* 5S RNA fractionated on the standard fingerprinting procedure with a diagram showing the structure of the nucleotides (*19*). B is the position of the blue marker on the second dimension.

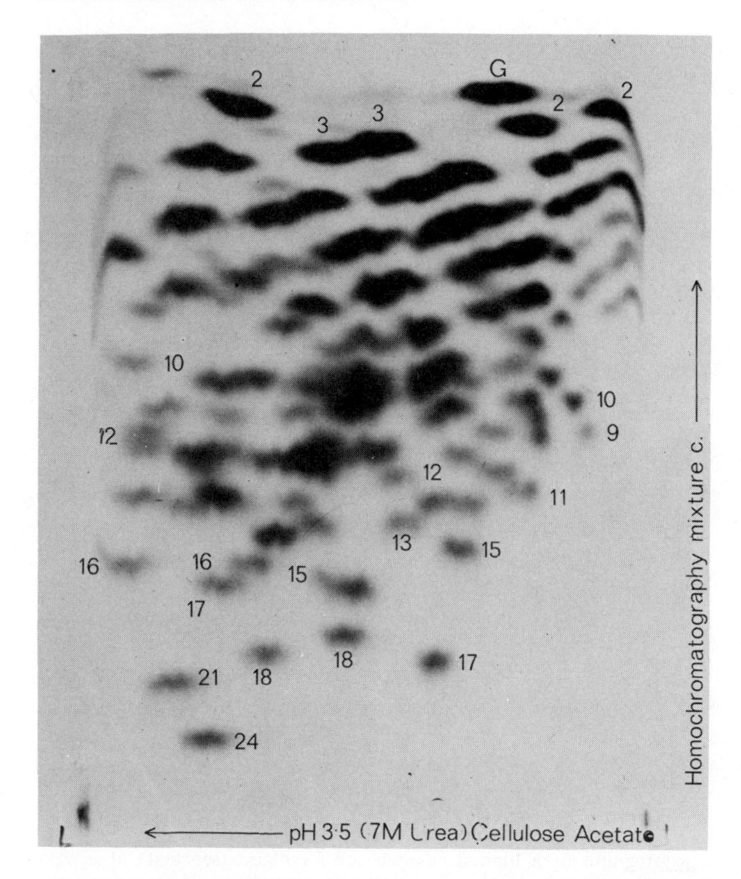

Figure 2. Radioautography of a complete ribonuclease T_1 digest of bacteriophage R17 RNA fingerprinted using the homochromatography procedure. The chain length of some of the nucleotides is given (P.G.N. Jeppesen, unpublished).

RNA fingerprinted by the homochromatography procedure. An example of a partial ribonuclease T_1 digest of *E. coli*^{Phe} tRNA fractionated by this procedure is shown in Fig. 3.

THE STANDARD FINGERPRINTING PROCEDURE

Enzymic Digestion

This procedure is normally used for fingerprinting ribonuclease T_1, ribonuclease A, and combined ribonuclease T_1 and bacterial alkaline phosphatase digests. An enzyme-to-substrate ratio of 1 : 20 (μg) in a buffer containing 0.01 M tris-chloride, pH 7.4, and neutralized 0.001 M EDTA, is

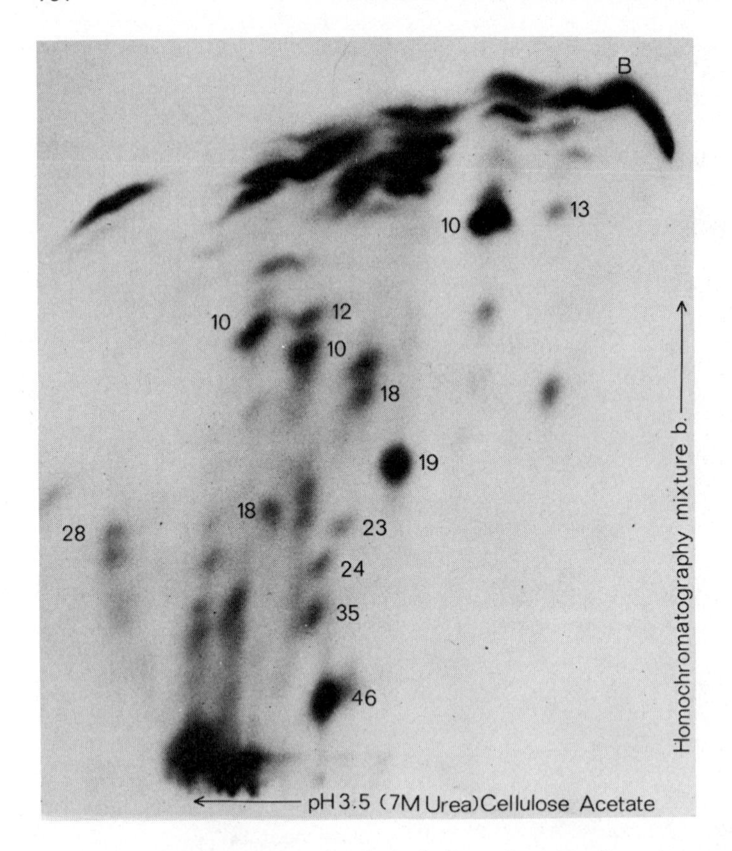

Figure 3. Radioautograph of a limited ribonuclease T_1 digest (see text) of *E. coli* $tRNA^{Phe}$ fractionated on the homochromatography system. The chain length of some of the partial products is given (B. G. Barrell unpublished). B is the position of the blue marker.

used for both ribonuclease T_1 and ribonuclease A digestion. Best results are obtained with under 100 μg of RNA in order not to overload the cellulose acetate; ideally 20 μg of RNA is used. Nonradioactive carrier RNA can be added when necessary to make the weight up to about 20 μg. Digestion of the ^{32}P-labeled RNA is ideally carried out in a volume of 3–5 μl in the tip of a drawn-out capillary tube (open ended melting point tubes 100 mm x 1.5–2.0 mm, MF388, Gallenkamp, London) which is placed tip down in a silicone treated test tube (½ x 2 in. is convenient) and incubated for 30 min at 37°C. For the combined ribonuclease T_1 and bacterial alkaline phosphatase digestion, the mixture is incubated for 60 min at 37°C using an enzyme-to-substrate ratio of 1 : 10 μg for the ribonuclease T_1 and 1 : 5 for the bacterial alkaline phosphatase, in 0.01 M tris-chloride, pH 8.0. The sample can then be directly loaded onto the cellulose-acetate.

Because of the very small amounts of RNA that we use, all test tubes are treated with "Repelcote" (Hopkin & Williams, Essex, England). This is poured into the tube and out again; the inverted tube is then allowed to dry on paper, and then washed very thoroughly with distilled water and dried in an oven.

Fractionation

The digest is fractionated on strips of cellulose acetate (available in sheets 95 x 25 cm from Oxoid, London, or in strips 55 x 3 cm from Schleicher & Schull, Dassel, Germany) using a freshly prepared pH 3.5 buffer containing 7 M urea (0.5% pyridine, 5% acetic acid, 7 M urea) by high-voltage ionophoresis in the apparatus described in Fig. 4. This kind of apparatus is

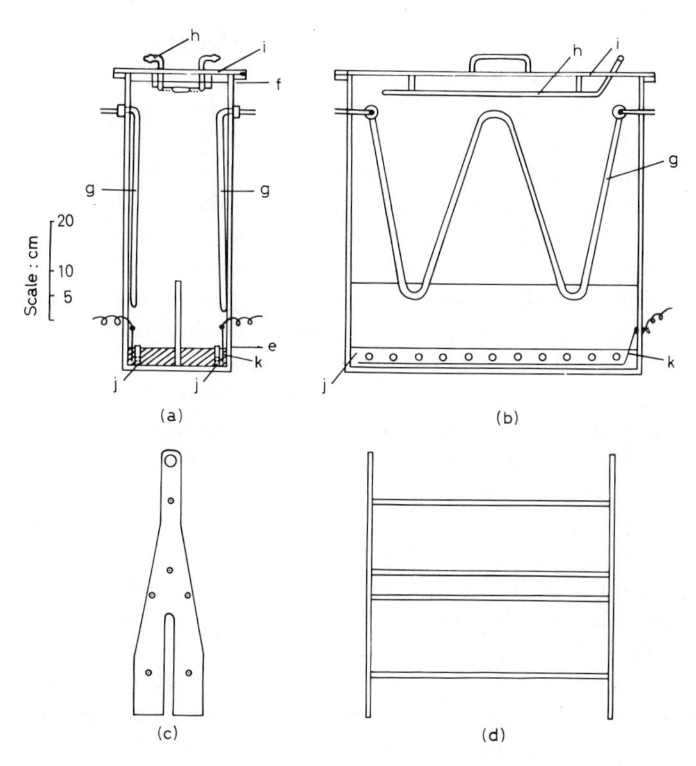

Figure 4. Ionophoresis tank and rack suitable for DEAE-paper and cellulose–acetate strips (*37*); (a) front of tank, (b) side of tank, (c) front of rack, (d) side of rack. The tank is made from 1-cm thick Perspex, (all joints are glued, screwed at 2-in. intervals) and filled with buffer to the level (e) and with white spirit 100, or Varsol, to the level (f). Tap water is passed through the stainless steel (g) and glass (h) cooling coils. The latter are attached to the lid (i) which is hinged to the tank. The small partitions (j) are drilled with holes (1.5 cm diameter). Electrodes (k) are of platinum coated wire. The rack is made of 1-cm thick Perspex (Lucite) and rods of 1-cm diameter. [From Ref. *37*. Reproduced with permission.]

also used for the DEAE-paper in the second dimension. Adequate fractionation can be obtained with 55 x 3-cm strips but, when necessary, strips 85 x 3 cm, cut from sheets, can be used in order to obtain higher resolution of the nucleotides. If strips are cut from the 95 x 25-cm sheets it is advisable not to use the first 3 cm of each of the long sides since edge effects are observed. The cellulose acetate is first wetted with the pH 3.5 buffer by floating it on the buffer contained in a petri dish to avoid inclusion of air bubbles. Excess buffer is blotted in the area of the point of application, which should be 10 cm from the cathode end of the strip. Two glass rods are placed under the strip at either side of the point of application. The digest is applied as a spot and is allowed to soak in. A dye mixture of Xylene cyanol F.F. (blue), acid fuchsin (red), and methyl orange (yellow) (all available from George T. Gurr, London) is applied to each side of the digest. Care must be taken to avoid drying out of the strip during sample application. Excess buffer on the strip is blotted and the strip is dipped in white spirit 100, or "Varsol," (Esso Petroleum Company) to avoid drying out while it is placed in the tank. The strip is placed on a rack (Fig. 4) and lowered into the tank. Ionophoresis is carried out at between 50 and 100 V/cm until the faster yellow marker reaches the anode buffer. The majority of the nucleotides will migrate slower than the pink markers but faster than the blue markers.

Fractionation in the second dimension is carried out on sheets of DEAE-paper (Whatman DE81). Normally sheets 85 x 46 cm cut from rolls are used, but sheets 57 x 46 cm are also satisfactory. Whereas the cellulose acetate is brittle when dry and easy to handle when wet, the DEAE-paper is very fragile when wet and difficult to handle. The material is transferred onto the DEAE-paper by the blotting procedure described below. Excess white spirit is allowed to drip off the strip and before the buffer dries the strip is placed on the DEAE-paper about 10 cm from the cathode end (i.e., from one of the short sides). A pad of three strips (46 x 3 cm) of Whatman 3MM paper that has been soaked in water is then put on top of the strip and a glass plate placed on top to press the strips together evenly. Water from the pad passes through the cellulose-acetate carrying the nucleotides with it and onto the DEAE-paper. Being acidic they are held on the DEAE-paper by ion exchange and remain in the position at which they are first washed on. To ensure complete transfer, more water can be added to the pad while it is still in position. The water is allowed to soak through until a 10-cm wide strip of the DEAE-paper is wetted. If this procedure is done carefully approximately 90% of the material is transferred.

For the second dimension 7% formic acid (v/v) is used, but for better separation of some of the smaller isomers a pH 1.9 buffer (2.5% formic acid, 8.7% acetic acid, v/v) can be used. After the transfer, this area of the DEAE-paper is soaked with ethanol in a trough to remove the urea. This is done three times, each time with fresh ethanol. The urea would otherwise cause streaking of some of the smaller nucleotides in the second dimension. Care must be taken because of the fragility of the wet DEAE-paper. The DEAE-paper is then dried and wetted with 7% formic acid in the following way. The paper is placed on a glass plate with glass rods under the paper at either side of the line where the nucleotides were applied. Wetting is started on either side of this line and the fronts of the solution allowed to meet along this line. After the rest of the cathode end of the paper is wetted (i.e., one third of the total length of the paper) the paper is put on a rack (Fig. 4). It is useful to prefold the paper in the center to facilitate this. The rest of the paper is wetted and the rack lowered into the ionophoresis tank (Fig. 4). The system has a high conductivity and there is considerable heating if high voltages are used. Normally ionophoresis is carried out overnight (16 hr) at 10 V/cm. The DEAE-paper is finally dried in air while still on the rack. Radioautographs are prepared by marking the paper with radioactive ink (^{35}S) and cutting the paper to a convenient size. The papers are then put in a folder with a sheet of 14 x 17 in. X-ray film on either side. Suitable X-ray films are Autoprocess 14 x 17 in. and Blue brand 14 x 17 in. and are processed in a Model M5AN Xomat, all available from Kodak Ltd. If hand processing is used Kodak Kodirex X-ray film is suitable. The folders are covered on one side with a thin sheet of lead (0.5 mm) with a sheet of aluminium for rigidity, so that they can be stacked together in a light-proof cabinet. When more than 1 μCi of ^{32}P-RNA is used the film can be developed after a day and there will be sufficient material to study the structure of the nucleotides. If only a fingerprint is required less material can be used, e.g., 0.01 μCi, and development will require 1–2 weeks.

The conditions for the DEAE-paper ionophoresis are rather acidic and some depurination of the nucleotides might be expected. Only occasionally after a long run at a high-voltage gradient the tank may become warm and streaking may be observed behind some of the spots, which could be due to depurination. It would seem that the effective pH on an ion-exchange material is not necessarily that of the buffer with which it is washed.

Position of Nucleotides on the Fingerprint

The position of a nucleotide on the fingerprint is determined largely by its composition. Figure 5 shows the relationship among the composition of nucleotides from a ribonuclease T_1 digest and their position on the standard fingerprinting procedure using 7% formic acid in the second dimension. Each point represents the "center of gravity" of a set of isomers. Lines are then drawn to link up those points representing nucleotides that differed only in the number of C residues they contained, e.g., among G, CG, CCG, CCCG and AG, (C,A)G, (C_2,A)G, (C_3,A)G, etc. Similarly, lines were drawn between

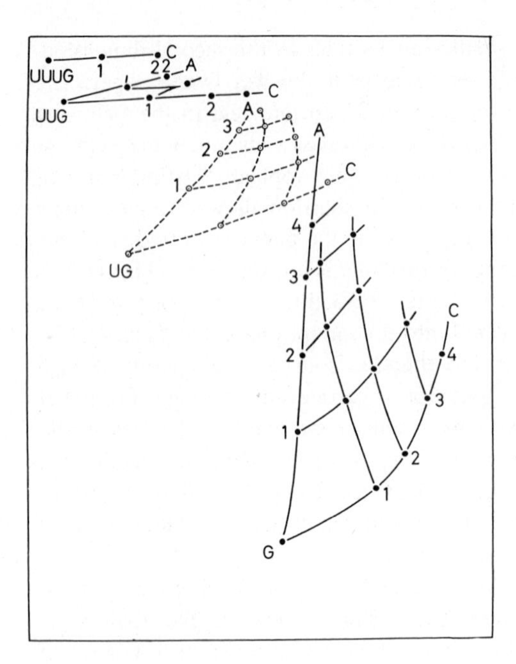

Figure 5. Diagram illustrating the relationship between the composition of nucleotides from a ribonuclease T_1 digest and their position on the standard fingerprinting system using ionophoresis with 7% formic acid in the DEAE-paper dimension. [From Ref. *37*. Reproduced with permission.]

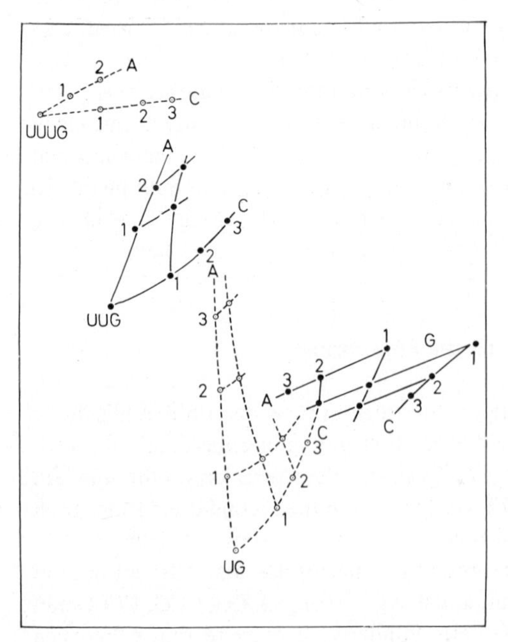

Figure 6. Same diagram as Fig. 5 but showing nucleotides from a ribonuclease T_1 and bacterial alkaline phosphatase digest. [From Ref. *37*. Reproduced with permission.]

points that differed only in A residues. The mononucleotide that has the greatest effect on mobility in the DEAE-paper dimension is U, so that the fingerprint may be regarded as being composed of three sections representing nucleotides with two, one, or no U residues, respectively. As we are considering a ribonuclease T_1 digest, the effect of G (which is approximately the same as U) is ignored because each nucleotide will have one G residue from the specificity of the enzyme. The lines joining the spots form three graticules corresponding to the different sections, and one axis on each graticule represents the number of A residues, whereas the other gives the number of C residues. In this way the probable composition of a nucleotide may be determined from its position on the map. Nucleotides containing two or more U residues move slowly on DEAE-paper ionophoresis and fractionation of these is relatively poor. Better resolution of this area can be achieved by combined ribonuclease T_1 and bacterial alkaline phosphatase digestion. The dephosphorylated nucleotides move faster than the corresponding phosphorylated ones (Fig. 6) on the DEAE-paper. Dephosphorylated nucleotides with two U residues, for example, run in the same region as the phosphorylated nucleotides with one U residue, but the dephosphorylated nucleotides with no U residue move more slowly on both dimensions than normal.

Figure 7 shows the position of nucleotides from a ribonuclease A digest. The simplicity in the composition of the nucleotides, and the smaller number of larger ones, makes the study of these digests much simpler than with ribonuclease T_1.

The fact that there is a separation of isomers indicates that the sequence affects the behavior on the two-dimensional system. Isomers which have an A residue as the 5′-terminus move more slowly in the second dimension than those having other 5′-terminal residues.

Elution

The position of the nucleotides is determined by lining up the radioactive ink marks on the film and the DEAE-paper on a light-box. The spots are cut out and eluted with 30% triethylamine carbonate, pH 10, prepared by passing CO_2 into a mixture of 70 ml water and 30 ml redistilled triethylamine until saturated and then adding more triethylamine to about pH 10. Elution is carried out according to Sanger and Tuppy (*3*) except that drawn-out, thin-walled capillary tubes are used (Fig. 8). Elution is very rapid and about 100 μl is collected. This is then transferred onto polythene sheets taped onto Perspex plates. These are placed in a rack in a desiccator, over concentrated sulfuric acid and sodium hydroxide pellets, and taken to dryness in a partial

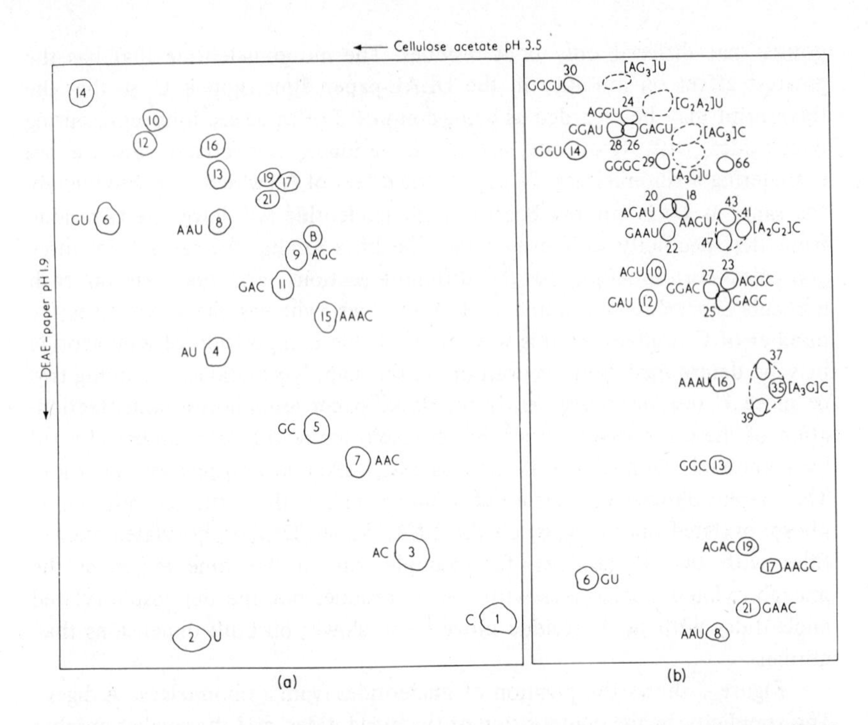

Figure 7. Diagram showing the position of nucleotides from a ribonuclease A digest fractionated on the standard fingerprinting system using pH 1.9 buffer in the DEAE-paper dimension. (a) A fractionation that has been run for a short time in the DEAE-paper dimension. (b) A fraction that has been run for a longer time. B marks the position of the blue marker. [From Ref. *37*. Reproduced with permission.]

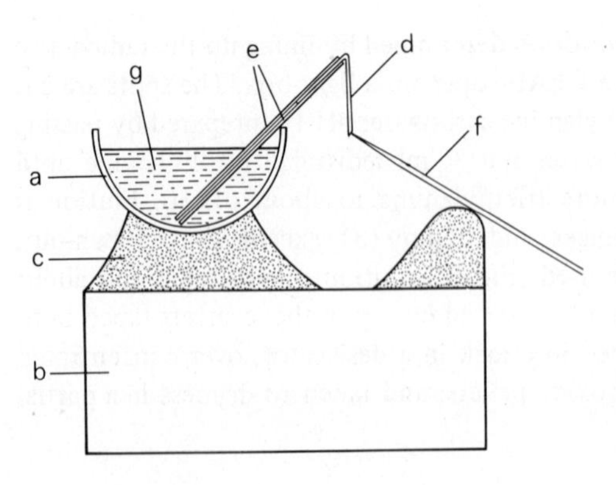

Figure 8. Cross section of the apparatus used for eluting nucleotides from paper. (a) Glass trough, (b) wooden base, (c) Plasticine, (d) paper to be eluted, (e) glass slide (3 x 1½ in), (f) drawn out thin walled capillary tubes, (g) eluting medium.

vacuum. If a good vacuum is put on the dessicator, the triethylamine bubbles badly. In order to ensure complete removal of the triethylamine carbonate, water is added to the spots and they are evaporated to dryness three times.

Nonradioactive nucleotides can be added to the elution medium to minimize losses. These are prepared by treating 1.0 gm yeast RNA with 25 ml 1.0 N KOH at 37°C for 20 min. The KOH is then neutralized with perchloric acid, the solution cooled to 0°C for 1 hr, and the precipitate removed by centrifugation. The supernatant is made up to 50 ml and thus contained 20 mg nucleotides per milliliter; 2 ml of this can be added to each 100 ml of elution medium.

Elution from ordinary paper is carried out in the same way but using water as the elution medium, except in the case of carbodiimide modified nucleotides which are eluted with 5% triethylamine carbonate pH 10.

THE THIN-LAYER HOMOCHROMATOGRAPHY FINGERPRINTING PROCEDURE

Preparation of Thin Layers

The Desaga basic thin-layer equipment and the absorbants were purchased from Camlab (Glass), Cambridge, England. A slurry of DEAE-cellulose (MN 300 DEAE) and cellulose (MN 300 cellulose) is prepared sufficient for two long glass plates (20 x 40 cm) or four short plates (20 x 20 cm). Normally 1 : 7.5 (DEAE-cellulose : cellulose) plates are prepared although 1 : 10 plates can be used for the separation of very large fragments that do not fractionate on the 1 : 7.5 plates. Plates (1 : 7.5) are prepared by mixing 1 gm DEAE-cellulose and 7.5 gm cellulose in 50 ml water until the mixture is fairly evenly dispersed and wetted. The slurry is homogenized for 5 min in a fast electric blender to break up aggregated particles. It is then heated for 2 min at 100°C and deaerated. The mixture is then added to the spreader (giving a fixed 250-μ layer), which is passed fairly rapidly across the plates on the template. For 1 : 10 plates the slurry is made from 0.75 gm DEAE-cellulose and 7.5 gm cellulose in 50 ml water. The plates are separated slightly and allowed to dry at room temperature. Finally a razor blade is used to define a sharp edge to the layers.

Enzymic Digestion and Fractionation

Complete ribonuclease T_1 digests are prepared as above. Partial digestion conditions obviously vary according to the amount of cleavage required and to the size and secondary structure of the molecule being analyzed. The example given below has been used by the author to obtain all the fragments necessary to order the complete ribonuclease T_1 and ribonuclease A products in deducing the structures of *E. coli* $tRNA^{Phe}$ and $tRNA_1^{Val}$. Suitable conditions are 1 : 500 (enzyme-to-substrate) for 10 min at 0°C in an ice bucket. The sample (of approximately 20 μl) in 2 μl of 0.02 M tris-chloride pH 7.5, 0.02 M $MgCl_2$, is incubated for 15 min at 37°C. This is then cooled to 0°C and 1 μl of cooled ribonuclease T_1 (0.05 mg/ml in the same buffer) is added and incubated for 10 min at 0°C. The sample is then rapidly applied to a 55 x 3 cm strip of cellulose acetate and ionophoresed for 45 min at 6 kV. The strip is removed from the tank, taking care not to let any buffer from the ends run along the strip, and monitored with a portable Geiger counter to find the position of the nucleotides. The strip is monitored from the anode end and the area of radioactivity, starting from this end and back toward the cathode for 20 cm, is taken. The fastest fragments ionophorese in the position of the pink marker. This part of the strip, which is not allowed to dry out, is then placed on a 40 x 20-cm thin-layer plate, 3 cm from one of the short ends, perpendicular to the spreading of the plate. Three moist, but not too wet, strips of Whatman 3MM paper are carefully placed on top followed by a glass plate to maintain efficient contact. This is then left for 15 min to allow the water to flow from the pad carrying the nucleotides on to the plate. Excess water or pressure applied to the glass plate tends to dislodge the layer. Transfer is not as good as onto DEAE-paper but at least 75% of the material is transferred.

Fractionation in the second dimension is carried out at 50–60°C using one of the *homomixtures* described below. For the digest in this example *homomixture b* is used. The thin-layer tank (formed by inverting one tank on top of another) is equilibrated with approximately 100 ml of the homomixture at 60°C in an oven; the plate is also brought to 60°C but not in the tank. Before starting the ascending chromatography the end of the plate in the region of the transferred nucleotides is sprayed with water to approximately 3 cm above the origin. This is to dissolve the urea so that during chromatography the urea moves ahead of the nucleotide front and does not interfere with the fractionation of the fast moving nucleotides. Alternatively, the plate can be washed very thoroughly with ethanol in a trough to remove the urea before equilibration. The plate is then transferred to the equilibrated tank, making sure an efficient seal is maintained with grease between the two tanks, and allowed to chromatograph until the front reaches the top of the plate – usually in 5–8 hr (2 hr for 20 x 20-cm plates).

The plate is then dried, marked with ^{35}S ink, and a radioautograph prepared by taping the plate in a lead-lined folder and exposing two X-ray films on top of each other, one for an elution template and the other for a record.

Preparation of Homomixtures

Homomixture a is prepared by dissolving 20 gm yeast RNA (British Drug Houses) in 200 ml water. It is then neutralized with 10 N KOH; 168 gm urea is added and when dissolved the volume is made up to 400 ml with water. The solution is finally filtered through filter paper.

Homomixture b is prepared as homomixture a but dialyzed against 2 liters of distilled water for 2 hr before adding the urea.

Homomixture c is a dialyzed and hydrolyzed homomixture prepared by dissolving 20 gm of yeast RNA in 200 ml of 1 N KOH and hydrolyzing for 15 min at room temperature. The solution is then neutralized to pH 7.5 with concentrated HCl and dialyzed against distilled water for 2 hr. Then 168 gm urea is added and the volume made up to 400 ml with distilled water to give a 5% mixture in 7 M urea. Homomixture c with only a 10-min hydrolysis may be diluted with 7 M urea to make a 3% mixture. This gives somewhat sharper spots but the R_f values are higher.

Homomixtures a and b give rather similar fractionations although mixture b gives better resolution. It may be that mixture a contains some salt (anions) which is an advantage for resolving very large fragments but gives very streaky separations. Thus its use has been confined to the purification of fragments in the region of 50 residues. Homomixture b is used for the separation of partial digests such as that of tRNA, giving fractionation in the range of 15–50 residues, depending on the complexity of the digest. Homomixture c can be used as an alternative procedure for fingerprinting digests of small molecules, separating the larger nucleotides (up to about 20 residues) better than the standard fingerprinting procedure but with less resolution of the smaller digestion products. The modified homomixture c is useful for fractionation of nucleotides in the range of 15 to 25 residues.

Elution

The spots that are to be eluted are marked on the radioautograph and a template is prepared by cutting out these spots on the radioautograph. This is then aligned on the thin-layer plate and the DEAE-cellulose containing the sample can be scraped off and collected on the glass sinter of the elution device shown in Fig. 9, which is attached to a vacuum line. The cellulose is sucked up through the orifice (b) and trapped on the sinter (c). The porosity of the sinter should be number 1 or 2. Urea is then removed by sucking 10–20 ml of 95% ethanol through the device while it is still connected to the vacuum line. The top section of the device is then removed and inverted and fixed over a small siliconed test tube. Approximately 200 μl of 30%

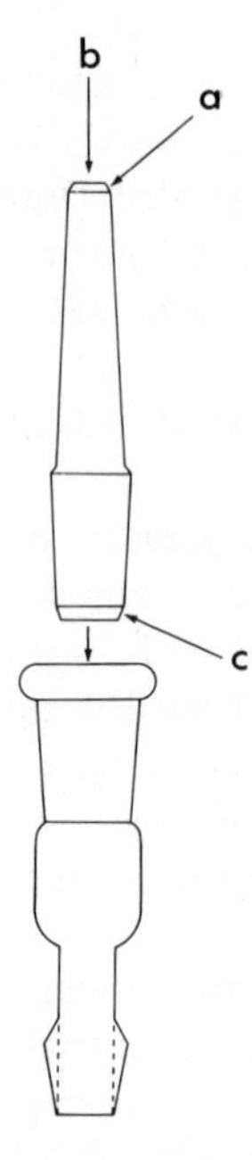

Figure 9. Diagram of a thin-layer elution device. This has a B10/19 ground glass joint which can be connected to a vacuum line. The upper part has an orifice (b) of approx. 2-mm internal diameter and is ground on its circumference to give a sharp edge (a) suitable for scraping and dislodging the layer. A sintered disc of porosity 1 or 2 is welded at (c). Device available from T. W. Wingent Ltd. (Milton, Cambridge, England). [From Ref. *2*. Reproduced with permission.]

triethylamine carbonate pH 10 is then added through the orifice (b) and allowed to drip through. This can be repeated and the eluate is then freeze dried in the test tube using a desiccator on an oil pump. The triethylamine carbonate is removed by three successive washes with water. Normally the sample is divided on the third wash and transferred to polythene sheets for analysis.

Fractionation by homochromatography is approximately dependant on chain length, but is also to a certain extent dependant on the purine content; i.e., with two ribonuclease T_1 products of the same chain length, the one containing more A residues will tend to move more slowly on this system. Because of the weight of RNA from the homomixture, stronger conditions have to be used for analysis of the nucleotides. Normally we assume an average spot from a homochromatograph to contain approximately 200 μg. In the section on sequence analysis these conditions are given after those for spots isolated on the standard fingerprinting procedure.

Sequence Analysis of the Nucleotides

Digestion of the nucleotides is carried out normally in drawn-out capillary tubes. A mouthpiece of polythene tubing is attached to the end of the capillary tube and the solution is blown onto the sample, stirring until it is dissolved. The solution, in the tip of the capillary tube, is then drawn into the middle of the tube by running one's thumbnail along the polythene tubing away from the capillary tube. The tip is sealed and the sample incubated. For long digestions, of 3 hr or more, both ends are sealed.

Composition

Alkaline Hydrolysis

The sample is incubated overnight (16 hr) with 10 μl of 0.2 N NaOH at 37°C. NaOH (0.5 N) is used for nucleotides from homochromatography fingerprints.

Ribonuclease T_2 Digestion

The sample is incubated for 2 hr at 37°C with 10 μl of 2 units/ml of ribonuclease T_2 (gift of Professor F. Egami) in a solution containing 0.05 mg/ml ribonuclease T_1, 0.05 mg/ml ribonuclease A, 0.05 M ammonium acetate, pH 4.5.

The mononucleotides are fractionated on either Whatman 540 paper or 3MM (57 x 46 cm), at pH 3.5 (no urea) in a hanging tank, shown in Fig. 10. Alkaline hydrolyzates of nucleotides from homochromatography fingerprints are fractionated on Whatman 3MM paper. Samples are applied as a streak 1-2 cm long with 1-cm gaps between them 10 cm from the cathode end. Ionophoresis is carried out at 60 V/cm. The four mononucleotides can be adequately resolved after 30 min although longer runs (up to 2 hr) can be used for separation of minor nucleotides, etc. With simple nucleotides it is usually possible to deduce how many residues of each mononucleotide were present by estimating the intensity of the bands on the radioautograph visually; however, this is usually checked by direct estimation on a scintillation counter. Areas of equal size for the various bands are cut out from the paper using the radioautograph as a guide. If the spots contain less than about 100 cpm the results are not very reliable, but above this about 10% accuracy can be achieved. Figure 11 shows the separation of mononucleotides on this system. Some minor nucleotides separate from the mononucleotides but they are normally further investigated by chromatography on Whatman No. 1 paper using the following systems. System a: propan-2-ol (680 ml), con HCl (176 ml), water to 1 liter (*4*). System b: propan-2-ol (70 ml), water (30 ml), con NH_3 (1 ml) (*5*). Table 1 shows the mobilities of some minor nucleotides on these three systems.

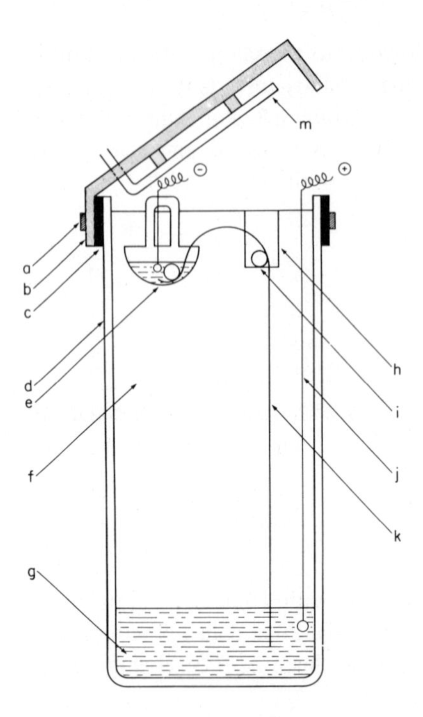

Figure 10. Side view of the "hanging tank" for high-voltage ionophoresis. The molded glass tank (d) (Panglass Chromatank Model 500 from Shandon Scientific Co., London) is attached to a piece of Bakelite (b), hinged to the Bakelite lid, over a strip of rubber (c) by means of a metal strip (a). The cooling coil (m) is tied with twine to the lid. The upper electrode compartment (e) is held at each end by two glass rods from a lip on the side of the tank. The ionophoresis paper (k) is held in the upper electrode compartment by means of a glass rod and hangs over another glass rod (i) (held in position by a piece of Bakelite (h) at each end from the other lip on the tank) into the lower electrode compartment (g). The tank is completely filled with white spirit 100 or Varsol (Esso Petroleum Co.). The electrodes enter at the back of the tank, one dipping into the upper trough and the other (j) passing down the back of the tank to the lower compartment. All parts, other than the tank itself, are available from T. W. Wingent Ltd. (Milton, Cambridge, England).

Digestion of Nucleotides with Enzymes

Generally the most useful enzymes for the preliminary analysis of ribonuclease T_1 products are ribonuclease A (pyrimidine specific), ribonuclease U_2 (purine specific), and ribonuclease A after modification of U (and G) residues with a water soluble carbodiimide reagent, which cleaves C residues. Partial digestion with exonucleases, i.e., snake venom phosphodiesterase and spleen phosphodiesterase, is used when the sequence is not obtained by the first three procedures. The sequence of ribonuclease A

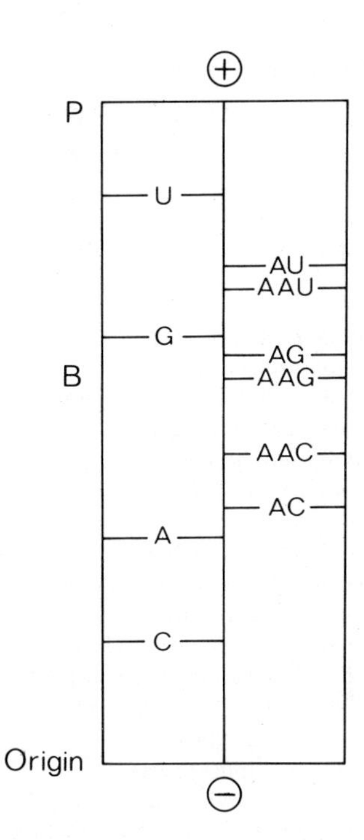

Figure 11. Diagram showing the separation of mononucleotides and some of the simpler products of ribonuclease A digestion by ionophoresis at pH 3.5 on Whatman 540 paper or 3 MM. B and P show the position of the blue and pink markers, respectively.

products can normally be obtained by the use of alkaline hydrolysis, ribonuclease T_1 digestion, and partial digestion with spleen phosphodiesterase.

Ribonuclease A

The sample is incubated with 10 μl of 0.1 mg/ml ribonuclease A (Worthington) and 1 mg/ml carrier RNA in 0.01 M tris chloride, 0.001 M EDTA, pH 7.5, for 30 min at 37°C. The nucleotides are fractionated on DEAE-paper at pH 3.5 in the apparatus shown in Fig. 4. Figure 12 shows the separation of the products of ribonuclease A digestion of nucleotides from a ribonuclease T_1 digest. These can be fractionated on Whatman 540 paper at pH 3.5 in a "hanging" tank but they are less well resolved (Fig. 11) and the system is only suitable for the more simple ribonuclease A products. If this system is used, the samples have to be treated with 0.1 N HCl to break down any cyclic phosphates which would complicate the fractionation as follows: 2 μl of 0.5 N HCl is put on a polythene sheet, the top end of the capillary is broken, and the digest blown out and mixed with the HCl. The solution is sucked back into the capillary, which is resealed, and incubated at 37°C for 1 hr.

Nucleotides from homochromatography fingerprints are digested with 0.2 mg/ml enzyme in the same buffer, but with the carrier RNA omitted, for 1 hr at 37°C.

TABLE 1. Ionophoretic and Chromatographic Properties of Some Minor Nucleotides Found in Radioactive Studies on tRNA and rRNA of *E. coli*

Compound	Symbol	Ionophoretic mobility (pH 3.5) relative to Up[a]	Chromatographic mobility[b]	
			System a	System b
β-Ureidopropionic acid	Y	1.07	0.93, 1.1	0.63
4-Thiouridylic acid	s^4U	~1.0		
2′-*O*-Methyluridine 3′-phosphate	U^m	1.0	1.12	1.3
5,6-Dihydrouridylic acid	D	1.0	0.9	
Ribothymidylic acid	T	0.98	1.1	1.7
Pseudo-uridylic acid	Ψ	0.98	0.79	0.63
Inosinic acid	I	0.90	0.54	0.68
2′-*O*-Methylguanylylguanylic acid	G^m–G	0.89	0.22	0.20
4-Amino-5(*N*-methyl)-formamido-isocytosine ribotide	"mG"	0.82	0.69	
2′-*O*-Methylcytidylyluridylic acid	C^m–U	0.79	0.67	~0.6
2′-*O*-Methylguanosine 3′-phosphate	G^m	0.78	0.82	0.90
N^2,N^2-Dimethylguanylic acid	m_2^2G	0.76	0.64	
N^1-Methylguanylic acid	m^1G	0.74	0.70	
Guanosine 3′-phosphate	G	0.74	0.50	0.66
N^2-Methylguanylic acid	m^2G	0.68	0.73	
N^6-Dimethyladenylyl-(N^6-dimethyl)	m_2^6A–m_2^6A	0.61		
2-Methylthio-6-isopentenyl adenylic acid[c]	ms^2i^6A	just ahead of A	1.2	near front
Isopentenyl adenylic acid	i^6A	just ahead of A	1.2	near front
N^6-Methyladenylic acid	m^6A	0.41	0.80	1.6
Adenylic acid	A	0.41	0.63	0.95
N^6-Dimethyladenylic acid	m_2^6A	0.40	0.9	2.1
N^4-Methyl(2′-*O*-methyl)cytidylyl-cytidylic acid	m^4C^mC	0.40, 0.38	0.63	1.15
2′-*O*-Methylcytidylylcytidylic acid	C^m–C	0.35	0.58	
Cytidylic acid	C	0.21	0.74	0.98
2′-*O*-Methylcytosine-3′-phosphate	C^m	0.20	0.94	1.15
5′-Methylcytidylic acid	m^5C	0.19	0.78	
N^7-Methylguanylic acid	m^7G	0.05	0.75	0.85, 0.69

[a] Ionophoretic mobilities are expressed relative to uridine 2′,3′-phosphate = 1.00 on Whatman 54 paper.

[b] Chromatographic mobility is for descending chromatography on Whatman 1 paper. System a propan-2-ol (680 ml), HCl (176 ml), and water to 1 liter (*4*). System b is propan-2-ol (70 ml), wa[ter] (30 ml), and ammonia (1 ml) (*5*).

[c] This compound is best distinguished from i^6A by ionophoresis on DEAE-paper at pH 3.5 (0.5 pyridine, 5% acetic acid) when it has an R_u of approximately 0.5 and i^6A has an R_u of approximate 0.9.

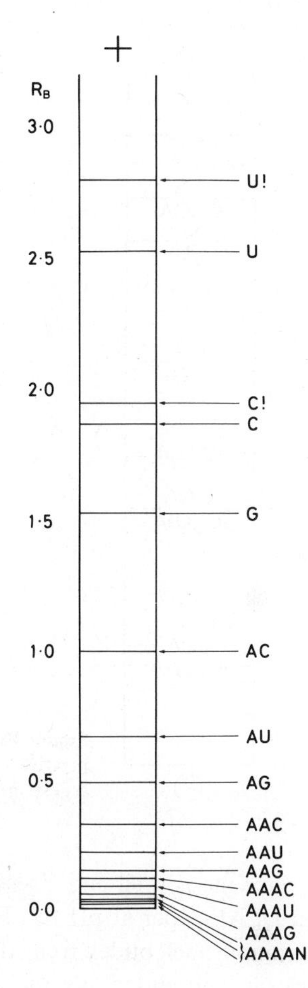

Figure 12. Fractionation of ribonuclease A digestion products on DEAE-paper at pH 3.5, relative to the position of the blue marker. N refers to any of the nucleotides C, U, or G. Cyclic nucleotides are indicated by (!). [From Ref. *29*. Reproduced with permission.]

Ribonuclease T_1

Nucleotides from ribonuclease A digests are digested with ribonuclease T_1 (Sankyo Co.) and fractionated in the same manner as for ribonuclease A above.

Ribonuclease U_2

Ribonuclease U_2 (preproduction batch from Sankyo Co.) is purine specific at low concentrations (*6*) and therefore with products from ribonuclease T_1 fingerprints, where G is 3′-terminal, splits specifically adjacent to A residues. Digestion is with 10 μl of 0.1 units/ml in 0.05 M sodium acetate, 0.002 M EDTA, pH 4.5, containing 0.1 mg/ml bovine serum albumin for 2 hr at 37°C. For nucleotides from homochromatography fingerprints, 1 unit/ml

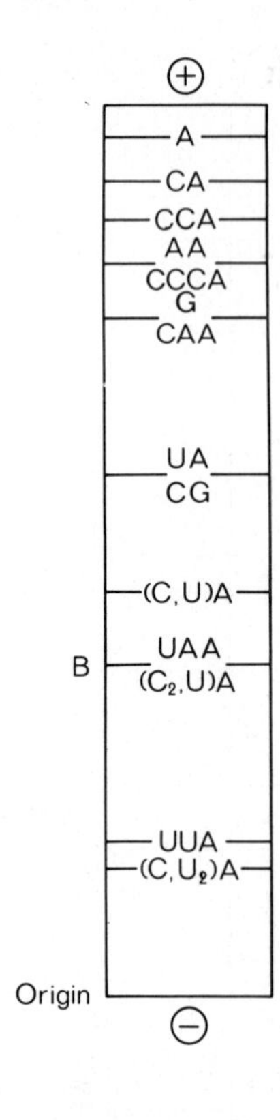

Figure 13. Fractionation of some of the simpler digestion products of ribonuclease U_2 of nucleotides from ribonuclease T_1 digests on DEAE-paper at pH 1.9.

of enzyme is used and digestion is for 4 hr at 37°C. Fractionation takes place on DEAE-paper at pH 1.9. Figure 13 shows the position of some of the more common nucleotides from ribonuclease U_2 digestion. Normally the spots are cut out and eluted and their composition determined by alkaline hydrolysis. This enzyme cleaves more slowly between two A residues and the conditions above are designed to preserve runs of A residues. For example, the nucleotide UACCAAACG will give the products UA, CCAAA, and CG after ribonuclease U_2 digestion under these conditions. For complete cleavage of all A residues incubation times can be extended.

Cleavage of Nucleotides Next to C Residues by Carbodiimide Blocking

N-Cyclohexyl-*N'*-(β-morpholinyl-(4)-ethyl) carbodiimide-methyl-*p*-toluene sulfonate (Fluka AG) reacts with uridine and to a lesser extent with guanosine (7). Ribonuclease A will then only cleave next to unmodified pyrimidines. This method enables specific cleavage of ribonuclease T_1 products next to C residues. The nucleotide is incubated overnight with 10 μl of the reagent at 100 mg/ml in 0.01 M tris-chloride, 0.001 M EDTA, pH 8.9, at 37°C. Then 10 μl of ribonuclease A (0.1 mg/ml) in 0.05 M tris-chloride, 0.001 M EDTA, pH 7.4, is then added and the solution incubated for 1 hr at 37°C. For spots from homochromatography fingerprints the modified nucleotides are digested with 0.2 mg/ml ribonuclease A for 1 hr at 37°C. Fractionation is on Whatman 3MM paper, placing the origin in the middle of the paper at pH 3.5, since the modified nucleotides are basic. These can be analyzed by alkaline hydrolysis or by unblocking with ammonia (0.2 N ammonia overnight at room temperature) and with further digestion using ribonuclease A as described above. Figure 14 shows the separation of some of the more simple nucleotides produced in this way. Results must be interpreted with caution, since in some cases not all of the U residues may be modified and/or not all of the C residues may be cleaved by ribonuclease.

Complete Digestion with Snake Venom Phosphodiesterase

Phosphorylated nucleotides are digested with 10 μl of 0.2 mg/ml snake venom phosphodiesterase (Worthington) in 0.01 M tris-chloride, 0.01 M $MgCl_2$, pH 8.9, for 2 hr at 37°C. The products are fractionated at 60 V/cm for 1½ hr on Whatman 540 paper at pH 3.5. The 5′-end is liberated as a

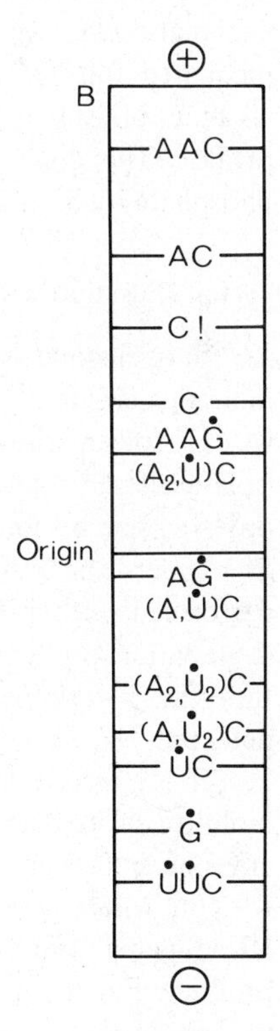

Figure 14. Fractionation of some products of ribonuclease A digestion of CMCT blocked nucleotides on Whatman 3MM paper at pH 3.5. (●) Indicates the modified mononucleotides. (!) Indicates cyclic nucleotides.

nucleoside and is thus not seen, and can be identified by a comparison with an alkaline digest. The 3′-end is liberated as a mononucleoside 3′,5′-diphosphate. The diphosphates have mobilities on Whatman 540 paper at pH 3.5 relative to uridine-5′-phosphate of pCp, 1.05; pAp, 1.1; pGp, 1.3; pGp!, 1.4; pUp, 1.8. If they are run on Whatman 3MM, EDTA, (0.002 M) in the pH 3.5 buffer prevents streaking of di-, tri- and tetraphosphates. However this reaction may not go to completion, since this digestion is normally carried out on dephosphorylated nucleotides.

Dephosphorylated nucleotides (i.e., those isolated by combined ribonuclease T_1 and bacterial alkaline phosphatase digestion, and nucleotides dephosphorylated as below) are digested with 10 μl of 0.1 mg/ml of enzyme in 0.05 M tris-chloride, 0.01 M $MgCl_2$, pH 8.9, and 1 mg/ml carrier RNA. The sample is incubated for 1 hr at 37°C and fractionated as described above. Mononucleotides with a 5′-phosphate migrate marginally more slowly than those with a 3′-phosphate at pH 3.5.

Dephosphorylation with Bacterial Alkaline Phosphatase

The sample is first heated with 10 μl of 0.1 N hydrochloric acid for 1 hr at 37°C to decyclicize any 2′,3′-cyclic phosphate which may be present after enzymatic cleavage. The acid is removed by lyophilization and the sample incubated for 1 hr at 37°C with 10 μl of a 0.1 mg/ml solution of bacterial alkaline phosphate (Worthington) in 0.05 M tris-chloride, 0.01 M $MgCl_2$, pH 8.9. The dephosphorylated product is freed of enzyme and inorganic phosphate by brief DEAE-paper ionophoresis at pH 1.9.

Partial Digestion with Snake Venom Phosphodiesterase

This enzyme is an exonuclease requiring a 3′-hydroxyl group removing mononucleoside 5′-phosphates sequentially from the 3′-end. The approach, as with a spleen phosphodiesterase, is to partially digest the nucleotide and isolate all the intermediary products and line these up as far as possible in order of size by ionophoresis on DEAE-paper. Normally pH 3.5 is used, but for larger nucleotides or those with many U residues ionophoresis is performed with pH 1.9 buffer or 7% formic acid. In the analysis of the degradation products with alkaline hydrolysis it is important to remember that the 3′-terminal residue is liberated as the nucleoside and is thus not detected.

It is possible to deduce the structure of a nucleotide directly from the mobility of its degradation products. A value of M is defined and its value for any one split is characteristic of the nucleotide released. M is defined as equal to x/y, where y is the distance of any nucleotide from the origin of the DEAE-paper and x is the distance between it and its first degradation product. For the nucleotides we have studied the values of M on DEAE-paper ionophoresis at pH 1.9 lie within the limits shown in Table 2 with the restriction that the values of M where the product of the split is faster than the blue marker are not included, these having M values that did not necessarily fall within these ranges. However this method of analysis should only be used as a guide and the degradation products should be eluted and analyzed by alkaline hydrolysis.

Digestion conditions vary according to the nucleotide and the conditions below serve only as a guide.

1. Nucleotides from combined ribonuclease T_1 and bacterial alkaline phosphatase digests: Digestion is performed with 10 μl of 0.005 mg/ml enzyme in 0.05 M tris-chloride, pH 8.9, 0.01 M $MgCl_2$ for 0 min (control), 5 min, and 10 min at room temperature.

2. Phosphorylated nucleotides: This procedure is adapted from that of Min Jou and Fiers (*8*). The nucleotide is incubated with bacterial alkaline phosphatase, with an enzyme-to-substrate ratio of 1 : 10, in 0.05 M tris-chloride, pH 8.9, 0.01 M $MgCl_2$ for 1 hr at 37°C. For nucleotides from the standard fingerprinting procedure 10–20 μg of carrier nucleotides are added, and for those from homochromatography fingerprints the weight of carrier nucleotides is assumed to be in the region of 100 μg. Snake venom phosphodiesterase in the same buffer is then added to give an enzyme to substrate ratio of about 1 : 10. Incubation times at 37°C for 0, 5, 10, and 20 min have been found to be the most useful. No attempt is made to destroy phosphatase activity, since the partial products lacking a 3′-phosphate are resistant and it may be an advantage to degrade the released mononucleoside 5′-phosphates, some of which have similar mobilities to those of the smaller degradation products.

Partial Digestion with Spleen Phosphodiesterase

Spleen phosphodiesterase (Worthington) releases mononucleoside 3′-phosphates sequentially from the 5′-terminus. The approach, as with snake venom phosphodiesterase, is to perform a partial digestion and isolate the various degradation products by ionophoresis at pH 1.9 on DEAE-paper. It has been found that on this system any phosphorylated nucleotide will move faster than a corresponding nucleotide having the same structure but with one extra nucleotide added to its 5′-terminus. Consequently the degradation products of spleen phosphodiesterase will be arranged in order of their size on fractionation in this system. Therefore we can define a value of M as with snake venom phosphodiesterase. These are shown in Table 2. This method can be applied to nucleotides from both ribonuclease T_1 and ribonuclease A digests. However one limitation is the rate at which different residues are released by the enzyme, U residues being released more rapidly than A and A more rapidly than C. Thus ribonuclease T_1 products with one or more C residues at the 5′-terminus are digested very slowly under normal conditions. Another difficulty with this procedure, and also with snake venom

TABLE 2A. Values of *M* on DEAE-Paper Ionophoresis (pH 3.5) for Nucleotides from Combined Ribonuclease T_1 and Alkaline Phosphatase Digests[a]

3′-Terminal residue	Range of *M* values
C	0.6 to 1.2
A	2.1 to 2.9
U	1.7 to 1.9
G	2.6 to 4.4

[a] The values of *M* for splits in which the product is faster than the blue marker are not included.

TABLE 2 B. Values of *M* on DEAE-Paper Ionophoresis (pH 1.9) for Nucleotides from Ribonuclease T_1 Digests and Ribonuclease A Digests

5′-Terminal residue	Range of *M* values
C	0.05 to 0.3
A	0.4 to 1.1
U	1.5 to 2.5
G	1.2 to 3.1

phosphodiesterase, is that internal splits are generated often between –Py–A– bonds so that results must be interpreted with caution and degradation products must be eluted and subjected to alkaline hydrolysis. This enzyme is very useful for sequencing ribonuclease A products. Here we are only concerned with G and A residues and these are both split off at similar rates. The *M* values associated with an A 5′-terminal residue vary from 0.4 to 1.1 and for a G 5′-terminal residue from 1.2 to 3.1. There is a wide spread and the limits are rather close; it is therefore advisable to do base compositions on the products. The extreme values of *M* are given by the smaller nucleotides.

Conditions, as before, vary according to the nucleotide and those given below serve only as a guide.

1. Analysis of ribonuclease T_1 nucleotides: Digestion is performed with 10 μl of 0.2 mg/ml of enzyme in 0.1 M ammonium acetate, pH 5.7, 0.002 M EDTA, 0.05% Tween 80 for 0 min (control), 30, min, and 60 min at room temperature.

2. Analysis of ribonuclease A nucleotides: Analysis is performed as

described above except lower concentrations of enzymes are required and 0.1 mg/ml is suitable.

3. Analysis of nucleotides from homochromatography fingerprints: 10 μl of 1.0 mg/ml enzyme in 0.1 M ammonium acetate pH 5.7, 0.002 M EDTA, 0.05% Tween 80, are incubated at 37°C for 0, 15, 30, and 60 min.

Fractionation at pH 1.9 on DEAE-paper is conveniently carried out for about 4 hr with 30 V/cm.

Digestion with Spleen Acid Ribonuclease

This is a useful enzyme with no absolute specificity which we have used on large ribonuclease T_1 products. The enzyme (8a) was a gift from Dr G. Bernardi. This enzyme seems to have a preference to –C–U– and –A–U– bonds (unpublished observations of B. G. Barrell and P. G. N. Jeppesen) although it must be emphasized that these are the main cleavage points and with longer digestion times other sites are cleaved. Conditions for digestion are given below.

1. Nucleotides from standard fingerprints: Digestion is performed with 10 μl of 0.2 mg/ml enzyme in 0.1 M sodium acetate, pH 5.0, 0.01 M EDTA at 37°C. Digestion times vary for different nucleotides and are in the range of 1 to 4 hr. DEAE-paper at pH 1.9 is used for fractionation.

2. Nucleotides from homochromatography fingerprints: Conditions are the same as described above except that 2.0 mg/ml enzyme is used.

Analysis of Nucleotides Produced by Limited Digestion with Ribonuclease T_1 or Ribonuclease A

Usually the eluted spot is divided in two, half being analyzed by further digestion with ribonuclease T_1 and half with ribonuclease A. The digests may be fractionated on DEAE-paper with 7% formic acid although the ribonuclease A digest may be fractionated at pH 1.9 on DEAE-paper. It is useful to run complete ribonuclease T_1 and ribonuclease A digests of the whole RNA alongside as markers. The products may be tentatively identified from their position and confirmed by analysis with alkali or by ribonuclease A digestion of the products of ribonuclease T_1 digestion and vice versa, the latter products being run on DEAE-paper at pH 3.5.

Although most of the digestion products of the various enzymes used above are analyzed by alkaline hydrolysis, some of the larger ones may have to be analyzed further. Ribonuclease U_2 and spleen acid ribonuclease products may be analyzed, for example, in any of the following ways: (1) with the carbodiimide reagent; (2) with dephosphorylation followed by complete digestion using snake venom phosphodiesterase, alkaline hydrolysis, or ribonuclease A digestion; (3) with partial digestion using spleen or snake venom phosphodiesterase.

APPLICATIONS OF THE "FINGERPRINTING" METHODS

The aim of this chapter has been to summarize the methods that have been developed in this laboratory for the sequence determination of ^{32}P-labeled RNA. This is a powerful and rapid approach and has been used for the sequence determination of several small RNA molecules and is now being applied to larger molecules such as ribosomal and bacteriophage RNAs. A brief survey of the applications of this method is given below so that the reader may refer to the literature on his particular sequence problem.

1. Transfer RNA

The method is particularly applicable to tRNAs and all the nucleotides in a ribonuclease T_1 digest are normally well separated. The main disadvantage is the identification of minor nucleotides and, except for those given in Table 1, unknowns have to be prepared on a larger, nonradioactive scale for identification.

The following *E. coli* tRNAs have been sequenced to date by this procedure: $tRNA_{I}^{Tyr}$, $tRNA_{II}^{Tyr}$, $tRNA_{Su+III}^{Tyr}$ (*9*); $tRNA^{fMet}$ (*10,-12*); $tRNA^{Met}$ (*13, 14*); $tRNA_{I}^{Val}$ (*15*); and $tRNA^{Phe}$ (*16*). The method has been particularly useful for fingerprinting and characterising mutants of the suppressor tyrosine tRNA (*17, 18*).

2. 5S RNA

The sequence of 5S rRNA has been determined (*19–21*) and all the spots on a ribonuclease T_1 fingerprint have been separated. Originally large fragments produced by limited digestion with nucleases were separated by homochromatography on DEAE-paper but this has been superceded by the thin-layer method which gives much better resolution. Forget and Weissman (*22*) have used the fingerprinting technique to sequence the 5S RNA from human carcinoma (KB) cells. Ohe and Weissman (*23*) have obtained the nucleotide sequence of an RNA from cells infected with Adenovirus 2 (this RNA is 156 residues long). Brownlee (personal communication) has obtained the sequence of an *E. coli* 6S component which is 184 residues long.

E. coli 16S and 23S Ribosomal RNA

These RNAs give distinctive fingerprints and differences even at the tetranucleotide level can be seen (*1*). The ribonuclease T_1 nucleotides containing methylated bases were located on the fingerprint and sequenced (*24, 25*). Fellner et al. (*26, 27*) have determined the sequence of

125 nucleotides from ribonuclease T_1 and bacterial alkaline phosphatase digests of 16 S rRNA, and are now examining larger fragments of this molecule.

Bacteriophage RNA

The RNA of the closely related small bacteriophages is approximately 3300 residues long. Dahlberg (*28*) has developed diagonal fractionation procedures, using ionophoresis on DEAE-paper, for the isolation of the 5′- and 3′-terminal sequences which were applied to f2, R17, and Qβ bacteriophage RNAs. Recently Adams et al. (*29*) have used initial fractionation on polyacrylamide gels of limited ribonuclease T_1 digests to isolate a 57-residue fragment from the coat protein cistron of bacteriophage R17 RNA. This fragment was further purified and sequenced using the homochromatography fingerprinting method. Work is well advanced on a 80 to 90-residue fragment containing the 5′-terminal pppGp (J. M. Adams, personal communication) and two other fragments from the coat protein cistron (P. G. N. Jeppesen, personal communication). Steitz (*30*) working on R17 RNA and Hindley and Staples (*31*) on Qβ RNA have isolated fragments corresponding to the initiation regions of the cistrons by ribonuclease A digestion of R17 RNA-ribosome complexes. These were fingerprinted and sequenced using the two dimensional procedures. The nucleotide sequence of the region between the end of the coat protein cistron and the beginning of the synthetase cistron has also been established by Nichols (*32*) using these procedures. Jeppesen et al. (*Nature*, in press) have used homochromatography fingerprints of the 40% and 60% fragments of R17 RNA produced by ribonuclease IV (*33*) to establish the gene order of bacteriophage R17 RNA.

Another approach has been used by Billeter et al. (*34*) for bacteriophage Qβ RNA. They have obtained synchronized synthesis, with a purified polymerase, of a segment of RNA from the 5′-terminal end extending about 160 residues into the molecule. The synthesized material has been analyzed using the standard fingerprinting procedure.

^{32}P-Labeling of Digests of Nonradioactive RNA using Polynucleotide Phosphokinase*

The RNA is degraded so as to yield nucleotides with free 5′-hydroxyl groups. These nucleotides are then phosphorylated with 5′-hydroxyl polynucleotide kinase using ($\gamma - {}^{32}$P) ATP as phosphate donor. The enzyme

* See also p. 780.

transfers the γ-phosphate group from ATP to the 5′-hydroxyl of each nucleotide in the digest, which can then be fingerprinted in the normal way. This method has been used to fingerprint digests of tRNA (*35*) and hemoglobin messenger RNA (*36*). This approach should be useful for sequence analysis of RNAs that are difficult to obtain highly labeled with ^{32}P.

DNA

The standard fingerprinting procedure can be used to fractionate digests of ^{32}P-labeled DNA (K. Murray, unpublished data). Székely and Sanger (*35*) have studied depurinated fd DNA and established several pyrimidine sequences by the use of polynucleotide kinase.

Compared with RNA there has been little progress in this field mainly because of the lack of specific nucleases and the absence of any small DNA molceule that could be used for the development of techniques.

References

1. Sanger, F., G. G. Brownlee, and B. G. Barrell. 1965. *J. Mol. Biol.,* **13**: 373.
2. Brownlee, G. G., and F. Sanger. 1969. *European J. Biochem.,* **11**: 395.
3. Sanger, F., and H. Tuppy. 1951. *Biochem. J.,* **49**: 463.
4. Wyatt, G. R. 1951. *Biochem. J.,* **48**: 584.
5. Markham, R., and J. D. Smith. 1952. *Biochem. J.,* **52**: 552, 558.
6. Arima, T., T. Uchida, and F. Egami. 1968. *Biochem. J.,* **106**: 609.
7. Gilham, P. T. 1962. *J. Am. Chem. Soc.,* **84**: 687.
8. Min Jou, W., and W. Fiers. 1969. *J. Mol. Biol.,* **40**: 187.

8a. Bernardi, G. 1967. In G. L. Cantoni and D. R. Davies (Eds). Procedures in nucleic acid research, Harper & Row, New York, Vol 1, p. 37.

9. Goodman, H. M., J. Abelson, A. Landy, S. Brenner, and J. D. Smith. 1968. *Nature,* **217**: 1019.
10. Dube, S. K., K. A. Marcker, B. F. C. Clark, and S. Cory. 1968. *Nature,* **218**: 232; *Biochem.,* **8**: 244.
11. Dube, S. K., K. A. Marcker, B. F. C. Clark, and S. Cory. 1969. *European J. Biochem.,* **8**: 244.
12. Dube, S. K., and K. A. Marcker. 1969. *European J. Biochem.,* **8**: 256.
13. Cory, S., K. A. Marcker, S. K. Dube, and B. F. C. Clark. 1968. *Nature,* **220**: 1039.
14. Cory, S., and K. A. Marcker. 1970. *European J. Biochem.,* **12**: 177.
15. Yaniv, M., and B. G. Barrell. 1969. *Nature,* **222**: 278.
16. Barrell, B. G., and F. Sanger. 1969. *FEBS Letters,* **3**: 275.
17. Abelson, J. N., L. Barnett, S. Brenner, M. L. Gefter, A. Landy, R. L. Russell, and J. D. Smith. 1969. *FEBS Letters,* **3**: 1.
18. Abelson, J. N., M. L. Gefter, L. Barnett, A. Landy, R. L. Russell, and J. D. Smith. 1970. *J. Mol. Biol.,* **47**: 15.
19. Brownlee, G. G., and F. Sanger. 1967. *J. Mol. Biol.,* **23**: 337.
20. Brownlee, G. G., F. Sanger, and B. G. Barrell. 1967. *Nature,* **215**: 735.

21. Brownlee, G. G., F. Sanger, and B. G. Barrell. 1968. *J. Mol. Biol.*, **34**: 379.
22. Forget, B. G., and S. H. Weissman. 1967. *Science*, **158**: 1695.
23. Ohe, K., and S. M. Weissman. 1970. *Science*, **167**: 879.
24. Fellner, P., and F. Sanger. 1968. *Nature*, **219**: 236.
25. Fellner, P. 1969. *European J. Biochem.*, **11**: 12.
26. Fellner, P., C. Ehresmann, and J. P. Ebel. 1970. *Nature*, **225**: 26.
27. Fellner, P., and J. P. Ebel. 1970. *FEBS Letters*, **6**: 102.
28. Dahlberg, J. E. 1968. *Nature*, **220**: 548.
29. Adams, J. M., P. G. N. Jeppesen, F. Sanger, and B. G. Barrell. 1969. *Nature*, **223**: 1009.
30. Steitz, J. A. 1969. *Nature*, **224**: 957.
31. Hindley, J., and D. H. Staples. 1969. *Nature*, **224**: 964.
32. Nichols, J. L. 1970. *Nature*, **225**: 147.
33. Spahr, P. F., and R. F. Gesteland. 1968. *Proc. Natl. Acad. Sci, U.S.* **59**: 876.
34. Billeter, M. A., J. E. Dahlberg, H. M. Goodman, J. Hindley, and C. Weissmann. 1969. *Nature*, **224**: 1083.
35. Székely, M., and F. Sanger. 1969. *J. Mol. Biol.*, **43**: 607.
36. Labrie, F. 1969. *Nature*, **221**: 1217.
37. Sanger, F., and G. G. Brownlee. 1967. In L. Grossman and K. Moldave (Eds.), Methods in enzymology. Academic Press, New York, Vol. XII, p. 361.

FINGERPRINTING NONRADIOACTIVE NUCLEIC ACIDS WITH THE AID OF POLYNUCLEOTIDE KINASE (*1*)

M. SZÉKELY*
MEDICAL RESEARCH COUNCIL
LABORATORY OF MOLECULAR BIOLOGY
CAMBRIDGE, ENGLAND

INTRODUCTION

5′-Hydroxylpolynucleotide kinase catalyzes the transfer of the γ-phosphate group of ATP to free 5′-hydroxyl of both ribo- and deoxyribonucleotides (*2–4*). With ATP labeled in the γ-phosphate, labeled nucleotides are produced. This reaction has been used for labeling the 5′-ends of RNAs (*5, 6*) and DNAs (*7–10*) and for determining their terminal nucleotides.

A digest of DNA or RNA can also be phosphorylated by polynucleotide kinase, using γ-^{32}P-ATP as phosphate donor. In this way labeled nucleotides are obtained from unlabeled nucleic acids. This *in vitro* labeling makes it possible to extend application of the fingerprinting technique of Sanger et al. (*11*) to nucleic acids that cannot be sufficiently labeled *in vivo*. The digest labeled with polynucleotide kinase can be fractionated by two-dimensional paper electrophoresis and some of the analytical techniques developed by Sanger et al. (*11–13*) can be applied to the separated nucleotides. Since these *in vitro* labeled nucleotides have a labeled phosphate group only at the 5′-end, techniques for further analysis are restricted to digestions which do not

* Present address: Department of Biochemistry, Imperial College of Science and Technology, London, England

remove this phosphate group. For studying the sequence of oligonucleotides, digestion with snake venom phosphodiesterase is therefore used, since this enzyme splits off nucleotides from the 3′-end and leaves the 5′-end intact.

In this paper the method of Székely and Sanger (*1*) is described for labeling and fractionating DNA and RNA digests and for determining sequences of oligonucleotides.

ISOLATION AND ASSAY OF POLYNUCLEOTIDE KINASE

Methods for the isolation of 5′-hydroxylpolynucleotide kinase have been described by Richardson (Ref. *3*, see also this volume, page 815), Novogrodsky and Hurwitz (*4*) and Takanami (*5*). In the studies described in this paper, the procedure of Richardson was used. The enzyme has been purified to stage 6 and was stored in ice or at –20°C in 50% glycerol. There is a slow loss of activity upon storage at 0°C.

In the assays of polynucleotide kinase according to Richardson (*3*) and to Novogrodsky and Hurwitz (*4*) DNA is phosphorylated with γ-^{32}P-ATP and the acid-insoluble ^{32}P is determined. It is more convenient for the present technique to assay the enzyme under conditions similar to those used for phosphorylation of nucleic acid digests. The semiquantitative assay described here is based on phosphorylation of dinucleoside monophosphate with polynucleotide kinase in the reaction mixture used for labeling nucleic acid digests. The phosphorylated products are detected by paper electrophoresis.

Five millimicromoles of UpG (or other dinucleoside monophosphates) are incubated for 30 min at 37°C with varying amounts of enzyme in the presence of 1 mμmole of γ-^{32}P-ATP (10^5 cpm/mμmole); 0.01 M Tris-HCl buffer, pH 8.1; 0.01 M $MgCl_2$; 0.02 M mercaptoethanol. Total volume is 5 μl. The dinucleotide produced is separated from ATP by ionophoresis on DEAE-paper at pH 3.5. Activity of the enzyme is estimated by determining the amount required for complete transfer of the labeled group. This assay was used to follow the course of purification of the enzyme. The presence of contaminating enzyme activities is indicated by the appearance of additional spots on the electrophoretograms; phosphatase gives rise to inorganic phosphate, some nucleases to mononucleotides.

PREPARATION OF LABELED ATP

Highly labeled ATP is required for this technique. Digests sufficiently labeled for sequence analysis are obtained when using ATP of a specific activity of 1–10 mCi/μmole. Labeled ATP preparations were either obtained from Radiochemical Centre, Amersham, or were prepared by the method of

Glynn and Chappell (*14*). A modification of this procedure used for preparing *small* quantities of ATP of *high* specific activity is described below.

Carrier-free $^{32}P_i$ (5 mCi) is dried *in vacuo* and is taken up in 100 μl reaction mixture containing 0.5 μmoles of ATP, 0.15 μmoles 3-phosphoglycerate, 20 μg glyceraldehyde phosphate dehydrogenase, 2 μg 3-phosphoglyceric phosphokinase, 0.01 M $MgCl_2$, 0.05 M Tris-HCl, pH 8.1, and 0.01 M mercaptoethanol. After 30 to 60 min incubation at 26°C the reaction mixture is applied to DEAE-paper and ATP and P_i are separated by ascending chromatography in 0.3 M ammonium formate. Inorganic phosphate moves about twice as fast as ATP. In some preparations some labeled material is found also at the startline. The yield of γ-^{32}P-ATP varies from one preparation to the other, it is usually between 30 and 75%. The ATP is eluted from the paper with alkaline triethylamine carbonate (*11*) and the eluant is removed by freeze drying. The residue is dissolved in water, freeze dried again, and this procedure is repeated twice. It is important to carry out elution and removal of triethylamine carbonate rather quickly, since some breakdown may occur during these steps.

γ-^{32}P-ATP preparations can be stored for several weeks but in old preparations nucleotides are produced which give additional spots upon electrophoresis. Such preparations can be purified by ionophoresis on DEAE-paper in 7% formic acid. Under these conditions ATP moves about two-thirds the distance of the blue marker (*11*). It is eluted with alkaline triethylamine carbonate and freeze dried as before.

FINGERPRINTING DNA

Degradation of DNA

DNA is degraded by depurination according to Burton (*15*), followed by treatment with bacterial alkaline phosphatase.

A 100-μg sample of DNA is dissolved in 100 μl water and is depurinated by incubating for 16 hr at 32°C with 200 μl of 3% diphenylamine in formic acid. After incubation the reagent is removed by ether extraction as follows. The reaction mixture is shaken with ether and the ether phase is reextracted five times with 300 μl water each. The ether phase is discarded and the whole procedure is repeated six times, using the same aqueous phases each time for reextracting the ether. After six extractions the aqueous phases are pooled and evaporated to dryness. The residue is dissolved in water and evaporated once more to remove traces of formic acid.

Five micrograms of the depurinated material is dissolved in 10 μl 0.01 M Tris-HCl, pH 8.1, containing 1 μg bacterial alkaline phosphatase and is incubated for 1hr at 37°C. The reaction is stopped by addition of acetate buffer, pH 4.5, to final concentration of 0.02 M. Phosphatase is immediately removed by adsorption to phosphocellulose paper. A small piece of phosphocellulose paper is washed with 0.01 M acetate buffer, pH 4.5 and the sample is applied to the wet paper. Nucleotides are eluted with 50 to 100 μl of 0.01 N acetic acid. The eluate is dried *in vacuo* onto a polythene sheet.

Phosphorylation of DNA Digest

The depurinated, dephosphorylated nucleotides are taken up in 10 μl of a reaction mixture containing a 10 to 20% excess of γ-^{32}P-ATP; an amount of polynucleotide kinase sufficient to transfer at least 50% more phosphate than present in the mixture; 0.01 M Tris-HCl, pH 8.1; 0.01 M $MgCl_2$; 0.02 M mercaptoethanol. The mixture is incubated at 37°C for 1 hr. An excess of ATP and enzyme is applied in order to ensure maximum phosphorylation.

Excess ATP may be removed after phosphorylation by further incubation for 10 min at 37°C with 1 mg/ml myosin. If this step is omitted, a strong spot of unreacted ATP appears on the fingerprint, it is, however, well separated from other nucleotides. Older ATP preparations give more spots.

Fractionation of DNA Digest by Two-Dimensional Paper Electrophoresis

The phosphorylated digest is applied directly to a cellulose-acetate strip and two-dimensional electrophoresis is carried out according to the techniques of Sanger et al. (*11, 12*).

A diagram showing the position of nucleotides from depurinated DNA of fd phage is seen in Fig 1. Since only pyrimidine nucleotides are present, the pattern is relatively simple. Isomers are not separated. Mobility at pH 3.5 is mainly dependent on the number of C residues in the nucleotide; mobility in 7% formic acid is mainly dependent on the number of T residues. Spots are arranged in five series: the first containing nucleotides with one T and different C residues, the second nucleotides with two T and different C residues, etc. C_2 is the only nucleotide of the series containing only C residues that was recovered in this fingerprint.

Nucleotides with no or one T residue move much faster in 7% formic acid than those containing more T residues. In order to obtain good separation of the latter, a relatively long run in the second dimension is required. (The blue marker (*11, 12*) moves 40 to 50 cm from the start). Under these conditions, however, the fast moving nucleotides run off the paper. Plate 1 shows a fingerprint of fd DNA obtained in this way. Only nucleotides containing two or more T residues are recovered. Nucleotides

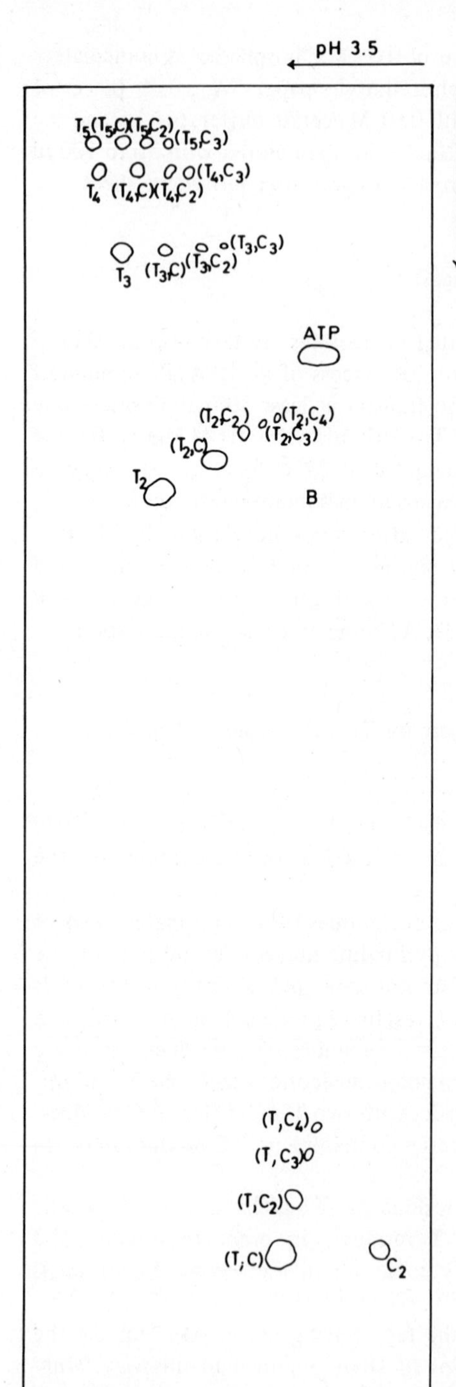

Figure 1. Diagram of the position of nucleotides on a fingerprint of depurinated DNA. B shows the position of the blue marker (*1*).

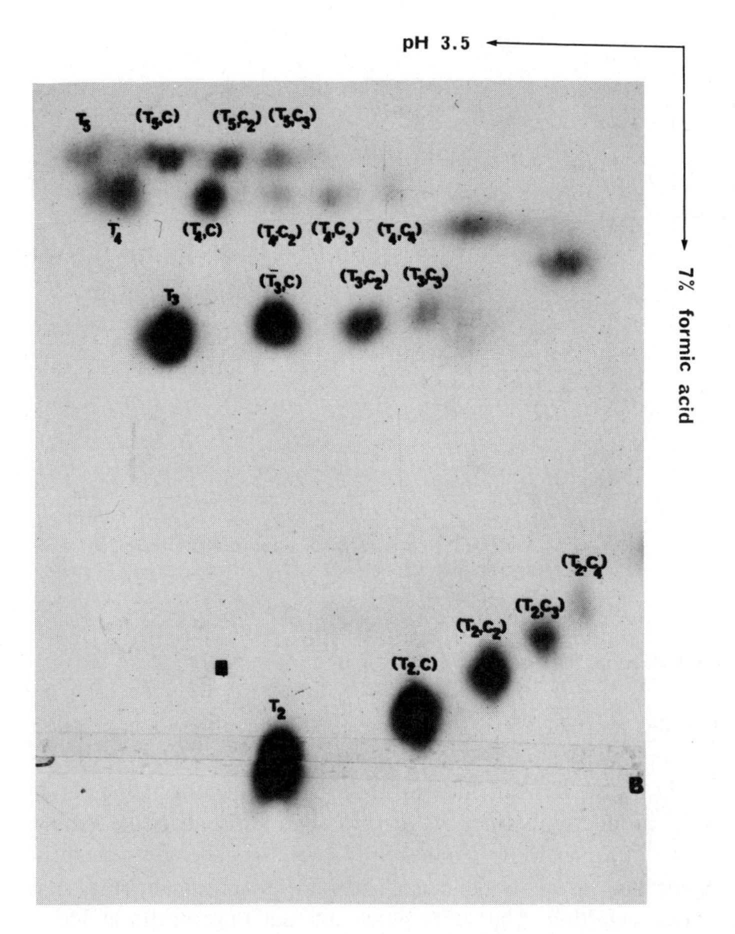

Plate 1. Two dimensional fractionation of digest of fd DNA. B shows position of the blue marker (*1*).

with less T residues can be detected in a shorter run. The following alternate procedure can be used to obtain good separation of both fast and slow-moving nucleotides in the same fingerprint. After a short run in the second dimension the paper is cut into two, between the series with one T and that with two T residues. (The blue marker moves with nucleotide T_2.) To the two halves of the paper other pieces of DEAE-paper are sewn and they are run separately for a longer time. The fingerprint in Fig.1 was obtained in this way.

For further analysis nucleotides are eluted as described by Sanger et al. (*11, 12*).

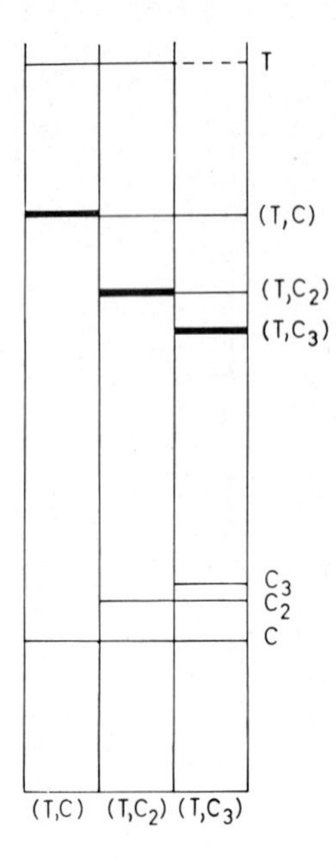

Figure 2. Diagram of the position of digestion products in partial phosphodiesterase digests of nucleotides (T,C), (T,C_2), and (T,C_3). The thick lines indicate the position of the undigested controls.

Composition of Nucleotides

As nucleotides are labeled in the 5′-phosphate only, their composition cannot be determined by complete digestion to mononucleotides. It can be established partly on the basis of their position on the fingerprint and partly on the basis of products obtained by partial digestion with snake venom phosphodiesterase. This method works well if several isomers are present in each spot, giving rise to all or almost all possible fragments upon partial phosphodiesterase digestion. The spots shown on the fingerprints in Fig. 1

TABLE 1. Determination of Composition of Nucleotides by Partial Digestion with Snake Venom Phosphodiesterase

Products of partial phosphodiesterase digestion	Composition of nucleotide
T, C	(T, C)
C_2, (T, C), T, C	(T, C_2)
C_3, (T, C_2), (T, C), C_2, T, C	(T, C_3)
T	T_2
(T, C), T_2, T, C	(T_2, C)
(T, C_2), (T_2, C), T_2, C_2, (T, C), T, C	(T_2, C_2)

and Plate 1 were identified as outlined in Table 1. For instance the nucleotide assigned the composition TC_3 could consist of a mixture of the following isomers TCCC, CTCC, CCTC, and CCCT. Partial snake venom digestion would give the following products: TCC, TC, T, CTC, CT, C, CCT, CC, C, and CCC, CC, C, that is C_3, (T, C_2), (T, C), C_2, T, C. Figure 2 shows a diagram of the position of products of partial phosphodiesterase digestion of (T, C), (T, C_2), and (T, C_3) separated by ionophoresis at pH 3.5 (as described below).

Sequence of Nucleotides

Sequence of nucleotides is studied by partial phosphodiesterase digestion. Nucleotides are incubated with 5 to 10 μl of 0.01 mg/ml snake venom phosphodiesterase in 0.02 M Tris-HCl, pH 8.5, 0.01 M $MgCl_2$, for 15 min at room temperature. Products are separated either by ionophoresis on paper at pH 3.5 or by two-dimensional electrophoresis as follows. Three samples are applied to a cellulose-acetate strip at a distance of about 10 cm from one another. After a short run at pH 3.5 (when the blue marker has moved 5–6 cm from the start) the nucleotides are transferred to DEAE-paper by the blotting procedure of Sanger et al. (*11, 12*) and subjected to electrophoresis at pH 1.9. The short run in the first dimension improves separation of nucleotides containing different numbers of C residues. This method is advantageous for fractionating digests of larger oligonucleotides. It gives good separation of fragments containing one, two, or three T and different C residues and a somewhat poorer separation of nucleotides with four T and different C residues. One-dimensional ionophoresis on paper at pH 3.5 gives better separation of nucleotides containing only C residues and can be applied also to nucleotides containing one or two T and several C residues. This technique was used mainly for analysis of smaller oligonucleotides. For separation of large fragments (4 or 5 T and several C residues) two-dimensional electrophoresis can be carried out in the same way as for fingerprinting the original digest. Isomers are not separated by any of these procedures. Figures 3 and 4 are diagrams of the position of nucleotides on ionophoresis at pH 3.5 and on two-dimensional ionophoresis, respectively. When electrophoresis is made in one dimension, an undigested control is also applied to the paper next to the digest.

Partial phosphodiesterase digestion gives information about the 5′-end as well as about the rest of the molecule and it is therefore usually not necessary to determine the 5′-end separately. It can be done, however, by complete digestion with phosphodiesterase (with 0.1 mg/ml enzyme, at 37°C for 15 min in 0.02 M Tris-HCl, pH 8.5, 0.01 M $MgCl_2$) followed by separation of T and C by ionophoresis at pH 3.5 on paper.

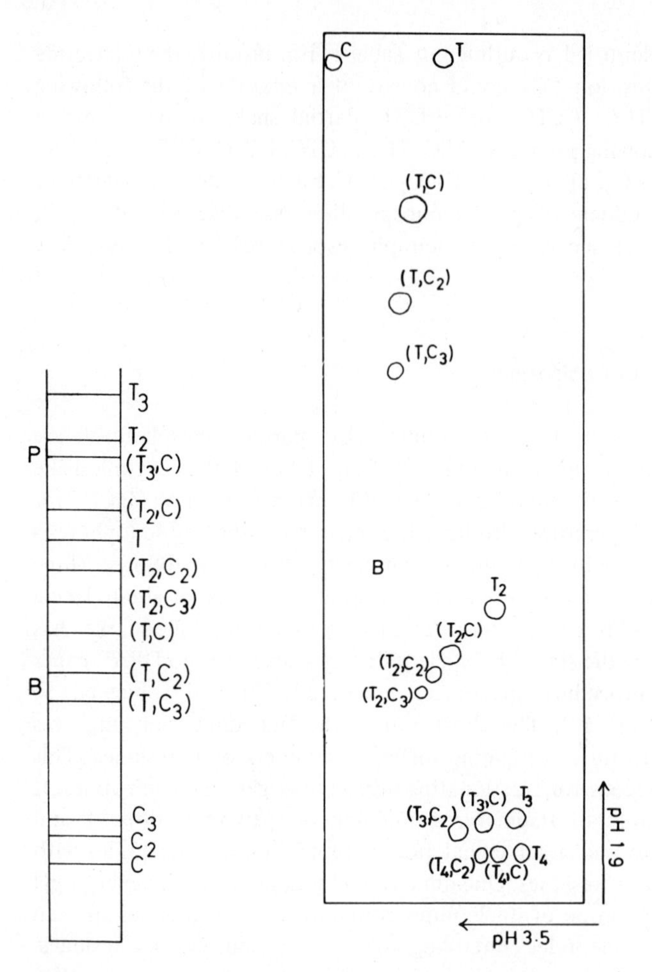

Figure 3. (Left) Diagram of the position of deoxyribonucleotides upon electrophoresis on paper at pH 3.5. B and P show the position of the blue and pink marker, respectively (*1*).

Figure 4. (Right) Diagram of the position of deoxyribonucleotides after two-dimensional electrophoresis. B shows the position of the blue marker. Data from two electrophoretograms combined.

Determination of Nucleotide Sequences in Mixtures of Isomeric Oligonucleotides

The sequence of a pure nucleotide can be deduced from the products of partial phosphodiesterase digestion. Since, however, nucleotides eluted from DNA fingerprints consist of mixtures of several isomers, one digestion does not give sufficient information to establish the structure of all isomers present. In the phosphodiesterase digest of nucleotide (T_2, C_2), for example, fragment T_2 proves the presence of the sequence TTCC and fragment C_2 proves the presence of CCTT. However, fragment (T_2, C) as well as (T, C_2) can be produced from three different isomers and (T, C) from four isomers (see Table 2). In order to identify all isomers in a mixture, products of the first phosphodiesterase digestion are eluted from the paper and digested again with phosphodiesterase under the same conditions. With some nucleotides it is necessary to repeat this procedure several times. In Table 2 an example of how repeated phosphodiesterase digestions are used to establish the sequences of all isomers of nucleotide (T_2, C_2) is given.

In a similar way sequences of several oligonucleotides in fd DNA have been determined. When studying the structure of longer nucleotides, this technique might become complicated if all isomers were present, since several repeated digestions would be needed to determine their sequences. Since fewer isomers of long oligonucleotides are present in the digest of DNA, sequence determination is somewhat simplified by the lack of digestion products corresponding to missing isomers. Thus the presence of some fragments in the phosphodiesterase digest and the absence of others both confirm a number of sequences. This was the case for example in the

TABLE 2. Determination of Sequences in a Mixture of Isomers of (T_2, C_2)

Products present in partial phosphodiesterase digests			
1st Digest	2nd Digest	3rd Digest	Sequence confirmed
T_2			TTCC
C_2			CCTT
(T, C)			
(T_2, C)	T		
	C		CTTC
	T_2		TTCC
	(T, C)	C	CTTC
		T	TCTC
(T, C_2)	T		TCCT
	C		
	C_2		CCTT
	(T, C)	T	TCCT
		C	CTCT

phosphodiesterase digest of nucleotide (T_5, C_3) from fd DNA. The first partial digestion yielded fragment (T_2, C_3), which in turn gave (T_2, C) and C but no T on second digestion. This confirms C as the 5′-end, CTT as the trinucleotide fragment, CTTCC as the pentanucleotide fragment, and CTTCCTTT as the sequence of the original octanucleotide. These data also eliminate the sequences with TTCCC and TCTCC at the 5′-end. Lack of some fragments from phosphodiesterase digests made it possible to eliminate a number of sequences in fd DNA.

APPLICATION OF THE TECHNIQUE TO RNA

RNA is digested with RNase T_1 and alkaline phosphatase and the nucleotides produced are phosphorylated with polynucleotide kinase and γ-^{32}P-ATP as in the case of DNA digests. Digestion and phosphorylation can be carried out in two different steps or in the same reaction mixture.

1. Digestion and phosphorylation made in two separate steps. Approximately 2 to 10 μg RNA is digested according to Brownlee and Sanger (*13*) with RNase T_1 and bacterial alkaline phosphatase in 0.01 M Tris-HCl, pH 8.1, using enzyme-to-substrate ratios of 1 : 20 and 1 : 5, respectively. The total volume is 5 μl. Incubation is performed for 30 min at 37°C. After incubation, enzymes are removed by phenol treatment (*16*) or by adsorption to phosphocellulose paper as described above.

The methods used for phosphorylation and two-dimensional fractionation of the digest are the same as those applied to DNA digests.

2. Digestion and phosphorylation in the same reaction mixture. When using this technique, smaller quantities of phosphatase are applied in order not to interfere with phosphorylation of the 5′-hydroxyls. It was found that polynucleotide kinase preparations are contaminated with a phosphatase that in itself is sufficient to remove the 3′-phosphate group and does not, at the same time, prevent phosphorylation of the 5′-hydroxyls. It is not known whether this enzyme does not attack 5′-phosphates or if these groups are removed as well, but are rapidly replaced by polynucleotide kinase. Although digestion can thus be carried out without the addition of phosphatase, a small amount of bacterial alkaline phosphatase (1 : 50 in proportion to RNA) was usually included in the reaction mixture.

Approximately 2 to 10 μg RNA is incubated with 1/20 amount of RNase T_1, 1/50 amount of bacterial alkaline phosphatase, 20% excess of γ-^{32}P-ATP, polynucleotide kinase in an amount sufficient to transfer at least 50% more phosphate than present in ATP. The reaction mixture also contains 0.01 M Tris-HCl, pH 8.1; 0.01 M $MgCl_2$; 0.02 M mercaptoethanol. The total volume is 10 μl. The mixture is incubated for 30 min at 37°C.

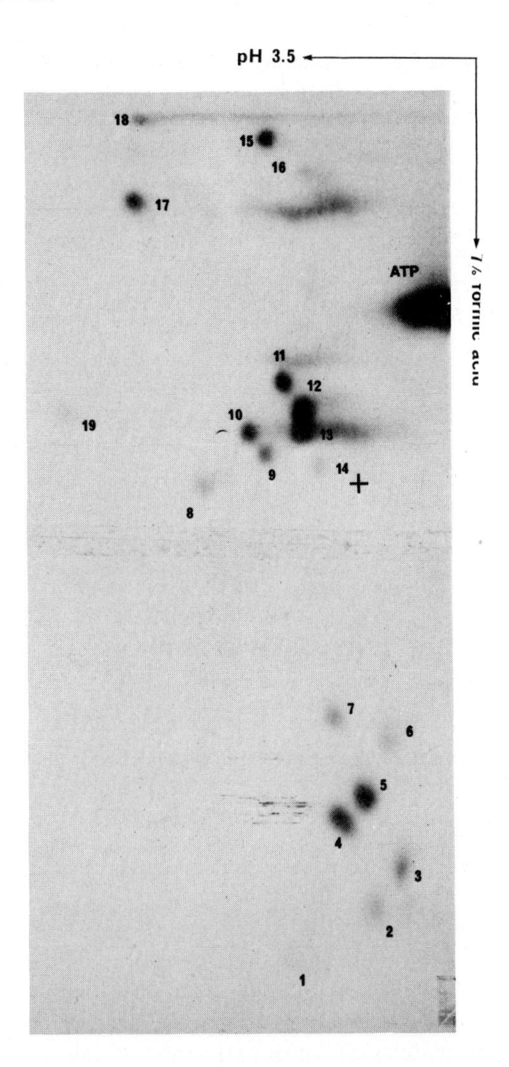

Plate 2. Two dimensional fractionation of a digest of $tRNA_f^{Met}$. The symbol (+) shows position of the blue marker (*1*). Structure of nucleotides is given in Table 3.

Two dimensional fractionation is carried out as with DNA digests. Plate 2 shows a fingerprint obtained in this way from unlabeled $tRNA_f^{Met}$.

Analysis of Nucleotides

Sequence of nucleotides is established by partial digestion with snake venom phosphodiesterase. Conditions for digestion are the same as for deoxyribonucleotides. The digest is subjected to electrophoresis at pH 1.9 on

TABLE 3. Structure of Nucleotides on Fingerprint in Plate 2[a]

No. of spot	Structure of nucleotide
1	G
2	CG
3	probably CAACCA[b]
4	AG
5	CAG
6	CCCCCG
7	AAG
8	UG
9	UCG
10	DAG
11	AUCG
12	CCUG
13	CUCG
14	7MeGUCG[b]
15	TψCAAAUCCG[b]
16	probably 2′O-MeCUCAUAACCCG[b]
19	pGp

[a] Sequences identical to those determined by Dube et al. (*17*) in *in vivo* labeled RNA.

[b] Only partial sequence determined, structure of nucleotides established by comparison with the data of Dube et al. (*17*).

DEAE-paper. An undigested control is applied to the paper next to the sample. Digestion products are identified on the basis of *M* values (*11*) where applicable and by comparing the position of fragments with that of nucleotides of known structure. *M* values determined by Sanger et al. for spleen phosphodiesterase digest of nucleotides containing 3′-phosphate can be applied to snake venom phosphodiesterase digest of nucleotides with 5′-phosphate. The method works well with smaller oligonucleotides but only partial sequences were obtained from long oligonucleotides. Fig. 5 shows the position of nucleotides in a partial phosphodiesterase digest of some oligonucleotides obtained from $tRNA_f^{Met}$.

Nucleotides from the 5′-end are also detected in partial phosphodiesterase digests. Better separation and identification is obtained, however, if the 5′-end is determined separately. This is done either by complete digestion with snake venom phosphodiesterase (as in the case of deoxyribonucleotides) or by alkaline hydrolysis (in 0.2 N NaOH at 37°C for 16 hr), followed by

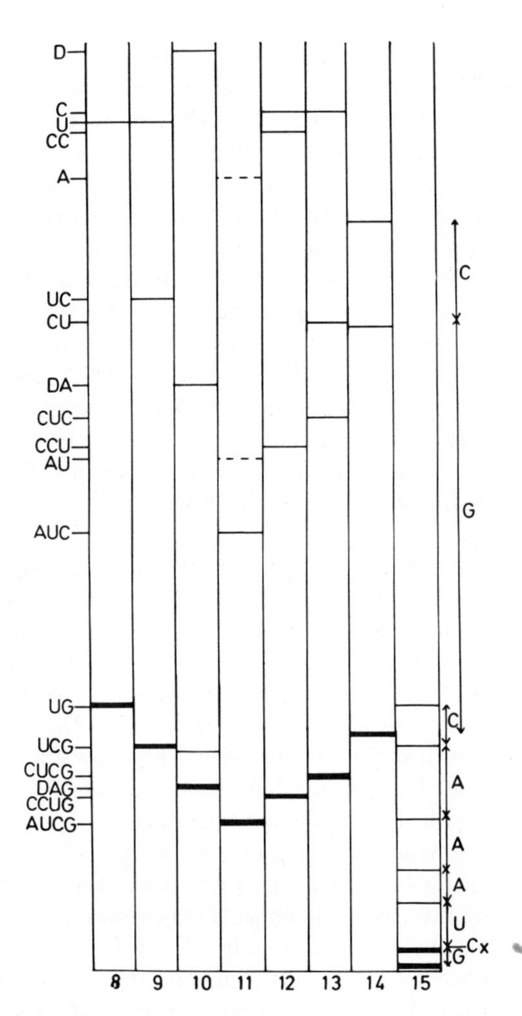

Figure 5. Diagram of partial phosphodiesterase digests of some nucleotides from $tRNA_f^{Met}$ fractionated on DEAE-paper at pH 1.9. Left: structure of digestion products of nucleotides 8 to 13; right: sequences at the 3′-end of nucleotides 14 and 15, determined on the basis of *M* values. The thick lines indicate the position of the undigested controls. Some faint bands behind the controls are not shown on the diagram.

ionophoresis on paper at pH 3.5. The latter technique yields nucleoside diphosphates the mobilities of which are, relative to pUp: pTp, 0.96; pGp, 0.79; pAp, 0.63; pCp, 0.59.

DISCUSSION

The use of polynucleotide kinase in labeling nucleic acid digests is advantageous in sequence studies of nucleic acids that can be obtained only in low yield and cannot be sufficiently labeled *in vivo*. With γ-^{32}P-ATP of high specific activity, fingerprints can be obtained from a few micrograms of nucleic acid. The technique has been applied to various DNAs and RNAs.

Fingerprints have been obtained from depurinated T_4 DNA, thymus DNA, and fd DNA. The method has also been used for phosphorylating enzymatic digests of DNA (K. Murray, unpublished results).

Fingerprints of *in vitro* and *in vivo* labeled RNAs have been compared in the case of 5 S RNA, $tRNA^{Phe}$ and $tRNA_f^{Met}$ The same technique has been applied in the study of different RNA fractions from reticulocytes (*16*). Its application is not restricted to paper electrophoretic techniques, it seems suitable also for obtaining fingerprints by thin-layer chromatography (T. Harrison and G. G. Brownlee, personal communication).

Only qualitative analyses of fingerprints have been made. Intensity of the spots shows that, especially in RNA digests, phosphorylation of some nucleotides is incomplete. These fingerprints seem therefore unsuited for quantitative estimations.

Comparison of fingerprints obtained from *in vivo* and *in vitro* labeled RNAs show only slight differences in the mobilities of nucleotides with a phosphate group at the 3′- or at the 5′-end. This causes slight differences in the separation of some nucleotides, e.g., separation of pAUG and pUAG is usually poor compared to AUGp and UAGp, whereas pCCUG and pCUCG separate better than CCUGp and CUCGp. A few additional spots appear on some RNA fingerprints obtained by the polynucleotide kinase technique. Spot 17 on the fingerprint in Plate 2 does not correspond to any nucleotide in the digest of *in vivo* labeled RNA. The position of such spots varies from one fingerprint to the other. They may arise from incomplete dephosphorylation, overdigestion, or from the action of nuclease contaminants in the polynucleotide kinase preparation.

Sequence determination by partial phosphodiesterase digestion works well with shorter oligonucleotides. It is an advantage that "secondary" splits by nuclease contaminants which sometimes give rise to anomalous spots in digests of uniformly labeled oligonucleotides do not interfere in this case, since only fragments containing the labeled 5′-end are detected in the digests. Difficulties arise, however, when sequencing long oligonucleotides. In order to work out long sequences, further techniques will be needed involving rephosphorylation of digestion products with polynucleotide kinase.

Acknowledgments

I am grateful to the Editors of the *Journal of Molecular Biology* for their permission to reuse Plates I and Vb. as well as Figures 1 and 2 of the paper M. Székely and F. Sanger, 1969. *J. Mol. Biol.*, **43**: 607.

This work has been made possible by an award of The Wellcome Trust which is gratefully acknowledged.

References

1. Székely, M., and F. Sanger. 1969. *J. Mol. Biol.,* **43**: 607.
2. Novogrodsky, A., and J. Hurwitz. 1965. *Federation Proc.,* **24**: 602.
3. Richardson, C. C. 1965. *Proc. Natl. Acad. Sci., U.S.,* **54**: 158.
4. Novogrodsky, A., and J. Hurwitz. 1966. *J. Biol. Chem.,* **241**: 2923.
5. Takanami, M. 1967. *J. Mol. Biol.,* **23**: 135.
6. Sugiura, M., and M. Takanami. 1967. *Proc. Natl. Acad. Sci., U.S.,* **58**: 1595.
7. Richardson, C. C. 1966. *J. Mol. Biol.,* **15**: 49.
8. Weiss, B., T. R. Live, and C. C. Richardson. 1968. *J. Biol. Chem.,* **243**: 4530.
9. Novogrodsky, A., M. Tal, A. Traub, and J. Hurwitz. 1966. *J. Biol. Chem.,* **241**: 2933.
10. Wu, R., and A. D. Kaiser. 1967. *Proc. Nat. Acad. Sci., U.S.,* **57**: 170.
11. Sanger, F., G. G. Brownlee, and B. G. Barrell. 1965. *J. Mol. Biol.,* **13**: 373.
12. Barrell, B. G. This volume, p. 751.
13. Brownlee, G. G., and F. Sanger. 1967. *J. Mol. Biol.,* **23**: 337.
14. Glynn, I. M., and J. B. Chappell. 1964. *Biochem. J.,* **90**: 147.
15. Burton, K. 1967. In L. Grossman and K. Moldave (Eds.), Methods in enzymology. Academic Press, New York, Vol. XIIA, p. 222.
16. Labrie, F., and F. Sanger. 1969. *Biochem. J.,* **114**: 29P.
17. Dube, S. K., K. A. Marcker, B. F. C. Clark, and S. Cory. 1969. *European J. Biochem.,* **8**: 244.

A TRITIUM DERIVATIVE METHOD FOR BASE ANALYSIS OF RIBONUCLEOTIDES AND RNA*†

KURT RANDERATH AND ERIKA RANDERATH

THE JOHN COLLINS WARREN LABORATORIES OF THE HUNTINGTON MEMORIAL HOSPITAL OF HARVARD UNIVERSITY AT THE MASSACHUSETTS GENERAL HOSPITAL
BOSTON, MASSACHUSETTS

AND

DEPARTMENT OF BIOLOGICAL CHEMISTRY
HARVARD MEDICAL SCHOOL
BOSTON, MASSACHUSETTS

INTRODUCTION

The base composition of ribopolynucleotides may be analyzed by a tritium derivative method (*1, 2*) after complete digestion of the polymers to the nucleoside level (Fig. 1). The method is very sensitive, less than 1μg of tRNA is, for example, required for a determination of the base composition including most minor constituents (*3*). In this chapter we shall describe the standard analytical procedure developed in our laboratory.

* Supported by USPHS Research Career Development Award 1 K04 CA42570-01 from the National Cancer Institute to K.R. and by Grant P-516 from the American Cancer Society. This is publication No. 1376 of the Cancer Commission of Harvard University.

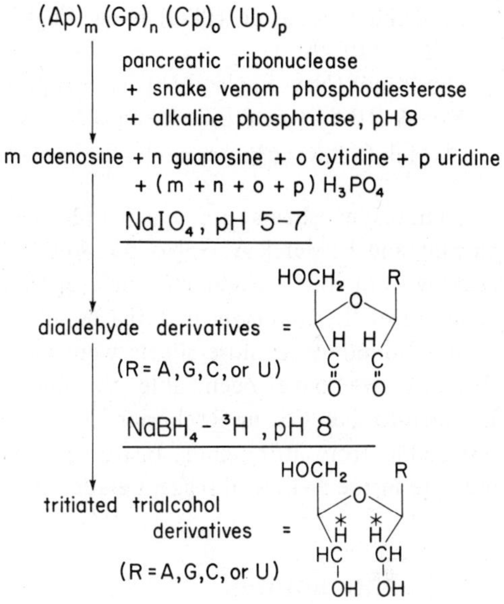

$(Ap)_m (Gp)_n (Cp)_o (Up)_p$

pancreatic ribonuclease
+ snake venom phosphodiesterase
+ alkaline phosphatase, pH 8

m adenosine + n guanosine + o cytidine + p uridine
+ (m + n + o + p) H_3PO_4

$NaIO_4$, pH 5–7

dialdehyde derivatives = HOCH$_2$ R, O, H H, C=O C=O

(R = A, G, C, or U)

$NaBH_4$ - ^3H, pH 8

tritiated trialcohol derivatives = HOCH$_2$ R, O, H* H*, HC CH, OH OH

(R = A, G, C, or U)

Figure 1. Quantitative analysis of ribopolynucleotides by enzymatic digestion and tritium labeling.

MATERIALS

Snake venom phosphodiesterase (Worthington Biochemical Corporation, Freehold, N.J., Code VPH)

E. coli alkaline phosphatase (Worthington Code BAPC), dialyzed exhaustively against distilled water

Ribonuclease A (Sigma Chemical Company, St. Louis, Mo., Type 1-A)

N,N-Bis(2-hydroxyethyl)glycine (bicine) (Calbiochem, Los Angeles, Calif., No. 391336)

Sodium metaperiodate, analytical reagent

Sodium borohydride and potassium borohydride pellets (Alfa Inorganics, Beverly, Mass.)

Tritium-labeled sodium borohydride (Amersham-Searle, Des Plaines, Ill., No. TRK-45) or potassium borohydride (Amersham-Searle, No. TRK-293)

† Abbreviations: A, adenosine; G, guanosine; C, cytidine; U, uridine; I, inosine; 1MeA, 1-methyladenosine; 6MeA, N^6-methyladenosine; 6DiMeA, N^6,N^6-dimethyladenosine; IPA, N^6-(Δ^2-isopentenyl)-adenosine; 1MeG, 1-methylguanosine; 2MeG, N^2-methylguanosine; 2DiMeG, N^2,N^2-dimethylguanosine; 7MeG, 7-methylguanosine; 1,7DiMeG, 1,7-dimethylguanosine; 3MeC, 3-methylcytidine; 5MeC, 5-methylcytidine; 4MeC, N^4-methylcytidine; 4DiMeC, N^4,N^4-dimethylcytidine; 3MeU, 3-methyluridine; rT, thymine riboside; ψ, pseudouridine; DiHU, 5, 6-dihydrouridine; 1 MeI, 1-methylinosine; 7MeI, 7-methylinosine; 4TU, 4-thiouridine; A′, etc., the trialcohol of adenosine, etc. PEI-cellulose, polyethyleneimine-cellulose, Tris, Tris(hydroxymethyl)aminomethane; bicine, *N, N-bis*(2-hydroxyethyl)glycine.

2,5-Diphenyloxazole (PPO) (New England Nuclear Corp., Boston, Mass., No. NEF-901)
Omnifluor (New England Nuclear Corp., No. NEF-906)
Kodak RB-54 Royal Blue Medical X-Ray Film
^{14}C-ink (Schwarz Bioresearch No. 7321-01).

Thin-layer sheets are Eastman-Kodak No. 6064 cellulose on plastic backing and E. Merck A. G. No. 5523/0025 silica gel/kieselguhr on aluminum backing. For the two-dimensional separations to be carried out a firm adhesion of the thin-layer material to the backing is essential. Some batches of the No. 6064 cellulose sheets were found unsatisfactory in this respect. Recently we have been able to obtain suitable sheets directly from Eastman-Kodak Co. in Rochester, N.Y. The silica gel/kieselguhr sheets are obtainable from Brinkmann Instruments, Westbury, N.Y. Organic solvents used are of the analytical reagent grade.

PROCEDURES

Preparation of Borotritiide Solution

All operations involving tritiated borohydride are carried out under a hood. Tritium-labeled sodium borohydride of high specific activity (5-20 Ci/mmole) is dissolved in 0.1 N NaOH (CO_2-free); unlabeled sodium borohydride (in 0.1 N NaOH) is added to give a final radioactivity of 2-4 Ci/mmole at 0.1 M total borohydride concentration. The solution is stored in portions of 20-50 μl at $-90°$.

We have recently observed that the use of a batch of potassium borotritiide resulted in a reduction of the radioactive background when compared with sodium borotritiide. The preparation of the solution was analogous to the procedure described above, except that potassium hydroxide and unlabeled potassium borohydride were used. According to the manufacturer, high specific activity potassium borotritiide is more stable than sodium borotritiide of similar radioactivity.

Preparation of Unlabeled Nucleoside Trialcohols

These compounds are prepared according to a modification of the procedure of Khym and Cohn (*4*). To a solution of 0.6 μmole nucleoside in 200 μl water is added 5 μl of freshly prepared 0.18 M aqueous $NaIO_4$. The solution is kept in the dark at room temperature for 30-40 min. Then 5 μl 0.72 M $NaBH_4$ in 0.1 N NaOH (fresh) is added and the reaction is allowed to proceed at room temperature for 30-40 min. Acetic acid (200 μl 1 N) is added and the solution is blown down to dryness in a stream of filtered air. The residue is taken up in 300 μl water (final concentration 2 nmole/μl). The solution is kept frozen.

Nucleosides which are poorly soluble in water may be treated in more dilute solution and the reaction times are extended to 2–3 hr.

Preparation of Tritium-Labeled Nucleoside Trialcohols

To a solution of 3 nmole nucleoside in 30 μl water is added 5 μl of an aqueous solution containing 6 nmole $NaIO_4$. The reaction is allowed to proceed for 60–100 min in the dark at room temperature. Then 0.5 μl of 0.1 M [^3H]-$NaBH_4$ (100 μCi) in 0.1 N NaOH is added and the solution is again incubated for 60–100 min in the dark at room temperature. Acetic acid (25 μl 1 N) is added to convert excess borotritiide to boric acid and tritium gas (*caution*). After 20–30 min in the open air, the solution is evaporated to dryness in a stream of filtered air at room temperature. Finally the residue is taken up in 60 μl water. The final radioactivity is about 0.05 μCi/μl indicating quantitative conversion of nucleoside to nucleoside trialcohol. The solution is kept frozen.

Some lots of borotritiide were found to require the presence of buffer during reduction of dialdehydes obtained from alkali-labile nucleosides. After periodate oxidation the sample is cooled on ice and 1 μl 0.1 M phosphate, pH 6.8, is added. The solution is thoroughly mixed and borotritiide is added immediately.

Enzymatic Hydrolysis of RNA

For complete digestion of RNA to nucleosides, the incubation mixture contains per microliter: 1 μg RNA, 0.25 μg snake venom phosphodiesterase, 0.25 μg ribonuclease A, 0.2 μg alkaline phosphatase, 30 nmole bicine-Na, and 10 nmole $MgCl_2$. The bicine buffer added is 0.6 M, pH 8.0. Incubation is at 37° for 6 hr. The enzymatic digest is appropriately diluted and used for subsequent periodate oxidation without removal of enzymes and electrolytes. A blank containing no RNA is treated identically and carried through all subsequent steps of the procedure.

Periodate Oxidation

The solution contains per microliter: RNA digest containing 0.5–1 nmole nucleoside and an approximately two-fold molar excess of $NaIO_4$. Incubation is at 22–25° for 2 hr in the dark.

Borotritiide Reduction

Following periodate oxidation, an aliquot of the borotritiide solution corresponding to about a 5-fold molar excess of borohydride over periodate is added and the reaction is allowed to proceed at 22–25° for 2 hr in the dark. As mentioned above, this reaction may require the presence of buffer. After periodate oxidation, the sample is cooled on ice and 1 M phosphate buffer, pH 6.8, is added to give a concentration of 15 nmole per microliter. The sample is mixed thoroughly and incubation with borotritiide is started immediately. The initial pH obtained this way is 7.0–7.4. During the reaction it rises to 7.6–8.0.

Decomposition of Borotritiide and Preparation of Solutions for Analysis

Following the reduction, a 300-fold molar excess of 1 N acetic acid over borohydride is added (*caution*: tritium gas is liberated). The solution is kept in the unstoppered tube for 20–30 min at room temperature. It is then evaporated to complete dryness in a stream of filtered air. The residue is taken up in 0.1 N formic acid. The final solution should contain about 0.5 nmole nucleoside per microliter. It is kept frozen at –10 to –20°.

Separation Procedures

The solvents are (*3*): A = *n*-butanol/isopropanol/7.5 N ammonia (3 : 3 : 2, by vol); B = *t*-amylalcohol/methyl ethyl ketone/water/formic acid, sp. gr. 1.2 (2 : 2 : 1 : 0.1); C = *n*-butanol/isopropanol/7.5 N ammonia (4 : 3 : 1); D = *n*-butanol/methyl ethyl ketone/water/formic acid, sp. gr. 1.2 (6 : 6 : 1 : 0.3); E = ethyl acetate/*n*-butanol/isopropanol/7.5 N ammonia (3 : 1 : 2 : 2).

For two-dimensional mapping, 1–15 μCi of the tritium-labeled final solution is applied to a thin-layer sheet (20 x 20 cm) at 2.5 cm each from the left-hand and the bottom edge. The solution is spotted in 1-μl portions with intermediate drying in a stream of cool air. Care should be taken not to damage the layer during this operation.

Development on the cellulose standard map is with solvent A to 16 cm from the origin (1st dimension) and twice with solvent B to 15 cm from the origin (2nd dimension). See also Note added in proof.

Development on silica gel/kieselguhr is with solvent C to 16 cm from the origin (1st dimension) and once with solvent D to 15 cm from the origin (2nd dimension).

There is no saturation of the tank atmosphere with solvent vapors. Chromatography (one chromatogram per tank) is started immediately after the solvent has been poured into the tank. Between developments the layer is thoroughly dried, first in a stream of cool air, then in a stream of warm (50–60°) air and the sheet is trimmed about 1 cm below the front of the first dimension and 1–2 cm above the bottom edge.

Because spots are less diffuse and separations generally more distinct on cellulose than on silica gel/kieselguhr (see below), we use the latter material mainly for resolving compounds that are difficult to separate on cellulose, see below in the section on rechromatography.

Unlabeled marker trialcohols of minor nucleosides may be cochromatographed in different systems with tritium-labeled RNA digests in order to help identify radioactive derivatives on the maps or to help locate derivatives of very low radioactivity. One should be aware of the possibility, however, that due to isotope effects, the [^{1}H]-derivatives do not necessarily completely coincide on chromatograms with the corresponding [^{3}H]-derivatives, particularly if borotritiide of relatively high specific activity (> 0.5 Ci/mmole) has been used for labeling. With borotritiide of 2 Ci/mmole specific activity, it has been our experience that the centers of the radioactive compound and its nonradioactive counterpart were rarely more than 3–4 mm apart. Due to isotope effects, the addition of unlabeled marker compounds cannot serve to locate unequivocally the major trialcohols in labeled RNA digests, cf. below in the section on fluorographic visualization of the tritium-labeled compounds.

If one keeps in mind that the behavior of the [^{1}H]-compound may differ slightly from that of its [^{3}H]-counterpart, the chromatographic similarity of the two compounds under a variety of conditions may help to establish the identity of the radioactive compound. For identification purposes, several (at least 5 or 6) different chromatographic conditions must be explored under which the radioactive compound and the added marker should almost coincide on the chromatogram. Additional solvents useful for this purpose (unpublished) are acetonitrile/H_2O (8 : 2 to 8 : 2.4 by volume) on PEI-cellulose (*5*) and on cellulose, acetonitrile/concentrated ammonia (8 : 2.25) on PEI-cellulose, and *t*-amylalcohol/methyl ethyl ketone/H_2O (3 : 3 : 2, by volume) on silica gel layers (*6*).

If the original RNA contains only the four major bases – adenine, guanine, cytosine, and uracil – the procedure can be simplified as follows (*2*). Labeling is carried out with borotritiide of < 200 mCi/mmole specific activity. An aliquot of the final labeled solution is mixed with an equal volume of an aqueous solution containing 2–4 nmole per microliter of unlabeled nucleoside trialcohols. For one-dimensional separation, 1-μl aliquots of this solution are applied to a cellulose thin layer at 2.5 cm from the bottom edge. The chromatogram is developed with solvent E to 12–14 cm above the origin. The compounds are located by examination under a short-wave (254 nm) ultraviolet light and circled with a pencil. R_f values of the nucleoside trialcohols are: 0.56 (A′), 0.45 (C′), 0.34 (U′), 0.23 (G′). These values may vary slightly from batch to batch of the layer. The areas corresponding to these compounds are cut out and processed as described below under elution of compounds and liquid scintillation counting. Count rates obtained by subjecting a blank without nucleotides to the entire procedure are subtracted from each trialcohol count rate.

Fluorographic Visualization of the Tritium-Labeled Compounds

Low temperature solid scintillation fluorography employing PPO as the scintillator (*6*) is used to visualize the labeled compounds. In this method photons emitted by the scintillator, not tritium β-particles, cause the blackening of the photographic emulsion (*6*). A 7% (w/v) solution of PPO (scintillation grade) in diethyl ether is poured from a small beaker over the entire chromatographic area as rapidly as possible. The volume of the PPO solution used for this treatment should be 35–40 $\mu l/cm^2$, i.e., 14–16 ml for a 20 x 20-cm layer. The scintillator is distributed evenly by tilting. This operation should take only 1–3 sec depending on the size of the layer. The treated chromatogram is then immediately brought into a vertical position and agitated until the ether has evaporated completely. All further operations are carried out in a darkroom under proper lighting conditions (7½ Watt bulb, Wratten 6B filter, distance $>$ 120 cm from the film). The layer is placed in contact with the emulsion of Kodak RB-54 Royal Blue Medical X-Ray Film. The thin-layer sheet is fastened to the film by stapling. Exposure is carried out in the dark between glass plates held together with adhesive tape. The sandwich is wrapped in aluminum foil and stored in an insulated container over Dry Ice (–78.5°) or in a freezer at –80 to –90° for a period of time required to visualize the labeled compounds. The X-ray film is subsequently developed, usually for 5 min at 23–25°, with Kodak Liquid X-Ray Developer, fixed with Kodak Liquid X-Ray Fixer for 5 min, washed under running water, and air dried.

The sensitivity of this procedure is 4–6 nCi $^3H/cm^2$/day or 1–5 nCi ^3H/day for an average thin-layer spot depending on the spot size. Sensitivity is higher on cellulose layers than on silica gel/kieselguhr layers because spots are more compact in the former medium. At a sensitivity of 2 nCi/day it is thus possible to detect 1 minor component in 1000 nucleotides after an exposure for 24 hr if a total of 2 μCi was applied to the origin of the map. To detect bases occurring in the RNA with a lower frequency, the total radioactivity applied or the exposure time (or both) has to be increased.

For locating the radioactive compounds on the original chromatograms, ^{14}C-ink is applied to a few points on the layer after the treatment with PPO. Following exposure these spots serve as reference points with which to line up the film and the chromatogram. The compounds are marked by perforating the superimposed film around the darkened areas with a needle.

Due to the exposure time required to visualize trace components in labeled RNA digests, the film is usually overexposed at the sites of the major derivatives. The film area blackened by A′, C′, U′, and G′ is thus considerably larger than the area actually occupied by the compound on the chromatogram. As pointed out above, one cannot rely on unlabeled marker compounds for localization. However, if one places two sheets of film on top of each other during the film exposure, the film that is not in direct contact with the thin layer indicates the exact location of the compound. Perforating with a punch may serve to line up films and chromatogram.

Elution of Compounds

The compound areas are cut from the sheet with scissors. The areas should be kept on the left side of the blades. If this is not possible, a line is scratched through the layer adjacent to the spot with a scalpel and a cut is made along this scratch line. For removing substance areas from silica gel/kieselguhr layers, each spot is circled with a needle and cut out along the line.

It is important to keep in mind that the blackened area usually does not completely coincide with the area actually occupied by the radioactive compound on the layer. The former may be larger than the latter due to overexposure (see above), or it may be smaller if the radioactivity of the compound is low. In this case the outer part of the area where the concentration is lower than in the center may fail to contribute to the blackening of the photographic emulsion and one should allow for this deficit by cutting a correspondingly larger area from the sheet. The question whether the area cut out contains all the radioactivity of the particular compound should always be evaluated in preliminary experiments.

Cellulose cut-outs are eluted layer side down in flat-bottom beakers with 1.0 ml 2 N aqueous ammonia. Silica gel/kieselguhr cut-outs are extracted layer side up with 1.0 or 2.0 ml 0.5 M aqueous LiCl solution* depending on

* Trialcohols carrying a positive charge at neutral pH (1MeA′, 3MeC′, 7MeG′) are not extracted quantitatively with water because they are bound by cation exchange to acidic groups present in the chromatographic media. Extraction for 1 hr with 2 N ammonia from cellulose or with 0.5 M LiCl solution from silica gel/kieselguhr is, however, quantitative for these and all other nucleoside derivatives mentioned in this article. Neutral nucleoside trialcohols may be extracted quantitatively with water from either medium.

the cut-out size. After the beakers have been sealed with Parafilm, extraction is carried out for one hour or longer at room temperature with occasional shaking. The somewhat hydrophobic silica gel/kieselguhr layer should be completely immersed in the eluant.

Liquid Scintillation Counting

Aliquots of the eluates (500 μl) are transferred to low-background scintillation vials. Scintillation fluid (10 ml) is added and after the vials have been shaken the solutions are counted in a liquid scintillation counter. For preparing the scintillation fluid, 6 gm Omnifluor and 80 gm naphthalene (scintillation grade) are dissolved in 500 ml of a mixture of 5 parts (by vol) of xylene, 5 parts of dioxane, and 3 parts of ethanol. The solution is made up to 1 liter with the xylene-dioxane-ethanol mixture. Using a Packard 2211 ambient temperature liquid scintillation spectrometer we obtained a counting efficiency for tritium of 36% under these conditions.

Calculation of the Base Composition

For the four major and most minor constituents of RNA (see examples and comments) the base composition f_i (expressed as nanomoles of an individual nucleoside/nanomoles of all nucleosides in the RNA) is reflected by the relative radioactivities of the corresponding tritium-labeled trialcohol derivatives (expressed as the count rate of each individual nucleoside derivative/total count rate of all derivatives):

$$f_i = \frac{\text{cpm}_i}{\sum_{i=1}^{N} \text{cpm}_i}$$

where N is the number of radioactive derivatives separated. If the total count rate is set equal to 100 the base composition may be expressed similarly as a percentage of the total.

Rechromatography

Compounds incompletely resolved on one map may be resolved by rechromatography in a different medium or with different solvents in the same medium. An aliquot of an ammonia or water eluate from the first chromatogram (see elution of compounds) is first freed from particles by low-speed centrifugation and then evaporated in a desiccator *in vacuo* over phosphorus pentoxide. The residue is taken up in a small volume of water and applied in 1-μl portions to a second chromatogram. We usually rechromatograph mixtures of compounds isolated from cellulose maps on silica gel/kieselguhr. If a minor component overlapping on cellulose with a major component, for example, in the A′-lMeG′ area (Fig. 2), is to be analyzed, rechromatography is usually performed two-dimensionally to avoid trailing of traces of the major component into the minor component. Two-dimensional rechromatography can frequently be carried out on "half sheet" maps, i.e., on a 20 x 10-cm piece, the second dimension being run for a shorter distance. An excellent separation of lMeG′ from A′ is thus achieved by rechromatography on such a half sheet of silica gel/kieselguhr in solvent C for 15–16 cm followed by solvent D for 7–8 cm in the second dimension. The compounds are assayed by liquid scintillation counting as described above.

In the previous example, by evaluating the 1MeG′-A′ area on the original cellulose map we obtain data pertaining to the composition with regard to the sum of these bases. After rechromatography on silica gel/kieselguhr, we are in a position to calculate the ratio 1MeG′/A′ and thus to correct the original data for the presence of 1MeG′.

EXAMPLES AND COMMENTS

Figures 2 and 3 depict schematically the distribution of the nucleoside derivatives on cellulose (Fig. 2) and silica gel/kieselguhr (Fig. 3), respectively. Although the overall patterns are similar in the two media there are distinct differences. For example, lMeG′ overlaps with A′ on the cellulose map (Fig. 2) while on silica gel/kieselguhr it overlaps with lMeI′ (Fig. 3). I′ and G′ are well separated on cellulose but not on silica gel/kieselguhr. 2MeG′ travels ahead of 2DiMeG′ in the first dimension on silica gel/kieselguhr but stays behind 2DiMeG′ in the first dimension on cellulose. 4MeC′ and 4DiMeC′ are resolved on cellulose but coincide on silica gel/kieselguhr. lMeI′ and 2DiMeG′ overlap with DiHU′ on cellulose only. Glycerol, a compound formed in trace amounts by overoxidation followed by reduction of positions 3′, 4′, and 5′, of the ribose moiety, coincides with DiHU′ on silica gel/kieselguhr only. IPA′ travels with the solvent fronts on cellulose only, (cf. also, Note added in proof). Spots are generally somewhat more diffuse on silica gel/kieselguhr.

The chromatographic behavior of U′, G′, and their derivatives strongly depends on the ammonia concentration of the first-dimension solvents; similarly, the formic acid concentration of the second-dimension solvents

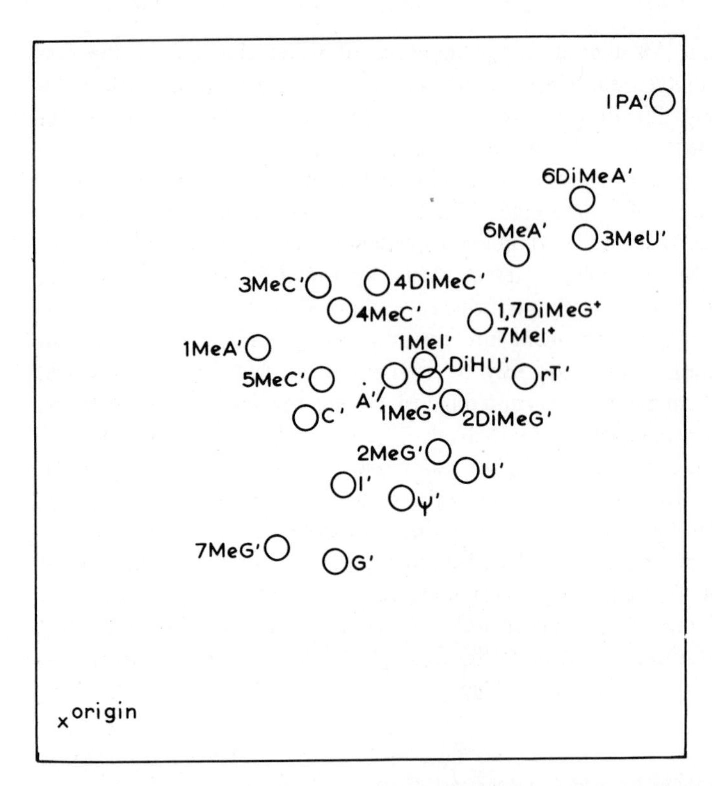

Figure 2. Cellulose map of ^{3}H-labeled nucleoside derivatives. First dimension from bottom to top. Second dimension from left to right. $7MeI^+$ and $1,7DiMeG^+$ represent glycerol, which is formed when 7MeI or 1,7DiMeG is subjected to $NaIO_4$–[^{3}H]-$NaBH_4$ treatment. (For the formation of glycerol from 7MeG, consult Table 1.)

strongly influences the mobility of A′ and its derivatives and, to a smaller degree, of G′ and its derivatives. This is due to the fact that positively or negatively charged species usually travel more slowly in partition chromatography than their neutral counterparts.

For complete conversion of the dialdehydes to trialcohols, a sufficiently large excess of borotritiide over periodate (>4 : 1) is required; otherwise partial reduction to ^{3}H-monoaldehydes (cf. Ref. *4*) may occur leading to spurious spots on the maps. On cellulose these monoaldehydes travel slightly slower in the ammonia system and slightly faster in the formic acid system than their completely reduced counterparts. They therefore are recognized as satellite spots located a little below and to the right of the corresponding trialcohol spots. Monoaldehyde formation was also observed when a sodium borotritiide solution of high specific activity (16 Ci/mmole) that had partly decomposed due to storage (–90° for 6 months) was used for reduction.

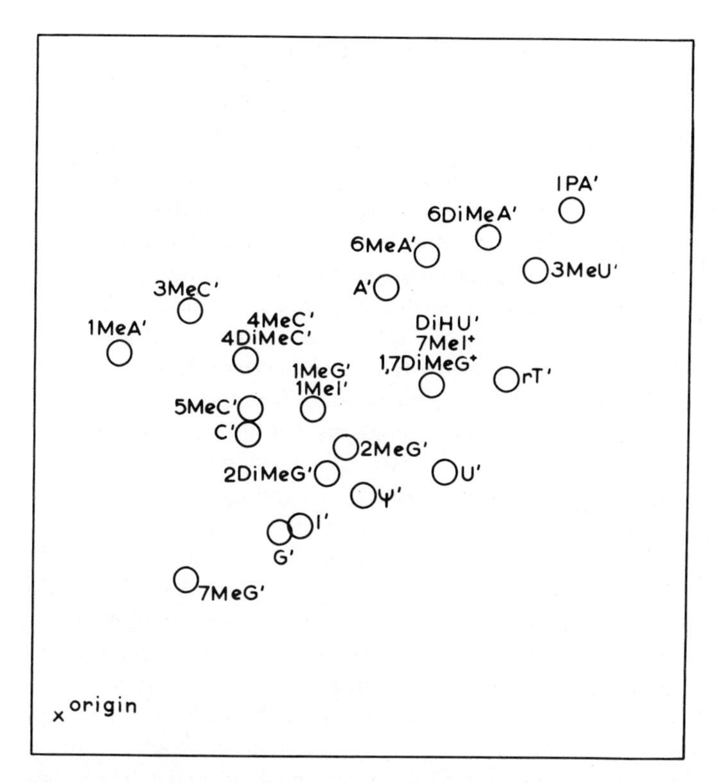

Figure 3. Silica gel/kieselguhr map of ^{3}H-labeled nucleoside derivatives. For explanation, see legend to Fig. 2.

Solutions of borotritiide of < 4 Ci/mmole specific activity could be stored at $-90°$ for many months without such a loss in reducing power. The presence of certain buffers also favors the formation of nucleoside monoaldehydes (see below).

The final tritium-labeled residue is taken up in 0.1 N formic acid rather than in water because CO_2 present in the solutions is reduced by borotritiide to tritium-labeled formic acid (*7, 8*), which may interfere with quantitative analysis under certain conditions. This effect is eliminated by dilution with unlabeled formic acid.

On some batches of the commercial cellulose sheets, U′ was found to form a double spot in the ammoniacal solvent. This phenomenon is without influence on the quantitative determination of uridine.

Some spots on the maps may be due to the introduction of tritium into compounds other than the nucleoside derivatives. In our experience this background has not presented any serious problems, provided the polynucleotide to be analyzed was not contaminated by carbohydrates or other substances susceptible to periodate oxidation. Only a small percentage of the total radioactivity of labeled digests is contributed by the background, and we have as yet seen no overlaps of background spots with spots derived from nucleosides. However, it appears advisable to carry a water blank (no nucleotide, no enzyme) and an enzyme blank (no nucleotide) through the entire procedure whenever a freshly prepared solution of borotritiide is used for the first time. We also test each freshly prepared solution of borotritiide by subjecting a mixture of adenosine, guanosine, cytidine, and uridine of known composition to labeling and quantitative evaluation (*1*).

The aromatic moiety of most ribonucleosides is not attacked by either periodate or borohydride under the conditions outlined above. Periodate oxidation, for example, of the base moiety of the major nucleosides requires higher concentration of the reagent, alkaline pH, and elevated temperature (*9*). More drastic conditions than those employed in the present procedure are also required to effect a reduction of susceptible nucleosides with borohydride (*10–14*). The ultraviolet spectra of the trialcohol derivatives usually closely resemble those of the parent nucleosides indicating the intactness of the original heterocyclic ring system. However, due to the possibility of side reactions, the reaction conditions specified above should be adhered to closely. For example, the replacement of bicine by *N*-2-hydroxyethylpiperazine-*N*′-2-ethanesulfonic acid (hepes) buffer of the same concentration and pH favors the formation of nucleoside monoaldehydes. The use of bicarbonate buffer (*2*) instead of bicine leads to side reactions of DiHU and possibly of other alkali-labile nucleosides, cf. below. Tris and similar buffers are attacked by periodate and cannot be used for this reason.

Pseudouridine reacts more sluggishly under the labeling conditions than "normal" nucleosides; too short reaction times or too low concentrations of the reagents may thus lead to incomplete labeling of this nucleoside. Pseudouridine also has a greater tendency to form monoaldehydes than other pyrimidine nucleosides. If the reduction is carried out in the presence of phosphate buffer about 5% of pseudouridine is converted to a decomposition product located between ψ' and G′ on cellulose.

With the exception of 4TU and some nucleosides carrying a positive charge at neutral pH (see below and Table 1), the radioactivity of each nucleoside derivative is directly proportional to the concentration of the corresponding parent nucleoside in the solution subjected to periodate-borotritiide treatment if the conditions specified are adhered to exactly. Statistical analysis of the data (cf. Ref. *1*) shows relative standard deviations to be usually $< \pm 2.5\%$.

4TU and possibly other thionucleosides cannot be analyzed directly by tritium labeling. The main product obtained from 4TU is $[^3H]$-U′. It

TABLE 1. Tritium Labeling of Alkali-Labile Nucleosides

Compounds	Products	Labeling only [a] (% recovered) [b]	Pretreatment enzymes, bicine, pH 8[c], followed by labeling (% recovered)[b]	Pretreatment NaOH, pH 13[d], followed by labeling (% recovered)[b]
DiHU	DiHU′	100	96	0
	Unidentified 1[e]	0	0	29
	Unidentified 2[e]	0	0	10
	Glycerol	0	0	15
	Total	100	96	54
7MeG	7MeG′	52	19	0
	Glycerol	8	21	8
	Unidentified[f]	22	12	0
	Total	82	52	8
1MeA	1MeA′	91	54	0
	6MeA′	3	33	100
	Total	94	87	100
3MeC	3MeC′	88	83	52
	Unidentified[g]	3	10	4
	Total	91	93	56

[a]Reaction conditions: 0.2 mM $NaIO_4$, 2 to 3-fold molar excess over nucleoside, 2 hr at room temperature; 1.1 mM [^{3}H]-KBH_4, 11 to 17-fold molar excess over nucleoside, 2 hr at room temperature. After addition of acetic acid and evaporation to dryness, the residue was taken up in 0.1 N formic acid. Chromatographic analysis on cellulose layers, fluorography, elution, and liquid scintillation counting were performed.

[b]As an internal standard a known amount of cytidine was added to each nucleoside at a molar concentration approximately equal to the concentration of the nucleoside to be analyzed.

[c]The same enzyme and buffer concentrations as described in the section on enzymatic hydrolysis of RNA were used, with the exception of a 13-fold greater molar excess of bicine over nucleoside. Incubation at 37° for 24 hr. Further treatment as under a. Due to the greater excess of buffer, a few percent of DiHU, 1MeA, 6MeA, and C[b] were recovered as the respective monoaldehydes whereas under standard conditions no monoaldehydes are obtained. The data are corrected for the monoaldehydes.

[d]Nucleoside (0.25 mM), 0.1 N NaOH, 22 hr at room temperature, neutralization with HCl, further treatment as under a.

[e]Unidentified products 1 and 2 are located closely together in the lower right quarter of the cellulose map. One of them may represent the expected trialcohol derivative of *N*-ribosylureidopropionic acid.

[f]The unidentified product is located in the lower right quarter of the cellulose map.

[g]The unidentified product is located in the upper right quarter of the cellulose map close to the front of the second dimension.

appears possible, however, to convert 4TU to 4MeC quantitatively at the macromolecular level (*15, 16*) and subsequently analyze the compound as [^{3}H]-4MeC′, which is well separated from all other trialcohols on the cellulose map (Fig. 2).

2′-*O*-Methyl nucleosides, not being susceptible to periodate oxidation, cannot be analyzed by tritium labeling. The behavior of N^4-acetylcytidine (*17*), [=N^6-(Δ^2-isopentenyl-)] adenosine (*17, 18*), and *N*-[9-(β-D-ribofuranosyl)purine-6-ylcarbamoyl]-threonine (*19*) has not been investigated in detail thus far.

Table 1 summarizes data obtained by labeling DiHU and nucleosides carrying a positive charge at physiological pH. These compounds are known to be alkali-labile (*20–26*). It was of interest therefore to evaluate whether the slightly alkaline conditions during enzymatic digestion and borotritiide treatment would result in side reactions. Pretreatment of 1MeA with 0.1 N NaOH led to quantitative conversion to 6MeA, a compound that is completely stable under the labeling conditions and is completely recovered therefore as [^{3}H]-6MeA′. Pretreatment of the other three nucleosides under the same conditions resulted in side reactions and an incomplete recovery of radioactive reaction products due to volatilization of 1- and 2-carbon fragments during the procedure. Under conditions similar to those described above for enzymatic RNA digestion, of the four compounds only 7MeG is degraded extensively so that a loss of about 50% in radioactivity occurs. 7MeI and 1,7DiMeG, which are not known to occur in RNA, behave similarly. DiHU appears completely stable, while 1MeA is partially converted to 6MeA. 3MeC undergoes an unknown side reaction but the total recoveries of radioactivity for this compound and 1MeA are approximately 90% of theory (cf. also Note added in proof).

In connection with the partial conversion of 1MeA to 6MeA it is noteworthy that plant and animal tRNA probably do not contain 6MeA (*27, 28*) so that the 1MeA content of the RNA can be obtained by adding the count rates from the 1MeA′ and 6MeA′ spots.

Periodate-borotritiide treatment may also be used for the characterization of diol end groups in ribopolynucleotides (*29, 30*).

It is obvious that the method presented can be easily adapted to the analysis of ribonucleoside mixtures obtained from other sources than ribopolynucleotides. For example, mononucleotides may be assayed by this procedure after enzymatic dephosphorylation. The method is therefore of potential value for pool size determinations of ribonucleotides and ribonucleosides. An analogous scheme involving periodate-borotritiide treatment can probably be developed for the direct analysis of 5′-ribonucleotides without prior dephosphorylation.

NOTE ADDED IN PROOF*

A two-dimensional combination of solvents recently developed in this laboratory allows on cellulose thin layers a complete resolution of the trialcohols of A, C, U, G, I, 1MeA, 6MeA, 6DiMeA, 3MeC, 4MeC, 4DiMeC, 5MeC, 3MeU, 5MeU (rT), DiHU, ψ, 1MeG, 2MeG, 2DiMeG, 7MeG, as well as glycerol. The solvents are: F = acetonitrile/ethyl acetate/*n*-butanol/isopropanol/6 N ammonia (7 : 2 : 1 : 1 : 2.7, by vol); G = *t*-amyl alcohol/methyl ethyl ketone/acetonitrile/ethyl acetate/water/formic acid, sp. gr. 1.2 (4 : 2 : 1.5 : 2 : 1.5 : 0.18). Development is with solvent F to 17 cm from the origin in the first dimension. For the second dimension, a 6-cm long paper wick (Whatman No. 1) is attached to the top of the sheet by stapling, and the chromatogram is developed with solvent G to 5 cm on the wick. To achieve a good resolution, it is essential that chromatography is begun immediately after the solvent has been poured into the tank. The overall chromatographic pattern under these conditions is similar to the one obtained with solvents A and B on cellulose. The area containing A′, 1MeG′, DiHU′, and 2DiMeG′ is, however, completely resolved in the new system.

IPA′, which travels close to the fronts of solvents A, B, F, and G, has an R_f of 0.74 in 10% ethanol on cellulose (R. M. Bock, personal communication).

Recoveries of alkali-labile nucleosides as [^{3}H]-trialcohols under the conditions described in the text (6 hr digestion, addition of phosphate buffer after periodate oxidation) are as follows: DiHU′, about 95%; 7MeG′, about 70%; 3MeC′, about 85%. Approximately 20% of 1MeA is converted to the N^6-methyl derivative (recovered as [^{3}H]-6MeA′).

* Randerath, K., unpublished data, April 1971.

References

1. Randerath, K., and E. Randerath. 1969. *Anal. Biochem.*, **28**: 110.
2. Randerath, E., J. W. Ten Broeke, and K. Randerath. 1968. *FEBS Letters*, **2**: 10.
3. Randerath, K., K. M. Flood, and E. Randerath. 1969. *FEBS Letters*, **5**: 31.
4. Khym, J. X., and W. E. Cohn. 1960. *J. Am. Chem. Soc.*, **82**. 6380.
5. Randerath, K., and E. Randerath. 1967. In L. Grossman and K. Moldave (Eds.), Methods in enzymology. Academic Press, New York. Vol. XIIA, p. 323.
6. Randerath, K., 1970. *Anal. Biochem.*, **34**: 188.
7. Wartik T., and R. K. Pearson. 1958. *J. Inorg. Nucl. Chem.*, **7**: 404.
8. Eisenberg, F., and A. H. Bolden. 1967. *Carbohydrate Res.*, **5**: 349.
9. Tomasz, M., Y. Sanno, and R. W. Chambers. 1965. *Biochemistry*, **4**: 1710.
10. Cerutti, P., K. Ikeda, and B. Witkop. 1965. *J. Am. Chem. Soc.*, **87**: 2505.

11. Ballé, G., P. Cerutti, and B. Witkop. 1966. *J. Am. Chem. Soc.*, **88**: 3946.
12. Cerutti, P. 1968. *Biochem. Biophys. Res. Commun.*, **30**: 434.
13. Cerutti, P., J. W. Holt, and N. Miller. 1968. *J. Mol. Biol.*, **34**: 505.
14. Macon, J. B., and R. Wolfenden. 1968. *Biochemistry*, **7**: 3453
15. Ziff, E. B., and J. R. Fresco. 1969. *Biochemistry*, **8**: 3242.
16. Burton, K. 1970. *FEBS Letters*, **6**: 77.
17. Feldmann, H., D. Dütting, and H. G. Zachau. 1966. *Hoppe-Seyler's Z. Physiol. Chem.*, **347**: 236.
18. Hall, R. H., M. J. Robins, L. Stasiuk, and R. Thedford. 1966. *J. Am. Chem. Soc.*, **88**: 2614.
19. Schweizer, M. P., G. B. Chheda, L. Baczynskyj, and R. H. Hall. 1969. *Biochemistry*, **8**: 3283.
20. Batt, R. D., J. K. Martin, J. M. Ploeser, and J. Murray. 1954. *J. Am. Chem. Soc.*, **76**: 3663.
21. Cohn, W. E., and D. G. Doherty. 1956. *J. Am. Chem. Soc.*, **78**: 2863.
22. Brookes, P., and P. D. Lawley. 1960. *J. Chem. Soc.*, p. 539.
23. Jones, J. W., and R. K. Robins. 1963. *J. Am. Chem. Soc.*, **85**: 193.
24. Haines, J. A., C. B. Reese, and A. R. Todd. 1962. *J. Chem. Soc.*, p. 5281.
25. Lawley, P. D., and P. Brookes. 1963. *Biochem. J.*, **89**: 127.
26. Townsend, L. B., and R. K. Robins. 1963. *J. Am. Chem. Soc.*, **85**: 242.
27. Dunn, D. B., 1959. *Biochim. Biophys. Acta*, **34**: 286.
28. Iwanami, Y., and G. M. Brown. 1968. *Arch. Biochem. Biophys.*, **124**: 472.
29. RajBhandary, U. L., 1968. *J. Biol. Chem.*, **243**: 556.
30. Leppla, S. H., B. Bjoraker, and R. M. Bock. 1968. L. Grossman and K. Moldave (Eds.), Methods in enzymology. Academic Press, New York, Vol. XIIB, p. 236.

SECTION

ENZYMES OF POLYNUCLEOTIDE SYNTHESIS

POLYNUCLEOTIDE KINASE FROM *Escherichia coli* INFECTED WITH BACTERIOPHAGE T4

CHARLES C. RICHARDSON

DEPARTMENT OF BIOLOGICAL CHEMISTRY

HARVARD MEDICAL SCHOOL

BOSTON, MASSACHUSETTS

Polynucleotide kinase (*1, 2*) has been isolated from *Escherichia coli* cells infected by bacteriophages T2 or T4. It catalyzes the transfer of the γ-phosphate group of ATP to the 5′-hydroxyl termini of deoxyribonucleic acid and ribonucleic acid (*1–3*), as shown in Fig. 1. The T2 and T4 phage-induced enzymes have similar properties and specificities. Unless otherwise stated, the following description refers to the studies carried out with the polynucleotide kinase found in T4 phage-infected cells.

ASSAY

Preparation of Substrates

5′-Hydroxyl-Terminated DNA

5′-Hydroxyl-terminated DNA is prepared by partially digesting native DNA with micrococcal nuclease (*4*). This enzyme hydrolyzes phosphodiester bonds of DNA to produce oligonucleotides terminated by 5′-hydroxyl and 3′-phosphoryl groups (*5–8*). The incubation mixture (20 ml) consists of 0.05 M glycine buffer (pH 9.2), 0.01 M $CaCl_2$, 20 μmoles of salmon sperm DNA, and 74 units of micrococcal nuclease. (A unit of enzyme is the amount catalyzing the production of 10 mμmoles of acid-soluble DNA nucleotide in

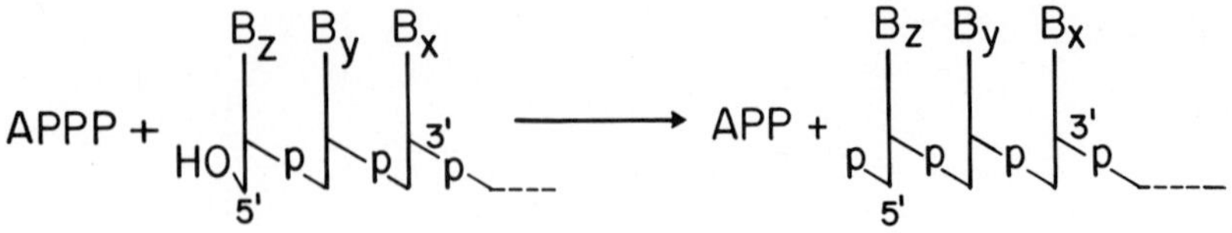

Figure 1. The transfer of phosphate from ATP to the 5′-hydroxyl terminus of a polynucleotide by polynucleotide kinase.

30 min.) The reaction mixture is incubated at 37°C until 20 to 30% of the DNA becomes acid-soluble. The assay consists of removing aliquots of the incubation mixture at intervals, adding perchloric acid to a final concentration of 0.4 N, centrifuging, and comparing the optical density (at 260 nm) of the supernatant fluid with that of an aliquot of the reaction mixture diluted into 1 N NaOH. The digestion is terminated by chilling the reaction mixture and dialyzing it against 100 volumes of 1 M KCl for 12 hr at 4°C to remove the dialyzable oligonucleotides. The dialysis procedure is repeated three times and the mixture is finally dialyzed against 0.02 M KCl. The extensively dialyzed solution represents the 5′-hydroxyl-terminated DNA. The solution is stored at –20°C.

γ-Labeled-^{32}P-ATP

ATP labeled with ^{32}P in the γ-phosphate is prepared by a modification (*9*) of the procedure of Glynn and Chappell (*10*). The incubation mixture (3.0 ml) contains 0.05 M Tris-HCl buffer (pH 8.0), 6 mM $MgCl_2$, 2 mM cysteine (free base), 0.33 mM 3-phosphoglycerate, 1 mM ATP, 25 mCi of $^{32}P_i$ (carrier-free, New England Nuclear), 0.2 mg of rabbit muscle 3-phosphoglyceraldehyde dehydrogenase (Calbiochem), and 0.02 mg of yeast 3-phosphoglycerate kinase (Calbiochem). The mixture is incubated at 26°C until 70–80% of the radioactivity is converted to Norit-adsorbable material. The assay consists of removing 0.01-ml aliquots before the addition of the enzymes and at intervals of 15 min. The aliquots are added to tubes containing 0.1 ml of 1 N HCl, 0.2 ml of bovine serum albumin (1 mg/ml), 0.2 ml of 5 mM P_i, 0.2 ml of 5 mM PP_i, 0.2 ml of Norit (20% packed volume), and 0.5 ml of H_2O, all at 0°C. After mixing, the Norit is removed by centrifugation for 5 min at 10,000 *g*, and the radioactivity in the supernatant fluid is measured by liquid scintillation counting. When the exchange reaction is complete (20–40 min), the incubation mixture is chilled to 0°C and diluted to a volume of 18 ml with H_2O. The solution is applied to a column of Dowex 1, Cl^-, 10% crosslinked (Bio-Rad AG 1-X10, 100 to 200 mesh; 1 x 3 cm) which has previously been washed with 50 ml of 1 N HCl and then with 1 liter of H_2O. The resin is washed first with 40 ml of 0.02 M

NH_4Cl–0.02 M HCl and then with 50 ml of H_2O. The ATP is eluted from the column with 70 ml of 0.25 N HCl. The ATP is concentrated by adsorption to, and elution from, Norit; the 70 ml of solution containing ATP is applied by suction to a column (0.5-cm diameter) containing 1 ml of a suspension of Norit and Celite 545 (Johns-Manville) (10% packed volume of each). The Norit is washed with 10 ml of cold 0.01 N HCl and the ATP is eluted with 6 ml of a solution of 50% ethanol–water containing 0.5 ml of 12 N NH_4OH per 100 ml of solution. Contaminating Norit particles are removed by centrifugation at 10,000 *g* for 10 min. The solution is reduced to a volume of approximately 1 ml under a stream of N_2 and is then diluted to yield a 1 mM solution of ATP. The specific radioactivity of the ^{32}P-ATP is approximately 1×10^{10} cpm per μmole. The overall recovery of both ATP and radioactivity is approximately 40%. ^{32}P-ATP having the specific radioactivity desired for the kinase assay is obtained by diluting a portion of the ^{32}P-ATP with unlabeled ATP.

Reagents

1. The reaction mixture (total volume 0.3 ml) contains 5′-hydroxyl-terminated salmon sperm DNA, 80 mμmoles of DNA-phosphorus
   0.066 mM γ-^{32}P-ATP, 5×10^6 cpm/μmole
   0.07 M Tris-HCl Buffer, pH 7.6
   0.01 M $MgCl_2$
   0.005 M dithiothreitol
   enzyme, 0.05 to 0.5 unit in a volume of 0.02 ml
2. Enzyme diluent: 0.05 M Tris-HCl buffer (pH 7.6), containing 0.01 M 2-mercaptoethanol and bovine plasma albumin (0.5 mg per ml)
3. Salmon sperm DNA, 0.25 mg/ml
4. 0.6 N trichloroacetic acid
5. 0.01 N HCl

Procedure

The standard assay for polynucleotide kinase measures the conversion of the radioactivity in ^{32}P-ATP into an acid-insoluble product.

After 30 min of incubation at 37°C, the reaction is stopped by the addition of 0.20 ml of a cold solution of salmon sperm DNA (0.25 mg/ml), 0.50 ml of cold 0.6 N trichloroacetic acid, and 2 ml of cold water. After centrifugation of 5 min at 10,000 *g*, the supernatant fluid is discarded. The precipitate is resuspended in 2 ml of 0.01 N HCl, recentrifuged, and the supernatant fluid discarded. The precipitate is dissolved in 0.5 ml of 0.1 N

NaOH, and reprecipitated by the addition of 0.5 ml of cold 0.6 N trichloroacetic acid. After the addition of 2 ml of cold water, the precipitate is collected on a glass filter (Whatman GF/C glass paper, 2.4-cm diameter) and washed three times with 2-ml portions of 0.01 N HCl. The paper is transferred to a planchet, dried, and the radioactivity is measured in a gas-flow counter. If the radioactivity is measured in a scintillation counter, the filter is transferred to a scintillation vial, dried, and covered with 10 ml of a scintillator solution consisting of 4 gm of 2,5-diphenyloxazole (PPO) and 50 mg of 1,4-*bis*-2-(4-methyl-5-phenyloxazolyl) benzene (dimethyl POPOP) per liter of toluene.

A control reaction mixture without enzyme is incubated and treated as described above. The control incubations contain 0.05–0.1% of the added radioactivity. A less time-consuming filtration assay, but one which yields higher control values, has been described (*9*). This latter procedure is useful when assaying a large number of fractions obtained during the purification of the enzyme.

A unit of polynucleotide kinase activity is defined as the amount catalyzing the production of 1 mμmole of acid-insoluble ^{32}P in 30 min. The radioactivity made acid-insoluble is proportional to enzyme concentrations at levels of from 0.05 to 0.5 unit of enzyme.

Enzyme Dilutions

Dilutions of the enzyme for the assay are made in a solution composed of 0.05 M Tris-HCl buffer (pH 7.6), bovine plasma albumin (0.5 mg/ml), and 0.01 M 2-mercaptoethanol. Samples of the diluted enzyme solutions are added to the reaction mixture within 30 min of dilution.

ISOLATION

Unless otherwise stated, all operations are carried out at 0–4°C. All centrifugations are performed at 15,000 *g* for 10 min. The purification procedure and the results of a typical preparation are summarized in Table 1. An alternative procedure for the purification of the enzyme from *E. coli* infected with phage T2*am*3 has been described (*2*). The following procedure is for the enzyme induced by phage T4 (*1*).

Growth of Cells

E. coli B is grown at 37° under forced aeration in a Fermocell (New Brunswick Scientific Company, New Brunswick, N. J.) in 100 liters of M-9 medium (*11*) containing casamino acids (2 gm per liter). At a cell density of

TABLE 1. Purification of Polynucleotide Kinase from *E. coli* Infected with Phage T4

Fraction and step	Total units ($\times 10^{-3}$)	Specific activity (units/mg)
I. Extract	95	40
II. Streptomycin	90	60
III. Autolysis	115	150
IV. Ammonium sulfate	105	420
V. DEAE-Cellulose	68	8,500
VI. Phosphocellulose	40	59,000
VII. Hydroxylapatite	27	– [a]

[a]Protein concentration insufficient to obtain an accurate specific activity.

2×10^9 per milliliter L-tryptophan is added to yield a final concentration of 1 μg per milliliter. T4r$^+$ phages are added at a multiplicity of 4. Twenty minutes after infection, the culture is quickly cooled to 5°C, and the cells are harvested by centrifugation at 35,000 *g* in a Sharples continuous-flow centrifuge at a flow rate of 0.5 liter per minute. The gummy paste (250 gm) is stored at –40° and shows no loss of activity after 1 yr.

Polynucleotide kinase has also been prepared from *E. coli* B (nonpermissive host) infected with T4*am*XFl defective in genes 41–45. Extracts prepared from *E. coli* B infected with this mutant are free of the T4 DNA polymerase (*12, 13*) and the T4 deoxycytidylate hydroxymethylase (*14, 15*). The possibility of contamination of the purified kinase by DNA ligase can be decreased by the use of T4*am*H39X (gene 30) infected cells (*16*), since gene 30 is the structural gene for the T4 DNA ligase (*17*). The same procedure for growth and infection as that described above for wild-type T4 is followed when the mutant phages are used, except that the cells are harvested 60 min after infection. Extracts prepared in this manner from the mutant-infected cells have a level of polynucleotide kinase activity 50 to 100% higher than that of wild-type extracts.

Ribonuclease activity can be greatly reduced in all fractions by using ribonuclease mutants of *E. coli. E. coli* AR19 (a mutant deficient in RNase 1) has been used for this purpose (*18*).

Preparation of Extracts

Frozen cells (25 gm) are suspended in 100 ml of 0.05 M Tris buffer (pH 7.4) containing 0.001 M glutathione. After disruption of the cells by sonication (Branson sonicator, Model S-57), cell debris is removed by centrifugation. The supernatant fluid is collected and sufficient buffer is added to yield a solution whose optical density at 260 nm is 105 (Fraction I). This fraction may be stored at 0°C for 1 week without loss of activity.

Streptomycin Precipitation

To 140 ml of extract are added, with stirring, 28 ml of 5% streptomycin sulfate over a 30-min period. After an additional 15 min, the suspension is centrifuged and the supernatant fluid discarded. The precipitate is suspended in 140 ml of 0.1 M potassium phosphate buffer (pH 7.5) containing 0.002 M glutathione (Fraction II). This fraction is immediately subjected to autolysis.

Autolysis

To 140 ml of Fraction II is added 0.40 ml of 1 M $MgCl_2$. The suspension is incubated at 37° for 1–2 hr, until 90% of the ultraviolet-absorbing material at 260 nm is rendered acid-soluble. The assay consists of removing 1-ml samples of the incubation mixture at intervals of 15 min, centrifuging, and determining the optical density of the supernatant fluid at 260 nm after a suitable dilution in 0.05 M Tris buffer (pH 7.4). A portion of the supernatant fluid is precipitated with an equal volume of cold 1 N perchloric acid, and the optical density of the acid-soluble fraction is determined after centrifugation. When the digestion is completed (60–120 min), the autolyzate is chilled to 0°C. The protein that precipitated out during the digestion is removed by centrifugation. The supernatant fluid (Fraction III) is immediately subjected to ammonium sulfate fractionation.

Ammonium Sulfate Fractionation

To 130 ml of Fraction III 13 gm of ammonium sulfate is added, with stirring, over a 30-min period. After an additional 20 min, the precipitate is removed by centrifugation and an additional 25 gm of ammonium sulfate is added to the supernatant fluid over a 30-min period. Twenty minutes later, the precipitate is collected by centrifugation and dissolved in 10 ml of 0.1 M potassium phosphate buffer (pH 7.5) containing 0.002 M glutathione (Fraction IV). Fraction IV may be stored for 4 years at –50° without loss of activity.

DEAE-Cellulose Fractionation

A column of DEAE-cellulose (8 cm^2 x 12 cm) is prepared and washed with 5 liters of 0.01 M potassium phosphate buffer (pH 7.5) containing 0.01

M 2-mercaptoethanol. Fraction IV (200 mg of protein) is diluted to 30 ml with the same buffer, dialyzed against 1 liter of the equilibrating buffer for 5 hr, and applied to the column. The adsorbent is washed with 200 ml of the same buffer and the enzyme is then eluted with 0.05 M potassium phosphate buffer (pH 7.5) containing 0.01 M 2-mercaptoethanol. The 0.05 M eluate is collected in 5-ml fractions. Of the enzyme applied to the adsorbent, 50–70% is obtained in 20 ml of the 0.05 M eluate (Fraction V). Fraction V has been stored at 0°C for 2 weeks without loss of activity.

Phosphocellulose Fractionation

A column of phosphocellulose (1 cm^2 x 10 cm) is prepared and washed with 1 liter of 0.05 M potassium phosphate buffer (pH 7.5) containing 0.01 M 2-mercaptoethanol. Fraction V (3 mg of protein) is applied to the column. After the resin is washed with 10 ml of the same buffer, elution of the enzyme is accomplished by 20-ml portions of 0.05 M potassium phosphate buffer (pH 7.5) containing 0.01 M 2-mercaptoethanol plus the following concentrations of KCl: 0.05 M, 0.1 M, and 0.25 M. Two-ml fractions are collected, and approximately 70% of the activity applied to the adsorbent is recovered in the 0.25 M eluate in a volume of 6 ml (Fraction VI). Fraction VI has been stored at 0°C for 8 months without loss of activity (less than 10%).

Hydroxylapatite Chromatography

An additional chromatography step has been introduced to remove traces of deoxyribonuclease or DNA ligase observed in some preparations of Fraction VI (*19*). A column of hydroxylapatite (1 cm^2 x 5 cm) is prepared and washed, first with 15 ml of 0.5 M potassium phosphate buffer (pH 7.0) containing 0.01 M 2-mercaptoethanol, and then with 175 ml of 0.02 M potassium phosphate buffer (pH 7.0) containing 0.01 M 2-mercaptoethanol. Fraction VI is diluted 4-fold with 0.02 M potassium phosphate buffer (pH 7.0), 0.01 M 2-mercaptoethanol and then applied to the column. The adsorbent is washed with 30 ml of the same buffer. The protein is eluted with 10-ml portions of potassium phosphate buffer (pH 7.0) containing 0.01 M 2-mercaptoethanol and having the following concentrations: 0.1 M, 0.2 M, 0.3 M, and finally 0.5 M. The eluate is collected in 5-ml fractions. More than 75% of the activity is found in the 0.3 M eluate. These fractions are pooled and dialyzed against 0.02 M potassium phosphate buffer (pH 7.0), 0.025 M potassium chloride, 0.01 M 2-mercaptoethanol in order to decrease the phosphate concentration in the enzyme fraction, since phosphate is inhibitory to the kinase activity.

Storage of the Purified Enzyme

Both the phosphocellulose (Fraction VI) and the hydroxylapatite fraction (Fraction VII) can be stored at 0°C for at least six months without significant loss in activity. However, in order to avoid bacterial growth, these fractions are usually made 50% with glycerol and stored at –20°C. Under these storage conditions there is no loss in activity (less than 10%) over a period of 12 months. These fractions should not be frozen, since freezing and thawing markedly reduces the activity.

PROPERTIES OF THE ENZYME

Contamination with Other Enzymes

In the majority of the preparations of polynucleotide kinase, Fraction VI is free of any contaminating deoxyribonuclease or DNA ligase. The absence of deoxyribonuclease is based on the following criteria: (1) no acid-soluble radioactivity is released from uniformly labeled double- or single-stranded DNA (*1*); (2) no single-strand breaks are introduced when the enzyme quantitatively labels the ends of dephosphorylated T7 DNA (*20*); and (3) alkaline sedimentation analysis of ϕX 174 RFI after treatment with the purified kinase fails to reveal the formation of any RFII. (Less than 1 mμmole of phosphodiester bonds is cleaved per 1000 units of enzyme.) There is no DNA ligase (less than 1 unit per 1000 units of kinase) measured in the standard T4 ligase assay (*21*) in the presence of ATP or DPN.

When *E. coli* A19 (a mutant deficient in RNase 1) is used as the host, the purified kinase is free of ribonuclease (*18*).

Fraction VI is free of phosphatase activity, as seen by the absence of hydrolysis of AMP, ADP, ATP, or 5′-^{32}P-phosphoryl-terminated DNA.

In approximately 10% of the preparations of polynucleotide kinase, Fraction VI has been contaminated with either an endonuclease or the T4 DNA ligase. Either activity can be decreased below an assayable level by chromatography on hydroxylapatite to yield Fraction VII as described above. The T4 ligase can also be eliminated from Fraction VI by using a ligase-negative mutant, T4*am*XFl, as the infecting phage.

Reaction Requirements

Effect of pH and Divalent Metals

Maximal activity is obtained at pH 7.6 in Tris-HCl buffer. In the absence of added $MgCl_2$, there is no detectable activity. The optimal Mg^{2+}

concentration at pH 7.6 is 1×10^{-2} M. Mn^{2+} can partially fulfill the metal requirement; at an optimal concentration of 3.3×10^{-3} M, only 50% of the maximal activity obtained with Mg^{2+} is observed.

Sulfhydryl Requirement

Maximal activity is obtained with 5 mM dithiothreitol in the reaction mixture. Protection is also achieved with 10 mM 2-mercaptoethanol and 10 mM glutathione, but only 80 and 70%, respectively, of that observed with dithiothreitol. In the absence of a sulfhydryl compound, there is 2% of the optimal activity. Although the phage T2 induced polynucleotide kinase is inhibited by *p*-hydroxymercuribenzoate, there is no requirement for a sulfhydryl compound in the standard reaction mixture (*2, 3*).

Inhibitors

Phosphate anion is inhibitory to the purified kinase. At pH 7.6 in either 0.07 M sodium or potassium phosphate buffer, 5% of the value observed in Tris buffer is obtained. Potassium or sodium phosphate at a concentration of 7 mM in the standard reaction mixture, containing Tris buffer, results in 50% inhibition. At an $(NH_4)_2SO_4$ concentration of 7 mM, the kinase activity is inhibited 75%. Sodium pyrophosphate at 3×10^{-4} M has been shown to inhibit the T2 polynucleotide kinase 50% (*3*), but has not been tested in the T4 kinase reaction.

Requirement for Nucleoside Triphosphate

A variety of nucleoside triphosphates can function as the phosphorylating agent. While ATP is used in the standard assay, CTP, UTP, GTP, dATP, and dTTP have all been found to function equally as well. The ATP concentration, 0.066 mM, used in the assay is not saturating. At a saturating level of ATP (0.3 mM), the rate of phosphorylation is approximately 3-fold higher than that observed with the concentration routinely used.

Requirement for Acceptor Activity of Nucleic Acid

A variety of nucleic acid compounds can be phosphorylated in the polynucleotide kinase reaction, provided they have a nucleotide bearing a free 5′-hydroxyl group with a phosphoryl group esterified at the 3′-position (Table 2). Thus, the substrates include DNA and RNA, oligonucleotides, and nucleoside 3′-monophosphates. There is no detectable phosphorylation of nucleosides, nucleoside 2′-phosphates, 3′-termini, or 5′-termini bearing phosphomonoesters. The enzyme will phosphorylate 5′-hydroxyl groups located at single-strand breaks in bihelical DNA molecules, but the rate of phosphorylation is reduced, and quantitative phosphorylation is difficult to

TABLE 2. Specificity of Polynucleotide Kinase

Substrate	Rate[a] (mμmoles/min/mg protein)
Acceptors	
5′-OH terminated salmon sperm DNA (micrococcal nuclease treated)	2200
5′-OH terminated ribosomal RNA (micrococcal nuclease treated)	2900
d(TpTpTpTp)	2100
d(ApApApAp)	2300
d(GpGpGp)	2400
d(CpCpCp)	2000
TpT	2400
ApA	2600
CpC	2700
GpG	2600
3′-AMP	2200
3′-dAMP	2900
Nonacceptors	
5′-P terminated T7 DNA (pancreatic DNase treated)	<20
5′-P terminated tRNA	<20
pApA	<5
2′-AMP	<50
Adenosine	<10

[a] The rate of reaction was measured in the standard reaction mixture with the replacement of the usual micrococcal nuclease treated salmon sperm DNA by the compounds listed.

achieve (*9*). All four of the nucleotides which are found in DNA, when terminally located, are phosphorylated at equal rates (Table 2).

Products of the Reaction

The reaction products are equimolar amounts of ADP and 5′-phosphoryl-terminated nucleic acid (Fig. 1). For each mole of phosphorylated product and ADP formed, one mole of ATP is lost from the reaction mixture. With the addition of excess enzyme or with prolonged incubation, the phosphorylation of nucleic acids is quantitative.

APPLICATIONS

Polynucleotide kinase can be used as a reagent in the study of nucleic acids. Not only is the reaction catalyzed by this enzyme highly specific, but the sensitivity of the reaction is limited only by the specific radioactivity of the ^{32}P-ATP used. The enzyme has been used (*1*) to characterize the 5′-termini of polynucleotides, (*2*) to prepare specifically labeled nucleic acid molecules for use in studying other enzymes, and (*3*) to radioactively label preformed polynucleotides that would otherwise be difficult to label *in vivo*. A summary of these applications follows.

End-Group Analysis

Unless otherwise stated, the detailed experimental procedure for end-group analysis is described in Ref. *9*. The removal of unreacted ^{32}P-ATP from the labeled product, a necessity for these procedures, can be accomplished by a number of methods, the choice of which depends on the properties of the product. For high-molecular-weight nucleic acids preparative sedimentation, dialysis (*20*), gel filtration (*22*), or a combination of acid precipitation and filtration (*2, 9*) can be used. However, in studies with small oligonucleotides, the ^{32}P-ATP and product must be separated by electrophoresis or chromatography (*1, 23, 24*). In some instances it is preferable to first degrade the ATP by chemical or enzymatic treatment (*3, 25*).

Total Number of 5′-End Groups

In order to determine the number of 5′-end groups, both external or internal (single-strand breaks), in a nucleic acid preparation, the nucleic acid is first denatured and then treated with *E. coli* alkaline phosphatase to remove all phosphomonoesters. After removal (*20*), inactivation (*26*), or inhibition (*9*) of the phosphatase, the nucleic acid is quantitatively phosphorylated with polynucleotide kinase and ^{32}P-ATP. The number of end groups is then determined by isolating the labeled nucleic acid free of unreacted ^{32}P-ATP, and measuring the radioactivity of the product.

Number of 5′-Hydroxyl and 5′-Phosphoryl End Groups

An analysis of the extent of phosphorylation of a denatured nucleic acid before and after phosphatase treatment permits the determination of the relative number of 5′-hydroxyl and 5′-phosphoryl end groups in a preparation.

Number of External and Internal 5′-End groups in DNA

At 37°C, under specific reaction conditions (*9*), *E. coli* alkaline phosphatase is specific for external phosphomonoester bonds in DNA. This fact permits the quantitative estimation of these end groups. The procedure for determining the number of external 5′-end groups is similar to that described above for the total number of end groups, except that the incubation with phosphatase at 37°C is performed prior to denaturation. The number of single-strand breaks can then be determined by subtracting the number of external 5′-end groups from the number of total 5′-end groups. The results are valid only if the DNA does not contain a significant number of internal 5′-hydroxyl end groups.

Identification of 5′-Nucleotides

In order to identify the 5′-terminal nucleotides of a nucleic acid, they are first radioactively labeled in the kinase reaction. After removal of the unreacted ^{32}P-ATP, the 5′-^{32}P-labeled polymer is enzymatically degraded to its constituent 5′-mononucleotides and the amount of radioactivity in each base is measured after fractionation of the nucleotides. By using the techniques described above, information can be obtained about the end groups present at single-strand breaks as well as at the ends of duplex DNA molecules.

Examples of End-Group Analysis

End-group analysis has been used to characterize the 5′-end groups of the DNA isolated from phages T7 (*20, 27*) and λ (*28, 29*). This technique has led to the identification of the 5′-end groups of these DNAs as well as providing an independent estimation of their molecular weights. The method has been used to characterize the interruptions found in T5 DNA (*19*) and in the replicative intermediates of φX 174 DNA (*22*). The enzyme has been used in studies on ribosomal RNA (*23*) and on Satellite Tobacco Necrosis Virus RNA (*24*). The technique has also been useful in characterizing phosphodiester bond cleavages introduced into DNA by sonic irradiation (*20*), by micrococcal nuclease (*1*), by pancreatic DNase (*9, 26*), and by T4 endonucleases II and IV (*30, 31*).

Preparation of Labeled Nucleic Acids

The ability of the kinase to specifically label the 5′-termini of nucleic acids permits the preparation of radioactively labeled molecules which can be used as substrates for other studies.

Several assays for DNA ligase make use of DNA substrates containing

^{32}P-phosphoryl groups at single-strand breaks. Such preparations include (1) DNA with internal 5′-^{32}P-phosphomonoesters (*9*), (2) a double-stranded homopolymer pair consisting of multiple dT units labeled with ^{32}P-phosphomonoesters hydrogen-bonded to a long dA chain (*32*), and (3) poly dAT circles bearing 5′-^{32}P-phosphomonoesters (*33*). These substrates make possible the rapid assay of joining, since the incorporation of the radioactivity into a phosphodiester bond renders it insusceptible to the action of alkaline phosphatase.

The characterization of synthetic polymers of known sequences and their joining to produce long polynucleotides has been facilitated by the introduction of ^{32}P at the 5′-termini (*34, 35*).

5′-^{32}P-labeled DNA and RNA have been used to purify and to characterize a 5′-phosphatase from T4-infected *E. coli* (*36*). DNA containing internal and external 5′-^{32}P-phosphomonoesters has been used to study the action of *E. coli* alkaline phosphatase on these end groups (*9*). A similar substrate has been useful in characterizing the 5′-hydrolytic activity of *E. coli* DNA polymerase (*37, 38*).

Fingerprinting Nonradioactive Nucleic Acids

Not all nucleic acids can be uniformly labeled *in vivo* or obtained in sufficient amounts to permit sequencing of their nucleotides by methods other than those involving radioactive labeling. However, if the oligomers isolated from digests of these nucleic acids can be radioactively labeled at their 5′-termini in the kinase reaction, the oligomer can then be fractionated by methods suitable for polymers labeled *in vivo*. This method has been applied to depurinated fd DNA and to RNase T_1 digests of formylmethionine tRNA (*25*), and is described more fully elsewhere in this volume (page 780).

References

1. Richardson, C. C. 1965. *Proc. Natl. Acad. Sci., U.S.*, **54**: 158.
2. Novogrodsky, A., and J. Hurwitz. 1966. *J. Biol. Chem.* **241**: 2923.
3. Novogrodsky, A., M. Tal, A. Traub, and J. Hurwitz. 1966. *J. Biol. Chem.*, **241**: 2933.
4. Richardson, C. C., and A. Kornberg. 1964. *J. Biol. Chem.*, **239**: 242.
5. Cunningham, L., B. W. Catlin, and M. J. Privat de Garilhe. 1956. *J. Am. Chem. Soc.*, **78**: 4642.
6. Cunningham, L. 1958. *J. Am. Chem. Soc.*, **80**: 2546.
7. Privat de Garilhe, M. J., L. Cunningham, U. R. Laurila, and M. Laskowski. 1957. *J. Biol. Chem.*, **224**: 751.
8. Alexander, M., L. A. Heppel, and J. Hurwitz. 1961. *J. Biol. Chem.*, **236**: 3014.
9. Weiss, B., T. R. Live, and C. C. Richardson. 1968. *J. Biol. Chem.*, **243**: 4530.

10. Glynn, I. M., and J. B. Chappell. 1964. *Biochem. J.*, **90**: 147.
11. Anderson, E. H. 1946. *Proc. Natl. Acad. Sci., U.S.*, **32**: 120.
12. De Waard, A., A. V. Paul, and I. R. Lehman. 1965. *Proc. Natl. Acad. Sci., U.S.*, **54**: 1241.
13. Warner, H. R., and J. E. Barnes. 1966. *Virology*, **28**: 100.
14. Wiberg, J. S., M. L. Dirksen, R. H. Epstein, S. E. Luria, and J. M. Buchanan. 1962. *Proc. Natl. Acad. Sci., U.S.*, **48**: 293.
15. Dirksen, M. L., J. C. Hutson, and J. M. Buchanan. 1963. *Proc. Natl. Acad. Sci., U.S.*, **50**: 507.
16. Lindahl, T., and G. M. Edelman. 1968. *Proc. Natl. Acad. Sci., U.S.*, **61**: 680.
17. Fareed, G. C., and C. C. Richardson. 1967. *Proc. Natl. Acad. Sci., U.S.*, **58**: 665.
18. Takanami, M. 1967. *J. Mol. Biol.*, **23**: 135.
19. Jacquemin-Sablon, A., and C. C. Richardson. 1970. *J. Mol. Biol.*, **47**: 477.
20. Richardson, C. C. 1966. *J. Mol. Biol.*, **15**: 49.
21. Weiss, B., A. Jacquemin-Sablon, T. R. Live, G. C. Fareed, and C. C. Richardson. 1968. *J. Biol. Chem.*, **243**: 4543.
22. Knippers, R., A. Razin, R. Davis, and R. L. Sinsheimer. 1969. *J. Mol. Biol.*, **45**: 237.
23. Takanami, M. 1967. *J. Mol. Biol.*, **29**: 323.
24. Reichmann, M. E., A. Y. Chang, L. Faiman, and J. M. Clark. 1966. *Cold Spring Harbor Symp. Quant. Biol.*, **31**: 139.
25. Székely, M., and F. Sanger. 1969. *J. Mol. Biol.*, **43**: 607.
26. Scheffler, I. E., E. L. Elson, and R. L. Baldwin. 1968. *J. Mol. Biol.*, **36**: 291.
27. Weiss, B., and C. C. Richardson. 1967. *J. Mol. Biol.*, **23**: 405.
28. Richardson, C. C., and B. Weiss. 1966. *J. Gen. Physiol.*, 49 (No. 6, Pt. 2): 81.
29. Wu, R., and A. D. Kaiser. 1967. *Proc. Natl. Acad. Sci., U.S.*, **57**: 170.
30. Sadowski, P. D., and J. Hurwitz. 1969. *J. Biol. Chem.*, **244**: 6182.
31. Sadowski. P. D., and J. Hurwitz. 1969. *J. Biol. Chem..*, **244**: 6192.
32. Olivera, B. M., and I. R. Lehman. 1967. *Proc. Natl. Acad. Sci., U.S.*, **57**: 1426.
33. Olivera, B. M., I. E. Scheffler, and I. R. Lehman. 1968. *J. Mol. Biol.*, **36**: 275.
34. Khorana, H. G., H. Büchi, M. H. Caruthers, S. H. Chang, N. K. Gupta, A. Kumar, E. Ohtsuka, V. Sgaramella, and H. Weber. 1968. *Cold Spring Harbor Symp. Quant. Biol.*, **33**: 35.
35. Narang, S. A., J. J. Michniewicz, and S. K. Dheer. 1969. *J. Am. Chem. Soc.*, **91**: 936.
36. Becker, A., and J. Hurwitz. 1967. *J. Biol. Chem.*, **242**: 936.
37. Klett, R. P., A. Cerami, and E. Reich. 1968. *Proc. Natl. Acad. Sci., U.S.*, **60**: 943.
38. Deutscher, M. P., and A. Kornberg. 1969. *J. Biol. Chem.*, **244**: 3029.

Qβ RNA POLYMERASE FROM PHAGE Qβ-INFECTED *E. coli*

LILLIAN EOYANG AND J. T. AUGUST
DEPARTMENT OF MOLECULAR BIOLOGY
DIVISION OF BIOLOGICAL SCIENCES
ALBERT EINSTEIN COLLEGE OF MEDICINE
BRONX, NEW YORK

Host cell infection by RNA viruses constitutes the only biological system in which the replication of RNA is known to occur. In plant, animal, and bacterial hosts, viral infection results in the appearance of enzymes which catalyze the synthesis of ribonucleic acid utilizing substrate ribonucleoside triphosphates and template RNA.

In 1966, Haruna and Spiegelman (*1*) reported the isolation and purification of a viral RNA polymerase induced upon infection of *E. coli* by the RNA bacteriophage Qβ. This enzyme has proven to be an exceedingly valuable tool in studies on RNA replication *in vitro* because of its apparent stability to extensive purification. The purification procedure described here is preferable to that previously utilized in this laboratory (*2*) in its operational ease and exceptional reproducibility. The Qβ RNA polymerase has been found to be catalytically and chemically homogeneous and capable of catalyzing the synthesis of biologically active RNA.

ASSAY

Reaction

The assay measures the amount of radioactive ribonucleoside monophosphate incorporated into acid insoluble material:

$$\begin{array}{c} nAPPP \\ + \\ nUPPP \\ + \\ nCPPP \\ + \\ nGPPP \end{array} \xrightarrow[\substack{Q\beta\ RNA \\ Mg^{2+}}]{\substack{Q\beta\ RNA\ polymerase \\ Factor\ I \\ Factor\ II}} \begin{bmatrix} AP \\ UP \\ CP \\ GP \end{bmatrix}_n + 4(n)PP$$

Reaction Mixture

The reaction mixture (volume 0.1 ml) contains the following:
100 mM Tris-HCl, pH 7.8
10 mM $MgCl_2$
4 mM 2-mercaptoethanol
0.8 mM each of ATP, UTP, CTP
0.8 mM ^{14}C-GTP (10^3 cpm/mμmole)
0.5 μg Qβ RNA
Factor I (*3*)
Factor II (*4*)
Enzyme

In crude extracts or where the enzyme fractions might be contaminated by DNA and the DNA-dependent RNA polymerase * or by polynucleotide phosphorylase † the following are also included in the assay mixture: 10 μg DNase (electrophoretically purified free from RNase, Worthington Biochemical Co.), 1 μm phosphoenolpyruvate (Calbiochem Co.), 10 μg pyruvate kinase (Calbiochem Co.)

With the purified enzyme, two additional protein fractions found in both Qβ-infected and uninfected *E. coli* are also required for replication of Qβ RNA. The purification of these factors is described in subsequent sections.

The reaction mixture is incubated for 20 min at 37°. It is then chilled in ice and approximately 3 ml of cold 5% trichloroacetic acid containing 20 mM sodium pyrophosphate added to stop the reaction. The reaction is filtered

* DNA-dependent RNA polymerase activity is detectable in crude extract and phase partition fractions and is inhibited by the addition of DNase. DNase is unnecessary when assaying fractions from subsequent steps in the purification, for the Qβ RNA polymerase is no longer contaminated with the DNA-dependent RNA polymerase. Despite the manufacturer's claims of purity the DNase is heavily contaminated with RNase and should not be used in assays of later fractions without purification.

† When normal *E. coli* is the host cell, polynucleotide phosphorylase activity is present in the crude extract, phase partition fractions, and the run-through of the phosphocellulose chromatography step. Although *E. coli* Q13 is a mutant deficient in polynucleotide phosphorylase activity, phosphoenolpyruvate and pyruvate kinase are included in assays of crude extracts to inhibit any residual activity.

through a nitrocellulose filter (Millipore HA, pore size 0.45 μ, 25 mm). The filter is washed 3 times with about 1 ml of the TCA-pyrophosphate solution, dried on a planchet, and counted in a gas-flow counter. One unit of activity is defined as one mμmole of GMP incorporated per 20 min at 37°.

ISOLATION OF Qβ RNA POLYMERASE

Growth of Infected Cells

E. coli Q13, a mutant strain (*5*) lacking RNase I (*6*) and polynucleotide phosphorylase, is grown in LB broth (*7*) * to a cell density of 5–8 x 10^8 cells/ml. Calcium chloride is then added to a concentration of 1 mM and the cells are infected with Qβ phage at a multiplicity of infection of 20–30. The culture is incubated at 37° for 15 min and then chilled for a total infection time of 20–30 min. The cells are harvested in a Sharples centrifuge and stored at –20°. The yield is 1.5–2.0 gm wet weight/liter.

Purification of Phage RNA (*8*)

Five milliliters of about a 1% suspension of purified Qβ phage † in SSC (150 mM NaCl, 15 mM sodium citrate) is adjusted to 2% sodium dodecyl sulfate. Fifty milligrams of Bentonite and an equal volume of freshly distilled phenol previously equilibrated with SSC are added. The tube is gently rotated for 10 min at room temperature. The phases are separated by centrifugation in a Sorvall SS-34 rotor at 10,000 rpm for 10 min. The top aqueous phase is removed and the phenol phase is reextracted with 1 ml of SSC. The combined aqueous phases are then extracted successively with 0.5 and 0.25 vol of phenol.

The final aqueous phase is mixed with 2 vol of cold 95% EtOH and allowed to stand at –20° for 1 hr. The RNA is centrifuged at 10,000 rpm for 10 min and dissolved in 2–3 ml of 0.1 mM EDTA, pH 7.0. The solution is adjusted to 0.5% potassium acetate and precipitated with 95% ethanol. The ethanol precipitation is twice repeated and the RNA is finally suspended in 0.1 mM EDTA.‡ Stock solutions of RNA are stored at –70°. For assays, solutions are routinely diluted to 5 OD/ml in 0.1 mM EDTA, and kept at

* The cells have more recently been grown in an enriched medium containing (per liter): 30 gm Bacto-tryptone, 20 gm yeast extract, and 10 gm sodium chloride adjusted to pH of 7.2 with NaOH.

† The specific absorption (OD_{260}/mg/ml) of Qβ phage is 8.02 (*15*).

‡ The OD_{260}/OD_{280} should be 2.0 to 2.1 and the OD_{220}/OD_{260} should be no greater than 0.8 for phenol free preparations.

−20°. Intact Qβ RNA is an essential requirement, since fragmented Qβ RNA will not serve as template.

Purification of Qβ RNA Polymerase

All procedures are carried out at 2–4°. Deionized water with a conductivity of less than 0.1 ppm as NaCl is used throughout. The standard buffer referred to below is the following: 50 mM Tris-HCl, pH 7.8, 5 mM $MgCl_2$, 5 mM 2-mercaptoethanol, 1 mM EDTA.

Preparation of Crude Extract

Two hundred grams of frozen cells are ground to a sticky paste with 400 gm of alumina (Alumina A301, Alcoa). The paste is suspended in 800 ml of standard buffer containing 50 mM $MgCl_2$. The suspension is centrifuged in a Sorvall GSA rotor at 10,000 rpm for 10 min and the supernatant decanted (Crude extract fraction, 810 ml).

Phase Partition (*9*)

With constant stirring, 150 ml of 50% w/w polyethylene glycol (Carbowax 6000, Union Carbide) * and 135 ml of 30% w/w Dextran 500 (Pharmacia Fine Chemicals, N.Y.) * are added to 810 ml of crude extract. The polymer concentrations are 92.5 mg/ml extract and 50 mg/ml extract, respectively. After stirring for 30 min, the suspension is centrifuged at 10,000 rpm for 10 min. Two phases are formed. The top (polyethylene glycol) phase (800 ml) containing proteins not associated with nucleic acid is removed. The bottom (Dextran) phase containing nucleic acid and associated proteins is retained and combined with fresh top phase (150 ml of 50% polyethylene glycol and 650 ml of standard buffer containing 50 mM $MgCl_2$). A portion of the buffer is used to remove the viscous bottom phase from the centrifuge bottles. Sodium chloride is then added with continuous stirring to a final concentration of 5.0 *M* (331.4 gm). The mixture is stirred until all the NaCl is dissolved (approximately 60 min) and centrifuged in a Sorvall GSA rotor for 10 min. The top phase, containing the Qβ RNA polymerase, is collected and dialyzed against 6 liters of standard buffer containing 20% v/v glycerol until the conductivity of a 1 to 50,000 dilution of the sample is less than 1 ppm as NaCl (approximately 3 hr with one change of dialysis buffer). The dialyzed fraction (910 ml) is adjusted to 35% saturation with 177 gm of ammonium sulfate ("Special Enzyme Grade," Schwarz-Mann) and centrifuged at 10,000 rpm in the Sorvall GSA rotor for 10 min to allow the phases to form.

* These polymer solutions are prepared by prolonged stirring at room temperature.

The preparation is then transferred to a separatory funnel and, after the phases have reformed, the bottom phase (870 ml) is collected. The bottom phase is brought to 55% saturation with 103 gm of ammonium sulfate and centrifuged in a Sorvall SS-34 rotor at 20,000 rpm for 15 min.* The pellet is dissolved in approximately 150 ml of standard buffer containing 20% glycerol.† Immediately prior to the next step, the ammonium sulfate fraction ‡ is dialyzed against 6 liters of buffer containing 50 mM Tris-HCl, pH 7.8, 5 mM 2-mercaptoethanol, 1 mM EDTA, and 20% glycerol (PC buffer) with 50 mM NaCl for 5 hr (phase partition fraction, 250 ml).

Phosphocellulose Chromatography

The dialyzed phase partition fraction is applied to a 4 x 20 cm column of phosphocellulose§ that has been previously equilibrated with PC buffer containing 50 mM NaCl. A linear gradient of 1600 ml total volume with concentration limits of 50 mM NaCl and 1.0 M NaCl in PC buffer is applied to the column. Fractions of 7.5 ml are collected overnight.|| Enzyme activity appears in fractions eluting at a NaCl concentration of about 500 mM (Fig. 1). The active fractions are pooled and dialyzed for 4 hr against 5 liters of standard buffer containing 100 mM ammonium sulfate (phosphocellulose fraction, 460 ml).

QAE-Sephadex Chromatography

The phosphocellulose fraction is applied to a 4 x 20 cm column of QAE-Sephadex A-50# (Pharmacia Fine Chemicals, N.Y.) previously equilibrated with standard buffer containing 100 mM ammonium sulfate. A linear gradient of 1 liter total volume with 100 mM ammonium sulfate and 300 mM ammonium sulfate in standard buffer as limiting concentrations is applied.|| Fractions of 14 ml are collected overnight. Enzyme activity is eluted between 200 mM and 250 mM ammonium sulfate (Fig. 2). The active fractions (160 ml) are pooled and adjusted to 85% saturation with ammonium sulfate

* This centrifugation time is essential due to the viscosity of the solution.

† The polymer layer floating at the top is decanted along with the supernatant. In dissolving the pellet, inclusion of the material adhering to the sides of the tubes is not critical to subsequent purification steps.

‡ This fraction is routinely stored overnight.

§ The phosphocellulose (Whatman PII) is precycled according to the manufacturer's manual. Prior to use, the resin is neutralized with 10 N NaOH. It is suspended in 2 M Tris-HCl, pH 7.8, poured into a column and washed with the buffer until a pH of 7.8 is reached. It is then equilibrated with PC buffer containing 50 mM NaCl.

|| The initial flow rate is approximately 1 ml/min. As the salt concentration increases, the flow rate increases to 2–3 ml/min.

The QAE-Sephadex is hydrated in standard buffer containing 100 mM ammonium sulfate.

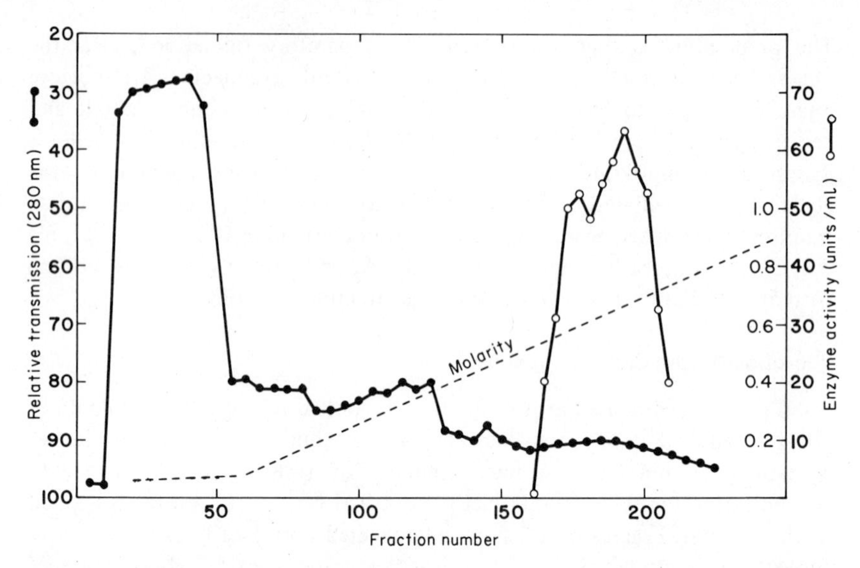

Figure 1. Chromatography on phosphocellulose. The dialyzed phase partition fraction (250 ml) was loaded on a 4 x 20-cm column equilibrated with PC buffer containing 50 mM NaCl. A linear gradient of 1600 ml total volume with concentration limits of 50 mM NaCl and 1 M NaCl was applied. Each fraction was 7.5 ml. Fractions 162–212 were pooled for further purification. The closed circles indicate relative transmission at 280 nm. The open circles indicate enzyme activity. The right hand inner scale indicates concentration of NaCl.

(90 gm) and centrifuged in a Spinco No. 30 rotor at 30,000 rpm for 20 min. The precipitate is dissolved in 3 ml of standard buffer containing 200 mM ammonium sulfate.* Immediately prior to the next step, this fraction is dialyzed† against 1 liter of standard buffer containing 50 mM ammonium sulfate for about 2.5 hr (QAE-Sephadex fraction, 5 ml).

Hydroxylapatite ‡

The QAE-Sephadex fraction is applied to a 0.9 x 12 cm-column of hydroxylapatite (Hypatite C, Clarkson Chemical Co., Williamsport, Pa.) equilibrated with standard buffer containing 50 mM ammonium sulfate. A stepwise elution with 20 ml each of 100 mM, 200 mM, 300 mM, 400 mM,

* See footnote (‡) p. 833.

† Dialysis tubing for this and subsequent steps is pretreated by boiling in 10% Na_2CO_3, 10 mM EDTA for 15 min followed by several rinses with H_2O. This procedure is repeated and the tubing is then autoclaved in 1 mM EDTA for 20 min. It is stored in 1 mM EDTA with a few drops of chloroform.

‡ This step may be omitted without appreciable loss in purification

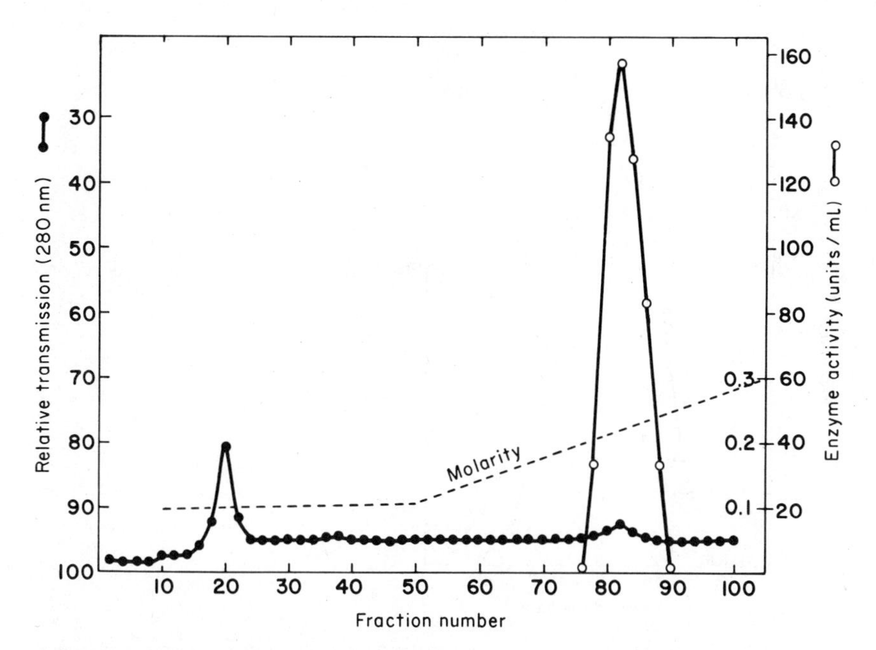

Figure 2. Chromatography on QAE-Sephadex A-50. The dialyzed phosphocellulose fraction (160 ml) was loaded on a 4 x 20-cm column equilibrated with standard buffer containing 100 mM ammonium sulfate. A linear gradient of 1 liter total volume with concentration limits of 100 mM to 300 mM ammonium sulfate was applied. Each fraction was 14 ml. Fractions 78–88 were pooled for further purification. The closed circles indicate relative transmission at 280 nm. The open circles indicate enzyme activity. The right hand inner scale indicates concentration of ammonium sulfate.

450 mM, and 500 mM ammonium sulfate in standard buffer is carried out with a flow rate of approximately 0.5 ml/min.* Enzyme activity is eluted at an ammonium sulfate concentration of 400–450 mM. The active fractions are pooled† (hydroxylapatite fraction, 31 ml).

Gel Filtration

The hydroxylapatite fraction is applied to a 3 x 90 cm-column of Bio-Gel A 0.5 m (Bio-Rad, New York, N.Y.)‡ equilibrated with a standard buffer containing 200 mM ammonium sulfate. Ten-milliliter fractions are collected

* Slow flow rates can be avoided by exhaustive de-fining of the hydroxylapatite prior to use.

† Concentration by ammonium sulfate precipitation should be avoided because of the low protein concentration.

‡ The Bio-Gel is not pretreated before equilibration.

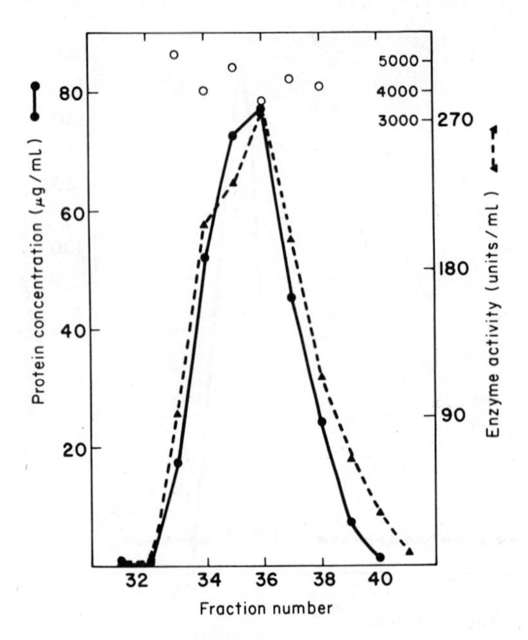

Figure 3. Gel filtration on Bio-Gel A 0.5 M. Thirty-one milliliters of the hydroxylapatite fraction were applied to a 3 x 90-cm column equilibrated with standard buffer containing 200 mM ammonium sulfate. Ten-milliliter fractions were collected. Assays of fractions eluting before and after the indicated peak showed no enzyme activity. The closed circles represent protein concentration. The closed triangles indicate enzyme activity. The open circles and the right hand inner scale indicate specific activities of fractions 33 through 39.

overnight.* The enzyme is found to elute at approximately 1.4 times the void volume (Bio-Gel fraction, 80 ml) (Fig. 3).

A summary of a purification is shown in Table 1.†

Properties of the Enzyme

Stability

The purified fraction is stable in standard buffer containing 200 mM ammonium sulfate and 20% glycerol at 4° for 2 years. In this solution, there

* The flow rate is approximately 0.3 ml/min.

† The figures in parentheses indicate reduced activities due to inhibitors. To test for inhibition, all fractions must be assayed at more than one protein level. The aliquots routinely used for enzyme assays are as follows (ml): Crude, 0.002–0.005; phase partition, 0.001–0.002; phosphocellulose, 0.002–0.005; QAE-Sephadex, 0.001–0.002; hydroxylapatite, 0.001–0.002; Bio-Gel, 0.002–0.02. Protein concentrations were determined by the method of Lowry (*16*) or Bücher (*17*).

TABLE 1. Summary of Purification of Qβ RNA Polymerase

Fraction	Specific activity (units/mg)	Units	Total protein (mg)
Crude	3.7	41,000	11,000
Phase partition	(7)	(17,000)	2,450
Phosphocellulose	(14.1)	(9,050)	640
QAE-Sephadex	1,320	17,290	13
Hydroxylapatite	(770)	(7,700)	10
Bio-Gel	4,200	12,205	2.9

is no significant loss of activity when the purified enzyme is incubated at 37° for 40 min. Stability is also observed in the presence of RNA and the other components of the standard reaction mixture.

Catalytic Purity

The purified fraction (Bio-Gel fraction) is completely free of DNA polymerase, DNA-dependent RNA polymerase, polynucleotide phosphorylase, and polyriboadenylate polymerase. In addition, there is no detectable RNase I or RNase II activity and the biological activity of Qβ RNA is not diminished by incubation with the enzyme at 37° for 20 min.

Physical Purity

When analyzed by polyacrylamide gel electrophoresis in the presence of sodium dodecyl sulfate (*10*), all pure preparations of active enzyme yield 4 protein bands which correspond to molecular weights of approximately 73,000, 66,000, 44,500, and 32,500. Although the sum of these components exceeds 200,000, estimations of molecular weight both by sedimentation equilibrium and zonal sedimentation centrifugation yield a value of 150,000. The reason for this discrepancy as well as the role of the four proteins in the reaction is as yet unclear.

Factor Requirement

At least two host cell proteins must be included in the reaction mixture when assaying fractions subsequent to the phase partition fraction. The isolation and use of these proteins are described elsewhere in this volume (*3, 4*). A crude preparation containing both factor activities may also be obtained from infected cells by DEAE-chromatography as described previously (*11*). This crude fraction, however, is unstable, gradually losing activity over several months. Final specific activities of the enzyme may vary, being dependent on the nature of the protein factors utilized.

The specific activity of the final fraction indicated in this purification was obtained with purified factor proteins. With some preparations of crude factors, specific activities as high as 10,000 units/mg protein have been observed.

Template Specificity

In addition to synthesizing Qβ RNA in the presence of template Qβ RNA, the Qβ RNA polymerase will also catalyze nucleotide incorporation in the presence of polyribocytidylic acid, synthetic polyribonucleotides containing cytidylate (*12*), the Qβ complementary strand (*13*), and Qβ 6S RNA (*14*). The protein factors are required only when Qβ RNA is used as template. Infectious RNA is synthesized with either the Qβ RNA or the Qβ complementary strand (*13*) as template.

Nucleotide Requirement

The reaction with Qβ RNA is dependent on the presence of all four ribonucleotides, ATP, UTP, CTP, GTP. The K_s for each is approximately 2×10^{-4} M. In the absence of one or more of these substrates, no significant incorporation is detected. In the presence of poly C, GTP alone serves as the substrate.

Acknowledgments

This work was supported in part by grants from the National Institute of General Medical Sciences, National Institutes of Health (GM 11301 and 11936). J. T. August is a Career Scientist of the Health Research Council of the City of New York. This is Communication No. 203 from the Joan and Lester Avnet Institute of Molecular Biology.

References

1. Haruna, I. and S. Spiegelman. 1965. *Proc. Natl. Acad. Sci., U.S.*, **54**: 579.
2. Eoyang, L. and J. T. August. 1968. In L. Grossman and K. Moldave (Eds.), Methods in enzymology. Academic Press, New York. Vol. XII B, p. 530.
3. Hayward, W. S. This volume, p. 340.
4. Kuo, C. H. This volume, p. 346.
5. Reiner, A. M. 1969. *J. Bacteriol.* **97**: 1431.
6. Gesteland, R. F. 1966. *J. Mol. Biol.*, **16**: 67.
7. Loeb, T. and N. D. Zinder. 1961. *Proc. Natl. Acad. Sci., U.S.*, **47**: 282.
8. Gesteland, R. F. 1964. *J. Mol. Biol.*, **8**: 496.

9. Albertsson, P-A. 1960. Partition of cell particles and macromolecules. Wiley, New York.
10. Shapiro, L., E. Vinuela, and J. V. Maizel. 1967. *Biochem. Biophys. Res. Commun.*, **28**: 815.
11. Franze de Fernandez, M. T., L. Eoyang, and J. T. August. 1968. *Nature,* **219**: 588.
12. Hori, K., A. K. Banerjee, L. Eoyang, and J. T. August. 1967. *Proc. Natl. Acad. Sci., U.S.,* **57**: 1790.
13. August, J. T., A. K. Banerjee, L. Eoyang, M. T. Franze de Fernandez, S. Hasegawa, K. Hori, C. H. Kuo, U. Rensing, and L. Shapiro. 1968. *Cold Spring Harbor Symp. Quant. Biol.,* **33**: 73.
14. Banerjee, A. K., C. H. Kuo, and J. T. August. 1969. *J. Mol. Biol.,* **40**: 445.
15. Overby, L. R., G. H. Barlow, R. H. Doi, M. Jacob, and S. Spiegelman. 1966. *J. Bacteriol.,* **91**: 442.
16. Lowry, O. H., H. S. Rosebrough, A. L. Farr, and R. J. Randall. 1951. *J. Biol. Chem.,* **193**: 265.
17. Bücher, T. 1947. *Biochim. Biophys. Acta,* **1**: 292.

HOST CELL FACTOR I FOR SYNTHESIS OF Qβ RNA

W. S. HAYWARD AND M. T. FRANZE DE FERNANDEZ*

DEPARTMENT OF MOLECULAR BIOLOGY

DIVISION OF BIOLOGICAL SCIENCES

ALBERT EINSTEIN COLLEGE OF MEDICINE

BRONX, NEW YORK

INTRODUCTION

Two host factors are required for the reaction catalyzed by Qβ RNA polymerase with Qβ RNA as template (*1–3*). The factors are not required with other RNA templates: the Qβ RNA, Qβ complementary strand, polycytidylate, or ribocopolymers containing cytidylate. The purification of one of the host cell components, termed "Factor I," is described in this chapter.

ASSAY

Reaction

The assay measures the amount of radioactive ribonucleoside monophosphate incorporated into acid insoluble material:

$$\begin{matrix} \text{ATP} \\ \text{GTP} \\ \text{UTP} \\ \text{CTP} \end{matrix} \xrightarrow[\substack{\text{Q}\beta\text{ RNA polymerase} \\ \text{Mg}^{2+} \\ \text{Q}\beta\text{ RNA}}]{\substack{\text{Factor I} \\ \text{Factor II}}} \begin{matrix} \text{Q}\beta\text{ Complementary} \\ \text{strand} \end{matrix} + \text{PP}_i$$

* Present address: Departamento de Quimica Biologica, Facultad de Farmacia y Biochemica, Universidad de Buenos Aires, Buenos Aires, Argentina

Reaction Mixture

Factor I activity is measured in the Qβ RNA polymerase reaction under conditions where this factor is limiting. The assay is described in the chapter by Eoyang and August (*4*) using a constant amount of RNA (0.7 μg), Qβ polymerase (1 unit), factor II (2 units), and 0.1-0.4 units of factor I. One unit of factor I corresponds to the incorporation of 1 mμmole ^{14}C–AMP in a 20-min reaction under the conditions described.*

Purification of the Qβ RNA polymerase and isolation of phage RNA were carried out as described earlier (*4*). The factor II used in this purification procedure was the partially purified fraction isolated from Qβ phage infected *E. coli* as previously described (*3*).

ISOLATION PROCEDURE

Growth of Cells

E. coli MRE 600 (*5*)† is grown in LB Broth (*6*) and harvested during late log phase. The packed cells are frozen and stored at –20°C.

Purification of Factor I

All operations are carried out at 2-4°C. Deionized water with a conductivity of less than 0.1 ppm as NaCl is used throughout. The standard buffer referred to below is the following: 0.05 *M* Tris, pH 7.8, and 0.005 M 2-mercaptoethanol. Centrifugations are at 10,000 rpm (12,000 *g*) for 10 min in the Sorvall GSA rotor.

Crude Extract

Frozen *E. coli* MRE 600 (200 gm) is ground to a sticky paste with 400 gm of cold alumina (Alumina A301, Alcoa). The cell paste is suspended in 400 ml of standard buffer containing 1 M NaCl, and centrifuged. The supernatant fraction is retained (crude supernatant, 455 ml).

* Factor I activity in the assay is strongly dependent on both RNA and Qβ polymerase concentrations. The assay conditions described give satisfactory, reproducible values.

† *E. coli* MRE 600 is an RNAse I female strain. Similar results have been obtained using either RNA phage infected or uninfected *E. coli* Q13.

Liquid Polymer Fractionation (*7*)

To 455 ml of crude supernatant are added 53 ml (0.117 vol) of 20% w/w Dextran 500, 141 ml (0.311 vol) of 30% w/w polyethylene glycol, and 80 gm NaCl (0.175 gm per ml crude supernatant). The mixture is stirred for 1 hr and centrifuged to separate the phases. The phase top (430 ml) is collected and dialyzed against 6 liters of standard buffer containing 0.05 M NaCl until the NaCl concentration is 0.05–0.07 M (approximately 6 hr with 2 changes of dialysis buffer).* Precipitated protein is removed by centrifugation (phase top, 750 ml).

Phosphocellulose Chromatography

The dialyzed phase top is applied overnight to a 4 x 30-cm phosphocellulose column† previously equilibrated with standard buffer containing 0.05 M NaCl. The column is then washed with 150 ml of standard buffer containing 0.05 M NaCl, and the factor is eluted with a linear gradient from 0.05 M to 0.7 M NaCl in 1200 ml (total volume) of standard buffer. Fractions of 10 ml are collected. The majority of the factor I activity elutes at approximately 0.5 M NaCl‡. Fractions from the major peak are pooled and dialyzed for 3 hr§ against 2 L standard buffer containing 0.07 M ammonium sulfate and 0.005 M $MgCl_2$ with one change of buffer solution (phosphocellulose fraction, 280 ml).

QAE–Sephadex Chromatography

The phosphocellulose fraction is applied to a column of QAE–Sephadex A-50‖ (3 x 20 cm) equilibrated with standard buffer containing 0.07 M

* Dialysis is slow due to the viscosity of the polymer.

† Phosphocellulose (Whatman P11) is precycled according to the manufacturer's manual. Prior to use it is neutralized with 10 N NaOH. The excess fluid is decanted and the phosphocellulose resuspended several times in standard buffer containing 0.05 M NaCl. The column is then poured and equilibrated with the same buffer until a pH of 7.8 is reached. The sample is normally applied overnight, since approximately 15 hr are required.

‡ Inhibition of the polymerase reaction is often observed in assays of phosphocellulose fractions. Satisfactory results are obtained by assaying small amounts, 1–2 μl per reaction, of this fraction.

§ Dialysis tubing in this and subsequent steps is pretreated by boiling in 10% Na_2CO_3 and 0.01 M EDTA for 15 min followed by several rinses with H_2O. This procedure is repeated and the tubing is then autoclaved in 0.001 M EDTA for 20 min. It is stored in 0.001 M EDTA with a few drops of chloroform.

‖ QAE-Sephadex is hydrated in standard buffer containing 0.005 $MgCl_2$, 0.07 M ammonium sulfate, and 20% glycerol. After the beads settle the buffer is decanted and the beads are resuspended an additional 5 times in the same buffer. The column is then poured, and washed with 6 bed volumes of the buffer.

ammonium sulfate and 0.005 M $MgCl_2$. The column is eluted with a linear gradient from 0.07 M to 0.4 M ammonium sulfate in 500 ml (total volume) of standard buffer containing 0.005 M $MgCl_2$. Fractions of 5 ml are collected overnight. Factor activity is found at 0.3 M ammonium sulfate. The active fractions are pooled and dialyzed for 3 hr against 2 L standard buffer containing 0.005 M $MgCl_2$, 0.05 M NaCl, and 20% glycerol with one change of buffer solution (QAE fraction, 120 ml).

SE–Sephadex Chromatography

The dialyzed QAE fraction is applied to a SE–Sephadex C-50 column* (2 x 20 cm) previously equilibrated with standard buffer containing 0.005 M $MgCl_2$, 0.05 M NaCl, and 20% glycerol. The column is eluted with a linear gradient from 0.05 M to 0.4 M NaCl in 500 ml (total volume) standard buffer containing 0.005 M $MgCl_2$ and 20% glycerol. Fractions of 5 ml are collected overnight. Factor I elutes at 0.3 M NaCl. Peak fractions are pooled and dialyzed for 3 hr against standard buffer containing 0.005 M $MgCl_2$, 20% glycerol, and 0.07 M ammonium sulfate with one change of buffer (SE–Sephadex fraction, 110 ml).

Gel Filtration

The SE-Sephadex fraction is concentrated by binding the factor to a QAE-Sephadex column (0.6 x 5 cm) previously equilibrated with the standard buffer containing 0.005 M $MgCl_2$, 0.07 M ammonium sulfate, and 20% glycerol. The column is eluted with 10 ml of standard buffer containing 0.005 M $MgCl_2$, 20% glycerol, and 0.5 M ammonium sulfate. Fractions of 1 ml are collected and factor I is recovered in 2–3 ml. The concentrated sample is applied to a column of Bio-Gel A-0.5 M† (3 x 70 cm) equilibrated with standard buffer containing 0.005 M $MgCl_2$, 20% glycerol, and 0.2 M ammonium sulfate. The column is eluted with the same buffer. Fractions of 5 ml are collected overnight. Factor I elutes at approximately 1.55 times the void volume (Bio-Gel fraction, 45 ml). A summary of the purification is shown in Table 1.

* SE-Sephadex is hydrated in standard buffer containing 0.005 M $MgCl_2$, 20% glycerol, and 0.05 M NaCl and resuspended several times to remove fines. The poured column is washed with several bed volumes of the same buffer.

† Bio-Gel is suspended in several volumes of standard buffer containing 0.005 M $MgCl_2$, 0.2 M ammonium sulfate, and 20% glycerol. After the beads have settled the buffer is decanted, and the beads are resuspended in the same buffer. The poured column is washed with 3 bed volumes of the buffer.

TABLE 1. Purification of Factor I

	Vol (ml)	Total units[a]	Total protein[b]	Specific activity (units/mg)	Recovery %
Crude supernatant	455	168,000	20.4 gm	8	100
Phase top	725	159,000	7.4 gm	21	95
Phosphocellulose fraction	280	98,000	110 mg	890	58
QAE fraction	120	51,000	12 mg	4,300	30
SE-Sephadex	110	32,700	1.9 mg	17,000	19
Bio-Gel fraction	45	27,000	1.3 mg	21,000	16

[a] Protein was determined by the method of Lowry (*9*) in steps 1 and 2. Subsequent determinations were performed by ninhydrin analysis (*10*).
[b] One unit of factor I corresponds to the incorporation of 1 mμmole ^{14}C-AMP in a 20-min reaction under the conditions described.

Properties of Factor I

Storage of Factor I

At concentrations greater than 10 μg/ml factor I is stable for at least 6 months at 2°–4°C in standard buffer containing 0.005 M $MgCl_2$, 0.2 M ammonium sulfate, and 20% glycerol.

Heat Stability

Factor I is extremely heat stable (*2*). No loss of activity was detected after incubation for 60 min at 100°C in the presence of 20% glycerol, 0.2 M ammonium sulfate, and 0.005 M $MgCl_2$.

Molecular Weight

Factor I elutes from Bio-Gel A-0.5 M at 1.55 x void volume, suggesting a molecular weight of approximately 75,000. A similar value has been obtained by zone centrifugation in glycerol gradients (*2, 8*). When analyzed by polyacrylamide gel electrophoresis in the presence of SDS, the purified protein migrates as a single polypeptide of molecular weight 12,500, suggesting that the active factor contains six identical subunits (*8*).

Binding to RNA

Factor I binds to single-stranded RNA from a variety of sources, but not to double stranded RNA or to DNA (*8*).

Role in the Qβ Polymerase Reaction

Analysis of synthesis of the 5′-terminal oligonucleotide, and incorporation of [γ-^{32}P] GTP suggests that host factor I is required for initiation of synthesis with Qβ RNA as template (*8*). One molecule of factor I (mol wt 75,000) is required per molecule of template RNA.

Acknowledgments

This work was supported in part by grants from the National Institute of General Medical Sciences, National Institutes of Health (GM 11301 and 11936). This is Communication no. 208 from the Joan and Lester Avnet Institute of Molecular Biology.

W. S. Hayward is a recipient of an American Cancer Society award #PF-596.

References

1. Hori, K., L. Eoyang, A. K. Banerjee, and J. T. August. 1967. *Proc. Natl. Acad. Sci., U.S.*, **57**: 1790.
2. Franze de Fernandez, M. T., L. Eoyang, and J. T. August. 1968. *Nature*, **219**: 588.
3. Shapiro, L., M. T. Franze de Fernandez, and J. T. August. 1968. *Nature*, **220**: 478.
4. Eoyang, L., and J. T. August. This volume, p. 829.
5. Cammack, R. A., and H. E. Wade. 1965. *Biochem. J.*, **96**: 671.
6. Loeb, T., and N. D. Zinder. 1961. *Proc. Natl. Acad. Sci., U.S.*, **47**: 282.
7. Albertsson, P-A. 1960. Partition of cell particles and macromolecules. Wiley, New York.
8. Franze de Fernandez, M. T., W. S. Hayward, and J. T. August. *J. Biol. Chem.*, in press.
9. Lowry, O. H., N. S. Rosebrough, A. L. Farr, and R. J. Randall. 1951. *J. Biol. Chem.*, **193**: 265.
10. Hill, R. L., and R. Delaney. 1967. In Methods in enzymology. Academic Press, New York, Vol. XI, p. 347.

HOST CELL FACTOR II FOR SYNTHESIS OF Qβ RNA

C. H. KUO
DEPARTMENT OF MOLECULAR BIOLOGY
DIVISION OF BIOLOGICAL SCIENCES
ALBERT EINSTEIN COLLEGE OF MEDICINE
BRONX, NEW YORK

INTRODUCTION

The requirement for a second protein factor (host factor II) in the reaction catalyzed by the Qβ RNA polymerase with Qβ RNA as template can be satisfied by any one of several basic proteins found in *E. coli*. These proteins are heterogeneous in size and charge and are acid soluble. They are purified in a manner analogous to the isolation of histone proteins.

ASSAY

Reaction

The assay measures the amount of radioactive ribonucleoside monophosphate incorporated into acid insoluble material:

$$\begin{matrix} \text{ATP} \\ \text{GTP} \\ \text{UTP} \\ \text{CTP} \end{matrix} \xrightarrow[\substack{\text{Q}\beta\text{ RNA Polymerase} \\ Mg^{2+} \\ \text{Q}\beta\text{ RNA}}]{\substack{\text{Factor I} \\ \text{Factor II}}} \begin{matrix} \text{Q}\beta\text{ Complementary} \\ \text{strand} \end{matrix} + PP_i$$

Reaction Mixture

Factor II activity is measured in the Qβ RNA polymerase reaction when this factor is limiting. The assay is as described in the chapter by Eoyang and August (*1*) using a constant amount of RNA (0.4 μg), Qβ RNA polymerase (0.5 μg), and an optimum amount of factor I. One unit of factor II corresponds to a minimum amount which would result in the incorporation of 1 mμmole ^{14}C-AMP in a 20-min reaction under the conditions described.*

Purification of the Qβ RNA polymerase and isolation of phage RNA were carried out as described in Ref. *1*. The factor I used in this purification procedure was the Bio-Gel Fraction described in the chapter by Hayward and Franze de Fernandez (*2*).

ISOLATION PROCEDURE

Growth of Cells

E. coli MRE 600 (*3*)† is grown in LB Broth (*4*) and harvested during late log phase. The packed cells are frozen and stored at –20°C.

Purification of Factor II

All operations are carried out at 2–4°C unless otherwise stated. Deionized water with conductivity of less than 0.1 ppm as NaCl is used throughout.

Crude Extract

Frozen *E. coli* MRE 600 (50 gm) is ground to a sticky paste with 100 gm of cold alumina (Alumina A301, Alcoa). The cell paste is suspended in 200 ml 10 mM Tris, pH 7.8, 1 mM dithiothreitol, and 1 mM EDTA, and centrifuged at 48,000 *g*‡ for 10 min (crude extract, 182 ml).

Perchloric Acid Extraction

Solid guanidine hydrochloride (Mann Research Laboratories) is added with mixing to the crude extract to a final concentration of 2 M and the mixing is continued for 20 min. Perchloric acid (Allied Chemicals), 20%, is

*Factor II activity must be measured at several different concentrations of the factor. With the purified fractions, activity is not proportional to concentration and in excess of factor II results in significant inhibition of nucleotide incorporation.
† Similar results have been obtained using *E. coli* Q-13.
‡ All centrifugations are with the Sorvall SS-2 rotor.

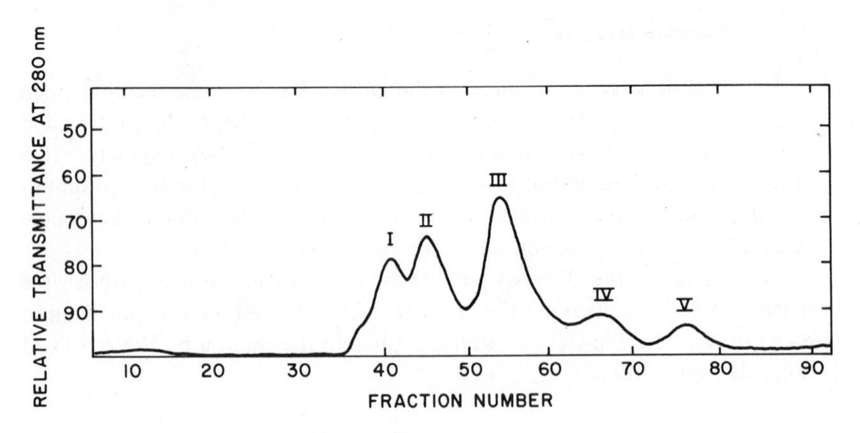

Figure 1. Sephadex G-50 gel filtration of host factor II.

then added dropwise to a final concentration of 2%. The solution is stirred for 20 min and the precipitate removed by centrifugation at 48,000 *g*. The supernatant is titrated to pH 4.2 with 10 N KOH and centrifuged at 12,000 *g* for 5 min. The supernatant is retained (acid extract, 230 ml).

CM-Cellulose Chromatography

The acid extract is desalted by passage through a 5 x 52 cm column of Bio-Gel P-2 (100–200 mesh) (Bio-Rad Laboratories) equilibrated and washed with 0.1 M sodium acetate buffer pH 4.2, 1 mM EDTA, and 1 mM dithiothreitol. The sample comprising the excluded volume is applied to a 2.5 x 25 cm column of CM-cellulose* (Whatman CM-11) previously equilibrated with 0.1 M sodium acetate buffer pH 4.2, 1 mM EDTA, and 1 mM dithiothreitol. The column is washed with 300 ml of starting buffer and then 300 ml of starting buffer containing 0.3 M NaCl. The factor is eluted with 300 ml of starting buffer containing 0.45 M NaCl and 300 ml of buffer containing 0.6 M NaCl. Fractions of 10 ml are collected at 3 ml per minute and the fraction containing factor activity are pooled (CM-cellulose fraction, 510 ml).

Sephadex G-50 Gel Filtration

The CM-cellulose fraction is brought to 85% saturation with solid ammonium sulfate. After stirring for 10 min the precipitate is collected by centrifugation at 48,000 *g* for 20 min and the pellet is dissolved in 16 ml of

* CM-cellulose (Whatman CM-11) is prepared according to the manufacturer's manual.

TABLE 1. Purification of Host Factor II

Fraction	Total units[a]	Total protein[b]	Specific activity (units/mg)
Crude	304,000	4.1 gm	74
Acid extract	322,000	780 mg	410
CM-cellulose fraction	288,000	147 mg	1,920
G-50 fraction I	15,400	14.6 mg	1,050
II	36,500	60.0 mg	610
III	113,000	46 mg	2,460
IV	6,500	16.9 mg	355
V	8,600	3.0 mg	2,870

[a]One unit of factor II corresponds to the incorporation of 1 mμmole, ^{14}C-AMP in a 20-min reaction under the condition described.

[b]Protein was determined by the method of Lowry (*5*) in step 1, 2, and 3. P-10 Fraction was measured by the procedure of Bücher (*6*).

0.1 M NH_4HCO_3, pH 8.0, 1 mM dithiothreitol. The sample then applied to a column of Sephadex G-50 (3 x 150 cm) previously equilibrated with the same buffer solution. The factor is eluted with the same buffer solution at 38 ml per hour and five discrete fractions with factor activity are collected (Fig. 1). A summary of the purification is shown in Table 1.

PROPERTIES

Storage

The G-50 fractions have been stored at 2–4° for 6 months without loss of activity. During purification all of the fractions appear to be stable under the conditions described.

Heat Stability

G-50 fractions I, II, IV, and V may be heated to 80° for 5 min without loss of activity, but the activity of fraction III is decreased by 70%.

Chemical and Physical Properties

The proteins isolated all of which satisfy the HF II requirement are similar to "histones" or their properties. They bind strongly to nucleic acids, are acid soluble, basic, rich in arginine and lysine, and heterogeneous in size and charge.

Purity

Each of the G-50 fractions contains several proteins as indicated by polyacrylamide gel electrophoresis in the presence of urea or SDS. After elution from the urea-gel, all of the proteins from G-50 fractions III and V are active as HF II, whereas fractions I, II and IV contain inactive proteins.

Acknowledgments

This work was supported in part by grants from the National Institute of General Medical Sciences, National Institutes of Health (GM 11301 and 11936). C. H. Kuo is a predoctoral trainee supported by the National Institutes of Health (GM 1191). This Communication No. 211 from the Joan and Lester Avnet Institute of Molecular Biology.

References

1. Eoyang, L. and J. T. August. This volume, p. 829
2. Hayward, W. S. and M. T. Franze de Fernandez. This volume, p. 840.
3. Cammack, R. A. and H. E. Wade. 1965. *Biochem. J.*, **96**: 671.
4. Loeb, T. and N. D. Zinder. 1961. *Proc. Natl. Acad. Sci., U.S.*, **47**: 282.
5. Lowry, O. H., H. S. Rosebrough, A. L. Farr, and R. J. Randall. 1951. *J. Biol. Chem.*, **193**: 265.
6. Bücher, T. 1947. *Biochim. Biophys. Acta,* **1**: 292.

DNA-DEPENDENT RNA POLYMERASE (EC 2.7.7.6)

RICHARD R. BURGESS
INSTITUT DE BIOLOGIE MOLECULAIRE
UNIVERSITÉ DE GENÈVE
GENÈVE, SWITZERLAND

ANDREW A. TRAVERS
MEDICAL RESEARCH COUNCIL
LABORATORY OF MOLECULAR BIOLOGY
CAMBRIDGE, ENGLAND

The procedure described below is but one of a number of satisfactory methods for the purification of RNA polymerase from *E. coli* (*1-8*). It provides enzyme, in good yield, of high activity and purity which contains full levels of sigma factor.

ASSAY

$$\begin{matrix} nATP \\ nGTP \\ nUTP \\ nCTP \end{matrix} + DNA \longrightarrow DNA + \begin{bmatrix} AMP \\ GMP \\ CMP \\ UMP \end{bmatrix}_n + 4nPP$$

Reagents

The assay mixture contains, in 0.25 ml, 0.04 M Tris-HCl, pH 7.9, at 25°, 0.01 M $MgCl_2$, 0.1 mM EDTA, 0.1 mM dithiothreitol, 0.15 M KCl, 0.4 mM potassium phosphate, 0.5 mg per ml of crystallized A Grade bovine serum albumin (Calbiochem), 0.15 mM each of UTP, GTP, and CTP (P-L Biochemicals) and 0.15 mM ^{14}C-ATP (Schwarz BioResearch) specific activity 1–2 mCi per mmole, and 0.15 mg per ml calf thymus DNA (Worthington).

Procedure

Assays are incubated for 10 min at 37°, chilled in ice, and precipitated with 3 ml of 5% trichloracetic acid, 0.01 M sodium pyrophosphate. After 15 min the precipitate is collected on a Millipore filter (0.65μ, pore size) or on a Whatman glass fibre filter (GF/C, 2.5 cm) and washed 4 times with 3 ml of 2% trichloroacetic acid, 0.01 M sodium pyrophosphate. The filter is affixed to an aluminium planchet with a drop of 1% bovine serum albumin, dried, and counted on an end window, low background, gas-flow counter.

One *activity unit of enzyme* incorporates 1 mμmole of AMP in 10 min of incubation under the conditions described above.

ISOLATION PROCEDURE

This purification is essentially the same as the alternative procedure described in the original purification paper (*8*). Slight modifications have been included to yield enzyme of higher purity while still retaining sigma. These modifications consist of a 30–47% saturated ammonium sulfate fractionation instead of 33–50%, elution of the DEAE-cellulose column with a linear salt gradient instead of a step gradient, and the use of agarose gel filtration at low ionic strength instead of a low salt glycerol gradient centrifugation when large quantities of enzyme are being purified. For a more detailed discussion of the steps involved in this purification one should consult the original paper (*8*).

A procedure involving chromatography on phosphocellulose, for producing RNA polymerase lacking sigma factor, is described elsewhere (*9, 10*).

Cells

E. coli K12 cells are grown on enriched medium at 37°, harvested in log phase when they reach about 3/4 of their maximal growth, and frozen without washing at –20° until use. Such cells may be grown with vigorous aeration in a medium containing 16 gm Difco Bacto-Tryptone, 10 gm Difco yeast extract, and 5 gm of NaCl per liter, or purchased from Grain Processing Corporation, Muscatine, Iowa.

Solutions

Stock solutions of 1 M Tris-HCl, pH 7.9, at 25°, 1 M Tris-HCl, pH 7.5 at 4°, 0.1 M EDTA, pH 7.0, and 1 M $MgCl_2$ are diluted with double distilled water to prepare the following buffers. Buffer G : 0.05 M Tris-HCl, pH 7.5, 0.01 M $MgCl_2$, 0.2 M KCl, 0.1 mM dithiothreitol, 0.1 mM EDTA, 5% (v/v) glycerol. Buffer A : 0.01 M Tris-HCl, pH 7.9, 0.01 M $MgCl_2$, 0.1 mM EDTA, 0.1 mM dithiothreitol, 5% glycerol. Buffer C : 0.05 M Tris-HCl, pH 7.9, 0.1 mM EDTA, 0.1 mM dithiothreitol, 5% glycerol.

Buffers containing added KCl, e.g., Buffer A + 0.13 M KCl, are conveniently prepared by mixing 100 ml of 10-times concentrated Buffer A, 0.13 mole of KCl, and enough water to make 1000 ml. Dithiothreitol is added just before use from a 0.1 M stock solution which has been stored at 4° in the dark. The percentage saturation of the ammonium sulfate solutions is calculated assuming that adding 70.0 gm of solid ammonium sulfate to 100 ml of water results in a 100% saturated solution and increases the volume by 35 ml.

Grinding and DNase Treatment

The frozen cells (1000 gm) are broken into small pieces with a mallet and placed in a Waring Blendor. Then 2.5 kg of acid-washed glass beads (Superbrite 100, 3 M Company) which were previously chilled to –20° and 1000 ml of Buffer G are added and the cells homogenized at low speed for 2–3 min and at high speed for 5 min. A 5-mg sample of bovine pancreatic DNase I (Electrophoretically pure, Worthington) dissolved in 5 ml of Buffer G is added and the extract blended for 30 sec longer. At this time the pH of the extract is adjusted by the addition of 1 N NaOH or 1 N HCl so that the pH of a one-to-ten dilution is 7.5. The extract is poured into a beaker and kept at 6–8°C for 30 min. The supernatant is decanted into a beaker and the beads poured into a large sintered glass funnel. Suction is applied to draw the bulk of the solution from the beads. The beads are washed with 200 ml of Buffer G and the filtrates are combined with the supernatant to give Fraction 1 (1800 ml).

High-Speed Centrifugation

Fraction 1 is centrifuged at 30,000 rpm for 2½ hours at 4° in a Spinco No. 30 rotor or at 40,000 rpm for 1½ hours in an International A-170 rotor or a Spinco type 42 rotor. This removes the cell debris and ribosomes in one step. About 1300 ml of clear amber supernatant results (Fraction 2).

Ammonium Sulfate Fractionation

To make a solution 30% saturated with ammonium sulfate, 21.0 gm of solid ammonium sulfate (Mann, enzyme grade) is added, slowly with stirring at 4°, per 100 ml of Fraction 2. The pH is prevented from dropping below 7 by the addition of 0.05 ml of 1 M NaOH per 10 gm of ammonium sulfate. The solution is stirred for 30 min at 4° and the precipitate removed by centrifuging at 9,000 rpm for 30 min in 250-ml polycarbonate bottles in a refrigerated Servall centrifuge. Then, 10.75 gm of solid ammonium sulfate per 100 ml is added to the 30% saturated supernatant to give a 47% saturated solution. The precipitate contains the RNA polymerase and is stirred and centrifuged as above. This precipitate is resuspended in 900 ml of 42% saturated ammonium sulfate in Buffer A (25.6 gm solid ammonium sulfate and 10 ml of 10-times concentrated Buffer A per 100 ml), stirred 45 min and centrifuged 45 min at 9000 rpm. The pellet is again resuspended in 900 ml 42% saturated ammonium sulfate in Buffer A, stirred, and centrifuged as

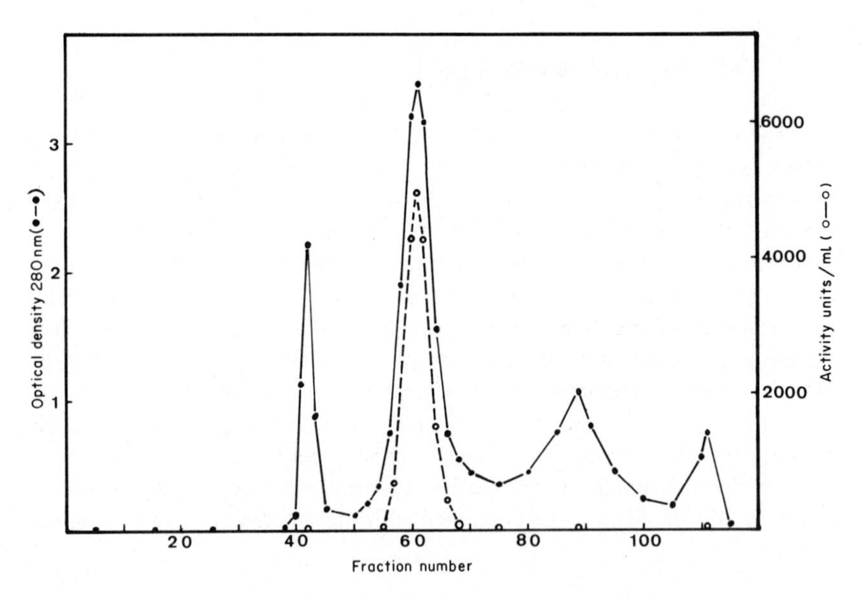

Figure 1. Agarose A-5m gel filtration at low ionic strength. A column (4 x 90 cm) containing agarose (Bio-Gel, A-5m, 200-400 mesh) was equilibrated at 4°C with Buffer C. The sample, containing precipitated Fraction 4 protein dissolved in 20 ml of Buffer C to give a solution 34 mg per ml with an $OD_{280\ nm}$ to $OD_{260\ nm}$ ratio of 1.46, was applied and the column eluted with Buffer C at 50 ml/hr. Fractions of 10 ml were collected, their optical densities read, and their activity assayed. The $OD_{280\ nm}$ to $OD_{260\ nm}$ ratios of fractions number 42, 61, 89, and 111 were 1.17, 1.62, 1.8, 0.98, respectively. Fractions number 57 through 65 were pooled and referred to as Fraction 5 material.

before. The twice-washed pellet, containing the enzyme, is dissolved in Buffer A and diluted until the solution (Fraction 3) has a conductivity equal to or just slightly less than Buffer A + 0.13 M KCl. For a preparation from 1000 gm of cells, diluting with 1000-1500 ml of Buffer A is usually sufficient if the precipitate has been pelleted hard and the supernatant has been carefully drained from the pellet.

DEAE-Cellulose Chromatography

DEAE-cellulose (Whatman, microgranular DE-52, 1.0 meq per gram, dry weight) is washed in 0.5 N HCl for 30 min, rinsed with distilled water on a large sintered glass funnel, washed in 0.5 N NaOH for 30 min, rinsed to pH 8, resuspended in 0.05 M Tris-HCl, pH 7.9, and titrated to pH 7.9 at 25°. The column (500 ml of column volume per 1000 gm of cells) is poured and equilibrated at 4° with Buffer A. Fraction 3 is applied at as great a rate as is possible and the column washed with 3-4 column volumes of Buffer A + 0.13 M KCl. A linear salt gradient is applied from Buffer A + 0.13 M KCl to Buffer A + 0.30 M KCl (total volume of 1500 ml). The peak of the polymerase activity elutes at 0.16-0.17 M KCl, just before a great deal of nucleic acid starts to elute. The fractions containing the bulk of the activity are pooled (Fraction 4). The $OD_{280\ nm}$ to $OD_{260\ nm}$ ratio of Fraction 4 should be greater than 1.1.

Agarose Gel Filtration Chromatography at Low Salt

Fraction 4 is precipitated by the addition of 35 gm of solid ammonium sulfate per 100 ml. After 30 min of stirring at 4°, the precipitate is centrifuged for 30 min at 9000 rpm, and the resulting pellet dissolved in 20 ml of Buffer C. A 4 x 90-cm column containing agarose (Bio-Gel A-5m, 200-400 mesh) is equilibrated with Buffer C at 4°. The enzyme is applied and eluted with Buffer C at a flow rate of 50 ml/hr. The elution profile is shown in Fig. 1. The polymerase peak (Fraction 5) elutes at 1.45 times the void volume. At this stage the enzyme is usually 90-95% pure as judged by electrophoresis on polyacrylamide gels containing sodium dodecyl sulfate, and has an $OD_{280\ nm}$ to $OD_{260\ nm}$ ratio of 1.6-1.7.

Glycerol Gradient Centrifugation at High Salt

For many purposes Fraction 5 material is sufficiently pure, but if complete removal of all impurities is desired, then sizing at high ionic strength is recommended. This is especially important if the $OD_{280\ nm}$ to $OD_{260\ nm}$ ratio of Fraction 5 is less than 1.4. The decreased ratio is due to small fragments of nucleic acid which bind to the enzyme at low ionic strength. The sizing at high salt can be accomplished by gel filtration chromatography on Bio-Gel A-1.5 m, 200–400 mesh, equilibrated with Buffer A + 1.0 M KCl. Under these conditions the polymerase elutes at 1.33 times the void volume and always has an $OD_{280\ nm}$ to $OD_{260\ nm}$ ratio of 1.6–1.7. Unfortunately, the minor protein impurities found in Fraction 5 cannot be removed by this column since it does not cleanly resolve the two major impurities (18 S and 8 S) from the 14 S polymerase. For this reason, the most satisfactory high ionic strength sizing technique is glycerol gradient centrifugation. Fraction 5 is precipitated by the addition of 35 gm of solid ammonium sulfate per 100 ml. After 30 min of stirring at 4°, the precipitate is centrifuged for 30 min at 9000 rpm, and the resulting pellet dissolved in 5 ml of Buffer A and dialyzed for 1 hr against 500 ml of Buffer A lacking 5% glycerol. Then 0.5–0.7 ml of this dialyzed enzyme is layered on a 12 ml 10–30% glycerol gradient containing Buffer A + 1.0 M KCl and centrifuged at 4°C for 22 hr at 40,000 rpm in a Spinco SW 41 rotor or an IEC SB 283 rotor. The polymerase activity sediments 2/3 of the way down the gradient and is free from detectable contamination (Fraction 6.)

Storage

It is best to store polymerase at a concentration of 5 mg per milliliter or greater at –20°C in a storage buffer containing 0.01 M Tris-HCl, pH 7.9, 0.01 M $MgCl_2$, 0.1 M KCl, 0.1 mM EDTA, 0.1 mM dithiothreitol, and 50% (v/v) glycerol.

If the protein concentration is above 3–4 mg per ml in the sample to be stored (for example, Fraction 5 or Fraction 6), then simple dialysis against storage buffer will reduce the volume and prepare it for storage. Since purified enzyme is quite susceptible to surface denaturation, precipitation by addition of solid ammonium sulfate with the accompanying stirring is avoided in the final stages if possible. Also, it should be noted that addition of solid ammonium sulfate to solutions containing 1.0 M KCl results in a large precipitate of KCl which is often difficult to handle. Therefore, if it is necessary to concentrate a sample before storage, it is convenient to dialyze the enzyme solution against several changes of 50% saturated ammonium sulfate in Buffer A. This will reduce the volume and prevent frothing. The precipitate is centrifuged and the pellet dissolved in, and then dialyzed against, storage buffer. Storage buffer will not freeze at –20°C and the enzyme is stable for at least 6 months.

Purity

A very convenient and reproducible way of monitoring purity throughout an enzyme purification is to analyze various fractions by electrophoresis on polyacrylamide gels containing 0.1 M sodium phosphate, pH 7.2, 0.1% sodium dodecyl sulfate, 5% acrylamide, and 0.135% *N,N'*-methylenebisacrylamide, as described by Shapiro et al. (*11*). The sample is applied in 100μl of buffer containing 0.1% sodium dodecyl sulfate, 0.01 M sodium phosphate, pH 7.2, 0.14 M 2-mercaptoethanol, 10% (v/v) glycerol, and 0.002% bromophenol blue. The band pattern is not altered by the presence, in the sample applied, of large amounts of nucleic acid, of up to 0.2 M ammonium sulfate, of 0.1–0.2 M KCl, of 1–2% sodium dodecyl sulfate, or of 4 M urea. In the presence of high ionic strength it will take longer for the protein and tracking dye to enter the gel.

The gels (0.5 x 6 cm) are run for 2½ hours at 8 mA per tube at room temperature (until the marker dye has migrated 5 cm), removed from the tubes, and stained for 2 hr in a filtered 0.2% coomassie brilliant blue solution in methanol–acetic acid–water (5 : 1 : 5). The gels are rinsed in 7.5% acetic acid, 5% methanol for 5 min and then destained electrophoretically in this same solution in a direction perpendicular to the axis of the gel at about 100 mA per gel. Figure 2 shows various fractions analyzed in such a fashion.

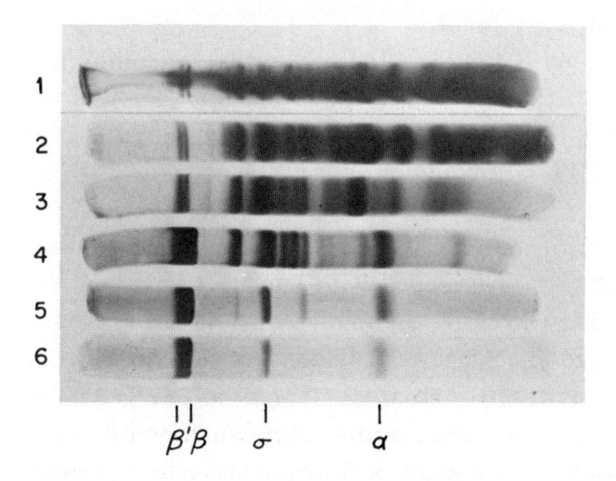

Figure 2. Electrophoretic analysis of fractions during purification. Electrophoresis of polyacrylamide gels containing 0.1% sodium dodecyl sulfate, 0.1 M sodium phosphate pH 7.2, and 5% acrylamide was performed as described in the text. Protein migration was from left to right. The smearing in gel 1 is due to cellular debris. The amount of protein and fraction for the various gels are: (1) 212 μg of Fraction 1; (2) 130 μg of Fraction 2; (3) 56 μg of Fraction 3; (4) 50 μg of Fraction 4; (5) 7 μg of Fraction 5; and (6) 45 μg of Fraction 6.

The most striking feature is that the β' and β bands of RNA polymerase can be clearly seen even in Fraction 1 where they only constitute 0.2-3% of the protein present. Due to their unusual size and equimolar amounts, they form an easily detectable doublet which allows one to recognize as little as 0.1 μg of polymerase on a gel. In this way it is easy to check fractions such as the 47% and 42% saturated ammonium sulfate supernatants and confirm that the precipitation has been complete.

Due to the ease with which sodium dodecyl sulfate containing gels may be run, and the increasing importance of knowing the purity of RNA polymerase, it is recommended that in the future all polymerase publications be accompanied by a gel pattern of the enzyme being used.

Yield

A summary of the purification described here is shown in Table 1.

Although every attempt has been made to maximize the yield of enzyme, there are several places where enzyme is lost. Since 25-30% of the volume of the crude extract is lost as pellet during the high-speed centrifugation, 15-20% of the enzyme is lost at this step. Even more will be lost if a low-speed centrifugation to remove the debris is employed because even more supernatant will be trapped in the loose pellet of debris and also the subsequent high-speed centrifugation will not effectively pellet the ribosomes in the given time unless the debris is present.

Often 10-15% of the enzyme is found in the 30% saturated ammonium sulfate pellet. While most of this is just trapped in the pellet, some polymerase does begin to precipitate slowly under these conditions and even more may precipitate if the 30% saturated solution is stirred more than 30 min.

Since such a large amount of material is applied to the DEAE column, the top of the column may become overloaded with nucleic acid. When this happens the polymerase fails to elute completely as a peak around 0.16-0.17 M KCl and trails into the higher salt region (up to 0.25 M KCl). This material, which can amount to 20-30% at worst, can be salvaged if the following principle is kept in mind: Do not fractionate polymerase by size at low ionic strength (agarose column) if a great deal of nucleic acid is present. Krakow et al. (*13*) have shown that the sigma factor from *A. vinelandii* is released when the enzyme binds to single stranded polynucleotides. This appears to be true for *E. coli* as well, and should be kept in mind when processing either the peak or the trailing material from the DEAE column. If the sample applied to the low salt agarose column has an OD_{280nm} to OD_{260nm} ratio of less than 1.0, considerable activity is lost and gels reveal

TABLE 1. Summary of *E. coli* K12 RNA Polymerase Purification[a]

Fraction Description	Total protein[b] (mg)	Total activity[c] (units)	Specific activity (units/mg)	Yield[d] (%)
1. DNase treated extract	86,000	200,000	2.3	100
2. High-speed supernatant	42,700	206,000	4.8	103
3. Ammonium sulfate fraction	11,200	203,000	18.1	101
4. Pooled DEAE-cellulose peak	680	268,000	412	134
5. Pooled agarose column peak	310	297,000	958	148
6. Pooled glycerol gradient peak	284	285,000	1000	142

[a] From 1000 gm of *E. coli* K12 grown at 37° with a generation time of 28 min.

[b] Protein in Fractions 1–4 determined by the method of Lowry et al. (*12*). Proteins in fractions 5 and 6 determined using $E_{280\ nm}^{1\%} = 6.6$ (5).

[c] Activity expressed as mμmoles of AMP incorporated in 10 min under assay conditions described in text.

[d] Yield is not very meaningful since it is difficult to get reproducible and reliable measurements of activity in Fractions 1–3 due to the presence of DNase I and the large amount of polynucleotides.

less than 30% as much sigma as is normally associated with polymerase. To prevent this loss of sigma, the material should be subjected to sizing at high salt prior to the low salt sizing. This is conveniently accomplished by passage through an agarose (Bio-Gel A-1.5m, 200-400 mesh) column equilibrated with Buffer A + 1.0 M KCl.

Taking these various losses of enzyme into account, a realistic estimate of the yield of purified enzyme under the best conditions would be 60–80%. The amount of enzyme in the cells might vary 2 to 4-fold depending on the conditions under which the cells are grown, however, and this of course affects the amount obtained.

PROPERTIES OF RNA POLYMERASE

For information about the RNA polymerase-catalyzed reaction the reviews of Richardson (*14*) and of Geiduschek and Haselkorn (*15*) and the Lepetit Colloquium on RNA Polymerase and Transcription (*16*) should be consulted.

Subunit Structure

Early investigators showed that RNA polymerase can be disrupted into material sedimenting more slowly than the intact enzyme (*4, 17*). This and the very large size of the enzyme suggested that it is composed of several, possibly different, subunits. Treatment of polymerase with various protein disrupting agents has allowed complete dissociation and separation of the polypeptide chain subunits. It has been shown that the enzyme is composed of at least four different polypeptide chains (*18, 19*). These chains are designated β' (165,000 ± 10% daltons), β (155,000 ± 10%), σ (95,000 ± 5%), and α (39,000 ± 5%) and they are found in the molar ratio of 1 : 1 : 1 : 2. The subunits are shown in gel 6 of Fig. 2. Another subunit, ω (9,000 ± 10%), is associated with purified enzyme but it is not yet known whether it is a part of the enzyme or merely a tightly binding impurity.

The subunits are not held together by disulfide linkages since they can be separated from each other without treatment with reducing agents such as 2-mercaptoethanol. Amino acid compositions have been determined for several of the separated subunits and α, β, and β' have been shown to have methionine N-terminal amino acids (*19*). Although these subunits appear to be intact polypeptide chains which cannot be further degraded without breaking peptide bonds (*19*), it has been reported that β' can be further dissociated to β and a small cationic subunit and that σ may be dissociated into several smaller pieces (*20*). It remains to be shown, however, that no peptide bond cleavage occurred in these studies.

Functions of the Subunits

At present the functions of the various subunits in the synthesis of RNA is only partially understood. It is possible by phosphocellulose chromatography (*9*), by complexing with polynucleotides (*13*), or by heat treatment (*21*) to dissociate the complete enzyme ($\alpha_2\beta\beta'\sigma$) into two main structural and functional components: (1) a core enzyme ($\alpha_2\beta\beta'$) which contains the machinery for synthesizing RNA but lacks the ability to initiate such synthesis accurately and efficiently, and (2) the sigma factor (σ) which has no synthetic activity itself but when added back to core enzyme restores the ability to initiate RNA synthesis at specific sites on the DNA template (*9, 22–25*).

It has been possible to study the function of the sigma subunit because it can be dissociated from the rest of the enzyme without denaturing it. It is necessary for specific initiation of RNA synthesis and is used catalytically and released after initiation has occurred (*22, 23*). The recent discovery of bacteriophage-coded sigma-like factors which direct *E. coli* core enzyme to initiate virus-specific RNA synthesis (*26–29*) has strengthened the model that such factors exert a general positive control on gene expression. A number of initiation factors with different initiation specificities would determine which genes the core enzyme could transcribe.

Less is known about the functions of the subunits in the core enzyme because these subunits must be denatured in order to be separated completely. A number of RNA synthesis mutants have been isolated in which the RNA polymerase is altered in such features as binding to DNA (*21*), response to antibiotics affecting chain initiation (*30–33*), and chain elongation (*34*). All of these mutations characterized so far have been shown to affect core enzyme and not sigma. In only one case has a mutation been correlated with an alteration in a particular subunit. The β subunit from a rifampicin-resistant enzyme has been shown to have an altered electrophoretic mobility (*35*). Since the antibiotic rifampicin has been demonstrated to inhibit initiation of RNA synthesis (*36*), the subunit presumably functions in initiation. Several different steps in RNA synthesis have been inhibited by chemical modification of the enzyme (*37*), but as yet the inhibition has not been correlated with the modification of a particular subunit.

The enzyme can be partially dissociated under mild dissociating conditions such as low urea or elevated pH into material intermediate between the native enzyme and the completely denatured subunits (*19*). With this technique it should be possible to isolate subassemblies of the enzyme which retain some partial function. Preliminary reports have suggested that $\alpha\beta$ and $\alpha\beta'$ (*19*), $\alpha\beta\sigma$ (*25*), and $\beta\sigma$ and possibly $\alpha\omega$ (*38*) may be subassemblies. Some success has been reported in reconstituting activity from such partially dissociated material (*39*).

Aggregation Properties

RNA polymerase undergoes a variety of complex aggregations and dissociations which are only beginning to be understood. Richardson (*5*) showed that the size of the enzyme is strongly dependent on the ionic conditions. More recently it has been recognized that sigma can also affect the aggregation of the enzyme and that enzyme preparations can vary in the amount of sigma they contain. A rapid equilibrium exists between the complete enzyme and the mixture of core enzyme plus sigma which strongly favors the former (*40*). As the ionic strength is lowered, the complete enzyme forms a dimer but does not appear to aggregate further, while the core enzyme undergoes extensive aggregation (*41*).

The molecular weight of the core enzyme monomer can be estimated from the molecular weights of its component subunits to be around 400,000

TABLE 2. The Multiple Forms of *E. coli* RNA Polymerase

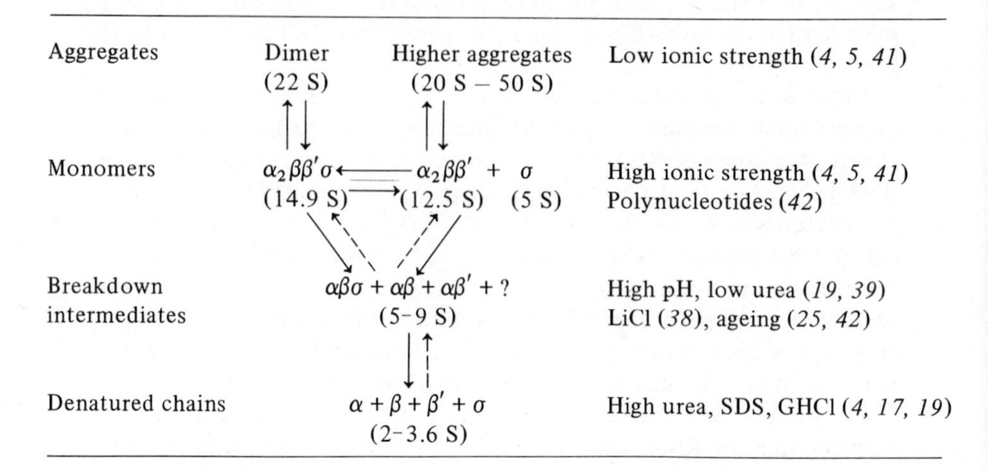

Aggregates	Dimer (22 S)	Higher aggregates (20 S – 50 S)	Low ionic strength (*4, 5, 41*)
Monomers	$\alpha_2\beta\beta'\sigma$ (14.9 S) ⇄	$\alpha_2\beta\beta'$ (12.5 S) + σ (5 S)	High ionic strength (*4, 5, 41*) Polynucleotides (*42*)
Breakdown intermediates	$\alpha\beta\sigma + \alpha\beta + \alpha\beta' + ?$ (5–9 S)		High pH, low urea (*19, 39*) LiCl (*38*), ageing (*25, 42*)
Denatured chains	$\alpha + \beta + \beta' + \sigma$ (2–3.6 S)		High urea, SDS, GHCl (*4, 17, 19*)

± 10% daltons (*19*). Recent direct measurements yield values around 350,000 for core enzyme (*41*).

The multiple forms of the enzyme at present inferred to exist are summarized in Table 2.

References

1. Chamberlin, M., and P. Berg. 1962. *Proc. Natl. Acad. Sci., U.S.*, **48**: 81.
2. Furth, J. J., J. Hurwitz, and M. Anders. 1962. *J. Biol. Chem.*, **237**: 2611.
3. Stevens, A., and J. Henry. 1964. *J. Biol. Chem.*, **239**: 196.
4. Zillig, W., E. Fuchs, and R. Millette. 1966. In G. L. Cantoni and D. R. Davies (Ed.), Procedures in nucleic acid research. Harper & Row, New York, Vol. I, p. 323.
5. Richardson, J. P. 1966. *Proc. Natl. Acad. Sci., U.S.*, **55**: 1616.
6. Babinet, C. 1967. *Biochem. Biophys. Res. Commun.*, **26**: 639.
7. Alberts, B. M., F. J. Amodio, M. Jenkins, E. D. Gutmann, and F. L. Ferris. 1968. *Cold Spring Harbor Symp. Quant. Biol.*, **33**: 289.
8. Burgess, R. R. 1969. *J. Biol. Chem.*, **244**: 6160.
9. Burgess, R. R., A. A. Travers, J. J. Dunn, and E. K. F. Bautz. 1969. *Nature*, **221**: 43.
10. Burgess, R. R., and A. A. Travers. 1970. In L. Grossman and K. Moldave (Eds.), Methods in enzymology. Academic Press, New York, in press.
11. Shapiro, A. L., E. Vinuela, and J. V. Maizel. 1967. *Biochem. Biophys. Res. Commun.*, **28**: 815.
12. Lowry, O. H., N. J. Rosebrough, A. L. Farr, and R. J. Randall. 1951. *J. Biol. Chem.*, **193**: 265.
13. Krakow, J. S., K. Daley, and M. Karstadt. 1969. *Proc. Nat. Acad. Sci., U.S.*, **62**: 432.
14. Richardson, J. P. 1969. Progress in nucleic acid research and molecular biology. Academic Press, New York, Vol. 9, p. 75.
15. Geiduschek, E. P., and R. Haselkorn. 1969. *Ann. Rev. Biochem.*, **38**: 647.

16. Silvestri, L. (Ed.). 1970. First Lepetit colloquium on RNA polymerase and transcription. North-Holland Publ. Co., Amsterdam.
17. Ishihama, A., and T. Kameyama. 1967. *Biochim. Biophys. Acta.,* **138**: 480.
18. Burgess, R. R. 1968. *Federation Proc.,* **27**: 295.
19. Burgess, R. R. 1969. *J. Biol. Chem.,* **244**: 6168.
20. Zillig, W., E. Fuchs, P. Palm. D. Rabussay, and K. Zechel. 1970. In L. Silvestri (Ed.), Lepetit colloquium on RNA polymerase and transcription, North-Holland Publ. Co., Amsterdam, p. 151.
21. Khesin, R. B. 1970. *ibid.,* p. 167.
22. Travers, A. A., and R. R. Burgess. 1969. *Nature,* **222**: 537.
23. Berg, D., K. Barrett, D. Hinkle, J. Mcgrath, and M. Chamberlin. 1969. *Federation Proc.,* **28**: 659.
24. Goff, C. G., and E. G. Minkley. 1970. In L. Silvestri, (Ed.), Lepetit colloquium on RNA polymerase and transcription. North-Holland Publ. Co., Amsterdam, p. 124.
25. Bautz, E. K. F., J. J. Dunn, F. A. Bautz, D. A. Schmidt, and J. Mazatis. 1970. *ibid.,* p. 90.
26. Travers, A. A. 1969. *Nature,* **223**: 1107.
27. Travers, A. A. 1970. *Nature,* **225**: in press.
28. Summers, W. C., and R. B. Siegel. 1969. *Nature,* **223**: 1111.
29. Summers, W. C. 1970. In L. Silvestri (Ed.), Lepetit colloquium on RNA polymerase and transcription. North-Holland Publ. Co., Amsterdam, p. 110.
30. Tocchini-Valentini, G. P., P. Marino, and A. J. Colvill. 1968. *Nature,* **220**: 275.
31. Ezekiel, D. H., and J. E. Hutchins. 1968. *Nature,* **220**: 276.
32. Babinet, C., and H. Condamine. 1968. *Compt. Rend. Acad. Sci. Paris,* **267**: 231.
33. Yura, T., and K. Igarishi. 1968. *Proc. Natl. Acad. Sci., U.S.,* **61**: 1313.
34. Schleif, R. F. 1969. *Nature,* **223**: 1068.
35. Rabussay, D., and W. Zillig. 1969. *FEBS Letters,* **5**: 104.
36. Sippel, A., and G. Hartmann. 1969. *Biochim. Biophys. Acta.,* **157**: 218.
37. Ishihama, A., and J. Hurwitz. 1969. *J. Biol. Chem.,* **244**: 6680.
38. Sethi, V. S., W. Zillig, and H. Bauer. 1970. *FEBS Letters,* **6**: 339.
39. Ishihama, A. 1969. *J. Cell Physiol.,* **74,** suppl. 1, p. 223.
40. Travers, A. A., in preparation.
41. Berg, D., and M. Chamberlin, in preparation.
42. Smith, D. A., A. M. Martinez, R. L. Ratliff, D. L. Williams, and F. N. Hayes. 1967. *Biochemistry,* **6**: 3057.

DNA POLYMERASE FROM

Escherichia coli

PAUL T. ENGLUND
DEPARTMENT OF PHYSIOLOGICAL CHEMISTRY
JOHNS HOPKINS SCHOOL OF MEDICINE
BALTIMORE, MARYLAND

DNA polymerase from *E. coli* was first purified to apparent homogeneity in 1964 by Richardson, Schildkraut, Aposhian, and Kornberg (*1*). Richardson described their purification scheme in Volume 1 (*2*). This article describes a modified isolation procedure which is both simpler and more suitable for operation at a large scale (*3*). The overall yield of the homogeneous polymerase is about 10 mg per kg of cell paste, and a by-product of the isolation is a purified preparation of exonuclease III.

ASSAY

DNA polymerase catalyzes the synthesis of DNA using as substrates the four deoxynucleoside 5′-triphosphates and a preexisting DNA. The assay measures the incorporation of radioactively labeled nucleotide precursor into a form which is insoluble in acid. In the early stages of purification (Fractions 1 through 4), the assay is dependent upon "activated" calf thymus DNA. Activation of this DNA, by mild treatment with pancreatic DNase, involves introduction of nicks which are the sites of action for polymerase (*4*). In later stages of purification (Fractions 4 through 7), the assay is dependent upon poly d(A-T), the alternating copolymer of deoxyadenylate and deoxythymidylate. Although this polymer is a superior substrate for DNA polymerase, it cannot be used in assaying the cruder fractions because it is extremely susceptible to degradation by nucleases which are present.

Reagents

1. The reaction mixtures (total volume 0.3 ml):

   A. Assay with "activated" calf thymus DNA:

   20 μmoles sodium glycinate, pH 9.2*
   2 μmoles $MgCl_2$
   0.3 μmoles mercaptoethanol
   10 nmoles each of dATP, dCTP, dGTP, and dTTP (one of which is labeled with approximately 3000 cpm/nmole)
   40 nmoles of "activated" calf thymus DNA †‡
   0.02-0.15 unit of polymerase, in a volume of 0.01 or 0.02 ml of polymerase diluent§

   B. Assay with poly d(A-T):

   20 μmoles potassium phosphate, pH 7.4
   2 μmoles $MgCl_2$
   0.3 μmoles mercaptoethanol
   10 nmoles each of dATP and dTTP (one of which is labeled with approximately 3000 cpm/nmole)
   6 nmoles poly d(A-T)||
   0.2 to 1.0 unit of exonuclease III#
   0.02 to 0.20 unit of polymerase, in a volume of 0.01 or 0.02 ml of polymerase diluent§

2. 0.2 M $Na_4P_2O_7$

3. 7% perchloric acid

4. 1.0 M HCl containing 0.1 M $Na_4P_2O_7$

* The pH of all buffers is determined at room temperature and at a concentration of 50 mM.

† DNA concentrations are expressed in moles of nucleotide.

‡ Calf thymus DNA is activated in a 10.0 ml reaction mixture containing 50 mM Tris-HCl, pH 7.5, 5 mM $MgCl_2$, 2.5 mg of calf thymus DNA, 5 mg of bovine serum albumin, and 5×10^{-3} μg of crystalline panereatic DNase. The solution is incubated for 15 min at 37°C, heated for 5 min at 77°C, and immediately chilled in an ice bath (*5*).

§ Polymerase diluant contains 50 mM Tris-HCl, pH 7.5, 0.1 M ammonium sulfate, 10 mM mercaptoethanol, and 1 mg/ml bovine serum albumin.

|| Poly d(A-T) is prepared enzymatically in a *de novo* or a primed reaction (*6*). Fraction 6 or 7 of the *E. coli* DNA polymerase has been routinely used in this synthesis.

Exonuclease III (*7, 8*) is added only in the poly d(A-T) dependent assay of Fraction 7. The characteristic specific activity of Fraction 7 (18,000 units/mg) is usually obtained only in the presence of this exonuclease. With some preparations of poly d(A-T) the specific activity of Fraction 7 is reduced by up to 50% when exonuclease III is omitted.

Procedure

Assays are generally made at two levels of enzyme (0.01 ml and 0.02 ml of diluted polymerase). Prior to the incubation, the reaction mixtures may be left at 0°C during addition of enzyme to all tubes; no synthesis occurs at this temperature. After incubation at 37°C for 30 min, the tubes are chilled in an ice bath. The reactions are terminated by addition to each tube of 0.2 ml of 0.2 M $Na_4P_2O_7$, which serves to reduce the blank radioactivity, and 0.5 ml of cold 7% perchloric acid. After about 5 min at 0°C, 3.0 ml of cold 1.0 M HCl containing 0.1 M $Na_4P_2O_7$ are added and the contents of each tube are filtered through a 2.4 cm diameter Whatman GF/C glass filter disk (presoaked in 0.2 M $Na_4P_2O_7$). Each tube and filter is washed with about ten 3.0-ml volumes of cold 1.0 M HCl containing 0.1 M $Na_4P_2O_7$. If the filters are to be counted in a scintillation counter, they are washed once with about 3 ml ethanol to remove any residual HCl which would act as a quenching agent. The filters are then dried under a heat lamp and counted in a gas-flow counter if the triphosphates were labeled in the α-position with ^{32}P or in a scintillation counter if they were labeled with ^{3}H or ^{14}C. Five milliliters of a scintillation fluid, containing 4 gm of 2,5-diphenyloxazole and 50 mg of 1,4-*bis*-2-(4-methyl-5-phenyloxazolyl)-benzene per liter of toluene, are used for each filter.

An alternative procedure has been routinely used in the assay of Fraction 7 (*9*). After chilling the reaction mixtures, 0.100 ml of each is transferred to a 2 cm square of DEAE-cellulose paper (Whatman DE81). The papers, which can be numbered by pencil, are then washed 5 times with occasional stirring in a beaker containing 0.3 M ammonium formate buffer, pH 8.0. This solvent elutes deoxynucleoside triphosphates, but not DNA (*10*). The papers are then washed twice in ethanol and once in ether. After drying under a heat lamp they are counted in a gas-flow counter or in a scintillation counter.

One unit of polymerase is defined as the amount catalyzing the incorporation of 10 nmoles of total nucleotide into DNA under the standard assay conditions. To calculate the total nucleotide incorporated it is necessary to take into account the base composition of the DNA used (43% G-C in the case of calf thymus DNA).

PURIFICATION OF POLYMERASE

The first 4 steps of the isolation procedure are practically identical with the earlier purification, and the description of these steps, with minor exceptions, is quoted directly from the previous publications (*1, 2*).

Unless otherwise indicated, all operations are carried out at 0°C to 4°C, and all centrifugations are at 15,000 *g* for 10 min. The results of a typical purification procedure are given in Table 1.

TABLE 1. Purification of Enzyme (*3*)

Fraction	Protein[a] (mg/ml)	Specific activity[b] (units/mg protein)	Yield (%)
1. Extract	20	3.5	100
2. Streptomycin	12	23	47
3. Autolysis	1.3	170	40
4. Ammonium sulfate	23	250 (DNA) 650 (poly d(A-T))	36
5. DEAE-cellulose[c]	11	750	36
6. Phosphocellulose[c]	3.7	10,000	28
7. Sephadex G-100[c]	4.0	18,000[d]	22

[a]The values for Steps 1 to 4 are averages for pools of different preparations. Deviations of 30% are observed.
[b]The assay utilizing activated calf thymus DNA is used in Steps 1 to 4. Deviations of up to 50% from the given values are observed. The poly d(A-T) primed assay is used in Steps 5 to 7 as well as in Step 4.
[c]Steps 5 to 7 have been successfully carried out on a scale 1/100th of that indicated by proportional adjustment of volumes and column cross-sectional areas.
[d]This assay is performed in the presence of exonuclease III. Variations of 10% from the given value are observed.

Step 1. Preparation of Extracts

E. coli strain B is grown in a medium containing 1.1% K_2HPO_4, 0.85% KH_2PO_4, 0.6% Difco yeast extract, and 1% glucose. The late log phase cells, which can be purchased from Grain Processing Corporation, Muscatine, Iowa, are frozen and stored at –20°C until used. Partially thawed cells (450 gm, wet weight) are mixed with 300 ml of 50 mM potassium glycylglycinate, pH 7.0, in a Waring Blendor (5-liter capacity) equipped with a cooling jacket and connected to a Variac transformer. Slow stirring is begun and after 5 min, 1350 gm of acid-washed glass beads (Superbrite, average diameter 200 μ, Minnesota Mining and Manufacturing Co.) are gradually added to the suspension.

When the mixture appears homogeneous, stirring is increased to approximately one-third of maximal speed. After 20 min an additional 1200 ml of the same buffer is added, and the homogenization is continued for 10 min at reduced speed to prevent excessive foaming. During the period of homogenization the temperature of the suspension is not permitted to rise above 12°C. The beads are then allowed to settle out and the broken cell suspension is decanted and saved. An additional 800 ml of buffer are added to the glass beads, and the residual broken cells are extracted by a 10-min homogenization at slow speed. The beads are again allowed to settle out, and the supernatant fluid is decanted and combined with the first supernatant fluid. After centrifugation to remove cell debris, the volume is approximately 2000 ml (Fraction 1, Table 1). This fraction may be stored at 0°C for at least 2 weeks without loss of activity.

Step 2. Streptomycin Precipitation

To 10 liters of extract are added 10 liters of 50 mM Tris-hydrochloride, pH 7.5, containing 1 mM EDTA; then, with constant stirring, 1440 ml of freshly prepared 5% streptomycin sulfate* are added over a 45-minute period. After 10 minutes, the suspension is centrifuged and the supernatant fluid is discarded. The thick, sticky precipitate is transferred to a beaker and 1000 ml of 50 mM potassium phosphate, pH 7.4, are added. The precipitate is suspended by slow mechanical stirring for approximately 12 hours, and the final volume is adjusted to 2500 ml by the addition of the same buffer (Fraction 2, Table 1). There is no detectable loss of activity after storage at 0°C for 1 month.

Step 3. Autolysis

Eleven liters of Fraction 2 are made 3 mM in $MgCl_2$ by the addition of 65 ml of 0.5 M $MgCl_2$. The suspension is incubated at 30°C for 7 to 12 hr, until 95% of the ultraviolet-absorbing material at 260 nm is rendered acid soluble. The percent-acid solubility is measured on 1-ml aliquots removed at 1-hr intervals. After centrifuging, the total absorption at 260 nm is determined after appropriate dilution in 50 mM Tris-hydrochloride, pH 7.4; to the supernatant is added an equal volume of 7% perchloric acid and the absorption of the acid soluble material is similarly determined. After autolysis the suspension is chilled to 0°C and the protein precipitate that formed during the digestion is removed by centrifugation. The supernatant (Fraction 3, Table 1) is stored at 0°C, and no loss in activity is observed over a 1-week period. During the several hours of nucleic acid digestion at 30°C, there is less than 10% loss of polymerase activity. However, if the incubation is allowed to continue beyond the complete acid solubilization of the nucleic acid, there is a further loss in enzymatic activity, i.e., 10% in 60 min.

* The amount of streptomycin sulfate required to precipitate the nucleic acids and DNA polymerase varies from one preparation of *E. coli* extract to another. Therefore, a trial precipitation on a small scale is required to determine the optimal amount of streptomycin sulfate.

Step 4. Ammonium Sulfate Fractionation

To 10 liters of Fraction 3 are added 50 ml of 0.2 M EDTA and 50 ml of 0.2 M reduced glutathione. Over a 60-min period, 3 kg of ammonium sulfate are added with stirring, and after 30 minutes at 4°C, the precipitate is removed by centrifugation. To the supernatant fluid an additional 1.15 kg of ammonium sulfate are added, with stirring, over a 60-min period. After 30 min at 4°C, the precipitate which forms is collected by centrifugation. This precipitate is dissolved in 220 ml of 20 mM potassium phosphate, pH 7.2 (Fraction 4, Table 1). Fraction 4 has been stored for 3 years at –20°C without loss of activity.

Step 5. DEAE-Cellulose

This treatment involves simply the removal of nucleic acids (without adsorption of the polymerase) which would otherwise prevent adsorption of the enzyme to the phosphocellulose column in the next step. The 12.6 cm^2 x 16 cm column of DEAE-cellulose (the resin can be purchased from Brown Co. and is washed in 0.25 M NaOH) is equilibrated with 0.2 M potassium phosphate-10 mM mercaptoethanol, pH 6.5. To 528 ml of Fraction 4 are added 25 ml of 1 M potassium phosphate buffer, pH 6.5, to lower the pH and increase the phosphate concentration to 0.07 M. The sample is applied to the column under moderate pressure, and is followed by 0.2 M potassium phosphate-10 mM mercaptoethanol, pH 6.5. The flow rate is 3 ml/min. After the first 100 ml of buffer have passed through the column, a single fraction (850 ml) containing the protein is collected and dialyzed for 10 hr against 20 liters of 0.02 M potassium phosphate-10 mM mercaptoethanol, pH 6.5. (The buffer is changed after 6 hr.) The slightly turbid solution, Fraction 5 (990 ml), has a pH of 6.6, a conductance corresponding to a phosphate molarity of 0.05 M, and an $A_{280} : A_{260}$ ratio of 1.3, compared to 0.6 before dialysis.* Fraction 5 is subjected to Step 6 within a few hours after its preparation.

Step 6. Phosphocellulose Chromatography

An 84 cm^2 x 25 cm column of Whatman phosphocellulose (the resin can be purchased from Brown Co. and is washed in 1.0 M NaOH) is equilibrated with

* It has been found in two experiments in which the A_{280}/A_{260} ratio was less than 1.3 that the polymerase did not adsorb to the phosphocellulose column in Step 6. However, if the dialyzed solution was passed through a second DEAE-cellulose column, the A_{280}/A_{260} ratio increased and the adsorption to phosphocellulose was successful.

0.02 M potassium phosphate–10 mM mercaptoethanol, pH 6.5. Two 500-ml volumes of Fraction 5 are successively diluted to 1 liter with 0.01 M mercaptoethanol, centrifuged to remove a slight precipitate, and applied to the column. The entire Fraction 5 is not diluted as a single unit to minimize the possibility of inactivation due to long exposure at low ionic strength. The column is washed with 5 liters of 0.02 M potassium phosphate–10 mM mercaptoethanol, pH 6.5, until the unadsorbed protein (68% of the applied load) is eluted. A 28-liter linear gradient from 0.02 M to 0.3 M potassium phosphate, pH 6.5, containing 10 mM mercaptoethanol is applied and, after the passage of 11 liters, 250-ml fractions are collected. Three major protein peaks are observed. In the central peak, eluted between 0.11 and 0.18 M phosphate, is 81% of the applied polymerase activity. The exonuclease III activity is eluted slightly earlier than the polymerase. Fractions which contain more than 700 polymerase units/ml are pooled (3.5 liters), and the protein is precipitated by the addition of 2360 gm of ammonium sulfate. After centrifugation in sterile 250-ml bottles, the precipitate is dissolved in sterile 0.1 M potassium phosphate–1 mM glutathione, pH 7.0. Recovery of activity in this concentration procedure is 95% or greater. Fraction 6 (188 ml) is stored in liquid nitrogen for up to 8 months without loss of activity.

Step 7. Sephadex G-100 Chromatography

Sephadex G-100 beads (600 gm) are equilibrated with 0.1 M potassium phosphate – 0.1 M ammonium sulfate – 1 mM mercaptoethanol, pH 7.0, and packed into two columns (45 cm^2 x 80 cm) constructed of Lucite tubing and machined fittings. The hydrostatic pressure across each column is not allowed to exceed 30 cm of water during the packing procedure. The columns, with a combined bed volume of 7.3 liters, are then connected in series and washed extensively with sterile pH 7.0 buffer. The void volume is 2.3 liters. To 100 ml of Fraction 6 are added 5 gm of sucrose to provide density stabilization, and the sample is applied to the column. A constant flow rate of about 2 ml/min is maintained by a peristaltic pump. Fractions of 79 ml are collected in sterile 100-ml prescription bottles mounted on a modified fraction collector. Of the applied polymerase activity, 85% (measured in the exonuclease III-augmented assay) is eluted at 35% of bed volume (Fig. 1). The nucleolytic activity of polymerase (previously designated exonuclease II (*12*)) elutes coincident with the polymerase activity. A second protein peak containing exonuclease III activity is eluted at 60% of bed volume. The polymerase fractions are pooled (237 ml) and precipitated by the addition of 142 gm of ammonium sulfate. The precipitate is dissolved in sterile 0.1 M potassium phosphate, pH 7.0, dialyzed exhaustively against that buffer, and clarified by centrifugation (Fraction 7, 37 ml). Recovery of polymerase activity in the concentration procedure is greater than 95%.

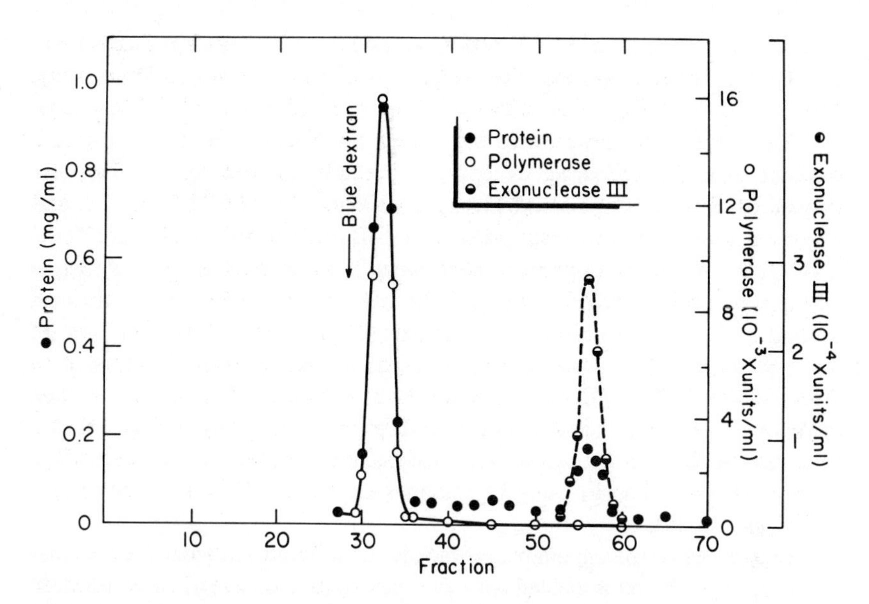

Figure 1. Sephadex G-100 filtration of Fraction 6 (*3*). Conditions are described in the text. The void volume, determined with Blue Dextran (Pharmacia), is 2300 ml. Over 95% of the applied protein is recovered. Exonuclease III is assayed as described by Richardson, Lehman, and Kornberg (*8*) except that 10 nmoles of ^{3}H-poly d(A-T) are substituted for the DNA substrate. Protein is measured according to Lowry et al. (*11*). Protein can also be measured spectrophotometrically; the absorbance at 280 nm of a 1 mg/ml solution in a cuvette of 1.0 cm path length is 0.85. (*3*).

Fraction 7 is stored in liquid nitrogen and is stable for at least 1½ years. It is also stable for at least 4 months at 0°C (under sterile conditions), frozen at −20°C, or in 50% (v/v) glycerol at −20°C. A sulfhydryl reagent such as dithiothreitol is not required for the preservation of activity.

The exonuclease III is processed in the same fashion, except that the ammonium sulfate precipitate is dissolved in 0.1 M potassium phosphate–1 mM dithiothreitol, pH 7.0. The concentration procedure yields 62% of the pooled activity. This exonuclease III fraction has a specific activity of 160,000 units/mg as compared to 16,000 units/mg for the purest enzyme reported previously (*8*), and is relatively free from endonuclease. At a concentration of 4,000 units/ml, the exonuclease III introduces less than one endonucleolytic scission per 30 min per 10,000 nucleotides of ϕX 174 RF I DNA under standard polymerase incubation conditions.

LARGE-SCALE PURIFICATION OF POLYMERASE

Steps 1 through 4 have been carried out on three 25-kg batches of cells at the New England Enzyme Center, Tufts University School of Medicine,

Boston. The average yield of Fraction 4 was 2.0×10^5 units/kg of cells (poly d(A-T) dependent assay) and the average specific activity was 670 units/mg. The work-up of 25 kg of cells, through Step 4, could be completed in 4 days.

The large scale procedures are scaled up directly from the standard method with the following exceptions. Partially thawed cells (2.5 kg) are suspended in 4 liters of 50 mM Tris-hydrochloride–1 mM EDTA, pH 7.6, and dispersed into a smooth suspension in a Gifford Wood Colloid Mill (Model VT-7-2BS). A pooled suspension containing 25 kg of cells is then diluted to 110 liters with the pH 7.6 buffer and the cells are broken by passage through a Manton Gaulin Laboratory Homogenizer (Model 15M–8TA). Only one of the two stages of the homogenizer is used, and the pressure is adjusted to 7000–8000 psi. The effluent from the Manton Gaulin Homogenizer is then centrifuged in the DeLaval continuous-flow centrifuge (Model PX-207) at a flow rate of about 3 liters/min. The supernatant from the DeLaval centrifuge is then centrifuged again using 11 Sharples Laboratory Presurtile centrifuges running in parallel.

Before the streptomycin precipitation, the supernatant from the Sharples centrifuges (96 liters) is diluted with an equal volume of water. After addition of 19.2 liters of 5% streptomycin sulfate, the precipitate is collected by centrifugation in the Sharples centrifuges. The sticky precipitate is suspended in 50 mM potassium phosphate, pH 7.4, using a Waring Blendor. The volume of the suspension is adjusted to 25 liters with 50 mM potassium phosphate, pH 7.4. Autolysis is carried out as in the standard procedure except that it is terminated, after the material absorbing at 260 nm is 95% acid-soluble, by the addition of 375 ml of 0.2 M disodium versenate. After chilling, the dense precipitate in the autolyzate is removed by centrifuging in the International PR-2 Centrifuge at 2400 RPM for 15 min. Ammonium sulfate fractionation is also carried out as in the standard procedure except that centrifugations are run in the Beckman Zonal Ultracentrifuge with a B-VIII rotor at 40,000 rpm and a flow rate of about 150 ml/min.

Further details of this large scale purification method are available from the author.

PROPERTIES OF THE PURIFIED POLYMERASE

Evidence for Purity

When assayed by polyacrylamide gel electrophoresis at pH 9.8, greater than 95% of the protein migrated as a single zone, and polymerase activity coincided with this zone. A similar degree of homogeneity was also found in gels run at pH 3.5 in the presence of 7 M urea, at pH 11, and at pH 7 in the presence of sodium dodecyl sulfate and mercaptoethanol (*3*).

When subjected to analytical centrifugation using either the sedimentation velocity or the sedimentation equilibrium method, the protein behaved as a single component (*3*).

Fraction 7 was free of significant levels of endonuclease, as shown in studies with the single-stranded, circular M13 DNA. The lack of ring opening in a 3-hr incubation with about 30 polymerase molecules per DNA molecule indicated the absence of endonucleases with specificity for either single or double-stranded DNA substrates, since in the course of replication the original single-stranded M13 DNA was converted to a double-stranded structure (*3*). There was evidence for possible contamination with polynucleotide ligase (*13*). dGTPase, which converts dGTP to deoxyguanosine and tripolyphosphate (*14*), and other phosphatase activities which convert dATP, dCTP, and dTTP to the corresponding diphosphates are present as trace contaminants, but these enzymes can be inhibited by adding $HgCl_2$ in molar excess of the polymerase (*15*). The $HgCl_2$ reacts with the single sulfhydryl group on polymerase, but has no detectable effect on polymerase activity (*9*).

The low content of phosphorus (0.6 phosphorus atom per molecule of enzyme) minimizes the possibility of contamination by nucleotide material (*3*).

Other Properties of DNA Polymerase

Besides the polymerization of nucleotides, the *E. coli* DNA polymerase catalyzes the hydrolysis of DNA from both the 3′- and the 5′- ends of a DNA strand (*12, 16, 17*), the pyrophosphorolysis of DNA (*18*), and the DNA-dependent exchange of inorganic pyrophosphate into triphosphate (*18*). The extensive studies on the specificity and mechanism of these reactions and the nature of the active site, as well as the physicochemical properties of the enzyme, have recently been reviewed and will not be discussed further here (*4, 19, 20*).

References

1. Richardson, C. C., C. L. Schildkraut, H. V. Aposhian, and A. Kornberg. 1964. *J. Biol. Chem.*, **239**: 222.
2. Richardson, C. C. 1966. In G. L. Cantoni and D. R. Davies (Eds.), Procedures in nucleic acid research. Harper & Row, New York, Vol. 1. p. 263.
3. Jovin, T. M., P. T. Englund, and L. L. Bertsch. 1969. *J. Biol. Chem.*, **244**: 2996.
4. Kornberg, A. 1969. *Science,* **163**: 1410.
5. Aposhian, H. V., and A. Kornberg. 1962. *J. Biol. Chem.*, **237**: 519.
6. Schachman, H. K., J. Adler, C. M. Radding, I. R. Lehman, and A. Kornberg. 1960. *J. Biol. Chem.*, **235**: 3242.

7. Richardson, C. C., and A. Kornberg. 1964. *J. Biol Chem.*, **239**: 242.
8. Richardson, C. C., I. R. Lehman, and A. Kornberg. 1964. *J. Biol. Chem.*, **239**: 251
9. Jovin, T. M., P. T. Englund, and A. Kornberg. 1969. *J. Biol. Chem.*, **244**: 3009.
10. Falaschi, A, J. Adler, and H. G. Khorana. 1963. *J. Biol. Chem.*, **238**: 3080.
11. Lowry, O. H., N. J. Rosebrough, A. L. Farr, and R. J. Randall. 1951. *J. Biol. Chem.*, **193**: 265.
12. Lehman, I. R., and C. C. Richardson. 1964. *J. Biol. Chem.*, **239**: 233.
13. Goulian, M., and A. Kornberg. 1967. *Proc. Natl. Acad. Sci., U.S.*, **58**: 1723.
14. Kornberg, S. R., I. R. Lehman, M. J. Bessman, E. S. Simms, and A. Kornberg. 1958. *J. Biol. Chem.*, **233**: 159.
15. Englund, P. T., J. A. Huberman, T. M. Jovin, and A. Kornberg. 1969. *J. Biol. Chem.*, **244**: 3038.
16. Klett, R. P., A. Cerami, and E. Reich. 1968. *Proc. Natl. Acad. Sci., U.S.*, **60**: 943.
17. Deutscher, M. P., and A. Kornberg. 1969. *J. Biol. Chem.*, **244**: 3029.
18. Deutscher, M. P., and A. Kornberg. 1969. *J. Biol. Chem.*, **244**: 3019.
19. Englund, P. T., M. P. Deutscher, T. M. Jovin, R. B. Kelly, N. R. Cozzarelli, and A. Kornberg. 1968. *Cold Spring Harbor Symp. Quant. Biol.*, **33**: 1.
20. Richardson, C. C. 1969. *Ann. Rev. Biochem.*, **38**: 795.

DNA LIGASE FROM *E. coli*

MARTIN GELLERT

LABORATORY OF MOLECULAR BIOLOGY

NATIONAL INSTITUTE OF ARTHRITIS

AND METABOLIC DISEASES

NATIONAL INSTITUTES OF HEALTH

BETHESDA, MARYLAND

DNA ligase joins single-strand breaks ("nicks") in double-stranded DNA. If the DNA substrate has 3′-hydroxyl and 5′-phosphoryl termini held in juxtaposition by the complementary strand, the enzyme catalyzes the formation of a 3′-5′ phosphodiester bond, thus reestablishing the covalent continuity of the DNA strand (*1-6*).

DNA-joining assays have been based on various properties of the reaction product:

1. Blocking of the terminal phosphoryl group, tested by its insensitivity to a phosphatase (*2, 3*).
2. Changes in physical structure; for example, the conversion of nicked to covalently circular DNA can be easily recognized by centrifugation (*1*).
3. Specific binding of the product to an adsorbent (*4, 6*).
4. Increase in length of covalently continuous segments, recognized in a transformation assay (*7*).
5. Insensitivity of a circular product to an exonuclease (*8*).

Two DNA-joining assays are described below. Assay I is convenient and sensitive, but requires the preparation of two additional enzymes. Assay II demands less preparation, but is less sensitive and is very laborious for large numbers of samples. It will be most useful for studies where relatively few assays are needed. In assaying crude extracts, it is more sensitive to interference by nucleases than is Assay I.

Another type of assay, independent of the DNA-joining reaction, is based on the reaction of the *E. coli* enzyme with DPN. Either the exchange of radioactive NMN into DPN or the adenylylation of the enzyme can be measured (*9, 10*). Assay III describes the adenylylation procedure.

ASSAY I

Formation of circles of Poly (dAT)

This assay measures the conversion of the alternating copolymer poly (dAT) to a circular form which is resistant to *E. coli* exonuclease III. The procedure is taken, with minor variations,* from that of Modrich and Lehman (*8*).

1. The reaction mixture contains (in 0.1 ml)†
   26 μM DPN
   0.03 M Tris-HCl, pH 8.0
   1 mM NH_4Cl
   4 mM $MgCl_2$
   1.2 mM EDTA
   ^3H-poly (dAT), 0.16 mM in nucleotides (1800 cpm per nmole of nucleotides): just before use, this is heated to 80° for 2 min and then quenched in ice.
   bovine plasma albumin (crystallized), 0.1 mg/ml
   DNA ligase, 0.002 to 0.02 units.
2. 0.1 M β-mercaptoethanol
3. *E. coli* exonuclease III (prepared as in (11)), 7500 units/ml.
4. 0.1 M Tris-HCl, pH 8.0
5. calf thymus DNA, 0.25 mM in nucleotides
6. 7% perchloric acid
7. 0.1 M sodium pyrophosphate
8. 1 M HCl
9. 95% ethanol

*Assay conditions have been altered in the following ways: doubling the concentration of albumin, adding NH_4Cl, and heating the poly (dAT) before assay.

† These conditions are satisfactory for all the fractions of *E. coli* DNA ligase to be described except the most highly purified. With Fraction VII, a 2-to 4-fold increase in apparent activity can be obtained by replacing Tris-HCl in both the reaction mixture and the enzyme diluent with potassium phosphate (pH 8.0) at the same molar concentration (10). In assaying DNA ligase from other sources, the correct cofactor must be used. For example with DNA ligase from T4-infected *E. coli*, 0.1 mM ATP is substituted for DPN, and 10 mM β-mercaptoethanol is added to the reaction.

Procedure

The reaction mixture is incubated at 30° for 30 min, then boiled for 2 min. Ten microliters of 0.1 M β-mercaptoethanol and 150 units of exonuclease III are added, and the mixture is incubated at 37° for 30 min. The reaction mixture is chilled in ice, and 0.2 ml of 0.1 M Tris-HCl (pH 8.0), and 50 μl of 0.25 mM calf thymus DNA are added, followed by 0.4 ml of cold 7% perchloric acid. After 10 min at 0°, the mixture is filtered on a Whatman GF/C 2.4-cm glass fiber filter previously soaked in 0.1 M sodium pyrophosphate. The filter is washed five times with 10 ml of cold 1 M HCl and three times with 10 ml of 95% ethanol. The filter is then dried and its radioactivity is determined in a scintillation counter with a toluene-base scintillation fluid (Liquifluor; New England Nuclear).

In the absence of DNA ligase, less than 0.5% of the added radioactivity remains acid-precipitable. One unit of DNA ligase is defined as the amount converting 100 nmoles of poly (dAT) to a form resistant to exonuclease III in 30 min under the conditions of the assay. The assay is linear over a tenfold range in enzyme concentration, and is reproducible as long as the same DNase-digested poly (dAT) sample is used. Different substrate samples give activities varying up to 40%.

Enzyme Dilutions

Enzyme is diluted in 0.05 M Tris-HCl (pH 8.0), 3 mM $MgCl_2$, 1 mM EDTA, and bovine plasma albumin (0.5 mg/ml).

Preparation of Substrate

^{3}H-poly (dAT) is made by the procedure of Modrich and Lehman (*8*). The reaction mixture contains 0.06 M potassium phosphate (pH 7.4), 6 mM $MgCl_2$, 1 mM β-mercaptoethanol, 0.5 mM dATP, ^{3}H-dTTP (0.5 mM and 8.0 Ci/mole), and 3.5 μM poly (dAT) previously treated with pancreatic DNase (see below).* DNA polymerase (Fraction VII (11); 10 units per milliliter of reaction volume) is added to start the reaction.

The reaction mixture is incubated at 37°. An aliquot is incubated in parallel in a 0.2-cm cuvette, whose absorbance at 260 nm is periodically measured (against a blank with an A_{260} of about 1) to follow the progress of the reaction. When the absorbance reaches a minimum (in 4-6 hr; the decrease in A_{260} is 0.55-0.60), the reaction is stopped by adding solid NaCl to a final concentration of 1.0 M and incubating at 70° for 25 min. To remove unreacted nucleotides, the product is dialyzed against several changes

* If one wants to vary the specific radioactivity of the poly (dAT) within a series of experiments, it is convenient to make unlabeled poly (dAT) and highly radioactive poly (dAT) separately and mix them as needed (*8*).

of 1.0 M NaCl and 1 mM EDTA until the A_{260} of the dialysis buffer is less than 0.002. The product is concentrated 3-fold by dialysis against solid polyethylene glycol (Carbowax 6000, Union Carbide), then dialyzed against 40 vol of 0.1 M Tris-HCl (pH 8.0) and 1 mM EDTA (2 changes). The final product is about 1 mM in nucleotides (using a molar extinction coefficient of 6700) with an overall yield of 30–40%. The number average molecular chain length of the poly (dAT) is about 5000 nucleotides.

This highly polymerized poly (dAT) is a poor substrate for DNA ligase and a relatively poor primer for DNA polymerase. The poly (dAT) is therefore partially digested with pancreatic DNase to produce a chain length which leads to the greatest apparent activity of DNA ligase in the joining assay. It is advisable to determine the optimal DNase concentration for each preparation of poly (dAT).

In a typical preparation of substrate, the reaction mixture (10 ml) contains 0.09 M Tris-HCl (pH 8.0), 0.01 M $MgCl_2$, 1 mM EDTA, 0.83 mM ^{3}H-poly (dAT) (1800 cpm per nmole) and pancreatic DNase, 4.5 ng/ml. After 35 min at 37°, the solution is heated to 75° for 30 min to inactivate the DNase, then rapidly chilled in ice to favor intramolecular helix formation. Such preparations have a number-average chain length of 700 to 1000 residues. The substrate can be stored frozen for some weeks.

ASSAY II

Formation of Covalently Circular λ DNA

In this assay, covalent circles (molecules in which both strands are covalently continuous) of phage λ DNA are formed by DNA ligase action and are detected by their rapid sedimentation in an alkaline sucrose gradient (*13*). λ DNA has complementary single-stranded regions at the ends of its double-stranded structure. At low DNA concentration, these regions can cohere intramolecularly to form hydrogen-bonded circular molecules (*14*). The structure so formed is double-stranded with single-strand interruptions (nicks) at the ends of the cohered region, and is a substrate for the action of DNA ligase. Covalent joining of the ends produces covalently circular molecules.

Reagents

1. λ DNA is cohered in a mixture containing:
   5 μg/ml ^{3}H-labeled λ DNA, 50-100 Ci/mole
   2 M NaCl
   0.01 M Tris-HCl, pH 8.0
   1 mM EDTA

2. The enzymatic reaction mixture (total volume 0.1 ml) contains:
   cohered ^{3}H-labeled λ DNA, 2 μg/ml
   0.02 M Tris-HCl, pH 8.0
   1 mM EDTA
   4 mM $MgSO_4$
   10 mM NH_4Cl
   50 μg/ml bovine plasma albumin (crystallized)
   5 μM DPN
   DNA ligase, 0.02 to 0.2 units (see Assay I for dilution medium and definition of units)
3. 0.2 M EDTA
4. 0.5 M NaOH
5. Sucrose solutions, 5% and 20% (w/v), containing 1 M NaCl and 1 mM EDTA; each solution adjusted to pH 12.3, using a low sodium error glass electrode (e.g., Radiometer G222B).*

Procedure

Preparation of Cohered λ DNA Circles

The mixture (usually 1 to 10 ml) is incubated for 2 hr at 50°, then chilled and dialyzed in the cold against a solution containing 0.01 M Tris-HCl, pH 8.0, 1 mM EDTA, and 2 mM $MgSO_4$. About 80% of the DNA is circularized. The low DNA concentration used in this step is necessary to avoid formation of intermolecular DNA aggregates. The dialyzed cohered DNA can be stored at 0° for at least a week.

Enzymatic Reaction

After incubation for 30 min at 30°, the reaction mixture is chilled, and 10 μl of 0.2 M EDTA and 10 μl of 0.5 M NaOH are added.

Sucrose Gradient Centrifugation

The reaction mixture (0.1 ml) is layered on a 5–20% alkaline sucrose gradient (total volume 4.4 ml) in a polyallomer tube and centrifuged for 70 min at 38,000 rpm in the SW-39 rotor of a Spinco model L centrifuge. The temperature is about 5°. The bottom of the tube is then punctured and 0.2-ml fractions are collected. Aliquots, usually 0.15 ml, are transferred to scintillation vials and counted. The scintillator solution contains: 3 liter toluene, 200 ml H_2O, 500 ml Biosolv BBS-3, and 32.2 gm Fluoralloy TLA (the last two reagents are from Beckman Instruments, Inc.). Ten milliliters of this solution will dissolve 0.15 ml of the sucrose gradient fractions.

* Sucrose is a strong buffer in this pH range. Merely adding a dilute alkaline buffer to adjust the pH of the solutions is not satisfactory, though sometimes done.

Covalent circles are found about 1 ml from the bottom of the gradient, forming a peak of material sedimenting 3.6 to 4.0 times as fast as linear λ DNA. The fraction of radioactivity sedimenting as covalent circles is determined.

With purified *E. coli* DNA ligase, the production of covalent circles is proportional to enzyme concentration up to at least 10% conversion to covalent circles. If the DNA cohesion step is omitted, or in the absence of ligase, less than 0.2% of the radioactivity is found at the position of covalent circles. Incubation with 0.2 units of *E. coli* ligase (as measured by Assay I above) catalyzes 10% conversion to covalent circles.

Even with large amounts of ligase, no more than 25 to 30% of the DNA becomes covalently closed. It has been suggested (*5*) that some molecules of λ DNA have nucleotides missing from their ends and thus cannot be joined unless the gaps are filled in by DNA polymerase action.

Preparation of Substrate

^{3}H-labeled λ phage is prepared by heat induction of the thymine-requiring strain *E. coli* CR 34 (λcI857)/λ in a synthetic medium containing ^{3}H-thymidine. DNA is extracted from the purified phage by shaking with phenol, and is stored in 0.01 M Tris-HCl (pH 8.0), 1 mM EDTA (see Ref. *1* for details).

ASSAY III

Adenylylation of DNA Ligase

E. coli DNA ligase reacts with DPN to form a stable complex, in which the AMP moiety of DPN is incorporated into an acid-insoluble form (*9*). Since the reaction is fast (the half-time in the assay mixture described below is less than 1 min even at 0°), the assay most conveniently measures the extent rather than the rate of adenylylation.

1. The reaction mixture (*10*) contains, in 0.1 ml:
   0.1 μM ^{14}C-DPN, adenosine-labeled (180 to 200 cpm/pmole)*
   0.1 M Tris-HCl (pH 8.0)
   4 mM $MgSO_4$
   1 mM dithiothreitol
   bovine plasma albumin (crystallized), 0.5 mg/ml
   DNA ligase, 0.1–2 units (see Assay I for dilution medium and definition of units).

* See Ref. *9* for preparation. DPN which is tritium-labeled in the adenosine moiety is now commercially available at a specific activity high enough for this assay.

2. Bovine plasma albumin, 10 mg/ml
3. 0.02 M EDTA (pH 8.0)
4. 10% trichloracetic acid
5. 2 M NH_4OH

Procedure

After 10 min of incubation at 37°, 0.4 ml of a solution containing 1 mg/ml bovine plasma albumin and 2 mM EDTA is added, and the tubes are chilled. Cold 10% trichloracetic acid (0.5 ml) is added, and after 5 min on ice the samples are centrifuged at 8000 *g* for 10 min at 5°. The supernatant fluid is discarded. The pellets are suspended in 2 ml of cold 5% trichloracetic acid, collected by centrifugation, dissolved in 2 ml of 2M NH_4OH, and plated on stainless steel planchets. After the solution is taken to dryness, the radioactivity is measured in a low-background Geiger counter. In control incubations without enzyme, 0.1% of the added radioactivity is in the final pellet.

This assay can be used for Fraction IV and later fractions of the purification procedure (see below). For crude fractions which contain DPN, it is first necessary to remove the DPN by passage through a Sephadex G-25 column or by dialysis. With this modification, crude *E. coli* extracts can be assayed reliably (*15*).

ISOLATION PROCEDURE

The procedure is that of Zimmerman and Oshinsky (*10*). Unless specified, all operations are carried out at 0–5°. Loading and elution of columns is at a flow rate of 1–2 ml per min per cm^2. The results of a typical preparation (*10*) are summarized in Table 1.

Preparation of Cell Extracts

Frozen cells (570 gm) of *E. coli* B (General Biochemicals, late log phase, Kornberg medium) suspended in 1140 ml of 0.05 M Tris-HCl buffer (pH 8.0) containing 20 μM EDTA are ruptured by two passages through a laboratory homogenizer (type 15M8TA, Manton-Gaulin Manufacturing Company, Everett, Mass.). After centrifuging for 3 hr at 13,000 *g*, the supernatant fluid is collected (Fraction 1). (Glass bead extraction (*4*) also provides suitable material for the following steps.)

Streptomycin Precipitation of Inactive Materials

Fraction I is diluted with an equal volume of 0.05 M Tris-HCl buffer (pH 8.0) containing 20 μM EDTA. An amount of freshly dissolved 10%

TABLE 1. Purification of *E. coli* DNA Ligase and Comparison of Assays (*10*)

Fraction and step	Volume (ml)	Protein (mg/ml)	DNA joining assay[a] Total activity units x 10^{-3}	Specific activity units/mg protein	Yield (%)	Enzyme adenylylation assay (pmoles AMP/mg protein)	Ratio of enzyme adenylylation to DNA joining activity (pmoles AMP/unit)
I. Crude extract	1,520	30	99	2.2	100		
II. Streptomycin	3,810	9.6	65	1.8	66		
III. Methanol	1,890	7.9	52	3.5	53	2	0.57
IV. Alumina Cγgel	226	11.5	21	7.9	21	5.6	0.71
V. DEAE-cellulose	175	0.89	9.5	60	10	39	0.65
VI. Ammonium sulfate	8.7	6.5	8.5	151	9	96	0.64
VII. Phosphocellulose	50	0.051	7.0	2750	7	2640	0.96

[a] Units of DNA-joining activity have been recalculated from those of Ref. *10* to the units defined for Assay 1. One unit of activity in the poly (dAT)-joining assay is equivalent to 20 units as defined in (*10*) (unpublished observations of the author). The ratio of activities in these two joining assays is constant throughout purification. Several of these steps were done repeatedly on a smaller scale as described in the text and the values given are calculated for the large scale procedure.

streptomycin sulfate (USP) equal to one-third of the volume of diluted extract is added over 5 min with stirring. After 15 min, the suspension is centrifuged for 20 min at 13,000 *g* and the supernatant fluid is collected (Fraction II).

Methanol Precipitation

This step is carried out with several smaller batches (e.g., 700 ml of Fraction II) to prevent excessively long time of exposure to methanol. Absolute methanol (0.45 volume, 0°) is added to Fraction II (at 0°) with efficient stirring over a ½-min period. After 5 min on ice, the suspension is centrifuged for 10 min at 27,000 *g*. The precipitate is promptly drained and dissolved in a volume of 0.1 M potassium phosphate buffer (pH 6.8) equal to 0.5 that of Fraction II (Fraction III).

Alumina Cγ Gel Fractionation

To 1890 ml of Fraction III are added 470 ml of a 1 : 5 dilution of alumina Cγ gel (aged gel, Bio-Rad; 40 mg of solids per ml before dilution) in 0.01 M potassium phosphate buffer (pH 6.8). The suspension is stirred 30 min and centrifuged for 10 min at 4000 *g*. The precipitate is resuspended in 1490 ml of 0.6 M potassium phosphate buffer (pH 6.8) and stirred 30 min. The supernatant fluid is then collected by centrifugation for 10 min at 10,000 *g*. The enzyme is concentrated by addition of ammonium sulfate (0.43 gm/ml), equilibration for 15 min, and centrifugation for 15 min at 13,000 *g*. The pellets are redissolved in 150 ml of 0.02 M potassium phosphate buffer (pH 8.0), centrifuged 10 min at 12,000 *g*, and dialyzed overnight against 6 liters of the same buffer. Inactive material is removed by centrifugation for 15 min at 8000 *g* (Fraction IV).

DEAE-Cellulose Chromatography

Fraction IV is reacted with DPN to form ligase-AMP before DEAE-cellulose chromatography, primarily because the yield of enzyme is up to 2-fold higher when the enzyme is in the adenylylated form. Also, the ligase-AMP can readily be made radioactive by using appropriately labeled DPN, and then assayed simply by the distribution of acid-insoluble radioactivity (see above). Accordingly, Fraction IV is brought to 2 mM $MgSO_4$ and 0.2 μM ^{14}C-DPN, adenosine-labeled (12 cpm/pmole; see footnote to Assay III). After 30 min at 0°, the mixture is brought to 4 mM EDTA and loaded onto a DEAE-cellulose column (DE52, Whatman; Reeve Angel and Co.) 4.5 cm^2 x 16 cm, equilibrated with 0.02 M potassium phosphate buffer (pH 8.0)). A linear gradient is applied from 0.05 M to 0.20 M potassium phosphate buffer (pH 8.0) with a total gradient volume of 1900 ml. Fractions of 25 ml are collected. Unreacted ^{14}C-DPN is marginally bound under these conditions and is completely eluted by the beginning of the gradient. Aliquots of 0.2 ml are precipitated as described for Assay III. A peak of ligase-^{14}C-AMP appears at 0.14 M buffer (Fraction V). An additional peak of ligase-^{14}C-AMP centered at 0.11 M buffer has appeared in approximately one-third of the fractionations and may contain up to 40% of the total ligase-^{14}C-AMP. This material has little if any DNA-joining activity (*10*) and is discarded.

Ammonium Sulfate Fractionation

Fraction V is equilibrated for 15 min at 0° with ammonium sulfate (0.262 gm/ml of Fraction V), centrifuged for 15 min at 11,000 *g*, and the supernatant fluid is further treated with ammonium sulfate (0.197 gm/ml of Fraction V) as above. The final precipitate is redissolved in 0.04 volume of 0.01 M potassium phosphate buffer (pH 6.8) (Fraction VI).

Phosphocellulose Chromatography

The ligase-AMP is discharged by reaction with NMN (*9*) before chromatography.* Fractionation on the following scale is repeated as needed. Fraction VI (1.50 ml) is incubated with 0.1 mM NMN, 2 mM $MgSO_4$, and H_2O in a total volume of 3.0 ml for 10 min at 37°, then chilled, and 0.24 ml of 0.1 M EDTA is added. After dialyzing overnight against 500 ml of 10 mM potassium phosphate buffer (pH 6.8) containing 0.1 mM EDTA and 0.1 mM dithiothreitol, the enzyme is brought to 20% glycerol, v/v, by addition of 0.67 volume of 50% glycerol, v/v, and applied to a column (0.9 cm^2 x 10 cm) of phosphocellulose equilibrated with 10 mM potassium phosphate buffer (pH 6.8) containing 20% glycerol, 0.1 mM EDTA, and 0.1 mM dithiothreitol. The column is eluted with a linear gradient (100 ml total volume) from 0.01 M to 0.15 M potassium phosphate buffer (pH 6.8) containing 20% glycerol, 0.1 mM EDTA, and 0.1 mM dithiothreitol. A peak containing both DNA-joining and AMP-accepting ability appears at about 0.07 M phosphate (Fraction VII).

Stability of Enzyme Fractions

Fraction I has been kept frozen for 6 months without change in activity. Fractions I through VI have been kept unfrozen at 0° from several weeks to several months without loss in activity. Fraction VII is relatively unstable to either frozen or cold storage, losing approximately 30% activity after a single freezing or after 2 weeks at 0°. Fraction VII, however, can be stabilized

*Ligase-AMP is retained to a variable extent on phosphocellulose, in contrast to the reproducible behavior of ligase.

greatly by adding bovine plasma albumin (0.1 vol of a 1% solution) shortly after collection; the activity is then stable to repeated freezings or to storage for several months at 0°.

If Fraction VII, or the corresponding adenylylated enzyme, is stored in buffers of low concentration (<0.05 M), a conversion to two less active forms of the enzyme occurs with a half-time of a few days (*10*). These forms, which are distinguishable by chromatography on DEAE-cellulose, are still able to accept AMP from labeled DPN, but have little or no DNA-joining activity. Storage in more concentrated salts prevents this inactivation.

PROPERTIES OF THE ENZYME

Specificity

The joining reaction does not appear to be specific with regard to the base of the nucleotide residue on either side of the break; formation of most of the 16 common internucleotide links has been reported by now (e.g., *5, 14*), although the influence of local sequence on the *rate* of joining has not been systematically studied.

Other requirements on the substrate structure are relatively stringent:

1. The juxtaposed strands must end in 5′-phosphoryl and 3′-hydroxyl groups. Dephosphorylated ends, or those with 3′-phosphoryl groups, cannot be joined (*4, 17*).

2. The apposed ends must be held in register by base-pairing to the complementary strand. If the strands are separated by denaturation, or if the complementary strand is also broken at the same point (resulting in a double strand break), joining does not take place. Similarly, joining is prevented by removal of a few nucleotides between the apposed ends (*5*).*

3. For joining by *E. coli* DNA ligase to occur, both strands must be deoxyribonucleotide polymers (*3, 21*). T4-induced ligase, however, can seal single-strand breaks in either strand of a ribonucleotide–deoxyribonucleotide hybrid (*21, 22*). The rate of this reaction is about 1% of that with a

* Several exceptions to this rule deserve mention. First, in synthetic DNA polymers where the repeating sequence allows the chains to slip relative to each other, the strands can be joined even if the most stable configuration does not have correctly apposed ends (as, for example, in poly (dAT) (*18*). In such systems, the rate of joining may be limited by the conformational mobility of the polymer rather than by the enzymatic reaction.

Second, a single mismatched base at the end of a base-paired region can be joined to a neighboring chain. This was shown with $dT_{11}dC$ bound to poly (dA); neighboring oligomers were joined through a CpT bond (*19*).

Third, a joining reaction whose final product is a dimer of double-strand DNA fragments has recently been reported. It is not known whether the first product of this reaction is joined intramolecularly across the end of a single fragment or intermolecularly (*20*). The latter two reactions have been demonstrated only with T4-induced DNA ligase and presumably with high levels of enzyme.

double-stranded deoxyribonucleotide polymer. No joining has been observed in double-stranded polyribonucleotides.

DNA ligase from *E. coli* requires DPN as cofactor (*4, 23*) and consumes it stoichiometrically in the DNA-joining reaction; one molecule of DPN is cleaved to AMP and NMN for each DNA break that is sealed (*4*). The K_m for DPN is 3–10 x 10^{-8} M (*4, 23*), a remarkably low value among enzymes which use DPN. Among DPN analogs tested, only thionicotinamide-DPN has substantial activity. In particular DPNH, TPN, and TPNH are totally inactive (*4*).

The T4-induced and mammalian ligases use ATP in an analogous reaction (*2, 24*), yielding AMP and pyrophosphate as products. The specificity of both groups of enzymes is strict; there is no stimulation of *E. coli* DNA ligase by ATP or of the T4 and mammalian enzymes by DPN.

Reaction Requirements

The optimal pH range for *E. coli* DNA ligase is 7.5–8.0 in Tris-HCl buffer and near 8.0 in sodium phosphate buffer. A divalent cation (Mg^{2+} or Mn^{2+}) is required for activity, the optimal Mg^{2+} concentration being 1–4 mM. Mn^{2+} at levels of 0.2–1.0 mM is slightly more effective than Mg^{2+}.

Mechanism – Isolation of Intermediates

The steps of the reaction catalyzed by *E. coli* DNA ligase are summarized below:

$$\text{DPN + ligase} \rightleftharpoons \text{AMP–ligase + NMN}$$
$$\text{AMP–ligase + nicked DNA} \longrightarrow \text{AMP–DNA + ligase}$$
$$\text{AMP–DNA} \xrightarrow{\text{ligase}} \text{joined DNA + AMP}$$

$$\text{DPN + nicked DNA} \xrightarrow{\text{ligase}} \text{joined DNA + AMP + NMN}$$

In the first step, which can take place in the absence of DNA, a covalent adenylyl-ligase complex is formed, releasing NMN (*9*). The equilibrium of this reaction strongly favors the complex, so that, without added NMN, essentially complete adenylylation of the enzyme is readily obtained. The reaction can be reversed by a sufficient excess of NMN (50% reversal by 10 μM NMN in the presence of 0.1 μM DPN).

Ligase is readily isolated in the adenylylated form (*9*). After reaction of the enzyme with DPN (see Assay II for typical conditions) excess EDTA is added and the mixture is passed through a Sephadex G-25 column in 0.10 M potassium phosphate (pH 8.0), 0.1 mM dithiothreitol, and 0.1 mM EDTA, or dialyzed against this buffer. AMP-ligase made from Fraction VII can be stored under the same conditions as specified above for Fraction VII, and has comparable stability. The complex is able to carry out the DNA-joining process without added cofactor.

In the next step of the reaction, the adenylyl group is transferred to the 5′-phosphoryl end of the single strand break on the DNA, forming an adenylyl-5′-phosphoryl linkage (*25*). For this reaction, as for joining, the proper DNA substrate configuration is essential. A 5′-phosphoryl end without an apposed 3′-hydroxyl end cannot be adenylylated.

In order to trap an appreciable fraction of adenylyl residues in AMP-DNA, it is necessary to choose conditions which slow the rate of the overall joining reaction. After incubation at pH 8 and 0° for a short time (30 sec), 5–10% of AMP originating from AMP-ligase can be found attached to DNA (*25*). Further incubation, even at 0°, leads to release of AMP from the DNA.

A larger fraction of adenylyl groups can be trapped in AMP-DNA if the pH is lowered to 6.5, where the joining reaction is drastically slowed. Since conditions for efficient isolation of AMP-DNA have not previously been published, they are given in some detail below (unpublished observations of the author). The reaction mixture contains pancreatic DNase-treated calf thymus DNA (about 8% acid-soluble), 0.6 mM in nucleotides, 0.07 M sodium phosphate (pH 6.5), 0.5 mM $MgSO_4$, 0.5 mM dithiothreitol, 50 μg/ml bovine plasma albumin, and 1–50 nM ^{14}C-AMP-ligase. After 5 min at 0°, 1/10 volume of a cold solution of 1 M sodium glycinate and 0.5 M EDTA, whose pH has been adjusted to 10.2 with NaOH, is added, and the mixture is shaken twice with phenol to remove unreacted AMP-ligase. The product is exhaustively dialyzed against 0.01 M Tris-HCl (pH 8.0) and 1 mM EDTA. By this procedure, 80–90% of radioactivity originating in AMP-ligase can be isolated in AMP-DNA.

If AMP-DNA is incubated with ligase under normal assay conditions, the final joining step then takes place, with liberation of AMP (*25*). In this step, the 3′-hydroxyl group presumably attacks the neighboring pyrophosphoryl linkage, forming a phosphodiester bond and displacing AMP. Unless AMP-DNA is artificially separated from ligase, the last two steps of the reaction probably occur in a concerted fashion, with ligase remaining bound to the DNA nick (*26*).

E. coli Strains with Altered DNA Ligase Activity

Mutants of *E. coli* with more or less than normal ligase activity have been isolated (*15, 27*). One mutant (*lop* 8) makes 4 to 5 times as much enzyme as wild-type *E. coli* (*15*). The enzyme itself appears to be unaltered, and can be purified by the procedure described above. This strain thus provides a convenient source of DNA ligase. Other strains, with a partially defective or temperature-sensitive DNA ligase, have also been described (*15, 27*).

References

1. Gellert, M. 1967. *Proc. Natl. Acad. Sci., U.S.*, **57**: 148.
2. Weiss, B., and C. C. Richardson. 1967. *Proc. Natl. Acad. Sci., U.S.*, **57**: 1021.
3. Olivera, B. M., and I. R. Lehman. 1967. *Proc. Natl. Acad. Sci., U.S.*, **57**: 1426.
4. Zimmerman, S. B., J. W. Little, C. K. Oshinsky, and M. Gellert. 1967. *Proc. Natl. Acad. Sci., U.S.*, **57**: 1841.
5. Gefter, M. L., A. Becker, and J. Hurwitz. 1967. *Proc. Natl. Acad. Sci., U.S.*, **58**: 240.
6. Cozzarelli, N. R., N. E. Melechen, T. M. Jovin, and A. Kornberg. 1967. *Biochem. Biophys. Res. Commun.*, **28**: 578.
7. Bautz, E. K. F. 1967. *Biochem. Biophys. Res. Commun.*, **28**: 641.
8. Modrich, P., and I. R. Lehman. 1970. *J. Biol. Chem.*, **245**: 3626.
9. Little, J. W., S. B. Zimmerman, C. K. Oshinsky, and M. Gellert. 1967. *Proc. Natl. Acad. Sci., U.S.*, **58**: 2004.
10. Zimmerman, S. B., and C. K. Oshinsky. 1969. *J. Biol. Chem.*, **244**: 4689.
11. Jovin, T. M., P. T. Englund, and L. L. Bertsch. 1969. *J. Biol. Chem.*, **244**: 2996.
12. Schachman, H. K., J. Adler, C. M. Radding, I. R. Lehman, and A. Kornberg. 1960. *J. Biol. Chem.*, **235**: 3242.
13. Bode, V. C., and A. D. Kaiser. 1965. *J. Mol. Biol.*, **14**: 399.
14. Hershey, A. D., E. Burgi, and L. Ingraham. 1963. *Proc. Natl. Acad. Sci., U.S.*, **49**: 748.
15. Gellert, M., and M. L. Bullock. 1970. *Proc. Natl. Acad. Sci., U.S.*, **67**: 1580.
16. Gupta, N. K., E. Ohtsuka, H. Weber, S. H. Chang, and H. G. Khorana. 1968. *Proc. Natl. Acad. Sci., U.S.*, **60**: 285.
17. Becker, A., G. Lyn, M. Gefter, and J. Hurwitz. 1967. *Proc. Natl. Acad. Sci., U.S.*, **58**: 1996.
18. Olivera, B. M., I. E. Scheffler, and I. R. Lehman. 1968. *J. Mol. Biol.*, **36**: 275.
19. Tsiapalis, C. M., and S. A. Narang. 1970. *Biochem. Biophys. Res. Commun.*, **39**: 631.
20. Sgaramella, V., J. H. Van de Sande, and H. G. Khorana. 1970. *Proc. Natl. Acad. Sci., U.S.*, **67**: 1468.
21. Fareed, G. C., E. M. Wilt, and C. C. Richardson. 1971. *J. Biol. Chem.*, **246**: 925.
22. Kleppe, K., J. H. Van de Sande, and H. G. Khorana. 1970. *Proc. Natl. Acad. Sci., U.S.*, **67**: 68.
23. Olivera, B. M., and I. R. Lehman. 1967. *Proc. Natl Acad. Sci., U.S.*, **57**: 1700.
24. Lindahl, T., and G. M. Edelman. 1968. *Proc. Natl. Acad. Sci., U.S.*, **61**: 680.
25. Olivera, B. M., Z. W. Hall, and I. R. Lehman. 1968. *Proc. Natl. Acad. Sci., U.S.*, **61**: 237.
26. Hall, Z. W., and I. R. Lehman. 1969. *J. Biol. Chem.*, **244**: 43.
27. Pauling, C., and L. Hamm. 1969. *Proc. Natl. Acad. Sci. U.S.*, **64**: 1195.

DNA-RESTRICTION ENZYME FROM *E. coli* K

MATTHEW MESELSON

AND

ROBERT YUAN

THE BIOLOGICAL LABORATORIES

HARVARD UNIVERSITY

CAMBRIDGE, MASSACHUSETTS

Most strains of *E. coli* are able to recognize and degrade DNA from foreign *E. coli* strains. Whether a foreign DNA molecule is degraded depends on certain nonheritable properties imparted by the cell from which it is obtained. These properties are called host-controlled modifications. For example, the result of infecting *E. coli* strain K with phage λ depends on the host in which the phages were last grown. Phages grown in bacteria possessing the modification character m_k multiply successfully, but phages from bacteria lacking m_k do not. Instead, their DNA is quickly degraded on entering cells of strain K. The ability of strain K to degrade or "restrict" DNA from cells lacking m_k is itself under genetic control, the responsible character being designated r_k. An endonuclease has been shown to be responsible for restriction in *E. coli* K. It is specifically active against λ DNA from strains lacking m_k and is called endonuclease R·K.

ASSAYS

Two assay methods will be described here: sedimentation analysis on neutral sucrose gradients and specific binding of enzyme-DNA complexes to nitrocellulose filters.

SEDIMENTATION ANALYSIS

Principle

^{3}H-Thymidine-labeled DNA is prepared from phage λ grown on *E. coli* strain K, designated λ·K DNA. ^{32}P-labeled λ DNA sensitive to endonuclease R·K is prepared from a $r_k^- m_k^-$ mutant of strain K and is referred to as λ·0 DNA. Restriction *in vitro* is observed by incubating a mixture of these two DNAs with enzyme in the presence of S-adenosylmethionine, ATP, and Mg^{2+}. Sedimentation analysis on a neutral sucrose gradient shows that λ·0 DNA is cleaved while λ·K DNA is not. The assay may be applied to crude cell extracts as well as to purified enzyme fractions.

Reagents

TES-NaOH buffer, 0.1 M, pH 8.0
Na-EDTA, 0.1 M, pH 8.0
$MgCl_2$, 0.01 M
Mercaptoethanol, 0.1 M
ATP, Sodium salt, 0.1 M, stored at $-10°$
S-adenosylmethionine (SAM), purified from the commercial product by elution from "Biorex 70" with acetic acid, ca. 10^{-2} M in 4 N acetic acid, stored at $-10°$.
Calf thymus DNA, 2.5% in 0.01 M Tris 10^{-4} M EDTA, pH 8.0
^{32}Pλ·0 DNA, approximately 10^{12} phage equivalents/ml in 0.01 M Tris, 0.01 M NaCl, 10^{-4} M EDTA adjusted to pH 8.0 (TNE)
^{3}Hλ·K DNA, approximately 10^{12} phage equivalents/ml in TNE.

The phage stocks are produced by ultraviolet light induction of C600.4(λ) and CR34(λ), respectively, in radioactive media. Phages are purified by two cycles of differential centrifugation followed by equilibrium sedimentation in CsCl solution. DNA is extracted with water-saturated phenol adjusted to pH 8.0 with NaOH and is extensively dialyzed against TNE. Before use in assays, DNA solutions are kept at 55° for 5 min in order to disassociate end-to-end aggregates.

Procedure

A reaction mixture of 0.25 ml contains: 25 μmole TES, 0.07 μmole EDTA, 1.6 μmole $MgCl_2$, 3 μmole mercaptoethanol, 0.1 μmole ATP, 0.02 μmole S-adenosylmethionine, approximately 10^{10} phage equivalents each of ^{32}P λ·0 DNA and ^{3}H λ·K DNA and, in the purification scheme described below, 10 μl of enzyme preparation.

After incubation for 20 min at 30°, 0.03 ml of 0.4 M EDTA, pH 9.5, is added. Each sample is layered on a 3.5 ml 6–20% exponential sucrose gradient containing 0.01 M Tris, pH 8.0, 0.001 M EDTA, 0.04% sodium dodecyl sulfate, and centrifuged for 1.6 hr at 55,000 rpm in an International SB-405 rotor at 20°. Fractions are collected on filter paper, dried, and counted by scintillation.

BINDING ASSAY

Principle

When λ·0 DNA and λ·K DNA are incubated with endonuclease R·K in the presence of SAM, ATP, and Mg^{2+}, and passed through a membrane filter, the λ·0 DNA is specifically retained. Apparently, λ·0 DNA forms a complex with endonuclease R·K which binds to the membrane. In all except the fractions from the final glycerol gradient, calf thymus DNA is added to suppress the nonspecific binding of λ DNA caused by other proteins.

Procedure

The reagents for the binding assay are the same as for the sedimentation analysis. The reaction mixture for binding is also the same except that only 2.5×10^9 phage equivalents of ^{32}P λ·0 DNA and of ^{3}H λ·K DNA are used and 0.01 ml of 2.5% calf thymus DNA solution is added. The amount of enzyme preparation is decreased to 5 μl. Incubation is carried out at 30° for 10 min in all assays except those for the final glycerol gradient fractions. The incubation for the assays of the glycerol fraction is shortened to 2 min, and the calf thymus DNA is omitted. Immediately after the incubation, 0.03 ml of 0.5 M EDTA, pH 8.0, is added. The membrane filters used are Schleicher and Schuell B6, 24-mm diameter. The filters are washed in boiling distilled water for 5 min, rinsed, and stored in 0.1 M Tris, pH 8.0, 10^{-4} M EDTA, and 10^{-5} M mercaptoethanol (TEM). Each filter is washed with 5 ml TEM prior to application of the sample. The sample is filtered slowly (approximately 1 ml/minute) with gentle suction. The filter is then washed with 5 ml TEM at the same flow rate. Filtration is done at room temperature. The filters are dried and counted by liquid scintillation. The percentage of ^{3}H DNA bound to the filter should be small. It is subtracted from the percentage of ^{32}P DNA bound to obtain the amount of specific binding.

This binding assay is quick and sensitive. Fractions that are shown to be active in the binding assay are further checked by using the sedimentation analysis assay.

Unit of Activity

A unit is defined as the amount of protein necessary to bind 30% of the input λ·0 DNA present in a standard reaction mixture in 2 min at 30°.

PURIFICATION PROCEDURE

The DNA endonuclease I deficient strain 1100 is grown to a concentration of 6–8 x 10^8 at 37° in tryptone broth supplemented with 1 μg/ml of vitamin B_1. (Although we have used only strain 1100, other K strains could probably be used as well.) The cells are harvested, washed with 0.01 M Tris, pH 7.4, 2 x 10^{-4} M $MgCl_2$, 10^{-4} M EDTA, and frozen at –20°.

Preparation of Crude Extract

A 120-gm portion of frozen cells is thawed in 70 ml of the above solution supplemented with 2 x 10^{-3} M mercaptoethanol. All operations are carried out at approximately 4°. The mixture is homogenized with 360 gm of acid washed "super-bright" beads in a water-jacketed blender for 5 min at low speed and then 5 min at top speed. The mixture is centrifuged to bring down beads and unbroken cells. The pellet is blended as described above with 100 ml of fresh buffer. The mixture is centrifuged and the second pellet blended once more in the same way. All supernatants are combined to give a volume of approximately 300 ml.

The combined supernatant is centrifuged at 35,000 rpm for 4 hr in the International A-170 rotor. The pellet is discarded.

Ammonium Sulfate Fractionation

The high-speed supernatant is fractionally precipitated by gradual addition of dry ammonium sulfate. Fractions are collected from 25 to 65% saturation at 10% intervals. If the pH falls below 7.0, it is restored by the addition of 0.1 M NaOH. The material precipitating between 35 and 55% saturation is dissolved in 100 ml of 0.02 M potassium phosphate, pH 7.0, 10^{-4} M EDTA, 2 x 10^{-3} M mercaptoethanol (PEM), and dialyzed against the same buffer to remove ammonium sulfate. Activity is sometimes found in other ammonium sulfate fractions. These should be saved until the activity is located.

Chromatography on DEAE-Cellulose

After dialysis, the volume of the ammonium sulfate fraction is adjusted to 200 ml with PEM. A column of Whatman DE-52 DEAE-Cellulose

6 (diam.) x 11 cm is washed with PEM. After the sample has been applied to the column, it is washed with 400 ml of the same buffer and eluted with a 1000 ml linear gradient of PEM running from 0.06 to 0.3 M phosphate. Fractions of 50 ml are collected. The activity emerges in approximately 100 ml at a mean phosphate concentration of approximately 0.15 M. The active fractions are dialyzed against PEM containing 0.02 M phosphate.

Chromatography on Phosphocellulose

The DE-52 fraction is applied to a 15-cm Whatman P-11 phosphocellulose column 1 centimenter in diameter prewashed with 150 ml PEM containing 0.02 M phosphate. The column is then washed with 40 ml PEM and eluted with a 300 ml linear gradient running from 0.05 M to 0.3 M phosphate. Fractions of 9 ml are collected. The activity comes off in approximately 30 ml at a mean phosphate concentration of 0.16 M. The active fractions are dialyzed against 0.01 M phosphate pH 7.5 containing 10^{-4} M EDTA and 5×10^{-4} M dithiothreitol. After concentration to 7.5 ml by dialysis against dry Sephadex G-200, the solution is dialyzed again against the same buffer.

Sedimentation on Glycerol Gradient

The concentrated phosphocellulose eluate is applied in 2.5-ml portions to 30 ml, 10–25% glycerol gradients made up in the dithiothreitol buffer. After centrifugation at 25,000 rpm for 20 hr at 2° in an International SB-110 rotor, the activity in each gradient is found in approximately 2 ml at a position corresponding to a sedimentation coefficient of 13 S. This enzyme fraction is free of contaminating endonuclease activity as shown by its inability to break supercoils of λ·K DNA. No exonuclease activity can be detected either. Table 1 summarizes the purification.

TABLE 1. Purification of the Enzyme

	Total Protein (mg)	Activity
Low-speed supernatant	8,000	~1
High-speed supernatant	5,700	
$(NH_4)_2SO_4$ precipitate	3,400	
DEAE eluate	260	
Phosphocellulose eluate	2	
Glycerol gradient fraction	⩽0.5	⪆5,000

PROPERTIES

Stability

The glycerol gradient fractions are stored at −10° and have been kept up to a year without significant loss of activity. Exposure to 55° for 5 min destroys the activity completely.

Cofactors and Inhibitors

ATP, SAM, and Mg^{2+} are essential for the reaction. The requirement for SAM cannot be satisfied by *l*-methionine, 5′-thiomethyladenosine, *S*-adenosyl-*dl*-homocysteine or *S*-adenosylethionine. ATP can be replaced by dATP. ATP cannot be replaced by the other three nucleoside triphosphates, ADP, AMP, adenosine, pyrophosphate, or phosphate. The enzyme is rapidly inactivated by exposure to ATP in the presence of Mg^{2+}. The only other triphosphate that shows the same effect is dATP.

Physicochemical Properties

The enzyme has a molecular weight of 390,000 as estimated by sedimentation on a glycerol gradient and gel filtration on Sephadex G-200. The optimal pH for the restriction reaction lies between 7.5 and 8.0.

Specificity

The nucleolytic activity is specific for DNA from λ grown on a m_k^- host, The action of the enzyme on λ·0 DNA liberates little or no nonsedimenting material, showing that degradation is endo- rather than exonucleolytic. The action of the enzyme on λ twisted circles, DNA molecules without ends, confirms this. The enzyme readily degrades λ·0 twisted circles but has no effect on λ·K twisted circles; neither double nor single chain scissions are produced. Denatured λ·0 DNA is not attacked by the enzyme.

The enzyme has been tested on DNA duplexes containing one strand that is modified and one that is not. Such "heteroduplexes" may be prepared by the annealing of separated strands from λ·0 and λ·K. The enzyme makes neither double- nor single-chain scissions in either of the two possible heteroduplexes.

Kinetics

During the first few seconds of incubation, the restriction enzyme makes only single-chain scissions. However, prolonged incubation breaks the DNA duplex into pieces ranging from approximately 15–40% of the λ genome. These pieces are free of single polynucleotide chain interruptions. Thus, the enzyme first makes only single-chain scissions and then, a few seconds later, breaks the complementary chain at a point directly or nearly opposite.

Binding

In the presence of Mg^{2+}, ATP, and SAM, the enzyme forms a specific complex with λ·0 DNA which can be detected by its retention on membrane filters. λ·K DNA is not retained. With purified enzyme, the amount of binding reaches a maximum at approximately two minutes of incubation and then falls. However, the addition of EDTA halts the reaction and stabilizes the complex.

References

1. Meselson, M., and R. Yuan. 1968. *Nature,* **217**:1110.
2. Yuan, R., and M. Meselson. 1970. *Proc. Natl. Acad. Sci., U.S.,* **65**: 357.

PRIMER-DEPENDENT POLYNUCLEOTIDE PHOSPHORYLASE

CLAUDE B. KLEE
LABORATORY OF BIOCHEMICAL PHARMACOLOGY
NATIONAL INSTITUTE OF ARTHRITIS
AND METABOLIC DISEASES
NATIONAL INSTITUTES OF HEALTH
BETHESDA, MARYLAND

Procedures for the preparation of highly purified polynucleotide phosphorylase from several sources were described in the preceding volume (*1*). A more recent purification procedure for the isolation of polynucleotide phosphorylase from *Escherichia coli* is now available (*2*). One of the striking differences among various preparations of this enzyme is the degree of dependence of the polymerization reaction on the presence of oligonucleotide primer (*1*). Although oligonucleotides serve as primers for the enzymes from *Azotobacter vinelandii* and *E. coli,* they have little effect on the rate of the polymerization reaction (*3*). The enzyme obtained from *Micrococcus luteus* (formerly *M. lysodeikticus*) has always been a better source for the enzyme dependent on primer for polymerization of nucleoside diphosphate (*4, 5*) than the enzymes from *E. coli* (*6*) and *A. vinelandii* (*7*).

More recently, highly purified preparations of the enzyme from *M. luteus* have been obtained in a form which does not depend on primer for the polymerization reaction (*8*). Limited proteolysis with trypsin is the most effective known means to convert the enzyme to a primer dependent form (*8–11*). The primer dependent polynucleotide phosphorylase obtained previously from *M. luteus* has the same electrophoretic mobility as the enzyme converted to primer dependence by limited proteolysis (*8*), and was probably the result of limited proteolysis occurring in the cells or during the early steps of the purification procedure. A preparation of a homogeneous, primer-dependent polynucleotide phosphorylase from *M. luteus* obtained by limited proteolysis with trypsin is described below. A preparation of partially purified enzyme, also obtained by limited proteolysis is usually suitable for the preparation of oligonucleotides of known sequence (*12*). This preparation is also outlined later in this chapter.

ASSAYS

Two assays can be used to follow the purification of polynucleotide phosphorylase: the phosphorolysis assay is used throughout the purification, and the polymerization assay is used in the late steps of the purification (Fractions VI through X, Table 1) and to follow the conversion of the enzyme to primer dependence.

Phosphorolysis Assay

The assay described here (*13*) is a modification of the one described previously (*1*). In this procedure the ^{32}P-labeled ADP is separated from $^{32}P_i$ by precipitation of the latter with ammonium molybdate according to the method of Sugino and Miyoshi (*14*).

Reagents

1. Tris buffer, pH 8.2 chloride salt, 1 M
2. $MgCl_2$, 0.1 M
3. $K_2H^{32}PO_4$, 0.1 M (containing 50–150 x 10 ^{5} cpm/ml)*
4. EDTA, 0.01 M
5. Polyadenylic acid, 10 μmoles of mononucleotide equivalent/ml
6. Bovine serum albumin, 1 mg/ml in water
7. Freshly prepared solution of 0.008 M triethylamine, pH 5.0, chloride salt, made 0.0064 M ammonium molybdate, and 0.16 N perchloric acid
8. Triton toluene scintillation fluid: 330 ml of Triton X-100 (Packard), mixed with 667 ml of toluene, 5.5 gm of 2,5-diphenyloxazole (PPO), and 100 mg of 1,4-*bis*[2-(5-phenyloxazolyl)] benzene (POPOP).

*The ^{32}P obtained from Tracerlab can be used without the purification procedure previously described (*1*). Less than 0.1% of the counts are adsorbed onto charcoal.

Procedure

The incubation mixture consists of 0.01 ml Tris, 0.005 ml $MgCl_2$, 0.01 ml K_2HPO_4, 0.01 ml polymer, 0.01 ml EDTA, 0.003 to 0.02 phosphorolysis unit of enzyme, and H_2O to a final volume of 0.1 ml. With purified fractions, 0.01 ml of bovine serum albumin is included in the assay. Bovine serum albumin is replaced by thiolated gelatin (obtained from Sigma) when pyrimidine-containing compounds are used since bovine serum albumin is contaminated with a pyrimidine-specific RNase (*15*). Since the rate of phosphorolysis varies with different batches of polymers, a more reproducible and accurate measurement of the specific activity of the enzyme can be obtained by measuring the phosphorolysis of the oligonucleotide ApApApA or ApApApApA, at a concentration of 0.67 mM (as oligonucleotide) in the assay mixture.*

After 15 min at 37°C, 1.25 ml of ice cold reagent 7 is added to precipitate $^{32}P_i$; the tubes are allowed to remain in ice for 5 to 10 min and are then centrifuged at 2,000 rpm for 15 min at 0–4°C. The supernatant solution (1 ml) is counted in a scintillation counter in 10 ml of triton scintillation fluid (*16*).

Polymerization Assay*†

This assay measures the formation of P_i from the polymerization of nucleoside diphosphate in the presence or absence of oligonucleotide primer (*4*).

Reagents

1. Tris buffer, pH 9.0, chloride salt, 1 M
2. $MgCl_2$, 0.05 M
3. EDTA, 0.004 M
4. ADP, 0.2 M
5. Bovine serum albumin, 1 mg/ml in water
6. ApApApA, 2.0 mM (as oligonucleotide)
7. A suspension containing 5 gm of washed Norite per 100 ml of 2.5% perchloric acid.‡

*M. F. Singer, personal communication.
†Method adapted from T. Godefroy, M. Cohn, and M. Grunberg Manago. 1970. *European J. Biochem.*, **12**: 236.
‡Acid washed Norite A (Pfanstiehl) is washed 3 times by decantation with 20 vol of 2.5% perchloric acid and then suspended in 20 vol of 2.5% perchloric acid to give a 5% solution. The A_{280} of 0.3 ml of the Norite-supernatant fluid in the Ames-Dubin procedure should be 0.06 or less.

Procedure

The incubation mixture contains 0.005 ml Tris, 0.005 ml $MgCl_2$, 0.005 ml EDTA, 0.005 ml ADP, 0.01 ml bovine serum albumin, 0.01 to 0.1 unit of enzyme, (measured by polymerization), with and without 0.005 ml of ApApApA, and H_2O to a final volume of 0.05 ml. After 15 min at 37°, 0.4 ml of perchloric acid–charcoal mixture is added. The tubes are allowed to stand in ice for 10 min with occasional shaking and are centrifuged at 2,000 rpm for 15 min. The P_i content in the supernatant is determined according to Chen, Toribara, and Warner (*17*).

Assay of Polynucleotide Phosphorylase on Polyacrylamide Gel

The assay described by Thang et al. (*18*) is used to follow the purification of polynucleotide phosphorylase and its conversion to primer dependence. (Primer-independent and primer-dependent polynucleotide phosphorylases have a different electrophoretic mobility as shown in Fig. 1.)

The polymer formed *in situ* is stained by acridine orange and allows the localization and identification of the enzyme.

Reagents

1. Incubation mixture containing 0.1 M Tris-HCl buffer, pH 9.0, 0.002 M $MgCl_2$, 0.005 M ADP.*
2. Solution of 1% acridine orange, 1% lanthanum acetate in 15% acetic acid.
3. Acetic acid, 7.5%

Procedure

The enzyme (1 to 6 polymerization units) is subjected to disc gel electrophoresis at pH 9.5 with a 7.5% gel according to the manual supplied by Canalco. At the end of the run, the gel is transferred to a tube containing 5 ml of incubation mixture, and incubated at 37° for 1–4 hr depending on the amount of enzyme present. The gel is then transferred to a tube containing 5 ml of the acridine orange solution and allowed to stand overnight at room temperature. The gel is then washed by repeated transfers in 7.5% acetic acid until no more color is left except for the activity bands. This procedure can be used with crude preparations of enzyme. The gels can also be split longitudinally, one half being used to stain for protein, the other for activity.

*Thiols or oligonucleotide primers can be added to the incubation mixture when necessary.

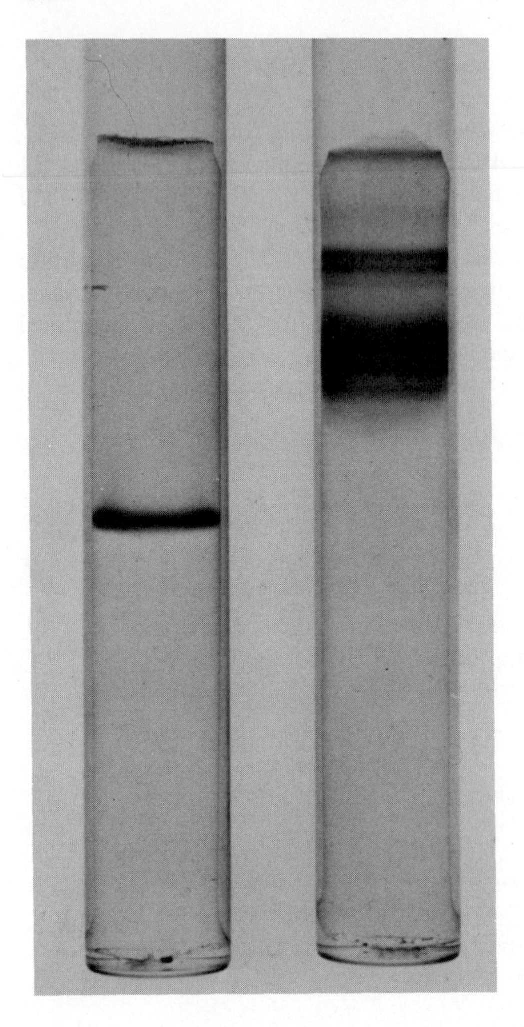

Figure 1. Electrophoretic pattern of primer independent (left) and primer dependent (right) polynucleotide phosphorylase.

PURIFICATION OF PRIMER-INDEPENDENT POLYNUCLEOTIDE PHOSPHORYLASE

Growth of Cells

Commercially available lyophilized or spray-dried cells of *Micrococcus luteus* can be used as a starting material as described previously (*1*). However, the crude enzyme from laboratory grown cells is more reproducible in respect to its specific activity and its electrophoretic properties. In this case *M. luteus* cells (ATCC 4698) were used; stock cultures of these cells to be used as inoculum, can be stored as a mixture of 0.4 ml of a broth culture at late or mid log phase with 0.6 ml of sterile 25% glycerol at –70°. For large scale growth the medium contained 5 gm of tryptone, 5 gm of yeast extract, 1 gm of sucrose, 1 gm of K_2HPO_4, in 1 liter of tap water.*

Preparation of Extract

For large scale preparations, the lysis of the cells can either be carried out repeatedly on 40-gm batches as described previously (*1*), or more conveniently on a large scale with similar recovery of enzyme.† In this case two batches of 250 gm of cells suspended in 2.5 liters each of 0.5% NaCl at 23°C are blended for 30 seconds in a precooled Waring Blendor at low speed at 0–4°C;[7] the residual lumps are broken up with a glass rod; and the mixture is then again blended for an additional 30 seconds. The resulting suspensions (23°) are poured into a 10 liter jar and the next step is carried out at room temperature. The pH, usually 5.4 to 5.6, is brought to 8.0 by dropwise addition of 1 M Tris base (about 250 ml) with constant stirring. A solution of crystalline lysozyme, 1.25 gm in 30 ml of water, is added at once and stirring is continued for about 1 hr at room temperature, after which the suspension becomes very viscous (as described previously (*1*)), and curd formation occurs usually in about 10 min. The lysis is then stopped by addition of 2.5 liters of cold saturated $(NH_4)_2SO_4$ with rapid mixing. The suspension is then centrifuged at 0–4°C for 15 min at 10,000 *g*. The clear or slightly turbid, bright yellow supernatant, called crude extract, is poured through glass wool. The volume is 5,000 ± 100 ml. The crude extract, Fraction I, is assayed as described previously after dialysis of a small sample against cold distilled water.

Purification

The results of a typical purification are shown in Table 1 (*16*). All operations after preparation of the crude extract were carried out between 0 and 4°C. The pH's were measured at room temperature except when otherwise stated. Protein was determined by the method of Lowry et al. in crude extracts (*19*) or by the absorption at 280 nm, for Fractions IX and X.

*C. Letendre and M. F. Singer, personal communication.
†When fresh cells are used, they are suspended in enough 0.5% NaCl to give an absorption at 620 nm of 300 measured in a cuvette of 1-cm path length.

TABLE 1. Purification of Primer-Independent Polynucleotide Phosphorylase (500 gm Cells)

Fraction	Protein (mg/ml)	Specific activity (units/mg)[a]		Total phosphoro-lysis units	Primer stimulation
		Phosphorolysis	Polymerization		
II. $(NH_4)_2SO_4$ (30-65%)[b]	50.0	0.28		7,580	
III. $(NH_4)_2SO_4$ (43-55%)	50.0	0.60		5,890	
IV. DEAE-cellulose I	8.5	0.60		5,930	
V. $(NH_4)_2SO_4$ (40-60%, pH 6.3)	84.0	1.04		6,100	
Sephadex G-100	13.0	2.50		3,700	
VI. Zn Sephadex G-100	0.9	22.2	112	3,650	1.5
VII. DEAE-cellulose II	1.0	45.0	330	2,000	1.1–1.7
VIII. Sephadex G-200	0.44[c]	62.0	690	825	1.1–1.4
IX. Microgranular DEAE-Cellulose IXa	0.20[c]	140.0	1,500	340	1.1–1.4
Microgranular DEAE-cellulose IXb	0.26[c]	79.0	660	210	1.6
X. Hydroxylapatite[d]	0.66[c]	138.0		175	

[a]One unit of enzyme catalyzes the formation of 1 μmole of ADP (phosphorolysis) or 1 μmole of P_i (polymerization) per 15 min in the conditions of the assays.
[b]Lysis of the cells was done on 40-gm batches.
[c]The protein content is estimated, assuming that a solution of 1 mg of protein/ml has an $A_{280\ nm}$ of 1.0.
[d]Fraction IXb was used for hydroxylapatite chromatography.

For pure enzyme (Fractions X or IXa), absorption at 280 nm can be converted to mg of protein/ml using an extinction coefficient $E_{1\%,\ 280\ nm}$ of 4.4 (*11*).

Precipitation by $(NH_4)_2SO_4$ (*1*)

When lysis is carried out on 40-gm batches, Fraction I is brought to 65% saturation of $(NH_4)_2SO_4$. The resulting batches of Fraction II are stored frozen at $-20°$ until all batches are collected. There is a 10 to 20% loss of activity over 3 months. After large scale lysis, Fraction I is brought to 65% saturation by the addition of 217 gm of solid $(NH_4)_2SO_4$ per liter. After stirring for an additional 20 min, the suspension is allowed to settle overnight at 0–4°C. A yellow turbid supernatant devoid of enzymatic activity is removed by suction as much as possible, the remainder is centrifuged for 15 min at 10,000 *g*, and the precipitate is dissolved in a minimum amount of 0.1 M Tris-HCl buffer (pH 8.1). This solution is dialyzed against distilled H_2O until free of $(NH_4)_2SO_4$.

Second Ammonium Sulfate Fractionation (*1*)

The pooled Fraction II obtained from the separate 40-gm batches or Fraction II from the large-scale lysis is diluted to 10 mg of protein/ml and simultaneously made 0.1 M in Tris-HCl buffer (pH 8.1). The solution is brought to 43% saturation with $(NH_4)_2SO_4$ by the addition of 244 gm of solid $(NH_4)_2SO_4$ per liter. The supernatant is collected by centrifugation at 20,000 *g* for 15 min and 74 gm of $(NH_4)_2SO_4$ is added per liter to obtain 55% saturation. The precipitate is dissolved in 0.1 M Tris-HCl buffer, pH 8.1 (about 100–200 ml of buffer for 500 gm of cells). The solution, Fraction III, is dialyzed against 10 liters of 0.1 M Tris-HCl buffer, pH 8.1, containing 1 mM EDTA for 4 hr. The dialysis fluid is changed every hour.

DEAE-Cellulose I

Fraction III is freed of nucleic acids by DEAE-cellulose chromatography (*20*). Fraction III is diluted to 30 mg of protein/ml with 0.1 M Tris-HCl buffer, pH 8.1, containing 1 mM EDTA. To this solution 2 vol of 0.015 M Tris-HCl buffer, pH 7.8, containing 1.5 mM β-mercaptoethanol, 1.5 mM $MgCl_2$, and 0.45 M NaCl are added. The enzyme solution is then applied to a column of DEAE-cellulose (Brown and Company, Selectacel type 20), (20 ml per gram of protein), which has been equilibrated with 0.01 M Tris-HCl buffer, pH 7.8, containing 1mM mercaptoethanol, 1 mM $MgCl_2$, and 0.3 M NaCl, at a flow rate of 2 ml/min. The enzyme is washed on the column with the above buffer, and 100 ml fractions are collected. The enzyme passes directly through the column, and all fractions showing a 280/260 ratio above 1.4 are pooled and stored overnight at 0–4°C before the next step.*

$(NH_4)_2SO_4$ Fractionation at pH 6.3 (*1*)

Fraction IV is adjusted to pH 6.3 with 1 M acetic acid, at 0°. The solution is then brought to 40% saturation with $(NH_4)_2SO_4$ by the addition of 226 gm of solid $(NH_4)_2SO_4$ per liter. After stirring for 10 min the supernatant fluid is collected by centrifugation and made 60% saturated with $(NH_4)_2SO_4$ by addition of 120 gm of solid salt per liter (pH maintained at 6.3). The precipitate is collected as above and dissolved in a minimum volume of 0.5 M Tris-HCl buffer, pH 8.1. This solution is freed of $(NH_4)_2SO_4$ by gel filtration through Sephadex G-100. The column (5 x 40 cm) is equilibrated and eluted with 0.01 M Tris-HCl buffer, pH 8.2, containing 0.001 M EDTA, 0.25 M NaCl, and 0.001 M mercaptoethanol. The flow rate is 1 ml/min and the fraction size is 10 ml. Only fractions with a specific activity greater than 1, with the electrophoretic behavior of primer-independent polynucleotide phosphorylase, are pooled. These fractions are eluted from the column just after the void volume prior to the bulk of the protein.

Zinc-Sephadex (*1*)

The $ZnCl_2$ step is carried out on the pooled material from the Sephadex G-100 column. Fraction V (110 ml) is diluted with water (40 ml) to 10 mg of protein/ml. Samples of 0.1 ml are treated with varying amounts of 0.1 M $ZnCl_2$ (from 0.03 to 0.2 volumes); the specific activity of the enzyme in the supernatant fluid is determined and the conditions giving the best purification and yield are applied to the rest of Fraction V. In the preparation described in Table 1, 7.2 ml of 0.1 M $ZnCl_2$ (0.05 volume) is added dropwise with mechanical stirring. The mixture is stirred for an additional 15 min and centrifuged for 10 min at 10,000 *g*. The supernatant fluid is freed of $ZnCl_2$ by gel filtration through the same Sephadex G-100 column used in the previous step. The enzyme is eluted just after the void volume, with most of

*The resulting Fraction IV corresponds to Fraction V described previously (*1*); therefore, the steps VI, VII, and VIII described by Singer correspond to steps V, VI, and VII, respectively, of this purification procedure.

the protein in a single peak. The peak fractions, containing 80 to 90% of the enzymatic activity, are pooled (Fraction VI). The tail part of the peak is discarded.

DEAE-Cellulose Chromatography II (*1*)

Fraction VI is subjected to DEAE-cellulose chromatography. Fraction VI is diluted three-fold with 0.01 M Tris-HCl buffer, pH 8.2, 0.001 M EDTA, and 0.001 M mercaptoethanol, and applied to a column (2.5 x 20 cm) of DEAE-cellulose (Brown and Company) which has been equilibrated with 0.01 M Tris-HCl, pH 8.2, 0.001 M $MgCl_2$, and 0.001 M mercaptoethanol (mixer solution). The enzyme is eluted with a nonlinear gradient, the constant volume mixing chamber containing 300 ml of mixer solution and the reservoir containing 300 ml of the same solution made 0.5 M NaCl. The enzyme fractions eluted between 230 and 270 ml of effluent, (50 to 60% of the total enzymatic activity), with a specific activity between 20 and 60 are pooled (Fraction VII). The fractions eluted later with most of the contaminating protein are discarded.

Fraction VII is concentrated on a small DEAE-cellulose column (*5*). It can also be concentrated in a collodion bag (Schleicher and Schuell) under vacuum, or in an Amincon ultrafiltration cell with a UM-1 membrane* under a pressure of nitrogen of 50 psi, according to the volume of the sample.

Sephadex G-200 Gel Filtration

The concentrated Fraction VII (10 ml) is applied to a column of Sephadex G-200 or preferably G-200 superfine (2.5 x 40 cm), equilibrated, and eluted with 0.01 M Tris-HCl buffer, pH 8.2, containing 1 mM $MgCl_2$, 0.05 M NaCl, and 1 mM mercaptoethanol. The flow rate is 0.1 to 0.3 ml/min, and the fraction size is 2 to 4 ml. The enzyme is excluded from the Sephadex G-200 and eluted in a total volume of 30 ml. Some contaminating protein eluted after the enzyme is removed.

Microgranular DEAE-Cellulose Chromatography

Fraction VIII is purified further on a column (1 x 5 cm) of microgranular DEAE-cellulose (DE 52) that has been equilibrated with 0.01 M Tris-HCl, pH 8.2, containing 1 mM $MgCl_2$, 0.05 M NaCl, and 1 mM mercaptoethanol. The

*The UM-1 membrane retains material with a molecular weight of 10,000.

enzyme is eluted with a linear NaCl gradient (0.1 to 0.3 M) in the above buffer mixture (100 ml of each buffer), the flow rate is 0.5 ml per minute, and the fraction size is 1 ml. The enzyme eluted at low ionic strength between 55 and 75 ml of effluent (Fraction IXa) usually has a constant specific activity and has been shown to be 95% pure. The only contaminant which is detected by disc gel electrophoresis can be removed easily by the next step, if necessary.

The enzyme eluted later, between 75 and 130 ml of effluent (Fraction IXb), should be pooled separately; it is contaminated by a yellow material which can also be removed by the next step, but may require repeated chromatography.

Hydroxylapatite Chromatography

When pure preparations are needed, Fraction IXa and IXb are purified separately by hydroxylapatite chromatography.* The preparation of hydroxylapatite is equilibrated just before use, with 0.01 M Tris-HC1 buffer, pH 7.4, containing 1 mM mercaptoethanol and 0.1 M NaCl. The column dimensions are calculated from the values described by Tiselius (*21*), 0.6 to 1 ml per milligram of protein. Immediately prior to application on the column, the enzyme solution is dialyzed for 3 hr against 500 vol of the above buffer solution with 2 changes. The enzyme is eluted from the column with a linear gradient of ammonium sulfate in the above buffer (*2*). The volume of the gradient is 15 ml per milliliter of column volume, and the flow rate is adjusted so that elution is complete in about 5 hr for small columns and 12 hr for larger columns. Fraction IXa, if more than 90% pure, is eluted with a gradient between 0.2 and 1 M $(NH_4)_2SO_4$, less purified enzyme is eluted with a gradient between 0 and 1.2 M $(NH_4)_2SO_4$. The pH of the 2 M ammonium sulfate stock solution is adjusted to 7.4 with NH_4OH prior to use. The cellulose nitrate receiving tubes contain a concentrated solution of Tris-HCl, pH 8.2, and $MgCl_2$ such that the final concentration in each collected fraction is 0.01 M Tris-HCl and 1 mM $MgCl_2$. Within 24 hr the eluted enzyme is pooled and concentrated by ultrafiltration under vacuum to a concentration of 0.4–2 mg of protein per milliliter. It can then be stored for more than a year at −20°.

Comments on the Purification

In large scale preparations Fraction III can be subjected to a DEAE-cellulose chromatography instead of being simply passed through the column for removal of nucleic acids. In this case a column of DEAE-cellulose (Brown

* Hydroxylapatite is made in the laboratory according to Tiselius and Eriksson-Quensel (*21*).

and Company, Selectacel type 20) of 10 x 25 cm, equilibrated with 0.01 M Tris-HCl buffer, pH 7.8, containing 0.1 M NaCl, 1 mM $MgCl_2$, and 1 mM mercaptoethanol is used. The enzyme is diluted with glass distilled water to bring the conductivity to that of the buffer used for the column (5 millimho at 0°C). The enzyme is washed onto the column with 2 liters of the above buffer and then eluted with a linear gradient formed of 4 liters of the equilibrating buffer and 4 liters of the same buffer containing 0.6 M NaCl. The enzyme is eluted as a sharp peak after about 4 liters of effluent. All fractions with a 280/260 ratio greater than 1.5 are pooled, and the following steps are carried out as described above with the exception of step VII, the DEAE-cellulose chromatography II, which can be omitted.*

The reproducibility of the purification procedure, as well as the conversion to primer dependence, is directly related to the properties of the starting material which should be composed primarily of primer independent polynucleotide phosphorylase with the electrophoretic pattern shown in Fig. 1. The relative proportion of the multiple components is variable. It is therefore important to check the starting material by disc gel electrophoresis and also in all subsequent steps to remove any enzyme with electrophoretic mobility greater than the fastest component of the primer-independent enzyme shown in Fig. 1. Such species are often present and will not separate as well from contaminating protein, and under the proteolysis conditions can lead to preparations with different properties (*9, 22, 23*).

PREPARATION OF PRIMER-DEPENDENT POLYNUCLEOTIDE PHOSPHORYLASE (*11, 12*)

Reagents

1. A fresh solution of trypsin 1 mg/ml (twice recrystallized from Worthington) in water.
2. Diisopropylfluorophosphate. Because of its toxicity this reagent should be used with great care, only in small amounts, in a well-ventilated hood. A beaker of NH_4OH should be kept handy in case of spillage.
3. NaOH, 0.05 N

Procedure

The reaction is usually followed in the pH stat as described below. A solution of primer-independent polynucleotide phosphorylase (Fraction IXa or X) containing about 1 mg of protein/ml is dialyzed for 3 hr at 0–4°C

*See footnote (*) on page 901.

against 500 vol of an unbuffered solution containing 1 mM $MgCl_2$, 1 mM mercaptoethanol, and 0.2 M NaCl, which had been adjusted to pH 8.2 with NaOH. The dialyzed solution (1 ml) is placed in the titration vessel of the pH stat. The temperature is maintained at 23° by circulating water from a thermostatted water bath. The pH of the reaction mixture is adjusted to 8.0 and automatically maintained at this value with 0.05 M NaOH. The reaction is started by addition of 1 μl of the trypsin solution. When uptake of NaOH ceases after 20 to 30 min of incubation, an additional 3 μl of trypsin solution are added, to ensure complete conversion of the enzyme to primer dependence, and the reaction is followed for 50 to 60 min. No further detectable uptake of NaOH should occur. The reaction is stopped by the addition of 0.1 ml of freshly prepared 1 M Tris-HCl buffer, pH 8.2, containing 0.2 μl of diisopropylfluorophosphate. The enzyme is freed of excess diisopropylfluorophosphate and small oligopeptides by dialysis at 0–4°C for 5 hr against 500 ml of 0.01 M Tris-HCl buffer, pH 8.2, 0.2 M NaCl, 1 mM $MgCl_2$, and 1 mM mercaptoethanol or by passage through a 1 x 30 cm Sephadex G-50 column equilibrated and eluted with the above buffer. The resulting preparation is then checked for primer dependence by the polymerization assay and by its electrophoretic mobility, shown in Fig. 1. Both forms of the enzyme should be kept concentrated (1 to 5 mg/ml) and diluted in a solution of bovine serum albumin, prior to assays.

When a pH stat is not available, the reaction is carried out as described above, but 0.01 M Tris-HCl buffer, pH 8.0, is present in the reaction mixture. Preliminary trials on small aliquots (*12*) are used to determine the trypsin concentration (between 1 and 5 μg/ml) and the time of incubation (30 to

TABLE 2. Comparison of Primer-Independent and Primer-Dependent Polynucleotide Phosphorylases

	Before trypsin	After trypsin
Polymerization	Primer independent	Primer dependent
Molecular weight	2.6×10^5	2.2×10^5
Subunit molecular weight (6 M guanidine hydrochloride, 0.1 M mercaptoethanol	6.7×10^4	5.9×10^4
$s_{20,w}$	9.1	9.0
Electrophoresis	Multiband	Single band
-SH content		
Before reduction	<0.05	0.2 - 0.5
After reduction		1.1 - 1.4

90 min) necessary to obtain primer dependence as measured by direct assay, and conversion to a single component with the electrophoretic mobility shown in Fig. 1. If the conversion is incomplete, enzymatically active species are observed on disc gel electrophoresis, between the starting material and the primer dependent enzyme. In contrast to trypsin, chymotrypsin leads to the formation of multiple products and should therefore be avoided (*22*).

The trypsin treated enzyme is included in Sephadex G-200 in contrast to the starting material which is eluted at the void volume. A convenient preparation of primer-dependent polynucleotide phosphorylase suitable for polynucleotide synthesis is based on this property. Fraction IV or lyophilized *M. luteus* polynucleotide phosphorylase from P-L Biochemicals, Inc., is first passed through a Sephadex G-200 column; the material eluted at the void volume with the electrophoretic mobility of the primer-independent enzyme (50 to 70% of the total activity) is treated with trypsin and purifed further on the same Sephadex G-200 column (*12*).

PROPERTIES OF PRIMER-DEPENDENT POLYNUCLEOTIDE PHOSPHORYLASE

Physical and Chemical Properties

Primer-dependent polynucleotide phosphorylase obtained by limited proteolysis with trypsin is a homogeneous protein with a molecular weight of 220,000 and an $s_{20,w}$ of 9.0 S. On disc gel electrophoresis a single component, with an R_f of 0.45, is observed. The enzyme can be dissociated into 4 subunits of molecular weight 59,000 in 6 M guanidine hydrochloride and mercaptoethanol (*11*). The properties of primer-independent and primer-dependent polynucleotide phosphorylase are summarized in Table 2. The conversion to primer dependence is accompanied by the breaking of between 5 and 6 peptide bonds per molecule of enzyme and results in the removal of a small portion of the peptide chain of each subunit (*11*).

The conversion to primer dependence is not the result of the preferential loss of 260 nm absorbing material. Both forms of the enzyme have very similar spectral characteristics and an $E_{1\%,\ 280\ \text{nm}}$ of 4.4 ± 0.6. Furthermore, the primer-dependent enzyme can be converted to primer independence by reduction with thiols, which results in the appearance of 1 to 2 titratable SH groups per mole of enzyme. Blocking of these SH groups with sulfhydryl inhibitors reinstates primer dependence. None of these treatments affects the ability of the enzyme to catalyze the phosphorolysis reaction. It has been suggested that the presence of a free SH group may not itself be correlated with primer independence, but that a structural change resulting from proteolysis and subsequent oxidation or reduction of a cysteine residue is responsible for the change in primer dependence (*11*). This interpretation is consistent with the recent finding of Gajda and Fitt that polynucleotide phosphorylase from *A. vinelandii* can undergo inactivation upon oxidation or treatment with sulfhydryl reagents without changes in primer dependence (*24*).

Catalytic Properties

The primer-dependent enzyme obtained by limited proteolysis with trypsin has a specific activity of 45 to 60 phosphorolysis units/mg * and 250 to 500 polymerization units/mg measured in the presence of primer; these values are the same as those of the pure primer-independent enzyme.

Primer-dependent polynucleotide phosphorylase catalyzes the polymerization of nucleoside diphosphate in the absence of primer at a very slow but linear rate (*23*). Autocatalytic kinetics previously observed are absent with the pure enzyme.

Primer dependence is observed for ADP, UDP, and CDP when the enzyme is prepared as indicated above. However, when less purified enzyme is used for the proteolysis, a preferential dependence on primer with CDP or UDP has sometimes been observed (*22, 23*).

It has been shown recently by Moses and Singer (*23*) that the primer-independent enzyme catalyzes *de novo* synthesis of long-chain polymers in the presence or absence of primer, in a processive fashion; the primer-dependent enzyme catalyzes *de novo* processive synthesis of long-chain polymers only when the primer is absent. When the primer is present, the primer-dependent enzyme catalyzes the elongation of the primers to short-chain polymers, in a nonprocessive fashion.

In the phosphorolysis of oligonucleotides no significant differences are observed between the two forms of the enzyme (*15*). They both catalyze a stepwise exonucleolytic cleavage, starting at the end bearing the 3′-hydroxyl group (*25*). This reaction proceeds by a processive mechanism with long-chain polymers (*16*) and by a random mechanism with oligonucleotides with chain lengths up to at least 6 (*13, 15*). The K_m for oligonucleotide substrates and K_i for oligonucleotide inhibitors decreases with increasing chain length and reaches a minimum with compounds containing 5 residues. In contrast the V_m is greater for oligonucleotides than long chain polymers (*15*). It has been proposed by Chou and Singer (*15*) that the enzyme has multiple binding sites for the interaction with polynucleotides.

* The protein was measured by absorption at 280 nm using an $E_{1\%\ 280\ nm}$ of 4.4.

Acknowledgments

I wish to express my gratitude to Dr. Maxine F. Singer for her constant and generous advice, and to thank Drs. R. E. Moses, J. Y. Chou, and C. Letendre for contributing much helpful information.

References

1. Singer, M. F. 1966. In G. L. Cantoni and D. R. Davies (Eds.), Procedures in nucleic acid research. Harper & Row, New York, Vol. 1, p. 245.
2. Kimhi, Y., and U. Z. Littauer. 1968. *J. Biol. Chem.*, **243**; 231.
3. Singer, M. F., L. A. Heppel, and R. J. Hilmoe. 1960. *J. Biol. Chem.*, **235**: 739.
4. Singer, M. F., and J. K. Guss. 1962. *J. Biol. Chem.*, **237**: 182.
5. Singer, M. F., and B. M. O'Brien. 1963. *J. Biol. Chem.*, **238**: 328.
6. Williams, F. R., and M. Grunberg Manago. 1964. *Biochim. Biophys. Acta,* **89**: 66.
7. Ochoa, S., and S. Mii. 1961. *J. Biol. Chem.*, **236**: 3303.
8. Klee, C. B. 1967. *J. Biol. Chem.*, **242**: 3579.
9. Fitt, P. S., and E. A. Fitt. 1967. *Biochem. J.*, **105**: 25.
10. Klee, C. B., and M. F. Singer. 1968. *J. Biol. Chem.*, **243**: 5094.
11. Klee, C. B. 1969. *J. Biol. Chem.*, **244**: 2558.
12. Klee, C. B., and M. F. Singer. 1967. *Biochem. Biophys. Res. Commun.*, **29**: 356.
13. Chou, J. Y., and M. F. Singer. 1970. *J. Biol. Chem.*, **245**: 995.
14. Sugino, Y., and Y. Miyoshi. 1964. *J. Biol. Chem.*, **239**: 2360.
15. Chou, J. Y., and M. F. Singer. 1970. *J. Biol. Chem.*, **245**: 1005.
16. Klee, C. B., and M. F. Singer. 1968. *J. Biol. Chem.*, **243**: 923.
17. Chen, P. S., T. Y. Toribara, and H. Warner. 1956. *Anal. Chem.*, **28**: 1756.
18. Thang, M. N., D. C. Thang, and J. Leautey. 1967. *Compt. Rend. Acad. Sci. Paris,* **265 D**: 1823.
19. Lowry, O. H., N. J. Rosebrough, A. L. Farr, and R. J. Randall. 1951. *J. Biol. Chem.*, **193**: 265.
20. Thanassi, N. M., and M. F. Singer. 1966. *J. Biol. Chem.*,**241**: 3639.
21. Tiselius, A., and I. B. Eriksson-Quensel. 1939. *Biochem. J.*, **33**: 1752.
22. Fitt, P. S., E. A. Fitt, and H. Wille. 1968. *Biochem. J.*, **110**: 475.
23. Moses, R. E., and M. F. Singer. 1970. *J. Biol. Chem.*, **245**: 2414.
24. Gajda, A. T., and P. S. Fitt. 1969. *Biochem. J.*, **112**: 381.
25. Singer, M. F., R. J. Hilmoe, and M. Grunberg Manago, 1960. *J. Biol. Chem.*, **235**: 2705.

INDEX